11TH EDITION

INTERMEDIATE ALGEBRA

Margaret L. Lial
American River College

John Hornsby
University of New Orleans

Terry McGinnis

Addison-Wesley

Boston   Columbus   Indianapolis   New York   San Francisco   Upper Saddle River
Amsterdam   Cape Town   Dubai   London   Madrid   Milan   Munich   Paris   Montreal   Toronto
Delhi   Mexico City   Sao Paulo   Sydney   Hong Kong   Seoul   Singapore   Taipei   Tokyo

Editorial Director: Christine Hoag
Editor-in-Chief: Maureen O'Connor
Executive Content Manager: Kari Heen
Content Editor: Courtney Slade
Assistant Editor: Mary St. Thomas
Editorial Assistant: Rachel Haskell
Senior Managing Editor: Karen Wernholm
Senior Production Project Manager: Kathleen A. Manley
Senior Author Support/Technology Specialist: Joe Vetere
Digital Assets Manager: Marianne Groth
Rights and Permissions Advisor: Michael Joyce
Image Manager: Rachel Youdelman
Media Producer: Lin Mahoney
Software Development: Kristina Evans and Mary Durnwald
Marketing Manager: Adam Goldstein
Marketing Assistant: Ashley Bryan
Design Manager: Andrea Nix
Cover Designer: Beth Paquin
Cover Art: *Beginning of Spring* by Gregory Packard Fine Art LLC, www.gregorypackard.com
Senior Manufacturing Buyer: Carol Melville
Senior Media Buyer: Ginny Michaud
Interior Design, Production Coordination, Composition, and Illustrations: Nesbitt Graphics, Inc.

For permission to use copyrighted material, grateful acknowledgment is made to the copyright holders on page G-8, which is hereby made part of this copyright page.

Many of the designations used by manufacturers and sellers to distinguish their products are claimed as trademarks. Where those designations appear in this book, and Addison-Wesley was aware of a trademark claim, the designations have been printed in initial caps or all caps.

Library of Congress Cataloging-in-Publication Data
Lial, Margaret L.
    Intermediate algebra/Margaret L. Lial, John Hornsby, Terry McGinnis.—11th ed.
        p. cm.
    ISBN-13: 978-0-321-71541-8
    (student edition)
    ISBN-10: 0-321-71541-1
    (student edition)
1. Algebra—Textbooks I. Hornsby, E. John. II. McGinnis, Terry. III. Title.
    QA152.3.L534 2012
    512.9—dc22

                                                2010002278

5 6 7 8 9 10—CKV—14 13 12

Addison-Wesley
is an imprint of

www.pearsonhighered.com

ISBN 13: 978-0-321-71541-8
ISBN 10: 0-321-71541-1

In memory of Frank Rickey, Paul Rees,
Earl Swokowski, and Chuck Miller
E.J.H.

To Papa
T.

Contents

4 Systems of Linear Equations   209

5 Exponents, Polynomials, and Polynomial Functions   263

6 Factoring   319

Rational Expressions and Functions   361

Roots, Radicals, and Root Functions   427

Quadratic Equations, Inequalities, and Functions   495

It is with pleasure that we offer the eleventh edition of *Intermediate Algebra*. With each new edition, the text has been shaped and adapted to meet the changing needs of both students and educators, and this edition faithfully continues that process. As always, we have taken special care to respond to the specific suggestions of users and reviewers through enhanced discussions, new and updated examples and exercises, helpful features, updated figures and graphs, and an extensive package of supplements and study aids. We believe the result is an easy-to-use, comprehensive text that is the best edition yet.

Students who have never studied algebra—as well as those who require further review of basic algebraic concepts before taking additional courses in mathematics, business, science, nursing, or other fields—will benefit from the text's student-oriented approach. Of particular interest to students and instructors will be the NEW Study Skills activities and Now Try Exercises.

This text is part of a series that also includes the following books:

▶ *Beginning Algebra,* Eleventh Edition, by Lial, Hornsby, and McGinnis

▶ *Beginning and Intermediate Algebra,* Fifth Edition, by Lial, Hornsby, and McGinnis

▶ *Algebra for College Students,* Seventh Edition, by Lial, Hornsby, and McGinnis

NEW IN THIS EDITION

We are pleased to offer the following new student-oriented features and study aids:

Lial Video Library This collection of video resources helps students navigate the road to success. It is available in MyMathLab and on Video Resources on DVD.

MyWorkBook This helpful guide provides extra practice exercises for every chapter of the text and includes the following resources for every section:

▶ Key vocabulary terms and vocabulary practice problems

▶ Guided Examples with step-by-step solutions and similar Practice Exercises, keyed to the text by Learning Objective

▶ References to textbook Examples and Section Lecture Videos for additional help

▶ Additional Exercises with ample space for students to show their work, keyed to the text by Learning Objective

Study Skills Poor study skills are a major reason why students do not succeed in mathematics. In these short activities, we provide helpful information, tips, and strategies on a variety of essential study skills, including *Reading Your Math Textbook, Tackling Your Homework, Taking Math Tests,* and *Managing Your Time.* While most of the activities are concentrated in the early chapters of the text, each has been designed independently to allow flexible use with individuals or small groups of students, or as a source of material for in-class discussions. (See pages 102 and 225.)

Now Try Exercises To actively engage students in the learning process, we now include a parallel margin exercise juxtaposed with each numbered example. These all-new exercises enable students to immediately apply and reinforce the concepts and skills presented in the corresponding examples. Answers are conveniently located on the same page so students can quickly check their results. (See pages 3 and 92.)

Revised Exposition As each section of the text was being revised, we paid special attention to the exposition, which has been tightened and polished. (See Section 5.2 Adding and Subtracting Polynomials, for example.) We believe this has improved discussions and presentations of topics.

Specific Content Changes These include the following:

▶ We gave the exercise sets special attention. There are approximately 1200 new and updated exercises, including problems that check conceptual understanding, focus on skill development, and provide review. We also worked to improve the even-odd pairing of exercises.

▶ Real-world data in over 165 applications in the examples and exercises have been updated.

▶ There is an increased emphasis on the difference between expressions and equations, including a new example at the beginning of Section 2.1, plus corresponding exercises. Throughout the text, we have reformatted many example solutions to use a "drop down" layout in order to further emphasize for students the difference between simplifying expressions and solving equations.

▶ We increased the emphasis on checking solutions and answers, as indicated by the new CHECK tag and ✓ in the exposition and examples.

▶ Section 2.2 has been expanded to include a new example and exercises on solving a linear equation in two variables for y. A new objective, example, and exercises on percent increase and decrease are also provided.

▶ Section 3.5 Introduction to Functions from the previous edition has been expanded and split into two sections.

▶ Key information about graphs is displayed prominently beside hand-drawn graphs for the various types of functions. (See Sections 5.3, 7.4, 8.1, 9.5, 9.6, 10.2, 10.3, and 11.1.)

▶ An objective, example, and exercises on using factoring to solve formulas for specified variables is included in Section 6.5.

▶ Presentations of the following topics have also been enhanced and expanded:

Solving three-part inequalities (Section 2.5)
Finding average rate of change (Section 3.2)
Writing equations of horizontal and vertical lines (Section 3.3)
Determining the number of solutions of a linear system (Section 4.1)
Solving systems of linear equations in three variables (Section 4.2)
Understanding the basic concepts and terminology of polynomials (Section 5.2)
Solving equations with rational expressions and graphing rational functions
   (Section 7.4)
Solving quadratic equations by factoring and the square root property
   (Section 9.1)

Solving quadratic equations by substitution (Section 9.3)
Evaluating expressions involving the greatest integer (Section 11.1)
Graphing hyperbolas (Section 11.3)
Evaluating factorials and binomial coefficients (Section 12.4)

HALLMARK FEATURES

We have included the following helpful features, each of which is designed to increase ease-of-use by students and/or instructors.

Annotated Instructor's Edition For convenient reference, we include answers to the exercises "on page" in the *Annotated Instructor's Edition,* using an enhanced, easy-to-read format. In addition, we have added approximately 15 new Teaching Tips and over 40 new and updated Classroom Examples.

Relevant Chapter Openers In the new and updated chapter openers, we feature real-world applications of mathematics that are relevant to students and tied to specific material within the chapters. Examples of topics include Americans' spending on pets, television ownership and viewing, and tourism. Each opener also includes a section outline. (See pages 1, 47, and 263.)

Helpful Learning Objectives We begin each section with clearly stated, numbered objectives, and the included material is directly keyed to these objectives so that students and instructors know exactly what is covered in each section. (See pages 2 and 48.)

Popular Cautions and Notes One of the most popular features of previous editions, we include information marked ⚠ CAUTION and NOTE to warn students about common errors and emphasize important ideas throughout the exposition. The updated text design makes them easy to spot. (See pages 53 and 140.)

Comprehensive Examples The new edition of this text features a multitude of step-by-step, worked-out examples that include pedagogical color, helpful side comments, and special pointers. We give increased attention to checking example solutions—more checks, designated using a special *CHECK* tag, are included than in past editions. (See pages 51 and 270.)

More Pointers Well received by both students and instructors in the previous edition, we incorporate more pointers in examples and discussions throughout this edition of the text. They provide students with important on-the-spot reminders and warnings about common pitfalls. (See pages 96 and 396.)

Updated Figures, Photos, and Hand-Drawn Graphs Today's students are more visually oriented than ever. As a result, we have made a concerted effort to include appealing mathematical figures, diagrams, tables, and graphs, including a "hand-drawn" style of graphs, whenever possible. (See pages 138 and 532.) Many of the graphs also use a style similar to that seen by students in today's print and electronic media. We have incorporated new photos to accompany applications in examples and exercises. (See pages 154 and 168.)

Relevant Real-Life Applications We include many new or updated applications from fields such as business, pop culture, sports, technology, and the life sciences that show the relevance of algebra to daily life. (See pages 76 and 244.)

Emphasis on Problem-Solving We introduce our six-step problem-solving method in Chapter 2 and integrate it throughout the text. The six steps, *Read, Assign a Variable, Write an Equation, Solve, State the Answer,* and *Check,* are emphasized in boldface type and repeated in examples and exercises to reinforce the problem-solving process for students. (See pages 69 and 234.) We also provide students with PROBLEM-SOLVING HINT boxes that feature helpful problem-solving tips and strategies. (See pages 81 and 233.)

Connections We include these to give students another avenue for making connections to the real world, graphing technology, or other mathematical concepts, as well as to provide historical background and thought-provoking questions for writing, class discussion, or group work. (See pages 117 and 143.)

Ample and Varied Exercise Sets One of the most commonly mentioned strengths of this text is its exercise sets. We include a wealth of exercises to provide students with opportunities to practice, apply, connect, review, and extend the algebraic concepts and skills they are learning. We also incorporate numerous illustrations, tables, graphs, and photos to help students visualize the problems they are solving. Problem types include writing ✐, graphing calculator 📷, multiple-choice, true/false, matching, and fill-in-the-blank problems, as well as the following:

▶ *Concept Check* exercises facilitate students' mathematical thinking and conceptual understanding. (See pages 108 and 413.)

▶ *WHAT WENT WRONG?* exercises ask students to identify typical errors in solutions and work the problems correctly. (See pages 274 and 502.)

▶ *Brain Busters* exercises challenge students to go beyond the section examples. (See pages 145 and 300.)

▶ RELATING CONCEPTS exercises help students tie together topics and develop problem-solving skills as they compare and contrast ideas, identify and describe patterns, and extend concepts to new situations. These exercises make great collaborative activities for pairs or small groups of students. (See pages 173 and 301.)

▶ TECHNOLOGY INSIGHTS exercises provide an opportunity for students to interpret typical results seen on graphing calculator screens. Actual screens from the TI-83/84 Plus graphing calculator are featured. (See pages 146 and 353.)

▶ PREVIEW EXERCISES allow students to *review* previously-studied concepts and *preview* skills needed for the upcoming section. These make good oral warm-up exercises to open class discussions. (See pages 283 and 371.)

Special Summary Exercises We include a set of these popular in-chapter exercises in selected chapters. They provide students with the all-important ***mixed* review problems** they need to master topics and often include summaries of solution methods and/or additional examples. (See pages 394 and 522.)

Extensive Review Opportunities We conclude each chapter with the following review components:

▶ A **Chapter Summary** that features a helpful list of **Key Terms,** organized by section, **New Symbols, Test Your Word Power** vocabulary quiz (with answers immediately following), and a **Quick Review** of each section's contents, complete with additional examples (See pages 483–486.)

▶ A comprehensive set of **Chapter Review Exercises,** keyed to individual sections for easy student reference, as well as a set of **Mixed Review Exercises** that helps students further synthesize concepts (See pages 487–490.)

▶ A **Chapter Test** that students can take under test conditions to see how well they have mastered the chapter material (See pages 490–491.)

▶ A set of **Cumulative Review Exercises** (beginning in Chapter 2) that covers material going back to Chapter 1 (See pages 492–493.)

Glossary For easy reference at the back of the book, we include a comprehensive glossary featuring key terms and definitions from throughout the text. (See pages G-1 to G-7.)

SUPPLEMENTS

For a comprehensive list of the supplements and study aids that accompany *Intermediate Algebra,* Eleventh Edition, see pages xv–xvii.

ACKNOWLEDGMENTS

The comments, criticisms, and suggestions of users, nonusers, instructors, and students have positively shaped this textbook over the years, and we are most grateful for the many responses we have received. Thanks to the following people for their review work, feedback, assistance at various meetings, and additional media contributions:

Barbara Aaker, *Community College of Denver*
Viola Lee Bean, *Boise State University*
Kim Bennekin, *Georgia Perimeter College*
Dixie Blackinton, *Weber State University*
Tim Caldwell, *Meridian Community College*
Sally Casey, *Shawnee Community College*
Callie Daniels, *St. Charles Community College*
Cheryl Davids, *Central Carolina Technical College*
Chris Diorietes, *Fayetteville Technical Community College*
Sylvia Dreyfus, *Meridian Community College*
Lucy Edwards, *Las Positas College*
LaTonya Ellis, *Bishop State Community College*
Jacqui Fields, *Wake Technical Community College*
Beverly Hall, *Fayetteville Technical Community College*
Sandee House, *Georgia Perimeter College*
Lynette King, *Gadsden State Community College*
Linda Kodama, *Windward Community College*
Ted Koukounas, *Suffolk Community College*
Karen McKarnin, *Allen County Community College*
James Metz, *Kapi´olani Community College*
Jean Millen, *Georgia Perimeter College*
Molly Misko, *Gadsden State Community College*
Jane Roads, *Moberly Area Community College*
Melanie Smith, *Bishop State Community College*

Linda Smoke, *Central Michigan University*
Erik Stubsten, *Chattanooga State Technical Community College*
Tong Wagner, *Greenville Technical College*
Sessia Wyche, *University of Texas at Brownsville*

Special thanks are due the many instructors at Broward College who provided insightful comments.

Over the years, we have come to rely on an extensive team of experienced professionals. Our sincere thanks go to these dedicated individuals at Addison-Wesley, who worked long and hard to make this revision a success: Chris Hoag, Maureen O'Connor, Michelle Renda, Adam Goldstein, Kari Heen, Courtney Slade, Kathy Manley, Lin Mahoney, and Mary St. Thomas.

We are especially grateful to Callie Daniels for her excellent work on the new Now Try Exercises. Abby Tanenbaum did a terrific job helping us revise real-data applications. Kathy Diamond provided expert guidance through all phases of production and rescued us from one snafu or another on multiple occasions. Marilyn Dwyer and Nesbitt Graphics, Inc., provided some of the highest quality production work we have experienced on the challenging format of these books.

Special thanks are due Jeff Cole, who continues to supply accurate, helpful solutions manuals; David Atwood, who wrote the comprehensive *Instructor's Resource Manual with Tests;* Beverly Fusfield, who provided the new MyWorkBook; Beth Anderson, who provided wonderful photo research; and Lucie Haskins, for yet another accurate, useful index. De Cook, Shannon d'Hemecourt, Paul Lorczak, and Sarah Sponholz did a thorough, timely job accuracy checking manuscript and page proofs. It has indeed been a pleasure to work with such an outstanding group of professionals.

As an author team, we are committed to providing the best possible text and supplements package to help instructors teach and students succeed. As we continue to work toward this goal, we would welcome any comments or suggestions you might have via e-mail to math@pearson.com.

Margaret L. Lial

John Hornsby

Terry McGinnis

STUDENT SUPPLEMENTS

Student's Solutions Manual

▶ By Jeffery A. Cole, *Anoka-Ramsey Community College*

▶ Provides detailed solutions to the odd-numbered, section-level exercises and to all Now Try Exercises, Relating Concepts, Summary, Chapter Review, Chapter Test, and Cumulative Review Exercises

ISBNs: 0-321-71582-9, 978-0-321-71582-1

NEW Video Resources on DVD featuring the Lial Video Library

▶ Provides a wealth of video resources to help students navigate the road to success

▶ Available in MyMathLab (with optional subtitles in English)

▶ Includes the following resources:

Section Lecture Videos that offer a new navigation menu for easy focus on key examples and exercises needed for review in each section (with optional subtitles in Spanish)

Solutions Clips that feature an instructor working through selected exercises marked in the text with a DVD icon 🌐

Quick Review Lectures that provide a short summary lecture of each key concept from Quick Reviews at the end of every chapter in the text

Chapter Test Prep Videos that include step-by-step solutions to all Chapter Test exercises and give guidance and support when needed most—the night before an exam. Also available on YouTube (searchable using author name and book title)

ISBNs: 0-321-71584-5, 978-0-321-71584-5

NEW MyWorkBook

▶ Provides Guided Examples and corresponding Now Try Exercises for each text objective

▶ Refers students to correlated Examples, Lecture Videos, and Exercise Solution Clips

▶ Includes extra practice exercises for every section of the text with ample space for students to show their work

▶ Lists the learning objectives and key vocabulary terms for every text section, along with vocabulary practice problems

ISBNs: 0-321-71586-1, 978-0-321-71586-9

INSTRUCTOR SUPPLEMENTS

Annotated Instructor's Edition

▶ Provides "on-page" answers to all text exercises in an easy-to-read margin format, along with Teaching Tips and extensive Classroom Examples

▶ Includes icons to identify writing 🖊 and calculator 🖩 exercises. These are in Student Edition also.

ISBNs: 0-321-71578-0, 978-0-321-71578-4

Instructor's Solutions Manual

▶ By Jeffery A. Cole, *Anoka-Ramsey Community College*

▶ Provides complete solutions to all text exercises, including all Classroom Examples and Now Try Exercises

ISBNs: 0-321-71580-2, 978-0-321-71580-7

Instructor's Resource Manual with Tests

▶ By David Atwood, *Rochester Community and Technical College*

▶ Contains two diagnostic pretests, four free-response and two multiple-choice test forms per chapter, and two final exams

▶ Includes a mini-lecture for each section of the text with objectives, key examples, and teaching tips

▶ Provides a correlation guide from the tenth to the eleventh edition

ISBNs: 0-321-71579-9, 978-0-321-71579-1

PowerPoint® Lecture Slides

▶ Present key concepts and definitions from the text

▶ Available for download at www.pearsonhighered.com/irc

ISBNs: 0-321-71585-3, 978-0-321-71585-2

TestGen® (www.pearsonhighered.com/testgen)

▶ Enables instructors to build, edit, print, and administer tests using a computerized bank of questions developed to cover all text objectives

▶ Allows instructors to create multiple but equivalent versions of the same question or test with the click of a button

▶ Allows instructors to modify test bank questions or add new questions

▶ Available for download from Pearson Education's online catalog

ISBNs: 0-321-71581-0, 978-0-321-71581-4

STUDENT SUPPLEMENTS

InterAct Math Tutorial Website
http://www.interactmath.com

▶ Provides practice and tutorial help online

▶ Provides algorithmically generated practice exercises that correlate directly to the exercises in the textbook

▶ Allows students to retry an exercise with new values each time for unlimited practice and mastery

▶ Includes an interactive guided solution for each exercise that gives helpful feedback when an incorrect answer is entered

▶ Enables students to view the steps of a worked-out sample problem similar to the one being worked on

INSTRUCTOR SUPPLEMENTS

Pearson Math Adjunct Support Center
(http://www.pearsontutorservices.com/math-adjunct.html)

▶ Staffed by qualified instructors with more than 50 years of combined experience at both the community college and university levels

Assistance is provided for faculty in the following areas:

▶ Suggested syllabus consultation

▶ Tips on using materials packed with your book

▶ Book-specific content assistance

▶ Teaching suggestions, including advice on classroom strategies

Available for Students and Instructors

MyMathLab® Online Course (Access code required.)

MyMathLab® is a text-specific, easily customizable online course that integrates interactive multimedia instruction with textbook content. MyMathLab gives instructors the tools they need to deliver all or a portion of their course online, whether their students are in a lab setting or working from home.

▶ **Interactive homework exercises,** correlated to the textbook at the objective level, are algorithmically generated for unlimited practice and mastery. Most exercises are free-response and provide guided solutions, sample problems, and tutorial learning aids for extra help.

▶ **Personalized homework** assignments can be designed to meet the needs of the class. MyMathLab tailors the assignment for each student based on their test or quiz scores so that each student's homework assignment contains only the problems they still need to master.

▶ **Personalized Study Plan,** generated when students complete a test or quiz or homework, indicates which topics have been mastered and links to tutorial exercises for topics students have not mastered. Instructors can customize the Study Plan so that the topics available match their course content.

▶ **Multimedia learning aids,** such as video lectures and podcasts, animations, and a complete multimedia textbook, help students independently improve their understanding and performance. Instructors can assign these multimedia learning aids as homework to help their students grasp the concepts.

▶ **Homework and Test Manager** lets instructors assign homework, quizzes, and tests that are automatically graded. They can select just the right mix of questions from the MyMathLab exercise bank, instructor-created custom exercises, and/or TestGen® test items.

▶ **Gradebook,** designed specifically for mathematics and statistics, automatically tracks students' results, lets instructors stay on top of student performance, and gives them control over how to calculate final grades. They can also add offline (paper-and-pencil) grades to the gradebook.

▶ **MathXL Exercise Builder** allows instructors to create static and algorithmic exercises for their online assignments. They can use the library of sample exercises as an easy starting point, or they can edit any course-related exercise.

▶ **Pearson Tutor Center** (www.pearsontutorservices.com) access is automatically included with MyMathLab. The Tutor Center is staffed by qualified math instructors who provide textbook-specific tutoring for students via toll-free phone, fax, email, and interactive Web sessions.

Students do their assignments in the Flash®-based MathXL Player, which is compatible with almost any browser (Firefox®, Safari™, or Internet Explorer®) on almost any platform (Macintosh® or Windows®). MyMathLab is powered by CourseCompass™, Pearson Education's online teaching and learning environment, and by MathXL®, our online homework, tutorial, and assessment system. MyMathLab is available to qualified adopters. For more information, visit our website at www.mymathlab.com or contact your Pearson representative.

MathXL® Online Course (access code required)

MathXL® is an online homework, tutorial, and assessment system that accompanies Pearson's textbooks in mathematics or statistics.

▶ **Interactive homework exercises,** correlated to the textbook at the objective level, are algorithmically generated for unlimited practice and mastery. Most exercises are free-response and provide guided solutions, sample problems, and learning aids for extra help.

▶ **Personalized homework** assignments are designed by the instructor to meet the needs of the class, and then personalized for each student based on their test or quiz results. As a result, each student receives a homework assignment that contains only the problems they still need to master.

▶ **Personalized Study Plan,** generated when students complete a test or quiz or homework, indicates which topics have been mastered and links to tutorial exercises for topics students have not mastered. Instructors can customize the available topics in the study plan to match their course concepts.

▶ **Multimedia learning aids,** such as video lectures and animations, help students independently improve their understanding and performance. These are assignable as homework, to further encourage their use.

▶ **Gradebook,** designed specifically for mathematics and statistics, automatically tracks students' results, lets instructors stay on top of student performance, and gives them control over how to calculate final grades.

▶ **MathXL Exercise Builder** allows instructors to create static and algorithmic exercises for their online assignments. They can use the library of sample exercises as an easy starting point or the Exercise Builder to edit any of the course-related exercises.

▶ **Homework and Test Manager** lets instructors create online homework, quizzes, and tests that are automatically graded. They can select just the right mix of questions from the MathXL exercise bank, instructor-created custom exercises, and/or TestGen test items.

The new, Flash®-based MathXL Player is compatible with almost any browser (Firefox®, Safari™, or Internet Explorer®) on almost any platform (Macintosh® or Windows®). MathXL is available to qualified adopters. For more information, visit our website at www.mathxl.com, or contact your Pearson representative.

STUDY SKILLS

Using Your Math Textbook

Your textbook is a valuable resource. You will learn more if you fully make use of the features it offers.

General Features

▶ **Table of Contents** Find this at the front of the text. Mark the chapters and sections you will cover, as noted on your course syllabus.

▶ **Answer Section** Tab this section at the back of the book so you can refer to it frequently when doing homework. Answers to odd-numbered section exercises are provided. Answers to ALL summary, chapter review, test, and cumulative review exercises are given.

▶ **Glossary** Find this feature after the answer section at the back of the text. It provides an alphabetical list of the key terms found in the text, with definitions and section references.

▶ **List of Formulas** Inside the back cover of the text is a helpful list of geometric formulas, along with review information on triangles and angles. Use these for reference throughout the course.

Specific Features

▶ **Objectives** The objectives are listed at the beginning of each section and again within the section as the corresponding material is presented. Once you finish a section, ask yourself if you have accomplished them.

▶ **Now Try Exercises** These margin exercises allow you to immediately practice the material covered in the examples and prepare you for the exercises. Check your results using the answers at the bottom of the page.

▶ **Pointers** These small shaded balloons provide on-the-spot warnings and reminders, point out key steps, and give other helpful tips.

▶ **Cautions** These provide warnings about common errors that students often make or trouble spots to avoid.

▶ **Notes** These provide additional explanations or emphasize important ideas.

▶ **Problem-Solving Hints** These green boxes give helpful tips or strategies to use when you work applications.

Find an example of each of these features in your textbook.

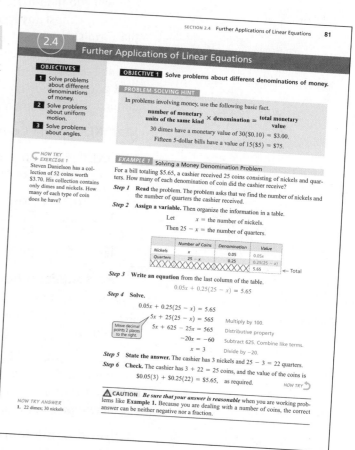

Review of the Real Number System

Americans love their pets. Over 71 million U.S. households owned pets in 2008. Combined, these households spent more than $44 billion pampering their animal friends. The fastest-growing segment of the pet industry is the high-end luxury area, which includes everything from gourmet pet foods, designer toys, and specialty furniture to groomers, dog walkers, boarding in posh pet hotels, and even pet therapists. (*Source:* American Pet Products Manufacturers Association.)

In **Exercise 101** of **Section 1.3,** we use an *algebraic expression*, one of the topics of this chapter, to determine how much Americans have spent annually on their pets in recent years.

1.1 Basic Concepts

OBJECTIVE 1 Write sets using set notation. A **set** is a collection of objects called the **elements** or **members** of the set. In algebra, the elements of a set are usually numbers. Set braces, { }, are used to enclose the elements.

For example, 2 is an element of the set $\{1, 2, 3\}$. Since we can count the number of elements in the set $\{1, 2, 3\}$, it is a **finite set.**

In our study of algebra, we refer to certain sets of numbers by name. The set

$$N = \{1, 2, 3, 4, 5, 6, \ldots\} \qquad \text{Natural (counting) numbers}$$

is called the **natural numbers,** or the **counting numbers.** The three dots (*ellipsis* points) show that the list continues in the same pattern indefinitely. We cannot list all of the elements of the set of natural numbers, so it is an **infinite set.**

Including 0 with the set of natural numbers gives the set of **whole numbers.**

$$W = \{0, 1, 2, 3, 4, 5, 6, \ldots\} \qquad \text{Whole numbers}$$

The set containing no elements, such as the set of whole numbers less than 0, is called the **empty set,** or **null set,** usually written $\emptyset$ or { }.

⚠ **CAUTION** Do not write $\{\emptyset\}$ for the empty set. $\{\emptyset\}$ is a set with one element: $\emptyset$. Use the notation $\emptyset$ or { } for the empty set.

To write the fact that 2 is an element of the set $\{1, 2, 3\}$, we use the symbol $\in$ (read "is an element of").

$$2 \in \{1, 2, 3\}$$

The number 2 is also an element of the set of natural numbers N.

$$2 \in N$$

To show that 0 is *not* an element of set N, we draw a slash through the symbol $\in$.

$$0 \notin N$$

Two sets are equal if they contain exactly the same elements. For example, $\{1, 2\} = \{2, 1\}$. (Order doesn't matter.) However, $\{1, 2\} \neq \{0, 1, 2\}$ ($\neq$ means "is not equal to"), since one set contains the element 0 while the other does not.

In algebra, letters called **variables** are often used to represent numbers or to define sets of numbers. For example,

$$\{x \mid x \text{ is a natural number between 3 and 15}\}$$

(read "the set of all elements x such that x is a natural number between 3 and 15") defines the set

$$\{4, 5, 6, 7, \ldots, 14\}.$$

The notation $\{x \mid x \text{ is a natural number between 3 and 15}\}$ is an example of **set-builder notation.**

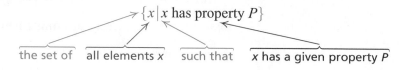

NOW TRY
EXERCISE 1
List the elements in

$\{p \mid p$ is a natural number less than 6$\}$.

EXAMPLE 1 Listing the Elements in Sets

List the elements in each set.

(a) $\{x \mid x$ is a natural number less than 4$\}$
The natural numbers less than 4 are 1, 2, and 3. This set is $\{1, 2, 3\}$.

(b) $\{x \mid x$ is one of the first five even natural numbers$\}$ is $\{2, 4, 6, 8, 10\}$.

(c) $\{x \mid x$ is a natural number greater than or equal to 7$\}$
The set of natural numbers greater than or equal to 7 is an infinite set, written with ellipsis points as

$$\{7, 8, 9, 10, \ldots\}.$$

NOW TRY

NOW TRY
EXERCISE 2
Use set-builder notation to describe the set.

$$\{9, 10, 11, 12\}$$

EXAMPLE 2 Using Set-Builder Notation to Describe Sets

Use set-builder notation to describe each set.

(a) $\{1, 3, 5, 7, 9\}$
There are often several ways to describe a set in set-builder notation. One way to describe the given set is

$$\{x \mid x \text{ is one of the first five odd natural numbers}\}.$$

(b) $\{5, 10, 15, \ldots\}$
This set can be described as $\{x \mid x$ is a multiple of 5 greater than 0$\}$. NOW TRY

OBJECTIVE 2 Use number lines.
A good way to get a picture of a set of numbers is to use a **number line.** See **FIGURE 1**.

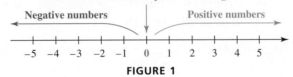

The number 0 is neither positive nor negative.

Negative numbers   Positive numbers

FIGURE 1

To draw a number line, choose any point on the line and label it 0. Then choose any point to the right of 0 and label it 1. Use the distance between 0 and 1 as the scale to locate, and then label, other points.

The set of numbers identified on the number line in **FIGURE 1**, including positive and negative numbers and 0, is part of the set of **integers.**

$$I = \{\ldots, -3, -2, -1, 0, 1, 2, 3, \ldots\} \quad \text{Integers}$$

Each number on a number line is called the **coordinate** of the point that it labels, while the point is the **graph** of the number. **FIGURE 2** shows a number line with several points graphed on it.

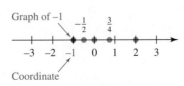

Graph of −1 $-\frac{1}{2}$ $\frac{3}{4}$

Coordinate

FIGURE 2

NOW TRY ANSWERS
1. $\{1, 2, 3, 4, 5\}$
2. $\{x \mid x$ is a natural number between 8 and 13$\}$

The fractions $-\frac{1}{2}$ and $\frac{3}{4}$, graphed on the number line in **FIGURE 2**, are *rational numbers*. A **rational number** can be expressed as the quotient of two integers, with denominator not 0. The set of all rational numbers is written as follows.

$$\left\{\frac{p}{q} \,\middle|\, p \text{ and } q \text{ are integers, } q \neq 0\right\} \qquad \text{Rational numbers}$$

The set of rational numbers includes the natural numbers, whole numbers, and integers, since these numbers can be written as fractions. For example,

$$14 = \frac{14}{1}, \quad -3 = \frac{-3}{1}, \quad \text{and} \quad 0 = \frac{0}{1}.$$

A rational number written as a fraction, such as $\frac{1}{8}$ or $\frac{2}{3}$, can also be expressed as a decimal by dividing the numerator by the denominator.

$$\begin{array}{r} 0.125 \\ 8\overline{)1.000} \\ \underline{8} \\ 20 \\ \underline{16} \\ 40 \\ \underline{40} \\ 0 \end{array}$$
← Terminating decimal (rational number)

← Remainder is 0.

$$\frac{1}{8} = 0.125$$

$$\begin{array}{r} 0.666\ldots \\ 3\overline{)2.000\ldots} \\ \underline{18} \\ 20 \\ \underline{18} \\ 20 \\ \underline{18} \\ 2 \end{array}$$
← Repeating decimal (rational number)

← Remainder is never 0.

$$\frac{2}{3} = 0.\overline{6}$$ ← A bar is written over the repeating digit(s).

Thus, terminating decimals, such as $0.125 = \frac{1}{8}$, $0.8 = \frac{4}{5}$, and $2.75 = \frac{11}{4}$, and repeating decimals, such as $0.\overline{6} = \frac{2}{3}$ and $0.\overline{27} = \frac{3}{11}$, are rational numbers.

Decimal numbers that neither terminate nor repeat, which include many square roots, are *irrational numbers.*

$$\sqrt{2} = 1.414213562\ldots \quad \text{and} \quad -\sqrt{7} = -2.6457513\ldots \qquad \text{Irrational numbers}$$

NOTE Some square roots, such as $\sqrt{16} = 4$ and $\sqrt{\frac{9}{25}} = \frac{3}{5}$, are rational.

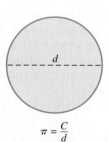

$$\pi = \frac{C}{d}$$

π is approximately 3.141592653....

FIGURE 3

Another irrational number is π, the ratio of the circumference of a circle to its diameter. See **FIGURE 3**.

Some rational and irrational numbers are graphed on the number line in **FIGURE 4**. The rational numbers together with the irrational numbers make up the set of **real numbers**. *Every point on a number line corresponds to a real number, and every real number corresponds to a point on the number line.*

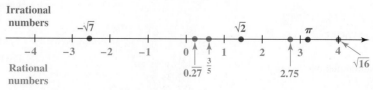

Real numbers

FIGURE 4

OBJECTIVE 3   Know the common sets of numbers.

Sets of Numbers	
Natural numbers, or counting numbers	$\{1, 2, 3, 4, 5, 6, \ldots\}$
Whole numbers	$\{0, 1, 2, 3, 4, 5, 6, \ldots\}$
Integers	$\{\ldots, -3, -2, -1, 0, 1, 2, 3, \ldots\}$
Rational numbers	$\left\{\frac{p}{q} \mid p \text{ and } q \text{ are integers}, q \neq 0\right\}$
	Examples: $\frac{4}{1}$ or 4, 1.3, $-\frac{9}{2}$ or $-4\frac{1}{2}$, $\frac{16}{8}$ or 2, $\sqrt{9}$ or 3, $0.\overline{6}$
Irrational numbers	$\{x \mid x \text{ is a real number that is not rational}\}$
	Examples: $\sqrt{3}, -\sqrt{2}, \pi$
Real numbers	$\{x \mid x \text{ is a rational number or an irrational number}\}^*$

FIGURE 5 shows the set of real numbers. Every real number is either rational or irrational. Notice that the integers are elements of the set of rational numbers and that the whole numbers and natural numbers are elements of the set of integers.

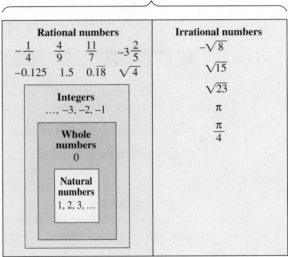

FIGURE 5

NOW TRY
EXERCISE 3

List the numbers in the following set that are elements of each set.

$$\left\{-2.4, -\sqrt{1}, -\frac{1}{2}, 0, 0.\overline{3}, \sqrt{5}, \pi, 5\right\}$$

(a) Whole numbers

(b) Rational numbers

NOW TRY ANSWERS

3. **(a)** $\{0, 5\}$
   (b) $\left\{-2.4, -\sqrt{1}, -\frac{1}{2}, 0, 0.\overline{3}, 5\right\}$

EXAMPLE 3   Identifying Examples of Number Sets

List the numbers in the following set that are elements of each set.

$$\left\{-8, -\sqrt{5}, -\frac{9}{64}, 0, 0.5, \frac{1}{3}, 1.\overline{12}, \sqrt{3}, 2, \pi\right\}$$

(a) Integers
-8, 0, and 2

(b) Rational numbers
$-8, -\frac{9}{64}, 0, 0.5, \frac{1}{3}, 1.\overline{12}$, and 2

(c) Irrational numbers
$-\sqrt{5}, \sqrt{3}$, and π

(d) Real numbers
All are real numbers.   NOW TRY

*An example of a number that is not real is $\sqrt{-1}$. This number, part of the *complex number system*, is discussed in **Chapter 8**.

NOW TRY
EXERCISE 4

Decide whether each statement is *true* or *false*. If it is false, tell why.

(a) All integers are irrational numbers.

(b) Every whole number is an integer.

EXAMPLE 4 Determining Relationships Between Sets of Numbers

Decide whether each statement is *true* or *false*.

(a) All irrational numbers are real numbers.

This is true. As shown in **FIGURE 5**, the set of real numbers includes all irrational numbers.

(b) Every rational number is an integer.

This statement is false. Although some rational numbers are integers, other rational numbers, such as $\frac{2}{3}$ and $-\frac{1}{4}$, are not.　NOW TRY

OBJECTIVE 4 **Find additive inverses.** Look at **FIGURE 6**. For each positive number, there is a negative number on the opposite side of 0 that lies the same distance from 0. These pairs of numbers are called *additive inverses, opposites,* or *negatives* of each other. For example, 3 and -3 are additive inverses.

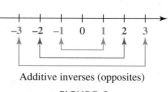
Additive inverses (opposites)
FIGURE 6

Additive Inverse

For any real number a, the number $-a$ is the **additive inverse** of a.

We change the sign of a number to find its additive inverse. As we shall see later, the sum of a number and its additive inverse is always 0.

Uses of the Symbol —

The symbol "$-$" is used to indicate any of the following:

1. **a negative number,** such as -9 or -15;

2. **the additive inverse of a number,** as in "-4 is the additive inverse of 4";

3. **subtraction,** as in $12 - 3$.

In the expression $-(-5)$, the symbol "$-$" is being used in two ways. The first $-$ indicates the additive inverse (or opposite) of -5, and the second indicates a negative number, -5. Since the additive inverse of -5 is 5, it follows that

$$-(-5) = 5.$$

Number	Additive Inverse
6	-6
-4	4
$\frac{2}{3}$	$-\frac{2}{3}$
-8.7	8.7
0	0

The number 0 is its own additive inverse.

$-(-a)$

For any real number a,　　$-(-a) = a.$

Numbers written with positive or negative signs, such as $+4$, $+8$, -9, and -5, are called **signed numbers.** A positive number can be called a signed number even though the positive sign is usually left off. The table in the margin shows the additive inverses of several signed numbers.

OBJECTIVE 5 **Use absolute value.** Geometrically, the **absolute value** of a number a, written $|a|$, is the distance on the number line from 0 to a. For example, the absolute value of 5 is the same as the absolute value of -5 because each number lies five units from 0. See **FIGURE 7** on the next page.

NOW TRY ANSWERS
4. (a) false; All integers are rational numbers.
　(b) true

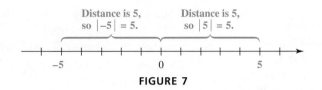

FIGURE 7

⚠ **CAUTION** *Because absolute value represents distance, and distance is never negative, the absolute value of a number is always positive or 0.*

The formal definition of absolute value follows.

Absolute Value

For any real number a, $|a| = \begin{cases} a & \text{if } a \text{ is positive or 0} \\ -a & \text{if } a \text{ is negative.} \end{cases}$

The second part of this definition, $|a| = -a$ if a is negative, requires careful thought. If a is a *negative* number, then $-a$, the additive inverse or opposite of a, is a positive number. Thus, $|a|$ is positive. For example, if $a = -3$, then

$$|a| = |-3| = -(-3) = 3. \quad |a| = -a \text{ if } a \text{ is negative.}$$

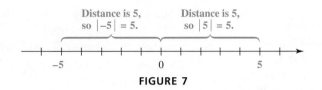

NOW TRY
EXERCISE 5
Simplify by finding each absolute value.
(a) $|-7|$ **(b)** $-|-15|$
(c) $|4| - |-4|$

EXAMPLE 5 Finding Absolute Value

Simplify by finding each absolute value.

(a) $|13| = 13$

(b) $|-2| = -(-2) = 2$

(c) $|0| = 0$

(d) $|-0.75| = 0.75$

(e) $-|8| = -(8) = -8$ Evaluate the absolute value.
Then find the additive inverse.

(f) $-|-8| = -(8) = -8$ Work as in part (e); $|-8| = 8$.

(g) $|-2| + |5| = 2 + 5 = 7$ Evaluate each absolute value, and then add.

(h) $-|5 - 2| = -|3| = -3$ Subtract inside the bars first.   NOW TRY ↺

EXAMPLE 6 Comparing Rates of Change in Industries

The projected total rates of change in employment (in percent) in some of the fastest-growing and in some of the most rapidly declining occupations from 2006 through 2016 are shown in the table.

Occupation (2006–2016)	Total Rate of Change (in percent)
Customer service representatives	24.8
Home health aides	48.7
Security guards	16.9
Word processors and typists	−11.6
File clerks	−41.3
Sewing machine operators	−27.2

Source: Bureau of Labor Statistics.

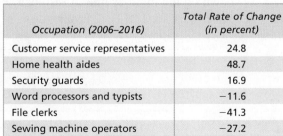

NOW TRY
EXERCISE 6

Refer to the table in **Example 6** on the preceding page. Of the security guards, file clerks, and customer service representatives, which occupation is expected to see the least change (without regard to sign)?

What occupation in the table on the preceding page is expected to see the greatest change?   The least change?

We want the greatest *change,* without regard to whether the change is an increase or a decrease. Look for the number in the table with the greatest absolute value. That number is for home health aides, since $|48.7| = 48.7$. Similarly, the least change is for word processors and typists: $|-11.6| = 11.6.$ NOW TRY

OBJECTIVE 6   **Use inequality symbols.** The statement

$$4 + 2 = 6$$

is an **equation**—a statement that two quantities are equal. The statement

$$4 \neq 6 \quad \text{(read "4 is not equal to 6")}$$

is an **inequality**—a statement that two quantities are *not* equal.

If two numbers are not equal, one must be less than the other. When reading from left to right, the symbol $<$ means "is less than."

$$8 < 9, \quad -6 < 15, \quad -6 < -1, \quad \text{and} \quad 0 < \frac{4}{3} \quad \text{All are true.}$$

Reading from left to right, the symbol $>$ means "is greater than."

$$12 > 5, \quad 9 > -2, \quad -4 > -6, \quad \text{and} \quad \frac{6}{5} > 0 \quad \text{All are true.}$$

In each case, the symbol "points" toward the lesser number.

The number line in **FIGURE 8** shows the graphs of the numbers 4 and 9. We know that $4 < 9$. On the graph, 4 is to the left of 9. *The lesser of two numbers is always to the left of the other on a number line.*

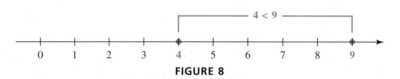

FIGURE 8

Inequalities on a Number Line

On a number line,

$a < b$ if a is to the left of b. $a > b$ if a is to the right of b.

NOW TRY
EXERCISE 7

Use a number line to determine whether each statement is *true* or *false.*

(a) $-5 > -1$

(b) $-7 < -6$

EXAMPLE 7   Determining Order on a Number Line

Use a number line to compare -6 and 1 and to compare -5 and -2.

As shown on the number line in **FIGURE 9**, -6 is located to the left of 1. For this reason, $-6 < 1$. Also, $1 > -6$. From **FIGURE 9**, $-5 < -2$, or $-2 > -5$.

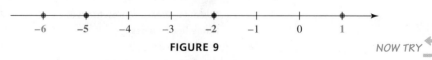

FIGURE 9   NOW TRY

⚠ CAUTION   *Be careful when ordering negative numbers.* Since -5 is to the left of -2 on the number line in **FIGURE 9**, $-5 < -2$, or $-2 > -5$. In each case, the symbol points to -5, the lesser number.

NOW TRY ANSWERS
6. security guards
7. (a) false **(b)** true

The following table summarizes results about positive and negative numbers in both words and symbols.

Words	Symbols
Every negative number is less than 0.	If a is negative, then $a < 0$.
Every positive number is greater than 0.	If a is positive, then $a > 0$.
0 is neither positive nor negative.	

In addition to the symbols $\neq$, $<$, and $>$, the symbols $\leq$ and $\geq$ are often used.

Inequality Symbols

Symbol	Meaning	Example
$\neq$	is not equal to	$3 \neq 7$
$<$	is less than	$-4 < -1$
$>$	is greater than	$3 > -2$
$\leq$	is less than or equal to	$6 \leq 6$
$\geq$	is greater than or equal to	$-8 \geq -10$

NOW TRY
EXERCISE 8
Simplify. Then tell whether the resulting statement is *true* or *false*.

$$-|-12| \geq 2 \cdot 5$$

| EXAMPLE 8 | Using Inequality Symbols |

The table shows some examples of uses of inequalities and why they are true.

Inequality	Why It Is True
$6 \leq 8$	$6 < 8$
$-2 \leq -2$	$-2 = -2$
$-9 \geq -12$	$-9 > -12$
$-3 \geq -3$	$-3 = -3$
$6 \cdot 4 \leq 5(5)$	$24 < 25$

Notice the reason that $-2 \leq -2$ is true. *With the symbol $\leq$, if either the $<$ part or the $=$ part is true, then the inequality is true. This is also the case with the $\geq$ symbol.*

In the last row of the table, recall that the dot in $6 \cdot 4$ indicates the product 6×4, or 24, and $5(5)$ means 5×5, or 25. Thus, the inequality $6 \cdot 4 \leq 5(5)$ becomes $24 \leq 25$, which is true.   NOW TRY

OBJECTIVE 7 **Graph sets of real numbers.** Inequality symbols and variables are used to write sets of real numbers. For example, the set

$$\{x \mid x > -2\}$$

consists of all the real numbers greater than -2. On a number line, we graph the elements of this set by drawing an arrow from -2 to the right. We use a parenthesis at -2 to indicate that -2 is *not* an element of the given set. See **FIGURE 10**.

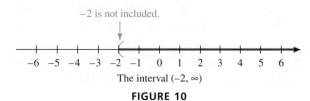

FIGURE 10

The set of numbers greater than -2 is an example of an **interval** on the number line. To write intervals, we use **interval notation.** We write the interval of all numbers greater than -2 as $(-2, \infty)$.

NOW TRY ANSWER
8. $-12 \geq 10$; false

In the interval $(-2, \infty)$, the **infinity symbol** ∞ does not indicate a number—it shows that the interval includes all real numbers greater than -2. The left parenthesis indicates that -2 is not included. *A parenthesis is always used next to the infinity symbol.*

The set of all real numbers is written in interval notation as $(-\infty, \infty)$.

NOW TRY
EXERCISE 9

Write in interval notation and graph.

$$\{x \mid x < -2\}$$

EXAMPLE 9   Graphing an Inequality Written in Interval Notation

Write $\{x \mid x < 4\}$ in interval notation and graph the interval.

The interval is written $(-\infty, 4)$. The graph is shown in **FIGURE 11**. Since the elements of the set are all real numbers *less than* 4, the graph extends to the left.

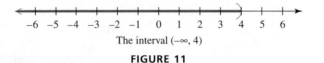

The interval $(-\infty, 4)$

FIGURE 11   *NOW TRY*

The set $\{x \mid x \le -6\}$ includes all real numbers less than or equal to -6. To show that -6 is part of the set, a square bracket is used at -6, as shown in **FIGURE 12**. In interval notation, this set is written $(-\infty, -6]$.

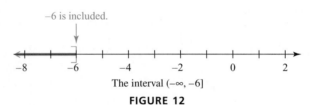

The interval $(-\infty, -6]$

FIGURE 12

NOW TRY
EXERCISE 10

Write in interval notation and graph.

$$\{x \mid x \ge 1\}$$

EXAMPLE 10   Graphing an Inequality Written in Interval Notation

Write $\{x \mid x \ge -4\}$ in interval notation and graph the interval.

This set is written in interval notation as $[-4, \infty)$. The graph is shown in **FIGURE 13**. We use a square bracket at -4, since -4 is part of the set.

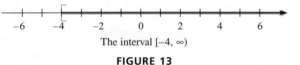

The interval $[-4, \infty)$

FIGURE 13   *NOW TRY*

We sometimes graph sets of numbers that are *between* two given numbers. For example, the set

$$\{x \mid -2 < x < 4\}$$

includes all real numbers between -2 and 4, but *not* the numbers -2 and 4 themselves. This set is written in interval notation as $(-2, 4)$. The graph has a heavy line between -2 and 4, with parentheses at -2 and 4. See **FIGURE 14**. The inequality $-2 < x < 4$, called a **three-part inequality,** is read "-2 is less than x and x is less than 4," or "x is between -2 and 4."

NOW TRY ANSWERS
9. $(-\infty, -2)$
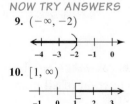
10. $[1, \infty)$

The interval $(-2, 4)$

FIGURE 14

NOW TRY EXERCISE 11

Write in interval notation and graph.

$\{x \mid -2 \leq x < 4\}$

NOW TRY ANSWER

11. $[-2, 4)$

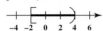

EXAMPLE 11 Graphing a Three-Part Inequality

Write $\{x \mid 3 < x \leq 10\}$ in interval notation and graph the interval.

Use a parenthesis at 3 and a square bracket at 10 to write the interval $(3, 10]$. The graph is shown in **FIGURE 15**. Read the inequality $3 < x \leq 10$ as "3 is less than x and x is less than or equal to 10," or "x is between 3 and 10, excluding 3 and including 10."

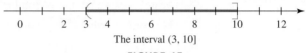

The interval (3, 10]

FIGURE 15

NOW TRY

1.1 EXERCISES

MyMathLab Math XL PRACTICE WATCH DOWNLOAD READ REVIEW

Complete solution available on the Video Resources on DVD

Write each set by listing its elements. ***See Example 1.***

1. $\{x \mid x$ is a natural number less than $6\}$

2. $\{m \mid m$ is a natural number less than $9\}$

3. $\{z \mid z$ is an integer greater than $4\}$

4. $\{y \mid y$ is an integer greater than $8\}$

5. $\{z \mid z$ is an integer less than or equal to $4\}$

6. $\{p \mid p$ is an integer less than $3\}$

7. $\{a \mid a$ is an even integer greater than $8\}$

8. $\{k \mid k$ is an odd integer less than $1\}$

9. $\{x \mid x$ is an irrational number that is also rational$\}$

10. $\{r \mid r$ is a number that is both positive and negative$\}$

11. $\{p \mid p$ is a number whose absolute value is $4\}$

12. $\{w \mid w$ is a number whose absolute value is $7\}$

Write each set using set-builder notation. ***See Example 2.*** *(More than one description is possible.)*

13. $\{2, 4, 6, 8\}$

14. $\{11, 12, 13, 14\}$

15. $\{4, 8, 12, 16, \ldots\}$

16. $\{\ldots, -6, -3, 0, 3, 6, \ldots\}$

Graph the elements of each set on a number line. ***See Objective 2.***

17. $\{-4, -2, 0, 3, 5\}$

18. $\{-3, -1, 0, 4, 6\}$

19. $\left\{-\frac{6}{5}, -\frac{1}{4}, 0, \frac{5}{6}, \frac{13}{4}, 5.2, \frac{11}{2}\right\}$

20. $\left\{-\frac{2}{3}, 0, \frac{4}{5}, \frac{12}{5}, \frac{9}{2}, 4.8\right\}$

Which elements of each set are ***(a)*** *natural numbers,* ***(b)*** *whole numbers,* ***(c)*** *integers,* ***(d)*** *rational numbers,* ***(e)*** *irrational numbers,* ***(f)*** *real numbers?* ***See Example 3.***

21. $\left\{-9, -\sqrt{6}, -0.7, 0, \frac{6}{7}, \sqrt{7}, 4.\overline{6}, 8, \frac{21}{2}, 13, \frac{75}{5}\right\}$

22. $\left\{-8, -\sqrt{5}, -0.6, 0, \frac{3}{4}, \sqrt{3}, \pi, 5, \frac{13}{2}, 17, \frac{40}{2}\right\}$

23. *Concept Check* A student claimed that $\{x \mid x$ is a natural number greater than $3\}$ and $\{y \mid y$ is a natural number greater than $3\}$ actually name the same set, even though different variables are used. Was this student correct?

24. *Concept Check* Give a real number that statisfies each condition.

(a) An integer between 6.75 and 7.75

(b) A rational number between $\frac{1}{4}$ and $\frac{3}{4}$

(c) A whole number that is not a natural number

(d) An integer that is not a whole number

(e) An irrational number between $\sqrt{4}$ and $\sqrt{9}$

Decide whether each statement is true *or* false. *If it is false, tell why.* **See Example 4.**

25. Every integer is a whole number.

26. Every natural number is an integer.

27. Every irrational number is an integer.

28. Every integer is a rational number.

29. Every natural number is a whole number.

30. Some rational numbers are irrational.

31. Some rational numbers are whole numbers.

32. Some real numbers are integers.

33. The absolute value of any number is the same as the absolute value of its additive inverse.

34. The absolute value of any nonzero number is positive.

35. *Concept Check*   Match each expression in parts (a)–(d) with its value in choices A–D. Choices may be used once, more than once, or not at all.

I		II	
(a) $-(-4)$ **(b)** $\lvert -4 \rvert$		**A.** 4	**B.** -4
(c) $-\lvert -4 \rvert$ **(d)** $-\lvert -(-4) \rvert$		**C.** Both A and B	**D.** Neither A nor B

36. *Concept Check*   For what value(s) of x is $\lvert x \rvert = 4$ true?

Give **(a)** *the additive inverse and* **(b)** *the absolute value of each number.* **See the discussion of additive inverses and Example 5.**

37. 6 **38.** 9 **39.** -12 **40.** -14 **41.** $\dfrac{6}{5}$ **42.** 0.16

Simplify by finding each absolute value. **See Example 5.**

43. $\lvert -8 \rvert$ **44.** $\lvert -19 \rvert$ **45.** $\left\lvert \dfrac{3}{2} \right\rvert$ **46.** $\left\lvert \dfrac{3}{4} \right\rvert$

47. $-\lvert 5 \rvert$ **48.** $-\lvert 12 \rvert$ **49.** $-\lvert -2 \rvert$ **50.** $-\lvert -6 \rvert$

51. $-\lvert 4.5 \rvert$ **52.** $-\lvert 12.4 \rvert$ **53.** $\lvert -2 \rvert + \lvert 3 \rvert$ **54.** $\lvert -16 \rvert + \lvert 14 \rvert$

55. $\lvert -9 \rvert - \lvert -3 \rvert$ **56.** $\lvert -10 \rvert - \lvert -7 \rvert$

57. $\lvert -1 \rvert + \lvert -2 \rvert - \lvert -3 \rvert$ **58.** $\lvert -7 \rvert + \lvert -3 \rvert - \lvert -10 \rvert$

Solve each problem. **See Example 6.**

59. The table shows the percent change in population from 2000 through 2008 for selected metropolitan areas.

Metropolitan Area	Percent Change
Las Vegas	35.6
San Francisco	3.7
Chicago	5.2
New Orleans	−13.9
Phoenix	31.7
Detroit	−0.6

Source: U.S. Census Bureau.

(a) Which metropolitan area had the greatest change in population? What was this change? Was it an increase or a decline?

(b) Which metropolitan area had the least change in population? What was this change? Was it an increase or a decline?

60. The table gives the net trade balance, in millions of U.S. dollars, for selected U.S. trade partners for October 2009.

Country	Trade Balance (in millions of dollars)
India	−493
China	−22,663
Netherlands	1314
France	−572
Turkey	400

Source: U.S. Census Bureau.

A negative balance means that imports to the U.S. exceeded exports from the U.S., while a positive balance means that exports exceeded imports.

(a) Which country had the greatest discrepancy between exports and imports?

(b) Which country had the least discrepancy between exports and imports?

Sea level refers to the surface of the ocean. The depth of a body of water such as an ocean or sea can be expressed as a negative number, representing average depth in feet below sea level. By contrast, the altitude of a mountain can be expressed as a positive number, indicating its height in feet above sea level. The table gives selected depths and heights.

Body of Water	Average Depth in Feet (as a negative number)	Mountain	Altitude in Feet (as a positive number)
Pacific Ocean	−12,925	McKinley	20,320
South China Sea	−4,802	Point Success	14,158
Gulf of California	−2,375	Matlalcueyetl	14,636
Caribbean Sea	−8,448	Rainier	14,410
Indian Ocean	−12,598	Steele	16,644

Source: World Almanac and Book of Facts.

61. List the bodies of water in order, starting with the deepest and ending with the shallowest.

62. List the mountains in order, starting with the shortest and ending with the tallest.

63. *True* or *false:* The absolute value of the depth of the Pacific Ocean is greater than the absolute value of the depth of the Indian Ocean.

64. *True* or *false:* The absolute value of the depth of the Gulf of California is greater than the absolute value of the depth of the Caribbean Sea.

Use the number line to answer true *or* false *to each statement.* ***See Example 7.***

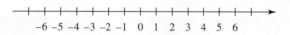

65. $-6 < -1$ **66.** $-4 < -2$ **67.** $-4 > -3$ **68.** $-3 > -1$

69. $3 > -2$ **70.** $6 > -3$ **71.** $-3 \geq -3$ **72.** $-5 \leq -5$

Rewrite each statement with > *so that it uses* < *instead. Rewrite each statement with* < *so that it uses* >. ***See Example 7.***

73. $6 > 2$ **74.** $5 > 1$ **75.** $-9 < 4$ **76.** $-6 < 1$

77. $-5 > -10$ **78.** $-7 > -12$ **79.** $0 < x$ **80.** $-3 < x$

Use an inequality symbol to write each statement. ***See Example 8.***

81. 7 is greater than y.

82. -4 is less than 10.

83. 5 is greater than or equal to 5.

84. -6 is less than or equal to -6.

85. $3t - 4$ is less than or equal to 10.

86. $5x + 4$ is greater than or equal to 21.

87. $5x + 3$ is not equal to 0.

88. $6x + 7$ is not equal to -9.

89. t is between -3 and 5.

90. r is between -5 and 12.

91. $3x$ is between -3 and 4, including -3 and excluding 4.

92. $6x$ is between -2 and 6, excluding -2 and including 6.

Simplify. Then tell whether the resulting statement is true *or* false. ***See Example 8.***

93. $-6 < 7 + 3$ **94.** $-7 < 4 + 1$ **95.** $2 \cdot 5 \geq 4 + 6$

96. $8 + 7 \leq 3 \cdot 5$ **97.** $-|-3| \geq -3$ **98.** $-|-4| \leq -4$

99. $-8 > -|-6|$ **100.** $-10 > -|-4|$

Write each set in interval notation and graph the interval. ***See Examples 9–11.***

🌐 **101.** $\{x \mid x > -1\}$ **102.** $\{x \mid x < 5\}$ 🌐 **103.** $\{x \mid x \leq 6\}$

104. $\{x \mid x \geq -3\}$ **105.** $\{x \mid 0 < x < 3.5\}$ **106.** $\{x \mid -4 < x < 6.1\}$

107. $\{x \mid 2 \le x \le 7\}$ **108.** $\{x \mid -3 \le x \le -2\}$ 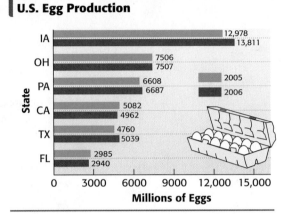 **109.** $\{x \mid -4 < x \le 3\}$

110. $\{x \mid 3 \le x < 6\}$ **111.** $\{x \mid 0 < x \le 3\}$ **112.** $\{x \mid -1 \le x < 6\}$

The graph shows egg production in millions of eggs in selected states for 2005 and 2006. Use this graph to work Exercises 113–116.

113. In 2006, which states had production greater than 6000 million eggs?

114. In which of the states was 2006 egg production less than 2005 egg production?

115. If x represents 2006 egg production for Texas (TX) and y represents 2006 egg production for Ohio (OH), which is true, $x < y$ or $x > y$?

116. If x represents 2005 egg production for Iowa (IA) and y represents 2005 egg production for Pennsylvania (PA), write two inequalities that compare the production in these two states.

U.S. Egg Production

IA 12,978 / 13,811
OH 7506 / 7507
PA 6608 / 6687
CA 5082 / 4962
TX 4760 / 5039
FL 2985 / 2940

State

0   3000   6000   9000   12,000   15,000
Millions of Eggs

2005
2006

Source: U.S. Department of Agriculture.

1.2 Operations on Real Numbers

OBJECTIVES

1 Add real numbers.

2 Subtract real numbers.

3 Find the distance between two points on a number line.

4 Multiply real numbers.

5 Find reciprocals and divide real numbers.

OBJECTIVE 1 **Add real numbers.** Recall that the answer to an addition problem is called the **sum.** The rules for adding real numbers follow.

Adding Real Numbers

Same sign   To add two numbers with the *same* sign, add their absolute values. The sum has the same sign as the given numbers.

Different signs   To add two numbers with *different* signs, find the absolute values of the numbers, and subtract the lesser absolute value from the greater. The sum has the same sign as the number with the greater absolute value.

EXAMPLE 1 Adding Two Negative Real Numbers

Find each sum.

(a) $-12 + (-8)$

First find the absolute values: $|-12| = 12$ and $|-8| = 8$.

Because -12 and -8 have the *same* sign, add their absolute values.

$$-12 + (-8) = -(12 + 8) \quad \text{Add the absolute values.}$$
$$= -(20)$$
$$= -20$$

Both numbers are negative, so the sum will be negative.

(b) $-6 + (-3) = -(|-6| + |-3|) \quad \text{Add the absolute values.}$
$$= -(6 + 3), \quad \text{or} \quad -9$$

NOW TRY
EXERCISE 1

Find each sum.

(a) $-4 + (-9)$

(b) $-7.25 + (-3.57)$

(c) $-\dfrac{2}{5} + \left(-\dfrac{3}{10}\right)$

(c) $-1.2 + (-0.4) = -(1.2 + 0.4),$ or -1.6

(d) $-\dfrac{5}{6} + \left(-\dfrac{1}{3}\right) = -\left(\dfrac{5}{6} + \dfrac{1}{3}\right)$ Add the absolute values. Both numbers are negative, so the sum will be negative.

$\qquad\qquad\quad = -\left(\dfrac{5}{6} + \dfrac{2}{6}\right)$ The least common denominator is 6; $\frac{1\,\cdot\,2}{3\,\cdot\,2} = \frac{2}{6}$

$\qquad\qquad\quad = -\dfrac{7}{6}$ Add numerators.
Keep the same denominator.   *NOW TRY*

NOW TRY
EXERCISE 2

Find each sum.

(a) $-15 + 7$

(b) $4.6 + (-2.8)$

(c) $-\dfrac{5}{9} + \dfrac{2}{7}$

EXAMPLE 2 Adding Real Numbers with Different Signs

Find each sum.

(a) $-17 + 11$

First find the absolute values: $|-17| = 17$ and $|11| = 11$.

Because -17 and 11 have *different* signs, subtract their absolute values.

$$17 - 11 = 6$$

The number -17 has a greater absolute value than 11, so the answer is negative.

$$-17 + 11 = -6 \quad\boxed{\text{The sum is negative because } |-17| > |11|.}$$

(b) $4 + (-1)$

Subtract the absolute values, 4 and 1. Because 4 has the greater absolute value, the sum must be positive.

$$4 + (-1) = 4 - 1 = 3 \quad\boxed{\text{The sum is positive because } |4| > |-1|.}$$

(c) $-9 + 17 = 17 - 9,$ or 8

(d) $-2.3 + 5.6 = 5.6 - 2.3,$ or 3.3

(e) $-16 + 12$

The absolute values are 16 and 12. Subtract the absolute values.

$$-16 + 12 = -(16 - 12) = -4 \quad\boxed{\text{The sum is negative because } |-16| > |12|.}$$

(f) $-\dfrac{4}{5} + \dfrac{2}{3} = -\dfrac{12}{15} + \dfrac{10}{15}$ The least common denominator is 15.
$-\frac{4\,\cdot\,3}{5\,\cdot\,3} = -\frac{12}{15}; \frac{2\,\cdot\,5}{3\,\cdot\,5} = \frac{10}{15}$

$\qquad\qquad = -\left(\dfrac{12}{15} - \dfrac{10}{15}\right)$ Subtract the absolute values. $-\frac{12}{15}$ has the greater absolute value, so the answer will be negative.

$\qquad\qquad = -\dfrac{2}{15}$ Subtract numerators.
Keep the same denominator.   *NOW TRY*

OBJECTIVE 2 **Subtract real numbers.** Recall that the answer to a subtraction problem is called the **difference.** Compare the following two statements.

$$6 - 4 = 2$$

$$6 + (-4) = 2$$

NOW TRY ANSWERS
1. (a) -13 **(b)** -10.82
(c) $-\frac{7}{10}$
2. (a) -8 **(b)** 1.8 **(c)** $-\frac{17}{63}$

Thus, $6 - 4 = 6 + (-4)$. To subtract 4 from 6, we add the additive inverse of 4 to 6.

The preceding example suggests the following definition of subtraction.

> **Subtraction**
>
> For all real numbers a and b,
>
> $$a - b = a + (-b).$$
>
> That is, to subtract b from a, add the additive inverse (or opposite) of b to a.

NOW TRY
EXERCISE 3

Find each difference.

(a) $-4 - 11$

(b) $-5.67 - (-2.34)$

(c) $\dfrac{4}{9} - \dfrac{3}{5}$

EXAMPLE 3 Subtracting Real Numbers

Find each difference.

Change to addition.

The additive inverse of 8 is -8.

(a) $6 - 8 = 6 + (-8) = -2$

Change to addition.

The additive inverse of 4 is -4.

(b) $-12 - 4 = -12 + (-4) = -16$

(c) $-10 - (-7) = -10 + 7$　The additive inverse of -7 is 7.

$= -3$

(d) $-2.4 - (-8.1)$

$= -2.4 + 8.1$

$= 5.7$

(e) $\dfrac{5}{6} - \left(-\dfrac{3}{8}\right) = \dfrac{5}{6} + \dfrac{3}{8}$　To subtract, add the additive inverse (opposite).

$= \dfrac{20}{24} + \dfrac{9}{24}$　Write each fraction with the least common denominator, 24.

$= \dfrac{29}{24}$　Add numerators.

Keep the same denominator.　NOW TRY

When working a problem that involves both addition and subtraction, add and subtract in order from left to right. Work inside brackets or parentheses first.

EXAMPLE 4 Adding and Subtracting Real Numbers

Perform the indicated operations.

(a) $15 - (-3) - 5 - 12$

$= (15 + 3) - 5 - 12$　Work from left to right.

$= 18 - 5 - 12$　Add inside parentheses.

$= 13 - 12$　Subtract from left to right.

$= 1$　Subtract.

NOW TRY ANSWERS

3. (a) -15　**(b)** -3.33　**(c)** $-\frac{7}{45}$

**⤺ NOW TRY
EXERCISE 4**

Perform the indicated operation.

$-4 - (-2 - 7) - 12$

(b) $-9 - [-8 - (-4)] + 6$

$= -9 - [-8 + 4] + 6$ Work inside brackets.

$= -9 - [-4] + 6$ Add.

$= -9 + 4 + 6$ Add the additive inverse.

$= -5 + 6$ Work from left to right.

$= 1$

NOW TRY ⤺

OBJECTIVE 3 **Find the distance between two points on a number line.**
The number line in **FIGURE 16** shows several points.

FIGURE 16

To find the distance between the points 4 and 7, we subtract $7 - 4 = 3$. Since distance is always positive (or 0), we must be careful to subtract in such a way that the answer is positive (or 0). To avoid this problem altogether, we can find the absolute value of the difference. Then the distance between 4 and 7 is found as follows.

$$|7 - 4| = |3| = 3 \quad \text{or} \quad |4 - 7| = |-3| = 3$$

Distance

The **distance** between two points on a number line is the absolute value of the difference between their coordinates.

**⤺ NOW TRY
EXERCISE 5**

Find the distance between the points -7 and 12.

EXAMPLE 5 **Finding Distance Between Points on the Number Line**

Find the distance between each pair of points. Refer to **FIGURE 16**.

(a) 8 and -4

Find the absolute value of the difference of the numbers, taken in either order.

$$|8 - (-4)| = 12, \quad \text{or} \quad |-4 - 8| = 12$$

(b) -4 and -6

$$|-4 - (-6)| = 2, \quad \text{or} \quad |-6 - (-4)| = 2 \quad \text{NOW TRY} ⤺$$

OBJECTIVE 4 **Multiply real numbers.** Recall that the answer to a multiplication problem is called the **product.**

Multiplying Real Numbers

Same sign The product of two numbers with the *same* sign is positive.

Different signs The product of two numbers with *different* signs is negative.

**NOW TRY
EXERCISE 6**

Find each product.

(a) $-3(-10)$

(b) $0.7(-1.2)$

(c) $-\dfrac{8}{11}(33)$

EXAMPLE 6 Multiplying Real Numbers

Find each product.

(a) $-3(-9) = 27$ Same sign; product is positive.

(b) $-0.5(-0.4) = 0.2$

(c) $-\dfrac{3}{4}\left(-\dfrac{5}{6}\right) = \dfrac{15}{24}$ Multiply numerators.
Multiply denominators.

$\qquad\qquad = \dfrac{5 \cdot 3}{8 \cdot 3}$ Factor to write in lowest terms.

$\qquad\qquad = \dfrac{5}{8}$ Divide out the common factor, 3.

(d) $6(-9) = -54$ Different signs; product is negative.

(e) $-0.05(0.3) = -0.015$ **(f)** $-\dfrac{5}{8}\left(\dfrac{12}{13}\right) = -\dfrac{15}{26}$ **(g)** $\dfrac{2}{3}(-6) = -4$

$\boxed{-6 = -\frac{6}{1}}$

NOW TRY

OBJECTIVE 5 **Find reciprocals and divide real numbers.** The definition of division depends on the idea of a **multiplicative inverse,** or *reciprocal.* Two numbers are *reciprocals* if they have a product of 1.

Reciprocal

The **reciprocal** of a nonzero number a is $\dfrac{1}{a}$.

The table gives several numbers and their reciprocals.

Number	Reciprocal
$-\frac{2}{5}$	$-\frac{5}{2}$
-6, or $-\frac{6}{1}$	$-\frac{1}{6}$
$\frac{7}{11}$	$\frac{11}{7}$
0.05	20
0	None

$-\dfrac{2}{5}\left(-\dfrac{5}{2}\right) = 1$
$-6\left(-\dfrac{1}{6}\right) = 1$ Reciprocals have a product of 1.
$\dfrac{7}{11}\left(\dfrac{11}{7}\right) = 1$
$0.05(20) = 1$

There is no reciprocal for 0 because there is no number that can be multiplied by 0 to give a product of 1.

⚠ **CAUTION** *A number and its additive inverse have opposite signs. However, a number and its reciprocal always have the same sign.*

The result of dividing one number by another is called the **quotient.** For example, we can write the quotient of 45 and 3 as $\frac{45}{3}$, which equals 15. The same answer will be obtained if 45 and $\frac{1}{3}$ are multiplied, as follows.

$$45 \div 3 = \frac{45}{3} = 45 \cdot \frac{1}{3} = 15$$

NOW TRY ANSWERS
6. (a) 30   **(b)** -0.84
   (c) -24

This suggests the following definition of division of real numbers.

> ### Division
>
> For all real numbers a and b (where $b \neq 0$),
>
> $$a \div b = \frac{a}{b} = a \cdot \frac{1}{b}.$$
>
> That is, multiply the first number (the **dividend**) by the reciprocal of the second number (the **divisor**).

There is no reciprocal for the number 0, so **division by 0 is undefined.** For example, $\frac{15}{0}$ is undefined and $-\frac{1}{0}$ is undefined.

⚠ **CAUTION**   Division by 0 is undefined. However, dividing 0 by a nonzero number gives the quotient 0. For example,

$$\frac{6}{0} \text{ is undefined,} \quad \text{but} \quad \frac{0}{6} = 0 \quad (\text{since } 0 \cdot 6 = 0).$$

Be careful when 0 is involved in a division problem.

Since division is defined as multiplication by the reciprocal, the rules for signs of quotients are the same as those for signs of products.

> ### Dividing Real Numbers
>
> ***Same sign***   The quotient of two nonzero real numbers with the *same* sign is positive.
>
> ***Different signs***   The quotient of two nonzero real numbers with *different* signs is negative.

EXAMPLE 7   Dividing Real Numbers

Find each quotient.

(a) $\dfrac{-12}{4} = -12 \cdot \dfrac{1}{4} = -3$ $\frac{a}{b} = a \cdot \frac{1}{b}$

(b) $\dfrac{6}{-3} = 6\left(-\dfrac{1}{3}\right) = -2$ The reciprocal of -3 is $-\frac{1}{3}$.

(c) $\dfrac{-\frac{2}{3}}{-\frac{5}{9}} = -\dfrac{2}{3} \cdot \left(-\dfrac{9}{5}\right) = \dfrac{6}{5}$ The reciprocal of $-\frac{5}{9}$ is $-\frac{9}{5}$.

This is a *complex fraction* (**Section 7.3**)—a quotient that has a fraction in the numerator, the denominator, or both.

NOW TRY
EXERCISE 7

Find each quotient.

(a) $\dfrac{-10}{-5}$ (b) $\dfrac{-\dfrac{10}{3}}{\dfrac{3}{8}}$

(c) $-\dfrac{6}{5} \div \left(-\dfrac{3}{7}\right)$

(d) $-\dfrac{9}{14} \div \dfrac{3}{7} = -\dfrac{9}{14} \cdot \dfrac{7}{3}$ Multiply by the reciprocal.

$= -\dfrac{63}{42}$ Multiply numerators and multiply denominators.

$= -\dfrac{7 \cdot 3 \cdot 3}{7 \cdot 3 \cdot 2}$ Factor.

$= -\dfrac{3}{2}$ Lowest terms

NOW TRY

Every fraction has three signs: the sign of the numerator, the sign of the denominator, and the sign of the fraction itself.

Equivalent Forms of a Fraction

The fractions $\dfrac{-x}{y}$, $\dfrac{x}{-y}$, and $-\dfrac{x}{y}$ are equivalent $(y \neq 0)$.

Example: $\dfrac{-4}{7} = \dfrac{4}{-7} = -\dfrac{4}{7}$

The fractions $\dfrac{x}{y}$ and $\dfrac{-x}{-y}$ are equivalent $(y \neq 0)$.

Example: $\dfrac{4}{7} = \dfrac{-4}{-7}$

NOW TRY ANSWERS
7. (a) 2   (b) $-\dfrac{80}{9}$ (c) $\dfrac{14}{5}$

1.2 EXERCISES

MyMathLab   Math XL PRACTICE   WATCH   DOWNLOAD   READ   REVIEW

◉ *Complete solution available on the Video Resources on DVD*

Concept Check   *Complete each statement and give an example.*

1. The sum of a positive number and a negative number is 0 if the numbers are _____.

2. The sum of two positive numbers is a _____ number.

3. The sum of two negative numbers is a _____ number.

4. The sum of a positive number and a negative number is negative if the negative number has the _____ absolute value.

5. The sum of a positive number and a negative number is positive if the positive number has the _____ absolute value.

6. The difference between two positive numbers is negative if _____.

7. The difference between two negative numbers is negative if _____.

8. The product of two numbers with the same sign is _____.

9. The product of two numbers with different signs is _____.

10. The quotient formed by any nonzero number divided by 0 is _____, and the quotient formed by 0 divided by any nonzero number is _____.

Add or subtract as indicated. **See Examples 1–3.**

◉ 11. $-6 + (-13)$ 12. $-8 + (-16)$ ◉ 13. $13 + (-4)$

14. $19 + (-18)$ 15. $-\dfrac{7}{3} + \dfrac{3}{4}$ 16. $-\dfrac{5}{6} + \dfrac{4}{9}$

17. $-2.3 + 0.45$ 18. $-0.238 + 4.58$ ◉ 19. $-6 - 5$

20. $-8 - 17$ 21. $8 - (-13)$ 22. $12 - (-22)$

23. $-16 - (-3)$ 24. $-21 - (-6)$ 25. $-12.31 - (-2.13)$

26. $-15.88 - (-9.42)$ **27.** $\dfrac{9}{10} - \left(-\dfrac{4}{3}\right)$ **28.** $\dfrac{3}{14} - \left(-\dfrac{3}{4}\right)$

29. $|-8 - 6|$ **30.** $|-7 - 15|$ **31.** $-|-4 + 9|$

32. $-|-5 + 6|$ **33.** $-2 - |-4|$ **34.** $16 - |-13|$

Perform the indicated operations. **See Example 4.**

35. $-7 + 5 - 9$ **36.** $-12 + 14 - 18$

37. $6 - (-2) + 8$ **38.** $7 - (-4) + 11$

39. $-9 - 4 - (-3) + 6$ **40.** $-10 - 6 - (-12) + 9$

41. $-8 - (-12) - (2 - 6)$ **42.** $-3 + (-14) + (-6 + 4)$

43. $-0.382 + 4 - 0.6$ **44.** $3 - 2.95 - (-0.63)$

45. $\left(-\dfrac{5}{4} - \dfrac{2}{3}\right) + \dfrac{1}{6}$ **46.** $\left(-\dfrac{5}{8} + \dfrac{1}{4}\right) - \left(-\dfrac{1}{4}\right)$

47. $-\dfrac{3}{4} - \left(\dfrac{1}{2} - \dfrac{3}{8}\right)$ **48.** $\dfrac{7}{5} - \left(\dfrac{9}{10} - \dfrac{3}{2}\right)$

49. $|-11| - |-5| - |7| + |-2|$ **50.** $|-6| + |-3| - |4| - |-8|$

The number line has several points labeled. Find the distance between each pair of points. **See Example 5.**

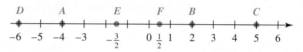

51. A and B **52.** A and C **53.** D and F **54.** E and C

55. A statement that is often heard is "Two negatives give a positive." When is this true? When is it not true? Give a more precise statement that conveys this message.

56. Explain why the reciprocal of a nonzero number must have the same sign as the number.

Multiply. **See Example 6.**

57. $5(-7)$ **58.** $6(-9)$ **59.** $-8(-5)$ **60.** $-20(-4)$

61. $-10\left(-\dfrac{1}{5}\right)$ **62.** $-\dfrac{1}{2}(-18)$ **63.** $\dfrac{3}{4}(-16)$ **64.** $\dfrac{3}{5}(-35)$

65. $-\dfrac{5}{2}\left(-\dfrac{12}{25}\right)$ **66.** $-\dfrac{9}{7}\left(-\dfrac{21}{36}\right)$ **67.** $-\dfrac{3}{8}\left(-\dfrac{24}{9}\right)$ **68.** $-\dfrac{2}{11}\left(-\dfrac{77}{4}\right)$

69. $-2.4(-2.45)$ **70.** $-3.45(-5.14)$ **71.** $3.4(-3.14)$ **72.** $5.66(-6.1)$

Divide where possible. **See Example 7.**

73. $\dfrac{-14}{2}$ **74.** $\dfrac{-39}{13}$ **75.** $\dfrac{-24}{-4}$ **76.** $\dfrac{-45}{-9}$ **77.** $\dfrac{100}{-25}$

78. $\dfrac{150}{-30}$ **79.** $\dfrac{0}{-8}$ **80.** $\dfrac{0}{-14}$ **81.** $\dfrac{5}{0}$ **82.** $\dfrac{13}{0}$

83. $-\dfrac{10}{17} \div \left(-\dfrac{12}{5}\right)$ **84.** $-\dfrac{22}{23} \div \left(-\dfrac{33}{5}\right)$ **85.** $\dfrac{\frac{12}{13}}{-\frac{4}{3}}$ **86.** $\dfrac{\frac{7}{6}}{-\frac{1}{30}}$

87. $\dfrac{-27.72}{13.2}$ **88.** $\dfrac{-162.9}{36.2}$ **89.** $\dfrac{-100}{-0.01}$ **90.** $\dfrac{-60}{-0.06}$

Exercises 91–116 provide more practice on operations with fractions and decimals. Perform the indicated operations.

91. $\dfrac{1}{6} - \left(-\dfrac{7}{9}\right)$

92. $\dfrac{7}{10} - \left(-\dfrac{1}{6}\right)$

93. $-\dfrac{1}{9} + \dfrac{7}{12}$

94. $-\dfrac{1}{12} + \dfrac{13}{16}$

95. $-\dfrac{3}{8} - \dfrac{5}{12}$

96. $-\dfrac{11}{15} - \dfrac{4}{9}$

97. $-\dfrac{7}{30} + \dfrac{2}{45} - \dfrac{3}{10}$

98. $-\dfrac{8}{15} - \dfrac{3}{20} + \dfrac{7}{6}$

99. $\dfrac{8}{25}\left(-\dfrac{5}{12}\right)$

100. $\dfrac{9}{20}\left(-\dfrac{7}{15}\right)$

101. $\dfrac{5}{6}\left(-\dfrac{9}{10}\right)\left(-\dfrac{4}{5}\right)$

102. $\dfrac{4}{3}\left(-\dfrac{9}{20}\right)\left(-\dfrac{5}{12}\right)$

103. $\dfrac{7}{6} \div \left(-\dfrac{9}{10}\right)$

104. $\dfrac{12}{5} \div \left(-\dfrac{18}{25}\right)$

105. $\dfrac{-\dfrac{8}{9}}{2}$

106. $\dfrac{-\dfrac{15}{16}}{5}$

107. $-8.6 - 3.751$

108. $-37.8 - 13.582$

109. $(-4.2)(1.4)(2.7)$

110. $(2.9)(-10.3)(0.04)$

111. $-24.84 \div 6$

112. $-32.84 \div 8$

113. $-2496 \div (-0.52)$

114. $-161.7 \div (-0.25)$

115. $-14.23 + 9.81 + 74.63 - 18.715$

116. $-89.416 + 21.32 - 478.91 + 298.212$

Solve each problem.

117. The highest temperature ever recorded in Juneau, Alaska, was 90°F. The lowest temperature ever recorded there was −22°F. What is the difference between these two temperatures? (*Source: World Almanac and Book of Facts.*)

118. On August 10, 1936, a temperature of 120°F was recorded in Ponds, Arkansas. On February 13, 1905, Ozark, Arkansas, recorded a temperature of −29°F. What is the difference between these two temperatures? (*Source: World Almanac and Book of Facts.*)

119. Andrew McGinnis has $48.35 in his checking account. He uses his debit card to make purchases of $35.99 and $20.00, which overdraws his account. His bank charges his account an overdraft fee of $28.50. He then deposits his paycheck for $66.27 from his part-time job at Arby's. What is the balance in his account?

120. Kayla Koolbeck has $37.60 in her checking account. She uses her debit card to make purchases of $25.99 and $19.34, which overdraws her account. Her bank charges her account an overdraft fee of $25.00. She then deposits her paycheck for $58.66 from her part-time job at Subway. What is the balance in her account?

121. Andrew Jauch owes $382.45 on his Visa account. He returns two items costing $25.10 and $34.50 for credit. Then he makes purchases of $45.00 and $98.17.

(a) How much should his payment be if he wants to pay off the balance on the account?

(b) Instead of paying off the balance, he makes a payment of $300 and then incurs a finance charge of $24.66. What is the balance on his account?

122. Charlene Macdowall owes $237.59 on her MasterCard account. She returns one item costing $47.25 for credit and then makes two purchases of $12.39 and $20.00.

(a) How much should her payment be if she wants to pay off the balance on the account?

(b) Instead of paying off the balance, she makes a payment of $75.00 and incurs a finance charge of $32.06. What is the balance on her account?

123. The graph shows profits and losses in thousands of dollars for a private company for the years 2007 through 2010.

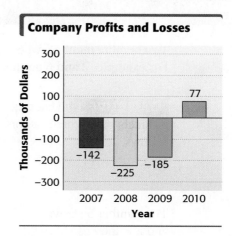

Company Profits and Losses

(a) What was the total profit or loss for the years 2007 through 2010?

(b) Find the difference between the profit or loss in 2010 and that in 2009.

(c) Find the difference between the profit or loss in 2008 and that in 2007.

124. The graph shows annual returns in percent for Class A shares of the AIM Charter Fund for the years 2002 through 2008.

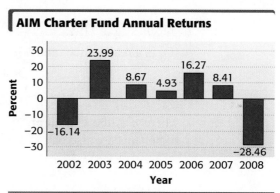

AIM Charter Fund Annual Returns

Source: Invesco Aim.

(a) Find the sum of the percents for the years shown in the graph.

(b) Find the difference between the returns in 2003 and 2002.

(c) Find the difference between the returns in 2008 and 2007.

125. The table shows Social Security finances (in billions of dollars).

Year	Tax Revenue	Cost of Benefits
2000	538	409
2010*	916	710
2020*	1479	1405
2030*	2041	2542

*Projected
Source: Social Security Board of Trustees.

(a) Find the difference between Social Security tax revenue and cost of benefits for each year shown in the table.

(b) Interpret your answer for 2030.

1.3

Exponents, Roots, and Order of Operations

Two or more numbers whose product is a third number are **factors** of that third number. For example, 2 and 6 are factors of 12, since $2 \cdot 6 = 12$.

OBJECTIVE 1   **Use exponents.**   In algebra, we use *exponents* as a way of writing products of repeated factors. For example, the product $2 \cdot 2 \cdot 2 \cdot 2 \cdot 2$ is written as follows.

$$\underbrace{2 \cdot 2 \cdot 2 \cdot 2 \cdot 2}_{5 \text{ factors of } 2} = 2^5$$

The number 5 shows that 2 is used as a factor 5 times. The number 5 is the *exponent*, and 2 is the *base*.

$$2^5 \longleftarrow \text{Exponent}$$
$$\uparrow$$
$$\text{Base}$$

Read 2^5 as "2 to the fifth power," or "2 to the fifth." Multiplying five 2s gives 32.

$$2^5 = 2 \cdot 2 \cdot 2 \cdot 2 \cdot 2 = 32$$

Exponential Expression

If a is a real number and n is a natural number, then

$$a^n = \underbrace{a \cdot a \cdot a \cdot \ldots \cdot a,}_{n \text{ factors of } a}$$

where n is the **exponent,** a is the **base,** and a^n is an **exponential expression.** Exponents are also called **powers.**

NOW TRY EXERCISE 1

Write using exponents.

(a) $(-3)(-3)(-3)$

(b) $t \cdot t \cdot t \cdot t \cdot t$

EXAMPLE 1   Using Exponential Notation

Write using exponents.

(a) $\underbrace{4 \cdot 4 \cdot 4}_{3 \text{ factors of } 4} = 4^3$

Read 4^3 as "4 **cubed.**"

(b) $\dfrac{3}{5} \cdot \dfrac{3}{5} = \left(\dfrac{3}{5}\right)^2$ 2 factors of $\frac{3}{5}$

Read $\left(\frac{3}{5}\right)^2$ as "$\frac{3}{5}$ **squared.**"

(c) $(-6)(-6)(-6)(-6) = (-6)^4$

Read $(-6)^4$ as "−6 to the fourth power," or "−6 to the fourth."

(d) $(0.3)(0.3)(0.3)(0.3)(0.3) = (0.3)^5$ **(e)** $x \cdot x \cdot x \cdot x \cdot x \cdot x = x^6$

NOW TRY

NOTE In **Example 1,** we used the terms *squared* and *cubed* to refer to powers of 2 and 3, respectively. The term *squared* comes from the figure of a square, which has the same measure for both length and width, as shown in **FIGURE 17(a)**. Similarly, the term *cubed* comes from the figure of a cube, where the length, width, and height have the same measure, as shown in **FIGURE 17(b)**.

NOW TRY ANSWERS
1. (a) $(-3)^3$ (b) t^5

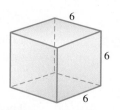

(a) 3 · 3 = 3 squared, or 3^2 **(b)** 6 · 6 · 6 = 6 cubed, or 6^3

FIGURE 17

NOW TRY
EXERCISE 2

Evaluate.

(a) 7^2 **(b)** $(-7)^2$

(c) -7^2

EXAMPLE 2 Evaluating Exponential Expressions

Evaluate.

(a) $5^2 = 5 \cdot 5 = 25$ 5 is used as a factor 2 times.

> 5^2 means 5 · 5, NOT 5 · 2.

(b) $\left(\dfrac{2}{3}\right)^3 = \dfrac{2}{3} \cdot \dfrac{2}{3} \cdot \dfrac{2}{3} = \dfrac{8}{27}$ $\frac{2}{3}$ is used as a factor 3 times.

(c) $2^6 = 2 \cdot 2 \cdot 2 \cdot 2 \cdot 2 \cdot 2 = 64$

(d) $(-3)^5 = (-3)(-3)(-3)(-3)(-3) = -243$ The base is -3.

(e) $(-2)^6 = (-2)(-2)(-2)(-2)(-2)(-2) = 64$ The base is -2.

(f) -2^6

There are no parentheses. The exponent 6 applies *only* to the number 2, not to -2.

$$-2^6 = -(2 \cdot 2 \cdot 2 \cdot 2 \cdot 2 \cdot 2) = -64$$ The base is 2.

NOW TRY

Examples **2(d)** and **(e)** suggest the following generalizations.

Sign of an Exponential Expression

The product of an *odd* number of negative factors is negative.

The product of an *even* number of negative factors is positive.

⚠ **CAUTION** As shown in **Examples 2(e)** and **(f),** it is important to distinguish between $-a^n$ and $(-a)^n$.

$-a^n$ means $-1\underbrace{(a \cdot a \cdot a \cdot \ldots \cdot a)}_{n \text{ factors of } a}$ The base is a.

$(-a)^n$ means $\underbrace{(-a)(-a) \cdot \ldots \cdot (-a)}_{n \text{ factors of } -a}$ The base is $-a$.

Be careful when evaluating an exponential expression with a negative sign.

OBJECTIVE 2 **Find square roots.** As we saw in **Example 2(a)**, $5^2 = 5 \cdot 5 = 25$, so 5 squared is 25. The opposite (inverse) of squaring a number is called taking its **square root.** For example, a square root of 25 is 5. Another square root of 25 is -5, since $(-5)^2 = 25$. Thus, 25 has two square roots: 5 and -5.

NOW TRY ANSWERS

2. (a) 49 **(b)** 49 **(c)** -49

We write the **positive** or **principal square root** of a number with the symbol $\sqrt{\ }$, called a **radical symbol.** For example, the positive or principal square root of 25 is written

$$\sqrt{25} = 5.$$

The **negative square root** of 25 is written

$$-\sqrt{25} = -5.$$

Since the square of any nonzero real number is positive, the square root of a negative number, such as $\sqrt{-25}$, is not a real number.

⌐ NOW TRY
⌐ EXERCISE 3
Find each square root that is a real number.

(a) $-\sqrt{144}$ **(b)** $\sqrt{\dfrac{100}{9}}$

(c) $\sqrt{-144}$

EXAMPLE 3 Finding Square Roots

Find each square root that is a real number.

(a) $\sqrt{36} = 6$, since 6 is positive and $6^2 = 36$.

(b) $\sqrt{0} = 0$, since $0^2 = 0$.

(c) $\sqrt{\dfrac{9}{16}} = \dfrac{3}{4}$, since $\left(\dfrac{3}{4}\right)^2 = \dfrac{9}{16}$.

(d) $\sqrt{0.16} = 0.4$, since $(0.4)^2 = 0.16$.

(e) $\sqrt{100} = 10$, since $10^2 = 100$.

(f) $-\sqrt{100} = -10$, since the negative sign is outside the radical symbol.

(g) $\sqrt{-100}$ is not a real number, because the negative sign is inside the radical symbol. No *real number* squared equals -100.

Notice the difference among the square roots in parts (e), (f), and (g). Part (e) is the positive or principal square root of 100, part (f) is the negative square root of 100, and part (g) is the square root of -100, which is not a real number. NOW TRY ⟳

⚠ **CAUTION** The symbol $\sqrt{\ }$ is used only for the *positive* square root, except that $\sqrt{0} = 0$. The symbol $-\sqrt{\ }$ is used for the negative square root.

OBJECTIVE 3 Use the order of operations. To simplify

$$5 + 2 \cdot 3,$$

what should we do first—add 5 and 2 or multiply 2 and 3? When an expression involves more than one operation symbol, we use the following **order of operations.**

Order of Operations

1. Work separately above and below any **fraction bar.**

2. If **grouping symbols** such as **parentheses ( )**, **brackets [ ]**, or **absolute value bars | |** are present, start with the innermost set and work outward.

3. Evaluate all **powers, roots,** and **absolute values.**

4. **Multiply** or **divide** in order from left to right.

5. **Add** or **subtract** in order from left to right.

NOW TRY ANSWERS
3. **(a)** -12 **(b)** $\frac{10}{3}$
   (c) not a real number

NOTE Some students like to use the mnemonic "Please Excuse My Dear Aunt Sally" to help remember the rules for order of operations.

Please	Excuse	My	Dear	Aunt	Sally
↑	↑	↑	↑	↑	↑
Parentheses	Exponents	Multiply	Divide	Add	Subtract

Be sure to multiply or divide in order from left to right. Then add or subtract in order from left to right.

**NOW TRY
EXERCISE 4**

Simplify.

$15 - 3 \cdot 4 + 2$

EXAMPLE 4 Using the Order of Operations

Simplify.

(a) $5 + 2 \cdot 3$

$\quad = 5 + 6 \qquad$ Multiply.

$\quad = 11 \qquad$ Add.

(b) $24 \div 3 \cdot 2 + 6$

$\quad = 8 \cdot 2 + 6 \qquad$ Divide.

$\quad = 16 + 6 \qquad$ Multiply.

$\quad = 22 \qquad$ Add.

Multiplications and divisions are done in the order in which they appear from left to right. So we divide first here.

NOW TRY

**NOW TRY
EXERCISE 5**

Simplify.

$-5^2 + 10 \div 5 - |3 - 7|$

EXAMPLE 5 Using the Order of Operations

Simplify.

(a) $10 \div 5 + 2|3 - 4| \qquad$ Work inside the absolute value bars first.

$\quad = 10 \div 5 + 2|-1| \qquad$ Subtract inside the absolute value bars.

$\quad = 10 \div 5 + 2 \cdot 1 \qquad$ Take the absolute value.

$\quad = 2 + 2 \qquad$ Divide first, and then multiply.

$\quad = 4 \qquad$ Add.

(b) $\qquad 4 \cdot 3^2 + 7 - (2 + 8) \qquad$ Work inside the parentheses first.

$\quad = 4 \cdot 3^2 + 7 - 10 \qquad$ Add inside the parentheses.

$\quad = 4 \cdot 9 + 7 - 10 \qquad$ Evaluate the power.

3^2 means $3 \cdot 3$, NOT $3 \cdot 2$.

$\quad = 36 + 7 - 10 \qquad$ Multiply.

$\quad = 43 - 10 \qquad$ Add.

$\quad = 33 \qquad$ Subtract.

(c) $\dfrac{1}{2} \cdot 4 + (6 \div 3 - 7) \qquad$ Work inside the parentheses first.

$\quad = \dfrac{1}{2} \cdot 4 + (2 - 7) \qquad$ Divide inside the parentheses.

$\quad = \dfrac{1}{2} \cdot 4 + (-5) \qquad$ Subtract inside the parentheses.

$\quad = 2 + (-5) \qquad$ Multiply.

$\quad = -3 \qquad$ Add.

NOW TRY

NOW TRY ANSWERS

4. 5   **5.** -27

NOW TRY
EXERCISE 6

Simplify.

$$\frac{\sqrt{36} - 4 \cdot 3^2}{-2^2 - 8 \cdot 3 + 13}$$

EXAMPLE 6   Using the Order of Operations

Simplify.

$$\frac{5 + (-2^3)(2)}{6 \cdot \sqrt{9} - 9 \cdot 4}$$ Work separately above and below the fraction bar.

$$= \frac{5 + (-8)(2)}{6 \cdot 3 - 9 \cdot 4}$$ Evaluate the power and the root.

$$= \frac{5 - 16}{18 - 36}$$ Multiply.

$$= \frac{-11}{-18}, \quad \text{or} \quad \frac{11}{18}$$ Subtract; $\frac{-a}{-b} = \frac{a}{b}$ NOW TRY

OBJECTIVE 4   Evaluate algebraic expressions for given values of variables.

Any sequence of numbers, variables, operation symbols, and/or grouping symbols formed in accordance with the rules of algebra is called an **algebraic expression.**

$$6ab, \qquad 5m - 9n, \qquad \text{and} \qquad -2(x^2 + 4y) \qquad \text{Algebraic expressions}$$

Algebraic expressions have different numerical values for different values of the variables. We evaluate such expressions by *substituting* given values for the variables.

For example, if movie tickets cost $9 each, the amount in dollars you pay for x tickets can be represented by the algebraic expression $9x$. We can substitute different numbers of tickets to get the costs of purchasing those tickets.

NOW TRY
EXERCISE 7

Evaluate the expression for $x = -4$, $y = 7$, and $z = 36$.

$$\frac{x^2 - \sqrt{z}}{-3xy}$$

EXAMPLE 7   Evaluating Algebraic Expressions

Evaluate each expression for $m = -4$, $n = 5$, $p = -6$, and $q = 25$.

(a) $5m - 9n$

> Use parentheses around substituted values to avoid errors.

$$= 5(-4) - 9(5)$$ Substitute $m = -4$ and $n = 5$.

$$= -20 - 45$$ Multiply.

$$= -65$$ Subtract.

(b) $\dfrac{m + 2n}{4p}$

$$= \frac{-4 + 2(5)}{4(-6)}$$ Substitute $m = -4$, $n = 5$, and $p = -6$.

$$= \frac{-4 + 10}{-24}$$ Work separately above and below the fraction bar.

$$= \frac{6}{-24}, \quad \text{or} \quad -\frac{1}{4}$$ Write in lowest terms; also, $\frac{a}{-b} = -\frac{a}{b}$.

(c) $-3m^3 - n^2\left(\sqrt{q}\right)$

$$= -3(-4)^3 - (5)^2\left(\sqrt{25}\right)$$ Substitute $m = -4$, $n = 5$, and $q = 25$.

$$= -3(-64) - 25(5)$$ Evaluate the powers and the root.

$$= 192 - 125$$ Multiply.

$$= 67$$ Subtract.   NOW TRY

NOW TRY ANSWERS
6. 2   7. $\frac{5}{42}$

1.3 EXERCISES

MyMathLab | Math XL PRACTICE | WATCH | DOWNLOAD | READ | REVIEW

🔘 *Complete solution available on the Video Resources on DVD*

Concept Check *Decide whether each statement is* true *or* false. *If it is false, correct the statement so that it is true.*

1. $-7^6 = (-7)^6$

2. $-5^7 = (-5)^7$

3. $\sqrt{25}$ is a positive number.

4. $3 + 5 \cdot 8 = 3 + (5 \cdot 8)$

5. $(-6)^7$ is a negative number.

6. $(-6)^8$ is a positive number.

7. The product of 10 positive factors and 10 negative factors is positive.

8. The product of 5 positive factors and 5 negative factors is positive.

9. In the exponential expression -8^5, -8 is the base.

10. $\sqrt{a}$ is positive for all positive numbers a.

Concept Check *In Exercises 11 and 12, evaluate each exponential expression.*

11. (a) 8^2 **(b)** -8^2

(c) $(-8)^2$ **(d)** $-(-8)^2$

12. (a) 4^3 **(b)** -4^3

(c) $(-4)^3$ **(d)** $-(-4)^3$

Write each expression by using exponents. ***See Example 1.***

🔘 **13.** $10 \cdot 10 \cdot 10 \cdot 10$

14. $8 \cdot 8 \cdot 8$

🔘 **15.** $\dfrac{3}{4} \cdot \dfrac{3}{4} \cdot \dfrac{3}{4} \cdot \dfrac{3}{4} \cdot \dfrac{3}{4}$

16. $\dfrac{3}{2} \cdot \dfrac{3}{2}$

🔘 **17.** $(-9)(-9)(-9)$

18. $(-9)(-9)(-9)(-9)$

🔘 **19.** $z \cdot z \cdot z \cdot z \cdot z \cdot z \cdot z$

20. $a \cdot a \cdot a \cdot a \cdot a$

Evaluate each expression. ***See Example 2.***

🔘 **21.** 4^2

22. 2^4

23. 0.28^3

24. 0.91^3

25. $\left(\dfrac{1}{5}\right)^3$

26. $\left(\dfrac{1}{6}\right)^4$

🔘 **27.** $\left(\dfrac{4}{5}\right)^4$

28. $\left(\dfrac{7}{10}\right)^3$

🔘 **29.** $(-5)^3$

30. $(-2)^5$

🔘 **31.** $(-2)^8$

32. $(-3)^6$

33. -3^6

34. -4^6

35. -8^4

36. -10^3

Find each square root. If it is not a real number, say so. ***See Example 3.***

🔘 **37.** $\sqrt{81}$

38. $\sqrt{64}$

39. $\sqrt{169}$

40. $\sqrt{225}$

🔘 **41.** $-\sqrt{400}$

42. $-\sqrt{900}$

🔘 **43.** $\sqrt{\dfrac{100}{121}}$

44. $\sqrt{\dfrac{225}{169}}$

45. $-\sqrt{0.49}$

46. $-\sqrt{0.64}$

🔘 **47.** $\sqrt{-36}$

48. $\sqrt{-121}$

49. *Concept Check* Match each square root with the appropriate value or description.

(a) $\sqrt{144}$ **(b)** $\sqrt{-144}$ **(c)** $-\sqrt{144}$

A. -12 **B.** 12   **C.** Not a real number

Concept Check *In Exercises 50 and 51, x represents a positive number.*

50. Is $-\sqrt{-x}$ positive, negative, or not a real number?

51. Is $-\sqrt{x}$ positive, negative, or not a real number?

52. *Concept Check* Frank Capek's grandson was asked to evaluate the following expression.

$$9 + 15 \div 3$$

Frank gave the answer as 8, but his grandson gave the answer as 14. The grandson explained that the answer is 14 because of the "Order of Process rule" which says that in a problem like this, you proceed from right to left rather than left to right. (*Note:* This is a true story.)

(a) Whose answer was correct for this expression, Frank's or his grandson's?

(b) Was the *reasoning* for the correct answer valid? Explain.

Simplify each expression. Use the order of operations. ***See Examples 4–6.***

53. $12 + 3 \cdot 4$

54. $15 + 5 \cdot 2$

55. $6 \cdot 3 - 12 \div 4$

56. $9 \cdot 4 - 8 \div 2$

57. $10 + 30 \div 2 \cdot 3$

58. $12 + 24 \div 3 \cdot 2$

59. $-3(5)^2 - (-2)(-8)$

60. $-9(2)^2 - (-3)(-2)$

61. $5 - 7 \cdot 3 - (-2)^3$

62. $-4 - 3 \cdot 5 + 6^2$

63. $-7\left(\sqrt{36}\right) - (-2)(-3)$

64. $-8\left(\sqrt{64}\right) - (-3)(-7)$

65. $6|4 - 5| - 24 \div 3$

66. $-4|2 - 4| + 8 \cdot 2$

67. $|-6 - 5|(-8) + 3^2$

68. $(-6 - 3)|-2 - 3| \div 9$

69. $6 + \dfrac{2}{3}(-9) - \dfrac{5}{8} \cdot 16$

70. $7 - \dfrac{3}{4}(-8) + 12 \cdot \dfrac{5}{6}$

71. $-14\left(-\dfrac{2}{7}\right) \div (2 \cdot 6 - 10)$

72. $-12\left(-\dfrac{3}{4}\right) - (6 \cdot 5 \div 3)$

73. $\dfrac{\left(-5 + \sqrt{4}\right)(-2^2)}{-5 - 1}$

74. $\dfrac{\left(-9 + \sqrt{16}\right)(-3^2)}{-4 - 1}$

75. $\dfrac{2(-5) + (-3)(-2)}{-8 + 3^2 - 1}$

76. $\dfrac{3(-4) + (-5)(-8)}{2^3 - 2 - 6}$

77. $\dfrac{5 - 3\left(\dfrac{-5 - 9}{-7}\right) - 6}{-9 - 11 + 3 \cdot 7}$

78. $\dfrac{-4\left(\dfrac{12 - (-8)}{3 \cdot 2 + 4}\right) - 5(-1 - 7)}{-9 - (-7) - [-5 - (-8)]}$

Evaluate each expression for $a = -3$, $b = 64$, and $c = 6$. ***See Example 7.***

79. $3a + \sqrt{b}$

80. $-2a - \sqrt{b}$

81. $\sqrt{b} + c - a$

82. $\sqrt{b} - c + a$

83. $4a^3 + 2c$

84. $-3a^4 - 3c$

85. $\dfrac{2c + a^3}{4b + 6a}$

86. $\dfrac{3c + a^2}{2b - 6c}$

Evaluate each expression for $w = 4$, $x = -\frac{3}{4}$, $y = \frac{1}{2}$, and $z = 1.25$. ***See Example 7.***

87. $wy - 8x$

88. $wz - 12y$

89. $xy + y^4$

90. $xy - x^2$

91. $-w + 2x + 3y + z$

92. $w - 6x + 5y - 3z$

93. $\dfrac{7x + 9y}{w}$

94. $\dfrac{7y - 5x}{2w}$

Solve each problem.

Residents of Linn County, Iowa, in the Cedar Rapids Community School District can use the expression

$$(v \times 0.5485 - 4850) \div 1000 \times 31.44$$

to determine their property taxes, where v is assessed home value. (*Source: The Gazette.*) Use the expression to calculate the amount of property taxes to the nearest dollar that the owner of a home with each of the following values would pay. Follow the order of operations.

95. $100,000   **96.** $150,000   **97.** $200,000

The Blood Alcohol Concentration (BAC) of a person who has been drinking is given by the expression

number of oz $\times$ % alcohol $\times$ 0.075 $\div$ body weight in lb $-$ hr of drinking $\times$ 0.015.

(*Source:* Lawlor, J., *Auto Math Handbook: Mathematical Calculations, Theory, and Formulas for Automotive Enthusiasts,* HP Books.)

98. Suppose a policeman stops a 190-lb man who, in 2 hr, has ingested four 12-oz beers (48 oz), each having a 3.2% alcohol content.

    (a) Substitute the values into the formula, and write the expression for the man's BAC.

    (b) Calculate the man's BAC to the nearest thousandth. Follow the order of operations.

99. Find the BAC to the nearest thousandth for a 135-lb woman who, in 3 hr, has drunk three 12-oz beers (36 oz), each having a 4.0% alcohol content.

100. (a) Calculate the BACs in **Exercises 98 and 99** if each person weighs 25 lb more and the rest of the variables stay the same. How does increased weight affect a person's BAC?

    (b) Predict how decreased weight would affect the BAC of each person in **Exercises 98 and 99.** Calculate the BACs if each person weighs 25 lb less and the rest of the variables stay the same.

101. An approximation of the amount in billions of dollars that Americans have spent on their pets from 1996 to 2008 can be obtained by substituting a given year for x in the expression

$$1.909x - 3791.$$

(*Source:* American Pet Products Manufacturers Association.) Approximate the amount spent in each year. Round answers to the nearest tenth.

    (a) 1996   **(b)** 2002   **(c)** 2008

    (d) How has the amount Americans have spent on their pets changed from 1996 to 2008?

102. An approximation of the average price of a theater ticket in the United States from 1977 through 2007 can be obtained by using the expression.

$$0.1399x - 274.4,$$

where x represents the year. (*Source:* National Association of Theater Owners.)

Year	Average Price (in dollars)
1977	
1987	3.58
1997	
2007	

    (a) Use the expression to complete the table. Round answers to the nearest cent.

    (b) How has the average price of a theater ticket in the United States changed from 1977 to 2007?

1.4 Properties of Real Numbers

OBJECTIVES

1 Use the distributive property.

2 Use the identity properties.

3 Use the inverse properties.

4 Use the commutative and associative properties.

5 Use the multiplication property of 0.

The basic properties of real numbers studied in this section reflect results that occur consistently in work with numbers. They have been generalized to apply to expressions with variables as well.

OBJECTIVE 1 Use the distributive property. Notice that

$$2(3 + 5) = 2 \cdot 8 = 16$$

and

$$2 \cdot 3 + 2 \cdot 5 = 6 + 10 = 16,$$

so

$$2(3 + 5) = 2 \cdot 3 + 2 \cdot 5.$$

This idea is illustrated by the divided rectangle in **FIGURE 18**. Similarly,

$$-4[5 + (-3)] = -4(2) = -8$$

and

$$-4(5) + (-4)(-3) = -20 + 12 = -8,$$

so

$$-4[5 + (-3)] = -4(5) + (-4)(-3).$$

These examples are generalized to *all* real numbers as the **distributive property of multiplication with respect to addition,** or simply the **distributive property.**

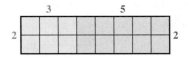

Area of left part is $2 \cdot 3 = 6$.
Area of right part is $2 \cdot 5 = 10$.
Area of total rectangle is $2(3 + 5) = 16$.

FIGURE 18

Distributive Property

For any real numbers a, b, and c, the following are true.

$$a(b + c) = ab + ac \qquad \text{and} \qquad (b + c)a = ba + ca$$

The distributive property can also be written in "reverse" as

$$ab + ac = a(b + c) \qquad \text{and} \qquad ba + ca = (b + c)a.$$

It can be extended to more than two numbers as well.

$$a(b + c + d) = ab + ac + ad$$

The distributive property provides a way to rewrite a product $a(b + c)$ as a sum $ab + ac$. It is also used to write a sum as a product.

NOTE When we rewrite $a(b + c)$ as $ab + ac$, we sometimes refer to the process as "removing" or "clearing" parentheses.

EXAMPLE 1 Using the Distributive Property

Use the distributive property to rewrite each expression.

(a) $3(x + y)$ 　　　　 Use the first form of the property to rewrite the given product as a sum.
$= 3x + 3y$

(b) $-2(5 + k)$

$= -2(5) + (-2)(k)$

$= -10 - 2k$

(c) $4x + 8x$ 　　　　 Use the distributive property in reverse to rewrite the given sum as a product.
$= (4 + 8)x$

$= 12x$

**NOW TRY
EXERCISE 1**
Use the distributive property
to rewrite each expression.
(a) $-2(3x - y)$
(b) $4k - 12k$

(d) $3r - 7r$

$$= 3r + (-7r) \qquad \text{Definition of subtraction}$$

$$= [3 + (-7)]r \qquad \text{Distributive property}$$

$$= -4r$$

(e) $5p + 7q$

Because there is no common number or variable here, we cannot use the distributive property to rewrite the expression.

(f) $6(x + 2y - 3z)$

$$= 6x + 6(2y) + 6(-3z)$$

$$= 6x + 12y - 18z$$

NOW TRY

The distributive property can also be used for subtraction **(Example 1(d))**, so

$$a(b - c) = ab - ac.$$

OBJECTIVE 2 **Use the identity properties.** The number 0 is the only number that can be added to any number to get that number, leaving the identity of the number unchanged. For this reason, 0 is called the **identity element for addition,** or the **additive identity.**

In a similar way, multiplying any number by 1 leaves the identity of the number unchanged. Thus, 1 is the **identity element for multiplication**, or the **multiplicative identity.**

The **identity properties** summarize this discussion and extend these properties from arithmetic to algebra.

Identity Properties

For any real number a, the following are true.

$$a + 0 = 0 + a = a$$

$$a \cdot 1 = 1 \cdot a = a$$

The identity properties leave the identity of a real number unchanged. Think of a child wearing a costume on Halloween. The child's appearance is changed, but his or her identity is unchanged.

EXAMPLE 2 Using the Identity Property $1 \cdot a = a$

Simplify each expression.

(a) $12m + m$

$$= 12m + 1m \qquad \text{Identity property; } m = 1 \cdot m, \text{ or } 1m$$

$$= (12 + 1)m \qquad \text{Distributive property}$$

$$= 13m \qquad \text{Add inside parentheses.}$$

(b) $y + y$

$$= 1y + 1y \qquad \text{Identity property}$$

$$= (1 + 1)y \qquad \text{Distributive property}$$

$$= 2y \qquad \text{Add inside parentheses.}$$

NOW TRY ANSWERS
1. (a) $-6x + 2y$ **(b)** $-8k$

NOW TRY
EXERCISE 2
Simplify each expression.

(a) $7x + x$ **(b)** $-(5p - 3q)$

(c)
$$-(m - 5n)$$
$$= -1(m - 5n) \qquad \text{Identity property}$$
$$= -1(m) + (-1)(-5n) \qquad \text{Distributive property}$$
$$= -m + 5n \qquad \text{Multiply.} \qquad \textit{NOW TRY}$$

> Multiply *each* term by -1. Be careful with signs.

OBJECTIVE 3 **Use the inverse properties.** The *additive inverse* (or *opposite*) of a number a is $-a$. Additive inverses have a sum of 0 (the additive identity).

$$5 \quad \text{and} \quad -5, \qquad -\frac{1}{2} \quad \text{and} \quad \frac{1}{2}, \qquad -34 \quad \text{and} \quad 34 \qquad \begin{array}{l}\text{Additive inverses}\\\text{(sum of 0)}\end{array}$$

The *multiplicative inverse (or reciprocal)* of a number a is $\frac{1}{a}$ (where $a \neq 0$), and multiplicative inverses have a product of 1 (the multiplicative identity).

$$5 \quad \text{and} \quad \frac{1}{5}, \qquad -\frac{1}{2} \quad \text{and} \quad -2, \qquad \frac{3}{4} \quad \text{and} \quad \frac{4}{3} \qquad \begin{array}{l}\text{Multiplicative inverses}\\\text{(product of 1)}\end{array}$$

This discussion leads to the **inverse properties** of addition and multiplication.

Inverse Properties

For any real number a, the following are true.

$$a + (-a) = 0 \qquad \text{and} \qquad -a + a = 0$$
$$a \cdot \frac{1}{a} = 1 \qquad \text{and} \qquad \frac{1}{a} \cdot a = 1 \quad (a \neq 0)$$

Term	Numerical Coefficient
$-7y$	-7
$34r^3$	34
$-26x^5yz^4$	-26
$-k = -1k$	-1
$r = 1r$	1
$\dfrac{3x}{8} = \dfrac{3}{8}x$	$\dfrac{3}{8}$
$\dfrac{x}{3} = \dfrac{1x}{3} = \dfrac{1}{3}x$	$\dfrac{1}{3}$

The inverse properties "undo" addition or multiplication. Think of putting on your shoes when you get up in the morning and then taking them off before you go to bed at night. These are inverse operations that undo each other.

Expressions such as $12m$ and $5n$ from **Example 2** are examples of *terms*. A **term** is a number or the product of a number and one or more variables raised to powers. The numerical factor in a term is called the **numerical coefficient,** or just the **coefficient.** Some examples are shown in the table in the margin.

Terms with exactly the same variables raised to exactly the same powers are called **like terms.**

$$5p \text{ and } -21p \qquad -6x^2 \text{ and } 9x^2 \qquad \text{Like terms}$$
$$3m \text{ and } 16x \qquad 7y^3 \text{ and } -3y^2 \qquad \text{Unlike terms}$$

OBJECTIVE 4 **Use the commutative and associative properties.** Simplifying expressions as in parts (a) and (b) of **Example 2** is called **combining like terms.** *Only like terms may be combined.* To combine like terms in an expression such as

$$-2m + 5m + 3 - 6m + 8,$$

we need two more properties. From arithmetic, we know that

$$3 + 9 = 12 \qquad \text{and} \qquad 9 + 3 = 12$$
$$3 \cdot 9 = 27 \qquad \text{and} \qquad 9 \cdot 3 = 27.$$

The order of the numbers being added or multiplied does not matter. The same answers result. Also,

$$(5 + 7) + 2 = 12 + 2 = 14$$
$$5 + (7 + 2) = 5 + 9 = 14,$$

NOW TRY ANSWERS
2. (a) $8x$ **(b)** $-5p + 3q$

and
$$(5 \cdot 7) \cdot 2 = 35 \cdot 2 = 70$$
$$5(7 \cdot 2) = 5 \cdot 14 = 70.$$

The way in which the numbers being added or multiplied are grouped does not matter. The same answers result.

These arithmetic examples can be extended to algebra.

Commutative and Associative Properties

For any real numbers a, b, and c, the following are true.

$$\left.\begin{array}{c} a + b = b + a \\ ab = ba \end{array}\right\} \text{ Commutative properties}$$

(The *order* of the two terms or factors changes.)

$$\left.\begin{array}{c} a + (b + c) = (a + b) + c \\ a(bc) = (ab)c \end{array}\right\} \text{ Associative properties}$$

(The *grouping* among the three terms or factors changes, but the order stays the same.)

The commutative properties are used to change the order of the terms or factors in an expression. Think of *commuting* from home to work and then from work to home. The associative properties are used to regroup the terms or factors of an expression. The grouped terms or factors are *associated*.

NOW TRY
EXERCISE 3
Simplify.

$$-7x + 10 - 3x - 4 + x$$

EXAMPLE 3   Using the Commutative and Associative Properties

Simplify.

$$-2m + 5m + 3 - 6m + 8$$

$= (-2m + 5m) + 3 - 6m + 8$	Order of operations
$= (-2 + 5)m + 3 - 6m + 8$	Distributive property
$= 3m + 3 - 6m + 8$	Add inside parentheses.

The next step would be to add $3m$ and 3, but they are unlike terms. To get $3m$ and $-6m$ together, we use the associative and commutative properties, inserting parentheses and brackets according to the order of operations.

$= [(3m + 3) - 6m] + 8$	
$= [3m + (3 - 6m)] + 8$	Associative property
$= [3m + (-6m + 3)] + 8$	Commutative property
$= [(3m + [-6m]) + 3] + 8$	Associative property
$= (-3m + 3) + 8$	Combine like terms.
$= -3m + (3 + 8)$	Associative property
$= -3m + 11$	Add.

In practice, many of these steps are not written down, but you should realize that the commutative and associative properties are used whenever the terms in an expression are rearranged to combine like terms.   NOW TRY

NOW TRY ANSWER
3. $-9x + 6$

NOW TRY
EXERCISE 4

Simplify each expression.

(a) $-3(t - 4) - t + 15$

(b) $5x(6y)$

EXAMPLE 4   Using the Properties of Real Numbers

Simplify each expression.

(a) $5y - 8y - 6y + 11y$

$= (5 - 8 - 6 + 11)y$ Distributive property

$= 2y$ Combine like terms.

(b) $3x + 4 - 5(x + 1) - 8$ ⟨ Be careful with signs. ⟩

$= 3x + 4 - 5x - 5 - 8$ Distributive property

$= 3x - 5x + 4 - 5 - 8$ Commutative property

$= -2x - 9$ Combine like terms.

(c) $8 - (3m + 2)$

$= 8 - 1(3m + 2)$ Identity property

$= 8 - 3m - 2$ Distributive property

$= 6 - 3m$ Combine like terms.

(d) $3x(5)(y)$

$= [3x(5)]y$ Order of operations

$= [3(x \cdot 5)]y$ Associative property

$= [3(5x)]y$ Commutative property

$= [(3 \cdot 5)x]y$ Associative property

$= (15x)y$ Multiply.

$= 15(xy)$ Associative property

$= 15xy$

As previously mentioned, many of these steps are not usually written out.

NOW TRY

⚠ **CAUTION**   Be careful. The distributive property does not apply in **Example 4(d),** because there is no addition involved.

$$(3x)(5)(y) \neq (3x)(5) \cdot (3x)(y)$$

OBJECTIVE 5   **Use the multiplication property of 0.** The additive identity property gives a special property of 0, namely, that

$$a + 0 = 0 + a = a, \quad \text{for any real number } a.$$

The **multiplication property of 0** gives another special property of 0, namely that the product of any real number and 0 is 0.

Multiplication Property of 0

For any real number a, the following are true.

$$a \cdot 0 = 0 \quad \text{and} \quad 0 \cdot a = 0$$

NOW TRY ANSWERS
4. (a) $-4t + 27$ (b) $30xy$

1.4 EXERCISES

MyMathLab Math XL
PRACTICE   WATCH   DOWNLOAD   READ   REVIEW

🌐 *Complete solution available
on the Video Resources on DVD*

Concept Check   Choose the correct response in Exercises 1–4.

1. The identity element for addition is

   A. $-a$ **B.** 0 **C.** 1 **D.** $\frac{1}{a}$.

2. The identity element for multiplication is

   A. $-a$ **B.** 0 **C.** 1 **D.** $\frac{1}{a}$.

3. The additive inverse of a is

   A. $-a$ **B.** 0 **C.** 1 **D.** $\frac{1}{a}$.

4. The multiplicative inverse of a, where $a \neq 0$, is

   A. $-a$ **B.** 0 **C.** 1 **D.** $\frac{1}{a}$.

Concept Check   Complete each statement.

5. The multiplication property of 0 says that the _____ of 0 and any real number is _____.

6. The commutative property is used to change the _____ of two terms or factors.

7. The associative property is used to change the _____ of three terms or factors.

8. Like terms are terms with the _____ variables raised to the _____ powers.

9. When simplifying an expression, only _____ terms can be combined.

10. The numerical coefficient in the term $-7yz^2$ is _____.

Simplify each expression. ***See Examples 1 and 2.***

🌐 **11.** $2(m + p)$ **12.** $3(a + b)$ **13.** $-12(x - y)$ **14.** $-10(p - q)$

15. $5k + 3k$ **16.** $6a + 5a$ **17.** $7r - 9r$ **18.** $4n - 6n$

19. $-8z + 4w$ **20.** $-12k + 3r$ 🌐 **21.** $a + 7a$ **22.** $s + 9s$

23. $-(2d - f)$ **24.** $-(3m - n)$ **25.** $-(-x - y)$ **26.** $-(-3x - 4y)$

Simplify each expression. ***See Examples 1–4.***

27. $-12y + 4y + 3 + 2y$ **28.** $-5r - 9r + 8r - 5$

🌐 **29.** $-6p + 5 - 4p + 6 + 11p$ **30.** $-8x - 12 + 3x - 5x + 9$

🌐 **31.** $3(k + 2) - 5k + 6 + 3$ **32.** $5(r - 3) + 6r - 2r + 4$

33. $-2(m + 1) - (m - 4)$ **34.** $6(a - 5) - (a + 6)$

35. $0.25(8 + 4p) - 0.5(6 + 2p)$ **36.** $0.4(10 - 5x) - 0.8(5 + 10x)$

37. $-(2p + 5) + 3(2p + 4) - 2p$ **38.** $-(7m - 12) - 2(4m + 7) - 8m$

39. $2 + 3(2z - 5) - 3(4z + 6) - 8$ **40.** $-4 + 4(4k - 3) - 6(2k + 8) + 7$

Concept Check   Complete each statement so that the indicated property is illustrated. Simplify each answer if possible.

41. $5x + 8x =$ _____
   (distributive property)

42. $9y - 6y =$ _____
   (distributive property)

43. $5(9r) =$ _____
   (associative property)

44. $-4 + (12 + 8) =$ _____
   (associative property)

45. $5x + 9y =$ _____
   (commutative property)

46. $-5 \cdot 7 =$ _____
   (commutative property)

47. $1 \cdot 7 =$ _____
(identity property)

48. $-12x + 0 =$ _____
(identity property)

49. $-\dfrac{1}{4}ty + \dfrac{1}{4}ty =$ _____
(inverse property)

50. $-\dfrac{9}{8}\left(-\dfrac{8}{9}\right) =$ _____
(inverse property)

51. $8(-4 + x) =$ _____
(distributive property)

52. $3(x - y + z) =$ _____
(distributive property)

53. $0(0.875x + 9y - 88z) =$ _____
(multiplication property of 0)

54. $0(35t^2 - 8t + 12) =$ _____
(multiplication property of 0)

55. *Concept Check*   Give an "everyday" example of a commutative operation and of an operation that is not commutative.

56. *Concept Check*   Give an "everyday" example of inverse operations.

The distributive property can be used to mentally perform calculations. For example, calculate $38 \cdot 17 + 38 \cdot 3$ *as follows.*

$$38 \cdot 17 + 38 \cdot 3 = 38(17 + 3) \qquad \text{Distributive property}$$
$$= 38(20) \qquad \text{Add inside the parentheses.}$$
$$= 760 \qquad \text{Multiply.}$$

Use the distributive property to calculate each value mentally.

57. $96 \cdot 19 + 4 \cdot 19$

58. $27 \cdot 60 + 27 \cdot 40$

59. $58 \cdot \dfrac{3}{2} - 8 \cdot \dfrac{3}{2}$

60. $8.75(15) - 8.75(5)$

61. $4.31(69) + 4.31(31)$

62. $\dfrac{8}{5}(17) + \dfrac{8}{5}(13)$

RELATING CONCEPTS   EXERCISES 63–68

FOR INDIVIDUAL OR GROUP WORK

When simplifying the expression $3x + 4 + 2x + 7$ *to* $5x + 11$, *some steps are usually done mentally.* **Work Exercises 63–68 in order,** *providing the property that justifies each statement in the given simplification. (These steps could be done in other orders.)*

$$3x + 4 + 2x + 7$$

63. $= (3x + 4) + (2x + 7)$ _____

64. $= 3x + (4 + 2x) + 7$ _____

65. $= 3x + (2x + 4) + 7$ _____

66. $= (3x + 2x) + (4 + 7)$ _____

67. $= (3 + 2)x + (4 + 7)$ _____

68. $= 5x + 11$ _____

---

69. By the distributive property, $a(b + c) = ab + ac$. This property is more completely named the distributive property of multiplication with respect to addition. Is there a distributive property of addition with respect to multiplication? In other words, is

$$a + (b \cdot c) = (a + b)(a + c)$$

true for all real numbers, a, b, and c? To find out, try some sample values of a, b, and c.

70. Explain how the distributive property is used to combine like terms. Give an example.

CHAPTER 1 SUMMARY

KEY TERMS

1.1
set
elements (members)
finite set
infinite set
empty set (null set)
variable
set-builder notation
number line
coordinate
graph
additive inverse
(opposite, negative)

signed numbers
absolute value
equation
inequality
interval
interval notation
three-part inequality

1.2
sum
difference
product
quotient

reciprocal
(multiplicative inverse)

1.3
factors
exponent (power)
base
exponential expression
square root
principal (positive)
square root
negative square root
algebraic expression

1.4
identity element for
addition
identity element for
multiplication
term
coefficient (numerical
coefficient)
like terms
combining like terms

NEW SYMBOLS

$\{a, b\}$	set containing the elements a and b	$\{x\|x \text{ has property } P\}$	set-builder notation	∞	infinity
$\emptyset$ or $\{\ \}$	empty set	$\|x\|$	absolute value of x	$-\infty$	negative infinity
$\in$	is an element of (a set)	$<$	is less than	$(-\infty, \infty)$	set of all real numbers
$\notin$	is not an element of	$\leq$	is less than or equal to	(a, ∞)	the interval $\{x\|x > a\}$
$\neq$	is not equal to	$>$	is greater than	$(-\infty, a)$	the interval $\{x\|x < a\}$
		$\geq$	is greater than or equal to		

$(a, b]$	the interval $\{x\|a < x \leq b\}$
a^m	m factors of a
$\sqrt{\ }$	radical symbol
$\sqrt{a}$	positive (or principal) square root of a

TEST YOUR WORD POWER

See how well you have learned the vocabulary in this chapter.

1. The **empty set** is a set
   A. with 0 as its only element
   B. with an infinite number of elements
   C. with no elements
   D. of ideas.

2. A **variable** is
   A. a symbol used to represent an unknown number
   B. a value that makes an equation true
   C. a solution of an equation
   D. the answer in a division problem.

3. The **absolute value** of a number is
   A. the graph of the number
   B. the reciprocal of the number
   C. the opposite of the number
   D. the distance between 0 and the number on a number line.

4. The **reciprocal** of a nonzero number a is
   A. a B. $\frac{1}{a}$ C. $-a$ D. 1.

5. A **factor** is
   A. the answer in an addition problem
   B. the answer in a multiplication problem
   C. one of two or more numbers that are added to get another number
   D. any number that divides evenly into a given number.

6. An **exponential expression** is
   A. a number that is a repeated factor in a product
   B. a number or a variable written with an exponent
   C. a number that shows how many times a factor is repeated in a product

   D. an expression that involves addition.

7. A **term** is
   A. a numerical factor
   B. a number or a product of a number and one or more variables raised to powers
   C. one of several variables with the same exponents
   D. a sum of numbers and variables raised to powers.

8. A **numerical coefficient** is
   A. the numerical factor in a term
   B. the number of terms in an expression
   C. a variable raised to a power
   D. the variable factor in a term.

QUICK REVIEW

CONCEPTS	EXAMPLES

1.1 Basic Concepts

Sets of Numbers

Natural Numbers {1, 2, 3, 4, . . .}

Whole Numbers {0, 1, 2, 3, 4, . . .}

Integers {. . . , −2, −1, 0, 1, 2, . . .}

Rational Numbers

$\left\{\frac{p}{q} \mid p \text{ and } q \text{ are integers}, q \neq 0\right\}$

(all terminating or repeating decimals)

Irrational Numbers

{x|x is a real number that is not rational}

(all nonterminating, nonrepeating decimals)

Real Numbers

{x|x is a rational or an irrational number}

Absolute Value $|a| = \begin{cases} a & \text{if } a \text{ is positive or } 0 \\ -a & \text{if } a \text{ is negative} \end{cases}$

Examples:

10, 25, 143 — Natural Numbers

0, 8, 47 — Whole Numbers

−22, −7, 0, 4, 9 — Integers

$-\frac{2}{3}, -0.14, 0, \frac{15}{8}, 6, 0.33333\ldots, \sqrt{4}$ — Rational Numbers

$-\sqrt{22}, \sqrt{3}, \pi$ — Irrational Numbers

$-3, -\frac{2}{7}, 0.7, \pi, \sqrt{11}$ — Real Numbers

$|12| = 12$

$|-12| = 12$

1.2 Operations on Real Numbers

Addition

Same Sign: Add the absolute values. The sum has the same sign as the given numbers.

Different Signs: Find the absolute values of the numbers, and subtract the lesser absolute value from the greater. The sum has the same sign as the number with the greater absolute value.

$-2 + (-7) = -(2 + 7) = -9$

$-5 + 8 = 8 - 5 = 3$
$-12 + 4 = -(12 - 4) = -8$

Subtraction

For all real numbers a and b,

$$a - b = a + (-b).$$

$-5 - (-3) = -5 + 3 = -2$

Multiplication and Division

Same Sign: The answer is positive when multiplying or dividing two numbers with the same sign.

$-3(-8) = 24 \qquad \dfrac{-15}{-5} = 3$

Different Signs: The answer is negative when multiplying or dividing two numbers with different signs.

$-7(5) = -35 \qquad \dfrac{-24}{12} = -2$

Division

For all real numbers a and b (where $b \neq 0$),

$$a \div b = \frac{a}{b} = a \cdot \frac{1}{b}.$$

$\dfrac{2}{3} \div \dfrac{5}{6} = \dfrac{2}{3} \cdot \dfrac{6}{5} = \dfrac{4}{5}$ — Multiply by the reciprocal of the divisor.

Note

(continued)

CONCEPTS	*EXAMPLES*

1.3 Exponents, Roots, and Order of Operations

The product of an even number of negative factors is positive. The product of an odd number of negative factors is negative.

$(-5)^2$ is positive: $(-5)^2 = (-5)(-5) = 25$

$(-5)^3$ is negative: $(-5)^3 = (-5)(-5)(-5) = -125$

Order of Operations

1. Work separately above and below any fraction bar.

$$\frac{12 + 3}{5 \cdot 2} = \frac{15}{10} = \frac{3}{2}$$

2. If parentheses, brackets, or absolute value bars are present, start with the innermost set and work outward.

3. Evaluate all exponents, roots, and absolute values.

4. Multiply or divide in order from left to right.

5. Add or subtract in order from left to right.

$$(-6)[2^2 - (3 + 4)] + 3$$
$$= (-6)[2^2 - 7] + 3$$
$$= (-6)[4 - 7] + 3$$
$$= (-6)[-3] + 3$$
$$= 18 + 3$$
$$= 21$$

1.4 Properties of Real Numbers

For real numbers a, b, and c, the following are true.

Distributive Property

$a(b + c) = ab + ac$ and $(b + c)a = ba + ca$

$12(4 + 2) = 12 \cdot 4 + 12 \cdot 2$

Identity Properties

$a + 0 = 0 + a = a$ and $a \cdot 1 = 1 \cdot a = a$

$-32 + 0 = -32$ $17.5 \cdot 1 = 17.5$

Inverse Properties

$a + (-a) = 0$ and $-a + a = 0$

$a \cdot \dfrac{1}{a} = 1$ and $\dfrac{1}{a} \cdot a = 1$

$5 + (-5) = 0$ $-12 + 12 = 0$

$5 \cdot \dfrac{1}{5} = 1$ $-\dfrac{1}{3}(-3) = 1$

Commutative Properties

$a + b = b + a$ and $ab = ba$

$9 + (-3) = -3 + 9$ $6(-4) = (-4)6$

Associative Properties

$a + (b + c) = (a + b) + c$ and $a(bc) = (ab)c$

$7 + (5 + 3) = (7 + 5) + 3$ $-4(6 \cdot 3) = (-4 \cdot 6)3$

Multiplication Property of 0

$a \cdot 0 = 0$ and $0 \cdot a = 0$

$4 \cdot 0 = 0$ $0(-3) = 0$

CHAPTER 1

REVIEW EXERCISES

1.1   *Graph the elements of each set on a number line.*

1. $\left\{-4, -1, 2, \dfrac{9}{4}, 4\right\}$

2. $\left\{-5, -\dfrac{11}{4}, -0.5, 0, 3, \dfrac{13}{3}\right\}$

Find the value of each expression.

3. $|-16|$

4. $-|-8|$

5. $|-8| - |-3|$

Let $S = \left\{-9, -\dfrac{4}{3}, -\sqrt{4}, -0.25, 0, 0.\overline{35}, \dfrac{5}{3}, \sqrt{7}, \sqrt{-9}, \dfrac{12}{3}\right\}$. *Simplify the elements of S as necessary, and then list those elements of S which belong to the specified set.*

6. Whole numbers

7. Integers

8. Rational numbers

9. Real numbers

Write each set by listing its elements.

10. $\{x \mid x$ is a natural number between 3 and 9$\}$

11. $\{y \mid y$ is a whole number less than 4$\}$

Write true *or* false *for each inequality.*

12. $4 \cdot 2 \le |12 - 4|$

13. $2 + |-2| > 4$

14. $4(3 + 7) > -|40|$

The graph shows the percent change in passenger car production at U.S. plants from 2006 to 2007 for various automakers. Use this graph to work Exercises 15–18.

15. Which automaker had the greatest change in sales? What was that change?

16. Which automaker had the least change in sales? What was that change?

17. *True* or *false:* The absolute value of the percent change for Chrysler was greater than the absolute value of the percent change for Hyundai.

18. *True* or *false:* The percent change for Subaru was more than twice the percent change for Chrysler.

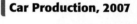

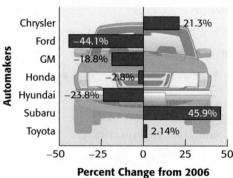

Car Production, 2007

Automakers:
- Chrysler: 21.3%
- Ford: −44.1%
- GM: −18.8%
- Honda: −2.8%
- Hyundai: −23.8%
- Subaru: 45.9%
- Toyota: 2.14%

Percent Change from 2006

Source: World Almanac and Book of Facts.

Write each set in interval notation and graph the interval.

19. $\{x \mid x < -5\}$

20. $\{x \mid -2 < x \le 3\}$

1.2   *Add or subtract as indicated.*

21. $-\dfrac{5}{8} - \left(-\dfrac{7}{3}\right)$

22. $-\dfrac{4}{5} - \left(-\dfrac{3}{10}\right)$

23. $-5 + (-11) + 20 - 7$

24. $-9.42 + 1.83 - 7.6 - 1.8$

25. $-15 + (-13) + (-11)$

26. $-1 - 3 - (-10) + (-6)$

27. $\dfrac{3}{4} - \left(\dfrac{1}{2} - \dfrac{9}{10}\right)$

28. $-|-12| - |-9| + (-4) - |10|$

29. Telescope Peak, altitude 11,049 ft, is next to Death Valley, 282 ft below sea level. Find the difference between these altitudes. (*Source: World Almanac and Book of Facts.*)

Multiply or divide as indicated.

30. $2(-5)(-3)(-3)$ **31.** $-\dfrac{3}{7}\left(-\dfrac{14}{9}\right)$ **32.** $\dfrac{75}{-5}$ **33.** $\dfrac{-2.3754}{-0.74}$

34. *Concept Check*   Which one of the following is undefined: $\dfrac{5}{7-7}$ or $\dfrac{7-7}{5}$?

`1.3`   *Evaluate each expression.*

35. 10^4 **36.** $\left(\dfrac{3}{7}\right)^3$ **37.** $(-5)^3$ **38.** -5^3

Find each square root. If it is not a real number, say so.

39. $\sqrt{400}$ **40.** $\sqrt{\dfrac{64}{121}}$ **41.** $-\sqrt{0.81}$ **42.** $\sqrt{-49}$

Simplify each expression.

43. $-14\left(\dfrac{3}{7}\right)+6\div3$ **44.** $-\dfrac{2}{3}[5(-2)+8-4^3]$ **45.** $\dfrac{-5(3^2)+9\left(\sqrt{4}\right)-5}{6-5(-2)}$

Evaluate each expression for $k=-4$, $m=2$, and $n=16$.

46. $4k-7m$ **47.** $-3\sqrt{n}+m+5k$ **48.** $\dfrac{4m^3-3n}{7k^2-10}$

49. The following expression for *body mass index* (BMI) can help determine ideal body weight.

$$704\times(\text{weight in pounds})\div(\text{height in inches})^2$$

A BMI of 19 to 25 corresponds to a healthy weight. (*Source: The Washington Post.*)

(a) Baseball player Grady Sizemore is 6 ft, 2 in., tall and weighs 200 lb. (*Source:* www.mlb.com) Find his BMI (to the nearest whole number).

(b) Calculate your BMI.

`1.4`   *Simplify each expression.*

50. $2q+18q$ **51.** $13z-17z$

52. $-m+4m$ **53.** $5p-p$

54. $-2(k+3)$ **55.** $6(r+3)$

56. $9(2m+3n)$ **57.** $-(-p+6q)-(2p-3q)$

58. $-3y+6-5+4y$ **59.** $2a+3-a-1-a-2$

60. $-3(4m-2)+2(3m-1)-4(3m+1)$

Complete each statement so that the indicated property is illustrated. Simplify each answer if possible.

61. $2x+3x=$ _____

(distributive property)

62. $-5\cdot1=$ _____

(identity property)

63. $2(4x)=$ _____

(associative property)

64. $-3+13=$ _____

(commutative property)

65. $-3+3=$ _____

(inverse property)

66. $6(x+z)=$ _____

(distributive property)

67. $0+7=$ _____

(identity property)

68. $4\cdot\dfrac{1}{4}=$ _____

(inverse property)

MIXED REVIEW EXERCISES*

The table gives U.S. exports and imports with Spain, in millions of U.S. dollars.

Year	Exports	Imports
2007	9766	10,498
2008	12,190	11,094
2009	7294	6495

Source: U.S. Census Bureau.

Determine the absolute value of the difference between imports and exports for each year. Is the balance of trade (exports minus imports) in each year positive or negative?

69. 2007   **70.** 2008   **71.** 2009

Perform the indicated operations.

72. $\left(-\dfrac{4}{5}\right)^4$ **73.** $-\dfrac{5}{8}(-40)$

74. $-25\left(-\dfrac{4}{5}\right) + 3^3 - 32 \div \sqrt{4}$ **75.** $-8 + |-14| + |-3|$

76. $\dfrac{6 \cdot \sqrt{4} - 3 \cdot \sqrt{16}}{-2 \cdot 5 + 7(-3) - 10}$ **77.** $-\sqrt{25}$ **78.** $-\dfrac{10}{21} \div \left(-\dfrac{5}{14}\right)$

79. $0.8 - 4.9 - 3.2 + 1.14$ **80.** -3^2 **81.** $\dfrac{-38}{-19}$

82. $-2(k - 1) + 3k - k$ **83.** $-\sqrt{-100}$ **84.** $-(3k - 6h)$

85. $-4.6(2.48)$ **86.** $-\dfrac{2}{3}(-15) + (2^4 - 8 \div 4)$

87. $-2x + 5 - 4x - 1$ **88.** $-\dfrac{2}{3} - \left(\dfrac{1}{6} - \dfrac{5}{9}\right)$

89. Evaluate $-m(3k^2 + 5m)$ for **(a)** $k = -4$ and $m = 2$ and **(b)** $k = \frac{1}{2}$ and $m = -\frac{3}{4}$.

90. *Concept Check*   To evaluate $(3 + 5)^2$, should you work within the parentheses first, or should you square 3 and square 5 and then add?

_____

*The order of exercises in this final group does not correspond to the order in which topics occur in the chapter. This random ordering should help you prepare for the chapter test in yet another way.

CHAPTER (1)   TEST

Step-by-step test solutions are found on the Chapter Test Prep Videos available via the Video Resources on DVD, in *MyMathLab*, or on *YouTube* (search "LialIntermediateAlg").

View the complete solutions to all Chapter Test exercises on the Video Resources on DVD.

1. Graph $\left\{-3, 0.75, \dfrac{5}{3}, 5, 6.3\right\}$ on a number line.

Let $A = \left\{-\sqrt{6}, -1, -0.5, 0, 3, \sqrt{25}, 7.5, \frac{24}{2}, \sqrt{-4}\right\}$. *Simplify the elements of A as necessary, and then list those elements of A which belong to the specified set.*

2. Whole numbers   **3.** Integers

4. Rational numbers   **5.** Real numbers

Write each set in interval notation and graph the interval.

6. $\{x \mid x < -3\}$

7. $\{x \mid -4 < x \leq 2\}$

Perform the indicated operations.

8. $-6 + 14 + (-11) - (-3)$

9. $10 - 4 \cdot 3 + 6(-4)$

10. $7 - 4^2 + 2(6) + (-4)^2$

11. $\dfrac{10 - 24 + (-6)}{\sqrt{16}(-5)}$

12. $\dfrac{-2[3 - (-1 - 2) + 2]}{\sqrt{9}(-3) - (-2)}$

13. $\dfrac{8 \cdot 4 - 3^2 \cdot 5 - 2(-1)}{-3 \cdot 2^3 + 1}$

The table shows the heights in feet of some selected mountains and the depths in feet (as negative numbers) of some selected ocean trenches.

Mountain	Height	Trench	Depth
Foraker	17,400	Philippine	−32,995
Wilson	14,246	Cayman	−24,721
Pikes Peak	14,110	Java	−23,376

Source: World Almanac and Book of Facts.

14. What is the difference between the height of Mt. Foraker and the depth of the Philippine Trench?

15. What is the difference between the height of Pikes Peak and the depth of the Java Trench?

16. How much deeper is the Cayman Trench than the Java Trench?

Find each square root. If the number is not real, say so.

17. $\sqrt{196}$

18. $-\sqrt{225}$

19. $\sqrt{-16}$

20. *Concept Check*   For the expression $\sqrt{a}$, under what conditions will its value be each of the following?

    (a) positive   **(b)** not real   **(c)** 0

21. Evaluate $\dfrac{8k + 2m^2}{r - 2}$ for $k = -3$, $m = -3$, and $r = 25$.

22. Simplify $-3(2k - 4) + 4(3k - 5) - 2 + 4k$.

23. How does the subtraction sign affect the terms $-4r$ and 6 when $(3r + 8) - (-4r + 6)$ is simplified? What is the simplified form?

Match each statement in Column I with the appropriate property in Column II. Answers may be used more than once.

I	II
24. $6 + (-6) = 0$	**A.** Distributive property
25. $-2 + (3 + 6) = (-2 + 3) + 6$	**B.** Inverse property
26. $5x + 15x = (5 + 15)x$	**C.** Identity property
27. $13 \cdot 0 = 0$	**D.** Associative property
28. $-9 + 0 = -9$	**E.** Commutative property
29. $4 \cdot 1 = 4$	**F.** Multiplication property of 0
30. $(a + b) + c = (b + a) + c$	

STUDY SKILLS

Reading Your Math Textbook

Take time to read each section and its examples before doing your homework. You will learn more and be better prepared to work the exercises your instructor assigns.

Approaches to Reading Your Math Textbook

Student A learns best by listening to her teacher explain things. She "gets it" when she sees the instructor work problems. She previews the section before the lecture, so she knows generally what to expect. **Student A carefully reads the section in her text *AFTER* she hears the classroom lecture on the topic.**

Student B learns best by reading on his own. He reads the section and works through the examples before coming to class. That way, he knows what the teacher is going to talk about and what questions he wants to ask. **Student B carefully reads the section in his text *BEFORE* he hears the classroom lecture on the topic.**

Which reading approach works best for you—that of Student A or Student B?

Tips for Reading Your Math Textbook

▶ **Turn off your cell phone.** You will be able to concentrate more fully on what you are reading.

▶ **Read slowly.** Read only one section—or even part of a section—at a sitting, with paper and pencil in hand.

▶ **Pay special attention to important information given in colored boxes or set in boldface type.**

▶ **Study the examples carefully.** Pay particular attention to the blue side comments and pointers.

▶ **Do the Now Try exercises in the margin on separate paper as you go.** These mirror the examples and prepare you for the exercise set. The answers are given at the bottom of the page.

▶ **Make study cards as you read.** (See **page 102.**) Make cards for new vocabulary, rules, procedures, formulas, and sample problems.

▶ **Mark anything you don't understand. *ASK QUESTIONS*** in class—everyone will benefit. Follow up with your instructor, as needed.

Select several reading tips to try this week.

Linear Equations, Inequalities, and Applications

Despite increasing competition from the Internet and video games, television remains a popular form of entertainment. In 2009, 114.9 million American households owned at least one TV set, and average viewing time for all viewers was almost 34 hours per week. During the 2008–2009 season, favorite prime-time television programs were *American Idol* and *Dancing with the Stars.* (*Source:* Nielsen Media Research.)

In **Section 2.2** we discuss the concept of *percent*—one of the most common everyday applications of mathematics—and use it in **Exercises 43–46** to determine additional information about television ownership and viewing in U.S. households.

2.1 Linear Equations in One Variable

OBJECTIVES

1. Distinguish between expressions and equations.
2. Identify linear equations, and decide whether a number is a solution of a linear equation.
3. Solve linear equations by using the addition and multiplication properties of equality.
4. Solve linear equations by using the distributive property.
5. Solve linear equations with fractions or decimals.
6. Identify conditional equations, contradictions, and identities.

OBJECTIVE 1 **Distinguish between expressions and equations.** In our work in **Chapter 1,** we reviewed *algebraic expressions.*

$$8x + 9, \quad y - 4, \quad \text{and} \quad \frac{x^3 y^8}{z} \qquad \text{Examples of algebraic expressions}$$

Equations and inequalities compare algebraic expressions, just as a balance scale compares the weights of two quantities. Recall from **Section 1.1** that an *equation* is a statement that two algebraic expressions are equal. *An equation always contains an equals symbol, while an expression does not.*

EXAMPLE 1 Distinguishing between Expressions and Equations

Decide whether each of the following is an *expression* or an *equation.*

(a) $3x - 7 = 2$ **(b)** $3x - 7$

In part (a) we have an equation, because there is an equals symbol. In part (b), there is no equals symbol, so it is an expression. See the diagram below.

$$\underbrace{3x - 7}_{\text{Left side}} \overset{\uparrow}{=} \underbrace{2}_{\text{Right side}} \qquad \qquad 3x - 7$$

Equation	Expression
(to solve)	(to simplify or evaluate)

NOW TRY

OBJECTIVE 2 **Identify linear equations, and decide whether a number is a solution of a linear equation.** A *linear equation in one variable* involves only real numbers and one variable raised to the first power.

$$x + 1 = -2, \quad x - 3 = 5, \quad \text{and} \quad 2k + 5 = 10 \qquad \text{Examples of linear equations}$$

Linear Equation in One Variable

A **linear equation in one variable** can be written in the form

$$Ax + B = C,$$

where A, B, and C are real numbers, with $A \neq 0$.

A linear equation is a **first-degree equation,** since the greatest power on the variable is 1. Some equations that are not linear (that is, *nonlinear*) follow.

$$x^2 + 3y = 5, \quad \frac{8}{x} = -22, \quad \text{and} \quad \sqrt{x} = 6 \qquad \text{Examples of nonlinear equations}$$

If the variable in an equation can be replaced by a real number that makes the statement true, then that number is a **solution** of the equation. For example, 8 is a solution of the equation $x - 3 = 5$, since replacing x with 8 gives a true statement. An equation is *solved* by finding its **solution set,** the set of all solutions. The solution set of the equation $x - 3 = 5$ is $\{8\}$.

NOW TRY
EXERCISE 1

Decide whether each of the following is an *expression* or an *equation.*

(a) $2x + 17 - 3x$

(b) $2x + 17 = 3x$

NOW TRY ANSWERS
1. **(a)** expression **(b)** equation

Equivalent equations are related equations that have the same solution set. To solve an equation, we usually start with the given equation and replace it with a series of simpler equivalent equations. For example,

$$5x + 2 = 17, \quad 5x = 15, \quad \text{and} \quad x = 3 \qquad \text{Equivalent equations}$$

are all equivalent, since each has the solution set $\{3\}$.

OBJECTIVE 3 **Solve linear equations by using the addition and multiplication properties of equality.** We use two important properties of equality to produce equivalent equations.

Addition and Multiplication Properties of Equality

Addition Property of Equality

For all real numbers A, B, and C, the equations

$$A = B \qquad \text{and} \qquad A + C = B + C \qquad \text{are equivalent.}$$

That is, *the same number may be added to each side of an equation without changing the solution set.*

Multiplication Property of Equality

For all real numbers A and B, and for $C \neq 0$, the equations

$$A = B \qquad \text{and} \qquad AC = BC \qquad \text{are equivalent.}$$

That is, *each side of an equation may be multiplied by the same nonzero number without changing the solution set.*

Because subtraction and division are defined in terms of addition and multiplication, respectively, the preceding properties can be extended.

The same number may be subtracted from each side of an equation, and each side of an equation may be divided by the same nonzero number, without changing the solution set.

EXAMPLE 2   Using the Properties of Equality to Solve a Linear Equation

Solve $4x - 2x - 5 = 4 + 6x + 3$.

The goal is to isolate x on one side of the equation.

$$4x - 2x - 5 = 4 + 6x + 3$$

$2x - 5 = 7 + 6x$	Combine like terms.
$2x - 5 + 5 = 7 + 6x + 5$	Add 5 to each side.
$2x = 12 + 6x$	Combine like terms.
$2x - 6x = 12 + 6x - 6x$	Subtract 6x from each side.
$-4x = 12$	Combine like terms.
$\dfrac{-4x}{-4} = \dfrac{12}{-4}$	Divide each side by -4.
$x = -3$	

Check by substituting -3 for x in the *original* equation.

NOW TRY
EXERCISE 2
Solve.

$5x + 11 = 2x - 13 - 3x$

CHECK

$$4x - 2x - 5 = 4 + 6x + 3$$ Original equation

$$4(-3) - 2(-3) - 5 \overset{?}{=} 4 + 6(-3) + 3$$ Let $x = -3$.

$$-12 + 6 - 5 \overset{?}{=} 4 - 18 + 3$$ Multiply.

$$-11 = -11 \checkmark$$ True

> Use parentheses around substituted values to avoid errors.

> This is *not* the solution.

The true statement indicates that $\{-3\}$ is the solution set. NOW TRY

⚠ **CAUTION** In **Example 2,** the equality symbols are aligned in a column. *Use only one equality symbol in a horizontal line of work when solving an equation.*

Solving a Linear Equation in One Variable

Step 1 **Clear fractions or decimals.** Eliminate fractions by multiplying each side by the least common denominator. Eliminate decimals by multiplying by a power of 10.

Step 2 **Simplify each side separately.** Use the distributive property to clear parentheses and combine like terms as needed.

Step 3 **Isolate the variable terms on one side.** Use the addition property to get all terms with variables on one side of the equation and all numbers on the other.

Step 4 **Isolate the variable.** Use the multiplication property to get an equation with just the variable (with coefficient 1) on one side.

Step 5 **Check.** Substitute the proposed solution into the original equation.

OBJECTIVE 4 **Solve linear equations by using the distributive property.**
In **Example 2,** we did not use Step 1 or the distributive property in Step 2 as given in the box. Many equations, however, will require one or both of these steps.

EXAMPLE 3 Using the Distributive Property to Solve a Linear Equation

Solve $2(x - 5) + 3x = x + 6$.

Step 1 Since there are no fractions in this equation, Step 1 does not apply.

Step 2 Use the distributive property to simplify and combine like terms on the left.

> Be sure to distribute over *all* terms within the parentheses.

$$2(x - 5) + 3x = x + 6$$

$$2x + 2(-5) + 3x = x + 6$$ Distributive property

$$2x - 10 + 3x = x + 6$$ Multiply.

$$5x - 10 = x + 6$$ Combine like terms.

Step 3 Next, use the addition property of equality.

$$5x - 10 + 10 = x + 6 + 10$$ Add 10.

$$5x = x + 16$$ Combine like terms.

$$5x - x = x + 16 - x$$ Subtract x.

$$4x = 16$$ Combine like terms.

NOW TRY ANSWER
2. $\{-4\}$

NOW TRY
EXERCISE 3
Solve.

$5(x - 4) - 9 = 3 - 2(x + 16)$

Step 4   Use the multiplication property of equality to isolate x on the left.

$$\frac{4x}{4} = \frac{16}{4} \qquad \text{Divide by 4.}$$

$$x = 4$$

Step 5   Check by substituting 4 for x in the original equation.

CHECK $\qquad 2(x - 5) + 3x = x + 6 \qquad$ Original equation

Always check your work.

$$2(4 - 5) + 3(4) \stackrel{?}{=} 4 + 6 \qquad \text{Let } x = 4.$$

$$2(-1) + 12 \stackrel{?}{=} 10 \qquad \text{Simplify.}$$

$$10 = 10 \ \checkmark \qquad \text{True}$$

The solution checks, so $\{4\}$ is the solution set.   NOW TRY

OBJECTIVE 5 **Solve linear equations with fractions or decimals.** When fractions or decimals appear as coefficients in equations, our work can be made easier if we multiply each side of the equation by the least common denominator (LCD) of all the fractions. This is an application of the multiplication property of equality.

NOW TRY
EXERCISE 4
Solve.

$$\frac{x - 4}{4} + \frac{2x + 4}{8} = 5$$

EXAMPLE 4 Solving a Linear Equation with Fractions

Solve $\frac{x + 7}{6} + \frac{2x - 8}{2} = -4$.

Step 1 $\qquad 6\left(\frac{x + 7}{6} + \frac{2x - 8}{2}\right) = 6(-4) \qquad$ Eliminate the fractions.
Multiply each side by the LCD, 6.

Step 2 $\quad 6\left(\frac{x + 7}{6}\right) + 6\left(\frac{2x - 8}{2}\right) = 6(-4) \qquad$ Distributive property

$$x + 7 + 3(2x - 8) = -24 \qquad \text{Multiply.}$$

$$x + 7 + 3(2x) + 3(-8) = -24 \qquad \text{Distributive property}$$

$$x + 7 + 6x - 24 = -24 \qquad \text{Multiply.}$$

$$7x - 17 = -24 \qquad \text{Combine like terms.}$$

Step 3 $\qquad\qquad 7x - 17 + 17 = -24 + 17 \qquad$ Add 17.

$$7x = -7 \qquad \text{Combine like terms.}$$

Step 4 $\qquad\qquad\qquad \frac{7x}{7} = \frac{-7}{7} \qquad$ Divide by 7.

$$x = -1$$

Step 5 CHECK $\qquad \frac{x + 7}{6} + \frac{2x - 8}{2} = -4$

$$\frac{-1 + 7}{6} + \frac{2(-1) - 8}{2} \stackrel{?}{=} -4 \qquad \text{Let } x = -1.$$

$$\frac{6}{6} + \frac{-10}{2} \stackrel{?}{=} -4 \qquad \text{Add and subtract in the numerators.}$$

$$1 - 5 \stackrel{?}{=} -4 \qquad \text{Simplify each fraction.}$$

$$-4 = -4 \ \checkmark \qquad \text{True}$$

NOW TRY ANSWERS
3. $\{0\}$ **4.** $\{11\}$

The solution checks, so the solution set is $\{-1\}$.   NOW TRY

Some equations have decimal coefficients. We can clear these decimals by multiplying by a power of 10, such as

$$10^1 = 10, \quad 10^2 = 100, \quad \text{and so on.}$$

This allows us to obtain integer coefficients.

NOW TRY
EXERCISE 5

Solve.

$$0.08x - 0.12(x - 4)$$
$$= 0.03(x - 5)$$

EXAMPLE 5 Solving a Linear Equation with Decimals

Solve $0.06x + 0.09(15 - x) = 0.07(15)$.

Because each decimal number is given in hundredths, multiply each side of the equation by 100. A number can be multiplied by 100 by moving the decimal point two places to the right.

$$0.06x + 0.09(15 - x) = 0.07(15)$$

$$0.06x + 0.09(15 - x) = 0.07(15) \qquad \text{Multiply each term by 100.}$$

> Move decimal points 2 places to the right.

$$6x + 9(15 - x) = 7(15)$$

$$6x + 9(15) - 9x = 7(15) \qquad \text{Distributive property}$$

$$-3x + 135 = 105 \qquad \text{Combine like terms and multiply.}$$

$$-3x + 135 - 135 = 105 - 135 \qquad \text{Subtract 135.}$$

$$-3x = -30 \qquad \text{Combine like terms.}$$

$$\frac{-3x}{-3} = \frac{-30}{-3} \qquad \text{Divide by } -3.$$

$$x = 10$$

CHECK
$$0.06x + 0.09(15 - x) = 0.07(15)$$

$$0.06(10) + 0.09(15 - 10) \overset{?}{=} 0.07(15) \qquad \text{Let } x = 10.$$

$$0.6 + 0.09(5) \overset{?}{=} 1.05 \qquad \text{Multiply and subtract.}$$

$$0.6 + 0.45 \overset{?}{=} 1.05 \qquad \text{Multiply.}$$

$$1.05 = 1.05 \ \checkmark \qquad \text{True}$$

The solution set is $\{10\}$.

NOW TRY

NOTE Because of space limitations, we will not always show the check when solving an equation. *To be sure that your solution is correct, you should always check your work.*

OBJECTIVE 6 **Identify conditional equations, contradictions, and identities.**
In **Examples 2–5,** all of the equations had solution sets containing *one* element, such as $\{10\}$ in **Example 5.** Some equations, however, have no solutions, while others have an infinite number of solutions. The table on the next page gives the names of these types of equations.

NOW TRY ANSWER
5. $\{9\}$

Type of Linear Equation	Number of Solutions	Indication when Solving
Conditional	One	Final line is $x = $ a number. (See **Example 6(a).**)
Identity	Infinite; solution set {all real numbers}	Final line is true, such as $0 = 0$. (See **Example 6(b).**)
Contradiction	None; solution set Ø	Final line is false, such as $-15 = -20$. (See **Example 6(c).**)

⌒ *NOW TRY*
↳ *EXERCISE 6*

Solve each equation. Decide whether it is a *conditional equation*, an *identity*, or a *contradiction*.

(a) $9x - 3(x + 4) = 6(x - 2)$

(b) $-3(2x - 1) - 2x = 3 + x$

(c) $10x - 21 = 2(x - 5) + 8x$

EXAMPLE 6   Recognizing Conditional Equations, Identities, and Contradictions

Solve each equation. Decide whether it is a *conditional equation*, an *identity*, or a *contradiction*.

(a)
$$5(2x + 6) - 2 = 7(x + 4)$$

$10x + 30 - 2 = 7x + 28$	Distributive property
$10x + 28 = 7x + 28$	Combine like terms.
$10x + 28 - 7x - 28 = 7x + 28 - 7x - 28$	Subtract 7x. Subtract 28.
$3x = 0$	Combine like terms.
$\dfrac{3x}{3} = \dfrac{0}{3}$	Divide by 3.
$x = 0$	

The solution set, $\{0\}$, has only one element, so $5(2x + 6) - 2 = 7(x + 4)$ is a conditional equation.

(b)
$$5x - 15 = 5(x - 3)$$

$5x - 15 = 5x - 15$	Distributive property
$5x - 15 - 5x + 15 = 5x - 15 - 5x + 15$	Subtract 5x. Add 15.
$0 = 0$	True

The final line, $0 = 0$, indicates that the solution set is {all real numbers}, and the equation $5x - 15 = 5(x - 3)$ is an identity. (The first step yielded $5x - 15 = 5x - 15$, which is *true* for all values of x. We could have identified the equation as an identity at that point.)

(c)
$$5x - 15 = 5(x - 4)$$

$5x - 15 = 5x - 20$	Distributive property
$5x - 15 - 5x = 5x - 20 - 5x$	Subtract 5x.
$-15 = -20$	False

Since the result, $-15 = -20$, is *false*, the equation has no solution. The solution set is Ø, so the equation $5x - 15 = 5(x - 4)$ is a contradiction.   *NOW TRY* ⤵

NOW TRY ANSWERS

6. (a) {all real numbers}; identity

(b) {0}; conditional equation

(c) Ø; contradiction

⚠ **CAUTION**   A common error in solving an equation like that in **Example 6(a)** is to think that the equation has no solution and write the solution set as Ø. This equation has one solution, the number 0, so it is a conditional equation with solution set {0}.

2.1 EXERCISES

MyMathLab Math XL PRACTICE WATCH DOWNLOAD READ REVIEW

Complete solution available on the Video Resources on DVD

1. *Concept Check* Which equations are linear equations in x?

A. $3x + x - 1 = 0$ **B.** $8 = x^2$ **C.** $6x + 2 = 9$ **D.** $\frac{1}{2}x - \frac{1}{x} = 0$

2. Which of the equations in **Exercise 1** are nonlinear equations in x? Explain why.

3. Decide whether 6 is a solution of $3(x + 4) = 5x$ by substituting 6 for x. If it is not a solution, explain why.

4. Use substitution to decide whether -2 is a solution of $5(x + 4) - 3(x + 6) = 9(x + 1)$. If it is not a solution, explain why.

Decide whether each of the following is an expression *or an* equation. *See Example 1.*

5. $-3x + 2 - 4 = x$

6. $-3x + 2 - 4 - x = 4$

7. $4(x + 3) - 2(x + 1) - 10$

8. $4(x + 3) - 2(x + 1) + 10$

9. $-10x + 12 - 4x = -3$

10. $-10x + 12 - 4x + 3 = 0$

Solve each equation, and check your solution. If applicable, tell whether the equation is an identity or a contradiction. *See Examples 2, 3, and 6.*

11. $7x + 8 = 1$

12. $5x - 4 = 21$

13. $5x + 2 = 3x - 6$

14. $9x + 1 = 7x - 9$

15. $7x - 5x + 15 = x + 8$

16. $2x + 4 - x = 4x - 5$

17. $12w + 15w - 9 + 5 = -3w + 5 - 9$

18. $-4x + 5x - 8 + 4 = 6x - 4$

19. $3(2t - 4) = 20 - 2t$

20. $2(3 - 2x) = x - 4$

21. $-5(x + 1) + 3x + 2 = 6x + 4$

22. $5(x + 3) + 4x - 5 = 4 - 2x$

23. $-2x + 5x - 9 = 3(x - 4) - 5$

24. $-6x + 2x - 11 = -2(2x - 3) + 4$

25. $2(x + 3) = -4(x + 1)$

26. $4(x - 9) = 8(x + 3)$

27. $3(2x + 1) - 2(x - 2) = 5$

28. $4(x - 2) + 2(x + 3) = 6$

29. $2x + 3(x - 4) = 2(x - 3)$

30. $6x - 3(5x + 2) = 4(1 - x)$

31. $6x - 4(3 - 2x) = 5(x - 4) - 10$

32. $-2x - 3(4 - 2x) = 2(x - 3) + 2$

33. $-2(x + 3) - x - 4 = -3(x + 4) + 2$

34. $4(2x + 7) = 2x + 25 + 3(2x + 1)$

35. $2[x - (2x + 4) + 3] = 2(x + 1)$

36. $4[2x - (3 - x) + 5] = -(2 + 7x)$

37. $-[2x - (5x + 2)] = 2 + (2x + 7)$

38. $-[6x - (4x + 8)] = 9 + (6x + 3)$

39. $-3x + 6 - 5(x - 1) = -5x - (2x - 4) + 5$

40. $4(x + 2) - 8x - 5 = -3x + 9 - 2(x + 6)$

41. $7[2 - (3 + 4x)] - 2x = -9 + 2(1 - 15x)$

42. $4[6 - (1 + 2x)] + 10x = 2(10 - 3x) + 8x$

43. $-[3x - (2x + 5)] = -4 - [3(2x - 4) - 3x]$

44. $2[-(x - 1) + 4] = 5 + [-(6x - 7) + 9x]$

45. *Concept Check* To solve the linear equation

$$\frac{8x}{3} - \frac{5x}{4} = -13,$$

we multiply each side by the least common denominator of all the fractions in the equation. What is this least common denominator?

46. Suppose that in solving the equation

$$\frac{1}{3}x + \frac{1}{2}x = \frac{1}{6}x,$$

we begin by multiplying each side by 12, rather than the *least* common denominator, 6. Would we get the correct solution? Explain.

47. *Concept Check* To solve a linear equation with decimals, we usually begin by multiplying by a power of 10 so that all coefficients are integers. What is the least power of 10 that will accomplish this goal in each equation?

   (a) $0.05x + 0.12(x + 5000) = 940$ **(Exercise 63)**

   (b) $0.006(x + 2) = 0.007x + 0.009$ **(Exercise 69)**

48. *Concept Check* The expression $0.06(10 - x)(100)$ is equivalent to which of the following?

   A. $0.06 - 0.06x$ **B.** $60 - 6x$ **C.** $6 - 6x$ **D.** $6 - 0.06x$

*Solve each equation, and check your solution. **See Examples 4 and 5.***

49. $-\dfrac{5}{9}x = 2$ **50.** $\dfrac{3}{11}x = -5$ **51.** $\dfrac{6}{5}x = -1$

52. $-\dfrac{7}{8}x = 6$ **53.** $\dfrac{x}{2} + \dfrac{x}{3} = 5$ **54.** $\dfrac{x}{5} - \dfrac{x}{4} = 1$

55. $\dfrac{3x}{4} + \dfrac{5x}{2} = 13$ **56.** $\dfrac{8x}{3} - \dfrac{x}{2} = -13$ **57.** $\dfrac{x - 10}{5} + \dfrac{2}{5} = -\dfrac{x}{3}$

58. $\dfrac{2x - 3}{7} + \dfrac{3}{7} = -\dfrac{x}{3}$ **59.** $\dfrac{3x - 1}{4} + \dfrac{x + 3}{6} = 3$ **60.** $\dfrac{3x + 2}{7} - \dfrac{x + 4}{5} = 2$

61. $\dfrac{4x + 1}{3} = \dfrac{x + 5}{6} + \dfrac{x - 3}{6}$ **62.** $\dfrac{2x + 5}{5} = \dfrac{3x + 1}{2} + \dfrac{-x + 7}{2}$

63. $0.05x + 0.12(x + 5000) = 940$ **64.** $0.09x + 0.13(x + 300) = 61$

65. $0.02(50) + 0.08x = 0.04(50 + x)$

66. $0.20(14{,}000) + 0.14x = 0.18(14{,}000 + x)$

67. $0.05x + 0.10(200 - x) = 0.45x$

68. $0.08x + 0.12(260 - x) = 0.48x$

69. $0.006(x + 2) = 0.007x + 0.009$

70. $0.004x + 0.006(50 - x) = 0.004(68)$

"Preview Exercises" are designed to review ideas introduced earlier, as well as preview ideas needed for the next section.

PREVIEW EXERCISES

*Use the given value(s) to evaluate each expression. **See Section 1.3.***

71. $2L + 2W$; $L = 10$, $W = 8$ **72.** rt; $r = 0.15, t = 3$

73. $\dfrac{1}{3}Bh$; $B = 27, h = 8$ **74.** prt; $p = 8000, r = 0.06, t = 2$

75. $\dfrac{5}{9}(F - 32)$; $F = 122$ **76.** $\dfrac{9}{5}C + 32$; $C = 60$

Tackling Your Homework

You are ready to do your homework **AFTER** you have read the corresponding textbook section and worked through the examples and Now Try exercises.

Homework Tips

▶ **Work problems neatly.** Use pencil and write legibly, so others can read your work. Skip lines between steps. Clearly separate problems from each other.

▶ **Show all your work.** It is tempting to take shortcuts. Include ALL steps.

▶ **Check your work frequently to make sure you are on the right track.** It is hard to unlearn a mistake. For all odd-numbered problems, answers are given in the back of the book.

▶ **If you have trouble with a problem, refer to the corresponding worked example in the section.** The exercise directions will often reference specific examples to review. Pay attention to every line of the worked example to see how to get from step to step.

▶ **If you are having trouble with an even-numbered problem, work the corresponding odd-numbered problem.** Check your answer in the back of the book, and apply the same steps to work the even-numbered problem.

▶ **Mark any problems you don't understand.** Ask your instructor about them.

Select several homework tips to try this week.

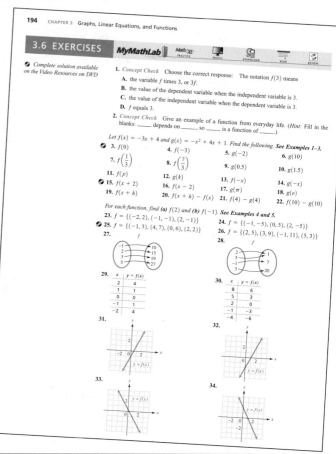

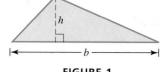

2.2 Formulas and Percent

OBJECTIVES

1. Solve a formula for a specified variable.
2. Solve applied problems by using formulas.
3. Solve percent problems.
4. Solve problems involving percent increase or decrease.

A **mathematical model** is an equation or inequality that describes a real situation. Models for many applied problems, called *formulas,* already exist. A formula is an equation in which variables are used to describe a relationship. For example, the formula for finding the area $\mathcal{A}$ of a triangle is

$$\mathcal{A} = \tfrac{1}{2}bh.$$

Here, b is the length of the base and h is the height. See **FIGURE 1.** A list of formulas used in algebra is given inside the covers of this book.

FIGURE 1

OBJECTIVE 1 **Solve a formula for a specified variable.** The formula $I = prt$ says that interest on a loan or investment equals principal (amount borrowed or invested) times rate (percent) times time at interest (in years). To determine how long it will take for an investment at a stated interest rate to earn a predetermined amount of interest, it would help to first solve the formula for t. This process is called **solving for a specified variable** or **solving a literal equation.**

When solving for a specified variable, the key is to treat that variable as if it were the only one. Treat all other variables like numbers (constants). The steps used in the following examples are very similar to those used in solving linear equations from **Section 2.1.**

NOW TRY
EXERCISE 1
Solve the formula $I = prt$
for p.

EXAMPLE 1 Solving for a Specified Variable

Solve the formula $I = prt$ for t.

We solve this formula for t by treating I, p, and r as constants (having fixed values) and treating t as the only variable.

$$prt = I$$ Our goal is to isolate t.

$$(pr)t = I$$ Associative property

$$\frac{(pr)t}{pr} = \frac{I}{pr}$$ Divide by pr.

$$t = \frac{I}{pr}$$

The result is a formula for t, time in years.   *NOW TRY*

Solving for a Specified Variable

Step 1   If the equation contains fractions, multiply both sides by the LCD to clear the fractions.

Step 2   Transform so that all terms containing the specified variable are on one side of the equation and all terms without that variable are on the other side.

Step 3   Divide each side by the factor that is the coefficient of the specified variable.

EXAMPLE 2 Solving for a Specified Variable

Solve the formula $P = 2L + 2W$ for W.

This formula gives the relationship between perimeter of a rectangle, P, length of the rectangle, L, and width of the rectangle, W. See **FIGURE 2.**

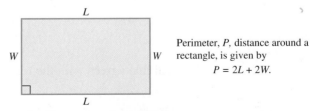

Perimeter, P, distance around a rectangle, is given by
$$P = 2L + 2W.$$

FIGURE 2

NOW TRY ANSWER
1. $p = \frac{I}{rt}$

We solve the formula for W by isolating W on one side of the equals symbol.

NOW TRY
EXERCISE 2
Solve the formula for b.

$$P = a + 2b + c$$

$$P = 2L + 2W \qquad \text{Solve for } W.$$

Step 1 is not needed here, since there are no fractions in the formula.

Step 2
$$P - 2L = 2L + 2W - 2L \qquad \text{Subtract } 2L.$$
$$P - 2L = 2W \qquad \text{Combine like terms.}$$

Step 3
$$\frac{P - 2L}{2} = \frac{2W}{2} \qquad \text{Divide by 2.}$$

$$\frac{P - 2L}{2} = W, \quad \text{or} \quad W = \frac{P - 2L}{2} \qquad \text{NOW TRY}$$

> ⚠ **CAUTION** In Step 3 of **Example 2,** we cannot simplify the fraction by dividing 2 into the term $2L$. The fraction bar serves as a grouping symbol. Thus, the subtraction in the numerator must be done before the division.
>
> $$\frac{P - 2L}{2} \neq P - L$$

NOW TRY
EXERCISE 3
Solve $P = 2(L + W)$ for L.

EXAMPLE 3   Solving a Formula Involving Parentheses

The formula for the perimeter of a rectangle is sometimes written in the equivalent form $P = 2(L + W)$. Solve this form for W.

One way to begin is to use the distributive property on the right side of the equation to get $P = 2L + 2W$, which we would then solve as in **Example 2.** Another way to begin is to divide by the coefficient 2.

$$P = 2(L + W)$$

$$\frac{P}{2} = L + W \qquad \text{Divide by 2.}$$

$$\frac{P}{2} - L = W, \quad \text{or} \quad W = \frac{P}{2} - L \qquad \text{Subtract } L.$$

We can show that this result is equivalent to our result in **Example 2** by rewriting L as $\frac{2}{2}L$.

$$\frac{P}{2} - L = W$$

$$\frac{P}{2} - \frac{2}{2}(L) = W \qquad \frac{2}{2} = 1, \text{ so } L = \frac{2}{2}(L).$$

$$\frac{P}{2} - \frac{2L}{2} = W$$

$$\frac{P - 2L}{2} = W \qquad \text{Subtract fractions.}$$

The final line agrees with the result in **Example 2.** NOW TRY

NOW TRY ANSWERS
2. $b = \frac{P - a - c}{2}$
3. $L = \frac{P}{2} - W$, or $L = \frac{P - 2W}{2}$

In **Examples 1–3,** we solved formulas for specified variables. In **Example 4,** we solve an equation with two variables for one of these variables. This process will be useful when we work with *linear equations in two variables* in **Chapter 4.**

**NOW TRY
EXERCISE 4**
Solve the equation for y.

$$5x - 6y = 12$$

EXAMPLE 4 Solving an Equation for One of the Variables

Solve the equation $3x - 4y = 12$ for y.

Our goal is to isolate y on one side of the equation.

$$3x - 4y = 12$$
$$3x - 4y - 3x = 12 - 3x \quad \text{Subtract } 3x.$$
$$-4y = 12 - 3x \quad \text{Combine like terms.}$$
$$\frac{-4y}{-4} = \frac{12 - 3x}{-4} \quad \text{Divide by } -4.$$
$$y = \frac{12 - 3x}{-4}$$

There are other equivalent forms of the final answer that are also correct. For example, since $\frac{a}{-b} = \frac{-a}{b}$ **(Section 1.2)**, we rewrite the fraction by moving the negative sign from the denominator to the numerator, taking care to distribute to both terms.

Multiply both terms of the numerator by -1.

$$y = \frac{12 - 3x}{-4} \quad \text{can be written as} \quad y = \frac{-(12 - 3x)}{4}, \quad \text{or} \quad y = \frac{3x - 12}{4}.$$

NOW TRY

OBJECTIVE 2 Solve applied problems by using formulas. The distance formula, $d = rt$, relates d, the distance traveled, r, the rate or speed, and t, the travel time.

**NOW TRY
EXERCISE 5**
It takes $\frac{1}{2}$ hr for Dorothy Easley to drive 21 mi to work each day. What is her average rate?

EXAMPLE 5 Finding Average Rate

Phyllis Koenig found that on average it took her $\frac{3}{4}$ hr each day to drive a distance of 15 mi to work. What was her average rate (or speed)?

Find the formula for rate r by solving $d = rt$ for r.

$$d = rt$$
$$\frac{d}{t} = \frac{rt}{t} \quad \text{Divide by } t.$$
$$\frac{d}{t} = r, \quad \text{or} \quad r = \frac{d}{t}$$

Notice that only Step 3 was needed to solve for r in this example. Now find the rate by substituting the given values of d and t into this formula.

$$r = \frac{15}{\frac{3}{4}} \quad \text{Let } d = 15, t = \frac{3}{4}.$$
$$r = 15 \cdot \frac{4}{3} \quad \text{Multiply by the reciprocal of } \frac{3}{4}.$$
$$r = 20$$

NOW TRY ANSWERS

4. $y = \frac{12 - 5x}{-6},$ or $y = \frac{5x - 12}{6}$

5. 42 mph

Her average rate was 20 mph. (That is, at times she may have traveled a little faster or slower than 20 mph, but overall her rate was 20 mph.)

NOW TRY

OBJECTIVE 3 **Solve percent problems.** An important everyday use of mathematics involves the concept of percent. Percent is written with the symbol %. The word **percent** means "per one hundred." One percent means "one per one hundred" or "one one-hundredth."

$$1\% = 0.01 \quad \text{or} \quad 1\% = \frac{1}{100}$$

> ### Solving a Percent Problem
>
> Let a represent a partial amount of b, the base, or whole amount. Then the following equation can be used to solve a percent problem.
>
> $$\frac{\textbf{partial amount } a}{\textbf{base } b} = \textbf{percent (represented as a decimal)}$$

For example, if a class consists of 50 students and 32 are males, then the percent of males in the class is found as follows.

$$\frac{\text{partial amount } a}{\text{base } b} = \frac{32}{50} \qquad \text{Let } a = 32, b = 50.$$

$$= \frac{64}{100} \qquad \frac{32}{50} \cdot \frac{2}{2} = \frac{64}{100}$$

$$= 0.64, \quad \text{or} \quad 64\% \qquad \text{Write as a decimal and then a percent.}$$

NOW TRY
EXERCISE 6

Solve each problem.

(a) A 5-L mixture of water and antifreeze contains 2 L of antifreeze. What is the percent of antifreeze in the mixture?

(b) If a savings account earns 2.5% interest on a balance of $7500 for one year, how much interest is earned?

EXAMPLE 6 Solving Percent Problems

(a) A 50-L mixture of acid and water contains 10 L of acid. What is the percent of acid in the mixture?

The given amount of the mixture is 50 L, and the part that is acid is 10 L. Let x represent the percent of acid in the mixture.

$$x = \frac{10}{50} \quad \begin{array}{l} \leftarrow \text{partial amount} \\ \leftarrow \text{whole amount (base)} \end{array}$$

$$x = 0.20, \quad \text{or} \quad 20\%$$

The mixture is 20% acid.

(b) If a savings account balance of $4780 earns 5% interest in one year, how much interest is earned?

Let x represent the amount of interest earned (that is, the part of the whole amount invested). Since 5% = 0.05, the equation is written as follows.

$$\frac{x}{4780} = 0.05 \qquad \frac{\text{partial amount } a}{\text{base } b} = \text{percent}$$

$$x = 0.05(4780) \qquad \text{Multiply by 4780.}$$

$$x = 239$$

NOW TRY ANSWERS
6. (a) 40% **(b)** $187.50

The interest earned is $239.

NOW TRY

**NOW TRY
EXERCISE 7**

Refer to **FIGURE 3**. How much was spent on vet care? Round your answer to the nearest tenth of a billion dollars.

EXAMPLE 7 Interpreting Percents from a Graph

In 2007, Americans spent about \$41.2 billion on their pets. Use the graph in **FIGURE 3** to determine how much of this amount was spent on pet food.

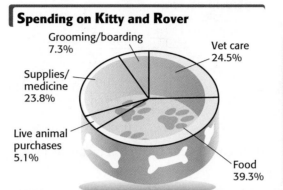

Spending on Kitty and Rover

Grooming/boarding 7.3%

Supplies/medicine 23.8%

Live animal purchases 5.1%

Vet care 24.5%

Food 39.3%

Source: American Pet Products Manufacturers Association Inc.

Pythagoras

FIGURE 3

Since 39.3% was spent on food, let $x =$ this amount in billions of dollars.

$$\frac{x}{41.2} = 0.393 \qquad \text{39.3\%} = 0.393$$

$$x = 41.2(0.393) \qquad \text{Multiply by 41.2.}$$

$$x \approx 16.2 \qquad \text{Nearest tenth}$$

Therefore, about \$16.2 billion was spent on pet food. NOW TRY

OBJECTIVE 4 **Solve problems involving percent increase or decrease.**
Percent is often used to express a change in some quantity. Buying an item that has been marked up and getting a raise at a job are applications of percent increase. Buying an item on sale and finding population decline are applications of percent decrease.

To solve problems of this type, we use the following form of the percent equation.

$$\text{percent change} = \frac{\text{amount of change}}{\text{base}}$$

Subtract to find this.

EXAMPLE 8 Solving Problems about Percent Increase or Decrease

(a) An electronics store marked up a laptop computer from their cost of \$1200 to a selling price of \$1464. What was the percent markup?

"Markup" is a name for an increase. Let $x =$ the percent increase (as a decimal).

$$\text{percent increase} = \frac{\text{amount of increase}}{\text{base}}$$

Subtract to find the *amount* of increase.

$$x = \frac{1464 - 1200}{1200} \qquad \text{Substitute the given values.}$$

Use the original cost.

$$x = \frac{264}{1200}$$

$$x = 0.22, \quad \text{or} \quad 22\% \qquad \text{Use a calculator.}$$

The computer was marked up 22%.

NOW TRY ANSWER
7. \$10.1 billion

NOW TRY
EXERCISE 8

(a) Jane Brand bought a jacket on sale for $56. The regular price of the jacket was $80. What was the percent markdown?

(b) When it was time for Horatio Loschak to renew the lease on his apartment, the landlord raised his rent from $650 to $689 a month. What was the percent increase?

(b) The enrollment at a community college declined from 12,750 during one school year to 11,350 the following year. Find the percent decrease to the nearest tenth. Let x = the percent decrease (as a decimal).

$$\text{percent decrease} = \frac{\text{amount of decrease}}{\text{base}}$$

Subtract to find the amount of decrease.

$$x = \frac{12,750 - 11,350}{12,750} \qquad \text{Substitute the given values.}$$

Use the original number.

$$x = \frac{1400}{12,750}$$

$$x \approx 0.11, \quad \text{or} \quad 11\% \qquad \text{Use a calculator.}$$

The college enrollment decreased by about 11%.

NOW TRY

> ⚠️ **CAUTION** When calculating a percent increase or decrease, be sure that you use the original number (*before* the increase or decrease) as the base. A common error is to use the final number (*after* the increase or decrease) in the denominator of the fraction.

NOW TRY ANSWERS
8. (a) 30% (b) 6%

(handwritten) 9/5/12

2.2 EXERCISES

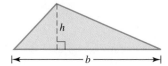

(handwritten notes in left margin)
perimeter of a triangle
6. P = a+b+c; for b
P = a+b+c
−a −a
P−a = b+c   answer
    −c   −c
b = [P−a−c] ✓

8. A = ½bh; for h
2A = 2(½)bh
2A = bh
2A/b = bh/b
h = 2A/b

⊙ Complete solution available on the Video Resources on DVD

Solve each formula for the specified variable. ***See Examples 1–3.***

1. $I = prt$ for r (simple interest)

2. $d = rt$ for t (distance)

⊙ **3.** $P = 2L + 2W$ for L
(perimeter of a rectangle)

4. $\mathscr{A} = bh$ for b (area of a parallelogram)*

5. $V = LWH$
(volume of a rectangular solid)
(a) for W **(b)** for H

6. $P = a + b + c$
(perimeter of a triangle)
(a) for b **(b)** for c

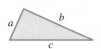

7. $C = 2\pi r$ for r
(circumference of a circle)

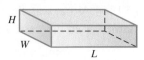

8. $\mathscr{A} = \dfrac{1}{2}bh$ for h
(area of a triangle)

*In this book, we use $\mathscr{A}$ to denote area.

9. $\mathcal{A} = \dfrac{1}{2}h(b + B)$ (area of a trapezoid)

(a) for h (b) for B

10. $S = 2\pi rh + 2\pi r^2$ for h
(surface area of a right circular cylinder)

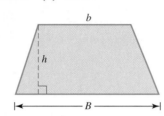

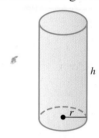

11. $F = \dfrac{9}{5}C + 32$ for C
(Celsius to Fahrenheit)

12. $C = \dfrac{5}{9}(F - 32)$ for F
(Fahrenheit to Celsius)

13. *Concept Check* When a formula is solved for a particular variable, several different equivalent forms may be possible. If we solve $\mathcal{A} = \frac{1}{2}bh$ for h, one possible correct answer is

$$h = \frac{2\mathcal{A}}{b}.$$

Which one of the following is *not* equivalent to this?

A. $h = 2\left(\dfrac{\mathcal{A}}{b}\right)$ **B.** $h = 2\mathcal{A}\left(\dfrac{1}{b}\right)$ **C.** $h = \dfrac{\mathcal{A}}{\frac{1}{2}b}$ **D.** $h = \dfrac{\frac{1}{2}\mathcal{A}}{b}$

14. *Concept Check* The answer to **Exercise 11** is given as $C = \frac{5}{9}(F - 32)$. Which one of the following is *not* equivalent to this?

A. $C = \dfrac{5}{9}F - \dfrac{160}{9}$ **B.** $C = \dfrac{5F}{9} - \dfrac{160}{9}$ **C.** $C = \dfrac{5F - 160}{9}$ **D.** $C = \dfrac{5}{9}F - 32$

Solve each equation for y. ***See Example 4.***

15. $4x + 9y = 11$ **16.** $7x + 8y = 11$ **17.** $-3x + 2y = 5$

18. $-5x + 3y = 12$ **19.** $6x - 5y = 7$ **20.** $8x - 3y = 4$

Solve each problem. ***See Example 5.***

21. Ryan Newman won the Daytona 500 (mile) race with a rate of 152.672 mph in 2008. Find his time to the nearest thousandth. (*Source:* www.daytona500.com)

22. In 2007, rain shortened the Indianapolis 500 race to 415 mi. It was won by Dario Franchitti, who averaged 151.774 mph. What was his time to the nearest thousandth? (*Source:* www.indy500.com)

23. Nora Demosthenes traveled from Kansas City to Louisville, a distance of 520 mi, in 10 hr. Find her rate in miles per hour.

24. The distance from Melbourne to London is 10,500 mi. If a jet averages 500 mph between the two cities, what is its travel time in hours?

25. As of 2009, the highest temperature ever recorded in Tennessee was 45°C. Find the corresponding Fahrenheit temperature. (*Source:* National Climatic Data Center.)

26. As of 2009, the lowest temperature ever recorded in South Dakota was $-58°$F. Find the corresponding Celsius temperature. (*Source:* National Climate Data Center.)

27. The base of the Great Pyramid of Cheops is a square whose perimeter is 920 m. What is the length of each side of this square? (*Source: Atlas of Ancient Archaeology.*)

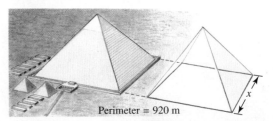
Perimeter = 920 m

28. Marina City in Chicago is a complex of two residential towers that resemble corncobs. Each tower has a concrete cylindrical core with a 35-ft diameter and is 588 ft tall. Find the volume of the core of one of the towers to the nearest whole number. (*Hint:* Use the π key on your calculator.) (*Source:* www.architechgallery.com; www.aviewoncities.com)

29. The circumference of a circle is 480π in. What is the radius? What is the diameter?

30. The radius of a circle is 2.5 in. What is the diameter? What is the circumference?

$r = 2.5$ in.

31. A sheet of standard-size copy paper measures 8.5 in. by 11 in. If a ream (500 sheets) of this paper has a volume of 187 in.^{3}, how thick is the ream?

32. Copy paper (**Exercise 31**) also comes in legal size, which has the same width, but is longer than standard size. If a ream of legal-size paper has the same thickness as standard-size paper and a volume of 238 in.^{3}, what is the length of a sheet of legal paper?

11 in.
8.5 in.

*Solve each problem. **See Example 6.***

33. A mixture of alcohol and water contains a total of 36 oz of liquid. There are 9 oz of pure alcohol in the mixture. What percent of the mixture is water? What percent is alcohol?

34. A mixture of acid and water is 35% acid. If the mixture contains a total of 40 L, how many liters of pure acid are in the mixture? How many liters of pure water are in the mixture?

35. A real-estate agent earned $6300 commission on a property sale of $210,000. What is her rate of commission?

36. A certificate of deposit for 1 yr pays $221 simple interest on a principal of $3400. What is the interest rate being paid on this deposit?

*When a consumer loan is paid off ahead of schedule, the finance charge is less than if the loan were paid off over its scheduled life. By one method, called the **rule of 78,** the amount of unearned interest (the finance charge that need not be paid) is given by*

$$u = f \cdot \frac{k(k+1)}{n(n+1)}.$$

In the formula, u is the amount of unearned interest (money saved) when a loan scheduled to run for n payments is paid off k payments ahead of schedule. The total scheduled finance charge is f. Use the formula for the rule of 78 to work Exercises 37–40.

37. Sondra Braeseker bought a new car and agreed to pay it off in 36 monthly payments. The total finance charge was $700. Find the unearned interest if she paid the loan off 4 payments ahead of schedule.

38. Donnell Boles bought a truck and agreed to pay it off in 36 monthly payments. The total finance charge on the loan was $600. With 12 payments remaining, he decided to pay the loan in full. Find the amount of unearned interest.

39. The finance charge on a loan taken out by Kha Le is $380.50. If 24 equal monthly installments were needed to repay the loan, and the loan is paid in full with 8 months remaining, find the amount of unearned interest.

40. Maky Manchola is scheduled to repay a loan in 24 equal monthly installments. The total finance charge on the loan is $450. With 9 payments remaining, he decides to repay the loan in full. Find the amount of unearned interest.

In baseball, winning percentage (Pct.) is commonly expressed as a decimal rounded to the nearest thousandth. To find the winning percentage of a team, divide the number of wins (W) by the total number of games played (W + L).

41. The final 2009 standings of the Eastern Division of the American League are shown in the table. Find the winning percentage of each team.

    (a) Boston   **(b)** Tampa Bay

    (c) Toronto   **(d)** Baltimore

	W	L	Pct.
New York Yankees	103	59	.636
Boston	95	67	
Tampa Bay	84	78	
Toronto	75	87	
Baltimore	64	98	

Source: World Almanac and Book of Facts.

42. Repeat **Exercise 41** for the following standings for the Eastern Division of the National League.

    (a) Philadelphia   **(b)** Atlanta

    (c) New York Mets   **(d)** Washington

	W	L	Pct.
Philadelphia	93	69	
Florida	87	75	.537
Atlanta	86	76	
New York Mets	70	92	
Washington	59	103	

Source: World Almanac and Book of Facts.

As mentioned in the chapter introduction, 114.9 million U.S. households owned at least one TV set in 2009. (Source: Nielsen Media Research.) Use this information to work Exercises 43–46. Round answers to the nearest percent in Exercises 43–44, and to the nearest tenth million in Exercises 45–46. ***See Example 6.***

43. About 62.0 million U.S. households owned 3 or more TV sets in 2009. What percent of those owning at least one TV set was this?

44. About 102.2 million households that owned at least one TV set in 2009 had a DVD player. What percent of those owning at least one TV set had a DVD player?

45. Of the households owning at least one TV set in 2009, 88% received basic cable. How many households received basic cable?

46. Of the households owning at least one TV set in 2009, 35% received premium cable. How many households received premium cable?

An average middle-income family will spend $221,190 to raise a child born in 2008 from birth through age 17. The graph shows the percents spent for various categories. Use the graph to answer Exercises 47–50. **See Example 7.**

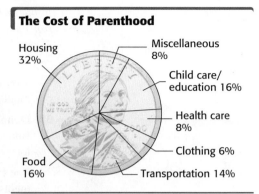

The Cost of Parenthood

Housing 32%

Miscellaneous 8%

Child care/ education 16%

Health care 8%

Clothing 6%

Transportation 14%

Food 16%

Source: U.S. Department of Agriculture.

47. To the nearest dollar, how much will be spent to provide housing for the child?

48. To the nearest dollar, how much will be spent for health care?

49. Use your answer from **Exercise 48** to find how much will be spent for child care and education.

50. About $35,000 will be spent for food. To the nearest percent, what percent of the cost of raising a child from birth through age 17 is this? Does your answer agree with the percent shown in the graph?

Solve each problem about percent increase or percent decrease. **See Example 8.**

51. After 1 yr on the job, Grady got a raise from $10.50 per hour to $11.34 per hour. What was the percent increase in his hourly wage?

52. Clayton bought a ticket to a rock concert at a discount. The regular price of the ticket was $70.00, but he only paid $59.50. What was the percent discount?

53. Between 2000 and 2007, the estimated population of Pittsfield, Massachusetts, declined from 134,953 to 129,798. What was the percent decrease to the nearest tenth? (*Source:* U.S. Census Bureau.)

54. Between 2000 and 2007, the estimated population of Anchorage, Alaska, grew from 320,391 to 362,340. What was the percent increase to the nearest tenth? (*Source:* U.S. Census Bureau.)

55. In April 2008, the audio CD of the Original Broadway Cast Recording of the musical *Wicked* was available for $9.97. The list price (full price) of this CD was $18.98. To the nearest tenth, what was the percent discount? (*Source:* www.amazon.com)

56. In April 2008, the DVD of the movie *Alvin and the Chipmunks* was released. This DVD had a list price of $29.99 and was on sale for $15.99. To the nearest tenth, what was the percent discount? (*Source:* www.amazon.com)

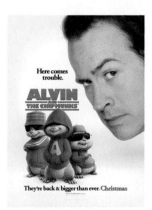

Solve each equation. **See Section 2.1.**

57. $4x + 4(x + 7) = 124$

58. $x + 0.20x = 66$

59. $2.4 + 0.4x = 0.25(6 + x)$

60. $0.07x + 0.05(9000 - x) = 510$

Evaluate. See Section 1.2.

61. The product of -3 and 5, divided by 1 less than

62. Half of -18, added to the reciprocal of $\frac{1}{5}$

63. The sum of 6 and -9, multiplied by the additiv

64. The product of -2 and 4, added to the product

Applications of Linear Equations

OBJECTIVES

1 Translate from words to mathematical expressions.

2 Write equations from given information.

3 Distinguish between simplifying expressions and solving equations.

4 Use the six steps in solving an applied problem.

5 Solve percent problems.

6 Solve investment problems.

7 Solve mixture problems.

OBJECTIVE 1 Translate from words to mathematical expressions.

PROBLEM-SOLVING HINT

There are usually key words and phrases in a verbal problem that translate into mathematical expressions involving addition, subtraction, multiplication, and division. Translations of some commonly used expressions follow.

Translating from Words to Mathematical Expressions

Verbal Expression	Mathematical Expression (where x and y are numbers)
Addition	
The **sum** of a number and 7	$x + 7$
6 **more than** a number	$x + 6$
3 **plus** a number	$3 + x$
24 **added to** a number	$x + 24$
A number **increased by** 5	$x + 5$
The **sum** of two numbers	$x + y$
Subtraction	
2 **less than** a number	$x - 2$
2 **less** a number	$2 - x$
12 **minus** a number	$12 - x$
A number **decreased by** 12	$x - 12$
A number **subtracted from** 10	$10 - x$
10 **subtracted from** a number	$x - 10$
The **difference between** two numbers	$x - y$
Multiplication	
16 **times** a number	$16x$
A number **multiplied by** 6	$6x$
$\frac{2}{3}$ **of** a number (used with fractions and percent)	$\frac{2}{3}x$
$\frac{3}{4}$ **as much as** a number	$\frac{3}{4}x$
Twice (2 times) a number	$2x$
The **product** of two numbers	xy
Division	
The **quotient** of 8 and a number	$\frac{8}{x}$ $(x \neq 0)$
A number **divided by** 13	$\frac{x}{13}$
The **ratio** of two numbers or the quotient of two numbers	$\frac{x}{y}$ $(y \neq 0)$

⚠ **CAUTION** *Because subtraction and division are not commutative operations, it is important to correctly translate expressions involving them.* For example,

"2 less than a number" is translated as $x - 2$, ***not*** $2 - x$.

"A number subtracted from 10" is expressed as $10 - x$, ***not*** $x - 10$.

For division, the number *by which* we are dividing is the denominator, and the number *into which* we are dividing is the numerator.

"A number divided by 13" and "13 divided into x" both translate as $\frac{x}{13}$.

"The quotient of x and y" is translated as $\frac{x}{y}$.

OBJECTIVE 2 Write equations from given information. The symbol for equality, $=$, is often indicated by the word *is*.

NOW TRY EXERCISE 1

Translate each verbal sentence into an equation, using x as the variable.

(a) The quotient of a number and 10 is twice the number.

(b) The product of a number and 5, decreased by 7, is zero.

EXAMPLE 1 Translating Words into Equations

Translate each verbal sentence into an equation.

Verbal Sentence	Equation
Twice a number, decreased by 3, is 42.	$2x - 3 = 42$
The product of a number and 12, decreased by 7, is 105.	$12x - 7 = 105$
The quotient of a number and the number plus 4 is 28.	$\dfrac{x}{x + 4} = 28$
The quotient of a number and 4, plus the number, is 10.	$\dfrac{x}{4} + x = 10$

Any words that indicate the idea of "sameness" translate as $=$.

NOW TRY

OBJECTIVE 3 Distinguish between simplifying expressions and solving equations. An expression translates as a phrase. An equation includes the $=$ symbol, with expressions on both sides, and translates as a sentence.

EXAMPLE 2 Distinguishing between Simplifying Expressions and Solving Equations

Decide whether each is an *expression* or an *equation*. Simplify any expressions, and solve any equations.

(a) $2(3 + x) - 4x + 7$

There is no equals symbol, so this is an expression.

$$2(3 + x) - 4x + 7$$
$$= 6 + 2x - 4x + 7 \qquad \text{Distributive property}$$
$$= -2x + 13 \qquad \text{Simplified expression}$$

(b) $2(3 + x) - 4x + 7 = -1$

Because there is an equals symbol with expressions on both sides, this is an equation.

$$2(3 + x) - 4x + 7 = -1$$
$$6 + 2x - 4x + 7 = -1 \qquad \text{Distributive property}$$

NOW TRY ANSWERS

1. (a) $\frac{x}{10} = 2x$ **(b)** $5x - 7 = 0$

NOW TRY
EXERCISE 2

Decide whether each is an
expression or an *equation*.
Simplify any expressions, and
solve any equations.

(a) $3(x - 5) + 2x - 1$

(b) $3(x - 5) + 2x = 1$

$$-2x + 13 = -1 \qquad \text{Combine like terms.}$$
$$-2x = -14 \qquad \text{Subtract 13.}$$
$$x = 7 \qquad \text{Divide by } -2.$$

The solution set is $\{7\}$.

OBJECTIVE 4 **Use the six steps in solving an applied problem.** While there
is no one method that allows us to solve all types of applied problems, the following
six steps are helpful.

Solving an Applied Problem

Step 1 **Read** the problem, several times if necessary. What information is
given? What is to be found?

Step 2 **Assign a variable** to represent the unknown value. Use a sketch, diagram, or table, as needed. Write down what the variable represents. If
necessary, express any other unknown values in terms of the variable.

Step 3 **Write an equation** using the variable expression(s).

Step 4 **Solve** the equation.

Step 5 **State the answer.** Label it appropriately. Does it seem reasonable?

Step 6 **Check** the answer in the words of the *original* problem.

EXAMPLE 3 Solving a Perimeter Problem

The length of a rectangle is 1 cm more than twice the width. The perimeter of the rectangle is 110 cm. Find the length and the width of the rectangle.

Step 1 **Read** the problem. What must be found? The
length and width of the rectangle. What is
given? The length is 1 cm more than twice
the width and the perimeter is 110 cm.

Step 2 **Assign a variable.** Let $W =$ the width. Then
$2W + 1 =$ the length. Make a sketch, as in
FIGURE 4.

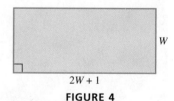

FIGURE 4

Step 3 **Write an equation.** Use the formula for the perimeter of a rectangle.

$$P = 2L + 2W \qquad \text{Perimeter of a rectangle}$$
$$110 = 2(2W + 1) + 2W \qquad \text{Let } L = 2W + 1 \text{ and } P = 110.$$

Step 4 **Solve** the equation obtained in Step 3.

$$110 = 4W + 2 + 2W \qquad \text{Distributive property}$$
$$110 = 6W + 2 \qquad \text{Combine like terms.}$$
$$110 - 2 = 6W + 2 - 2 \qquad \text{Subtract 2.}$$
$$108 = 6W \qquad \text{Combine like terms.}$$
$$\frac{108}{6} = \frac{6W}{6} \qquad \text{Divide by 6.}$$
$$18 = W$$

We also
need to find
the length.

NOW TRY ANSWERS

2. (a) expression; $5x - 16$

  (b) equation; $\left\{\frac{16}{5}\right\}$

NOW TRY
EXERCISE 3
The length of a rectangle is 2 ft more than twice the width. The perimeter is 34 ft. Find the length and width of the rectangle.

Step 5   **State the answer.** The width of the rectangle is 18 cm and the length is

$$2(18) + 1 = 37 \text{ cm}.$$

Step 6   **Check.** The length, 37 cm, is 1 cm more than $2(18)$ cm (twice the width). The perimeter is

$$2(37) + 2(18) = 74 + 36 = 110 \text{ cm}, \quad \text{as required.} \qquad \textit{NOW TRY} \curvearrowright$$

NOW TRY
EXERCISE 4
During the 2008 regular NFL football season, Drew Brees of the New Orleans Saints threw 4 more touchdown passes than Kurt Warner of the Arizona Cardinals. Together, these two quarterbacks completed a total of 64 touchdown passes. How many touchdown passes did each player complete? (*Source:* www.nfl.com)

Tim Lincecum

EXAMPLE 4   Finding Unknown Numerical Quantities

During the 2009 regular season, Justin Verlander of the Detroit Tigers and Tim Lincecum of the San Francisco Giants were the top major league pitchers in strikeouts. The two pitchers had a total of 530 strikeouts. Verlander had 8 more strikeouts than Lincecum. How many strikeouts did each pitcher have? (*Source:* www.mlb.com)

Step 1   **Read** the problem. We are asked to find the number of strikeouts each pitcher had.

Step 2   **Assign a variable** to represent the number of strikeouts for one of the men.

Let s = the number of strikeouts for Tim Lincecum.

We must also find the number of strikeouts for Justin Verlander. Since he had 8 more strikeouts than Lincecum,

$s + 8$ = the number of strikeouts for Verlander.

Step 3   **Write an equation.** The sum of the numbers of strikeouts is 530.

Lincecum's strikeouts	+	Verlander's strikeouts	=	Total
↓		↓		↓
s	$+$	$(s + 8)$	$=$	530

Step 4   **Solve** the equation.

$$s + (s + 8) = 530$$

$$2s + 8 = 530 \qquad \text{Combine like terms.}$$

$$2s + 8 - 8 = 530 - 8 \qquad \text{Subtract 8.}$$

$$2s = 522 \qquad \text{Combine like terms.}$$

$$\frac{2s}{2} = \frac{522}{2} \qquad \text{Divide by 2.}$$

> Don't stop here.

$$s = 261$$

Step 5   **State the answer.** We let s represent the number of strikeouts for Lincecum, so Lincecum had 261. Then Verlander had

$$s + 8 = 261 + 8 = 269 \text{ strikeouts.}$$

Step 6   **Check.** 269 is 8 more than 261, and $261 + 269 = 530$. The conditions of the problem are satisfied, and our answer checks. $\qquad \textit{NOW TRY} \curvearrowright$

⚠ **CAUTION**   Be sure to answer all the questions asked in the problem. In **Example 4,** we were asked for the number of strikeouts for *each* player, so there was extra work in Step 5 in order to find Verlander's number.

NOW TRY ANSWERS
3. width: 5 ft; length: 12 ft
4. Drew Brees: 34; Kurt Warner: 30

OBJECTIVE 5 **Solve percent problems.** Recall from **Section 2.2** that percent means "per one hundred," so 5% means 0.05, 14% means 0.14, and so on.

NOW TRY
EXERCISE 5

In the fall of 2009, there were 96 Introductory Statistics students at a certain community college, an increase of 700% over the number of Introductory Statistics students in the fall of 1992. How many Introductory Statistics students were there in the fall of 1992?

EXAMPLE 5   Solving a Percent Problem

In 2006, total annual health expenditures in the United States were about $2000 billion (or $2 trillion). This was an increase of 180% over the total for 1990. What were the approximate total health expenditures in billions of dollars in the United States in 1990? (*Source:* U.S. Centers for Medicare & Medicaid Services.)

Step 1   **Read** the problem. We are given that the total health expenditures increased by 180% from 1990 to 2006, and $2000 billion was spent in 2006. We must find the expenditures in 1990.

Step 2   **Assign a variable.** Let x represent the total health expenditures for 1990.

$$180\% \ = \ 180(0.01) \ = \ 1.8,$$

so $1.8x$ represents the additional expenditures since 1990.

Step 3   **Write an equation** from the given information.

the expenditures in 1990 + the increase = 2000

$$x \ + \ 1.8x \ = 2000$$

Note the x in $1.8x$.

Step 4   **Solve** the equation.

$1x + 1.8x = 2000$	Identity property
$2.8x = 2000$	Combine like terms.
$x \approx 714$	Divide by 2.8.

Step 5   **State the answer.** Total health expenditures in the United States for 1990 were about $714 billion.

Step 6   **Check** that the increase, $2000 - 714 = 1286$, is about 180% of 714.

NOW TRY

⚠ **CAUTION**   Avoid two common errors that occur in solving problems like the one in **Example 5.**

1. Do not try to find 180% of 2000 and subtract that amount from 2000. The 180% should be applied to *the amount in 1990, not the amount in 2006.*

2. Do not write the equation as

$$x + 1.8 = 2000. \quad \text{Incorrect}$$

The percent must be multiplied by some number. In this case, the number is the amount spent in 1990, giving $1.8x$.

OBJECTIVE 6 **Solve investment problems.** The investment problems in this chapter deal with *simple interest.* In most real-world applications, *compound interest* (covered in a later chapter) is used.

NOW TRY ANSWER
5. 12

NOW TRY
EXERCISE 6
Gary Jones received a $20,000 inheritance from his grandfather. He invested some of the money in an account earning 3% annual interest and the remaining amount in an account earning 2.5% annual interest. If the total annual interest earned is $575, how much is invested at each rate?

EXAMPLE 6 Solving an Investment Problem

Thomas Flanagan has $40,000 to invest. He will put part of the money in an account paying 4% interest and the remainder into stocks paying 6% interest. The total annual income from these investments should be $2040. How much should he invest at each rate?

Step 1   **Read** the problem again. We must find the two amounts.

Step 2   **Assign a variable.**

$$\text{Let} \qquad x = \text{the amount to invest at 4\%;}$$
$$40,000 - x = \text{the amount to invest at 6\%.}$$

The formula for interest is $I = prt$. Here the time t is 1 yr. Use a table to organize the given information.

Principal	Rate (as a decimal)	Interest	
x	0.04	$0.04x$	Multiply principal, rate, and time (here, 1 yr) to get interest.
$40,000 - x$	0.06	$0.06(40,000 - x)$	
40,000	✕✕✕✕	2040	← Total

Step 3   **Write an equation.** The last column of the table gives the equation.

$$\text{interest at 4\%} \quad + \quad \text{interest at 6\%} \quad = \quad \text{total interest}$$
$$0.04x \qquad + \quad 0.06(40,000 - x) \quad = \qquad 2040$$

Step 4   **Solve** the equation.

$$0.04x + 0.06(40,000) - 0.06x = 2040 \qquad \text{Distributive property.}$$
$$0.04x + 2400 - 0.06x = 2040 \qquad \text{Multiply.}$$
$$-0.02x + 2400 = 2040 \qquad \text{Combine like terms.}$$
$$-0.02x = -360 \qquad \text{Subtract 2400.}$$
$$x = 18,000 \qquad \text{Divide by } -0.02.$$

Step 5   **State the answer.** Thomas should invest $18,000 of the money at 4%. At 6%, he should invest

$$\$40,000 - \$18,000 = \$22,000.$$

Step 6   **Check.** Find the annual interest at each rate. The sum of these two amounts should total $2040.

$$0.04(\$18,000) = \$720 \quad \text{and} \quad 0.06(\$22,000) = \$1320$$
$$\$720 + \$1320 = \$2040, \quad \text{as required.} \qquad \text{NOW TRY}$$

PROBLEM-SOLVING HINT

In **Example 6,** we chose to let the variable represent the amount invested at 4%. Students often ask, "Can I let the variable represent the other unknown?" The answer is yes. The equation will be different, but in the end the answers will be the same.

NOW TRY ANSWER
6. $15,000 at 3%; $5000 at 2.5%

OBJECTIVE 7 Solve mixture problems.

NOW TRY
EXERCISE 7
How many liters of a 20% acid solution must be mixed with 5 L of a 30% acid solution to get a 24% acid solution?

EXAMPLE 7 Solving a Mixture Problem

A chemist must mix 8 L of a 40% acid solution with some 70% solution to get a 50% solution. How much of the 70% solution should be used?

Step 1 **Read** the problem. The problem asks for the amount of 70% solution to be used.

Step 2 **Assign a variable.** Let $x =$ the number of liters of 70% solution to be used. The information in the problem is illustrated in **FIGURE 5** and organized in the table.

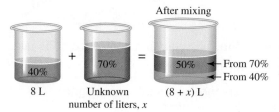

Number of Liters	Percent (as a decimal)	Liters of Pure Acid	
8	0.40	$0.40(8) = 3.2$	⎤ Sum must
x	0.70	$0.70x$	⎬ equal
$8 + x$	0.50	$0.50(8 + x)$	⎦

FIGURE 5

The numbers in the last column of the table were found by multiplying the strengths by the numbers of liters. The number of liters of pure acid in the 40% solution plus the number of liters in the 70% solution must equal the number of liters in the 50% solution.

Step 3 **Write an equation.**

$$3.2 + 0.70x = 0.50(8 + x)$$

Step 4 **Solve.**

$3.2 + 0.70x = 4 + 0.50x$	Distributive property
$0.20x = 0.8$	Subtract 3.2 and 0.50x.
$x = 4$	Divide by 0.20.

Step 5 **State the answer.** The chemist should use 4 L of the 70% solution.

Step 6 **Check.** 8 L of 40% solution plus 4 L of 70% solution is

$$8(0.40) + 4(0.70) = 6 \text{ L of acid.}$$

Similarly, $8 + 4$ or 12 L of 50% solution has

$$12(0.50) = 6 \text{ L of acid.}$$

The total amount of pure acid is 6 L both before and after mixing, so the answer checks. NOW TRY

NOW TRY ANSWER
7. $7\frac{1}{2}$ L

PROBLEM-SOLVING HINT

Remember that when pure water is added to a solution, water is 0% of the chemical (acid, alcohol, etc.). Similarly, pure chemical is 100% chemical.

NOW TRY
EXERCISE 8

How much pure antifreeze must be mixed with 3 gal of a 30% antifreeze solution to get a 40% antifreeze solution?

EXAMPLE 8   Solving a Mixture Problem When One Ingredient Is Pure

The octane rating of gasoline is a measure of its antiknock qualities. For a standard fuel, the octane rating is the percent of isooctane. How many liters of pure isooctane should be mixed with 200 L of 94% isooctane, referred to as 94 octane, to get a mixture that is 98% isooctane?

Step 1   **Read** the problem. The problem asks for the amount of pure isooctane.

Step 2   **Assign a variable.** Let x = the number of liters of pure (100%) isooctane. Complete a table. Recall that $100\% = 100(0.01) = 1$.

Number of Liters	Percent (as a decimal)	Liters of Pure Isooctane
x	1	x
200	0.94	0.94(200)
$x + 200$	0.98	0.98(x + 200)

Step 3   **Write an equation.** The equation comes from the last column of the table.

$$x + 0.94(200) = 0.98(x + 200)$$

Step 4   **Solve.**

$x + 0.94(200) = 0.98x + 0.98(200)$	Distributive property
$x + 188 = 0.98x + 196$	Multiply.
$0.02x = 8$	Subtract 0.98x and 188.
$x = 400$	Divide by 0.02.

Step 5   **State the answer.** 400 L of isooctane is needed.

NOW TRY ANSWER

8. $\frac{1}{2}$ gal

Step 6   **Check** by showing that $400 + 0.94(200) = 0.98(400 + 200)$ is true.

NOW TRY

2.3 EXERCISES

MyMathLab   Math XL PRACTICE   WATCH   DOWNLOAD   READ   REVIEW

Complete solution available on the Video Resources on DVD

Concept Check   In each of the following, **(a)** translate as an expression and **(b)** translate as an equation or inequality. Use x to represent the number.

1. **(a)** 15 more than a number
   (b) 15 is more than a number.

2. **(a)** 5 greater than a number
   (b) 5 is greater than a number.

3. **(a)** 8 less than a number
   (b) 8 is less than a number.

4. **(a)** 6 less than a number
   (b) 6 is less than a number.

5. *Concept Check*   Which one of the following is *not* a valid translation of "40% of a number," where x represents the number?

   A. $0.40x$ **B.** $0.4x$ **C.** $\dfrac{2x}{5}$ **D.** $40x$

6. Explain why $13 - x$ is *not* a correct translation of "13 less than a number."

Translate each verbal phrase into a mathematical expression. Use x to represent the unknown number. **See Example 1.**

7. Twice a number, decreased by 13

8. The product of 6 and a number, decreased by 14

9. 12 increased by four times a number

10. 15 more than one-half of a number

11. The product of 8 and 16 less than a number

12. The product of 8 more than a number and 5 less than the number

13. The quotient of three times a number and 10

14. The quotient of 9 and five times a nonzero number

*Use the variable x for the unknown, and write an equation representing the verbal sentence. Then solve the problem. **See Example 1.***

15. The sum of a number and 6 is -31. Find the number.

16. The sum of a number and -4 is 18. Find the number.

17. If the product of a number and -4 is subtracted from the number, the result is 9 more than the number. Find the number.

18. If the quotient of a number and 6 is added to twice the number, the result is 8 less than the number. Find the number.

19. When $\frac{2}{3}$ of a number is subtracted from 14, the result is 10. Find the number.

20. When 75% of a number is added to 6, the result is 3 more than the number. Find the number.

Decide whether each is an expression *or an* equation. *Simplify any expressions, and solve any equations. **See Example 2.***

21. $5(x + 3) - 8(2x - 6)$

22. $-7(x + 4) + 13(x - 6)$

23. $5(x + 3) - 8(2x - 6) = 12$

24. $-7(x + 4) + 13(x - 6) = 18$

25. $\frac{1}{2}x - \frac{1}{6}x + \frac{3}{2} - 8$

26. $\frac{1}{3}x + \frac{1}{5}x - \frac{1}{2} + 7$

Concept Check *Complete the six suggested problem-solving steps to solve each problem.*

27. In 2008, the corporations securing the most U.S. patents were IBM and Samsung. Together, the two corporations secured a total of 7671 patents, with Samsung receiving 667 fewer patents than IBM. How many patents did each corporation secure? (*Source:* U.S. Patent and Trademark Office.)

> *Step 1* **Read** the problem carefully. We are asked to find _____.
>
> *Step 2* **Assign a variable.** Let $x =$ the number of patents that IBM secured. Then $x - 667 =$ the number of _____.
>
> *Step 3* **Write an equation.** _____ $+$ _____ $= 7671$
>
> *Step 4* **Solve** the equation. $x =$ _____
>
> *Step 5* **State the answer.** IBM secured _____ patents, and Samsung secured _____ patents.
>
> *Step 6* **Check.** The number of Samsung patents was _____ fewer than the number of _____, and the total number of patents was $4169 +$ _____ $=$ _____.

28. In 2008, 7.8 million more U.S. residents traveled to Mexico than to Canada. There was a total of 32.8 million U.S. residents traveling to these two countries. How many traveled to each country? (*Source:* U.S. Department of Commerce.)

> *Step 1* **Read** the problem carefully. We are asked to find _____.
>
> *Step 2* **Assign a variable.** Let $x =$ the number of travelers to Mexico (in millions). Then $x - 7.8 =$ the number of _____.
>
> *Step 3* **Write an equation.** _____ $+$ _____ $= 32.8$
>
> *Step 4* **Solve** the equation. $x =$ _____
>
> *Step 5* **State the answer.** There were _____ travelers to Mexico and _____ travelers to Canada.
>
> *Step 6* **Check.** The number of _____ was _____ more than the number of _____, and the total number of these travelers was $20.3 +$ _____ $=$ _____.

Solve each problem. See Examples 3 and 4.

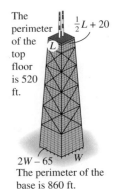

The perimeter of the top floor is 520 ft.

29. The John Hancock Center in Chicago has a rectangular base. The length of the base measures 65 ft less than twice the width. The perimeter of the base is 860 ft. What are the dimensions of the base?

30. The John Hancock Center (**Exercise 29**) tapers as it rises. The top floor is rectangular and has perimeter 520 ft. The width of the top floor measures 20 ft more than one-half its length. What are the dimensions of the top floor?

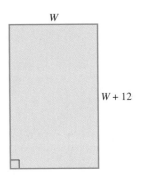

$\frac{1}{2}L + 20$

$2W - 65$

The perimeter of the base is 860 ft.

31. Grant Wood painted his most famous work, *American Gothic,* in 1930 on composition board with perimeter 108.44 in. If the painting is 5.54 in. taller than it is wide, find the dimensions of the painting. (*Source: The Gazette.*)

American Gothic by Grant Wood. © Figge Art Museum/Estate of Nan Wood Graham/VAGA, NY

32. The perimeter of a certain rectangle is 16 times the width. The length is 12 cm more than the width. Find the length and width of the rectangle.

W

$W + 12$

33. The Bermuda Triangle supposedly causes trouble for aircraft pilots. It has a perimeter of 3075 mi. The shortest side measures 75 mi less than the middle side, and the longest side measures 375 mi more than the middle side. Find the lengths of the three sides.

34. The Vietnam Veterans Memorial in Washington, DC, is in the shape of two sides of an isosceles triangle. If the two walls of equal length were joined by a straight line of 438 ft, the perimeter of the resulting triangle would be 931.5 ft. Find the lengths of the two walls. (*Source:* Pamphlet obtained at Vietnam Veterans Memorial.)

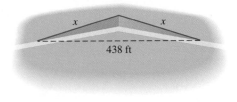

x x

438 ft

35. The two companies with top revenues in the Fortune 500 list for 2009 were Exxon Mobil and Wal-Mart. Their revenues together totaled $848.5 billion. Wal-Mart revenues were $37.3 billion less than Exxon Mobil revenues. What were the revenues of each corporation? (*Source:* www.money.cnn.com)

36. Two of the longest-running Broadway shows were *Cats,* which played from 1982 through 2000, and *Les Misérables,* which played from 1987 through 2003. Together, there were 14,165 performances of these two shows during their Broadway runs. There were 805 fewer performances of *Les Misérables* than of *Cats.* How many performances were there of each show? (*Source:* The Broadway League.)

37. Galileo Galilei conducted experiments involving Italy's famous Leaning Tower of Pisa to investigate the relationship between an object's speed of fall and its weight. The Leaning Tower is 880 ft shorter than the Eiffel Tower in Paris, France. The two towers have a total height of 1246 ft. How tall is each tower? (*Source:* www.leaned.org, www.tour-eiffel.fr.)

38. In 2009, the New York Yankees and the New York Mets had the highest payrolls in Major League Baseball. The Mets' payroll was $65.6 million less than the Yankees' payroll, and the two payrolls totaled $337.2 million. What was the payroll for each team? (*Source:* Associated Press.)

39. In the 2008 presidential election, Barack Obama and John McCain together received 538 electoral votes. Obama received 192 more votes than McCain. How many votes did each candidate receive? (*Source: World Almanac and Book of Facts.*)

40. Ted Williams and Rogers Hornsby were two great hitters in Major League Baseball. Together, they got 5584 hits in their careers. Hornsby got 276 more hits than Williams. How many base hits did each get? (*Source:* Neft, D. S., and R. M. Cohen, *The Sports Encyclopedia: Baseball,* St. Martins Griffin; New York, 2007.)

Solve each percent problem. **See Example 5.**

41. In 2009, the number of graduating seniors taking the ACT exam was 1,480,469. In 2000, a total of 1,065,138 graduating seniors took the exam. By what percent did the number increase over this period of time, to the nearest tenth of a percent? (*Source:* ACT.)

42. Composite scores on the ACT exam rose from 20.8 in 2002 to 21.1 in 2009. What percent increase was this, to the nearest tenth of a percent? (*Source:* ACT.)

43. In 1995, the average cost of tuition and fees at public four-year universities in the United States was $2811 for full-time students. By 2009, it had risen approximately 150%. To the nearest dollar, what was the approximate cost in 2009? (*Source:* The College Board.)

44. In 1995, the average cost of tuition and fees at private four-year universities in the United States was $12,216 for full-time students. By 2009, it had risen approximately 115.1%. To the nearest dollar, what was the approximate cost in 2009? (*Source:* The College Board.)

45. In 2009, the average cost of a traditional Thanksgiving dinner for 10, featuring turkey, stuffing, cranberries, pumpkin pie, and trimmings, was $42.91, a decrease of 3.8% over the cost in 2008. What was the cost, to the nearest cent, in 2008? (*Source:* American Farm Bureau.)

46. Refer to **Exercise 45.** The cost of a traditional Thanksgiving dinner in 2009 was $42.91, an increase of 60.4% over the cost in 1987 when data was first collected. What was the cost, to the nearest cent, in 1987? (*Source:* American Farm Bureau.)

47. At the end of a day, Lawrence Hawkins found that the total cash register receipts at the motel where he works amounted to $2725. This included the 9% sales tax charged. Find the amount of the tax.

48. David Ruppel sold his house for $159,000. He got this amount knowing that he would have to pay a 6% commission to his agent. What amount did he have after the agent was paid?

Solve each investment problem. **See Example 6.**

49. Mario Toussaint earned $12,000 last year by giving tennis lessons. He invested part of the money at 3% simple interest and the rest at 4%. In one year, he earned a total of $440 in interest. How much did he invest at each rate?

Principal	Rate (as a decimal)	Interest
x	0.03	
	0.04	✕✕✕✕✕

50. Sheryl Zavertnik won $60,000 on a slot machine in Las Vegas. She invested part of the money at 2% simple interest and the rest at 3%. In one year, she earned a total of $1600 in interest. How much was invested at each rate?

Principal	Rate (as a decimal)	Interest
x	0.02	
		✕✕✕✕✕

51. Jennifer Siegel invested some money at 4.5% simple interest and $1000 less than twice this amount at 3%. Her total annual income from the interest was $1020. How much was invested at each rate?

52. Piotr Galkowski invested some money at 3.5% simple interest, and $5000 more than three times this amount at 4%. He earned $1440 in annual interest. How much did he invest at each rate?

53. Dan Abbey has invested $12,000 in bonds paying 6%. How much additional money should he invest in a certificate of deposit paying 3% simple interest so that the total return on the two investments will be 4%?

54. Mona Galland received a year-end bonus of $17,000 from her company and invested the money in an account paying 6.5%. How much additional money should she deposit in an account paying 5% so that the return on the two investments will be 6%?

Solve each problem involving rates of concentration and mixtures. **See Examples 7 and 8.**

55. Ten liters of a 4% acid solution must be mixed with a 10% solution to get a 6% solution. How many liters of the 10% solution are needed?

Liters of Solution	Percent (as a decimal)	Liters of Pure Acid
10	0.04	
x	0.10	
	0.06	

56. How many liters of a 14% alcohol solution must be mixed with 20 L of a 50% solution to get a 30% solution?

Liters of Solution	Percent (as a decimal)	Liters of Pure Alcohol
x	0.14	
	0.50	

57. In a chemistry class, 12 L of a 12% alcohol solution must be mixed with a 20% solution to get a 14% solution. How many liters of the 20% solution are needed?

58. How many liters of a 10% alcohol solution must be mixed with 40 L of a 50% solution to get a 40% solution?

59. How much pure dye must be added to 4 gal of a 25% dye solution to increase the solution to 40%? (*Hint:* Pure dye is 100% dye.)

60. How much water must be added to 6 gal of a 4% insecticide solution to reduce the concentration to 3%? (*Hint:* Water is 0% insecticide.)

61. Randall Albritton wants to mix 50 lb of nuts worth $2 per lb with some nuts worth $6 per lb to make a mixture worth $5 per lb. How many pounds of $6 nuts must he use?

Pounds of Nuts	Cost per Pound	Total Cost

62. Lee Ann Spahr wants to mix tea worth 2¢ per oz with 100 oz of tea worth 5¢ per oz to make a mixture worth 3¢ per oz. How much 2¢ tea should be used?

Ounces of Tea	Cost per Ounce	Total Cost

63. Why is it impossible to mix candy worth $4 per lb and candy worth $5 per lb to obtain a final mixture worth $6 per lb?

64. Write an equation based on the following problem, solve the equation, and explain why the problem has no solution:

How much 30% acid should be mixed with 15 L of 50% acid to obtain a mixture that is 60% acid?

RELATING CONCEPTS EXERCISES 65–68

FOR INDIVIDUAL OR GROUP WORK

Consider each problem.

Problem A Jack has $800 invested in two accounts. One pays 5% interest per year and the other pays 10% interest per year. The amount of yearly interest is the same as he would get if the entire $800 was invested at 8.75%. How much does he have invested at each rate?

Problem B Jill has 800 L of acid solution. She obtained it by mixing some 5% acid with some 10% acid. Her final mixture of 800 L is 8.75% acid. How much of each of the 5% and 10% solutions did she use to get her final mixture?

In Problem A, let x represent the amount invested at 5% interest, and in Problem B, let y represent the amount of 5% acid used. **Work Exercises 65–68 in order.**

65. (a) Write an expression in x that represents the amount of money Jack invested at 10% in Problem A.

(b) Write an expression in y that represents the amount of 10% acid solution Jill used in Problem B.

66. (a) Write expressions that represent the amount of interest Jack earns per year at 5% and at 10%.

(b) Write expressions that represent the amount of pure acid in Jill's 5% and 10% acid solutions.

67. (a) The sum of the two expressions in part (a) of **Exercise 66** must equal the total amount of interest earned in one year. Write an equation representing this fact.

(b) The sum of the two expressions in part (b) of **Exercise 66** must equal the amount of pure acid in the final mixture. Write an equation representing this fact.

68. (a) Solve Problem A.   (b) Solve Problem B.

(c) Explain the similarities between the processes used in solving Problems A and B.

PREVIEW EXERCISES

Solve each problem. ***See Section 2.2.***

69. Use $d = rt$ to find d if $r = 50$ and $t = 4$.

70. Use $P = 2L + 2W$ to find P if $L = 10$ and $W = 6$.

71. Use $P = a + b + c$ to find a if $b = 13$, $c = 14$, and $P = 46$.

72. Use $\mathcal{A} = \frac{1}{2}h(b + B)$ to find h if $\mathcal{A} = 156$, $b = 12$, and $B = 14$.

STUDY **SKILLS**

Taking Lecture Notes

Study the set of sample math notes given here.

▶ **Use a new page** for each day's lecture.

▶ **Include the date and title** of the day's lecture topic.

▶ **Skip lines and write neatly** to make reading easier.

▶ **Include cautions and warnings** to emphasize common errors to avoid.

▶ **Mark important concepts with stars, underlining, circling, boxes, etc.**

▶ **Use two columns,** which allows an example and its explanation to be close together.

▶ **Use brackets and arrows** to clearly show steps, related material, etc.

With a partner or in a small group, compare lecture notes.

1. What are you doing to show main points in your notes (such as boxing, using stars or capital letters, etc.)?

2. In what ways do you set off explanations from worked problems and subpoints (such as indenting, using arrows, circling, etc.)?

3. What new ideas did you learn by examining your classmates' notes?

4. What new techniques will you try in your note taking?

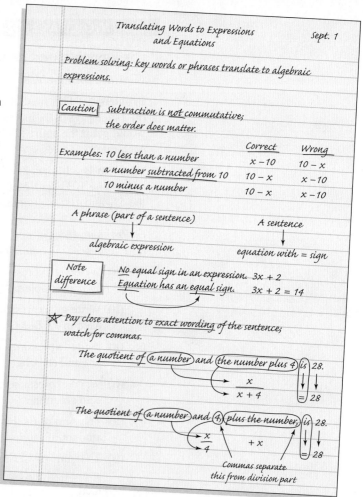

2.4 Further Applications of Linear Equations

OBJECTIVES

1 Solve problems about different denominations of money.
2 Solve problems about uniform motion.
3 Solve problems about angles.

NOW TRY EXERCISE 1

Steven Danielson has a collection of 52 coins worth $3.70. His collection contains only dimes and nickels. How many of each type of coin does he have?

OBJECTIVE 1 Solve problems about different denominations of money.

PROBLEM-SOLVING HINT

In problems involving money, use the following basic fact.

$$\text{number of monetary units of the same kind} \times \text{denomination} = \text{total monetary value}$$

30 dimes have a monetary value of $30(\$0.10) = \$3.00$.

Fifteen 5-dollar bills have a value of $15(\$5) = \$75$.

EXAMPLE 1 Solving a Money Denomination Problem

For a bill totaling $5.65, a cashier received 25 coins consisting of nickels and quarters. How many of each denomination of coin did the cashier receive?

Step 1 **Read** the problem. The problem asks that we find the number of nickels and the number of quarters the cashier received.

Step 2 **Assign a variable.** Then organize the information in a table.

Let $x =$ the number of nickels.

Then $25 - x =$ the number of quarters.

	Number of Coins	Denomination	Value
Nickels	x	0.05	$0.05x$
Quarters	$25 - x$	0.25	$0.25(25 - x)$
			5.65	← Total

Step 3 **Write an equation** from the last column of the table.

$$0.05x + 0.25(25 - x) = 5.65$$

Step 4 **Solve.**

$$0.05x + 0.25(25 - x) = 5.65$$

$$5x + 25(25 - x) = 565 \quad \text{Multiply by 100.}$$

Move decimal points 2 places to the right.

$$5x + 625 - 25x = 565 \quad \text{Distributive property}$$

$$-20x = -60 \quad \text{Subtract 625. Combine like terms.}$$

$$x = 3 \quad \text{Divide by } -20.$$

Step 5 **State the answer.** The cashier has 3 nickels and $25 - 3 = 22$ quarters.

Step 6 **Check.** The cashier has $3 + 22 = 25$ coins, and the value of the coins is

$$\$0.05(3) + \$0.25(22) = \$5.65, \quad \text{as required.} \qquad \text{NOW TRY}$$

⚠ CAUTION *Be sure that your answer is reasonable* when you are working problems like **Example 1.** Because you are dealing with a number of coins, the correct answer can be neither negative nor a fraction.

NOW TRY ANSWER
1. 22 dimes; 30 nickels

> OBJECTIVE 2   **Solve problems about uniform motion.**

> **PROBLEM-SOLVING HINT**
>
> Uniform motion problems use the distance formula $d = rt$. ***When rate (or speed) is given in miles per hour, time must be given in hours. Draw a sketch*** to illustrate what is happening. ***Make a table*** to summarize given information.

*NOW TRY
EXERCISE 2*

Two trains leave a city traveling in opposite directions. One travels at a rate of 80 km per hr and the other at a rate of 75 km per hr. How long will it take before they are 387.5 km apart?

> EXAMPLE 2   Solving a Motion Problem (Motion in Opposite Directions)

Two cars leave the same place at the same time, one going east and the other west. The eastbound car averages 40 mph, while the westbound car averages 50 mph. In how many hours will they be 300 mi apart?

Step 1   **Read** the problem. We are looking for the time it takes for the two cars to be 300 mi apart.

Step 2   **Assign a variable.** A sketch shows what is happening in the problem. The cars are going in *opposite* directions. See **FIGURE 6**.

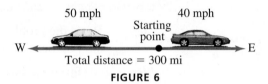

FIGURE 6

Let x represent the time traveled by each car, and summarize the information of the problem in a table.

	Rate	Time	Distance
Eastbound Car	40	x	$40x$
Westbound Car	50	x	$50x$
			300

Fill in each distance by multiplying rate by time, using the formula $d = rt$. The sum of the two distances is 300.

Step 3   **Write an equation.** The sum of the two distances is 300.

$$40x + 50x = 300$$

Step 4   **Solve.** $90x = 300$ Combine like terms.

$$x = \frac{300}{90} = \frac{10}{3} \qquad \text{Divide by 90; lowest terms}$$

Step 5   **State the answer.** The cars travel $\frac{10}{3} = 3\frac{1}{3}$ hr, or 3 hr, 20 min.

Step 6   **Check.** The eastbound car traveled $40\left(\frac{10}{3}\right) = \frac{400}{3}$ mi. The westbound car traveled $50\left(\frac{10}{3}\right) = \frac{500}{3}$ mi, for a total distance of $\frac{400}{3} + \frac{500}{3} = \frac{900}{3} = 300$ mi, as required.

             NOW TRY

⚠ **CAUTION**   It is a common error to write 300 as the distance traveled by each car in **Example 2**. Three hundred miles is the *total* distance traveled.

     As in **Example 2,** in general, the equation for a problem involving motion in *opposite* directions is of the following form.

partial distance + partial distance = total distance

NOW TRY ANSWER
2. $2\frac{1}{2}$ hr

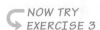

NOW TRY
EXERCISE 3

Michael Good can drive to work in $\frac{1}{2}$ hr. When he rides his bicycle, it takes $1\frac{1}{2}$ hours. If his average rate while driving to work is 30 mph faster than his rate while bicycling to work, determine the distance that he lives from work.

EXAMPLE 3 Solving a Motion Problem (Motion in the Same Direction)

Jeff can bike to work in $\frac{3}{4}$ hr. When he takes the bus, the trip takes $\frac{1}{4}$ hr. If the bus travels 20 mph faster than Jeff rides his bike, how far is it to his workplace?

Step 1 **Read** the problem. We must find the distance between Jeff's home and his workplace.

Step 2 **Assign a variable.** Although the problem asks for a distance, it is easier here to let x be Jeff's rate when he rides his bike to work. Then the rate of the bus is $x + 20$.

For the trip by bike, $\qquad d = rt = x \cdot \dfrac{3}{4} = \dfrac{3}{4}x.$

For the trip by bus, $\qquad d = rt = (x + 20) \cdot \dfrac{1}{4} = \dfrac{1}{4}(x + 20).$

Summarize this information in a table.

	Rate	Time	Distance
Bike	x	$\frac{3}{4}$	$\frac{3}{4}x$
Bus	$x + 20$	$\frac{1}{4}$	$\frac{1}{4}(x + 20)$

Same

Step 3 **Write an equation.** The key to setting up the correct equation is to understand that the distance in each case is the same. See **FIGURE 7**.

Home   Workplace

FIGURE 7

$\dfrac{3}{4}x = \dfrac{1}{4}(x + 20)$ The distance is the same in each case.

Step 4 **Solve.** $4\left(\dfrac{3}{4}x\right) = 4\left(\dfrac{1}{4}\right)(x + 20)$ Multiply by 4.

$\qquad\qquad 3x = x + 20$ Multiply; $1x = x$

$\qquad\qquad 2x = 20$ Subtract x.

$\qquad\qquad x = 10$ Divide by 2.

Step 5 **State the answer.** The required distance is

$$d = \frac{3}{4}x = \frac{3}{4}(10) = \frac{30}{4} = 7.5 \text{ mi.}$$

Step 6 **Check** by finding the distance using

$$d = \frac{1}{4}(x + 20) = \frac{1}{4}(10 + 20) = \frac{30}{4} = 7.5 \text{ mi.}$$

The same result

NOW TRY

As in **Example 3,** the equation for a problem involving motion in the *same* direction is usually of the following form.

one distance = other distance

NOW TRY ANSWER
3. 22.5 mi

> **PROBLEM-SOLVING HINT**
>
> In **Example 3,** it was easier to let the variable represent a quantity other than the one that we were asked to find. It takes practice to learn when this approach works best.

OBJECTIVE 3 **Solve problems about angles.** An important result of Euclidean geometry (the geometry of the Greek mathematician Euclid) is that *the sum of the angle measures of any triangle is* **180°.** This property is used in the next example.

NOW TRY
EXERCISE 4

Find the value of x, and determine the measure of each angle.

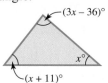

EXAMPLE 4   Finding Angle Measures

Find the value of x, and determine the measure of each angle in **FIGURE 8**.

Step 1   **Read** the problem. We are asked to find the measure of each angle.

Step 2   **Assign a variable.**

Let x = the measure of one angle.

Step 3   **Write an equation.** The sum of the three measures shown in the figure must be 180°.

$$x + (x + 20) + (210 - 3x) = 180$$

Step 4   **Solve.** $-x + 230 = 180$ Combine like terms.

                              $-x = -50$ Subtract 230.

                               $x = 50$ Multiply by -1.

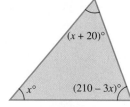

FIGURE 8

Step 5   **State the answer.** One angle measures 50°. The other two angles measure

$$x + 20 = 50 + 20 = 70°$$

and $210 - 3x = 210 - 3(50) = 60°.$

Step 6   **Check.** Since $50° + 70° + 60° = 180°$, the answers are correct.

NOW TRY

NOW TRY ANSWER
4. 41°, 52°, 87°

2.4 EXERCISES

🌐 *Complete solution available on the Video Resources on DVD*

Concept Check   *Solve each problem.*

1. What amount of money is found in a coin hoard containing 14 dimes and 16 quarters?

2. The distance between Cape Town, South Africa, and Miami is 7700 mi. If a jet averages 550 mph between the two cities, what is its travel time in hours?

3. Tri Phong traveled from Chicago to Des Moines, a distance of 300 mi, in 10 hr. What was his rate in miles per hour?

4. A square has perimeter 80 in. What would be the perimeter of an equilateral triangle whose sides each measure the same length as the side of the square?

✒ *Concept Check*   *Answer the questions in Exercises 5–8.*

5. Read over **Example 3** in this section. The solution of the equation is 10. Why is *10 mph* not the answer to the problem?

6. Suppose that you know that two angles of a triangle have equal measures and the third angle measures 36°. How would you find the measures of the equal angles without actually writing an equation?

7. In a problem about the number of coins of different denominations, would an answer that is a fraction be reasonable? Would a negative answer be reasonable?

8. In a motion problem the rate is given as *x* mph and the time is given as 10 min. What variable expression represents the distance in miles?

Solve each problem. **See Example 1.**

9. Otis Taylor has a box of coins that he uses when he plays poker with his friends. The box currently contains 44 coins, consisting of pennies, dimes, and quarters. The number of pennies is equal to the number of dimes, and the total value is $4.37. How many of each denomination of coin does he have in the box?

Number of Coins	Denomination	Value
x	0.01	0.01x
x		
	0.25	
XXXXXXXXXX		4.37	← Total

10. Nana Nantambu found some coins while looking under her sofa pillows. There were equal numbers of nickels and quarters and twice as many half-dollars as quarters. If she found $2.60 in all, how many of each denomination of coin did she find?

Number of Coins	Denomination	Value
x	0.05	0.05x
x		
2x	0.50	
XXXXXXXXXX		2.60	← Total

11. In Canada, $1 and $2 bills have been replaced by coins. The $1 coins are called "loonies" because they have a picture of a loon (a well-known Canadian bird) on the reverse, and the $2 coins are called "toonies." When Marissa returned home to San Francisco from a trip to Vancouver, she found that she had acquired 37 of these coins, with a total value of 51 Canadian dollars. How many coins of each denomination did she have?

12. Dan Ulmer works at an ice cream shop. At the end of his shift, he counted the bills in his cash drawer and found 119 bills with a total value of $347. If all of the bills are $5 bills and $1 bills, how many of each denomination were in his cash drawer?

13. Dave Bowers collects U.S. gold coins. He has a collection of 41 coins. Some are $10 coins, and the rest are $20 coins. If the face value of the coins is $540, how many of each denomination does he have?

14. In the 19th century, the United States minted two-cent and three-cent pieces. Frances Steib has three times as many three-cent pieces as two-cent pieces, and the face value of these coins is $2.42. How many of each denomination does she have?

15. In 2010, general admission to the Art Institute of Chicago cost $18 for adults and $12 for children and seniors. If $22,752 was collected from the sale of 1460 general admission tickets, how many adult tickets were sold? (*Source:* www.artic.edu)

16. For a high school production of *Annie Get Your Gun*, student tickets cost $5 each while nonstudent tickets cost $8. If 480 tickets were sold for the Saturday night show and a total of $2895 was collected, how many tickets of each type were sold?

In Exercises 17–20, find the rate on the basis of the information provided. Use a calculator and round your answers to the nearest hundredth. All events were at the 2008 Summer Olympics in Beijing, China. (Source: World Almanac and Book of Facts.)

	Event	Participant	Distance	Time
17.	100-m hurdles, women	Dawn Harper, USA	100 m	12.54 sec
18.	400-m hurdles, women	Melanie Walker, Jamaica	400 m	52.64 sec
19.	400-m hurdles, men	Angelo Taylor, USA	400 m	47.25 sec
20.	400-m run, men	LaShawn Merritt, USA	400 m	43.75 sec

Solve each problem. ***See Examples 2 and 3.***

21. Two steamers leave a port on a river at the same time, traveling in opposite directions. Each is traveling 22 mph. How long will it take for them to be 110 mi apart?

	Rate	Time	Distance
First Steamer		t	
Second Steamer	22		
			110

22. A train leaves Kansas City, Kansas, and travels north at 85 km per hr. Another train leaves at the same time and travels south at 95 km per hr. How long will it take before they are 315 km apart?

	Rate	Time	Distance
First Train	85	t	
Second Train			
			315

23. Mulder and Scully are driving to Georgia to investigate "Big Blue," a giant reptile reported in one of the local lakes. Mulder leaves the office at 8:30 A.M. averaging 65 mph. Scully leaves at 9:00 A.M., following the same path and averaging 68 mph. At what time will Scully catch up with Mulder?

	Rate	Time	Distance
Mulder			
Scully			

24. Lois and Clark, two elderly reporters, are covering separate stories and have to travel in opposite directions. Lois leaves the *Daily Planet* building at 8:00 A.M. and travels at 35 mph. Clark leaves at 8:15 A.M. and travels at 40 mph. At what time will they be 140 mi apart?

	Rate	Time	Distance
Lois			
Clark			

25. It took Charmaine 3.6 hr to drive to her mother's house on Saturday morning for a weekend visit. On her return trip on Sunday night, traffic was heavier, so the trip took her 4 hr. Her average rate on Sunday was 5 mph slower than on Saturday. What was her average rate on Sunday?

	Rate	Time	Distance
Saturday			
Sunday			

26. Sharon Kobrin commutes to her office by train. When she walks to the train station, it takes her 40 min. When she rides her bike, it takes her 12 min. Her average walking rate is 7 mph less than her average biking rate. Find the distance from her house to the train station.

	Rate	Time	Distance
Walking			
Biking			

27. Johnny leaves Memphis to visit his cousin, Anne Hoffman, who lives in the town of Hornsby, Tennessee, 80 mi away. He travels at an average rate of 50 mph. One-half hour later, Anne leaves to visit Johnny, traveling at an average rate of 60 mph. How long after Anne leaves will it be before they meet?

28. On an automobile trip, Laura Iossi maintained a steady rate for the first two hours. Rush-hour traffic slowed her rate by 25 mph for the last part of the trip. The entire trip, a distance of 125 mi, took $2\frac{1}{2}$ hr. What was her rate during the first part of the trip?

Find the measure of each angle in the triangles shown. ***See Example 4.***

29.

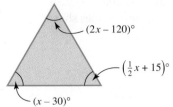

30.

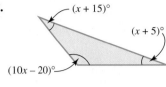
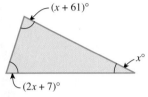

31.

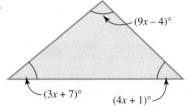

32.

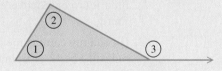

FOR INDIVIDUAL OR GROUP WORK

Consider the following two figures. ***Work Exercises 33–36 in order.***

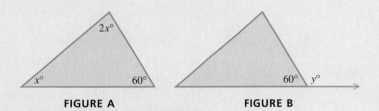

FIGURE A   **FIGURE B**

33. Solve for the measures of the unknown angles in **FIGURE A**.

34. Solve for the measure of the unknown angle marked $y°$ in **FIGURE B**.

35. Add the measures of the two angles you found in **Exercise 33**. How does the sum compare to the measure of the angle you found in **Exercise 34?**

36. Based on the answers to **Exercises 33–35,** make a conjecture (an educated guess) about the relationship among the angles marked ①, ②, and ③ in the figure shown below.

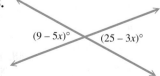

*In Exercises 37 and 38, the angles marked with variable expressions are called **vertical angles.** It is shown in geometry that vertical angles have equal measures. Find the measure of each angle.*

37.

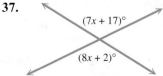

38.

39. Two angles whose sum is 90° are called **complementary angles**. Find the measures of the complementary angles shown in the figure.

40. Two angles whose sum is 180° are called **supplementary angles**. Find the measures of the supplementary angles shown in the figure.

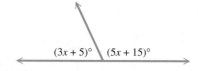

Consecutive Integer Problems

Consecutive integers are integers that follow each other in counting order, such as 8, 9, and 10. Suppose we wish to solve the following problem:

Find three consecutive integers such that the sum of the first and third, increased by 3, is 50 more than the second.

Let $x =$ the first of the unknown integers, $x + 1 =$ the second, and $x + 2 =$ the third. We solve the following equation.

$$
\underset{\substack{\text{Sum of the}\\\text{first and third}}}{\underbrace{x + (x + 2)}} \quad \underset{\substack{\text{increased}\\\text{by 3}}}{\underbrace{+\ 3}} \quad \underset{\text{is}}{=} \quad \underset{\substack{\text{50 more than}\\\text{the second.}}}{\underbrace{(x + 1) + 50}}
$$

$$2x + 5 = x + 51$$

$$x = 46$$

The solution of this equation is 46, so the first integer is $x = 46$, the second is $x + 1 = 47$, and the third is $x + 2 = 48$. The three integers are 46, 47, and 48. Check by substituting these numbers back into the words of the original problem.

Solve each problem involving consecutive integers.

41. Find three consecutive integers such that the sum of the first and twice the second is 17 more than twice the third.

42. Find four consecutive integers such that the sum of the first three is 54 more than the fourth.

43. If I add my current age to the age I will be next year on this date, the sum is 103 yr. How old will I be 10 yr from today?

44. Two pages facing each other in this book have 193 as the sum of their page numbers. What are the two page numbers?

45. Find three consecutive *even* integers such that the sum of the least integer and the middle integer is 26 more than the greatest integer.

46. Find three consecutive *even* integers such that the sum of the least integer and the greatest integer is 12 more than the middle integer.

47. Find three consecutive *odd* integers such that the sum of the least integer and the middle integer is 19 more than the greatest integer.

48. Find three consecutive *odd* integers such that the sum of the least integer and the greatest integer is 13 more than the middle integer.

*Graph each interval. **See Section 1.1.***

49. $(4, \infty)$ **50.** $(-\infty, -2]$ **51.** $(-2, 6)$ **52.** $[-1, 6]$

SUMMARY EXERCISES on Solving Applied Problems

Solve each problem.

1. The length of a rectangle is 3 in. more than its width. If the length were decreased by 2 in. and the width were increased by 1 in., the perimeter of the resulting rectangle would be 24 in. Find the dimensions of the original rectangle.

$x + 3$

x

2. A farmer wishes to enclose a rectangular region with 210 m of fencing in such a way that the length is twice the width and the region is divided into two equal parts, as shown in the figure. What length and width should be used?

Width

Length

3. After a discount of 46%, the sale price for a *Harry Potter* Paperback Boxed Set (Books 1–7) by J. K. Rowling was $46.97. What was the regular price of the set of books to the nearest cent? (*Source:* www.amazon.com)

4. An electronics store offered a Blu-ray player for $255, the sale price after the regular price was discounted 40%. What was the regular price?

5. An amount of money is invested at 4% annual simple interest, and twice that amount is invested at 5%. The total annual interest is $112. How much is invested at each rate?

6. An amount of money is invested at 3% annual simple interest, and $2000 more than that amount is invested at 4%. The total annual interest is $920. How much is invested at each rate?

7. LeBron James of the Cleveland Cavaliers was the leading scorer in the NBA for the 2007–2008 season, and Dwyane Wade was the leading scorer for the 2008–2009 season. Together, they scored 4636 points, with James scoring 136 points fewer than Wade. How many points did each of them score?

8. Before being overtaken by *Avatar*, the two all-time top-grossing American movies were *Titanic* and *The Dark Knight*. *Titanic* grossed $67.5 million more than *The Dark Knight*. Together, the two films brought in $1134.1 million. How much did each movie gross? (*Source:* www.imdb.com)

(continued)

9. Atlanta and Cincinnati are 440 mi apart. John leaves Cincinnati, driving toward Atlanta at an average rate of 60 mph. Pat leaves Atlanta at the same time, driving toward Cincinnati in her antique auto, averaging 28 mph. How long will it take them to meet?

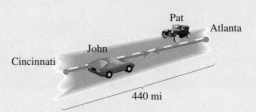

440 mi

10. Deriba Merga from Ethiopia won the 2009 men's Boston Marathon with a winning time of 2 hr, 8 min, 42 sec, or 2.145 hr. The women's race was won by Salina Kosgei from Kenya, whose winning time was 2 hr, 32 min, 16 sec, or 2.538 hr. Kosgei's average rate was 1.9 mph slower than Merga's. Find the average rate for each runner, to the nearest hundredth. (*Source: World Almanac and Book of Facts.*)

11. A pharmacist has 20 L of a 10% drug solution. How many liters of 5% solution must be added to get a mixture that is 8%?

12. A certain metal is 20% tin. How many kilograms of this metal must be mixed with 80 kg of a metal that is 70% tin to get a metal that is 50% tin?

13. A cashier has a total of 126 bills in fives and tens. The total value of the money is $840. How many of each denomination of bill does he have?

14. The top-grossing domestic movie in 2008 was *The Dark Knight*. On the opening weekend, one theater showing this movie took in $20,520 by selling a total of 2460 tickets, some at $9 and the rest at $7. How many tickets were sold at each price? (*Source: Variety.*)

15. Find the measure of each angle.   **16.** Find the measure of each marked angle.

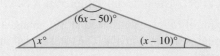

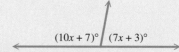

17. The sum of the least and greatest of three consecutive integers is 32 more than the middle integer. What are the three integers?

18. If the lesser of two consecutive odd integers is doubled, the result is 7 more than the greater of the two integers. Find the two integers.

19. The perimeter of a triangle is 34 in. The middle side is twice as long as the shortest side. The longest side is 2 in. less than three times the shortest side. Find the lengths of the three sides.

x inches

20. The perimeter of a rectangle is 43 in. more than the length. The width is 10 in. Find the length of the rectangle.

2.5 Linear Inequalities in One Variable

In **Section 1.1,** we used interval notation to write solution sets of inequalities.

- A parenthesis indicates that an endpoint is *not* included.
- A square bracket indicates that an endpoint is included.

We summarize the various types of intervals here.

Type of Interval	Set-Builder Notation	Interval Notation	Graph
Open interval	$\{x \mid a < x < b\}$	(a, b)	
Closed interval	$\{x \mid a \leq x \leq b\}$	$[a, b]$	
Half-open (or half-closed) interval	$\{x \mid a \leq x < b\}$	$[a, b)$	
	$\{x \mid a < x \leq b\}$	$(a, b]$	
Disjoint interval*	$\{x \mid x < a \text{ or } x > b\}$	$(-\infty, a) \cup (b, \infty)$	
Infinite interval	$\{x \mid x > a\}$	(a, ∞)	
	$\{x \mid x \geq a\}$	$[a, \infty)$	
	$\{x \mid x < a\}$	$(-\infty, a)$	
	$\{x \mid x \leq a\}$	$(-\infty, a]$	
	$\{x \mid x \text{ is a real number}\}$	$(-\infty, \infty)$	

NOTE A parenthesis is *always* used next to an infinity symbol, $-\infty$ or ∞.

An **inequality** says that two expressions are *not* equal. Solving inequalities is similar to solving equations.

Linear Inequality in One Variable

A **linear inequality in one variable** can be written in the form

$$Ax + B < C, \quad Ax + B \leq C, \quad Ax + B > C, \quad \text{or} \quad Ax + B \geq C,$$

where A, B, and C are real numbers, with $A \neq 0$.

$x + 5 < 2$, $\quad x - 3 \geq 5$, $\quad$ and $\quad 2k + 5 \leq 10$ $\qquad$ Examples of linear inequalities

---

*We will work with disjoint intervals in **Section 2.6** when we study *set operations* and *compound inequalities*.

OBJECTIVE 1 **Solve linear inequalities by using the addition property.** We solve an inequality by finding all numbers that make the inequality true. Usually, an inequality has an infinite number of solutions. These solutions, like solutions of equations, are found by producing a series of simpler related equivalent inequalities. **Equivalent inequalities** are inequalities with the same solution set.

We use two important properties to produce equivalent inequalities. The first is the *addition property of inequality*.

Addition Property of Inequality

For all real numbers A, B, and C, the inequalities

$$A < B \qquad \text{and} \qquad A + C < B + C \qquad \text{are equivalent.}$$

That is, adding the same number to each side of an inequality does not change the solution set.

NOW TRY
EXERCISE 1
Solve $x - 10 > -7$, and graph the solution set.

EXAMPLE 1 Using the Addition Property of Inequality

Solve $x - 7 < -12$, and graph the solution set.

$$x - 7 < -12$$
$$x - 7 + 7 < -12 + 7 \qquad \text{Add 7.}$$
$$x < -5 \qquad \text{Combine like terms.}$$

CHECK Substitute -5 for x in the *equation* $x - 7 = -12$.

$$x - 7 = -12 \qquad \text{Related equation}$$
$$-5 - 7 \overset{?}{=} -12 \qquad \text{Let } x = -5.$$
$$-12 = -12 \; \checkmark \quad \text{True}$$

This shows that -5 is the boundary point. Now test a number on each side of -5 to verify that numbers *less than* -5 make the inequality true. We choose -4 and -6.

$$x - 7 < -12$$

$-4 - 7 \overset{?}{<} -12$ Let $x = -4$.	$-6 - 7 \overset{?}{<} -12$ Let $x = -6$.
$-11 < -12$ False	$-13 < -12 \; \checkmark$ True
-4 is not in the solution set.	-6 is in the solution set.

The check confirms that $(-\infty, -5)$, graphed in **FIGURE 9**, is the correct solution set.

FIGURE 9

NOW TRY

NOW TRY ANSWER
1. $(3, \infty)$

As with equations, the addition property can be used to *subtract* the same number from each side of an inequality.

NOW TRY
EXERCISE 2
Solve $4x + 1 \geq 5x$, and graph the solution set.

EXAMPLE 2 Using the Addition Property of Inequality

Solve $14 + 2x \leq 3x$ and graph the solution set.

$$14 + 2x \leq 3x$$

$$14 + 2x - 2x \leq 3x - 2x \qquad \text{Subtract } 2x.$$

$$14 \leq x \qquad \text{Combine like terms.}$$

Be careful. $\qquad x \geq 14 \qquad \text{Rewrite.}$

The inequality $14 \leq x$ (14 is less than or equal to x) can also be written $x \geq 14$ (x is greater than or equal to 14). *Notice that in each case the inequality symbol points to the lesser number,* **14.**

CHECK $\qquad\qquad 14 + 2x = 3x \qquad \text{Related equation}$

$$14 + 2(14) \overset{?}{=} 3(14) \qquad \text{Let } x = 14.$$

$$42 = 42 \quad \checkmark \qquad \text{True}$$

So 14 satisfies the equality part of $\leq$. Choose 10 and 15 as test values.

$$14 + 2x < 3x$$

$$14 + 2(10) \overset{?}{<} 3(10) \quad \text{Let } x = 10. \qquad\qquad 14 + 2(15) \overset{?}{<} 3(15) \quad \text{Let } x = 15.$$

$$34 < 30 \qquad \text{False} \qquad\qquad\qquad\qquad 44 < 45 \quad \checkmark \quad \text{True}$$

10 is not in the solution set. $\qquad\qquad\qquad$ 15 is in the solution set.

The check confirms that $[14, \infty)$ is the correct solution set. See **FIGURE 10.**

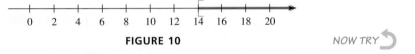

FIGURE 10 $\qquad\qquad$ NOW TRY

OBJECTIVE 2 **Solve linear inequalities by using the multiplication property.** Solving an inequality such as $3x \leq 15$ requires dividing each side by 3, using the *multiplication property of inequality.*

Consider the following true statement.

$$-2 < 5$$

Multiply each side by, say, 8.

$$-2(8) < 5(8) \qquad \text{Multiply by 8.}$$

$$-16 < 40 \qquad \text{True}$$

This gives a true statement. Start again with $-2 < 5$, and multiply each side by -8.

$$-2(-8) < 5(-8) \qquad \text{Multiply by } -8.$$

$$16 < -40 \qquad \text{False}$$

The result, $16 < -40$, is false. To make it true, we must change the direction of the inequality symbol.

$$16 > -40 \qquad \text{True}$$

NOW TRY ANSWER
2. $(-\infty, 1]$

As these examples suggest, multiplying each side of an inequality by a *negative* number requires reversing the direction of the inequality symbol. The same is true for dividing by a negative number, since division is defined in terms of multiplication.

Multiplication Property of Inequality

For all real numbers A, B, and C, with $C \neq 0$,

(a) the inequalities

$$A < B \quad \text{and} \quad AC < BC \quad \text{are equivalent if } C > 0;$$

(b) the inequalities

$$A < B \quad \text{and} \quad AC > BC \quad \text{are equivalent if } C < 0.$$

That is, each side of an inequality may be multiplied (or divided) by a *positive* number without changing the direction of the inequality symbol. ***Multiplying (or dividing) by a negative number requires that we reverse the inequality symbol.***

NOW TRY
EXERCISE 3

Solve each inequality and graph the solution set.

(a) $8x \geq -40$

(b) $-20x > -60$

EXAMPLE 3   Using the Multiplication Property of Inequality

Solve each inequality, and graph the solution set.

(a) $5x \leq -30$

Divide each side by 5. ***Since 5 > 0, do not reverse the direction of the inequality symbol.***

$$5x \leq -30$$

$$\frac{5x}{5} \leq \frac{-30}{5} \qquad \text{Divide by 5.}$$

$$x \leq -6$$

Check that the solution set is the interval $(-\infty, -6]$, graphed in **FIGURE 11**.

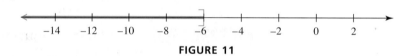

FIGURE 11

(b) $-4x \leq 32$

Divide each side by -4. ***Since $-4 < 0$, reverse the direction of the inequality symbol.***

$$-4x \leq 32$$

$$\frac{-4x}{-4} \geq \frac{32}{-4} \qquad \begin{array}{l}\text{Divide by } -4. \\ \text{Reverse the direction of the symbol.}\end{array}$$

Reverse the inequality symbol when dividing by a *negative* number.

$$x \geq -8$$

Check the solution set. **FIGURE 12** shows the graph of the solution set, $[-8, \infty)$.

NOW TRY ANSWERS

3. (a) $[-5, \infty)$

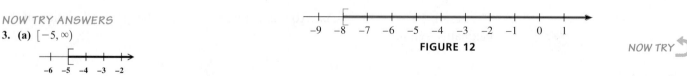

(b) $(-\infty, 3)$

FIGURE 12   NOW TRY

⚠ **CAUTION** *Reverse the direction of the inequality symbol when multiplying or dividing each side of an inequality by a negative number.*

Solving a Linear Inequality

Step 1 **Simplify each side separately.** Clear parentheses, fractions, and decimals using the distributive property, and combine like terms.

Step 2 **Isolate the variable terms on one side.** Use the addition property of inequality to get all terms with variables on one side of the inequality and all numbers on the other side.

Step 3 **Isolate the variable.** Use the multiplication property of inequality to change the inequality to one of these forms.

$$x < k, \quad x \leq k, \quad x > k, \quad \text{or} \quad x \geq k$$

NOW TRY
EXERCISE 4

Solve and graph the solution set.

$$5 - 2(x - 4) \leq 11 - 4x$$

EXAMPLE 4 Solving a Linear Inequality by Using the Distributive Property

Solve $-3(x + 4) + 2 \geq 7 - x$, and graph the solution set.

Step 1	$-3(x + 4) + 2 \geq 7 - x$	
	$-3x - 12 + 2 \geq 7 - x$	Distributive property
	$-3x - 10 \geq 7 - x$	Combine like terms.
Step 2	$-3x - 10 + x \geq 7 - x + x$	Add x.
	$-2x - 10 \geq 7$	Combine like terms.
	$-2x - 10 + 10 \geq 7 + 10$	Add 10.
	$-2x \geq 17$	Combine like terms.
Step 3	$\dfrac{-2x}{-2} \leq \dfrac{17}{-2}$	Divide by -2. Change $\geq$ to $\leq$.

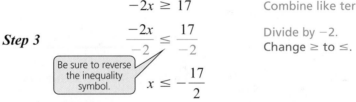

Be sure to reverse the inequality symbol.

$$x \leq -\frac{17}{2}$$

FIGURE 13 shows the graph of the solution set, $\left(-\infty, -\frac{17}{2}\right]$.

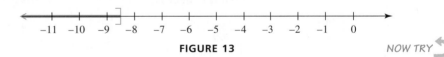

FIGURE 13

NOW TRY

NOTE In Step 2 of **Example 4**, if we add $3x$ (instead of x) to both sides of the inequality, we have the following sequence of equivalent inequalities.

$-3x - 10 + 3x \geq 7 - x + 3x$	Add $3x$.
$-10 \geq 2x + 7$	Combine like terms.
$-10 - 7 \geq 2x + 7 - 7$	Subtract 7.
$-17 \geq 2x$	Combine like terms.
$-\dfrac{17}{2} \geq x$	Divide by 2; $\dfrac{-a}{b} = -\dfrac{a}{b}$

NOW TRY ANSWER
4. $(-\infty, -1]$

The result, "$-\frac{17}{2}$ is greater than or equal to x," means the same as "x is less than or equal to $-\frac{17}{2}$." Thus, the solution set is the same.

NOW TRY
EXERCISE 5

Solve and graph the solution set.

$$\frac{3}{4}(x - 2) + \frac{1}{2} > \frac{1}{5}(x - 8)$$

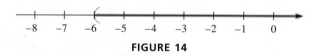

EXAMPLE 5 Solving a Linear Inequality with Fractions

Solve $-\frac{2}{3}(x - 3) - \frac{1}{2} < \frac{1}{2}(5 - x)$, and graph the solution set.

To clear fractions, multiply each side by the least common denominator, 6.

$$-\frac{2}{3}(x - 3) - \frac{1}{2} < \frac{1}{2}(5 - x)$$

$$6\left[-\frac{2}{3}(x - 3) - \frac{1}{2}\right] < 6\left[\frac{1}{2}(5 - x)\right] \qquad \text{Multiply by 6, the LCD.}$$

Be careful here.

$$6\left[-\frac{2}{3}(x - 3)\right] - 6\left(\frac{1}{2}\right) < 6\left[\frac{1}{2}(5 - x)\right] \qquad \text{Distributive property}$$

$$-4(x - 3) - 3 < 3(5 - x) \qquad \text{Multiply.}$$

Step 1 $-4x + 12 - 3 < 15 - 3x$ Distributive property

$$-4x + 9 < 15 - 3x$$

Step 2 $-4x + 9 + 3x < 15 - 3x + 3x$ Add 3x.

$$-x + 9 < 15$$

$$-x + 9 - 9 < 15 - 9 \qquad \text{Subtract 9.}$$

$$-x < 6$$

Step 3 $-1(-x) > -1(6)$ Multiply by -1.
Change $<$ to $>$.

Reverse the inequality symbol when multiplying by a negative number.

$$x > -6$$

Check that the solution set is $(-6, \infty)$. See the graph in **FIGURE 14**.

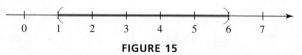

FIGURE 14

NOW TRY

OBJECTIVE 3 **Solve linear inequalities with three parts.** For some applications, it is necessary to work with a **three-part inequality** such as

$$3 < x + 2 < 8,$$

where $x + 2$ is *between* 3 and 8.

NOW TRY
EXERCISE 6

Solve and graph the solution set.

$$-1 < x - 2 < 3$$

EXAMPLE 6 Solving a Three-Part Inequality

Solve $3 < x + 2 < 8$, and graph the solution set.

To solve this inequality, we subtract 2 from *each* of the three parts of the inequality.

$$3 < \quad x + 2 \quad < 8$$

$$3 - 2 < x + 2 - 2 < 8 - 2 \qquad \text{Subtract 2 from all three parts.}$$

$$1 < \quad x \quad < 6$$

Thus, x must be between 1 and 6 so that $x + 2$ will be between 3 and 8. The solution set, $(1, 6)$, is graphed in **FIGURE 15**.

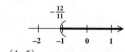
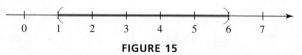

FIGURE 15

NOW TRY

⚠ **CAUTION** *In three-part inequalities, the order of the parts is important.* For example, do not write $8 < x + 2 < 3$, since this would imply that $8 < 3$, a false statement. *Write three-part inequalities so that the symbols point in the same direction and both point toward the lesser number.*

NOW TRY
EXERCISE 7

Solve and graph the solution set.

$$-2 < -4x - 5 \le 7$$

EXAMPLE 7 Solving a Three-Part Inequality

Solve $-2 \le -3x - 1 \le 5$ and graph the solution set.

$$-2 \le \quad -3x - 1 \quad \le 5$$

$$-2 + 1 \le -3x - 1 + 1 \le 5 + 1 \qquad \text{Add 1 to each part.}$$

$$-1 \le \quad -3x \quad \le 6$$

$$\frac{-1}{-3} \ge \quad \frac{-3x}{-3} \quad \ge \frac{6}{-3} \qquad \begin{array}{l}\text{Divide each part by } -3.\\ \text{Reverse the inequality symbols.}\end{array}$$

$$\frac{1}{3} \ge \quad x \quad \ge -2$$

$$-2 \le \quad x \quad \le \frac{1}{3} \quad \leftarrow \boxed{\begin{array}{l}\text{Rewrite in the}\\\text{order on the}\\\text{number line.}\end{array}}$$

Check that the solution set is $\left[-2, \frac{1}{3}\right]$, as shown in **FIGURE 16**.

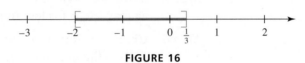

FIGURE 16

NOW TRY

Types of solution sets for linear equations and inequalities are summarized here.

Equation or Inequality	Typical Solution Set	Graph of Solution Set
Linear equation $5x + 4 = 14$	$\{2\}$	● 2
Linear inequality $5x + 4 < 14$ or $5x + 4 > 14$	$(-\infty, 2)$ $(2, \infty)$	) 2 (2
Three-part inequality $-1 \le 5x + 4 \le 14$	$[-1, 2]$	[———] -1 2

OBJECTIVE 4 **Solve applied problems by using linear inequalities.** The table gives some common words and phrases that suggest inequality.

Word Expression	Interpretation
a exceeds b	$a > b$
a is at least b	$a \ge b$
a is no less than b	$a \ge b$
a is at most b	$a \le b$
a is no more than b	$a \le b$

NOW TRY ANSWER

7. $\left[-3, -\frac{3}{4}\right)$

$$\begin{array}{c} \quad -\frac{3}{4} \\ \text{[———)} \\ -4 \ -3 \ -2 \ -1 \quad 0 \quad 1 \end{array}$$

In **Example 8,** we use the six problem-solving steps from **Section 2.3,** changing Step 3 from

"Write an equation"   to   "Write an inequality."

NOW TRY
EXERCISE 8

A local health club charges a $40 one-time enrollment fee, plus $35 per month for a membership. Sara can spend no more than $355 on this exercise expense. What is the *maximum* number of months that Sara can belong to this health club?

EXAMPLE 8   Using a Linear Inequality to Solve a Rental Problem

A rental company charges $15 to rent a chain saw, plus $2 per hr. Tom Ruhberg can spend no more than $35 to clear some logs from his yard. What is the *maximum* amount of time he can use the rented saw?

Step 1   **Read** the problem again.

Step 2   **Assign a variable.** Let x = the number of hours he can rent the saw.

Step 3   **Write an inequality.** He must pay $15, plus $2x$, to rent the saw for x hours, and this amount must be *no more than* $35.

$$\underset{\text{renting}}{\underbrace{\text{Cost of}}} \quad \underset{\text{more than}}{\underbrace{\text{is no}}} \quad \underset{}{\underbrace{\text{35 dollars.}}}$$
$$15 + 2x \qquad \leq \qquad 35$$

Step 4   **Solve.**
$$2x \leq 20 \qquad \text{Subtract 15.}$$
$$x \leq 10 \qquad \text{Divide by 2.}$$

Step 5   **State the answer.** He can use the saw for a maximum of 10 hr. (Of course, he may use it for less time, as indicated by the inequality $x \leq 10$.)

Step 6   **Check.** If Tom uses the saw for 10 hr, he will spend $15 + 2(10) = 35$ dollars, the maximum amount.

NOW TRY

NOW TRY
EXERCISE 9

Joel has scores of 82, 97, and 93 on his first three exams. What score must he earn on the fourth exam to keep an average of at least 90?

EXAMPLE 9   Finding an Average Test Score

Martha has scores of 88, 86, and 90 on her first three algebra tests. An average score of at least 90 will earn an A in the class. What possible scores on her fourth test will earn her an A average?

Let x = the score on the fourth test. Her average score must be at least 90. To find the average of four numbers, add them and then divide by 4.

$$\underset{}{\underbrace{\text{Average}}} \qquad \underset{\text{least}}{\underbrace{\text{is at}}} \quad \underset{}{\underbrace{90.}}$$
$$\frac{88 + 86 + 90 + x}{4} \qquad \geq \qquad 90$$

$$\frac{264 + x}{4} \geq 90 \qquad \text{Add the scores.}$$
$$264 + x \geq 360 \qquad \text{Multiply by 4.}$$
$$x \geq 96 \qquad \text{Subtract 264.}$$

She must score 96 or more on her fourth test.

CHECK $\dfrac{88 + 86 + 90 + 96}{4} = \dfrac{360}{4} = 90,$ the minimum score. ✓

NOW TRY ANSWERS
8. 9 months
9. at least 88

A score of 96 or more will give an average of at least 90, as required.   NOW TRY

2.5 EXERCISES

Complete solution available on the Video Resources on DVD

 Concept Check Match each inequality in Column I with the correct graph or interval in Column II.

I

II

1. $x \le 3$

A.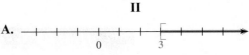

2. $x > 3$

B.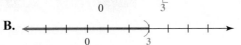

3. $x < 3$ **C.** $(3, \infty)$

4. $x \ge 3$ **D.** $(-\infty, 3]$

5. $-3 \le x \le 3$ **E.** $(-3, 3)$

6. $-3 < x < 3$ **F.** $[-3, 3]$

7. *Concept Check* A student solved the following inequality as shown.

$$4x \ge -64$$
$$\frac{4x}{4} \le \frac{-64}{4}$$
$$x \le -16$$

Solution set: $(-\infty, -16]$

WHAT WENT WRONG? Give the correct solution set.

8. *Concept Check* Dr. Paul Donohue writes a syndicated column in which readers question him on a variety of health topics. Reader C. J. wrote, "Many people say they can weigh more because they have a large frame. How is frame size determined?" Here is Dr. Donohue's response:

> *"For a man, a wrist circumference between 6.75 and 7.25 in. [inclusive] indicates a medium frame. Anything above is a large frame and anything below, a small frame."*

Using x to represent wrist circumference in inches, write an inequality or a three-part inequality that represents wrist circumference for a male with the following.

(a) a small frame   **(b)** a medium frame   **(c)** a large frame

(*Source: The Gazette.*)

Solve each inequality. Give the solution set in both interval and graph form. ***See Examples 1–5.***

9. $x - 4 \ge 12$ **10.** $x - 3 \ge 7$ **11.** $3k + 1 > 22$

12. $5x + 6 < 76$ **13.** $4x < -16$ **14.** $2x > -10$

15. $-\frac{3}{4}x \ge 30$ **16.** $-\frac{2}{3}x \le 12$ **17.** $-1.3x \ge -5.2$

18. $-2.5x \le -1.25$ **19.** $5x + 2 \le -48$ **20.** $4x + 1 \le -31$

21. $\frac{5x - 6}{8} < 8$ **22.** $\frac{3x - 1}{4} > 5$ **23.** $\frac{2x - 5}{-4} > 5$

24. $\frac{3x - 2}{-5} < 6$ **25.** $6x - 4 \ge -2x$ **26.** $2x - 8 \ge -2x$

27. $x - 2(x - 4) \le 3x$ **28.** $x - 3(x + 1) \le 4x$

29. $-(4 + r) + 2 - 3r < -14$ **30.** $-(9 + x) - 5 + 4x \ge 4$

31. $-3(x - 6) > 2x - 2$ **32.** $-2(x + 4) \le 6x + 16$

33. $\frac{2}{3}(3x - 1) \ge \frac{3}{2}(2x - 3)$ **34.** $\frac{7}{5}(10x - 1) < \frac{2}{3}(6x + 5)$

35. $-\dfrac{1}{4}(p + 6) + \dfrac{3}{2}(2p - 5) < 10$

36. $\dfrac{3}{5}(t - 2) - \dfrac{1}{4}(2t - 7) \le 3$

37. $3(2x - 4) - 4x < 2x + 3$

38. $7(4 - x) + 5x < 2(16 - x)$

39. $8\left(\dfrac{1}{2}x + 3\right) < 8\left(\dfrac{1}{2}x - 1\right)$

40. $10\left(\dfrac{1}{5}x + 2\right) < 10\left(\dfrac{1}{5}x + 1\right)$

RELATING CONCEPTS EXERCISES 41–45

FOR INDIVIDUAL OR GROUP WORK

Work Exercises 41–45 in order.

41. Solve the linear equation $5(x + 3) - 2(x - 4) = 2(x + 7)$, and graph the solution set on a number line.

42. Solve the linear inequality $5(x + 3) - 2(x - 4) > 2(x + 7)$, and graph the solution set on a number line.

43. Solve the linear inequality $5(x + 3) - 2(x - 4) < 2(x + 7)$, and graph the solution set on a number line.

44. Graph all the solution sets of the equation and inequalities in **Exercises 43–45** on the same number line. What set do you obtain?

45. Based on the results of **Exercises 41–43,** complete the following, using a conjecture (educated guess): The solution set of $-3(x + 2) = 3x + 12$ is $\{-3\}$, and the solution set of $-3(x + 2) < 3x + 12$ is $(-3, \infty)$. Therefore the solution set of $-3(x + 2) > 3x + 12$ is _____ .

46. *Concept Check* Which is the graph of $-2 < x$?

A. **B.**

C. **D.**

Solve each inequality. Give the solution set in both interval and graph form. ***See Examples 6 and 7.***

47. $-4 < x - 5 < 6$

48. $-1 < x + 1 < 8$

49. $-9 \le x + 5 \le 15$

50. $-4 \le x + 3 \le 10$

51. $-6 \le 2x + 4 \le 16$

52. $-15 < 3x + 6 < -12$

53. $-19 \le 3x - 5 \le 1$

54. $-16 < 3x + 2 < -10$

55. $-1 \le \dfrac{2x - 5}{6} \le 5$

56. $-3 \le \dfrac{3x + 1}{4} \le 3$

57. $4 \le -9x + 5 < 8$

58. $4 \le -2x + 3 < 8$

Give, in interval notation, the unknown numbers in each description.

59. Six times a number is between -12 and 12.

60. Half a number is between -3 and 2.

61. When 1 is added to twice a number, the result is greater than or equal to 7.

62. If 8 is subtracted from a number, then the result is at least 5.

63. One third of a number is added to 6, giving a result of at least 3.

64. Three times a number, minus 5, is no more than 7.

*Solve each problem. **See Examples 8 and 9.***

65. Faith Varnado earned scores of 90 and 82 on her first two tests in English literature. What score must she make on her third test to keep an average of 84 or greater?

66. Greg Tobin scored 92 and 96 on his first two tests in "Methods in Teaching Mathematics." What score must he make on his third test to keep an average of 90 or greater?

67. Amber is signing up for cell phone service. She is trying to decide between Plan A, which costs $54.99 a month with a free phone included, and Plan B, which costs $49.99 a month, but would require her to buy a phone for $129. Under either plan, Amber does not expect to go over the included number of monthly minutes. After how many months would Plan B be a better deal?

68. Newlyweds Bryce and Lauren Tomlin need to rent a truck to move their belongings to their new apartment. They can rent a truck of the size they need from U-Haul for $29.95 a day plus 28 cents per mile or from Budget Truck Rentals for $34.95 a day plus 25 cents per mile. After how many miles would the Budget rental be a better deal than the U-Haul one?

A product will produce a profit only when the revenue R from selling the product exceeds the cost C of producing it. In Exercises 69 and 70, find the least whole number of units x that must be sold for the business to show a profit for the item described.

69. Peripheral Visions, Inc., finds that the cost of producing x studio-quality DVDs is $C = 20x + 100$, while the revenue produced from them is $R = 24x$ (C and R in dollars).

70. Speedy Delivery finds that the cost of making x deliveries is $C = 3x + 2300$, while the revenue produced from them is $R = 5.50x$ (C and R in dollars).

71. A body mass index (BMI) between 19 and 25 is considered healthy. Use the formula

$$\text{BMI} = \frac{704 \times (\text{weight in pounds})}{(\text{height in inches})^2}$$

to find the weight range w, to the nearest pound, that gives a healthy BMI for each height. (*Source: Washington Post.*)

(a) 72 in.   **(b)** 63 in.   **(c)** Your height in inches

72. To achieve the maximum benefit from exercising, the heart rate, in beats per minute, should be in the target heart rate (THR) zone. For a person aged A, the formula is as follows.

$$0.7(220 - A) \leq \text{THR} \leq 0.85(220 - A)$$

Find the THR to the nearest whole number for each age. (*Source:* Hockey, Robert V., *Physical Fitness: The Pathway to Healthful Living,* Times Mirror/Mosby College Publishing.)

(a) 35   **(b)** 55   **(c)** Your age

PREVIEW EXERCISES

Each exercise requires the graph of two inequalities. Graph them and respond to the statement that follows. **See Section 1.1.**

73. (a) Graph $x > 4$.

     (b) Graph $x < 5$.

     (c) Describe the set of numbers belonging to *both* of these sets.

74. (a) Graph $x < 7$.

     (b) Graph $x < 9$.

     (c) Describe the set of numbers belonging to *either* of these sets.

STUDY SKILLS

Using Study Cards

You may have used "flash cards" in other classes. In math, "study cards" can help you remember terms and definitions, procedures, and concepts. Use study cards to

▶ Quickly review when you have a few minutes;

▶ Review before a quiz or test.

One of the advantages of study cards is that you learn while you are making them.

Vocabulary Cards

Put the word and a page reference on the front of the card. On the back, write the definition, an example, any related words, and a sample problem (if appropriate).

Procedure ("Steps") Cards

Write the name of the procedure on the front of the card. Then write each step in words. On the back of the card, put an example showing each step.

Make a vocabulary card and a procedure card for material you are learning now.

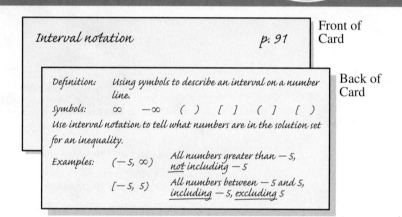

Interval notation   p. 91   Front of Card

Definition: Using symbols to describe an interval on a number line.   Back of Card

Symbols: ∞ $-\infty$ ( ) [ ] ( ] [ )

Use interval notation to tell what numbers are in the solution set for an inequality.

Examples: $(-5, \infty)$ All numbers greater than -5, not including -5

     $[-5, 5)$ All numbers between -5 and 5, including -5, excluding 5

Solving a Linear Inequality   p. 95   Front of Card

1. Simplify each side separately. (clear parentheses and combine like terms.)

2. Isolate variable terms on one side. (Add or subtract the same number from both sides.)

3. Isolate the variable. (Divide both sides by the same number; if dividing by a *negative* number, *reverse direction* of inequality.)

Solve $-3(x + 4) + 2 \geq 7 - x$ and graph the solution set.   Back of Card

$-3(x + 4) + 2 \geq 7 - x$ Clear parentheses.

$-3x - 12 + 2 \geq 7 - x$ Combine like terms.

$-3x - 10 \geq 7 - x$ Both sides are simplified.

$-3x - 10 + x \geq 7 - x + x$ Add x to both sides.

$-2x - 10 \geq 7$ Variable term still not isolated.

$-2x - 10 + 10 \geq 7 + 10$ Add 10 to both sides.

$\frac{-2x}{-2} \leq \frac{17}{-2}$ Divide both sides by -2; dividing by negative, *reverse* direction of inequality symbol.

$x \leq -\frac{17}{2}$

$-\frac{17}{2} = -8\frac{1}{2}$

2.6 Set Operations and Compound Inequalities

OBJECTIVES

1 Find the intersection of two sets.

2 Solve compound inequalities with the word *and*.

3 Find the union of two sets.

4 Solve compound inequalities with the word *or*.

Consider the two sets A and B defined as follows.

$$A = \{1, 2, 3\}, \qquad B = \{2, 3, 4\}$$

The set of all elements that belong to both A **and** B, called their *intersection* and symbolized $A \cap B$, is given by

$$A \cap B = \{2, 3\}. \qquad \text{Intersection}$$

The set of all elements that belong to either A **or** B, or both, called their *union* and symbolized $A \cup B$, is given by

$$A \cup B = \{1, 2, 3, 4\}. \qquad \text{Union}$$

We discuss the use of the words *and* and *or* as they relate to sets and inequalities.

OBJECTIVE 1 **Find the intersection of two sets.** The intersection of two sets is defined with the word *and*.

Intersection of Sets

For any two sets A and B, the **intersection** of A and B, symbolized $A \cap B$, is defined as follows.

$$A \cap B = \{x \mid x \text{ is an element of } A \textbf{ and } x \text{ is an element of } B\}$$

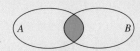

NOW TRY
EXERCISE 1

Let $A = \{2, 4, 6, 8\}$ and $B = \{0, 2, 6, 8\}$. Find $A \cap B$.

EXAMPLE 1 Finding the Intersection of Two Sets

Let $A = \{1, 2, 3, 4\}$ and $B = \{2, 4, 6\}$. Find $A \cap B$.

The set $A \cap B$ contains those elements that belong to both A *and* B: the numbers 2 and 4. Therefore,

$$A \cap B = \{1, 2, 3, 4\} \cap \{2, 4, 6\}$$
$$= \{2, 4\}. \qquad \text{NOW TRY}$$

A **compound inequality** consists of two inequalities linked by a connective word.

$$x + 1 \le 9 \quad \text{and} \quad x - 2 \ge 3 \qquad \text{Examples of compound inequalities}$$
$$2x > 4 \quad \text{or} \quad 3x - 6 < 5 \qquad \text{linked by } \textit{and} \text{ or } \textit{or}$$

OBJECTIVE 2 **Solve compound inequalities with the word *and*.** We use the following steps to solve a compound inequality such as "$x + 1 \le 9$ and $x - 2 \ge 3$."

Solving a Compound Inequality with *and*

Step 1 Solve each inequality individually.

Step 2 Since the inequalities are joined with *and*, the solution set of the compound inequality will include all numbers that satisfy both inequalities in Step 1 (the intersection of the solution sets).

NOW TRY ANSWER
1. $\{2, 6, 8\}$

⌐ NOW TRY
 ↳ EXERCISE 2
Solve the compound inequality, and graph the solution set.

$$x - 2 \le 5 \quad \text{and} \quad x + 5 \ge 9$$

EXAMPLE 2   Solving a Compound Inequality with *and*

Solve the compound inequality, and graph the solution set.

$$x + 1 \le 9 \quad \text{and} \quad x - 2 \ge 3$$

Step 1   Solve each inequality individually.

$$x + 1 \le 9 \qquad \text{and} \qquad x - 2 \ge 3$$
$$x + 1 - 1 \le 9 - 1 \quad \text{and} \quad x - 2 + 2 \ge 3 + 2$$
$$x \le 8 \qquad \text{and} \qquad x \ge 5$$

Step 2   Because of the word *and,* the solution set will include all numbers that satisfy both inequalities in Step 1 at the same time. The compound inequality is true whenever $x \le 8$ and $x \ge 5$ are both true. See the graphs in **FIGURE 17**.

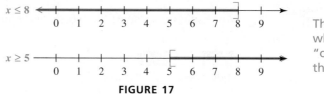

The set of points where the graphs "overlap" represents the intersection.

FIGURE 17

The intersection of the two graphs is the solution set of the compound inequality. **FIGURE 18** shows that the solution set, in interval notation, is $[5, 8]$.

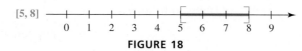

FIGURE 18

NOW TRY ↻

⌐ NOW TRY
 ↳ EXERCISE 3
Solve and graph.

$$-4x - 1 < 7 \quad \text{and}$$
$$3x + 4 \ge -5$$

EXAMPLE 3   Solving a Compound Inequality with *and*

Solve the compound inequality, and graph the solution set.

$$-3x - 2 > 5 \quad \text{and} \quad 5x - 1 \le -21$$

Step 1   Solve each inequality individually.

$$-3x - 2 > 5 \qquad \text{and} \quad 5x - 1 \le -21$$

Remember to reverse the inequality symbol.

$$-3x > 7 \qquad \text{and} \qquad 5x \le -20$$
$$x < -\frac{7}{3} \quad \text{and} \qquad x \le -4$$

The graphs of $x < -\frac{7}{3}$ and $x \le -4$ are shown in **FIGURE 19**.

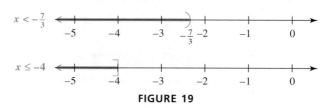

FIGURE 19

Step 2   Now find all values of x that are less than $-\frac{7}{3}$ and also less than or equal to -4. As shown in **FIGURE 20**, the solution set is $(-\infty, -4]$.

NOW TRY ANSWERS
2. $[4, 7]$

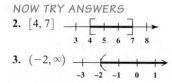

3. $(-2, \infty)$

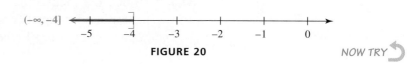

FIGURE 20

NOW TRY ↻

NOW TRY
EXERCISE 4
Solve and graph.

$x - 7 < -12$ and
$\qquad 2x + 1 > 5$

EXAMPLE 4 Solving a Compound Inequality with *and*

Solve the compound inequality, and graph the solution set.

$$x + 2 < 5 \quad \text{and} \quad x - 10 > 2$$

Step 1 Solve each inequality individually.

$$x + 2 < 5 \quad \text{and} \quad x - 10 > 2$$
$$x < 3 \quad \text{and} \qquad x > 12$$

The graphs of $x < 3$ and $x > 12$ are shown in **FIGURE 21**.

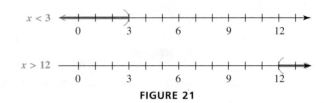

FIGURE 21

Step 2 There is no number that is both less than 3 *and* greater than 12, so the given compound inequality has no solution. The solution set is $\emptyset$. See **FIGURE 22**.

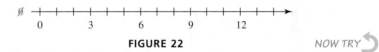

FIGURE 22　　　　　　　　　NOW TRY

OBJECTIVE 3 **Find the union of two sets.** The union of two sets is defined with the word *or*.

Union of Sets

For any two sets A and B, the **union** of A and B, symbolized $A \cup B$, is defined as follows.

$$A \cup B = \{x \mid x \text{ is an element of } A \text{ or } x \text{ is an element of } B\}$$

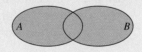

NOW TRY
EXERCISE 5
Let $A = \{5, 10, 15, 20\}$
and $B = \{5, 15, 25\}$.
Find $A \cup B$.

EXAMPLE 5 Finding the Union of Two Sets

Let $A = \{1, 2, 3, 4\}$ and $B = \{2, 4, 6\}$. Find $A \cup B$.

Begin by listing all the elements of set A: 1, 2, 3, 4. Then list any additional elements from set B. In this case the elements 2 and 4 are already listed, so the only additional element is 6.

$$A \cup B = \{1, 2, 3, 4\} \cup \{2, 4, 6\}$$
$$= \{1, 2, 3, 4, 6\}$$

The union consists of all elements in either A *or* B (or both).　　　NOW TRY

NOTE In **Example 5,** notice that although the elements 2 and 4 appeared in both sets A and B, they are written only once in $A \cup B$.

NOW TRY ANSWERS
4. $\emptyset$　**5.** $\{5, 10, 15, 20, 25\}$

OBJECTIVE 4 **Solve compound inequalities with the word *or*.** Use the following steps to solve a compound inequality such as "$6x - 4 < 2x$ or $-3x \leq -9$."

Solving a Compound Inequality with *or*

Step 1 Solve each inequality individually.

Step 2 Since the inequalities are joined with *or,* the solution set of the compound inequality includes all numbers that satisfy either one of the two inequalities in Step 1 (the union of the solution sets).

NOW TRY
EXERCISE 6
Solve and graph.

$-12x \leq -24$ or $x + 9 < 8$

EXAMPLE 6 **Solving a Compound Inequality with *or***

Solve the compound inequality, and graph the solution set.

$$6x - 4 < 2x \quad \text{or} \quad -3x \leq -9$$

Step 1 Solve each inequality individually.

$$6x - 4 < 2x \quad \text{or} \quad -3x \leq -9$$
$$4x < 4$$
$$x < 1 \quad \text{or} \quad x \geq 3$$

> Remember to reverse the inequality symbol.

The graphs of these two inequalities are shown in **FIGURE 23**.

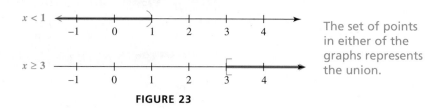

The set of points in either of the graphs represents the union.

FIGURE 23

Step 2 Since the inequalities are joined with *or,* find the union of the two solution sets. The union is shown in **FIGURE 24** and is written

$$(-\infty, 1) \cup [3, \infty).$$

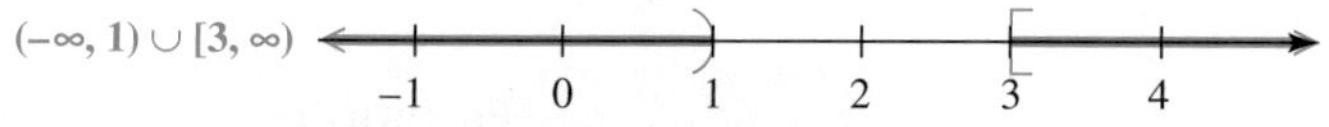

FIGURE 24

NOW TRY

⚠ CAUTION When inequalities are used to write the solution set in **Example 6,** it *must* be written as

$$x < 1 \quad \text{or} \quad x \geq 3,$$

which keeps the numbers 1 and 3 in their order on the number line. Writing $3 \leq x < 1$, which translates using *and,* would imply that $3 \leq 1$, which is ***FALSE.*** There is no other way to write the solution set of such a union.

NOW TRY ANSWER
6. $(-\infty, -1) \cup [2, \infty)$

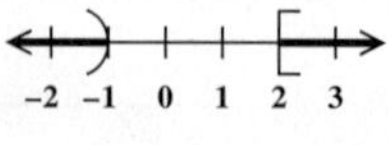

NOW TRY
EXERCISE 7

Solve and graph.

$-x + 2 < 6$ or $6x - 8 \geq 10$

EXAMPLE 7   Solving a Compound Inequality with *or*

Solve the compound inequality, and graph the solution set.

$$-4x + 1 \geq 9 \quad \text{or} \quad 5x + 3 \leq -12$$

Step 1   Solve each inequality individually.

$$-4x + 1 \geq 9 \quad \text{or} \quad 5x + 3 \leq -12$$
$$-4x \geq 8 \quad \text{or} \quad 5x \leq -15$$
$$x \leq -2 \quad \text{or} \quad x \leq -3$$

The graphs of these two inequalities are shown in **FIGURE 25**.

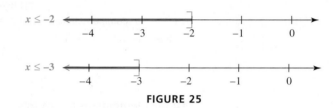

FIGURE 25

Step 2   By taking the union, we obtain the interval $(-\infty, -2]$. See **FIGURE 26**.

FIGURE 26   NOW TRY

NOW TRY
EXERCISE 8

Solve and graph.

$8x - 4 \geq 20$ or
       $-2x + 1 > -9$

EXAMPLE 8   Solving a Compound Inequality with *or*

Solve the compound inequality, and graph the solution set.

$$-2x + 5 \geq 11 \quad \text{or} \quad 4x - 7 \geq -27$$

Step 1   Solve each inequality separately.

$$-2x + 5 \geq 11 \quad \text{or} \quad 4x - 7 \geq -27$$
$$-2x \geq 6 \quad \text{or} \quad 4x \geq -20$$
$$x \leq -3 \quad \text{or} \quad x \geq -5$$

The graphs of these two inequalities are shown in **FIGURE 27**.

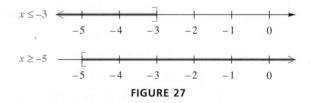

FIGURE 27

Step 2   By taking the union, we obtain every real number as a solution, since every real number satisfies at least one of the two inequalities. The set of all real numbers is written in interval notation as $(-\infty, \infty)$ and graphed as in **FIGURE 28**.

NOW TRY ANSWERS

7. $(-4, \infty)$

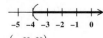

8. $(-\infty, \infty)$

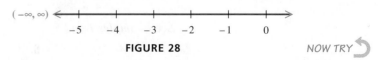

FIGURE 28   NOW TRY

NOW TRY
EXERCISE 9

In **Example 9,** list the elements that satisfy each set.

(a) The set of films with admissions greater than 140,000,000 and gross income less than $1,200,000,000

(b) The set of films with admissions less than 200,000,000 or gross income less than $1,200,000,000

EXAMPLE 9 Applying Intersection and Union

The five highest-grossing domestic films (adjusted for inflation) as of 2009 are listed in the table.

Five All-Time Highest-Grossing Domestic Films

Film	Admissions	Gross Income
Gone with the Wind	202,044,600	$1,450,680,400
Star Wars	178,119,600	$1,278,898,700
The Sound of Music	142,415,400	$1,022,542,400
E.T.	141,854,300	$1,018,514,100
The Ten Commandments	131,000,000	$ 940,580,000

Source: boxofficemojo.com.

List the elements of the following sets.

(a) The set of the top five films with admissions greater than 180,000,000 *and* gross income greater than $1,000,000,000

The only film that satisfies both conditions is *Gone with the Wind,* so the set is

$$\{Gone\ with\ the\ Wind\}.$$

(b) The set of the top five films with admissions less than 140,000,000 *or* gross income greater than $1,000,000,000

Here, any film that satisfies at least one of the conditions is in the set. This set includes all five films:

$$\{Gone\ with\ the\ Wind,\ Star\ Wars,\ The\ Sound\ of\ Music,\ E.T.,\ The\ Ten\ Commandments\}.$$

NOW TRY

NOW TRY ANSWERS
9. (a) {*The Sound of Music, E.T.*}
   (b) {*Star Wars, The Sound of Music, E.T., The Ten Commandments*}

2.6 EXERCISES

MyMathLab · Math XL PRACTICE · WATCH · DOWNLOAD · READ · REVIEW

⊕ *Complete solution available on the Video Resources on DVD*

Concept Check *Decide whether each statement is* true *or* false. *If it is false, explain why.*

1. The union of the solution sets of $x + 1 = 6$, $x + 1 < 6$, and $x + 1 > 6$ is $(-\infty, \infty)$.

2. The intersection of the sets $\{x \mid x \geq 9\}$ and $\{x \mid x \leq 9\}$ is $\emptyset$.

3. The union of the sets $(-\infty, 7)$ and $(7, \infty)$ is $\{7\}$.

4. The intersection of the sets $(-\infty, 7]$ and $[7, \infty)$ is $\{7\}$.

5. The intersection of the set of rational numbers and the set of irrational numbers is $\{0\}$.

6. The union of the set of rational numbers and the set of irrational numbers is the set of real numbers.

Let $A = \{1, 2, 3, 4, 5, 6\}$, $B = \{1, 3, 5\}$, $C = \{1, 6\}$ and $D = \{4\}$. *Specify each set.* **See Examples 1 and 5.**

⊕ **7.** $B \cap A$ **8.** $A \cap B$ **9.** $A \cap D$ **10.** $B \cap C$

11. $B \cap \emptyset$ **12.** $A \cap \emptyset$ ⊕ **13.** $A \cup B$ **14.** $B \cup D$

Concept Check   *Two sets are specified by graphs. Graph the intersection of the two sets.*

15.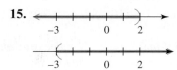

16.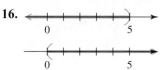

17.

18.

For each compound inequality, give the solution set in both interval and graph form. **See Examples 2–4.**

19. $x < 2$ and $x > -3$

20. $x < 5$ and $x > 0$

21. $x \leq 2$ and $x \leq 5$

22. $x \geq 3$ and $x \geq 6$

23. $x \leq 3$ and $x \geq 6$

24. $x \leq -1$ and $x \geq 3$

25. $x - 3 \leq 6$ and $x + 2 \geq 7$

26. $x + 5 \leq 11$ and $x - 3 \geq -1$

27. $-3x > 3$ and $x + 3 > 0$

28. $-3x < 3$ and $x + 2 < 6$

29. $3x - 4 \leq 8$ and $-4x + 1 \geq -15$

30. $7x + 6 \leq 48$ and $-4x \geq -24$

Concept Check   *Two sets are specified by graphs. Graph the union of the two sets.*

31.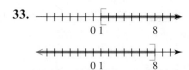

32.

33.

34.

For each compound inequality, give the solution set in both interval and graph form. **See Examples 6–8.**

35. $x \leq 1$ or $x \leq 8$

36. $x \geq 1$ or $x \geq 8$

37. $x \geq -2$ or $x \geq 5$

38. $x \leq -2$ or $x \leq 6$

39. $x \geq -2$ or $x \leq 4$

40. $x \geq 5$ or $x \leq 7$

41. $x + 2 > 7$ or $1 - x > 6$

42. $x + 1 > 3$ or $x + 4 < 2$

43. $x + 1 > 3$ or $-4x + 1 > 5$

44. $3x < x + 12$ or $x + 1 > 10$

45. $4x + 1 \geq -7$ or $-2x + 3 \geq 5$

46. $3x + 2 \leq -7$ or $-2x + 1 \leq 9$

Concept Check   *Express each set in the simplest interval form. (Hint: Graph each set and look for the intersection or union.)*

47. $(-\infty, -1] \cap [-4, \infty)$

48. $[-1, \infty) \cap (-\infty, 9]$

49. $(-\infty, -6] \cap [-9, \infty)$

50. $(5, 11] \cap [6, \infty)$

51. $(-\infty, 3) \cup (-\infty, -2)$

52. $[-9, 1] \cup (-\infty, -3)$

53. $[3, 6] \cup (4, 9)$

54. $[-1, 2] \cup (0, 5)$

For each compound inequality, decide whether intersection *or* union *should be used. Then give the solution set in both interval and graph form.* **See Examples 2–4 and 6–8.**

55. $x < -1$ and $x > -5$

56. $x > -1$ and $x < 7$

57. $x < 4$ or $x < -2$

58. $x < 5$ or $x < -3$

59. $-3x \leq -6$ or $-3x \geq 0$ **60.** $2x - 6 \leq -18$ and $2x \geq -18$

61. $x + 1 \geq 5$ and $x - 2 \leq 10$ **62.** $-8x \leq -24$ or $-5x \geq 15$

Average expenses for full-time resident college students at 4-year institutions during the 2007–2008 academic year are shown in the table.

College Expenses (in Dollars), 4-Year Institutions

Type of Expense	Public Schools (in-state)	Private Schools
Tuition and fees	5950	21,588
Board rates	3402	3993
Dormitory charges	4072	4812

Source: National Center for Education Statistics.

Refer to the table on college expenses. List the elements of each set. **See Example 9.**

63. The set of expenses that are less than $6500 for public schools *and* are greater than $10,000 for private schools

64. The set of expenses that are greater than $3000 for public schools *and* are less than $4000 for private schools

65. The set of expenses that are less than $6500 for public schools *or* are greater than $10,000 for private schools

66. The set of expenses that are greater than $12,000 *or* are between $5000 and $6000

RELATING CONCEPTS EXERCISES 67–72

FOR INDIVIDUAL OR GROUP WORK

The figures represent the backyards of neighbors Luigi, Maria, Than, and Joe. Find the area and the perimeter of each yard. Suppose that each resident has 150 ft of fencing and enough sod to cover 1400 ft² of lawn. Give the name or names of the residents whose yards satisfy each description. **Work Exercises 67–72 in order.**

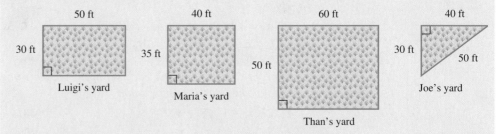

67. The yard can be fenced *and* the yard can be sodded.

68. The yard can be fenced *and* the yard cannot be sodded.

69. The yard cannot be fenced *and* the yard can be sodded.

70. The yard cannot be fenced *and* the yard cannot be sodded.

71. The yard can be fenced *or* the yard can be sodded.

72. The yard cannot be fenced *or* the yard can be sodded.

Solve each inequality. ***See Section 2.5.***

73. $2x - 4 \le 3x + 2$

74. $5x - 8 < 6x - 7$

75. $-5 < 2x + 1 < 5$

76. $-7 \le 3x - 2 < 7$

Evaluate. ***See Sections 1.1 and 1.2.***

77. $-|6| - |-11| + (-4)$

78. $(-5) - |-9| + |5 - 4|$

79. *True* or *false?*   The absolute value of a number is always positive.

80. *True* or *false?*   If $a < 0$, then $|a| = -a$.

STUDY SKILLS

Using Study Cards Revisited

We introduced study cards on **page 102**. Another type of study card follows.

Practice Quiz Cards

Write a problem with direction words (like *solve, simplify*) on the front of the card, and work the problem on the back. Make one for each type of problem you learn. Use the card when you review for a quiz or test.

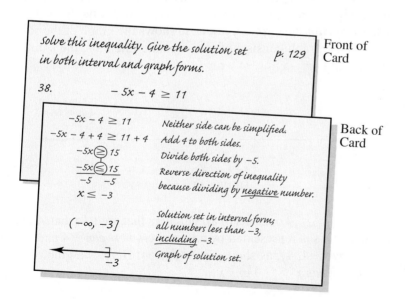

Make a practice quiz card for material you are learning now.

2.7 Absolute Value Equations and Inequalities

Suppose that the government of a country decides that it will comply with a certain restriction on greenhouse gas emissions *within* 3 years of 2020. This means that the *difference* between the year it will comply and 2020 is less than 3, *without regard to sign*. We state this mathematically as

$$|x - 2020| < 3, \quad \text{Absolute value inequality}$$

where x represents the year in which it complies.

Reasoning tells us that the year must be between 2017 and 2023, and thus $2017 < x < 2023$ makes this inequality true. But what general procedure is used to solve such an inequality? We now investigate how to solve absolute value equations and inequalities.

OBJECTIVE 1 **Use the distance definition of absolute value.** In **Section 1.1,** we saw that the absolute value of a number x, written $|x|$, represents the distance from x to 0 on the number line. For example, the solutions of $|x| = 4$ are 4 and -4, as shown in **FIGURE 29**.

$$x = -4 \quad \text{or} \quad x = 4$$

FIGURE 29

Because absolute value represents distance from 0, we interpret the solutions of $|x| > 4$ to be all numbers that are *more* than four units from 0. The set $(-\infty, -4) \cup (4, \infty)$ fits this description. **FIGURE 30** shows the graph of the solution set of $|x| > 4$. Because the graph consists of two separate intervals, the solution set is described using the word *or*: $x < -4$ or $x > 4$.

$$x < -4 \quad \text{or} \quad x > 4$$

FIGURE 30

The solution set of $|x| < 4$ consists of all numbers that are *less* than 4 units from 0 on the number line. This is represented by all numbers *between* -4 and 4. This set of numbers is given by $(-4, 4)$, as shown in **FIGURE 31**. Here, the graph shows that $-4 < x < 4$, which means $x > -4$ *and* $x < 4$.

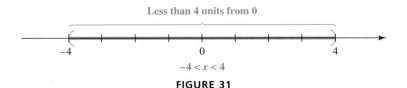

$$-4 < x < 4$$

FIGURE 31

The equation and inequalities just described are examples of **absolute value equations and inequalities.** They involve the absolute value of a variable expression and generally take the form

$$|ax + b| = k, \qquad |ax + b| > k, \qquad \text{or} \qquad |ax + b| < k,$$

where k is a positive number. From **FIGURES 29–31**, we see that

$$|x| = 4 \quad \text{has the same solution set as} \quad x = -4 \quad \text{or} \quad x = 4,$$

$$|x| > 4 \quad \text{has the same solution set as} \quad x < -4 \quad \text{or} \quad x > 4,$$

$$|x| < 4 \quad \text{has the same solution set as} \quad x > -4 \quad \text{and} \quad x < 4.$$

Thus, we solve an absolute value equation or inequality by solving the appropriate compound equation or inequality.

Solving Absolute Value Equations and Inequalities

Let k be a positive real number and p and q be real numbers.

Case 1 To solve $|ax + b| = k$, solve the following compound equation.

$$ax + b = k \quad \text{or} \quad ax + b = -k$$

The solution set is usually of the form $\{p, q\}$, which includes two numbers.

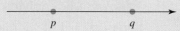

Case 2 To solve $|ax + b| > k$, solve the following compound inequality.

$$ax + b > k \quad \text{or} \quad ax + b < -k$$

The solution set is of the form $(-\infty, p) \cup (q, \infty)$, which is a disjoint interval.

Case 3 To solve $|ax + b| < k$, solve the following three-part inequality.

$$-k < ax + b < k$$

The solution set is of the form (p, q), a single interval.

NOTE Some people prefer to write the compound statements in Cases 1 and 2 of the preceding box as follows.

$$ax + b = k \quad \text{or} \quad -(ax + b) = k \qquad \text{Alternative for Case 1}$$

and $\qquad ax + b > k \quad \text{or} \quad -(ax + b) > k \qquad$ Alternative for Case 2

These forms produce the same results.

OBJECTIVE 2 **Solve equations of the form $|ax + b| = k$, for $k > 0$.** *Remember that because absolute value refers to distance from the origin, an absolute value equation will have two parts.*

NOW TRY
EXERCISE 1
Solve $|4x - 1| = 11$.

EXAMPLE 1 Solving an Absolute Value Equation

Solve $|2x + 1| = 7$. Graph the solution set.

For $|2x + 1|$ to equal 7, $2x + 1$ must be 7 units from 0 on the number line. This can happen only when $2x + 1 = 7$ or $2x + 1 = -7$. This is Case 1 in the preceding box. Solve this compound equation as follows.

$$2x + 1 = 7 \quad \text{or} \quad 2x + 1 = -7$$

$$2x = 6 \quad \text{or} \qquad 2x = -8 \qquad \text{Subtract 1.}$$

$$x = 3 \quad \text{or} \qquad x = -4 \qquad \text{Divide by 2.}$$

Check by substituting 3 and then -4 into the original absolute value equation to verify that the solution set is $\{-4, 3\}$. The graph is shown in **FIGURE 32**.

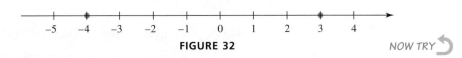

FIGURE 32   NOW TRY

OBJECTIVE 3 Solve inequalities of the form $|ax + b| < k$ and of the form $|ax + b| > k$, for $k > 0$.

NOW TRY
EXERCISE 2
Solve $|4x - 1| > 11$.

EXAMPLE 2 Solving an Absolute Value Inequality with $>$

Solve $|2x + 1| > 7$. Graph the solution set.

By Case 2 described in the previous box, this absolute value inequality is rewritten as

$$2x + 1 > 7 \quad \text{or} \quad 2x + 1 < -7,$$

because $2x + 1$ must represent a number that is *more* than 7 units from 0 on either side of the number line. Now, solve the compound inequality.

$$2x + 1 > 7 \quad \text{or} \quad 2x + 1 < -7$$

$$2x > 6 \quad \text{or} \qquad 2x < -8 \qquad \text{Subtract 1.}$$

$$x > 3 \quad \text{or} \qquad x < -4 \qquad \text{Divide by 2.}$$

Check these solutions. The solution set is $(-\infty, -4) \cup (3, \infty)$. See **FIGURE 33**. Notice that the graph is a disjoint interval.

FIGURE 33   NOW TRY

EXAMPLE 3 Solving an Absolute Value Inequality with $<$

Solve $|2x + 1| < 7$. Graph the solution set.

The expression $2x + 1$ must represent a number that is less than 7 units from 0 on either side of the number line. That is, $2x + 1$ must be between -7 and 7. As Case 3 in the previous box shows, that relationship is written as a three-part inequality.

$$-7 < 2x + 1 < 7$$

$$-8 < \quad 2x \quad < 6 \qquad \text{Subtract 1 from each part.}$$

$$-4 < \quad x \quad < 3 \qquad \text{Divide each part by 2.}$$

NOW TRY ANSWERS
1. $\left\{-\frac{5}{2}, 3\right\}$
2. $\left(-\infty, -\frac{5}{2}\right) \cup (3, \infty)$

NOW TRY
EXERCISE 3
Solve $|4x - 1| < 11$.

Check that the solution set is $(-4, 3)$. The graph consists of the single interval shown in **FIGURE 34**.

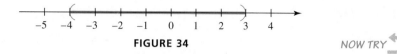

FIGURE 34

NOW TRY

Look back at **FIGURES 32, 33, AND 34**, with the graphs of

$$|2x + 1| = 7, \quad |2x + 1| > 7, \quad \text{and} \quad |2x + 1| < 7,$$

respectively. If we find the union of the three sets, we get the set of all real numbers. This is because, for any value of x, $|2x + 1|$ will satisfy one and only one of the following: It is equal to 7, greater than 7, or less than 7.

⚠ **CAUTION**   When solving absolute value equations and inequalities of the types in **Examples 1, 2, and 3,** remember the following.

1. The methods described apply when the constant is alone on one side of the equation or inequality and is *positive.*

2. Absolute value equations and absolute value inequalities of the form $|ax + b| > k$ translate into "or" compound statements.

3. Absolute value inequalities of the form $|ax + b| < k$ translate into "and" compound statements, which may be written as three-part inequalities.

4. An "or" statement *cannot* be written in three parts. It would be incorrect to write $-7 > 2x + 1 > 7$ in **Example 2,** because this would imply that $-7 > 7$, which is *false.*

OBJECTIVE 4   **Solve absolute value equations that involve rewriting.**

NOW TRY
EXERCISE 4
Solve $|10x - 2| - 2 = 12$.

EXAMPLE 4   Solving an Absolute Value Equation That Requires Rewriting

Solve $|x + 3| + 5 = 12$.

First isolate the absolute value expression on one side of the equals symbol.

$$|x + 3| + 5 = 12$$

$$|x + 3| + 5 - 5 = 12 - 5 \qquad \text{Subtract 5.}$$

$$|x + 3| = 7 \qquad \text{Combine like terms.}$$

Now use the method shown in **Example 1** to solve $|x + 3| = 7$.

$$x + 3 = 7 \quad \text{or} \quad x + 3 = -7$$

$$x = 4 \quad \text{or} \qquad x = -10 \qquad \text{Subtract 3.}$$

Check these solutions by substituting each one in the original equation.

CHECK $\qquad\qquad\qquad\qquad |x + 3| + 5 = 12$

$$|4 + 3| + 5 \overset{?}{=} 12 \qquad \text{Let } x = 4. \qquad\quad |-10 + 3| + 5 \overset{?}{=} 12 \qquad \text{Let } x = -10.$$

$$|7| + 5 \overset{?}{=} 12 \qquad\qquad\qquad\qquad\quad |-7| + 5 \overset{?}{=} 12$$

$$12 = 12 \checkmark \text{ True} \qquad\qquad\qquad\qquad 12 = 12 \checkmark \text{ True}$$

The check confirms that the solution set is $\{-10, 4\}$.

NOW TRY

NOW TRY ANSWERS
3. $\left(-\frac{5}{2}, 3\right)$
4. $\left\{-\frac{6}{5}, \frac{8}{5}\right\}$

NOW TRY
EXERCISE 5

Solve each inequality.

(a) $|x - 1| - 4 \leq 2$

(b) $|x - 1| - 4 \geq 2$

EXAMPLE 5 Solving Absolute Value Inequalities That Require Rewriting

Solve each inequality.

(a)
$$|x + 3| + 5 \geq 12$$
$$|x + 3| \geq 7$$
$$x + 3 \geq 7 \quad \text{or} \quad x + 3 \leq -7$$
$$x \geq 4 \quad \text{or} \quad x \leq -10$$

Solution set: $(-\infty, -10] \cup [4, \infty)$

(b)
$$|x + 3| + 5 \leq 12$$
$$|x + 3| \leq 7$$
$$-7 \leq x + 3 \leq 7$$
$$-10 \leq \quad x \quad \leq 4$$

Solution set: $[-10, 4]$

NOW TRY

OBJECTIVE 5 Solve equations of the form $|ax + b| = |cx + d|$. *If two expressions have the same absolute value, they must either be equal or be negatives of each other.*

> **Solving $|ax + b| = |cx + d|$**
>
> To solve an absolute value equation of the form
> $$|ax + b| = |cx + d|,$$
> solve the following compound equation.
> $$ax + b = cx + d \quad \text{or} \quad ax + b = -(cx + d)$$

NOW TRY
EXERCISE 6

Solve

$$|3x - 4| = |5x + 12|.$$

EXAMPLE 6 Solving an Equation with Two Absolute Values

Solve $|x + 6| = |2x - 3|$.

This equation is satisfied either if $x + 6$ and $2x - 3$ are equal to each other or if $x + 6$ and $2x - 3$ are negatives of each other.

$$x + 6 = 2x - 3 \quad \text{or} \quad x + 6 = -(2x - 3)$$
$$x + 9 = 2x \quad \text{or} \quad x + 6 = -2x + 3$$
$$9 = x \quad \text{or} \quad 3x = -3$$
$$x = -1$$

Check that the solution set is $\{-1, 9\}$.

NOW TRY

OBJECTIVE 6 Solve special cases of absolute value equations and inequalities. When an absolute value equation or inequality involves a *negative constant or 0* alone on one side, use the properties of absolute value to solve the equation or inequality.

> **Special Cases of Absolute Value**
>
> **Case 1** The absolute value of an expression can never be negative. That is, $|a| \geq 0$ for all real numbers a.
>
> **Case 2** The absolute value of an expression equals 0 only when the expression is equal to 0.

NOW TRY ANSWERS
5. (a) $[-5, 7]$
   (b) $(-\infty, -5] \cup [7, \infty)$
6. $\{-8, -1\}$

**NOW TRY
EXERCISE 7**

Solve each equation.

(a) $|3x - 8| = -2$

(b) $|7x + 12| = 0$

EXAMPLE 7 Solving Special Cases of Absolute Value Equations

Solve each equation.

(a) $|5x - 3| = -4$

See Case 1 in the preceding box. ***The absolute value of an expression can never be negative,*** so there are no solutions for this equation. The solution set is $\emptyset$.

(b) $|7x - 3| = 0$

See Case 2 in the preceding box. The expression $|7x - 3|$ will equal 0 *only* if

$$7x - 3 = 0$$

$$7x = 3 \qquad \text{Add 3.}$$

> Check by substituting in the original equation.

$$x = \frac{3}{7}. \qquad \text{Divide by 7.}$$

The solution of this equation is $\frac{3}{7}$. Thus, the solution set is $\left\{\frac{3}{7}\right\}$, with just one element.

NOW TRY

**NOW TRY
EXERCISE 8**

Solve each inequality.

(a) $|x| > -10$

(b) $|4x + 1| + 5 < 4$

(c) $|x - 2| - 3 \le -3$

EXAMPLE 8 Solving Special Cases of Absolute Value Inequalities

Solve each inequality.

(a) $|x| \ge -4$

The absolute value of a number is always greater than or equal to 0. Thus, $|x| \ge -4$ is true for *all* real numbers. The solution set is $(-\infty, \infty)$.

(b)
$$|x + 6| - 3 < -5$$

$$|x + 6| < -2 \qquad \text{Add 3 to each side.}$$

There is no number whose absolute value is less than -2, so this inequality has no solution. The solution set is $\emptyset$.

(c)
$$|x - 7| + 4 \le 4$$

$$|x - 7| \le 0 \qquad \text{Subtract 4 from each side.}$$

The value of $|x - 7|$ will never be less than 0. However, $|x - 7|$ will equal 0 when $x = 7$. Therefore, the solution set is $\{7\}$.

NOW TRY

CONNECTIONS

Absolute value is used to find the *relative error* of a measurement. If x_t represents the expected measurement and x represents the actual measurement, then the relative error in x equals the absolute value of the difference between x_t and x, divided by x_t.

$$\text{relative error in } x = \left|\frac{x_t - x}{x_t}\right|$$

In quality control situations, the relative error often must be less than some predetermined amount. For example, suppose a machine filling *quart* milk cartons is set for a relative error *no greater than* 0.05. Here $x_t = 32$ oz, the relative error = 0.05 oz, and we must find x, given the following condition.

$$\left|\frac{32 - x}{32}\right| \le 0.05 \qquad \textit{No greater than translates as } \le.$$

For Discussion or Writing

With this tolerance level, how many *ounces* may a carton contain?

NOW TRY ANSWERS

7. (a) $\emptyset$ (b) $\left\{-\frac{12}{7}\right\}$

8. (a) $(-\infty, \infty)$ (b) $\emptyset$ (c) $\{2\}$

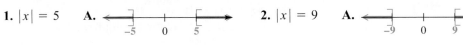

Complete solution available on the Video Resources on DVD

Concept Check *Match each absolute value equation or inequality in Column I with the graph of its solution set in Column II.*

I | II | I | II

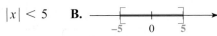

1. $|x| = 5$ **A.**
2. $|x| = 9$ **A.**

$|x| < 5$ **B.**
$|x| > 9$ **B.**

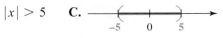

$|x| > 5$ **C.**
$|x| \geq 9$ **C.**

$|x| \leq 5$ **D.**
$|x| < 9$ **D.**

$|x| \geq 5$ **E.**
$|x| \leq 9$ **E.**

3. *Concept Check*   How many solutions will $|ax + b| = k$ have for each situation?

   (a) $k = 0$ **(b)** $k > 0$ **(c)** $k < 0$

4. Explain when to use *and* and when to use *or* if you are solving an absolute value equation or inequality of the form $|ax + b| = k$, $|ax + b| < k$, or $|ax + b| > k$, where k is a positive number.

Solve each equation. **See Example 1.**

5. $|x| = 12$ **6.** $|x| = 14$ **7.** $|4x| = 20$

8. $|5x| = 30$ **9.** $|x - 3| = 9$ **10.** $|x - 5| = 13$

11. $|2x - 1| = 11$ **12.** $|2x + 3| = 19$ **13.** $|4x - 5| = 17$

14. $|5x - 1| = 21$ **15.** $|2x + 5| = 14$ **16.** $|2x - 9| = 18$

17. $\left|\dfrac{1}{2}x + 3\right| = 2$ **18.** $\left|\dfrac{2}{3}x - 1\right| = 5$ **19.** $\left|1 + \dfrac{3}{4}x\right| = 7$

20. $\left|2 - \dfrac{5}{2}x\right| = 14$ **21.** $|0.02x - 1| = 2.50$ **22.** $|0.04x - 3| = 5.96$

Solve each inequality, and graph the solution set. **See Example 2.**

23. $|x| > 3$ **24.** $|x| > 5$ **25.** $|x| \geq 4$

26. $|x| \geq 6$ **27.** $|r + 5| \geq 20$ **28.** $|3r - 1| \geq 8$

29. $|x + 2| > 10$ **30.** $|4x + 1| \geq 21$ **31.** $|3 - x| > 5$

32. $|5 - x| > 3$ **33.** $|-5x + 3| \geq 12$ **34.** $|-2x - 4| \geq 5$

35. *Concept Check*   The graph of the solution set of $|2x + 1| = 9$ is given here.

Without actually doing the algebraic work, graph the solution set of each inequality, referring to the graph shown.

   (a) $|2x + 1| < 9$ **(b)** $|2x + 1| > 9$

36. *Concept Check* The graph of the solution set of $|3x - 4| < 5$ is given here.

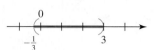

Without actually doing the algebraic work, graph the solution set of the following, referring to the graph shown.

(a) $|3x - 4| = 5$ (b) $|3x - 4| > 5$

Solve each inequality, and graph the solution set. See Example 3. (Hint: Compare your answers with those in Exercises 23–34.)

37. $|x| \leq 3$ **38.** $|x| \leq 5$ **39.** $|x| < 4$

40. $|x| < 6$ **41.** $|r + 5| < 20$ **42.** $|3r - 1| < 8$

43. $|x + 2| \leq 10$ **44.** $|4x + 1| < 21$ **45.** $|3 - x| \leq 5$

46. $|5 - x| \leq 3$ **47.** $|-5x + 3| < 12$ **48.** $|-2x - 4| < 5$

In Exercises 49–66, decide which method of solution applies, and find the solution set. In Exercises 49–60, graph the solution set. See Examples 1–3.

49. $|-4 + x| > 9$ **50.** $|-3 + x| > 8$ **51.** $|x + 5| > 20$

52. $|2x - 1| < 7$ **53.** $|7 + 2x| = 5$ **54.** $|9 - 3x| = 3$

55. $|3x - 1| \leq 11$ **56.** $|2x - 6| \leq 6$ **57.** $|-6x - 6| \leq 1$

58. $|-2x - 6| \leq 5$ **59.** $|2x - 1| \geq 7$ **60.** $|-4 + x| \leq 9$

61. $|x + 2| = 3$ **62.** $|x + 3| = 10$ **63.** $|x - 6| = 3$

64. $|x - 4| = 1$ **65.** $|2 - 0.2x| = 2$ **66.** $|5 - 0.5x| = 4$

Solve each equation or inequality. See Examples 4 and 5.

67. $|x| - 1 = 4$ **68.** $|x| + 3 = 10$ **69.** $|x + 4| + 1 = 2$

70. $|x + 5| - 2 = 12$ **71.** $|2x + 1| + 3 > 8$ **72.** $|6x - 1| - 2 > 6$

73. $|x + 5| - 6 \leq -1$ **74.** $|x - 2| - 3 \leq 4$

75. $\left|\frac{1}{2}x + \frac{1}{3}\right| + \frac{1}{4} = \frac{3}{4}$ **76.** $\left|\frac{2}{3}x + \frac{1}{6}\right| + \frac{1}{2} = \frac{5}{2}$

77. $|0.1x - 2.5| + 0.3 \geq 0.8$ **78.** $|0.5x - 3.5| + 0.2 \geq 0.6$

Solve each equation. See Example 6.

79. $|3x + 1| = |2x + 4|$ **80.** $|7x + 12| = |x - 8|$ **81.** $\left|x - \frac{1}{2}\right| = \left|\frac{1}{2}x - 2\right|$

82. $\left|\frac{2}{3}x - 2\right| = \left|\frac{1}{3}x + 3\right|$ **83.** $|6x| = |9x + 1|$ **84.** $|13x| = |2x + 1|$

85. $|2x - 6| = |2x + 11|$ **86.** $|3x - 1| = |3x + 9|$

Solve each equation or inequality. See Examples 7 and 8.

87. $|x| \geq -10$ **88.** $|x| \geq -15$ **89.** $|12t - 3| = -8$

90. $|13x + 1| = -3$ **91.** $|4x + 1| = 0$ **92.** $|6x - 2| = 0$

93. $|2x - 1| = -6$ **94.** $|8x + 4| = -4$ **95.** $|x + 5| > -9$

96. $|x + 9| > -3$ **97.** $|7x + 3| \leq 0$ **98.** $|4x - 1| \leq 0$

99. $|5x - 2| = 0$ **100.** $|7x + 4| = 0$ **101.** $|x - 2| + 3 \geq 2$

102. $|x - 4| + 5 \geq 4$ **103.** $|10x + 7| + 3 < 1$ **104.** $|4x + 1| - 2 < -5$

105. The recommended daily intake (RDI) of calcium for females aged 19–50 is 1000 mg. Actual needs vary from person to person. Write this statement as an absolute value inequality, with x representing the RDI, to express the RDI plus or minus 100 mg, and solve the inequality. (*Source:* National Academy of Sciences—Institute of Medicine.)

106. The average clotting time of blood is 7.45 sec, with a variation of plus or minus 3.6 sec. Write this statement as an absolute value inequality, with x representing the time, and solve the inequality.

RELATING CONCEPTS EXERCISES 107–110

FOR INDIVIDUAL OR GROUP WORK

The 10 tallest buildings in Houston, Texas, as of 2009 are listed, along with their heights.

Building	Height (in feet)
JPMorgan Chase Tower	1002
Wells Fargo Plaza	992
Williams Tower	901
Bank of America Center	780
Texaco Heritage Plaza	762
Enterprise Plaza	756
Centerpoint Energy Plaza	741
Continental Center I	732
Fulbright Tower	725
One Shell Plaza	714

Source: World Almanac and Book of Facts.

Use this information to **work Exercises 107–110 in order.**

107. To find the average of a group of numbers, we add the numbers and then divide by the number of numbers added. Use a calculator to find the average of the heights.

108. Let k represent the average height of these buildings. If a height x satisfies the inequality

$$|x - k| < t,$$

then the height is said to be within t feet of the average. Using your result from **Exercise 107,** list the buildings that are within 50 ft of the average.

109. Repeat **Exercise 108,** but list the buildings that are within 95 ft of the average.

110. **(a)** Write an absolute value inequality that describes the height of a building that is *not* within 95 ft of the average.

(b) Solve the inequality you wrote in part (a).

(c) Use the result of part (b) to list the buildings that are not within 95 ft of the average.

(d) Confirm that your answer to part (c) makes sense by comparing it with your answer to **Exercise 109.**

PREVIEW EXERCISES

*For the equations **(a)** $3x + 2y = 24$, and **(b)** $-2x + 5y = 20$, find y for the given value of x. **See Section 1.3.***

111. $x = 0$ **112.** $x = -2$ **113.** $x = 8$ **114.** $x = 1.5$

SUMMARY EXERCISES on Solving Linear and Absolute Value Equations and Inequalities

Solve each equation or inequality. Give the solution set in set notation for equations and in interval notation for inequalities.

1. $4x + 1 = 49$

2. $|x - 1| = 6$

3. $6x - 9 = 12 + 3x$

4. $3x + 7 = 9 + 8x$

5. $|x + 3| = -4$

6. $2x + 1 \leq x$

7. $8x + 2 \geq 5x$

8. $4(x - 11) + 3x = 20x - 31$

9. $2x - 1 = -7$

10. $|3x - 7| - 4 = 0$

11. $6x - 5 \leq 3x + 10$

12. $|5x - 8| + 9 \geq 7$

13. $9x - 3(x + 1) = 8x - 7$

14. $|x| \geq 8$

15. $9x - 5 \geq 9x + 3$

16. $13x - 5 > 13x - 8$

17. $|x| < 5.5$

18. $4x - 1 = 12 + x$

19. $\dfrac{2}{3}x + 8 = \dfrac{1}{4}x$

20. $-\dfrac{5}{8}x \geq -20$

21. $\dfrac{1}{4}x < -6$

22. $7x - 3 + 2x = 9x - 8x$

23. $\dfrac{3}{5}x - \dfrac{1}{10} = 2$

24. $|x - 1| < 7$

25. $x + 9 + 7x = 4(3 + 2x) - 3$

26. $6 - 3(2 - x) < 2(1 + x) + 3$

27. $|2x - 3| > 11$

28. $\dfrac{x}{4} - \dfrac{2x}{3} = -10$

29. $|5x + 1| \leq 0$

30. $5x - (3 + x) \geq 2(3x + 1)$

31. $-2 \leq 3x - 1 \leq 8$

32. $-1 \leq 6 - x \leq 5$

33. $|7x - 1| = |5x + 3|$

34. $|x + 2| = |x + 4|$

35. $|1 - 3x| \geq 4$

36. $\dfrac{1}{2} \leq \dfrac{2}{3}x \leq \dfrac{5}{4}$

37. $-(x + 4) + 2 = 3x + 8$

38. $\dfrac{x}{6} - \dfrac{3x}{5} = x - 86$

39. $-6 \leq \dfrac{3}{2} - x \leq 6$

40. $|5 - x| < 4$

41. $|x - 1| \geq -6$

42. $|2x - 5| = |x + 4|$

43. $8x - (1 - x) = 3(1 + 3x) - 4$

44. $8x - (x + 3) = -(2x + 1) - 12$

45. $|x - 5| = |x + 9|$

46. $|x + 2| < -3$

47. $2x + 1 > 5$ or $3x + 4 < 1$

48. $1 - 2x \geq 5$ and $7 + 3x \geq -2$

Reviewing a Chapter

Your textbook provides material to help you prepare for quizzes or tests in this course. Refer to a **Chapter Summary** as you read through the following techniques.

Chapter Reviewing Techniques

▶ **Review the Key Terms and New Symbols.** Make a study card for each. Include a definition, an example, a sketch (if appropriate), and a section or page reference.

▶ **Take the Test Your Word Power quiz** to check your understanding of new vocabulary. The answers immediately follow.

▶ **Read the Quick Review.** Pay special attention to the headings. Study the explanations and examples given for each concept. Try to think about the whole chapter.

▶ **Reread your lecture notes.** Focus on what your instructor has emphasized in class, and review that material in your text.

▶ **Work the Review Exercises.** They are grouped by section.

✓ Pay attention to direction words, such as *simplify*, *solve*, and *estimate*.

✓ After you've done each section of exercises, check your answers in the answer section.

✓ Are your answers exact and complete? Did you include the correct labels, such as $, cm², ft, etc.?

✓ Make study cards for difficult problems.

▶ **Work the Mixed Review Exercises.** They are in mixed-up order. Check your answers in the answer section.

▶ **Take the Chapter Test under test conditions.**

✓ Time yourself.

✓ Use a calculator or notes (if your instructor permits them on tests).

✓ Take the test in one sitting.

✓ Show all your work.

✓ Check your answers in the back of the book. Section references are provided.

Reviewing a chapter will take some time. Avoid rushing through your review in one night. Use the suggestions over a few days or evenings to better understand the material and remember it longer.

Follow these reviewing techniques for your next test. Evaluate how they worked for you.

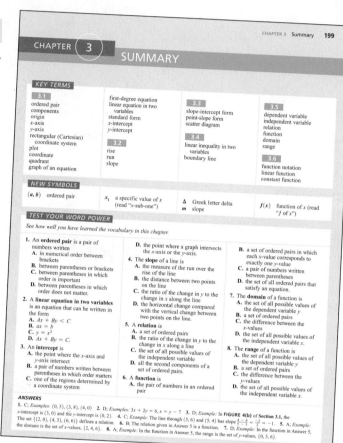

CHAPTER 2 SUMMARY

KEY TERMS

2.1
linear (first-degree) equation in one variable
solution
solution set
equivalent equations
conditional equation
contradiction
identity

2.2
mathematical model
formula
percent

2.5
inequality
linear inequality in one variable
equivalent inequalities
three-part inequality

2.6
intersection
compound inequality
union

2.7
absolute value equation
absolute value inequality

NEW SYMBOLS

$1°$ one degree

∞ infinity
$-\infty$ negative infinity

$(-\infty, \infty)$ the set of real numbers

$\cap$ set intersection
$\cup$ set union

TEST YOUR WORD POWER

See how well you have learned the vocabulary in this chapter.

1. An **algebraic expression** is
  A. an expression that uses any of the four basic operations or the operation of raising to powers or taking roots on any collection of variables and numbers formed according to the rules of algebra
  B. an equation that uses any of the four basic operations or the operation of raising to powers or taking roots on any collection of variables and numbers formed according to the rules of algebra
  C. an equation in algebra
  D. an expression that contains fractions.

2. An **equation** is
  A. an algebraic expression
  B. an expression that contains fractions
  C. an expression that uses any of the four basic operations or the operation of raising to powers or taking roots on any collection of variables and numbers formed according to the rules of algebra
  D. a statement that two algebraic expressions are equal.

3. The **intersection** of two sets A and B is the set of elements that belong
  A. to both A and B
  B. to either A or B, or both
  C. to either A or B, but not both
  D. to just A.

4. The **union** of two sets A and B is the set of elements that belong
  A. to both A and B
  B. to either A or B, or both
  C. to either A or B, but not both
  D. to just B.

ANSWERS

1. A; *Examples:* $\frac{3y-1}{2}$, $6 + \sqrt{2x}$, $4a^3b - c$ **2.** D; *Examples:* $2a + 3 = 7$, $3y = -8$, $x^2 = 4$ **3.** A; *Example:* If $A = \{2, 4, 6, 8\}$ and $B = \{1, 2, 3\}$, then $A \cap B = \{2\}$.   **4.** B; *Example:* Using the sets A and B from Answer 3, $A \cup B = \{1, 2, 3, 4, 6, 8\}$.

QUICK REVIEW

CONCEPTS	EXAMPLES

2.1 Linear Equations in One Variable

Solving a Linear Equation in One Variable

Step 1   Clear fractions.

Step 2   Simplify each side separately.

Step 3   Isolate the variable terms on one side.

Step 4   Isolate the variable.

Step 5   Check.

Solve $4(8 - 3x) = 32 - 8(x + 2)$.

$$32 - 12x = 32 - 8x - 16 \qquad \text{Distributive property}$$

$$32 - 12x = 16 - 8x$$

$$32 - 12x + 12x = 16 - 8x + 12x \qquad \text{Add } 12x.$$

$$32 = 16 + 4x$$

$$32 - 16 = 16 + 4x - 16 \qquad \text{Subtract 16.}$$

$$16 = 4x$$

$$\frac{16}{4} = \frac{4x}{4} \qquad \text{Divide by 4.}$$

$$4 = x$$

The solution set is $\{4\}$. This can be checked by substituting 4 for x in the original equation.

2.2 Formulas and Percent

Solving a Formula for a Specified Variable (Solving a Literal Equation)

Step 1   If the equation contains fractions, multiply both sides by the LCD to clear the fractions.

Step 2   Transform so that all terms with the specified variable are on one side and all terms without that variable are on the other side.

Step 3   Divide each side by the factor that is the coefficient of the specified variable.

Solve $\mathcal{A} = \frac{1}{2}bh$ for h.

$$\mathcal{A} = \frac{1}{2}bh$$

$$2\mathcal{A} = 2\left(\frac{1}{2}bh\right) \qquad \text{Multiply by 2.}$$

$$2\mathcal{A} = bh$$

$$\frac{2\mathcal{A}}{b} = h, \quad \text{or} \quad h = \frac{2\mathcal{A}}{b} \qquad \text{Divide by } b.$$

2.3 Applications of Linear Equations

Solving an Applied Problem

Step 1   Read the problem.

Step 2   Assign a variable.

How many liters of 30% alcohol solution and 80% alcohol solution must be mixed to obtain 100 L of 50% alcohol solution?

Let $\quad x =$ number of liters of 30% solution needed.

Then $100 - x =$ number of liters of 80% solution needed.

Liters of Solution	Percent (as a decimal)	Liters of Pure Alcohol
x	0.30	$0.30x$
$100 - x$	0.80	$0.80(100 - x)$
100	0.50	$0.50(100)$

Step 3   Write an equation.

Step 4   Solve the equation.

Step 5   State the answer.

Step 6   Check.

The equation is $0.30x + 0.80(100 - x) = 0.50(100)$.

The solution of the equation is 60. Thus, 60 L of 30% solution and $100 - 60 = 40$ L of 80% solution are needed.

$$0.30(60) + 0.80(100 - 60) = 50 \text{ is true.}$$

(continued)

CONCEPTS	EXAMPLES

2.4 **Further Applications of Linear Equations**

To solve a uniform motion problem, draw a sketch and make a table. Use the formula

$$d = rt.$$

Two cars start from towns 400 mi apart and travel toward each other. They meet after 4 hr. Find the rate of each car if one travels 20 mph faster than the other.

Let $\quad x$ = rate of the slower car in miles per hour.

Then $x + 20$ = rate of the faster car.

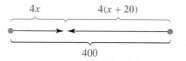

	Rate	Time	Distance
Slower Car	x	4	$4x$
Faster Car	$x + 20$	4	$4(x + 20)$
			400

Use the information in the problem and $d = rt$ to complete a table.

←Total

A sketch shows that the sum of the distances, $4x$ and $4(x + 20)$, must be 400.

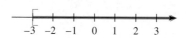

The equation is $\quad 4x + 4(x + 20) = 400$.

Solving this equation gives $x = 40$. The slower car travels 40 mph, and the faster car travels $40 + 20 = 60$ mph.

2.5 **Linear Inequalities in One Variable**

Solving a Linear Inequality in One Variable

Step 1 Simplify each side of the inequality by clearing parentheses and combining like terms.

Step 2 Use the addition property of inequality to get all terms with variables on one side and all terms without variables on the other side.

Step 3 Use the multiplication property of inequality to write the inequality in one of these forms.

$$x < k, \quad x \le k, \quad x > k, \quad \text{or} \quad x \ge k$$

If an inequality is multiplied or divided by a negative number, the inequality symbol must be reversed.

To solve a three-part inequality, work with all three parts at the same time.

Solve $3(x + 2) - 5x \le 12$.

$$3x + 6 - 5x \le 12 \qquad \text{Distributive property}$$
$$-2x + 6 \le 12$$
$$-2x + 6 - 6 \le 12 - 6 \qquad \text{Subtract 6.}$$
$$-2x \le 6$$
$$\frac{-2x}{-2} \ge \frac{6}{-2} \qquad \begin{array}{l}\text{Divide by } -2.\\ \text{Change } \le \text{ to } \ge.\end{array}$$
$$x \ge -3$$

The solution set, $[-3, \infty)$, is graphed here.

Solve $-4 < 2x + 3 \le 7$.

$$-4 - 3 < 2x + 3 - 3 \le 7 - 3 \qquad \text{Subtract 3.}$$
$$-7 < \quad 2x \quad \le 4$$
$$\frac{-7}{2} < \quad \frac{2x}{2} \quad \le \frac{4}{2} \qquad \text{Divide by 2.}$$
$$-\frac{7}{2} < \quad x \quad \le 2$$

The solution set, $\left(-\frac{7}{2}, 2\right]$, is graphed here.

(continued)

CONCEPTS	EXAMPLES

2.6 Set Operations and Compound Inequalities

Solving a Compound Inequality

Step 1 Solve each inequality in the compound inequality individually.

Step 2 If the inequalities are joined with *and*, then the solution set is the intersection of the two individual solution sets.

If the inequalities are joined with *or*, then the solution set is the union of the two individual solution sets.

Solve $x + 1 > 2$ and $2x < 6$.

$$x + 1 > 2 \quad \text{and} \quad 2x < 6$$
$$x > 1 \quad \text{and} \quad x < 3$$

The solution set is $(1, 3)$.

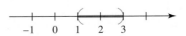

Solve $x \geq 4$ or $x \leq 0$.

The solution set is $(-\infty, 0] \cup [4, \infty)$.

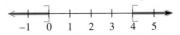

2.7 Absolute Value Equations and Inequalities

Solving Absolute Value Equations and Inequalities
Let k be a positive number.

To solve $|ax + b| = k$, solve the following compound equation.

$$ax + b = k \quad \text{or} \quad ax + b = -k$$

Solve $|x - 7| = 3$.

$$x - 7 = 3 \quad \text{or} \quad x - 7 = -3$$
$$x = 10 \quad \text{or} \quad x = 4 \qquad \text{Add 7.}$$

The solution set is $\{4, 10\}$.

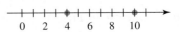

To solve $|ax + b| > k$, solve the following compound inequality.

$$ax + b > k \quad \text{or} \quad ax + b < -k$$

Solve $|x - 7| > 3$.

$$x - 7 > 3 \quad \text{or} \quad x - 7 < -3$$
$$x > 10 \quad \text{or} \quad x < 4 \qquad \text{Add 7.}$$

The solution set is $(-\infty, 4) \cup (10, \infty)$.

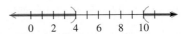

To solve $|ax + b| < k$, solve the following compound inequality.

$$-k < ax + b < k$$

Solve $|x - 7| < 3$.

$$-3 < x - 7 < 3$$
$$4 < x < 10 \qquad \text{Add 7.}$$

The solution set is $(4, 10)$.

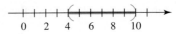

To solve an absolute value equation of the form

$$|ax + b| = |cx + d|,$$

solve the following compound equation.

$$ax + b = cx + d \quad \text{or} \quad ax + b = -(cx + d)$$

Solve $|x + 2| = |2x - 6|$.

$$x + 2 = 2x - 6 \quad \text{or} \quad x + 2 = -(2x - 6)$$
$$x = 8 \qquad\qquad x + 2 = -2x + 6$$
$$3x = 4$$
$$x = \frac{4}{3}$$

The solution set is $\left\{\frac{4}{3}, 8\right\}$.

CHAPTER **2**   REVIEW EXERCISES

2.1   *Solve each equation.*

1. $-(8 + 3x) + 5 = 2x + 6$

2. $-\dfrac{3}{4}x = -12$

3. $\dfrac{2x + 1}{3} - \dfrac{x - 1}{4} = 0$

4. $5(2x - 3) = 6(x - 1) + 4x$

Solve each equation. Then tell whether the equation is conditional, *an* identity, *or a* contradiction.

5. $7x - 3(2x - 5) + 5 + 3x = 4x + 20$

6. $8x - 4x - (x - 7) + 9x + 6 = 12x - 7$

7. $-2x + 6(x - 1) + 3x - (4 - x) = -(x + 5) - 5$

2.2   *Solve each formula for the specified variable.*

8. $V = LWH$ for L

9. $A = \dfrac{1}{2}h(b + B)$ for b

Solve each equation for x.

10. $M = -\dfrac{1}{4}(x + 3y)$

11. $P = \dfrac{3}{4}x - 12$

12. Give the steps you would use to solve $-2x + 5 = 7$.

2.2, 2.3   *Solve each problem.*

13. A rectangular solid has a volume of 180 ft³. Its length is 6 ft and its width is 5 ft. Find its height.

14. The number of students attending college in the United States in 2000 was about 15.3 million. By 2007, the number had increased to about 18.2 million. To the nearest tenth, what was the percent increase? (*Source:* National Center for Education Statistics.)

15. Find the simple-interest rate that Halina Adamska is earning if the principal of $30,000 earns $7800 interest in 4 yr.

16. If the Fahrenheit temperature is 77°, what is the corresponding Celsius temperature?

For 2008, total U.S. government spending was about $2980 billion (or $2.98 trillion). The circle graph shows how the spending was divided.

17. About how much was spent on Social Security?

18. About how much did the U.S. government spend on education and social services in 2008?

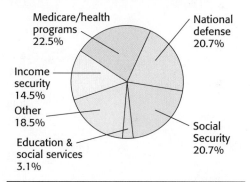

2008 U.S. Government Spending

Medicare/health programs 22.5%

National defense 20.7%

Income security 14.5%

Other 18.5%

Education & social services 3.1%

Social Security 20.7%

Source: U.S. Office of Management and Budget.

Write each phrase as a mathematical expression, using x as the variable.

19. One-third of a number, subtracted from 9

20. The product of 4 and a number, divided by 9 more than the number

Solve each problem.

21. The length of a rectangle is 3 m less than twice the width. The perimeter of the rectangle is 42 m. Find the length and width of the rectangle.

22. In a triangle with two sides of equal length, the third side measures 15 in. less than the sum of the two equal sides. The perimeter of the triangle is 53 in. Find the lengths of the three sides.

23. A candy clerk has three times as many kilograms of chocolate creams as peanut clusters. The clerk has 48 kg of the two candies altogether. How many kilograms of peanut clusters does the clerk have?

24. How many liters of a 20% solution of a chemical should be mixed with 15 L of a 50% solution to get a 30% mixture?

25. How much water should be added to 30 L of a 40% acid solution to reduce it to a 30% solution?

Liters of Solution	Percent (as a decimal)	Liters of Pure Acid
	0.40	
x		
	0.30	

26. Eric Gorenstein invested some money at 6% and $4000 less than that amount at 4%. Find the amount invested at each rate if his total annual interest income is $840.

Principal	Rate (as a decimal)	Interest
x	0.06	
	0.04	

2.4

27. A grocery store clerk has $3.50 in dimes and quarters in her cash drawer. The number of dimes is 1 less than twice the number of quarters. How many of each denomination are there?

28. When Jim emptied his pockets one evening, he found he had 19 nickels and dimes with a total value of $1.55. How many of each denomination did he have?

29. *Concept Check*   Which choice is the best *estimate* for the average rate for a trip of 405 mi that lasted 8.2 hr?

 A. 50 mph   **B.** 30 mph   **C.** 60 mph   **D.** 40 mph

30. (a) A driver averaged 53 mph and took 10 hr to travel from Memphis to Chicago. What is the distance between Memphis and Chicago?

  (b) A small plane traveled from Warsaw to Rome, averaging 164 mph. The trip took 2 hr. What is the distance from Warsaw to Rome?

31. A passenger train and a freight train leave a town at the same time and go in opposite directions. They travel at rates of 60 mph and 75 mph, respectively. How long will it take for them to be 297 mi apart?

	Rate	Time	Distance
Passenger Train	60	x	
Freight Train	75	x	

32. Two cars leave small towns 230 km apart at the same time, traveling directly toward one another. One car travels 15 km per hr slower than the other car. They pass one another 2 hr later. What are their rates?

	Rate	Time	Distance
Faster car	x	2	
Slower car	x − 15	2	

33. An automobile averaged 45 mph for the first part of a trip and 50 mph for the second part. If the entire trip took 4 hr and covered 195 mi, for how long was the rate 45 mph?

34. An 85-mi trip to the beach took Susan Frandsen 2 hr. During the second hour, a rainstorm caused her to average 7 mph less than she traveled during the first hour. Find her average rate for the first hour.

35. Find the measure of each angle in the triangle.

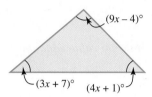

36. Find the measure of each marked angle.

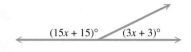

2.5 *Solve each inequality. Express the solution set in interval form.*

37. $-\dfrac{2}{3}x < 6$

38. $-5x - 4 \geq 11$

39. $\dfrac{6x + 3}{-4} < -3$

40. $5 - (6 - 4x) \geq 2x - 7$

41. $8 \leq 3x - 1 < 14$

42. $\dfrac{5}{3}(x - 2) + \dfrac{2}{5}(x + 1) > 1$

Solve each problem.

43. The perimeter of a rectangular playground must be no greater than 120 m. One dimension of the playground must be 22 m. Find the possible lengths of the other dimension of the playground.

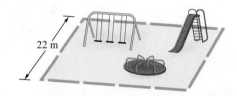

44. A group of college students wants to buy tickets to attend a performance of *The Lion King* at the Overture Center for the Arts in Madison, Wisconsin. They selected mid-balcony seats and were able to get a group rate of $48 per ticket for 15 tickets or more. If they have $1600 available to spend on tickets and they qualify for a $50 group discount along with the group ticket price, how many tickets can they purchase?

45. To pass algebra, a student must have an average of at least 70 on five tests. On the first four tests, a student has scores of 75, 79, 64, and 71. What possible scores on the fifth test would guarantee the student a passing grade in the class?

46. *Concept Check* While solving the following inequality, a student did all the work correctly and obtained the statement $-8 < -13$. The student did not know what to do at this point, because the variable "disappeared," so he gave $\{-8, -13\}$ as the solution set. **WHAT WENT WRONG?**

$$10x + 2(x - 4) < 12x - 13$$

2.6 *Let $A = \{a, b, c, d\}$, $B = \{a, c, e, f\}$, and $C = \{a, e, f, g\}$. Find each set.*

47. $A \cap B$ **48.** $A \cap C$ **49.** $B \cup C$ **50.** $A \cup C$

Solve each compound inequality. Give the solution set in both interval and graph form.

51. $x > 6$ and $x < 9$

52. $x + 4 > 12$ and $x - 2 < 12$

53. $x > 5$ or $x \le -3$

54. $x \ge -2$ or $x < 2$

55. $x - 4 > 6$ and $x + 3 \le 10$

56. $-5x + 1 \ge 11$ or $3x + 5 \ge 26$

Express each union or intersection in simplest interval form.

57. $(-3, \infty) \cap (-\infty, 4)$

58. $(-\infty, 6) \cap (-\infty, 2)$

59. $(4, \infty) \cup (9, \infty)$

60. $(1, 2) \cup (1, \infty)$

2.7 *Solve each absolute value equation.*

61. $|x| = 7$

62. $|x + 2| = 9$

63. $|3x - 7| = 8$

64. $|x - 4| = -12$

65. $|2x - 7| + 4 = 11$

66. $|4x + 2| - 7 = -3$

67. $|3x + 1| = |x + 2|$

68. $|2x - 1| = |2x + 3|$

Solve each absolute value inequality. Give the solution set in interval form.

69. $|x| < 14$

70. $|-x + 6| \le 7$

71. $|2x + 5| \le 1$

72. $|x + 1| \ge -3$

MIXED REVIEW EXERCISES

Solve.

73. $5 - (6 - 4x) > 2x - 5$

74. $ak + bt = 6r$ for k

75. $x < 3$ and $x \ge -2$

76. $\dfrac{4x + 2}{4} + \dfrac{3x - 1}{8} = \dfrac{x + 6}{16}$

77. $|3x + 6| \ge 0$

78. $-5x \ge -10$

79. A newspaper recycling collection bin is in the shape of a box 1.5 ft wide and 5 ft long. If the volume of the bin is 75 ft^3, find the height.

80. The sum of the first and third of three consecutive integers is 47 more than the second integer. What are the integers?

81. $|3x + 2| + 4 = 9$

82. $0.05x + 0.03(1200 - x) = 42$

83. $|x + 3| \le 13$

84. $\dfrac{3}{4}(x - 2) - \dfrac{1}{3}(5 - 2x) < -2$

85. $-4 < 3 - 2x < 9$

86. $-0.3x + 2.1(x - 4) \le -6.6$

87. The complement of an angle measures 10° less than one-fifth of its supplement. Find the measure of the angle.

88. To qualify for a company pension plan, an employee must average at least $1000 per month in earnings. During the first four months of the year, an employee made $900, $1200, $1040, and $760. What possible amounts earned during the fifth month will qualify the employee?

89. $|5x - 1| > 14$

90. $x \ge -2$ or $x < 4$

91. How many liters of a 20% solution of a chemical should be mixed with 10 L of a 50% solution to get a 40% mixture?

92. $|x - 1| = |2x + 3|$

93. $\dfrac{3x}{5} - \dfrac{x}{2} = 3$

94. $|x + 3| \le 1$

95. $|3x - 7| = 4$ **96.** $5(2x - 7) = 2(5x + 3)$

97. *Concept Check*   If $k < 0$, what is the solution set of each of the following?

    (a) $|5x + 3| < k$ **(b)** $|5x + 3| > k$ **(c)** $|5x + 3| = k$

In Exercises 98 and 99, sketch the graph of each solution set.

98. $x > 6$ and $x < 8$ **99.** $-5x + 1 \geq 11$ or $3x + 5 \geq 26$

100. The numbers of civilian workers (to the nearest thousand) for several states in 2008 are shown in the table.

Number of Workers

State	Female	Male
Illinois	2,918,000	3,345,000
Maine	320,000	349,000
North Carolina	2,016,000	2,242,000
Oregon	869,000	976,000
Utah	571,000	755,000
Wisconsin	1,427,000	1,526,000

Source: U.S. Bureau of Labor Statistics.

List the elements of each set.

(a) The set of states with less than 3 million female workers *and* more than 3 million male workers

(b) The set of states with less than 1 million female workers *or* more than 2 million male workers

(c) The set of states with a total of more than 7 million civilian workers

---

CHAPTER 2 TEST

CHAPTER Test Prep VIDEOS

Step-by-step test solutions are found on the Chapter Test Prep Videos available via the Video Resources on DVD, in *MyMathLab*, or on YouTube (search "LialIntermediateAlg").

View the complete solutions to all Chapter Test exercises on the Video Resources on DVD.

Solve each equation.

1. $3(2x - 2) - 4(x + 6) = 3x + 8 + x$

2. $0.08x + 0.06(x + 9) = 1.24$

3. $\dfrac{x + 6}{10} + \dfrac{x - 4}{15} = \dfrac{x + 2}{6}$

4. Solve each equation. Then tell whether the equation is a *conditional equation,* an *identity,* or a *contradiction.*

    (a) $3x - (2 - x) + 4x + 2 = 8x + 3$

    (b) $\dfrac{x}{3} + 7 = \dfrac{5x}{6} - 2 - \dfrac{x}{2} + 9$

    (c) $-4(2x - 6) = 5x + 24 - 7x$

5. Solve $V = \frac{1}{3}bh$ for h.

6. Solve $-16t^2 + vt - S = 0$ for v.

Solve each problem.

7. The 2009 Indianapolis 500 (mile) race was won by Helio Castroneves, who averaged 150.318 mph. What was his time to the nearest thousandth of an hour? (*Source: World Almanac and Book of Facts.*)

8. A certificate of deposit pays $2281.25 in simple interest for 1 yr on a principal of $36,500. What is the rate of interest?

9. In 2008, there were 36,723 offices, stations, and branches of the U.S. Postal Service, of which 27,232 were actually classified as post offices. What percent, to the nearest tenth, were classified as post offices? (*Source:* U.S. Postal Service.)

10. Tyler McGinnis invested some money at 3% simple interest and some at 5% simple interest. The total amount of his investments was $28,000, and the interest he earned during the first year was $1240. How much did he invest at each rate?

11. Two cars leave from the same point at the same time, traveling in opposite directions. One travels 15 mph slower than the other. After 6 hr, they are 630 mi apart. Find the rate of each car.

12. Find the measure of each angle.

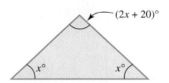

Solve each inequality. Give the solution set in both interval and graph forms.

13. $4 - 6(x + 3) \leq -2 - 3(x + 6) + 3x$ **14.** $-\dfrac{4}{7}x > -16$

15. $-1 < 3x - 4 < 2$ **16.** $-6 \leq \dfrac{4}{3}x - 2 \leq 2$

17. *Concept Check*   Which one of the following inequalities is equivalent to $x < -3$?

A. $-3x < 9$ **B.** $-3x > -9$ **C.** $-3x > 9$ **D.** $-3x < -9$

18. A student must have an average of at least 80 on the four tests in a course to get a B. The student had scores of 83, 76, and 79 on the first three tests. What minimum score on the fourth test would guarantee the student a B in the course?

19. Let $A = \{1, 2, 5, 7\}$ and $B = \{1, 5, 9, 12\}$. Write each of the following sets.

(a) $A \cap B$ **(b)** $A \cup B$

Solve each compound or absolute value inequality.

20. $3x \geq 6$ and $x - 4 < 5$ **21.** $-4x \leq -24$ or $4x - 2 < 10$

22. $|4x + 3| \leq 7$ **23.** $|5 - 6x| > 12$

24. $|7 - x| \leq -1$ **25.** $|-3x + 4| - 4 < -1$

Solve each absolute value equation.

26. $|3x - 2| + 1 = 8$ **27.** $|3 - 5x| = |2x + 8|$

28. *Concept Check*   If $k < 0$, what is the solution set of each of the following?

(a) $|8x - 5| < k$ **(b)** $|8x - 5| > k$ **(c)** $|8x - 5| = k$

Let $A = \left\{-8, -\frac{2}{3}, -\sqrt{6}, 0, \frac{4}{5}, 9, \sqrt{36}\right\}$. *Simplify the elements of A as necessary and then list the elements that belong to each set.*

1. Natural numbers   **2.** Whole numbers   **3.** Integers

4. Rational numbers   **5.** Irrational numbers   **6.** Real numbers

Add or subtract, as indicated.

7. $-\dfrac{4}{3} - \left(-\dfrac{2}{7}\right)$ **8.** $|-4| - |2| + |-6|$

9. $(-2)^4 + (-2)^3$ **10.** $\sqrt{25} - 5(-1)^0$

Evaluate each expression.

11. $(-3)^5$ **12.** $\left(\dfrac{6}{7}\right)^3$ **13.** $\left(-\dfrac{2}{3}\right)^3$ **14.** -4^6

Evaluate for $a = 2$, $b = -3$, and $c = 4$.

15. $-3a + 2b - c$ **16.** $-8(a^2 + b^3)$ **17.** $\dfrac{3a^3 - b}{4 + 3c}$

Simplify each expression.

18. $-7r + 5 - 13r + 12$

19. $-(3k + 8) - 2(4k - 7) + 3(8k + 12)$

Identify the property of real numbers illustrated by each equation.

20. $(a + b) + 4 = 4 + (a + b)$ **21.** $4x + 12x = (4 + 12)x$

Solve each equation.

22. $-4x + 7(2x + 3) = 7x + 36$ **23.** $-\dfrac{3}{5}x + \dfrac{2}{3}x = 2$

24. $0.06x + 0.03(100 + x) = 4.35$ **25.** $P = a + b + c$ for b

26. $4(2x - 6) + 3(x - 2) = 11x + 1$ **27.** $\dfrac{2}{3}x + \dfrac{5}{8}x = \dfrac{31}{24}x$

Solve each inequality. Give the solution set in both interval and graph forms.

28. $3 - 2(x + 7) \le -x + 3$ **29.** $-4 < 5 - 3x \le 0$

30. $2x + 1 > 5$ or $2 - x > 2$ **31.** $|-7k + 3| \ge 4$

Solve each problem.

32. Shamil Mamedov invested some money at 5% simple interest and $2000 more than that amount at 6%. The interest for the year totaled $670. How much was invested at each rate?

33. How much pure alcohol should be added to 7 L of 10% alcohol to increase the concentration to 30% alcohol?

34. Clark's rule, a formula used in reducing drug dosage according to weight from the recommended adult dosage to a child dosage, is

$$\frac{\text{weight of child in pounds}}{150} \times \text{adult dose} = \text{child's dose.}$$

Find a child's dosage if the child weighs 55 lb and the recommended adult dosage is 120 mg.

35. Since 1975, the number of daily newspapers in the United States has steadily declined, as shown in the table.

(a) By how many did the number of daily newspapers decrease between 1975 and 2008?

(b) By what *percent* did the number of daily newspapers decrease from 1975 to 2008? Round to the nearest tenth.

Year	Number of Daily Newspapers
1975	1756
1980	1745
1985	1676
1990	1611
1995	1533
2000	1480
2005	1452
2006	1437
2007	1422
2008	1408

Source: Editorial & Publisher Co.

Graphs, Linear Equations, and Functions

The two most common measures of temperature are Fahrenheit (F) and Celsius (C). It is fairly common knowledge that water freezes at 32°F, or 0°C, and boils at 212°F, or 100°C. Because there is a *linear* relationship between the Fahrenheit and Celsius temperature scales, using these two equivalences we can derive the familiar formulas for converting from one temperature scale to the other, as seen in **Section 3.3, Exercises 93–100.**

Graphs are widely used in the media because they present a great deal of information in a concise form. In this chapter, we see how information such as the relationship between the two temperature scales can be depicted by graphs.

OBJECTIVES

1. Interpret a line graph.
2. Plot ordered pairs.
3. Find ordered pairs that satisfy a given equation.
4. Graph lines.
5. Find x- and y-intercepts.
6. Recognize equations of horizontal and vertical lines and lines passing through the origin.
7. Use the midpoint formula.

OBJECTIVE 1 **Interpret a line graph.** The line graph in **FIGURE 1** shows personal spending (in billions of dollars) on medical care in the United States from 2000 through 2007. About how much was spent on medical care in 2006? (We will answer this question shortly.)

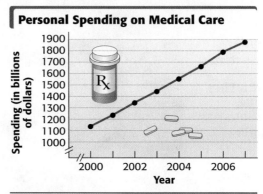

Personal Spending on Medical Care

Source: U.S. Department of Commerce.

FIGURE 1

The line graph in **FIGURE 1** presents information based on a method for locating a point in a plane developed by René Descartes, a 17th-century French mathematician. Legend has it that Descartes, who was lying in bed ill, was watching a fly crawl about on the ceiling near a corner of the room. It occurred to him that the location of the fly could be described by determining its distances from the two adjacent walls. See the figure.

Locating a fly on a ceiling

We use this insight to plot points and graph linear equations in two variables whose graphs are straight lines.

René Descartes (1596–1650)

OBJECTIVE 2 **Plot ordered pairs.** Each of the pairs of numbers

$$(3, 2), \quad (-5, 6), \quad \text{and} \quad (4, -1)$$

is an example of an **ordered pair**—that is, a pair of numbers written within parentheses, consisting of a **first component** and a **second component**. We graph an ordered pair by using two perpendicular number lines that intersect at their 0 points, as shown in the plane in **FIGURE 2**. The common 0 point is called the **origin.**

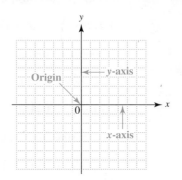

Rectangular coordinate system

FIGURE 2

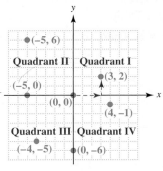

FIGURE 3

The position of any point in this plane is determined by referring to the horizontal number line, or **x-axis,** and the vertical number line, or **y-axis.** The *x*-axis and the *y*-axis make up a **rectangular** (or **Cartesian,** for Descartes) **coordinate system.**

In an ordered pair, the first component indicates position relative to the *x*-axis, and the second component indicates position relative to the *y*-axis. For example, to locate, or **plot,** the point on the graph that corresponds to the ordered pair $(3, 2)$, we move three units from 0 to the right along the *x*-axis and then two units up parallel to the *y*-axis. See **FIGURE 3.** The numbers in an ordered pair are called the **coordinates** of the corresponding point.

We can apply this method of locating ordered pairs to the line graph in **FIGURE 1.** We move along the horizontal axis to a year and then up parallel to the vertical axis to approximate spending for that year. Thus, the ordered pair $(2006, 1800)$ indicates that in 2006 personal spending on medical care was about \$1800 billion.

⚠ **CAUTION**   The parentheses used to represent an ordered pair are also used to represent an open interval (**Section 1.1**). The context of the discussion tells whether ordered pairs or open intervals are being represented.

The four regions of the graph, shown in **FIGURE 3,** are called **quadrants I, II, III,** and **IV,** reading counterclockwise from the upper right quadrant. *The points on the x-axis and y-axis do not belong to any quadrant.*

OBJECTIVE 3   Find ordered pairs that satisfy a given equation. Each solution of an equation with two variables, such as

$$2x + 3y = 6,$$

includes two numbers, one for each variable. To keep track of which number goes with which variable, we write the solutions as ordered pairs. *(If x and y are used as the variables, the x-value is given first.)* For example, we can show that $(6, -2)$ is a solution of $2x + 3y = 6$ by substitution.

$$2x + 3y = 6$$
$$2(6) + 3(-2) \overset{?}{=} 6 \qquad \text{Let } x = 6, y = -2.$$
$$12 - 6 \overset{?}{=} 6 \qquad \text{Multiply.}$$
$$6 = 6 \ \checkmark \quad \text{True}$$

Use parentheses to avoid errors.

Because the ordered pair $(6, -2)$ makes the equation true, it is a solution. On the other hand, $(5, 1)$ is *not* a solution of the equation $2x + 3y = 6$.

$$2x + 3y = 6$$
$$2(5) + 3(1) \overset{?}{=} 6 \qquad \text{Let } x = 5, y = 1.$$
$$10 + 3 \overset{?}{=} 6 \qquad \text{Multiply.}$$
$$13 = 6 \qquad \text{False}$$

To find ordered pairs that satisfy an equation, select a number for one of the variables, substitute it into the equation for that variable, and solve for the other variable.

Since any real number could be selected for one variable and would lead to a real number for the other variable, linear equations in two variables have an infinite number of solutions.

NOW TRY
EXERCISE 1
Complete the table of ordered pairs for $2x - y = 4$.

x	y
0	
	0
4	
	2

EXAMPLE 1 Completing Ordered Pairs and Making a Table

In parts (a)–(d), complete each ordered pair for $2x + 3y = 6$. Then, in part (e), write the results as a table of ordered pairs.

(a) $(0, \underline{\quad})$

$2x + 3y = 6$

$2(0) + 3y = 6$ Let $x = 0$.

$3y = 6$ Multiply, and then add.

$y = 2$ Divide by 3.

The ordered pair is $(0, 2)$.

(b) $(\underline{\quad}, 0)$

$2x + 3y = 6$

$2x + 3(0) = 6$ Let $y = 0$.

$2x = 6$ Multiply, and then add.

$x = 3$ Divide by 2.

The ordered pair is $(3, 0)$.

(c) $(-3, \underline{\quad})$

$2x + 3y = 6$

$2(-3) + 3y = 6$ Let $x = -3$.

$-6 + 3y = 6$ Multiply.

$3y = 12$ Add 6.

$y = 4$ Divide by 3.

The ordered pair is $(-3, 4)$.

(d) $(\underline{\quad}, -4)$

$2x + 3y = 6$

$2x + 3(-4) = 6$ Let $y = -4$.

$2x - 12 = 6$ Multiply.

$2x = 18$ Add 12.

$x = 9$ Divide by 2.

The ordered pair is $(9, -4)$.

(e) We write a table of these ordered pairs as shown.

x	y	
0	2	← Represents the ordered pair (0, 2)
3	0	← Represents the ordered pair (3, 0)
-3	4	← Represents the ordered pair (-3, 4)
9	-4	← Represents the ordered pair (9, -4)

NOW TRY

OBJECTIVE 4 **Graph lines.** The **graph of an equation** is the set of points corresponding to *all* ordered pairs that satisfy the equation. It gives a "picture" of the equation.

To graph $2x + 3y = 6$, we plot the ordered pairs found in **Objective 3** and **Example 1.** These points, plotted in **FIGURE 4(a)**, appear to lie on a straight line. If all the ordered pairs that satisfy the equation $2x + 3y = 6$ were graphed, they would form the straight line shown in **FIGURE 4(b)**.

NOW TRY ANSWER
1.

x	y
0	-4
2	0
4	4
3	2

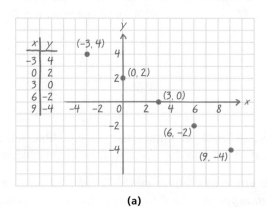

(a)

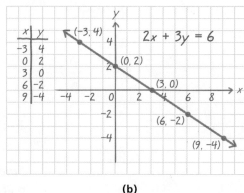

(b)

FIGURE 4

The equation $2x + 3y = 6$ is called a **first-degree equation,** because it has no term with a variable to a power greater than 1.

The graph of any first-degree equation in two variables is a straight line.

Since first-degree equations with two variables have straight-line graphs, they are called *linear equations in two variables.*

Linear Equation in Two Variables

A **linear equation in two variables** can be written in the form

$$Ax + By = C,$$

where A, B, and C are real numbers and A and B are not both 0. This form is called **standard form.**

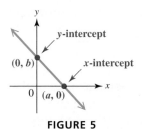

FIGURE 5

OBJECTIVE 5 **Find** ***x-*** **and** ***y-*** **intercepts.** A straight line is determined if any two different points on the line are known. Therefore, finding two different points is enough to graph the line.

Two useful points for graphing are the *x-* and *y-* intercepts. The ***x-*intercept** is the point (if any) where the line intersects the *x-*axis. The ***y-*intercept** is the point (if any) where the line intersects the *y-*axis.* See **FIGURE 5**.

The *y-*value of the point where the line intersects the *x-*axis is 0. Similarly, the *x-*value of the point where the line intersects the *y-*axis is 0. This suggests a method for finding the *x-* and *y-*intercepts.

Finding Intercepts

When graphing the equation of a line, find the intercepts as follows.

Let $y = 0$ to find the *x-*intercept.

Let $x = 0$ to find the *y-*intercept.

EXAMPLE 2 Finding Intercepts

Find the *x-* and *y-*intercepts of $4x - y = -3$ and graph the equation.

To find the *x-*intercept, let $y = 0$.

$$4x - y = -3$$
$$4x - 0 = -3 \qquad \text{Let } y = 0.$$
$$4x = -3$$
$$x = -\frac{3}{4} \qquad \text{x-intercept is } \left(-\tfrac{3}{4}, 0\right).$$

To find the *y-*intercept, let $x = 0$.

$$4x - y = -3$$
$$4(0) - y = -3 \qquad \text{Let } x = 0.$$
$$-y = -3$$
$$y = 3 \qquad \text{y-intercept is } (0, 3).$$

---

*Some texts define an intercept as a number, not a point. For example, "*y-*intercept $(0, 4)$" would be given as "*y-*intercept 4."

NOW TRY
EXERCISE 2
Find the *x*- and *y*-intercepts, and graph the equation.
$$x - 2y = 4$$

The intercepts of $4x - y = -3$ are the points $\left(-\frac{3}{4}, 0\right)$ and $(0, 3)$. Verify by substitution that $(-2, -5)$ also satisfies the equation. We use these ordered pairs to draw the graph in **FIGURE 6**.

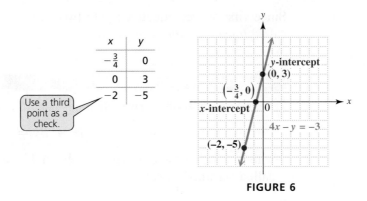

x	y
$-\frac{3}{4}$	0
0	3
-2	-5

Use a third point as a check.

FIGURE 6

NOW TRY

NOTE While two points, such as the two intercepts in **FIGURE 6**, are sufficient to graph a straight line, *it is a good idea to use a third point to guard against errors.*

OBJECTIVE 6 **Recognize equations of horizontal and vertical lines and lines passing through the origin.** A line parallel to the *x*-axis will not have an *x*-intercept. Similarly, a line parallel to the *y*-axis will not have a *y*-intercept. We graph these types of lines in the next two examples.

NOW TRY
EXERCISE 3
Graph $y = -2$.

EXAMPLE 3 Graphing a Horizontal Line

Graph $y = 2$.

Writing $y = 2$ as $0x + 1y = 2$ shows that any value of *x*, including $x = 0$, gives $y = 2$. Thus, the *y*-intercept is $(0, 2)$. Since *y* is always 2, there is no value of *x* corresponding to $y = 0$, so the graph has no *x*-intercept. The graph is shown with a table of ordered pairs in **FIGURE 7**. It is a horizontal line.

NOW TRY ANSWERS
2. *x*-intercept: $(4, 0)$;
  y-intercept: $(0, -2)$

3.

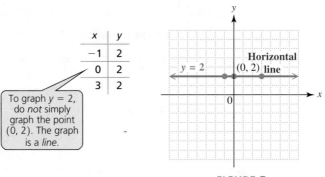

x	y
-1	2
0	2
3	2

To graph $y = 2$, do *not* simply graph the point $(0, 2)$. The graph is a *line*.

FIGURE 7

NOW TRY

NOTE The horizontal line $y = 0$ is the *x*-axis.

NOW TRY
EXERCISE 4

Graph $x + 3 = 0$.

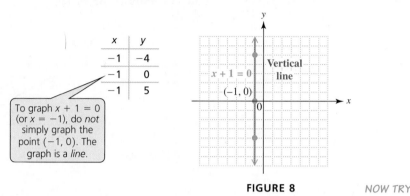

EXAMPLE 4 Graphing a Vertical Line

Graph $x + 1 = 0$.

The form $1x + 0y = -1$ shows that every value of y leads to $x = -1$, making the x-intercept $(-1, 0)$. No value of y makes $x = 0$, so the graph has no y-intercept. A straight line that has no y-intercept is vertical. See **FIGURE 8**.

x	y
−1	−4
−1	0
−1	5

To graph $x + 1 = 0$ (or $x = -1$), do *not* simply graph the point $(-1, 0)$. The graph is a *line*.

FIGURE 8

NOW TRY

NOTE The vertical line $x = 0$ is the y-axis.

NOW TRY
EXERCISE 5

Graph $2x + 3y = 0$.

EXAMPLE 5 Graphing a Line That Passes through the Origin

Graph $x + 2y = 0$.

Find the x-intercept.

$x + 2y = 0$

$x + 2(0) = 0$ Let $y = 0$.

$x + 0 = 0$ Multiply.

$x = 0$ x-intercept is $(0, 0)$.

Find the y-intercept.

$x + 2y = 0$

$0 + 2y = 0$ Let $x = 0$.

$2y = 0$ Add.

$y = 0$ y-intercept is $(0, 0)$.

Both intercepts are the same point, $(0, 0)$, which means that the graph passes through the origin. To find another point, choose any nonzero number for x or y and solve for the other variable. We choose $x = 4$.

$x + 2y = 0$

$4 + 2y = 0$ Let $x = 4$.

$2y = -4$ Subtract 4.

$y = -2$ Divide by 2.

This gives the ordered pair $(4, -2)$. As a check, verify that $(-2, 1)$ also lies on the line. The graph is shown in **FIGURE 9**.

NOW TRY ANSWERS

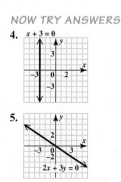

4. $x + 3 = 0$

5. $2x + 3y = 0$

x	y
−2	1
0	0
4	−2

x-intercept and y-intercept

$(-2, 1)$ $(0, 0)$

$(4, -2)$

$x + 2y = 0$

FIGURE 9

NOW TRY

OBJECTIVE 7 **Use the midpoint formula.** If the coordinates of the endpoints of a line segment are known, then the coordinates of the *midpoint* of the segment can be found.

FIGURE 10 shows a line segment PQ with endpoints $P(-8, 4)$ and $Q(3, -2)$. R is the point with the same x-coordinate as P and the same y-coordinate as Q. So the coordinates of R are $(-8, -2)$.

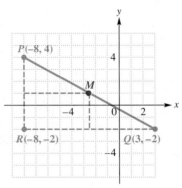

FIGURE 10

The x-coordinate of the midpoint M of PQ is the same as the x-coordinate of the midpoint of RQ. Since RQ is horizontal, the x-coordinate of its midpoint is the *average* of the x-coordinates of its endpoints.

$$\frac{1}{2}(-8 + 3) = -2.5$$

The y-coordinate of M is the average of the y-coordinates of the midpoint of PR.

$$\frac{1}{2}(4 + (-2)) = 1$$

The midpoint of PQ is $M(-2.5, 1)$. This discussion leads to the *midpoint formula.*

Midpoint Formula

If the endpoints of a line segment PQ are (x_1, y_1) and (x_2, y_2), its midpoint M is

$$\left(\frac{x_1 + x_2}{2}, \frac{y_1 + y_2}{2}\right).$$

The small numbers 1 and 2 in the ordered pairs above are called **subscripts.** Read (x_1, y_1) as "**x-sub-one, y-sub-one.**"

NOW TRY
EXERCISE 6

Find the coordinates of the midpoint of the line segment PQ with endpoints $P(2, -5)$ and $Q(-4, 7)$.

NOW TRY ANSWER
6. $(-1, 1)$

EXAMPLE 6 Finding the Coordinates of a Midpoint

Find the coordinates of the midpoint of line segment PQ with endpoints $P(4, -3)$ and $Q(6, -1)$.

Use the midpoint formula with $x_1 = 4, x_2 = 6, y_1 = -3,$ and $y_2 = -1$.

$$\left(\frac{4 + 6}{2}, \frac{-3 + (-1)}{2}\right) = \left(\frac{10}{2}, \frac{-4}{2}\right) = (5, -2) \longleftarrow \text{Midpoint}$$

NOW TRY

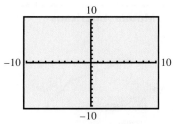

10

−10 ....|........ 10

−10

Standard viewing window
FIGURE 11

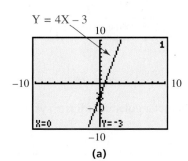

Y = 4X − 3

(a)

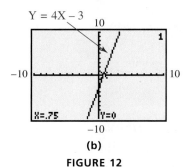

Y = 4X − 3

(b)

FIGURE 12

NOTE When finding the coordinates of the midpoint of a line segment, we are finding the *average* of the x-coordinates and the *average* of the y-coordinates of the endpoints of the segment. In both cases, add the corresponding coordinates and divide the sum by 2.

CONNECTIONS

When graphing with a graphing calculator, we must tell the calculator how to set up a rectangular coordinate system. In the screen in **FIGURE 11**, we chose minimum x- and y-values of -10 and maximum x- and y-values of 10. The **scale** on each axis determines the distance between the tick marks. In the screen shown, the scale is 1 for both axes. We refer to this screen as the **standard viewing window.**

To graph an equation such as $4x - y = 3$, we use the intercepts to determine an appropriate window. Here, the x-intercept is $(0.75, 0)$ and the y-intercept is $(0, -3)$. Although many choices are possible, we choose the standard viewing window.

We must solve the equation for y to enter it into the calculator.

$$4x - y = 3$$
$$-y = -4x + 3 \qquad \text{Subtract } 4x.$$
$$y = 4x - 3 \qquad \text{Multiply by } -1.$$

The graph in **FIGURE 12** also gives the intercepts at the bottoms of the screens. Some calculators have the capability of locating the x-intercept (called "Root" or "Zero"). Consult your owner's manual.

For Discussion or Writing

1. The graphing calculator screens in **Exercise 73** on **page 146** show the graph of a linear equation. What are the intercepts?

Graph each equation with a graphing calculator. Use the standard viewing window.

2. $4x - y = -3$ **(Example 2)**   3. $x + 2y = 0$ **(Example 5)**

3.1 EXERCISES

MyMathLab | Math XL PRACTICE | WATCH | DOWNLOAD | READ | REVIEW

🌐 *Complete solution available on the Video Resources on DVD*

*Solve each problem by locating ordered pairs on the graphs. **See Objective 1.***

1. The graph indicates higher education financial aid in billions of dollars.

   (a) If the ordered pair (x, y) represents a point on the graph, what does x represent? What does y represent?

   (b) Estimate higher education aid in 2007.

   (c) Write an ordered pair (x, y) that gives approximate aid in 2007.

   ✍ **(d)** What does the ordered pair $(1997, 75)$ mean in the context of this graph?

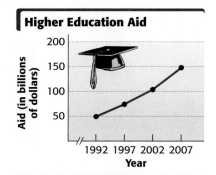

Higher Education Aid

Aid (in billions of dollars)

200
150
100
50

1992 1997 2002 2007
Year

Source: The College Board.

2. The graph shows the percentage of Americans who moved in selected years.

(a) If the ordered pair (x, y) represents a point on the graph, what does x represent? What does y represent?

(b) Estimate the percentage of Americans who moved in 2008.

(c) Write an ordered pair (x, y) that gives the approximate percentage of Americans who moved in 2008.

(d) What does the ordered pair $(1951, 21)$ mean in the context of this graph?

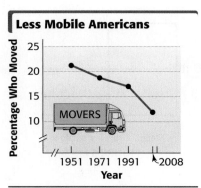

Less Mobile Americans

Source: U.S. Census Bureau.

Concept Check   *Fill in each blank with the correct response.*

3. The point with coordinates $(0, 0)$ is called the _____ of a rectangular coordinate system.

4. For any value of x, the point $(x, 0)$ lies on the _____-axis.

5. To find the x-intercept of a line, we let _____ equal 0 and solve for _____. To find the y-intercept, we let _____ equal 0 and solve for _____.

6. The equation _____ = 4 has a horizontal line as its graph.
$(x \text{ or } y)$

7. To graph a straight line, we must find a minimum of _____ points.

8. The point (_____ , 4) is on the graph of $2x - 3y = 0$.

Name the quadrant, if any, in which each point is located. ***See Objective 2.***

9. (a) $(1, 6)$ **(b)** $(-4, -2)$ **10. (a)** $(-2, -10)$ **(b)** $(4, 8)$

   (c) $(-3, 6)$ **(d)** $(7, -5)$ **(c)** $(-9, 12)$ **(d)** $(3, -9)$

   (e) $(-3, 0)$ **(f)** $(0, -0.5)$ **(e)** $(0, -8)$ **(f)** $(2.3, 0)$

11. *Concept Check*   Use the given information to determine the quadrants in which the point (x, y) may lie.

   (a) $xy > 0$ **(b)** $xy < 0$ **(c)** $\dfrac{x}{y} < 0$ **(d)** $\dfrac{x}{y} > 0$

12. *Concept Check*   What must be true about the value of at least one of the coordinates of any point that lies along an axis?

Plot each point in a rectangular coordinate system. ***See Objective 2.***

13. $(2, 3)$ **14.** $(-1, 2)$ **15.** $(-3, -2)$ **16.** $(1, -4)$ **17.** $(0, 5)$

18. $(-2, -4)$ **19.** $(-2, 4)$ **20.** $(3, 0)$ **21.** $(-2, 0)$ **22.** $(3, -3)$

In Exercises 23–32, **(a)** *complete the given table for each equation and then* **(b)** *graph the equation.* ***See Example 1 and*** **FIGURE 4.**

23. $y = x - 4$

x	y
0	
1	
2	
3	
4	

24. $y = x + 3$

x	y
0	
1	
2	
3	
4	

25. $x - y = 3$

x	y
0	
	0
5	
2	

26. $x - y = 5$

x	y
0	
	0
1	
3	

27. $x + 2y = 5$

x	y
0	
	0
2	
	2

28. $x + 3y = -5$

x	y
0	
	0
1	
	-1

29. $4x - 5y = 20$

x	y
0	
	0
2	
	-3

30. $6x - 5y = 30$

x	y
0	
	0
3	
	-2

31. $y = -2x + 3$

x	y
0	
1	
2	
3	

32. $y = -3x + 1$

x	y
0	
1	
2	
3	

33. *Concept Check* Consider the patterns formed in the tables for **Exercises 23 and 31.** Fill in each blank with the appropriate number.

   (a) In **Exercise 23,** for every increase in x by 1 unit, y increases by _____ unit(s).

   (b) In **Exercise 31,** for every increase in x by 1 unit, y decreases by _____ unit(s).

   (c) On the basis of your observations in parts (a) and (b), make a conjecture about a similar pattern for $y = 2x + 4$. Then test your conjecture.

34. *Concept Check* What is the equation of the x-axis? The equation of the y-axis?

*Find the x- and y-intercepts. Then graph each equation. **See Examples 2–5.***

35. $2x + 3y = 12$ **36.** $5x + 2y = 10$ **37.** $x - 3y = 6$

38. $x - 2y = -4$ **39.** $5x + 6y = -10$ **40.** $3x - 7y = 9$

41. $\dfrac{2}{3}x - 3y = 7$ **42.** $\dfrac{5}{7}x + \dfrac{6}{7}y = -2$ **43.** $y = 5$

44. $y = -3$ **45.** $x = 2$ **46.** $x = -3$

47. $x + 4 = 0$ **48.** $x - 4 = 0$ **49.** $y + 2 = 0$

50. $y - 5 = 0$ **51.** $x + 5y = 0$ **52.** $x - 3y = 0$

53. $2x = 3y$ **54.** $4y = 3x$ **55.** $-\dfrac{2}{3}y = x$ **56.** $-\dfrac{9}{4}y = x$

*Find the midpoint of each segment with the given endpoints. **See Example 6.***

57. $(-8, 4)$ and $(-2, -6)$ **58.** $(5, 2)$ and $(-1, 8)$

59. $(3, -6)$ and $(6, 3)$ **60.** $(-10, 4)$ and $(7, 1)$

61. $(-9, 3)$ and $(9, 8)$ **62.** $(4, -3)$ and $(-1, 3)$

63. $(2.5, 3.1)$ and $(1.7, -1.3)$ **64.** $(6.2, 5.8)$ and $(1.4, -0.6)$

Brain Busters *Find the midpoint of each segment with the given endpoints.*

65. $\left(\dfrac{1}{2}, \dfrac{1}{3}\right)$ and $\left(\dfrac{3}{2}, \dfrac{5}{3}\right)$ **66.** $\left(\dfrac{21}{4}, \dfrac{2}{5}\right)$ and $\left(\dfrac{7}{4}, \dfrac{3}{5}\right)$

67. $\left(-\dfrac{1}{3}, \dfrac{2}{7}\right)$ and $\left(-\dfrac{1}{2}, \dfrac{1}{14}\right)$ **68.** $\left(\dfrac{3}{5}, -\dfrac{1}{3}\right)$ and $\left(\dfrac{1}{2}, -\dfrac{7}{2}\right)$

Segment PQ has the given coordinates for one endpoint P and for its midpoint M. Find the coordinates of the other endpoint Q. (Hint: Represent Q by (x, y) and write two equations using the midpoint formula, one involving x and the other involving y. Then solve for x and y.)

69. $P(5, 8), M(8, 2)$

70. $P(7, 10), M(5, 3)$

71. $P(1.5, 1.25), M(3, 1)$

72. $P(2.5, 1.75), M(3, 2)$

TECHNOLOGY INSIGHTS EXERCISES 73–75

73. The screens show the graph of one of the equations in A–D. Which equation is it?

A. $3x + 2y = 6$ **B.** $-3x + 2y = 6$

C. $-3x - 2y = 6$ **D.** $3x - 2y = 6$

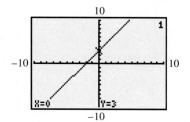

74. The table of ordered pairs was generated by a graphing calculator.

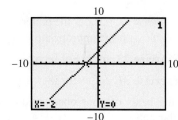

(a) What is the x-intercept?   **(b)** What is the y-intercept?

(c) Which equation corresponds to this table of values?

A. $Y_1 = 2X - 3$ **B.** $Y_1 = -2X - 3$

C. $Y_1 = 2X + 3$ **D.** $Y_1 = -2X + 3$

75. The screens each show the graph of $x + y = 15$ (which was entered as $y = -x + 15$). However, different viewing windows are used. Which window would be more useful for this graph? Why?

A. 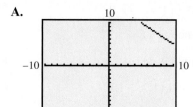   **B.**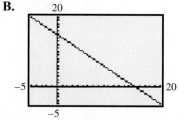

Graph each equation with a graphing calculator. Use the standard viewing window. See the Connections box.

76. $3.6x - y = -5.8$ **77.** $5x + 2y = -10$ **78.** $0.3x + 0.4y = -0.6$

PREVIEW EXERCISES

Find each quotient. **See Section 1.2.**

79. $\dfrac{6-2}{5-3}$ **80.** $\dfrac{5-7}{-4-2}$ **81.** $\dfrac{-5-(-5)}{3-2}$ **82.** $\dfrac{7-(-2)}{-3-(-3)}$

Solve each equation for y. **See Section 2.2.**

83. $4x - y = 10$ **84.** $2y - 3x = 6$ **85.** $3x + 4y = 12$

Graphing lines using the slope intercept forms

STUDY SKILLS

Managing Your Time

Many college students juggle a difficult schedule and multiple responsibilities, including school, work, and family demands.

Time Management Tips

▶ **Read the syllabus for each class.** Understand class policies, such as attendance, late homework, and make-up tests. Find out how you are graded.

▶ **Make a semester or quarter calendar.** Put test dates and major due dates for *all* your classes on the *same* calendar. Try using a different color pen for each class.

▶ **Make a weekly schedule.** After you fill in your classes and other regular responsibilities, block off some study periods. Aim for 2 hours of study for each 1 hour in class.

▶ **Choose a regular study time and place** (such as the campus library). Routine helps.

▶ **Make "to-do" lists.** Number tasks in order of importance. Cross off tasks as you complete them.

▶ **Break big assignments into smaller chunks.** Make deadlines for each smaller chunk so that you stay on schedule.

▶ **Take breaks when studying.** Do not try to study for hours at a time. Take a 10-minute break each hour or so.

▶ **Ask for help when you need it.** Talk with your instructor during office hours. Make use of the learning center, tutoring center, counseling office, or other resources available at your school.

Select several tips to help manage your time this week.

3.2 The Slope of a Line

Slope (steepness) is used in many practical ways. The slope of a highway (sometimes called the *grade*) is often given as a percent. For example, a 10% $\left(\text{or } \frac{10}{100} = \frac{1}{10}\right)$ slope means that the highway rises 1 unit for every 10 horizontal units. Stairs and roofs have slopes too, as shown in **FIGURE 13**.

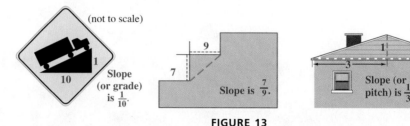

FIGURE 13

Slope is the ratio of vertical change, or **rise**, to horizontal change, or **run**. A simple way to remember this is to think, *"Slope is rise over run."*

OBJECTIVE 1 **Find the slope of a line, given two points on the line.** To get a formal definition of the slope of a line, we designate two different points on the line. To differentiate between the points, we write them as (x_1, y_1) and (x_2, y_2). See **FIGURE 14**.

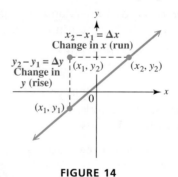

FIGURE 14

As we move along the line in **FIGURE 14** from (x_1, y_1) to (x_2, y_2), the y-value changes (vertically) from y_1 to y_2, an amount equal to $y_2 - y_1$. As y changes from y_1 to y_2, the value of x changes (horizontally) from x_1 to x_2 by the amount $x_2 - x_1$.

NOTE The Greek letter **delta**, Δ, is used in mathematics to denote "change in," so Δy and Δx represent the change in y and the change in x, respectively.

The ratio of the change in y to the change in x (the rise over the run) is called the *slope* of the line, with the letter m traditionally used for slope.

Slope Formula

The **slope** m of the line through the distinct points (x_1, y_1) and (x_2, y_2) is

$$m = \frac{\text{rise}}{\text{run}} = \frac{\text{change in } y}{\text{change in } x} = \frac{\Delta y}{\Delta x} = \frac{y_2 - y_1}{x_2 - x_1} \quad (x_1 \neq x_2).$$

NOW TRY
EXERCISE 1

Find the slope of the line through the points $(2, -6)$ and $(-3, 5)$.

EXAMPLE 1   Finding the Slope of a Line

Find the slope of the line through the points $(2, -1)$ and $(-5, 3)$.

We let $(2, -1) = (x_1, y_1)$ and $(-5, 3) = (x_2, y_2)$ in the slope formula.

$$m = \frac{y_2 - y_1}{x_2 - x_1} = \frac{3 - (-1)}{-5 - 2} = \frac{4}{-7} = -\frac{4}{7}$$

Thus, the slope is $-\frac{4}{7}$. See **FIGURE 15**.

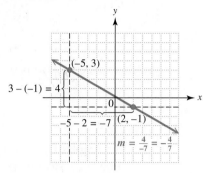

FIGURE 15

If we interchange the ordered pairs so that $(-5, 3) = (x_1, y_1)$ and $(2, -1) = (x_2, y_2)$ in the slope formula, the slope is the same.

> *y*-values are in the *numerator, x*-values in the *denominator.*

$$m = \frac{-1 - 3}{2 - (-5)} = \frac{-4}{7} = -\frac{4}{7}$$

NOW TRY

Example 1 suggests the following important ideas regarding slope:

1. The slope is the same no matter which point we consider first.

2. Using similar triangles from geometry, we can show that the slope is the same no matter which two different points on the line we choose.

⚠ **CAUTION**   *In calculating slope, be careful to subtract the y-values and the x-values in the same order.*

Correct		**Incorrect**

$$\frac{y_2 - y_1}{x_2 - x_1} \quad \text{or} \quad \frac{y_1 - y_2}{x_1 - x_2} \qquad \frac{y_2 - y_1}{x_1 - x_2} \;\text{or}\; \frac{y_1 - y_2}{x_2 - x_1}$$

The change in y is the numerator and the change in x is the denominator.

OBJECTIVE 2   **Find the slope of a line, given an equation of the line.**   When an equation of a line is given, one way to find the slope is to first find two different points on the line and then use the slope formula.

NOW TRY
EXERCISE 2

Find the slope of the line $3x - 7y = 21$.

EXAMPLE 2   Finding the Slope of a Line

Find the slope of the line $4x - y = -8$.

The intercepts can be used as the two different points needed to find the slope. Let $y = 0$ to find that the *x*-intercept is $(-2, 0)$. Then let $x = 0$ to find that the *y*-intercept is $(0, 8)$. Use these two points in the slope formula.

NOW TRY ANSWERS
1. $-\frac{11}{5}$
2. $\frac{3}{7}$

$$m = \frac{\text{rise}}{\text{run}} = \frac{8 - 0}{0 - (-2)} = \frac{8}{2} = 4$$

NOW TRY

NOW TRY
EXERCISE 3
Find the slope of each line.

(a) $y - 6 = 0$ **(b)** $x = 4$

EXAMPLE 3 Finding Slopes of Horizontal and Vertical Lines

Find the slope of each line.

(a) $y = 2$

The graph of $y = 2$ is a horizontal line. See **FIGURE 16**. To find the slope, select two different points on the line, such as $(3, 2)$ and $(-1, 2)$, and use the slope formula.

$$m = \frac{\text{rise}}{\text{run}} = \frac{2 - 2}{3 - (-1)} = \frac{0}{4} = 0$$

In this case, the *rise* is 0, so the slope is 0.

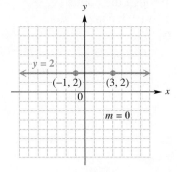

FIGURE 16

(b) $x + 1 = 0$

The graph of $x + 1 = 0$, or $x = -1$, is a vertical line. See **FIGURE 17**. Two points that satisfy the equation $x = -1$ are $(-1, 5)$ and $(-1, -4)$. We use these two points and the slope formula.

$$m = \frac{\text{rise}}{\text{run}} = \frac{-4 - 5}{-1 - (-1)} = \frac{-9}{0}$$

Since division by 0 is undefined, the slope is undefined. This is why the definition of slope includes the restriction that $x_1 \neq x_2$.

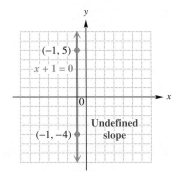

FIGURE 17

NOW TRY

Example 3 illustrates the following important concepts.

> ## Horizontal and Vertical Lines
>
> - An equation of the form $y = b$ always intersects the y-axis at the point $(0, b)$. The line with that equation is horizontal and has slope 0.
>
> - An equation of the form $x = a$ always intersects the x-axis at the point $(a, 0)$. The line with that equation is vertical and has undefined slope.

The slope of a line can also be found directly from its equation. Look again at the equation $4x - y = -8$ from **Example 2.** Solve this equation for y.

$$4x - y = -8 \qquad \text{Equation from Example 2}$$
$$-y = -4x - 8 \qquad \text{Subtract } 4x.$$
$$y = 4x + 8 \qquad \text{Multiply by } -1.$$

Notice that the slope, 4, found with the slope formula in **Example 2** is the same number as the coefficient of x in the equation $y = 4x + 8$. We will see in the next section that this always happens, *as long as the equation is solved for y.*

NOW TRY ANSWERS

3. **(a)** 0 **(b)** undefined

NOW TRY
EXERCISE 4

Find the slope of the graph
of $5x - 4y = 7$.

EXAMPLE 4   Finding the Slope from an Equation

Find the slope of the graph of $3x - 5y = 8$.
  Solve the equation for y.

$$3x - 5y = 8$$

$$-5y = -3x + 8 \qquad \text{Subtract } 3x.$$

$$\boxed{\tfrac{-3x}{-5} = \tfrac{-3}{-5} \cdot \tfrac{x}{1} = \tfrac{3}{5}x} \qquad y = \frac{3}{5}x - \frac{8}{5} \qquad \text{Divide } each \text{ term by } -5.$$

The slope is given by the coefficient of x, so the slope is $\frac{3}{5}$.   NOW TRY

OBJECTIVE 3   Graph a line, given its slope and a point on the line.

NOW TRY
EXERCISE 5

Graph the line passing
through $(-4, 1)$ that has
slope $-\frac{2}{3}$.

EXAMPLE 5   Using the Slope and a Point to Graph Lines

Graph each line described.

(a) With slope $\frac{2}{3}$ and y-intercept $(0, -4)$
  Begin by plotting the point $P(0, -4)$, as shown in **FIGURE 18**. Then use the slope to find a second point.

$$m = \frac{\text{change in } y}{\text{change in } x} = \frac{2}{3}$$

We move 2 units *up* from $(0, -4)$ and then 3 units to the *right* to locate another point on the graph, $R(3, -2)$. The line through $P(0, -4)$ and R is the required graph.

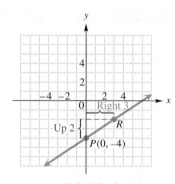

FIGURE 18

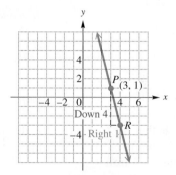

FIGURE 19

(b) Through $(3, 1)$ with slope -4
  Start by locating the point $P(3, 1)$, as shown in **FIGURE 19**. Find a second point R on the line by writing the slope -4 as $\frac{-4}{1}$ and using the slope formula.

$$m = \frac{\text{change in } y}{\text{change in } x} = \frac{-4}{1}$$

We move 4 units *down* from $(3, 1)$ and then 1 unit to the *right* to locate this second point $R(4, -3)$. The line through $P(3, 1)$ and R is the required graph.
  The slope also could be written as

$$m = \frac{\text{change in } y}{\text{change in } x} = \frac{4}{-1}.$$

In this case, the second point R is located 4 units *up* and 1 unit to the *left*. Verify that this approach also produces the line in **FIGURE 19**.   NOW TRY

NOW TRY ANSWERS
4. $\frac{5}{4}$
5.

In **Example 5(a),** the slope of the line is the *positive* number $\frac{2}{3}$. The graph of the line in **FIGURE 18** slants up (rises) from left to right. The line in **Example 5(b)** has *negative* slope -4. As **FIGURE 19** shows, its graph slants down (falls) from left to right. These facts illustrate the following generalization.

Orientation of a Line in the Plane

A positive slope indicates that the line slants *up* (rises) from left to right.

A negative slope indicates that the line slants *down* (falls) from left to right.

FIGURE 20 shows lines of positive, 0, negative, and undefined slopes.

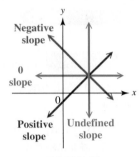

FIGURE 20

OBJECTIVE 4   **Use slopes to determine whether two lines are parallel, perpendicular, or neither.** Recall that the slope of a line measures the steepness of the line and that parallel lines have equal steepness.

Slopes of Parallel Lines

Two nonvertical lines with the same slope are parallel.

Two nonvertical parallel lines have the same slope.

NOW TRY
EXERCISE 6

Determine whether the line through $(2, 5)$ and $(4, 8)$ is parallel to the line through $(2, 0)$ and $(-1, -2)$.

EXAMPLE 6   Determining Whether Two Lines Are Parallel

Determine whether the lines L_1, through $(-2, 1)$ and $(4, 5)$, and L_2, through $(3, 0)$ and $(0, -2)$, are parallel.

Find the slope of L_1.   Find the slope of L_2.

$$m_1 = \frac{5 - 1}{4 - (-2)} = \frac{4}{6} = \frac{2}{3} \qquad\qquad m_2 = \frac{-2 - 0}{0 - 3} = \frac{-2}{-3} = \frac{2}{3}$$

Because the slopes are equal, the two lines are parallel.   *NOW TRY*

To see how the slopes of perpendicular lines are related, consider a nonvertical line with slope $\frac{a}{b}$. If this line is rotated $90°$, the vertical change and the horizontal change are interchanged and the slope is $-\frac{b}{a}$, since the horizontal change is now negative. See **FIGURE 21**. Thus, the slopes of perpendicular lines have product -1 and are negative reciprocals of each other.

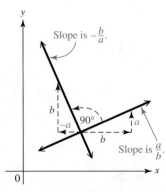

NOW TRY ANSWER
6. no

FIGURE 21

> ### Slopes of Perpendicular Lines
>
> If neither is vertical, perpendicular lines have slopes that are negative reciprocals—that is, their product is -1. Also, lines with slopes that are negative reciprocals are perpendicular.
>
> A line with 0 slope is perpendicular to a line with undefined slope.

NOW TRY
EXERCISE 7
Are the lines with these equations perpendicular?

$$x + 2y = 7$$
$$2x = y - 4$$

EXAMPLE 7   Determining Whether Two Lines Are Perpendicular

Are the lines with equations $2y = 3x - 6$ and $2x + 3y = -6$ perpendicular?
   Find the slope of each line by solving each equation for y.

$2y = 3x - 6$	$2x + 3y = -6$
$y = \dfrac{3}{2}x - 3$ Divide by 2.	$3y = -2x - 6$ Subtract 2x.
$\uparrow$ Slope	$y = -\dfrac{2}{3}x - 2$ Divide by 3.
	$\uparrow$ Slope

Since the product of the slopes is $\frac{3}{2}\left(-\frac{2}{3}\right) = -1$, the lines are perpendicular.

NOW TRY

NOTE In **Example 7,** alternatively, we could have found the slope of each line by using intercepts and the slope formula. For the graph of the equation $2y = 3x - 6$, the x-intercept is $(2, 0)$ and the y-intercept is $(0, -3)$.

$$m = \frac{0 - (-3)}{2 - 0} = \frac{3}{2} \qquad \text{See \textbf{Example 7}.}$$

Find the intercepts of the graph of $2x + 3y = -6$ and use them to confirm the slope $-\frac{2}{3}$. Since the slopes are negative reciprocals, the lines are perpendicular.

NOW TRY
EXERCISE 8
Determine whether the lines with these equations are *parallel, perpendicular,* or *neither.*

$$2x - y = 4$$
$$2x + y = 6$$

EXAMPLE 8   Determining Whether Two Lines Are Parallel, Perpendicular, or Neither

Determine whether the lines with equations $2x - 5y = 8$ and $2x + 5y = 8$ are *parallel, perpendicular,* or *neither.*
   Find the slope of each line by solving each equation for y.

$2x - 5y = 8$	$2x + 5y = 8$
$-5y = -2x + 8$ Subtract 2x.	$5y = -2x + 8$ Subtract 2x.
$y = \dfrac{2}{5}x - \dfrac{8}{5}$ Divide by -5.	$y = -\dfrac{2}{5}x + \dfrac{8}{5}$ Divide by 5.
$\uparrow$ Slope	$\uparrow$ Slope

The slopes, $\frac{2}{5}$ and $-\frac{2}{5}$, are not equal, and they are not negative reciprocals because their product is $-\frac{4}{25}$, not -1. Thus, the two lines are neither parallel nor perpendicular.

NOW TRY

NOW TRY ANSWERS
7. yes
8. neither

OBJECTIVE 5 **Solve problems involving average rate of change.** The slope formula applied to any two points on a line gives the **average rate of change** in *y* per unit change in *x*, where the value of *y* depends on the value of *x*.

For example, suppose the height of a boy increased from 60 to 68 in. between the ages of 12 and 16, as shown in **FIGURE 22**.

Growth Rate

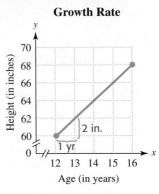

FIGURE 22

$$\underset{\text{Change in age } x \longrightarrow}{\overset{\text{Change in height } y \longrightarrow}{}} \frac{68 - 60}{16 - 12} = \frac{8}{4} = 2 \text{ in.}$$ Boy's average growth rate (or average change in height) *per year*

The boy may actually have grown more than 2 in. during some years and less than 2 in. during other years. If we plotted ordered pairs (age, height) for those years and drew a line connecting any two of the points, the average rate of change would likely be slightly different than that found above. However using the data for ages 12 and 16, the boy's *average* change in height was 2 in. per year over these years.

NOW TRY
EXERCISE 9

Americans spent an average of 828 hr in 2002 watching cable and satellite TV. Using this number for 2002 and the number for 2000 from the graph in **FIGURE 23**, find the average rate of change from 2000 to 2002. How does it compare with the average rate of change found in **Example 9**?

EXAMPLE 9 Interpreting Slope as Average Rate of Change

The graph in **FIGURE 23** approximates the average number of hours per year spent watching cable and satellite TV for each person in the United States from 2000 to 2005. Find the average rate of change in number of hours per year.

Watching Cable and Satellite TV

(2005, 980)

(2000, 690)

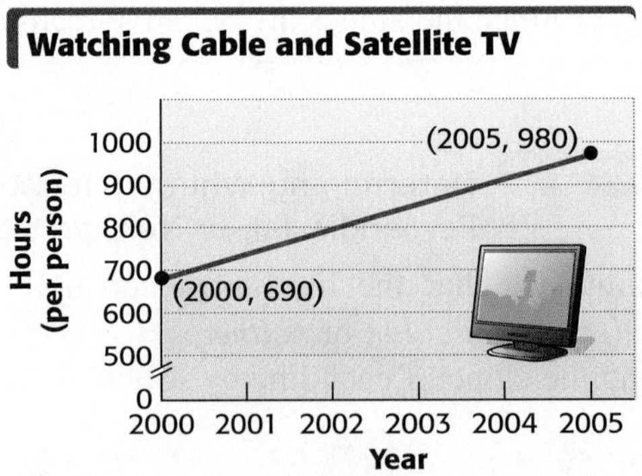

Source: Veronis Suhler Stevenson.

FIGURE 23

To find the average rate of change, we need two pairs of data. From the graph, we have the ordered pairs (2000, 690) and (2005, 980). We use the slope formula.

$$\text{average rate of change} = \frac{980 - 690}{2005 - 2000} = \frac{290}{5} = 58$$

A positive slope indicates an increase.

NOW TRY ANSWER
9. 69 hr per yr; It is greater than the average rate of change from 2000 to 2005.

This means that the average time per person spent watching cable and satellite TV *increased* by 58 hr per year from 2000 to 2005.

NOW TRY

NOW TRY
EXERCISE 10

In 2000, sales of digital cam-corders in the United States totaled $2838 million. In 2008, sales totaled $1885 million. Find the average rate of change in sales of digital camcorders per year, to the nearest million dollars. (*Source:* Consumer Electronics Association.)

EXAMPLE 10 Interpreting Slope as Average Rate of Change

During the year 2000, the average person in the United States spent 812 hr watching broadcast TV. In 2005, the average number of hours per person spent watching broadcast TV was 679. Find the average rate of change in number of hours per year. (*Source:* Veronis Suhler Stevenson.)

To use the slope formula, we let one ordered pair be (2000, 812) and the other be (2005, 679).

$$\text{average rate of change} = \frac{679 - 812}{2005 - 2000} = \frac{-133}{5} = -26.6$$

> A negative slope indicates a decrease.

The graph in **FIGURE 24** confirms that the line through the ordered pairs falls from left to right and therefore has negative slope. Thus, the average time per person spent watching broadcast TV *decreased* by about 27 hr per year from 2000 to 2005.

NOW TRY ANSWER
10. −$119 million per yr

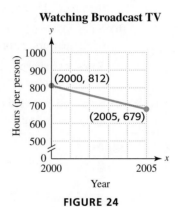

Watching Broadcast TV

FIGURE 24

NOW TRY

3.2 EXERCISES

MyMathLab Math XL PRACTICE WATCH DOWNLOAD READ REVIEW

⊕ *Complete solution available on the Video Resources on DVD*

Concept Check Answer each question about slope in Exercises 1–4.

1. A hill rises 30 ft for every horizontal 100 ft. Which of the following express its slope (or grade)? (There are several correct choices.)

A. 0.3 **B.** $\frac{3}{10}$ **C.** $3\frac{1}{3}$ **D.** $\frac{30}{100}$

E. $\frac{10}{3}$ **F.** 30 **G.** 30% **H.** $-\frac{10}{3}$

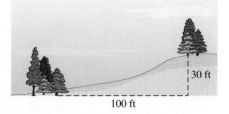

2. If a walkway rises 2 ft for every 10 ft on the horizontal, which of the following express its slope (or grade)? (There are several correct choices.)

A. 0.2 **B.** $\frac{2}{10}$ **C.** $\frac{1}{5}$ **D.** 20%

E. 5 **F.** $\frac{20}{100}$ **G.** $\frac{10}{2}$ **H.** −5

3. A ladder leaning against a wall has slope 3. How many feet in the horizontal direction correspond to a rise of 15 ft?

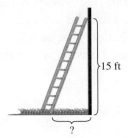

15 ft

?

4. A hill has slope 0.05. How many feet in the vertical direction correspond to a run of 50 ft?

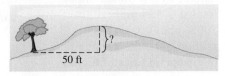

50 ft

NOT TO SCALE

5. *Concept Check* Match each situation in (a)–(d) with the most appropriate graph in A–D.

(a) Sales rose sharply during the first quarter, leveled off during the second quarter, and then rose slowly for the rest of the year.

(b) Sales fell sharply during the first quarter and then rose slowly during the second and third quarters before leveling off for the rest of the year.

(c) Sales rose sharply during the first quarter and then fell to the original level during the second quarter before rising steadily for the rest of the year.

(d) Sales fell during the first two quarters of the year, leveled off during the third quarter, and rose during the fourth quarter.

A.

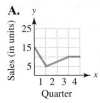

Quarter

B.

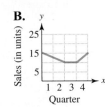

Quarter

C.

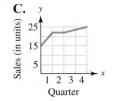

Quarter

D.

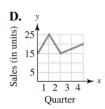

Quarter

6. *Concept Check* Using the given axes, draw a graph that illustrates the following description:

Profits for a business were $10 million in 2000. They rose sharply from 2000 through 2004, remained constant from 2004 through 2008, and then fell slowly from 2008 through 2010.

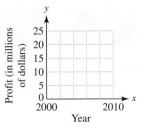

Concept Check Determine the slope of each line segment in the given figure.

7. *AB* **8.** *BC* **9.** *CD*

10. *DE* **11.** *EF* **12.** *FG*

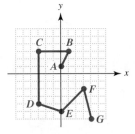

13. *Concept Check* If *A* and *F* were joined by a line segment in the figure for **Exercises 7–12**, what would be the slope of the segment?

14. *Concept Check* If *B* and *D* were joined by a line segment in the figure for **Exercises 7–12**, what would be the slope of the segment?

*Calculate the value of each slope m, if possible, by using the slope formula. **See Example 1.***

15. $m = \dfrac{6 - 2}{5 - 3}$

16. $m = \dfrac{5 - 7}{-4 - 2}$

17. $m = \dfrac{4 - (-1)}{-3 - (-5)}$

18. $m = \dfrac{-6 - 0}{0 - (-3)}$

19. $m = \dfrac{-5 - (-5)}{3 - 2}$

20. $m = \dfrac{-2 - (-2)}{4 - (-3)}$

21. $m = \dfrac{3 - 8}{-2 - (-2)}$

22. $m = \dfrac{5 - 6}{-8 - (-8)}$

23. *Concept Check*   On the basis of the figure shown here, determine which line satisfies the given description.

(a) The line has positive slope.

(b) The line has negative slope.

(c) The line has slope 0.

(d) The line has undefined slope.

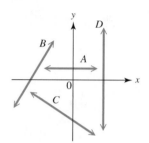

24. Which of the following forms of the slope formula are correct? Explain.

A. $m = \dfrac{y_1 - y_2}{x_2 - x_1}$ **B.** $m = \dfrac{y_1 - y_2}{x_1 - x_2}$ **C.** $m = \dfrac{x_2 - x_1}{y_2 - y_1}$ **D.** $m = \dfrac{y_2 - y_1}{x_2 - x_1}$

*For Exercises 25–36, **(a)** find the slope of the line through each pair of points, if possible, and **(b)** based on the slope, indicate whether the line through the points* rises *from left to right,* falls *from left to right, is* horizontal, *or is* vertical. ***See Examples 1 and 3 and* FIGURE 20.**

25. $(-2, -3)$ and $(-1, 5)$

26. $(-4, 1)$ and $(-3, 4)$

27. $(-4, 1)$ and $(2, 6)$

28. $(-3, -3)$ and $(5, 6)$

29. $(2, 4)$ and $(-4, 4)$

30. $(-6, 3)$ and $(2, 3)$

31. $(-2, 2)$ and $(4, -1)$

32. $(-3, 1)$ and $(6, -2)$

33. $(5, -3)$ and $(5, 2)$

34. $(4, -1)$ and $(4, 3)$

35. $(1.5, 2.6)$ and $(0.5, 3.6)$

36. $(3.4, 4.2)$ and $(1.4, 10.2)$

Brain Busters   *Find the slope of the line through each pair of points.* $\left(\text{Hint: } \dfrac{\frac{a}{b}}{\frac{c}{d}} = \dfrac{a}{b} \div \dfrac{c}{d}\right)$

37. $\left(\dfrac{1}{6}, \dfrac{1}{2}\right)$ and $\left(\dfrac{5}{6}, \dfrac{9}{2}\right)$

38. $\left(\dfrac{3}{4}, \dfrac{1}{3}\right)$ and $\left(\dfrac{5}{4}, \dfrac{10}{3}\right)$

39. $\left(-\dfrac{2}{9}, \dfrac{5}{18}\right)$ and $\left(\dfrac{1}{18}, -\dfrac{5}{9}\right)$

40. $\left(-\dfrac{4}{5}, \dfrac{9}{10}\right)$ and $\left(-\dfrac{3}{10}, \dfrac{1}{5}\right)$

Find the slope of each line.

41.

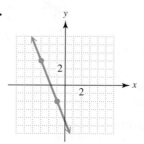

42.

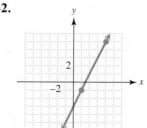

43.

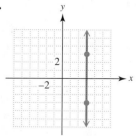

44. *Concept Check* Let k be the number of letters in your last name. Sketch the graph of $y = k$. What is the slope of this line?

Find the slope of the line and sketch the graph. **See Examples 1–4.**

45. $x + 2y = 4$ **46.** $x + 3y = -6$ **47.** $5x - 2y = 10$

48. $4x - y = 4$ **49.** $y = 4x$ **50.** $y = -3x$

51. $x - 3 = 0$ **52.** $x + 2 = 0$ **53.** $y = -5$

54. $y = -4$ **55.** $2y = 3$ **56.** $3x = 4$

Graph each line described. **See Example 5.**

57. Through $(-4, 2)$; $m = \frac{1}{2}$ **58.** Through $(-2, -3)$; $m = \frac{5}{4}$

59. y-intercept $(0, -2)$; $m = -\frac{2}{3}$ **60.** y-intercept $(0, -4)$; $m = -\frac{3}{2}$

61. Through $(-1, -2)$; $m = 3$ **62.** Through $(-2, -4)$; $m = 4$

63. $m = 0$; through $(2, -5)$ **64.** $m = 0$; through $(5, 3)$

65. Undefined slope; through $(-3, 1)$ **66.** Undefined slope; through $(-4, 1)$

67. *Concept Check* If a line has slope $-\frac{4}{9}$, then any line parallel to it has slope _____, and any line perpendicular to it has slope _____.

68. *Concept Check* If a line has slope 0.2, then any line parallel to it has slope _____, and any line perpendicular to it has slope _____.

Decide whether each pair of lines is parallel, perpendicular, *or* neither. **See Examples 6–8.**

69. The line through $(15, 9)$ and $(12, -7)$ and the line through $(8, -4)$ and $(5, -20)$

70. The line through $(4, 6)$ and $(-8, 7)$ and the line through $(-5, 5)$ and $(7, 4)$

71. $x + 4y = 7$ and $4x - y = 3$ **72.** $2x + 5y = -7$ and $5x - 2y = 1$

73. $4x - 3y = 6$ and $3x - 4y = 2$ **74.** $2x + y = 6$ and $x - y = 4$

75. $x = 6$ and $6 - x = 8$ **76.** $3x = y$ and $2y - 6x = 5$

77. $4x + y = 0$ and $5x - 8 = 2y$ **78.** $2x + 5y = -8$ and $6 + 2x = 5y$

79. $2x = y + 3$ and $2y + x = 3$ **80.** $4x - 3y = 8$ and $4y + 3x = 12$

Concept Check *Find and interpret the average rate of change illustrated in each graph.*

81.
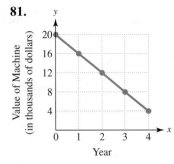

82.
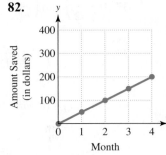

83.
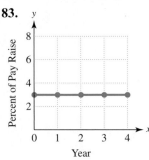

84. *Concept Check* If the graph of a linear equation rises from left to right, then the average rate of change is _____ . If the graph of a linear equation falls from
                                             (positive/negative)

left to right, then the average rate of change is _____ .
                                                  (positive/negative)

Concept Check   *Solve each problem.*

85. When designing the arena now known as TD Banknorth Garden in Boston, architects designed the ramps leading up to the entrances so that circus elephants would be able to walk up the ramps. The maximum grade (or slope) that an elephant will walk on is 13%. Suppose that such a ramp was constructed with a horizontal run of 150 ft. What would be the maximum vertical rise the architects could use?

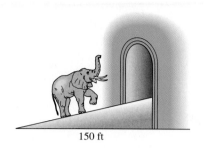

150 ft

86. The upper deck at U.S. Cellular Field in Chicago has produced, among other complaints, displeasure with its steepness. It is 160 ft from home plate to the front of the upper deck and 250 ft from home plate to the back. The top of the upper deck is 63 ft above the bottom. What is its slope? (Consider the slope as a positive number here.)

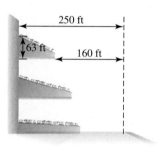

250 ft

63 ft   160 ft

Solve each problem. ***See Examples 9 and 10.***

87. The graph shows the number of cellular phone subscribers (in millions) in the United States from 2005 to 2008.

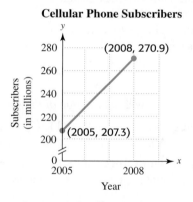

Cellular Phone Subscribers

(2008, 270.9)

(2005, 207.3)

Subscribers (in millions)

Year

Source: CTIA: The Wireless Association.

(a) Use the given ordered pairs to find the slope of the line.

(b) Interpret the slope in the context of this problem.

88. The graph shows spending on personal care products (in billions of dollars) in the United States from 2005 to 2008.

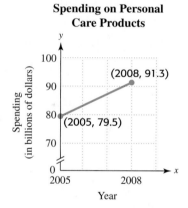

Spending on Personal Care Products

(2008, 91.3)

(2005, 79.5)

Spending (in billions of dollars)

Year

Source: U.S. Department of Commerce.

(a) Use the given ordered pairs to find the slope of the line to the nearest tenth.

(b) Interpret the slope in the context of this problem.

89. The graph provides a good approximation of the number of drive-in theaters in the United States from 2000 through 2007.

(a) Use the given ordered pairs to find the average rate of change in the number of drive-in theaters per year during this period. Round your answer to the nearest whole number.

(b) Explain how a negative slope is interpreted in this situation.

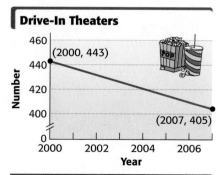

Drive-In Theaters

(2000, 443)

(2007, 405)

Number

Year

Source: www.drive-ins.com

90. The graph provides a good approximation of the number of mobile homes (in thousands) placed in use in the United States from 2000 through 2008.

(a) Use the given ordered pairs to find the average rate of change in the number of mobile homes per year during this period.

🖉 (b) Explain how a negative slope is interpreted in this situation.

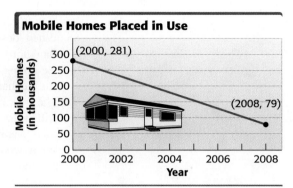

Mobile Homes Placed in Use

Source: U.S. Census Bureau.

🖉 **91.** The total amount spent on plasma TVs in the United States changed from $1590 million in 2003 to $5705 million in 2006. Find and interpret the average rate of change in sales, in millions of dollars per year. Round your answer to the nearest hundredth. (*Source:* Consumer Electronics Association.)

🖉 **92.** The total amount spent on analog TVs in the United States changed from $5836 million in 2003 to

$1424 million in 2006. Find and interpret the average rate of change in sales, in millions of dollars per year. Round your answer to the nearest hundredth. (*Source:* Consumer Electronics Association.)

Brain Busters *Use your knowledge of the slopes of parallel and perpendicular lines.*

93. Show that $(-13, -9)$, $(-11, -1)$, $(2, -2)$, and $(4, 6)$ are the vertices of a parallelogram. (*Hint:* A parallelogram is a four-sided figure with opposite sides parallel.)

94. Is the figure with vertices at $(-11, -5)$, $(-2, -19)$, $(12, -10)$, and $(3, 4)$ a parallelogram? Is it a rectangle? (*Hint:* A rectangle is a parallelogram with a right angle.)

RELATING CONCEPTS EXERCISES 95–100

FOR INDIVIDUAL OR GROUP WORK

*Three points that lie on the same straight line are said to be **collinear**. Consider the points $A(3, 1)$, $B(6, 2)$, and $C(9, 3)$.* ***Work Exercises 95–100 in order.***

95. Find the slope of segment AB.

96. Find the slope of segment BC.

97. Find the slope of segment AC.

98. If slope of segment AB = slope of segment BC = slope of segment AC, then A, B, and C are collinear. Use the results of **Exercises 95–97** to show that this statement is satisfied.

99. Use the slope formula to determine whether the points $(1, -2)$, $(3, -1)$, and $(5, 0)$ are collinear.

100. Repeat **Exercise 99** for the points $(0, 6)$, $(4, -5)$, and $(-2, 12)$.

Solve each equation for y. ***See Section 2.2.***

101. $3x + 2y = 8$ **102.** $4x + 3y = 0$ **103.** $y - 2 = 4(x + 3)$

Write each equation in the form Ax + By = C. ***See Section 2.1.***

104. $y - (-2) = \dfrac{3}{2}(x - 5)$ **105.** $y - (-1) = -\dfrac{1}{2}[x - (-2)]$

3.3   Linear Equations in Two Variables

OBJECTIVES

1 Write an equation of a line, given its slope and *y*-intercept.

2 Graph a line, using its slope and *y*-intercept.

3 Write an equation of a line, given its slope and a point on the line.

4 Write equations of horizontal and vertical lines.

5 Write an equation of a line, given two points on the line.

6 Write an equation of a line parallel or perpendicular to a given line.

7 Write an equation of a line that models real data.

OBJECTIVE 1   **Write an equation of a line, given its slope and *y*-intercept.** In **Section 3.2,** we found the slope of a line from its equation by solving the equation for *y*. For example, we found that the slope of the line with equation

$$y = 4x + 8$$

is 4, the coefficient of *x*. What does the number 8 represent?

To find out, suppose a line has slope *m* and *y*-intercept $(0, b)$. We can find an equation of this line by choosing another point (x, y) on the line, as shown in **FIGURE 25**, and using the slope formula.

$$m = \frac{y - b}{x - 0}$$

$$m = \frac{y - b}{x}$$

$$mx = y - b \qquad \text{Multiply by } x.$$

$$mx + b = y \qquad \text{Add } b.$$

$$y = mx + b \qquad \text{Rewrite.}$$

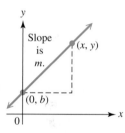

FIGURE 25

This last equation is called the *slope-intercept form* of the equation of a line, because we can identify the slope *m* and *y*-intercept $(0, b)$ at a glance. Thus, in the line with equation $y = 4x + 8$, the number 8 indicates that the *y*-intercept is $(0, 8)$.

Slope-Intercept Form

The **slope-intercept form** of the equation of a line with slope *m* and *y*-intercept $(0, b)$ is

$$\boldsymbol{y = mx + b.}$$

      ↑   ↑

    Slope   *y*-intercept is $(0, b)$.

NOW TRY
EXERCISE 1

Write an equation of the line with slope $\frac{2}{3}$ and *y*-intercept $(0, 1)$.

NOW TRY ANSWER
1. $y = \frac{2}{3}x + 1$

EXAMPLE 1   Writing an Equation of a Line

Write an equation of the line with slope $-\frac{4}{5}$ and *y*-intercept $(0, -2)$.

Here, $m = -\frac{4}{5}$ and $b = -2$. Substitute these values into the slope-intercept form.

$$y = mx + b \qquad \text{Slope-intercept form}$$

$$y = -\frac{4}{5}x - 2 \qquad m = -\frac{4}{5}; b = -2 \qquad \text{NOW TRY}$$

NOTE Every linear equation (of a nonvertical line) has a *unique* (one and only one) slope-intercept form. In **Section 3.6,** we study *linear functions,* which are defined using slope-intercept form. Also, this is the form we use when graphing a line with a graphing calculator.

OBJECTIVE 2 Graph a line, using its slope and *y*-intercept. We first saw this approach in **Example 5(a)** of **Section 3.2.**

NOW TRY
EXERCISE 2
Graph the line, using the slope and *y*-intercept.

$$4x + 3y = 6$$

EXAMPLE 2 Graphing Lines Using Slope and *y*-Intercept

Graph each line, using the slope and *y*-intercept.

(a) $y = 3x - 6$

Here, $m = 3$ and $b = -6$. Plot the *y*-intercept $(0, -6)$. The slope 3 can be interpreted as

$$m = \frac{\text{rise}}{\text{run}} = \frac{\text{change in } y}{\text{change in } x} = \frac{3}{1}.$$

From $(0, -6)$, move 3 units *up* and 1 unit to the *right,* and plot a second point at $(1, -3)$. Join the two points with a straight line. See **FIGURE 26.**

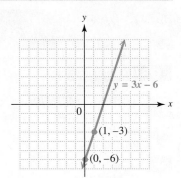

FIGURE 26

(b) $3y + 2x = 9$

Write the equation in slope-intercept form by solving for *y*.

$$3y + 2x = 9$$

$$3y = -2x + 9 \qquad \text{Subtract } 2x.$$

$$y = -\frac{2}{3}x + 3 \qquad \text{Divide by 3.}$$

Slope ↗   ↖ *y*-intercept is $(0, 3)$.

Plot the *y*-intercept $(0, 3)$. The slope can be interpreted as either $\frac{-2}{3}$ or $\frac{2}{-3}$. Using $\frac{-2}{3}$, begin at $(0, 3)$ and move 2 units *down* and 3 units to the *right* to locate the point $(3, 1)$. The line through these two points is the required graph. See **FIGURE 27.** (Verify that the point obtained with $\frac{2}{-3}$ as the slope is also on this line.)

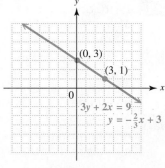

FIGURE 27

NOW TRY ↩

OBJECTIVE 3 Write an equation of a line, given its slope and a point on the line. Let *m* represent the slope of a line and (x_1, y_1) represent a given point on the line. Let (x, y) represent any other point on the line. See **FIGURE 28.**

$$m = \frac{y - y_1}{x - x_1} \qquad \text{Slope formula}$$

$$m(x - x_1) = y - y_1 \qquad \text{Multiply each side by } x - x_1.$$

$$y - y_1 = m(x - x_1) \qquad \text{Rewrite.}$$

This last equation is the *point-slope form* of the equation of a line.

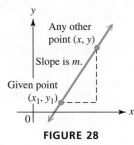

FIGURE 28

NOW TRY ANSWER
2.
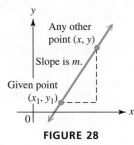

Point-Slope Form

The **point-slope form** of the equation of a line with slope m passing through the point (x_1, y_1) is

$$\overset{\text{Slope}}{\underset{\text{Given point}}{y - y_1 = m(x - x_1).}}$$

NOW TRY
EXERCISE 3

Write an equation of the line with slope $-\frac{1}{5}$ and passing through the point $(5, -3)$.

EXAMPLE 3 Writing an Equation of a Line, Given the Slope and a Point

Write an equation of the line with slope $\frac{1}{3}$ and passing through the point $(-2, 5)$.

Method 1 Use the point-slope form of the equation of a line, with $(x_1, y_1) = (-2, 5)$ and $m = \frac{1}{3}$.

$$y - y_1 = m(x - x_1) \qquad \text{Point-slope form}$$

$$y - 5 = \frac{1}{3}[x - (-2)] \qquad \text{Substitute for } y_1, m, \text{ and } x_1.$$

$$y - 5 = \frac{1}{3}(x + 2) \qquad \text{Definition of subtraction}$$

$$3y - 15 = x + 2 \qquad \text{(*) Multiply by 3.}$$

$$3y = x + 17 \qquad \text{Add 15.}$$

Slope-intercept form $\longrightarrow$ $y = \dfrac{1}{3}x + \dfrac{17}{3}$ Divide by 3.

Method 2 An alternative method for finding this equation uses slope-intercept form, with $(x, y) = (-2, 5)$ and $m = \frac{1}{3}$.

$$y = mx + b \qquad \text{Slope-intercept form}$$

$$5 = \frac{1}{3}(-2) + b \qquad \text{Substitute for } y, m, \text{ and } x.$$

Solve for *b.* $5 = -\dfrac{2}{3} + b$ Multiply.

$$\frac{17}{3} = b, \quad \text{or} \quad b = \frac{17}{3} \qquad 5 = \tfrac{15}{3}; \text{Add } \tfrac{2}{3}. \quad \boxed{\text{Same equation found in Method 1}}$$

Since $m = \frac{1}{3}$ and $b = \frac{17}{3}$, the equation is $y = \frac{1}{3}x + \frac{17}{3}$.   NOW TRY

OBJECTIVE 4 **Write equations of horizontal and vertical lines.** A horizontal line has slope 0. Using point-slope form, we can find the equation of a horizontal line through the point (a, b).

$$y - y_1 = m(x - x_1) \qquad \text{Point-slope form}$$

$$y - b = 0(x - a) \qquad y_1 = b, m = 0, x_1 = a$$

$$y - b = 0 \qquad \text{Multiplication property of 0}$$

$$y = b \qquad \text{Add } b.$$

NOW TRY ANSWER
3. $y = -\frac{1}{5}x - 2$

Point-slope form does not apply to a vertical line, since the slope of a vertical line is undefined. A vertical line through the point (a, b) has equation $x = a$.

Equations of Horizontal and Vertical Lines

The horizontal line through the point (a, b) has equation $y = b$.

The vertical line through the point (a, b) has equation $x = a$.

NOW TRY
EXERCISE 4

Write an equation of the line passing through the point $(4, -4)$ that satisfies the given condition.

(a) Undefined slope

(b) Slope 0

EXAMPLE 4 Writing Equations of Horizontal and Vertical Lines

Write an equation of the line passing through the point $(-3, 3)$ that satisfies the given condition.

(a) Slope 0

Since the slope is 0, this is a horizontal line. A horizontal line through the point (a, b) has equation $y = b$. Here the y-coordinate is 3, so the equation is $y = 3$.

(b) Undefined slope

This is a vertical line, since the slope is undefined. A vertical line through the point (a, b) has equation $x = a$. Here the x-coordinate is -3, so the equation is $x = -3$.

Both lines are graphed in **FIGURE 29**.

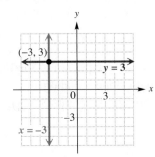

FIGURE 29

OBJECTIVE 5 Write an equation of a line, given two points on the line.

In **Section 3.1,** we defined *standard form* for a linear equation as

$$Ax + By = C, \quad \text{Standard form}$$

where A, B, and C are real numbers and A and B are not both 0. (In most cases, A, B, and C are rational numbers.) For consistency in this book, we give answers so that A, B, and C are integers with greatest common factor 1, and $A \geq 0$. For example, the equation in **Example 3** is written in standard form as follows.

$$3y - 15 = x + 2 \qquad \text{Equation (*) from \textbf{Example 3}}$$

$$-x + 3y = 17 \qquad \text{Subtract } x \text{ and add 15.}$$

$$\text{Standard form} \longrightarrow x - 3y = -17 \qquad \text{Multiply by } -1.$$

NOTE The definition of "standard form" is not standard among texts. A linear equation can be written in many different, equally correct ways. For example, the equation $2x + 3y = 8$ can be written as

$$2x = 8 - 3y, \quad 3y = 8 - 2x, \quad x + \frac{3}{2}y = 4, \quad \text{and} \quad 4x + 6y = 16.$$

We prefer the standard form $2x + 3y = 8$ over any multiples of each side, such as $4x + 6y = 16$. (To write $4x + 6y = 16$ in this preferred form, divide each side by 2.)

EXAMPLE 5 Writing an Equation of a Line, Given Two Points

Write an equation of the line passing through the points $(-4, 3)$ and $(5, -7)$. Give the final answer in standard form.

First find the slope by the slope formula.

$$m = \frac{-7 - 3}{5 - (-4)} = -\frac{10}{9}$$

NOW TRY ANSWERS
4. (a) $x = 4$ **(b)** $y = -4$

NOW TRY
EXERCISE 5

Write an equation of the line passing through the points $(3, -4)$ and $(-2, -1)$. Give the final answer in standard form.

Use either $(-4, 3)$ or $(5, -7)$ as (x_1, y_1) in the point-slope form of the line. We choose $(-4, 3)$, so $-4 = x_1$ and $3 = y_1$.

$$y - y_1 = m(x - x_1) \qquad \text{Point-slope form}$$

$$y - 3 = -\frac{10}{9}[x - (-4)] \qquad y_1 = 3,\ m = -\tfrac{10}{9},\ x_1 = -4$$

$$y - 3 = -\frac{10}{9}(x + 4) \qquad \text{Definition of subtraction}$$

$$9y - 27 = -10(x + 4) \qquad \text{Multiply by 9 to clear the fraction.}$$

$$9y - 27 = -10x - 40 \qquad \text{Distributive property}$$

$$\text{Standard form} \longrightarrow 10x + 9y = -13 \qquad \text{Add } 10x. \text{ Add } 27.$$

Verify that if $(5, -7)$ were used, the same equation would result. **NOW TRY**

OBJECTIVE 6 **Write an equation of a line parallel or perpendicular to a given line.** As mentioned in **Section 3.2,** parallel lines have the same slope and perpendicular lines have slopes that are negative reciprocals of each other.

EXAMPLE 6 Writing Equations of Parallel or Perpendicular Lines

Write an equation of the line passing through the point $(-3, 6)$ and **(a)** parallel to the line $2x + 3y = 6$; **(b)** perpendicular to the line $2x + 3y = 6$. Give final answers in slope-intercept form.

(a) We find the slope of the line $2x + 3y = 6$ by solving for y.

$$2x + 3y = 6$$

$$3y = -2x + 6 \qquad \text{Subtract } 2x.$$

$$y = -\frac{2}{3}x + 2 \qquad \text{Divide by 3.}$$

$$\uparrow \text{Slope}$$

The slope of the line is given by the coefficient of x, so $m = -\frac{2}{3}$. See **FIGURE 30.**

The required equation of the line through $(-3, 6)$ and parallel to $2x + 3y = 6$ must also have slope $-\frac{2}{3}$. To find this equation, we use the point-slope form, with $(x_1, y_1) = (-3, 6)$ and $m = -\frac{2}{3}$.

$$y - 6 = -\frac{2}{3}[x - (-3)] \qquad y_1 = 6,\ m = -\tfrac{2}{3},\ x_1 = -3$$

$$y - 6 = -\frac{2}{3}(x + 3) \qquad \text{Definition of subtraction}$$

$$y - 6 = -\frac{2}{3}x - 2 \qquad \text{Distributive property}$$

$$y = -\frac{2}{3}x + 4 \qquad \text{Add 6.}$$

We did not clear the fraction here because we want the equation in slope-intercept form—that is, solved for y. Both lines are shown in **FIGURE 31.**

FIGURE 30

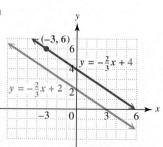

FIGURE 31

NOW TRY ANSWER
5. $3x + 5y = -11$

Give final answers in slope-intercept form.

(b) To be perpendicular to the line $2x + 3y = 6$, a line must have a slope that is the negative reciprocal of $-\frac{2}{3}$, which is $\frac{3}{2}$. We use $(-3, 6)$ and slope $\frac{3}{2}$ in the point-slope form to find the equation of the perpendicular line shown in **FIGURE 32**.

$$y - 6 = \frac{3}{2}[x - (-3)] \qquad y_1 = 6,\ m = \frac{3}{2},\ x_1 = -3$$

$$y - 6 = \frac{3}{2}(x + 3) \qquad \text{Definition of subtraction}$$

$$y - 6 = \frac{3}{2}x + \frac{9}{2} \qquad \text{Distributive property}$$

$$y = \frac{3}{2}x + \frac{21}{2} \qquad \text{Add } 6 = \frac{12}{2}.$$

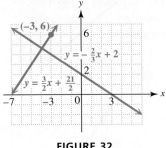

FIGURE 32

NOW TRY

A summary of the various forms of linear equations follows.

Forms of Linear Equations

Equation	Description	When to Use
$y = mx + b$	**Slope-Intercept Form** Slope is m. y-intercept is $(0, b)$.	The slope and y-intercept can be easily identified and used to quickly graph the equation.
$y - y_1 = m(x - x_1)$	**Point-Slope Form** Slope is m. Line passes through (x_1, y_1).	This form is ideal for finding the equation of a line if the slope and a point on the line or two points on the line are known.
$Ax + By = C$	**Standard Form** (A, B, and C integers, $A \geq 0$) Slope is $-\frac{A}{B}$ ($B \neq 0$). x-intercept is $\left(\frac{C}{A}, 0\right)$ ($A \neq 0$). y-intercept is $\left(0, \frac{C}{B}\right)$ ($B \neq 0$).	The x- and y-intercepts can be found quickly and used to graph the equation. The slope must be calculated.
$y = b$	**Horizontal Line** Slope is 0. y-intercept is $(0, b)$.	If the graph intersects only the y-axis, then y is the only variable in the equation.
$x = a$	**Vertical Line** Slope is undefined. x-intercept is $(a, 0)$.	If the graph intersects only the x-axis, then x is the only variable in the equation.

OBJECTIVE 7 **Write an equation of a line that models real data.** If a given set of data changes at a fairly constant rate, the data may fit a linear pattern, where the rate of change is the slope of the line.

EXAMPLE 7 Determining a Linear Equation to Describe Real Data

A local gasoline station is selling 89-octane gas for $3.20 per gal.

(a) Write an equation that describes the cost y to buy x gallons of gas.

The total cost is determined by the number of gallons we buy multiplied by the price per gallon (in this case, $3.20). As the gas is pumped, two sets of numbers spin by: the number of gallons pumped and the cost of that number of gallons.

NOW TRY ANSWERS
6. (a) $y = \frac{3}{5}x - \frac{23}{5}$
   (b) $y = -\frac{5}{3}x + 9$

**NOW TRY
EXERCISE 7**

A cell phone plan costs $100 for the telephone plus $85 per month for service. Write an equation that gives the cost y in dollars for x months of cell phone service using this plan.

The table illustrates this situation.

Number of Gallons Pumped	Cost of This Number of Gallons
0	0($3.20) = $ 0.00
1	1($3.20) = $ 3.20
2	2($3.20) = $ 6.40
3	3($3.20) = $ 9.60
4	4($3.20) = $12.80

If we let x denote the number of gallons pumped, then the total cost y in dollars can be found using the following linear equation.

Total cost ⌐   ⌐ Number of gallons
$$y = 3.20x$$

Theoretically, there are infinitely many ordered pairs (x, y) that satisfy this equation, but here we are limited to nonnegative values for x, since we cannot have a negative number of gallons. In this situation, there is also a practical maximum value for x that varies from one car to another. What determines this maximum value?

(b) A car wash at this gas station costs an additional $3.00. Write an equation that defines the cost of gas and a car wash.

The cost will be $3.20x + 3.00$ dollars for x gallons of gas and a car wash.

$$y = 3.2x + 3 \qquad \text{Delete unnecessary zeros.}$$

(c) Interpret the ordered pairs $(5, 19)$ and $(10, 35)$ in relation to the equation from part (b).

The ordered pair $(5, 19)$ indicates that 5 gal of gas and a car wash costs $19.00. Similarly, $(10, 35)$ indicates that 10 gal of gas and a car wash costs $35.00.

NOW TRY ⟳

NOTE In **Example 7(a),** the ordered pair $(0, 0)$ satisfied the equation, so the linear equation has the form $y = mx$, where $b = 0$. If a realistic situation involves an initial charge plus a charge per unit, as in **Example 7(b),** the equation has the form $y = mx + b$, where $b \neq 0$.

EXAMPLE 8   **Finding an Equation of a Line That Models Data**

Average annual tuition and fees for in-state students at public two-year colleges are shown in the table for selected years and graphed as ordered pairs of points in the **scatter diagram** in **FIGURE 33**, where $x = 0$ represents 2004, $x = 1$ represents 2005, and so on, and y represents the cost in dollars.

Year	Cost (in dollars)
2004	2079
2005	2182
2006	2266
2007	2292
2008	2372
2009	2544

Source: The College Board.

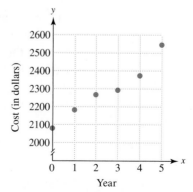

FIGURE 33

NOW TRY ANSWER
7. $y = 85x + 100$

NOW TRY
EXERCISE 8
Refer to **Example 8.**

(a) Using the data values for the years 2004 and 2008, write an equation that models the data.

(b) Use the equation from part (a) to approximate the cost of tuition and fees in 2009.

(a) Write an equation that models the data.

Since the points in **FIGURE 33** lie approximately on a straight line, we can write a linear equation that models the relationship between year x and cost y. We choose two data points, $(0, 2079)$ and $(5, 2544)$, to find the slope of the line.

$$m = \frac{2544 - 2079}{5 - 0} = \frac{465}{5} = 93$$

The slope 93 indicates that the cost of tuition and fees increased by about \$93 per year from 2004 to 2009. We use this slope and the y-intercept $(0, 2079)$ to write an equation of the line in slope-intercept form.

$$y = 93x + 2079$$

(b) Use the equation from part (a) to approximate the cost of tuition and fees in 2010. The value $x = 6$ corresponds to the year 2010.

$y = 93x + 2079$	Equation from part (a)
$y = 93(6) + 2079$	Substitute 6 for x.
$y = 2637$	Multiply, and then add.

According to the model, average tuition and fees for in-state students at public two-year colleges in 2010 were about \$2637. NOW TRY

EXAMPLE 9 Writing an Equation of a Line That Models Data

Retail spending (in billions of dollars) on prescription drugs in the United States is shown in the graph in **FIGURE 34**.

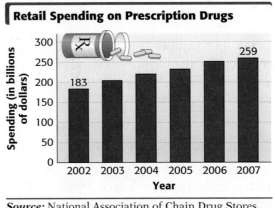

Retail Spending on Prescription Drugs

Source: National Association of Chain Drug Stores.

FIGURE 34

(a) Write an equation that models the data.

The data increase linearly—that is, a straight line through the tops of any two bars in the graph would be close to the top of each bar. To model the relationship between year x and spending on prescription drugs y, we let $x = 2$ represent 2002, $x = 3$ represent 2003, and so on. The given data for 2002 and 2007 can be written as the ordered pairs $(2, 183)$ and $(7, 259)$.

$$m = \frac{259 - 183}{7 - 2} = \frac{76}{5} = 15.2 \qquad \begin{array}{l}\text{Find the slope of the line} \\ \text{through } (2, 183) \text{ and } (7, 259).\end{array}$$

NOW TRY ANSWERS
8. (a) $y = 73.25x + 2079$
   (b) about \$2445

Thus, spending increased by about \$15.2 billion per year. To write an equation, we substitute this slope and one of the points, say, $(2, 183)$, into the point-slope form.

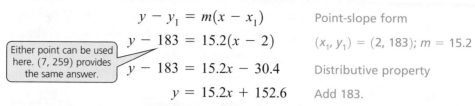

$$y - y_1 = m(x - x_1) \qquad \text{Point-slope form}$$

$$y - 183 = 15.2(x - 2) \qquad (x_1, y_1) = (2, 183); \; m = 15.2$$

Either point can be used here. (7, 259) provides the same answer.

$$y - 183 = 15.2x - 30.4 \qquad \text{Distributive property}$$

$$y = 15.2x + 152.6 \qquad \text{Add 183.}$$

NOW TRY
EXERCISE 9

Refer to **Example 9.**

(a) Use the ordered pairs (2, 183) and (6, 251) to write an equation that models the data.

(b) Use the equation from part (a) to estimate retail spending on prescription drugs in 2011.

Retail spending y (in billions of dollars) on prescription drugs in the United States in year x can be approximated by the equation $y = 15.2x + 152.6$.

(b) Use the equation from part (a) to estimate retail spending on prescription drugs in the United States in 2010. (Assume a constant rate of change.)

Since $x = 2$ represents 2002 and 2010 is 8 yr after 2002, $x = 10$ represents 2010.

$$y = 15.2x + 152.6 \qquad \text{Equation from part (a)}$$

$$y = 15.2(10) + 152.6 \qquad \text{Substitute 10 for } x.$$

$$y = 304.6 \qquad \text{Multiply, and then add.}$$

About \$305 billion was spent on prescription drugs in 2010.   NOW TRY

 CONNECTIONS

 We can use a graphing calculator to solve a linear equation in one variable, such as

$$-2x - 4(2 - x) = 3x + 4.$$

We must write the equation as an equivalent equation with 0 on one side.

$$-2x - 4(2 - x) - 3x - 4 = 0 \qquad \text{Subtract 3x and 4.}$$

Then we graph

$$Y = -2X - 4(2 - X) - 3X - 4$$

to find the x-intercept. (The word "Zero" indicates that the x-intercept has been located.) The standard viewing window *cannot* be used here because the x-intercept does not lie in the interval $[-10, 10]$. See **FIGURE 35**. The x-intercept of the graph is the point $(-12, 0)$. Thus, the solution of the equation is -12, and the solution set is $\{-12\}$.

$Y = -2X - 4(2 - X) - 3X - 4$

FIGURE 35

For Discussion or Writing

Use a graphing calculator to solve each equation.

1. $7x - 2x + 4 - 5 = 3x - 1$
2. $3(x - 7) + 4x = -3x - 11$
3. $3(2x + 1) - 2(x - 2) = 5$
4. $4(x - 3) - x = x - 6$

NOW TRY ANSWERS
9. **(a)** $y = 17x + 149$
   (b) \$336 billion

 3.3 EXERCISES   **MyMathLab**   Math XL PRACTICE   WATCH   DOWNLOAD READ   REVIEW

◉ *Complete solution available on the Video Resources on DVD*

Concept Check   *In Exercises 1–6, provide the appropriate response.*

1. The following equations all represent the same line. Which one is in standard form as defined in the text?

   A. $3x - 2y = 5$ **B.** $2y = 3x - 5$ **C.** $\dfrac{3}{5}x - \dfrac{2}{5}y = 1$ **D.** $3x = 2y + 5$

2. Which equation is in point-slope form?

   A. $y = 6x + 2$ **B.** $4x + y = 9$ **C.** $y - 3 = 2(x - 1)$ **D.** $2y = 3x - 7$

3. Which equation in **Exercise 2** is in slope-intercept form?

4. Write the equation $y + 2 = -3(x - 4)$ in slope-intercept form.

5. Write the equation from **Exercise 4** in standard form.

6. Write the equation $10x - 7y = 70$ in slope-intercept form.

Concept Check Match each equation with the graph that it most closely resembles. *(Hint: Determine the signs of m and b to help you make your decision.)*

7. $y = 2x + 3$

8. $y = -2x + 3$

9. $y = -2x - 3$

10. $y = 2x - 3$

11. $y = 2x$

12. $y = -2x$

13. $y = 3$

14. $y = -3$

A.

B.

C.

D.

E.

F.

G.

H.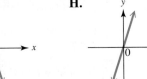

I.

Write the equation in slope-intercept form of the line satisfying the given conditions. **See Example 1.**

15. $m = 5; b = 15$

16. $m = 2; b = 12$

17. $m = -\frac{2}{3}; b = \frac{4}{5}$

18. $m = -\frac{5}{8}; b = -\frac{1}{3}$

19. Slope 1; y-intercept $(0, -1)$

20. Slope -1; y-intercept $(0, -3)$

21. Slope $\frac{2}{5}$; y-intercept $(0, 5)$

22. Slope $-\frac{3}{4}$; y-intercept $(0, 7)$

Concept Check Write an equation in slope-intercept form of the line shown in each graph. *(Hint: Use the indicated points to find the slope.)*

23.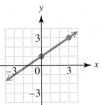

24.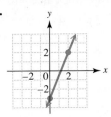

25.

26.
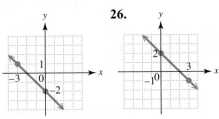

For each equation, **(a)** *write it in slope-intercept form,* **(b)** *give the slope of the line,* **(c)** *give the y-intercept, and* **(d)** *graph the line.* **See Example 2.**

27. $-x + y = 4$

28. $-x + y = 6$

29. $6x + 5y = 30$

30. $3x + 4y = 12$

31. $4x - 5y = 20$

32. $7x - 3y = 3$

33. $x + 2y = -4$

34. $x + 3y = -9$

Find an equation of the line that satisfies the given conditions. **(a)** *Write the equation in standard form.* **(b)** *Write the equation in slope-intercept form.* **See Example 3.**

35. Through $(5, 8)$; slope -2

36. Through $(12, 10)$; slope 1

37. Through $(-2, 4)$; slope $-\frac{3}{4}$

38. Through $(-1, 6)$; slope $-\frac{5}{6}$

39. Through $(-5, 4)$; slope $\frac{1}{2}$

40. Through $(7, -2)$; slope $\frac{1}{4}$

41. x-intercept $(3, 0)$; slope 4

42. x-intercept $(-2, 0)$; slope -5

43. Through $(2, 6.8)$; slope 1.4

44. Through $(6, -1.2)$; slope 0.8

*Find an equation of the line that satisfies the given conditions. **See Example 4.***

45. Through $(9, 5)$; slope 0

46. Through $(-4, -2)$; slope 0

47. Through $(9, 10)$; undefined slope

48. Through $(-2, 8)$; undefined slope

49. Through $\left(-\frac{3}{4}, -\frac{3}{2}\right)$; slope 0

50. Through $\left(-\frac{5}{8}, -\frac{9}{2}\right)$; slope 0

51. Through $(-7, 8)$; horizontal

52. Through $(2, -7)$; horizontal

53. Through $(0.5, 0.2)$; vertical

54. Through $(0.1, 0.4)$; vertical

*Find an equation of the line passing through the given points. **(a)** Write the equation in standard form. **(b)** Write the equation in slope-intercept form if possible. **See Example 5.***

55. $(3, 4)$ and $(5, 8)$

56. $(5, -2)$ and $(-3, 14)$

57. $(6, 1)$ and $(-2, 5)$

58. $(-2, 5)$ and $(-8, 1)$

59. $(2, 5)$ and $(1, 5)$

60. $(-2, 2)$ and $(4, 2)$

61. $(7, 6)$ and $(7, -8)$

62. $(13, 5)$ and $(13, -1)$

63. $\left(\frac{1}{2}, -3\right)$ and $\left(-\frac{2}{3}, -3\right)$

64. $\left(-\frac{4}{9}, -6\right)$ and $\left(\frac{12}{7}, -6\right)$

65. $\left(-\frac{2}{5}, \frac{2}{5}\right)$ and $\left(\frac{4}{3}, \frac{2}{3}\right)$

66. $\left(\frac{3}{4}, \frac{8}{3}\right)$ and $\left(\frac{2}{5}, \frac{2}{3}\right)$

*Find an equation of the line that satisfies the given conditions. **(a)** Write the equation in slope-intercept form. **(b)** Write the equation in standard form. **See Example 6.***

67. Through $(7, 2)$; parallel to $3x - y = 8$

68. Through $(4, 1)$; parallel to $2x + 5y = 10$

69. Through $(-2, -2)$; parallel to $-x + 2y = 10$

70. Through $(-1, 3)$; parallel to $-x + 3y = 12$

71. Through $(8, 5)$; perpendicular to $2x - y = 7$

72. Through $(2, -7)$; perpendicular to $5x + 2y = 18$

73. Through $(-2, 7)$; perpendicular to $x = 9$

74. Through $(8, 4)$; perpendicular to $x = -3$

*Write an equation in the form $y = mx$ for each situation. Then give the three ordered pairs associated with the equation for x-values 0, 5, and 10. **See Example 7(a).***

75. x represents the number of hours traveling at 45 mph, and y represents the distance traveled (in miles).

76. x represents the number of t-shirts sold at $26 each, and y represents the total cost of the t-shirts (in dollars).

77. x represents the number of gallons of gas sold at $3.10 per gal, and y represents the total cost of the gasoline (in dollars).

78. x represents the number of days a DVD movie is rented at $4.50 per day, and y represents the total charge for the rental (in dollars).

79. x represents the number of credit hours taken at Kirkwood Community College at $111 per credit hour, and y represents the total tuition paid for the credit hours (in dollars). (*Source:* www.kirkwood.edu)

80. x represents the number of tickets to a performance of *Jersey Boys* at the Des Moines Civic Center purchased at $125 per ticket, and y represents the total paid for the tickets (in dollars). (*Source:* Ticketmaster.)

For each situation, **(a)** *write an equation in the form* $y = mx + b$, **(b)** *find and interpret the ordered pair associated with the equation for* $x = 5$, *and* **(c)** *answer the question.* **See Examples 7(b) and 7(c).**

81. A ticket for the 2010 Troubadour Reunion, featuring James Taylor and Carole King, costs $112.50. A parking pass costs $12. (*Source:* Ticketmaster.) Let x represent the number of tickets and y represent the cost. How much does it cost for 2 tickets and a parking pass?

82. Resident tuition at Broward College is $87.95 per credit hour. There is also a $20 health science application fee. (*Source:* www.broward.edu) Let x represent the number of credit hours and y represent the cost. How much does it cost for a student in health science to take 15 credit hours?

83. A membership in the Midwest Athletic Club costs $99, plus $41 per month. (*Source:* Midwest Athletic Club.) Let x represent the number of months and y represent the cost. How much does the first year's membership cost?

84. For a family membership, the athletic club in **Exercise 83** charges a membership fee of $159, plus $60 for each additional family member after the first. Let x represent the number of additional family members and y represent the cost. What is the membership fee for a four-person family?

85. A cell phone plan includes 900 anytime minutes for $60 per month, plus a one-time activation fee of $36. A Nokia 6650 cell phone is included at no additional charge. (*Source:* AT&T.) Let x represent the number of months of service and y represent the cost. If you sign a 1-yr contract, how much will this cell phone plan cost? (Assume that you never use more than the allotted number of minutes.)

86. Another cell phone plan includes 450 anytime minutes for $40 per month, plus $50 for a Nokia 2320 cell phone and $36 for a one-time activation fee. (*Source:* AT&T.) Let x represent the number of months of service and y represent the cost. If you sign a 1-yr contract, how much will this cell phone plan cost? (Assume that you never use more than the allotted number of minutes.)

87. There is a $30 fee to rent a chain saw, plus $6 per day. Let x represent the number of days the saw is rented and y represent the charge to the user in dollars. If the total charge is $138, for how many days is the saw rented?

88. A rental car costs $50 plus $0.20 per mile. Let x represent the number of miles driven and y represent the total charge to the renter. How many miles was the car driven if the renter paid $84.60?

Solve each problem. In part (a), give equations in slope-intercept form. (Round the slope to the nearest tenth.) **See Examples 8 and 9.**

89. Total sales of digital cameras in the United States (in millions of dollars) are shown in the graph, where the year 2003 corresponds to $x = 0$.

(a) Use the ordered pairs from the graph to write an equation that models the data. What does the slope tell us in the context of this problem?

(b) Use the equation from part (a) to approximate the sales of digital cameras in the United States in 2007.

Digital Camera Sales

Source: Consumer Electronics Association.

90. Total sales of fax machines in the United States (in millions of dollars) are shown in the graph, where the year 2003 corresponds to $x = 0$.

(a) Use the ordered pairs from the graph to write an equation that models the data. What does the slope tell us in the context of this problem?

(b) Use the equation from part (a) to approximate the sales of fax machines in the United States in 2007.

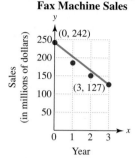

Fax Machine Sales

Source: Consumer Electronics Association.

91. Expenditures for home health care in the United States are shown in the graph.

(a) Use the information given for the years 2003 and 2007, letting $x = 3$ represent 2003, $x = 7$ represent 2007, and y represent the amount (in billions of dollars) to write an equation that models home health care spending.

(b) Use the equation from part (a) to approximate the amount spent on home health care in 2005. How does your result compare with the actual value, $48.1 billion?

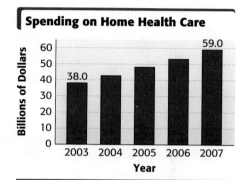

Spending on Home Health Care

Source: U.S. Centers for Medicare & Medicaid Services.

92. The number of post offices in the United States is shown in the graph.

(a) Use the information given for the years 2003 and 2008, letting $x = 3$ represent 2003, $x = 8$ represent 2008, and y represent the number of post offices, to write an equation that models the data.

(b) Use the equation to approximate the number of post offices in 2006. How does this result compare with the actual value, 27,318?

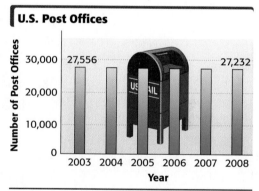

U.S. Post Offices

Source: U.S. Postal Service.

RELATING CONCEPTS   EXERCISES 93–100

FOR INDIVIDUAL OR GROUP WORK

*In **Section 2.2**, we worked with formulas. **Work Exercises 93–100 in order,** to see how the formula that relates Celsius and Fahrenheit temperatures is derived.*

93. There is a linear relationship between Celsius and Fahrenheit temperatures. When $C = 0°$, $F =$ _____°, and when $C = 100°$, $F =$ _____°.

94. Think of ordered pairs of temperatures (C, F), where C and F represent corresponding Celsius and Fahrenheit temperatures. The equation that relates the two scales has a straight-line graph that contains the two points determined in **Exercise 93**. What are these two points?

95. Find the slope of the line described in **Exercise 94**.

(continued)

96. Use the slope found in **Exercise 95** and one of the two points determined earlier, and write an equation that gives F in terms of C. (*Hint:* Use the point-slope form, with C replacing x and F replacing y.)

97. To obtain another form of the formula, use the equation found in **Exercise 96** and solve for C in terms of F.

98. Use the equation from **Exercise 96** to find the Fahrenheit temperature when $C = 30$.

99. Use the equation from **Exercise 97** to find the Celsius temperature when $F = 50$.

100. For what temperature is $F = C$? (Use the photo on the previous page to confirm your answer.)

PREVIEW EXERCISES

Solve each inequality. ***See Section 2.5.***

101. $2x + 5 < 9$ **102.** $-x + 4 > 3$ **103.** $5 - 3x \geq 9$ **104.** $-x \leq 0$

SUMMARY EXERCISES on Slopes and Equations of Lines

Find the slope of each line, if possible.

1. $3x + 5y = 9$ **2.** $4x + 7y = 3$ **3.** $y = 2x - 5$

4. $5x - 2y = 4$ **5.** $x - 4 = 0$ **6.** $y = 0.5$

*For each line described, write an equation of the line **(a)** in slope-intercept form and **(b)** in standard form.*

7. Through the points $(-2, 6)$ and $(4, 1)$

8. Through $(-2, 5)$ and parallel to the graph of $3x - y = 4$

9. Through the origin and perpendicular to the graph of $2x - 5y = 6$

10. Through $(5, -8)$ and parallel to the graph of $y = 4$

11. Through $\left(\frac{3}{4}, -\frac{7}{9}\right)$ and perpendicular to the graph of $x = \frac{2}{3}$

12. Through $(4, -2)$ with slope -3

13. Through $(-4, 2)$ and parallel to the line through $(3, 9)$ and $(6, 11)$

14. Through $(4, -2)$ and perpendicular to the line through $(3, 7)$ and $(5, 6)$

15. Through the points $(4, -8)$ and $(-4, 12)$

16. Through $(-3, 6)$ with slope $\frac{2}{3}$

17. Through $(0, 3)$ and the midpoint of the segment with endpoints $(2, 8)$ and $(-4, 12)$

18. *Concept Check* Match the description in Column I with its equation in Column II.

I	II
(a) Slope -0.5, $b = -2$	**A.** $y = -\frac{1}{2}x$
(b) x-intercept $(4, 0)$, y-intercept $(0, 2)$	**B.** $y = -\frac{1}{2}x - 2$
(c) Passes through $(4, -2)$ and $(0, 0)$	**C.** $x - 2y = 2$
(d) $m = \frac{1}{2}$, passes through $(-2, -2)$	**D.** $x + 2y = 4$
(e) $m = \frac{1}{2}$, passes through the origin	**E.** $x = 2y$

3.4 Linear Inequalities in Two Variables

OBJECTIVES

1. Graph linear inequalities in two variables.
2. Graph the intersection of two linear inequalities.
3. Graph the union of two linear inequalities.

OBJECTIVE 1 **Graph linear inequalities in two variables.** In **Chapter 2,** we graphed linear inequalities in one variable on the number line. In this section, we graph linear inequalities in two variables on a rectangular coordinate system.

Linear Inequality in Two Variables

An inequality that can be written as

$$Ax + By < C, \quad Ax + By \le C, \quad Ax + By > C, \quad \text{or} \quad Ax + By \ge C,$$

where A, B, and C are real numbers and A and B are not both 0, is a **linear inequality in two variables.**

Consider the graph in **FIGURE 36**. The graph of the line $x + y = 5$ divides the points in the rectangular coordinate system into three sets:

1. Those points that lie on the line itself and satisfy the equation $x + y = 5$ [like $(0, 5)$, $(2, 3)$, and $(5, 0)$];
2. Those that lie in the half-plane above the line and satisfy the inequality $x + y > 5$ [like $(5, 3)$ and $(2, 4)$];
3. Those that lie in the half-plane below the line and satisfy the inequality $x + y < 5$ [like $(0, 0)$ and $(-3, -1)$].

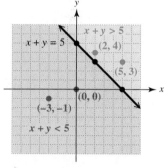

FIGURE 36

The graph of the line $x + y = 5$ is called the **boundary line** for the inequalities $x + y > 5$ and $x + y < 5$. Graphs of linear inequalities in two variables are *regions* in the real number plane that may or may not include boundary lines.

To graph a linear inequality in two variables, follow these steps.

Graphing a Linear Inequality

Step 1 **Draw the graph of the straight line that is the boundary.** Make the line solid if the inequality involves $\le$ or $\ge$. Make the line dashed if the inequality involves $<$ or $>$.

Step 2 **Choose a test point.** Choose any point not on the line, and substitute the coordinates of that point in the inequality.

Step 3 **Shade the appropriate region.** Shade the region that includes the test point if it satisfies the original inequality. Otherwise, shade the region on the other side of the boundary line.

⚠ **CAUTION** When drawing the boundary line in Step 1, be careful to draw a solid line if the inequality includes equality ($\le$, $\ge$) or a dashed line if equality is not included ($<$, $>$).

NOW TRY
EXERCISE 1
Graph $-x + 2y \geq 4$.

EXAMPLE 1 Graphing a Linear Inequality

Graph $3x + 2y \geq 6$.

Step 1 First graph the boundary line $3x + 2y = 6$, as shown in **FIGURE 37**.

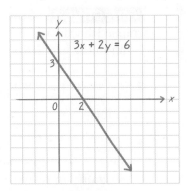

FIGURE 37

Step 2 The graph of the inequality $3x + 2y \geq 6$ includes the points of the line $3x + 2y = 6$ and either the points *above* that line or the points *below* it. To decide which, select any point not on the boundary line to use as a test point. Substitute the values from the test point, here $(0, 0)$, for x and y in the inequality.

$$3x + 2y \geq 6 \qquad \text{Original inequality}$$

(0, 0) is a convenient test point. $\longrightarrow 3(0) + 2(0) \overset{?}{\geq} 6 \qquad \text{Let } x = 0 \text{ and } y = 0.$

$$0 \geq 6 \qquad \text{False}$$

Step 3 Because the result is false, $(0, 0)$ does *not* satisfy the inequality. The solution set includes all points on the other side of the line. See **FIGURE 38**.

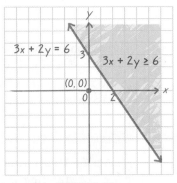

FIGURE 38

NOW TRY

If the inequality is written in the form $y > mx + b$ or $y < mx + b$, then the inequality symbol indicates which half-plane to shade.

If $y > mx + b$, then shade above the boundary line.

If $y < mx + b$, then shade below the boundary line.

This method works only if the inequality is solved for y.

NOW TRY ANSWER
1.

> ⚠ **CAUTION** A common error in using the method just described is to use the original inequality symbol when deciding which half-plane to shade. Be sure to use the inequality symbol found in the inequality *after* it is solved for y.

NOW TRY
EXERCISE 2
Graph $3x - y < 6$.

EXAMPLE 2 Graphing a Linear Inequality

Graph $x - 3y < 4$.

First graph the boundary line, shown in **FIGURE 39**. The points of the boundary line do not belong to the inequality $x - 3y < 4$ (because the inequality symbol is $<$, not $\leq$). For this reason, the line is dashed. Now solve the inequality for y.

$$x - 3y < 4$$

$$-3y < -x + 4 \qquad \text{Subtract } x.$$

$$y > \frac{1}{3}x - \frac{4}{3} \qquad \text{Multiply by } -\frac{1}{3}. \text{ Change } < \text{ to } >.$$

Because of the *is greater than* symbol that occurs **when the inequality is solved for y,** shade *above* the line.

CHECK Choose a test point not on the line, say, $(0, 0)$.

$$x - 3y < 4$$

$$0 - 3(0) \overset{?}{<} 4 \qquad \text{Let } x = 0 \text{ and } y = 0.$$

$$0 < 4 \ \checkmark \quad \text{True}$$

This result agrees with the decision to shade above the line. The solution set, graphed in **FIGURE 39**, includes only those points in the shaded half-plane (not those on the line).

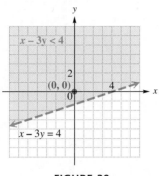

FIGURE 39

NOW TRY

OBJECTIVE 2 **Graph the intersection of two linear inequalities.** A pair of inequalities joined with the word **and** is interpreted as the intersection of the solution sets of the inequalities. *The graph of the intersection of two or more inequalities is the region of the plane where all points satisfy all of the inequalities at the same time.*

NOW TRY
EXERCISE 3
Graph $x + y < 3$ and $y \leq 2$.

EXAMPLE 3 Graphing the Intersection of Two Inequalities

Graph $2x + 4y \geq 5$ and $x \geq 1$.

To begin, we graph each of the two inequalities $2x + 4y \geq 5$ and $x \geq 1$ separately, as shown in **FIGURES 40(a) AND (b)**. Then we use heavy shading to identify the intersection of the graphs, as shown in **FIGURE 40(c)**.

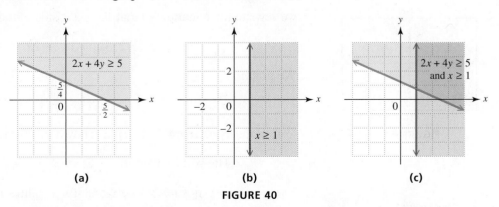

(a)   (b)   (c)

FIGURE 40

NOW TRY ANSWERS

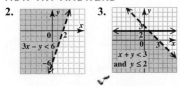

In practice, the graphs in **FIGURES 40(a) AND (b)** are graphed on the same axes.

CHECK Using **FIGURE 40(c)**, choose a test point from each of the four regions formed by the intersection of the boundary lines. Verify that only ordered pairs in the heavily shaded region satisfy *both* inequalities.

NOW TRY

OBJECTIVE 3 Graph the union of two linear inequalities. When two inequalities are joined by the word *or,* we must find the union of the graphs of the inequalities. *The graph of the union of two inequalities includes all of the points that satisfy either inequality.*

NOW TRY
EXERCISE 4

Graph

$3x - 5y < 15$ or $x > 4$.

EXAMPLE 4 Graphing the Union of Two Inequalities

Graph $2x + 4y \geq 5$ or $x \geq 1$.

The graphs of the two inequalities are shown in **FIGURES 40(a) AND (b)** in **Example 3** on the preceding page. The graph of the union is shown in **FIGURE 41**.

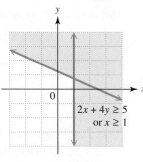

$2x + 4y \geq 5$
or $x \geq 1$

FIGURE 41

NOW TRY

CONNECTIONS

Recall from **Section 3.3** that the *x*-intercept of the graph of the line $y = mx + b$ indicates the solution of the equation $mx + b = 0$. We can extend this observation to find solutions of the associated inequalities $mx + b > 0$ and $mx + b < 0$.

For example, to solve the equation

$$-2(3x + 1) = -2x + 18$$

and the associated inequalities

$$-2(3x + 1) > -2x + 18 \quad \text{and} \quad -2(3x + 1) < -2x + 18,$$

we rewrite the equation so that the right side equals 0.

$$-2(3x + 1) + 2x - 18 = 0$$

We graph

$$Y = -2(3X + 1) + 2X - 18$$

to find the *x*-intercept $(-5, 0)$, as shown in **FIGURE 42**.

The solution set of $-2(3x + 1) = -2x + 18$ is $\{-5\}$.

The graph of Y lies *above* the *x*-axis for *x*-values less than -5.

Thus, the solution set of $-2(3x + 1) > -2x + 18$ is $(-\infty, -5)$.

The graph of Y lies *below* the *x*-axis for *x*-values greater than -5.

Thus, the solution set of $-2(3x + 1) < -2x + 18$ is $(-5, \infty)$.

Y = −2(3X + 1) + 2X − 18

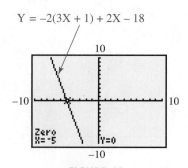

FIGURE 42

NOW TRY ANSWER

4.

For Discussion or Writing

Solve the equation in part (a) and the associated inequalities in parts (b) and (c), by graphing the left side as y in the standard viewing window of a graphing calculator. Explain your answers using the graph.

1. (a) $5x + 3 = 0$ **(b)** $5x + 3 > 0$ **(c)** $5x + 3 < 0$

2. (a) $6x + 3 = 0$ **(b)** $6x + 3 > 0$ **(c)** $6x + 3 < 0$

3. (a) $-8x - (2x + 12) = 0$ **(b)** $-8x - (2x + 12) \geq 0$
   (c) $-8x - (2x + 12) \leq 0$

4. (a) $-4x - (2x + 18) = 0$ **(b)** $-4x - (2x + 18) \geq 0$
   (c) $-4x - (2x + 18) \leq 0$

3.4 EXERCISES

MyMathLab   Math XL PRACTICE   WATCH   DOWNLOAD   READ   REVIEW

Complete solution available on the Video Resources on DVD

Concept Check   In Exercises 1–4, fill in the first blank with either solid *or* dashed. *Fill in the second blank with either* above *or* below.

1. The boundary of the graph of $y \leq x + 2$ will be a _____ line, and the shading will be _____ the line.

2. The boundary of the graph of $y < -x + 2$ will be a _____ line, and the shading will be _____ the line.

3. The boundary of the graph of $y > -x + 2$ will be a _____ line, and the shading will be _____ the line.

4. The boundary of the graph of $y \geq -x + 2$ will be a _____ line, and the shading will be _____ the line.

5. How is the boundary line $Ax + By = C$ used in graphing either $Ax + By < C$ or $Ax + By > C$?

6. Describe the two methods discussed in the text for deciding which region is the solution set of a linear inequality in two variables.

*Graph each linear inequality in two variables. **See Examples 1 and 2.***

7. $x + y \leq 2$ **8.** $x + y \leq -3$ **9.** $4x - y < 4$

10. $3x - y < 3$ **11.** $x + 3y \geq -2$ **12.** $x + 4y \geq -3$

13. $2x + 3y \geq 6$ **14.** $3x + 4y \geq 12$ **15.** $5x - 3y > 15$

16. $4x - 5y > 20$ **17.** $x + y > 0$ **18.** $x + 2y > 0$

19. $x - 3y \leq 0$ **20.** $x - 5y \leq 0$ **21.** $y < x$ **22.** $y \leq 4x$

*Graph each compound inequality. **See Example 3.***

23. $x + y \leq 1$ and $x \geq 1$ **24.** $x - y \geq 2$ and $x \geq 3$

25. $2x - y \geq 2$ and $y < 4$ **26.** $3x - y \geq 3$ and $y < 3$

27. $x + y > -5$ and $y < -2$ **28.** $6x - 4y < 10$ and $y > 2$

Use the method described in Section 2.7 to write each inequality as a compound inequality, and graph its solution set in the rectangular coordinate plane.

29. $|x| < 3$ **30.** $|y| < 5$

31. $|x + 1| < 2$ **32.** $|y - 3| < 2$

Graph each compound inequality. **See Example 4.**

 33. $x - y \geq 1$ or $y \geq 2$ **34.** $x + y \leq 2$ or $y \geq 3$

35. $x - 2 > y$ or $x < 1$ **36.** $x + 3 < y$ or $x > 3$

37. $3x + 2y < 6$ or $x - 2y > 2$ **38.** $x - y \geq 1$ or $x + y \leq 4$

TECHNOLOGY INSIGHTS   EXERCISES 39–46

Match each inequality with its calculator graph. (Hint: Use the slope, y-intercept, and inequality symbol in making your choice.)

39. $y \leq 3x - 6$ **40.** $y \geq 3x - 6$

41. $y \leq -3x - 6$ **42.** $y \geq -3x - 6$

A.
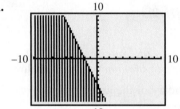

B.
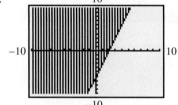

C.
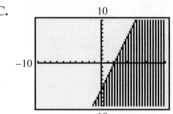

D.
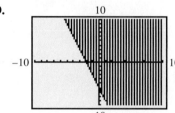

The graph of a linear equation $y = mx + b$ is shown on a graphing calculator screen, along with the x-value of the x-intercept of the line. Use the screen to solve (a) $y = 0$, (b) $y < 0$, and (c) $y > 0$. **See the Connections box.**

 43.
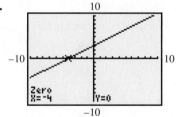

44.
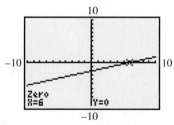

45.
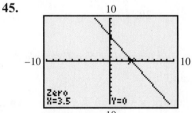

46.
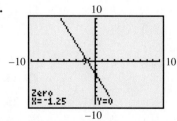

RELATING CONCEPTS EXERCISES 47–52

FOR INDIVIDUAL OR GROUP WORK

Suppose a factory can have no more than 200 *workers on a shift, but must have* at least 100 *and must manufacture* at least 3000 *units at minimum cost. The managers need to know how many workers should be on a shift in order to produce the required units at minimal cost.* **Linear programming** *is a method for finding the optimal (best possible) solution that meets all the conditions for such problems.*

Let x represent the number of workers and y represent the number of units manufactured. **Work Exercises 47–52 in order.**

47. Write three inequalities expressing the conditions given in the problem.

48. Graph the inequalities from **Exercise 47** and shade the intersection.

49. The cost per worker is $50 per day and the cost to manufacture 1 unit is $100. Write an equation in x, y, and C representing the total daily cost C.

50. Find values of x and y for several points in or on the boundary of the shaded region. Include any "corner points." These are the points that maximize or minimize C.

51. Of the values of x and y that you chose in **Exercise 50,** which gives the least value when substituted in the cost equation from **Exercise 49?**

 52. What does your answer in **Exercise 51** mean in terms of the given problem?

PREVIEW EXERCISES

Write each inequality or compound inequality using interval notation. **See Sections 1.1 and 2.6.**

53. $x \geq 0$ 54. $x \leq 0$ 55. $x < 1$ or $x > 1$ 56. $-4 \leq x \leq 4$

3.5 Introduction to Relations and Functions

OBJECTIVES

1. Distinguish between independent and dependent variables.
2. Define and identify relations and functions.
3. Find the domain and range.
4. Identify functions defined by graphs and equations.

OBJECTIVE 1 Distinguish between independent and dependent variables.
We often describe one quantity in terms of another. Consider the following:

- The amount of a paycheck for an hourly employee depends on the number of hours worked.
- The cost at a gas station depends on the number of gallons of gas pumped.
- The distance traveled by a car moving at a constant rate depends on the time traveled.

We can use ordered pairs to represent these corresponding quantities. We indicate the relationship between hours worked and paycheck amount as follows.

$(5, 40)$ Working 5 hr results in a $40 paycheck.

Number of hours worked ⌐↑   ↑⌐ Paycheck amount in dollars

Similarly, the ordered pair $(10, 80)$ indicates that working 10 hr results in an $80 paycheck. In this example, what would the ordered pair $(20, 160)$ indicate?

Since paycheck amount *depends* on number of hours worked, paycheck amount is called the *dependent variable,* and number of hours worked is called the *independent variable.* Generalizing, if the value of the variable y depends on the value of the variable x, then y is the **dependent variable** and x is the **independent variable.**

Independent variable ⌐ ⌐ Dependent variable

$$(x, y)$$

OBJECTIVE 2 **Define and identify relations and functions.** Since we can write related quantities as ordered pairs, a set of ordered pairs such as

$$\{(5, 40), (10, 80), (20, 160), (40, 320)\}$$

is called a *relation.*

Relation

A **relation** is any set of ordered pairs.

A *function* is a special kind of relation.

Function

A **function** is a relation in which, for each value of the first component of the ordered pairs, there is *exactly one value* of the second component.

NOW TRY
EXERCISE 1

Determine whether each relation defines a function.

(a) $\{(1, 5), (3, 5), (5, 5)\}$

(b) $\{(-1, -3), (0, 2), (-1, 6)\}$

EXAMPLE 1 Determining Whether Relations Are Functions

Determine whether each relation defines a function.

(a) $F = \{(1, 2), (-2, 4), (3, -1)\}$

For $x = 1$, there is only one value of y, 2.

For $x = -2$, there is only one value of y, 4.

For $x = 3$, there is only one value of y, -1.

Thus, relation F is a function, because for each different x-value, there is exactly one y-value.

(b) $G = \{(-2, -1), (-1, 0), (0, 1), (1, 2), (2, 2)\}$

Relation G is also a function. Although the last two ordered pairs have the same y-value (1 is paired with 2 and 2 is paired with 2), this does not violate the definition of a function. The first components (x-values) are different, and each is paired with only one second component (y-value).

(c) $H = \{(-4, 1), (-2, 1), (-2, 0)\}$

In relation H, the last two ordered pairs have the *same x*-value paired with *two different y*-values (-2 is paired with both 1 and 0), so H is a relation, but *not* a function.

Different y-values

$$H = \{(-4, 1), (-2, 1), (-2, 0)\} \quad \text{Not a function}$$

Same x-value

In a function, no two ordered pairs can have the same first component and different second components.

NOW TRY

Relations and functions can be defined in several different ways.

- **As a set of ordered pairs** (See Example 1.)
- **As a correspondence or *mapping***

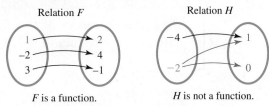

Relation *F*   Relation *H*

F is a function.   *H* is not a function.

FIGURE 43

See **FIGURE 43**. In the mapping for relation *F* from **Example 1(a),** 1 is mapped to 2, −2 is mapped to 4, and 3 is mapped to −1. Thus, *F* is a function, since each first component is paired with exactly one second component. In the mapping for relation *H* from **Example 1(c),** which is not a function, the first component −2 is paired with two different second components.

- **As a table**
- **As a graph**

  FIGURE 44 includes a table and graph for relation *F* from **Example 1(a).**

x	y
1	2
−2	4
3	−1

Table for relation *F*

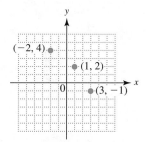

Graph of relation *F*

FIGURE 44

- **As an equation (or rule)**

  An equation (or rule) can tell how to determine the dependent variable for a specific value of the independent variable. For example, if the value of *y* is twice the value of *x*, the equation is

  $$y = 2x.$$

  Dependent   Independent
  variable   variable

  The solutions of this equation define an infinite set of ordered pairs that can be represented by the graph in **FIGURE 45**.

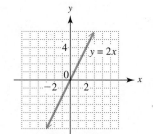

Graph of the relation defined by $y = 2x$

FIGURE 45

NOTE Another way to think of a function relationship is to think of the independent variable as an input and the dependent variable as an output. This is illustrated by the input-output (function) machine for the function defined by

$$y = 2x.$$

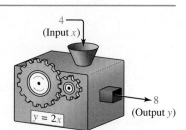

Function machine

In a function, there is exactly one value of the dependent variable, the second component, for each value of the independent variable, the first component.

OBJECTIVE 3 **Find the domain and range.** For every relation, there are two important sets of elements called the *domain* and *range*.

> ### Domain and Range
>
> In a relation, the set of all values of the independent variable (x) is the **domain.** The set of all values of the dependent variable (y) is the **range.**

NOW TRY
EXERCISE 2
Give the domain and range of each relation. Tell whether the relation defines a function.

(a) $\{(2, 2), (2, 5), (4, 8), (6, 5)\}$

(b)

x	y
1	$ 1.39
10	$10.39
15	$20.85

EXAMPLE 2 Finding Domains and Ranges of Relations

Give the domain and range of each relation. Tell whether the relation defines a function.

(a) $\{(3, -1), (4, 2), (4, 5), (6, 8)\}$

The domain, the set of x-values, is $\{3, 4, 6\}$. The range, the set of y-values, is $\{-1, 2, 5, 8\}$. This relation is not a function because the same x-value 4 is paired with two different y-values, 2 and 5.

(b)

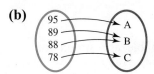

(c)

x	y
-5	2
0	2
5	2

The domain of the relation represented by this mapping is $\{95, 89, 88, 78\}$, and the range is $\{A, B, C\}$. The mapping defines a function—each domain value corresponds to exactly one range value.

In this table, the domain is the set of x-values $\{-5, 0, 5\}$ and the range is the set of y-values $\{2\}$. The table defines a function—each x-value corresponds to exactly one y-value (even though it is the same y-value).

NOW TRY

A graph gives a "picture" of a relation and can be used to determine its domain and range.

EXAMPLE 3 Finding Domains and Ranges from Graphs

Give the domain and range of each relation.

(a)

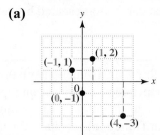

(b)

NOW TRY ANSWERS
2. **(a)** domain: $\{2, 4, 6\}$;
range: $\{2, 5, 8\}$;
not a function
(b) domain: $\{1, 10, 15\}$;
range: $\{\$1.39, \$10.39, \$20.85\}$;
function

This relation includes the four ordered pairs that are graphed. The domain is the set of x-values,

$$\{-1, 0, 1, 4\}.$$

The range is the set of y-values,

$$\{-3, -1, 1, 2\}.$$

The x-values of the points on the graph include all numbers between -4 and 4, inclusive. The y-values include all numbers between -6 and 6, inclusive.

The domain is $[-4, 4]$. Use interval
The range is $[-6, 6]$. notation.

Give the domain and range of the relation.

(c)

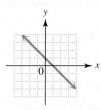

The arrowheads indicate that the line extends indefinitely left and right, as well as up and down. Therefore, both the domain and the range include all real numbers, written $(-\infty, \infty)$.

(d)

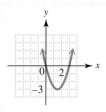

The graph extends indefinitely left and right, as well as upward. The domain is $(-\infty, \infty)$. Because there is a least y-value, -3, the range includes all numbers greater than or equal to -3, written $[-3, \infty)$.

NOW TRY

OBJECTIVE 4   Identify functions defined by graphs and equations. Since each value of x in a function corresponds to only one value of y, any vertical line drawn through the graph of a function must intersect the graph in at most one point.

Vertical Line Test

If every vertical line intersects the graph of a relation in no more than one point, then the relation is a function.

FIGURE 46 illustrates the vertical line test with the graphs of two relations.

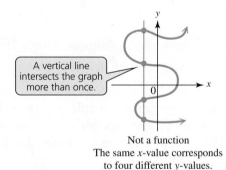

A vertical line intersects the graph more than once.

Any vertical line intersects the graph only once.

Not a function
The same x-value corresponds to four different y-values.

Function
Each x-value corresponds to only one y-value.

FIGURE 46

Use the vertical line test to determine whether the relation is a function.

EXAMPLE 4   Using the Vertical Line Test

Use the vertical line test to determine whether each relation graphed in **Example 3** is a function. (We repeat the graphs here.)

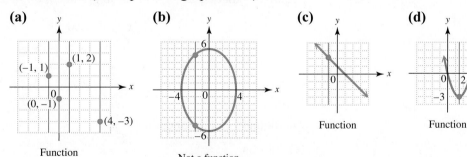

(a)

$(-1, 1)$
$(1, 2)$
$(0, -1)$
$(4, -3)$

Function

(b)

Not a function

(c)

Function

(d)

Function

The graphs in (a), (c), and (d) satisfy the vertical line test and represent functions. The graph in (b) fails the vertical line test, since the same x-value corresponds to two different y-values, and is not the graph of a function.

NOW TRY

NOTE Graphs that do not represent functions are still relations. *All equations and graphs represent relations, and all relations have a domain and range.*

Relations are often defined by equations. If a relation is defined by an equation, keep the following guidelines in mind when finding its domain.

1. **Exclude from the domain any values that make the denominator of a fraction equal to 0.**

   Example: The function defined by $y = \frac{1}{x}$ has all real numbers except 0 as its domain, since division by 0 is undefined.

2. **Exclude from the domain any values that result in an even root of a negative number.**

   Example: The function defined by $y = \sqrt{x}$ has all *nonnegative* real numbers as its domain, since the square root of a negative number is not real.

In this book, we assume the following agreement on the domain of a relation.

Agreement on Domain

Unless specified otherwise, the domain of a relation is assumed to be all real numbers that produce real numbers when substituted for the independent variable.

EXAMPLE 5 Identifying Functions from Their Equations

Decide whether each relation defines y as a function of x, and give the domain.

(a) $y = x + 4$

In the defining equation (or rule) $y = x + 4$, y is always found by adding 4 to x. Thus, each value of x corresponds to just one value of y, and the relation defines a function. Since x can be any real number, the domain is

$$\{x \mid x \text{ is a real number}\}, \quad \text{or} \quad (-\infty, \infty).$$

(b) $y = \sqrt{2x - 1}$

For any choice of x in the domain, there is exactly one corresponding value for y (the radical is a nonnegative number), so this equation defines a function. Since the equation involves a square root, the quantity under the radical symbol cannot be negative—that is, $2x - 1$ *must be greater than or equal to* 0.

$$2x - 1 \geq 0$$

$$2x \geq 1 \qquad \text{Add 1.}$$

$$x \geq \frac{1}{2} \qquad \text{Divide by 2.}$$

The domain of the function is $\left[\frac{1}{2}, \infty\right)$.

(c) $y^2 = x$

The ordered pairs $(16, 4)$ and $(16, -4)$ both satisfy this equation. Since one value of x, 16, corresponds to two values of y, 4 and -4, this equation does not define a function. Because x is equal to the square of y, the values of x must always be nonnegative. The domain of the relation is $[0, \infty)$.

NOW TRY
EXERCISE 5

Decide whether each relation defines y as a function of x, and give the domain.

(a) $y = 4x - 3$

(b) $y = \sqrt{2x - 4}$

(c) $y = \dfrac{1}{x - 2}$

(d) $y < 3x + 1$

(d) $y \leq x - 1$

By definition, y is a function of x if every value of x leads to exactly one value of y. Here, a particular value of x, say, 1, corresponds to many values of y. The ordered pairs

$$(1, 0), \quad (1, -1), \quad (1, -2), \quad (1, -3), \quad \text{and so on}$$

all satisfy the inequality. Thus, this relation does not define a function. Any number can be used for x, so the domain is the set of all real numbers, $(-\infty, \infty)$.

(e) $y = \dfrac{5}{x - 1}$

Given any value of x in the domain, we find y by subtracting 1 and then dividing the result into 5. This process produces exactly one value of y for each value in the domain, so the given equation defines a function.

The domain includes all real numbers except those which make the denominator 0. We find these numbers by setting the denominator equal to 0 and solving for x.

$$x - 1 = 0$$

$$x = 1 \quad \text{Add 1.}$$

The domain includes all real numbers *except* 1, written $(-\infty, 1) \cup (1, \infty)$.

NOW TRY

In summary, we give three variations of the definition of a function.

Variations of the Definition of a Function

1. A **function** is a relation in which, for each value of the first component of the ordered pairs, there is exactly one value of the second component.

2. A **function** is a set of distinct ordered pairs in which no first component is repeated.

3. A **function** is a correspondence or rule that assigns exactly one range value to each domain value.

NOW TRY ANSWERS
5. **(a)** yes; $(-\infty, \infty)$
   (b) yes; $[2, \infty)$
   (c) yes; $(-\infty, 2) \cup (2, \infty)$
   (d) no; $(-\infty, \infty)$

3.5 EXERCISES

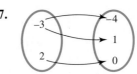

● *Complete solution available on the Video Resources on DVD*

✎ **1.** In your own words, define a function and give an example.

✎ **2.** In your own words, define the domain of a function and give an example.

3. *Concept Check* In an ordered pair of a relation, is the first element the independent or the dependent variable?

4. *Concept Check* Give an example of a relation that is not a function and that has domain $\{-3, 2, 6\}$ and range $\{4, 6\}$. (There are many possible correct answers.)

Concept Check Express each relation using a different form. There is more than one correct way to do this. **See Objective 2.**

5. $\{(0, 2), (2, 4), (4, 6)\}$

6.

x	y
-1	-3
0	-1
1	1
3	3

7.

8. *Concept Check* Does the relation given in **Exercise 7** define a function? Why or why not?

Decide whether each relation defines a function, and give the domain and range. **See Examples 1–4.**

9. $\{(5, 1), (3, 2), (4, 9), (7, 6)\}$

10. $\{(8, 0), (5, 4), (9, 3), (3, 8)\}$

11. $\{(2, 4), (0, 2), (2, 5)\}$

12. $\{(9, -2), (-3, 5), (9, 2)\}$

13. $\{(-3, 1), (4, 1), (-2, 7)\}$

14. $\{(-12, 5), (-10, 3), (8, 3)\}$

15. $\{(1, 1), (1, -1), (0, 0), (2, 4), (2, -4)\}$

16. $\{(2, 5), (3, 7), (4, 9), (5, 11)\}$

17.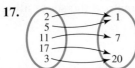

18.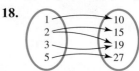

19.

x	y
1	5
1	2
1	-1
1	-4

20.

x	y
-4	-4
-4	0
-4	4
-4	8

21.

x	y
4	-3
2	-3
0	-3
-2	-3

22.

x	y
-3	-6
-1	-6
1	-6
3	-6

23.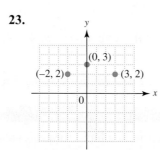

24.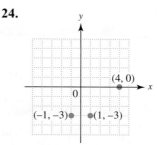

25.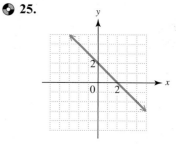

26.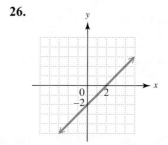

27.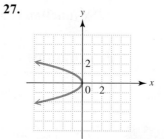

28.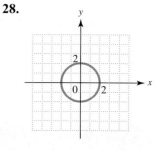

29.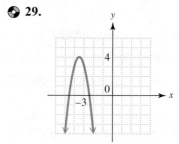

30.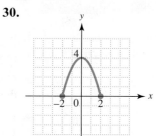

31.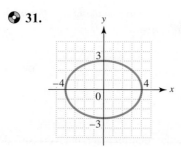

32.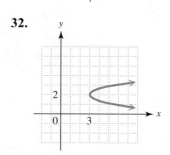

*Decide whether each relation defines y as a function of x. (Solve for y first if necessary.) Give the domain. **See Example 5.***

33. $y = -6x$

34. $y = -9x$

35. $y = 2x - 6$

36. $y = 6x + 8$

37. $y = x^2$

38. $y = x^3$

39. $x = y^6$

40. $x = y^4$

41. $x + y < 4$

42. $x - y < 3$

43. $y = \sqrt{x}$

44. $y = -\sqrt{x}$

45. $y = \sqrt{x - 3}$

46. $y = \sqrt{x - 7}$

47. $y = \sqrt{4x + 2}$

48. $y = \sqrt{2x + 9}$

49. $y = \dfrac{x + 4}{5}$

50. $y = \dfrac{x - 3}{2}$

51. $y = -\dfrac{2}{x}$

52. $y = -\dfrac{6}{x}$

53. $y = \dfrac{2}{x - 4}$

54. $y = \dfrac{7}{x - 2}$

55. $xy = 1$

56. $xy = 3$

Solve each problem.

57. The table shows the percentage of students at 4-year public colleges who graduated within 5 years.

(a) Does the table define a function?

(b) What are the domain and range?

(c) Call this function f. Give two ordered pairs that belong to f.

Year	Percentage
2004	42.3
2005	42.3
2006	42.8
2007	43.7
2008	43.8

Source: ACT.

58. The table shows the percentage of full-time college freshmen who said they had discussed politics in election years.

(a) Does the table define a function?

(b) What are the domain and range?

(c) Call this function g. Give two ordered pairs that belong to g.

Year	Percentage
1992	83.7
1996	73.0
2000	69.6
2004	77.4
2008	85.9

Source: Cooperative Institutional Research Program.

PREVIEW EXERCISES

*Evaluate y for x = 3. **See Section 3.1.***

59. $y = -7x + 12$

60. $y = -5x - 4$

61. $y = 3x - 8$

*Solve for y. **See Section 2.2.***

62. $3x - 7y = 8$

63. $2x - 4y = 7$

64. $\dfrac{3}{4}x + 2y = 9$

3.6 Function Notation and Linear Functions

OBJECTIVES

1 Use function notation.

2 Graph linear and constant functions.

OBJECTIVE 1 **Use function notation.** When a function f is defined with a rule or an equation using x and y for the independent and dependent variables, we say, "*y is a function of x*" to emphasize that y *depends on* x. We use the notation

$$y = f(x),$$

The parentheses here do *not* indicate multiplication.

called **function notation,** to express this and read $f(x)$ as "**f of x.**" The letter f is a name for this particular function. For example, if $y = 9x - 5$, we can name this function f and write

$$f(x) = 9x - 5.$$

f is the name of the function.
x is a value from the domain.
$f(x)$ is the function value (or y-value) that corresponds to x.

$f(x)$ *is just another name for the dependent variable y.*

We can evaluate a function at different values of x by substituting x-values from the domain into the function.

NOW TRY
EXERCISE 1

Let $f(x) = 4x + 3$. Find the value of the function f for $x = -2$.

EXAMPLE 1 Evaluating a Function

Let $f(x) = 9x - 5$. Find the value of the function f for $x = 2$.

$$f(x) = 9x - 5$$

Read $f(2)$ as "f of 2" or "f at 2."

$$f(2) = 9 \cdot 2 - 5 \qquad \text{Replace } x \text{ with 2.}$$

$$f(2) = 18 - 5 \qquad \text{Multiply.}$$

$$f(2) = 13 \qquad \text{Add.}$$

Thus, for $x = 2$, the corresponding function value (or y-value) is 13. $f(2) = 13$ is an abbreviation for the statement "If $x = 2$ in the function f, then $y = 13$" and is represented by the ordered pair $(2, 13)$.　　　　　　　　　　　　　**NOW TRY**

⚠ **CAUTION** The symbol $f(x)$ *does not* indicate "f times x," but represents the y-value associated with the indicated x-value. As just shown, $f(2)$ is the y-value that corresponds to the x-value 2 in the function.

These ideas can be illustrated as follows.

Name of the function

Defining expression

$$y = f(x) = 9x - 5$$

Value of the function　　Name of the independent variable

NOW TRY
EXERCISE 2
Let $f(x) = 2x^2 - 4x + 1$.
Find the following.

(a) $f(-2)$ **(b)** $f(a)$

EXAMPLE 2   Evaluating a Function

Let $f(x) = -x^2 + 5x - 3$. Find the following.

(a) $f(4)$

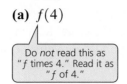

Do *not* read this as
"*f* times 4." Read it as
"*f* of 4."

$f(x) = -x^2 + 5x - 3$ The base in $-x^2$ is *x*, not $(-x)$.

$f(4) = -4^2 + 5 \cdot 4 - 3$ Replace *x* with 4.

$f(4) = -16 + 20 - 3$ Apply the exponent. Multiply.

$f(4) = 1$ Add and subtract.

Thus, $f(4) = 1$, and the ordered pair $(4, 1)$ belongs to f.

(b) $f(q)$

$$f(x) = -x^2 + 5x - 3$$
$$f(q) = -q^2 + 5q - 3 \quad \text{Replace } x \text{ with } q.$$

The replacement of one variable with another is important in later courses.

NOW TRY

Sometimes letters other than f, such as g, h, or capital letters F, G, and H are used to name functions.

NOW TRY
EXERCISE 3
Let $g(x) = 8x - 5$. Find and
simplify $g(a - 2)$.

EXAMPLE 3   Evaluating a Function

Let $g(x) = 2x + 3$. Find and simplify $g(a + 1)$.

$$g(x) = 2x + 3$$
$$g(a + 1) = 2(a + 1) + 3 \quad \text{Replace } x \text{ with } a + 1.$$
$$g(a + 1) = 2a + 2 + 3 \quad \text{Distributive property}$$
$$g(a + 1) = 2a + 5 \quad \text{Add.} \qquad \text{NOW TRY}$$

NOW TRY
EXERCISE 4
Find $f(-1)$ for each function.

(a) $f = \{(-5, -1), (-3, 2), (-1, 4)\}$

(b) $f(x) = x^2 - 12$

EXAMPLE 4   Evaluating Functions

For each function, find $f(3)$.

(a) $f(x) = 3x - 7$

$f(3) = 3(3) - 7$ Replace *x* with 3.

$f(3) = 9 - 7$ Multiply.

$f(3) = 2$ Subtract.

(b)

x	$y = f(x)$
6	-12
3	-6	←Here, $f(3) = -6$.
0	0
-3	6

(c) $f = \{(-3, 5), (0, 3), (3, 1), (6, -1)\}$

We want $f(3)$, the *y*-value of the ordered pair whose first component is $x = 3$. As indicated by the ordered pair $(3, 1)$, for $x = 3$, $y = 1$. Thus, $f(3) = 1$.

(d)

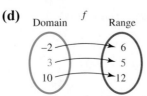

The domain element 3 is paired with 5 in the range, so $f(3) = 5$.

NOW TRY ANSWERS
2. (a) 17   **(b)** $2a^2 - 4a + 1$
3. $8a - 21$
4. (a) 4   **(b)** -11

NOW TRY

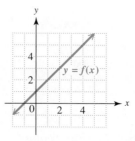

FIGURE 47

NOW TRY
EXERCISE 5

Refer to the function graphed
in **FIGURE 47**.

(a) Find $f(-1)$.

(b) For what value of x is
$f(x) = 2$?

EXAMPLE 5 Finding Function Values from a Graph

Refer to the function graphed in **FIGURE 47**.

(a) Find $f(3)$.

Locate 3 on the x-axis. See **FIGURE 48**. Moving up to the graph of f and over to the y-axis gives 4 for the corresponding y-value. Thus, $f(3) = 4$, which corresponds to the ordered pair $(3, 4)$.

(b) Find $f(0)$.

Refer to **FIGURE 48** to see that $f(0) = 1$.

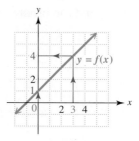

FIGURE 48   **FIGURE 49**

(c) For what value of x is $f(x) = 5$?

Since $f(x) = y$, we want the value of x that corresponds to $y = 5$. Locate 5 on the y-axis. See **FIGURE 49**. Moving across to the graph of f and down to the x-axis gives $x = 4$. Thus, $f(4) = 5$, which corresponds to the ordered pair $(4, 5)$.   NOW TRY

If a function f is defined by an equation with x and y, and y is not solved for x, use the following steps to find $f(x)$.

Finding an Expression for $f(x)$

Step 1   Solve the equation for y.

Step 2   Replace y with $f(x)$.

EXAMPLE 6 Writing Equations Using Function Notation

Rewrite each equation using function notation $f(x)$. Then find $f(-2)$ and $f(a)$.

(a) $y = x^2 + 1$

This equation is already solved for y, so we replace y with $f(x)$.

$$f(x) = x^2 + 1 \qquad y = f(x)$$

To find $f(-2)$, let $x = -2$.

$$f(x) = x^2 + 1$$
$$f(-2) = (-2)^2 + 1 \qquad \text{Let } x = -2.$$
$$f(-2) = 4 + 1 \qquad (-2)^2 = -2(-2)$$
$$f(-2) = 5 \qquad \text{Add.}$$

Find $f(a)$ by letting $x = a$: $f(a) = a^2 + 1$.

(b) $x - 4y = 5$ Solve for y. (Step 1)

$x - 5 = 4y$ Add $4y$. Subtract 5.

$y = \dfrac{x - 5}{4}$,   so $f(x) = \dfrac{1}{4}x - \dfrac{5}{4}$
$\boxed{\frac{a - b}{c} = \frac{a}{c} - \frac{b}{c}}$ $y = f(x)$ (Step 2)

NOW TRY ANSWERS
5. **(a)** 0 **(b)** 1

NOW TRY
EXERCISE 6
Rewrite the equation using function notation $f(x)$. Then find $f(-3)$ and $f(h)$.
$$-4x^2 + y = 5$$

Now find $f(-2)$ and $f(a)$.

$$f(-2) = \frac{1}{4}(-2) - \frac{5}{4} = -\frac{7}{4} \qquad \text{Let } x = -2.$$

$$f(a) = \frac{1}{4}a - \frac{5}{4} \qquad \text{Let } x = a. \qquad \text{NOW TRY}$$

OBJECTIVE 2 **Graph linear and constant functions.** Linear equations (except for vertical lines with equations $x = a$) define *linear functions*.

Linear Function

A function that can be defined by

$$f(x) = ax + b$$

for real numbers a and b is a **linear function.** The value of a is the slope m of the graph of the function. The domain of any linear function is $(-\infty, \infty)$.

A linear function whose graph is a horizontal line is defined by

$$f(x) = b \qquad \text{Constant function}$$

and is sometimes called a **constant function.** While the range of any nonconstant linear function is $(-\infty, \infty)$, the range of a constant function defined by $f(x) = b$ is $\{b\}$.

NOW TRY
EXERCISE 7
Graph the function. Give the domain and range.
$$g(x) = \frac{1}{3}x - 2$$

EXAMPLE 7 **Graphing Linear and Constant Functions**

Graph each function. Give the domain and range.

(a) $f(x) = \dfrac{1}{4}x - \dfrac{5}{4}$ (from **Example 6(b)**)

Slope ⸻   ⸻ y-intercept is $\left(0, -\frac{5}{4}\right)$.

The graph of $y = \frac{1}{4}x - \frac{5}{4}$ has slope $m = \frac{1}{4}$ and y-intercept $\left(0, -\frac{5}{4}\right)$. To graph this function, plot the y-intercept $\left(0, -\frac{5}{4}\right)$ and use the definition of slope as $\frac{\text{rise}}{\text{run}}$ to find a second point on the line. Since the slope is $\frac{1}{4}$, move 1 unit up from $\left(0, -\frac{5}{4}\right)$ and 4 units to the right to find this second point. Draw the straight line through the points to obtain the graph shown in **FIGURE 50**. The domain and range are both $(-\infty, \infty)$.

NOW TRY ANSWERS
6. $f(x) = 4x^2 + 5; f(-3) = 41;$
    $f(h) = 4h^2 + 5$

7.

domain: $(-\infty, \infty)$;
range: $(-\infty, \infty)$

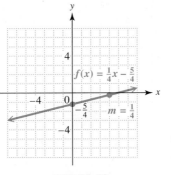

FIGURE 50

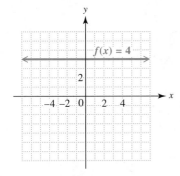

FIGURE 51

(b) $f(x) = 4$

The graph of this constant function is the horizontal line containing all points with y-coordinate 4. See **FIGURE 51**. The domain is $(-\infty, \infty)$ and the range is $\{4\}$.

NOW TRY

3.6 EXERCISES

Complete solution available on the Video Resources on DVD

1. *Concept Check*   Choose the correct response: The notation $f(3)$ means

A. the variable f times 3, or $3f$.

B. the value of the dependent variable when the independent variable is 3.

C. the value of the independent variable when the dependent variable is 3.

D. f equals 3.

2. *Concept Check*   Give an example of a function from everyday life. (*Hint:* Fill in the blanks: _____ depends on _____, so _____ is a function of _____.)

Let $f(x) = -3x + 4$ and $g(x) = -x^2 + 4x + 1$. Find the following. ***See Examples 1–3.***

3. $f(0)$ **4.** $f(-3)$ **5.** $g(-2)$ **6.** $g(10)$

7. $f\left(\dfrac{1}{3}\right)$ **8.** $f\left(\dfrac{7}{3}\right)$ **9.** $g(0.5)$ **10.** $g(1.5)$

11. $f(p)$ **12.** $g(k)$ **13.** $f(-x)$ **14.** $g(-x)$

15. $f(x + 2)$ **16.** $f(x - 2)$ **17.** $g(\pi)$ **18.** $g(e)$

19. $f(x + h)$ **20.** $f(x + h) - f(x)$ **21.** $f(4) - g(4)$ **22.** $f(10) - g(10)$

For each function, find **(a)** $f(2)$ *and* **(b)** $f(-1)$. ***See Examples 4 and 5.***

23. $f = \{(-2, 2), (-1, -1), (2, -1)\}$ **24.** $f = \{(-1, -5), (0, 5), (2, -5)\}$

25. $f = \{(-1, 3), (4, 7), (0, 6), (2, 2)\}$ **26.** $f = \{(2, 5), (3, 9), (-1, 11), (5, 3)\}$

27.   **28.**

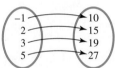   

29.

x	$y = f(x)$
2	4
1	1
0	0
−1	1
−2	4

30.

x	$y = f(x)$
8	6
5	3
2	0
−1	−3
−4	−6

31.   **32.**

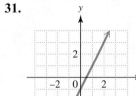   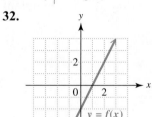

33.   **34.**

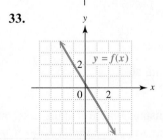   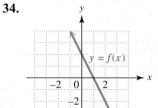

35. Refer to **Exercise 31.** Find the value of x for each value of $f(x)$. **See Example 5(c).**

  (a) $f(x) = 3$ (b) $f(x) = -1$ (c) $f(x) = -3$

36. Refer to **Exercise 32.** Find the value of x for each value of $f(x)$. **See Example 5(c).**

  (a) $f(x) = 4$ (b) $f(x) = -2$ (c) $f(x) = 0$

An equation that defines y as a function f of x is given. **(a)** *Solve for y in terms of x, and re-place y with the function notation f(x).* **(b)** *Find f(3).* **See Example 6.**

37. $x + 3y = 12$ **38.** $x - 4y = 8$ **39.** $y + 2x^2 = 3$

40. $y - 3x^2 = 2$ **41.** $4x - 3y = 8$ **42.** $-2x + 5y = 9$

43. *Concept Check*   Fill in each blank with the correct response.

The equation $2x + y = 4$ has a straight _____ as its graph. One point that lies on the graph is $(3, \underline{\quad})$. If we solve the equation for y and use function notation, we obtain $f(x) = \underline{\quad}$. For this function, $f(3) = \underline{\quad}$, meaning that the point $(\underline{\quad}, \underline{\quad})$ lies on the graph of the function.

44. *Concept Check*   Which of the following defines y as a linear function of x?

  A. $y = \dfrac{1}{4}x - \dfrac{5}{4}$ **B.** $y = \dfrac{1}{x}$ **C.** $y = x^2$ **D.** $y = \sqrt{x}$

Graph each linear function. Give the domain and range. **See Example 7.**

45. $f(x) = -2x + 5$ **46.** $g(x) = 4x - 1$ **47.** $h(x) = \dfrac{1}{2}x + 2$

48. $F(x) = -\dfrac{1}{4}x + 1$ **49.** $G(x) = 2x$ **50.** $H(x) = -3x$

51. $g(x) = -4$ **52.** $f(x) = 5$ **53.** $f(x) = 0$ **54.** $f(x) = -2.5$

55. *Concept Check*   What is the name that is usually given to the graph in **Exercise 53?**

56. Can the graph of a linear function have an undefined slope? Explain.

Solve each problem.

57. A package weighing x pounds costs $f(x)$ dollars to mail to a given location, where

$$f(x) = 3.75x.$$

  (a) Evaluate $f(3)$.

  (b) Describe what 3 and the value $f(3)$ mean in part (a), using the terminology *independent variable* and *dependent variable*.

  (c) How much would it cost to mail a 5-lb package? Interpret this question and its answer, using function notation.

58. A taxicab driver charges $2.50 per mile.

  (a) Fill in the table with the correct response for the price $f(x)$ he charges for a trip of x miles.

x	$f(x)$
0	
1	
2	
3	

  (b) The linear function that gives a rule for the amount charged is $f(x) = \underline{\quad}$.

  (c) Graph this function for the domain $\{0, 1, 2, 3\}$.

59. Forensic scientists use the lengths of certain bones to calculate the height of a person. Two bones often used are the tibia (t), the bone from the ankle to the knee, and the femur (r), the bone from the knee to the hip socket. A person's height (h) in centimeters is determined from the lengths of these bones by using functions defined by the following formulas.

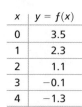

For men: $h(r) = 69.09 + 2.24r$ or $h(t) = 81.69 + 2.39t$

For women: $h(r) = 61.41 + 2.32r$ or $h(t) = 72.57 + 2.53t$

Femur

Tibia

(a) Find the height of a man with a femur measuring 56 cm.

(b) Find the height of a man with a tibia measuring 40 cm.

(c) Find the height of a woman with a femur measuring 50 cm.

(d) Find the height of a woman with a tibia measuring 36 cm.

60. Federal regulations set standards for the size of the quarters of marine mammals. A pool to house sea otters must have a volume of "the square of the sea otter's average adult length (in meters) multiplied by 3.14 and by 0.91 meter." If x represents the sea otter's average adult length and $f(x)$ represents the volume (in cubic meters) of the corresponding pool size, this formula can be written as

$$f(x) = 0.91(3.14)x^2.$$

Find the volume of the pool for each adult sea otter length (in meters). Round answers to the nearest hundredth.

(a) 0.8   (b) 1.0   (c) 1.2   (d) 1.5

61. To print t-shirts, there is a $100 set-up fee, plus a $12 charge per t-shirt. Let x represent the number of t-shirts printed and $f(x)$ represent the total charge.

(a) Write a linear function that models this situation.

(b) Find $f(125)$. Interpret your answer in the context of this problem.

(c) Find the value of x if $f(x) = 1000$. Express this situation using function notation, and interpret it in the context of this problem.

62. Rental on a car is $150, plus $0.20 per mile. Let x represent the number of miles driven and $f(x)$ represent the total cost to rent the car.

(a) Write a linear function that models this situation.

(b) How much would it cost to drive 250 mi? Interpret this question and answer, using function notation.

(c) Find the value of x if $f(x) = 230$. Interpret your answer in the context of this problem.

63. The table represents a linear function.

(a) What is $f(2)$?

(b) If $f(x) = -2.5$, what is the value of x?

(c) What is the slope of the line?

(d) What is the y-intercept of the line?

(e) Using your answers from parts (c) and (d), write an equation for $f(x)$.

x	$y = f(x)$
0	3.5
1	2.3
2	1.1
3	−0.1
4	−1.3
5	−2.5

64. The table represents a linear function.

x	$y = f(x)$
−1	−3.9
0	−2.4
1	−0.9
2	0.6
3	2.1

  (a) What is $f(2)$?

  (b) If $f(x) = 2.1$, what is the value of x?

  (c) What is the slope of the line?

  (d) What is the y-intercept of the line?

  (e) Using your answers from parts (c) and (d), write an equation for $f(x)$.

65. Refer to the graph to answer each of the questions.

Gallons of Water in a Pool at Time t

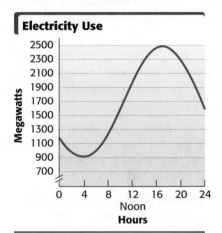

  (a) What numbers are possible values of the independent variable? The dependent variable?

  (b) For how long is the water level increasing? Decreasing?

  (c) How many gallons of water are in the pool after 90 hr?

  (d) Call this function f. What is $f(0)$? What does it mean?

  (e) What is $f(25)$? What does it mean?

66. The graph shows megawatts of electricity used on a summer day.

Electricity Use

Source: Sacramento Municipal Utility District.

  (a) Why is this the graph of a function?

  (b) What is the domain?

  (c) Estimate the number of megawatts used at 8 A.M.

  (d) At what time was the most electricity used? The least electricity?

  (e) Call this function f. What is $f(12)$? What does it mean?

PREVIEW EXERCISES

*Find y if $x = -3$. **See Section 2.1.***

67. $6x + 5y = 2$

68. $\dfrac{4}{3}x + y = 9$

69. $1.5x + 2.5y = 5.5$

*Solve each equation. **See Section 2.1.***

70. $-2(3x + 1) + 5x = -20$ **71.** $2x - 4(x - 3) = 8$ **72.** $2\left(\dfrac{3x + 1}{2}\right) - x = -4$

STUDY SKILLS

Taking Math Tests

Techniques To Improve Your Test Score	Comments
Come prepared with a pencil, eraser, paper, and calculator, if allowed.	Working in pencil lets you erase, keeping your work neat and readable.
Scan the entire test, note the point values of different problems, and plan your time accordingly.	To do 20 problems in 50 minutes, allow $50 \div 20 = 2.5$ minutes per problem. Spend less time on easy problems.
Do a "knowledge dump" when you get the test. Write important notes to yourself in a corner of the test, such as formulas.	Writing down tips and things that you've memorized at the beginning allows you to relax later.
Read directions carefully, and circle any significant words. When you finish a problem, read the directions again to make sure you did what was asked.	Pay attention to announcements written on the board or made by your instructor. Ask if you don't understand.
Show all your work. Many teachers give partial credit if some steps are correct, even if the final answer is wrong. ***Write neatly.***	If your teacher can't read your writing, you won't get credit for it. If you need more space to work, ask to use extra paper.
Write down anything that might help solve a problem: a formula, a diagram, etc. If you can't get it, circle the problem and come back to it later. Do *not* erase anything you wrote down.	If you know even a little bit about the problem, write it down. The answer may come to you as you work on it, or you may get partial credit. Don't spend too long on any one problem.
If you can't solve a problem, make a guess. Do not change it unless you find an obvious mistake.	Have a good reason for changing an answer. Your first guess is usually your best bet.
Check that the answer to an application problem is reasonable and makes sense. Read the problem again to make sure that you have answered the question.	Use common sense. Can the father really be seven years old? Would a month's rent be $32,140? Label your answer: $, years, inches, etc.
Check for careless errors. Rework the problem without looking at your previous work. Compare the two answers.	Reworking the problem from the beginning forces you to rethink it. If possible, use a different method to solve the problem.

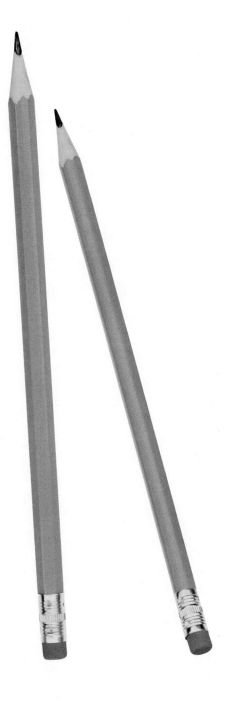

Select several tips to try when you take your next math test.

CHAPTER 3 SUMMARY

KEY TERMS

3.1
ordered pair
components
origin
x-axis
y-axis
rectangular (Cartesian)
   coordinate system
plot
coordinate
quadrant
graph of an equation

first-degree equation
linear equation in two
   variables
standard form
x-intercept
y-intercept

3.2
rise
run
slope

3.3
slope-intercept form
point-slope form
scatter diagram

3.4
linear inequality in two
   variables
boundary line

3.5
dependent variable
independent variable
relation
function
domain
range

3.6
function notation
linear function
constant function

NEW SYMBOLS

(a, b) ordered pair	x_1 a specific value of x (read "*x*-sub-one")	Δ Greek letter delta m slope

$f(x)$ function of x (read "f of x")

TEST YOUR WORD POWER

See how well you have learned the vocabulary in this chapter.

1. An **ordered pair** is a pair of numbers written
   A. in numerical order between brackets
   B. between parentheses or brackets
   C. between parentheses in which order is important
   D. between parentheses in which order does not matter.

2. A **linear equation in two variables** is an equation that can be written in the form
   A. $Ax + By < C$
   B. $ax = b$
   C. $y = x^2$
   D. $Ax + By = C$.

3. An **intercept** is
   A. the point where the *x*-axis and *y*-axis intersect
   B. a pair of numbers written between parentheses in which order matters
   C. one of the regions determined by a coordinate system

   D. the point where a graph intersects the *x*-axis or the *y*-axis.

4. The **slope** of a line is
   A. the measure of the run over the rise of the line
   B. the distance between two points on the line
   C. the ratio of the change in *y* to the change in *x* along the line
   D. the horizontal change compared with the vertical change between two points on the line.

5. A **relation** is
   A. a set of ordered pairs
   B. the ratio of the change in *y* to the change in *x* along a line
   C. the set of all possible values of the independent variable
   D. all the second components of a set of ordered pairs.

6. A **function** is
   A. the pair of numbers in an ordered pair

   B. a set of ordered pairs in which each *x*-value corresponds to exactly one *y*-value
   C. a pair of numbers written between parentheses
   D. the set of all ordered pairs that satisfy an equation.

7. The **domain** of a function is
   A. the set of all possible values of the dependent variable *y*
   B. a set of ordered pairs
   C. the difference between the *x*-values
   D. the set of all possible values of the independent variable *x*.

8. The **range** of a function is
   A. the set of all possible values of the dependent variable *y*
   B. a set of ordered pairs
   C. the difference between the *y*-values
   D. the set of all possible values of the independent variable *x*.

ANSWERS

1. C; *Examples:* $(0, 3)$, $(3, 8)$, $(4, 0)$ **2.** D; *Examples:* $3x + 2y = 6$, $x = y - 7$ **3.** D; *Example:* In **FIGURE 4(b)** of **Section 3.1,** the *x*-intercept is $(3, 0)$ and the *y*-intercept is $(0, 2)$.   **4.** C; *Example:* The line through $(3, 6)$ and $(5, 4)$ has slope $\frac{4 - 6}{5 - 3} = \frac{-2}{2} = -1$.   **5.** A; *Example:* The set $\{(2, 0), (4, 3), (6, 6)\}$ defines a relation.   **6.** B; The relation given in Answer 5 is a function.   **7.** D; *Example:* In the function in Answer 5, the domain is the set of *x*-values, $\{2, 4, 6\}$.   **8.** A; *Example:* In the function in Answer 5, the range is the set of *y*-values, $\{0, 3, 6\}$.

EXAMPLES

The Rectangular Coordinate System

Finding Intercepts

To find the *x*-intercept, let $y = 0$ and solve for *x*.

To find the *y*-intercept, let $x = 0$ and solve for *y*.

Find the intercepts of the graph of $2x + 3y = 12$.

$$2x + 3(0) = 12 \qquad\qquad 2(0) + 3y = 12$$
$$2x = 12 \qquad\qquad\qquad 3y = 12$$
$$x = 6 \qquad\qquad\qquad\quad y = 4$$

The *x*-intercept is $(6, 0)$. | The *y*-intercept is $(0, 4)$.

Midpoint Formula

If the endpoints of a line segment PQ are $P(x_1, y_1)$ and $Q(x_2, y_2)$, then its midpoint M is

$$\left(\frac{x_1 + x_2}{2}, \frac{y_1 + y_2}{2}\right).$$

Find the midpoint of the segment with endpoints $(4, -7)$ and $(-10, -13)$.

$$\left(\frac{4 + (-10)}{2}, \frac{-7 + (-13)}{2}\right) = (-3, -10)$$

3.2 The Slope of a Line

If $x_2 \neq x_1$, then

$$\textbf{slope } \boldsymbol{m} = \frac{\textbf{rise}}{\textbf{run}} = \frac{\textbf{change in } \boldsymbol{y}}{\textbf{change in } \boldsymbol{x}} = \frac{\Delta y}{\Delta x} = \frac{y_2 - y_1}{x_2 - x_1}.$$

Find the slope of the graph of $2x + 3y = 12$.
  Use the intercepts $(6, 0)$ and $(0, 4)$ and the slope formula.

$$m = \frac{4 - 0}{0 - 6} = \frac{4}{-6} = -\frac{2}{3} \qquad x_1 = 6, y_1 = 0, x_2 = 0, y_2 = 4$$

A vertical line has undefined slope.

A horizontal line has 0 slope.

Parallel lines have equal slopes.

The graph of the line $x = 3$ has undefined slope.

The graph of the line $y = -5$ has slope $m = 0$.

The lines $y = 2x + 3$ and $4x - 2y = 6$ are parallel—both have $m = 2$.

$$y = 2x + 3 \qquad\qquad\quad 4x - 2y = 6$$
$$m = 2 \qquad\qquad\qquad -2y = -4x + 6$$
$$y = 2x - 3$$
$$m = 2$$

The slopes of perpendicular lines, neither of which is vertical, are negative reciprocals with a product of -1.

The lines $y = 3x - 1$ and $x + 3y = 4$ are perpendicular—their slopes are negative reciprocals.

$$y = 3x - 1 \qquad\qquad x + 3y = 4$$
$$m = 3 \qquad\qquad\qquad 3y = -x + 4$$
$$y = -\frac{1}{3}x + \frac{4}{3}$$
$$m = -\frac{1}{3}$$

(continued)

CONCEPTS	EXAMPLES

3.3 Linear Equations in Two Variables

Slope-Intercept Form
$y = mx + b$

Point-Slope Form
$y - y_1 = m(x - x_1)$

Standard Form
$Ax + By = C$

Horizontal Line
$y = b$

Vertical Line
$x = a$

$y = 2x + 3$ $m = 2$, y-intercept is $(0, 3)$.

$y - 3 = 4(x - 5)$ $(5, 3)$ is on the line, $m = 4$.

$2x - 5y = 8$ Standard form

$y = 4$ Horizontal line

$x = -1$ Vertical line

3.4 Linear Inequalities in Two Variables

Graphing a Linear Inequality

Step 1 Draw the graph of the line that is the boundary. Make the line solid if the inequality involves $\leq$ or $\geq$. Make the line dashed if the inequality involves $<$ or $>$.

Step 2 Choose any point not on the line as a test point. Substitute the coordinates into the inequality.

Step 3 Shade the region that includes the test point if the test point satisfies the original inequality. Otherwise, shade the region on the other side of the boundary line.

Graph $2x - 3y \leq 6$.

Draw the graph of $2x - 3y = 6$. Use a solid line because of the inclusion of equality in the symbol $\leq$.

Choose $(0, 0)$, for example.

$2(0) - 3(0) \overset{?}{\leq} 6$

     $0 \leq 6$ True

Shade the side of the line that includes $(0, 0)$.

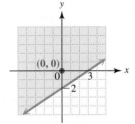

3.5 Introduction to Relations and Functions

A **function** is a set of ordered pairs such that, for each first component, there is one and only one second component. The set of first components is called the **domain,** and the set of second components is called the **range.**

$f = \{(-1, 4), (0, 6), (1, 4)\}$ defines a function f with domain, the set of x-values, $\{-1, 0, 1\}$ and range, the set of y-values, $\{4, 6\}$.

$y = x^2$ defines a function with domain $(-\infty, \infty)$ and range $[0, \infty)$.

3.6 Function Notation and Linear Functions

To evaluate a function f, where $f(x)$ defines the range value for a given value of x in the domain, substitute the value wherever x appears.

To write an equation that defines a function f in function notation, follow these steps.

Step 1 Solve the equation for y.

Step 2 Replace y with $f(x)$.

If $f(x) = x^2 - 7x + 12$, then

$$f(1) = 1^2 - 7(1) + 12 = 6.$$

Write $2x + 3y = 12$ using notation for a function f.

$3y = -2x + 12$ Subtract $2x$.

$y = -\dfrac{2}{3}x + 4$ Divide by 3.

$f(x) = -\dfrac{2}{3}x + 4$ $y = f(x)$

CHAPTER 3 REVIEW EXERCISES

3.1 *Complete the table of ordered pairs for each equation. Then graph the equation.*

1. $3x + 2y = 10$

x	y
0	
	0
2	
	−2

2. $x − y = 8$

x	y
2	
	−3
3	
	−2

Find the x- and y-intercepts and then graph each equation.

3. $4x − 3y = 12$

4. $5x + 7y = 28$

5. $2x + 5y = 20$

6. $x − 4y = 8$

Use the midpoint formula to find the midpoint of each segment with the given endpoints.

7. $(−8, −12)$ and $(8, 16)$

8. $(0, −5)$ and $(−9, 8)$

3.2 *Find the slope of each line.*

9. Through $(−1, 2)$ and $(4, −5)$

10. Through $(0, 3)$ and $(−2, 4)$

11. $y = 2x + 3$

12. $3x − 4y = 5$

13. $x = 5$

14. Parallel to $3y = 2x + 5$

15. Perpendicular to $3x − y = 4$

16. Through $(−1, 5)$ and $(−1, −4)$

17.

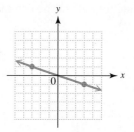

18.

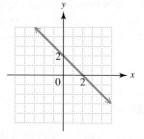

Tell whether each line has positive, negative, 0, *or* undefined *slope.*

19.

20.

21.

22.

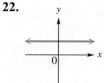

23. *Concept Check* If the pitch of a roof is $\frac{1}{4}$, how many feet in the horizontal direction correspond to a rise of 3 ft?

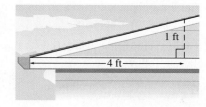

24. Family income in the United States has increased steadily for many years (primarily due to inflation). In 1980, the median family income was about $21,000 per year. In 2007, it was about $61,400 per year. Find the average rate of change of median family income to the nearest dollar over that period. (*Source:* U.S. Census Bureau.)

3.3 *Find an equation for each line.* **(a)** *Write the equation in slope-intercept form.* **(b)** *Write the equation in standard form.*

25. Slope $-\frac{1}{3}$; y-intercept $(0, -1)$

26. Slope 0; y-intercept $(0, -2)$

27. Slope $-\frac{4}{3}$; through $(2, 7)$

28. Slope 3; through $(-1, 4)$

29. Vertical; through $(2, 5)$

30. Through $(2, -5)$ and $(1, 4)$

31. Through $(-3, -1)$ and $(2, 6)$

32. The line pictured in **Exercise 18**

33. Parallel to $4x - y = 3$ and through $(7, -1)$

34. Perpendicular to $2x - 5y = 7$ and through $(4, 3)$

35. The Midwest Athletic Club offers two special membership plans. (*Source:* Midwest Athletic Club.) For each plan, write a linear equation in slope-intercept form and give the cost y in dollars of a 1-yr membership. Let x represent the number of months.

(a) Executive VIP/Gold membership: $159 fee, plus $57 per month

(b) Executive Regular/Silver membership: $159 fee, plus $47 per month

36. Revenue for skiing facilities in the United States is shown in the graph.

(a) Use the information given for the years 2003 and 2007, letting $x = 3$ represent 2003, $x = 7$ represent 2007, and y represent revenue (in millions of dollars) to find a linear equation that models the data. Write the equation in slope-intercept form. Interpret the slope.

(b) Use your equation from part (a) to estimate revenue for skiing facilities in 2008, to the nearest million.

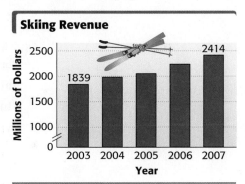

Skiing Revenue

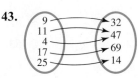

2500

2414

2000 1839

1500

1000

0

2003 2004 2005 2006 2007

Year

Millions of Dollars

Source: U.S. Census Bureau.

3.4 *Graph the solution set of each inequality or compound inequality.*

37. $3x - 2y \le 12$

38. $5x - y > 6$

39. $2x + y \le 1$ and $x \ge 2y$

40. $x \ge 2$ or $y \ge 2$

41. *Concept Check*   Which one of the following has as its graph a dashed boundary line and shading below the line?

A. $y \ge 4x + 3$ **B.** $y > 4x + 3$ **C.** $y \le 4x + 3$ **D.** $y < 4x + 3$

3.5 *In Exercises 42–45, give the domain and range of each relation. Identify any functions.*

42. $\{(-4, 2), (-4, -2), (1, 5), (1, -5)\}$

43.

9

11

4

17

25

32

47

69

14

44.

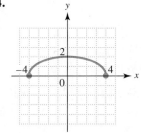

45.

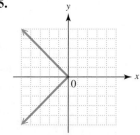

3.5–3.6 *Determine whether each equation or inequality defines y as a function of x. Give the domain in each case. Identify any linear functions.*

46. $y = 3x - 3$

47. $y < x + 2$

48. $y = |x|$

49. $y = \sqrt{4x + 7}$

50. $x = y^2$

51. $y = \dfrac{7}{x - 6}$

3.6 *Given* $f(x) = -2x^2 + 3x - 6$, *find each function value or expression.*

52. $f(0)$

53. $f(2.1)$

54. $f\left(-\dfrac{1}{2}\right)$

55. $f(k)$

56. The equation $2x^2 - y = 0$ defines y as a function f of x. Write it using function notation, and find $f(3)$.

57. *Concept Check* Suppose that $2x - 5y = 7$ defines y as a function f of x. If $y = f(x)$, which one of the following defines the same function?

A. $f(x) = -\dfrac{2}{5}x + \dfrac{7}{5}$ **B.** $f(x) = -\dfrac{2}{5}x - \dfrac{7}{5}$

C. $f(x) = \dfrac{2}{5}x - \dfrac{7}{5}$ **D.** $f(x) = \dfrac{2}{5}x + \dfrac{7}{5}$

58. The table shows life expectancy at birth in the United States for selected years.

(a) Does the table define a function?

(b) What are the domain and range?

(c) Call this function f. Give two ordered pairs that belong to f.

(d) Find $f(1980)$. What does this mean?

(e) If $f(x) = 76.8$, what does x equal?

Year	Life Expectancy at Birth (years)
1960	69.7
1970	70.8
1980	73.7
1990	75.4
2000	76.8
2009	78.1

Source: National Center for Health Statistics.

RELATING CONCEPTS EXERCISES 59–70

FOR INDIVIDUAL OR GROUP WORK

Refer to the straight-line graph and **work Exercises 59–70 in order.**

59. By just looking at the graph, how can you tell whether the slope is positive, negative, 0, or undefined?

60. Use the slope formula to find the slope of the line.

61. What is the slope of any line parallel to the line shown? Perpendicular to the line shown?

62. Find the x-intercept of the graph.

63. Find the y-intercept of the graph.

64. Use function notation to write the equation of the line. Use f to designate the function.

65. Find $f(8)$.

66. If $f(x) = -8$, what is the value of x?

67. Graph the solution set of $f(x) \geq 0$.

68. What is the solution set of $f(x) = 0$?

69. What is the solution set of $f(x) < 0$? (Use the graph and the result of **Exercise 68.**)

70. What is the solution set of $f(x) > 0$? (Use the graph and the result of **Exercise 68.**)

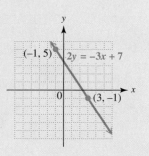

🌐 *View the complete solutions to all Chapter Test exercises on the Video Resources on DVD.*

1. Complete the table of ordered pairs for the equation $2x - 3y = 12$.

x	y
1	
3	
	-4

Find the x- and y-intercepts, and graph each equation.

2. $3x - 2y = 20$ **3.** $y = 5$ **4.** $x = 2$

5. Find the slope of the line through the points $(6, 4)$ and $(-4, -1)$.

✏️ **6.** Describe how the graph of a line with undefined slope is situated in a rectangular coordinate system.

Determine whether each pair of lines is parallel, perpendicular, *or* neither.

7. $5x - y = 8$ and $5y = -x + 3$

8. $2y = 3x + 12$ and $3y = 2x - 5$

✏️ **9.** In 1980, there were 119,000 farms in Iowa. As of 2008, there were 93,000. Find and interpret the average rate of change in the number of farms per year, to the nearest whole number. (*Source:* U.S. Department of Agriculture.)

*Find an equation of each line, and write it in **(a)** slope-intercept form if possible and **(b)** standard form.*

10. Through $(4, -1)$; $m = -5$ **11.** Through $(-3, 14)$; horizontal

12. Through $(-2, 3)$ and $(6, -1)$ **13.** Through $(5, -6)$; vertical

14. Through $(-7, 2)$ and parallel to $3x + 5y = 6$

15. Through $(-7, 2)$ and perpendicular to $y = 2x$

16. *Concept Check* Which line has positive slope and negative y-coordinate for its y-intercept?

A. 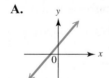   **B.**   **C.** 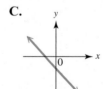   **D.**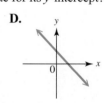

17. The bar graph shows median household income for Asians and Pacific Islanders in the United States.

   (a) Use the information for the years 2001 and 2007 to find an equation that models the data. Let $x = 1$ represent 2001, $x = 7$ represent 2007, and y represent the median income. Write the equation in slope-intercept form.

   (b) Use the equation from part (a) to approximate median household income for 2005 to the nearest dollar. How does your result compare against the actual value, $61,094?

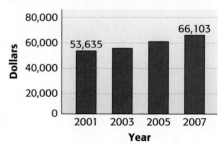

Median Household Income for Asians and Pacific Islanders

Source: U.S. Census Bureau.

Graph each inequality or compound inequality.

18. $3x - 2y > 6$

19. $y < 2x - 1$ and $x - y < 3$

20. Which one of the following is the graph of a function?

A.

B.

C.

D.

21. Which of the following does not define y as a function of x?

A. $\{(0, 1), (-2, 3), (4, 8)\}$ B. $y = 2x - 6$ C. $y = \sqrt{x + 2}$ D.

x	y
0	1
3	2
0	2
6	3

Give the domain and range of the relation shown in each of the following.

22. Choice A of **Exercise 20**

23. Choice A of **Exercise 21**

24. For $f(x) = -x^2 + 2x - 1$, find **(a)** $f(1)$, and **(b)** $f(a)$.

25. Graph the linear function defined by $f(x) = \frac{2}{3}x - 1$. What is its domain and range?

CHAPTERS **1–3**

CUMULATIVE REVIEW EXERCISES

Decide whether each statement is always true, sometimes true, *or* never true. *If the statement is* sometimes true, *give examples in which it is true and in which it is false.*

1. The absolute value of a negative number equals the additive inverse of the number.

2. The sum of two negative numbers is positive.

3. The sum of a positive number and a negative number is 0.

Perform each operation.

4. $-|-2| - 4 + |-3| + 7$ **5.** $(-0.8)^2$ **6.** $\sqrt{-64}$

Simplify.

7. $-(-4m + 3)$ **8.** $3x^2 - 4x + 4 + 9x - x^2$ **9.** $\dfrac{(4^2 - 4) - (-1)7}{4 + (-6)}$

10. Write $-3 < x \leq 5$ in interval notation.

Evaluate each expression for $p = -4$, $q = \frac{1}{2}$, *and* $r = 16$.

11. $-3(2q - 3p)$ **12.** $\dfrac{\sqrt{r}}{8p + 2r}$

Solve.

13. $2z - 5 + 3z = 2 - z$ **14.** $\dfrac{3x - 1}{5} + \dfrac{x + 2}{2} = -\dfrac{3}{10}$

Solve each problem.

15. If each side of a square were increased by 4 in., the perimeter would be 8 in. less than twice the perimeter of the original square. Find the length of a side of the original square.

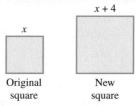

Original square   New square

16. Two planes leave the Dallas-Fort Worth airport at the same time. One travels east at 550 mph, and the other travels west at 500 mph. Assuming no wind, how long will it take for the planes to be 2100 mi apart?

West ← ✈ Airport ✈ → East

Solve. Write each solution set in interval notation and graph it.

17. $-4 < 3 - 2k < 9$

18. $-0.3x + 2.1(x - 4) \le -6.6$

19. $\dfrac{1}{2}x > 3$ and $\dfrac{1}{3}x < \dfrac{8}{3}$

20. $-5x + 1 \ge 11$ or $3x + 5 > 26$

Solve.

21. $|2k - 7| + 4 = 11$

22. $|3m + 6| \ge 0$

23. Find the x- and y-intercepts of the line with equation $3x + 5y = 12$, and graph the line.

24. Consider the points $A(-2, 1)$ and $B(3, -5)$.

   (a) Find the slope of the line AB.

   (b) Find the slope of a line perpendicular to line AB.

25. Graph the inequality $-2x + y < -6$.

*Write an equation for each line. Express the equation **(a)** in slope-intercept form if possible and **(b)** in standard form.*

26. Slope $-\dfrac{3}{4}$; y-intercept $(0, -1)$

27. Through $(4, -3)$ and $(1, 1)$

28. Give the domain and range of the relation. Does it define a function? Explain.

```
14 ──────→ 9
91 ──────→ 70
75 ══════→ 56
23 ──────→ 5
```

29. Consider the function defined by
$$f(x) = -4x + 10.$$

   (a) Find the domain and range.

   (b) Evaluate $f(-3)$.

   (c) If $f(x) = 6$, find the value of x.

30. Use the information in the graph to find and interpret the average rate of change in the per capita consumption of potatoes in the United States from 2003 to 2008.

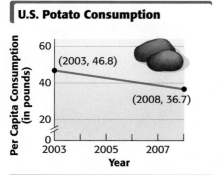

U.S. Potato Consumption

(2003, 46.8)

(2008, 36.7)

Per Capita Consumption (in pounds)

Year

Source: U.S. Department of Agriculture.

Systems of Linear Equations

In the early 1970s, the NBC television network presented *The Bill Cosby Show,* in which the popular comedian played Chet Kincaid, a Los Angeles high school physical education teacher. In the episode "Let *x* Equal a Lousy Weekend," Chet must substitute for the algebra teacher. He and the entire class are stumped by the following problem:

> *How many pounds of candy that sells for $0.75 per lb must be mixed with candy that sells for $1.25 per lb to obtain 9 lb of a mixture that should sell for $0.96 per lb?*

The smartest student in the class eventually helps Chet solve this problem. In **Exercise 31** of **Section 4.3,** we ask you to use a *system of linear equations,* the topic of this chapter, to do so.

4.1 Systems of Linear Equations in Two Variables

In recent years, the sale of digital cameras has increased, while that of digital camcorders has decreased. See **FIGURE 1**. The two straight-line graphs intersect at the point in time when the two products had the *same* sales.

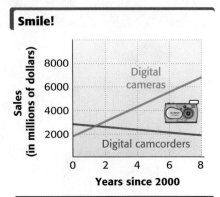

Smile!

Source: Consumer Electronics Association.

FIGURE 1

With regard to **FIGURE 1**, we can use a linear equation to model the graph of digital camera sales (the blue equation below) and another linear equation to model the graph of digital camcorder sales (the red equation below).

$$624x - y = -1823$$
$$119x + y = 2838$$

(Here, $x = 0$ represents 2000, $x = 1$ represents 2001, and so on; y represents sales in millions of dollars.)

Such a set of equations is called a **system of equations**—in this case, a **linear system of equations.** The point where the graphs in **FIGURE 1** intersect is a solution of each of the individual equations. It is also the solution of the linear system of equations.

OBJECTIVE 1 Decide whether an ordered pair is a solution of a linear system. The **solution set of a linear system** of equations contains all ordered pairs that satisfy all the equations of the system *at the same time.*

EXAMPLE 1 Deciding Whether an Ordered Pair Is a Solution

Decide whether the given ordered pair is a solution of the given system.

(a) $\begin{aligned} x + y &= 6 \\ 4x - y &= 14 \end{aligned}$; $(4, 2)$

Replace x with 4 and y with 2 in each equation of the system.

$x + y = 6$	$4x - y = 14$
$4 + 2 \overset{?}{=} 6$	$4(4) - 2 \overset{?}{=} 14$
$6 = 6$ ✓ True	$14 = 14$ ✓ True

Since $(4, 2)$ makes *both* equations true, $(4, 2)$ is a solution of the system.

NOW TRY
EXERCISE 1
Is the ordered pair $(2, 5)$ a solution of the system?
$$-x + 2y = 8$$
$$3x - 2y = 0$$

(b) $3x + 2y = 11$
$x + 5y = 36$; $(-1, 7)$

$$3x + 2y = 11$$
$$3(-1) + 2(7) \stackrel{?}{=} 11$$
$$-3 + 14 \stackrel{?}{=} 11$$
$$11 = 11 \ \checkmark \ \text{True}$$

$$x + 5y = 36$$
$$-1 + 5(7) \stackrel{?}{=} 36$$
$$-1 + 35 \stackrel{?}{=} 36$$
$$34 = 36 \ \checkmark \ \text{False}$$

The ordered pair $(-1, 7)$ is not a solution of the system, since it does not make *both* equations true.   NOW TRY

OBJECTIVE 2 **Solve linear systems by graphing.** One way to find the solution set of a linear system of equations is to graph each equation and find the point where the graphs intersect.

NOW TRY
EXERCISE 2
Solve the system of equations by graphing.
$$3x - 2y = 6$$
$$x - y = 1$$

EXAMPLE 2 Solving a System by Graphing

Solve the system of equations by graphing.

$$x + y = 5 \quad (1)$$
$$2x - y = 4 \quad (2)$$

To graph these linear equations, we plot several points for each line.

$x + y = 5$

x	y
0	5
5	0
2	3

The intercepts are a convenient choice.

$2x - y = 4$

x	y
0	-4
2	0
4	4

Find a third ordered pair as a check.

As shown in **FIGURE 2**, the graph suggests that the point of intersection is the ordered pair $(3, 2)$.

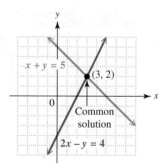

FIGURE 2

To be sure that $(3, 2)$ is a solution of *both* equations, we check by substituting 3 for x and 2 for y in each equation.

CHECK $x + y = 5 \quad (1)$
$$3 + 2 \stackrel{?}{=} 5$$
$$5 = 5 \ \checkmark \ \text{True}$$

$2x - y = 4 \quad (2)$
$$2(3) - 2 \stackrel{?}{=} 4$$
$$6 - 2 \stackrel{?}{=} 4$$
$$4 = 4 \ \checkmark \ \text{True}$$

Since $(3, 2)$ makes both equations true, $\{(3, 2)\}$ is the solution set of the system.
                                        NOW TRY

NOW TRY ANSWERS
1. no
2. $\{(4, 3)\}$

There are three possibilities for the number of elements in the solution set of a linear system in two variables.

Graphs of Linear Systems in Two Variables

Case 1 **The two graphs intersect in a single point.** The coordinates of this point give the only solution of the system. Since the system has a solution, it is **consistent.** The equations are *not* equivalent, so they are **independent.** See **FIGURE 3(a)**.

Case 2 **The graphs are parallel lines.** There is no solution common to both equations, so the solution set is $\emptyset$ and the system is **inconsistent.** Since the equations are *not* equivalent, they are **independent.** See **FIGURE 3(b)**.

Case 3 **The graphs are the same line.** Since any solution of one equation of the system is a solution of the other, the solution set is an infinite set of ordered pairs representing the points on the line. This type of system is **consistent** because there is a solution. The equations are equivalent, so they are **dependent.** See **FIGURE 3(c)**.

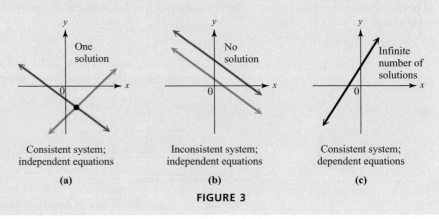

Consistent system; independent equations	Inconsistent system; independent equations	Consistent system; dependent equations
(a)	**(b)**	**(c)**

FIGURE 3

OBJECTIVE 3 **Solve linear systems (with two equations and two variables) by substitution.** Since it can be difficult to read exact coordinates, especially if they are not integers, from a graph, we usually use algebraic methods to solve systems. One such method, the **substitution method,** is most useful for solving linear systems in which one equation is solved or can be easily solved for one variable in terms of the other.

EXAMPLE 3 Solving a System by Substitution

Solve the system.

$$2x - y = 6 \quad \text{(1)}$$
$$x = y + 2 \quad \text{(2)}$$

Since equation (2) is solved for x, substitute $y + 2$ for x in equation (1).

$$2x - y = 6 \quad \text{(1)}$$
$$2(y + 2) - y = 6 \qquad \text{Let } x = y + 2.$$
$$2y + 4 - y = 6 \qquad \text{Distributive property}$$
$$y + 4 = 6 \qquad \text{Combine like terms.}$$
$$y = 2 \qquad \text{Subtract 4.}$$

> Be sure to use parentheses here.

**NOW TRY
EXERCISE 3**
Solve the system.

$$x = 3 + 2y$$
$$4x - 3y = 32$$

We found y. Now solve for x by substituting 2 for y in equation (2).

$$x = y + 2 = 2 + 2 = 4 \quad \boxed{\text{Write the } x\text{-value first in the ordered pair.}}$$

Thus, $x = 4$ and $y = 2$, giving the ordered pair $(4, 2)$. Check this solution in both equations of the original system.

CHECK

$2x - y = 6$ (1)	$x = y + 2$ (2)
$2(4) - 2 \overset{?}{=} 6$	$4 \overset{?}{=} 2 + 2$
$8 - 2 \overset{?}{=} 6$	$4 = 4$ ✓ True
$6 = 6$ ✓ True	

Since $(4, 2)$ makes both equations true, the solution set is $\{(4, 2)\}$. NOW TRY ↻

Solving a Linear System by Substitution

Step 1 **Solve one of the equations for either variable.** If one of the equations has a variable term with coefficient 1 or -1, choose it, since the substitution method is usually easier this way.

Step 2 **Substitute** for that variable in the other equation. The result should be an equation with just one variable.

Step 3 **Solve** the equation from Step 2.

Step 4 **Find the other value.** Substitute the result from Step 3 into the equation from Step 1 to find the value of the other variable.

Step 5 **Check** the ordered-pair solution in *both* of the *original* equations. Then write the solution set.

EXAMPLE 4 Solving a System by Substitution

Solve the system.

$$3x + 2y = 13 \quad (1)$$
$$4x - y = -1 \quad (2)$$

Step 1 Solve one of the equations for either x or y. Since the coefficient of y in equation (2) is -1, it is easiest to solve for y in equation (2).

$$4x - y = -1 \quad (2)$$
$$-y = -1 - 4x \quad \text{Subtract } 4x.$$
$$y = 1 + 4x \quad \text{Multiply by } -1.$$

Step 2 Substitute $1 + 4x$ for y in equation (1).

$$3x + 2y = 13 \quad (1)$$
$$3x + 2(1 + 4x) = 13 \quad \text{Let } y = 1 + 4x.$$

Step 3 Solve for x.

$$3x + 2 + 8x = 13 \quad \text{Distributive property}$$
$$11x = 11 \quad \text{Combine like terms. Subtract 2.}$$
$$x = 1 \quad \text{Divide by 11.}$$

NOW TRY ANSWER
3. $\{(11, 4)\}$

NOW TRY
EXERCISE 4
Solve the system.

$$5x + y = 7$$
$$3x - 2y = 25$$

Step 4 Now solve for y. From Step 1, $y = 1 + 4x$, so if $x = 1$, then

$$y = 1 + 4(1) = 5. \quad \text{Let } x = 1.$$

Step 5 Check the solution $(1, 5)$ in both equations (1) and (2).

CHECK

$3x + 2y = 13$ (1)	$4x - y = -1$ (2)
$3(1) + 2(5) \overset{?}{=} 13$	$4(1) - 5 \overset{?}{=} -1$
$3 + 10 \overset{?}{=} 13$	$4 - 5 \overset{?}{=} -1$
$13 = 13$ ✓ True	$-1 = -1$ ✓ True

The solution set is $\{(1, 5)\}$.

NOW TRY

EXAMPLE 5 Solving a System with Fractional Coefficients

Solve the system.

$$\frac{2}{3}x - \frac{1}{2}y = \frac{7}{6} \quad (1)$$

$$3x - y = 6 \quad (2)$$

This system will be easier to solve if we clear the fractions in equation (1).

$$6\left(\frac{2}{3}x - \frac{1}{2}y\right) = 6\left(\frac{7}{6}\right) \quad \text{Multiply (1) by the LCD, 6.}$$

Remember to multiply *each* term by 6.

$$6 \cdot \frac{2}{3}x - 6 \cdot \frac{1}{2}y = 6 \cdot \frac{7}{6} \quad \text{Distributive property}$$

$$4x - 3y = 7 \quad (3)$$

Now the system consists of equations (2) and (3).

This equation is equivalent to equation (1).

$$3x - y = 6 \quad (2)$$
$$4x - 3y = 7 \quad (3)$$

To use the substitution method, we solve equation (2) for y.

$$3x - y = 6 \quad (2)$$
$$-y = 6 - 3x \quad \text{Subtract } 3x.$$
$$y = 3x - 6 \quad \text{Multiply by } -1. \text{ Rewrite.}$$

Substitute $3x - 6$ for y in equation (3).

$$4x - 3y = 7 \quad (3)$$
$$4x - 3(3x - 6) = 7 \quad \text{Let } y = 3x - 6.$$
$$4x - 9x + 18 = 7 \quad \text{Distributive property}$$

Be careful with signs.

$$-5x + 18 = 7 \quad \text{Combine like terms.}$$
$$-5x = -11 \quad \text{Subtract 18.}$$
$$x = \frac{11}{5} \quad \text{Divide by } -5.$$

NOW TRY ANSWER
4. $\{(3, -8)\}$

NOW TRY
EXERCISE 5

Solve the system.

$$\frac{1}{10}x - \frac{3}{5}y = \frac{2}{5}$$
$$-2x + 3y = 1$$

Since $y = 3x - 6$ and $x = \frac{11}{5}$,

$$y = 3\left(\frac{11}{5}\right) - 6 = \frac{33}{5} - \frac{30}{5} = \frac{3}{5}.$$

A check verifies that the solution set is $\left\{\left(\frac{11}{5}, \frac{3}{5}\right)\right\}$.

NOW TRY

NOTE If an equation in a system contains decimal coefficients, it is best to first clear the decimals by multiplying by 10, 100, or 1000, depending on the number of decimal places. Then solve the system. For example, we multiply *each side* of the equation

$$0.5x + 0.75y = 3.25$$

by 10^2, or 100, to get the equivalent equation

$$50x + 75y = 325.$$

OBJECTIVE 4 **Solve linear systems (with two equations and two variables) by elimination.** Another algebraic method, the **elimination method,** involves combining the two equations in a system so that one variable is eliminated. This is done using the following logic.

If $a = b$ and $c = d$,   then $a + c = b + d$.

NOW TRY
EXERCISE 6

Solve the system.

$$8x - 2y = 5$$
$$5x + 2y = -18$$

EXAMPLE 6 Solving a System by Elimination

Solve the system.

$$2x + 3y = -6 \quad (1)$$
$$4x - 3y = 6 \quad (2)$$

Notice that adding the equations together will eliminate the variable y.

$$2x + 3y = -6 \quad (1)$$
$$\underline{4x - 3y = 6} \quad (2)$$
$$6x = 0 \quad \text{Add.}$$
$$x = 0 \quad \text{Solve for } x.$$

To find y, substitute 0 for x in either equation (1) or equation (2).

$$2x + 3y = -6 \quad (1)$$
$$2(0) + 3y = -6 \quad \text{Let } x = 0.$$
$$0 + 3y = -6 \quad \text{Multiply.}$$
$$3y = -6 \quad \text{Add.}$$
$$y = -2 \quad \text{Divide by 3.}$$

The solution is $(0, -2)$. Check by substituting 0 for x and -2 for y in both equations of the original system. The solution set is $\{(0, -2)\}$.

NOW TRY

By adding the equations in **Example 6,** we eliminated the variable y because the coefficients of the y-terms were opposites. In many cases the coefficients will *not* be opposites, and we must transform one or both equations so that the coefficients of one pair of variable terms are opposites.

NOW TRY ANSWERS
5. $\{(-2, -1)\}$
6. $\left\{\left(-1, -\frac{13}{2}\right)\right\}$

Solving a Linear System by Elimination

Step 1 **Write both equations in standard form** $Ax + By = C$.

Step 2 **Make the coefficients of one pair of variable terms opposites.** Multiply one or both equations by appropriate numbers so that the sum of the coefficients of either the x- or y-terms is 0.

Step 3 **Add** the new equations to eliminate a variable. The sum should be an equation with just one variable.

Step 4 **Solve** the equation from Step 3 for the remaining variable.

Step 5 **Find the other value.** Substitute the result of Step 4 into either of the original equations and solve for the other variable.

Step 6 **Check** the ordered-pair solution in *both* of the *original* equations. Then write the solution set.

NOW TRY
EXERCISE 7
Solve the system.

$$2x - 5y = 4$$
$$5x + 2y = 10$$

EXAMPLE 7 Solving a System by Elimination

Solve the system.

$$5x - 2y = 4 \quad (1)$$
$$2x + 3y = 13 \quad (2)$$

Step 1 Both equations are in standard form.

Step 2 Suppose that you wish to eliminate the variable x. One way to do this is to multiply equation (1) by 2 and equation (2) by -5.

The goal is to have *opposite* coefficients.

$$10x - 4y = 8 \quad \text{2 times each side of equation (1)}$$
$$-10x - 15y = -65 \quad \text{-5 times each side of equation (2)}$$

Step 3 Now add.

$$10x - 4y = 8$$
$$\underline{-10x - 15y = -65}$$
$$-19y = -57 \quad \text{Add.}$$

Step 4 Solve for y. $\qquad\qquad y = 3 \qquad$ Divide by -19.

Step 5 To find x, substitute 3 for y in either equation (1) or equation (2).

$$2x + 3y = 13 \quad (2)$$
$$2x + 3(3) = 13 \quad \text{Let } y = 3.$$
$$2x + 9 = 13 \quad \text{Multiply.}$$
$$2x = 4 \quad \text{Subtract 9.}$$
$$x = 2 \quad \text{Divide by 2.}$$

Step 6 To check, substitute 2 for x and 3 for y in both equations (1) and (2).

CHECK	$5x - 2y = 4$	(1)		$2x + 3y = 13$	(2)
	$5(2) - 2(3) \overset{?}{=} 4$			$2(2) + 3(3) \overset{?}{=} 13$	
	$10 - 6 \overset{?}{=} 4$			$4 + 9 \overset{?}{=} 13$	
	$4 = 4$ ✓ True			$13 = 13$ ✓ True	

NOW TRY ANSWER
7. $\{(2, 0)\}$

The solution set is $\{(2, 3)\}$.

NOW TRY

OBJECTIVE 5 Solve special systems. As we saw in **FIGURES 3(b) AND (c)**, some systems of linear equations have no solution or an infinite number of solutions.

NOW TRY
EXERCISE 8
Solve the system.

$$x - 3y = 7$$
$$-3x + 9y = -21$$

EXAMPLE 8   Solving a System of Dependent Equations

Solve the system.

$$2x - y = 3 \quad \text{(1)}$$
$$6x - 3y = 9 \quad \text{(2)}$$

We multiply equation (1) by −3 and then add the result to equation (2).

$$-6x + 3y = -9 \qquad \text{times each side of equation (1)}$$
$$\underline{6x - 3y = 9} \quad \text{(2)}$$
$$0 = 0 \qquad \text{True}$$

Adding these equations gives the true statement $0 = 0$. In the original system, we could get equation (2) from equation (1) by multiplying equation (1) by 3. Because of this, equations (1) and (2) are equivalent and have the same graph, as shown in **FIGURE 4**. The equations are dependent.

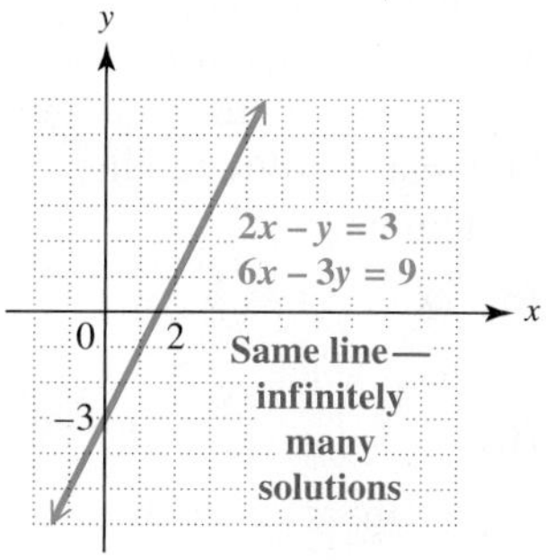

The solution set is the set of all points on the line with equation $2x - y = 3$, written in set-builder notation **(Section 1.1)** as

$$\{(x, y) \mid 2x - y = 3\}$$

and read "the set of all ordered pairs (x, y), such that $2x - y = 3$."   *NOW TRY*

FIGURE 4

NOTE When a system has dependent equations and an infinite number of solutions, as in **Example 8,** either equation of the system could be used to write the solution set. *In this book, we use the equation in standard form with coefficients that are integers having greatest common factor 1 and positive coefficient of x.*

NOW TRY
EXERCISE 9
Solve the system.

$$-2x + 5y = 6$$
$$6x - 15y = 4$$

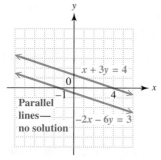

FIGURE 5

NOW TRY ANSWERS
8. $\{(x, y) \mid x - 3y = 7\}$
9. $\emptyset$

EXAMPLE 9   Solving an Inconsistent System

Solve the system.

$$x + 3y = 4 \quad \text{(1)}$$
$$-2x - 6y = 3 \quad \text{(2)}$$

Multiply equation (1) by 2, and then add the result to equation (2).

$$2x + 6y = 8 \qquad \text{Equation (1) multiplied by 2}$$
$$\underline{-2x - 6y = 3} \quad \text{(2)}$$
$$0 = 11 \qquad \text{False}$$

The result of the addition step is a false statement, which indicates that the system is inconsistent. As shown in **FIGURE 5**, the graphs of the equations of the system are parallel lines.

There are no ordered pairs that satisfy both equations, so there is no solution for the system. The solution set is $\emptyset$.   *NOW TRY*

The results of **Examples 8 and 9** are generalized as follows.

> **Special Cases of Linear Systems**
>
> If both variables are eliminated when a system of linear equations is solved, then the solution sets are determined as follows.
>
> **Case 1** There are infinitely many solutions if the resulting statement is *true*.
>
> **Case 2** There is no solution if the resulting statement is *false*.

Slopes and y-intercepts can be used to decide whether the graphs of a system of equations are parallel lines or whether they coincide.

NOW TRY
EXERCISE 10

Write each equation in slope-intercept form and then tell how many solutions the system has.

(a) $2x - 3y = 3$
$\quad 4x - 6y = -6$

(b) $\quad 5y = -x - 4$
$\quad -10y = 2x + 8$

EXAMPLE 10 Using Slope-Intercept Form to Determine the Number of Solutions

Refer to **Examples 8 and 9.** Write each pair of equations in slope-intercept form, and use the results to tell how many solutions the system has.

Solve each equation from **Example 8** for y.

$2x - y = 3$	(1)	$6x - 3y = 9$	(2)
$-y = -2x + 3$ Subtract $2x$.		$2x - y = 3$ Divide by 3.	
$y = 2x - 3$ Multiply by -1.		This leads to the same result as on the left.	

Slope y-intercept $(0, -3)$

The lines have the same slope and same y-intercept, meaning that they coincide. There are infinitely many solutions.

Solve each equation from **Example 9** for y.

$x + 3y = 4$	(1)	$-2x - 6y = 3$	(2)
$3y = -x + 4$ Subtract x.		$-6y = 2x + 3$ Add $2x$.	
$y = -\dfrac{1}{3}x + \dfrac{4}{3}$ Divide by 3.		$y = -\dfrac{1}{3}x - \dfrac{1}{2}$ Divide by -6.	

Slope y-intercept $\left(0, \frac{4}{3}\right)$ Slope y-intercept $\left(0, -\frac{1}{2}\right)$

The lines have the same slope, but different y-intercepts, indicating that they are parallel. Thus, the system has no solution.   NOW TRY

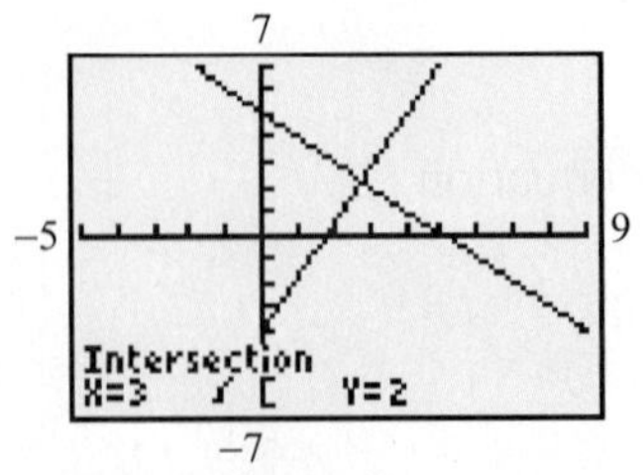

CONNECTIONS

In **Example 2**, we showed how to solve the system

$$x + y = 5 \quad (1)$$
$$2x - y = 4 \quad (2)$$

by graphing the two lines and finding their point of intersection. We can also do this with a graphing calculator, as shown in **FIGURE 6**. The two lines were graphed by solving each equation for y.

$$y = 5 - x \qquad \text{Equation (1) solved for } y$$
$$y = 2x - 4 \qquad \text{Equation (2) solved for } y$$

FIGURE 6

NOW TRY ANSWERS
10. (a) $y = \frac{2}{3}x - 1$; $y = \frac{2}{3}x + 1$; no solution
(b) Both are $y = -\frac{1}{5}x - \frac{4}{5}$; infinitely many solutions

The coordinates of their point of intersection are displayed at the bottom of the screen, indicating that the solution set is $\{(3, 2)\}$. (Compare this graph with the one in **FIGURE 2** on **page 211**.)

For Discussion or Writing

Consider this system.

$$2x - 3y = 3 \quad \text{(1)}$$
$$2x + 2y = 8 \quad \text{(2)}$$

Give the slope-intercept form of each equation. Then graph them both in the same standard viewing window, and give the solution set of the system.

4.1 EXERCISES

MyMathLab | Math XL PRACTICE | WATCH | DOWNLOAD | READ | REVIEW

🌐 *Complete solution available on the Video Resources on DVD*

Concept Check *Fill in the blanks with the correct responses.*

1. If $(4, -3)$ is a solution of a linear system in two variables, then substituting _____ for x and _____ for y leads to true statements in *both* equations.

2. A solution of a system of independent linear equations in two variables is an ordered _____.

3. If solving a system leads to a false statement such as $0 = 3$, the solution set is _____.

4. If solving a system leads to a true statement such as $0 = 0$, the system has _____ equations.

5. If the two lines forming a system have the same slope and different y-intercepts, the system has _____ solution(s).
   (how many?)

6. If the two lines forming a system have different slopes, the system has _____ solution(s).
   (how many?)

7. *Concept Check* Which ordered pair could be a solution of the graphed system of equations? Why?

   A. $(4, 4)$
   B. $(-4, 4)$
   C. $(-4, -4)$
   D. $(4, -4)$

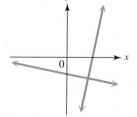

8. *Concept Check* Which ordered pair could be a solution of the graphed system of equations? Why?

   A. $(4, 0)$
   B. $(-4, 0)$
   C. $(0, 4)$
   D. $(0, -4)$

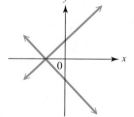

9. *Concept Check* Match each system with the correct graph.

   (a) $x + y = 6$
   $x - y = 0$

   (b) $x + y = -6$
   $x - y = 0$

   (c) $x + y = 0$
   $x - y = -6$

   (d) $x + y = 0$
   $x - y = 6$

   A.
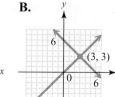

   B.
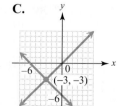

   C.
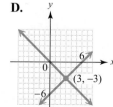

   D.

10. *Concept Check* Make up a system that has $(0, 0)$ as its only solution.

Decide whether the given ordered pair is a solution of the given system. ***See Example 1.***

11. $x - y = 17$
$x + y = -1$; $(8, -9)$

12. $x + y = 6$
$x - y = 4$; $(5, 1)$

13. $3x - 5y = -12$
$x - y = 1$; $(-1, 2)$

14. $2x - y = 8$
$3x + 2y = 20$; $(5, 2)$

Solve each system by graphing. ***See Example 2.***

15. $x + y = -5$
$-2x + y = 1$

16. $x - 4y = -4$
$3x + y = 1$

17. $x + y = 4$
$2x - y = 2$

18. $2x + y = 4$
$3x - y = 6$

Solve each system by substitution. If the system is inconsistent or has dependent equations, say so. ***See Examples 3–5, 8, and 9.***

19. $4x + y = 6$
$y = 2x$

20. $2x - y = 6$
$y = 5x$

21. $-x - 4y = -14$
$y = 2x - 1$

22. $-3x - 5y = -17$
$y = 4x + 8$

23. $3x - 4y = -22$
$-3x + y = 0$

24. $-3x + y = -5$
$x + 2y = 0$

25. $5x - 4y = 9$
$3 - 2y = -x$

26. $6x - y = -9$
$4 + 7x = -y$

27. $4x - 5y = -11$
$x + 2y = 7$

28. $3x - y = 10$
$2x + 5y = 1$

29. $x = 3y + 5$
$x = \dfrac{3}{2}y$

30. $x = 6y - 2$
$x = \dfrac{3}{4}y$

31. $\dfrac{1}{2}x + \dfrac{1}{3}y = 3$
$-3x + y = 0$

32. $\dfrac{1}{4}x - \dfrac{1}{5}y = 9$
$5x - y = 0$

33. $y = 2x$
$4x - 2y = 0$

34. $x = 3y$
$3x - 9y = 0$

35. $x = 5y$
$5x - 25y = 5$

36. $y = -4x$
$8x + 2y = 4$

37. $y = 0.5x$
$1.5x - 0.5y = 5.0$

38. $y = 1.4x$
$0.5x + 1.5y = 26.0$

39. *Concept Check* If a system of the form

$$Ax + By = 0$$
$$Cx + Dy = 0$$

has a single solution, what must that solution be?

40. Assuming that you want to minimize the amount of work required, tell whether you would use the substitution or elimination method to solve each system. Explain your answers. *Do not actually solve.*

(a) $3x + y = -7$
$x - y = -5$

(b) $6x - y = 5$
$y = 11x$

(c) $3x - 2y = 0$
$9x + 8y = 7$

Solve each system by elimination. If the system is inconsistent or has dependent equations, say so. **See Examples 6–9.**

41. $-2x + 3y = -16$
$2x - 5y = 24$

42. $6x + 5y = -7$
$-6x - 11y = 1$

43. $2x - 5y = 11$
$3x + y = 8$

44. $-2x + 3y = 1$
$-4x + y = -3$

45. $3x + 4y = -6$
$5x + 3y = 1$

46. $4x + 3y = 1$
$3x + 2y = 2$

47. $7x + 2y = 6$
$-14x - 4y = -12$

48. $x - 4y = 2$
$4x - 16y = 8$

49. $3x + 3y = 0$
$4x + 2y = 3$

50. $8x + 4y = 0$
$4x - 2y = 2$

51. $5x - 5y = 3$
$x - y = 12$

52. $2x - 3y = 7$
$-4x + 6y = 14$

53. $x + y = 0$
$2x - 2y = 0$

54. $3x + 3y = 0$
$-2x - y = 0$

55. $x - \dfrac{1}{2}y = 2$
$-x + \dfrac{2}{5}y = -\dfrac{8}{5}$

56. $\dfrac{3}{2}x + y = 3$
$\dfrac{2}{3}x + \dfrac{1}{3}y = 1$

57. $\dfrac{1}{2}x + \dfrac{1}{3}y = \dfrac{49}{18}$
$\dfrac{1}{2}x + 2y = \dfrac{4}{3}$

58. $\dfrac{1}{5}x + \dfrac{1}{7}y = \dfrac{12}{5}$
$\dfrac{1}{10}x + \dfrac{1}{3}y = \dfrac{5}{6}$

Write each equation in slope-intercept form and then tell how many solutions the system has. Do not actually solve. **See Example 10.**

59. $3x + 7y = 4$
$6x + 14y = 3$

60. $-x + 2y = 8$
$4x - 8y = 1$

61. $2x = -3y + 1$
$6x = -9y + 3$

62. $5x = -2y + 1$
$10x = -4y + 2$

Solve each system by the method of your choice. **See Examples 3–9.** *(For Exercises 63–65, see your answers to* **Exercise 40.***)*

63. $3x + y = -7$
$x - y = -5$

64. $6x - y = 5$
$y = 11x$

65. $3x - 2y = 0$
$9x + 8y = 7$

66. $3x - 5y = 7$
$2x + 3y = 30$

67. $2x + 3y = 10$
$-3x + y = 18$

68. $\dfrac{1}{6}x + \dfrac{1}{3}y = 8$
$\dfrac{1}{4}x + \dfrac{1}{2}y = 12$

69. $\dfrac{1}{2}x - \dfrac{1}{8}y = -\dfrac{1}{4}$
$4x - y = -2$

70. $0.2x + 0.5y = 6$
$0.4x + y = 9$

71. $0.3x + 0.2y = 0.4$
$0.5x + 0.4y = 0.7$

72. Explain why the system $\begin{array}{l}3x + 7y = 19\\ 8x - 5y = 4\end{array}$ would be more difficult to solve by substitution than $\begin{array}{l}x + 7y = 8\\ 3x - 2y = 4\end{array}$.

TECHNOLOGY INSIGHTS EXERCISES 73–76

73. The table shown was generated by a graphing calculator. The functions defined by Y_1 and Y_2 are linear. Based on the table, what are the coordinates of the point of intersection of the graphs?

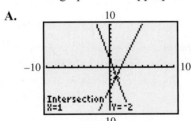

74. The functions defined by Y_1 and Y_2 in the table are linear.

(a) Use the methods of **Chapter 3** to find an equation for Y_1 and an equation for Y_2.

(b) Solve the system of equations formed by Y_1 and Y_2.

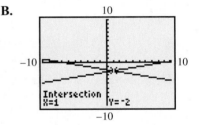

75. The solution set of the system

$$y_1 = 3x - 5$$
$$y_2 = -4x + 2$$

is $\{(1, -2)\}$. Using slopes and y-intercepts, determine which one of the two calculator graphs is the appropriate one for this system.

A.

B.

76. Which one of the ordered pairs listed is the only possible solution of the system whose graphs are shown in the standard viewing window of a graphing calculator?

A. $(15, -15)$ **B.** $(15, 15)$

C. $(-15, 15)$ **D.** $(-15, -15)$

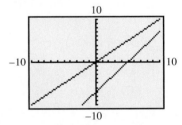

---

*For each system, **(a)** solve by elimination or substitution and **(b)** use a graphing calculator to support your result. In part (b), be sure to solve each equation for y first.*

77. $x + y = 10$
$2x - y = 5$

78. $6x + y = 5$
$-x + y = -9$

Use each graph provided to work Exercises 79–82.

79. The figure shows graphs that represent supply and demand for a certain brand of low-fat frozen yogurt at various prices per half-gallon (in dollars).

(a) At what price does supply equal demand?

(b) For how many half-gallons does supply equal demand?

(c) What are the supply and demand at a price of $2 per half-gallon?

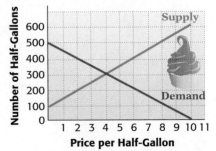

The Fortunes of Frozen Yogurt

80. Sharon Kobrin compared the monthly payments she would incur for two types of mortgages: fixed rate and variable rate. Her observations led to the following graphs.

(a) For which years would the monthly payment be more for the fixed-rate mortgage than for the variable-rate mortgage?

(b) In what year would the payments be the same, and what would those payments be?

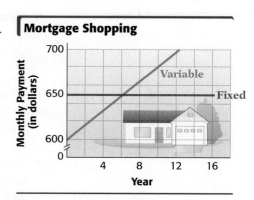

81. The graph shows sales (in millions of dollars) in the United States of three types of television displays from 2003 through 2008.

(a) During what years did sales of plasma flat panel displays exceed those of front projection displays?

(b) Between what two consecutive years did LCD flat panel first become the most popular type of display?

(c) When were sales of LCD flat panel and plasma flat panel displays approximately equal? What were sales at that time?

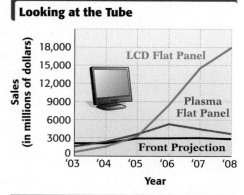

Source: Consumer Electronics Association.

(d) Write your answer for part (c) as an ordered pair, rounding down to the nearest year.

(e) Describe the trends in sales of the three types of displays during the years shown.

82. The graph shows the numbers of non-Hispanic whites and African-Americans (in thousands) living with AIDS in the United States during the period 1995 through 2007.

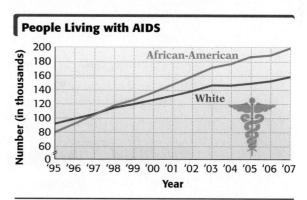

Source: U.S. Centers for Disease Control and Prevention.

(a) In what year were the numbers approximately the same?

(b) How many people of each race were living with AIDS in that year?

(c) Express the point of intersection of the graphs for whites and African-Americans living with AIDS as an ordered pair of the form (year, number).

(d) During what two consecutive years did the number of whites living with AIDS first stabilize and remain fairly constant?

(e) Describe the trend in the number of African-Americans living with AIDS from 1995 through 2007. If a straight line were used to approximate its graph, would the line have a positive, negative, or 0 slope?

Use the graph shown in **FIGURE 1** *at the beginning of this section (and repeated here) to work Exercises 83–86.*

83. For which years during the period 2000–2008 were sales of digital cameras less than sales of digital camcorders?

84. Estimate the year in which sales for the two products were the same. About what was this sales figure?

85. If $x = 0$ represents 2000 and $x = 8$ represents 2008, sales (y) in millions of units can be modeled by the linear equations in the following system.

$$624x - y = -1823 \quad \text{Digital cameras}$$
$$119x + y = 2838 \quad \text{Digital camcorders}$$

Solve this system. Express values to the nearest tenth. Write the solution as an ordered pair of the form (year, sales).

86. Interpret your answer for **Exercise 85.** How does that answer compare with your estimate from **Exercise 84?**

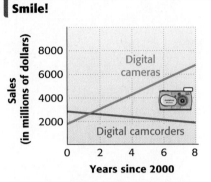

Smile!

Source: Consumer Electronics Association.

Systems such as those in Exercises 87–90 can be solved by elimination. One way to do this is to let $p = \frac{1}{x}$ and $q = \frac{1}{y}$. Substitute, solve for p and q, and then find x and y. $\left(\text{For example, in Exercise 87, } \frac{3}{x} = 3 \cdot \frac{1}{x} = 3p.\right)$ Use this method to solve each system.

87. $\dfrac{3}{x} + \dfrac{4}{y} = \dfrac{5}{2}$

$\dfrac{5}{x} - \dfrac{3}{y} = \dfrac{7}{4}$

88. $\dfrac{2}{x} - \dfrac{5}{y} = \dfrac{3}{2}$

$\dfrac{4}{x} + \dfrac{1}{y} = \dfrac{4}{5}$

89. $\dfrac{2}{x} + \dfrac{3}{y} = \dfrac{11}{2}$

$-\dfrac{1}{x} + \dfrac{2}{y} = -1$

90. $\dfrac{4}{x} - \dfrac{9}{y} = -1$

$-\dfrac{7}{x} + \dfrac{6}{y} = -\dfrac{3}{2}$

Brain Busters *Solve by any method. Assume that a and b represent nonzero constants.*

91. $ax + by = c$
$ax - 2by = c$

92. $ax + by = 2$
$-ax + 2by = 1$

93. $2ax - y = 3$
$y = 5ax$

94. $3ax + 2y = 1$
$-ax + y = 2$

PREVIEW EXERCISES

Multiply both sides of each equation by the given number. ***See Section 2.1.***

95. $2x - 3y + z = 5$ by 4

96. $-3x + 8y - z = 0$ by -3

Solve for z if $x = 1$ and $y = -2$. ***See Section 2.1.***

97. $x + 2y + 3z = 9$

98. $-3x - y + z = 1$

By what number must the first equation be multiplied so that x is eliminated when the two equations are added? ***See Section 4.1.***

99. $x + 2y - z = 0$
$3x - 4y + 2z = 6$

100. $x - 2y + 5z = -7$
$-2x - 3y + 4z = -14$

Analyzing Your Test Results

An exam is a learning opportunity—learn from your mistakes. After a test is returned, do the following:

▶ **Note what you got wrong and why you had points deducted.**

▶ **Figure out how to solve the problems you missed.** Check your textbook or notes, or ask your instructor. Rework the problems correctly.

▶ **Keep all quizzes and tests that are returned to you.** Use them to study for future tests and the final exam.

Typical Reasons for Errors on Math Tests

These are test taking errors. They are easy to correct if you read carefully, show all your work, proofread, and double-check units and labels.

1. You read the directions wrong.
2. You read the question wrong or skipped over something.
3. You made a computation error.
4. You made a careless error. (For example, you incorrectly copied a correct answer onto a separate answer sheet.)
5. Your answer is not complete.
6. You labeled your answer wrong. (For example, you labeled an answer "ft" instead of "ft^2.")
7. You didn't show your work.

These are test preparation errors. You must practice the kinds of problems that you will see on tests.

8. You didn't understand a concept.
9. You were unable to set up the problem (in an application).
10. You were unable to apply a procedure.

Below are sample charts for tracking your test taking progress. Use them to find out if you tend to make certain kinds of errors on tests. Check the appropriate box when you've made an error in a particular category.

Test Taking Errors

Test	Read directions wrong	Read question wrong	Computation error	Not exact or accurate	Not complete	Labeled wrong	Didn't show work
1							
2							
3							

Test Preparation Errors

Test	Didn't understand concept	Didn't set up problem correctly	Couldn't apply concept to new situation
1			
2			
3			

What will you do to avoid these kinds of errors on your next test?

4.2 Systems of Linear Equations in Three Variables

A solution of an equation in three variables, such as

$$2x + 3y - z = 4, \qquad \text{Linear equation in three variables}$$

is called an **ordered triple** and is written (x, y, z). For example, the ordered triple $(0, 1, -1)$ is a solution of the preceding equation, because

$$2(0) + 3(1) - (-1) = 4$$

is a true statement. Verify that another solution of this equation is $(10, -3, 7)$.

We now extend the term *linear equation* to equations of the form

$$Ax + By + Cz + \cdots + Dw = K,$$

where not all the coefficients $A, B, C, \ldots, D$ equal 0. For example,

$$2x + 3y - 5z = 7 \quad \text{and} \quad x - 2y - z + 3w = 8$$

are linear equations, the first with three variables and the second with four.

OBJECTIVE 1 Understand the geometry of systems of three equations in three variables. Consider the solution of a system such as the following.

$$
\begin{aligned}
4x + 8y + z &= 2 \\
x + 7y - 3z &= -14 \qquad \text{System of linear equations} \\
2x - 3y + 2z &= 3 \qquad \text{in three variables}
\end{aligned}
$$

Theoretically, a system of this type can be solved by graphing. However, the graph of a linear equation with three variables is a *plane,* not a line. Since visualizing a plane requires three-dimensional graphing, the method of graphing is not practical with these systems. However, it does illustrate the number of solutions possible for such systems, as shown in **FIGURE 7.**

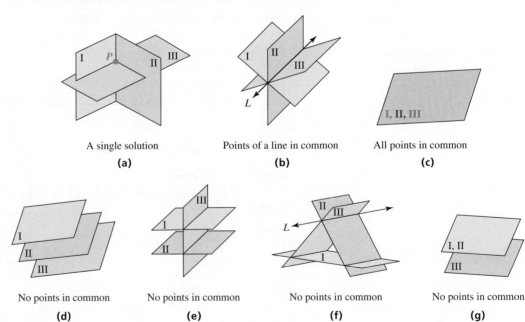

A single solution **(a)**   Points of a line in common **(b)**   All points in common **(c)**

No points in common **(d)**   No points in common **(e)**   No points in common **(f)**   No points in common **(g)**

FIGURE 7

FIGURE 7 illustrates the following cases.

Graphs of Linear Systems in Three Variables

Case 1   **The three planes may meet at a single, common point** that is the solution of the system. See **FIGURE 7(a)**.

Case 2   **The three planes may have the points of a line in common,** so that the infinite set of points that satisfy the equation of the line is the solution of the system. See **FIGURE 7(b)**.

Case 3   **The three planes may coincide,** so that the solution of the system is the set of all points on a plane. See **FIGURE 7(c)**.

Case 4   **The planes may have no points common to all three,** so that there is no solution of the system. See **FIGURES 7(d)–(g)**.

OBJECTIVE 2 **Solve linear systems (with three equations and three variables) by elimination.** Since graphing to find the solution set of a system of three equations in three variables is impractical, these systems are solved with an extension of the elimination method from **Section 4.1.**

In the steps that follow, we use the term **focus variable** to identify the first variable to be eliminated in the process. The focus variable will always be present in the **working equation,** which will be used twice to eliminate this variable.

Solving a Linear System in Three Variables*

Step 1   **Select a variable and an equation.** A good choice for the variable, which we call the *focus variable,* is one that has coefficient 1 or -1. Then select an equation, one that contains the focus variable, as the *working equation*.

Step 2   **Eliminate the focus variable.** Use the working equation and one of the other two equations of the original system. The result is an equation in two variables.

Step 3   **Eliminate the focus variable again.** Use the working equation and the remaining equation of the original system. The result is another equation in two variables.

Step 4   **Write the equations in two variables that result from Steps 2 and 3 as a system, and solve it.** Doing this gives the values of two of the variables.

Step 5   **Find the value of the remaining variable.** Substitute the values of the two variables found in Step 4 into the working equation to obtain the value of the focus variable.

Step 6   **Check** the ordered-triple solution in *each* of the *original* equations of the system. Then write the solution set.

---

*The authors wish to thank Christine Heinecke Lehmann of Purdue University North Central for her suggestions here.

EXAMPLE 1 Solving a System in Three Variables

Solve the system.

$$4x + 8y + z = 2 \qquad (1)$$
$$x + 7y - 3z = -14 \qquad (2)$$
$$2x - 3y + 2z = 3 \qquad (3)$$

Step 1 Since z in equation (1) has coefficient 1, we choose z as the focus variable and (1) as the working equation. (Another option would be to choose x as the focus variable, since it also has coefficient 1, and use (2) as the working equation.)

Focus variable
$$\underset{\downarrow}{}$$
$$4x + 8y + z = 2 \qquad (1) \leftarrow \text{Working equation}$$

Step 2 Multiply working equation (1) by 3 and add the result to equation (2) to eliminate focus variable z.

$$\begin{array}{ll} 12x + 24y + 3z = 6 & \text{Multiply each side of (1) by 3.} \\ \underline{x + 7y - 3z = -14} & (2) \\ 13x + 31y = -8 & \text{Add. (4)} \end{array}$$

Step 3 Multiply working equation (1) by -2 and add the result to remaining equation (3) to again eliminate focus variable z.

$$\begin{array}{ll} -8x - 16y - 2z = -4 & \text{Multiply each side of (1) by } -2. \\ \underline{2x - 3y + 2z = 3} & (3) \\ -6x - 19y = -1 & \text{Add. (5)} \end{array}$$

Step 4 Write the equations in two variables that result in Steps 2 and 3 as a system.

> Make sure these equations have the same variables.

$$13x + 31y = -8 \qquad (4) \quad \text{The result from Step 2}$$
$$-6x - 19y = -1 \qquad (5) \quad \text{The result from Step 3}$$

Now solve this system. We choose to eliminate x.

$$\begin{array}{ll} 78x + 186y = -48 & \text{Multiply each side of (4) by 6.} \\ \underline{-78x - 247y = -13} & \text{Multiply each side of (5) by 13.} \\ -61y = -61 & \text{Add.} \\ y = 1 & \text{Divide by } -61. \end{array}$$

Substitute 1 for y in either equation (4) or (5) to find x.

$$\begin{array}{ll} -6x - 19y = -1 & (5) \\ -6x - 19(1) = -1 & \text{Let } y = 1. \\ -6x - 19 = -1 & \text{Multiply.} \\ -6x = 18 & \text{Add 19.} \\ x = -3 & \text{Divide by } -6. \end{array}$$

Step 5 Now substitute the two values we found in Step 4 in working equation (1) to find the value of the remaining variable, focus variable z.

$$\begin{array}{ll} 4x + 8y + z = 2 & (1) \\ 4(-3) + 8(1) + z = 2 & \text{Let } x = -3 \text{ and } y = 1. \\ -4 + z = 2 & \text{Multiply, and then add.} \\ z = 6 & \text{Add 4.} \end{array}$$

NOW TRY
EXERCISE 1

Solve the system.

$$x - y + 2z = 1$$
$$3x + 2y + 7z = 8$$
$$-3x - 4y + 9z = -10$$

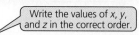

Write the values of x, y, and z in the correct order.

Step 6 It appears that the ordered triple $(-3, 1, 6)$ is the only solution of the system. We must check that the solution satisfies all three original equations of the system. We begin with equation (1).

CHECK

$$4x + 8y + z = 2 \qquad (1)$$
$$4(-3) + 8(1) + 6 \overset{?}{=} 2 \qquad \text{Substitute.}$$
$$-12 + 8 + 6 \overset{?}{=} 2 \qquad \text{Multiply.}$$
$$2 = 2 \;\checkmark\; \text{True}$$

In **Exercise 2** you are asked to show that $(-3, 1, 6)$ also satisfies equations (2) and (3). The solution set is $\{(-3, 1, 6)\}$.

NOW TRY

OBJECTIVE 3 **Solve linear systems (with three equations and three variables) in which some of the equations have missing terms.** If a linear system has an equation missing a term or terms, one elimination step can be omitted.

EXAMPLE 2 Solving a System of Equations with Missing Terms

Solve the system.

$$6x - 12y = -5 \qquad (1) \quad \text{Missing } z$$
$$8y + z = 0 \qquad (2) \quad \text{Missing } x$$
$$9x - z = 12 \qquad (3) \quad \text{Missing } y$$

Since equation (3) is missing the variable y, one way to begin is to eliminate y again, using equations (1) and (2).

$$12x - 24y \qquad\quad = -10 \qquad \text{Multiply each side of (1) by 2.}$$

Leave space for the missing term.

$$\underline{\qquad\quad 24y + 3z = \quad 0} \qquad \text{Multiply each side of (2) by 3.}$$
$$12x \qquad\quad + 3z = -10 \qquad \text{Add.} \quad (4)$$

Use the resulting equation (4) in x and z, together with equation (3), $9x - z = 12$, to eliminate z. Multiply equation (3) by 3.

$$27x - 3z = \quad 36 \qquad \text{Multiply each side of (3) by 3.}$$
$$\underline{12x + 3z = -10} \qquad (4)$$
$$39x \qquad = \quad 26 \qquad \text{Add.}$$

$$x = \frac{26}{39}, \quad \text{or} \quad \frac{2}{3} \qquad \text{Divide by 39; lowest terms}$$

We can find z by substituting this value for x in equation (3).

$$9x - z = 12 \qquad (3)$$
$$9\left(\frac{2}{3}\right) - z = 12 \qquad \text{Let } x = \tfrac{2}{3}.$$
$$6 - z = 12 \qquad \text{Multiply.}$$
$$z = -6 \qquad \text{Subtract 6. Multiply by } -1.$$

NOW TRY ANSWER
1. $\{(2, 1, 0)\}$

NOW TRY
EXERCISE 2
Solve the system.
$$3x - z = -10$$
$$4y + 5z = 24$$
$$x - 6y = -8$$

We can find y by substituting -6 for z in equation (2).

$$8y + z = 0 \qquad \text{(2)}$$
$$8y - 6 = 0 \qquad \text{Let } z = -6.$$
$$8y = 6 \qquad \text{Add 6.}$$
$$y = \frac{6}{8}, \quad \text{or} \quad \frac{3}{4} \qquad \text{Divide by 8; lowest terms}$$

Check to verify that the solution set is $\left\{\left(\frac{2}{3}, \frac{3}{4}, -6\right)\right\}$.

NOW TRY

OBJECTIVE 4 Solve special systems.

NOW TRY
EXERCISE 3
Solve the system.
$$x - 5y + 2z = 4$$
$$3x + y - z = 6$$
$$-2x + 10y - 4z = 7$$

EXAMPLE 3 Solving an Inconsistent System with Three Variables

Solve the system.

$$2x - 4y + 6z = 5 \qquad \text{(1)}$$
$$-x + 3y - 2z = -1 \qquad \text{(2)}$$
$$x - 2y + 3z = 1 \qquad \text{(3)}$$

> Use as the working equation, with focus variable x.

Eliminate the focus variable, x, using equations (1) and (3).

$$-2x + 4y - 6z = -2 \qquad \text{Multiply each side of (3) by } -2.$$
$$\underline{2x - 4y + 6z = 5} \qquad \text{(1)}$$
$$0 = 3 \qquad \text{Add; false}$$

The resulting false statement indicates that equations (1) and (3) have no common solution. Thus, the system is inconsistent and the solution set is $\emptyset$. The graph of this system would show the two planes parallel to one another.

NOW TRY

NOTE If a false statement results when adding as in **Example 3**, it is not necessary to go any further with the solution. Since two of the three planes are parallel, it is not possible for the three planes to have any points in common.

NOW TRY
EXERCISE 4
Solve the system.
$$x - 3y + 2z = 10$$
$$-2x + 6y - 4z = -20$$
$$\frac{1}{2}x - \frac{3}{2}y + z = 5$$

EXAMPLE 4 Solving a System of Dependent Equations with Three Variables

Solve the system.

$$2x - 3y + 4z = 8 \qquad \text{(1)}$$
$$-x + \frac{3}{2}y - 2z = -4 \qquad \text{(2)}$$
$$6x - 9y + 12z = 24 \qquad \text{(3)}$$

Multiplying each side of equation (1) by 3 gives equation (3). Multiplying each side of equation (2) by -6 also gives equation (3). Because of this, the equations are dependent. All three equations have the same graph, as illustrated in **FIGURE 7(c)**. The solution set is written as follows.

$$\{(x, y, z) \mid 2x - 3y + 4z = 8\} \qquad \text{Set-builder notation}$$

NOW TRY ANSWERS
2. $\{(-2, 1, 4)\}$
3. $\emptyset$
4. $\{(x, y, z) \mid x - 3y + 2z = 10\}$

Although any one of the three equations could be used to write the solution set, we use the equation in standard form with coefficients that are integers with greatest common factor 1, as we did in **Section 4.1**.

NOW TRY

NOW TRY
EXERCISE 5

Solve the system.

$$x - 3y + 2z = 4$$
$$\frac{1}{3}x - y + \frac{2}{3}z = 7$$
$$\frac{1}{2}x - \frac{3}{2}y + z = 2$$

EXAMPLE 5   Solving Another Special System

Solve the system.

$$2x - y + 3z = 6 \quad (1)$$
$$x - \frac{1}{2}y + \frac{3}{2}z = 3 \quad (2)$$
$$4x - 2y + 6z = 1 \quad (3)$$

Multiplying each side of equation (2) by 2 gives equation (1), so these two equations are dependent. Equations (1) and (3) are not equivalent, however. Multiplying equation (3) by $\frac{1}{2}$ does *not* give equation (1). Instead, we obtain two equations with the same coefficients, but with different constant terms.

The graphs of equations (1) and (3) have no points in common (that is, the planes are parallel). Thus, the system is inconsistent and the solution set is $\emptyset$, as illustrated in **FIGURE 7(g)**.

NOW TRY

NOW TRY ANSWER
5. $\emptyset$

4.2 EXERCISES

Complete solution available on the Video Resources on DVD

1. *Concept Check*   The two equations $\begin{matrix} x + y + z = 6 \\ 2x - y + z = 3 \end{matrix}$ have a common solution of $(1, 2, 3)$. Which equation would complete a system of three linear equations in three variables having solution set $\{(1, 2, 3)\}$?

A. $3x + 2y - z = 1$ **B.** $3x + 2y - z = 4$

C. $3x + 2y - z = 5$ **D.** $3x + 2y - z = 6$

2. Complete the work of **Example 1** and show that the ordered triple $(-3, 1, 6)$ is also a solution of equations (2) and (3).

$$x + 7y - 3z = -14 \qquad \text{Equation (2)}$$
$$2x - 3y + 2z = 3 \qquad \text{Equation (3)}$$

Solve each system of equations. ***See Example 1.***

3. $2x - 5y + 3z = -1$
$x + 4y - 2z = 9$
$x - 2y - 4z = -5$

4. $x + 3y - 6z = 7$
$2x - y + z = 1$
$x + 2y + 2z = -1$

5. $3x + 2y + z = 8$
$2x - 3y + 2z = -16$
$x + 4y - z = 20$

6. $-3x + y - z = -10$
$-4x + 2y + 3z = -1$
$2x + 3y - 2z = -5$

7. $2x + 5y + 2z = 0$
$4x - 7y - 3z = 1$
$3x - 8y - 2z = -6$

8. $5x - 2y + 3z = -9$
$4x + 3y + 5z = 4$
$2x + 4y - 2z = 14$

9. $x + 2y + z = 4$
$2x + y - z = -1$
$x - y - z = -2$

10. $x - 2y + 5z = -7$
$-2x - 3y + 4z = -14$
$-3x + 5y - z = -7$

11. $-x + 2y + 6z = 2$
$3x + 2y + 6z = 6$
$x + 4y - 3z = 1$

12. $2x + y + 2z = 1$
$x + 2y + z = 2$
$x - y - z = 0$

13. $x + y - z = -2$
$2x - y + z = -5$
$-x + 2y - 3z = -4$

14. $x + 2y + 3z = 1$
$-x - y + 3z = 2$
$-6x + y + z = -2$

15. $\dfrac{1}{3}x + \dfrac{1}{6}y - \dfrac{2}{3}z = -1$

$-\dfrac{3}{4}x - \dfrac{1}{3}y - \dfrac{1}{4}z = 3$

$\dfrac{1}{2}x + \dfrac{3}{2}y + \dfrac{3}{4}z = 21$

16. $\dfrac{2}{3}x - \dfrac{1}{4}y + \dfrac{5}{8}z = 0$

$\dfrac{1}{5}x + \dfrac{2}{3}y - \dfrac{1}{4}z = -7$

$-\dfrac{3}{5}x + \dfrac{4}{3}y - \dfrac{7}{8}z = -5$

17. $5.5x - 2.5y + 1.6z = 11.83$
$2.2x + 5.0y - 0.1z = -5.97$
$3.3x - 7.5y + 3.2z = 21.25$

18. $6.2x - 1.4y + 2.4z = -1.80$
$3.1x + 2.8y - 0.2z = 5.68$
$9.3x - 8.4y - 4.8z = -34.20$

Solve each system of equations. **See Example 2.**

19. $2x - 3y + 2z = -1$
$x + 2y + z = 17$
$2y - z = 7$

20. $2x - y + 3z = 6$
$x + 2y - z = 8$
$2y + z = 1$

21. $4x + 2y - 3z = 6$
$x - 4y + z = -4$
$-x + 2z = 2$

22. $2x + 3y - 4z = 4$
$x - 6y + z = -16$
$-x + 3z = 8$

23. $2x + y = 6$
$3y - 2z = -4$
$3x - 5z = -7$

24. $4x - 8y = -7$
$4y + z = 7$
$-8x + z = -4$

25. $-5x + 2y + z = 5$
$-3x - 2y - z = 3$
$-x + 6y = 1$

26. $-4x + 3y - z = 4$
$-5x - 3y + z = -4$
$-2x - 3z = 12$

27. $7x - 3z = -34$
$2y + 4z = 20$
$\dfrac{3}{4}x + \dfrac{1}{6}y = -2$

28. $5x - 2z = 8$
$4y + 3z = -9$
$\dfrac{1}{2}x + \dfrac{2}{3}y = -1$

29. $4x - z = -6$
$\dfrac{3}{5}y + \dfrac{1}{2}z = 0$
$\dfrac{1}{3}x + \dfrac{2}{3}z = -5$

30. $5x - z = 38$
$\dfrac{2}{3}y + \dfrac{1}{4}z = -17$
$\dfrac{1}{5}y + \dfrac{5}{6}z = 4$

Solve each system of equations. If the system is inconsistent or has dependent equations, say so. **See Examples 1, 3, 4, and 5.**

31. $2x + 2y - 6z = 5$
$-3x + y - z = -2$
$-x - y + 3z = 4$

32. $-2x + 5y + z = -3$
$5x + 14y - z = -11$
$7x + 9y - 2z = -5$

33. $-5x + 5y - 20z = -40$
$x - y + 4z = 8$
$3x - 3y + 12z = 24$

34. $x + 4y - z = 3$
$-2x - 8y + 2z = -6$
$3x + 12y - 3z = 9$

35. $x + 5y - 2z = -1$
$-2x + 8y + z = -4$
$3x - y + 5z = 19$

36. $x + 3y + z = 2$
$4x + y + 2z = -4$
$5x + 2y + 3z = -2$

37. $2x + y - z = 6$
$4x + 2y - 2z = 12$
$-x - \dfrac{1}{2}y + \dfrac{1}{2}z = -3$

38. $2x - 8y + 2z = -10$
$-x + 4y - z = 5$
$\dfrac{1}{8}x - \dfrac{1}{2}y + \dfrac{1}{8}z = -\dfrac{5}{8}$

39. $x + y - 2z = 0$
$3x - y + z = 0$
$4x + 2y - z = 0$

40. $2x + 3y - z = 0$
$x - 4y + 2z = 0$
$3x - 5y - z = 0$

41. $x - 2y + \dfrac{1}{3}z = 4$

$3x - 6y + z = 12$

$-6x + 12y - 2z = -3$

42. $4x + y - 2z = 3$

$x + \dfrac{1}{4}y - \dfrac{1}{2}z = \dfrac{3}{4}$

$2x + \dfrac{1}{2}y - z = 1$

Brain Busters   *Extend the method of this section to solve each system. Express the solution in the form (x, y, z, w).*

43.
$$\begin{aligned} x + y + z - w &= 5 \\ 2x + y - z + w &= 3 \\ x - 2y + 3z + w &= 18 \\ -x - y + z + 2w &= 8 \end{aligned}$$

44.
$$\begin{aligned} 3x + y - z + 2w &= 9 \\ x + y + 2z - w &= 10 \\ x - y - z + 3w &= -2 \\ -x + y - z + w &= -6 \end{aligned}$$

45.
$$\begin{aligned} 3x + y - z + w &= -3 \\ 2x + 4y + z - w &= -7 \\ -2x + 3y - 5z + w &= 3 \\ 5x + 4y - 5z + 2w &= -7 \end{aligned}$$

46.
$$\begin{aligned} x - 3y + 7z + w &= 11 \\ 2x + 4y + 6z - 3w &= -3 \\ 3x + 2y + z + 2w &= 19 \\ 4x + y - 3z + w &= 22 \end{aligned}$$

PREVIEW EXERCISES

*Solve each problem. **See Sections 2.3 and 2.4.***

47. The perimeter of a triangle is 323 in. The shortest side measures five-sixths the length of the longest side, and the medium side measures 17 in. less than the longest side. Find the lengths of the sides of the triangle.

48. The sum of the three angles of a triangle is 180°. The largest angle is twice the measure of the smallest, and the third angle measures 10° less than the largest. Find the measures of the three angles.

49. The sum of three numbers is 16. The greatest number is -3 times the least, while the middle number is four less than the greatest. Find the three numbers.

50. Witny Librun has a collection of pennies, dimes, and quarters. The number of dimes is one less than twice the number of pennies. If there are 27 coins in all worth a total of $4.20, how many of each denomination of coin is in the collection?

4.3 Applications of Systems of Linear Equations

OBJECTIVES

1 Solve geometry problems by using two variables.

2 Solve money problems by using two variables.

3 Solve mixture problems by using two variables.

4 Solve distance-rate-time problems by using two variables.

5 Solve problems with three variables by using a system of three equations.

Although some problems with two unknowns can be solved by using just one variable, it is often easier to use two variables and a system of equations. The following problem, which can be solved with a system, appeared in a Hindu work that dates back to about A.D. 850. (See **Exercise 35.**)

The mixed price of 9 citrons (a lemonlike fruit) and 7 fragrant wood apples is 107; again, the mixed price of 7 citrons and 9 fragrant wood apples is 101. O you arithmetician, tell me quickly the price of a citron and the price of a wood apple here, having distinctly separated those prices well.

PROBLEM-SOLVING HINT

When solving an applied problem using two variables, it is a good idea to pick letters that correspond to the descriptions of the unknown quantities. In the example above, we could choose c to represent the number of citrons, and w to represent the number of wood apples.

The following steps are based on the problem-solving method of **Section 2.3.**

Solving an Applied Problem by Writing a System of Equations

Step 1 **Read** the problem, several times if necessary. What information is given? What is to be found? This is often stated in the last sentence.

Step 2 **Assign variables** to represent the unknown values. Use a sketch, diagram, or table, as needed.

Step 3 **Write a system of equations** using the variable expressions.

Step 4 **Solve** the system of equations.

Step 5 **State the answer** to the problem. Label it appropriately. Does it seem reasonable?

Step 6 **Check** the answer in the words of the *original* problem.

OBJECTIVE 1 Solve geometry problems by using two variables.

EXAMPLE 1 Finding the Dimensions of a Soccer Field

A rectangular soccer field may have a width between 50 and 100 yd and a length between 100 and 130 yd. One particular soccer field has a perimeter of 320 yd. Its length measures 40 yd more than its width. What are the dimensions of this field? (*Source:* www.soccer-training-guide.com)

Step 1 **Read** the problem again. We are asked to find the dimensions of the field.

Step 2 **Assign variables.** Let L = the length and W = the width. See **FIGURE 8**.

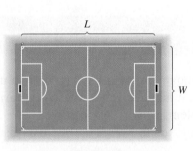

FIGURE 8

Step 3 **Write a system of equations.** Because the perimeter is 320 yd, we find one equation by using the perimeter formula.

$$2L + 2W = 320 \qquad \text{\small 2L + 2W = P}$$

We write a second equation using the fact that the length is 40 yd more than the width.

$$L = W + 40$$

These two equations form a system of equations.

$$2L + 2W = 320 \qquad (1)$$
$$L = W + 40 \qquad (2)$$

NOW TRY
EXERCISE 1

A rectangular parking lot has a length that is 10 ft more than twice its width. The perimeter of the parking lot is 620 ft. What are the dimensions of the parking lot?

Step 4 **Solve** the system of equations. Since equation (2), $L = W + 40$, is solved for L, we can substitute $W + 40$ for L in equation (1) and solve for W.

$$2L + 2W = 320 \quad \text{(1)}$$
$$2(W + 40) + 2W = 320 \quad \text{Let } L = W + 40.$$
$$2W + 80 + 2W = 320 \quad \text{Distributive property}$$
$$4W + 80 = 320 \quad \text{Combine like terms.}$$
$$4W = 240 \quad \text{Subtract 80.}$$
$$W = 60 \quad \text{Divide by 4.}$$

Be sure to use parentheses around $W + 40$.

Don't stop here.

Let $W = 60$ in the equation $L = W + 40$ to find L.

$$L = 60 + 40 = 100$$

Step 5 **State the answer.** The length is 100 yd, and the width is 60 yd. Both dimensions are within the ranges given in the problem.

Step 6 **Check.** Calculate the perimeter and the difference between the length and the width.

$$2(100) + 2(60) = 320 \quad \text{The perimeter is 320 yd, as required.}$$
$$100 - 60 = 40 \quad \text{Length is 40 yd more than width, as required.}$$

The answer is correct.

NOW TRY

OBJECTIVE 2 **Solve money problems by using two variables.**

EXAMPLE 2 Solving a Problem about Ticket Prices

For the 2008–2009 National Hockey League and National Basketball Association seasons, two hockey tickets and one basketball ticket purchased at their average prices would have cost $148.79. One hockey ticket and two basketball tickets would have cost $148.60. What were the average ticket prices for the two sports? (*Source:* Team Marketing Report.)

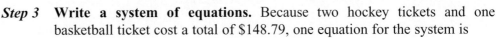

Step 1 **Read** the problem again. There are two unknowns.

Step 2 **Assign variables.**

Let $h =$ the average price for a hockey ticket

and $b =$ the average price for a basketball ticket.

Step 3 **Write a system of equations.** Because two hockey tickets and one basketball ticket cost a total of $148.79, one equation for the system is

$$2h + b = 148.79.$$

By similar reasoning, the second equation is

$$h + 2b = 148.60.$$

These two equations form a system of equations.

$$2h + b = 148.79 \quad \text{(1)}$$
$$h + 2b = 148.60 \quad \text{(2)}$$

NOW TRY ANSWER
1. length: 210 ft; width: 100 ft

NOW TRY
EXERCISE 2

For the 2009–2010 season at Six Flags St. Louis, two general admission tickets and three tickets for children under 48 in. tall cost $172.98. One general admission ticket and four tickets for children under 48 in. tall cost $163.99. Determine the ticket prices for general admission and for children under 48 in. tall. (*Source:* www.sixflags.com)

Step 4 **Solve** the system. To eliminate h, multiply equation (2), $h + 2b = 148.60$, by -2 and add.

$$
\begin{array}{rl}
2h + b = 148.79 & \text{(1)} \\
\underline{-2h - 4b = -297.20} & \text{Multiply each side of (2) by } -2. \\
-3b = -148.41 & \text{Add.} \\
b = 49.47 & \text{Divide by } -3.
\end{array}
$$

To find the value of h, let $b = 49.47$ in equation (2).

$$
\begin{array}{rl}
h + 2b = 148.60 & \text{(2)} \\
h + 2(49.47) = 148.60 & \text{Let } b = 49.47. \\
h + 98.94 = 148.60 & \text{Multiply.} \\
h = 49.66 & \text{Subtract } 98.94.
\end{array}
$$

Step 5 **State the answer.** The average price for one basketball ticket was $49.47. For one hockey ticket, the average price was $49.66.

Step 6 **Check** that these values satisfy the conditions stated in the problem.

NOW TRY

OBJECTIVE 3 **Solve mixture problems by using two variables.** We solved mixture problems in **Section 2.3** using one variable. Many mixture problems can also be solved using more than one variable and a system of equations.

EXAMPLE 3 Solving a Mixture Problem

How many ounces each of 5% hydrochloric acid and 20% hydrochloric acid must be combined to get 10 oz of solution that is 12.5% hydrochloric acid?

Step 1 **Read** the problem. Two solutions of different strengths are being mixed together to get a specific amount of a solution with an "in-between" strength.

Step 2 **Assign variables.**

Let x = the number of ounces of 5% solution

and y = the number of ounces of 20% solution.

Use a table to summarize the information from the problem.

Ounces of Solution	Percent (as a decimal)	Ounces of Pure Acid
x	5% = 0.05	0.05x
y	20% = 0.20	0.20y
10	12.5% = 0.125	(0.125)10

Gives equation (1) ⌣   Gives equation (2) ⌣

Multiply the amount of each solution (given in the first column) by its concentration of acid (given in the second column) to find the amount of acid in that solution (given in the third column).

FIGURE 9 illustrates what is happening in the problem.

Ounces of solution

Ounces of pure acid

FIGURE 9

NOW TRY ANSWER
2. general admission: $39.99;
children under 48 in. tall: $31.00

⌐ *NOW TRY*
↳ *EXERCISE 3*
How many liters each of a 15% acid solution and a 25% acid solution should be mixed to get 30 L of an 18% acid solution?

Step 3 **Write a system of equations.** When the x ounces of 5% solution and the y ounces of 20% solution are combined, the total number of ounces is 10, giving the following equation.

$$x + y = 10 \qquad (1)$$

The number of ounces of acid in the 5% solution $(0.05x)$ plus the number of ounces of acid in the 20% solution $(0.20y)$ should equal the total number of ounces of acid in the mixture, which is $(0.125)10$, or 1.25.

$$0.05x + 0.20y = 1.25 \qquad (2)$$

Notice that these equations can be quickly determined by reading down the table or using the labels in FIGURE 9.

Step 4 **Solve** the system of equations (1) and (2) by eliminating x.

$$
\begin{array}{ll}
-5x - 5y = -50 & \text{Multiply each side of (1) by } -5. \\
\underline{5x + 20y = 125} & \text{Multiply each side of (2) by 100.} \\
15y = 75 & \text{Add.} \\
y = 5 & \text{Divide by 15.}
\end{array}
$$

Substitute $y = 5$ in equation (1) to find that x is also 5.

Step 5 **State the answer.** The desired mixture will require 5 oz of the 5% solution and 5 oz of the 20% solution.

Step 6 **Check.**

Total amount of solution: $\quad x + y = 5 \text{ oz} + 5 \text{ oz}$
$$= 10 \text{ oz}, \quad \text{as required.}$$

Total amount of acid: $\quad$ 5% of 5 oz + 20% of 5 oz
$$= 0.05(5) + 0.20(5)$$
$$= 1.25 \text{ oz}$$

Percent of acid in solution:

Total acid $\longrightarrow \dfrac{1.25}{10} = 0.125,$ or 12.5%, as required.
Total solution $\longrightarrow$

NOW TRY ↻

OBJECTIVE 4 **Solve distance-rate-time problems by using two variables.** Motion problems require the distance formula $d = rt$, where d is distance, r is rate (or speed), and t is time.

EXAMPLE 4 Solving a Motion Problem

A car travels 250 km in the same time that a truck travels 225 km. If the rate of the car is 8 km per hr faster than the rate of the truck, find both rates.

Step 1 **Read** the problem again. Given the distances traveled, we need to find the rate of each vehicle.

Step 2 **Assign variables.**

Let $x =$ the rate of the car,
and $y =$ the rate of the truck.

NOW TRY ANSWER
3. 9 L of the 25% acid solution;
21 L of the 15% acid solution

NOW TRY
EXERCISE 4

Vann and Ivy Sample are planning a bicycle ride to raise money for cancer research. Vann can travel 50 mi in the same amount of time that Ivy can travel 40 mi. Determine both bicyclists' rates, if Vann's rate is 2 mph faster than Ivy's.

As in **Example 3,** a table helps organize the information. Fill in the distance for each vehicle, and the variables for the unknown rates.

	d	r	t	
Car	250	x	$\frac{250}{x}$	← To find the expressions for time, we solved the distance formula
Truck	225	y	$\frac{225}{y}$	← $d = rt$ for t. Thus, $\frac{d}{r} = t$.

Step 3 **Write a system of equations.** The car travels 8 km per hr faster than the truck. Since the two rates are x and y,

$$x = y + 8. \quad (1)$$

Both vehicles travel for the *same* time, so the times must be equal.

Time for car ⟶ $\dfrac{250}{x} = \dfrac{225}{y}$ ← Time for truck

This is not a linear equation. However, multiplying each side by xy gives

$$250y = 225x, \quad (2)$$

which is linear. The system to solve consists of equations (1) and (2).

$$x = y + 8 \quad (1)$$
$$250y = 225x \quad (2)$$

Step 4 **Solve** the system by substitution. Replace x with $y + 8$ in equation (2).

$$250y = 225x \qquad (2)$$
$$250y = 225(y + 8) \qquad \text{Let } x = y + 8.$$

> Be sure to use parentheses around $y + 8$.

$$250y = 225y + 1800 \qquad \text{Distributive property}$$
$$25y = 1800 \qquad \text{Subtract } 225y.$$
$$y = 72 \qquad \text{Divide by 25.}$$

Because $x = y + 8$, the value of x is $72 + 8 = 80$.

Step 5 **State the answer.** The rate of the car is 80 km per hr, and the rate of the truck is 72 km per hr.

Step 6 **Check.**

$$\text{Car:} \quad t = \frac{d}{r} = \frac{250}{80} = 3.125 \;←$$
$$\text{Truck:} \quad t = \frac{d}{r} = \frac{225}{72} = 3.125 \;←$$

Times are equal.

Since 80 is 8 greater than 72, the conditions of the problem are satisfied.

NOW TRY

OBJECTIVE 5 Solve problems with three variables by using a system of three equations.

PROBLEM-SOLVING HINT

If an application requires finding *three* unknown quantities, we can use a system of *three* equations to solve it. We extend the method used for two unknowns.

NOW TRY ANSWER
4. Vann: 10 mph; Ivy: 8 mph

NOW TRY
EXERCISE 5

At a concession stand, a bottle of Gatorade costs $2.00, a pretzel costs $1.50, and candy items cost $1.00. Workers sold four times as many pretzels as candy items. The number of Gatorades sold exceeded the number of pretzels sold by 310. Sales for these items totaled $1520. How many of each item was sold?

EXAMPLE 5   Solving a Problem Involving Prices

At Panera Bread, a loaf of honey wheat bread costs $2.95, a loaf of sunflower bread costs $2.99, and a loaf of French bread costs $5.79. On a recent day, three times as many loaves of honey wheat bread were sold as sunflower bread. The number of loaves of French bread sold was 5 less than the number of loaves of honey wheat bread sold. Total receipts for these breads were $87.89. How many loaves of each type of bread were sold? (*Source:* Panera Bread menu.)

Step 1 **Read** the problem again. There are three unknowns in this problem.

Step 2 **Assign variables** to represent the three unknowns.

$$\text{Let} \quad x = \text{the number of loaves of honey wheat bread,}$$
$$y = \text{the number of loaves of sunflower bread,}$$
$$\text{and} \quad z = \text{the number of loaves of French bread.}$$

Step 3 **Write a system of three equations.** Since three times as many loaves of honey wheat bread were sold as sunflower bread,

$$x = 3y, \quad \text{or} \quad x - 3y = 0. \quad \text{Subtract } 3y. \quad (1)$$

Also, we have the information needed for another equation.

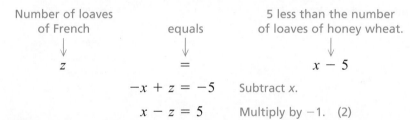

$$-x + z = -5 \qquad \text{Subtract } x.$$
$$x - z = 5 \qquad \text{Multiply by } -1. \quad (2)$$

Multiplying the cost of a loaf of each kind of bread by the number of loaves of that kind sold and adding gives the total receipts.

$$2.95x + 2.99y + 5.79z = 87.89$$

Multiply each side of this equation by 100 to clear it of decimals.

$$295x + 299y + 579z = 8789 \qquad (3)$$

Step 4 **Solve** the system of three equations.

$$x - 3y = 0 \qquad (1)$$
$$x - z = 5 \qquad (2)$$
$$295x + 299y + 579z = 8789 \qquad (3)$$

Using the method shown in **Section 4.2,** we will find that $x = 12$, $y = 4$, and $z = 7$.

Step 5 **State the answer.** The solution set is $\{(12, 4, 7)\}$, meaning that 12 loaves of honey wheat bread, 4 loaves of sunflower bread, and 7 loaves of French bread were sold.

Step 6 **Check.** Since $12 = 3 \cdot 4$, the number of loaves of honey wheat bread is three times the number of loaves of sunflower bread. Also, $12 - 7 = 5$, so the number of loaves of French bread is 5 less than the number of loaves of honey wheat bread. Multiply the appropriate cost per loaf by the number of loaves sold and add the results to check that total receipts were $87.89. *NOW TRY*

NOW TRY ANSWER
5. 60 candy items, 240 pretzels, 550 bottles of Gatorade

NOW TRY
EXERCISE 6

Katherine has a quilting shop and makes three kinds of quilts: the lone star quilt, the bandana quilt, and the log cabin quilt.

- Each lone star quilt requires 8 hr of piecework, 4 hr of machine quilting, and 2 hr of finishing.
- Each bandana quilt requires 2 hr of piecework, 2 hr of machine quilting, and 2 hr of finishing.
- Each log cabin quilt requires 10 hr of piecework, 5 hr of machine quilting, and 2 hr of finishing.

Katherine allocates 74 hr for piecework, 42 hr for machine quilting, and 24 hr for finishing quilts each month. How many of each type of quilt should be made each month if all available time must be used?

NOW TRY ANSWER

6. lone star quilts: 3;
bandana quilts: 5;
log cabin quilts: 4

EXAMPLE 6 Solving a Business Production Problem

A company produces three flat screen television sets: models X, Y, and Z.

- Each model X set requires 2 hr of electronics work, 2 hr of assembly time, and 1 hr of finishing time.
- Each model Y requires 1 hr of electronics work, 3 hr of assembly time, and 1 hr of finishing time.
- Each model Z requires 3 hr of electronics work, 2 hr of assembly time, and 2 hr of finishing time.

There are 100 hr available for electronics, 100 hr available for assembly, and 65 hr available for finishing per week. How many of each model should be produced each week if all available time must be used?

Step 1 **Read** the problem again. There are three unknowns.

Step 2 **Assign variables.** Then organize the information in a table.

$$\text{Let} \quad x = \text{the number of model X produced per week,}$$
$$y = \text{the number of model Y produced per week,}$$
$$\text{and} \quad z = \text{the number of model Z produced per week.}$$

	Each Model X	Each Model Y	Each Model Z	Totals
Hours of Electronics Work	2	1	3	100
Hours of Assembly Time	2	3	2	100
Hours of Finishing Time	1	1	2	65

Step 3 **Write a system of three equations.** The x model X sets require $2x$ hours of electronics, the y model Y sets require $1y$ (or y) hours of electronics, and the z model Z sets require $3z$ hours of electronics. Since 100 hr are available for electronics, we write the first equation.

$$2x + y + 3z = 100 \quad (1)$$

Since 100 hr are available for assembly, we can write another equation.

$$2x + 3y + 2z = 100 \quad (2)$$

The fact that 65 hr are available for finishing leads to this equation.

$$x + y + 2z = 65 \quad (3)$$

Notice that by reading across the table, we can easily determine the coefficients and constants in the equations of the system.

Step 4 **Solve** the system of equations (1), (2), and (3).

$$2x + y + 3z = 100 \quad (1)$$
$$2x + 3y + 2z = 100 \quad (2)$$
$$x + y + 2z = 65 \quad (3)$$

We find that $x = 15$, $y = 10$, and $z = 20$.

Step 5 **State the answer.** The company should produce 15 model X, 10 model Y, and 20 model Z sets per week.

Step 6 **Check** that these values satisfy the conditions of the problem. NOW TRY

4.3 EXERCISES

MyMathLab

🌀 *Complete solution available on the Video Resources on DVD*

Solve each problem. ***See Example 1.***

1. During the 2009 Major League Baseball season, the Los Angeles Dodgers played 162 games. They won 28 more games than they lost. What was their win-loss record that year?

2. Refer to **Exercise 1.** During the same 162-game season, the Arizona Diamondbacks lost 22 more games than they won. What was the team's win-loss record?

Team	W	L
L.A. Dodgers	—	—
Colorado	92	70
San Francisco	88	74
San Diego	75	87
Arizona	—	—

Source: World Almanac and Book of Facts.

🌀 **3.** Venus and Serena measured a tennis court and found that it was 42 ft longer than it was wide and had a perimeter of 228 ft. What were the length and the width of the tennis court?

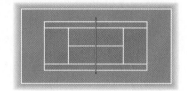

4. LeBron and Shaq measured a basketball court and found that the width of the court was 44 ft less than the length. If the perimeter was 288 ft, what were the length and the width of the basketball court?

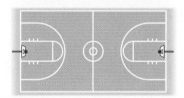

5. In 2009, the two American telecommunication companies with the greatest revenues were AT&T and Verizon. The two companies had combined revenues of $221.4 billion. AT&T's revenue was $26.6 more than that of Verizon. What was the revenue for each company? (*Source: Fortune* magazine.)

6. In 2008, U.S. exports to Canada were $110 billion more than exports to Mexico. Together, exports to these two countries totaled $412 billion. How much were exports to each country? (*Source:* U.S. Census Bureau.)

In Exercises 7 and 8, find the measures of the angles marked x and y. Remember that **(1)** *the sum of the measures of the angles of a triangle is* 180°, **(2)** *supplementary angles have a sum of* 180°, *and* **(3)** *vertical angles have equal measures.*

7.

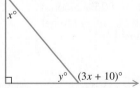

8.

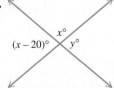

The Fan Cost Index (FCI) represents the cost of four average-price tickets, four small soft drinks, two small beers, four hot dogs, parking for one car, two game programs, and two souvenir caps to a sporting event. (*Source:* Team Marketing Report.)

Use the concept of FCI in Exercises 9 and 10. ***See Example 2.***

9. For the 2008–2009 season, the FCI prices for the National Hockey League and the National Basketball Association totaled $580.16. The hockey FCI was $3.70 less than that of basketball. What were the FCIs for these sports?

10. In 2009, the FCI prices for Major League Baseball and the National Football League totaled $609.53. The football FCI was $215.75 more than that of baseball. What were the FCIs for these sports?

Solve each problem. ***See Example 2.***

11. Andrew McGinnis works at Arby's. During one particular day he sold 15 Junior Roast Beef sandwiches and 10 Big Montana sandwiches, totaling $75.25. Another day he sold 30 Junior Roast Beef sandwiches and 5 Big Montana sandwiches, totaling $84.65. How much did each type of sandwich cost? (*Source:* Arby's menu.)

12. New York City and Washington, D.C., were the two most expensive cities for business travel in 2009. On the basis of the average total costs per day for each city (which include a hotel room, car rental, and three meals), 2 days in New York and 3 days in Washington cost $2772, while 4 days in New York and 2 days in Washington cost $3488. What was the average cost per day in each city? (*Source: Business Travel News.*)

Concept Check   *The formulas* $p = br$ *(percentage = base × rate) and* $I = prt$ *(simple interest = principal × rate × time) are used in the applications in* **Exercises 17–24.** *To prepare to use these formulas, answer the questions in Exercises 13 and 14.*

13. If a container of liquid contains 60 oz of solution, what is the number of ounces of pure acid if the given solution contains the following acid concentrations?

    (a) 10%   **(b)** 25%   **(c)** 40%   **(d)** 50%

14. If $5000 is invested in an account paying simple annual interest, how much interest will be earned during the first year at the following rates?

    (a) 2%   **(b)** 3%   **(c)** 4%   **(d)** 3.5%

15. *Concept Check*   If 1 pound of turkey costs $2.29, give an expression for the cost of x pounds.

16. *Concept Check*   If 1 ticket to the movie *Avatar* costs $9 and y tickets are sold, give an expression for the amount collected from the sale.

Solve each problem. ***See Example 3.***

17. How many gallons each of 25% alcohol and 35% alcohol should be mixed to get 20 gal of 32% alcohol?

18. How many liters each of 15% acid and 33% acid should be mixed to get 120 L of 21% acid?

Gallons of Solution	Percent (as a decimal)	Gallons of Pure Alcohol
x	25% = 0.25	
y	35% = 0.35	
20	32% =	

Liters of Solution	Percent (as a decimal)	Liters of Pure Acid
x	15% = 0.15	
y	33% =	
120	21% =	

19. Pure acid is to be added to a 10% acid solution to obtain 54 L of a 20% acid solution. What amounts of each should be used?

20. A truck radiator holds 36 L of fluid. How much pure antifreeze must be added to a mixture that is 4% antifreeze to fill the radiator with a mixture that is 20% antifreeze?

21. A party mix is made by adding nuts that sell for $2.50 per kg to a cereal mixture that sells for $1 per kg. How much of each should be added to get 30 kg of a mix that will sell for $1.70 per kg?

	Number of Kilograms	Price per Kilogram	Value
Nuts	x	2.50	
Cereal	y	1.00	
Mixture		1.70	

22. A fruit drink is made by mixing fruit juices. Such a drink with 50% juice is to be mixed with another drink that is 30% juice to get 200 L of a drink that is 45% juice. How much of each should be used?

	Liters of Drink	Percent (as a decimal)	Liters of Pure Juice
50% Juice	x	0.50	
30% Juice	y	0.30	
Mixture		0.45	

23. A total of $3000 is invested, part at 2% simple interest and part at 4%. If the total annual return from the two investments is $100, how much is invested at each rate?

Principal	Rate (as a decimal)	Interest
x	0.02	0.02x
y	0.04	0.04y
3000		100

24. An investor will invest a total of $15,000 in two accounts, one paying 4% annual simple interest and the other 3%. If he wants to earn $550 annual interest, how much should he invest at each rate?

Principal	Rate (as a decimal)	Interest
x	0.04	
y	0.03	
15,000		

Concept Check *The formula d = rt (distance = rate × time) is used in the applications in* **Exercises 27–30.** *To prepare to use this formula, work Exercises 25 and 26.*

25. If the rate of a boat in still water is 10 mph, and the rate of the current of a river is *x* mph, what is the rate of the boat

  (a) going upstream (that is, against the current, which slows the boat down);

  (b) going downstream (that is, with the current, which speeds the boat up)?

Upstream (against the current)

Downstream (with the current)

26. If the rate of a killer whale is 25 mph and the whale swims for *y* hours, give an expression for the number of miles the whale travels.

Solve each problem. **See Example 4.**

27. A train travels 150 km in the same time that a plane covers 400 km. If the rate of the plane is 20 km per hr less than 3 times the rate of the train, find both rates.

	r	t	d
Train	x		150
Plane	y		400

28. A freight train and an express train leave towns 390 km apart, traveling toward one another. The freight train travels 30 km per hr slower than the express train. They pass one another 3 hr later. What are their rates?

	r	t	d
Freight Train	x	3	
Express Train	y	3	

29. In his motorboat, Bill Ruhberg travels upstream at top speed to his favorite fishing spot, a distance of 36 mi, in 2 hr. Returning, he finds that the trip downstream, still at top speed, takes only 1.5 hr. Find the rate of Bill's boat and the rate of the current. Let x = the rate of the boat and y = the rate of the current.

	r	t	d
Upstream	$x - y$	2	
Downstream	$x + y$		

30. Traveling for 3 hr into a steady head wind, a plane flies 1650 mi. The pilot determines that flying *with* the same wind for 2 hr, he could make a trip of 1300 mi. Find the rate of the plane and the wind speed.

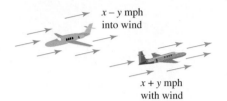

$x - y$ mph
into wind

$x + y$ mph
with wind

Solve each problem by using two variables. ***See Examples 1–4.***

31. (See the Chapter Introduction.) How many pounds of candy that sells for $0.75 per lb must be mixed with candy that sells for $1.25 per lb to obtain 9 lb of a mixture that should sell for $0.96 per lb?

32. The top-grossing tour on the North American concert circuit for 2009 was U2, followed in second place by Bruce Springsteen and the E Street Band. Together, they took in $217.5 million from ticket sales. If Springsteen took in $28.5 million less than U2, how much did each band generate? (*Source:* Pollstar.)

33. Tickets to a production of *A Midsummer Night's Dream* at Broward College cost $5 for general admission or $4 with a student ID. If 184 people paid to see a performance and $812 was collected, how many of each type of ticket were sold?

34. At a business meeting at Panera Bread, the bill for two cappuccinos and three house lattes was $14.55. At another table, the bill for one cappuccino and two house lattes was $8.77. How much did each type of beverage cost? (*Source:* Panera Bread menu.)

35. The mixed price of 9 citrons and 7 fragrant wood apples is 107; again, the mixed price of 7 citrons and 9 fragrant wood apples is 101. O you arithmetician, tell me quickly the price of a citron and the price of a wood apple here, having distinctly separated those prices well. (*Source:* Hindu work, A.D. 850.)

36. Braving blizzard conditions on the planet Hoth, Luke Skywalker sets out in his snow speeder for a rebel base 4800 mi away. He travels into a steady head wind and makes the trip in 3 hr. Returning, he finds that the trip back, now with a tailwind, takes only 2 hr. Find the rate of Luke's snow speeder and the speed of the wind.

	r	t	d
Into Head Wind			
With Tailwind			

Solve each problem by using three variables. **See Examples 5 and 6.** *(In Exercises 37–40, remember that the sum of the measures of the angles of a triangle is 180°.)*

37. In the figure, $z = x + 10$ and $x + y = 100$. Determine a third equation involving x, y, and z, and then find the measures of the three angles.

38. In the figure, x is 10 less than y and 20 less than z. Write a system of equations and find the measures of the three angles.

39. In a certain triangle, the measure of the second angle is 10° greater than three times the first. The third angle measure is equal to the sum of the measures of the other two. Find the measures of the three angles.

40. The measure of the largest angle of a triangle is 12° less than the sum of the measures of the other two. The smallest angle measures 58° less than the largest. Find the measures of the angles.

41. The perimeter of a triangle is 70 cm. The longest side is 4 cm less than the sum of the other two sides. Twice the shortest side is 9 cm less than the longest side. Find the length of each side of the triangle.

42. The perimeter of a triangle is 56 in. The longest side measures 4 in. less than the sum of the other two sides. Three times the shortest side is 4 in. more than the longest side. Find the lengths of the three sides.

43. In the 2008 Summer Olympics in Beijing, China, Russia earned 5 fewer gold medals than bronze. The number of silver medals earned was 35 less than twice the number of bronze medals. Russia earned a total of 72 medals. How many of each kind of medal did Russia earn? (*Source: World Almanac and Book of Facts.*)

44. In a random sample of Americans of voting age conducted in 2010, 8% more people identified themselves as Independents than as Republicans, while 6% fewer people identified themselves as Republicans than as Democrats. Of those sampled, 2% did not identify with any of the three categories. What percent of the people in the sample identified themselves with each of the three political affiliations? (*Source:* Gallup, Inc.)

45. Tickets for the Harlem Globetrotters show at Michigan State University in 2010 cost $16, $23, or, for VIP seats, $40. If nine times as many $16 tickets were sold as VIP tickets, and the number of $16 tickets sold was 55 more than the sum of the number of $23 tickets and VIP tickets, sales of all three kinds of tickets would total $46,575. How many of each kind of ticket would have been sold? (*Source:* Breslin Student Events Center.)

46. Three kinds of tickets are available for a *Cowboy Mouth* concert: "up close," "in the middle," and "far out." "Up close" tickets cost $10 more than "in the middle" tickets, while "in the middle" tickets cost $10 more than "far out" tickets. Twice the cost of an "up close" ticket is $20 more than 3 times the cost of a "far out" ticket. Find the price of each kind of ticket.

47. A wholesaler supplies college T-shirts to three college bookstores: A, B, and C. The wholesaler recently shipped a total of 800 T-shirts to the three bookstores. In order to meet student demand at the three colleges, twice as many T-shirts were shipped to bookstore B as to bookstore A, and the number shipped to bookstore C was 40 less than the sum of the numbers shipped to the other two bookstores. How many T-shirts were shipped to each bookstore?

48. An office supply store sells three models of computer desks: A, B, and C. In January, the store sold a total of 85 computer desks. The number of model B desks was five more than the number of model C desks, and the number of model A desks was four more than twice the number of model C desks. How many of each model did the store sell in January?

49. A plant food is to be made from three chemicals. The mix must include 60% of the first and second chemicals. The second and third chemicals must be in the ratio of 4 to 3 by weight. How much of each chemical is needed to make 750 kg of the plant food?

50. How many ounces of 5% hydrochloric acid, 20% hydrochloric acid, and water must be combined to get 10 oz of solution that is 8.5% hydrochloric acid if the amount of water used must equal the total amount of the other two solutions?

Starting with the 2005–2006 season, the National Hockey League adopted a new system for awarding points used to determine team standings. A team is awarded 2 points for a win (W), 0 points for a loss in regulation play (L), and 1 point for an overtime loss (OTL). Use this information in Exercises 51 and 52.

51. During the 2008–2009 NHL regular season, the Boston Bruins played 82 games. Their wins and overtime losses resulted in a total of 116 points. They had 9 more losses in regulation play than overtime losses. How many wins, losses, and overtime losses did they have that year?

Team	GP	W	L	OTL	Points
Boston	82	___	___	___	116
Montreal	82	41	30	11	93
Buffalo	82	41	32	9	91
Ottawa	82	36	35	11	83
Toronto	82	34	35	13	81

Source: World Almanac and Book of Facts.

52. During the 2008–2009 NHL regular season, the Los Angeles Kings played 82 games. Their wins and overtime losses resulted in a total of 79 points. They had 14 more total losses (in regulation play and overtime) than wins. How many wins, losses, and overtime losses did they have that year?

Team	GP	W	L	OTL	Points
San Jose	82	53	18	11	117
Anaheim	82	42	33	7	91
Dallas	82	36	35	11	83
Phoenix	82	36	39	7	79
Los Angeles	82	___	___	___	79

Source: World Almanac and Book of Facts.

PREVIEW EXERCISES

Give **(a)** the additive inverse and **(b)** the multiplicative inverse (reciprocal) of each number. *See Sections 1.1 and 1.2.*

53. -6 **54.** 0.2 **55.** $\dfrac{7}{8}$ **56.** 2.25

Solving Systems of Linear Equations by Matrix Methods

OBJECTIVE 1 Define a matrix. An ordered array of numbers such as

$$\text{Rows} \left[\begin{array}{ccc} 2 & 3 & 5 \\ 7 & 1 & 2 \end{array} \right] \quad \text{Matrix}$$

Columns

is called a **matrix.** The numbers are called **elements** of the matrix. *Matrices* (the plural of *matrix*) are named according to the number of **rows** and **columns** they contain. The rows are read horizontally, and the columns are read vertically. This matrix is a 2×3 (read "two by three") matrix, because it has 2 rows and 3 columns. The number of rows followed by the number of columns gives the **dimensions** of the matrix.

$$\begin{bmatrix} -1 & 0 \\ 1 & -2 \end{bmatrix} \quad \begin{array}{c} 2 \times 2 \\ \text{matrix} \end{array} \qquad \begin{bmatrix} 8 & -1 & -3 \\ 2 & 1 & 6 \\ 0 & 5 & -3 \\ 5 & 9 & 7 \end{bmatrix} \quad \begin{array}{c} 4 \times 3 \\ \text{matrix} \end{array}$$

A **square matrix** is a matrix that has the same number of rows as columns. The 2×2 matrix above is a square matrix.

FIGURE 10 shows how a graphing calculator displays the preceding two matrices. Consult your owner's manual for details for using matrices.

In this section, we discuss a matrix method of solving linear systems that is a structured way of using the elimination method. The advantage of this new method is that it can be done by a graphing calculator or a computer.

```
[A]
        [[-1  0 ]
         [1  -2]]
```

```
[B]
        [[8  -1  -3]
         [2   1   6]
         [0   5  -3]
         [5   9   7]]
```

FIGURE 10

OBJECTIVE 2 Write the augmented matrix of a system. To solve a linear system using matrices, we begin by writing an *augmented matrix* of the system. An **augmented matrix** has a vertical bar that separates the columns of the matrix into two groups. For example, to solve the system

$$x - 3y = 1$$
$$2x + y = -5,$$

start by writing the augmented matrix

$$\left[\begin{array}{cc|c} 1 & -3 & 1 \\ 2 & 1 & -5 \end{array} \right]. \quad \text{Augmented matrix}$$

Place the coefficients of the variables to the left of the bar, and the constants to the right. The bar separates the coefficients from the constants. *The matrix is just a shorthand way of writing the system of equations, so the rows of the augmented matrix can be treated the same as the equations of a system of equations.*

Exchanging the positions of two equations in a system does not change the system. Also, multiplying any equation in a system by a nonzero number does not change the system. Comparable changes to the augmented matrix of a system of equations produce new matrices that correspond to systems with the same solutions as the original system.

The following **row operations** produce new matrices that lead to systems having the same solutions as the original system.

> ### Matrix Row Operations
>
> 1. Any two rows of the matrix may be interchanged.
> 2. The elements of any row may be multiplied by any nonzero real number.
> 3. Any row may be changed by adding to the elements of the row the product of a real number and the corresponding elements of another row.

Examples of these row operations follow.

Row operation 1

$$\begin{bmatrix} 2 & 3 & 9 \\ 4 & 8 & -3 \\ 1 & 0 & 7 \end{bmatrix} \text{ becomes } \begin{bmatrix} 1 & 0 & 7 \\ 4 & 8 & -3 \\ 2 & 3 & 9 \end{bmatrix}$$

Interchange row 1 and row 3.

Row operation 2

$$\begin{bmatrix} 2 & 3 & 9 \\ 4 & 8 & -3 \\ 1 & 0 & 7 \end{bmatrix} \text{ becomes } \begin{bmatrix} 6 & 9 & 27 \\ 4 & 8 & -3 \\ 1 & 0 & 7 \end{bmatrix}$$

Multiply the numbers in row 1 by 3.

Row operation 3

$$\begin{bmatrix} 2 & 3 & 9 \\ 4 & 8 & -3 \\ 1 & 0 & 7 \end{bmatrix} \text{ becomes } \begin{bmatrix} 0 & 3 & -5 \\ 4 & 8 & -3 \\ 1 & 0 & 7 \end{bmatrix}$$

Multiply the numbers in row 3 by -2. Add them to the corresponding numbers in row 1.

The third row operation corresponds to the way we eliminated a variable from a pair of equations in previous sections.

OBJECTIVE 3 **Use row operations to solve a system with two equations.** Row operations can be used to rewrite a matrix until it is the matrix of a system whose solution is easy to find. The goal is a matrix in the form

$$\left[\begin{array}{cc|c} 1 & a & b \\ 0 & 1 & c \end{array}\right] \quad \text{or} \quad \left[\begin{array}{ccc|c} 1 & a & b & c \\ 0 & 1 & d & e \\ 0 & 0 & 1 & f \end{array}\right]$$

for systems with two and three equations, respectively. Notice that there are 1's down the diagonal from upper left to lower right and 0's below the 1's. A matrix written this way is said to be in **row echelon form.**

EXAMPLE 1 **Using Row Operations to Solve a System with Two Variables**

Use row operations to solve the system.

$$x - 3y = 1$$
$$2x + y = -5$$

We start by writing the augmented matrix of the system.

$$\left[\begin{array}{cc|c} 1 & -3 & 1 \\ 2 & 1 & -5 \end{array}\right]$$

Write the augmented matrix.

**NOW TRY
EXERCISE 1**
Use row operations to solve
the system.

$$x + 3y = 3$$
$$2x - 3y = -12$$

*Our goal is to use the various row operations to change this matrix into one that
leads to a system that is easier to solve. It is best to work by columns.*

We start with the first column and make sure that there is a 1 in the first row, first
column, position. There already is a 1 in this position.

Next, we get 0 in every position below the first. To get a 0 in row two, column
one, we add to the numbers in row two the result of multiplying each number in row
one by -2. (We abbreviate this as $-2R_1 + R_2$.) Row one remains unchanged.

$$\begin{bmatrix} 1 & -3 & \bigm| & 1 \\ 2 + 1(-2) & 1 + (-3)(-2) & \bigm| & -5 + 1(-2) \end{bmatrix}$$

Original number -2 times number
from row two   from row one

1 in the first position of column one $\longrightarrow$ $\begin{bmatrix} 1 & -3 & \bigm| & 1 \\ 0 & 7 & \bigm| & -7 \end{bmatrix}$ $-2R_1 + R_2$
0 in every position below the first $\longrightarrow$

Now we go to column two. The number 1 is needed in row two, column two. We
use the second row operation, multiplying each number of row two by $\frac{1}{7}$.

Stop here—this
matrix is in row
echelon form. $\begin{bmatrix} 1 & -3 & \bigm| & 1 \\ 0 & 1 & \bigm| & -1 \end{bmatrix}$ $\frac{1}{7}R_2$

This augmented matrix leads to the system of equations

$$1x - 3y = 1 \qquad\qquad x - 3y = 1$$
$$0x + 1y = -1, \qquad \text{or} \qquad y = -1.$$

From the second equation, $y = -1$, we substitute -1 for y in the first equation to
find x.

$$x - 3y = 1$$
$$x - 3(-1) = 1 \qquad \text{Let } y = -1.$$
$$x + 3 = 1 \qquad \text{Multiply.}$$
$$x = -2 \qquad \text{Subtract 3.}$$

The solution set of the system is $\{(-2, -1)\}$. Check this solution by substitution in
both equations of the system.

Write the values of x
and y in the correct order.

NOW TRY

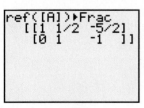

(a)

(b)

FIGURE 11

NOTE If the augmented matrix of the system in **Example 1** is entered as matrix [A]
in a graphing calculator (**FIGURE 11(a)**) and the row echelon form of the matrix is
found (**FIGURE 11(b)**), then the system becomes the following.

$$x + \frac{1}{2}y = -\frac{5}{2}$$

$$y = -1$$

NOW TRY ANSWER
1. $\{(-3, 2)\}$

While this system looks different from the one we obtained in **Example 1,** it is equivalent, since its solution set is also $\{(-2, -1)\}$.

OBJECTIVE 4 **Use row operations to solve a system with three equations.**

EXAMPLE 2 **Using Row Operations to Solve a System with Three Variables**

Use row operations to solve the system.

$$x - y + 5z = -6$$
$$3x + 3y - z = 10$$
$$x + 3y + 2z = 5$$

Start by writing the augmented matrix of the system.

$$\begin{bmatrix} 1 & -1 & 5 & | & -6 \\ 3 & 3 & -1 & | & 10 \\ 1 & 3 & 2 & | & 5 \end{bmatrix}$$ Write the augmented matrix.

This matrix already has 1 in row one, column one. Next get 0's in the rest of column one. First, add to row two the results of multiplying each number of row one by -3.

$$\begin{bmatrix} 1 & -1 & 5 & | & -6 \\ 0 & 6 & -16 & | & 28 \\ 1 & 3 & 2 & | & 5 \end{bmatrix}$$ $-3R_1 + R_2$

Now add to the numbers in row three the results of multiplying each number of row one by -1.

$$\begin{bmatrix} 1 & -1 & 5 & | & -6 \\ 0 & 6 & -16 & | & 28 \\ 0 & 4 & -3 & | & 11 \end{bmatrix}$$ $-1R_1 + R_3$

Introduce 1 in row two, column two, by multiplying each number in row two by $\frac{1}{6}$.

$$\begin{bmatrix} 1 & -1 & 5 & | & -6 \\ 0 & 1 & -\frac{8}{3} & | & \frac{14}{3} \\ 0 & 4 & -3 & | & 11 \end{bmatrix}$$ $\frac{1}{6}R_2$

To obtain 0 in row three, column two, add to row three the results of multiplying each number in row two by -4.

$$\begin{bmatrix} 1 & -1 & 5 & | & -6 \\ 0 & 1 & -\frac{8}{3} & | & \frac{14}{3} \\ 0 & 0 & \frac{23}{3} & | & -\frac{23}{3} \end{bmatrix}$$ $-4R_2 + R_3$

Obtain 1 in row three, column three, by multiplying each number in row three by $\frac{3}{23}$.

This matrix is in row echelon form. $\longrightarrow$ $$\begin{bmatrix} 1 & -1 & 5 & | & -6 \\ 0 & 1 & -\frac{8}{3} & | & \frac{14}{3} \\ 0 & 0 & 1 & | & -1 \end{bmatrix}$$ $\frac{3}{23}R_3$

The final matrix gives this system of equations.

$$x - y + 5z = -6$$
$$y - \frac{8}{3}z = \frac{14}{3}$$
$$z = -1$$

NOW TRY
EXERCISE 2

Use row operations to solve the system.

$$x + y - 2z = -5$$
$$-x + 2y + z = -1$$
$$2x - y + 3z = 14$$

Substitute -1 for z in the second equation, $y - \frac{8}{3}z = \frac{14}{3}$, to find that $y = 2$. Finally, substitute 2 for y and -1 for z in the first equation,

$$x - y + 5z = -6,$$

to determine that $x = 1$. The solution set of the original system is $\{(1, 2, -1)\}$. Check by substitution. **NOW TRY**

OBJECTIVE 5 **Use row operations to solve special systems.**

NOW TRY
EXERCISE 3

Use row operations to solve each system.

(a) $3x - y = 8$
$-6x + 2y = 4$

(b) $x + 2y = 7$
$-x - 2y = -7$

EXAMPLE 3 Recognizing Inconsistent Systems or Dependent Equations

Use row operations to solve each system.

(a) $2x - 3y = 8$
$-6x + 9y = 4$

$$\begin{bmatrix} 2 & -3 & | & 8 \\ -6 & 9 & | & 4 \end{bmatrix}$$ Write the augmented matrix.

$$\begin{bmatrix} 1 & -\frac{3}{2} & | & 4 \\ -6 & 9 & | & 4 \end{bmatrix}$$ $\frac{1}{2}R_1$

$$\begin{bmatrix} 1 & -\frac{3}{2} & | & 4 \\ 0 & 0 & | & 28 \end{bmatrix}$$ $6R_1 + R_2$

The corresponding system of equations is

$$x - \frac{3}{2}y = 4$$

$$0 = 28, \quad \text{False}$$

which has no solution and is inconsistent. The solution set is $\emptyset$.

(b) $-10x + 12y = 30$
$5x - 6y = -15$

$$\begin{bmatrix} -10 & 12 & | & 30 \\ 5 & -6 & | & -15 \end{bmatrix}$$ Write the augmented matrix.

$$\begin{bmatrix} 1 & -\frac{6}{5} & | & -3 \\ 5 & -6 & | & -15 \end{bmatrix}$$ $-\frac{1}{10}R_1$

$$\begin{bmatrix} 1 & -\frac{6}{5} & | & -3 \\ 0 & 0 & | & 0 \end{bmatrix}$$ $-5R_1 + R_2$

The corresponding system is

$$x - \frac{6}{5}y = -3$$

$$0 = 0, \quad \text{True}$$

which has dependent equations. We use the second equation of the given system, which is in standard form, to express the solution set.

$$\{(x, y) \mid 5x - 6y = -15\}$$ **NOW TRY**

NOW TRY ANSWERS
2. $\{(2, -1, 3)\}$
3. (a) $\emptyset$
   (b) $\{(x, y) \mid x + 2y = 7\}$

4.4 EXERCISES

MyMathLab

PRACTICE WATCH DOWNLOAD READ REVIEW

⊕ *Complete solution available on the Video Resources on DVD*

1. *Concept Check* Consider the matrix $\begin{bmatrix} -2 & 3 & 1 \\ 0 & 5 & -3 \\ 1 & 4 & 8 \end{bmatrix}$ and answer the following.

    (a) What are the elements of the second row?

    (b) What are the elements of the third column?

    (c) Is this a square matrix? Explain why or why not.

    (d) Give the matrix obtained by interchanging the first and third rows.

    (e) Give the matrix obtained by multiplying the first row by $-\frac{1}{2}$.

    (f) Give the matrix obtained by multiplying the third row by 3 and adding to the first row.

2. Repeat **Exercise 1** for the matrix $\begin{bmatrix} -7 & 0 & 1 \\ 3 & 2 & -2 \\ 0 & 1 & 6 \end{bmatrix}$.

Concept Check   *Give the dimensions of each matrix.*

3. $\begin{bmatrix} 3 & -7 \\ 4 & 5 \\ -1 & 0 \end{bmatrix}$

4. $\begin{bmatrix} 4 & 9 & 0 \\ -1 & 2 & -4 \end{bmatrix}$

5. $\begin{bmatrix} 6 & 3 \\ -2 & 5 \\ 4 & 10 \\ 1 & -1 \end{bmatrix}$

6. $\begin{bmatrix} 8 & 4 & 3 & 2 \end{bmatrix}$

Use row operations to solve each system. ***See Examples 1 and 3.***

7. $\begin{aligned} x + y &= 5 \\ x - y &= 3 \end{aligned}$

8. $\begin{aligned} x + 2y &= 7 \\ x - y &= -2 \end{aligned}$

⊕ **9.** $\begin{aligned} 2x + 4y &= 6 \\ 3x - y &= 2 \end{aligned}$

10. $\begin{aligned} 4x + 5y &= -7 \\ x - y &= 5 \end{aligned}$

11. $\begin{aligned} 3x + 4y &= 13 \\ 2x - 3y &= -14 \end{aligned}$

12. $\begin{aligned} 5x + 2y &= 8 \\ 3x - y &= 7 \end{aligned}$

⊕ **13.** $\begin{aligned} -4x + 12y &= 36 \\ x - 3y &= 9 \end{aligned}$

14. $\begin{aligned} 2x - 4y &= 8 \\ -3x + 6y &= 5 \end{aligned}$

15. $\begin{aligned} 2x + y &= 4 \\ 4x + 2y &= 8 \end{aligned}$

16. $\begin{aligned} -3x - 4y &= 1 \\ 6x + 8y &= -2 \end{aligned}$

17. $\begin{aligned} -3x + 2y &= 0 \\ x - y &= 0 \end{aligned}$

18. $\begin{aligned} -5x + 3y &= 0 \\ 7x + 2y &= 0 \end{aligned}$

Use row operations to solve each system. ***See Examples 2 and 3.***

⊕ **19.** $\begin{aligned} x + y - 3z &= 1 \\ 2x - y + z &= 9 \\ 3x + y - 4z &= 8 \end{aligned}$

20. $\begin{aligned} 2x + 4y - 3z &= -18 \\ 3x + y - z &= -5 \\ x - 2y + 4z &= 14 \end{aligned}$

21. $\begin{aligned} x + y - z &= 6 \\ 2x - y + z &= -9 \\ x - 2y + 3z &= 1 \end{aligned}$

22. $\begin{aligned} x + 3y - 6z &= 7 \\ 2x - y + 2z &= 0 \\ x + y + 2z &= -1 \end{aligned}$

23. $\begin{aligned} x - y &= 1 \\ y - z &= 6 \\ x + z &= -1 \end{aligned}$

24. $\begin{aligned} x + y &= 1 \\ 2x - z &= 0 \\ y + 2z &= -2 \end{aligned}$

25. $\begin{aligned} x - 2y + z &= 4 \\ 3x - 6y + 3z &= 12 \\ -2x + 4y - 2z &= -8 \end{aligned}$

26. $\begin{aligned} x + 3y + z &= 1 \\ 2x + 6y + 2z &= 2 \\ 3x + 9y + 3z &= 3 \end{aligned}$

27. $\begin{aligned} x + 2y + 3z &= -2 \\ 2x + 4y + 6z &= -5 \\ x - y + 2z &= 6 \end{aligned}$

28. $\begin{aligned} 4x + 8y + 4z &= 9 \\ x + 3y + 4z &= 10 \\ 5x + 10y + 5z &= 12 \end{aligned}$

The augmented matrix of the system $\begin{array}{l} 4x + 8y = 44 \\ 2x - y = -3 \end{array}$ *is shown in the graphing calculator screen on the left as matrix* [A]. *The screen in the middle shows the row echelon form for* [A]. *The screen on the right shows the "reduced" row echelon form, and from this it can be determined by inspection that the solution set of the system is* $\{(1, 5)\}$.

```
[A]
      [[4 8  44]
       [2 -1 -3]]
```

```
ref([A])
      [[1 2 11]
       [0 1 5 ]]
```

```
rref([A])
      [[1 0 1]
       [0 1 5]]
```

Use a graphing calculator and either matrix method illustrated to solve each system.

29. $4x + y = 5$
$\quad 2x + y = 3$

30. $5x + 3y = 7$
$\quad 7x - 3y = -19$

31. $\quad 5x + y - 3z = -6$
$\quad\ \ 2x + 3y + z = 5$
$\quad -3x - 2y + 4z = 3$

32. $x + y + z = 3$
$\quad 3x - 3y - 4z = -1$
$\quad\ x + y + 3z = 11$

33. $x + z = -3$
$\quad y + z = 3$
$\quad x + y = 8$

34. $\quad x - y = -1$
$\quad -y + z = -2$
$\quad\ x + z = -2$

PREVIEW EXERCISES

*Evaluate each exponential expression. **See Section 1.3.***

35. 2^6

36. $(-4)^5$

37. $(-5)^4$

38. -5^4

39. $\left(\dfrac{3}{4}\right)^4$

40. $\left(-\dfrac{3}{2}\right)^5$

CHAPTER (4)

SUMMARY

KEY TERMS

4.1

system of equations
system of linear
   equations
solution set of a linear
   system

consistent system
independent equations
inconsistent system
dependent equations
elimination method
substitution method

4.2

ordered triple
focus variable
working equation

4.4

matrix
element of a matrix
row
column
square matrix
augmented matrix
row operations
row echelon form

NEW SYMBOLS

(x, y, z) ordered triple

$\begin{bmatrix} a & b & c \\ d & e & f \end{bmatrix}$ matrix with two rows, three columns

TEST YOUR WORD POWER

See how well you have learned the vocabulary in this chapter.

1. A **system of equations** consists of
   A. at least two equations with different variables
   B. two or more equations that have an infinite number of solutions
   C. two or more equations that are to be solved at the same time
   D. two or more inequalities that are to be solved.

2. The **solution set of a system of equations** is
   A. all ordered pairs that satisfy one equation of the system
   B. all ordered pairs that satisfy all the equations of the system at the same time

C. any ordered pair that satisfies one or more equations of the system
D. the set of values that make all the equations of the system false.

3. An **inconsistent system** is a system of equations
   A. with one solution
   B. with no solution
   C. with an infinite number of solutions
   D. that have the same graph.

4. **Dependent equations**
   A. have different graphs
   B. have no solution

C. have one solution
D. are different forms of the same equation.

5. A **matrix** is
   A. an ordered pair of numbers
   B. an array of numbers with the same number of rows and columns
   C. a pair of numbers written between brackets
   D. a rectangular array of numbers.

ANSWERS

1. C; *Example:* $\begin{array}{l} 3x - y = 3 \\ 2x + y = 7 \end{array}$ 2. B; *Example:* The ordered pair $(2, 3)$ satisfies both equations of the system in Answer 1, so $\{(2, 3)\}$ is the solution set of the system.   3. B; *Example:* The equations of two parallel lines form an inconsistent system. Their graphs never intersect, so the system has no solution.   4. D; *Example:* The equations $4x - y = 8$ and $8x - 2y = 16$ are dependent because their graphs are the same line.

5. D; *Examples:* $\begin{bmatrix} 3 & -1 & 0 \\ 4 & 2 & 1 \end{bmatrix}, \begin{bmatrix} 1 & 2 \\ 4 & 3 \end{bmatrix}$

QUICK REVIEW

CONCEPTS	EXAMPLES
4.1 Systems of Linear Equations in Two Variables	
Solving a Linear System by Substitution	Solve by substitution. $\begin{array}{ll} 4x - y = 7 & (1) \\ 3x + 2y = 30 & (2) \end{array}$
Step 1 Solve one of the equations for either variable.	Solve for y in equation (1): $y = 4x - 7$.
Step 2 Substitute for that variable in the other equation. The result should be an equation with just one variable.	Substitute $4x - 7$ for y in equation (2), and solve for x. $\begin{array}{lll} 3x + 2y = 30 & (2) & \\ 3x + 2(4x - 7) = 30 & & \text{Let } y = 4x - 7. \\ 3x + 8x - 14 = 30 & & \text{Distributive property} \end{array}$
Step 3 Solve the equation from Step 2.	$\begin{array}{lll} 11x = 44 & & \text{Combine like terms. Add 14.} \\ x = 4 & & \text{Divide by 11.} \end{array}$
Step 4 Find the value of the other variable by substituting the result from Step 3 into the equation from Step 1.	Substitute 4 for x in the equation $y = 4x - 7$ to find that $y = 4(4) - 7 = 9$.
Step 5 Check the ordered-pair solution in *both* of the *original* equations. Then write the solution set.	Check to see that $\{(4, 9)\}$ is the solution set.

(continued)

CONCEPTS	EXAMPLES

Solving a Linear System by Elimination

Step 1 Write both equations in standard form.

Step 2 Make the coefficients of one pair of variable terms opposites.

Step 3 Add the new equations. The sum should be an equation with just one variable.

Step 4 Solve the equation from Step 3.

Step 5 Find the value of the other variable by substituting the result of Step 4 into either of the original equations.

Step 6 Check the ordered-pair solution in *both* of the *original* equations. Then write the solution set.

If the result of the addition step (Step 3) is a false statement, such as $0 = 4$, the graphs are parallel lines and *there is no solution. The solution set is* $\emptyset$.

If the result is a true statement, such as $0 = 0$, the graphs are the same line, and an *infinite number of ordered pairs are solutions. The solution set is written in set-builder notation as*

$$\{(x, y) \,|\, \underline{\quad\quad}\},$$

where a form of the equation is written in the blank.

Solve by elimination.
$$5x + y = 2 \quad (1)$$
$$2x - 3y = 11 \quad (2)$$

To eliminate y, multiply equation (1) by 3 and add the result to equation (2).

$$
\begin{aligned}
15x + 3y &= 6 & \text{3 times equation (1)} \\
2x - 3y &= 11 & (2) \\
\hline
17x &= 17 & \text{Add.} \\
x &= 1 & \text{Divide by 17.}
\end{aligned}
$$

Let $x = 1$ in equation (1), and solve for y.
$$5(1) + y = 2$$
$$y = -3$$

Check to verify that $\{(1, -3)\}$ is the solution set.

$$
\begin{aligned}
x - 2y &= 6 \\
-x + 2y &= -2 \\
\hline
0 &= 4 & \text{Solution set: } \emptyset
\end{aligned}
$$

$$
\begin{aligned}
x - 2y &= 6 \\
-x + 2y &= -6 \\
\hline
0 &= 0 & \text{Solution set: } \{(x, y) \,|\, x - 2y = 6\}
\end{aligned}
$$

4.2 Systems of Linear Equations in Three Variables

Solving a Linear System in Three Variables

Step 1 Select a focus variable, preferably one with coefficient 1 or -1, and a working equation.

Step 2 Eliminate the focus variable, using the working equation and one of the equations of the system.

Step 3 Eliminate the focus variable again, using the working equation and the remaining equation of the system.

Step 4 Solve the system of two equations in two variables formed by the equations from Steps 2 and 3.

Solve the system.
$$x + 2y - z = 6 \quad (1)$$
$$x + y + z = 6 \quad (2)$$
$$2x + y - z = 7 \quad (3)$$

We choose z as the focus variable and (2) as the working equation.

Add equations (1) and (2).
$$2x + 3y = 12 \quad (4)$$

Add equations (2) and (3).
$$3x + 2y = 13 \quad (5)$$

Use equations (4) and (5) to eliminate x.

$$
\begin{aligned}
-6x - 9y &= -36 & \text{Multiply (4) by } -3. \\
6x + 4y &= 26 & \text{Multiply (5) by 2.} \\
\hline
-5y &= -10 & \text{Add.} \\
y &= 2 & \text{Divide by } -5.
\end{aligned}
$$

(continued)

CONCEPTS	EXAMPLES
	To find x, substitute 2 for y in equation (4).

$$2x + 3(2) = 12 \qquad \text{Let } y = 2 \text{ in (4)}.$$
$$2x + 6 = 12 \qquad \text{Multiply.}$$
$$2x = 6 \qquad \text{Subtract 6.}$$
$$x = 3 \qquad \text{Divide by 2.}$$

Step 5   Find the value of the remaining variable.

Substitute 3 for x and 2 for y in working equation (2).

$$x + y + z = 6 \qquad (2)$$
$$3 + 2 + z = 6$$
$$z = 1$$

Step 6   Check the ordered-triple solution in each of the original equations of the system. Then write the solution set.

A check of the solution $(3, 2, 1)$ confirms that the solution set is $\{(3, 2, 1)\}$.

4.3   Applications of Systems of Linear Equations

Use the six-step problem-solving method.

Step 1   Read the problem carefully.

Step 2   Assign variables.

Step 3   Write a system of equations that relates the unknowns.

Step 4   Solve the system.

Step 5   State the answer.

Step 6   Check.

The perimeter of a rectangle is 18 ft. The length is 3 ft more than twice the width. What are the dimensions of the rectangle?

Let x represent the length and y represent the width. From the perimeter formula, one equation is $2x + 2y = 18$. From the problem, another equation is $x = 2y + 3$. Solve the system

$$2x + 2y = 18$$
$$x = 2y + 3$$

to get $x = 7$ and $y = 2$. The length is 7 ft, and the width is 2 ft. Since the perimeter is

$$2(7) + 2(2) = 18, \quad \text{and} \quad 2(2) + 3 = 7,$$

the solution checks.

4.4   Solving Systems of Linear Equations by Matrix Methods

Matrix Row Operations

1. Any two rows of the matrix may be interchanged.

$$\begin{bmatrix} 1 & 5 & 7 \\ 3 & 9 & -2 \\ 0 & 6 & 4 \end{bmatrix} \text{ becomes } \begin{bmatrix} 3 & 9 & -2 \\ 1 & 5 & 7 \\ 0 & 6 & 4 \end{bmatrix} \quad \begin{array}{l}\text{Interchange}\\ \text{R}_1 \text{ and R}_2.\end{array}$$

2. The elements of any row may be multiplied by any nonzero real number.

$$\begin{bmatrix} 1 & 5 & 7 \\ 3 & 9 & -2 \\ 0 & 6 & 4 \end{bmatrix} \text{ becomes } \begin{bmatrix} 1 & 5 & 7 \\ 1 & 3 & -\frac{2}{3} \\ 0 & 6 & 4 \end{bmatrix} \quad \frac{1}{3}\text{R}_2$$

3. Any row may be changed by adding to the elements of the row the product of a real number and the elements of another row.

$$\begin{bmatrix} 1 & 5 & 7 \\ 3 & 9 & -2 \\ 0 & 6 & 4 \end{bmatrix} \text{ becomes } \begin{bmatrix} 1 & 5 & 7 \\ 0 & -6 & -23 \\ 0 & 6 & 4 \end{bmatrix} \quad -3\text{R}_1 + \text{R}_2$$

(continued)

CONCEPTS	EXAMPLES
A system can be solved by matrix methods. Write the augmented matrix and use row operations to obtain a matrix in row echelon form.	Solve using row operations. $\begin{aligned} x + 3y &= 7 \\ 2x + y &= 4 \end{aligned}$ $\begin{bmatrix} 1 & 3 & \mid & 7 \\ 2 & 1 & \mid & 4 \end{bmatrix}$ Write the augmented matrix. $\begin{bmatrix} 1 & 3 & \mid & 7 \\ 0 & -5 & \mid & -10 \end{bmatrix}$ $-2R_1 + R_2$ $\begin{bmatrix} 1 & 3 & \mid & 7 \\ 0 & 1 & \mid & 2 \end{bmatrix}$ $-\frac{1}{5}R_2$ implies $\begin{aligned} x + 3y &= 7 \\ y &= 2 \end{aligned}$ When $y = 2$, $x + 3(2) = 7$, so $x = 1$. The solution set is $\{(1, 2)\}$.

CHAPTER 4 REVIEW EXERCISES

4.1

1. The graph shows the trends during the years 1975 through 2007 relating to bachelor's degrees awarded in the United States.

   (a) Between what years shown on the horizontal axis did the number of degrees for men equal that for women?

   (b) When the number of degrees for men was equal to that for women, what was that number (approximately)?

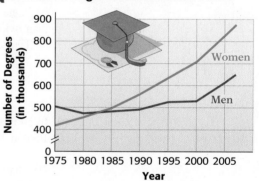

Bachelor's Degrees in the United States

Source: U.S. National Center for Education Statistics.

2. Solve the system by graphing.

$$\begin{aligned} x + 3y &= 8 \\ 2x - y &= 2 \end{aligned}$$

3. *Concept Check* Which ordered pair is *not* a solution of the equation $3x + 2y = 6$?

   A. $(2, 0)$ **B.** $(0, 3)$ **C.** $(4, -3)$ **D.** $(3, -2)$

4. *Concept Check* Suppose that two linear equations are graphed on the same set of coordinate axes. Sketch what the graph might look like if the system has the given description.

   (a) The system has a single solution.

   (b) The system has no solution.

   (c) The system has infinitely many solutions.

Solve each system by the substitution method. If a system is inconsistent or has dependent equations, say so.

5. $3x + y = -4$
$x = \dfrac{2}{3}y$

6. $9x - y = -4$
$y = x + 4$

7. $-5x + 2y = -2$
$x + 6y = 26$

Solve each system of equations by the elimination method. If a system is inconsistent or has dependent equations, say so.

8. $5x + y = 12$
$2x - 2y = 0$

9. $x - 4y = -4$
$3x + y = 1$

10. $6x + 5y = 4$
$-4x + 2y = 8$

11. $\dfrac{1}{6}x + \dfrac{1}{6}y = -\dfrac{1}{2}$
$x - y = -9$

12. $-3x + y = 6$
$y = 6 + 3x$

13. $5x - 4y = 2$
$-10x + 8y = 7$

14. Without doing any algebraic work, but answering on the basis of your knowledge of the graphs of the two lines, explain why the following system has $\emptyset$ as its solution set.

$$y = 3x + 2$$
$$y = 3x - 4$$

4.2   *Solve each system. If a system is inconsistent or has dependent equations, say so.*

15. $2x + 3y - z = -16$
$x + 2y + 2z = -3$
$-3x + y + z = -5$

16. $4x - y = 2$
$3y + z = 9$
$x + 2z = 7$

17. $3x - y - z = -8$
$4x + 2y + 3z = 15$
$-6x + 2y + 2z = 10$

4.3   *Solve each problem by using a system of equations.*

18. A regulation National Hockey League ice rink has perimeter 570 ft. The length of the rink is 30 ft longer than twice the width. What are the dimensions of an NHL ice rink? (*Source:* www.nhl.com)

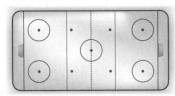

19. In 2009, the New York Yankees and the Boston Red Sox had the most expensive ticket prices in Major League Baseball. Two Yankees tickets and three Red Sox tickets purchased at their average prices cost $296.66, while three Yankees tickets and two Red Sox tickets cost $319.39. Find the average ticket price for a Yankees ticket and a Red Sox ticket. (*Source:* Team Marketing Report.)

20. A plane flies 560 mi in 1.75 hr traveling with the wind. The return trip later against the same wind takes the plane 2 hr. Find the speed of the plane and the speed of the wind. Let x = the speed of the plane and y = the speed of the wind.

	r	t	d
With Wind	$x + y$	1.75	
Against Wind		2	

21. For Valentine's Day, Ms. Sweet will mix some $2-per-lb nuts with some $1-per-lb chocolate candy to get 100 lb of mix, which she will sell at $1.30 per lb. How many pounds of each should she use?

	Number of Pounds	Price per Pound	Value
Nuts	x		
Chocolate	y		
Mixture	100		

22. The sum of the measures of the angles of a triangle is 180°. The largest angle measures 10° less than the sum of the other two. The measure of the middle-sized angle is the average of the other two. Find the measures of the three angles.

23. Noemi Alfonso-Triana sells real estate. On three recent sales, she made 10% commission, 6% commission, and 5% commission. Her total commissions on these sales were $17,000, and she sold property worth $280,000. If the 5% sale amounted to the sum of the other two, what were the three sales prices?

24. How many liters each of 8%, 10%, and 20% hydrogen peroxide should be mixed together to get 8 L of 12.5% solution if the amount of 8% solution used must be 2 L more than the amount of 20% solution used?

25. In the great baseball year of 1961, Yankee teammates Mickey Mantle, Roger Maris, and Yogi Berra combined for 137 home runs. Mantle hit 7 fewer than Maris. Maris hit 39 more than Berra. What were the home run totals for each player? (*Source:* Neft, David S., Richard M. Cohen, and Michael Lo Neft, *The Sports Encyclopedia: Baseball 2006.*)

4.4 *Solve each system of equations by using row operations.*

26. $2x + 5y = -4$
$4x - y = 14$

27. $6x + 3y = 9$
$-7x + 2y = 17$

28. $x + 2y - z = 1$
$3x + 4y + 2z = -2$
$-2x - y + z = -1$

29. $x + 3y = 7$
$3x + z = 2$
$y - 2z = 4$

MIXED REVIEW EXERCISES

30. *Concept Check* Which system, A or B, would be easier to solve using the substitution method? Why?

A. $5x - 3y = 7$
$2x + 8y = 3$

B. $7x + 2y = 4$
$y = -3x + 1$

Solve by any method.

31. $\frac{2}{3}x + \frac{1}{6}y = \frac{19}{2}$
$\frac{1}{3}x - \frac{2}{9}y = 2$

32. $2x + 5y - z = 12$
$-x + y - 4z = -10$
$-8x - 20y + 4z = 31$

33. $x = 7y + 10$
$2x + 3y = 3$

34. $x + 4y = 17$
$-3x + 2y = -9$

35. $-7x + 3y = 12$
$5x + 2y = 8$

36. $2x - 5y = 8$
$3x + 4y = 10$

37. To make a 10% acid solution, Jeffrey Guild wants to mix some 5% solution with 10 L of 20% solution. How many liters of 5% solution should he use?

38. In the 2010 Winter Olympics, Germany, the United States, and Canada won a combined total of 93 medals. Germany won seven fewer medals than the United States, while Canada won 11 fewer medals than the United States. How many medals did each country win? (*Source:* www.vancouver2010.com/olympic-medals)

CHAPTER 4 TEST

Step-by-step test solutions are found on the Chapter Test Prep Videos available via the Video Resources on DVD, in *MyMathLab*, or on YouTube (search "LialIntermediateAlg").

View the complete solutions to all Chapter Test exercises on the Video Resources on DVD.

If the rates of growth between 1990 and 2000 continue, the populations of Houston, Phoenix, Dallas, and Philadelphia will follow the trends indicated in the graph. Use the graph to work Exercises 1 and 2.

The Growth Game

Size of cities if the rate of population growth from 1990 to 2000 continues:

PROJECTED
Houston
Phoenix
Dallas
Philadelphia

Population (in millions): 3.5, 3.0, 2.5, 2.0, 1.5, 1.0

Year: 1990 2000 2010 2020 2030

Source: U.S. Census Bureau, *Chronicle* research.

1. (a) Which of these cities will experience population growth?

(b) Which city will experience population decline?

(c) Rank the city populations from least to greatest for the year 2000.

2. (a) In which year will the population of Dallas equal that of Philadelphia? About what will this population be?

(b) Write as an ordered pair (year, population in millions) the point at which Houston and Phoenix will have the same population.

3. Use a graph to solve the system.
$$x + y = 7$$
$$x - y = 5$$

Solve each system by substitution or elimination. If a system is inconsistent or has dependent equations, say so.

4. $2x - 3y = 24$
$y = -\dfrac{2}{3}x$

5. $3x - y = -8$
$2x + 6y = 3$

6. $12x - 5y = 8$
$3x = \dfrac{5}{4}y + 2$

7. $3x + y = 12$
$2x - y = 3$

8. $-5x + 2y = -4$
$6x + 3y = -6$

9. $3x + 4y = 8$
$8y = 7 - 6x$

10. $3x + 5y + 3z = 2$
$6x + 5y + z = 0$
$3x + 10y - 2z = 6$

11. $4x + y + z = 11$
$x - y - z = 4$
$y + 2z = 0$

Solve each problem using a system of equations.

12. Harrison Ford is a box-office star. As of January 2010, his two top-grossing domestic films, *Star Wars Episode IV: A New Hope* and *Indiana Jones and the Kingdom of the Crystal Skull,* earned $778.0 million together. If *Indiana Jones and the Kingdom of the Crystal Skull* grossed $144.0 million less than *Star Wars Episode IV: A New Hope,* how much did each film gross? (*Source:* www.the-numbers.com)

13. Two cars start from points 420 mi apart and travel toward each other. They meet after 3.5 hr. Find the average rate of each car if one travels 30 mph slower than the other.

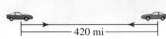

420 mi

14. A chemist needs 12 L of a 40% alcohol solution. She must mix a 20% solution and a 50% solution. How many liters of each will be required to obtain what she needs?

15. A local electronics store will sell seven AC adaptors and two rechargeable flashlights for $86, or three AC adaptors and four rechargeable flashlights for $84. What is the price of a single AC adaptor and a single rechargeable flashlight?

16. The owner of a tea shop wants to mix three kinds of tea to make 100 oz of a mixture that will sell for $0.83 per oz. He uses Orange Pekoe, which sells for $0.80 per oz, Irish Breakfast, for $0.85 per oz, and Earl Grey, for $0.95 per oz. If he wants to use twice as much Orange Pekoe as Irish Breakfast, how much of each kind of tea should he use?

Solve each system using row operations.

17. $3x + 2y = 4$
$5x + 5y = 9$

18. $x + 3y + 2z = 11$
$3x + 7y + 4z = 23$
$5x + 3y - 5z = -14$

CHAPTERS (1–4)

CUMULATIVE REVIEW EXERCISES

Perform each operation.

1. $-\dfrac{3}{4} - \dfrac{2}{5}$

2. $\dfrac{8}{15} \div \left(-\dfrac{12}{5}\right)$

Evaluate each expression if possible.

3. $(-3)^4$

4. -3^4

5. $-(-3)^4$

6. $\sqrt{0.49}$

7. $-\sqrt{0.49}$

8. $\sqrt{-0.49}$

Evaluate for $x = -4$, $y = 3$, and $z = 6$.

9. $|2x| + 3y - z^3$

10. $-5(x^3 - y^3)$

11. Which property of real numbers justifies the statement $5 + (3 \cdot 6) = 5 + (6 \cdot 3)$?

Solve each equation.

12. $7(2x + 3) - 4(2x + 1) = 2(x + 1)$

13. $|6x - 8| = 4$

14. $ax + by = d$ for x

15. $0.04x + 0.06(x - 1) = 1.04$

Solve each inequality.

16. $\dfrac{2}{3}x + \dfrac{5}{12}x \le 20$

17. $|3x + 2| \le 4$

18. $|12t + 7| \ge 0$

19. $2x + 3 > 5$ or $x - 1 \le 6$

20. A survey measured public recognition of some popular contemporary advertising slogans. Complete the results shown in the table if 2500 people were surveyed.

Slogan (product or company)	Percent Recognition (nearest tenth of a percent)	Actual Number Who Recognized Slogan (nearest whole number)
Please Don't Squeeze the . . . (Charmin)	80.4%	
The Breakfast of Champions (Wheaties)	72.5%	
The King of Beers (Budweiser)		1570
Like a Good Neighbor (State Farm)		1430

Source: Department of Integrated Marketing Communications, Northwestern University.

Solve each problem.

21. A jar contains only pennies, nickels, and dimes. The number of dimes is one more than the number of nickels, and the number of pennies is six more than the number of nickels. How many of each denomination are in the jar, if the total value is $4.80?

22. Two angles of a triangle have the same measure. The measure of the third angle is 4° less than twice the measure of each of the equal angles. Find the measures of the three angles.

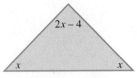

Measures are in degrees.

In Exercises 23–27, point A has coordinates $(-2, 6)$ and point B has coordinates $(4, -2)$.

23. What is the equation of the horizontal line through A?

24. What is the equation of the vertical line through B?

25. What is the slope of line AB?

26. What is the slope of a line perpendicular to line AB?

27. What is the standard form of the equation of line AB?

28. Graph the line having slope $\frac{2}{3}$ and passing through the point $(-1, -3)$.

29. Graph the inequality $-3x - 2y \leq 6$.

30. Given that $f(x) = x^2 + 3x - 6$, find **(a)** $f(-3)$ and   **(b)** $f(a)$.

Solve by any method.

31. $-2x + 3y = -15$
$\quad 4x - \ y = 15$

32. $\ x - 3y = 7$
$\quad 2x - 6y = 14$

33. $\quad x + y + z = 10$
$\quad\ x - y - z = 0$
$\quad -x + y - z = -4$

34. Ten years after the original Tickle Me Elmo became a must-have toy, a new version, called T.M.X., was released in the fall of 2006. The original Tickle Me Elmo's average cost was $12.37 less than the recommended cost of T.M.X., and one of each cost $67.63. Find the average cost of Tickle Me Elmo and the recommended cost of T.M.X. (*Source:* NPD Group, Inc.; *USA Today*.)

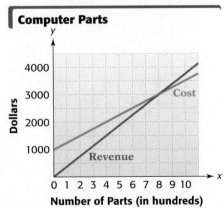

35. The graph shows a company's costs to produce computer parts and the revenue from the sale of computer parts.

(a) At what production level does the cost equal the revenue? What is the revenue at that point?

(b) Profit is revenue less cost. Estimate the profit on the sale of 1100 parts.

Exponents, Polynomials, and Polynomial Functions

One of the most popular tourist sites in the world is the magnificent Taj Mahal (or "the Taj"), at Agra, India, which attracts about 2.4 million visitors per year. The Taj, a domed white marble tomb built by Emperor Shah Jahan in memory of his wife, was completed in the year 1648. (*Source:* www.greatbuildings.com, www.forbestraveler.com)

We introduced the concept of function in **Section 3.5** and extend our work to include *polynomial functions* in this chapter. In **Exercise 11** of **Section 5.3**, we use a polynomial function to model the amount Americans spend on foreign travel.

5.1 Integer Exponents and Scientific Notation

Recall that we use exponents to write products of repeated factors. For example,

$$2^5 \text{ is defined as } 2 \cdot 2 \cdot 2 \cdot 2 \cdot 2 = 32.$$

The number 5, the **exponent**, shows that the **base** 2 appears as a factor five times. The quantity 2^5 is called an **exponential** or a **power**. We read 2^5 as **"2 to the fifth power"** or **"2 to the fifth."**

OBJECTIVE 1 **Use the product rule for exponents.** Consider the product $2^5 \cdot 2^3$, which can be simplified as follows.

$$\overbrace{2^5 \cdot 2^3 = (2 \cdot 2 \cdot 2 \cdot 2 \cdot 2)(2 \cdot 2 \cdot 2) = 2^8}^{5 + 3 = 8}$$

This result, that products of exponential expressions with the *same base* are found by adding exponents, is generalized as the **product rule for exponents.**

Product Rule for Exponents

If m and n are natural numbers and a is any real number, then

$$a^m \cdot a^n = a^{m+n}.$$

That is, when multiplying powers of like bases, keep the same base and add the exponents.

To see that the product rule is true, use the definition of an exponent.

$$a^m = \underbrace{a \cdot a \cdot a \cdot \ldots \cdot a}_{a \text{ appears as a factor } m \text{ times.}} \qquad a^n = \underbrace{a \cdot a \cdot a \cdot \ldots \cdot a}_{a \text{ appears as a factor } n \text{ times.}}$$

$$a^m \cdot a^n = \underbrace{a \cdot a \cdot a \cdot \ldots \cdot a}_{m \text{ factors}} \cdot \underbrace{a \cdot a \cdot a \cdot \ldots \cdot a}_{n \text{ factors}}$$

$$= \underbrace{a \cdot a \cdot a \cdot \ldots \cdot a}_{(m+n) \text{ factors}}$$

$$a^m \cdot a^n = a^{m+n}$$

NOW TRY
EXERCISE 1

Apply the product rule, if possible, in each case.

(a) $8^5 \cdot 8^4$

(b) $(5x^4y^7)(-7xy^3)$

(c) $p^2 \cdot q^2$

EXAMPLE 1 Using the Product Rule for Exponents

Apply the product rule for exponents, if possible, in each case.

(a) $3^4 \cdot 3^7 = 3^{4+7} = 3^{11}$ — Do *not* multiply the bases. Keep the same base.

(b) $5^3 \cdot 5 = 5^3 \cdot 5^1 = 5^{3+1} = 5^4$

(c) $y^3 \cdot y^8 \cdot y^2 = y^{3+8+2} = y^{13}$

(d) $(5y^2)(-3y^4)$

$= 5(-3)y^2y^4$ Commutative property

$= -15y^{2+4}$ Multiply; product rule

$= -15y^6$

(e) $(7p^3q)(2p^5q^2)$

$= 7(2)p^3p^5q^1q^2$

$= 14p^{3+5}q^{1+2}$

$= 14p^8q^3$

(f) $x^2 \cdot y^4$ The product rule does not apply because the bases are not the same.

NOW TRY

NOW TRY ANSWERS

1. (a) 8^9 **(b)** $-35x^5y^{10}$
(c) The product rule does not apply.

⚠️ **CAUTION**   Be careful not to multiply the bases. In **Example 1(a)**, $3^4 \cdot 3^7 = 3^{11}$, *not* 9^{11}. *Keep the same base and add the exponents.*

OBJECTIVE 2 **Define 0 and negative exponents.** Consider the following, where the product rule is applied to an exponent that is not a natural number.

$$4^2 \cdot 4^0 = 4^{2+0} = 4^2$$

For the product rule to hold, 4^0 must equal 1, so we define a^0 this way for any nonzero real number a.

Zero Exponent

If a is any nonzero real number, then

$$a^0 = 1.$$

---

*The expression 0^0 is undefined.**

NOW TRY
EXERCISE 2
Evaluate.

(a) 5^0 **(b)** $(-5x)^0$, $x \neq 0$

(c) -5^0 **(d)** $10^0 - 9^0$

EXAMPLE 2 Using 0 as an Exponent

Evaluate.

(a) $6^0 = 1$

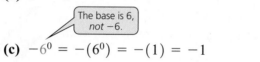
The base is 6, not −6.

(b) $(-6)^0 = 1$

Here the base is −6.

(c) $-6^0 = -(6^0) = -(1) = -1$

(d) $-(-6)^0 = -1$

(e) $5^0 + 12^0$

$= 1 + 1$

$= 2$

(f) $(8k)^0 = 1$, $k \neq 0$

NOW TRY

To define a negative exponent, we extend the product rule, as follows.

$$8^2 \cdot 8^{-2} = 8^{2+(-2)} = 8^0 = 1$$

Here, 8^{-2} is the reciprocal of 8^2. But $\frac{1}{8^2}$ is the reciprocal of 8^2, and a number can have only one reciprocal. Therefore, $8^{-2} = \frac{1}{8^2}$. We can generalize this result.

Negative Exponent

For any natural number n and any nonzero real number a,

$$a^{-n} = \frac{1}{a^n}.$$

With this definition, the expression a^n is meaningful for any integer exponent n and any nonzero real number a.

NOW TRY ANSWERS
2. (a) 1   (b) 1   (c) −1   (d) 0

*In advanced treatments, 0^0 is called an *indeterminate form.*

⚠️ **CAUTION** *A negative exponent does not indicate that an expression repre-sents a negative number.* Negative exponents lead to reciprocals.

$$3^{-2} = \frac{1}{3^2} = \frac{1}{9} \quad \text{Not negative} \quad \Big| \quad -3^{-2} = -\frac{1}{3^2} = -\frac{1}{9} \quad \text{Negative}$$

⤷ **NOW TRY**
EXERCISE 3

Write with only positive exponents.

(a) 9^{-4}

(b) $(3y)^{-6}, \quad y \neq 0$

(c) $-4k^{-3}, \quad k \neq 0$

(d) Evaluate $4^{-1} + 6^{-1}$.

EXAMPLE 3 Using Negative Exponents

In parts (a)–(f), write with only positive exponents.

(a) $2^{-3} = \frac{1}{2^3} \qquad a^{-n} = \frac{1}{a^n}$

(b) $6^{-1} = \frac{1}{6^1} = \frac{1}{6} \qquad a^{-n} = \frac{1}{a^n}$

(c) $(5z)^{-3} = \frac{1}{(5z)^3}, \quad z \neq 0$
↑
Base is $5z$.

(d) $5z^{-3} = 5\left(\frac{1}{z^3}\right) = \frac{5}{z^3}, \quad z \neq 0$
↑
Base is z.

(e) $-m^{-2} = -\frac{1}{m^2}, \quad m \neq 0$
(What is the base here?)

(f) $(-m)^{-2} = \frac{1}{(-m)^2}, \quad m \neq 0$
(What is the base here?)

In parts (g) and (h), evaluate.

(g) $3^{-1} + 4^{-1}$

$= \frac{1}{3} + \frac{1}{4}$ Definition of negative exponent

$= \frac{4}{12} + \frac{3}{12}$ $\frac{1}{3} \cdot \frac{4}{4} = \frac{4}{12}; \frac{1}{4} \cdot \frac{3}{3} = \frac{3}{12}$

$= \frac{7}{12}$ $\frac{a}{c} + \frac{b}{c} = \frac{a+b}{c}$

(h) $5^{-1} - 2^{-1}$

$= \frac{1}{5} - \frac{1}{2}$ Definition of negative exponent

$= \frac{2}{10} - \frac{5}{10}$ Get a common denominator.

$= -\frac{3}{10}$ $\frac{a}{c} - \frac{b}{c} = \frac{a-b}{c}$

NOW TRY ⤻

⚠️ **CAUTION** In Example 3(g), note that $3^{-1} + 4^{-1} \neq (3 + 4)^{-1}$. The expres-sion $3^{-1} + 4^{-1}$ simplifies to $\frac{7}{12}$, as shown in the example, while the expression $(3 + 4)^{-1}$ simplifies to 7^{-1}, or $\frac{1}{7}$. Similar reasoning can be applied to part (h).

⤷ **NOW TRY**
EXERCISE 4

Evaluate.

(a) $\frac{1}{5^{-3}}$ **(b)** $\frac{10^{-2}}{2^{-5}}$

EXAMPLE 4 Using Negative Exponents

Evaluate.

(a) $\dfrac{1}{2^{-3}} = \dfrac{1}{\frac{1}{2^3}} = 1 \div \frac{1}{2^3} = 1 \cdot \frac{2^3}{1} = 2^3 = 8$

Multiply by the reciprocal of the divisor.

(b) $\dfrac{2^{-3}}{3^{-2}} = \dfrac{\frac{1}{2^3}}{\frac{1}{3^2}} = \frac{1}{2^3} \div \frac{1}{3^2} = \frac{1}{2^3} \cdot \frac{3^2}{1} = \frac{3^2}{2^3} = \frac{9}{8}$

NOW TRY ⤻

Example 4 suggests the following generalizations.

Special Rules for Negative Exponents

If $a \neq 0$ and $b \neq 0$, then $\qquad \dfrac{1}{a^{-n}} = a^n \qquad$ and $\qquad \dfrac{a^{-n}}{b^{-m}} = \dfrac{b^m}{a^n}.$

OBJECTIVE 3   Use the quotient rule for exponents. We simplify a quotient, such as $\frac{a^8}{a^3}$, in much the same way as a product. (In all quotients of this type, assume that the denominator is not 0.) Consider this example.

$$\frac{a^8}{a^3} = \frac{a \cdot a \cdot a \cdot a \cdot a \cdot a \cdot a \cdot a}{a \cdot a \cdot a} = a \cdot a \cdot a \cdot a \cdot a = a^5$$

Notice that $8 - 3 = 5$. In the same way, we simplify $\frac{a^3}{a^8}$.

$$\frac{a^3}{a^8} = \frac{a \cdot a \cdot a}{a \cdot a \cdot a \cdot a \cdot a \cdot a \cdot a \cdot a} = \frac{1}{a^5} = a^{-5}$$

Here $3 - 8 = -5$. These examples suggest the **quotient rule for exponents.**

Quotient Rule for Exponents

If a is any nonzero real number and m and n are integers, then

$$\frac{a^m}{a^n} = a^{m-n}.$$

That is, when dividing powers of like bases, keep the same base and subtract the exponent of the denominator from the exponent of the numerator.

NOW TRY
EXERCISE 5

Apply the quotient rule, if possible, and write each result with only positive exponents.

(a) $\dfrac{t^8}{t^2},\quad t \neq 0$ (b) $\dfrac{4^5}{4^{-2}}$

(c) $\dfrac{m^4}{n^3},\quad n \neq 0$

NOW TRY ANSWERS
5. (a) t^6 **(b)** 4^7
   (c) The quotient rule does not apply.

EXAMPLE 5   Using the Quotient Rule for Exponents

Apply the quotient rule for exponents, if possible, and write each result with only positive exponents.

Numerator exponent
Denominator exponent

(a) $\dfrac{3^7}{3^2} = 3^{7-2} = 3^5$ (b) $\dfrac{p^6}{p^2} = p^{6-2} = p^4,\quad p \neq 0$

Subtraction symbol

Use parentheses to avoid errors.

(c) $\dfrac{k^7}{k^{12}} = k^{7-12} = k^{-5} = \dfrac{1}{k^5},\quad k \neq 0$ (d) $\dfrac{2^7}{2^{-3}} = 2^{7-(-3)} = 2^{7+3} = 2^{10}$

(e) $\dfrac{8^{-2}}{8^5} = 8^{-2-5} = 8^{-7} = \dfrac{1}{8^7}$ (f) $\dfrac{6}{6^{-1}} = \dfrac{6^1}{6^{-1}} = 6^{1-(-1)} = 6^2$

(g) $\dfrac{z^{-5}}{z^{-8}} = z^{-5-(-8)} = z^3,\quad z \neq 0$ (h) $\dfrac{a^3}{b^4},\quad b \neq 0$ This expression cannot be simplified further.

Be careful with signs.

The quotient rule does not apply because the bases are different.

NOW TRY

⚠️ **CAUTION** Be careful when working with quotients that involve negative exponents in the denominator. Write the numerator exponent, then a subtraction symbol, and then the denominator exponent. Use parentheses.

OBJECTIVE 4 **Use the power rules for exponents.** We can simplify $(3^4)^2$ as follows.

$$(3^4)^2 = 3^4 \cdot 3^4 = 3^{4+4} = 3^8$$

Notice that $4 \cdot 2 = 8$. This example suggests the first **power rule for exponents.** The other two power rules can be demonstrated with similar examples.

Power Rules for Exponents

If a and b are real numbers and m and n are integers, then

(a) $(a^m)^n = a^{mn}$, **(b)** $(ab)^m = a^m b^m$, and **(c)** $\left(\dfrac{a}{b}\right)^m = \dfrac{a^m}{b^m}$ $(b \neq 0)$.

That is,

(a) To raise a power to a power, multiply exponents.

(b) To raise a product to a power, raise each factor to that power.

(c) To raise a quotient to a power, raise the numerator and the denominator to that power.

NOW TRY
EXERCISE 6

Simplify, using the power rules.

(a) $(-2m^3)^4$

(b) $\left(\dfrac{3x^2}{y^3}\right)^3$, $y \neq 0$

EXAMPLE 6 Using the Power Rules for Exponents

Simplify, using the power rules.

(a) $(p^8)^3$
$= p^{8 \cdot 3}$ Power rule (a)
$= p^{24}$

(b) $(3y)^4$
$= 3^4 y^4$ Power rule (b)
$= 81y^4$

(c) $\left(\dfrac{2}{3}\right)^4$
$= \dfrac{2^4}{3^4}$ Power rule (c)
$= \dfrac{16}{81}$

(d) $(6p^7)^2$
$= 6^2 p^{7 \cdot 2}$ Power rule (b)
$= 6^2 p^{14}$ Multiply exponents.
$= 36p^{14}$ Square 6.

(e) $\left(\dfrac{-2m^5}{z}\right)^3$
$= \dfrac{(-2m^5)^3}{z^3}$ Power rule (c)
$= \dfrac{(-2)^3 m^{5 \cdot 3}}{z^3}$ Power rule (b)
$= \dfrac{-8m^{15}}{z^3}$, $z \neq 0$ Simplify.
NOW TRY

The reciprocal of a^n is $\dfrac{1}{a^n} = \left(\dfrac{1}{a}\right)^n$. Also, a^n and a^{-n} are reciprocals.

$$a^n \cdot a^{-n} = a^n \cdot \dfrac{1}{a^n} = 1$$

NOW TRY ANSWERS

6. (a) $16m^{12}$ **(b)** $\dfrac{27x^6}{y^9}$

Thus, since both $\left(\frac{1}{a}\right)^n$ and a^{-n} are reciprocals of a^n, the follo

$$a^{-n} = \left(\frac{1}{a}\right)^n$$

Some examples of this result are

$$6^{-3} = \left(\frac{1}{6}\right)^3 \quad \text{and} \quad \left(\frac{1}{3}\right)^{-2} = 3^2.$$

Special Rules for Negative Exponents, Continued

If $a \neq 0$ and $b \neq 0$ and n is an integer, then

$$a^{-n} = \left(\frac{1}{a}\right)^n \quad \text{and} \quad \left(\frac{a}{b}\right)^{-n} = \left(\frac{b}{a}\right)^n.$$

That is, any nonzero number raised to the negative nth power is equal to the reciprocal of that number raised to the nth power.

NOW TRY
EXERCISE 7

Write with only positive exponents, and then evaluate.

$$\left(\frac{5}{3}\right)^{-3}$$

EXAMPLE 7 Using Negative Exponents with Fractions

Write with only positive exponents and then evaluate.

(a) $\left(\frac{3}{7}\right)^{-2}$

$= \left(\frac{7}{3}\right)^2$ — Change the fraction to its reciprocal and change the sign of the exponent.

$= \dfrac{49}{9}$

(b) $\left(\frac{4x}{5}\right)^{-3}, \quad x \neq 0$

$= \left(\frac{5}{4x}\right)^3$

$= \dfrac{5^3}{4^3 x^3}, \quad$ or $\quad \dfrac{125}{64x^3}$ *NOW TRY*

The definitions and rules of this section are summarized here.

Definitions and Rules for Exponents

For all integers m and n and all real numbers a and b, the following rules apply.

Product Rule $\quad a^m \cdot a^n = a^{m+n}$

Quotient Rule $\quad \dfrac{a^m}{a^n} = a^{m-n} \quad (a \neq 0)$

Zero Exponent $\quad a^0 = 1 \quad (a \neq 0)$

Negative Exponent $\quad a^{-n} = \dfrac{1}{a^n} \quad (a \neq 0)$

Power Rules $\quad$ **(a)** $(a^m)^n = a^{mn}$ $\qquad$ **(b)** $(ab)^m = a^m b^m$

$\qquad\qquad$ **(c)** $\left(\dfrac{a}{b}\right)^m = \dfrac{a^m}{b^m} \quad (b \neq 0)$

Special Rules $\quad \dfrac{1}{a^{-n}} = a^n \quad (a \neq 0) \qquad \dfrac{a^{-n}}{b^{-m}} = \dfrac{b^m}{a^n} \quad (a, b \neq 0)$

$\qquad\qquad\qquad a^{-n} = \left(\dfrac{1}{a}\right)^n \quad (a \neq 0) \qquad \left(\dfrac{a}{b}\right)^{-n} = \left(\dfrac{b}{a}\right)^n \quad (a, b \neq 0)$

NOW TRY ANSWER

7. $\frac{27}{125}$

OBJECTIVE 5 Simplify exponential expressions.

NOW TRY
EXERCISE 8

Simplify. Assume that all variables represent nonzero real numbers.

(a) $x^{-8} \cdot x \cdot x^4$

(b) $(5^{-3})^2$

(c) $\dfrac{p^2 q^{-4}}{p^{-2} q^{-1}}$

(d) $\left(\dfrac{2x^2}{y^2}\right)^3 \left(\dfrac{-5x^{-2}}{y}\right)^{-2}$

EXAMPLE 8 Using the Definitions and Rules for Exponents

Simplify each exponential expression so that no negative exponents appear in the final result. Assume that all variables represent nonzero real numbers.

(a) $3^2 \cdot 3^{-5}$

$= 3^{2+(-5)}$ Product rule

$= 3^{-3}$ Add exponents.

$= \dfrac{1}{3^3}$,   or $\dfrac{1}{27}$ $a^{-n} = \frac{1}{a^n}$

(b) $x^{-3} \cdot x^{-4} \cdot x^2$

$= x^{-3+(-4)+2}$ Product rule

$= x^{-5}$ Add exponents.

$= \dfrac{1}{x^5}$ $a^{-n} = \frac{1}{a^n}$

(c) $(4^{-2})^{-5}$

$= 4^{(-2)(-5)}$ Power rule (a)

$= 4^{10}$ Multiply exponents.

(d) $(x^{-4})^6$

$= x^{(-4)6}$ Power rule (a)

$= x^{-24}$ Multiply exponents.

$= \dfrac{1}{x^{24}}$ $a^{-n} = \frac{1}{a^n}$

(e) $\dfrac{x^{-4}y^2}{x^2 y^{-5}}$

$= \dfrac{x^{-4}}{x^2} \cdot \dfrac{y^2}{y^{-5}}$

$= x^{-4-2} \cdot y^{2-(-5)}$ Quotient rule

$= x^{-6} y^7$

$= \dfrac{y^7}{x^6}$ $a^{-n} = \frac{1}{a^n}$

(f) $(2^3 x^{-2})^{-2}$

$= (2^3)^{-2} \cdot (x^{-2})^{-2}$ Power rule (b)

$= 2^{-6} x^4$

$= \dfrac{x^4}{2^6}$,   or $\dfrac{x^4}{64}$ $a^{-n} = \frac{1}{a^n}$

(g) $\left(\dfrac{3x^2}{y}\right)^2 \left(\dfrac{4x^3}{y^{-2}}\right)^{-1}$

$= \dfrac{3^2(x^2)^2}{y^2} \cdot \dfrac{y^{-2}}{4x^3}$ Combination of rules

$= \dfrac{9x^4}{y^2} \cdot \dfrac{y^{-2}}{4x^3}$ Power rule (a)

$= \dfrac{9}{4} x^{4-3} y^{-2-2}$ Quotient rule

$= \dfrac{9}{4} x^1 y^{-4}$ Subtract exponents.

$= \dfrac{9x}{4y^4}$ $a^{-n} = \frac{1}{a^n}$

(h) $\left(\dfrac{-4m^5 n^4}{24mn^{-7}}\right)^{-2}$

$= \left(\dfrac{m^{5-1} n^{4-(-7)}}{-6}\right)^{-2}$ Quotient rule; divide coefficients.

$= \left(\dfrac{m^4 n^{11}}{-6}\right)^{-2}$ Subtract exponents.

$= \dfrac{(m^4)^{-2}(n^{11})^{-2}}{(-6)^{-2}}$ Power rules (b) and (c)

$= \dfrac{m^{-8} n^{-22}}{(-6)^{-2}}$ Power rule (a)

The sign on -6 does *not* change in this step.

$= \dfrac{(-6)^2}{m^8 n^{22}}$ $\frac{a^{-n}}{b^{-m}} = \frac{b^m}{a^n}$

$= \dfrac{36}{m^8 n^{22}}$ $(-6)^2 = 36$

NOW TRY

NOW TRY ANSWERS
8. (a) $\dfrac{1}{x^3}$ (b) $\dfrac{1}{5^6}$
(c) $\dfrac{p^4}{q^3}$ (d) $\dfrac{8x^{10}}{25y^4}$

NOTE There is often more than one way to simplify expressions involving negative exponents.

In **Example 8(e),** we began by using the quotient rule. At the right, we simplify the same expression by using one of the special rules for exponents. The final result is the same.

$$\frac{x^{-4}y^2}{x^2y^{-5}}$$ Example 8(e)

$$= \frac{y^5 y^2}{x^4 x^2}$$ Use $\frac{a^{-n}}{b^{-m}} = \frac{b^m}{a^n}$.

$$= \frac{y^7}{x^6}$$ Product rule

OBJECTIVE 6 **Use the rules for exponents with scientific notation.** The number of one-celled organisms that will sustain a whale for a few hours is 400,000,000,000,000, and the shortest wavelength of visible light is approximately 0.0000004 m. It is often simpler to write such numbers in *scientific notation.*

In scientific notation, a number is written with the decimal point after the first nonzero digit and multiplied by a power of 10.

Scientific Notation

A number is written in **scientific notation** when it is expressed in the form

$$a \times 10^n, \quad \text{where } 1 \le |a| < 10 \text{ and } n \text{ is an integer.}$$

> It is customary to use $\times$ rather than $\cdot$ for multiplication.

$$8500 = 8.5 \times 1000 = 8.5 \times 10^3 \quad \text{In scientific notation}$$

$$0.230 \times 10^4 \qquad 46.5 \times 10^{-3} \qquad \textit{Not in scientific notation}$$

0.230 is less than 1.   46.5 is greater than 10.

To write a number in scientific notation, use the following steps. (If the number is negative, ignore the negative sign, go through these steps, and then attach a negative sign to the result.)

Converting a Positive Number to Scientific Notation

Step 1 **Position the decimal point.** Place a caret, ^, to the right of the first nonzero digit, where the decimal point will be placed.

Step 2 **Determine the numeral for the exponent.** Count the number of digits from the decimal point to the caret. This number gives the absolute value of the exponent on 10.

Step 3 **Determine the sign for the exponent.** Decide whether multiplying by 10^n should make the result of Step 1 greater or less. The exponent should be positive to make the result greater. It should be negative to make the result less.

It is helpful to remember that, for **$n \ge 1$, $10^{-n} < 1$ and $10^n \ge 10$.**

NOW TRY
EXERCISE 9
Write each number in scientific notation.

(a) 7,560,000,000

(b) −0.000000245

EXAMPLE 9   Writing Numbers in Scientific Notation

Write each number in scientific notation.

(a) 820,000

> **Step 1** Place a caret to the right of the 8 (the first nonzero digit) to mark the new location of the decimal point.
>
> $$8_\wedge 20,000$$
>
> **Step 2** Count from the decimal point, which is understood to be after the last 0, to the caret.
>
> $$8_\wedge 20,000. \leftarrow \text{Decimal point}$$
>
> Count 5 places.
>
> **Step 3** Since 8.2 is to be made greater, the exponent on 10 is positive.
>
> $$820,000 = 8.2 \times 10^5$$

(b) 0.0000072

$$0.000007_\wedge 2 \qquad \text{Count from left to right.}$$

6 places

Since the number 7.2 is to be made less, the exponent on 10 is negative.

$$0.0000072 = 7.2 \times 10^{-6}$$

(c) $-0.000462 = -4.62 \times 10^{-4}$ Remember the negative sign.

Count 4 places.

NOW TRY

Converting a Positive Number from Scientific Notation

Multiplying a positive number by a positive power of 10 makes the number greater, so move the decimal point to the right if n is positive in 10^n.

Multiplying a positive number by a negative power of 10 makes a number less, so move the decimal point to the left if n is negative.

If n is 0, leave the decimal point where it is.

NOW TRY
EXERCISE 10
Write each number in standard notation.

(a) 4.45×10^{10}

(b) -5.9×10^{-5}

EXAMPLE 10   Converting from Scientific Notation to Standard Notation

Write each number in standard notation.

(a) 6.93×10^7

$$6.9300000_\wedge \qquad \text{Attach 0's as necessary.}$$

7 places

We moved the decimal point 7 places to the right. (We had to attach five 0's.)

$$6.93 \times 10^7 = 69,300,000$$

(b) 4.7×10^{-3}

$$_\wedge 004.7 \qquad \text{Attach 0's as necessary.}$$

3 places

We moved the decimal point 3 places to the left.

> Add a leading 0, since the decimal is between 0 and 1.

$$4.7 \times 10^{-3} = .0047, \quad \text{or} \quad 0.0047$$

(c) $-1.083 \times 10^0 = -1.083 \times 1 = -1.083$

NOW TRY

NOW TRY ANSWERS
9. (a) 7.56×10^9
   (b) -2.45×10^{-7}
10. (a) 44,500,000,000
    (b) −0.000059

NOTE When converting from scientific notation to standard notation, *use the exponent to determine the number of places and the direction in which to move the decimal point.*

NOW TRY
EXERCISE 11

Evaluate.

$$\frac{0.00063 \times 400,000}{1400 \times 0.000003}$$

EXAMPLE 11 Using Scientific Notation in Computation

Evaluate.

$$\frac{1,920,000 \times 0.0015}{0.000032 \times 45,000}$$

$$= \frac{1.92 \times 10^6 \times 1.5 \times 10^{-3}}{3.2 \times 10^{-5} \times 4.5 \times 10^4} \qquad \text{Express all numbers in scientific notation.}$$

$$= \frac{1.92 \times 1.5 \times 10^6 \times 10^{-3}}{3.2 \times 4.5 \times 10^{-5} \times 10^4} \qquad \text{Commutative property}$$

$$= \frac{1.92 \times 1.5 \times 10^3}{3.2 \times 4.5 \times 10^{-1}} \qquad \text{Product rule}$$

$$= \frac{1.92 \times 1.5}{3.2 \times 4.5} \times 10^4 \qquad \text{Quotient rule}$$

$$= 0.2 \times 10^4 \qquad \text{Simplify.}$$

Don't stop here! $= (2 \times 10^{-1}) \times 10^4 \qquad \text{Write 0.2 in scientific notation.}$

$$= 2 \times 10^3 \qquad \text{Product rule}$$

$$= 2000 \qquad \text{Standard notation} \qquad \text{NOW TRY}$$

NOW TRY
EXERCISE 12

The speed of light is approximately 30,000,000,000 cm per sec. How long will it take light to travel 1.2×10^{15} cm?

EXAMPLE 12 Using Scientific Notation to Solve Problems

In 1990, private health care expenditures in the United States were \$427 billion. By 2006, this figure had risen by a factor of 2.7—that is, it had risen by over $2\frac{1}{2}$ times in only 16 yr. (*Source:* U.S. Centers for Medicare & Medicaid Services.)

(a) Write the 1990 health care expenditure in scientific notation.

$$427 \text{ billion}$$

$$= 427 \times 10^9 \qquad \text{1 billion} = 1{,}000{,}000{,}000 = 10^9$$

$$= (4.27 \times 10^2) \times 10^9 \qquad \text{Write 427 in scientific notation.}$$

$$= 4.27 \times 10^{11} \qquad \text{Product rule}$$

In 1990, the expenditure was $\$4.27 \times 10^{11}$.

(b) What were expenditures in 2006?

$$(4.27 \times 10^{11}) \times 2.7 \qquad \text{Multiply the result in part (a) by 2.7.}$$

$$= (2.7 \times 4.27) \times 10^{11} \qquad \text{Commutative and associative properties}$$

$$= 11.529 \times 10^{11} \qquad \text{Multiply.}$$

$$= (1.1529 \times 10^1) \times 10^{11} \qquad \text{Write 11.529 in scientific notation.}$$

$$= 1.1529 \times 10^{12} \qquad \text{Product rule}$$

NOW TRY ANSWERS
11. 60,000
12. 4.0×10^4 sec, or 40,000 sec

Expenditures in 2006 were about \$1,152,900,000,000 (over \$1.1 trillion).

NOW TRY

5.1 EXERCISES

MyMathLab PRACTICE WATCH DOWNLOAD READ REVIEW

⊕ *Complete solution available on the Video Resources on DVD*

Concept Check Decide whether each expression has been simplified correctly. If not, correct it.

1. $(ab)^2 = ab^2$

2. $(5x)^3 = 5^3x^3$

3. $\left(\dfrac{3}{a}\right)^4 = \dfrac{3^4}{a}$ $(a \neq 0)$

4. $y^2 \cdot y^4 = y^8$

5. $x^3 \cdot x^4 = x^7$

6. $xy^0 = 0$ $(y \neq 0)$

7. *Concept Check* Your friend evaluated $4^5 \cdot 4^2$ as 16^7. **WHAT WENT WRONG?** Give the correct answer.

8. *Concept Check* Another friend evaluated $\dfrac{6^5}{3^2}$ as 2^3, or 8. **WHAT WENT WRONG?** Give the correct answer.

Apply the product rule for exponents, if possible, in each case. **See Example 1.**

⊕ **9.** $13^4 \cdot 13^8$

10. $11^6 \cdot 11^4$

11. $8^9 \cdot 8$

12. $12 \cdot 12^6$

13. $x^3 \cdot x^5 \cdot x^9$

14. $y^4 \cdot y^5 \cdot y^6$

15. $(-3w^5)(9w^3)$

16. $(-5x^2)(3x^4)$

17. $(2x^2y^5)(9xy^3)$

18. $(8s^4t)(3s^3t^5)$

19. $r^2 \cdot s^4$

20. $p^3 \cdot q^2$

In Exercises 21 and 22, match the expression in Column I with its equivalent expression in Column II. Choices may be used once, more than once, or not at all. * **See Example 2.**

	I	II
⊕ **21.**	**(a)** 9^0	**A.** 0
	(b) -9^0	**B.** 1
	(c) $(-9)^0$	**C.** -1
	(d) $-(-9)^0$	**D.** 9
		E. -9

	I	II
22.	**(a)** $8x^0$	**A.** 0
	(b) $-8x^0$	**B.** 1
	(c) $(8x)^0$	**C.** -1
	(d) $(-8x)^0$	**D.** 8
		E. -8

Evaluate. Assume that all variables represent nonzero real numbers. * **See Example 2.**

23. 15^0

24. 19^0

25. -8^0

26. -10^0

27. $(-25)^0$

28. $(-30)^0$

29. $3^0 + (-3)^0$

30. $5^0 + (-5)^0$

31. $-3^0 + 3^0$

32. $-5^0 + 5^0$

33. $-4^0 - m^0$

34. $-8^0 - k^0$

In Exercises 35 and 36, match the expression in Column I with its equivalent expression in Column II. Choices may be used once, more than once, or not at all. * **See Example 3.**

	I	II
35.	**(a)** 5^{-2}	**A.** 25
	(b) -5^{-2}	**B.** $\dfrac{1}{25}$
	(c) $(-5)^{-2}$	**C.** -25
	(d) $-(-5)^{-2}$	**D.** $-\dfrac{1}{25}$

	I	II
36.	**(a)** 4^{-3}	**A.** 64
	(b) -4^{-3}	**B.** -64
	(c) $(-4)^{-3}$	**C.** $\dfrac{1}{64}$
	(d) $-(-4)^{-3}$	**D.** $-\dfrac{1}{64}$

Write each expression with only positive exponents. Assume that all variables represent nonzero real numbers. In Exercises 49–52, simplify each expression. **See Example 3.**

⊕ **37.** 5^{-4}

38. 7^{-2}

39. 9^{-1}

40. 14^{-1}

41. $(4x)^{-2}$

42. $(5t)^{-3}$

43. $4x^{-2}$

44. $5t^{-3}$

---

*The authors thank Mitchel Levy of Broward College for his suggestions for these exercises.

45. $-a^{-3}$ **46.** $-b^{-4}$ **47.** $(-a)^{-4}$ **48.** $(-b)^{-6}$

49. $5^{-1} + 6^{-1}$ **50.** $2^{-1} + 8^{-1}$ **51.** $8^{-1} - 3^{-1}$ **52.** $6^{-1} - 4^{-1}$

Evaluate each expression. ***See Examples 4 and 7.***

⊙ 53. $\dfrac{1}{4^{-2}}$ **54.** $\dfrac{1}{3^{-3}}$ **55.** $\dfrac{2^{-2}}{3^{-3}}$ **56.** $\dfrac{3^{-3}}{2^{-2}}$

⊙ 57. $\left(\dfrac{2}{3}\right)^{-3}$ **58.** $\left(\dfrac{3}{2}\right)^{-3}$ **59.** $\left(\dfrac{4}{5}\right)^{-2}$ **60.** $\left(\dfrac{5}{4}\right)^{-2}$

In Exercises 61 and 62, match the expression in Column I with its equivalent expression in Column II. Choices may be used once, more than once, or not at all. * ***See Example 7.***

	I	**II**		**I**	**II**
61. (a)	$\left(\dfrac{1}{3}\right)^{-1}$	**A.** $\dfrac{1}{3}$	**62. (a)**	$\left(\dfrac{2}{5}\right)^{-2}$	**A.** $\dfrac{25}{4}$
(b)	$\left(-\dfrac{1}{3}\right)^{-1}$	**B.** 3	**(b)**	$\left(-\dfrac{2}{5}\right)^{-2}$	**B.** $-\dfrac{25}{4}$
(c)	$-\left(\dfrac{1}{3}\right)^{-1}$	**C.** $-\dfrac{1}{3}$	**(c)**	$-\left(\dfrac{2}{5}\right)^{-2}$	**C.** $\dfrac{4}{25}$
(d)	$-\left(-\dfrac{1}{3}\right)^{-1}$	**D.** -3	**(d)**	$-\left(-\dfrac{2}{5}\right)^{-2}$	**D.** $-\dfrac{4}{25}$

Write each result with only positive exponents. Assume that all variables represent nonzero real numbers. ***See Example 5.***

⊙ 63. $\dfrac{4^8}{4^6}$ **64.** $\dfrac{5^9}{5^7}$ **65.** $\dfrac{x^{12}}{x^8}$ **66.** $\dfrac{y^{14}}{y^{10}}$

67. $\dfrac{r^7}{r^{10}}$ **68.** $\dfrac{y^8}{y^{12}}$ **69.** $\dfrac{6^4}{6^{-2}}$ **70.** $\dfrac{7^5}{7^{-3}}$

71. $\dfrac{6^{-3}}{6^7}$ **72.** $\dfrac{5^{-4}}{5^2}$ **73.** $\dfrac{7}{7^{-1}}$ **74.** $\dfrac{8}{8^{-1}}$

75. $\dfrac{r^{-3}}{r^{-6}}$ **76.** $\dfrac{s^{-4}}{s^{-8}}$ **77.** $\dfrac{x^3}{y^2}$ **78.** $\dfrac{y^5}{t^3}$

Simplify each expression. Assume that all variables represent nonzero real numbers. ***See Example 6.***

⊙ 79. $(x^3)^6$ **80.** $(y^5)^4$ **81.** $\left(\dfrac{3}{5}\right)^3$ **82.** $\left(\dfrac{4}{3}\right)^2$

83. $(4t)^3$ **84.** $(5t)^4$ **85.** $(-6x^2)^3$ **86.** $(-2x^5)^5$

87. $\left(\dfrac{-4m^2}{t}\right)^3$ **88.** $\left(\dfrac{-5n^4}{r^2}\right)^3$ **89.** $\left(\dfrac{-s^3}{t^5}\right)^4$ **90.** $\left(\dfrac{-2a^4}{b^5}\right)^6$

Simplify each expression so that no negative exponents appear in the final result. Assume that all variables represent nonzero real numbers. ***See Examples 1–8.***

91. $3^5 \cdot 3^{-6}$ **92.** $4^4 \cdot 4^{-6}$ **⊙ 93.** $a^{-3}a^2a^{-4}$

94. $k^{-5}k^{-3}k^4$ **95.** $(k^2)^{-3}k^4$ **96.** $(x^3)^{-4}x^5$

---

*The authors thank Mitchel Levy of Broward College for his suggestions for these exercises.

97. $-4r^{-2}(r^4)^2$

98. $-2m^{-1}(m^3)^2$

99. $(5a^{-1})^4(a^2)^{-3}$

100. $(3p^{-4})^2(p^3)^{-1}$

101. $(z^{-4}x^3)^{-1}$

102. $(y^{-2}z^4)^{-3}$

103. $7k^2(-2k)(4k^{-5})^0$

104. $3a^2(-5a^{-6})(-2a)^0$

105. $\dfrac{(p^{-2})^0}{5p^{-4}}$

106. $\dfrac{(m^4)^0}{9m^{-3}}$

107. $\dfrac{(3pq)q^2}{6p^2q^4}$

108. $\dfrac{(-8xy)y^3}{4x^5y^4}$

109. $\dfrac{4a^5(a^{-1})^3}{(a^{-2})^{-2}}$

110. $\dfrac{12k^{-2}(k^{-3})^{-4}}{6k^5}$

111. $\dfrac{(-y^{-4})^2}{6(y^{-5})^{-1}}$

112. $\dfrac{2(-m^{-1})^{-4}}{9(m^{-3})^2}$

113. $\dfrac{(2k)^2m^{-5}}{(km)^{-3}}$

114. $\dfrac{(3rs)^{-2}}{3^2r^2s^{-4}}$

115. $\dfrac{(2k)^2k^3}{k^{-1}k^{-5}}(5k^{-2})^{-3}$

116. $\dfrac{(3r^2)^2r^{-5}}{r^{-2}r^3}(2r^{-6})^2$

117. $\left(\dfrac{3k^{-2}}{k^4}\right)^{-1}\cdot\dfrac{2}{k}$

118. $\left(\dfrac{7m^{-2}}{m^{-3}}\right)^{-2}\cdot\dfrac{m^3}{4}$

119. $\left(\dfrac{2p}{q^2}\right)^3\left(\dfrac{3p^4}{q^{-4}}\right)^{-1}$

120. $\left(\dfrac{5z^3}{2a^2}\right)^{-3}\left(\dfrac{8a^{-1}}{15z^{-2}}\right)^{-3}$

121. $\dfrac{2^2y^4(y^{-3})^{-1}}{2^5y^{-2}}$

122. $\dfrac{3^{-1}m^4(m^2)^{-1}}{3^2m^{-2}}$

123. $\left(\dfrac{5m^4n^{-3}}{m^{-5}n^2}\right)^{-2}$

124. $\left(\dfrac{8x^{-6}y^3}{x^4y^{-4}}\right)^{-2}$

125. $\left(\dfrac{-3x^4y^6}{15x^{-6}y^7}\right)^{-3}$

126. $\left(\dfrac{-4a^3b^2}{12a^5b^{-4}}\right)^{-3}$

Brain Busters *Simplify each expression. Assume that all variables represent nonzero real numbers.*

127. $\dfrac{(2m^2p^3)^2(4m^2p)^{-2}}{(-3mp^4)^{-1}(2m^3p^4)^3}$

128. $\dfrac{(-5y^3z^4)^2(2yz^5)^{-2}}{10(y^4z)^3(3y^3z^2)^{-1}}$

129. $\dfrac{(-3y^3x^3)(-4y^4x^2)(x^2)^{-4}}{18x^3y^2(y^3)^3(x^3)^{-2}}$

130. $\dfrac{(2m^3x^2)^{-1}(3m^4x)^{-3}}{(m^2x^3)^3(m^2x)^{-5}}$

131. $\left(\dfrac{p^2q^{-1}}{2p^{-2}}\right)^2\cdot\left(\dfrac{p^3\cdot4q^{-2}}{3q^{-5}}\right)^{-1}\cdot\left(\dfrac{pq^{-5}}{q^{-2}}\right)^3$

132. $\left(\dfrac{a^6b^{-2}}{2a^{-2}}\right)^{-1}\cdot\left(\dfrac{6a^{-2}}{5b^{-4}}\right)^2\cdot\left(\dfrac{2b^{-1}a^2}{3b^{-2}}\right)^{-1}$

Write each number in scientific notation. ***See Example 9.***

133. 530

134. 1600

135. 0.830

136. 0.0072

137. 0.00000692

138. 0.875

139. $-38,500$

140. $-976,000,000$

Write each number in standard notation. ***See Example 10.***

141. 7.2×10^4

142. 8.91×10^2

143. 2.54×10^{-3}

144. 5.42×10^{-4}

145. -6×10^4

146. -9×10^3

147. 1.2×10^{-5}

148. 2.7×10^{-6}

Evaluate. Express answers in standard notation. ***See Example 11.***

149. $\dfrac{12\times10^4}{2\times10^6}$

150. $\dfrac{16\times10^5}{4\times10^8}$

151. $\dfrac{3\times10^{-2}}{12\times10^3}$

152. $\dfrac{5\times10^{-3}}{25\times10^2}$

153. $\dfrac{0.05\times1600}{0.0004}$

154. $\dfrac{0.003\times40,000}{0.00012}$

155. $\dfrac{20,000\times0.018}{300\times0.0004}$

156. $\dfrac{840,000\times0.03}{0.00021\times600}$

Solve each problem. ***See Example 12.***

157. The U.S. budget first passed **$1,000,000,000** in 1917. Seventy years later, in 1987, it exceeded **$1,000,000,000,000** for the first time. The budget request for fiscal-year 2009 was **$3,100,000,000,000**. If stacked in dollar bills, this amount would stretch **210,385** mi, almost 90% of the distance to the moon. Write the four boldfaced numbers in scientific notation. (*Source:* www.gpoaccess.gov, *The New York Times.*)

158. By area, the largest of the fifty United States is Alaska, with land area of about **365,482,000** acres, while the smallest is Rhode Island, with land area of about **677,000** acres. The total land area of the United States is about **2,271,343,000** acres. Write these three numbers in scientific notation. (*Source:* General Services Administration.)

159. In January 2010, the population of the United States was approximately 308.4 million. (*Source:* U.S. Census Bureau.)

   (a) Write the January 2010 population using scientific notation.

   (b) Write $1 trillion, that is, $1,000,000,000,000, using scientific notation.

   (c) Using your answers from parts (a) and (b), calculate how much each person in the United States in the year 2010 would have had to contribute in order to make someone a trillionaire. Write this amount in standard notation to the nearest dollar.

160. In February 2010, the U.S. House of Representatives voted to allow the government to go $1.9 trillion deeper in debt. Calculate how much additional debt this is for every U.S. resident. Use the 2010 population of 308.4 million from **Exercise 159,** and write the amount in standard notation to the nearest dollar. (*Source: The Gazette.*)

161. In 2009, the national debt of the U.S. government was about $11.5 trillion. Using 307.0 million as the population for that year, about how much was this per American? Write the amount in standard notation to the nearest dollar. (*Source: The Gazette.*)

162. In the early years of the Powerball® Lottery, a player would choose five numbers from 1 through 49 and one number from 1 through 42. It can be shown that there are about 8.009×10^7 different ways to do this. Suppose that a group of 2000 persons decided to purchase tickets for all these numbers and each ticket cost $1.00. How much should each person have expected to pay? (*Source:* www.powerball.com)

163. The speed of light is approximately 3×10^{10} cm per sec. How long will it take light to travel 9×10^{12} cm?

164. The average distance from Earth to the sun is 9.3×10^7 mi. How long would it take a rocket, traveling at 2.9×10^3 mph, to reach the sun?

165. A *light-year* is the distance that light travels in one year. Find the number of miles in a light-year if light travels 1.86×10^5 mi per sec.

166. Use the information given in the previous two exercises to find the number of minutes necessary for light from the sun to reach Earth.

167. In 2009, the estimated population of Luxembourg was 4.92×10^5. The population density was 493 people per square mile. What is the area of Luxembourg, to the nearest square mile? (*Source: The World Factbook.*)

168. In 2009, the population of Costa Rica was approximately 4.25×10^6. The population density was 83.3 people per square kilometer. (*Source: The World Factbook.*)

   (a) Write the population density in scientific notation.

   (b) To the nearest square kilometer, what is the area of Costa Rica?

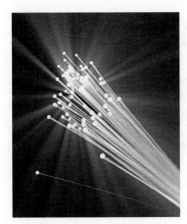

The screen on the left shows how a graphing calculator displays 250,000 and 0.000000034 in scientific notation. When put in scientific mode, it will calculate and display results as shown in the screen on the right.

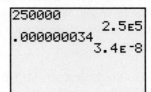

```
250000
           2.5E5
.000000034
          3.4E-8
```

```
(2.5E5)*(2E-3)
            5E2
(1.25E-4)/(5E-9)

          2.5E4
```

Predict the result the calculator will give for each screen. (Use the usual scientific notation to write your answers.)

169.
```
(1.5E12)*(5E-3)
```

170.
```
(3.2E-5)*(3E12)
```

171.
```
(8.4E14)/(2.1E-3
)
```

172.
```
(2.5E10)/(2E-3)
```

PREVIEW EXERCISES

*Simplify each expression. **See Section 1.4.***

173. $9x + 5x - x + 8x - 12x$

174. $6 - 4(3 - z) + 5(2 - 3z)$

175. $3(5 + q) - 2(6 - q)$

176. $7x - (5 + 5x) + 3$

5.2 Adding and Subtracting Polynomials

OBJECTIVES

1 Know the basic definitions for polynomials.

2 Add and subtract polynomials.

OBJECTIVE 1 Know the basic definitions for polynomials. Recall from **Chapter 1** that a **term** is a number (constant), a variable, or the product of a number and one or more variables raised to powers. A term or a sum of two or more terms is an **algebraic expression.** The simplest kind of algebraic expression is a *polynomial.*

> **Polynomial**
>
> A **polynomial** in x is a term or a finite sum of terms of the form ax^n, where a is a real number and the exponent n is a whole number.

$$12x^9, \quad 3t - 5, \quad \text{and} \quad 4m^3 - 5m^2 + 8 \qquad \text{Polynomials in } x, t, \text{ and } m$$

Even though the expression $3x - 5$ involves subtraction, it is a sum of terms, since it could be written as $3x + (-5)$.

For each term ax^n of a polynomial, a is called the **numerical coefficient,** or just the **coefficient,** and n is the **degree of the term.** The table on the next page gives examples.

Term ax^n	Numerical Coefficient	Degree
$12x^9$	12	9
$3x$, or $3x^1$	3	1
-6, or $-6x^0$	-6	0
$-x^4$, or $-1x^4$	-1	4
$\dfrac{x^2}{3} = \dfrac{1x^2}{3} = \dfrac{1}{3}x^2$	$\dfrac{1}{3}$	2

> Any nonzero constant has degree 0.

NOTE The number 0 has no degree, since 0 times a variable to any power is 0.

A polynomial containing only the variable x is called a **polynomial in x.** (Other variables may be used.) A polynomial in one variable is written in **descending powers** of the variable if the exponents on the variable decrease from left to right.

$$x^5 - 6x^2 + 12x - 5$$

> Think of $12x$ as $12x^1$ and -5 as $-5x^0$.

Descending powers of x

When written in descending powers of the variable, the greatest-degree term is written first and is called the **leading term** of the polynomial. Its coefficient is the **leading coefficient.**

NOW TRY
EXERCISE 1
Write the polynomial in descending powers of the variable. Then give the leading term and the leading coefficient.

$$-2x^3 - 2x^5 + 4x^2 + 7 - x$$

EXAMPLE 1 Writing Polynomials in Descending Powers

Write each polynomial in descending powers of the variable. Then give the leading term and the leading coefficient.

(a) $y - 6y^3 + 8y^5 - 9y^4 + 12$ is written as $8y^5 - 9y^4 - 6y^3 + y + 12$.

(b) $-2 + m + 6m^2 - m^3$ is written as $-m^3 + 6m^2 + m - 2$.

Each leading term is shown in color. In part (a), the leading coefficient is 8, and in part (b) it is -1.

NOW TRY

Some polynomials with a specific number of terms are given special names.

- A polynomial with exactly three terms is a **trinomial.**
- A two-term polynomial is a **binomial.**
- A single-term polynomial is a **monomial.**

Although many polynomials contain only one variable, polynomials may have more than one variable. The degree of a term with more than one variable is defined to be the sum of the exponents on the variables. The **degree of a polynomial** is the greatest degree of all of its terms. The table gives examples.

Type of Polynomial	Example	Degree
Monomial	7	0 $(7 = 7x^0)$
	$5x^3y^7$	10 $(3 + 7 = 10)$
Binomial	$6 + 2x^3$	3
	$11y + 8$	1 $(y = y^1)$
Trinomial	$t^2 + 11t + 4$	2
	$-3 + 2k^5 + 9z^4$	5
	$x^3y^9 + 12xy^4 + 7xy$	12 (The terms have degrees 12, 5, and 2, and 12 is the greatest.)

NOW TRY ANSWER
1. $-2x^5 - 2x^3 + 4x^2 - x + 7$;
$-2x^5$; -2

NOTE If a polynomial in a single variable is written in descending powers of that variable, the degree of the polynomial will be the degree of the leading term.

NOW TRY
EXERCISE 2
Identify the polynomial as a *monomial, binomial, trinomial,* or *none of these.* Also, give the degree.

$$-4x^3 + 10x^5 + 7$$

EXAMPLE 2 Classifying Polynomials

Identify each polynomial as a *monomial, binomial, trinomial,* or *none of these.* Also, give the degree.

(a) $-x^2 + 5x + 1$ This is a trinomial of degree 2.

(b) $\frac{3}{4}xy^4$ $\left(\text{or } \frac{3}{4}x^1y^4\right)$ This is a monomial of degree 5 (because $1 + 4 = 5$).

(c) $7m^9 + 18m^{14}$ This is a binomial of degree 14.

(d) $p^4 - p^2 - 6p - 5$ Polynomials of four terms or more do not have special names, so *none of these* is the answer that applies here. This polynomial has degree 4.   **NOW TRY**

OBJECTIVE 2 **Add and subtract polynomials.** We use the distributive property to simplify polynomials by combining like terms.

$$x^3 + 4x^2 + 5x^2 - 1$$
$$= x^3 + (4 + 5)x^2 - 1 \qquad \text{Distributive property}$$
$$= x^3 + 9x^2 - 1 \qquad \text{Add.}$$

Notice that the terms in a polynomial such as $4x + 5x^2$ cannot be combined. Only terms containing exactly the same variables to the same powers may be combined. Recall that such terms are called **like terms.**

⚠ **CAUTION** *Only like terms can be combined.*

NOW TRY
EXERCISE 3
Combine like terms.
(a) $3p^3 - 2q + p^3 - 5q$
(b) $-x^2t + 4x^2t + 3xt^2 - 7xt^2$

EXAMPLE 3 Combining Like Terms

Combine like terms.

(a) $-5y^3 + 8y^3 - y^3$
$$= (-5 + 8 - 1)y^3 \qquad \text{Distributive property}$$
$$= 2y^3 \qquad \text{Add and subtract.}$$

(b) $6x + 5y - 9x + 2y$
$$= 6x - 9x + 5y + 2y \qquad \text{Commutative property}$$
$$= -3x + 7y \qquad \text{Combine like terms.}$$

Since $-3x$ and $7y$ are unlike terms, no further simplification is possible.

(c) $5x^2y - 6xy^2 + 9x^2y + 13xy^2$
$$= 5x^2y + 9x^2y - 6xy^2 + 13xy^2 \qquad \text{Commutative property}$$
$$= 14x^2y + 7xy^2 \qquad \text{Combine like terms.} \qquad \textit{NOW TRY}$$

Adding Polynomials

To add two polynomials, combine like terms.

NOW TRY ANSWERS
2. trinomial; 5
3. (a) $4p^3 - 7q$ **(b)** $3x^2t - 4xt^2$

NOW TRY
EXERCISE 4

Add.

$(7x^2 - 9x + 4) +$
$(x^3 - 3x^2 - 5)$

EXAMPLE 4 Adding Polynomials

Add $(3a^5 - 9a^3 + 4a^2) + (-8a^5 + 8a^3 + 2)$.

$$(3a^5 - 9a^3 + 4a^2) + (-8a^5 + 8a^3 + 2)$$
$$= 3a^5 - 8a^5 - 9a^3 + 8a^3 + 4a^2 + 2 \quad \text{Commutative and associative properties}$$
$$= -5a^5 - a^3 + 4a^2 + 2 \quad \text{Combine like terms.}$$

Alternatively, we can add these two polynomials vertically.

$$3a^5 - 9a^3 + 4a^2$$
$$\underline{-8a^5 + 8a^3 \qquad + 2} \quad \text{Place like terms in columns.}$$
$$-5a^5 - a^3 + 4a^2 + 2$$

NOW TRY

In **Section 1.2,** we defined subtraction of real numbers as follows.

$$a - b = a + (-b)$$

That is, we add the first number and the negative (or opposite) of the second. We define the **negative of a polynomial** as that polynomial with the sign of every coefficient changed.

Subtracting Polynomials

To subtract two polynomials, add the first polynomial (minuend) and the negative (or opposite) of the *second* polynomial (subtrahend).

NOW TRY
EXERCISE 5

Subtract.

$(2y^2 - 7y - 4) -$
$(8y^2 - 2y + 10)$

EXAMPLE 5 Subtracting Polynomials

Subtract $(-6m^2 - 8m + 5) - (-5m^2 + 7m - 8)$.

Change every sign in the second polynomial (subtrahend) and add.

$$(-6m^2 - 8m + 5) - (-5m^2 + 7m - 8)$$
$$= -6m^2 - 8m + 5 + 5m^2 - 7m + 8 \quad \text{Definition of subtraction}$$
$$= -6m^2 + 5m^2 - 8m - 7m + 5 + 8 \quad \text{Rearrange terms.}$$
$$= -m^2 - 15m + 13 \quad \text{Combine like terms.}$$

CHECK $\quad \underset{\uparrow \atop \text{Answer}}{(-m^2 - 15m + 13)} + \underset{\uparrow \atop \text{Subtrahend}}{(-5m^2 + 7m - 8)} = \underset{\uparrow \atop \text{Minuend}}{-6m^2 - 8m + 5}$ ✓

Alternatively, we can subtract these two polynomials vertically.

$$-6m^2 - 8m + 5 \qquad \text{Write the subtrahend below the minuend, lining up like terms in columns.}$$
$$\underline{-5m^2 + 7m - 8}$$

Change all the signs in the subtrahend and add.

$$-6m^2 - 8m + 5$$
$$\underline{+5m^2 - 7m + 8} \qquad \text{Change all signs.}$$
$$-m^2 - 15m + 13 \qquad \text{Add in columns.}$$

The answer is the same.

NOW TRY

NOW TRY ANSWERS
4. $x^3 + 4x^2 - 9x - 1$
5. $-6y^2 - 5y - 14$

5.2 EXERCISES

● *Complete solution available on the Video Resources on DVD*

Give the numerical coefficient and the degree of each term. **See Objective 1.**

1. $7z$ **2.** $3r$ **3.** $-15p^2$ **4.** $-27k^3$ **5.** x^4 **6.** y^6

7. $\dfrac{t}{6}$ **8.** $\dfrac{m}{4}$ **9.** 8 **10.** 2 **11.** $-x^3$ **12.** $-y^9$

Write each polynomial in descending powers of the variable. Then give the leading term and the leading coefficient. **See Example 1.**

● **13.** $2x^3 + x - 3x^2 + 4$ **14.** $q^2 + 3q^4 - 2q + 1$ **15.** $4p^3 - 8p^5 + p^7$

16. $3y^2 + y^4 - 2y^3$ **17.** $10 - m^3 - 3m^4$ **18.** $4 - x - 8x^2$

Identify each polynomial as a monomial, binomial, trinomial, *or* none of these. *Also, give the degree.* **See Example 2.**

19. 25 **20.** 15 **21.** $7m - 22$ **22.** $6x + 15$

● **23.** $-7y^6 + 11y^8$ **24.** $12k^2 - 9k^5$ **25.** $-mn^5$ **26.** $-a^3b$

27. $-5m^3 + 6m - 9m^2$ **28.** $4z^2 - 11z + 2$

29. $-6p^4q - 3p^3q^2 + 2pq^3 - q^4$ **30.** $8s^3t - 4s^2t^2 + 2st^3 + 9$

31. *Concept Check* Which one of the following is a trinomial in descending powers, having degree 6?

A. $5x^6 - 4x^5 + 12$ **B.** $6x^5 - x^6 + 4$

C. $2x + 4x^2 - x^6$ **D.** $4x^6 - 6x^4 + 9x^2 - 8$

32. *Concept Check* Give an example of a polynomial of four terms in the variable x, having degree 5, written in descending powers, and lacking a fourth-degree term.

Combine like terms. **See Example 3.**

33. $5z^4 + 3z^4$ **34.** $8r^5 - 2r^5$ ● **35.** $-m^3 + 2m^3 + 6m^3$

36. $3p^4 + 5p^4 - 2p^4$ **37.** $x + x + x + x + x$ **38.** $z - z - z + z$

39. $m^4 - 3m^2 + m$ **40.** $5a^5 + 2a^4 - 9a^3$ **41.** $5t + 4s - 6t + 9s$

42. $8p - 9q - 3p + q$ **43.** $2k + 3k^2 + 5k^2 - 7$ **44.** $4x^2 + 2x - 6x^2 - 6$

45. $n^4 - 2n^3 + n^2 - 3n^4 + n^3$ **46.** $2q^3 + 3q^2 - 4q - q^3 + 5q^2$

47. $3ab^2 + 7a^2b - 5ab^2 + 13a^2b$ **48.** $6m^2n - 8mn^2 + 3mn^2 - 7m^2n$

49. $4 - (2 + 3m) + 6m + 9$ **50.** $8a - (3a + 4) - (5a - 3)$

51. $6 + 3p - (2p + 1) - (2p + 9)$ **52.** $4x - 8 - (-1 + x) - (11x + 5)$

Add or subtract as indicated. **See Examples 4 and 5.**

● **53.** $(5x^2 + 7x - 4) + (3x^2 - 6x + 2)$ **54.** $(4k^3 + k^2 + k) + (2k^3 - 4k^2 - 3k)$

55. $(6t^2 - 4t^4 - t) + (3t^4 - 4t^2 + 5)$ **56.** $(3p^2 + 2p - 5) + (7p^2 - 4p^3 + 3p)$

57. $(y^3 + 3y + 2) + (4y^3 - 3y^2 + 2y - 1)$ **58.** $(2x^5 - 2x^4 + x^3 - 1) + (x^4 - 3x^3 + 2)$

59. $(3r + 8) - (2r - 5)$ **60.** $(2d + 7) - (3d - 1)$

61. $(2a^2 + 3a - 1) - (4a^2 + 5a + 6)$ **62.** $(q^4 - 2q^2 + 10) - (3q^4 + 5q^2 - 5)$

● **63.** $(z^5 + 3z^2 + 2z) - (4z^5 + 2z^2 - 5z)$ **64.** $(5t^3 - 3t^2 + 2t) - (4t^3 + 2t^2 + 3t)$

65. Add.
$$21p - 8$$
$$\underline{-9p + 4}$$

66. Add.
$$15m - 9$$
$$\underline{4m + 12}$$

67. Add.
$$-12p^2 + 4p - 1$$
$$\underline{3p^2 + 7p - 8}$$

68. Add.
$$-6y^3 + 8y + 5$$
$$\underline{9y^3 + 4y - 6}$$

69. Subtract.
$$12a + 15$$
$$\underline{7a - 3}$$

70. Subtract.
$$-3b + 6$$
$$\underline{2b - 8}$$

71. Subtract.
$$6m^2 - 11m + 5$$
$$\underline{-8m^2 + 2m - 1}$$

72. Subtract.
$$-4z^2 + 2z - 1$$
$$\underline{3z^2 - 5z + 2}$$

73. Add.
$$12z^2 - 11z + 8$$
$$5z^2 + 16z - 2$$
$$\underline{-4z^2 + 5z - 9}$$

74. Add.
$$-6m^3 + 2m^2 + 5m$$
$$8m^3 + 4m^2 - 6m$$
$$\underline{-3m^3 + 2m^2 - 7m}$$

75. Add.
$$6y^3 - 9y^2 \quad\quad + 8$$
$$\underline{4y^3 + 2y^2 + 5y}$$

76. Add.
$$-7r^8 + 2r^6 - r^5$$
$$\underline{3r^6 \quad\quad + 5}$$

77. Subtract.
$$-5a^4 \quad + 8a^2 - 9$$
$$\underline{6a^3 - a^2 + 2}$$

78. Subtract.
$$-2m^3 + 8m^2$$
$$\underline{m^4 - m^3 \quad\quad + 2m}$$

The following problems are of mixed variety. Perform the indicated operations. ***See Examples 3–5.***

79. Subtract $4y^2 - 2y + 3$ from $7y^2 - 6y + 5$.

80. Subtract $-(-4x + 2z^2 + 3m)$ from $[(2z^2 - 3x + m) + (z^2 - 2m)]$.

81. $(-4m^2 + 3n^2 - 5n) - [(3m^2 - 5n^2 + 2n) + (-3m^2) + 4n^2]$

82. $[-(4m^2 - 8m + 4m^3) - (3m^2 + 2m + 5m^3)] + m^2$

83. $[-(y^4 - y^2 + 1) - (y^4 + 2y^2 + 1)] + (3y^4 - 3y^2 - 2)$

84. $[2p - (3p - 6)] - [(5p - (8 - 9p)) + 4p]$

85. $-[3z^2 + 5z - (2z^2 - 6z)] + [(8z^2 - [5z - z^2]) + 2z^2]$

86. $5k - (5k - [2k - (4k - 8k)]) + 11k - (9k - 12k)$

PREVIEW EXERCISES

Decide whether each relation is the graph of a function. Give its domain and range. ***See Section 3.5.***

87.

88.

89.

Evaluate ***(a)*** $f(-1)$ *and* ***(b)*** $f(2)$ *for each function.* ***See Section 3.6.***

90. $f(x) = 3x + 1$

91. $f(x) = x^2 + 2$

92. $f(x) = x^3 - 8$

Polynomial Functions, Graphs, and Composition

OBJECTIVES

1 Recognize and evaluate polynomial functions.

2 Use a polynomial function to model data.

3 Add and subtract polynomial functions.

4 Find the composition of functions.

5 Graph basic polynomial functions.

OBJECTIVE 1 **Recognize and evaluate polynomial functions.** In **Chapter 3,** we studied linear (first-degree polynomial) functions, defined as $f(x) = ax + b$. Now we consider more general polynomial functions.

Polynomial Function

A **polynomial function of degree n** is defined by

$$f(x) = a_n x^n + a_{n-1} x^{n-1} + \cdots + a_1 x + a_0,$$

for real numbers $a_n, a_{n-1}, \ldots, a_1$, and a_0, where $a_n \neq 0$ and n is a whole number.

Another way of describing a polynomial function is to say that it is a function defined by a polynomial in one variable, consisting of one or more terms. It is usually written in descending powers of the variable, and its degree is the degree of the polynomial that defines it.

We can evaluate a polynomial function $f(x)$ at different values of the variable x.

⟲ NOW TRY
  EXERCISE 1

Let $f(x) = x^3 - 2x^2 + 7$.
Find $f(-3)$.

EXAMPLE 1 Evaluating Polynomial Functions

Let $f(x) = 4x^3 - x^2 + 5$. Find each value.

(a) $f(3)$

Read this as "f of 3," not "f times 3."	$f(x) = 4x^3 - x^2 + 5$ — Given function
	$f(3) = 4(3)^3 - 3^2 + 5$ — Substitute 3 for x.
	$f(3) = 4(27) - 9 + 5$ — Apply the exponents.
	$f(3) = 108 - 9 + 5$ — Multiply.
	$f(3) = 104$ — Subtract, and then add.

Thus, $f(3) = 104$ and the ordered pair $(3, 104)$ belongs to f.

(b) $f(-4)$

$f(x) = 4x^3 - x^2 + 5$

$f(-4) = 4 \cdot (-4)^3 - (-4)^2 + 5$ ⟵ Use parentheses. Let $x = -4$.

$f(-4) = 4 \cdot (-64) - 16 + 5$ ⟵ Be careful with signs.

$f(-4) = -256 - 16 + 5$ — Multiply.

$f(-4) = -267$ — Subtract, and then add.

So, $f(-4) = -267$. The ordered pair $(-4, -267)$ belongs to f. NOW TRY ⟳

While f is the most common letter used to represent functions, recall that other letters, such as g and h, are also used. *The capital letter P is often used for polynomial functions.* The function defined as

$$P(x) = 4x^3 - x^2 + 5$$

yields the same ordered pairs as the function f in **Example 1.**

NOW TRY ANSWER
1. -38

OBJECTIVE 2 Use a polynomial function to model data.

EXAMPLE 2 Using a Polynomial Model to Approximate Data

The number of students enrolled in public schools (grades pre-K–12) in the United States during the years 1990 through 2006 can be modeled by the polynomial function defined by

$$P(x) = -0.01774x^2 + 0.7871x + 41.26,$$

where $x = 0$ corresponds to the year 1990, $x = 1$ corresponds to 1991, and so on, and $P(x)$ is in millions. Use this function to approximate the number of public school students in 2006. (*Source:* Department of Education.)

Since $x = 16$ corresponds to 2006, we must find $P(16)$.

$$P(x) = -0.01774x^2 + 0.7871x + 41.26$$
$$P(16) = -0.01774(16)^2 + 0.7871(16) + 41.26 \qquad \text{Let } x = 16.$$
$$P(16) \approx 49.3 \qquad \text{Evaluate.}$$

There were about 49.3 million public school students in 2006.   NOW TRY

NOW TRY
EXERCISE 2
Use the function in **Example 2** to approximate the number of public school students in 2002.

OBJECTIVE 3 Add and subtract polynomial functions. The operations of addition, subtraction, multiplication, and division are also defined for functions. For example, businesses use the equation "profit equals revenue minus cost," which can be written in function notation.

$$P(x) = R(x) - C(x) \qquad \begin{array}{l} x \text{ is the number of items} \\ \text{produced and sold.} \end{array}$$
$$\uparrow \qquad \uparrow \qquad \uparrow$$
$$\begin{array}{ccc} \text{Profit} & \text{Revenue} & \text{Cost} \\ \text{function} & \text{function} & \text{function} \end{array}$$

The profit function is found by subtracting the cost function from the revenue function. We define the following **operations on functions.**

Adding and Subtracting Functions

If $f(x)$ and $g(x)$ define functions, then

$$(f + g)(x) = f(x) + g(x) \qquad \text{Sum function}$$

and

$$(f - g)(x) = f(x) - g(x). \qquad \text{Difference function}$$

In each case, the domain of the new function is the intersection of the domains of $f(x)$ and $g(x)$.

EXAMPLE 3 Adding and Subtracting Functions

Find each of the following for the polynomial functions defined by

$$f(x) = x^2 - 3x + 7 \quad \text{and} \quad g(x) = -3x^2 - 7x + 7.$$

(a) $(f + g)(x)$ This notation does *not* indicate the distributive property.

$$= f(x) + g(x) \qquad \text{Use the definition.}$$
$$= (x^2 - 3x + 7) + (-3x^2 - 7x + 7) \qquad \text{Substitute.}$$
$$= -2x^2 - 10x + 14 \qquad \text{Add the polynomials.}$$

NOW TRY ANSWER
2. about 48.2 million

NOW TRY
EXERCISE 3
For $f(x) = x^3 - 3x^2 + 4$
and $g(x) = -2x^3 + x^2 - 12$,
find each of the following.

(a) $(f + g)(x)$

(b) $(f - g)(x)$

(b) $(f - g)(x)$

$= f(x) - g(x)$	Use the definition.
$= (x^2 - 3x + 7) - (-3x^2 - 7x + 7)$	Substitute.
$= (x^2 - 3x + 7) + (3x^2 + 7x - 7)$	Change subtraction to addition.
$= 4x^2 + 4x$	Add. NOW TRY

NOW TRY
EXERCISE 4
For $f(x) = x^2 - 4$
and $g(x) = -6x^2$,
find each of the following.

(a) $(f + g)(x)$

(b) $(f - g)(-4)$

EXAMPLE 4 Adding and Subtracting Functions

Find each of the following for the functions defined by
$$f(x) = 10x^2 - 2x \quad \text{and} \quad g(x) = 2x.$$

(a) $(f + g)(2)$

$= f(2) + g(2)$	Use the definition.

$\overbrace{f(x) = 10x^2 - 2x}^{} \quad \overbrace{g(x) = 2x}^{}$

$= [10(2)^2 - 2(2)] + 2(2)$	Substitute.

This is a key step.

$= [40 - 4] + 4$	Order of operations
$= 40$	Subtract, and then add.

Alternatively, we could first find $(f + g)(x)$.

$(f + g)(x)$

$= f(x) + g(x)$	Use the definition.
$= (10x^2 - 2x) + 2x$	Substitute.
$= 10x^2$	Combine like terms.

Then, $(f + g)(2)$

$= 10(2)^2$	Substitute.
$= 40.$	The result is the same.

(b) $(f - g)(x)$ and $(f - g)(1)$

$(f - g)(x)$

$= f(x) - g(x)$	Use the definition.
$= (10x^2 - 2x) - 2x$	Substitute.
$= 10x^2 - 4x$	Combine like terms.

Then, $(f - g)(1)$

$= 10(1)^2 - 4(1)$	Substitute.
$= 6.$	Simplify.

Confirm that $f(1) - g(1)$ gives the same result.   NOW TRY

OBJECTIVE 4 **Find the composition of functions.** The diagram in **FIGURE 1** on the next page shows a function f that assigns, to each element x of set X, some element y of set Y. Suppose that a function g takes each element of set Y and assigns a value z of set Z. Then f and g together assign an element x in X to an element z in Z. The result of this process is a new function h that takes an element x in X and assigns it an element z in Z.

NOW TRY ANSWERS
3. (a) $-x^3 - 2x^2 - 8$
   (b) $3x^3 - 4x^2 + 16$
4. (a) $-5x^2 - 4$
   (b) 108

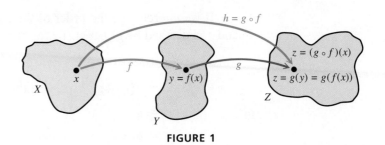

FIGURE 1

This function h is called the *composition* of functions g and f, written $\boldsymbol{g \circ f}$.

Composition of Functions

If f and g are functions, then the **composite function**, or **composition**, of g and f is defined by

$$(g \circ f)(x) = g(f(x))$$

for all x in the domain of f such that $f(x)$ is in the domain of g.

Read $g \circ f$ as "g of f".

As a real-life example of how composite functions occur, consider the following retail situation.

> *A $40 pair of blue jeans is on sale for 25% off. If you purchase the jeans before noon, the retailer offers an additional 10% off. What is the final sale price of the blue jeans?*

You might be tempted to say that the blue jeans are 25% + 10% = 35% off and calculate $40(0.35) = $14, giving a final sale price of

$$\$40 - \$14 = \$26. \quad \text{This is not correct.}$$

To find the correct final sale price, we must first find the price after taking 25% off, and then take an additional 10% off that price.

$40(0.25) = $10, giving a sale price of $40 − $10 = $30.   Take 25% off original price.

$30(0.10) = $3, giving a **final sale price** of $30 − $3 = $27.   Take additional 10% off.

This is the idea behind composition of functions.

NOW TRY
EXERCISE 5

Let $f(x) = 3x + 7$
and $g(x) = x - 2$.
Find $(f \circ g)(7)$.

EXAMPLE 5   Evaluating a Composite Function

Let $f(x) = x^2$ and $g(x) = x + 3$. Find $(f \circ g)(4)$.

$(f \circ g)(4)$ Evaluate the "inside" function value first.

$= f(g(4))$ Definition

$= f(4 + 3)$ Use the rule for $g(x)$; $g(4) = 4 + 3$.

$= f(7)$ Add.

Now evaluate the "outside" function.

$= 7^2$ Use the rule for $f(x)$; $f(7) = 7^2$.

$= 49$ Square 7.   NOW TRY

NOW TRY ANSWER
5. 22

If we interchange the order of the functions in **Example 5**, the composition of g and f is defined by $g(f(x))$. To find $(g \circ f)(4)$, we let $x = 4$.

$(g \circ f)(4)$

$\quad = g(f(4))$ Definition

$\quad = g(4^2)$ Use the rule for $f(x)$; $f(4) = 4^2$.

$\quad = g(16)$ Square 4.

$\quad = 16 + 3$ Use the rule for $g(x)$; $g(16) = 16 + 3$.

$\quad = 19$ Add.

Here we see that $(f \circ g)(4) \neq (g \circ f)(4)$ because $49 \neq 19$. In general,

$$(f \circ g)(x) \neq (g \circ f)(x).$$

⌐ *NOW TRY*
  EXERCISE 6
Let $f(x) = x - 5$
and $g(x) = -x^2 + 2$.
Find the following.
(a) $(g \circ f)(-1)$
(b) $(f \circ g)(x)$

EXAMPLE 6   Finding Composite Functions

Let $f(x) = 4x - 1$ and $g(x) = x^2 + 5$. Find the following.

(a) $(f \circ g)(2)$

$\quad = f(g(2))$

$\quad = f(2^2 + 5)$ $g(x) = x^2 + 5$

$\quad = f(9)$ Work inside the parentheses.

$\quad = 4(9) - 1$ $f(x) = 4x - 1$

$\quad = 35$ Multiply, and then subtract.

(b) $(f \circ g)(x)$

$\quad = f(g(x))$ Use $g(x)$ as the input for the function f.

$\quad = 4(g(x)) - 1$ Use the rule for $f(x)$; $f(x) = 4x - 1$.

$\quad = 4(x^2 + 5) - 1$ $g(x) = x^2 + 5$

$\quad = 4x^2 + 20 - 1$ Distributive property

$\quad = 4x^2 + 19$ Combine like terms.

(c) Find $(f \circ g)(2)$ again, this time using the rule obtained in part (b).

$\quad (f \circ g)(x) = 4x^2 + 19$ From part (b)

$\quad (f \circ g)(2) = 4(2)^2 + 19$ Let $x = 2$.

$\quad\quad\quad\quad\quad = 4(4) + 19$ Square 2.

$\quad\quad\quad\quad\quad = 16 + 19$ Multiply.

Same result as in part (a) ⟶ $= 35$ Add.   *NOW TRY* ↻

OBJECTIVE 5   **Graph basic polynomial functions.** Recall from **Section 3.5** that each input (or x-value) of a function results in one output (or y-value). The set of input values (for x) defines the domain of the function, and the set of output values (for y) defines the range.

    The simplest polynomial function is the **identity function,** defined by $f(x) = x$ and graphed in **FIGURE 2** on the next page. This function pairs each real number with itself.

NOW TRY ANSWERS
6. (a) -34 **(b)** $-x^2 - 3$

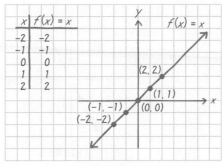

Identity function

$$f(x) = x$$

Domain: $(-\infty, \infty)$

Range: $(-\infty, \infty)$

FIGURE 2

NOTE A *linear function* **(Section 3.6)** is a specific kind of polynomial function.

Another polynomial function, defined by $f(x) = x^2$ and graphed in **FIGURE 3**, is the **squaring function.** For this function, every real number is paired with its square. The graph of the squaring function is a *parabola.*

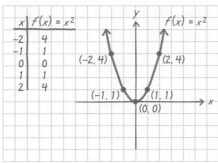

Squaring function

$$f(x) = x^2$$

Domain: $(-\infty, \infty)$

Range: $[0, \infty)$

FIGURE 3

The **cubing function** is defined by $f(x) = x^3$ and graphed in **FIGURE 4**. This function pairs every real number with its cube.

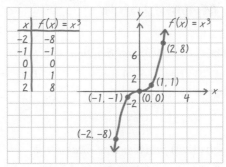

Cubing function

$$f(x) = x^3$$

Domain: $(-\infty, \infty)$

Range: $(-\infty, \infty)$

FIGURE 4

EXAMPLE 7 Graphing Variations of Polynomial Functions

Graph each function by creating a table of ordered pairs. Give the domain and range of each function by observing its graph.

(a) $f(x) = 2x$

To find each range value, multiply the domain value by 2. Plot the points and join them with a straight line. See **FIGURE 5** on the next page. Both the domain and the range are $(-\infty, \infty)$.

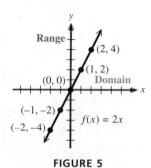

FIGURE 5

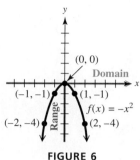

FIGURE 6

NOW TRY
EXERCISE 7
Graph $f(x) = x^2 - 4$. Give
the domain and range.

(b) $f(x) = -x^2$

For each input x, square it and then take its opposite. Plotting and joining the points gives a parabola that opens down. It is a *reflection* of the graph of the squaring function across the x-axis. See the table and **FIGURE 6**. The domain is $(-\infty, \infty)$ and the range is $(-\infty, 0]$.

(c) $f(x) = x^3 - 2$

For this function, cube the input and then subtract 2 from the result. The graph is that of the cubing function *shifted* 2 units down. See the table and **FIGURE 7**. The domain and range are both $(-\infty, \infty)$.

NOW TRY ANSWER
7.
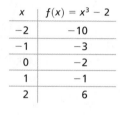

domain: $(-\infty, \infty)$;
range: $[-4, \infty)$

x	$f(x) = x^3 - 2$
-2	-10
-1	-3
0	-2
1	-1
2	6

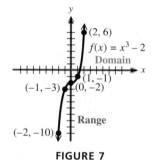

FIGURE 7

NOW TRY

5.3 EXERCISES

MyMathLab Math XL PRACTICE WATCH DOWNLOAD READ REVIEW

⊕ *Complete solution available
on the Video Resources on DVD*

*For each polynomial function, find **(a)** $f(-1)$ and **(b)** $f(2)$. **See Example 1.***

1. $f(x) = 6x - 4$
2. $f(x) = -2x + 5$
3. $f(x) = x^2 - 3x + 4$

4. $f(x) = 3x^2 + x - 5$
5. $f(x) = 5x^4 - 3x^2 + 6$
6. $f(x) = -4x^4 + 2x^2 - 1$

⊕ **7.** $f(x) = -x^2 + 2x^3 - 8$
8. $f(x) = -x^2 - x^3 + 11x$

*Solve each problem. **See Example 2.***

9. Imports of Fair Trade Certified™ coffee into the United States during the years 2000 through 2006 can be modeled by the polynomial function defined by

$$P(x) = 1667x^2 + 22.78x + 4300,$$

where $x = 0$ corresponds to the year 2000, $x = 1$ corresponds to 2001, and so on, and $P(x)$ is in thousands of pounds. Use this function to approximate the amount (to the nearest whole number) of Fair Trade coffee imported into the United States in each given year. (*Source:* TransFair USA.)

(a) 2000　　**(b)** 2003　　**(c)** 2006

10. The total number of airports (public and private) in the United States during the years 1980 through 2007 can be approximated by the polynomial function defined by

$$f(x) = -2.576x^2 + 253.1x + 15{,}160,$$

where $x = 0$ represents 1980, $x = 1$ represents 1981, and so on. Use this function to approximate the number of airports (to the nearest whole number) in each given year. (*Source:* U.S. Bureau of Transportation Statistics.)

(a) 1980　　　**(b)** 2000　　　**(c)** 2007

11. The amount spent by Americans on foreign travel during the years 1985 through 2006 can be modeled by the polynomial function defined by

$$P(x) = -0.00189x^3 + 0.1193x^2 + 2.027x + 28.19,$$

where $x = 0$ represents 1985, $x = 1$ represents 1986, and so on, and $P(x)$ is in billions of dollars. Use this function to approximate the amount spent by Americans on foreign travel in each given year. Round answers to the nearest tenth. (*Source:* U.S. Department of Commerce.)

(a) 1985　　　**(b)** 2000　　　**(c)** 2006

12. The percent of births to unmarried women during the years 1990 through 2007 can be approximated by the polynomial function defined by

$$f(x) = 0.0077x^3 - 0.1945x^2 + 1.857x + 26.64,$$

where $x = 0$ represents 1990, $x = 1$ represents 1991, and so on. Use this function to approximate the percent (to the nearest tenth) of births to unmarried women in each given year. (*Source:* National Center for Health Statistics.)

(a) 1990　　　**(b)** 1997　　　**(c)** 2007

*For each pair of functions, find **(a)** $(f + g)(x)$ and **(b)** $(f - g)(x)$. **See Example 3.***

13. $f(x) = 5x - 10, \quad g(x) = 3x + 7$

14. $f(x) = -4x + 1, \quad g(x) = 6x + 2$

15. $f(x) = 4x^2 + 8x - 3, \quad g(x) = -5x^2 + 4x - 9$

16. $f(x) = 3x^2 - 9x + 10, \quad g(x) = -4x^2 + 2x + 12$

*Let $f(x) = x^2 - 9, g(x) = 2x,$ and $h(x) = x - 3$. Find each of the following. **See Example 4.***

17. $(f + g)(x)$　　　**18.** $(f - g)(x)$　　　**19.** $(f + g)(3)$　　　**20.** $(f - g)(-3)$

21. $(f - h)(x)$　　　**22.** $(f + h)(x)$　　　**23.** $(f - h)(-3)$

24. $(f + h)(-2)$　　　**25.** $(g + h)(-10)$　　　**26.** $(g - h)(10)$

27. $(g - h)(-3)$　　　**28.** $(g + h)(1)$　　　**29.** $(g + h)\left(\dfrac{1}{4}\right)$

30. $(g + h)\left(\dfrac{1}{3}\right)$　　　**31.** $(g + h)\left(-\dfrac{1}{2}\right)$　　　**32.** $(g + h)\left(-\dfrac{1}{4}\right)$

33. Construct two functions defined by $f(x)$, a polynomial of degree 3, and $g(x)$, a polynomial of degree 4. Find $(f - g)(x)$ and $(g - f)(x)$. Use your answers to decide whether subtraction of functions is a commutative operation. Explain.

34. *Concept Check* Find two polynomial functions defined by $f(x)$ and $g(x)$ such that

$$(f + g)(x) = 3x^3 - x + 3.$$

Let $f(x) = x^2 + 4$, $g(x) = 2x + 3$, and $h(x) = x - 5$. Find each value or expression. ***See Examples 5 and 6.***

35. $(h \circ g)(4)$ **36.** $(f \circ g)(4)$ **37.** $(g \circ f)(6)$ **38.** $(h \circ f)(6)$

39. $(f \circ h)(-2)$ **40.** $(h \circ g)(-2)$ **41.** $(f \circ g)(0)$ **42.** $(f \circ h)(0)$

43. $(g \circ f)(x)$ **44.** $(g \circ h)(x)$ **45.** $(h \circ g)(x)$ **46.** $(h \circ f)(x)$

47. $(f \circ h)\left(\dfrac{1}{2}\right)$ **48.** $(h \circ f)\left(\dfrac{1}{2}\right)$ **49.** $(f \circ g)\left(-\dfrac{1}{2}\right)$ **50.** $(g \circ f)\left(-\dfrac{1}{2}\right)$

Solve each problem.

51. The function defined by $f(x) = 12x$ computes the number of inches in x feet, and the function defined by $g(x) = 5280x$ computes the number of feet in x miles. What is $(f \circ g)(x)$ and what does it compute?

52. The perimeter x of a square with sides of length s is given by the formula $x = 4s$.

    (a) Solve for s in terms of x.

    (b) If y represents the area of this square, write y as a function of the perimeter x.

    (c) Use the composite function of part (b) to find the area of a square with perimeter 6.

53. When a thermal inversion layer is over a city (as happens often in Los Angeles), pollutants cannot rise vertically, but are trapped below the layer and must disperse horizontally. Assume that a factory smokestack begins emitting a pollutant at 8 A.M. Assume that the pollutant disperses horizontally over a circular area. Suppose that t represents the time, in hours, since the factory began emitting pollutants ($t = 0$ represents 8 A.M.), and assume that the radius of the circle of pollution is $r(t) = 2t$ miles. Let $\mathcal{A}(r) = \pi r^2$ represent the area of a circle of radius r. Find and interpret $(\mathcal{A} \circ r)(t)$.

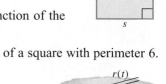

54. An oil well is leaking, with the leak spreading oil over the surface as a circle. At any time t, in minutes, after the beginning of the leak, the radius of the circular oil slick on the surface is $r(t) = 4t$ feet. Let $\mathcal{A}(r) = \pi r^2$ represent the area of a circle of radius r. Find and interpret $(\mathcal{A} \circ r)(t)$.

Graph each function. Give the domain and range. ***See Example 7.***

55. $f(x) = -2x + 1$ **56.** $f(x) = 3x + 2$

57. $f(x) = -3x^2$ **58.** $f(x) = \dfrac{1}{2}x^2$

59. $f(x) = x^3 + 1$ **60.** $f(x) = -x^3 + 2$

PREVIEW EXERCISES

Find each product. ***See Section 5.1.***

61. $3m^3(4m^2)$ **62.** $5z^2(7z^4)$

63. $-3b^5(2a^3b^4)$ **64.** $-4k^2(-3k^5t^3)$

65. $12x^2y(5xy^3)$ **66.** $6mn^3(3m^4n)$

5.4 Multiplying Polynomials

OBJECTIVES

1. Multiply terms.
2. Multiply any two polynomials.
3. Multiply binomials.
4. Find the product of the sum and difference of two terms.
5. Find the square of a binomial.
6. Multiply polynomial functions.

OBJECTIVE 1 Multiply terms.

EXAMPLE 1 Multiplying Monomials

Find each product.

(a) $(3x^4)(5x^3)$

$= 3 \cdot 5 \cdot x^4 \cdot x^3$ Commutative and associative properties

$= 15x^{4+3}$ Multiply; product rule for exponents

$= 15x^7$ Add the exponents.

(b) $-4a^3(3a^5)$

$= -4(3)a^3 \cdot a^5$

$= -12a^8$

(c) $2m^2z^4(8m^3z^2)$

$= 2(8)m^2 \cdot m^3 \cdot z^4 \cdot z^2$

$= 16m^5z^6$ *NOW TRY*

OBJECTIVE 2 Multiply any two polynomials.

EXAMPLE 2 Multiplying Polynomials

Find each product.

(a) $-2(8x^3 - 9x^2)$ Be careful with signs.

$= -2(8x^3) - 2(-9x^2)$ Distributive property

$= -16x^3 + 18x^2$ Multiply.

(b) $5x^2(-4x^2 + 3x - 2)$

$= 5x^2(-4x^2) + 5x^2(3x) + 5x^2(-2)$ Distributive property

$= -20x^4 + 15x^3 - 10x^2$ Multiply.

(c) $(3x - 4)(2x^2 + x)$

$(3x - 4)(2x^2 + x)$ Distributive property;

Treat $3x - 4$ as a single expression.

   Multiply each term of $2x^2 + x$ by $3x - 4$.

$= (3x - 4)(2x^2) + (3x - 4)(x)$

$= 3x(2x^2) + (-4)(2x^2) + (3x)(x) + (-4)(x)$

     Distributive property

$= 6x^3 - 8x^2 + 3x^2 - 4x$ Multiply.

$= 6x^3 - 5x^2 - 4x$ Combine like terms.

(d) $2x^2(x + 1)(x - 3)$

$= 2x^2[(x + 1)(x) + (x + 1)(-3)]$ Distributive property

$= 2x^2[x^2 + x - 3x - 3]$ Distributive property

$= 2x^2(x^2 - 2x - 3)$ Combine like terms.

$= 2x^4 - 4x^3 - 6x^2$ Distributive property   *NOW TRY*

NOW TRY EXERCISE 1

Find the product.

$-3s^2t(15s^3t^4)$

NOW TRY EXERCISE 2

Find each product.

(a) $3k^3(-2k^5 + 3k^2 - 4)$

(b) $5x(2x - 1)(x + 4)$

NOW TRY ANSWERS

1. $-45s^5t^5$
2. **(a)** $-6k^8 + 9k^5 - 12k^3$
   (b) $10x^3 + 35x^2 - 20x$

NOW TRY
EXERCISE 3
Find the product.

$$3t^2 - 5t + 4$$
$$\underline{ t - 3}$$

EXAMPLE 3 Multiplying Polynomials Vertically

Find each product.

(a) $(5a - 2b)(3a + b)$

$$
\begin{array}{r}
5a \;-\; 2b \\
3a \;+\; b \\
\hline
5ab \;-\; 2b^2 \\
15a^2 \;-\; 6ab \\
\hline
15a^2 \;-\; ab \;-\; 2b^2
\end{array}
$$

Write the factors vertically.

← Multiply $b(5a - 2b)$.

← Multiply $3a(5a - 2b)$.

Combine like terms.

(b) $(3m^3 - 2m^2 + 4)(3m - 5)$

Be sure to write like terms in columns.

$$
\begin{array}{r}
3m^3 \;-\; 2m^2 \;+\; 4 \\
3m \;-\; 5 \\
\hline
-15m^3 \;+\; 10m^2 \;-\; 20 \\
9m^4 \;-\; 6m^3 \;+\; 12m \\
\hline
9m^4 \;-\; 21m^3 \;+\; 10m^2 \;+\; 12m \;-\; 20
\end{array}
$$

$-5(3m^3 - 2m^2 + 4)$

$3m(3m^3 - 2m^2 + 4)$

Combine like terms.

NOW TRY

NOTE We can use a rectangle to model polynomial multiplication.

$$(5a - 2b)(3a + b) \qquad \text{Example 3(a)}$$

Label a rectangle with each term, as shown below on the left. Then put the product of each pair of monomials in the appropriate box, as shown on the right.

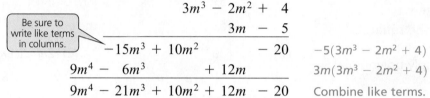

$$(5a - 2b)(3a + b)$$

$$= 15a^2 + 5ab - 6ab - 2b^2 \qquad \text{Add the four monomial products.}$$

$$= 15a^2 - ab - 2b^2 \qquad \text{Same result as in **Example 3(a)**}$$

OBJECTIVE 3 **Multiply binomials.** There is a shortcut method for finding the product of two binomials.

$$(3x - 4)(2x + 3)$$

$$= 3x(2x + 3) - 4(2x + 3) \qquad \text{Distributive property}$$

$$= 3x(2x) + 3x(3) - 4(2x) - 4(3) \qquad \text{Distributive property again}$$

$$= 6x^2 + 9x - 8x - 12 \qquad \text{Multiply.}$$

Before combining like terms to find the simplest form of the answer, we check the origin of each of the four terms in the sum

$$6x^2 + 9x - 8x - 12.$$

NOW TRY ANSWER
3. $3t^3 - 14t^2 + 19t - 12$

First, $6x^2$ is the product of the two *first* terms.

$$(3x - 4)(2x + 3) \qquad 3x(2x) = 6x^2 \qquad \text{First terms}$$

To get $9x$, the *outer* terms are multiplied.

$$(3x - 4)(2x + 3) \qquad 3x(3) = 9x \qquad \text{Outer terms}$$

The term $-8x$ comes from the *inner* terms.

$$(3x - 4)(2x + 3) \qquad -4(2x) = -8x \qquad \text{Inner terms}$$

Finally, -12 comes from the *last* terms.

$$(3x - 4)(2x + 3) \qquad -4(3) = -12 \qquad \text{Last terms}$$

The product is found by adding these four results.

$$(3x - 4)(2x + 3)$$
$$= 6x^2 + 9x + (-8x) + (-12) \qquad \text{FOIL}$$
$$= 6x^2 + x - 12 \qquad\qquad \text{Combine like terms.}$$

To keep track of the order of multiplying terms, we use the initials FOIL (First, Outer, Inner, Last). This procedure can be written in compact form as follows.

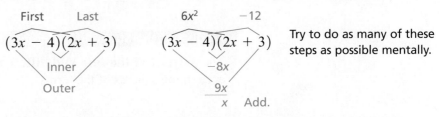

Try to do as many of these steps as possible mentally.

⚠ **CAUTION**   The FOIL method *applies only to multiplying two binomials.*

EXAMPLE 4   Using the FOIL Method

Use the FOIL method to find each product.

(a) $(4m - 5)(3m + 1)$

First terms	$(4m - 5)(3m + 1)$	$4m(3m) = 12m^2$
Outer terms	$(4m - 5)(3m + 1)$	$4m(1) = 4m$
Inner terms	$(4m - 5)(3m + 1)$	$-5(3m) = -15m$
Last terms	$(4m - 5)(3m + 1)$	$-5(1) = -5$

$$(4m - 5)(3m + 1)$$

$$ \ \ \text{F} \qquad \text{O} \qquad \text{I} \qquad \text{L}$$
$$= 12m^2 + 4m - 15m - 5$$
$$= 12m^2 - 11m - 5 \qquad \text{Combine like terms.}$$

**NOW TRY
EXERCISE 4**
Use the FOIL method to find the product.

$$(3p - k)(5p + 4k)$$

Compact form:

$$\underbrace{(4m - 5)(3m + 1)} = 12m^2 - 11m - 5$$

$12m^2 \qquad -5$

$-15m$

$4m$

$-11m$ Add.

(b) $(6a - 5b)(3a + 4b)$

    First   Outer   Inner   Last

$$= 18a^2 + 24ab - 15ab - 20b^2$$

$$= 18a^2 + 9ab - 20b^2 \qquad \text{Combine like terms.}$$

(c) $(2k + 3z)(5k - 3z)$

$$= 10k^2 - 6kz + 15kz - 9z^2 \qquad \text{FOIL}$$

$$= 10k^2 + 9kz - 9z^2 \qquad \text{Combine like terms.} \qquad \text{NOW TRY}$$

OBJECTIVE 4 **Find the product of the sum and difference of two terms.**
The product of the sum and difference of the same two terms occurs frequently.

$$(x + y)(x - y)$$

$$= x^2 - xy + xy - y^2 \qquad \text{FOIL}$$

$$= x^2 - y^2 \qquad \text{Combine like terms.}$$

Product of the Sum and Difference of Two Terms

The **product of the sum and difference of the two terms x and y** is the difference of the squares of the terms.

$$(x + y)(x - y) = x^2 - y^2$$

**NOW TRY
EXERCISE 5**
Find each product.

(a) $(3x - 7y)(3x + 7y)$

(b) $5k(2k - 3)(2k + 3)$

EXAMPLE 5 Multiplying the Sum and Difference of Two Terms

Find each product.

(a) $(p + 7)(p - 7)$

$$= p^2 - 7^2$$

$$= p^2 - 49$$

(b) $(2r + 5)(2r - 5)$

$$= (2r)^2 - 5^2$$

$$= 2^2 r^2 - 25$$

$$= 4r^2 - 25$$

(c) $(6m + 5n)(6m - 5n)$

$$= (6m)^2 - (5n)^2$$

$$= 36m^2 - 25n^2$$

(d) $2x^3(x + 3)(x - 3)$

$$= 2x^3(x^2 - 9)$$

$$= 2x^5 - 18x^3 \qquad \text{NOW TRY}$$

OBJECTIVE 5 **Find the square of a binomial.** To find the *square of a binomial* $x + y$, or $(x + y)^2$, multiply $x + y$ by itself.

$$(x + y)(x + y)$$

$$= x^2 + xy + xy + y^2 \qquad \text{FOIL}$$

$$= x^2 + 2xy + y^2 \qquad \text{Combine like terms.}$$

A similar result is true for the square of a difference.

NOW TRY ANSWERS
4. $15p^2 + 7kp - 4k^2$
5. (a) $9x^2 - 49y^2$ **(b)** $20k^3 - 45k$

> ### Square of a Binomial
>
> The **square of a binomial** is the sum of the square of the first term, twice the product of the two terms, and the square of the last term.
>
> $$(x + y)^2 = x^2 + 2xy + y^2$$
> $$(x - y)^2 = x^2 - 2xy + y^2$$

**NOW TRY
EXERCISE 6**

Find each product.

(a) $(y - 10)^2$ **(b)** $(4x + 5y)^2$

EXAMPLE 6 Squaring Binomials

Find each product.

(a) $(m + 7)^2$

$$= m^2 + 2 \cdot m \cdot 7 + 7^2 \qquad (x + y)^2 = x^2 + 2xy + y^2$$

$$= m^2 + 14m + 49$$

(b) $(p - 4)^2$

$$= p^2 - 2 \cdot p \cdot 4 + 4^2 \qquad (x - y)^2 = x^2 - 2xy + y^2$$

$$= p^2 - 8p + 16$$

(c) $(2p + 3v)^2$

$$= (2p)^2 + 2(2p)(3v) + (3v)^2$$

$$= 4p^2 + 12pv + 9v^2$$

(d) $(3r - 5s)^2$

$$= (3r)^2 - 2(3r)(5s) + (5s)^2$$

$$= 9r^2 - 30rs + 25s^2$$

NOW TRY ⟲

⚠ **CAUTION**   As the products in the formula for the square of a binomial show,

$$(x + y)^2 \neq x^2 + y^2.$$

More generally, $(x + y)^n \neq x^n + y^n \quad (n \neq 1).$

EXAMPLE 7 Multiplying More Complicated Binomials

Find each product.

(a) $[(3p - 2) + 5q][(3p - 2) - 5q]$

$$= (3p - 2)^2 - (5q)^2 \qquad \text{Product of sum and difference of terms}$$

$$= 9p^2 - 12p + 4 - 25q^2 \qquad \text{Square both quantities.}$$

(b) $[(2z + r) + 1]^2$

$$= (2z + r)^2 + 2(2z + r)(1) + 1^2 \qquad \text{Square of a binomial}$$

$$= 4z^2 + 4zr + r^2 + 4z + 2r + 1 \qquad \text{Square again. Use the distributive property.}$$

(c) $(x + y)^3$

$$= (x + y)^2(x + y) \qquad a^3 = a^2 \cdot a$$

This does not equal $x^3 + y^3$.

$$= (x^2 + 2xy + y^2)(x + y) \qquad \text{Square } x + y.$$

$$= x^3 + 2x^2y + xy^2 + x^2y + 2xy^2 + y^3 \qquad \text{Distributive property}$$

$$= x^3 + 3x^2y + 3xy^2 + y^3 \qquad \text{Combine like terms.}$$

NOW TRY ANSWERS

6. **(a)** $y^2 - 20y + 100$
    (b) $16x^2 + 40xy + 25y^2$

NOW TRY
EXERCISE 7
Find each product.
(a) $[(4x - y) + 2][(4x - y) - 2]$
(b) $(y - 3)^4$

(d) $(2a + b)^4$

$= (2a + b)^2 (2a + b)^2$

$= (4a^2 + 4ab + b^2)(4a^2 + 4ab + b^2)$ Square $2a + b$ twice.

$= 4a^2(4a^2 + 4ab + b^2) + 4ab(4a^2 + 4ab + b^2)$ Distributive property
$+ b^2(4a^2 + 4ab + b^2)$

$= 16a^4 + 16a^3b + 4a^2b^2 + 16a^3b + 16a^2b^2 + 4ab^3$ Distributive property
$+ 4a^2b^2 + 4ab^3 + b^4$ again

$= 16a^4 + 32a^3b + 24a^2b^2 + 8ab^3 + b^4$ Combine like terms.

NOW TRY

OBJECTIVE 6 **Multiply polynomial functions.** In **Section 5.3**, we added and subtracted functions. Functions can also be multiplied.

> **Multiplying Functions**
>
> If $f(x)$ and $g(x)$ define functions, then
> $$(fg)(x) = f(x) \cdot g(x). \quad \text{Product function}$$
> The domain of the product function is the intersection of the domains of $f(x)$ and $g(x)$.

⚠ **CAUTION** Write the product $f(x) \cdot g(x)$ as $(fg)(x)$, **not** $f(g(x))$, which indicates the composition of functions f and g. (See **Section 5.3**.)

NOW TRY
EXERCISE 8
For $f(x) = 3x^2 - 1$
and $g(x) = 8x + 7$,
find $(fg)(x)$ and $(fg)(-2)$.

EXAMPLE 8 Multiplying Polynomial Functions

For $f(x) = 3x + 4$ and $g(x) = 2x^2 + x$, find $(fg)(x)$ and $(fg)(-1)$.

$(fg)(x)$

$= f(x) \cdot g(x)$ Use the definition.

$= (3x + 4)(2x^2 + x)$ Substitute.

$= 6x^3 + 3x^2 + 8x^2 + 4x$ FOIL

$= 6x^3 + 11x^2 + 4x$ Combine like terms.

$(fg)(-1)$

$= 6(-1)^3 + 11(-1)^2 + 4(-1)$ Let $x = -1$ in $(fg)(x)$.

$= -6 + 11 - 4$ Be careful with signs.

$= 1$ Add and subtract.

Confirm that $f(-1) \cdot g(-1)$ is equal to $(fg)(-1)$.   NOW TRY

NOW TRY ANSWERS
7. (a) $16x^2 - 8xy + y^2 - 4$
   (b) $y^4 - 12y^3 + 54y^2 -$
       $108y + 81$
8. $24x^3 + 21x^2 - 8x - 7; -99$

5.4 EXERCISES

MyMathLab | Math XL PRACTICE | WATCH | DOWNLOAD | READ | REVIEW

Complete solution available on the Video Resources on DVD

Concept Check Match each product in Column I with the correct polynomial in Column II.

I	II
1. $(2x - 5)(3x + 4)$	**A.** $6x^2 + 23x + 20$
2. $(2x + 5)(3x + 4)$	**B.** $6x^2 + 7x - 20$
3. $(2x - 5)(3x - 4)$	**C.** $6x^2 - 7x - 20$
4. $(2x + 5)(3x - 4)$	**D.** $6x^2 - 23x + 20$

Find each product. **See Examples 1–3.**

5. $-8m^3(3m^2)$ **6.** $-4p^2(5p^4)$ **7.** $14x^2y^3(-2x^5y)$

8. $5m^3n^4(-4m^2n^5)$ **9.** $3x(-2x + 5)$ **10.** $5y(-6y + 1)$

11. $-q^3(2 + 3q)$ **12.** $-3a^4(4 + a)$ **13.** $6k^2(3k^2 + 2k + 1)$

14. $5r^3(2r^2 + 3r + 4)$ **15.** $(2t + 3)(3t^2 - 4t - 1)$ **16.** $(4z + 2)(z^2 - 3z - 5)$

17. $m(m + 5)(m - 8)$ **18.** $p(p + 4)(p - 6)$ **19.** $4z(2z + 1)(3z - 4)$

20. $2y(2y + 1)(8y - 3)$ **21.** $4x^3(x - 3)(x + 2)$ **22.** $2y^5(y - 8)(y + 2)$

23. $(2y + 3)(3y - 4)$ **24.** $(2m + 6)(5m - 3)$ **25.** $-b^2 + 3b + 3$
$$\underline{2b + 4}$$

26. $-r^2 - 4r + 8$
$$\underline{3r - 2}$$
    27. $5m - 3n$
$$\underline{5m + 3n}$$
    28. $2k + 6q$
$$\underline{2k - 6q}$$

29. $2z^3 - 5z^2 + 8z - 1$
$$\underline{4z + 3}$$
    30. $3z^4 - 2z^3 + z - 5$
$$\underline{2z - 5}$$

31. $2p^2 + 3p + 6$
$$\underline{3p^2 - 4p - 1}$$
    32. $5y^2 - 2y + 4$
$$\underline{2y^2 + y + 3}$$

Use the FOIL method to find each product. **See Example 4.**

33. $(m + 5)(m - 8)$ **34.** $(p + 4)(p - 6)$ **35.** $(4k + 3)(3k - 2)$

36. $(5w + 2)(2w - 5)$ **37.** $(z - w)(3z + 4w)$ **38.** $(s - t)(2s + 5t)$

39. $(6c - d)(2c + 3d)$ **40.** $(2m - n)(3m + 5n)$

41. Describe the FOIL method in your own words.

42. Explain why the product of the sum and difference of two terms is not a trinomial.

Find each product. **See Example 5.**

43. $(x + 9)(x - 9)$ **44.** $(z + 6)(z - 6)$ **45.** $(2p - 3)(2p + 3)$

46. $(3x - 8)(3x + 8)$ **47.** $(5m - 1)(5m + 1)$ **48.** $(6y - 3)(6y + 3)$

49. $(3a + 2c)(3a - 2c)$ **50.** $(5r + 4s)(5r - 4s)$ **51.** $(4m + 7n^2)(4m - 7n^2)$

52. $(2k^2 + 6h)(2k^2 - 6h)$ **53.** $3y(5y^3 + 2)(5y^3 - 2)$ **54.** $4x(3x^3 + 4)(3x^3 - 4)$

Find each product. **See Example 6.**

55. $(y - 5)^2$ **56.** $(a - 3)^2$ **57.** $(x + 1)^2$ **58.** $(t + 2)^2$

59. $(2p + 7)^2$ **60.** $(3z + 8)^2$ **61.** $(4n - 3m)^2$ **62.** $(5r - 7s)^2$

63. *Concept Check* Use the special product rule $(x + y)(x - y) = x^2 - y^2$ to find the product $101 \cdot 99$.

64. *Concept Check* Repeat **Exercise 63** for the product $202 \cdot 198$.

In Exercises 65–74, the factors involve fractions or decimals. Apply the methods of this section, and find each product.

65. $(0.2x + 1.3)(0.5x - 0.1)$

66. $(0.1y + 2.1)(0.5y - 0.4)$

67. $\left(3w + \dfrac{1}{4}z\right)(w - 2z)$

68. $\left(5r + \dfrac{2}{3}y\right)(r - 5y)$

69. $\left(4x - \dfrac{2}{3}\right)\left(4x + \dfrac{2}{3}\right)$

70. $\left(3t - \dfrac{5}{4}\right)\left(3t + \dfrac{5}{4}\right)$

71. $\left(k - \dfrac{5}{7}p\right)^2$

72. $\left(q - \dfrac{3}{4}r\right)^2$

73. $(0.2x - 1.4y)^2$

74. $(0.3x - 1.6y)^2$

Find each product. ***See Example 7.***

75. $[(5x + 1) + 6y]^2$

76. $[(3m - 2) + p]^2$

77. $[(2a + b) - 3]^2$

78. $[(4k + h) - 4]^2$

79. $[(2a + b) - 3][(2a + b) + 3]$

80. $[(m + p) - 5][(m + p) + 5]$

81. $[(2h - k) + j][(2h - k) - j]$

82. $[(3m - y) + z][(3m - y) - z]$

83. $(y + 2)^3$

84. $(z - 3)^3$

85. $(5r - s)^3$

86. $(x + 3y)^3$

87. $(q - 2)^4$

88. $(r + 3)^4$

Brain Busters *Find each product.*

89. $(2a + b)(3a^2 + 2ab + b^2)$

90. $(m - 5p)(m^2 - 2mp + 3p^2)$

91. $(4z - x)(z^3 - 4z^2x + 2zx^2 - x^3)$

92. $(3r + 2s)(r^3 + 2r^2s - rs^2 + 2s^3)$

93. $(m^2 - 2mp + p^2)(m^2 + 2mp - p^2)$

94. $(3 + x + y)(-3 + x - y)$

95. $ab(a + b)(a + 2b)(a - 3b)$

96. $mp(m - p)(m - 2p)(2m + p)$

In Exercises 97–100, two expressions are given. Replace x with 3 and y with 4 to show that, in general, the two expressions are not equivalent.

97. $(x + y)^2$; $\quad x^2 + y^2$

98. $(x + y)^3$; $\quad x^3 + y^3$

99. $(x + y)^4$; $\quad x^4 + y^4$

100. $(x + y)^5$; $\quad x^5 + y^5$

Find the area of each figure. Express it as a polynomial in descending powers of the variable x. Refer to the formulas on the inside covers of this book if necessary.

101.

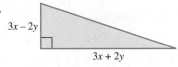

$3x - 2y$

$3x + 2y$

102.

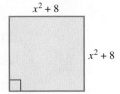

$x^2 + 8$

$x^2 + 8$

103.

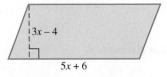

$3x - 4$

$5x + 6$

104.

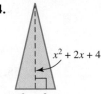

$x^2 + 2x + 4$

$2x + 3$

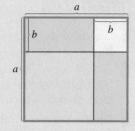

RELATING CONCEPTS EXERCISES 105–112

FOR INDIVIDUAL OR GROUP WORK

*Consider the figure. **Work Exercises 105–112 in order.***

105. What is the length of each side of the blue square in terms of a and b?

106. What is the formula for the area of a square? Use the formula to write an expression, in the form of a product, for the area of the blue square.

107. Each green rectangle has an area of _____. Therefore, the total area in green is represented by the polynomial _____.

108. The yellow square has an area of _____.

109. The area of the entire colored region is represented by _____, because each side of the entire colored region has length _____.

110. The area of the blue square is equal to the area of the entire colored region, minus the total area of the green squares, minus the area of the yellow square. Write this as a simplified polynomial in a and b.

111. (a) What must be true about the expressions for the area of the blue square found in **Exercises 106 and 110?**

   (b) Write an equation based on your answer in part (a).

112. Draw a figure and give a similar proof for $(a + b)^2 = a^2 + 2ab + b^2$.

For each pair of functions, find the product $(fg)(x)$. ***See Example 8.***

113. $f(x) = 2x, \quad g(x) = 5x - 1$

114. $f(x) = 3x, \quad g(x) = 6x - 8$

115. $f(x) = x + 1, \quad g(x) = 2x - 3$

116. $f(x) = x - 7, \quad g(x) = 4x + 5$

117. $f(x) = 2x - 3, \quad g(x) = 4x^2 + 6x + 9$

118. $f(x) = 3x + 4, \quad g(x) = 9x^2 - 12x + 16$

Let $f(x) = x^2 - 9, g(x) = 2x,$ and $h(x) = x - 3$. Find each of the following. ***See Example 8.***

119. $(fg)(x)$

120. $(fh)(x)$

121. $(fg)(2)$

122. $(fh)(1)$

123. $(gh)(x)$

124. $(fh)(-1)$

125. $(gh)(-3)$

126. $(fg)(-2)$

127. $(fg)\left(-\dfrac{1}{2}\right)$

128. $(fg)\left(-\dfrac{1}{3}\right)$

129. $(fh)\left(-\dfrac{1}{4}\right)$

130. $(fh)\left(-\dfrac{1}{5}\right)$

PREVIEW EXERCISES

Perform the indicated operations. ***See Sections 5.1 and 5.2.***

131. $\dfrac{12p^7}{6p^3}$

132. $\dfrac{-9y^{11}}{3y}$

133. $\dfrac{-8a^3b^7}{6a^5b}$

134. $\dfrac{-20r^3s^5}{15rs^9}$

135. Subtract.
$$-3a^2 + 4a - 5$$
$$5a^2 + 3a - 9$$

136. Subtract.
$$-4p^2 - 8p + 5$$
$$3p^2 + 2p + 9$$

5.5 Dividing Polynomials

OBJECTIVES

1 Divide a polynomial by a monomial.
2 Divide a polynomial by a polynomial of two or more terms.
3 Divide polynomial functions.

OBJECTIVE 1   Divide a polynomial by a monomial. Recall that a monomial is a single term, such as 3, $5m^2$, or x^2y^2.

Dividing a Polynomial by a Monomial

To divide a polynomial by a monomial, divide each term in the polynomial by the monomial, and then write each quotient in lowest terms.

NOW TRY
EXERCISE 1
Divide.
$$\frac{81y^4 - 54y^3 + 18y}{9y^3}$$

EXAMPLE 1   Dividing a Polynomial by a Monomial

Divide.

(a) $\dfrac{15x^2 - 12x + 6}{3}$

$$= \frac{15x^2}{3} - \frac{12x}{3} + \frac{6}{3} \qquad \text{Divide each term by 3.}$$

$$= 5x^2 - 4x + 2 \qquad \text{Write in lowest terms.}$$

CHECK $3\underbrace{(5x^2 - 4x + 2)}_{} = \underbrace{15x^2 - 12x + 6}_{}$ ✓

Divisor Quotient   Original polynomial (Dividend)

(b) $\dfrac{5m^3 - 9m^2 + 10m}{5m^2}$

Think: $\frac{10m}{5m^2} = \frac{10}{5}m^{1-2} = 2m^{-1} = \frac{2}{m}$

$$= \frac{5m^3}{5m^2} - \frac{9m^2}{5m^2} + \frac{10m}{5m^2} \qquad \text{Divide each term by } 5m^2.$$

$$= m - \frac{9}{5} + \frac{2}{m} \qquad \begin{array}{l}\text{Simplify each term.}\\ \text{Use the quotient rule for exponents.}\end{array}$$

CHECK $5m^2\left(m - \dfrac{9}{5} + \dfrac{2}{m}\right) = 5m^3 - 9m^2 + 10m$ ✓   Divisor × Quotient = Original polynomial

The result $m - \frac{9}{5} + \frac{2}{m}$ is not a polynomial. (Why?) The quotient of two polynomials need not be a polynomial.

(c) $\dfrac{8xy^2 - 9x^2y + 6x^2y^2}{x^2y^2}$

$$= \frac{8xy^2}{x^2y^2} - \frac{9x^2y}{x^2y^2} + \frac{6x^2y^2}{x^2y^2} \qquad \text{Divide each term by } x^2y^2.$$

$$= \frac{8}{x} - \frac{9}{y} + 6 \qquad \frac{a^m}{a^n} = a^{m-n}$$

NOW TRY

NOW TRY ANSWER
1. $9y - 6 + \dfrac{2}{y^2}$

OBJECTIVE 2 Divide a polynomial by a polynomial of two or more terms.
This process is similar to that for dividing whole numbers.

NOW TRY
EXERCISE 2
Divide.
$$\frac{3x^2 - 4x - 15}{x - 3}$$

EXAMPLE 2 Dividing a Polynomial by a Polynomial

Divide $\dfrac{2m^2 + m - 10}{m - 2}$.

Make sure each polynomial is written in descending powers of the variables.

$$m - 2\overline{)2m^2 + m - 10} \quad \boxed{\text{Write as if dividing whole numbers.}}$$

Divide the first term of the dividend $2m^2 + m - 10$ by the first term of the divisor $m - 2$. Here $\frac{2m^2}{m} = 2m$.

$$\begin{array}{r} 2m \quad \leftarrow \text{Result of } \frac{2m^2}{m} \\ m - 2\overline{)2m^2 + m - 10} \end{array}$$

Multiply $m - 2$ and $2m$, and write the result below $2m^2 + m - 10$.

$$\begin{array}{r} 2m \\ m - 2\overline{)2m^2 + m - 10} \\ \underline{2m^2 - 4m} \quad \leftarrow 2m(m-2) = 2m^2 - 4m \end{array}$$

Now subtract by mentally changing the signs on $2m^2 - 4m$ and *adding*.

$$\begin{array}{r} 2m \\ m - 2\overline{)2m^2 + m - 10} \\ \boxed{\text{To subtract, add the opposite.}}\quad \underline{2m^2 - 4m} \\ 5m \quad \leftarrow \text{Subtract. The difference is } 5m. \end{array}$$

Bring down -10 and continue by dividing $5m$ by m.

$$\begin{array}{r} 2m + 5 \leftarrow \frac{5m}{m} = 5\\ m - 2\overline{)2m^2 + m - 10} \\ \underline{2m^2 - 4m} \\ 5m - 10 \leftarrow \text{Bring down } -10.\\ \underline{5m - 10} \leftarrow 5(m-2) = 5m - 10\\ 0 \leftarrow \text{Subtract. The difference is 0.} \end{array}$$

CHECK Multiply $m - 2$ (the divisor) and $2m + 5$ (the quotient). The result is $2m^2 + m - 10$ (the dividend). ✓   NOW TRY

EXAMPLE 3 Dividing a Polynomial with a Missing Term

Divide $3x^3 - 2x + 5$ by $x - 3$.

Make sure that $3x^3 - 2x + 5$ is in descending powers of the variable. Add a term with 0 coefficient as a placeholder for the missing x^2-term.

Missing term

$$x - 3\overline{)3x^3 + 0x^2 - 2x + 5}$$

NOW TRY ANSWER
2. $3x + 5$

NOW TRY
EXERCISE 3

Divide
$2x^3 - 12x - 10$ by $x - 4$.

Start with $\frac{3x^3}{x} = 3x^2$.

$$\begin{array}{r} 3x^2 \\ x - 3 \overline{)\ 3x^3 + 0x^2 - 2x + 5} \\ \underline{3x^3 - 9x^2 } \end{array}$$

$\leftarrow \frac{3x^3}{x} = 3x^2$

$\leftarrow 3x^2(x - 3)$

Subtract by mentally changing the signs on $3x^3 - 9x^2$ and adding.

$$\begin{array}{r} 3x^2 \\ x - 3 \overline{)\ 3x^3 + 0x^2 - 2x + 5} \\ \underline{3x^3 - 9x^2 } \\ 9x^2 \end{array}$$

$\leftarrow$ Subtract.

Bring down the next term.

$$\begin{array}{r} 3x^2 \\ x - 3 \overline{)\ 3x^3 + 0x^2 - 2x + 5} \\ \underline{3x^3 - 9x^2 } \\ 9x^2 - 2x \end{array}$$

$\leftarrow$ Bring down $-2x$.

In the next step, $\frac{9x^2}{x} = 9x$.

$$\begin{array}{r} 3x^2 + 9x \\ x - 3 \overline{)\ 3x^3 + 0x^2 - 2x + 5} \\ \underline{3x^3 - 9x^2 } \\ 9x^2 - 2x \\ \underline{9x^2 - 27x } \\ 25x + 5 \end{array}$$

$\leftarrow \frac{9x^2}{x} = 9x$

$\leftarrow 9x(x - 3)$

$\leftarrow$ Subtract. Bring down 5.

Finally, $\frac{25x}{x} = 25$.

$$\begin{array}{r} 3x^2 + 9x + 25 \\ x - 3 \overline{)\ 3x^3 + 0x^2 - 2x + 5} \\ \underline{3x^3 - 9x^2 } \\ 9x^2 - 2x \\ \underline{9x^2 - 27x } \\ 25x + 5 \\ \underline{25x - 75 } \\ 80 \end{array}$$

$\leftarrow \frac{25x}{x} = 25$

$\leftarrow 25(x - 3)$

$\leftarrow$ Remainder

Write the remainder, 80, as the numerator of the fraction $\frac{80}{x - 3}$.

$$\frac{3x^3 - 2x + 5}{x - 3} = 3x^2 + 9x + 25 + \frac{80}{x - 3}$$

Be sure to add $\frac{\text{remainder}}{\text{divisor}}$.
Don't forget the $+$ sign.

CHECK   Multiply $x - 3$ (the divisor) and $3x^2 + 9x + 25$ (the quotient), and then add 80 (the remainder). The result is $3x^3 - 2x + 5$. ✓   NOW TRY

> ⚠ **CAUTION** Remember to include $\frac{\text{remainder}}{\text{divisor}}$ as part of the answer. ***Don't forget to insert a plus sign between the polynomial quotient and this fraction.***

NOW TRY ANSWER
3. $2x^2 + 8x + 20 + \frac{70}{x - 4}$

NOW TRY
EXERCISE 4

Divide

$2x^4 + 8x^3 + 2x^2 - 5x - 3$

by $2x^2 - 2$.

EXAMPLE 4 Dividing by a Polynomial with a Missing Term

Divide $6r^4 + 9r^3 + 2r^2 - 8r + 7$ by $3r^2 - 2$.

$$
\begin{array}{r}
2r^2 + 3r + 2 \\
3r^2 + 0r - 2 \overline{)\ 6r^4 + 9r^3 + 2r^2 - 8r + 7} \\
6r^4 + 0r^3 - 4r^2 \\
\hline
9r^3 + 6r^2 - 8r \\
9r^3 + 0r^2 - 6r \\
\hline
6r^2 - 2r + 7 \\
6r^2 + 0r - 4 \\
\hline
-2r + 11
\end{array}
$$

Missing term ⟶

Stop when the degree of the remainder is less than the degree of the divisor.

⟵ Remainder

The degree of the remainder, $-2r + 11$, is less than the degree of the divisor, $3r^2 - 2$, so the division process is finished. The result is written as follows.

$$2r^2 + 3r + 2 + \frac{-2r + 11}{3r^2 - 2} \qquad \text{Quotient} + \frac{\text{remainder}}{\text{divisor}}$$

NOW TRY

⚠ **CAUTION** When dividing a polynomial by a polynomial of two or more terms:

1. Be sure the terms in both polynomials are written in descending powers.

2. Write any missing terms with 0 placeholders.

NOW TRY
EXERCISE 5

Divide

$6m^3 - 8m^2 - 5m - 6$

by $3m - 6$.

EXAMPLE 5 Finding a Quotient with a Fractional Coefficient

Divide $2p^3 + 5p^2 + p - 2$ by $2p + 2$.

$$\frac{3p^2}{2p} = \frac{3}{2}p$$

$$
\begin{array}{r}
p^2 + \dfrac{3}{2}p - 1 \\
2p + 2 \overline{)\ 2p^3 + 5p^2 + p - 2} \\
2p^3 + 2p^2 \\
\hline
3p^2 + p \\
3p^2 + 3p \\
\hline
-2p - 2 \\
-2p - 2 \\
\hline
0
\end{array}
$$

Since the remainder is 0, the quotient is $p^2 + \frac{3}{2}p - 1$.

NOW TRY

OBJECTIVE 3 **Divide polynomial functions.** We now define the quotient of two functions.

Dividing Functions

If $f(x)$ and $g(x)$ define functions, then

$$\left(\frac{f}{g}\right)(x) = \frac{f(x)}{g(x)}. \qquad \text{Quotient function}$$

The domain of the quotient function is the intersection of the domains of $f(x)$ and $g(x)$, excluding any values of x for which $g(x) = 0$.

NOW TRY ANSWERS

4. $x^2 + 4x + 2 + \dfrac{3x + 1}{2x^2 - 2}$

5. $2m^2 + \dfrac{4}{3}m + 1$

**NOW TRY
EXERCISE 6**

For $f(x) = 8x^2 + 2x - 3$
and $g(x) = 2x - 1$,

find $\left(\frac{f}{g}\right)(x)$ and $\left(\frac{f}{g}\right)(8)$.

EXAMPLE 6 Dividing Polynomial Functions

For $f(x) = 2x^2 + x - 10$ and $g(x) = x - 2$, find $\left(\frac{f}{g}\right)(x)$ and $\left(\frac{f}{g}\right)(-3)$. What value of x is not in the domain of the quotient function?

$$\left(\frac{f}{g}\right)(x) = \frac{f(x)}{g(x)} = \frac{2x^2 + x - 10}{x - 2}$$

This quotient was found in **Example 2,** with m replacing x. The result here is $2x + 5$, so

$$\left(\frac{f}{g}\right)(x) = 2x + 5, \quad x \neq 2.$$

The number 2 is not in the domain because it causes the denominator $g(x) = x - 2$ to equal 0. Then

$$\left(\frac{f}{g}\right)(-3) = 2(-3) + 5 = -1. \quad \text{Let } x = -3.$$

NOW TRY ANSWER

6. $4x + 3, \quad x \neq \frac{1}{2}; 35$

Verify that the same value is found by evaluating $\frac{f(-3)}{g(-3)}$.

NOW TRY

5.5 EXERCISES

MyMathLab | Math XL PRACTICE | WATCH | DOWNLOAD | READ | REVIEW

● *Complete solution available on the Video Resources on DVD*

Concept Check *Complete each statement with the correct word(s).*

1. We find the quotient of two monomials by using the _____ rule for _____.

2. When dividing polynomials that are not monomials, first write them in _____ powers.

3. If a polynomial in a division problem has a missing term, insert a term with coefficient equal to _____ as a placeholder.

4. To check a division problem, multiply the _____ by the quotient. Then add the _____.

Divide. *See Example 1.*

● **5.** $\dfrac{15x^3 - 10x^2 + 5}{5}$

6. $\dfrac{27m^4 - 18m^3 + 9m}{9}$

7. $\dfrac{9y^2 + 12y - 15}{3y}$

8. $\dfrac{80r^2 - 40r + 10}{10r}$

9. $\dfrac{15m^3 + 25m^2 + 30m}{5m^3}$

10. $\dfrac{64x^3 - 72x^2 + 12x}{8x^3}$

11. $\dfrac{4m^2n^2 - 21mn^3 + 18mn^2}{14m^2n^3}$

12. $\dfrac{24h^2k + 56hk^2 - 28hk}{16h^2k^2}$

13. $\dfrac{8wxy^2 + 3wx^2y + 12w^2xy}{4wx^2y}$

14. $\dfrac{12ab^2c + 10a^2bc + 18abc^2}{6a^2bc}$

Complete the division.

15.
$$
\begin{array}{r}
r^2 \\
3r - 1 \overline{)3r^3 - 22r^2 + 25r - 6} \\
\underline{3r^3 - r^2 } \\
-21r^2
\end{array}
$$

16.
$$
\begin{array}{r}
3b^2 \\
2b - 5 \overline{)6b^3 - 7b^2 - 4b - 40} \\
\underline{6b^3 - 15b^2 } \\
8b^2
\end{array}
$$

Divide. See Examples 2–5.

17. $\dfrac{y^2 + y - 20}{y + 5}$

18. $\dfrac{y^2 + 3y - 18}{y + 6}$

19. $\dfrac{q^2 + 4q - 32}{q - 4}$

20. $\dfrac{q^2 + 2q - 35}{q - 5}$

21. $\dfrac{3t^2 + 17t + 10}{3t + 2}$

22. $\dfrac{2k^2 - 3k - 20}{2k + 5}$

23. $\dfrac{p^2 + 2p + 20}{p + 6}$

24. $\dfrac{x^2 + 11x + 16}{x + 8}$

25. $\dfrac{3m^3 + 5m^2 - 5m + 1}{3m - 1}$

26. $\dfrac{8z^3 - 6z^2 - 5z + 3}{4z + 3}$

27. $\dfrac{m^3 - 2m^2 - 9}{m - 3}$

28. $\dfrac{p^3 + 3p^2 - 4}{p + 2}$

29. $(2z^3 - 5z^2 + 6z - 15) \div (2z - 5)$

30. $(3p^3 + p^2 + 18p + 6) \div (3p + 1)$

31. $(4x^3 + 9x^2 - 10x + 3) \div (4x + 1)$

32. $(10z^3 - 26z^2 + 17z - 13) \div (5z - 3)$

33. $\dfrac{6x^3 - 19x^2 + 14x - 15}{3x^2 - 2x + 4}$

34. $\dfrac{8m^3 - 18m^2 + 37m - 13}{2m^2 - 3m + 6}$

35. $(x^3 + 2x - 3) \div (x - 1)$

36. $(x^3 + 5x^2 - 18) \div (x + 3)$

37. $(2x^3 - 11x^2 + 25) \div (x - 5)$

38. $(2x^3 + 3x^2 - 5) \div (x - 1)$

39. $(3x^3 - x + 4) \div (x - 2)$

40. $(3k^3 + 9k - 14) \div (k - 2)$

41. $\dfrac{4k^4 + 6k^3 + 3k - 1}{2k^2 + 1}$

42. $\dfrac{9k^4 + 12k^3 - 4k - 1}{3k^2 - 1}$

43. $\dfrac{6y^4 + 4y^3 + 4y - 6}{3y^2 + 2y - 3}$

44. $\dfrac{8t^4 + 6t^3 + 12t - 32}{4t^2 + 3t - 8}$

45. $(x^4 - 4x^3 + 5x^2 - 3x + 2) \div (x^2 + 3)$

46. $(3t^4 + 5t^3 - 8t^2 - 13t + 2) \div (t^2 - 5)$

47. $(2p^3 + 7p^2 + 9p + 3) \div (2p + 2)$

48. $(3x^3 + 4x^2 + 7x + 4) \div (3x + 3)$

49. $(3a^2 - 11a + 17) \div (2a + 6)$

50. $(5t^2 + 19t + 7) \div (4t + 12)$

51. $\dfrac{p^3 - 1}{p - 1}$

52. $\dfrac{8a^3 + 1}{2a + 1}$

Brain Busters   *Divide.*

53. $\left(2x^2 - \dfrac{7}{3}x - 1 \right) \div (3x + 1)$

54. $\left(m^2 + \dfrac{7}{2}m + 3 \right) \div (2m + 3)$

55. $\left(3a^2 - \dfrac{23}{4}a - 5 \right) \div (4a + 3)$

56. $\left(3q^2 + \dfrac{19}{5}q - 3 \right) \div (5q - 2)$

Solve each problem.

57. Suppose that the volume of a box is $(2p^3 + 15p^2 + 28p)$. The height is p and the length is $(p + 4)$. Give an expression in p that represents the width.

58. Suppose that a minivan travels a distance of $(2m^3 + 15m^2 + 35m + 36)$ miles in $(2m + 9)$ hours. Give an expression in m that represents the rate of the van in mph.

59. For $P(x) = x^3 - 4x^2 + 3x - 5$, find $P(-1)$. Then divide $P(x)$ by $D(x) = x + 1$. Compare the remainder with $P(-1)$. What do these results suggest?

60. *Concept Check*　Let $P(x) = 4x^3 - 8x^2 + 13x - 2$ and $D(x) = 2x - 1$. Use division to find polynomials $Q(x)$ and $R(x)$ such that

$$P(x) = Q(x) \cdot D(x) + R(x).$$

For each pair of functions, find the quotient $\left(\dfrac{f}{g}\right)(x)$ and give any x-values that are not in the domain of the quotient function. See Example 6.

61. $f(x) = 10x^2 - 2x, \quad g(x) = 2x$　　**62.** $f(x) = 18x^2 - 24x, \quad g(x) = 3x$

63. $f(x) = 2x^2 - x - 3, \quad g(x) = x + 1$　　**64.** $f(x) = 4x^2 - 23x - 35, \quad g(x) = x - 7$

65. $f(x) = 8x^3 - 27, \quad g(x) = 2x - 3$　　**66.** $f(x) = 27x^3 + 64, \quad g(x) = 3x + 4$

Let $f(x) = x^2 - 9, g(x) = 2x,$ and $h(x) = x - 3$. Find each of the following. See Example 6.

67. $\left(\dfrac{f}{g}\right)(x)$　　**68.** $\left(\dfrac{f}{h}\right)(x)$　　**69.** $\left(\dfrac{f}{g}\right)(2)$　　**70.** $\left(\dfrac{f}{h}\right)(1)$

71. $\left(\dfrac{h}{g}\right)(x)$　　**72.** $\left(\dfrac{g}{h}\right)(x)$　　**73.** $\left(\dfrac{h}{g}\right)(3)$　　**74.** $\left(\dfrac{g}{h}\right)(-1)$

75. $\left(\dfrac{f}{g}\right)\left(\dfrac{1}{2}\right)$　　**76.** $\left(\dfrac{f}{g}\right)\left(\dfrac{3}{2}\right)$　　**77.** $\left(\dfrac{h}{g}\right)\left(-\dfrac{1}{2}\right)$　　**78.** $\left(\dfrac{h}{g}\right)\left(-\dfrac{3}{2}\right)$

PREVIEW EXERCISES

Use the distributive property to rewrite each expression. See Section 1.4.

79. $8(y - 5)$　　**80.** $-(2x - 11)$　　**81.** $4p(2p + 1)$

82. $9 \cdot 6 + 9 \cdot r^2$　　**83.** $7(2x) - 7(3z)$　　**84.** $3x(x + 1) + 4(x + 1)$

CHAPTER 5 SUMMARY

KEY TERMS

5.1

exponent
base
exponential (power)

5.2

term
algebraic expression

polynomial
numerical coefficient
　(coefficient)
degree of a term
polynomial in x
descending powers
leading term
leading coefficient

trinomial
binomial
monomial
degree of a polynomial
like terms
negative of a polynomial

5.3

polynomial function
composition of functions
identity function
squaring function
cubing function

NEW SYMBOLS

$(f \circ g)(x) = f(g(x))$　composite function

TEST YOUR WORD POWER

See how well you have learned the vocabulary in this chapter.

1. A **polynomial** is an algebraic expression made up of
   A. a term or a finite product of terms with positive coefficients and exponents
   B. the sum of two or more terms with whole number coefficients and exponents
   C. the product of two or more terms
   D. a term or a finite sum of terms with real number coefficients and whole number exponents.

2. A **monomial** is a polynomial with
   A. only one term
   B. exactly two terms
   C. exactly three terms
   D. more than three terms.

3. A **binomial** is a polynomial with
   A. only one term
   B. exactly two terms
   C. exactly three terms
   D. more than three terms.

4. A **trinomial** is a polynomial with
   A. only one term
   B. exactly two terms
   C. exactly three terms
   D. more than three terms.

5. The **FOIL** method is used for
   A. adding two binomials
   B. adding two trinomials
   C. multiplying two binomials
   D. multiplying two trinomials.

ANSWERS

1. D; *Example:* $5x^3 + 2x^2 - 7$ **2.** A; *Examples:* $-4, 2x^3, 15a^2b$ **3.** B; *Example:* $3t^3 + 5t$ **4.** C; *Example:* $2a^2 - 3ab + b^2$
5. C; *Example:* $(m + 4)(m - 3) = m(m) - 3m + 4m + 4(-3) = m^2 + m - 12$
                    F   O   I   L

QUICK REVIEW

CONCEPTS

EXAMPLES

5.1   Integer Exponents and Scientific Notation

Definitions and Rules for Exponents
For all integers m and n and all real numbers a and b,

Product Rule $a^m \cdot a^n = a^{m+n}$

Quotient Rule $\dfrac{a^m}{a^n} = a^{m-n}$ $(a \neq 0)$

Zero Exponent $a^0 = 1$ $(a \neq 0)$

Negative Exponent $a^{-n} = \dfrac{1}{a^n}$ $(a \neq 0)$

Power Rules   **(a)** $(a^m)^n = a^{mn}$ **(b)** $(ab)^m = a^m b^m$

   (c) $\left(\dfrac{a}{b}\right)^n = \dfrac{a^n}{b^n}$ $(b \neq 0)$

Special Rules for Negative Exponents

$\dfrac{1}{a^{-n}} = a^n$ $(a \neq 0)$ $\dfrac{a^{-n}}{b^{-m}} = \dfrac{b^m}{a^n}$ $(a, b \neq 0)$

$a^{-n} = \left(\dfrac{1}{a}\right)^n$ $(a \neq 0)$ $\left(\dfrac{a}{b}\right)^{-n} = \left(\dfrac{b}{a}\right)^n$ $(a, b \neq 0)$

Apply the rules for exponents.

$$3^4 \cdot 3^2 = 3^6$$

$$\frac{2^5}{2^3} = 2^2$$

$$27^0 = 1, \quad (-5)^0 = 1$$

$$5^{-2} = \frac{1}{5^2}$$

$$(6^3)^4 = 6^{12}, \quad (5p)^4 = 5^4 p^4$$

$$\left(\frac{2}{3}\right)^5 = \frac{2^5}{3^5}$$

$$\frac{1}{3^{-2}} = 3^2, \qquad \frac{5^{-3}}{4^{-6}} = \frac{4^6}{5^3}$$

$$4^{-3} = \left(\frac{1}{4}\right)^3, \quad \left(\frac{4}{7}\right)^{-2} = \left(\frac{7}{4}\right)^2$$

Scientific Notation
A number is in scientific notation when it is expressed in the form

$a \times 10^n$,   where $1 \leq |a| < 10$ and n is an integer.

Write 23,500,000,000 in scientific notation.

$$23{,}500{,}000{,}000 = 2.35 \times 10^{10}$$

Write 4.3×10^{-6} in standard notation.

$$4.3 \times 10^{-6} = 0.0000043$$

5.2   Adding and Subtracting Polynomials

Add or subtract polynomials by combining like terms.

$$(5x^4 + 3x^2) - (7x^4 + x^2 - x)$$
$$= -2x^4 + 2x^2 + x$$

(continued)

CONCEPTS	EXAMPLES

5.3 Polynomial Functions, Graphs, and Composition

Adding and Subtracting Functions

If $f(x)$ and $g(x)$ define functions, then

$$(f + g)(x) = f(x) + g(x)$$

and

$$(f - g)(x) = f(x) - g(x).$$

Let $f(x) = x^2$ and $g(x) = 2x + 1$.

$$
\begin{aligned}
(f + g)(x) && (f - g)(x) \\
&= f(x) + g(x) && = f(x) - g(x) \\
&= x^2 + 2x + 1 && = x^2 - (2x + 1) \\
&&&= x^2 - 2x - 1
\end{aligned}
$$

Composition of f and g

$$(f \circ g)(x) = f(g(x))$$

Let $f(x) = x^2$ and $g(x) = 2x + 1$.

$$
\begin{aligned}
(f \circ g)(x) = f(g(x)) && (g \circ f)(x) = g(f(x)) \\
= f(2x + 1) && = g(x^2) \\
= (2x + 1)^2 && = 2x^2 + 1
\end{aligned}
$$

Graphs of Basic Polynomial Functions

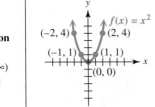

Identity function
$f(x) = x$
Domain: $(-\infty, \infty)$
Range: $(-\infty, \infty)$

$f(x) = x^2$ Squaring function
$f(x) = x^2$
Domain: $(-\infty, \infty)$
Range: $[0, \infty)$

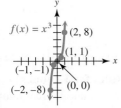

Cubing function
$f(x) = x^3$
Domain: $(-\infty, \infty)$
Range: $(-\infty, \infty)$

5.4 Multiplying Polynomials

To multiply two polynomials, multiply each term of one by each term of the other.

$$
\begin{aligned}
(x^3 + 3x)&(4x^2 - 5x + 2) \\
&= 4x^5 + 12x^3 - 5x^4 - 15x^2 + 2x^3 + 6x \\
&= 4x^5 - 5x^4 + 14x^3 - 15x^2 + 6x
\end{aligned}
$$

To multiply two binomials, use the **FOIL method.** Multiply the **First** terms, the **Outer** terms, the **Inner** terms, and the **Last** terms. Then add these products.

$$
\begin{aligned}
(2x + 3)&(x - 7) \\
&= 2x(x) + 2x(-7) + 3x + 3(-7) \quad \text{FOIL} \\
&= 2x^2 - 14x + 3x - 21 \\
&= 2x^2 - 11x - 21
\end{aligned}
$$

Special Products

$$(x + y)(x - y) = x^2 - y^2$$
$$(x + y)^2 = x^2 + 2xy + y^2$$
$$(x - y)^2 = x^2 - 2xy + y^2$$

$$
\begin{aligned}
(3m + 8)&(3m - 8) \\
&= 9m^2 - 64
\end{aligned}
$$

$$
\begin{aligned}
(5a + 3b)^2 && (2k - 1)^2 \\
= 25a^2 + 30ab + 9b^2 && = 4k^2 - 4k + 1
\end{aligned}
$$

Multiplying Functions

If $f(x)$ and $g(x)$ define functions, then

$$(fg)(x) = f(x) \cdot g(x).$$

Let $f(x) = x^2$ and $g(x) = 2x + 1$.

$$
\begin{aligned}
(fg)(x) \\
&= f(x) \cdot g(x) \\
&= x^2(2x + 1) \\
&= 2x^3 + x^2
\end{aligned}
$$

(continued)

CONCEPTS	EXAMPLES

5.5 Dividing Polynomials

Dividing by a Monomial
To divide a polynomial by a monomial, divide each term in the polynomial by the monomial, and then write each fraction in lowest terms.

$$\frac{2x^3 - 4x^2 + 6x - 8}{2x}$$
$$= \frac{2x^3}{2x} - \frac{4x^2}{2x} + \frac{6x}{2x} - \frac{8}{2x}$$
$$= x^2 - 2x + 3 - \frac{4}{x}$$

Dividing by a Polynomial
Use the "long division" process. The process ends when the remainder is 0 or when the degree of the remainder is less than the degree of the divisor.

Divide $\dfrac{m^3 - m^2 + 2m + 5}{m + 1}$.

$$
\begin{array}{r}
m^2 - 2m + 4 \\
m + 1\overline{)m^3 - m^2 + 2m + 5} \\
\underline{m^3 + m^2} \\
-2m^2 + 2m \\
\underline{-2m^2 - 2m} \\
4m + 5 \\
\underline{4m + 4} \\
1 \leftarrow \text{Remainder}
\end{array}
$$

The answer is
$$m^2 - 2m + 4 + \frac{1}{m + 1}.$$

Dividing Functions
If $f(x)$ and $g(x)$ define functions, then
$$\left(\frac{f}{g}\right)(x) = \frac{f(x)}{g(x)}, \quad g(x) \neq 0.$$

Let $f(x) = x^2$ and $g(x) = 2x + 1$.
$$\left(\frac{f}{g}\right)(x) = \frac{f(x)}{g(x)} = \frac{x^2}{2x + 1}, \quad x \neq -\frac{1}{2}$$

CHAPTER 5 REVIEW EXERCISES

5.1 *Simplify. Write answers with only positive exponents. Assume that all variables represent nonzero real numbers.*

1. 4^3

2. $\left(\frac{1}{3}\right)^4$

3. $(-5)^3$

4. $\dfrac{2}{(-3)^{-2}}$

5. $\left(\frac{2}{3}\right)^{-4}$

6. $\left(\frac{5}{4}\right)^{-2}$

7. $5^{-1} - 6^{-1}$

8. $2^{-1} + 4^{-1}$

9. $-3^0 + 3^0$

10. $(3^{-4})^2$

11. $(x^{-4})^{-2}$

12. $(xy^{-3})^{-2}$

13. $(z^{-3})^3 z^{-6}$

14. $(5m^{-3})^2(m^4)^{-3}$

15. $\dfrac{(3r)^2 r^4}{r^{-2} r^{-3}}(9r^{-3})^{-2}$

16. $\left(\dfrac{5z^{-3}}{z^{-1}}\right)\dfrac{5}{z^2}$

17. $\left(\dfrac{6m^{-4}}{m^{-9}}\right)^{-1}\left(\dfrac{m^{-2}}{16}\right)$

18. $\left(\dfrac{3r^5}{5r^{-3}}\right)^{-2}\left(\dfrac{9r^{-1}}{2r^{-5}}\right)^3$

19. $(-3x^4y^3)(4x^{-2}y^5)$

20. $\dfrac{6m^{-4}n^3}{-3mn^2}$

21. $\dfrac{(5p^{-2}q)(4p^5q^{-3})}{2p^{-5}q^5}$

22. Explain the difference between the expressions $(-6)^0$ and -6^0.

23. *Concept Check* Is $\left(\dfrac{a}{b}\right)^{-1} = \dfrac{a^{-1}}{b^{-1}}$ true for all $a, b \neq 0$? If not, explain.

24. *Concept Check* Is $(ab)^{-1} = ab^{-1}$ true for all $a, b \neq 0$? If not, explain.

Write in scientific notation.

25. 13,450

26. 0.0000000765

27. 0.138

28. In July 2008, the total resident population of the United States was **304,060,000.** Of this amount, **92,000** Americans were centenarians—that is, age **100** or older. Write the three boldfaced numbers in scientific notation. (*Source:* U.S. Census Bureau.)

Write each number in standard notation.

29. 1.21×10^6

30. 5.8×10^{-3}

Find each value. Give answers in both scientific notation and standard notation.

31. $\dfrac{16 \times 10^4}{8 \times 10^8}$

32. $\dfrac{6 \times 10^{-2}}{4 \times 10^{-5}}$

33. $\dfrac{0.0000000164}{0.0004}$

34. $\dfrac{0.0009 \times 12,000,000}{400,000}$

35. (a) The planet Mercury has an average distance from the sun of 3.6×10^7 mi, while the average distance from Venus to the sun is 6.7×10^7 mi. How long would it take a spacecraft traveling at 1.55×10^3 mph to travel from Venus to Mercury? (Give your answer in hours, in standard notation.)

    (b) Use the information from part (a) to find the number of days it would take the spacecraft to travel from Venus to Mercury. Round your answer to the nearest whole number of days.

5.2 *Give the numerical coefficient of each term.*

36. $14p^5$

37. $-z$

38. $\dfrac{x}{10}$

39. $504p^3r^5$

For each polynomial, (a) write it in descending powers, (b) identify it as a monomial, binomial, trinomial, or none of these, and (c) give its degree.

40. $9k + 11k^3 - 3k^2$

41. $14m^6 + 9m^7$

42. $-5y^4 + 3y^3 + 7y^2 - 2y$

43. $-7q^5r^3$

44. *Concept Check* Give an example of a polynomial in the variable x such that the polynomial has degree 5, is lacking a third-degree term, and is in descending powers of the variable.

Add or subtract as indicated.

45. Add.
$$3x^2 - 5x + 6$$
$$-4x^2 + 2x - 5$$

46. Subtract.
$$-5y^3 \qquad + 8y - 3$$
$$4y^2 + 2y + 9$$

47. $(4a^3 - 9a + 15) - (-2a^3 + 4a^2 + 7a)$ **48.** $(3y^2 + 2y - 1) + (5y^2 - 11y + 6)$

49. Find the perimeter of the triangle.

$4x^2 + 2$ $6x^2 + 5x + 2$

$2x^2 + 3x + 1$

5.3

50. Find each of the following for the polynomial function defined by

$$f(x) = -2x^2 + 5x + 7.$$

  (a) $f(-2)$ **(b)** $f(3)$

51. Find each of the following for the polynomial functions defined by

$$f(x) = 2x + 3 \quad \text{and} \quad g(x) = 5x^2 - 3x + 2.$$

  (a) $(f + g)(x)$ **(b)** $(f - g)(x)$ **(c)** $(f + g)(-1)$ **(d)** $(f - g)(-1)$

52. Find each of the following for the polynomial functions defined by

$$f(x) = 3x^2 + 2x - 1 \quad \text{and} \quad g(x) = 5x + 7.$$

  (a) $(g \circ f)(3)$ **(b)** $(f \circ g)(3)$ **(c)** $(f \circ g)(-2)$

  (d) $(g \circ f)(-2)$ **(e)** $(f \circ g)(x)$ **(f)** $(g \circ f)(x)$

53. The number of twin births in the United States during the years 1990 through 2006 can be modeled by the polynomial function defined by

$$f(x) = -21.307x^3 + 603.07x^2$$
$$- 1576.7x + 94{,}319,$$

where $x = 0$ corresponds to 1990, $x = 1$ corresponds to 1991, and so on. Use this model to approximate the number of twin births in each given year. (*Source:* National Center for Health Statistics.)

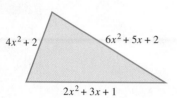

  (a) 1990   **(b)** 2000   **(c)** 2006

Graph each polynomial function defined as follows.

54. $f(x) = -2x + 5$ **55.** $f(x) = x^2 - 6$ **56.** $f(x) = -x^3 + 1$

5.4 *Find each product.*

57. $-6k(2k^2 + 7)$ **58.** $(3m - 2)(5m + 1)$

59. $(3w - 2t)(2w - 3t)$ **60.** $(2p^2 + 6p)(5p^2 - 4)$

61. $(3q^2 + 2q - 4)(q - 5)$ **62.** $(6r^2 - 1)(6r^2 + 1)$

63. $(4m + 3)^2$ **64.** $t(3t + 2)^2$

5.5 *Divide.*

65. $\dfrac{4y^3 - 12y^2 + 5y}{4y}$ **66.** $\dfrac{x^3 - 9x^2 + 26x - 30}{x - 5}$

67. $\dfrac{2p^3 + 9p^2 + 27}{2p - 3}$ **68.** $\dfrac{5p^4 + 15p^3 - 33p^2 - 9p + 18}{5p^2 - 3}$

MIXED REVIEW EXERCISES

69. *Concept Check* Match each expression (a)–(j) in Column I with its equivalent expression A–J in Column II. Choices may be used once, more than once, or not at all.

I

(a) 4^{-2} **(b)** -4^2

(c) 4^0 **(d)** $(-4)^0$

(e) $(-4)^{-2}$ **(f)** -4^0

(g) $-4^0 + 4^0$ **(h)** $-4^0 - 4^0$

(i) $4^{-2} + 4^{-1}$ **(j)** 4^2

II

A. $\dfrac{1}{16}$ **B.** 0

C. 1 **D.** $-\dfrac{1}{16}$

E. -1 **F.** $\dfrac{5}{16}$

G. -16 **H.** -2

I. 16 **J.** none of these

In Exercises 70–81, perform the indicated operations and then simplify. Write answers with only positive exponents. Assume that all variables represent nonzero real numbers.

70. $\dfrac{6^{-1}y^3(y^2)^{-2}}{6y^{-4}(y^{-1})}$

71. 5^{-3}

72. $-(-3)^2$

73. $7p^5(3p^4 + p^3 + 2p^2)$

74. $(2x - 9)^2$

75. $\dfrac{(-z^{-2})^3}{5(z^{-3})^{-1}}$

76. $(y^6)^{-5}(2y^{-3})^{-4}$

77. $\dfrac{8x^2 - 23x + 2}{x - 3}$

78. $\dfrac{(5z^2x^3)^2(2zx^2)^{-1}}{(-10zx^{-3})^{-2}(3z^{-1}x^{-4})^2}$

79. $[(3m - 5n) + p][(3m - 5n) - p]$

80. $\dfrac{20y^3x^3 + 15y^4x + 25yx^4}{10yx^2}$

81. $(2k - 1) - (3k^2 - 2k + 6)$

82. In 2009, the estimated population of New Zealand was 4.2134×10^6. The population density was 40.76 people per square mile. Based on this information, what is the area of New Zealand, to the nearest square mile? (*Source: The World Factbook.*)

CHAPTER 5 TEST

CHAPTER Test Prep VIDEOS

Step-by-step test solutions are found on the Chapter Test Prep Videos available via the Video Resources on DVD, in *MyMathLab*, or on YouTube (search "LialIntermediateAlg").

View the complete solutions to all Chapter Test exercises on the Video Resources on DVD.

1. Match each expression (a)–(j) in Column I with its equivalent expression A–J in Column II. Choices may be used once, more than once, or not at all.

I

(a) 7^{-2} **(b)** 7^0

(c) -7^0 **(d)** $(-7)^0$

(e) -7^2 **(f)** $7^{-1} + 2^{-1}$

(g) $(7 + 2)^{-1}$ **(h)** $\dfrac{7^{-1}}{2^{-1}}$

(i) 7^2 **(j)** $(-7)^{-2}$

II

A. 1 **B.** $\dfrac{1}{9}$

C. $\dfrac{1}{49}$ **D.** -1

E. -49 **F.** $\dfrac{9}{14}$

G. $\dfrac{2}{7}$ **H.** 0

I. 49 **J.** none of these

Simplify. Write answers with only positive exponents. Assume that all variables represent nonzero real numbers.

2. $(3x^{-2}y^3)^{-2}(4x^3y^{-4})$

3. $\dfrac{36r^{-4}(r^2)^{-3}}{6r^4}$

4. $\left(\dfrac{4p^2}{q^4}\right)^3\left(\dfrac{6p^8}{q^{-8}}\right)^{-2}$

5. $(-2x^4y^{-3})^0(-4x^{-3}y^{-8})^2$

6. Write 9.1×10^{-7} in standard form.

7. Use scientific notation to simplify

$$\frac{2,500,000 \times 0.00003}{0.05 \times 5,000,000}.$$

Write the answer in both scientific notation and standard notation.

8. Find each of the following for the functions defined by

$$f(x) = -2x^2 + 5x - 6 \quad \text{and} \quad g(x) = 7x - 3.$$

(a) $f(4)$ **(b)** $(f + g)(x)$ **(c)** $(f - g)(x)$ **(d)** $(f - g)(-2)$

9. Find each of the following for the functions defined by

$$f(x) = 3x + 5 \quad \text{and} \quad g(x) = x^2 + 2.$$

(a) $(f \circ g)(-2)$ **(b)** $(f \circ g)(x)$ **(c)** $(g \circ f)(x)$

Graph each polynomial function.

10. $f(x) = -2x^2 + 3$

11. $f(x) = -x^3 + 3$

12. The number of medical doctors, in thousands, in the United States during the years from 1980 through 2005 can be modeled by the polynomial function defined by

$$f(x) = 0.139x^2 + 14.16x + 465.9,$$

where $x = 0$ corresponds to 1980, $x = 1$ corresponds to 1981, and so on. Use this model to approximate the number of doctors to the nearest thousand in each given year. (*Source: American Medical Association.*)

(a) 1980 **(b)** 1995 **(c)** 2005

Perform the indicated operations.

13. $(4x^3 - 3x^2 + 2x - 5) - (3x^3 + 11x + 8) + (x^2 - x)$

14. $(5x - 3)(2x + 1)$

15. $(2m - 5)(3m^2 + 4m - 5)$

16. $(6x + y)(6x - y)$

17. $(3k + q)^2$

18. $[2y + (3z - x)][2y - (3z - x)]$

19. $\dfrac{16p^3 - 32p^2 + 24p}{4p^2}$

20. $(x^3 + 3x^2 - 4) \div (x - 1)$

21. If $f(x) = x^2 + 3x + 2$ and $g(x) = x + 1$, find each of the following.

(a) $(fg)(x)$ **(b)** $(fg)(-2)$

22. Use $f(x)$ and $g(x)$ from **Exercise 21** to find each of the following.

(a) $\left(\dfrac{f}{g}\right)(x)$ **(b)** $\left(\dfrac{f}{g}\right)(-2)$

CUMULATIVE REVIEW EXERCISES

1. Match each number in Column I with the choice or choices of sets of numbers in Column II to which the number belongs.

I		**II**	
(a) 34	**(b)** 0	**A.** Natural numbers	**B.** Whole numbers
(c) 2.16	**(d)** $-\sqrt{36}$	**C.** Integers	**D.** Rational numbers
(e) $\sqrt{13}$	**(f)** $-\dfrac{4}{5}$	**E.** Irrational numbers	**F.** Real numbers

Evaluate.

2. $9 \cdot 4 - 16 \div 4$

3. $\left(\dfrac{1}{3}\right)^2 - \left(\dfrac{1}{2}\right)^3$

4. $-|8 - 13| - |-4| + |-9|$

Solve.

5. $-5(8 - 2z) + 4(7 - z) = 7(8 + z) - 3$

6. $3(x + 2) - 5(x + 2) = -2x - 4$

7. $A = p + prt$ for t

8. $2(m + 5) - 3m + 1 > 5$

9. $|3x - 1| = 2$

10. $|3z + 1| \geq 7$

11. A recent survey polled teens about the most important inventions of the 20th century. Complete the results shown in the table if 1500 teens were surveyed.

Most Important Invention	Percent	Actual Number
Personal computer		480
Pacemaker	26%	
Wireless communication	18%	
Television		150

Source: Lemelson–MIT Program.

12. Find the measure of each angle of the triangle.

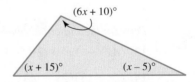

$(6x + 10)°$

$(x + 15)°$ $(x - 5)°$

Find the slope of each line described.

13. Through $(-4, 5)$ and $(2, -3)$

14. Horizontal, through $(4, 5)$

*Find an equation of each line. Write the equation in **(a)** slope-intercept form and in **(b)** standard form.*

15. Through $(4, -1)$, $m = -4$

16. Through $(0, 0)$ and $(1, 4)$

Graph each equation or inequality.

17. $-3x + 4y = 12$

18. $y \leq 2x - 6$

19. $3x + 2y < 0$

20. The graph shows the number of pounds of shrimp caught in the United States (in thousands of pounds) in selected years.

(a) Use the information given in the graph to find and interpret the average rate of change in the number of pounds of shrimp caught per year.

(b) If $x = 0$ represents the year 2000, $x = 1$ represents 2001, and so on, use your answer from part (a) to write an equation of the line in slope-intercept form that models the annual amount of shrimp caught (in thousands of pounds, to the nearest whole number) for the years 2000 through 2005.

(c) Use the equation from part (b) to approximate the amount of shrimp caught in 2003.

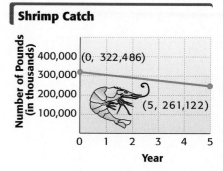

Shrimp Catch

(0, 322,486)

(5, 261,122)

Source: National Oceanic and Atmospheric Administration.

21. Give the domain and range of the relation

$$\{(-4, -2), (-1, 0), (2, 0), (5, 2)\}.$$

Does this relation define a function?

22. If $g(x) = -x^2 - 2x + 6$, find $g(3)$.

Solve each system.

23. $3x - 4y = 1$
$2x + 3y = 12$

24. $3x - 2y = 4$
$-6x + 4y = 7$

25. $x + 3y - 6z = 7$
$2x - y + z = 1$
$x + 2y + 2z = -1$

Use a system of equations to solve each problem.

26. The Star-Spangled Banner that flew over Fort McHenry during the War of 1812 had a perimeter of 144 ft. Its length measured 12 ft more than its width. Find the dimensions of this flag, which is displayed in the Smithsonian Institution's Museum of American History in Washington, DC. (*Source:* National Park Service brochure.)

27. Agbe Asiamigbe needs 9 L of a 20% solution of alcohol. Agbe has a 15% solution on hand, as well as a 30% solution. How many liters of the 15% solution and the 30% solution should Agbe mix to get the 20% solution needed?

Simplify. Write answers with only positive exponents. Assume that all variables represent positive real numbers.

28. $\left(\dfrac{2m^3n}{p^2}\right)^3$

29. $\dfrac{x^{-6}y^3z^{-1}}{x^7y^{-4}z}$

30. $(2m^{-2}n^3)^{-3}$

Perform the indicated operations.

31. $(3x^2 - 8x + 1) - (x^2 - 3x - 9)$

32. $(x + 2y)(x^2 - 2xy + 4y^2)$

33. $(3x + 2y)(5x - y)$

34. $\dfrac{16x^3y^5 - 8x^2y^2 + 4}{4x^2y}$

35. $\dfrac{m^3 - 3m^2 + 5m - 3}{m - 1}$

Factoring

In our number system, whole numbers greater than 1 are categorized as either *prime* or *composite*. A composite number such as 10 can be *factored* into its prime factors in one and only one way: $10 = 2 \cdot 5$. Similarly, polynomials can be factored into their prime factors.

In algebra, *factoring* can be used to solve *quadratic equations*. One important application is to express the distance a falling or projected object travels in a specific time. In **Section 6.5,** we use these concepts to find the heights of objects after they are projected or dropped.

6.1   Greatest Common Factors and Factoring by Grouping

OBJECTIVES

1 Factor out the greatest common factor.

2 Factor by grouping.

Writing a polynomial as the product of two or more simpler polynomials is called **factoring** the polynomial. For example, the product of $3x$ and $5x - 2$ is $15x^2 - 6x$, and $15x^2 - 6x$ can be factored as the product $3x(5x - 2)$.

$$3x(5x - 2) = 15x^2 - 6x \qquad \text{Multiplying}$$
$$15x^2 - 6x = 3x(5x - 2) \qquad \text{Factoring}$$

Notice that both multiplying and factoring use the distributive property, but in opposite directions. ***Factoring "undoes," or reverses, multiplying.***

OBJECTIVE 1   **Factor out the greatest common factor.** The first step in factoring a polynomial is to find the *greatest common factor* for the terms of the polynomial. The **greatest common factor (GCF)** is the largest term that is a factor of all terms in the polynomial.

For example, the greatest common factor for $8x + 12$ is 4, since 4 is the largest term that is a factor of (divides into) both $8x$ and 12.

$$8x + 12$$
$$= 4(2x) + 4(3) \qquad \text{Factor 4 from each term.}$$
$$= 4(2x + 3) \qquad \text{Distributive property}$$

As a check, multiply 4 and $2x + 3$. The result should be $8x + 12$. Using the distributive property this way is called **factoring out the greatest common factor.**

NOW TRY EXERCISE 1

Factor out the greatest common factor.

(a) $54m - 45$

(b) $2k - 7$

EXAMPLE 1   Factoring Out the Greatest Common Factor

Factor out the greatest common factor.

(a) $9z - 18$

Since 9 is the GCF, factor 9 from each term.

$$9z - 18$$
$$= 9 \cdot z - 9 \cdot 2 \qquad \text{GCF} = 9$$
$$= 9(z - 2) \qquad \text{Distributive property}$$

CHECK   Multiply $9(z - 2)$ to obtain $9z - 18. \leftarrow$ Original polynomial

(b) $56m + 35p$

$$= 7(8m + 5p)$$

(c) $2y + 5$ There is no common factor other than 1.

(d) $\qquad 12 + 24z$

> Remember to write the 1.

$$= 12 \cdot 1 + 12 \cdot 2z \qquad \text{Identity property}$$
$$= 12(1 + 2z) \qquad \text{12 is the GCF.}$$

CHECK $12(1 + 2z)$

$$= 12(1) + 12(2z) \qquad \text{Distributive property}$$
$$= 12 + 24z \checkmark \qquad \text{Original polynomial} \qquad\qquad \text{NOW TRY}$$

NOW TRY ANSWERS

1. **(a)** $9(6m - 5)$
    (b) There is no common factor other than 1.

NOW TRY
EXERCISE 2

Factor out the greatest common factor.

$20m^3n^3 - 15m^3n^2 + 10m^2n$

EXAMPLE 2 Factoring Out the Greatest Common Factor

Factor out the greatest common factor.

(a) $9x^2 + 12x^3$

The numerical part of the GCF is 3, the largest number that divides into both 9 and 12. The least exponent that appears on x is 2. The GCF is $3x^2$.

$$9x^2 + 12x^3$$
$$= 3x^2(3) + 3x^2(4x) \qquad \text{GCF} = 3x^2$$
$$= 3x^2(3 + 4x) \qquad \text{Distributive property}$$

(b) $32p^4 - 24p^3 + 40p^5$
$$= 8p^3(4p) + 8p^3(-3) + 8p^3(5p^2) \qquad \text{GCF} = 8p^3$$
$$= 8p^3(4p - 3 + 5p^2) \qquad \text{Distributive property}$$

(c) $\qquad 3k^4 - 15k^7 + 24k^9$

Remember the 1.
$$= 3k^4(1 - 5k^3 + 8k^5) \qquad \text{GCF} = 3k^4$$

(d) $24m^3n^2 - 18m^2n + 6m^4n^3$
$$= 6m^2n(4mn) + 6m^2n(-3) + 6m^2n(m^2n^2)$$
$$= 6m^2n(4mn - 3 + m^2n^2)$$

(e) $25x^2y^3 + 30y^5 - 15x^4y^7$
$$= 5y^3(5x^2 + 6y^2 - 3x^4y^4)$$

In each case, remember to check the factored form by multiplying. *NOW TRY*

NOW TRY
EXERCISE 3

Factor out the greatest common factor.

(a) $(3x - 2)(x + 1) +$
$\qquad (3x - 2)(2x - 5)$

(b) $z(a - b)^3 - 2w(a - b)^2$

EXAMPLE 3 Factoring Out a Binomial Factor

Factor out the greatest common factor.

(a) $(x - 5)(x + 6) + (x - 5)(2x + 5)$ The greatest common factor is $x - 5$.
$$= (x - 5)[(x + 6) + (2x + 5)] \qquad \text{Factor out } x - 5.$$
$$= (x - 5)(x + 6 + 2x + 5)$$
$$= (x - 5)(3x + 11) \qquad \text{Combine like terms.}$$

(b) $z^2(m + n)^2 + x^2(m + n)^2$
$$= (m + n)^2(z^2 + x^2) \qquad \text{Factor out } (m + n)^2.$$

(c) $p(r + 2s)^2 - q(r + 2s)^3$
$$= (r + 2s)^2[p - q(r + 2s)] \qquad \text{Factor out the common factor.}$$
$$= (r + 2s)^2(p - qr - 2qs) \qquad \text{Be careful with signs.}$$

(d) $(p - 5)(p + 2) - (p - 5)(3p + 4)$
$$= (p - 5)[(p + 2) - (3p + 4)] \qquad \text{Factor out } p - 5.$$
$$= (p - 5)[p + 2 - 3p - 4] \qquad \text{Distributive property}$$
$$= (p - 5)[-2p - 2] \qquad \text{Combine like terms.}$$
$$= (p - 5)[-2(p + 1)] \qquad \text{Look for a common factor.}$$
$$= -2(p - 5)(p + 1) \qquad \text{Commutative property} \qquad \text{NOW TRY}$$

NOW TRY ANSWERS
2. $5m^2n(4mn^2 - 3mn + 2)$
3. (a) $(3x - 2)(3x - 4)$
   (b) $(a - b)^2(za - zb - 2w)$

NOW TRY
EXERCISE 4
Factor $-4y^5 - 3y^3 + 8y$ in two ways.

EXAMPLE 4 Factoring Out a Negative Common Factor

Factor $-a^3 + 3a^2 - 5a$ in two ways.

First, a could be used as the common factor.

$$-a^3 + 3a^2 - 5a$$
$$= a(-a^2) + a(3a) + a(-5) \quad \text{Factor out } a.$$
$$= a(-a^2 + 3a - 5)$$

Because of the leading negative sign, $-a$ could also be used as the common factor.

$$-a^3 + 3a^2 - 5a$$
$$= -a(a^2) + (-a)(-3a) + (-a)(5) \quad \text{Factor out } -a.$$
$$= -a(a^2 - 3a + 5)$$

Sometimes there may be a reason to prefer one of these forms, but either is correct.

NOW TRY

NOTE The answer section in this book will usually give the factored form where the common factor has a positive coefficient.

OBJECTIVE 2 **Factor by grouping.** Sometimes the *individual terms* of a polynomial have a greatest common factor of 1, but it still may be possible to factor the polynomial by using a process called *factoring by grouping*. **We usually factor by grouping when a polynomial has more than three terms.**

NOW TRY
EXERCISE 5
Factor.

$$3m - 3n + xm - xn$$

EXAMPLE 5 Factoring by Grouping

Factor $ax - ay + bx - by$.

Group the terms as follows.

Terms with common factor a Terms with common factor b

$$(ax - ay) + (bx - by)$$

Then factor $ax - ay$ as $a(x - y)$ and factor $bx - by$ as $b(x - y)$.

$$ax - ay + bx - by$$
$$= (ax - ay) + (bx - by) \quad \text{Group the terms.}$$
$$= a(x - y) + b(x - y) \quad \text{Factor each group.}$$
$$= (x - y)(a + b) \quad \text{The common factor is } x - y.$$

CHECK $(x - y)(a + b)$

$$= xa + xb - ya - yb \quad \text{Multiply using the FOIL method.}$$
$$= ax + bx - ay - by \quad \text{Commutative property}$$
$$= ax - ay + bx - by \ \checkmark \quad \text{Original polynomial} \qquad \text{NOW TRY}$$

NOW TRY ANSWERS
4. $y(-4y^4 - 3y^2 + 8)$;
   $-y(4y^4 + 3y^2 - 8)$
5. $(m - n)(3 + x)$

NOW TRY
EXERCISE 6

Factor.

$$ab - 7a - 5b + 35$$

EXAMPLE 6 Factoring by Grouping

Factor $3x - 3y - ax + ay$.

$$3x - 3y - ax + ay$$ Pay close attention here.

$$= (3x - 3y) + (-ax + ay) \quad \text{Group the terms.}$$

$$= 3(x - y) + a(-x + y) \quad \text{Factor out 3, and factor out } a.$$

The factors $(x - y)$ and $(-x + y)$ are opposites. If we factor out $-a$ instead of a in the second group, we get the common binomial factor $(x - y)$. So we start over.

$$(3x - 3y) + (-ax + ay)$$

$$= 3(x - y) - a(x - y) \quad \text{Be careful with signs.}$$

$$= (x - y)(3 - a) \quad \text{Factor out } x - y.$$

CHECK $(x - y)(3 - a)$

$$= 3x - ax - 3y + ay \quad \text{Multiply using the FOIL method.}$$

$$= 3x - 3y - ax + ay \;\checkmark \quad \text{Original polynomial} \qquad \textit{NOW TRY}$$

NOTE In **Example 6,** a different grouping would lead to the factored form $(a - 3)(y - x)$. Verify by multiplying that this is also correct.

Factoring by Grouping

Step 1 **Group terms.** Collect the terms into groups so that each group has a common factor.

Step 2 **Factor within the groups.** Factor out the common factor in each group.

Step 3 **Factor the entire polynomial.** If each group now has a common factor, factor it out. If not, try a different grouping.

Always check the factored form by multiplying.

NOW TRY
EXERCISE 7

Factor.

$$3ax - 6xy - a + 2y$$

EXAMPLE 7 Factoring by Grouping

Factor $6ax + 12bx + a + 2b$.

$$6ax + 12bx + a + 2b$$

$$= (6ax + 12bx) + (a + 2b) \quad \text{Group the terms.}$$

Now factor $6x$ from the first group, and use the identity property of multiplication to introduce the factor 1 in the second group.

Remember to write the 1.

$$= 6x(a + 2b) + 1(a + 2b) \quad \text{Factor each group.}$$

$$= (a + 2b)(6x + 1) \quad \text{Factor out } a + 2b.$$

CHECK $(a + 2b)(6x + 1)$

$$= 6ax + a + 12bx + 2b \quad \text{FOIL}$$

$$= 6ax + 12bx + a + 2b \;\checkmark \quad \text{Original polynomial} \qquad \textit{NOW TRY}$$

NOW TRY ANSWERS
6. $(b - 7)(a - 5)$
7. $(a - 2y)(3x - 1)$

NOW TRY
EXERCISE 8

Factor.

$$2kp^2 + 6 - 3p^2 - 4k$$

EXAMPLE 8 Rearranging Terms before Factoring by Grouping

Factor $p^2q^2 - 10 - 2q^2 + 5p^2$.

Neither the first two terms nor the last two terms have a common factor except 1. We can rearrange and group the terms as follows.

$$p^2q^2 - 10 - 2q^2 + 5p^2$$

$$= (p^2q^2 - 2q^2) + (5p^2 - 10) \qquad \text{Rearrange and group the terms.}$$

$$= q^2(p^2 - 2) + 5(p^2 - 2) \qquad \text{Factor out the common factors.}$$

Don't stop here.

$$= (p^2 - 2)(q^2 + 5) \qquad \text{Factor out } p^2 - 2. \text{ Use parentheses.}$$

CHECK $\qquad (p^2 - 2)(q^2 + 5)$

$$= p^2q^2 + 5p^2 - 2q^2 - 10 \qquad \text{FOIL}$$

$$= p^2q^2 - 10 - 2q^2 + 5p^2 \; \checkmark \qquad \text{Original polynomial} \quad \text{NOW TRY}$$

⚠ **CAUTION** In **Example 8,** do not stop at the step

$$q^2(p^2 - 2) + 5(p^2 - 2).$$

NOW TRY ANSWER

8. $(p^2 - 2)(2k - 3)$

This expression is *not in factored form,* because it is a *sum* of two terms, $q^2(p^2 - 2)$ and $5(p^2 - 2)$, not a *product.*

6.1 EXERCISES

MyMathLab | Math XL PRACTICE | WATCH | DOWNLOAD | READ | REVIEW

⊙ *Complete solution available on the Video Resources on DVD*

Factor out the greatest common factor. Simplify the factors, if possible. ***See Examples 1–4.***

1. $12m - 60$ **2.** $15r - 45$ **3.** $4 + 20z$

4. $9 + 27x$ **5.** $8y - 15$ **6.** $7x - 40$

7. $8k^3 + 24k$ **8.** $9z^4 + 81z$ **9.** $-4p^3q^4 - 2p^2q^5$

10. $-3z^5w^2 - 18z^3w^4$ **11.** $7x^3 + 35x^4 - 14x^5$ **12.** $6k^3 - 36k^4 - 48k^5$

13. $10t^5 - 2t^3 - 4t^4$ **14.** $6p^3 - 3p^2 - 9p^4$

15. $15a^2c^3 - 25ac^2 + 5ac$ **16.** $15y^3z^3 - 27y^2z^4 + 3yz^3$

17. $16z^2n^6 + 64zn^7 - 32z^3n^3$ **18.** $5r^3s^5 + 10r^2s^2 - 15r^4s^2$

19. $14a^3b^2 + 7a^2b - 21a^5b^3 + 42ab^4$ **20.** $12km^3 - 24k^3m^2 + 36k^2m^4 - 60k^4m^3$

21. $(m - 4)(m + 2) + (m - 4)(m + 3)$ **22.** $(z - 5)(z + 7) + (z - 5)(z + 10)$

23. $(2z - 1)(z + 6) - (2z - 1)(z - 5)$ **24.** $(3x + 2)(x - 4) - (3x + 2)(x + 8)$

25. $5(2 - x)^2 - 2(2 - x)^3$ **26.** $2(5 - x)^3 - 3(5 - x)^2$

27. $4(3 - x)^2 - (3 - x)^3 + 3(3 - x)$ **28.** $2(t - s) + 4(t - s)^2 - (t - s)^3$

29. $15(2z + 1)^3 + 10(2z + 1)^2 - 25(2z + 1)$

30. $6(a + 2b)^2 - 4(a + 2b)^3 + 12(a + 2b)^4$

31. $5(m + p)^3 - 10(m + p)^2 - 15(m + p)^4$

32. $-9a^2(p + q) - 3a^3(p + q)^2 + 6a(p + q)^3$

Factor each polynomial twice. First use a common factor with a positive coefficient, and then use a common factor with a negative coefficient. **See Example 4.**

🌀 **33.** $-r^3 + 3r^2 + 5r$

34. $-t^4 + 8t^3 - 12t$

35. $-12s^5 + 48s^4$

36. $-16y^4 + 64y^3$

37. $-2x^5 + 6x^3 + 4x^2$

38. $-5a^3 + 10a^4 - 15a^5$

Factor by grouping. **See Examples 5–8.**

🌀 **39.** $mx + qx + my + qy$

40. $2k + 2h + jk + jh$

41. $10m + 2n + 5mk + nk$

42. $3ma + 3mb + 2ab + 2b^2$

🌀 **43.** $4 - 2q - 6p + 3pq$

44. $20 + 5m + 12n + 3mn$

45. $p^2 - 4zq + pq - 4pz$

46. $r^2 - 9tw + 3rw - 3rt$

🌀 **47.** $2xy + 3y + 2x + 3$

48. $7ab + 35bc + a + 5c$

49. $m^3 + 4m^2 - 6m - 24$

50. $2a^3 + a^2 - 14a - 7$

51. $-3a^3 - 3ab^2 + 2a^2b + 2b^3$

52. $-16m^3 + 4m^2p^2 - 4mp + p^3$

🌀 **53.** $4 + xy - 2y - 2x$

54. $10ab - 21 - 6b + 35a$

55. $8 + 9y^4 - 6y^3 - 12y$

56. $x^3y^2 - 3 - 3y^2 + x^3$

57. $1 - a + ab - b$

58. $2ab^2 - 8b^2 + a - 4$

Brain Busters *Factor out the variable that is raised to the lesser exponent. (For example, in* **Exercise 59,** *factor out m^{-5}.)*

59. $3m^{-5} + m^{-3}$

60. $k^{-2} + 2k^{-4}$

61. $3p^{-3} + 2p^{-2}$

62. $-5q^{-3} + 8q^{-2}$

63. *Concept Check* When directed to completely factor the polynomial $4x^2y^5 - 8xy^3$, a student wrote

$$2xy^3(2xy^2 - 4).$$

When the teacher did not give him full credit, he complained because when his answer is multiplied out, the result is the original polynomial. *WHAT WENT WRONG?* Give the correct answer.

64. *Concept Check* Refer to **Exercise 58.** One form of the answer is $(2b^2 + 1)(a - 4)$. Give two other acceptable factored forms of $2ab^2 - 8b^2 + a - 4$.

65. *Concept Check* Which choice is an example of a polynomial in factored form?

A. $3x^2y^3 + 6x^2(2x + y)$

B. $5(x + y)^2 - 10(x + y)^3$

C. $(-2 + 3x)(5y^2 + 4y + 3)$

D. $(3x + 4)(5x - y) - (3x + 4)(2x - 1)$

66. *Concept Check* Refer to the answer in **Exercise 65.** When expressed in polynomial form, what is the degree of the polynomial?

PREVIEW EXERCISES

Find each product. **See Section 5.4.**

67. $(k + 7)(k - 1)$

68. $(4y - 2)(5y + 6)$

69. $(5x - 2t)(5x + 2t)$

70. $(2x^2 - 5)(x^2 - 6)$

71. $(3y^3 - 4)(2y^3 + 3)$

72. $5t(3t + 2)(t - 8)$

6.2 Factoring Trinomials

OBJECTIVE 1 Factor trinomials when the coefficient of the quadratic term is 1. We begin by finding the product of $x + 3$ and $x - 5$.

$$(x + 3)(x - 5)$$
$$= x^2 - 5x + 3x - 15 \qquad \text{FOIL}$$
$$= x^2 - 2x - 15 \qquad \text{Combine like terms.}$$

By this result, the factored form of $x^2 - 2x - 15$ is $(x + 3)(x - 5)$.

$$\text{Factored form} \longrightarrow (x + 3)(x - 5) = x^2 - 2x - 15 \longleftarrow \text{Product}$$

Multiplication / Factoring

Since multiplying and factoring are operations that "undo" each other, factoring trinomials involves using FOIL in reverse. As shown here, the x^2-term comes from multiplying x and x, and -15 comes from multiplying 3 and -5.

Product of x and x is x^2.

$$(x + 3)(x - 5) = x^2 - 2x - 15$$

Product of 3 and -5 is -15.

We find the $-2x$ in $x^2 - 2x - 15$ by multiplying the outer terms, multiplying the inner terms, and adding.

Outer terms: $x(-5) = -5x$

$$(x + 3)(x - 5) \qquad \text{Add to get } -2x.$$

Inner terms: $3 \cdot x = 3x$

Based on this example, use the following steps to factor a trinomial $x^2 + bx + c$, where 1 is the coefficient of the quadratic term.

Factoring $x^2 + bx + c$

Step 1 **Find pairs whose product is c.** Find all pairs of integers whose product is c, the third term of the trinomial.

Step 2 **Find the pair whose sum is b.** Choose the pair whose sum is b, the coefficient of the middle term.

If there are no such integers, the trinomial cannot be factored.

A polynomial that cannot be factored with integer coefficients is a **prime polynomial.**

$$x^2 + x + 2, \quad x^2 - x - 1, \quad 2x^2 + x + 7 \qquad \text{Examples of prime polynomials}$$

**NOW TRY
EXERCISE 1**
Factor each trinomial.

(a) $t^2 - t - 30$

(b) $w^2 + 12w + 32$

EXAMPLE 1 Factoring Trinomials in $x^2 + bx + c$ Form

Factor each trinomial.

(a) $x^2 + 2x - 35$

Step 1 Find pairs of integers whose product is -35.

$$-35(1)$$
$$35(-1)$$
$$7(-5)$$
$$5(-7)$$

Step 2 Write sums of those integers.

$$-35 + 1 = -34$$
$$35 + (-1) = 34$$
$$7 + (-5) = 2 \leftarrow \text{Coefficient of the middle term}$$
$$5 + (-7) = -2$$

The integers 7 and -5 have the necessary product and sum, so

$$x^2 + 2x - 35 \quad \text{factors as} \quad (x + 7)(x - 5).$$ Multiply to check.

(b) $r^2 + 8r + 12$

Look for two integers with a product of 12 and a sum of 8. Of all pairs having a product of 12, only the pair 6 and 2 has a sum of 8.

$$r^2 + 8r + 12 \quad \text{factors as} \quad (r + 6)(r + 2).$$

Because of the commutative property, it would be equally correct to write $(r + 2)(r + 6)$. ***Check by using FOIL to multiply the factored form.*** NOW TRY

**NOW TRY
EXERCISE 2**
Factor $m^2 + 12m - 11$.

EXAMPLE 2 Recognizing a Prime Polynomial

Factor $m^2 + 6m + 7$.

Look for two integers whose product is 7 and whose sum is 6. Only two pairs of integers, 7 and 1 and -7 and -1, give a product of 7. Neither of these pairs has a sum of 6, so $m^2 + 6m + 7$ cannot be factored with integer coefficients and is prime. NOW TRY

**NOW TRY
EXERCISE 3**
Factor $a^2 + ab - 20b^2$.

EXAMPLE 3 Factoring a Trinomial in Two Variables

Factor $x^2 + 6ax - 16a^2$.

This trinomial is in the form $x^2 + bx + c$, where $b = 6a$ and $c = -16a^2$.

Step 1 Find pairs of expressions whose product is $-16a^2$.

$$16a(-a)$$
$$-16a(a)$$
$$8a(-2a)$$
$$-8a(2a)$$
$$-4a(4a)$$

Step 2 Write sums of the pairs of expressions from Step 1, looking for a sum of $6a$.

$$16a + (-a) = 15a$$
$$-16a + a = -15a$$
$$8a + (-2a) = 6a$$
$$-8a + 2a = -6a$$
$$-4a + 4a = 0$$

The expressions $8a$ and $-2a$ have the necessary product and sum, so

$$x^2 + 6ax - 16a^2 \quad \text{factors as} \quad (x + 8a)(x - 2a).$$

CHECK $(x + 8a)(x - 2a)$

$$= x^2 - 2ax + 8ax - 16a^2 \quad \text{FOIL}$$
$$= x^2 + 6ax - 16a^2 \ \checkmark \quad \text{Original polynomial} \quad \text{NOW TRY}$$

NOW TRY ANSWERS
1. **(a)** $(t - 6)(t + 5)$
   (b) $(w + 4)(w + 8)$
2. prime
3. $(a + 5b)(a - 4b)$

NOW TRY
EXERCISE 4
Factor $5w^3 - 40w^2 + 60w$.

EXAMPLE 4 Factoring a Trinomial with a Common Factor

Factor $16y^3 - 32y^2 - 48y$.

$$16y^3 - 32y^2 - 48y$$
$$= 16y(y^2 - 2y - 3) \qquad \text{Factor out the GCF, } 16y.$$

To factor $y^2 - 2y - 3$, look for two integers whose product is -3 and whose sum is -2. The necessary integers are -3 and 1.

$$= 16y(y - 3)(y + 1) \qquad \text{Factor the trinomial.}$$

Remember to include the GCF, $16y$.

NOW TRY

⚠ **CAUTION** *When factoring, always look for a common factor first. Remember to write the common factor as part of the answer.*

OBJECTIVE 2 Factor trinomials when the coefficient of the quadratic term is not 1. We can use a generalization of the method shown in **Objective 1** to factor a trinomial of the form $ax^2 + bx + c$, where $a \neq 1$. To factor $3x^2 + 7x + 2$, for example, we first identify the values a, b, and c.

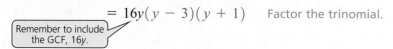

$$ax^2 + bx + c$$
$$3x^2 + 7x + 2, \qquad \text{so} \qquad a = 3, \quad b = 7, \quad c = 2$$

The product ac is $3 \cdot 2 = 6$, so we must find integers having a product of 6 and a sum of 7 (since the middle term has coefficient 7). The necessary integers are 1 and 6, so we write $7x$ as $1x + 6x$, or $x + 6x$.

$$3x^2 + 7x + 2$$
$$= 3x^2 + \underline{x + 6x} + 2$$
$$\qquad\qquad x + 6x = 7x$$
$$= (3x^2 + x) + (6x + 2) \qquad \text{Group the terms.}$$
$$= x(3x + 1) + 2(3x + 1) \qquad \text{Factor by grouping.}$$
$$= (3x + 1)(x + 2) \qquad \text{Factor out the common factor.}$$

Check by multiplying.

NOW TRY
EXERCISE 5
Factor $8y^2 - 10y - 3$.

EXAMPLE 5 Factoring a Trinomial in $ax^2 + bx + c$ Form

Factor $12r^2 - 5r - 2$.

Since $a = 12$, $b = -5$, and $c = -2$, the product ac is -24. The two integers whose product is -24 and whose sum b is -5, are 3 and -8.

$$12r^2 - 5r - 2$$
$$= 12r^2 + 3r - 8r - 2 \qquad \text{Write } -5r \text{ as } 3r - 8r.$$
$$= 3r(4r + 1) - 2(4r + 1) \qquad \text{Factor by grouping.}$$
$$= (4r + 1)(3r - 2) \qquad \text{Factor out the common factor.}$$

Check by multiplying.

NOW TRY

NOW TRY ANSWERS
4. $5w(w - 6)(w - 2)$
5. $(2y - 3)(4y + 1)$

OBJECTIVE 3 Use an alternative method for factoring trinomials. This method involves trying repeated combinations and using FOIL.

NOW TRY
EXERCISE 6
Factor $10r^2 + 19r + 6$.

EXAMPLE 6 Factoring Trinomials in $ax^2 + bx + c$ Form

Factor each trinomial.

(a) $3x^2 + 7x + 2$

To factor this trinomial, we use an alternative method. The goal is to find the correct numbers to put in the blanks.

$$3x^2 + 7x + 2 = (\underline{\quad}x + \underline{\quad})(\underline{\quad}x + \underline{\quad})$$

Addition signs are used, since all the signs in the polynomial indicate addition. The first two expressions have a product of $3x^2$, so they must be $3x$ and x.

$$3x^2 + 7x + 2 = (3x + \underline{\quad})(x + \underline{\quad})$$

The product of the two last terms must be 2, so the numbers must be 2 and 1. There is a choice. The 2 could be placed with the $3x$ or with the x. Only one of these choices will give the correct middle term, $7x$. We use the FOIL method to check each one.

$$\overset{3x}{(3x + 2)(x + 1)}$$
$$\underset{2x}{}$$

$3x + 2x = 5x$
Wrong middle term

$$\overset{6x}{(3x + 1)(x + 2)}$$
$$\underset{x}{}$$

$6x + x = 7x$
Correct middle term

Therefore, $3x^2 + 7x + 2$ factors as $(3x + 1)(x + 2)$. (Compare with the solution obtained by factoring by grouping on the preceding page.)

(b) $12r^2 - 5r - 2$

To reduce the number of trials, we note that the terms of the trinomial have greatest common factor 1. This means that neither of its factors can have a common factor except 1. We should keep this in mind as we choose factors. We try 4 and 3 for the two first terms.

$$12r^2 - 5r - 2 = (4r \underline{\quad})(3r \underline{\quad})$$

The factors of -2 are -2 and 1 or -1 and 2. We try both possibilities.

$$(4r - 2)(3r + 1)$$

Wrong: $4r - 2$ has a factor of 2, which cannot be correct, since 2 is not a factor of $12r^2 - 5r - 2$.

$$\overset{8r}{(4r - 1)(3r + 2)}$$
$$\underset{-3r}{}$$

$8r - 3r = 5r$
Wrong middle term

The middle term on the right is $5r$, instead of the $-5r$ that is needed. We get $-5r$ by interchanging the signs of the second terms in the factors.

$$\overset{-8r}{(4r + 1)(3r - 2)}$$
$$\underset{3r}{}$$

$-8r + 3r = -5r$
Correct middle term

Thus, $12r^2 - 5r - 2$ factors as $(4r + 1)(3r - 2)$. (Compare with **Example 5.**)

NOW TRY

NOW TRY ANSWER
6. $(2r + 3)(5r + 2)$

NOTE As shown in **Example 6(b),** if the terms of a polynomial have greatest common factor 1, then none of the terms of its factors can have a common factor (except 1). Remembering this will eliminate some potential factors.

Factoring $ax^2 + bx + c$ (GCF of a, b, c is 1)

Step 1 **Find pairs whose product is *a*.** Write all pairs of integer factors of a, the coefficient of the second-degree term.

Step 2 **Find pairs whose product is *c*.** Write all pairs of integer factors of c, the last term.

Step 3 **Choose inner and outer terms.** Use FOIL and various combinations of the factors from Steps 1 and 2 until the necessary middle term is found.

If no such combinations exist, the trinomial is prime.

NOW TRY
EXERCISE 7

Factor $12a^2 - 19ab - 21b^2$.

EXAMPLE 7 Factoring a Trinomial in Two Variables

Factor $18m^2 - 19mx - 12x^2$.

The terms of the polynomial have greatest common factor 1. Follow the steps for factoring a trinomial. There are many possible factors of both 18 and -12. Try 6 and 3 for 18 and -3 and 4 for -12.

$$(6m - 3x)(3m + 4x) \qquad (6m + 4x)(3m - 3x)$$
Wrong: common factor $\qquad$ Wrong: common factors

Since 6 and 3 do not work as factors of 18, try 9 and 2 instead, with 3 and -4 as factors of -12.

$$(9m + 3x)(2m - 4x) \qquad \overset{\overset{\displaystyle 27mx}{\overbrace{\qquad\qquad}}}{(9m - 4x)(2m + 3x)}$$
Wrong: common factors $\qquad \underset{-8mx}{\underbrace{\quad}}$

$$27mx + (-8mx) = 19mx$$
Wrong middle term

The result on the right differs from the correct middle term only in sign, so interchange the signs of the second terms in the factors.

$$18m^2 - 19mx - 12x^2 \quad \text{factors as} \quad (9m + 4x)(2m - 3x) \longleftarrow \boxed{\text{Check by multiplying.}}$$

NOW TRY

NOW TRY
EXERCISE 8

Factor $-8x^2 + 22x - 15$.

EXAMPLE 8 Factoring a Trinomial in $ax^2 + bx + c$ Form ($a < 0$)

Factor $-3x^2 + 16x + 12$.

While we could factor directly, it is helpful to first factor out -1 so that the coefficient of the x^2-term is positive.

$$-3x^2 + 16x + 12$$
$$= -1(3x^2 - 16x - 12) \qquad \text{Factor out } -1.$$
$$= -1(3x + 2)(x - 6) \qquad \text{Factor the trinomial.}$$
$$= -(3x + 2)(x - 6) \qquad -1a = -a$$

NOW TRY

NOW TRY ANSWERS
7. $(4a + 3b)(3a - 7b)$
8. $-(4x - 5)(2x - 3)$

NOTE The factored form in **Example 8** can be written in other ways, such as

$$(-3x - 2)(x - 6) \quad \text{and} \quad (3x + 2)(-x + 6).$$

Verify that these both give the original trinomial when multiplied.

**NOW TRY
EXERCISE 9**
Factor $12y^3 + 33y^2 - 9y$.

EXAMPLE 9 Factoring a Trinomial with a Common Factor

Factor $16y^3 + 24y^2 - 16y$.

$$16y^3 + 24y^2 - 16y$$

> Remember the common factor.

$$= 8y(2y^2 + 3y - 2) \qquad \text{GCF} = 8y$$

$$= 8y(2y - 1)(y + 2) \qquad \text{Factor the trinomial.} \qquad \text{NOW TRY}$$

OBJECTIVE 4 Factor by substitution.

**NOW TRY
EXERCISE 10**
Factor.

$$3(a + 2)^2 - 11(a + 2) - 4$$

EXAMPLE 10 Factoring a Polynomial by Substitution

Factor $2(x + 3)^2 + 5(x + 3) - 12$.

Since the binomial $x + 3$ appears to powers 2 and 1, we let a substitution variable represent $x + 3$. We may choose any letter we wish except x. We choose t.

$$2(x + 3)^2 + 5(x + 3) - 12$$

$$= 2t^2 + 5t - 12 \qquad \text{Let } t = x + 3.$$

$$= (2t - 3)(t + 4) \qquad \text{Factor.}$$

$$= [2(x + 3) - 3][(x + 3) + 4] \qquad \text{Replace } t \text{ with } x + 3.$$

$$= (2x + 6 - 3)(x + 7) \qquad \text{Simplify.}$$

$$= (2x + 3)(x + 7) \qquad \text{NOW TRY}$$

⚠ **CAUTION** *Remember to make the final substitution* of $x + 3$ for t in **Example 10**.

**NOW TRY
EXERCISE 11**
Factor $6x^4 + 11x^2 + 3$.

EXAMPLE 11 Factoring a Trinomial in $ax^4 + bx^2 + c$ Form

Factor $6y^4 + 7y^2 - 20$.

The variable y appears to powers in which the greater exponent is twice the lesser exponent. We can let a substitution variable represent the variable to the lesser power. Here, we let $t = y^2$.

$$6y^4 + 7y^2 - 20$$

$$= 6(y^2)^2 + 7y^2 - 20 \qquad y^4 = (y^2)^2$$

$$= 6t^2 + 7t - 20 \qquad \text{Substitute } t \text{ for } y^2.$$

> Don't stop here. Replace t with y^2.

$$= (3t - 4)(2t + 5) \qquad \text{Factor.}$$

$$= (3y^2 - 4)(2y^2 + 5) \qquad t = y^2 \qquad \text{NOW TRY}$$

NOW TRY ANSWERS
9. $3y(4y - 1)(y + 3)$
10. $(3a + 7)(a - 2)$
11. $(3x^2 + 1)(2x^2 + 3)$

NOTE Some students feel comfortable factoring polynomials like the one in **Example 11** directly, without using the substitution method.

6.2 EXERCISES

Complete solution available on the Video Resources on DVD

1. *Concept Check*   Which is *not* a valid way of starting the process of factoring the trinomial $12x^2 + 29x + 10$?

    A. $(12x \quad)(x \quad)$ **B.** $(4x \quad)(3x \quad)$

    C. $(6x \quad)(2x \quad)$ **D.** $(8x \quad)(4x \quad)$

2. *Concept Check*   Which is the completely factored form of $2x^6 - 5x^5 - 3x^4$?

    A. $x^4(2x + 1)(x - 3)$ **B.** $x^4(2x - 1)(x + 3)$

    C. $(2x^5 + x^4)(x - 3)$ **D.** $x^3(2x^2 + x)(x - 3)$

3. *Concept Check*   Which is *not* a factored form of $-x^2 + 16x - 60$?

    A. $(x - 10)(-x + 6)$ **B.** $(-x - 10)(x + 6)$

    C. $(-x + 10)(x - 6)$ **D.** $-(x - 10)(x - 6)$

4. *Concept Check*   Which is the completely factored form of $4x^2 - 4x - 24$?

    A. $4(x - 2)(x + 3)$ **B.** $4(x + 2)(x + 3)$

    C. $4(x + 2)(x - 3)$ **D.** $4(x - 2)(x - 3)$

Factor each trinomial. **See Examples 1–9.**

5. $y^2 + 7y - 30$ **6.** $z^2 + 2z - 24$ **7.** $p^2 + 15p + 56$

8. $k^2 - 11k + 30$ **9.** $m^2 - 11m + 60$ **10.** $p^2 - 12p - 27$

11. $a^2 - 2ab - 35b^2$ **12.** $z^2 + 8zw + 15w^2$ **13.** $a^2 - 9ab + 18b^2$

14. $k^2 - 11hk + 28h^2$ **15.** $x^2y^2 + 12xy + 18$ **16.** $p^2q^2 - 5pq - 18$

17. $-6m^2 - 13m + 15$ **18.** $-15y^2 + 17y + 18$ **19.** $10x^2 + 3x - 18$

20. $8k^2 + 34k + 35$ **21.** $20k^2 + 47k + 24$ **22.** $27z^2 + 42z - 5$

23. $15a^2 - 22ab + 8b^2$ **24.** $14c^2 - 17cd - 6d^2$ **25.** $36m^2 - 60m + 25$

26. $25r^2 - 90r + 81$ **27.** $40x^2 + xy + 6y^2$ **28.** $15p^2 + 24pq + 8q^2$

29. $6x^2z^2 + 5xz - 4$ **30.** $8m^2n^2 - 10mn + 3$ **31.** $24x^2 + 42x + 15$

32. $36x^2 + 18x - 4$ **33.** $-15a^2 - 70a + 120$ **34.** $-12a^2 - 10a + 42$

35. $-11x^3 + 110x^2 - 264x$ **36.** $-9k^3 - 36k^2 + 189k$

37. $2x^3y^3 - 48x^2y^4 + 288xy^5$ **38.** $6m^3n^2 - 60m^2n^3 + 150mn^4$

39. $6a^3 + 12a^2 - 90a$ **40.** $3m^4 + 6m^3 - 72m^2$

41. $13y^3 + 39y^2 - 52y$ **42.** $4p^3 + 24p^2 - 64p$

43. $12p^3 - 12p^2 + 3p$ **44.** $45t^3 + 60t^2 + 20t$

45. *Concept Check*   When a student was given the polynomial $4x^2 + 2x - 20$ to factor completely on a test, the student lost some credit when her answer was $(4x + 10)(x - 2)$. She complained to her teacher that when we multiply $(4x + 10)(x - 2)$, we get the original polynomial. **WHAT WENT WRONG?** Give the correct answer.

46. When factoring the polynomial $-4x^2 - 29x + 24$, Terry obtained $(-4x + 3)(x + 8)$, while John got $(4x - 3)(-x - 8)$. Who is correct? Explain your answer.

Factor each trinomial. **See Example 10.**

47. $12p^6 - 32p^3r + 5r^2$ **48.** $2y^6 + 7xy^3 + 6x^2$

49. $10(k + 1)^2 - 7(k + 1) + 1$ **50.** $4(m - 5)^2 - 4(m - 5) - 15$

51. $3(m + p)^2 - 7(m + p) - 20$ **52.** $4(x - y)^2 - 23(x - y) - 6$

Factor each trinomial. (Hint: *Factor out the GCF first.*)

53. $a^2(a + b)^2 - ab(a + b)^2 - 6b^2(a + b)^2$

54. $m^2(m - p)^2 - mp(m - p)^2 - 12p^2(m - p)^2$

55. $p^2(p + q) + 4pq(p + q) + 3q^2(p + q)$

56. $2k^2(5 - y) - 7k(5 - y) + 5(5 - y)$

57. $z^2(z - x) - zx(x - z) - 2x^2(z - x)$

58. $r^2(r - s) - 5rs(s - r) - 6s^2(r - s)$

Factor each trinomial. ***See Example 11.***

59. $p^4 - 10p^2 + 16$ **60.** $k^4 + 10k^2 + 9$ **61.** $2x^4 - 9x^2 - 18$

62. $6z^4 + z^2 - 1$ **63.** $16x^4 + 16x^2 + 3$ **64.** $9r^4 + 9r^2 + 2$

PREVIEW EXERCISES

Find each product. ***See Section 5.4.***

65. $(3x - 5)(3x + 5)$ **66.** $(8m + 3)(8m - 3)$

67. $(p + 3q)^2$ **68.** $(2z - 7)^2$

69. $(y + 3)(y^2 - 3y + 9)$ **70.** $(3m - 1)(9m^2 + 3m + 1)$

6.3   Special Factoring

OBJECTIVES

1   Factor a difference of squares.

2   Factor a perfect square trinomial.

3   Factor a difference of cubes.

4   Factor a sum of cubes.

OBJECTIVE 1 **Factor a difference of squares.** The special products introduced in **Section 5.4** are used in reverse when factoring. Recall that the product of the sum and difference of two terms leads to a **difference of squares.**

Difference of Squares

$$x^2 - y^2 = (x + y)(x - y)$$

EXAMPLE 1   Factoring Differences of Squares

Factor each polynomial.

(a) $t^2 - 36$

$\quad = t^2 - 6^2$ $36 = 6^2$

$\quad = (t + 6)(t - 6)$ Factor the difference of squares.

(b) $4a^2 - 64$

$\quad = 4(a^2 - 16)$ Factor out the common factor, 4.

$\quad = 4(a + 4)(a - 4)$ Factor the difference of squares.

$$x^2 \quad - \quad y^2 \quad = \quad (x \ + \ y) \ (x \ - \ y)$$
$$\downarrow \qquad \downarrow \qquad\qquad \downarrow \quad\ \downarrow \quad \downarrow \quad\ \downarrow$$

(c) $16m^2 - 49p^2 = (4m)^2 - (7p)^2 = (4m + 7p)(4m - 7p)$

NOW TRY
EXERCISE 1

Factor each polynomial.

(a) $4m^2 - 25n^2$

(b) $9x^2 - 729$

(c) $(a + b)^2 - 25$

(d) $v^4 - 1$

$$\underset{\downarrow}{x^2} \quad - \quad \underset{\downarrow}{y^2} \quad = \quad (\underset{\downarrow}{x} \; + \; \underset{\downarrow}{y}) \; (\underset{\downarrow}{x} \; - \; \underset{\downarrow}{y})$$

(d) $81k^2 - (a + 2)^2 = (9k)^2 - (a + 2)^2 = (9k + a + 2)(9k - (a + 2))$

$$= (9k + a + 2)(9k - a - 2)$$

We could have used the method of substitution here.

(e) $x^4 - 81$

$= (x^2 + 9)(x^2 - 9)$ Factor the difference of squares.

$= (x^2 + 9)(x + 3)(x - 3)$ Factor the difference of squares again.

NOW TRY

⚠ **CAUTION** *Assuming that the greatest common factor is 1, it is not possible to factor (with real numbers) a sum of squares* such as $x^2 + 9$ in **Example 1(e)**. In particular, $x^2 + y^2 \neq (x + y)^2$, as shown next.

OBJECTIVE 2 **Factor a perfect square trinomial.** Two other special products from **Section 5.4** lead to the following rules for factoring.

Perfect Square Trinomial

$$x^2 + 2xy + y^2 = (x + y)^2$$
$$x^2 - 2xy + y^2 = (x - y)^2$$

Because the trinomial $x^2 + 2xy + y^2$ is the square of $x + y$, it is called a **perfect square trinomial.** In this pattern, both the first and the last terms of the trinomial must be perfect squares. In the factored form $(x + y)^2$, twice the product of the first and the last terms must give the middle term of the trinomial.

$$4m^2 + 20m + 25 \qquad\qquad p^2 - 8p + 64$$

Perfect square trinomial; $\qquad$ Not a perfect square trinomial;
$4m^2 = (2m)^2$, $25 = 5^2$, $\qquad$ middle term would have to be
and $2(2m)(5) = 20m$. $\qquad\qquad$ $16p$ or $-16p$.

EXAMPLE 2 **Factoring Perfect Square Trinomials**

Factor each polynomial.

(a) $144p^2 - 120p + 25$

Here, $144p^2 = (12p)^2$ and $25 = 5^2$. The sign on the middle term is $-$, so if $144p^2 - 120p + 25$ is a perfect square trinomial, the factored form will have to be

$$(12p - 5)^2.$$

Determine twice the product of the two terms to see if this is correct.

$$2(12p)(-5) = -120p$$

This is the middle term of the given trinomial.

$$144p^2 - 120p + 25 \quad \text{factors as} \quad (12p - 5)^2.$$

NOW TRY ANSWERS
1. **(a)** $(2m + 5n)(2m - 5n)$
   (b) $9(x + 9)(x - 9)$
   (c) $(a + b + 5)(a + b - 5)$
   (d) $(v^2 + 1)(v + 1)(v - 1)$

NOW TRY
EXERCISE 2
Factor each polynomial.

(a) $a^2 + 12a + 36$

(b) $16x^2 - 56xy + 49y^2$

(c) $y^2 - 16y + 64 - z^2$

(b) $4m^2 + 20mn + 49n^2$

If this is a perfect square trinomial, it will equal $(2m + 7n)^2$. By the pattern in the box, if multiplied out, this squared binomial has a middle term of

$$2(2m)(7n) = 28mn,$$

which *does not equal* $20mn$. Verify that this trinomial cannot be factored by the methods of the previous section either. It is prime.

(c) $(r + 5)^2 + 6(r + 5) + 9$

$\qquad = [(r + 5) + 3]^2 \qquad\qquad 2(r + 5)(3) = 6(r + 5),$ the middle term.

$\qquad = (r + 8)^2$

(d) $m^2 - 8m + 16 - p^2$

Since there are four terms, use factoring by grouping. The first three terms here form a perfect square trinomial. Group them together, and factor as follows.

$$m^2 - 8m + 16 - p^2$$

$$= (m^2 - 8m + 16) - p^2$$

$$= (m - 4)^2 - p^2 \qquad\qquad \text{Factor the perfect square trinomial.}$$

$$= (m - 4 + p)(m - 4 - p) \qquad \text{Factor the difference of squares.}$$

NOW TRY

NOTE Perfect square trinomials can be factored by the general methods shown earlier for other trinomials. The patterns given here provide "shortcuts."

OBJECTIVE 3 **Factor a difference of cubes.** A **difference of cubes,** $x^3 - y^3$, can be factored as follows.

Difference of Cubes

$$x^3 - y^3 = (x - y)(x^2 + xy + y^2)$$

Check by showing that the product of $x - y$ and $x^2 + xy + y^2$ is $x^3 - y^3$.

EXAMPLE 3   Factoring Differences of Cubes

Factor each polynomial.

$$x^3 - y^3 = (x - y)(x^2 + x \cdot y + y^2)$$

$$\downarrow \quad \downarrow \quad\quad \downarrow \quad \downarrow \downarrow \quad \downarrow \downarrow \quad \downarrow$$

(a) $m^3 - 8 = m^3 - 2^3 = (m - 2)(m^2 + m \cdot 2 + 2^2)$

$$= (m - 2)(m^2 + 2m + 4)$$

CHECK $(m - 2)(m^2 + 2m + 4)$

$$= m^3 + 2m^2 + 4m - 2m^2 - 4m - 8 \qquad \text{Distributive property}$$

$$= m^3 - 8 \ \checkmark \qquad\qquad\qquad\qquad\quad \text{Combine like terms.}$$

NOW TRY ANSWERS

2. (a) $(a + 6)^2$

   (b) $(4x - 7y)^2$

   (c) $(y - 8 + z)(y - 8 - z)$

Factor each polynomial.

(a) $t^3 - 1$

(b) $125a^3 - 8b^3$

(b) $27x^3 - 8y^3$

$\qquad = (3x)^3 - (2y)^3$ Difference of cubes

$\qquad = (3x - 2y)[(3x)^2 + (3x)(2y) + (2y)^2]$ Factor.

$\qquad = (3x - 2y)(9x^2 + 6xy + 4y^2)$ $(3x)^2 = 3^2x^2, \text{ not } 3x^2.$
$\qquad\qquad\qquad\qquad\qquad\qquad\qquad\qquad\quad (2y)^2 = 2^2y^2, \text{ not } 2y^2.$

(c) $1000k^3 - 27n^3$

$\qquad = (10k)^3 - (3n)^3$ Difference of cubes

$\qquad = (10k - 3n)[(10k)^2 + (10k)(3n) + (3n)^2]$ Factor.

$\qquad = (10k - 3n)(100k^2 + 30kn + 9n^2)$ Multiply.   *NOW TRY*

OBJECTIVE 4 Factor a sum of cubes. While the binomial $x^2 + y^2$ cannot be factored with real numbers, a **sum of cubes,** such as $x^3 + y^3$, is factored as follows.

Sum of Cubes

$$x^3 + y^3 = (x + y)(x^2 - xy + y^2)$$

NOTE The sign of the second term in the binomial factor of a sum or difference of cubes is *always the same* as the sign in the original polynomial. In the trinomial factor, the first and last terms are *always positive*. The sign of the middle term is *the opposite of* the sign of the second term in the binomial factor.

EXAMPLE 4 Factoring Sums of Cubes

Factor each polynomial.

(a) $r^3 + 27$

$\qquad = r^3 + 3^3$ Sum of cubes

$\qquad = (r + 3)(r^2 - 3r + 3^2)$ Factor.

$\qquad = (r + 3)(r^2 - 3r + 9)$ $3^2 = 9$

(b) $27z^3 + 125$

$\qquad = (3z)^3 + 5^3$ Sum of cubes

$\qquad = (3z + 5)[(3z)^2 - (3z)(5) + 5^2]$ Factor.

$\qquad = (3z + 5)(9z^2 - 15z + 25)$ Multiply.

(c) $125t^3 + 216s^6$

$\qquad = (5t)^3 + (6s^2)^3$ Sum of cubes

$\qquad = (5t + 6s^2)[(5t)^2 - (5t)(6s^2) + (6s^2)^2]$ Factor.

$\qquad = (5t + 6s^2)(25t^2 - 30ts^2 + 36s^4)$ Multiply.

(d) $3x^2 + 192$

$\qquad = 3(x^3 + 64)$ Factor out the common factor.

$\qquad = 3(x^3 + 4^3)$ Write as a sum of cubes.

   Remember the common factor.

$\qquad = 3(x + 4)(x^2 - 4x + 16)$ Factor.

NOW TRY ANSWERS
3. (a) $(t - 1)(t^2 + t + 1)$
 (b) $(5a - 2b)(25a^2 + 10ab + 4b^2)$

NOW TRY
EXERCISE 4
Factor each polynomial.

(a) $1000 + z^3$

(b) $81a^6 + 3b^3$

(c) $(x - 3)^3 + y^3$

(e) $(x + 2)^3 + t^3$

$= [(x + 2) + t][(x + 2)^2 - (x + 2)t + t^2]$ Sum of cubes

$= (x + 2 + t)(x^2 + 4x + 4 - xt - 2t + t^2)$ Multiply. **NOW TRY**

⚠ **CAUTION** A common error when factoring $x^3 + y^3$ or $x^3 - y^3$ is to think that the xy-term has a coefficient of 2. Since there is no coefficient of 2, expressions of the form $x^2 + xy + y^2$ and $x^2 - xy + y^2$ usually cannot be factored further.

The special types of factoring are summarized here. *These should be memorized.*

NOW TRY ANSWERS

4. **(a)** $(10 + z) \cdot$
$(100 - 10z + z^2)$

(b) $3(3a^2 + b) \cdot$
$(9a^4 - 3a^2b + b^2)$

(c) $(x - 3 + y) \cdot$
$(x^2 - 6x + 9 - xy + 3y + y^2)$

Special Types of Factoring	
Difference of Squares	$x^2 - y^2 = (x + y)(x - y)$
Perfect Square Trinomial	$x^2 + 2xy + y^2 = (x + y)^2$
	$x^2 - 2xy + y^2 = (x - y)^2$
Difference of Cubes	$x^3 - y^3 = (x - y)(x^2 + xy + y^2)$
Sum of Cubes	$x^3 + y^3 = (x + y)(x^2 - xy + y^2)$

6.3 EXERCISES

MyMathLab | *Math* XL PRACTICE WATCH DOWNLOAD READ REVIEW

🌐 *Complete solution available on the Video Resources on DVD*

Concept Check Work each problem.

1. Which of the following binomials are differences of squares?

   A. $64 - k^2$ **B.** $2x^2 - 25$ **C.** $k^2 + 9$ **D.** $4z^4 - 49$

2. Which of the following binomials are sums or differences of cubes?

   A. $64 + r^3$ **B.** $125 - p^6$ **C.** $9x^3 + 125$ **D.** $(x + y)^3 - 1$

3. Which of the following trinomials are perfect squares?

   A. $x^2 - 8x - 16$ **B.** $4m^2 + 20m + 25$

   C. $9z^4 + 30z^2 + 25$ **D.** $25p^2 - 45p + 81$

4. Of the 12 polynomials listed in **Exercises 1–3,** which ones can be factored by the methods of this section?

5. The binomial $4x^2 + 64$ is an example of a sum of two squares that can be factored. Under what conditions can the sum of two squares be factored?

6. Insert the correct signs in the blanks.

   (a) $8 + m^3 = (2 \underline{\quad} m)(4 \underline{\quad} 2m \underline{\quad} m^2)$

   (b) $n^3 - 1 = (n \underline{\quad} 1)(n^2 \underline{\quad} n \underline{\quad} 1)$

Factor each polynomial. ***See Examples 1–4.***

🌐 **7.** $p^2 - 16$ **8.** $k^2 - 9$ **9.** $25x^2 - 4$

10. $36m^2 - 25$ **11.** $18a^2 - 98b^2$ **12.** $32c^2 - 98d^2$

13. $64m^4 - 4y^4$ **14.** $243x^4 - 3t^4$ **15.** $(y + z)^2 - 81$

16. $(h + k)^2 - 9$ **17.** $16 - (x + 3y)^2$ **18.** $64 - (r + 2t)^2$

19. $p^4 - 256$ **20.** $a^4 - 625$ **21.** $k^2 - 6k + 9$

22. $x^2 + 10x + 25$ 🌐 **23.** $4z^2 + 4zw + w^2$ **24.** $9y^2 + 6yz + z^2$

25. $16m^2 - 8m + 1 - n^2$ **26.** $25c^2 - 20c + 4 - d^2$ **27.** $4r^2 - 12r + 9 - s^2$

28. $9a^2 - 24a + 16 - b^2$ **29.** $x^2 - y^2 + 2y - 1$ **30.** $-k^2 - h^2 + 2kh + 4$

31. $98m^2 + 84mn + 18n^2$ **32.** $80z^2 - 40zw + 5w^2$

33. $(p + q)^2 + 2(p + q) + 1$ **34.** $(x + y)^2 + 6(x + y) + 9$

35. $(a - b)^2 + 8(a - b) + 16$ **36.** $(m - n)^2 + 4(m - n) + 4$

🌐 **37.** $x^3 - 27$ **38.** $y^3 - 64$ **39.** $216 - t^3$

40. $512 - m^3$ 🌐 **41.** $x^3 + 64$ **42.** $r^3 + 343$

43. $1000 + y^3$ **44.** $729 + x^3$ **45.** $8x^3 + 1$

46. $27y^3 + 1$ **47.** $125x^3 - 216$ **48.** $8w^3 - 125$

49. $x^3 - 8y^3$ **50.** $z^3 - 125p^3$ **51.** $64g^3 - 27h^3$

52. $27a^3 - 8b^3$ **53.** $343p^3 + 125q^3$ **54.** $512t^3 + 27s^3$

55. $24n^3 + 81p^3$ **56.** $250x^3 + 16y^3$ **57.** $(y + z)^3 + 64$

58. $(p - q)^3 + 125$ **59.** $m^6 - 125$ **60.** $27r^6 + 1$

61. $27 - 1000x^9$ **62.** $64 - 729p^9$ **63.** $125y^6 + z^3$

64. *Concept Check*   Consider $(x - y)^2 - 25$. To factor this polynomial, is the first step $x^2 - 2xy + y^2 - 25$ correct?

RELATING CONCEPTS   **EXERCISES 65–70**

FOR INDIVIDUAL OR GROUP WORK

The binomial $x^6 - y^6$ may be considered either as a difference of squares or a difference of cubes. **Work Exercises 65–70 in order.**

65. Factor $x^6 - y^6$ by first factoring as a difference of squares. Then factor further by considering one of the factors as a sum of cubes and the other factor as a difference of cubes.

66. Based on your answer in **Exercise 65**, fill in the blank with the correct factors so that $x^6 - y^6$ is factored completely.

$$x^6 - y^6 = (x - y)(x + y)\underline{\hspace{4cm}}$$

67. Factor $x^6 - y^6$ by first factoring as a difference of cubes. Then factor further by considering one of the factors as a difference of squares.

68. Based on your answer in **Exercise 67**, fill in the blank with the correct factor so that $x^6 - y^6$ is factored.

$$x^6 - y^6 = (x - y)(x + y)\underline{\hspace{4cm}}$$

69. Notice that the factor you wrote in the blank in **Exercise 68** is a fourth-degree polynomial, while the two factors you wrote in the blank in **Exercise 66** are both second-degree polynomials. What must be true about the product of the two factors you wrote in the blank in **Exercise 66**? Verify this.

70. If you have a choice of factoring as a difference of squares or a difference of cubes, how should you start to more easily obtain the completely factored form of the polynomial? Base the answer on your results in **Exercises 65–69**.

In some cases, the method of factoring by grouping can be combined with the methods of special factoring discussed in this section. Consider this example.

$8x^3 + 4x^2 + 27y^3 - 9y^2$

$= (8x^3 + 27y^3) + (4x^2 - 9y^2)$ Associative and commutative properties

$= (2x + 3y)(4x^2 - 6xy + 9y^2) + (2x + 3y)(2x - 3y)$ Factor within groups.

$= (2x + 3y)[(4x^2 - 6xy + 9y^2) + (2x - 3y)]$ Factor out the greatest common factor, $2x + 3y$.

$= (2x + 3y)(4x^2 - 6xy + 9y^2 + 2x - 3y)$ Combine terms.

In problems such as this, how we choose to group in the first step is essential to factoring correctly. If we reach a "dead end," then we should group differently and try again.

Use the method just described to factor each polynomial.

71. $125p^3 + 25p^2 + 8q^3 - 4q^2$ **72.** $27x^3 + 9x^2 + y^3 - y^2$

73. $27a^3 + 15a - 64b^3 - 20b$ **74.** $1000k^3 + 20k - m^3 - 2m$

75. $8t^4 - 24t^3 + t - 3$ **76.** $y^4 + y^3 + y + 1$

77. $64m^2 - 512m^3 - 81n^2 + 729n^3$ **78.** $10x^2 + 5x^3 - 10y^2 + 5y^3$

PREVIEW EXERCISES

Factor completely. **See Sections 6.1 and 6.2.**

79. $2ax + ay - 2bx - by$ **80.** $y^2 - y - 2$

81. $p^2 + 4p - 21$ **82.** $6t^2 + 19ts - 7s^2$

6.4   A General Approach to Factoring

OBJECTIVES

1   Factor out any common factor.

2   Factor binomials.

3   Factor trinomials.

4   Factor polynomials of more than three terms.

A polynomial is completely factored when *both* conditions are satisfied.

1. It is written as a *product* of prime polynomials with integer coefficients.

2. None of the polynomial factors can be factored further.

Factoring a Polynomial

Step 1   **Factor out any common factor.**

Step 2   **If the polynomial is a binomial,** check to see if it is a difference of squares, a difference of cubes, or a sum of cubes.

        If the polynomial is a trinomial, check to see if it is a perfect square trinomial. If it is not, factor as in **Section 6.2.**

        If the polynomial has more than three terms, try to factor by grouping.

Step 3   **Check the factored form by multiplying.**

OBJECTIVE 1 Factor out any common factor. *This step is always the same, regardless of the number of terms in the polynomial.*

NOW TRY
EXERCISE 1
Factor each polynomial.

(a) $21x^3y^2 - 27x^2y^4$

(b) $8y(m - n) - 5(m - n)$

EXAMPLE 1 Factoring Out a Common Factor

Factor each polynomial.

(a) $9p + 45$

$= 9(p + 5)$ GCF = 9

(b) $8m^2p^2 + 4mp$

$= 4mp(2mp + 1)$

(c) $5x(a + b) - y(a + b)$

$= (a + b)(5x - y)$ Factor out $a + b$.

NOW TRY

OBJECTIVE 2 Factor binomials. For binomials, use one of the following rules.

Factoring a Binomial

For a **binomial** (two terms), check for the following patterns.

Difference of squares	$x^2 - y^2 = (x + y)(x - y)$
Difference of cubes	$x^3 - y^3 = (x - y)(x^2 + xy + y^2)$
Sum of cubes	$x^3 + y^3 = (x + y)(x^2 - xy + y^2)$

NOW TRY
EXERCISE 2
Factor each binomial if possible.

(a) $4a^2 - 49b^2$

(b) $9x^2 + 100$

(c) $27v^3 - 1000$

EXAMPLE 2 Factoring Binomials

Factor each binomial if possible.

(a) $64m^2 - 9n^2$

$= (8m)^2 - (3n)^2$ Difference of squares

$= (8m + 3n)(8m - 3n)$ $x^2 - y^2 = (x + y)(x - y)$

(b) $8p^3 - 27$

$= (2p)^3 - 3^3$ Difference of cubes

$= (2p - 3)[(2p)^2 + (2p)(3) + 3^2]$ $x^3 - y^3 = (x - y)(x^2 + xy + y^2)$

$= (2p - 3)(4p^2 + 6p + 9)$ $(2p)^2 = 2^2p^2 = 4p^2$

(c) $1000m^3 + 1$

$= (10m)^3 + 1^3$ Sum of cubes

$= (10m + 1)[(10m)^2 - (10m)(1) + 1^2]$ $x^3 + y^3 = (x + y)(x^2 - xy + y^2)$

$= (10m + 1)(100m^2 - 10m + 1)$ $(10m)^2 = 10^2m^2 = 100m^2$

(d) $25m^2 + 121$ is prime. It is a *sum* of squares.

NOW TRY

NOTE The binomial $25m^2 + 625$ is a sum of squares. It can be factored, however, because the greatest common factor of the terms is not 1.

$$25m^2 + 625 = 25(m^2 + 25)$$ Factor out the common factor 25.

This sum of squares cannot be factored further.

NOW TRY ANSWERS
1. (a) $3x^2y^2(7x - 9y^2)$
  (b) $(m - n)(8y - 5)$
2. (a) $(2a + 7b)(2a - 7b)$
  (b) prime
  (c) $(3v - 10)(9v^2 + 30v + 100)$

OBJECTIVE 3 **Factor trinomials.** Take the following into consideration.

Factoring a Trinomial

For a **trinomial** (three terms), decide whether it is a perfect square trinomial of either of these forms.

$$x^2 + 2xy + y^2 = (x + y)^2 \quad \text{or} \quad x^2 - 2xy + y^2 = (x - y)^2$$

If not, use the methods of **Section 6.2.**

> ⌐ *NOW TRY*
> *EXERCISE 3*
>
> Factor each trinomial.
> **(a)** $25x^2 - 90x + 81$
> **(b)** $7x^2 - 7xy - 84y^2$
> **(c)** $12m^2 + 5m - 28$

EXAMPLE 3 Factoring Trinomials

Factor each trinomial.

(a) $p^2 + 10p + 25$

$\quad = (p + 5)^2$ Perfect square trinomial

(b) $49z^2 - 42z + 9$

$\quad = (7z - 3)^2$ Perfect square trinomial

(c) $y^2 - 5y - 6$

$\quad = (y - 6)(y + 1)$ The numbers -6 and 1 have a product of -6 and a sum of -5.

(d) $2k^2 - k - 6$

$\quad = (2k + 3)(k - 2)$ Use either method from **Section 6.2.**

(e) $\quad\quad\quad 28z^2 + 6z - 10$

> Remember the common factor.

$\quad = 2(14z^2 + 3z - 5)$ Factor out the common factor.

$\quad = 2(7z + 5)(2z - 1)$ Factor the trinomial.   *NOW TRY* ↻

OBJECTIVE 4 **Factor polynomials of more than three terms.** Consider factoring by grouping in this case.

EXAMPLE 4 Factoring Polynomials with More Than Three Terms

Factor each polynomial.

(a) $xy^2 - y^3 + x^3 - x^2y$

$\quad = (xy^2 - y^3) + (x^3 - x^2y)$ Group the terms.

$\quad = y^2(x - y) + x^2(x - y)$ Factor each group.

$\quad = (x - y)(y^2 + x^2)$ $x - y$ is a common factor.

(b) $20k^3 + 4k^2 - 45k - 9$ > Be careful with signs.

$\quad = (20k^3 + 4k^2) - (45k + 9)$

$\quad = 4k^2(5k + 1) - 9(5k + 1)$ Factor each group.

$\quad = (5k + 1)(4k^2 - 9)$ $5k + 1$ is a common factor.

$\quad = (5k + 1)(2k + 3)(2k - 3)$ Difference of squares

NOW TRY ANSWERS
3. **(a)** $(5x - 9)^2$
   (b) $7(x + 3y)(x - 4y)$
   (c) $(4m + 7)(3m - 4)$

NOW TRY
EXERCISE 4

Factor each polynomial.

(a) $5a^3 + 5a^2b - ab^2 - b^3$

(b) $9u^2 - 48u + 64 - v^2$

(c) $x^3 - 9y^2 - 27y^3 + x^2$

(c) $4a^2 + 4a + 1 - b^2$

$\qquad = (4a^2 + 4a + 1) - b^2$ Associative property

$\qquad = (2a + 1)^2 - b^2$ Perfect square trinomial

$\qquad = (2a + 1 + b)(2a + 1 - b)$ Difference of squares

(d) $8m^3 + 4m^2 - n^3 - n^2$

$\qquad = \underbrace{(8m^3 - n^3)}_{\substack{\text{Difference} \\ \text{of cubes}}} + \underbrace{(4m^2 - n^2)}_{\substack{\text{Difference} \\ \text{of squares}}}$ Rearrange and group the terms.

$\qquad = (2m - n)(4m^2 + 2mn + n^2) + (2m - n)(2m + n)$ Factor each group.

$\qquad = (2m - n)(4m^2 + 2mn + n^2 + 2m + n)$ Factor out the common factor $2m - n$.

NOW TRY

NOW TRY ANSWERS

4. (a) $(a + b)(5a^2 - b^2)$
  (b) $(3u - 8 + v)(3u - 8 - v)$
  (c) $(x - 3y)(x^2 + 3xy + 9y^2 + x + 3y)$

6.4 EXERCISES

MyMathLab   Math XL PRACTICE   WATCH   DOWNLOAD   READ   REVIEW

⊙ *Complete solution available on the Video Resources on DVD*

The following exercises are of mixed variety. Factor each polynomial. **See Examples 1–4.**

1. $100a^2 - 9b^2$

2. $10r^2 + 13r - 3$

3. $3p^4 - 3p^3 - 90p^2$

4. $k^4 - 16$

5. $3a^2pq + 3abpq - 90b^2pq$

6. $49z^2 - 16$

⊙ **7.** $225p^2 + 256$

8. $18m^3n + 3m^2n^2 - 6mn^3$

⊙ **9.** $6b^2 - 17b - 3$

10. $k^2 - 6k - 16$

11. $x^3 - 1000$

12. $6t^2 + 19tu - 77u^2$

⊙ **13.** $4(p + 2) + m(p + 2)$

14. $40p - 32r$

15. $9m^2 - 45m + 18m^3$

16. $4k^2 + 28kr + 49r^2$

17. $54m^3 - 2000$

18. $mn - 2n + 5m - 10$

19. $9m^2 - 30mn + 25n^2$

20. $2a^2 - 7a - 4$

⊙ **21.** $kq - 9q + kr - 9r$

22. $56k^3 - 875$

23. $16z^3x^2 - 32z^2x$

24. $9r^2 + 100$

25. $x^2 + 2x - 35$

26. $9 - a^2 + 2ab - b^2$

27. $625 - x^4$

28. $2m^2 - mn - 15n^2$

29. $p^3 + 1$

30. $48y^2z^3 - 28y^3z^4$

31. $64m^2 - 625$

32. $14z^2 - 3zk - 2k^2$

33. $12z^3 - 6z^2 + 18z$

34. $225k^2 - 36r^2$

35. $256b^2 - 400c^2$

36. $z^2 - zp - 20p^2$

37. $512 + 1000z^3$

38. $64m^2 - 25n^2$

39. $10r^2 + 23rs - 5s^2$

40. $12k^2 - 17kq - 5q^2$

41. $24p^3q + 52p^2q^2 + 20pq^3$

42. $32x^2 + 16x^3 - 24x^5$

43. $48k^4 - 243$

44. $14x^2 - 25xq - 25q^2$

45. $m^3 + m^2 - n^3 - n^2$

46. $64x^3 + y^3 - 16x^2 + y^2$

47. $x^2 - 4m^2 - 4mn - n^2$

48. $4r^2 - s^2 - 2st - t^2$

49. $18p^5 - 24p^3 + 12p^6$

50. $k^2 - 6k + 16$

51. $2x^2 - 2x - 40$

52. $27x^3 - 3y^3$

53. $(2m + n)^2 - (2m - n)^2$

54. $(3k + 5)^2 - 4(3k + 5) + 4$

55. $50p^2 - 162$

56. $y^2 + 3y - 10$

57. $12m^2rx + 4mnrx + 40n^2rx$

58. $18p^2 + 53pr - 35r^2$

59. $21a^2 - 5ab - 4b^2$

60. $x^2 - 2xy + y^2 - 4$

61. $x^2 - y^2 - 4$

62. $(5r + 2s)^2 - 6(5r + 2s) + 9$

63. $(p + 8q)^2 - 10(p + 8q) + 25$

64. $z^4 - 9z^2 + 20$

65. $21m^4 - 32m^2 - 5$

66. $(x - y)^3 - (27 - y)^3$

67. $(r + 2t)^3 + (r - 3t)^3$

68. $16x^3 + 32x^2 - 9x - 18$

69. $x^5 + 3x^4 - x - 3$

70. $1 - x^{16}$

71. $m^2 - 4m + 4 - n^2 + 6n - 9$

72. $x^2 + 4 + x^2y + 4y$

PREVIEW EXERCISES

Solve each equation. ***See Section 2.1.***

73. $3x + 2 = 0$

74. $-2x + 7 = 0$

75. $5x = 0$

76. $-8x = 0$

77. $\frac{1}{2}x + 5 = 0$

78. $-\frac{3}{4}x - 6 = 0$

6.5 Solving Equations by Factoring

OBJECTIVES

1 Learn and use the zero-factor property.

2 Solve applied problems that require the zero-factor property.

3 Solve a formula for a specified variable, where factoring is necessary.

In **Chapter 2,** we developed methods for solving linear, or first-degree, equations. Solving higher degree polynomial equations requires other methods, one of which involves factoring.

OBJECTIVE 1 **Learn and use the zero-factor property.** Solving equations by factoring depends on a special property of the number 0, called the **zero-factor property.**

Zero-Factor Property

If two numbers have a product of 0, then at least one of the numbers must be 0.

If $ab = 0$, then either $a = 0$ or $b = 0$.

To prove the zero-factor property, we first assume that $a \neq 0$. (If a does equal 0, then the property is proved already.) If $a \neq 0$, then $\frac{1}{a}$ exists, and both sides of $ab = 0$ can be multiplied by $\frac{1}{a}$.

$$\frac{1}{a} \cdot ab = \frac{1}{a} \cdot 0$$

$$b = 0$$

Thus, if $a \neq 0$, then $b = 0$, and the property is proved.

⚠ **CAUTION** If $ab = 0$, then $a = 0$ or $b = 0$. However, if $ab = 6$, for example, it is not necessarily true that $a = 6$ or $b = 6$. In fact, it is very likely that *neither $a = 6$ nor $b = 6$. **The zero-factor property works only for a product equal to 0.***

NOW TRY
EXERCISE 1
Solve.

$(x + 5)(4x - 7) = 0$

EXAMPLE 1 Using the Zero-Factor Property to Solve an Equation

Solve $(x + 6)(2x - 3) = 0$.

Here, the product of $x + 6$ and $2x - 3$ is 0. By the zero-factor property, the following must hold true.

$$x + 6 = 0 \quad \text{or} \quad 2x - 3 = 0 \qquad \text{Zero-factor property}$$
$$x = -6 \quad \text{or} \qquad 2x = 3 \qquad \text{Solve each of these equations.}$$
$$x = \frac{3}{2}$$

CHECK $\quad (x + 6)(2x - 3) = 0 \qquad\qquad\qquad (x + 6)(2x - 3) = 0$

$(-6 + 6)[2(-6) - 3] \overset{?}{=} 0 \qquad\qquad \left(\frac{3}{2} + 6\right)\left(2 \cdot \frac{3}{2} - 3\right) \overset{?}{=} 0$

Let $x = 6$.

$0(-15) \overset{?}{=} 0 \qquad\qquad\qquad\qquad\qquad$ Let $x = \frac{3}{2}$.

$0 = 0 \; \checkmark \; \text{True} \qquad\qquad\qquad\qquad \frac{15}{2}(0) \overset{?}{=} 0$

$0 = 0 \; \checkmark \; \text{True}$

Both solutions check, so the solution set is $\left\{-6, \frac{3}{2}\right\}$. NOW TRY ↻

Since the product $(x + 6)(2x - 3)$ equals $2x^2 + 9x - 18$, the equation of **Example 1** has a term with a squared variable and is an example of a *quadratic equation. A quadratic equation has degree 2.*

Quadratic Equation

An equation that can be written in the form

$$ax^2 + bx + c = 0,$$

where a, b, and c are real numbers, with $a \neq 0$, is a **quadratic equation**. This form is called **standard form.**

NOW TRY ANSWER
1. $\left\{-5, \frac{7}{4}\right\}$

The steps for solving a quadratic equation by factoring are summarized here.

> ### Solving a Quadratic Equation by Factoring
>
> *Step 1* **Write in standard form.** Rewrite the equation if necessary so that one side is 0.
>
> *Step 2* **Factor** the polynomial.
>
> *Step 3* **Use the zero-factor property.** Set each variable factor equal to 0.
>
> *Step 4* **Find the solution(s).** Solve each equation formed in Step 3.
>
> *Step 5* **Check** each solution in the *original* equation.

NOW TRY
EXERCISE 2

Solve each equation.

(a) $7x = 3 - 6x^2$

(b) $16x^2 + 40x + 25 = 0$

EXAMPLE 2   Solving Quadratic Equations by Factoring

Solve each equation.

(a) $2x^2 + 3x = 2$

Step 1
$$2x^2 + 3x = 2$$
$$2x^2 + 3x - 2 = 0 \qquad \text{Standard form}$$

Step 2
$$(2x - 1)(x + 2) = 0 \qquad \text{Factor.}$$

Step 3
$$2x - 1 = 0 \quad \text{or} \quad x + 2 = 0 \qquad \text{Zero-factor property}$$

Step 4
$$2x = 1 \quad \text{or} \quad x = -2 \qquad \text{Solve each equation.}$$
$$x = \frac{1}{2}$$

Step 5 *Check* each solution in the original equation.

CHECK
$$2x^2 + 3x = 2$$
$$2\left(\frac{1}{2}\right)^2 + 3\left(\frac{1}{2}\right) \overset{?}{=} 2 \qquad \text{Let } x = \tfrac{1}{2}.$$
$$2\left(\frac{1}{4}\right) + \frac{3}{2} \overset{?}{=} 2$$
$$\frac{1}{2} + \frac{3}{2} \overset{?}{=} 2$$
$$2 = 2 \;\checkmark\; \text{True}$$

$$2x^2 + 3x = 2$$
$$2(-2)^2 + 3(-2) \overset{?}{=} 2 \qquad \text{Let } x = -2.$$
$$2(4) - 6 \overset{?}{=} 2$$
$$8 - 6 \overset{?}{=} 2$$
$$2 = 2 \;\checkmark\; \text{True}$$

Because both solutions check, the solution set is $\left\{-2, \frac{1}{2}\right\}$.

(b)
$$4x^2 - 4x + 1 = 0 \qquad \text{Standard form}$$

We could factor as $(2x - 1)(2x - 1)$.
$$(2x - 1)^2 = 0 \qquad \text{Factor.}$$
$$2x - 1 = 0 \qquad \text{Zero-factor property}$$
$$2x = 1 \qquad \text{Add 1.}$$
$$x = \frac{1}{2} \qquad \text{Divide by 2.}$$

There is only one solution, called a **double solution,** because the trinomial is a perfect square. The solution set is $\left\{\frac{1}{2}\right\}$.

NOW TRY

NOW TRY ANSWERS
2. (a) $\left\{-\frac{3}{2}, \frac{1}{3}\right\}$ (b) $\left\{-\frac{5}{4}\right\}$

**NOW TRY
EXERCISE 3**
Solve $3x^2 + 12x = 0$.

EXAMPLE 3 Solving a Quadratic Equation with a Missing Constant Term

Solve $4x^2 - 20x = 0$.

This quadratic equation has a missing term. Comparing it with the standard form $ax^2 + bx + c = 0$ shows that $c = 0$. The zero-factor property can still be used.

$$4x^2 - 20x = 0$$

$$4x(x - 5) = 0 \quad \text{Factor.}$$

Set each *variable* factor equal to 0. $\quad 4x = 0 \quad \text{or} \quad x - 5 = 0 \quad$ Zero-factor property

$$x = 0 \quad \text{or} \quad x = 5 \quad \text{Solve each equation.}$$

CHECK
$$4x^2 - 20x = 0$$
$$4(0)^2 - 20(0) \stackrel{?}{=} 0 \quad \text{Let } x = 0.$$
$$0 - 0 = 0 \checkmark \text{ True}$$

$$4x^2 - 20x = 0$$
$$4(5)^2 - 20(5) \stackrel{?}{=} 0 \quad \text{Let } x = 5.$$
$$100 - 100 = 0 \checkmark \text{ True}$$

The solution set is $\{0, 5\}$.

NOW TRY

⚠ **CAUTION** Remember to include 0 as a solution in **Example 3.**

**NOW TRY
EXERCISE 4**
Solve $4x^2 - 100 = 0$.

EXAMPLE 4 Solving a Quadratic Equation with a Missing Linear Term

Solve $3x^2 - 108 = 0$.

$$3x^2 - 108 = 0$$

$$3(x^2 - 36) = 0 \quad \text{Factor out 3.}$$

The factor 3 does *not* lead to a solution. $\quad 3(x + 6)(x - 6) = 0 \quad$ Factor $x^2 - 36$.

$$x + 6 = 0 \quad \text{or} \quad x - 6 = 0 \quad \text{Zero-factor property}$$

$$x = -6 \quad \text{or} \quad x = 6$$

Check that the solution set is $\{-6, 6\}$.

NOW TRY

⚠ **CAUTION** The factor 3 in **Example 4** is not a *variable* factor, so it does *not* lead to a solution of the equation. In **Example 3,** however, the factor x is a variable factor and leads to the solution 0.

EXAMPLE 5 Solving an Equation That Requires Rewriting

Solve $(2x + 1)(x + 1) = 2(1 - x) + 6$.

$$(2x + 1)(x + 1) = 2(1 - x) + 6$$

$$2x^2 + 3x + 1 = 2 - 2x + 6 \quad \text{Multiply on each side.}$$

Write in standard form. $\quad 2x^2 + 5x - 7 = 0 \quad$ Add 2x. Subtract 6.

$$(2x + 7)(x - 1) = 0 \quad \text{Factor.}$$

$$2x + 7 = 0 \quad \text{or} \quad x - 1 = 0 \quad \text{Zero-factor property}$$

$$x = -\frac{7}{2} \quad \text{or} \quad x = 1 \quad \text{Solve each equation.}$$

NOW TRY ANSWERS
3. $\{0, -4\}$ **4.** $\{-5, 5\}$

SECTION 6.5 Solving Equations by Factoring 347

NOW TRY
EXERCISE 5
Solve.

$(x + 3)(2x - 1)$
$= 4(x + 4) - 4$

CHECK

$$(2x + 1)(x + 1) = 2(1 - x) + 6$$

$$\left[2\left(-\frac{7}{2}\right) + 1\right]\left(-\frac{7}{2} + 1\right) \stackrel{?}{=} 2\left[1 - \left(-\frac{7}{2}\right)\right] + 6 \qquad \text{Let } x = -\frac{7}{2}.$$

$$(-7 + 1)\left(-\frac{5}{2}\right) \stackrel{?}{=} 2\left(\frac{9}{2}\right) + 6 \qquad \text{Simplify; } 1 = \frac{2}{2}.$$

$$(-6)\left(-\frac{5}{2}\right) \stackrel{?}{=} 9 + 6$$

$$15 = 15 \checkmark \qquad \text{True}$$

Check that 1 is a solution. The solution set is $\left\{-\frac{7}{2}, 1\right\}$.   NOW TRY

The zero-factor property can be extended to solve certain polynomial equations of degree 3 or greater, as shown in the next example.

NOW TRY
EXERCISE 6
Solve $12x = 2x^3 + 5x^2$.

EXAMPLE 6 Solving an Equation of Degree 3

Solve $-x^3 + x^2 = -6x$.

$$-x^3 + x^2 = -6x$$

$$-x^3 + x^2 + 6x = 0 \qquad \text{Add } 6x \text{ to each side.}$$

$$x^3 - x^2 - 6x = 0 \qquad \text{Multiply each side by } -1.$$

$$x(x^2 - x - 6) = 0 \qquad \text{Factor out } x.$$

$$x(x - 3)(x + 2) = 0 \qquad \text{Factor the trinomial.}$$

$x = 0$ or $x - 3 = 0$ or $x + 2 = 0$ Extend the zero-factor property to the three *variable* factors.

Remember to set x equal to 0.

$x = 3$ or $x = -2$

Check that the solution set is $\{-2, 0, 3\}$.   NOW TRY

OBJECTIVE 2 Solve applied problems that require the zero-factor property.
An application may lead to a quadratic equation.

EXAMPLE 7 Using a Quadratic Equation in an Application

A piece of sheet metal is in the shape of a parallelogram. The longer sides of the parallelogram are each 8 m longer than the distance between them. The area of the piece is 48 m². Find the length of the longer sides and the distance between them.

Step 1 **Read** the problem again. There will be two answers.

Step 2 **Assign a variable.**

Let $x = $ the distance between the longer sides

and $x + 8 = $ the length of each longer side. (See **FIGURE 1**.)

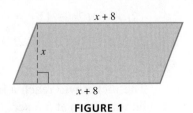

FIGURE 1

NOW TRY ANSWERS
5. $\left\{-3, \frac{5}{2}\right\}$ **6.** $\left\{-4, 0, \frac{3}{2}\right\}$

NOW TRY
EXERCISE 7
The height of a triangle is 1 ft less than twice the length of the base. The area is 14 ft². What are the measures of the base and the height?

Step 3 **Write an equation.** The area of a parallelogram is given by $\mathcal{A} = bh$, where b is the length of the longer side and h is the distance between the longer sides. Here, $b = x + 8$ and $h = x$.

$$\mathcal{A} = bh$$
$$48 = (x + 8)x \qquad \text{Let } \mathcal{A} = 48,\ b = x + 8,\ h = x.$$

Step 4 **Solve.**
$$48 = x^2 + 8x \qquad \text{Distributive property}$$
$$x^2 + 8x - 48 = 0 \qquad \text{Standard form}$$
$$(x + 12)(x - 4) = 0 \qquad \text{Factor.}$$
$$x + 12 = 0 \quad \text{or} \quad x - 4 = 0 \qquad \text{Zero-factor property}$$
$$x = -12 \quad \text{or} \qquad x = 4 \qquad \text{Solve each equation.}$$

Step 5 **State the answer.** *A distance cannot be negative, so reject −12 as an answer.* The only possible answer is 4, so the distance between the longer sides is 4 m. The length of the longer sides is $4 + 8 = 12$ m.

Step 6 **Check.** The length of the longer sides is 8 m more than the distance between them, and the area is $4 \cdot 12 = 48$ m², as required, so the answer checks.

NOW TRY

⚠ **CAUTION** When applications lead to quadratic equations, a solution of the equation may not satisfy the physical requirements of the problem, as in **Example 7.** Reject such solutions as answers.

A function defined by a quadratic polynomial is called a *quadratic function*. (See **Chapter 9.**) The next example uses such a function.

NOW TRY
EXERCISE 8
Refer to **Example 8.** After how many seconds will the rocket be 192 ft above the ground?

EXAMPLE 8 Using a Quadratic Function in an Application

Quadratic functions are used to describe the height a falling object or a projected object reaches in a specific time. For example, if a small rocket is launched vertically upward from ground level with an initial velocity of 128 ft per sec, then its height in feet after t seconds is a function defined by

$$h(t) = -16t^2 + 128t$$

if air resistance is neglected. After how many seconds will the rocket be 220 ft above the ground?

We must let $h(t) = 220$ and solve for t.

$$220 = -16t^2 + 128t \qquad \text{Let } h(t) = 220.$$
$$16t^2 - 128t + 220 = 0 \qquad \text{Standard form}$$
$$4t^2 - 32t + 55 = 0 \qquad \text{Divide by 4.}$$
$$(2t - 5)(2t - 11) = 0 \qquad \text{Factor.}$$
$$2t - 5 = 0 \quad \text{or} \quad 2t - 11 = 0 \qquad \text{Zero-factor property}$$
$$t = 2.5 \quad \text{or} \qquad t = 5.5 \qquad \text{Solve each equation.}$$

NOW TRY ANSWERS
7. base: 4 ft; height: 7 ft
8. 2 sec and 6 sec

The rocket will reach a height of 220 ft twice: on its way up at 2.5 sec and again on its way down at 5.5 sec.

NOW TRY

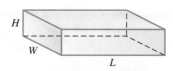

Rectangular solid
$\mathcal{A} = 2HW + 2LW + 2LH$
FIGURE 2

OBJECTIVE 3 **Solve a formula for a specified variable, where factoring is necessary.** In **Section 2.2** we solved certain formulas for variables. In some cases, factoring is required to accomplish this.

A rectangular solid has the shape of a box, but is solid. See **FIGURE 2**. The surface area of any solid three-dimensional figure is the total area of its surface. For a rectangular solid, the surface area $\mathcal{A}$ is

$$\mathcal{A} = 2HW + 2LW + 2LH.$$ *H, W, and L represent height, width, and length.*

**NOW TRY
EXERCISE 9**

Solve the formula for H.

$\mathcal{A} = 2HW + 2LW + 2LH$

EXAMPLE 9 Using Factoring to Solve for a Specified Variable

Solve the formula $\mathcal{A} = 2HW + 2LW + 2LH$ for L.

To solve for the length L, treat L as the only variable and treat all other variables as constants.

We must isolate the L-terms.

$$\mathcal{A} = 2HW + 2LW + 2LH$$

$$\mathcal{A} - 2HW = 2LW + 2LH \qquad \text{Subtract } 2HW.$$

$$\mathcal{A} - 2HW = L(2W + 2H) \qquad \text{Factor out } L.$$

This is a key step.

$$\frac{\mathcal{A} - 2HW}{2W + 2H} = L, \quad \text{or} \quad L = \frac{\mathcal{A} - 2HW}{2W + 2H} \qquad \text{Divide by } 2W + 2H.$$

NOW TRY

⚠ **CAUTION** *In Example 9, we must write the expression so that the specified variable is a factor. Then we can divide by its coefficient in the final step.*

CONNECTIONS

In **Section 5.3**, we saw that the graph of $f(x) = x^2$ is a parabola. In general, the graph of

$$f(x) = ax^2 + bx + c, \quad a \neq 0,$$

is a parabola, and the x-intercepts of its graph give the real number solutions of the equation $ax^2 + bx + c = 0$.

A graphing calculator can locate these x-intercepts (called **zeros** of the function) for $f(x) = 2x^2 + 3x - 2$. Notice that this quadratic expression was found on the left side of the equation in **Example 2(a)** earlier in this section, where the equation was written in standard form. The x-intercepts (zeros) given with the graphs in **FIGURE 3** are the same as the solutions found in **Example 2(a)**.

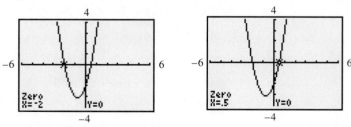

FIGURE 3

For Discussion or Writing

Solve each quadratic equation using the zero-factor property. Then support the solution(s) with a graphing calculator.

1. $x^2 - 6x - 7 = 0$ 2. $x^2 - 6x + 9 = 0$ 3. $x^2 = 4$

NOW TRY ANSWER

9. $H = \frac{\mathcal{A} - 2LW}{2W + 2L}$

Complete solution available on the Video Resources on DVD

1. Explain in your own words how the zero-factor property is used in solving a quadratic equation.

2. *Concept Check* One of the following equations is *not* in proper form for using the zero-factor property. Which one is it? Tell why it is not in proper form.

A. $(x + 2)(x - 6) = 0$ **B.** $x(3x - 7) = 0$

C. $3x(x + 8)(x - 9) = 0$ **D.** $x(x - 3) + 6(x - 3) = 0$

Solve each equation. **See Examples 1–5.**

3. $(x + 10)(x - 5) = 0$ **4.** $(x + 7)(x + 3) = 0$

5. $(2k - 5)(3k + 8) = 0$ **6.** $(3q - 4)(2q + 5) = 0$

7. $x^2 - 3x - 10 = 0$ **8.** $x^2 + x - 12 = 0$

9. $x^2 + 9x + 18 = 0$ **10.** $x^2 - 18x + 80 = 0$

11. $2x^2 = 7x + 4$ **12.** $2x^2 = 3 - x$

13. $15x^2 - 7x = 4$ **14.** $3x^2 + 3 = -10x$

15. $2x^2 - 12 - 4x = x^2 - 3x$ **16.** $3x^2 + 9x + 30 = 2x^2 - 2x$

17. $(5x + 1)(x + 3) = -2(5x + 1)$ **18.** $(3x + 1)(x - 3) = 2 + 3(x + 5)$

19. $4p^2 + 16p = 0$ **20.** $2t^2 - 8t = 0$

21. $6x^2 - 36x = 0$ **22.** $-3x^2 + 27x = 0$

23. $4p^2 - 16 = 0$ **24.** $9z^2 - 81 = 0$

25. $-3x^2 + 27 = 0$ **26.** $-2x^2 + 8 = 0$

27. $-x^2 = 9 - 6x$ **28.** $-x^2 - 8x = 16$

29. $9x^2 + 24x + 16 = 0$ **30.** $4x^2 - 20x + 25 = 0$

31. $(x - 3)(x + 5) = -7$ **32.** $(x + 8)(x - 2) = -21$

33. $(2x + 1)(x - 3) = 6x + 3$ **34.** $(3x + 2)(x - 3) = 7x - 1$

35. $(x + 3)(x - 6) = (2x + 2)(x - 6)$ **36.** $(2x + 1)(x + 5) = (x + 11)(x + 3)$

Solve each equation. **See Example 6.**

37. $2x^3 - 9x^2 - 5x = 0$ **38.** $6x^3 - 13x^2 - 5x = 0$

39. $x^3 - 2x^2 = 3x$ **40.** $x^3 - 6x^2 = -8x$

41. $9x^3 = 16x$ **42.** $25x^3 = 64x$

43. $2x^3 + 5x^2 - 2x - 5 = 0$ **44.** $2x^3 + x^2 - 98x - 49 = 0$

45. $x^3 - 6x^2 - 9x + 54 = 0$ **46.** $x^3 - 3x^2 - 4x + 12 = 0$

47. *Concept Check* A student tried to solve the equation in **Exercise 41** by first dividing each side by x, obtaining $9x^2 = 16$. She then solved the resulting equation by the zero-factor property to get the solution set $\left\{-\frac{4}{3}, \frac{4}{3}\right\}$. ***WHAT WENT WRONG?*** Give the correct solution set.

48. *Concept Check* Without actually solving each equation, determine which one of the following has 0 in its solution set.

A. $4x^2 - 25 = 0$ **B.** $x^2 + 2x - 3 = 0$

C. $6x^2 + 9x + 1 = 0$ **D.** $x^3 + 4x^2 = 3x$

Brain Busters *Solve each equation.*

49. $2(x - 1)^2 - 7(x - 1) - 15 = 0$

50. $4(2x + 3)^2 - (2x + 3) - 3 = 0$

51. $5(3x - 1)^2 + 3 = -16(3x - 1)$

52. $2(x + 3)^2 = 5(x + 3) - 2$

53. $(2x - 3)^2 = 16x^2$

54. $9x^2 = (5x + 2)^2$

Solve each problem. ***See Examples 7 and 8.***

55. A garden has an area of 320 ft². Its length is 4 ft more than its width. What are the dimensions of the garden? (*Hint:* $320 = 16 \cdot 20$)

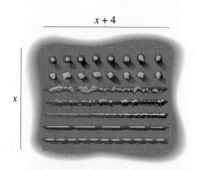

56. A square mirror has sides measuring 2 ft less than the sides of a square painting. If the difference between their areas is 32 ft², find the lengths of the sides of the mirror and the painting.

57. The base of a parallelogram is 7 ft more than the height. If the area of the parallelogram is 60 ft², what are the measures of the base and the height?

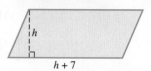

58. A sign has the shape of a triangle. The length of the base is 3 m less than the height. What are the measures of the base and the height if the area is 44 m²?

59. A farmer has 300 ft of fencing and wants to enclose a rectangular area of 5000 ft². What dimensions should she use? (*Hint:* $5000 = 50 \cdot 100$)

60. A rectangular landfill has an area of 30,000 ft². Its length is 200 ft more than its width. What are the dimensions of the landfill? (*Hint:* $30,000 = 300 \cdot 100$)

61. Find two consecutive integers such that the sum of their squares is 61.

62. Find two consecutive integers such that their product is 72.

63. A box with no top is to be constructed from a piece of cardboard whose length measures 6 in. more than its width. The box is to be formed by cutting squares that measure 2 in. on each side from the four corners and then folding up the sides. If the volume of the box will be 110 in.³, what are the dimensions of the piece of cardboard?

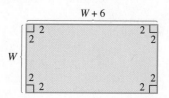

64. The surface area of the box with open top shown in the figure is 161 in.^{2}. Find the dimensions of the base. (*Hint:* The surface area of the box is a function defined by $S(x) = x^2 + 16x$.)

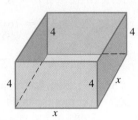

65. If an object is projected upward with an initial velocity of 64 ft per sec from a height of 80 ft, then its height in feet t seconds after it is projected is a function defined by

$$f(t) = -16t^2 + 64t + 80.$$

How long after it is projected will it hit the ground? (*Hint:* When it hits the ground, its height is 0 ft.)

66. Refer to **Example 8.** After how many seconds will the rocket be

(a) 240 ft above the ground?

(b) 112 ft above the ground?

67. If a baseball is dropped from a helicopter 625 ft above the ground, then its distance in feet from the ground t seconds later is a function defined by

$$f(t) = -16t^2 + 625.$$

How long after it is dropped will it hit the ground?

68. If a rock is dropped from a building 576 ft high, then its distance in feet from the ground t seconds later is a function defined by

$$f(t) = -16t^2 + 576.$$

How long after it is dropped will it hit the ground?

*Solve each equation for the specified variable. **See Example 9.***

69. $2k + ar = r - 3y$ for r

70. $4s + 7p = tp - 7$ for p

71. $w = \dfrac{3y - x}{y}$ for y

72. $c = \dfrac{-2t + 4}{t}$ for t

73. *Concept Check* Suppose a student solved the formula $\mathcal{A} = 2HW + 2LW + 2LH$ for L as follows.

$$\mathcal{A} = 2HW + 2LW + 2LH$$
$$\mathcal{A} - 2LW - 2HW = 2LH$$
$$\frac{\mathcal{A} - 2LW - 2HW}{2H} = L$$

WHAT WENT WRONG?

74. *Concept Check* Which one of the following is the correct result for solving the following equation for t?

$$2t + c = kt$$

A. $t = \dfrac{-c}{2 - k}$
B. $t = \dfrac{c - kt}{-2}$
C. $t = \dfrac{2t + c}{k}$
D. $t = \dfrac{kt - c}{2}$

TECHNOLOGY INSIGHTS   EXERCISES 75–76

The solutions of the quadratic equation $ax^2 + bx + c = 0$ ($a \neq 0$) are represented on the graph of the quadratic function $f(x) = ax^2 + bx + c$ by the x-intercepts.

Use the zero-factor property to solve each equation, and confirm that your solutions correspond to the x-intercepts (zeros) shown on the accompanying graphing calculator screens. **See the Connections box.**

75. $2x^2 - 7x - 4 = 0$

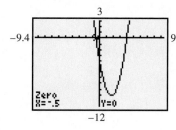
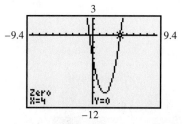

76. $-x^2 + 3x = -10$

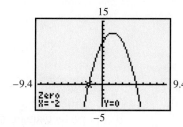
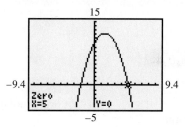

RELATING CONCEPTS   EXERCISES 77–82

If air resistance is neglected, the height $f(x)$ (in feet) of an object projected directly upward from an initial height s_0 feet with initial velocity (speed) v_0 feet per second is

$$f(x) = -16x^2 + v_0 x + s_0,$$

where x is the number of seconds after the object is projected.

Suppose that a ball is projected directly upward from an initial height of 100 ft with an initial velocity of 80 ft per sec. Use this information and a graphing calculator to answer the following. **Work Exercises 77–82 in order.**

77. Define a function that describes the height of the ball in terms of time x.

78. Use a graphing calculator to graph the function from **Exercise 77**. Use domain $[0, 10]$ and range $[-100, 300]$.

79. Use the graph from **Exercise 78** and the tracing capability of the calculator to estimate the maximum height of the ball. When does it reach that height?

80. After how many seconds will the ball reach the ground (that is, have height 0 ft)? Estimate the answer from the graph and check it in the equation.

81. Use the graph to estimate the time interval during which the height of the ball is greater than 150 ft. (*Hint:* Also graph $g(x) = 150$.) Check your estimate by substituting it into the function from **Exercise 77**.

82. Define a function that describes the height of an object if the initial velocity is 100 ft per sec and the object is projected from ground level.

PREVIEW EXERCISES

Simplify. **See Section 5.1.**

83. $\dfrac{12p^2}{3p}$

84. $\dfrac{-50a^4b^5}{150a^6b^4}$

85. $\dfrac{-27m^2n^5}{36m^6n^8}$

Write each fraction with the indicated denominator. **See Section 7.2.**

86. $\dfrac{5}{8} = \dfrac{?}{24}$

87. $\dfrac{12}{25} = \dfrac{?}{75}$

88. $\dfrac{8}{3} = \dfrac{?}{15}$

CHAPTER 6 SUMMARY

KEY TERMS

6.1
factoring
greatest common factor
(GCF)

6.2
prime polynomial

6.3
difference of squares
perfect square trinomial
difference of cubes
sum of cubes

6.5
quadratic equation
standard form of a quadratic
equation
double solution

TEST YOUR WORD POWER

See how well you have learned the vocabulary in this chapter.

1. **Factoring** is
   A. a method of multiplying polynomials
   B. the process of writing a polynomial as a product
   C. the answer in a multiplication problem
   D. a way to add the terms of a polynomial.

2. A **difference of squares** is a binomial
   A. that can be factored as the difference of two cubes

   B. that cannot be factored
   C. that is squared
   D. that can be factored as the product of the sum and difference of two terms.

3. A **perfect square trinomial** is a trinomial
   A. that can be factored as the square of a binomial
   B. that cannot be factored
   C. that is multiplied by a binomial
   D. where all terms are perfect squares.

4. A **quadratic equation** is a polynomial equation of
   A. degree one
   B. degree two
   C. degree three
   D. degree four.

5. The **zero-factor property** is used to
   A. factor a perfect square trinomial
   B. factor by grouping
   C. solve a polynomial equation of degree 2 or more
   D. solve a linear equation.

ANSWERS
1. B; *Example:* $x^2 - 5x - 14$ factors as $(x - 7)(x + 2)$. **2.** D; *Example:* $b^2 - 49$ is the difference of the squares b^2 and 7^2. It can be factored as $(b + 7)(b - 7)$. **3.** A; *Example:* $a^2 + 2a + 1$ is a perfect square trinomial. Its factored form is $(a + 1)^2$. **4.** B; *Examples:* $x^2 - 3x + 2 = 0$, $x^2 - 9 = 0, 2x^2 = 6x + 8$ **5.** C; *Example:* Use the zero-factor property to write $(x + 4)(x - 2) = 0$ as $x + 4 = 0$ or $x - 2 = 0$, and then solve each linear equation to find the solution set $\{-4, 2\}$.

QUICK REVIEW

CONCEPTS	EXAMPLES

6.1 Greatest Common Factors and Factoring by Grouping

The Greatest Common Factor

The product of the largest common numerical factor and each common variable raised to the least exponent that appears on that variable in any term is the greatest common factor of the terms of the polynomial.

Factor $4x^2y - 50xy^2$.

$$4x^2y - 50xy^2$$

$$= 2xy(2x - 25y) \qquad \text{The greatest common factor is } 2xy.$$

Factoring by Grouping

Step 1 Group the terms so that each group has a common factor.

Step 2 Factor out the common factor in each group.

Step 3 If the groups now have a common factor, factor it out. If not, try a different grouping.

Always check the factored form by multiplying.

Factor by grouping.

$$5a - 5b - ax + bx$$

$$= (5a - 5b) + (-ax + bx) \qquad \text{Group the terms.}$$

$$= 5(a - b) - x(a - b) \qquad \text{Factor out 5 and } -x.$$

$$= (a - b)(5 - x) \qquad \text{Factor out } a - b.$$

6.2 Factoring Trinomials

To factor a trinomial, choose factors of the first term and factors of the last term. Then place them within a pair of parentheses of this form.

$$(\qquad)(\qquad)$$

Try various combinations of the factors until the correct middle term of the trinomial is found.

Factor $15x^2 + 14x - 8$.

The factors of 15 are 5 and 3, and 15 and 1.

The factors of -8 are -4 and 2, 4 and -2, -1 and 8, and 1 and -8.

Various combinations lead to the correct factorization.

$$15x^2 + 14x - 8$$

$$= (5x - 2)(3x + 4) \qquad \text{Check by multiplying.}$$

6.3 Special Factoring

Difference of Squares

$$x^2 - y^2 = (x + y)(x - y)$$

Factor.

$$4m^2 - 25n^2$$

$$= (2m)^2 - (5n)^2$$

$$= (2m + 5n)(2m - 5n)$$

Perfect Square Trinomials

$$x^2 + 2xy + y^2 = (x + y)^2$$

$$x^2 - 2xy + y^2 = (x - y)^2$$

$$9y^2 + 6y + 1 \qquad\qquad 16p^2 - 56p + 49$$

$$= (3y + 1)^2 \qquad\qquad = (4p - 7)^2$$

Difference of Cubes

$$x^3 - y^3 = (x - y)(x^2 + xy + y^2)$$

$$8 - 27a^3$$

$$= (2 - 3a)(4 + 6a + 9a^2)$$

Sum of Cubes

$$x^3 + y^3 = (x + y)(x^2 - xy + y^2)$$

$$64z^3 + 1$$

$$= (4z + 1)(16z^2 - 4z + 1)$$

(continued)

CONCEPTS	EXAMPLES
6.4 **A General Approach to Factoring**	Factor.
Step 1 Factor out any common factors.	$ak^3 + 2ak^2 - 9ak - 18a$
Step 2 For a binomial, check for the difference of squares, the difference of cubes, or the sum of cubes.	$\quad = a(k^3 + 2k^2 - 9k - 18)$ Factor out the common factor.
For a trinomial, see if it is a perfect square. If not, factor as in **Section 6.2.**	$\quad = a[(k^3 + 2k^2) - (9k + 18)]$ Factor by grouping.
For more than three terms, try factoring by grouping.	$\quad = a[k^2(k + 2) - 9(k + 2)]$
	$\quad = a[(k + 2)(k^2 - 9)]$
Step 3 Check the factored form by multiplying.	$\quad = a(k + 2)(k - 3)(k + 3)$ Factor the difference of squares.

6.5 **Solving Equations by Factoring**	Solve. $\qquad 2x^2 + 5x = 3$
Step 1 Rewrite the equation if necessary so that one side is 0.	$2x^2 + 5x - 3 = 0$ Standard form
Step 2 Factor the polynomial.	$(2x - 1)(x + 3) = 0$ Factor.
Step 3 Set each factor equal to 0.	$2x - 1 = 0 \quad$ or $\quad x + 3 = 0$ Zero-factor property
Step 4 Solve each equation from Step 3.	$2x = 1 \quad$ or $\qquad x = -3$
	$x = \dfrac{1}{2}$
Step 5 Check each solution.	A check verifies that the solution set is $\left\{-3, \frac{1}{2}\right\}$.

CHAPTER ⑥ REVIEW EXERCISES

6.1 *Factor out the greatest common factor.*

1. $12p^2 - 6p$

2. $21x^2 + 35x$

3. $12q^2b + 8qb^2 - 20q^3b^2$

4. $6r^3t - 30r^2t^2 + 18rt^3$

5. $(x + 3)(4x - 1) - (x + 3)(3x + 2)$

6. $(z + 1)(z - 4) + (z + 1)(2z + 3)$

Factor by grouping.

7. $4m + nq + mn + 4q$

8. $x^2 + 5y + 5x + xy$

9. $2m + 6 - am - 3a$

10. $x^2 + 3x - 3y - xy$

6.2 *Factor completely.*

11. $3p^2 - p - 4$

12. $6k^2 + 11k - 10$

13. $12r^2 - 5r - 3$

14. $10m^2 + 37m + 30$

15. $10k^2 - 11kh + 3h^2$

16. $9x^2 + 4xy - 2y^2$

17. $24x - 2x^2 - 2x^3$

18. $6b^3 - 9b^2 - 15b$

19. $y^4 + 2y^2 - 8$

20. $2k^4 - 5k^2 - 3$

21. $p^2(p + 2)^2 + p(p + 2)^2 - 6(p + 2)^2$

22. $3(r + 5)^2 - 11(r + 5) - 4$

23. *Concept Check*   When asked to factor $x^2y^2 - 6x^2 + 5y^2 - 30$, a student gave the following incorrect answer: $x^2(y^2 - 6) + 5(y^2 - 6)$. *WHAT WENT WRONG?* What is the correct answer?

24. If the area of this rectangle is represented by $4p^2 + 3p - 1$, what is the width in terms of p?

$4p - 1$

6.3   *Factor completely.*

25. $16x^2 - 25$

26. $9t^2 - 49$

27. $36m^2 - 25n^2$

28. $x^2 + 14x + 49$

29. $9k^2 - 12k + 4$

30. $r^3 + 27$

31. $125x^3 - 1$

32. $m^6 - 1$

33. $x^8 - 1$

34. $x^2 + 6x + 9 - 25y^2$

35. $(a + b)^3 - (a - b)^3$

36. $x^5 - x^3 - 8x^2 + 8$

6.5   *Solve each equation.*

37. $x^2 - 8x + 16 = 0$

38. $(5x + 2)(x + 1) = 0$

39. $x^2 - 5x + 6 = 0$

40. $x^2 + 2x = 8$

41. $6x^2 = 5x + 50$

42. $6x^2 + 7x = 3$

43. $8x^2 + 14x + 3 = 0$

44. $-4x^2 + 36 = 0$

45. $6x^2 + 9x = 0$

46. $(2x + 1)(x - 2) = -3$

47. $(x + 2)(x - 2) = (x - 2)(x + 3) - 2$

48. $2x^3 - x^2 - 28x = 0$

49. $-x^3 - 3x^2 + 4x + 12 = 0$

50. $(x + 2)(5x^2 - 9x - 18) = 0$

Solve each problem.

51. A triangular wall brace has the shape of a right triangle. One of the perpendicular sides is 1 ft longer than twice the other. The area enclosed by the triangle is 10.5 ft². Find the shorter of the perpendicular sides.

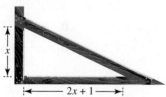

The area is 10.5 ft².

52. A rectangular parking lot has a length 20 ft more than its width. Its area is 2400 ft². What are the dimensions of the lot?

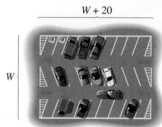

The area is 2400 ft².

A rock is projected directly upward from ground level. After t seconds, its height (if air resistance is neglected) is given by

$$f(t) = -16t^2 + 256t.$$

53. After how many seconds will the rock return to the ground?

54. After how many seconds will it be 240 ft above the ground?

55. Why does the question in **Exercise 54** have two answers?

56. After how many seconds does the rock reach its maximum height, 1024 ft?

Solve each equation for the specified variable.

57. $3s + bk = k - 2t$ for k

58. $z = \dfrac{3w + 7}{w}$ for w

MIXED REVIEW EXERCISES

Factor completely.

59. $16 - 81k^2$　　　　**60.** $30a + am - am^2$　　　　**61.** $9x^2 + 13xy - 3y^2$

62. $8 - a^3$　　　　**63.** $25z^2 - 30zm + 9m^2$　　　　**64.** $15y^3 + 20y^2$

Solve.

65. $5x^2 - 17x = 12$　　　　　　**66.** $x^3 - x = 0$

67. The length of a rectangular picture frame is 2 in. longer than its width. The area enclosed by the frame is 48 in.^{2}. What is the width?

68. When Europeans arrived in America, many Native Americans of the Northeast lived in *longhouses* that sheltered several related families. The rectangular floor area of a typical Huron longhouse was about 2750 ft^2. The length was 85 ft greater than the width. What were the dimensions of the floor?

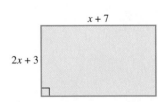

CHAPTER **6**

TEST

Step-by-step test solutions are found on the Chapter Test Prep Videos available via the Video Resources on DVD, in *MyMathLab* , or on YouTube (search "LialIntermediateAlg").

View the complete solutions to all Chapter Test exercises on the Video Resources on DVD.

Factor.

1. $11z^2 - 44z$　　　　　　　　**2.** $10x^2y^5 - 5x^2y^3 - 25x^5y^3$

3. $3x + by + bx + 3y$　　　　　**4.** $-2x^2 - x + 36$

5. $6x^2 + 11x - 35$　　　　**6.** $4p^2 + 3pq - q^2$　　　**7.** $16a^2 + 40ab + 25b^2$

8. $x^2 + 2x + 1 - 4z^2$　　　**9.** $a^3 + 2a^2 - ab^2 - 2b^2$　　**10.** $9k^2 - 121j^2$

11. $y^3 - 216$　　　　　**12.** $6k^4 - k^2 - 35$　　　　**13.** $27x^6 + 1$

14. *Concept Check*　Which one of the following is *not* a factored form of $-x^2 - x + 12$?

　　A. $(3 - x)(x + 4)$　　　　**B.** $-(x - 3)(x + 4)$

　　C. $(-x + 3)(x + 4)$　　　**D.** $(x - 3)(-x + 4)$

Solve each equation.

15. $3x^2 + 8x = -4$　　　　　　**16.** $3x^2 - 5x = 0$

17. $5m(m - 1) = 2(1 - m)$　　　**18.** $ar + 2 = 3r - 6t$ for r

Solve each problem.

19. The area of the rectangle shown is 40 in.^{2}. Find the length and the width of the rectangle.

The area is 40 in.^{2}.

20. A ball is projected upward from ground level. After t seconds, its height in feet is given by the function f defined by

$$f(t) = -16t^2 + 96t.$$

After how many seconds will it reach a height of 128 ft?

CHAPTERS (1–6)
CUMULATIVE REVIEW EXERCISES

Use the properties of real numbers to simplify.

1. $-2(m - 3)$

2. $3x^2 - 4x + 4 + 9x - x^2$

Evaluate for $p = -4$, $q = -2$, and $r = 5$.

3. $\dfrac{5p + 6r^2}{p^2 + q - 1}$

4. $\dfrac{\sqrt{r}}{-p + 2q}$

Solve.

5. $2x - 5 + 3x = 4 - (x + 2)$

6. $\dfrac{3x - 1}{5} + \dfrac{x + 2}{2} = -\dfrac{3}{10}$

7. $3 - 2(x + 3) < 4x$

8. $2x + 4 < 10$ and $3x - 1 > 5$

9. $2x + 4 > 10$ or $3x - 1 < 5$

10. $|5x + 3| - 10 = 3$

11. $|x + 2| < 9$

12. $|2x - 5| \geq 9$

13. Two planes leave the Dallas-Fort Worth airport at the same time. One travels east at 550 mph, and the other travels west at 500 mph. Assuming no wind, how long will it take for the planes to be 2100 mi apart?

Plane	r	t	d
Eastbound	550	x	
Westbound	500	x	
			← Total

14. Graph $4x + 2y = -8$.

15. Find the slope of the line passing through the points $(-4, 8)$ and $(-2, 6)$.

16. What is the slope of the line shown here?

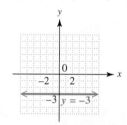

Use the function defined by $f(x) = 2x + 7$ to find each of the following.

17. $f(-4)$

18. The x-intercept of its graph

19. The y-intercept of its graph

Solve each system.

20. $3x - 2y = -7$
$2x + 3y = 17$

21. $2x + 3y - 6z = 5$
$8x - y + 3z = 7$
$3x + 4y - 3z = 7$

Perform the indicated operations. In Exercise 22, assume that variables represent nonzero real numbers.

22. $(3x^2y^{-1})^{-2}(2x^{-3}y)^{-1}$

23. $(7x + 3y)^2$

24. $(3x^3 + 4x^2 - 7) - (2x^3 - 8x^2 + 3x)$

Factor.

25. $16w^2 + 50wz - 21z^2$

26. $4x^2 - 4x + 1 - y^2$

27. $100x^4 - 81$

28. $8p^3 + 27$

Solve.

29. $(x - 1)(2x + 3)(x + 4) = 0$

30. $9x^2 = 6x - 1$

31. A sign is to have the shape of a triangle with a height 3 ft greater than the length of the base. How long should the base be if the area is to be 14 ft²?

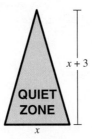

QUIET ZONE

$x + 3$

x

32. A game board has the shape of a rectangle. The longer sides are each 2 in. longer than the distance between them. The area of the board is 288 in.². Find the length of the longer sides and the distance between them.

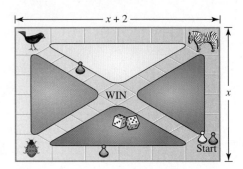

Rational Expressions and Functions

The year 1908 was momentous in automobile history—General Motors was founded and the first Ford Model T was sold. To celebrate the centennials of these landmark events, MotorCities created "2008: Year of the Car," a summer-long auto-tourism festival in southeastern Michigan. A highlight of the festival was Autopalooza August, which was so popular that it has continued on an annual basis. Classic vehicle lovers especially enjoy the Meadowbrook Concours d'Elegance. (*Source:* www.motorcities.org)

In **Exercise 91** of **Section 7.2,** we use a *rational expression* to determine the cost of restoring a vintage automobile to compete in a concours d'elegance.

Rational Expressions and Functions; Multiplying and Dividing

OBJECTIVE 1 Define rational expressions. In arithmetic, a rational number is the quotient of two integers, with the denominator not 0. In algebra, a **rational expression,** or *algebraic fraction,* is the quotient of two polynomials, again with the denominator not 0.

$$\frac{x}{y}, \quad \frac{-a}{4}, \quad \frac{m + 4}{m - 2}, \quad \frac{8x^2 - 2x + 5}{4x^2 + 5x}, \quad \text{and} \quad x^5 \left(\text{or } \frac{x^5}{1} \right) \qquad \text{Examples of rational expressions}$$

Rational expressions are elements of the set

$$\left\{ \frac{P}{Q} \,\middle|\, P \text{ and } Q \text{ are polynomials, with } Q \neq 0 \right\}.$$

OBJECTIVE 2 Define rational functions and describe their domains. A function that is defined by a quotient of polynomials is called a **rational function** and has the form

$$f(x) = \frac{P(x)}{Q(x)}, \quad \text{where } Q(x) \neq 0.$$

The domain of a rational function consists of all real numbers except those that make $Q(x)$—that is, the denominator—equal to 0. For example, the domain of

$$f(x) = \frac{2}{\underbrace{x - 5}_{\text{Cannot equal 0}}}$$

includes all real numbers except 5, because 5 would make the denominator equal to 0. **FIGURE 1** shows a graph of the function defined by $f(x) = \frac{2}{x - 5}$. Notice that the graph does not exist when $x = 5$. (It does not intersect the dashed vertical line whose equation is $x = 5$.) We discuss graphs of rational functions in **Section 7.4.**

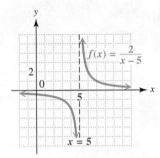

FIGURE 1

EXAMPLE 1 Finding Domains of Rational Functions

For each rational function, find all numbers that are not in the domain. Then give the domain, using set-builder notation.

(a) $f(x) = \dfrac{3}{7x - 14}$

The only values that cannot be used for x are those that make the denominator 0. To find these values, set the denominator equal to 0 and solve the resulting equation.

$$7x - 14 = 0$$
$$7x = 14 \qquad \text{Add 14.}$$
$$x = 2 \qquad \text{Divide by 7.}$$

The number 2 cannot be used as a replacement for x. The domain of f includes all real numbers except 2, written using set-builder notation as $\{x \mid x \neq 2\}$.

NOW TRY
EXERCISE 1

For each rational function, find all numbers that are not in the domain. Then give the domain, using set-builder notation.

(a) $\dfrac{2x - 1}{x^2 - 4x - 5}$

(b) $\dfrac{15}{2x^2 + 1}$

(b) $g(x) = \dfrac{3 + x}{x^2 - 4x + 3}$ ⟵ *Values that make the denominator 0 must be excluded.*

$$x^2 - 4x + 3 = 0 \qquad \text{Set the denominator equal to 0.}$$

$$(x - 1)(x - 3) = 0 \qquad \text{Factor.}$$

$$x - 1 = 0 \quad \text{or} \quad x - 3 = 0 \qquad \text{Zero-factor property}$$

$$x = 1 \quad \text{or} \qquad x = 3 \qquad \text{Solve each equation.}$$

The domain of g includes all real numbers except 1 and 3, written $\{x \mid x \neq 1, 3\}$.

(c) $h(x) = \dfrac{8x + 2}{3}$

The denominator, 3, can never be 0, so the domain of h includes all real numbers, written $\{x \mid x \text{ is a real number}\}$.

(d) $f(x) = \dfrac{2}{x^2 + 4}$

Setting $x^2 + 4$ equal to 0 leads to $x^2 = -4$. There is no real number whose square is -4. Therefore, any real number can be used as a replacement for x. As in part (c), the domain of f is $\{x \mid x \text{ is a real number}\}$. NOW TRY

OBJECTIVE 3 **Write rational expressions in lowest terms.** In arithmetic, we write the fraction $\frac{15}{20}$ in lowest terms by dividing the numerator and denominator by 5 to get $\frac{3}{4}$. We write rational expressions in lowest terms in a similar way, using the **fundamental property of rational numbers.**

Fundamental Property of Rational Numbers

If $\frac{a}{b}$ is a rational number and if c is any nonzero real number, then

$$\frac{a}{b} = \frac{ac}{bc}.$$

That is, the numerator and denominator of a rational number may either be multiplied or divided by the same *nonzero number* without changing the value of the rational number.

Because $\frac{c}{c}$ is equivalent to 1, the fundamental property is based on the identity property of multiplication.

NOTE A rational expression is a quotient of two polynomials. Since the value of a polynomial is a real number for every value of the variable for which it is defined, any statement that applies to rational numbers will also apply to rational expressions.

We use the following steps to write rational expressions in lowest terms.

Writing a Rational Expression in Lowest Terms

Step 1 **Factor** both numerator and denominator to find their greatest common factor (GCF).

Step 2 **Apply the fundamental property.** Divide out common factors.

NOW TRY ANSWERS
1. (a) $-1, 5$; $\{x \mid x \neq -1, 5\}$
  (b) none; $\{x \mid x \text{ is a real number}\}$

EXAMPLE 2 Writing Rational Expressions in Lowest Terms

Write each rational expression in lowest terms.

(a) $\dfrac{(x + 5)(x + 2)}{(x + 2)(x - 3)}$

$= \dfrac{(x + 5)(x + 2)}{(x - 3)(x + 2)}$ Commutative property

$= \dfrac{x + 5}{x - 3} \cdot 1$ Fundamental property

$= \dfrac{x + 5}{x - 3}$ Lowest terms

(b) $\dfrac{a^2 - a - 6}{a^2 + 5a + 6}$

$= \dfrac{(a - 3)(a + 2)}{(a + 3)(a + 2)}$ Factor the numerator.
    Factor the denominator.

$= \dfrac{a - 3}{a + 3} \cdot 1$ Fundamental property

$= \dfrac{a - 3}{a + 3}$ Lowest terms

(c) $\dfrac{y^2 - 4}{2y + 4}$

$= \dfrac{(y + 2)(y - 2)}{2(y + 2)}$ Factor the difference of squares in the numerator. Factor the denominator.

$= \dfrac{y - 2}{2}$ Lowest terms

(d) $\dfrac{x^3 - 27}{x - 3}$

$= \dfrac{(x - 3)(x^2 + 3x + 9)}{x - 3}$ Factor the difference of cubes in the numerator.

$= x^2 + 3x + 9$ Lowest terms

(e) $\dfrac{pr + qr + ps + qs}{pr + qr - ps - qs}$

$= \dfrac{(pr + qr) + (ps + qs)}{(pr + qr) - (ps + qs)}$ Group the terms.

$= \dfrac{r(p + q) + s(p + q)}{r(p + q) - s(p + q)}$ Factor within the groups.

$= \dfrac{(p + q)(r + s)}{(p + q)(r - s)}$ Factor by grouping.

$= \dfrac{r + s}{r - s}$ Lowest terms

**NOW TRY
EXERCISE 2**

Write each rational
expression in lowest terms.

(a) $\dfrac{3a^2 - 7a + 2}{a^2 + 2a - 8}$

(b) $\dfrac{t^3 + 8}{t + 2}$

(c) $\dfrac{am - bm + an - bn}{am + bm + an + bn}$

(f) $\dfrac{8 + k}{16}$ Be careful. The numerator cannot be factored.

This expression cannot be simplified further and is in lowest terms.   **NOW TRY**

⚠ **CAUTION**   Be careful! ***When using the fundamental property of rational numbers, only common factors may be divided.*** For example,

$$\dfrac{y - 2}{2} \neq y \quad \text{and} \quad \dfrac{y - 2}{2} \neq y - 1$$

because the 2 in $y - 2$ is not a *factor* of the numerator. ***Remember to factor before writing a fraction in lowest terms.***

Look again at the rational expression from **Example 2(b).**

$$\dfrac{a^2 - a - 6}{a^2 + 5a + 6}, \quad \text{or} \quad \dfrac{(a - 3)(a + 2)}{(a + 3)(a + 2)}$$

In this expression, a can take any value except -3 or -2, since these values make the denominator 0. In the simplified expression $\frac{a - 3}{a + 3}$, a cannot equal -3. Thus,

$$\dfrac{a^2 - a - 6}{a^2 + 5a + 6} = \dfrac{a - 3}{a + 3}, \quad \text{for all values of } a \text{ except } -3 \text{ or } -2.$$

From now on, such statements of equality will be made with the understanding that they apply only to those real numbers which make neither denominator equal 0. We will no longer state such restrictions.

**NOW TRY
EXERCISE 3**

Write each rational expression
in lowest terms.

(a) $\dfrac{a - 10}{10 - a}$ (b) $\dfrac{81 - y^2}{y - 9}$

EXAMPLE 3   Writing Rational Expressions in Lowest Terms

Write each rational expression in lowest terms.

(a) $\dfrac{m - 3}{3 - m}$ ⟶ Here, the numerator and denominator are opposites.

To write this expression in lowest terms, write the denominator as $-1(m - 3)$.

$$\dfrac{m - 3}{3 - m} = \dfrac{m - 3}{-1(m - 3)} = \dfrac{1}{-1} = -1$$

The numerator could have been rewritten instead to get the same result. (Try this.)

(b) $\dfrac{r^2 - 16}{4 - r}$

$= \dfrac{(r + 4)(r - 4)}{4 - r}$ Factor the difference of squares in the numerator.

$= \dfrac{(r + 4)(r - 4)}{-1(r - 4)}$ Write $4 - r$ as $-1(r - 4)$.

$= \dfrac{r + 4}{-1}$ Fundamental property

$= -(r + 4), \quad \text{or} \quad -r - 4$ Lowest terms   **NOW TRY**

NOW TRY ANSWERS

2. (a) $\dfrac{3a - 1}{a + 4}$
    (b) $t^2 - 2t + 4$
    (c) $\dfrac{a - b}{a + b}$
3. (a) -1
    (b) $-(y + 9)$, or $-y - 9$

As shown in **Example 3,** the quotient $\frac{a}{-a}$ $(a \neq 0)$ can be simplified.

$$\frac{a}{-a} = \frac{a}{-1(a)} = \frac{1}{-1} = -1$$

The following statement summarizes this result.

Quotient of Opposites

In general, if the numerator and the denominator of a rational expression are opposites, then the expression equals -1.

Based on this result, the following are true statements.

$$\frac{q - 7}{7 - q} = -1 \quad \text{and} \quad \frac{-5a + 2b}{5a - 2b} = -1$$

Numerator and denominator in each expression are opposites.

However, the following expression cannot be simplified further.

$$\frac{r - 2}{r + 2}$$ Numerator and denominator are *not* opposites.

OBJECTIVE 4 **Multiply rational expressions.** To multiply rational expressions, follow these steps. (In practice, we usually simplify before multiplying.)

Multiplying Rational Expressions

Step 1 **Factor** all numerators and denominators as completely as possible.

Step 2 **Apply the fundamental property.**

Step 3 **Multiply** the numerators and multiply the denominators.

Step 4 **Check** to be sure that the product is in lowest terms.

EXAMPLE 4 **Multiplying Rational Expressions**

Multiply.

(a) $\dfrac{5p - 5}{p} \cdot \dfrac{3p^2}{10p - 10}$

$= \dfrac{5(p - 1)}{p} \cdot \dfrac{3p \cdot p}{2 \cdot 5(p - 1)}$ Factor.

$= \dfrac{5(p - 1)}{5(p - 1)} \cdot \dfrac{p}{p} \cdot \dfrac{3p}{2}$ Commutative property

$= \dfrac{1}{1} \cdot \dfrac{1}{1} \cdot \dfrac{1}{1} \cdot \dfrac{3p}{2}$ Fundamental property

$= \dfrac{3p}{2}$ Lowest terms

Multiply.

(a) $\dfrac{8t^2}{t^2 - 4} \cdot \dfrac{3t + 6}{9t}$

(b) $\dfrac{m^2 + 2m - 15}{m^2 - 5m + 6} \cdot \dfrac{m^2 - 4}{m^2 + 5m}$

(b) $\dfrac{k^2 + 2k - 15}{k^2 - 4k + 3} \cdot \dfrac{k^2 - k}{k^2 + k - 20}$

$= \dfrac{(k + 5)(k - 3)}{(k - 3)(k - 1)} \cdot \dfrac{k(k - 1)}{(k + 5)(k - 4)}$ Factor.

$= \dfrac{k}{k - 4}$ Fundamental property

(c) $(p - 4) \cdot \dfrac{3}{5p - 20}$

$= \dfrac{p - 4}{1} \cdot \dfrac{3}{5p - 20}$ Write $p - 4$ as $\frac{p-4}{1}$.

$= \dfrac{p - 4}{1} \cdot \dfrac{3}{5(p - 4)}$ Factor.

$= \dfrac{3}{5}$ Fundamental property

(d) $\dfrac{x^2 + 2x}{x + 1} \cdot \dfrac{x^2 - 1}{x^3 + x^2}$

$= \dfrac{x(x + 2)}{x + 1} \cdot \dfrac{(x + 1)(x - 1)}{x^2(x + 1)}$ Factor.

$= \dfrac{(x + 2)(x - 1)}{x(x + 1)}$ Multiply; lowest terms

(e) $\dfrac{x - 6}{x^2 - 12x + 36} \cdot \dfrac{x^2 - 3x - 18}{x^2 + 7x + 12}$

$= \dfrac{x - 6}{(x - 6)^2} \cdot \dfrac{(x + 3)(x - 6)}{(x + 3)(x + 4)}$ Factor.

$= \dfrac{1}{x + 4}$ Lowest terms   NOW TRY

> Remember to include 1 in the numerator when all other factors are eliminated.

OBJECTIVE 5 Find reciprocals of rational expressions. The rational numbers $\frac{a}{b}$ and $\frac{c}{d}$ are reciprocals of each other if they have a product of 1. The **reciprocal** of a rational expression is defined in the same way: *Two rational expressions are reciprocals of each other if they have a product of 1. Recall that 0 has no reciprocal.*

The table shows several rational expressions and their reciprocals.

Rational Expression	Reciprocal
3, or $\frac{3}{1}$	$\frac{1}{3}$
$\frac{5}{k}$	$\frac{k}{5}$
$\frac{m^2 - 9m}{2}$	$\frac{2}{m^2 - 9m}$
$\frac{0}{4}$	undefined

Reciprocals have a product of 1.

Finding the Reciprocal

To find the reciprocal of a nonzero rational expression, interchange the numerator and denominator of the expression.

OBJECTIVE 6 Divide rational expressions.

Dividing Rational Expressions

To divide two rational expressions, *multiply* the first (the *dividend*) by the reciprocal of the second (the *divisor*).

NOW TRY ANSWERS

4. (a) $\dfrac{8t}{3(t - 2)}$ (b) $\dfrac{m + 2}{m}$

NOW TRY
EXERCISE 5

Divide.

(a) $\dfrac{2p^2q}{3pq^4} \div \dfrac{pq}{6p^2q^2}$

(b) $\dfrac{3k^2 + 5k - 2}{9k^2 - 1} \div \dfrac{4k^2 + 8k}{k^2 - 7k}$

EXAMPLE 5 Dividing Rational Expressions

Divide.

(a) $\dfrac{2z}{9} \div \dfrac{5z^2}{18}$

$= \dfrac{2z}{9} \cdot \dfrac{18}{5z^2}$ Multiply by the reciprocal.

$= \dfrac{2z}{9} \cdot \dfrac{2 \cdot 9}{5z \cdot z}$ Factor.

$= \dfrac{4}{5z}$ Multiply; lowest terms

(b) $\dfrac{m^2pq^3}{mp^4} \div \dfrac{m^5p^2q}{mpq^2}$

$= \dfrac{m^2pq^3}{mp^4} \cdot \dfrac{mpq^2}{m^5p^2q}$ Multiply by the reciprocal.

$= \dfrac{m^3p^2q^5}{m^6p^6q}$ Properties of exponents

$= \dfrac{q^4}{m^3p^4}$ Properties of exponents

(c) $\dfrac{8k - 16}{3k} \div \dfrac{3k - 6}{4k^2}$

$= \dfrac{8k - 16}{3k} \cdot \dfrac{4k^2}{3k - 6}$ Multiply by the reciprocal.

$= \dfrac{8(k - 2)}{3k} \cdot \dfrac{4k \cdot k}{3(k - 2)}$ Factor.

$= \dfrac{32k}{9}$ Multiply; lowest terms

(d) $\dfrac{5m^2 + 17m - 12}{3m^2 + 7m - 20} \div \dfrac{5m^2 + 2m - 3}{15m^2 - 34m + 15}$

$= \dfrac{5m^2 + 17m - 12}{3m^2 + 7m - 20} \cdot \dfrac{15m^2 - 34m + 15}{5m^2 + 2m - 3}$ Definition of division

$= \dfrac{(5m - 3)(m + 4)}{(m + 4)(3m - 5)} \cdot \dfrac{(3m - 5)(5m - 3)}{(5m - 3)(m + 1)}$ Factor.

$= \dfrac{5m - 3}{m + 1}$ Lowest terms

NOW TRY

NOW TRY ANSWERS

5. (a) $\dfrac{4p^2}{q^2}$ **(b)** $\dfrac{k - 7}{4(3k + 1)}$

7.1 EXERCISES

MyMathLab PRACTICE WATCH DOWNLOAD READ REVIEW

🌐 *Complete solution available on the Video Resources on DVD*

For each rational function, find all numbers that are not in the domain. Then give the domain, using set-builder notation. **See Example 1.**

1. $f(x) = \dfrac{x}{x - 7}$

2. $f(x) = \dfrac{x}{x + 3}$

🌐 **3.** $f(x) = \dfrac{6x - 5}{7x + 1}$

4. $f(x) = \dfrac{8x - 3}{2x + 7}$

5. $f(x) = \dfrac{12x + 3}{x}$

6. $f(x) = \dfrac{9x + 8}{x}$

7. $f(x) = \dfrac{3x + 1}{2x^2 + x - 6}$

8. $f(x) = \dfrac{2x + 4}{3x^2 + 11x - 42}$

9. $f(x) = \dfrac{x + 2}{14}$

10. $f(x) = \dfrac{x - 9}{26}$

11. $f(x) = \dfrac{2x^2 - 3x + 4}{3x^2 + 8}$

12. $f(x) = \dfrac{9x^2 - 8x + 3}{4x^2 + 1}$

Concept Check As review, multiply or divide the rational numbers as indicated. Write answers in lowest terms.

13. $\dfrac{4}{21} \cdot \dfrac{7}{10}$

14. $\dfrac{5}{9} \cdot \dfrac{12}{25}$

15. $\dfrac{3}{8} \div \dfrac{5}{12}$

16. $\dfrac{5}{6} \div \dfrac{14}{15}$

17. $\dfrac{2}{3} \div \dfrac{8}{9}$

18. $\dfrac{3}{8} \div \dfrac{9}{14}$

19. *Concept Check* Rational expressions often can be written in lowest terms in seemingly different ways. For example,

$$\frac{y-3}{-5} \quad \text{and} \quad \frac{-y+3}{5}$$

look different, but we get the second quotient by multiplying the first by -1 in both the numerator and denominator. To practice recognizing equivalent rational expressions, match the expressions in parts (a)–(f) with their equivalents in choices A–F.

(a) $\dfrac{x-3}{x+4}$ **(b)** $\dfrac{x+3}{x-4}$ **(c)** $\dfrac{x-3}{x-4}$ **(d)** $\dfrac{x+3}{x+4}$ **(e)** $\dfrac{3-x}{x+4}$ **(f)** $\dfrac{x+3}{4-x}$

A. $\dfrac{-x-3}{4-x}$ **B.** $\dfrac{-x-3}{-x-4}$ **C.** $\dfrac{3-x}{-x-4}$ **D.** $\dfrac{-x+3}{-x+4}$ **E.** $\dfrac{x-3}{-x-4}$ **F.** $\dfrac{-x-3}{x-4}$

20. *Concept Check* Which rational expressions equal $-\dfrac{x}{y}$?

A. $\dfrac{-x}{-y}$ **B.** $\dfrac{x}{-y}$ **C.** $\dfrac{x}{y}$ **D.** $-\dfrac{x}{-y}$ **E.** $\dfrac{-x}{y}$ **F.** $-\dfrac{-x}{-y}$

21. *Concept Check* Identify the two *terms* in the numerator and the two *terms* in the denominator of the rational expression $\dfrac{x^2+4x}{x+4}$, and write it in lowest terms.

22. *Concept Check* Which rational expression can be simplified?

A. $\dfrac{x^2+2}{x^2}$ **B.** $\dfrac{x^2+2}{2}$ **C.** $\dfrac{x^2+y^2}{y^2}$ **D.** $\dfrac{x^2-5x}{x}$

23. *Concept Check* Which rational expression is *not* equivalent to $\dfrac{x-3}{4-x}$?

A. $\dfrac{3-x}{x-4}$ **B.** $\dfrac{x+3}{4+x}$ **C.** $-\dfrac{3-x}{4-x}$ **D.** $-\dfrac{x-3}{x-4}$

24. *Concept Check* Which two rational expressions equal -1?

A. $\dfrac{2x+3}{2x-3}$ **B.** $\dfrac{2x-3}{3-2x}$ **C.** $\dfrac{2x+3}{3+2x}$ **D.** $\dfrac{2x+3}{-2x-3}$

Write each rational expression in lowest terms. **See Example 2.**

25. $\dfrac{x^2(x+1)}{x(x+1)}$

26. $\dfrac{y^3(y-4)}{y^2(y-4)}$

27. $\dfrac{(x+4)(x-3)}{(x+5)(x+4)}$

28. $\dfrac{(2x+7)(x-1)}{(2x+3)(2x+7)}$

29. $\dfrac{4x(x+3)}{8x^2(x-3)}$

30. $\dfrac{5y^2(y+8)}{15y(y-8)}$

31. $\dfrac{3x+7}{3}$

32. $\dfrac{4x-9}{4}$

33. $\dfrac{6m+18}{7m+21}$

34. $\dfrac{5r-20}{3r-12}$

35. $\dfrac{3z^2+z}{18z+6}$

36. $\dfrac{2x^2-5x}{16x-40}$

37. $\dfrac{t^2-9}{3t+9}$

38. $\dfrac{m^2-25}{4m-20}$

39. $\dfrac{2t+6}{t^2-9}$

40. $\dfrac{5s - 25}{s^2 - 25}$

41. $\dfrac{x^2 + 2x - 15}{x^2 + 6x + 5}$

42. $\dfrac{y^2 - 5y - 14}{y^2 + y - 2}$

43. $\dfrac{8x^2 - 10x - 3}{8x^2 - 6x - 9}$

44. $\dfrac{12x^2 - 4x - 5}{8x^2 - 6x - 5}$

45. $\dfrac{a^3 + b^3}{a + b}$

46. $\dfrac{r^3 - s^3}{r - s}$

47. $\dfrac{2c^2 + 2cd - 60d^2}{2c^2 - 12cd + 10d^2}$

48. $\dfrac{3s^2 - 9st - 54t^2}{3s^2 - 6st - 72t^2}$

49. $\dfrac{ac - ad + bc - bd}{ac - ad - bc + bd}$

50. $\dfrac{2xy + 2xw + y + w}{2xy + y - 2xw - w}$

*Write each rational expression in lowest terms. **See Example 3.***

51. $\dfrac{7 - b}{b - 7}$

52. $\dfrac{r - 13}{13 - r}$

53. $\dfrac{x^2 - y^2}{y - x}$

54. $\dfrac{m^2 - n^2}{n - m}$

55. $\dfrac{(a - 3)(x + y)}{(3 - a)(x - y)}$

56. $\dfrac{(8 - p)(x + 2)}{(p - 8)(x - 2)}$

57. $\dfrac{5k - 10}{20 - 10k}$

58. $\dfrac{7x - 21}{63 - 21x}$

59. $\dfrac{a^2 - b^2}{a^2 + b^2}$

60. $\dfrac{p^2 + q^2}{p^2 - q^2}$

*Multiply or divide as indicated. **See Examples 4 and 5.***

61. $\dfrac{x^3}{3y} \cdot \dfrac{9y^2}{x^5}$

62. $\dfrac{a^4}{5b^2} \cdot \dfrac{25b^4}{a^3}$

63. $\dfrac{5a^4b^2}{16a^2b} \div \dfrac{25a^2b}{60a^3b^2}$

64. $\dfrac{s^3t^2}{10s^2t^4} \div \dfrac{8s^4t^2}{5t^6}$

65. $\dfrac{(-3mn)^2 \cdot 64(m^2n)^3}{16m^2n^4(mn^2)^3} \div \dfrac{24(m^2n^2)^4}{(3m^2n^3)^2}$

66. $\dfrac{(-4a^2b^3)^2 \cdot 9(a^2b^4)^2}{(2a^2b^3)^4 \cdot (3a^3b)^2} \div \dfrac{(ab)^4}{(a^2b^3)^2}$

67. $\dfrac{(x + 2)(x + 1)}{(x + 3)(x - 2)} \cdot \dfrac{(x + 3)(x + 4)}{(x + 2)(x + 1)}$

68. $\dfrac{(x + 3)(x - 4)}{(x - 4)(x + 2)} \cdot \dfrac{(x + 5)(x - 6)}{(x + 3)(x - 6)}$

69. $\dfrac{(2x + 3)(x - 4)}{(x + 8)(x - 4)} \div \dfrac{(x - 4)(x + 2)}{(x - 4)(x + 8)}$

70. $\dfrac{(6x + 5)(x - 3)}{(x - 1)(x - 3)} \div \dfrac{(2x + 7)(x + 9)}{(x - 1)(x + 9)}$

71. $\dfrac{4x}{8x + 4} \cdot \dfrac{14x + 7}{6}$

72. $\dfrac{12x - 20}{5x} \cdot \dfrac{6}{9x - 15}$

73. $\dfrac{p^2 - 25}{4p} \cdot \dfrac{2}{5 - p}$

74. $\dfrac{a^2 - 1}{4a} \cdot \dfrac{2}{1 - a}$

75. $(7k + 7) \div \dfrac{4k + 4}{5}$

76. $(8y - 16) \div \dfrac{3y - 6}{10}$

77. $(z^2 - 1) \cdot \dfrac{1}{1 - z}$

78. $(y^2 - 4) \div \dfrac{2 - y}{8y}$

79. $\dfrac{4x - 20}{5x} \div \dfrac{2x - 10}{7x^3}$

80. $\dfrac{k^2 - 4}{3k^2} \div \dfrac{2 - k}{11k}$

81. $\dfrac{12x - 10y}{3x + 2y} \cdot \dfrac{6x + 4y}{10y - 12x}$

82. $\dfrac{9s - 12t}{2s + 2t} \cdot \dfrac{3s + 3t}{4t - 3s}$

83. $\dfrac{x^2 - 25}{x^2 + x - 20} \cdot \dfrac{x^2 + 7x + 12}{x^2 - 2x - 15}$

84. $\dfrac{t^2 - 49}{t^2 + 4t - 21} \cdot \dfrac{t^2 + 8t + 15}{t^2 - 2t - 35}$

85. $\dfrac{a^3 - b^3}{a^2 - b^2} \div \dfrac{2a - 2b}{2a + 2b}$

86. $\dfrac{x^3 + y^3}{2x + 2y} \div \dfrac{x^2 - y^2}{2x - 2y}$

87. $\dfrac{8x^3 - 27}{2x^2 - 18} \cdot \dfrac{2x + 6}{8x^2 + 12x + 18}$

88. $\dfrac{64x^3 + 1}{4x^2 - 100} \cdot \dfrac{4x + 20}{64x^2 - 16x + 4}$

89. $\dfrac{a^3 - 8b^3}{a^2 - ab - 6b^2} \cdot \dfrac{a^2 + ab - 12b^2}{a^2 + 2ab - 8b^2}$

90. $\dfrac{p^3 - 27q^3}{p^2 + pq - 12q^2} \cdot \dfrac{p^2 - 2pq - 24q^2}{p^2 - 5pq - 6q^2}$

91. $\dfrac{6x^2 + 5x - 6}{12x^2 - 11x + 2} \div \dfrac{4x^2 - 12x + 9}{8x^2 - 14x + 3}$

92. $\dfrac{8a^2 - 6a - 9}{6a^2 - 5a - 6} \div \dfrac{4a^2 + 11a + 6}{9a^2 + 12a + 4}$

Brain Busters   *Multiply or divide as indicated.*

93. $\dfrac{3k^2 + 17kp + 10p^2}{6k^2 + 13kp - 5p^2} \div \dfrac{6k^2 + kp - 2p^2}{6k^2 - 5kp + p^2}$

94. $\dfrac{16c^2 + 24cd + 9d^2}{16c^2 - 16cd + 3d^2} \div \dfrac{16c^2 - 9d^2}{16c^2 - 24cd + 9d^2}$

95. $\left(\dfrac{6k^2 - 13k - 5}{k^2 + 7k} \div \dfrac{2k - 5}{k^3 + 6k^2 - 7k} \right) \cdot \dfrac{k^2 - 5k + 6}{3k^2 - 8k - 3}$

96. $\left(\dfrac{2x^3 + 3x^2 - 2x}{3x - 15} \div \dfrac{2x^3 - x^2}{x^2 - 3x - 10} \right) \cdot \dfrac{5x^2 - 10x}{3x^2 + 12x + 12}$

PREVIEW EXERCISES

Add or subtract as indicated. ***See Section 1.2.***

97. $-\dfrac{3}{4} + \dfrac{1}{12}$

98. $\dfrac{9}{10} - \left(-\dfrac{1}{3} \right)$

99. $\dfrac{4}{7} + \dfrac{1}{3} - \dfrac{1}{2}$

100. $-\dfrac{2}{3} + \left(-\dfrac{4}{7} \right) - \dfrac{1}{8}$

7.2   Adding and Subtracting Rational Expressions

OBJECTIVES

1. Add and subtract rational expressions with the same denominator.
2. Find a least common denominator.
3. Add and subtract rational expressions with different denominators.

OBJECTIVE 1   **Add and subtract rational expressions with the same denominator.** We do this as we would with rational numbers.

Adding or Subtracting Rational Expressions

Step 1   **If the denominators are the same,** add or subtract the numerators. Place the result over the common denominator.

If the denominators are different, first find the least common denominator. Write all rational expressions with this least common denominator, and then add or subtract the numerators. Place the result over the common denominator.

Step 2   **Simplify.** Write all answers in lowest terms.

NOW TRY
EXERCISE 1
Add or subtract as indicated.

(a) $\dfrac{5}{3x} + \dfrac{2}{3x}$

(b) $\dfrac{x^2}{x-3} - \dfrac{9}{x-3}$

(c) $\dfrac{2}{x^2+x-2} + \dfrac{x}{x^2+x-2}$

EXAMPLE 1 Adding and Subtracting Rational Expressions (Same Denominators)

Add or subtract as indicated.

(a) $\dfrac{3y}{5} + \dfrac{x}{5}$

$= \dfrac{3y + x}{5}$ ← Add the numerators.
  ← Keep the common denominator.

(b) $\dfrac{7}{2r^2} - \dfrac{11}{2r^2}$

$= \dfrac{7-11}{2r^2}$ Subtract the numerators.
   Keep the common denominator.

$= \dfrac{-4}{2r^2}$

$= -\dfrac{2}{r^2}$ Lowest terms

(c)

$\dfrac{m}{m^2-p^2} + \dfrac{p}{m^2-p^2}$

$= \dfrac{m+p}{m^2-p^2}$ Add the numerators.
   Keep the common denominator.

$= \dfrac{m+p}{(m+p)(m-p)}$ Factor.

Remember to write 1 in the numerator. $= \dfrac{1}{m-p}$ Lowest terms

(d) $\dfrac{4}{x^2+2x-8} + \dfrac{x}{x^2+2x-8}$

$= \dfrac{4+x}{x^2+2x-8}$ Add.

$= \dfrac{4+x}{(x-2)(x+4)}$ Factor.

$= \dfrac{1}{x-2}$ Lowest terms   NOW TRY

OBJECTIVE 2 **Find a least common denominator.** We add or subtract rational expressions with different denominators by first writing them with a common denominator, usually the **least common denominator (LCD).**

Finding the Least Common Denominator

Step 1 **Factor** each denominator.

Step 2 **Find the least common denominator.** The LCD is the product of all of the different factors from each denominator, with each factor raised to the *greatest* power that occurs in any denominator.

NOW TRY ANSWERS
1. (a) $\frac{7}{3x}$ (b) $x+3$ (c) $\frac{1}{x-1}$

**NOW TRY
EXERCISE 2**

Find the LCD for each group of denominators.

(a) $15m^3n$, $10m^2n$

(b) t, $t - 8$

(c) $3x^2 + 9x - 30$, $x^2 - 4$, $x^2 + 10x + 25$

EXAMPLE 2 Finding Least Common Denominators

Suppose that the given expressions are denominators of fractions. Find the LCD for each group of denominators.

(a) $5xy^2$, $2x^3y$

$$5xy^2 = 5 \cdot x \cdot y^2 \qquad \text{Each denominator is already factored.}$$
$$2x^3y = 2 \cdot x^3 \cdot y$$

Greatest exponent on x is 3.

$$\text{LCD} = 5 \cdot 2 \cdot x^3 \cdot y^2 \leftarrow \text{Greatest exponent on } y \text{ is 2.}$$
$$= 10x^3y^2$$

(b) $k - 3$, k

Each denominator is already factored. The LCD must be divisible by *both* $k - 3$ and k.

$$\text{LCD} = k(k - 3) \quad \boxed{\text{Don't forget the factor } k.}$$

It is usually best to leave a least common denominator in factored form.

(c) $y^2 - 2y - 8$, $y^2 + 3y + 2$

$$\left. \begin{array}{l} y^2 - 2y - 8 = (y - 4)(y + 2) \\ y^2 + 3y + 2 = (y + 2)(y + 1) \end{array} \right\} \text{Factor.}$$
$$\text{LCD} = (y - 4)(y + 2)(y + 1)$$

(d) $8z - 24$, $5z^2 - 15z$

$$\left. \begin{array}{l} 8z - 24 = 8(z - 3) \\ 5z^2 - 15z = 5z(z - 3) \end{array} \right\} \text{Factor.}$$
$$\text{LCD} = 8 \cdot 5z \cdot (z - 3) = 40z(z - 3)$$

(e) $m^2 + 5m + 6$, $m^2 + 4m + 4$, $2m^2 + 4m - 6$

$$\left. \begin{array}{l} m^2 + 5m + 6 = (m + 3)(m + 2) \\ m^2 + 4m + 4 = (m + 2)^2 \\ 2m^2 + 4m - 6 = 2(m + 3)(m - 1) \end{array} \right\} \text{Factor.}$$
$$\text{LCD} = 2(m + 3)(m + 2)^2(m - 1) \qquad \textit{NOW TRY}$$

OBJECTIVE 3 **Add and subtract rational expressions with different denominators.** First we must write each expression with the least common denominator by multiplying its numerator and denominator by the factors needed to get the LCD. This is valid because we are multiplying by a form of 1, the identity element for multiplication.

Consider the sum $\frac{7}{15} + \frac{5}{12}$.

$$\frac{7}{15} + \frac{5}{12} \qquad \text{The LCD for 15 and 12 is 60.}$$

$$= \frac{7 \cdot 4}{15 \cdot 4} + \frac{5 \cdot 5}{12 \cdot 5} \qquad \text{Fundamental property}$$

$$= \frac{28}{60} + \frac{25}{60} \qquad \begin{array}{l} \text{Write each fraction with} \\ \text{the common denominator.} \end{array}$$

$$= \frac{28 + 25}{60} \qquad \begin{array}{l} \text{Add the numerators.} \\ \text{Keep the common denominator.} \end{array}$$

$$= \frac{53}{60}$$

NOW TRY ANSWERS

2. (a) $30m^3n$ (b) $t(t - 8)$
   (c) $3(x - 2)(x + 2)(x + 5)^2$

**NOW TRY
EXERCISE 3**
Add or subtract as indicated.

(a) $\dfrac{2}{3x} + \dfrac{7}{4x}$

(b) $\dfrac{3}{z} - \dfrac{6}{z-5}$

EXAMPLE 3 Adding and Subtracting Rational Expressions (Different Denominators)

Add or subtract as indicated.

(a) $\dfrac{5}{2p} + \dfrac{3}{8p}$ The LCD for $2p$ and $8p$ is $8p$.

$= \dfrac{5 \cdot 4}{2p \cdot 4} + \dfrac{3}{8p}$ Fundamental property

$= \dfrac{20}{8p} + \dfrac{3}{8p}$ Write the first fraction with the common denominator.

$= \dfrac{20 + 3}{8p}$ Add the numerators. Keep the common denominator.

$= \dfrac{23}{8p}$

(b) $\dfrac{6}{r} - \dfrac{5}{r-3}$ The LCD is $r(r-3)$.

$= \dfrac{6(r-3)}{r(r-3)} - \dfrac{r \cdot 5}{r(r-3)}$ Fundamental property

$= \dfrac{6r - 18}{r(r-3)} - \dfrac{5r}{r(r-3)}$ Distributive and commutative properties

$= \dfrac{6r - 18 - 5r}{r(r-3)}$ Subtract numerators.

$= \dfrac{r - 18}{r(r-3)}$ Combine like terms in the numerator.

NOW TRY

⚠ **CAUTION** Sign errors occur easily when a rational expression with two or more terms in the numerator is being subtracted. ***In this situation, the subtraction sign must be distributed to every term in the numerator of the fraction that follows it.*** Study **Example 4** carefully to see how this is done.

EXAMPLE 4 Subtracting Rational Expressions

Subtract.

(a) $\dfrac{7x}{3x+1} - \dfrac{x-2}{3x+1}$

The denominators are the same. ***The subtraction sign must be applied to both terms in the numerator of the second rational expression.***

$$\dfrac{7x}{3x+1} - \dfrac{x-2}{3x+1}$$ Use parentheses to avoid errors.

$$= \dfrac{7x - (x-2)}{3x+1}$$ Subtract the numerators. Keep the common denominator.

Be careful with signs. $$= \dfrac{7x - x + 2}{3x+1}$$ Distributive property

NOW TRY ANSWERS

3. (a) $\dfrac{29}{12x}$ (b) $\dfrac{-3z-15}{z(z-5)}$

NOW TRY
EXERCISE 4
Subtract.

(a) $\dfrac{18x + 7}{4x + 5} - \dfrac{2x - 13}{4x + 5}$

(b) $\dfrac{5}{x - 3} - \dfrac{5}{x + 3}$

$= \dfrac{6x + 2}{3x + 1}$ Combine like terms in the numerator.

$= \dfrac{2(3x + 1)}{3x + 1}$ Factor the numerator.

$= 2$ Lowest terms

(b) $\dfrac{1}{q - 1} - \dfrac{1}{q + 1}$ The LCD is $(q - 1)(q + 1)$.

$= \dfrac{1(q + 1)}{(q - 1)(q + 1)} - \dfrac{1(q - 1)}{(q + 1)(q - 1)}$ Fundamental property

$= \dfrac{(q + 1) - (q - 1)}{(q - 1)(q + 1)}$ Subtract the numerators.

$= \dfrac{q + 1 - q + 1}{(q - 1)(q + 1)}$ Be careful with signs. Distributive property

$= \dfrac{2}{(q - 1)(q + 1)}$ Combine like terms in the numerator.

NOW TRY

NOW TRY
EXERCISE 5
Add.

$\dfrac{t}{t - 9} + \dfrac{2}{9 - t}$

EXAMPLE 5 **Adding Rational Expressions (Denominators Are Opposites)**

Add.

$\dfrac{y}{y - 2} + \dfrac{8}{2 - y}$ ⟵ Denominators are opposites.

$= \dfrac{y}{y - 2} + \dfrac{8(-1)}{(2 - y)(-1)}$ Multiply the second expression by $\frac{-1}{-1}$.

$= \dfrac{y}{y - 2} + \dfrac{-8}{y - 2}$ The LCD is $y - 2$.

$= \dfrac{y - 8}{y - 2}$ Add the numerators.

We could use $2 - y$ as the common denominator and rewrite the first expression.

$\dfrac{y}{y - 2} + \dfrac{8}{2 - y}$

$= \dfrac{y(-1)}{(y - 2)(-1)} + \dfrac{8}{2 - y}$ Multiply the first expression by $\frac{-1}{-1}$.

$= \dfrac{-y}{2 - y} + \dfrac{8}{2 - y}$ The LCD is $2 - y$.

$= \dfrac{-y + 8}{2 - y}$, or $\dfrac{8 - y}{2 - y}$ This is an equivalent form of the answer.

NOW TRY

NOW TRY ANSWERS
4. (a) 4 (b) $\frac{30}{(x - 3)(x + 3)}$
5. $\frac{t - 2}{t - 9}$, or $\frac{2 - t}{9 - t}$

NOW TRY
EXERCISE 6
Add and subtract as indicated.

$$\frac{6}{y} - \frac{1}{y-3} + \frac{3}{y^2 - 3y}$$

EXAMPLE 6 Adding and Subtracting Three Rational Expressions

Add and subtract as indicated.

$$\frac{3}{x-2} + \frac{5}{x} - \frac{6}{x^2 - 2x}$$

$$= \frac{3}{x-2} + \frac{5}{x} - \frac{6}{x(x-2)} \qquad \text{Factor the third denominator.}$$

$$= \frac{3x}{x(x-2)} + \frac{5(x-2)}{x(x-2)} - \frac{6}{x(x-2)} \qquad \text{The LCD is } x(x-2); \\ \text{fundamental property}$$

$$= \frac{3x + 5(x-2) - 6}{x(x-2)} \qquad \text{Add and subtract the numerators.}$$

$$= \frac{3x + 5x - 10 - 6}{x(x-2)} \qquad \text{Distributive property}$$

$$= \frac{8x - 16}{x(x-2)} \qquad \text{Combine like terms in the numerator.}$$

$$= \frac{8(x-2)}{x(x-2)} \qquad \text{Factor the numerator.}$$

$$= \frac{8}{x} \qquad \text{Lowest terms}$$

NOW TRY

NOW TRY
EXERCISE 7
Subtract.

$$\frac{t-1}{t^2 - 2t - 8} - \frac{2t+3}{t^2 + 3t + 2}$$

EXAMPLE 7 Subtracting Rational Expressions

Subtract.

$$\frac{m+4}{m^2 - 2m - 3} - \frac{2m-3}{m^2 - 5m + 6}$$

$$= \frac{m+4}{(m-3)(m+1)} - \frac{2m-3}{(m-3)(m-2)} \qquad \text{Factor each denominator.}$$

$$= \frac{(m+4)(m-2)}{(m-3)(m+1)(m-2)} - \frac{(2m-3)(m+1)}{(m-3)(m-2)(m+1)} \qquad \text{The LCD is } (m-3) \cdot \\ (m+1)(m-2).$$

$$= \frac{(m+4)(m-2) - (2m-3)(m+1)}{(m-3)(m+1)(m-2)} \qquad \text{Subtract the numerators.}$$

$$= \frac{m^2 + 2m - 8 - (2m^2 - m - 3)}{(m-3)(m+1)(m-2)} \quad \boxed{\text{Note the careful use of parentheses.}} \qquad \text{Multiply in the numerator.}$$

$$= \frac{m^2 + 2m - 8 - 2m^2 + m + 3}{(m-3)(m+1)(m-2)} \quad \boxed{\text{Be careful with signs.}} \qquad \text{Distributive property}$$

$$= \frac{-m^2 + 3m - 5}{(m-3)(m+1)(m-2)} \qquad \text{Combine like terms in the numerator.}$$

NOW TRY ANSWERS
6. $\dfrac{5}{y}$

7. $\dfrac{-t^2 + 5t + 11}{(t+2)(t-4)(t+1)}$

If we try to factor the numerator, we find that this rational expression is in lowest terms.

NOW TRY

NOW TRY
EXERCISE 8

Add.

$$\frac{2}{m^2 - 6m + 9} + \frac{4}{m^2 + m - 12}$$

NOW TRY ANSWER

8. $\dfrac{6m - 4}{(m - 3)^2(m + 4)}$

EXAMPLE 8 Adding Rational Expressions

Add.

$$\frac{5}{x^2 + 10x + 25} + \frac{2}{x^2 + 7x + 10}$$

$$= \frac{5}{(x + 5)^2} + \frac{2}{(x + 5)(x + 2)} \qquad \text{Factor each denominator.}$$

$$= \frac{5(x + 2)}{(x + 5)^2(x + 2)} + \frac{2(x + 5)}{(x + 5)^2(x + 2)} \qquad \begin{array}{l}\text{The LCD is } (x + 5)^2 (x + 2);\\ \text{fundamental property}\end{array}$$

$$= \frac{5(x + 2) + 2(x + 5)}{(x + 5)^2(x + 2)} \qquad \text{Add.}$$

$$= \frac{5x + 10 + 2x + 10}{(x + 5)^2(x + 2)} \qquad \text{Distributive property}$$

$$= \frac{7x + 20}{(x + 5)^2(x + 2)} \qquad \begin{array}{l}\text{Combine like terms in}\\ \text{the numerator.}\end{array} \quad \textit{NOW TRY}$$

7.2 EXERCISES

MyMathLab Math XL PRACTICE WATCH DOWNLOAD READ REVIEW

⊕ *Complete solution available on the Video Resources on DVD*

Concept Check *As review, add or subtract the rational numbers as indicated. Write answers in lowest terms.*

1. $\dfrac{8}{15} + \dfrac{4}{15}$

2. $\dfrac{5}{16} + \dfrac{9}{16}$

3. $\dfrac{5}{6} - \dfrac{8}{9}$

4. $\dfrac{3}{4} - \dfrac{5}{6}$

5. $\dfrac{5}{18} + \dfrac{7}{12}$

6. $\dfrac{3}{10} + \dfrac{7}{15}$

Add or subtract as indicated. Write all answers in lowest terms. ***See Example 1.***

7. $\dfrac{7}{t} + \dfrac{2}{t}$

8. $\dfrac{5}{r} + \dfrac{9}{r}$

⊕ **9.** $\dfrac{6x}{7} + \dfrac{y}{7}$

10. $\dfrac{12t}{5} + \dfrac{s}{5}$

11. $\dfrac{11}{5x} - \dfrac{1}{5x}$

12. $\dfrac{7}{4y} - \dfrac{3}{4y}$

13. $\dfrac{9}{4x^3} - \dfrac{17}{4x^3}$

14. $\dfrac{6}{5y^4} - \dfrac{21}{5y^4}$

15. $\dfrac{5x + 4}{6x + 5} + \dfrac{x + 1}{6x + 5}$

16. $\dfrac{6y + 12}{4y + 3} + \dfrac{2y - 6}{4y + 3}$

17. $\dfrac{x^2}{x + 5} - \dfrac{25}{x + 5}$

18. $\dfrac{y^2}{y + 6} - \dfrac{36}{y + 6}$

19. $\dfrac{-3p + 7}{p^2 + 7p + 12} + \dfrac{8p + 13}{p^2 + 7p + 12}$

20. $\dfrac{5x + 6}{x^2 + x - 20} + \dfrac{4 - 3x}{x^2 + x - 20}$

21. $\dfrac{a^3}{a^2 + ab + b^2} - \dfrac{b^3}{a^2 + ab + b^2}$

22. $\dfrac{p^3}{p^2 - pq + q^2} + \dfrac{q^3}{p^2 - pq + q^2}$

Suppose that the expressions given are denominators of fractions. Find the least common denominator (LCD) for each group. ***See Example 2.***

⊕ **23.** $18x^2y^3, \quad 24x^4y^5$

24. $24a^3b^4, \quad 18a^5b^2$

25. $z - 2, \quad z$

26. $k + 3, \quad k$

27. $2y + 8, \quad y + 4$

28. $3r - 21, \quad r - 7$

29. $x^2 - 81, \quad x^2 + 18x + 81$

30. $y^2 - 16, \quad y^2 - 8y + 16$

31. $m + n, \quad m - n, \quad m^2 - n^2$

32. $r + s, \quad r - s, \quad r^2 - s^2$

33. $x^2 - 3x - 4, \quad x + x^2$

34. $y^2 - 8y + 12, \quad y^2 - 6y$

35. $2t^2 + 7t - 15, \quad t^2 + 3t - 10$

36. $s^2 - 3s - 4, \quad 3s^2 + s - 2$

37. $2y + 6, \quad y^2 - 9, \quad y$

38. $9x + 18, \quad x^2 - 4, \quad x$

39. $2x - 6, \quad x^2 - x - 6, \quad (x + 2)^2$

40. $3a - 3b, \quad a^2 + ab - 2b^2, \quad (a - b)^2$

41. *Concept Check* Consider the following *incorrect* work. **WHAT WENT WRONG?**

$$\frac{x}{x + 2} - \frac{4x - 1}{x + 2}$$

$$= \frac{x - 4x - 1}{x + 2}, \quad \text{or} \quad \frac{-3x - 1}{x + 2}$$

42. One student added two rational expressions and obtained the answer $\frac{3}{5 - y}$. Another student obtained the answer $\frac{-3}{y - 5}$ for the same problem. Both are correct. Explain.

Add or subtract as indicated. Write all answers in lowest terms. **See Examples 3–6.**

43. $\dfrac{8}{t} + \dfrac{7}{3t}$

44. $\dfrac{5}{x} + \dfrac{9}{4x}$

45. $\dfrac{5}{12x^2y} - \dfrac{11}{6xy}$

46. $\dfrac{7}{18a^3b^2} - \dfrac{2}{9ab}$

47. $\dfrac{4}{15a^4b^5} + \dfrac{3}{20a^2b^6}$

48. $\dfrac{5}{12x^5y^2} + \dfrac{5}{18x^4y^5}$

49. $\dfrac{2r}{7p^3q^4} + \dfrac{3s}{14p^4q}$

50. $\dfrac{4t}{9a^8b^7} + \dfrac{5s}{27a^4b^3}$

51. $\dfrac{1}{a^3b^2} - \dfrac{2}{a^4b} + \dfrac{3}{a^5b^7}$

52. $\dfrac{5}{t^4u^7} - \dfrac{3}{t^5u^9} + \dfrac{6}{t^{10}u}$

53. $\dfrac{1}{x - 1} - \dfrac{1}{x}$

54. $\dfrac{3}{x - 3} - \dfrac{1}{x}$

55. $\dfrac{3a}{a + 1} + \dfrac{2a}{a - 3}$

56. $\dfrac{2x}{x + 4} + \dfrac{3x}{x - 7}$

57. $\dfrac{17y + 3}{9y + 7} - \dfrac{-10y - 18}{9y + 7}$

58. $\dfrac{7x + 8}{3x + 2} - \dfrac{x + 4}{3x + 2}$

59. $\dfrac{11x - 13}{2x - 3} - \dfrac{3x - 1}{2x - 3}$

60. $\dfrac{13x - 5}{4x - 1} - \dfrac{x - 2}{4x - 1}$

61. $\dfrac{2}{4 - x} + \dfrac{5}{x - 4}$

62. $\dfrac{3}{2 - t} + \dfrac{1}{t - 2}$

63. $\dfrac{w}{w - z} - \dfrac{z}{z - w}$

64. $\dfrac{a}{a - b} - \dfrac{b}{b - a}$

65. $\dfrac{1}{x + 1} - \dfrac{1}{x - 1}$

66. $\dfrac{-2}{x - 1} + \dfrac{2}{x + 1}$

67. $\dfrac{4x}{x - 1} - \dfrac{2}{x + 1} - \dfrac{4}{x^2 - 1}$

68. $\dfrac{4}{x + 3} - \dfrac{x}{x - 3} - \dfrac{18}{x^2 - 9}$

69. $\dfrac{15}{y^2 + 3y} + \dfrac{2}{y} + \dfrac{5}{y + 3}$

70. $\dfrac{7}{t - 2} - \dfrac{6}{t^2 - 2t} - \dfrac{3}{t}$

71. $\dfrac{5}{x - 2} + \dfrac{1}{x} + \dfrac{2}{x^2 - 2x}$

72. $\dfrac{5x}{x - 3} + \dfrac{2}{x} + \dfrac{6}{x^2 - 3x}$

73. $\dfrac{3x}{x + 1} + \dfrac{4}{x - 1} - \dfrac{6}{x^2 - 1}$

74. $\dfrac{5x}{x + 3} + \dfrac{x + 2}{x} - \dfrac{6}{x^2 + 3x}$

75. $\dfrac{4}{x + 1} + \dfrac{1}{x^2 - x + 1} - \dfrac{12}{x^3 + 1}$

76. $\dfrac{5}{x + 2} + \dfrac{2}{x^2 - 2x + 4} - \dfrac{60}{x^3 + 8}$

77. $\dfrac{2x + 4}{x + 3} + \dfrac{3}{x} - \dfrac{6}{x^2 + 3x}$

78. $\dfrac{4x + 1}{x + 5} - \dfrac{2}{x} + \dfrac{10}{x^2 + 5x}$

79. $\dfrac{3}{(p-2)^2} - \dfrac{5}{p-2} + 4$

80. $\dfrac{8}{(3r-1)^2} + \dfrac{2}{3r-1} - 6$

Add or subtract as indicated. Write all answers in lowest terms. ***See Examples 7 and 8.****

81. $\dfrac{3}{x^2 - 5x + 6} - \dfrac{2}{x^2 - 4x + 4}$

82. $\dfrac{2}{m^2 - 4m + 4} + \dfrac{3}{m^2 + m - 6}$

83. $\dfrac{5x}{x^2 + xy - 2y^2} - \dfrac{3x}{x^2 + 5xy - 6y^2}$

84. $\dfrac{6x}{6x^2 + 5xy - 4y^2} - \dfrac{2y}{9x^2 - 16y^2}$

85. $\dfrac{5x - y}{x^2 + xy - 2y^2} - \dfrac{3x + 2y}{x^2 + 5xy - 6y^2}$

86. $\dfrac{6x + 5y}{6x^2 + 5xy - 4y^2} - \dfrac{x + 2y}{9x^2 - 16y^2}$

87. $\dfrac{r + s}{3r^2 + 2rs - s^2} - \dfrac{s - r}{6r^2 - 5rs + s^2}$

88. $\dfrac{3y}{y^2 + yz - 2z^2} + \dfrac{4y - 1}{y^2 - z^2}$

89. $\dfrac{3}{x^2 + 4x + 4} + \dfrac{7}{x^2 + 5x + 6}$

90. $\dfrac{5}{x^2 + 6x + 9} - \dfrac{2}{x^2 + 4x + 3}$

Work each problem.

91. A **concours d'elegance** is a competition in which a maximum of 100 points is awarded to a car on the basis of its general attractiveness. The function defined by the rational expression

$$c(x) = \dfrac{1010}{49(101 - x)} - \dfrac{10}{49}$$

approximates the cost, in thousands of dollars, of restoring a car so that it will win x points.

(a) Simplify the expression for $c(x)$ by performing the indicated subtraction.

(b) Use the simplified expression to determine how much it would cost to win 95 points.

92. A **cost-benefit model** expresses the cost of an undertaking in terms of the benefits received. One cost-benefit model gives the cost in thousands of dollars to remove x percent of a certain pollutant as

$$c(x) = \dfrac{6.7x}{100 - x}.$$

Another model produces the relationship

$$c(x) = \dfrac{6.5x}{102 - x}.$$

(a) What is the cost found by averaging the two models? (*Hint:* The average of two quantities is half their sum.)

(b) Using the two given models and your answer to part (a), find the cost to the nearest dollar to remove 95% ($x = 95$) of the pollutant.

(c) Average the two costs in part (b) from the given models. What do you notice about this result compared with the cost obtained by using the average of the two models?

PREVIEW EXERCISES

Simplify. ***See Section 1.2.***

93. $\dfrac{\frac{2}{3}}{\frac{3}{4}}$

94. $\dfrac{\frac{3}{5}}{\frac{5}{6}}$

95. $\dfrac{\frac{3}{4}}{\frac{5}{12}}$

96. $\dfrac{\frac{5}{6}}{\frac{4}{9}}$

*The authors wish to thank Joyce Nemeth of Broward College for her suggestions regarding some of these exercises.

7.3 Complex Fractions

OBJECTIVES

1 Simplify complex fractions by simplifying the numerator and denominator (Method 1).

2 Simplify complex fractions by multiplying by a common denominator (Method 2).

3 Compare the two methods of simplifying complex fractions.

4 Simplify rational expressions with negative exponents.

A **complex fraction** is a quotient having a fraction in the numerator, denominator, or both.

$$\dfrac{1 + \dfrac{1}{x}}{2}, \qquad \dfrac{\dfrac{4}{y}}{6 - \dfrac{3}{y}}, \qquad \text{and} \qquad \dfrac{\dfrac{m^2 - 9}{m + 1}}{\dfrac{m + 3}{m^2 - 1}} \qquad \text{Examples of complex fractions}$$

OBJECTIVE 1 Simplify complex fractions by simplifying the numerator and denominator (Method 1).

> **Simplifying a Complex Fraction (Method 1)**
>
> **Step 1** Simplify the numerator and denominator separately.
>
> **Step 2** Divide by multiplying the numerator by the reciprocal of the denominator.
>
> **Step 3** Simplify the resulting fraction if possible.

Before performing Step 2, be sure that both numerator and denominator are single fractions.

EXAMPLE 1 Simplifying Complex Fractions (Method 1)

Use Method 1 to simplify each complex fraction.

(a) $\dfrac{\dfrac{x + 1}{x}}{\dfrac{x - 1}{2x}}$ Both the numerator and the denominator are already simplified. (Step 1)

$= \dfrac{x + 1}{x} \div \dfrac{x - 1}{2x}$ Write as a division problem.

$= \dfrac{x + 1}{x} \cdot \dfrac{2x}{x - 1}$ Multiply by the reciprocal of $\frac{x - 1}{2x}$. (Step 2)

$= \dfrac{2x(x + 1)}{x(x - 1)}$ Multiply.

$= \dfrac{2(x + 1)}{x - 1}$ Simplify. (Step 3)

(b) $\dfrac{2 + \dfrac{1}{y}}{3 - \dfrac{2}{y}}$ Simplify the numerator and denominator separately. (Step 1)

$= \dfrac{\dfrac{2y}{y} + \dfrac{1}{y}}{\dfrac{3y}{y} - \dfrac{2}{y}}$ Prepare to write the numerator and denominator as single fractions.

NOW TRY
EXERCISE 1
Use Method 1 to simplify
each complex fraction.

(a) $\dfrac{\frac{t+4}{3t}}{\frac{2t+1}{9t}}$ **(b)** $\dfrac{5-\frac{2}{y}}{4+\frac{1}{y}}$

$= \dfrac{\frac{2y+1}{y}}{\frac{3y-2}{y}}$ The numerator is a single fraction. So is the denominator.

$= \dfrac{2y+1}{y} \div \dfrac{3y-2}{y}$ Write as a division problem.

$= \dfrac{2y+1}{y} \cdot \dfrac{y}{3y-2}$ Multiply by the reciprocal of $\frac{3y-2}{y}$. (Step 2)

$= \dfrac{2y+1}{3y-2}$ Multiply and simplify. (Step 3)

NOW TRY

OBJECTIVE 2   Simplify complex fractions by multiplying by a common denominator (Method 2). This method uses the identity property for multiplication.

Simplifying a Complex Fraction (Method 2)

Step 1   Multiply the numerator and denominator of the complex fraction by the least common denominator of the fractions in the numerator and the fractions in the denominator of the complex fraction.

Step 2   Simplify the resulting fraction if possible.

EXAMPLE 2   Simplifying Complex Fractions (Method 2)

Use Method 2 to simplify each complex fraction.

(a) $\dfrac{2+\frac{1}{y}}{3-\frac{2}{y}}$ This is the same fraction as in **Example 1(b)**. Compare the solution methods.

$= \dfrac{2+\frac{1}{y}}{3-\frac{2}{y}} \cdot 1$ Identity property of multiplication

$= \dfrac{\left(2+\frac{1}{y}\right) \cdot y}{\left(3-\frac{2}{y}\right) \cdot y}$ The LCD of all the fractions is y. Multiply the numerator and denominator by y, since $\frac{y}{y}=1$. (Step 1)

$= \dfrac{2 \cdot y + \frac{1}{y} \cdot y}{3 \cdot y - \frac{2}{y} \cdot y}$ Distributive property (Step 2)

$= \dfrac{2y+1}{3y-2}$ Multiply.

NOW TRY ANSWERS
1. (a) $\frac{3(t+4)}{2t+1}$ **(b)** $\frac{5y-2}{4y+1}$

NOW TRY
EXERCISE 2
Use Method 2 to simplify each complex fraction.

(a) $\dfrac{5 - \dfrac{2}{y}}{4 + \dfrac{1}{y}}$ (b) $\dfrac{x + \dfrac{5}{x}}{4x - \dfrac{1}{x-2}}$

(b) $\dfrac{2p + \dfrac{5}{p-1}}{3p - \dfrac{2}{p}}$

$= \dfrac{\left(2p + \dfrac{5}{p-1}\right) \cdot p(p-1)}{\left(3p - \dfrac{2}{p}\right) \cdot p(p-1)}$

Multiply the numerator and denominator by the LCD, $p(p-1)$. (Step 1)

$= \dfrac{2p[p(p-1)] + \dfrac{5}{p-1} \cdot p(p-1)}{3p[p(p-1)] - \dfrac{2}{p} \cdot p(p-1)}$

Distributive property (Step 2)

$= \dfrac{2p[p(p-1)] + 5p}{3p[p(p-1)] - 2(p-1)}$

Multiply.

$= \dfrac{2p^3 - 2p^2 + 5p}{3p^3 - 3p^2 - 2p + 2}$

Multiply again.

This rational expression is in lowest terms.

NOW TRY

OBJECTIVE 3 **Compare the two methods of simplifying complex fractions.** In the next example, we illustrate how to simplify a complex fraction by both methods. Some students prefer one method over the other, while other students feel comfortable with both methods and rely on practice with many examples to determine which method they will use on a particular problem.

EXAMPLE 3 Simplifying Complex Fractions (Both Methods)

Use both Method 1 and Method 2 to simplify each complex fraction.

Method 1	**Method 2**
(a) $\dfrac{\dfrac{2}{x-3}}{\dfrac{5}{x^2-9}}$	(a) $\dfrac{\dfrac{2}{x-3}}{\dfrac{5}{x^2-9}}$
$= \dfrac{\dfrac{2}{x-3}}{\dfrac{5}{(x-3)(x+3)}}$	$= \dfrac{\dfrac{2}{x-3}}{\dfrac{5}{(x-3)(x+3)}}$
$= \dfrac{2}{x-3} \div \dfrac{5}{(x-3)(x+3)}$	$= \dfrac{\dfrac{2}{x-3} \cdot (x-3)(x+3)}{\dfrac{5}{(x-3)(x+3)} \cdot (x-3)(x+3)}$
$= \dfrac{2}{x-3} \cdot \dfrac{(x-3)(x+3)}{5}$	
$= \dfrac{2(x+3)}{5}$	$= \dfrac{2(x+3)}{5}$

NOW TRY ANSWERS

2. (a) $\dfrac{5y-2}{4y+1}$

   (b) $\dfrac{x^3 - 2x^2 + 5x - 10}{4x^3 - 8x^2 - x}$

NOW TRY
EXERCISE 3
Simplify each complex fraction by both methods.

(a) $\dfrac{\dfrac{1}{p-6}}{\dfrac{5}{p^2-36}}$ **(b)** $\dfrac{\dfrac{1}{m^2}-\dfrac{1}{n^2}}{\dfrac{1}{m}+\dfrac{1}{n}}$

(b) $\dfrac{\dfrac{1}{x}+\dfrac{1}{y}}{\dfrac{1}{x^2}-\dfrac{1}{y^2}}$

$= \dfrac{\dfrac{y}{xy}+\dfrac{x}{xy}}{\dfrac{y^2}{x^2y^2}-\dfrac{x^2}{x^2y^2}}$

$= \dfrac{\dfrac{y+x}{xy}}{\dfrac{y^2-x^2}{x^2y^2}}$

$= \dfrac{y+x}{xy} \div \dfrac{y^2-x^2}{x^2y^2}$

$= \dfrac{y+x}{xy} \cdot \dfrac{x^2y^2}{(y-x)(y+x)}$

$= \dfrac{xy}{y-x}$

(b) $\dfrac{\dfrac{1}{x}+\dfrac{1}{y}}{\dfrac{1}{x^2}-\dfrac{1}{y^2}}$

$= \dfrac{\left(\dfrac{1}{x}+\dfrac{1}{y}\right)\cdot x^2y^2}{\left(\dfrac{1}{x^2}-\dfrac{1}{y^2}\right)\cdot x^2y^2}$

$= \dfrac{\left(\dfrac{1}{x}\right)x^2y^2+\left(\dfrac{1}{y}\right)x^2y^2}{\left(\dfrac{1}{x^2}\right)x^2y^2-\left(\dfrac{1}{y^2}\right)x^2y^2}$

$= \dfrac{xy^2+x^2y}{y^2-x^2}$

$= \dfrac{xy(y+x)}{(y+x)(y-x)}$

$= \dfrac{xy}{y-x}$ NOW TRY

OBJECTIVE 4 **Simplify rational expressions with negative exponents.** To simplify, we begin by rewriting the expressions with only positive exponents.

EXAMPLE 4 **Simplifying Rational Expressions with Negative Exponents**

Simplify each expression, using only positive exponents in the answer.

(a) $\dfrac{m^{-1}+p^{-2}}{2m^{-2}-p^{-1}}$ $a^{-n}=\dfrac{1}{a^n}$ **(Section 5.1)**

$= \dfrac{\dfrac{1}{m}+\dfrac{1}{p^2}}{\dfrac{2}{m^2}-\dfrac{1}{p}}$ Write with positive exponents.
$2m^{-2}=2\cdot m^{-2}=\dfrac{2}{1}\cdot\dfrac{1}{m^2}=\dfrac{2}{m^2}$

> The base of $2m^{-2}$ is m, not $2m$: $2m^{-2}=\dfrac{2}{m^2}$.

$= \dfrac{m^2p^2\left(\dfrac{1}{m}+\dfrac{1}{p^2}\right)}{m^2p^2\left(\dfrac{2}{m^2}-\dfrac{1}{p}\right)}$ Simplify by Method 2, multiplying the numerator and denominator by the LCD, m^2p^2.

$= \dfrac{m^2p^2\cdot\dfrac{1}{m}+m^2p^2\cdot\dfrac{1}{p^2}}{m^2p^2\cdot\dfrac{2}{m^2}-m^2p^2\cdot\dfrac{1}{p}}$ Distributive property

$= \dfrac{mp^2+m^2}{2p^2-m^2p}$ Lowest terms

NOW TRY ANSWERS
3. (a) $\dfrac{p+6}{5}$ (b) $\dfrac{n-m}{mn}$

NOW TRY
EXERCISE 4

Simplify the expression, using only positive exponents in the answer.

$$\frac{2y^{-1} - 3y^{-2}}{y^{-2} + 3x^{-1}}$$

(b) $\dfrac{x^{-2} - 2y^{-1}}{y - 2x^2}$

> The 2 does *not* go in the denominator of this fraction.

$$= \frac{\dfrac{1}{x^2} - \dfrac{2}{y}}{y - 2x^2}$$ Write with positive exponents.

$$= \frac{\left(\dfrac{1}{x^2} - \dfrac{2}{y}\right)x^2y}{(y - 2x^2)x^2y}$$ Use Method 2. Multiply by the LCD, x^2y.

$$= \frac{y - 2x^2}{(y - 2x^2)x^2y}$$ Use the distributive property in the numerator.

$$= \frac{1}{x^2y}$$ Lowest terms

> Remember to write 1 in the numerator.

NOW TRY ANSWER

4. $\dfrac{2xy - 3x}{x + 3y^2}$

NOW TRY ↻

7.3 EXERCISES

MyMathLab | Math XL PRACTICE | WATCH | DOWNLOAD | READ | REVIEW

⊙ *Complete solution available on the Video Resources on DVD*

Concept Check *Simplify.*

1. $\dfrac{\dfrac{5}{9} - \dfrac{1}{3}}{\dfrac{2}{3} + \dfrac{1}{6}}$

2. $\dfrac{\dfrac{7}{8} - \dfrac{3}{2}}{-\dfrac{1}{4} - \dfrac{3}{8}}$

3. $\dfrac{2 - \dfrac{1}{4}}{\dfrac{5}{4} + 3}$

4. $\dfrac{\dfrac{4}{3} - 2}{1 - \dfrac{3}{8}}$

Use either method to simplify each complex fraction. ***See Examples 1–3.***

5. $\dfrac{\dfrac{12}{x - 1}}{\dfrac{6}{x}}$

6. $\dfrac{\dfrac{24}{t + 4}}{\dfrac{6}{t}}$

⊙ **7.** $\dfrac{\dfrac{k + 1}{2k}}{\dfrac{3k - 1}{4k}}$

8. $\dfrac{\dfrac{1 - r}{4r}}{\dfrac{1 + r}{8r}}$

9. $\dfrac{\dfrac{4z^2x^4}{9}}{\dfrac{12x^2z^5}{15}}$

10. $\dfrac{\dfrac{3y^2x^3}{8}}{\dfrac{9y^3x^4}{16}}$

⊙ **11.** $\dfrac{6 + \dfrac{1}{x}}{7 - \dfrac{3}{x}}$

12. $\dfrac{4 - \dfrac{1}{p}}{9 + \dfrac{5}{p}}$

13. $\dfrac{\dfrac{3}{x} + \dfrac{3}{y}}{\dfrac{3}{x} - \dfrac{3}{y}}$

14. $\dfrac{\dfrac{4}{t} - \dfrac{4}{s}}{\dfrac{4}{t} + \dfrac{4}{s}}$

15. $\dfrac{\dfrac{8x - 24y}{10}}{\dfrac{x - 3y}{5x}}$

16. $\dfrac{\dfrac{20x - 10y}{12y}}{\dfrac{2x - y}{6y^2}}$

17. $\dfrac{\dfrac{x^2 - 16y^2}{xy}}{\dfrac{1}{y} - \dfrac{4}{x}}$

18. $\dfrac{\dfrac{4t^2 - 9s^2}{st}}{\dfrac{2}{s} - \dfrac{3}{t}}$

⊙ **19.** $\dfrac{\dfrac{6}{y - 4}}{\dfrac{12}{y^2 - 16}}$

20. $\dfrac{\dfrac{8}{t + 7}}{\dfrac{24}{t^2 - 49}}$

21. $\dfrac{\dfrac{1}{b^2} - \dfrac{1}{a^2}}{\dfrac{1}{b} - \dfrac{1}{a}}$

22. $\dfrac{\dfrac{1}{x^2} - \dfrac{1}{y^2}}{\dfrac{1}{x} + \dfrac{1}{y}}$

23. $\dfrac{x + y}{\dfrac{1}{y} + \dfrac{1}{x}}$

24. $\dfrac{s - r}{\dfrac{1}{r} - \dfrac{1}{s}}$

25. $\dfrac{y - \dfrac{y - 3}{3}}{\dfrac{4}{9} + \dfrac{2}{3y}}$

26. $\dfrac{p - \dfrac{p + 2}{4}}{\dfrac{3}{4} - \dfrac{5}{2p}}$

27. $\dfrac{\dfrac{x + 2}{x} + \dfrac{1}{x + 2}}{\dfrac{5}{x} + \dfrac{x}{x + 2}}$

28. $\dfrac{\dfrac{y + 3}{y} - \dfrac{4}{y - 1}}{\dfrac{y}{y - 1} + \dfrac{1}{y}}$

RELATING CONCEPTS   EXERCISES 29–34

FOR INDIVIDUAL OR GROUP WORK

Simplifying a complex fraction by Method 1 is a good way to review the methods of adding, subtracting, multiplying, and dividing rational expressions. Method 2 gives a good review of the fundamental principle of rational expressions. Refer to the following complex fraction and **work Exercises 29–34 in order.**

$$\dfrac{\dfrac{4}{m} + \dfrac{m + 2}{m - 1}}{\dfrac{m + 2}{m} - \dfrac{2}{m - 1}}$$

29. Add the fractions in the numerator.

30. Subtract as indicated in the denominator.

31. Divide your answer from **Exercise 29** by your answer from **Exercise 30.**

32. Go back to the original complex fraction and find the LCD of all denominators.

33. Multiply the numerator and denominator of the complex fraction by your answer from **Exercise 32.**

34. Your answers for **Exercises 31 and 33** should be the same. Which method do you prefer? Explain why.

Simplify each expression, using only positive exponents in your answer. **See Example 4.**

35. $\dfrac{1}{x^{-2} + y^{-2}}$

36. $\dfrac{1}{p^{-2} - q^{-2}}$

37. $\dfrac{x^{-2} + y^{-2}}{x^{-1} + y^{-1}}$

38. $\dfrac{x^{-1} - y^{-1}}{x^{-2} - y^{-2}}$

39. $\dfrac{x^{-1} + 2y^{-1}}{2y + 4x}$

40. $\dfrac{a^{-2} - 4b^{-2}}{3b - 6a}$

41. (a) Start with the complex fraction $\dfrac{\dfrac{3}{mp} - \dfrac{4}{p} + \dfrac{8}{m}}{2m^{-1} - 3p^{-1}}$, and write it so that there are no negative exponents in your expression.

(b) Explain why $\dfrac{\dfrac{3}{mp} - \dfrac{4}{p} + \dfrac{8}{m}}{\dfrac{1}{2m} - \dfrac{1}{3p}}$ would *not* be a correct response in part (a).

(c) Simplify the complex fraction in part (a).

42. Are $\dfrac{m^{-1} + n^{-1}}{m^{-2} + n^{-2}}$ and $\dfrac{m^2 + n^2}{m + n}$ equivalent? Explain why or why not.

Solve each equation. *See Section 2.1.*

43. $\dfrac{1}{2}x + \dfrac{1}{4}x = -9$

44. $\dfrac{x}{3} - \dfrac{x}{8} = -5$

45. $\dfrac{x-6}{5} = \dfrac{x+4}{10}$

For each rational function, find all numbers that are not in the domain. Then give the domain, using set-builder notation. *See Section 7.1.*

46. $f(x) = \dfrac{1}{x^2 - 16}$

47. $f(x) = \dfrac{1}{x}$

48. $f(x) = \dfrac{6}{x^2 + 4}$

7.4 Equations with Rational Expressions and Graphs

OBJECTIVES

1 Determine the domain of the variable in a rational equation.

2 Solve rational equations.

3 Recognize the graph of a rational function.

In **Section 7.1,** we defined the domain of a rational expression as the set of all possible values of the variable. (We also refer to this as "the domain of the variable.") Any value that makes the denominator 0 is excluded.

OBJECTIVE 1   **Determine the domain of the variable in a rational equation.** The **domain of the variable in a rational equation** is the intersection of the domains of the rational expressions in the equation.

EXAMPLE 1   Determining the Domains of the Variables in Rational Equations

Find the domain of the variable in each equation.

(a) $\dfrac{2}{x} - \dfrac{3}{2} = \dfrac{7}{2x}$

The domains of the three rational expressions in the equation are, in order,

$$\{x \mid x \neq 0\}, \quad \{x \mid x \text{ is a real number}\}, \quad \text{and} \quad \{x \mid x \neq 0\}.$$

The intersection of these three domains is all real numbers except 0, written using set-builder notation as $\{x \mid x \neq 0\}$.

(b) $\dfrac{2}{x-3} - \dfrac{3}{x+3} = \dfrac{12}{x^2 - 9}$

The domains of the three expressions are, respectively,

$$\{x \mid x \neq 3\}, \quad \{x \mid x \neq -3\}, \quad \text{and} \quad \{x \mid x \neq \pm 3\}.$$

> ± is read "positive or negative," or "plus or minus."

The domain of the variable is the intersection of the three domains, all real numbers except 3 and −3, written $\{x \mid x \neq \pm 3\}$.

NOW TRY

NOW TRY EXERCISE 1

Find the domain of the variable in each equation.

(a) $\dfrac{1}{3x} - \dfrac{3}{4x} = \dfrac{1}{3}$

(b) $\dfrac{1}{x-7} + \dfrac{2}{x+7} = \dfrac{14}{x^2 - 49}$

NOW TRY ANSWERS

1. (a) $\{x \mid x \neq 0\}$

   (b) $\{x \mid x \neq \pm 7\}$

OBJECTIVE 2   **Solve rational equations.** To solve rational equations, we usually multiply all terms in the equation by the least common denominator to clear the fractions. *We can do this only with equations, not expressions.*

> **Solving an Equation with Rational Expressions**
>
> **Step 1** Determine the domain of the variable.
>
> **Step 2** Multiply each side of the equation by the LCD to clear the fractions.
>
> **Step 3** Solve the resulting equation.
>
> **Step 4** Check that each proposed solution is in the domain, and discard any values that are not. Check the remaining proposed solution(s) in the original equation.

NOW TRY
EXERCISE 2
Solve.

$$\frac{1}{3x} - \frac{3}{4x} = \frac{1}{3}$$

EXAMPLE 2 Solving a Rational Equation

Solve $\dfrac{2}{x} - \dfrac{3}{2} = \dfrac{7}{2x}$.

Step 1 The domain, which excludes 0, was found in **Example 1(a).**

Step 2 $\qquad 2x\left(\dfrac{2}{x} - \dfrac{3}{2}\right) = 2x\left(\dfrac{7}{2x}\right) \qquad$ Multiply by the LCD, $2x$.

Step 3 $\quad 2x\left(\dfrac{2}{x}\right) - 2x\left(\dfrac{3}{2}\right) = 2x\left(\dfrac{7}{2x}\right) \qquad$ Distributive property

$$4 - 3x = 7 \qquad \text{Multiply.}$$

$$-3x = 3 \qquad \text{Subtract 4.}$$

Proposed solution $\rightarrow x = -1 \qquad$ Divide by -3.

Step 4 CHECK $\qquad \dfrac{2}{x} - \dfrac{3}{2} = \dfrac{7}{2x} \qquad$ Original equation

$$\dfrac{2}{-1} - \dfrac{3}{2} \overset{?}{=} \dfrac{7}{2(-1)} \qquad \text{Let } x = -1.$$

$$-\dfrac{7}{2} = -\dfrac{7}{2} \ \checkmark \qquad \text{True}$$

The solution set is $\{-1\}$.

NOW TRY

⚠ **CAUTION** When each side of an equation is multiplied by a *variable* expression, the resulting "solutions" may not satisfy the original equation. *You must either determine and observe the domain or check all proposed solutions in the original equation. It is wise to do both.*

EXAMPLE 3 Solving a Rational Equation with No Solution

Solve $\dfrac{2}{x-3} - \dfrac{3}{x+3} = \dfrac{12}{x^2 - 9}$.

Step 1 From **Example 1(b)**, we know that the domain excludes 3 and -3.

Step 2 Factor $x^2 - 9$. Then multiply each side by the LCD, $(x+3)(x-3)$.

NOW TRY ANSWER
2. $\left\{-\frac{5}{4}\right\}$

$$(x+3)(x-3)\left(\dfrac{2}{x-3} - \dfrac{3}{x+3}\right) = (x+3)(x-3)\left[\dfrac{12}{(x+3)(x-3)}\right]$$

**NOW TRY
EXERCISE 3**
Solve.

$$\frac{1}{t-7} + \frac{2}{t+7} = \frac{14}{t^2 - 49}$$

Step 3 $(x+3)(x-3)\left(\dfrac{2}{x-3}\right) - (x+3)(x-3)\left(\dfrac{3}{x+3}\right)$

$$= (x+3)(x-3)\left[\frac{12}{(x+3)(x-3)}\right]$$

Distributive property

$2(x+3) - 3(x-3) = 12$ Multiply.

$2x + 6 - 3x + 9 = 12$ Distributive property

$-x + 15 = 12$ Combine like terms.

$-x = -3$ Subtract 15.

Proposed solution $\rightarrow x = 3$ Divide by -1.

Step 4 Since the proposed solution, 3, is not in the domain, it cannot be a solution of the equation. Substituting 3 into the original equation shows why.

CHECK $\dfrac{2}{x-3} - \dfrac{3}{x+3} = \dfrac{12}{x^2 - 9}$ Original equation

$\dfrac{2}{3-3} - \dfrac{3}{3+3} \overset{?}{=} \dfrac{12}{3^2 - 9}$ Let $x = 3$.

$\dfrac{2}{0} - \dfrac{3}{6} \overset{?}{=} \dfrac{12}{0}$ Division by 0 is undefined.

The equation has no solution. The solution set is $\emptyset$.   NOW TRY

**NOW TRY
EXERCISE 4**
Solve.

$$\frac{2}{t^2 - 2t - 8} - \frac{4}{t^2 + 6t + 8} = \frac{2}{t^2 - 16}$$

EXAMPLE 4   Solving a Rational Equation

Solve $\dfrac{3}{p^2 + p - 2} - \dfrac{1}{p^2 - 1} = \dfrac{7}{2(p^2 + 3p + 2)}$.

Factor each denominator to find the domain and the LCD.

$$\frac{3}{(p-1)(p+2)} - \frac{1}{(p+1)(p-1)}$$

$$= \frac{7}{2(p+2)(p+1)}$$ Factor the denominators.

The domain excludes 1, -2, and -1. Multiply each side of the equation by the LCD, $2(p-1)(p+2)(p+1)$.

$$2(p-1)(p+2)(p+1)\left[\frac{3}{(p-1)(p+2)} - \frac{1}{(p+1)(p-1)}\right]$$

$$= 2(p-1)(p+2)(p+1)\left[\frac{7}{2(p+2)(p+1)}\right]$$

$2 \cdot 3(p+1) - 2(p+2) = 7(p-1)$ Distributive property

$6p + 6 - 2p - 4 = 7p - 7$ $2 \cdot 3 = 6$; Distributive property

$4p + 2 = 7p - 7$ Combine like terms.

$9 = 3p$ Subtract $4p$. Add 7.

Proposed solution $\rightarrow 3 = p$ Divide by 3.

NOW TRY ANSWERS
3. $\emptyset$ **4.** $\{5\}$

Note that 3 is in the domain. Substitute 3 for p in the original equation to check that the solution set is $\{3\}$.   NOW TRY

NOW TRY
EXERCISE 5
Solve.

$$\frac{2x}{x - 2} = \frac{-3}{x} + \frac{4}{x - 2}$$

EXAMPLE 5 Solving a Rational Equation

Solve $\dfrac{2}{3x + 1} = \dfrac{1}{x} - \dfrac{6x}{3x + 1}$.

> We must exclude $-\frac{1}{3}$ and 0 from the domain.

$$x(3x + 1)\left(\frac{2}{3x + 1}\right) = x(3x + 1)\left(\frac{1}{x} - \frac{6x}{3x + 1}\right) \qquad \text{Multiply by the LCD, } x(3x + 1).$$

$$x(3x + 1)\left(\frac{2}{3x + 1}\right) = x(3x + 1)\left(\frac{1}{x}\right) - x(3x + 1)\left(\frac{6x}{3x + 1}\right)$$

Distributive property

$$2x = 3x + 1 - 6x^2 \qquad \text{Multiply.}$$

$$6x^2 - x - 1 = 0 \qquad \text{Standard form}$$

$$(3x + 1)(2x - 1) = 0 \qquad \text{Factor.}$$

$$3x + 1 = 0 \quad \text{or} \quad 2x - 1 = 0 \qquad \text{Zero-factor property}$$

$$x = -\frac{1}{3} \quad \text{or} \quad x = \frac{1}{2} \qquad \text{Proposed solutions}$$

Because $-\frac{1}{3}$ is not in the domain, it is not a solution. Check that the solution set is $\left\{\frac{1}{2}\right\}$.

NOW TRY

OBJECTIVE 3 Recognize the graph of a rational function. A function defined by a quotient of polynomials is a **rational function**. Because one or more values of x may be excluded from the domain of a rational function, their graphs are often **discontinuous**. That is, there will be one or more breaks in the graph.

One simple rational function, defined by $f(x) = \frac{1}{x}$ and graphed in **FIGURE 2**, is the **reciprocal function**. The domain of this function includes all real numbers except 0. Thus, this function pairs every real number except 0 with its reciprocal.

| The closer negative values of x are to 0, the smaller ("more negative") y is. | | | | | | The closer positive values of x are to 0, the larger y is. | | | | | |

x	-3	-2	-1	-0.5	-0.25	-0.1	0.1	0.25	0.5	1	2	3
y	$-\frac{1}{3}$	$-\frac{1}{2}$	-1	-2	-4	-10	10	4	2	1	$\frac{1}{2}$	$\frac{1}{3}$

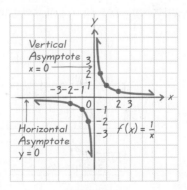

Reciprocal function

$$f(x) = \frac{1}{x}$$

Domain: $\{x \mid x \neq 0\}$

Range: $\{x \mid x \neq 0\}$

FIGURE 2

Since the domain of this function includes all real numbers except 0, there is no point on the graph with $x = 0$. The vertical line with equation $x = 0$ is called a **vertical asymptote** of the graph. Also, the horizontal line with equation $y = 0$ is called a **horizontal asymptote**.

NOW TRY ANSWER
5. $\left\{-\frac{3}{2}\right\}$

NOTE In general, if the y-values of a rational function approach ∞ or $-\infty$ as the x-values approach a real number a, the vertical line $x = a$ is a vertical asymptote of the graph. Also, if the x-values approach a real number b as $|x|$ increases without bound, the horizontal line $y = b$ is a horizontal asymptote of the graph.

NOW TRY
EXERCISE 6

Graph, and give the equations of the vertical and horizontal asymptotes.

$$f(x) = \frac{1}{x + 1}$$

EXAMPLE 6 Graphing a Rational Function

Graph, and give the equations of the vertical and horizontal asymptotes.

$$g(x) = \frac{-2}{x - 3}$$

Some ordered pairs that belong to the function are listed in the table.

x	-2	-1	0	1	2	2.5	2.75	3.25	3.5	4	5	6	7
y	$\frac{2}{5}$	$\frac{1}{2}$	$\frac{2}{3}$	1	2	4	8	-8	-4	-2	-1	$-\frac{2}{3}$	$-\frac{1}{2}$

There is no point on the graph, shown in **FIGURE 3**, for $x = 3$, because 3 is excluded from the domain of the rational function. The dashed line $x = 3$ represents the vertical asymptote and is not part of the graph. The graph gets closer to the vertical asymptote as the x-values get closer to 3. Again, $y = 0$ is a horizontal asymptote.

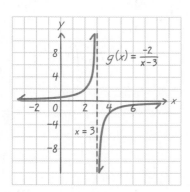

FIGURE 3   NOW TRY

CONNECTIONS

We can solve rational equations with a graphing calculator by finding the x-intercepts of the graph of the corresponding rational function.

FIGURE 4 shows two views of the graph of the following rational function.

$$f(x) = \frac{1}{x^2} + \frac{1}{x} - \frac{3}{2}$$

The x-intercepts (or zeros) of the graph determine the solution set of $f(x) = 0$.

$$\frac{1}{x^2} + \frac{1}{x} - \frac{3}{2} = 0$$

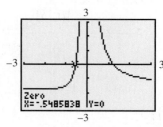 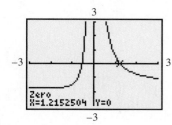

FIGURE 4

NOW TRY ANSWER
6. vertical asymptote: $x = -1$; horizontal asymptote: $y = 0$

The bottom of each calculator screen gives approximations of the x-values for which $y = 0$, here -0.5485838 and 1.2152504. These are the solutions of the equation. Thus, the solution set is $\{-0.5485838, 1.2152504\}$.

For Discussion or Writing

Use each graph to determine the solution set of the equation $f(x) = 0$.

1. $f(x) = \dfrac{1}{x^3} + \dfrac{1}{x} + 2$

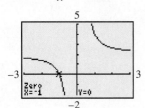

2. $f(x) = \dfrac{2}{x} - \dfrac{1}{x^2}$

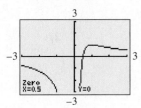

7.4 EXERCISES   *MyMathLab* Math XL PRACTICE   WATCH   DOWNLOAD   READ   REVIEW

🌐 *Complete solution available on the Video Resources on DVD*

As explained in this section, any values that would cause a denominator to equal 0 must be excluded from the domain and, consequently, as solutions of an equation that has variable expressions in the denominators.

(a) *Without actually solving each equation, list all possible values that would have to be rejected if they appeared as proposed solutions.*

(b) *Then give the domain, using set-builder notation.*

See Example 1.

🌐 **1.** $\dfrac{1}{3x} + \dfrac{1}{2x} = \dfrac{x}{3}$

2. $\dfrac{5}{6x} - \dfrac{8}{2x} = \dfrac{x}{4}$

3. $\dfrac{1}{x+1} - \dfrac{1}{x-2} = 0$

4. $\dfrac{3}{x+4} - \dfrac{2}{x-9} = 0$

5. $\dfrac{1}{x^2-16} - \dfrac{2}{x-4} = \dfrac{1}{x+4}$

6. $\dfrac{2}{x^2-25} - \dfrac{1}{x+5} = \dfrac{1}{x-5}$

7. $\dfrac{2}{x^2-x} + \dfrac{1}{x+3} = \dfrac{4}{x-2}$

8. $\dfrac{3}{x^2+x} - \dfrac{1}{x+5} = \dfrac{2}{x-7}$

9. $\dfrac{6}{4x+7} - \dfrac{3}{x} = \dfrac{5}{6x-13}$

10. $\dfrac{4}{3x-5} + \dfrac{2}{x} = \dfrac{9}{4x+13}$

11. $\dfrac{3x+1}{x-4} = \dfrac{6x+5}{2x-7}$

12. $\dfrac{4x-1}{2x+3} = \dfrac{12x-25}{6x-2}$

✏ **13.** Suppose that in solving the equation

$$\frac{x+7}{4} - \frac{x+3}{3} = \frac{x}{12},$$

all of your algebraic steps are correct. Is there a possibility that your proposed solution will have to be rejected? Explain.

14. Consider the equation in **Exercise 13.**

(a) Solve it.   (b) Check your solution, showing all steps.

Solve each equation. ***See Examples 2–5.***

15. $\dfrac{3}{4x} = \dfrac{5}{2x} - \dfrac{7}{4}$

16. $\dfrac{2}{3x} = \dfrac{6}{5x} + \dfrac{8}{45}$

17. $x - \dfrac{24}{x} = -2$

18. $p + \dfrac{15}{p} = -8$

19. $\dfrac{x}{4} - \dfrac{21}{4x} = -1$

20. $\dfrac{x}{2} - \dfrac{12}{x} = 1$

21. $\dfrac{x - 4}{x + 6} = \dfrac{2x + 3}{2x - 1}$

22. $\dfrac{5x - 8}{x + 2} = \dfrac{5x - 1}{x + 3}$

23. $\dfrac{3x + 1}{x - 4} = \dfrac{6x + 5}{2x - 7}$

24. $\dfrac{4x - 1}{2x + 3} = \dfrac{12x - 25}{6x - 2}$

25. $\dfrac{1}{y - 1} + \dfrac{5}{12} = \dfrac{-2}{3y - 3}$

26. $\dfrac{4}{m + 2} - \dfrac{11}{9} = \dfrac{1}{3m + 6}$

27. $\dfrac{7}{6x + 3} - \dfrac{1}{3} = \dfrac{2}{2x + 1}$

28. $\dfrac{3}{4m + 2} = \dfrac{17}{2} - \dfrac{7}{2m + 1}$

29. $\dfrac{6}{x - 4} + \dfrac{5}{x} = \dfrac{-20}{x^2 - 4x}$

30. $\dfrac{7}{x - 4} + \dfrac{3}{x} = \dfrac{-12}{x^2 - 4x}$

31. $\dfrac{3}{x + 2} - \dfrac{2}{x^2 - 4} = \dfrac{1}{x - 2}$

32. $\dfrac{3}{x - 2} + \dfrac{21}{x^2 - 4} = \dfrac{14}{x + 2}$

33. $\dfrac{1}{y + 2} + \dfrac{3}{y + 7} = \dfrac{5}{y^2 + 9y + 14}$

34. $\dfrac{1}{t + 3} + \dfrac{4}{t + 5} = \dfrac{2}{t^2 + 8t + 15}$

35. $\dfrac{9}{x} + \dfrac{4}{6x - 3} = \dfrac{2}{6x - 3}$

36. $\dfrac{5}{n} + \dfrac{4}{6 - 3n} = \dfrac{2n}{6 - 3n}$

37. $\dfrac{1}{x - 2} + \dfrac{1}{4} = \dfrac{1}{4(x^2 - 4)}$

38. $\dfrac{1}{x + 4} + \dfrac{1}{3} = \dfrac{-10}{3(x^2 - 16)}$

39. $\dfrac{6}{w + 3} + \dfrac{-7}{w - 5} = \dfrac{-48}{w^2 - 2w - 15}$

40. $\dfrac{2}{r - 5} + \dfrac{3}{2r + 1} = \dfrac{22}{2r^2 - 9r - 5}$

41. $\dfrac{x}{x - 3} + \dfrac{4}{x + 3} = \dfrac{18}{x^2 - 9}$

42. $\dfrac{2x}{x - 3} + \dfrac{4}{x + 3} = \dfrac{-24}{x^2 - 9}$

43. $\dfrac{1}{x + 4} + \dfrac{x}{x - 4} = \dfrac{-8}{x^2 - 16}$

44. $\dfrac{5}{x - 4} - \dfrac{3}{x - 1} = \dfrac{x^2 - 1}{x^2 - 5x + 4}$

45. $\dfrac{2}{k^2 + k - 6} + \dfrac{1}{k^2 - k - 2} = \dfrac{4}{k^2 + 4k + 3}$

46. $\dfrac{5}{p^2 + 3p + 2} - \dfrac{3}{p^2 - 4} = \dfrac{1}{p^2 - p - 2}$

47. $\dfrac{5x + 14}{x^2 - 9} = \dfrac{-2x^2 - 5x + 2}{x^2 - 9} + \dfrac{2x + 4}{x - 3}$

48. $\dfrac{4x - 7}{4x^2 - 9} = \dfrac{-2x^2 + 5x - 4}{4x^2 - 9} + \dfrac{x + 1}{2x + 3}$

Graph each rational function. Give the equations of the vertical and horizontal asymptotes. ***See Example 6.***

49. $f(x) = \dfrac{2}{x}$

50. $f(x) = \dfrac{3}{x}$

51. $g(x) = -\dfrac{1}{x}$

52. $g(x) = -\dfrac{2}{x}$

53. $f(x) = \dfrac{1}{x - 2}$

54. $f(x) = \dfrac{1}{x + 2}$

Solve each problem.

55. The average number of vehicles waiting in line to enter a parking area is modeled by the function defined by

$$w(x) = \frac{x^2}{2(1-x)},$$

where x is a quantity between 0 and 1 known as the **traffic intensity.** (*Source:* Mannering, F., and W. Kilareski, *Principles of Highway Engineering and Traffic Control,* John Wiley and Sons.) For each traffic intensity, find the average number of vehicles waiting (to the nearest tenth).

(a) 0.1 **(b)** 0.8 **(c)** 0.9

(d) What happens to waiting time as traffic intensity increases?

56. The percent of deaths caused by smoking is modeled by the rational function defined by

$$p(x) = \frac{x-1}{x},$$

where x is the number of times a smoker is more likely to die of lung cancer than a nonsmoker is. This is called the **incidence rate.** (*Source:* Walker, A., *Observation and Inference: An Introduction to the Methods of Epidemiology,* Epidemiology Resources Inc.) For example, $x = 10$ means that a smoker is 10 times more likely than a nonsmoker to die of lung cancer.

(a) Find $p(x)$ if x is 10.

(b) For what values of x is $p(x) = 80\%$? (*Hint:* Change 80% to a decimal.)

(c) Can the incidence rate equal 0? Explain.

57. The force required to keep a 2000-lb car going 30 mph from skidding on a curve, where r is the radius of the curve in feet, is given by

$$F(r) = \frac{225{,}000}{r}.$$

(a) What radius must a curve have if a force of 450 lb is needed to keep the car from skidding?

(b) As the radius of the curve is lengthened, how is the force affected?

58. The amount of heating oil produced (in gallons per day) by an oil refinery is modeled by the rational function defined by

$$f(x) = \frac{125{,}000 - 25x}{125 + 2x},$$

where x is the amount of gasoline produced (in hundreds of gallons per day). Suppose the refinery must produce 300 gal of heating oil per day to meet the needs of its customers.

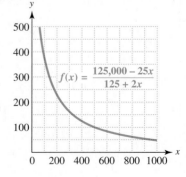

(a) How much gasoline will be produced per day?

(b) The graph of f is shown in the figure. Use it to decide what happens to the amount of gasoline (x) produced as the amount of heating oil (y) produced increases.

Two views of the graph of $f(x) = \frac{7}{x^4} - \frac{8}{x^2} + 1$ are shown. Use the graphs to respond to Exercises 59 and 60. **See the Connections box.**

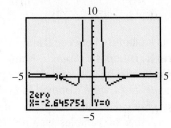
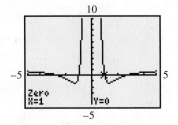

59. How many solutions does the equation $f(x) = 0$ have?

60. Use the display to give the solution set of $f(x) = 0$.

PREVIEW EXERCISES

Solve each formula for the specified variable. **See Section 2.2.**

61. $d = rt$ for t

62. $I = prt$ for r

63. $P = a + b + c$ for c

64. $\mathcal{A} = \frac{1}{2}h(b + B)$ for B

SUMMARY EXERCISES on Rational Expressions and Equations

A common student error is to confuse an equation, such as $\frac{x}{2} + \frac{x}{3} = -5$, with an expression involving an operation, such as $\frac{x}{2} + \frac{x}{3}$. **Equations are solved for a numerical answer, while problems involving operations result in simplified expressions.**

Solving an Equation	Simplifying an Expression Involving an Operation
Solve: $\frac{x}{2} + \frac{x}{3} = -5$ Look for the equals symbol.	**Add:** $\frac{x}{2} + \frac{x}{3}$
Multiply each side by the LCD, 6.	Write both fractions with the LCD, 6.
$6\left(\dfrac{x}{2} + \dfrac{x}{3}\right) = 6(-5)$	$\dfrac{x}{2} + \dfrac{x}{3}$
$6\left(\dfrac{x}{2}\right) + 6\left(\dfrac{x}{3}\right) = 6(-5)$	$= \dfrac{x \cdot 3}{2 \cdot 3} + \dfrac{x \cdot 2}{3 \cdot 2}$
$3x + 2x = -30$	$= \dfrac{3x}{6} + \dfrac{2x}{6}$
$5x = -30$	$= \dfrac{3x + 2x}{6}$
$x = -6$	$= \dfrac{5x}{6}$
Check that the solution set is $\{-6\}$.	

Identify each exercise as an equation *or an* expression. *Then simplify the expression by performing the indicated operation, or solve the equation, as appropriate.*

1. $\dfrac{x}{2} - \dfrac{x}{4} = 5$

2. $\dfrac{4x - 20}{x^2 - 25} \cdot \dfrac{(x + 5)^2}{10}$

3. $\dfrac{6}{7x} - \dfrac{4}{x}$

4. $\dfrac{\dfrac{1}{x} + \dfrac{1}{y}}{\dfrac{1}{x} - \dfrac{1}{y}}$

5. $\dfrac{5}{7t} = \dfrac{52}{7} - \dfrac{3}{t}$

6. $\dfrac{x - 5}{3} + \dfrac{1}{3} = \dfrac{x - 2}{5}$

7. $\dfrac{7}{6x} + \dfrac{5}{8x}$

8. $\dfrac{4}{x} - \dfrac{8}{x + 1} = 0$

9. $\dfrac{\dfrac{6}{x + 1} - \dfrac{1}{x}}{\dfrac{2}{x} - \dfrac{4}{x + 1}}$

10. $\dfrac{8}{r + 2} - \dfrac{7}{4r + 8}$

11. $\dfrac{x}{x + y} + \dfrac{2y}{x - y}$

12. $\dfrac{3p^2 - 6p}{p + 5} \div \dfrac{p^2 - 4}{8p + 40}$

13. $\dfrac{x - 2}{9} \cdot \dfrac{5}{8 - 4x}$

14. $\dfrac{a - 4}{3} + \dfrac{11}{6} = \dfrac{a + 1}{2}$

15. $\dfrac{b^2 + b - 6}{b^2 + 2b - 8} \cdot \dfrac{b^2 + 8b + 16}{3b + 12}$

16. $\dfrac{10z^2 - 5z}{3z^3 - 6z^2} \div \dfrac{2z^2 + 5z - 3}{z^2 + z - 6}$

17. $\dfrac{5}{x^2 - 2x} - \dfrac{3}{x^2 - 4}$

18. $\dfrac{6}{t + 1} + \dfrac{4}{5t + 5} = \dfrac{34}{15}$

19. $\dfrac{\dfrac{5}{x} - \dfrac{3}{y}}{\dfrac{9x^2 - 25y^2}{x^2 y}}$

20. $\dfrac{-2}{a^2 + 2a - 3} - \dfrac{5}{3 - 3a} = \dfrac{4}{3a + 9}$

21. $\dfrac{4y^2 - 13y + 3}{2y^2 - 9y + 9} \div \dfrac{4y^2 + 11y - 3}{6y^2 - 5y - 6}$

22. $\dfrac{8}{3k + 9} - \dfrac{8}{15} = \dfrac{2}{5k + 15}$

23. $\dfrac{3r}{r - 2} = 1 + \dfrac{6}{r - 2}$

24. $\dfrac{6z^2 - 5z - 6}{6z^2 + 5z - 6} \cdot \dfrac{12z^2 - 17z + 6}{12z^2 - z - 6}$

25. $\dfrac{-1}{3 - x} - \dfrac{2}{x - 3}$

26. $\dfrac{\dfrac{t}{4} - \dfrac{1}{t}}{1 + \dfrac{t + 4}{t}}$

27. $\dfrac{2}{y + 1} - \dfrac{3}{y^2 - y - 2} = \dfrac{3}{y - 2}$

28. $\dfrac{7}{2x^2 - 8x} + \dfrac{3}{x^2 - 16}$

29. $\dfrac{3}{y - 3} - \dfrac{3}{y^2 - 5y + 6} = \dfrac{2}{y - 2}$

30. $\dfrac{2k + \dfrac{5}{k - 1}}{3k - \dfrac{2}{k}}$

7.5 Applications of Rational Expressions

OBJECTIVES

1. Find the value of an unknown variable in a formula.
2. Solve a formula for a specified variable.
3. Solve applications by using proportions.
4. Solve applications about distance, rate, and time.
5. Solve applications about work rates.

OBJECTIVE 1 **Find the value of an unknown variable in a formula.** Formulas may contain rational expressions, as does $t = \frac{d}{r}$ and $\frac{1}{f} = \frac{1}{p} + \frac{1}{q}$.

EXAMPLE 1 Finding the Value of a Variable in a Formula

In physics, the focal length f of a lens is given by the formula

$$\frac{1}{f} = \frac{1}{p} + \frac{1}{q}.$$

In the formula, p is the distance from the object to the lens and q is the distance from the lens to the image. See **FIGURE 5**. Find q if $p = 20$ cm and $f = 10$ cm.

Focal Length of Camera Lens

FIGURE 5

NOW TRY EXERCISE 1

Use the formula in **Example 1** to find f if $p = 50$ cm and $q = 10$ cm.

$$\frac{1}{f} = \frac{1}{p} + \frac{1}{q}$$

$$\frac{1}{10} = \frac{1}{20} + \frac{1}{q} \qquad \text{Solve this equation for } q. \qquad \text{Let } f = 10, \, p = 20.$$

$$20q \cdot \frac{1}{10} = 20q\left(\frac{1}{20} + \frac{1}{q}\right) \qquad \text{Multiply by the LCD, } 20q.$$

$$20q \cdot \frac{1}{10} = 20q\left(\frac{1}{20}\right) + 20q\left(\frac{1}{q}\right) \qquad \text{Distributive property}$$

$$2q = q + 20 \qquad \text{Multiply.}$$

$$q = 20 \qquad \text{Subtract } q.$$

The distance from the lens to the image is 20 cm. **NOW TRY**

OBJECTIVE 2 **Solve a formula for a specified variable.** *Recall that the goal in solving for a specified variable is to isolate it on one side of the equals symbol.*

NOW TRY EXERCISE 2

Solve for m.

$$\frac{1}{m} - \frac{2}{n} = 5$$

EXAMPLE 2 Solving a Formula for a Specified Variable

Solve $\frac{1}{f} = \frac{1}{p} + \frac{1}{q}$ for p.

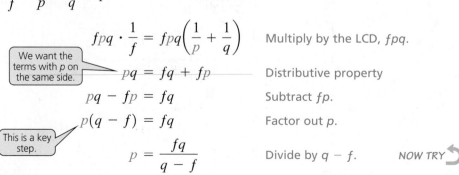

$$fpq \cdot \frac{1}{f} = fpq\left(\frac{1}{p} + \frac{1}{q}\right) \qquad \text{Multiply by the LCD, } fpq.$$

$$pq = fq + fp \qquad \text{We want the terms with } p \text{ on the same side.} \qquad \text{Distributive property}$$

$$pq - fp = fq \qquad \text{Subtract } fp.$$

$$p(q - f) = fq \qquad \text{This is a key step.} \qquad \text{Factor out } p.$$

$$p = \frac{fq}{q - f} \qquad \text{Divide by } q - f. \qquad \text{NOW TRY}$$

NOW TRY ANSWERS

1. $\frac{25}{3}$ cm   2. $m = \frac{n}{5n + 2}$

NOW TRY
EXERCISE 3
Solve for N.

$$\frac{NR}{N - n} = t$$

EXAMPLE 3 Solving a Formula for a Specified Variable

Solve $I = \dfrac{nE}{R + nr}$ for n.

$$(R + nr)I = (R + nr)\frac{nE}{R + nr} \qquad \text{Multiply by } R + nr.$$

$$RI + nrI = nE \qquad \text{Distributive property on the left}$$

$$RI = nE - nrI \qquad \text{Subtract } nrI.$$

$$RI = n(E - rI) \qquad \text{Factor out } n.$$

$$\frac{RI}{E - rI} = n, \quad \text{or} \quad n = \frac{RI}{E - rI} \qquad \text{Divide by } E - rI. \qquad \text{NOW TRY}$$

⚠ **CAUTION** Refer to the steps in **Examples 2 and 3** that factor out the desired variable. This variable *must* be a factor on only one side of the equation, so that each side can be divided by the remaining factor in the last step.

OBJECTIVE 3 **Solve applications by using proportions.** A **ratio** is a comparison of two quantities. The ratio of a to b may be written in any of the following ways.

$$a \text{ to } b, \quad a:b, \quad \text{or} \quad \frac{a}{b} \qquad \text{Ratio of } a \text{ to } b$$

Ratios are usually written as quotients in algebra. A **proportion** is a statement that two ratios are equal.

$$\frac{a}{b} = \frac{c}{d} \qquad \text{Proportion}$$

NOW TRY
EXERCISE 4
In 2008, about 13 of every 100 Americans lived in poverty. The population at that time was about 302 million. How many Americans lived in poverty in 2008? (*Source:* U.S. Census Bureau.)

EXAMPLE 4 Solving a Proportion

In 2008, about 15 of every 100 Americans had no health insurance. The population at that time was about 302 million. How many million Americans had no health insurance? (*Source:* U.S. Census Bureau.)

Step 1 **Read** the problem.

Step 2 **Assign a variable.** Let $x =$ the number of Americans (in millions) who had no health insurance.

Step 3 **Write an equation.** The ratio 15 to 100 should equal the ratio x to 302.

$$\frac{15}{100} = \frac{x}{302} \qquad \text{Write a proportion.}$$

Step 4 **Solve.** $30{,}200\left(\dfrac{15}{100}\right) = 30{,}200\left(\dfrac{x}{302}\right)$ Multiply by a common denominator, $100 \cdot 302 = 30{,}200$.

$$4530 = 100x \qquad \text{Simplify.}$$

$$x = 45.3 \qquad \text{Divide by 100.}$$

Step 5 **State the answer.** There were 45.3 million Americans with no health insurance in 2008.

NOW TRY ANSWERS
3. $N = \dfrac{nt}{t - R}$
4. 39.26 million

Step 6 **Check** that the ratio of 45.3 million to 302 million equals $\frac{15}{100}$. NOW TRY

NOW TRY
EXERCISE 5

Clayton's SUV uses 28 gal of gasoline to drive 500 mi. He has 10 gal of gasoline in the SUV, and wants to know how much more gasoline he will need to drive 400 mi. If we assume the car continues to use gasoline at the same rate, how many more gallons will he need?

EXAMPLE 5 Solving a Proportion Involving Rates

Jodie Fry's car uses 10 gal of gasoline to travel 210 mi. She has 5 gal of gasoline in the car, and she wants to know how much more gasoline she will need to drive 640 mi. If we assume the car continues to use gasoline at the same rate, how many more gallons will she need?

Step 1 **Read** the problem.

Step 2 **Assign a variable.** Let $x =$ the additional number of gallons of gas needed.

Step 3 **Write an equation.** To get an equation, set up a proportion.

$$\text{gallons} \rightarrow \frac{10}{210} = \frac{5 + x}{640} \leftarrow \text{gallons} \atop \leftarrow \text{miles}$$

Step 4 **Solve.** We could multiply both sides by the LCD $10 \cdot 21 \cdot 64$. Instead we use an alternative method that involves **cross products:**

For $\frac{a}{b} = \frac{c}{d}$ to be true, the cross products ad and bc must be equal.

$$a \cdot d = b \cdot c$$

$$10 \cdot 640 = 210(5 + x) \qquad \text{If } \tfrac{a}{b} = \tfrac{c}{d}, \text{ then } ad = bc.$$

$$6400 = 1050 + 210x \qquad \text{Multiply; distributive property}$$

$$5350 = 210x \qquad \text{Subtract 1050.}$$

$$25.5 \approx x \qquad \text{Divide by 210. Round to the nearest tenth.}$$

Step 5 **State the answer.** Jodie will need about 25.5 more gallons of gas.

Step 6 **Check.** The 25.5 gal plus the 5 gal equals 30.5 gal.

$$\frac{30.5}{640} \approx 0.048 \quad \text{and} \quad \frac{10}{210} \approx 0.048$$

Since the rates are equal, the solution is correct. NOW TRY

OBJECTIVE 4 **Solve applications about distance, rate, and time.** We introduced the distance formula $d = rt$ in **Section 2.2.** Using this formula,

Rate is the ratio of distance to time, or $r = \dfrac{d}{t}.$

Time is the ratio of distance to rate, or $t = \dfrac{d}{r}.$

EXAMPLE 6 Solving a Problem about Distance, Rate, and Time

A paddle wheeler goes 10 mi against the current in a river in the same time that it goes 15 mi with the current. If the rate of the current is 3 mph, find the rate of the boat in still water.

Step 1 **Read** the problem.

Step 2 **Assign a variable.** Let $x =$ the rate of the boat in still water.

Traveling ***against*** the current slows the boat down, so the rate of the boat is the ***difference*** between its rate in still water and the rate of the current—that is, $(x - 3)$ mph.

NOW TRY ANSWER
5. 12.4 more gallons

NOW TRY
EXERCISE 6

A small fishing boat goes 36 mi against the current in a river in the same time that it goes 44 mi with the current. If the rate of the boat in still water is 20 mph, find the rate of the current.

Traveling **with** the current speeds the boat up, so the rate of the boat is the **sum** of its rate in still water and the rate of the current—that is, $(x + 3)$ mph.

Thus, $x - 3 =$ the rate of the boat *against* the current,

and $x + 3 =$ the rate of the boat *with* the current.

Because the time is the same going against the current as with the current, find time in terms of distance and rate for each situation. Against the current, the distance is 10 mi and the rate is $(x - 3)$ mph.

$$t = \frac{d}{r} = \frac{10}{x - 3} \qquad \text{Time } \textit{against} \text{ the current}$$

With the current, the distance is 15 mi and the rate is $(x + 3)$ mph.

$$t = \frac{d}{r} = \frac{15}{x + 3} \qquad \text{Time } \textit{with} \text{ the current}$$

	Distance	Rate	Time
Against Current	10	$x - 3$	$\dfrac{10}{x - 3}$
With Current	15	$x + 3$	$\dfrac{15}{x + 3}$

We summarize the information in a table.

Times are equal.

Step 3 **Write an equation,** using the fact that the times are equal.

$$\frac{10}{x - 3} = \frac{15}{x + 3}$$

Step 4 **Solve.**

$$(x + 3)(x - 3)\left(\frac{10}{x - 3}\right) = (x + 3)(x - 3)\left(\frac{15}{x + 3}\right) \qquad \begin{array}{l}\text{Multiply by the LCD,}\\ (x + 3)(x - 3).\end{array}$$

$$10(x + 3) = 15(x - 3)$$

$$10x + 30 = 15x - 45 \qquad \text{Distributive property}$$

$$75 = 5x \qquad \text{Subtract } 10x. \text{ Add } 45.$$

$$15 = x \qquad \text{Divide by 5.}$$

Step 5 **State the answer.** The rate of the boat in still water is 15 mph.

Step 6 **Check** the answer: $\frac{10}{15 - 3} = \frac{15}{15 + 3}$ is true.   NOW TRY

EXAMPLE 7   Solving a Problem about Distance, Rate, and Time

At O'Hare International Airport in Chicago, Cheryl and Bill are walking to the gate at the same rate to catch their flight to Akron, Ohio. Since Bill wants a window seat, he steps onto the moving sidewalk and continues to walk while Cheryl uses the stationary sidewalk. If the sidewalk moves at 1 m per sec and Bill saves 50 sec covering the 300-m distance, what is their walking rate?

Step 1 **Read** the problem. We must find their walking rate.

Step 2 **Assign a variable.** Let x represent their walking rate in meters per second. Thus, Cheryl travels at a rate of x meters per second and Bill travels at a rate of $(x + 1)$ meters per second.

NOW TRY ANSWER
6. 2 mph

NOW TRY
EXERCISE 7

James and Pat are driving from Atlanta to Jacksonville, a distance of 310 mi. James, whose average rate is 5 mph faster than Pat's, will drive the first 130 mi of the trip and then Pat will drive the rest of the way to their destination. If the total driving time is 5 hr, determine the average rate of each driver.

Express their times in terms of the known distances and the variable rates, as in **Example 6.** Cheryl travels 300 m at a rate of x meters per second.

$$t = \frac{d}{r} = \frac{300}{x} \qquad \text{Cheryl's time}$$

Bill travels 300 m at a rate of $(x + 1)$ meters per second.

$$t = \frac{d}{r} = \frac{300}{x + 1} \qquad \text{Bill's time}$$

	Distance	Rate	Time
Cheryl	300	x	$\dfrac{300}{x}$
Bill	300	$x+1$	$\dfrac{300}{x+1}$

← Bill's time is 50 sec less than Cheryl's time.

Step 3   **Write an equation,** using the times from the table.

$$\underset{\substack{\text{Bill's} \\ \text{time}}}{\underbrace{\frac{300}{x+1}}} \;\; \underset{\text{is}}{=} \;\; \underset{\substack{\text{Cheryl's} \\ \text{time}}}{\underbrace{\frac{300}{x}}} \;\; \underset{\substack{\text{less 50} \\ \text{seconds.}}}{\underbrace{- \quad 50}}$$

Step 4   **Solve.**

$$x(x+1)\left(\frac{300}{x+1}\right) = x(x+1)\left(\frac{300}{x} - 50\right) \qquad \text{Multiply by the LCD, } x(x+1).$$

$$x(x+1)\left(\frac{300}{x+1}\right) = x(x+1)\left(\frac{300}{x}\right) - x(x+1)(50) \qquad \text{Distributive property}$$

$$300x = 300(x+1) - 50x(x+1) \qquad \text{Multiply.}$$

$$300x = 300x + 300 - 50x^2 - 50x \qquad \text{Distributive property}$$

$$50x^2 + 50x - 300 = 0 \qquad \text{Standard form}$$

$$x^2 + x - 6 = 0 \qquad \text{Divide by 50.}$$

$$(x+3)(x-2) = 0 \qquad \text{Factor.}$$

$$x + 3 = 0 \quad \text{or} \quad x - 2 = 0 \qquad \text{Zero-factor property}$$

$$x = -3 \quad \text{or} \quad x = 2 \qquad \text{Solve each equation.}$$

Discard the negative answer, since rate (speed) cannot be negative.

Step 5   **State the answer.** Their walking rate is 2 m per sec.

Step 6   **Check** the solution in the words of the original problem.   NOW TRY

⚠ **CAUTION**   We cannot solve the rational equation in **Example 7,**

$$\frac{300}{x+1} = \frac{300}{x} - 50, \qquad \begin{array}{l}\text{To solve, multiply} \\ \text{by the LCD.}\end{array}$$

by using cross products. *That method can be used only when there is a single rational expression on each side,* such as in the equation in **Example 6,**

$$\frac{10}{x-3} = \frac{15}{x+3}. \qquad \begin{array}{l}\text{To solve, multiply by the} \\ \text{LCD **or** use cross products.}\end{array}$$

OBJECTIVE 5 Solve applications about work rates.

PROBLEM-SOLVING HINT

People work at different rates. If the letters r, t, and A represent the rate at which work is done, the time required, and the amount of work accomplished, respectively, then $A = rt$. Notice the similarity to the distance formula, $d = rt$.

Amount of work can be measured in terms of jobs accomplished. Thus, if 1 job is completed, then $A = 1$, and the formula gives the rate as

$$1 = rt, \quad \text{or} \quad r = \frac{1}{t}.$$

To solve a work problem, we use this fact to express all rates of work.

Rate of Work

If a job can be accomplished in t units of time, then the rate of work is

$$\frac{1}{t} \text{ job per unit of time.}$$

NOW TRY
EXERCISE 8

Using her zero-turn riding lawn mower, Gina can mow her lawn in 2 hr. Using a traditional mower, Matt can mow the same lawn in 3 hr. How long will it take them to mow the lawn working together?

EXAMPLE 8 Solving a Problem about Work

Letitia and Kareem are working on a neighborhood cleanup. Kareem can clean up all the trash in the area in 7 hr, while Letitia can do the same job in 5 hr. How long will it take them if they work together?

Let $x =$ the number of hours it will take the two people working together.

We use $A = rt$. Since $A = 1$, the rate for each person will be $\frac{1}{t}$, where t is the time it takes the person to complete the job alone. Kareem can clean up all the trash in 7 hr, so his rate is $\frac{1}{7}$ of the job per hour. Similarly, Letitia's rate is $\frac{1}{5}$ of the job per hour.

	Rate	Time Working Together	Fractional Part of the Job Done
Kareem	$\frac{1}{7}$	x	$\frac{1}{7}x$
Letitia	$\frac{1}{5}$	x	$\frac{1}{5}x$

Part done by Kareem	+	part done by Letitia	is	1 whole job.
$\frac{1}{7}x$	$+$	$\frac{1}{5}x$	$=$	1

Together they complete 1 job. The sum of the fractional parts should equal 1.

$$35\left(\frac{1}{7}x + \frac{1}{5}x\right) = 35 \cdot 1 \qquad \text{Multiply by the LCD, 35.}$$

$$5x + 7x = 35 \qquad \text{Distributive property}$$

$$12x = 35 \qquad \text{Combine like terms.}$$

$$x = \frac{35}{12} \qquad \text{Divide by 12.}$$

NOW TRY ANSWER
8. $\frac{6}{5}$ hr, or 1 hr, 12 min

Working together, Kareem and Letitia can do the job in $\frac{35}{12}$ hr, or 2 hr, 55 min. Check this result in the original problem.

NOW TRY

NOTE There is another way to approach problems about rates of work. In **Example 8,** x represents the number of hours it will take the two people working together to complete the entire job. In 1 hr, $\frac{1}{x}$ of the entire job will be completed. In 1 hr, Kareem completes $\frac{1}{7}$ of the job and Letitia completes $\frac{1}{5}$ of the job.

$$\frac{1}{7} + \frac{1}{5} = \frac{1}{x} \qquad \text{The sum of their rates equals } \frac{1}{x}.$$

When each side of this equation is multiplied by $35x$, the result is $5x + 7x = 35$, the same equation we got in **Example 8** in the third line from the bottom. Thus, the solution of the equation is the same with either approach.

7.5 EXERCISES

MyMathLab | Math XL PRACTICE | WATCH | DOWNLOAD | READ | REVIEW

🌐 *Complete solution available on the Video Resources on DVD*

Concept Check *Exercises 1–4 present a familiar formula. Give the letter of the choice that is an equivalent form of the formula.*

1. $p = br$ (percent)

   A. $b = \dfrac{p}{r}$ **B.** $r = \dfrac{b}{p}$ **C.** $b = \dfrac{r}{p}$ **D.** $p = \dfrac{r}{b}$

2. $V = LWH$ (geometry)

   A. $H = \dfrac{LW}{V}$ **B.** $L = \dfrac{V}{WH}$ **C.** $L = \dfrac{WH}{V}$ **D.** $W = \dfrac{H}{VL}$

3. $m = \dfrac{F}{a}$ (physics)

   A. $a = mF$ **B.** $F = \dfrac{m}{a}$ **C.** $F = \dfrac{a}{m}$ **D.** $F = ma$

4. $I = \dfrac{E}{R}$ (electricity)

   A. $R = \dfrac{I}{E}$ **B.** $R = IE$ **C.** $E = \dfrac{I}{R}$ **D.** $E = RI$

Solve each problem. ***See Example 1.***

🌐 **5.** In work with electric circuits, the formula

$$\frac{1}{a} = \frac{1}{b} + \frac{1}{c}$$

occurs. Find b if $a = 8$ and $c = 12$.

6. A gas law in chemistry says that

$$\frac{PV}{T} = \frac{pv}{t}.$$

Suppose that $T = 300$, $t = 350$, $V = 9$, $P = 50$, and $v = 8$. Find p.

7. A formula from anthropology says that

$$c = \frac{100b}{L}.$$

Find L if $c = 80$ and $b = 5$.

8. The gravitational force between two masses is given by

$$F = \frac{GMm}{d^2}.$$

Find M if $F = 10$, $G = 6.67 \times 10^{-11}$, $m = 1$, and $d = 3 \times 10^{-6}$.

Solve each formula for the specified variable. **See Examples 2 and 3.**

9. $F = \dfrac{GMm}{d^2}$ for G (physics)

10. $F = \dfrac{GMm}{d^2}$ for M (physics)

11. $\dfrac{1}{a} = \dfrac{1}{b} + \dfrac{1}{c}$ for a (electricity)

12. $\dfrac{1}{a} = \dfrac{1}{b} + \dfrac{1}{c}$ for b (electricity)

13. $\dfrac{PV}{T} = \dfrac{pv}{t}$ for v (chemistry)

14. $\dfrac{PV}{T} = \dfrac{pv}{t}$ for T (chemistry)

15. $I = \dfrac{nE}{R + nr}$ for r (engineering)

16. $a = \dfrac{V - v}{t}$ for V (physics)

17. $A = \dfrac{1}{2}h(b + B)$ for b (mathematics)

18. $S = \dfrac{n}{2}(a + \ell)d$ for n (mathematics)

19. $\dfrac{E}{e} = \dfrac{R + r}{r}$ for r (engineering)

20. $y = \dfrac{x + z}{a - x}$ for x

21. *Concept Check*   In solving the equation $m = \dfrac{ab}{a - b}$ for a, what is the first step?

22. *Concept Check*   To solve the following equation for r, what is the first step?

$$rp - rq = p + q$$

Solve each problem mentally. Use proportions in Exercises 23 and 24.

23. In a mathematics class, 3 of every 4 students are girls. If there are 28 students in the class, how many are girls? How many are boys?

24. In a certain southern state, sales tax on a purchase of $1.50 is $0.12. What is the sales tax on a purchase of $9.00?

25. If Marin can mow her yard in 3 hr, what is her rate (in terms of the proportion of the job per hour)?

26. A van traveling from Atlanta to Detroit averages 50 mph and takes 14 hr to make the trip. How far is it from Atlanta to Detroit?

The water content of snow is affected by the temperature, wind speed, and other factors present when the snow is falling. The average snow-to-liquid ratio is 10 in. of snow to 1 in. of liquid precipitation. This means that if 10 in. of snow fell and was melted, it would produce 1 in. of liquid precipitation in a rain gauge. (Source: www.theweatherprediction.com)

27. A dry snow might have a snow-to-liquid ratio of 18 to 1. Using this ratio, how much liquid precipitation would produce 31.5 in. of snow?

28. A wet, sticky snow good for making a snow man might have a snow-to-liquid ratio of 5 to 1. How many inches of fresh snow would 3.25 in. of liquid precipitation produce using this ratio?

Use a proportion to solve each problem. **See Examples 4 and 5.**

29. On a map of the United States, the distance between Seattle and Durango is 4.125 in. The two cities are actually 1238 miles apart. On this same map, what would be the distance between Chicago and El Paso, two cities that are actually 1606 mi apart? Round your answer to the nearest tenth. (*Source:* Universal Map Atlas.)

30. On a map of the United States, the distance between Reno and Phoenix is 2.5 in. The two cities are actually 768 miles apart. On this same map, what would be the distance between St. Louis and Jacksonville, two cities that are actually 919 mi apart? Round your answer to the nearest tenth. (*Source:* Universal Map Atlas.)

31. On a world globe, the distance between New York and Cairo, two cities that are actually 5619 mi apart, is 8.5 in. On this same globe, how far apart are Madrid and Rio de Janeiro, two cities that are actually 5045 mi apart? Round your answer to the nearest tenth. (*Sources:* Author's globe, *The World Almanac and Book of Facts.*)

32. On a world globe, the distance between San Francisco and Melbourne, two cities that are actually 7856 mi apart, is 11.875 in. On this same globe, how far apart are Mexico City and Singapore, two cities that are actually 10,327 mi apart? Round your answer to the nearest tenth. (*Sources:* Author's globe, *The World Almanac and Book of Facts.*)

33. During the 2008–2009 school year, the average ratio of teachers to students in public elementary and secondary schools was approximately 1 to 15. If a public school had 846 students, how many teachers would be at the school if this ratio was valid for that school? Round your answer to the nearest whole number. (*Source:* U.S. National Center for Education Statistics.)

34. On January 6, 2010, the eventual NBA champion L.A. Lakers were in first place in the Western Conference of the NBA, having won 28 of their first 34 regular season games. If the team continued to win the same fraction of its games, how many games would the Lakers win for the complete 82-game season? (*Source:* www.nba.com)

35. To estimate the deer population of a forest preserve, wildlife biologists caught, tagged, and then released 42 deer. A month later, they returned and caught a sample of 75 deer and found that 15 of them were tagged. Based on this experiment, approximately how many deer lived in the forest preserve?

36. Suppose that in the experiment in **Exercise 35,** only 5 of the previously tagged deer were collected in the sample of 75. What would be the estimate of the deer population?

37. Biologists tagged 500 fish in a lake on January 1. On February 1, they returned and collected a random sample of 400 fish, 8 of which had been previously tagged. On the basis of this experiment, approximately how many fish does the lake have?

38. Suppose that in the experiment of **Exercise 37,** 10 of the previously tagged fish were collected on February 1. What would be the estimate of the fish population?

39. Bruce Johnston's Shelby Cobra uses 5 gal of gasoline to drive 156 mi. He has 3 gal of gasoline in the car, and he wants to know how much more gasoline he will need to drive 300 mi. If we assume that the car continues to use gasoline at the same rate, how many more gallons, to the nearest tenth, will he need?

40. Mike Love's T-bird uses 6 gal of gasoline to drive 141 mi. He has 4 gal of gasoline in the car, and he wants to know how much more gasoline he will need to drive 275 mi. If we assume that the car continues to use gasoline at the same rate, how many more gallons, to the nearest tenth, will he need?

*In geometry, two triangles with corresponding angle measures equal, called **similar triangles,** have corresponding sides proportional. For example, in the figure, angle A = angle D, angle B = angle E, and angle C = angle F, so the triangles are similar. Then the following ratios of corresponding sides are equal.*

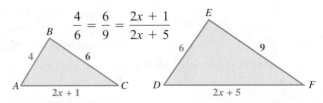

$$\frac{4}{6} = \frac{6}{9} = \frac{2x+1}{2x+5}$$

41. Solve for x, using the given proportion to find the lengths of the third sides of the triangles.

42. Suppose the following triangles are similar. Find y and the lengths of the two unknown sides of each triangle.

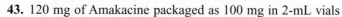

Nurses use proportions to determine the amount of a drug to administer when the dose is measured in milligrams, but the drug is packaged in a diluted form in milliliters. (Source: Hoyles, Celia, Richard Noss, and Stefano Pozzi, "Proportional Reasoning in Nursing Practice," Journal for Research in Mathematics Education.) For example, to find the number of milliliters of fluid needed to administer 300 mg of a drug that comes packaged as 120 mg in 2 mL of fluid, a nurse sets up the proportion

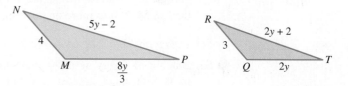

$$\frac{120 \text{ mg}}{2 \text{ mL}} = \frac{300 \text{ mg}}{x \text{ mL}},$$

where x represents the amount to administer in milliliters. Use this method to find the correct dose for each prescription.

43. 120 mg of Amakacine packaged as 100 mg in 2-mL vials

44. 1.5 mg of morphine packaged as 20 mg ampules diluted in 10 mL of fluid

Solve each problem. ***See Examples 6 and 7.***

45. Kellen's boat goes 12 mph. Find the rate of the current of the river if she can go 6 mi upstream in the same amount of time she can go 10 mi downstream.

46. Kasey can travel 8 mi upstream in the same time it takes her to go 12 mi downstream. Her boat goes 15 mph in still water. What is the rate of the current?

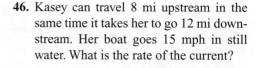

	Distance	Rate	Time
Downstream	10	12 + x	
Upstream	6	12 − x	

	Distance	Rate	Time
Downstream			
Upstream			

47. On his drive from Montpelier, Vermont, to Columbia, South Carolina, Victor Samuels averaged 51 mph. If he had been able to average 60 mph, he would have reached his destination 3 hr earlier. What is the driving distance between Montpelier and Columbia?

48. Kelli Hammer is a college professor who lives in an off-campus apartment. On days when she rides her bike to campus, she gets to her first class 36 min faster than when she walks. If her average walking rate is 3 mph and her average biking rate is 12 mph, how far is it from her apartment to her first class?

49. A plane averaged 500 mph on a trip going east, but only 350 mph on the return trip. The total flying time in both directions was 8.5 hr. What was the one-way distance?

50. Mary Smith drove from her apartment to her parents' house on a Saturday morning. She was able to average 60 mph because traffic was light. However, returning on Sunday night, she was able to average only 45 mph on the same route. The drive on Sunday took her 1.5 hr longer than the drive on Saturday. What is the distance between Mary's apartment and her parents' house?

51. On the first part of a trip to Carmel traveling on the freeway, Marge averaged 60 mph. On the rest of the trip, which was 10 mi longer than the first part, she averaged 50 mph. Find the total distance to Carmel if the second part of the trip took 30 min more than the first part.

52. During the first part of a trip on the highway, Jim and Annie averaged 60 mph. In Houston, traffic caused them to average only 30 mph. The distance they drove in Houston was 100 mi less than their distance on the highway. What was their total driving distance if they spent 50 min more on the highway than they did in Houston?

Solve each problem. **See Example 8.**

53. Butch and Peggy want to pick up the mess that their grandson, Grant, has made in his playroom. Butch can do it in 15 min working alone. Peggy, working alone, can clean it in 12 min. How long will it take them if they work together?

	Rate	Time Working Together	Fractional Part of the Job Done
Butch	$\frac{1}{15}$	x	
Peggy	$\frac{1}{12}$	x	

54. Lou can groom Randy Smith's dogs in 8 hr, but it takes his business partner, Janet, only 5 hr to groom the same dogs. How long will it take them to groom Randy's dogs if they work together?

	Rate	Time Working Together	Fractional Part of the Job Done
Lou	$\frac{1}{8}$	x	
Janet	$\frac{1}{5}$	x	

55. Jerry and Kuba are laying a hardwood floor. Working alone, Jerry can do the job in 20 hr. If the two of them work together, they can complete the job in 12 hr. How long would it take Kuba to lay the floor working alone?

56. Mrs. Disher is a high school mathematics teacher. She can grade a set of chapter tests in 5 hr working alone. If her student teacher Mr. Howes helps her, it will take 3 hr to grade the tests. How long would it take Mr. Howes to grade the tests if he worked alone?

57. Dixie can paint a room in 3 hr working alone. Trixie can paint the same room in 6 hr working alone. How long after Dixie starts to paint the room will it be finished if Trixie joins her 1 hr later?

58. Sheldon can wash a car in 30 min working alone. Leonard can do the same job in 45 min working alone. How long after Sheldon starts to wash the car will it be finished if Leonard joins him 5 min later?

59. If a vat of acid can be filled by an inlet pipe in 10 hr and emptied by an outlet pipe in 20 hr, how long will it take to fill the vat if both pipes are open?

60. A winery has a vat to hold Zinfandel. An inlet pipe can fill the vat in 9 hr, while an outlet pipe can empty it in 12 hr. How long will it take to fill the vat if both the outlet and the inlet pipes are open?

61. Suppose that Hortense and Mort can clean their entire house in 7 hr, while their toddler, Mimi, can mess it up in only 2 hr. If Hortense and Mort clean the house while Mimi is at her grandma's and then start cleaning up after Mimi the minute she gets home, how long does it take from the time Mimi gets home until the whole place is a shambles?

62. An inlet pipe can fill an artificial lily pond in 60 min, while an outlet pipe can empty it in 80 min. Through an error, both pipes are left open. How long will it take for the pond to fill?

PREVIEW EXERCISES

Find k, given that y $= 1$ *and x* $= 3$. **See Section 2.1.**

63. $y = kx$ **64.** $y = kx^2$ **65.** $y = \dfrac{k}{x}$ **66.** $y = \dfrac{k}{x^2}$

7.6 Variation

OBJECTIVES

1. Write an equation expressing direct variation.
2. Find the constant of variation, and solve direct variation problems.
3. Solve inverse variation problems.
4. Solve joint variation problems.
5. Solve combined variation problems.

Functions in which *y depends on a multiple of x* or *y depends on a number divided by x* are common in business and the physical sciences.

OBJECTIVE 1 **Write an equation expressing direct variation.** The circumference of a circle is given by the formula $C = 2\pi r$, where r is the radius of the circle. See **FIGURE 6**. The circumference is always a constant multiple of the radius. (C is always found by multiplying r by the constant 2π.)

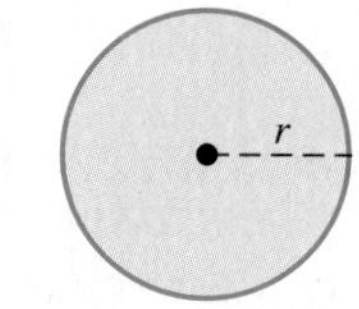

$C = 2\pi r$

FIGURE 6

As the ***radius increases,*** the ***circumference increases.***

As the ***radius decreases,*** the ***circumference decreases.***

Because of these relationships, the circumference is said to *vary directly* as the radius.

Direct Variation

y **varies directly as** x if there exists a real number k such that

$$y = kx.$$

y is said to be **proportional to** x. The number k is called the **constant of variation.**

In direct variation, for $k > 0$, as the value of x increases, the value of y increases. Similarly, as x decreases, y decreases.

OBJECTIVE 2 **Find the constant of variation, and solve direct variation problems.** ***The direct variation equation $y = kx$ defines a linear function, where the constant of variation k is the slope of the line.*** For example, we wrote the following equation to describe the cost y to buy x gallons of gasoline.

$$y = 3.20x \qquad \text{See Section 3.3, Example 7.}$$

The cost varies directly as, or is proportional to, the number of gallons of gasoline purchased.

As the *number of gallons* of gasoline *increases,* the *cost increases.*

As the *number of gallons* of gasoline *decreases,* the *cost decreases.*

The constant of variation k is 3.20, the cost of 1 gal of gasoline.

EXAMPLE 1 **Finding the Constant of Variation and the Variation Equation**

Eva Lutchman is paid an hourly wage. One week she worked 43 hr and was paid \$795.50. How much does she earn per hour?

Let h represent the number of hours she works and P represent her corresponding pay. Write the variation equation.

$$P = kh \qquad P \text{ varies directly as } h.$$

Here, k represents Eva's hourly wage.

NOW TRY
EXERCISE 1
One week Morgan sold 8 dozen eggs for $20. How much does she charge for one dozen eggs?

$$P = kh \qquad \textit{P varies directly as h.}$$

$$795.50 = 43k \qquad \text{Substitute 795.50 for } P \text{ and 43 for } h.$$

 This is the constant of variation.

$$k = 18.50 \qquad \text{Use a calculator.}$$

Her hourly wage is $18.50, and P and h are related by

$$P = 18.50h.$$

We can use this equation to find her pay for any number of hours worked.

NOW TRY

NOW TRY
EXERCISE 2
For a constant height, the area of a parallelogram is directly proportional to its base. If the area is 20 cm² when the base is 4 cm, find the area when the base is 7 cm.

EXAMPLE 2 Solving a Direct Variation Problem

Hooke's law for an elastic spring states that the distance a spring stretches is directly proportional to the force applied. If a force of 150 newtons* stretches a certain spring 8 cm, how much will a force of 400 newtons stretch the spring? See **FIGURE 7**.

If d is the distance the spring stretches and f is the force applied, then $d = kf$ for some constant k. Since a force of 150 newtons stretches the spring 8 cm, we use these values to find k.

FIGURE 7

$$d = kf \qquad \text{Variation equation}$$

$$8 = k \cdot 150 \qquad \text{Let } d = 8 \text{ and } f = 150.$$

$$k = \frac{8}{150} \qquad \text{Solve for } k.$$

$$k = \frac{4}{75} \qquad \text{Lowest terms}$$

Substitute $\frac{4}{75}$ for k in the variation equation $d = kf$.

$$d = \frac{4}{75}f \qquad \text{Here, } k = \frac{4}{75}.$$

For a force of 400 newtons, substitute 400 for f.

$$d = \frac{4}{75}(400) = \frac{64}{3} \qquad \text{Let } f = 400.$$

The spring will stretch $\frac{64}{3}$ cm, or $21\frac{1}{3}$ cm, if a force of 400 newtons is applied.

NOW TRY

Solving a Variation Problem

Step 1 Write the variation equation.

Step 2 Substitute the initial values and solve for k.

Step 3 Rewrite the variation equation with the value of k from Step 2.

Step 4 Substitute the remaining values, solve for the unknown, and find the required answer.

NOW TRY ANSWERS
1. $2.50   **2.** 35 cm²

*A newton is a unit of measure of force used in physics.

One variable can be proportional to a power of another variable.

> ### Direct Variation as a Power
>
> **y varies directly as the *n*th power of *x*** if there exists a real number k such that
> $$y = kx^n.$$

$A = \pi r^2$

FIGURE 8

The formula for the area of a circle, $A = \pi r^2$, is an example. See **FIGURE 8**. Here, π is the constant of variation, and the area varies directly as the *square* of the radius.

NOW TRY
EXERCISE 3

Suppose y varies directly as the square of x, and $y = 200$ when $x = 5$. Find y when $x = 7$.

> **EXAMPLE 3** Solving a Direct Variation Problem

The distance a body falls from rest varies directly as the square of the time it falls (disregarding air resistance). If a skydiver falls 64 ft in 2 sec, how far will she fall in 8 sec?

Step 1 If d represents the distance the skydiver falls and t the time it takes to fall, then d is a function of t for some constant k.

$$d = kt^2 \qquad \text{d varies directly as the square of t.}$$

Step 2 To find the value of k, use the fact that the skydiver falls 64 ft in 2 sec.

$$d = kt^2 \qquad \text{Variation equation}$$
$$64 = k(2)^2 \qquad \text{Let } d = 64 \text{ and } t = 2.$$
$$k = 16 \qquad \text{Find } k.$$

Step 3 Now we rewrite the variation equation $d = kt^2$ using 16 for k.

$$d = 16t^2 \qquad \text{Here, } k = 16.$$

Step 4 Let $t = 8$ to find the number of feet the skydiver will fall in 8 sec.

$$d = 16(8)^2 = 1024 \qquad \text{Let } t = 8.$$

The skydiver will fall 1024 ft in 8 sec. NOW TRY

As pressure on trash increases, volume of trash decreases.

FIGURE 9

OBJECTIVE 3 **Solve inverse variation problems.** Another type of variation is *inverse variation*. **With inverse variation, where $k > 0$, as one variable increases, the other variable decreases.**

For example, in a closed space, volume decreases as pressure increases, which can be illustrated by a trash compactor. See **FIGURE 9**. As the compactor presses down, the pressure on the trash increases, and in turn, the trash occupies a smaller space.

> ### Inverse Variation
>
> **y varies inversely as x** if there exists a real number k such that
> $$y = \frac{k}{x}.$$
>
> Also, **y varies inversely as the *n*th power of x** if there exists a real number k such that
> $$y = \frac{k}{x^n}.$$

NOW TRY ANSWER
3. 392

The inverse variation equation defines a rational function. Another example of inverse variation comes from the distance formula.

$$d = rt \qquad \text{Distance formula}$$

$$t = \frac{d}{r} \qquad \text{Divide each side by } r.$$

Here, t (time) varies inversely as r (rate or speed), with d (distance) serving as the constant of variation. For example, if the distance between Chicago and Des Moines is 300 mi, then

$$t = \frac{300}{r},$$

and the values of r and t might be any of the following.

$$\left.\begin{array}{l} r = 50, t = 6 \\ r = 60, t = 5 \\ r = 75, t = 4 \end{array}\right\} \begin{array}{l}\text{As } r \text{ increases,} \\ t \text{ decreases.}\end{array} \qquad \left.\begin{array}{l} r = 30, t = 10 \\ r = 25, t = 12 \\ r = 20, t = 15 \end{array}\right\} \begin{array}{l}\text{As } r \text{ decreases,} \\ t \text{ increases.}\end{array}$$

If we *increase* the rate (speed) at which we drive, time *decreases*. If we *decrease* the rate (speed) at which we drive, time *increases*.

**NOW TRY
EXERCISE 4**

For a constant area, the height of a triangle varies inversely as the base. If the height is 7 cm when the base is 8 cm, find the height when the base is 14 cm.

EXAMPLE 4 Solving an Inverse Variation Problem

In the manufacture of a certain medical syringe, the cost of producing the syringe varies inversely as the number produced. If 10,000 syringes are produced, the cost is $2 per syringe. Find the cost per syringe of producing 25,000 syringes.

Let $\quad x =$ the number of syringes produced,

and $\quad c =$ the cost per syringe.

Here, as production increases, cost decreases, and as production decreases, cost increases. We write a variation equation using the variables c and x and the constant k.

$$c = \frac{k}{x} \qquad c \text{ varies inversely as } x.$$

To find k, we replace c with 2 and x with 10,000.

$$2 = \frac{k}{10,000} \qquad \text{Substitute in the variation equation.}$$

$$20,000 = k \qquad \text{Multiply by 10,000.}$$

Since $c = \frac{k}{x}$,

$$c = \frac{20,000}{25,000} = 0.80. \qquad \text{Here, } k = 20,000. \text{ Let } x = 25,000.$$

The cost per syringe to make 25,000 syringes is $0.80. NOW TRY

EXAMPLE 5 Solving an Inverse Variation Problem

The weight of an object above Earth varies inversely as the square of its distance from the center of Earth. A space shuttle in an elliptical orbit has a maximum distance from the center of Earth (*apogee*) of 6700 mi. Its minimum distance from the center of Earth (*perigee*) is 4090 mi. See **FIGURE 10** on the next page. If an astronaut in the shuttle weighs 57 lb at its apogee, what does the astronaut weigh at its perigee?

NOW TRY ANSWER
4. 4 cm

**NOW TRY
EXERCISE 5**

The weight of an object above Earth varies inversely as the square of its distance from the center of Earth. If an object weighs 150 lb on the surface of Earth, and the radius of Earth is about 3960 mi, how much does it weigh when it is 1000 mi above Earth's surface?

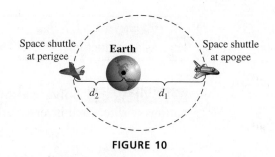

FIGURE 10

Let w = the weight and d = the distance from the center of Earth, for some constant k.

$$w = \frac{k}{d^2} \qquad \text{\textit{w varies inversely as the square of d.}}$$

At the apogee, the astronaut weighs 57 lb, and the distance from the center of Earth is 6700 mi. Use these values to find k.

$$57 = \frac{k}{(6700)^2} \qquad \text{Let } w = 57 \text{ and } d = 6700.$$

$$k = 57(6700)^2 \qquad \text{Solve for } k.$$

Substitute $k = 57(6700)^2$ and $d = 4090$ to find the weight at the perigee.

$$w = \frac{57(6700)^2}{(4090)^2} \approx 153 \text{ lb} \qquad \text{Use a calculator.} \qquad \text{NOW TRY}$$

OBJECTIVE 4 **Solve joint variation problems.** If one variable varies directly as the *product* of several other variables (perhaps raised to powers), the first variable is said to *vary jointly* as the others.

Joint Variation

y varies jointly as x and z if there exists a real number k such that

$$y = kxz.$$

**NOW TRY
EXERCISE 6**

The volume of a right pyramid varies jointly as the height and the area of the base. If the volume is 100 ft³ when the area of the base is 30 ft² and the height is 10 ft, find the volume when the area of the base is 90 ft² and the height is 20 ft.

EXAMPLE 6 Solving a Joint Variation Problem

The interest on a loan or an investment is given by the formula $I = prt$. Here, for a given principal p, the interest earned, I, varies jointly as the interest rate r and the time t the principal is left earning interest. If an investment earns \$100 interest at 5% for 2 yr, how much interest will the same principal earn at 4.5% for 3 yr?

We use the formula $I = prt$, where p is the constant of variation because it is the same for both investments.

$$I = prt$$

$$100 = p(0.05)(2) \qquad \text{Let } I = 100, r = 0.05, \text{ and } t = 2.$$

$$100 = 0.1p$$

$$p = 1000 \qquad \text{Divide by 0.1. Rewrite.}$$

Now we find I when $p = 1000$, $r = 0.045$, and $t = 3$.

$$I = 1000(0.045)(3) = 135 \qquad \text{Here, } p = 1000. \text{ Let } r = 0.045 \text{ and } t = 3.$$

The interest will be \$135. NOW TRY

NOW TRY ANSWERS
5. about 96 lb   **6.** 600 ft³

⚠ **CAUTION** Note that *and* in the expression "*y* varies directly as *x and z*" translates as a product in $y = kxz$. The word *and* does not indicate addition here.

OBJECTIVE 5 **Solve combined variation problems.** There are many combinations of direct and inverse variation, typically called **combined variation.**

EXAMPLE 7 Solving a Combined Variation Problem

Body mass index, or BMI, is used to assess a person's level of fatness. A BMI from 19 through 25 is considered desirable. BMI varies directly as an individual's weight in pounds and inversely as the square of the individual's height in inches.

A person who weighs 116.5 lb and is 64 in. tall has a BMI of 20. (The BMI is rounded to the nearest whole number.) Find the BMI of a man who weighs 165 lb and is 70 in. tall. (*Source: Washington Post.*)

Let B represent the BMI, w the weight, and h the height.

$$B = \frac{kw}{h^2} \quad \longleftarrow \text{BMI varies directly as the weight.}$$
$$\phantom{B = \frac{kw}{h^2}} \longleftarrow \text{BMI varies inversely as the square of the height.}$$

To find k, let $B = 20$, $w = 116.5$, and $h = 64$.

$$20 = \frac{k(116.5)}{64^2} \qquad B = \frac{kw}{h^2}$$

$$k = \frac{20(64^2)}{116.5} \qquad \text{Multiply by } 64^2. \text{ Divide by } 116.5.$$

$$k \approx 703 \qquad \text{Use a calculator.}$$

Now find B when $k = 703$, $w = 165$, and $h = 70$.

$$B = \frac{703(165)}{70^2} \approx 24 \qquad \text{Nearest whole number}$$

The man's BMI is 24.

NOW TRY ↻

NOW TRY
EXERCISE 7
In statistics, the sample size used to estimate a population mean varies directly as the variance and inversely as the square of the maximum error of the estimate. If the sample size is 200 when the variance is 25 m² and the maximum error of the estimate is 0.5 m, find the sample size when the variance is 25 m² and the maximum error of the estimate is 0.1 m.

NOW TRY ANSWER
7. 5000

7.6 EXERCISES

MyMathLab | Math XL PRACTICE | WATCH | DOWNLOAD | READ | REVIEW

🌐 *Complete solution available on the Video Resources on DVD*

Concept Check Use personal experience or intuition to determine whether the situation suggests direct *or* inverse *variation.*

1. The number of lottery tickets you buy and your probability of winning that lottery

2. The rate and the distance traveled by a pickup truck in 3 hr

3. The amount of pressure put on the accelerator of a car and the speed of the car

4. The number of days from now until December 25 and the magnitude of the frenzy of Christmas shopping

5. Your age and the probability that you believe in Santa Claus

6. The surface area of a balloon and its diameter

7. The number of days until the end of the baseball season and the number of home runs that Albert Pujols has

8. The amount of gasoline you pump and the amount you pay

Concept Check *Determine whether each equation represents* direct, inverse, joint, *or* combined *variation.*

9. $y = \dfrac{3}{x}$ **10.** $y = \dfrac{8}{x}$ **11.** $y = 10x^2$ **12.** $y = 2x^3$

13. $y = 3xz^4$ **14.** $y = 6x^3z^2$ **15.** $y = \dfrac{4x}{wz}$ **16.** $y = \dfrac{6x}{st}$

17. *Concept Check* For $k > 0$, if y varies directly as x, then when x increases, y _____, and when x decreases, y _____.

18. *Concept Check* For $k > 0$, if y varies inversely as x, then when x increases, y _____, and when x decreases, y _____.

Concept Check *Write each formula using the "language" of variation. For example, the formula for the circumference of a circle,* $C = 2\pi r$, *can be written as*

"*The circumference of a circle varies directly as the length of its radius.*"

19. $P = 4s$, where P is the perimeter of a square with side of length s

20. $d = 2r$, where d is the diameter of a circle with radius r

21. $S = 4\pi r^2$, where S is the surface area of a sphere with radius r

22. $V = \frac{4}{3}\pi r^3$, where V is the volume of a sphere with radius r

23. $A = \frac{1}{2}bh$, where A is the area of a triangle with base b and height h

24. $V = \frac{1}{3}\pi r^2 h$, where V is the volume of a cone with radius r and height h

25. *Concept Check* What is the constant of variation in each of the variation equations in **Exercises 19–24?**

26. What is meant by the constant of variation in a direct variation problem? If you were to graph the linear equation $y = kx$ for some nonnegative constant k, what role would k play in the graph?

Solve each problem. See Examples 1–7.

27. If x varies directly as y, and $x = 9$ when $y = 3$, find x when $y = 12$.

28. If x varies directly as y, and $x = 10$ when $y = 7$, find y when $x = 50$.

29. If a varies directly as the square of b, and $a = 4$ when $b = 3$, find a when $b = 2$.

30. If h varies directly as the square of m, and $h = 15$ when $m = 5$, find h when $m = 7$.

31. If z varies inversely as w, and $z = 10$ when $w = 0.5$, find z when $w = 8$.

32. If t varies inversely as s, and $t = 3$ when $s = 5$, find s when $t = 5$.

33. If m varies inversely as p^2, and $m = 20$ when $p = 2$, find m when $p = 5$.

34. If a varies inversely as b^2, and $a = 48$ when $b = 4$, find a when $b = 7$.

35. p varies jointly as q and r^2, and $p = 200$ when $q = 2$ and $r = 3$. Find p when $q = 5$ and $r = 2$.

36. f varies jointly as g^2 and h, and $f = 50$ when $g = 4$ and $h = 2$. Find f when $g = 3$ and $h = 6$.

Solve each problem. See Examples 1–7.

37. Ben bought 15 gal of gasoline and paid \$43.79. To the nearest tenth of a cent, what is the price of gasoline per gallon?

38. Sara gives horseback rides at Shadow Mountain Ranch. A 2.5-hr ride costs \$50.00. What is the price per hour?

39. The weight of an object on Earth is directly proportional to the weight of that same object on the moon. A 200-lb astronaut would weigh 32 lb on the moon. How much would a 50-lb dog weigh on the moon?

40. The pressure exerted by a certain liquid at a given point is directly proportional to the depth of the point beneath the surface of the liquid. The pressure at 30 m is 80 newtons. What pressure is exerted at 50 m?

41. The volume of a can of tomatoes is directly proportional to the height of the can. If the volume of the can is 300 cm³ when its height is 10.62 cm, find the volume of a can with height 15.92 cm.

42. The force required to compress a spring is directly proportional to the change in length of the spring. If a force of 20 newtons is required to compress a certain spring 2 cm, how much force is required to compress the spring from 20 cm to 8 cm?

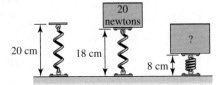

43. For a body falling freely from rest (disregarding air resistance), the distance the body falls varies directly as the square of the time. If an object is dropped from the top of a tower 576 ft high and hits the ground in 6 sec, how far did it fall in the first 4 sec?

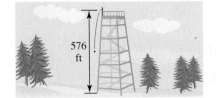

44. The amount of water emptied by a pipe varies directly as the square of the diameter of the pipe. For a certain constant water flow, a pipe emptying into a canal will allow 200 gal of water to escape in an hour. The diameter of the pipe is 6 in. How much water would a 12-in. pipe empty into the canal in an hour, assuming the same water flow?

45. Over a specified distance, rate varies inversely with time. If a Dodge Viper on a test track goes a certain distance in one-half minute at 160 mph, what rate is needed to go the same distance in three-fourths minute?

46. For a constant area, the length of a rectangle varies inversely as the width. The length of a rectangle is 27 ft when the width is 10 ft. Find the width of a rectangle with the same area if the length is 18 ft.

47. The frequency of a vibrating string varies inversely as its length. That is, a longer string vibrates fewer times in a second than a shorter string. Suppose a piano string 2 ft long vibrates 250 cycles per sec. What frequency would a string 5 ft long have?

48. The current in a simple electrical circuit varies inversely as the resistance. If the current is 20 amps when the resistance is 5 ohms, find the current when the resistance is 7.5 ohms.

49. The amount of light (measured in foot-candles) produced by a light source varies inversely as the square of the distance from the source. If the illumination produced 1 m from a light source is 768 foot-candles, find the illumination produced 6 m from the same source.

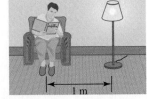

50. The force with which Earth attracts an object above Earth's surface varies inversely as the square of the distance of the object from the center of Earth. If an object 4000 mi from the center of Earth is attracted with a force of 160 lb, find the force of attraction if the object were 6000 mi from the center of Earth.

51. For a given interest rate, simple interest varies jointly as principal and time. If $2000 left in an account for 4 yr earned interest of $280, how much interest would be earned in 6 yr?

52. The collision impact of an automobile varies jointly as its mass and the square of its speed. Suppose a 2000-lb car traveling at 55 mph has a collision impact of 6.1. What is the collision impact of the same car at 65 mph?

53. The weight of a bass varies jointly as its girth and the square of its length. (**Girth** is the distance around the body of the fish.) A prize-winning bass weighed in at 22.7 lb and measured 36 in. long with a 21-in. girth. How much would a bass 28 in. long with an 18-in. girth weigh?

54. The weight of a trout varies jointly as its length and the square of its girth. One angler caught a trout that weighed 10.5 lb and measured 26 in. long with an 18-in. girth. Find the weight of a trout that is 22 in. long with a 15-in. girth.

55. The force needed to keep a car from skidding on a curve varies inversely as the radius of the curve and jointly as the weight of the car and the square of the speed. If 242 lb of force keeps a 2000-lb car from skidding on a curve of radius 500 ft at 30 mph, what force would keep the same car from skidding on a curve of radius 750 ft at 50 mph?

56. The maximum load that a cylindrical column with a circular cross section can hold varies directly as the fourth power of the diameter of the cross section and inversely as the square of the height. A 9-m column 1 m in diameter will support 8 metric tons. How many metric tons can be supported by a column 12 m high and $\frac{2}{3}$ m in diameter?

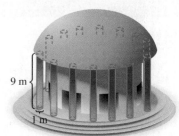

Load = 8 metric tons

57. The number of long-distance phone calls between two cities during a certain period varies jointly as the populations of the cities, p_1 and p_2, and inversely as the distance between them. If 80,000 calls are made between two cities 400 mi apart, with populations of 70,000 and 100,000, how many calls are made between cities with populations of 50,000 and 75,000 that are 250 mi apart?

58. In 2007, 51.2% of the homes in the United States used natural gas as the primary heating fuel. (*Source:* U.S. Census Bureau.) The volume of gas varies inversely as the pressure and directly as the temperature. (Temperature must be measured in *Kelvin* (K), a unit of measurement used in physics.) If a certain gas occupies a volume of 1.3 L at 300 K and a pressure of 18 newtons, find the volume at 340 K and a pressure of 24 newtons.

59. A body mass index from 27 through 29 carries a slight risk of weight-related health problems, while one of 30 or more indicates a great increase in risk. Use your own height and weight and the information in **Example 7** to determine your BMI and whether you are at risk.

60. The maximum load of a horizontal beam that is supported at both ends varies directly as the width and the square of the height and inversely as the length between the supports. A beam 6 m long, 0.1 m wide, and 0.06 m high supports a load of 360 kg. What is the maximum load supported by a beam 16 m long, 0.2 m wide, and 0.08 m high?

PREVIEW EXERCISES

Find each square root. If it is not a real number, say so. ***See Section 1.3.***

61. $\sqrt{64}$ **62.** $\sqrt{100}$ **63.** $-\sqrt{16}$ **64.** $-\sqrt{49}$ **65.** $\sqrt{-25}$

66. Does the relation $y = \sqrt{x - 1}$ define y as a function of x? Give its domain using interval notation. **See Section 3.5.**

CHAPTER 7 SUMMARY

KEY TERMS

7.1
rational expression
rational function

7.2
least common denominator
(LCD)

7.3
complex fraction

7.4
domain of the variable in a
rational equation
discontinuous

reciprocal function
vertical asymptote
horizontal asymptote

7.5
ratio
proportion

7.6
vary directly
proportional
constant of variation
vary inversely
vary jointly
combined variation

TEST YOUR WORD POWER

See how well you have learned the vocabulary in this chapter.

1. A **rational expression** is
   A. an algebraic expression made up of a term or the sum of a finite number of terms with real coefficients and integer exponents
   B. a polynomial equation of degree 2
   C. an expression with one or more fractions in the numerator, denominator, or both
   D. the quotient of two polynomials with denominator not 0.

2. In a given set of fractions, the **least common denominator** is
   A. the smallest denominator of all the denominators
   B. the smallest expression that is divisible by all the denominators
   C. the largest integer that evenly divides the numerator and denominator of all the fractions
   D. the largest denominator of all the denominators.

3. A **complex fraction** is
   A. an algebraic expression made up of a term or the sum of a finite number of terms with real coefficients and integer exponents
   B. a polynomial equation of degree 2
   C. an expression with one or more fractions in the numerator, denominator, or both

4. A **ratio**
   A. compares two quantities by using a quotient
   B. says that two quotients are equal
   C. is a product of two quantities
   D. is a difference between two quantities.

5. A **proportion**
   A. compares two quantities by using a quotient
   B. says that two quotients are equal
   C. is a product of two quantities
   D. is a difference between two quantities.

ANSWERS

1. D; *Examples:* $-\dfrac{3}{4y^2}, \dfrac{5x^3}{x+2}, \dfrac{a+3}{a^2-4a-5}$ 2. B; *Example:* The LCD of $\dfrac{1}{x}, \dfrac{2}{3}$, and $\dfrac{5}{x+1}$ is $3x(x+1)$.   3. C; *Examples:* $\dfrac{\frac{2}{3}}{\frac{4}{7}}, \dfrac{x-\frac{1}{x}}{x+\frac{1}{y}}, \dfrac{\frac{2}{a+1}}{a^2-1}$

4. A; *Example:* $\dfrac{7 \text{ in.}}{12 \text{ in.}}$ compares two quantities.   5. B; *Example:* The proportion $\dfrac{2}{3} = \dfrac{8}{12}$ states that the two ratios are equal.

QUICK REVIEW

CONCEPTS	EXAMPLES

7.1 Rational Expressions and Functions; Multiplying and Dividing

Rational Function

A function of the form

$$f(x) = \frac{P(x)}{Q(x)}, \quad \text{where } Q(x) \neq 0,$$

is a rational function. Its domain consists of all real numbers except those that make $Q(x) = 0$.

Find the domain.

$$f(x) = \frac{2x + 1}{3x + 6}$$

Solve $3x + 6 = 0$ to find $x = -2$. This is the only real number excluded from the domain. The domain is $\{x \mid x \neq -2\}$.

Fundamental Property of Rational Numbers

If $\frac{a}{b}$ is a rational number and if c is any nonzero real number, then

$$\frac{a}{b} = \frac{ac}{bc}.$$

$$\frac{3}{4} = \frac{3 \cdot 5}{4 \cdot 5} = \frac{15}{20}$$

Writing a Rational Expression in Lowest Terms

Step 1 Factor the numerator and the denominator completely.

Step 2 Apply the fundamental property. Divide out common factors.

Write in lowest terms.

$$\frac{2x + 8}{x^2 - 16}$$

$$= \frac{2(x + 4)}{(x - 4)(x + 4)} \qquad \text{Factor.}$$

$$= \frac{2}{x - 4} \qquad \text{Lowest terms}$$

Multiplying Rational Expressions

Step 1 Factor numerators and denominators.

Step 2 Apply the fundamental property.

Step 3 Multiply the numerators and multiply the denominators.

Step 4 Check that the product is in lowest terms.

Multiply.

$$\frac{x^2 + 2x + 1}{x^2 - 1} \cdot \frac{5}{3x + 3}$$

$$= \frac{(x + 1)^2}{(x - 1)(x + 1)} \cdot \frac{5}{3(x + 1)} \qquad \text{Factor.}$$

$$= \frac{5}{3(x - 1)} \qquad \text{Multiply; lowest terms}$$

Dividing Rational Expressions

Multiply the first rational expression (the dividend) by the reciprocal of the second (the divisor).

Divide.

$$\frac{2x + 5}{x - 3} \div \frac{2x^2 + 3x - 5}{x^2 - 9}$$

$$= \frac{2x + 5}{x - 3} \cdot \frac{x^2 - 9}{2x^2 + 3x - 5} \qquad \text{Multiply by the reciprocal.}$$

$$= \frac{2x + 5}{x - 3} \cdot \frac{(x + 3)(x - 3)}{(2x + 5)(x - 1)} \qquad \text{Factor.}$$

$$= \frac{x + 3}{x - 1} \qquad \text{Multiply; lowest terms}$$

(continued)

CONCEPTS	EXAMPLES

7.2 Adding and Subtracting Rational Expressions

Adding or Subtracting Rational Expressions

Step 1 **If the denominators are the same,** add or subtract the numerators. Place the result over the common denominator.

If the denominators are different, write all rational expressions with the LCD. Then add or subtract the numerators, and place the result over the common denominator.

Step 2 Make sure that the answer is in lowest terms.

Subtract.

$$\frac{1}{x + 6} - \frac{3}{x + 2}$$

$$= \frac{x + 2}{(x + 6)(x + 2)} - \frac{3(x + 6)}{(x + 6)(x + 2)}$$

$$= \frac{x + 2 - 3(x + 6)}{(x + 6)(x + 2)}$$

$$= \frac{x + 2 - 3x - 18}{(x + 6)(x + 2)}$$

$$= \frac{-2x - 16}{(x + 6)(x + 2)}$$

7.3 Complex Fractions

Simplifying a Complex Fraction

Method 1

Step 1 Simplify the numerator and denominator separately, as much as possible.

Step 2 Multiply the numerator by the reciprocal of the denominator.

Step 3 Then simplify the result.

Method 2

Step 1 Multiply the numerator and denominator of the complex fraction by the least common denominator of all fractions appearing in the complex fraction.

Step 2 Then simplify the result.

Method 1

$$\frac{\dfrac{1}{x^2} - \dfrac{1}{y^2}}{\dfrac{1}{x} + \dfrac{1}{y}}$$

$$= \frac{\dfrac{y^2}{x^2y^2} - \dfrac{x^2}{x^2y^2}}{\dfrac{y}{xy} + \dfrac{x}{xy}}$$

$$= \frac{\dfrac{y^2 - x^2}{x^2y^2}}{\dfrac{y + x}{xy}}$$

$$= \frac{y^2 - x^2}{x^2y^2} \div \frac{y + x}{xy}$$

$$= \frac{(y + x)(y - x)}{x^2y^2} \cdot \frac{xy}{y + x}$$

$$= \frac{y - x}{xy}$$

Method 2

$$\frac{\dfrac{1}{x^2} - \dfrac{1}{y^2}}{\dfrac{1}{x} + \dfrac{1}{y}}$$

$$= \frac{x^2y^2\left(\dfrac{1}{x^2} - \dfrac{1}{y^2}\right)}{x^2y^2\left(\dfrac{1}{x} + \dfrac{1}{y}\right)}$$

$$= \frac{y^2 - x^2}{xy^2 + x^2y}$$

$$= \frac{(y - x)(y + x)}{xy(y + x)}$$

$$= \frac{y - x}{xy}$$

(continued)

CONCEPTS	EXAMPLES

7.4 Equations with Rational Expressions and Graphs

Solving an Equation with Rational Expressions

Step 1 Determine the domain of the variable.

Step 2 Multiply each side of the equation by the least common denominator.

Step 3 Solve the resulting equation.

Solve.

$$\frac{3x + 2}{x - 2} + \frac{2}{x(x - 2)} = \frac{-1}{x}$$ Note that 0 and 2 are excluded from the domain.

$$x(3x + 2) + 2 = -(x - 2)$$ Multiply by the LCD, $x(x - 2)$.

$$3x^2 + 2x + 2 = -x + 2$$ Distributive property

$$3x^2 + 3x = 0$$ Add x. Subtract 2.

$$3x(x + 1) = 0$$ Factor.

$$3x = 0 \quad \text{or} \quad x + 1 = 0$$ Zero-factor property

$$x = 0 \quad \text{or} \quad x = -1$$ Solve each equation.

Step 4 Check that each proposed solution is in the domain, and discard any values that are not. Check the remaining proposed solutions in the original equation.

Of the two proposed solutions, 0 must be discarded because it is not in the domain. The solution set is $\{-1\}$.

Graphing a Rational Function

The graph of a rational function (written in lowest terms) may have one or more breaks. At such points, the graph will approach an asymptote.

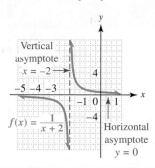

7.5 Applications of Rational Expressions

To solve a motion problem, use the formula

$$d = rt$$

or one of its equivalents,

$$t = \frac{d}{r} \quad \text{or} \quad r = \frac{d}{t}.$$

To solve a work problem, use the fact that if a complete job is done in t units of time, the rate of work is $\frac{1}{t}$ job per unit of time.

Solve.

A canal has a current of 2 mph. Find the rate of Amy's boat in still water if it goes 11 mi downstream in the same time that it goes 8 mi upstream.

Let x represent the rate of the boat in still water.

	Distance	Rate	Time
Downstream	11	$x + 2$	$\dfrac{11}{x + 2}$
Upstream	8	$x - 2$	$\dfrac{8}{x - 2}$

Times are equal.

$$\frac{11}{x + 2} = \frac{8}{x - 2}$$ Write an equation.

The LCD is $(x + 2)(x - 2)$.

$$11(x - 2) = 8(x + 2)$$ Multiply by the LCD.

$$11x - 22 = 8x + 16$$ Distributive property

$$3x = 38$$ Subtract $8x$. Add 22.

$$x = 12\frac{2}{3}$$ Divide by 3.

The rate in still water is $12\frac{2}{3}$ mph.

(continued)

CONCEPTS	EXAMPLES
7.6 Variation Let k be a real number. If $y = kx^n$, then y varies directly as x^n. If $y = \dfrac{k}{x^n}$, then y varies inversely as x^n. If $y = kxz$, then y varies jointly as x and z.	The area of a circle varies directly as the square of the radius. $$\mathcal{A} = kr^2 \qquad \text{Here, } k = \pi.$$ Pressure varies inversely as volume. $$p = \dfrac{k}{V}$$ For a given principal, interest varies jointly as interest rate and time. $$I = krt \qquad k \text{ is the given principal.}$$

CHAPTER 7 REVIEW EXERCISES

7.1 *(a) Find all real numbers that are excluded from the domain. (b) Give the domain, using set-builder notation.*

1. $f(x) = \dfrac{-7}{3x + 18}$

2. $f(x) = \dfrac{5x + 17}{x^2 - 7x + 10}$

3. $f(x) = \dfrac{9}{x^2 - 18x + 81}$

Write in lowest terms.

4. $\dfrac{12x^2 + 6x}{24x + 12}$

5. $\dfrac{25m^2 - n^2}{25m^2 - 10mn + n^2}$

6. $\dfrac{r - 2}{4 - r^2}$

7. What is meant by the reciprocal of a rational expression?

Multiply or divide. Write the answer in lowest terms.

8. $\dfrac{(2y + 3)^2}{5y} \cdot \dfrac{15y^3}{4y^2 - 9}$

9. $\dfrac{w^2 - 16}{w} \cdot \dfrac{3}{4 - w}$

10. $\dfrac{z^2 - z - 6}{z - 6} \cdot \dfrac{z^2 - 6z}{z^2 + 2z - 15}$

11. $\dfrac{m^3 - n^3}{m^2 - n^2} \div \dfrac{m^2 + mn + n^2}{m + n}$

7.2 *Suppose that each expression is the denominator of a rational expression. Find the least common denominator for each group.*

12. $32b^3, \quad 24b^5$

13. $9r^2, \quad 3r + 1, \quad 9$

14. $6x^2 + 13x - 5, \quad 9x^2 + 9x - 4$

15. $3x - 12, \quad x^2 - 2x - 8, \quad x^2 - 8x + 16$

Add or subtract as indicated.

16. $\dfrac{5}{3x^6y^5} - \dfrac{8}{9x^4y^7}$

17. $\dfrac{5y + 13}{y + 1} - \dfrac{1 - 7y}{y + 1}$

18. $\dfrac{6}{5a + 10} + \dfrac{7}{6a + 12}$

19. $\dfrac{3r}{10r^2 - 3rs - s^2} + \dfrac{2r}{2r^2 + rs - s^2}$

20. *Concept Check* Two students worked the following problem. One student got a sum of $\dfrac{1}{y - x}$ and the other got a sum of $\dfrac{-1}{x - y}$. Which student(s) got the correct answer?

$$\dfrac{2}{y - x} + \dfrac{1}{x - y}$$

7.3 *Simplify each complex fraction.*

21. $\dfrac{\dfrac{3}{t} + 2}{\dfrac{4}{t} - 7}$

22. $\dfrac{\dfrac{2}{m - 3n}}{\dfrac{1}{3n - m}}$

23. $\dfrac{\dfrac{3}{p} - \dfrac{2}{q}}{\dfrac{9q^2 - 4p^2}{qp}}$

24. $\dfrac{x^{-2} - y^{-2}}{x^{-1} - y^{-1}}$

7.4 *Solve each equation.*

25. $\dfrac{1}{t + 4} + \dfrac{1}{2} = \dfrac{3}{2t + 8}$

26. $\dfrac{-5m}{m + 1} + \dfrac{m}{3m + 3} = \dfrac{56}{6m + 6}$

27. $\dfrac{2}{k - 1} - \dfrac{4k + 1}{k^2 - 1} = \dfrac{-1}{k + 1}$

28. $\dfrac{5}{x + 2} + \dfrac{3}{x + 3} = \dfrac{x}{x^2 + 5x + 6}$

29. *Concept Check* Professor Dan Abbey asked the following question on a test: What is the solution set of $\frac{x + 3}{x + 3} = 1$? Only one student answered it correctly.

 (a) What is the solution set?

 (b) Many students answered $(-\infty, \infty)$. Why is this not correct?

30. Explain the difference between simplifying the expression $\frac{4}{x} + \frac{1}{2} - \frac{1}{3}$ and solving the equation $\frac{4}{x} + \frac{1}{2} = \frac{1}{3}$.

31. *Concept Check* Which graph has vertical and horizontal asymptotes? What are their equations?

A.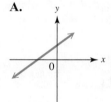

B.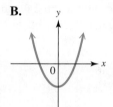

C.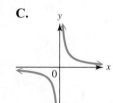

D.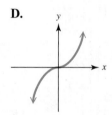

7.5

32. An equation of a law from physics is given in the form $\frac{1}{A} = \frac{1}{B} + \frac{1}{C}$. Find A if $B = 30$ and $C = 10$.

Solve each formula for the specified variable.

33. $F = \dfrac{GMm}{d^2}$ for m (physics)

34. $\mu = \dfrac{Mv}{M + m}$ for M (electronics)

Solve each problem.

35. An article in *Scientific American* predicted that, in the year 2050, about 23,200 of the 58,000 passenger-km per day in North America will be provided by high-speed trains. If the traffic volume in a typical region of North America is 15,000, how many passenger-kilometers per day will high-speed trains provide there? (*Source:* Schafer, Andreas, and David Victor, "The Past and Future of Global Mobility," *Scientific American.*)

36. A river has a current of 4 km per hr. Find the rate of Herby Assad's boat in still water if it goes 40 km downstream in the same time that it takes to go 24 km upstream.

	d	r	t
Upstream	24	x − 4	
Downstream	40		

37. A sink can be filled by a cold-water tap in 8 min and by a hot-water tap in 12 min. How long would it take to fill the sink with both taps open?

38. Jane Riggs and Jessica Ross need to sort a pile of bottles at the recycling center. Working alone, Jane could do the entire job in 9 hr, while Jessica could do the entire job in 6 hr. How long will it take them if they work together?

7.6

39. *Concept Check* In which one of the following does y vary inversely as x?

A. $y = 2x$ **B.** $y = \dfrac{x}{3}$ **C.** $y = \dfrac{3}{x}$ **D.** $y = x^2$

Solve each problem.

40. For the subject in a photograph to appear in the same perspective in the photograph as in real life, the viewing distance must be properly related to the amount of enlargement. For a particular camera, the viewing distance varies directly as the amount of enlargement. A picture that is taken with this camera and enlarged 5 times should be viewed from a distance of 250 mm. Suppose a print 8.6 times the size of the negative is made. From what distance should it be viewed?

41. The frequency (number of vibrations per second) of a vibrating guitar string varies inversely as its length. That is, a longer string vibrates fewer times in a second than a shorter string. Suppose a guitar string 0.65 m long vibrates 4.3 times per sec. What frequency would a string 0.5 m long have?

42. The volume of a rectangular box of a given height is proportional to its width and length. A box with width 2 ft and length 4 ft has volume 12 ft³. Find the volume of a box with the same height, but that is 3 ft wide and 5 ft long.

MIXED REVIEW EXERCISES

Write in lowest terms.

43. $\dfrac{x + 2y}{x^2 - 4y^2}$

44. $\dfrac{x^2 + 2x - 15}{x^2 - x - 6}$

Perform the indicated operations.

45. $\dfrac{2}{m} + \dfrac{5}{3m^2}$

46. $\dfrac{9}{3 - x} - \dfrac{2}{x - 3}$

47. $\dfrac{\dfrac{-3}{x} + \dfrac{x}{2}}{1 + \dfrac{x + 1}{x}}$

48. $\dfrac{\dfrac{3}{x} - 5}{6 + \dfrac{1}{x}}$

49. $\dfrac{4y + 16}{30} \div \dfrac{2y + 8}{5}$

50. $\dfrac{t^{-2} + s^{-2}}{t^{-1} - s^{-1}}$

51. $\dfrac{k^2 - 6k + 9}{1 - 216k^3} \cdot \dfrac{6k^2 + 17k - 3}{9 - k^2}$

52. $\dfrac{9x^2 + 46x + 5}{3x^2 - 2x - 1} \div \dfrac{x^2 + 11x + 30}{x^3 + 5x^2 - 6x}$

53. $\dfrac{4a}{a^2 - ab - 2b^2} - \dfrac{6b - a}{a^2 + 4ab + 3b^2}$

54. $\dfrac{a}{b} + \dfrac{b}{c} + \dfrac{c}{d}$

Solve.

55. $\dfrac{x + 3}{x^2 - 5x + 4} - \dfrac{1}{x} = \dfrac{2}{x^2 - 4x}$

56. $A = \dfrac{Rr}{R + r}$ for r

57. $1 - \dfrac{5}{r} = \dfrac{-4}{r^2}$

58. $\dfrac{3x}{x - 4} + \dfrac{2}{x} = \dfrac{48}{x^2 - 4x}$

Solve each problem.

59. The strength of a contact lens is given in units called diopters and also in millimeters of arc. As the diopters increase, the millimeters of arc decrease. The rational function defined by

$$f(x) = \frac{337}{x}$$

relates the arc measurement $f(x)$ to the diopter measurement x. (*Source:* Bausch and Lomb.)

   (a) What arc measurement will correspond to 40.5-diopter lenses?

   (b) A lens with an arc measurement of 7.51 mm will provide what diopter strength?

60. The hot-water tap can fill a tub in 20 min. The cold-water tap takes 15 min to fill the tub. How long would it take to fill the tub with both taps open?

61. At a certain gasoline station, 3 gal of unleaded gasoline cost $8.37. How much would 13 gal of the same gasoline cost?

62. The area of a triangle varies jointly as the lengths of the base and height. A triangle with base 10 ft and height 4 ft has area 20 ft^2. Find the area of a triangle with base 3 ft and height 8 ft.

63. If Dr. Dawson rides his bike to his office, he averages 12 mph. If he drives his car, he averages 36 mph. His time driving is $\frac{1}{4}$ hr less than his time riding his bike. How far is his office from home?

64. A private plane traveled from San Francisco to a secret rendezvous. It averaged 200 mph. On the return trip, the average rate was 300 mph. If the total traveling time was 4 hr, how far from San Francisco was the secret rendezvous?

65. Johnny averages 30 mph when he drives on the old highway to his favorite fishing hole, and he averages 50 mph when he takes the interstate. If both routes are the same length, and he saves 2 hr by traveling on the interstate, how far away is the fishing hole?

CHAPTER 7

TEST

Step-by-step test solutions are found on the Chapter Test Prep Videos available via the Video Resources on DVD, in *MyMathLab*, or on YouTube (search "LialIntermediateAlg").

View the complete solutions to all Chapter Test exercises on the Video Resources on DVD.

1. Find all real numbers excluded from the domain of $f(x) = \dfrac{x + 3}{3x^2 + 2x - 8}$. Then give the domain, using set-builder notation.

2. Write $\dfrac{6x^2 - 13x - 5}{9x^3 - x}$ in lowest terms.

Multiply or divide.

3. $\dfrac{(x + 3)^2}{4} \cdot \dfrac{6}{2x + 6}$

4. $\dfrac{y^2 - 16}{y^2 - 25} \cdot \dfrac{y^2 + 2y - 15}{y^2 - 7y + 12}$

5. $\dfrac{3 - t}{5} \div \dfrac{t - 3}{10}$

6. $\dfrac{x^2 - 9}{x^3 + 3x^2} \div \dfrac{x^2 + x - 12}{x^3 + 9x^2 + 20x}$

7. Find the least common denominator for the following group of denominators: $t^2 + t - 6$, $t^2 + 3t$, t^2.

Add or subtract as indicated.

8. $\dfrac{7}{6t^2} - \dfrac{1}{3t}$

9. $\dfrac{3}{7a^4b^3} + \dfrac{5}{21a^5b^2}$

10. $\dfrac{9}{x^2 - 6x + 9} + \dfrac{2}{x^2 - 9}$

11. $\dfrac{6}{x + 4} + \dfrac{1}{x + 2} - \dfrac{3x}{x^2 + 6x + 8}$

Simplify each complex fraction.

12. $\dfrac{\dfrac{12}{r + 4}}{\dfrac{11}{6r + 24}}$

13. $\dfrac{\dfrac{1}{a} - \dfrac{1}{b}}{\dfrac{a}{b} - \dfrac{b}{a}}$

14. $\dfrac{2x^{-2} + y^{-2}}{x^{-1} - y^{-1}}$

15. Identify each of the following as an expression to be simplified or an equation to be solved. Then simplify the expression and solve the equation.

   (a) $\dfrac{2x}{3} + \dfrac{x}{4} - \dfrac{11}{2}$

   (b) $\dfrac{2x}{3} + \dfrac{x}{4} = \dfrac{11}{2}$

Solve each equation.

16. $\dfrac{1}{x} - \dfrac{4}{3x} = \dfrac{1}{x - 2}$

17. $\dfrac{y}{y + 2} - \dfrac{1}{y - 2} = \dfrac{8}{y^2 - 4}$

18. Solve for the variable ℓ in this formula from mathematics: $S = \dfrac{n}{2}(a + \ell)$.

19. Graph the function defined by $f(x) = \dfrac{-2}{x + 1}$. Give the equations of its vertical and horizontal asymptotes.

Solve each problem.

20. Chris Dada can do a job in 9 hr, while Dana Bortes can do the same job in 5 hr. How long would it take them to do the job if they worked together?

21. The rate of the current in a stream is 3 mph. Nana's boat can go 36 mi downstream in the same time that it takes to go 24 mi upstream. Find the rate of her boat in still water.

22. Biologists collected a sample of 600 fish from West Okoboji Lake on May 1 and tagged each of them. When they returned on June 1, a new sample of 800 fish was collected and 10 of these had been previously tagged. Use this experiment to determine the approximate fish population of West Okoboji Lake.

23. In biology, the function defined by

$$g(x) = \dfrac{5x}{2 + x}$$

gives the growth rate of a population for x units of available food. (*Source:* Smith, J. Maynard, *Models in Ecology,* Cambridge University Press.)

   (a) What amount of food (in appropriate units) would produce a growth rate of 3 units of growth per unit of food?

   (b) What is the growth rate if no food is available?

24. The current in a simple electrical circuit is inversely proportional to the resistance. If the current is 80 amps when the resistance is 30 ohms, find the current when the resistance is 12 ohms.

25. The force of the wind blowing on a vertical surface varies jointly as the area of the surface and the square of the velocity. If a wind blowing at 40 mph exerts a force of 50 lb on a surface of 500 ft², how much force will a wind of 80 mph place on a surface of 2 ft²?

CHAPTERS (1–7) CUMULATIVE REVIEW EXERCISES

1. Evaluate $|2x| + 3y - z^3$ for $x = -4$, $y = 3$, and $z = 6$.

Solve each equation or inequality.

2. $7(2x + 3) - 4(2x + 1) = 2(x + 1)$

3. $|6x - 8| - 4 = 0$

4. $\dfrac{2}{3}y + \dfrac{5}{12}y \le 20$

5. $|3x + 2| \ge 4$

Solve each problem.

6. Abushieba Ibrahim invested some money at 4% interest and twice as much at 3% interest. His interest for the first year was $400. How much did he invest at each rate?

7. A triangle has area 42 m². The base is 14 m long. Find the height of the triangle.

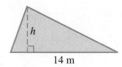

14 m

8. Graph $-4x + 2y = 8$ and give the intercepts.

Find the slope of each line described in Exercises 9 and 10.

9. Through $(-5, 8)$ and $(-1, 2)$

10. Perpendicular to $4x - 3y = 12$

11. Write an equation of the line in **Exercise 9.** Give the equation in the form $y = mx + b$.

Graph the solution set of each inequality.

12. $2x + 5y > 10$

13. $x - y \ge 3$ and $3x + 4y \le 12$

14. Decide whether the relation $y = -\sqrt{x + 2}$ is a function, and give its domain and range.

15. Suppose that $y = f(x)$ and $5x - 3y = 8$.

(a) Find the equation that defines $f(x)$. That is, $f(x) = $ _____.

(b) Find $f(1)$.

Solve each system.

16. $4x - y = -7$
$5x + 2y = 1$

17. $x + y - 2z = -1$
$2x - y + z = -6$
$3x + 2y - 3z = -3$

18. $x + 2y + z = 5$
$x - y + z = 3$
$2x + 4y + 2z = 11$

19. With traffic taken into account, an automobile can travel 7 km, on the average, in the same time that an airplane can travel 100 km. The average rate of an airplane is 558 km per hr greater than that of an automobile. Find both rates. (*Source:* Schafer, Andreas, and David Victor, "The Past and Future of Global Mobility," *Scientific American.*)

20. Simplify $\left(\dfrac{m^{-4}n^2}{m^2n^{-3}}\right) \cdot \left(\dfrac{m^5n^{-1}}{m^{-2}n^5}\right)$. Write the answer with only positive exponents. Assume that all variables represent nonzero real numbers.

Perform the indicated operations.

21. $(3y^2 - 2y + 6) - (-y^2 + 5y + 12)$

22. $(4f + 3)(3f - 1)$

23. $\left(\dfrac{1}{4}x + 5\right)^2$

24. $(3x^3 + 13x^2 - 17x - 7) \div (3x + 1)$

25. For the polynomial functions defined by

$$f(x) = x^2 + 2x - 3, \quad g(x) = 2x^3 - 3x^2 + 4x - 1, \quad \text{and} \quad h(x) = x^2,$$

find **(a)** $(f + g)(x)$, **(b)** $(g - f)(x)$, **(c)** $(f + g)(-1)$, and **(d)** $(f \circ h)(x)$.

Factor each polynomial completely.

26. $2x^2 - 13x - 45$ **27.** $100t^4 - 25$ **28.** $8p^3 + 125$

29. Solve the equation $3x^2 + 4x = 7$.

30. Write $\dfrac{y^2 - 16}{y^2 - 8y + 16}$ in lowest terms.

Perform the indicated operations. Express the answer in lowest terms.

31. $\dfrac{2a^2}{a + b} \cdot \dfrac{a - b}{4a}$ **32.** $\dfrac{x^2 - 9}{2x + 4} \div \dfrac{x^3 - 27}{4}$ **33.** $\dfrac{x + 4}{x - 2} + \dfrac{2x - 10}{x - 2}$

34. Solve the equation $\dfrac{-3x}{x + 1} + \dfrac{4x + 1}{x} = \dfrac{-3}{x^2 + x}$.

Solve each problem.

35. Machine A can complete a certain job in 2 hr. To speed up the work, Machine B, which could complete the job alone in 3 hr, is brought in to help. How long will it take the two machines to complete the job working together?

36. The cost of a pizza varies directly as the square of its radius. If a pizza with a 7-in. radius costs $6.00, how much should a pizza with a 9-in. radius cost?

Roots, Radicals, and Root Functions

The Pythagorean theorem states that in any right triangle with perpendicular sides a and b and longest side c,

$$a^2 + b^2 = c^2.$$

It is used in surveying, drafting, engineering, navigation, and many other fields. Although attributed to Pythagoras, it was known to surveyors from Egypt to China for a thousand years before Pythagoras.

In the 1939 movie *The Wizard of Oz,* the Scarecrow asks the Wizard for a brain. When the Wizard presents him with a diploma granting him a Th.D. (Doctor of Thinkology), the Scarecrow declares the following.

The sum of the square roots of any two sides of an isosceles triangle is equal to the square root of the remaining side. . . . Oh joy! Rapture! I've got a brain.

Did the Scarecrow state the Pythagorean theorem correctly? We will investigate this in **Exercises 133–134** of **Section 8.3.**

8.1 Radical Expressions and Graphs

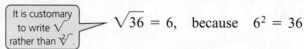

OBJECTIVES

1. Find roots of numbers.
2. Find principal roots.
3. Graph functions defined by radical expressions.
4. Find nth roots of nth powers.
5. Use a calculator to find roots.

OBJECTIVE 1 **Find roots of numbers.** Recall that $6^2 = 36$. We say that 6 *squared* is 36. The opposite (or inverse) of *squaring* a number is taking its *square root*.

> It is customary to write $\sqrt{}$ rather than $\sqrt[2]{}$.

$\sqrt{36} = 6$, because $6^2 = 36$

We extend this discussion to *cube roots* $\sqrt[3]{}$, *fourth roots* $\sqrt[4]{}$, and higher roots.

$\sqrt[n]{a}$

The nth root of a, written $\sqrt[n]{a}$, is a number whose nth power equals a. That is,

$$\sqrt[n]{a} = b \quad \text{means} \quad b^n = a.$$

The number a is the **radicand**, n is the **index** or **order,** and the expression $\sqrt[n]{a}$ is a **radical.**

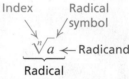

NOW TRY EXERCISE 1

Simplify.

(a) $\sqrt[3]{1000}$ **(b)** $\sqrt[4]{625}$

(c) $\sqrt[4]{\dfrac{1}{256}}$ **(d)** $\sqrt[3]{0.027}$

EXAMPLE 1 Simplifying Higher Roots

Simplify.

(a) $\sqrt[3]{64} = 4$, because $4^3 = 64$. **(b)** $\sqrt[3]{125} = 5$, because $5^3 = 125$.

(c) $\sqrt[4]{16} = 2$, because $2^4 = 16$. **(d)** $\sqrt[5]{32} = 2$, because $2^5 = 32$.

(e) $\sqrt[3]{\dfrac{8}{27}} = \dfrac{2}{3}$, because $\left(\dfrac{2}{3}\right)^3 = \dfrac{8}{27}$.

(f) $\sqrt[4]{0.0016} = 0.2$, because $(0.2)^4 = 0.0016$. NOW TRY

OBJECTIVE 2 **Find principal roots.** If n is even, positive numbers have two nth roots. For example, both 4 and -4 are square roots of 16, and 2 and -2 are fourth roots of 16. In such cases, the notation $\sqrt[n]{a}$ represents the positive root, called the **principal root,** and $-\sqrt[n]{a}$ represents the negative root.

nth Root

Case 1 If n is **even** and a is **positive or 0,** then

$$\sqrt[n]{a} \text{ represents the \textbf{principal }}n\textbf{th root of } a,$$

and $-\sqrt[n]{a}$ represents the **negative nth root** of a.

Case 2 If n is **even** and a is **negative,** then

$$\sqrt[n]{a} \text{ is not a real number.}$$

Case 3 If n is **odd,** then

there is exactly one real nth root of a, written $\sqrt[n]{a}$.

NOW TRY ANSWERS

1. **(a)** 10 **(b)** 5 **(c)** $\frac{1}{4}$ **(d)** 0.3

If n is even, the two nth roots of a are often written together as $\pm\sqrt[n]{a}$, with $\pm$ read "positive or negative," or "plus or minus."

NOW TRY
EXERCISE 2

Find each root.

(a) $-\sqrt{25}$ **(b)** $\sqrt[4]{-625}$

(c) $\sqrt[5]{-32}$ **(d)** $-\sqrt[3]{64}$

EXAMPLE 2 Finding Roots

Find each root.

(a) $\sqrt{100} = 10$

Because the radicand, 100, is *positive*, there are two square roots: 10 and -10. We want the principal square root, which is 10.

(b) $-\sqrt{100} = -10$

Here, we want the negative square root, -10.

(c) $\sqrt[4]{81} = 3$ Principal 4th root   **(d)** $-\sqrt[4]{81} = -3$ Negative 4th root

Parts (a)–(d) illustrate Case 1 in the preceding box.

(e) $\sqrt[4]{-81}$

The index is *even* and the radicand is *negative*, so $\sqrt[4]{-81}$ is not a real number. This is Case 2 in the preceding box.

(f) $\sqrt[3]{8} = 2$, because $2^3 = 8$.   **(g)** $\sqrt[3]{-8} = -2$, because $(-2)^3 = -8$.

Parts (f) and (g) illustrate Case 3 in the box. The index is *odd*, so each radical represents exactly one nth root (regardless of whether the radicand is positive, negative, or 0).

NOW TRY

OBJECTIVE 3 **Graph functions defined by radical expressions.** A **radical expression** is an algebraic expression that contains radicals.

$$3 - \sqrt{x}, \quad \sqrt[3]{x}, \quad \text{and} \quad \sqrt{2x - 1} \qquad \text{Examples of radical expressions}$$

In earlier chapters, we graphed functions defined by polynomial and rational expressions. Now we examine the graphs of functions defined by the basic radical expressions $f(x) = \sqrt{x}$ and $f(x) = \sqrt[3]{x}$.

FIGURE 1 shows the graph of the **square root function,** together with a table of selected points. Only nonnegative values can be used for x, so the domain is $[0, \infty)$. Because $\sqrt{x}$ is the principal square root of x, it always has a nonnegative value, so the range is also $[0, \infty)$.

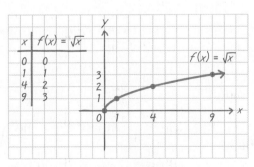

Square root function

$$f(x) = \sqrt{x}$$

Domain: $[0, \infty)$
Range: $[0, \infty)$

FIGURE 1

NOW TRY ANSWERS
2. (a) -5
  (b) It is not a real number.
  (c) -2 **(d)** -4

FIGURE 2 shows the graph of the **cube root function.** Since any real number (positive, negative, or 0) can be used for x in the cube root function, $\sqrt[3]{x}$ can be positive, negative, or 0. Thus, both the domain and the range of the cube root function are $(-\infty, \infty)$.

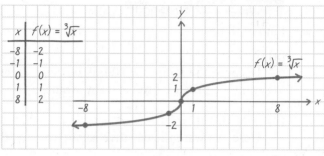

Cube root function

$$f(x) = \sqrt[3]{x}$$

Domain: $(-\infty, \infty)$
Range: $(-\infty, \infty)$

FIGURE 2

NOW TRY
EXERCISE 3

Graph each function. Give the domain and range.

(a) $f(x) = \sqrt{x + 1}$

(b) $f(x) = \sqrt[3]{x} - 1$

EXAMPLE 3 Graphing Functions Defined with Radicals

Graph each function by creating a table of values. Give the domain and range.

(a) $f(x) = \sqrt{x - 3}$

A table of values is given with the graph in **FIGURE 3**. The x-values were chosen in such a way that the function values are all integers. For the radicand to be nonnegative, we must have

$$x - 3 \geq 0, \quad \text{or} \quad x \geq 3.$$

Therefore, the domain of this function is $[3, \infty)$. Function values are positive or 0, so the range is $[0, \infty)$.

x	$f(x) = \sqrt{x - 3}$
3	$\sqrt{3 - 3} = 0$
4	$\sqrt{4 - 3} = 1$
7	$\sqrt{7 - 3} = 2$

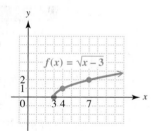

FIGURE 3

(b) $f(x) = \sqrt[3]{x} + 2$

See **FIGURE 4**. Both the domain and range are $(-\infty, \infty)$.

NOW TRY ANSWERS

3. (a)

domain: $[-1, \infty)$;
range: $[0, \infty)$

(b)

domain: $(-\infty, \infty)$;
range: $(-\infty, \infty)$

x	$f(x) = \sqrt[3]{x} + 2$
-8	$\sqrt[3]{-8} + 2 = 0$
-1	$\sqrt[3]{-1} + 2 = 1$
0	$\sqrt[3]{0} + 2 = 2$
1	$\sqrt[3]{1} + 2 = 3$
8	$\sqrt[3]{8} + 2 = 4$

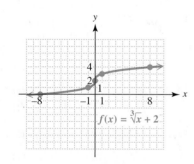

FIGURE 4

NOW TRY

OBJECTIVE 4 **Find *n*th roots of *n*th powers.** Consider the expression $\sqrt{a^2}$. At first glance, you may think that it is equivalent to *a*. However, this is not necessarily true. For example, consider the following.

If $a = 6$, then $\sqrt{a^2} =$ $\sqrt{6^2}$ $= \sqrt{36} = 6$.

If $a = -6$, then $\sqrt{a^2} = \sqrt{(-6)^2} = \sqrt{36} = 6.$ ⟵ Instead of **−6**, we get **6**, the *absolute value* of −6.

Since the symbol $\sqrt{a^2}$ represents the *nonnegative* square root, we express $\sqrt{a^2}$ with absolute value bars, as $|a|$, because *a* may be a negative number.

$\sqrt{a^2}$

For any real number *a*, $\sqrt{a^2} = |a|.$

That is, the principal square root of a^2 is the absolute value of *a*.

NOW TRY
EXERCISE 4

Find each square root.

(a) $\sqrt{11^2}$ (b) $\sqrt{(-11)^2}$

(c) $\sqrt{z^2}$ (d) $\sqrt{(-z)^2}$

EXAMPLE 4 Simplifying Square Roots by Using Absolute Value

Find each square root.

(a) $\sqrt{7^2} = |7| = 7$ (b) $\sqrt{(-7)^2} = |-7| = 7$

(c) $\sqrt{k^2} = |k|$ (d) $\sqrt{(-k)^2} = |-k| = |k|$ NOW TRY

We can generalize this idea to any *n*th root.

$\sqrt[n]{a^n}$

If *n* is an **even** positive integer, then $\sqrt[n]{a^n} = |a|.$

If *n* is an **odd** positive integer, then $\sqrt[n]{a^n} = a.$

That is, use the absolute value symbol when *n* is even. Absolute value is not used when *n* is odd.

NOW TRY
EXERCISE 5

Simplify each root.

(a) $\sqrt[8]{(-2)^8}$ (b) $\sqrt[3]{(-9)^3}$

(c) $-\sqrt[4]{(-10)^4}$ (d) $-\sqrt{m^8}$

(e) $\sqrt[3]{x^{18}}$ (f) $\sqrt[4]{t^{20}}$

EXAMPLE 5 Simplifying Higher Roots by Using Absolute Value

Simplify each root.

(a) $\sqrt[6]{(-3)^6} = |-3| = 3$ *n* is even. Use absolute value.

(b) $\sqrt[5]{(-4)^5} = -4$ *n* is odd.

(c) $-\sqrt[4]{(-9)^4} = -|-9| = -9$ *n* is even. Use absolute value.

(d) $-\sqrt{m^4} = -|m^2| = -m^2$ For all *m*, $|m^2| = m^2$.
    No absolute value bars are needed here, because m^2 is nonnegative for any real number value of *m*.

(e) $\sqrt[3]{a^{12}} = a^4$, because $a^{12} = (a^4)^3$.

(f) $\sqrt[4]{x^{12}} = |x^3|$

NOW TRY ANSWERS
4. (a) 11 (b) 11 (c) $|z|$
   (d) $|z|$
5. (a) 2 (b) −9 (c) −10
   (d) $-m^4$ (e) x^6 (f) $|t^5|$

We use absolute value to guarantee that the result is not negative (because x^3 is negative when *x* is negative). If desired $|x^3|$ can be written as $x^2 \cdot |x|$. NOW TRY

OBJECTIVE 5   Use a calculator to find roots. While numbers such as $\sqrt{9}$ and $\sqrt[3]{-8}$ are rational, radicals are often irrational numbers. To find approximations of such radicals, we usually use a scientific or graphing calculator. For example,

$$\sqrt{15} \approx 3.872983346, \quad \sqrt[3]{10} \approx 2.15443469, \quad \text{and} \quad \sqrt[4]{2} \approx 1.189207115,$$

where the symbol $\approx$ means "is approximately equal to." In this book, we often show approximations rounded to three decimal places. Thus,

$$\sqrt{15} \approx 3.873, \quad \sqrt[3]{10} \approx 2.154, \quad \text{and} \quad \sqrt[4]{2} \approx 1.189.$$

FIGURE 5 shows how the preceding approximations are displayed on a TI-83/84 Plus graphing calculator.

There is a simple way to check that a calculator approximation is "in the ballpark." For example, because 16 is a little larger than 15, $\sqrt{16} = 4$ should be a little larger than $\sqrt{15}$. Thus, 3.873 is reasonable as an approximation for $\sqrt{15}$.

```
√(15)
            3.873
³√(10)
            2.154
4 ˣ√2
            1.189
```

FIGURE 5

NOTE The methods for finding approximations differ among makes and models of calculators. *You should always consult your owner's manual for keystroke instructions.* Be aware that graphing calculators often differ from scientific calculators in the order in which keystrokes are made.

⤷ **NOW TRY**
  EXERCISE 6

Use a calculator to approximate each radical to three decimal places.

(a) $-\sqrt{92}$ **(b)** $\sqrt[4]{39}$

(c) $\sqrt[5]{33}$

EXAMPLE 6   Finding Approximations for Roots

Use a calculator to verify that each approximation is correct.

(a) $\sqrt{39} \approx 6.245$ **(b)** $-\sqrt{72} \approx -8.485$

(c) $\sqrt[3]{93} \approx 4.531$ **(d)** $\sqrt[4]{39} \approx 2.499$ *NOW TRY*

⤷ **NOW TRY**
  EXERCISE 7

Use the formula in **Example 7** to approximate f to the nearest thousand if

$$L = 7 \times 10^{-5}$$

and $C = 3 \times 10^{-9}.$

EXAMPLE 7   Using Roots to Calculate Resonant Frequency

In electronics, the resonant frequency f of a circuit may be found by the formula

$$f = \frac{1}{2\pi\sqrt{LC}},$$

where f is in cycles per second, L is in henrys, and C is in farads. (Henrys and farads are units of measure in electronics.) Find the resonant frequency f if $L = 5 \times 10^{-4}$ henry and $C = 3 \times 10^{-10}$ farad. Give your answer to the nearest thousand.

Find the value of f when $L = 5 \times 10^{-4}$ and $C = 3 \times 10^{-10}$.

$$f = \frac{1}{2\pi\sqrt{LC}} \qquad \text{Given formula}$$

$$= \frac{1}{2\pi\sqrt{(5 \times 10^{-4})(3 \times 10^{-10})}} \qquad \text{Substitute for } L \text{ and } C.$$

$$\approx 411,000 \qquad \text{Use a calculator.}$$

NOW TRY ANSWERS
6. **(a)** -9.592 **(b)** 2.499
   (c) 2.012
7. 347,000 cycles per sec

The resonant frequency f is approximately 411,000 cycles per sec.   *NOW TRY*

⊙ *Complete solution available on the Video Resources on DVD*

Match each expression from Column I with the equivalent choice from Column II. Answers may be used more than once. **See Examples 1 and 2.**

I

1. $-\sqrt{25}$ **2.** $\sqrt{-25}$

3. $\sqrt[3]{-27}$ **4.** $\sqrt[5]{-32}$

⊙ **5.** $\sqrt[4]{81}$ **6.** $\sqrt[3]{8}$

II

A. 3   **B.** -2

C. 2   **D.** -3

E. -5 **F.** Not a real number

Find each root that is a real number. **See Examples 1 and 2.**

⊙ **7.** $-\sqrt{81}$ **8.** $-\sqrt{121}$ **9.** $\sqrt[3]{216}$ **10.** $\sqrt[3]{343}$

11. $\sqrt[3]{-64}$ **12.** $\sqrt[3]{-125}$ **13.** $-\sqrt[3]{512}$ **14.** $-\sqrt[3]{1000}$

15. $\sqrt[4]{1296}$ **16.** $\sqrt[4]{625}$ **17.** $-\sqrt[4]{16}$ **18.** $-\sqrt[4]{256}$

19. $\sqrt[4]{-625}$ **20.** $\sqrt[4]{-256}$ **21.** $\sqrt[6]{64}$ **22.** $\sqrt[6]{729}$

23. $\sqrt[6]{-32}$ **24.** $\sqrt[8]{-1}$ **25.** $\sqrt{\dfrac{64}{81}}$ **26.** $\sqrt{\dfrac{100}{9}}$

27. $\sqrt[3]{\dfrac{64}{27}}$ **28.** $\sqrt[4]{\dfrac{81}{16}}$ **29.** $-\sqrt[6]{\dfrac{1}{64}}$ **30.** $-\sqrt[5]{\dfrac{1}{32}}$

31. $-\sqrt[3]{-27}$ **32.** $-\sqrt[3]{-64}$ **33.** $\sqrt{0.25}$ **34.** $\sqrt{0.36}$

35. $-\sqrt{0.49}$ **36.** $-\sqrt{0.81}$ **37.** $\sqrt[3]{0.001}$ **38.** $\sqrt[3]{0.125}$

39. *Concept Check*   Consider the expression $-\sqrt{-a}$. Decide whether it is *positive, negative,* 0, or *not a real number* if

   (a) $a > 0$,   **(b)** $a < 0$,   **(c)** $a = 0$.

40. *Concept Check*   If n is odd, under what conditions is $\sqrt[n]{a}$

   (a) positive,   **(b)** negative,   **(c)** 0?

Graph each function and give its domain and range. **See Example 3.**

⊙ **41.** $f(x) = \sqrt{x + 3}$ **42.** $f(x) = \sqrt{x - 5}$ **43.** $f(x) = \sqrt{x - 2}$

44. $f(x) = \sqrt{x + 4}$ **45.** $f(x) = \sqrt[3]{x} - 3$ **46.** $f(x) = \sqrt[3]{x} + 1$

47. $f(x) = \sqrt[3]{x - 3}$ **48.** $f(x) = \sqrt[3]{x + 1}$

Simplify each root. **See Examples 4 and 5.**

⊙ **49.** $\sqrt{12^2}$ **50.** $\sqrt{19^2}$ **51.** $\sqrt{(-10)^2}$ **52.** $\sqrt{(-13)^2}$

⊙ **53.** $\sqrt[6]{(-2)^6}$ **54.** $\sqrt[6]{(-4)^6}$ **55.** $\sqrt[5]{(-9)^5}$ **56.** $\sqrt[5]{(-8)^5}$

57. $-\sqrt[6]{(-5)^6}$ **58.** $-\sqrt[6]{(-7)^6}$ **59.** $\sqrt{x^2}$ **60.** $-\sqrt{x^2}$

61. $\sqrt{(-z)^2}$ **62.** $\sqrt{(-q)^2}$ **63.** $\sqrt[3]{x^3}$ **64.** $-\sqrt[3]{x^3}$

65. $\sqrt[3]{x^{15}}$ **66.** $\sqrt[3]{m^9}$ **67.** $\sqrt[6]{x^{30}}$ **68.** $\sqrt[4]{k^{20}}$

Concept Check *Choose the closest approximation of each square root.*

69. $\sqrt{123.5}$

    A. 9   **B.** 10   **C.** 11   **D.** 12

70. $\sqrt{67.8}$

    A. 7   **B.** 8   **C.** 9   **D.** 10

Find a decimal approximation for each radical. Round the answer to three decimal places.
See Example 6.

71. $\sqrt{9483}$ **72.** $\sqrt{6825}$ **73.** $\sqrt{284.361}$ **74.** $\sqrt{846.104}$

75. $-\sqrt{82}$ **76.** $-\sqrt{91}$ **77.** $\sqrt[3]{423}$ **78.** $\sqrt[3]{555}$

79. $\sqrt[4]{100}$ **80.** $\sqrt[4]{250}$ **81.** $\sqrt[5]{23.8}$ **82.** $\sqrt[5]{98.4}$

Refer to the rectangle to answer the questions in Exercises 83 and 84.

83. Which one of the following is the best estimate of its area?

    A. 50   **B.** 250   **C.** 2500   **D.** 100

84. Which one of the following is the best estimate of its perimeter?

    A. 15   **B.** 250   **C.** 100   **D.** 30

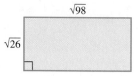

Solve each problem. See Example 7.

85. Use the formula in **Example 7** to calculate the resonant frequency of a circuit to the nearest thousand if $L = 7.237 \times 10^{-5}$ henry and $C = 2.5 \times 10^{-10}$ farad.

86. The threshold weight T for a person is the weight above which the risk of death increases greatly. The threshold weight in pounds for men aged 40–49 is related to height h in inches by the formula

$$h = 12.3\sqrt[3]{T}.$$

What height corresponds to a threshold weight of 216 lb for a 43-year-old man? Round your answer to the nearest inch and then to the nearest tenth of a foot.

87. According to an article in *The World Scanner Report,* the distance D, in miles, to the horizon from an observer's point of view over water or "flat" earth is given by

$$D = \sqrt{2H},$$

where H is the height of the point of view, in feet. If a person whose eyes are 6 ft above ground level is standing at the top of a hill 44 ft above "flat" earth, approximately how far to the horizon will she be able to see?

88. The time t in seconds for one complete swing of a simple pendulum, where L is the length of the pendulum in feet, and g, the acceleration due to gravity, is about 32 ft per sec², is

$$t = 2\pi\sqrt{\frac{L}{g}}.$$

Find the time of a complete swing of a 2-ft pendulum to the nearest tenth of a second.

89. **Heron's formula** gives a method of finding the area of a triangle if the lengths of its sides are known. Suppose that a, b, and c are the lengths of the sides. Let s denote one-half of the perimeter of the triangle (called the *semiperimeter*); that is, $s = \frac{1}{2}(a + b + c)$. Then the area of the triangle is

$$\mathcal{A} = \sqrt{s(s - a)(s - b)(s - c)}.$$

Find the area of the Bermuda Triangle, to the nearest thousand square miles, if the "sides" of this triangle measure approximately 850 mi, 925 mi, and 1300 mi.

90. Use Heron's formula from **Exercise 89** to find the area of a triangle with sides of lengths $a = 11$ m, $b = 60$ m, and $c = 61$ m.

91. The coefficient of self-induction L (in henrys), the energy P stored in an electronic circuit (in joules), and the current I (in amps) are related by this formula.

$$I = \sqrt{\frac{2P}{L}}$$

(a) Find I if $P = 120$ and $L = 80$.   **(b)** Find I if $P = 100$ and $L = 40$.

92. The Vietnam Veterans Memorial in Washington, DC, is in the shape of an unenclosed isosceles triangle with equal sides of length 246.75 ft. If the triangle were enclosed, the third side would have length 438.14 ft. Use Heron's formula from **Exercise 89** to find the area of this enclosure to the nearest hundred square feet. (*Source:* Information pamphlet obtained at the Vietnam Veterans Memorial.)

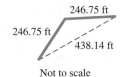

246.75 ft

246.75 ft

438.14 ft

Not to scale

PREVIEW EXERCISES

Apply the rules for exponents. Write each result with only positive exponents. Assume that all variables represent nonzero real numbers. ***See Section 5.1.***

93. $x^5 \cdot x^{-1} \cdot x^{-3}$ **94.** $(4x^2y^3)(2^3x^5y)$ **95.** $\left(\dfrac{2}{3}\right)^{-3}$ **96.** $\dfrac{5}{5^{-1}}$

8.2 Rational Exponents

OBJECTIVES

1 Use exponential notation for nth roots.

2 Define and use expressions of the form $a^{m/n}$.

3 Convert between radicals and rational exponents.

4 Use the rules for exponents with rational exponents.

OBJECTIVE 1   Use exponential notation for nth roots.   Consider the product $(3^{1/2})^2 = 3^{1/2} \cdot 3^{1/2}$. Using the rules of exponents from **Section 5.1,** we can simplify this product as follows.

$$
\begin{aligned}
(3^{1/2})^2 &= 3^{1/2} \cdot 3^{1/2} \\
&= 3^{1/2+1/2} && \text{Product rule: } a^m \cdot a^n = a^{m+n} \\
&= 3^1 && \text{Add exponents.} \\
&= 3 && a^1 = a
\end{aligned}
$$

Also, by definition,

$$\left(\sqrt{3}\right)^2 = \sqrt{3} \cdot \sqrt{3} = 3.$$

Since both $(3^{1/2})^2$ and $\left(\sqrt{3}\right)^2$ are equal to 3, it seems reasonable to define

$$3^{1/2} = \sqrt{3}.$$

This suggests the following generalization.

$a^{1/n}$

If $\sqrt[n]{a}$ is a real number, then $a^{1/n} = \sqrt[n]{a}.$

$$4^{1/2} = \sqrt{4}, \quad 8^{1/3} = \sqrt[3]{8}, \quad \text{and} \quad 16^{1/4} = \sqrt[4]{16} \qquad \text{Examples of } a^{1/n}$$

Notice that the denominator of the rational exponent is the index of the radical.

NOW TRY
EXERCISE 1

Evaluate each exponential.

(a) $81^{1/2}$ (b) $125^{1/3}$

(c) $-625^{1/4}$ (d) $(-625)^{1/4}$

(e) $(-125)^{1/3}$ (f) $\left(\dfrac{1}{16}\right)^{1/4}$

EXAMPLE 1 Evaluating Exponentials of the Form $a^{1/n}$

Evaluate each exponential.

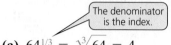

The denominator is the index.

The denominator is the index. $\sqrt{}$ means $\sqrt[2]{}$.

(a) $64^{1/3} = \sqrt[3]{64} = 4$

(b) $100^{1/2} = \sqrt{100} = 10$

(c) $-256^{1/4} = -\sqrt[4]{256} = -4$

(d) $(-256)^{1/4} = \sqrt[4]{-256}$ is not a real number, because the radicand, -256, is negative and the index is even.

(e) $(-32)^{1/5} = \sqrt[5]{-32} = -2$

(f) $\left(\dfrac{1}{8}\right)^{1/3} = \sqrt[3]{\dfrac{1}{8}} = \dfrac{1}{2}$ NOW TRY

⚠ **CAUTION** Notice the difference between **Examples 1(c) and (d).** The radical in part (c) is the *negative fourth root of a positive number,* while the radical in part (d) is the *principal fourth root of a negative number, which is not a real number.*

OBJECTIVE 2 Define and use expressions of the form $a^{m/n}$. We know that $8^{1/3} = \sqrt[3]{8}$. We can define a number like $8^{2/3}$, where the numerator of the exponent is not 1. For past rules of exponents to be valid,

$$8^{2/3} = 8^{(1/3)2} = (8^{1/3})^2.$$

Since $8^{1/3} = \sqrt[3]{8}$,

$$8^{2/3} = \left(\sqrt[3]{8}\right)^2 = 2^2 = 4.$$

Generalizing from this example, we define $a^{m/n}$ as follows.

$a^{m/n}$

If m and n are positive integers with m/n in lowest terms, then

$$a^{m/n} = (a^{1/n})^m,$$

provided that $a^{1/n}$ is a real number. If $a^{1/n}$ is not a real number, then $a^{m/n}$ is not a real number.

EXAMPLE 2 Evaluating Exponentials of the Form $a^{m/n}$

Evaluate each exponential.

Think:
$36^{1/2} = \sqrt{36} = 6$

Think:
$125^{1/3} = \sqrt[3]{125} = 5$

(a) $36^{3/2} = (36^{1/2})^3 = 6^3 = 216$

(b) $125^{2/3} = (125^{1/3})^2 = 5^2 = 25$

Be careful. The base is 4.

(c) $-4^{5/2} = -(4^{5/2}) = -(4^{1/2})^5 = -(2)^5 = -32$

Because the base here is 4, the negative sign is *not* affected by the exponent.

○ NOW TRY
● EXERCISE 2
Evaluate each exponential.
(a) $32^{2/5}$ **(b)** $8^{5/3}$
(c) $-100^{3/2}$ **(d)** $(-121)^{3/2}$
(e) $(-125)^{4/3}$

(d) $(-27)^{2/3} = [(-27)^{1/3}]^2 = (-3)^2 = 9$

Notice in part (c) that we first evaluate the exponential and then find its negative. In part (d), the $-$ sign is part of the base, -27.

(e) $(-100)^{3/2} = [(-100)^{1/2}]^3$, which is not a real number, since $(-100)^{1/2}$, or $\sqrt{-100}$, is not a real number.   *NOW TRY* ↻

When a rational exponent is negative, the earlier interpretation of negative exponents is applied.

$a^{-m/n}$

If $a^{m/n}$ is a real number, then

$$a^{-m/n} = \frac{1}{a^{m/n}} \quad (a \neq 0).$$

○ NOW TRY
● EXERCISE 3
Evaluate each exponential.
(a) $243^{-3/5}$ **(b)** $4^{-5/2}$
(c) $\left(\dfrac{216}{125}\right)^{-2/3}$

EXAMPLE 3 Evaluating Exponentials with Negative Rational Exponents

Evaluate each exponential.

(a) $16^{-3/4} = \dfrac{1}{16^{3/4}} = \dfrac{1}{(16^{1/4})^3} = \dfrac{1}{(\sqrt[4]{16})^3} = \dfrac{1}{2^3} = \dfrac{1}{8}$

> The denominator of 3/4 is the index and the numerator is the exponent.

(b) $25^{-3/2} = \dfrac{1}{25^{3/2}} = \dfrac{1}{(25^{1/2})^3} = \dfrac{1}{(\sqrt{25})^3} = \dfrac{1}{5^3} = \dfrac{1}{125}$

(c) $\left(\dfrac{8}{27}\right)^{-2/3} = \dfrac{1}{\left(\dfrac{8}{27}\right)^{2/3}} = \dfrac{1}{\left(\sqrt[3]{\dfrac{8}{27}}\right)^2} = \dfrac{1}{\left(\dfrac{2}{3}\right)^2} = \dfrac{1}{\dfrac{4}{9}} = \dfrac{9}{4}$

> $\dfrac{1}{\frac{4}{9}} = 1 \div \dfrac{4}{9} = 1 \cdot \dfrac{9}{4}$

We can also use the rule $\left(\dfrac{b}{a}\right)^{-m} = \left(\dfrac{a}{b}\right)^m$ here, as follows.

$$\left(\frac{8}{27}\right)^{-2/3} = \left(\frac{27}{8}\right)^{2/3} = \left(\sqrt[3]{\frac{27}{8}}\right)^2 = \left(\frac{3}{2}\right)^2 = \frac{9}{4}$$

> Take the reciprocal only of the base, **not** the exponent.

NOW TRY ↻

⚠ **CAUTION** Be careful to distinguish between exponential expressions like the following.

$16^{-1/4}$, which equals $\dfrac{1}{2}$, $-16^{1/4}$, which equals -2, and $-16^{-1/4}$, which equals $-\dfrac{1}{2}$

A negative exponent does not necessarily lead to a negative result. Negative exponents lead to reciprocals, which may be positive.

NOW TRY ANSWERS
2. (a) 4 **(b)** 32 **(c)** -1000
  (d) It is not a real number.
  (e) 625
3. (a) $\frac{1}{27}$ **(b)** $\frac{1}{32}$ **(c)** $\frac{25}{36}$

We obtain an alternative definition of $a^{m/n}$ by using the power rule for exponents differently than in the earlier definition. If all indicated roots are real numbers,

then $\qquad a^{m/n} = a^{m(1/n)} = (a^m)^{1/n}, \quad$ so $\quad a^{m/n} = (a^m)^{1/n}.$

$a^{m/n}$

If all indicated roots are real numbers, then

$$a^{m/n} = (a^{1/n})^m = (a^m)^{1/n}.$$

We can now evaluate an expression such as $27^{2/3}$ in two ways.

$$27^{2/3} = (27^{1/3})^2 = 3^2 = 9$$

or $\qquad 27^{2/3} = (27^2)^{1/3} = 729^{1/3} = 9 \qquad$ The result is the same.

In most cases, it is easier to use $(a^{1/n})^m$.

Radical Form of $a^{m/n}$

If all indicated roots are real numbers, then

$$a^{m/n} = \sqrt[n]{a^m} = (\sqrt[n]{a})^m.$$

That is, raise a to the mth power and then take the nth root, or take the nth root of a and then raise to the mth power.

For example,

$$8^{2/3} = \sqrt[3]{8^2} = \sqrt[3]{64} = 4, \quad \text{and} \quad 8^{2/3} = (\sqrt[3]{8})^2 = 2^2 = 4,$$

so $\qquad 8^{2/3} = \sqrt[3]{8^2} = (\sqrt[3]{8})^2.$

OBJECTIVE 3 **Convert between radicals and rational exponents.** Using the definition of rational exponents, we can simplify many problems involving radicals by converting the radicals to numbers with rational exponents. After simplifying, we can convert the answer back to radical form if required.

EXAMPLE 4 Converting between Rational Exponents and Radicals

Write each exponential as a radical. Assume that all variables represent positive real numbers. Use the definition that takes the root first.

(a) $13^{1/2} = \sqrt{13}$ **(b)** $6^{3/4} = (\sqrt[4]{6})^3$ **(c)** $9m^{5/8} = 9(\sqrt[8]{m})^5$

(d) $6x^{2/3} - (4x)^{3/5} = 6(\sqrt[3]{x})^2 - (\sqrt[5]{4x})^3$

(e) $r^{-2/3} = \dfrac{1}{r^{2/3}} = \dfrac{1}{(\sqrt[3]{r})^2}$

(f) $(a^2 + b^2)^{1/2} = \sqrt{a^2 + b^2} \quad \boxed{\sqrt{a^2+b^2} \neq a+b}$

NOW TRY
EXERCISE 4

Write each exponential as a radical. Assume that all variables represent positive real numbers.

(a) $21^{1/2}$ **(b)** $17^{5/4}$

(c) $4t^{3/5} + (4t)^{2/3}$

(d) $w^{-2/5}$ **(e)** $(a^2 - b^2)^{1/4}$

In parts (f)–(h), write each radical as an exponential. Simplify. Assume that all variables represent positive real numbers.

(f) $\sqrt[3]{15}$ **(g)** $\sqrt[4]{4^2}$

(h) $\sqrt[4]{x^4}$

In parts (g)–(i), write each radical as an exponential. Simplify. Assume that all variables represent positive real numbers.

(g) $\sqrt{10} = 10^{1/2}$

(h) $\sqrt[4]{3^8} = 3^{8/4} = 3^2 = 9$

(i) $\sqrt[6]{z^6} = z$, since z is positive.

NOW TRY

NOTE In **Example 4(i)**, it is not necessary to use absolute value bars, since the directions specifically state that the variable represents a positive real number. Because the absolute value of the positive real number z is z itself, the answer is simply z.

OBJECTIVE 4 **Use the rules for exponents with rational exponents.** The definition of rational exponents allows us to apply the rules for exponents from **Section 5.1.**

Rules for Rational Exponents

Let r and s be rational numbers. For all real numbers a and b for which the indicated expressions exist, the following are true.

$$a^r \cdot a^s = a^{r+s} \qquad a^{-r} = \frac{1}{a^r} \qquad \frac{a^r}{a^s} = a^{r-s} \qquad \left(\frac{a}{b}\right)^{-r} = \frac{b^r}{a^r}$$

$$(a^r)^s = a^{rs} \qquad (ab)^r = a^r b^r \qquad \left(\frac{a}{b}\right)^r = \frac{a^r}{b^r} \qquad a^{-r} = \left(\frac{1}{a}\right)^r$$

EXAMPLE 5   Applying Rules for Rational Exponents

Write with only positive exponents. Assume that all variables represent positive real numbers.

(a) $2^{1/2} \cdot 2^{1/4}$

$\qquad = 2^{1/2+1/4}$ Product rule

$\qquad = 2^{3/4}$ Add exponents.

(b) $\dfrac{5^{2/3}}{5^{7/3}}$

$\qquad = 5^{2/3-7/3}$ Quotient rule

$\qquad = 5^{-5/3}$ Subtract exponents.

$\qquad = \dfrac{1}{5^{5/3}}$ $a^{-r} = \frac{1}{a^r}$

(c) $\dfrac{(x^{1/2}y^{2/3})^4}{y}$

$\qquad = \dfrac{(x^{1/2})^4(y^{2/3})^4}{y}$ Power rule

$\qquad = \dfrac{x^2 y^{8/3}}{y^1}$ Power rule

$\qquad = x^2 y^{8/3-1}$ Quotient rule

$\qquad = x^2 y^{5/3}$ $\frac{8}{3} - 1 = \frac{8}{3} - \frac{3}{3} = \frac{5}{3}$

NOW TRY ANSWERS

4. **(a)** $\sqrt{21}$ **(b)** $\left(\sqrt[4]{17}\right)^5$

(c) $4\left(\sqrt[5]{t}\right)^3 + \left(\sqrt[3]{4t}\right)^2$

(d) $\dfrac{1}{\left(\sqrt[5]{w}\right)^2}$ **(e)** $\sqrt[4]{a^2 - b^2}$

(f) $15^{1/3}$ **(g)** 2 **(h)** x

NOW TRY
EXERCISE 5
Write with only positive exponents. Assume that all variables represent positive real numbers.

(a) $5^{1/4} \cdot 5^{2/3}$ **(b)** $\dfrac{9^{3/5}}{9^{7/5}}$

(c) $\dfrac{(r^{2/3}t^{1/4})^8}{t}$

(d) $\left(\dfrac{2x^{1/2}y^{-2/3}}{x^{-3/5}y^{-1/5}}\right)^{-3}$

(e) $y^{2/3}(y^{1/3} + y^{5/3})$

(d) $\left(\dfrac{x^4y^{-6}}{x^{-2}y^{1/3}}\right)^{-2/3}$

$= \dfrac{(x^4)^{-2/3}(y^{-6})^{-2/3}}{(x^{-2})^{-2/3}(y^{1/3})^{-2/3}}$ Power rule

$= \dfrac{x^{-8/3}y^4}{x^{4/3}y^{-2/9}}$ Power rule

$= x^{-8/3-4/3}y^{4-(-2/9)}$ Quotient rule

$= x^{-4}y^{38/9}$ Use parentheses to avoid errors. $4 - \left(-\frac{2}{9}\right) = \frac{36}{9} + \frac{2}{9} = \frac{38}{9}$

$= \dfrac{y^{38/9}}{x^4}$ Definition of negative exponent

The same result is obtained if we simplify within the parentheses first.

$\left(\dfrac{x^4y^{-6}}{x^{-2}y^{1/3}}\right)^{-2/3}$

$= (x^{4-(-2)}y^{-6-1/3})^{-2/3}$ Quotient rule

$= (x^6y^{-19/3})^{-2/3}$ $-6 - \frac{1}{3} = -\frac{18}{3} - \frac{1}{3} = -\frac{19}{3}$

$= (x^6)^{-2/3}(y^{-19/3})^{-2/3}$ Power rule

$= x^{-4}y^{38/9}$ Power rule

$= \dfrac{y^{38/9}}{x^4}$ Definition of negative exponent

(e) $m^{3/4}(m^{5/4} - m^{1/4})$

Do not make the common mistake of multiplying exponents in the first step.

$= m^{3/4}(m^{5/4}) - m^{3/4}(m^{1/4})$ Distributive property

$= m^{3/4+5/4} - m^{3/4+1/4}$ Product rule

$= m^{8/4} - m^{4/4}$ Add exponents.

$= m^2 - m$ Lowest terms in exponents

NOW TRY

> ⚠ **CAUTION** Use the rules of exponents in problems like those in **Example 5**. Do not convert the expressions to radical form.

EXAMPLE 6 Applying Rules for Rational Exponents

Write all radicals as exponentials, and then apply the rules for rational exponents. Leave answers in exponential form. Assume that all variables represent positive real numbers.

(a) $\sqrt[3]{x^2} \cdot \sqrt[4]{x}$

$= x^{2/3} \cdot x^{1/4}$ Convert to rational exponents.

$= x^{2/3+1/4}$ Product rule

$= x^{8/12+3/12}$ Write exponents with a common denominator.

$= x^{11/12}$ Add exponents.

NOW TRY ANSWERS

5. (a) $5^{11/12}$ **(b)** $\dfrac{1}{9^{4/5}}$

(c) $r^{16/3}t$ **(d)** $\dfrac{y^{7/5}}{8x^{33/10}}$

(e) $y + y^{7/3}$

NOW TRY
EXERCISE 6

Write all radicals as exponentials, and then apply the rules for rational exponents. Leave answers in exponential form. Assume that all variables represent positive real numbers.

(a) $\sqrt[5]{y^3} \cdot \sqrt[3]{y}$ **(b)** $\dfrac{\sqrt[4]{y^3}}{\sqrt{y^5}}$

(c) $\sqrt{\sqrt[3]{y}}$

NOW TRY ANSWERS

6. **(a)** $y^{14/15}$ **(b)** $\dfrac{1}{y^{7/4}}$ **(c)** $y^{1/6}$

(b) $\dfrac{\sqrt{x^3}}{\sqrt[3]{x^2}}$

$= \dfrac{x^{3/2}}{x^{2/3}}$ Convert to rational exponents.

$= x^{3/2 - 2/3}$ Quotient rule

$= x^{5/6}$ $\dfrac{3}{2} - \dfrac{2}{3} = \dfrac{9}{6} - \dfrac{4}{6} = \dfrac{5}{6}$

(c) $\sqrt{\sqrt[4]{z}}$

$= \sqrt{z^{1/4}}$ Convert the inside radical to rational exponents.

$= (z^{1/4})^{1/2}$ Convert to rational exponents.

$= z^{1/8}$ Power rule

NOW TRY

NOTE The ability to convert between radicals and rational exponents is important in the study of exponential and logarithmic functions in **Chapter 10**.

8.2 EXERCISES

● *Complete solution available on the Video Resources on DVD*

Concept Check Match each expression from Column I with the equivalent choice from Column II.

I		II	
1. $3^{1/2}$	**2.** $(-27)^{1/3}$	**A.** -4	**B.** 8
3. $-16^{1/2}$	**4.** $(-25)^{1/2}$	**C.** $\sqrt{3}$	**D.** $-\sqrt{6}$
5. $(-32)^{1/5}$	**6.** $(-32)^{2/5}$	**E.** -3	**F.** $\sqrt{6}$
7. $4^{3/2}$	**8.** $6^{2/4}$	**G.** 4	**H.** -2
9. $-6^{2/4}$	**10.** $36^{0.5}$	**I.** 6	**J.** Not a real number

Evaluate each exponential. ***See Examples 1–3.***

● 11. $169^{1/2}$ **12.** $121^{1/2}$ **13.** $729^{1/3}$ **14.** $512^{1/3}$

15. $16^{1/4}$ **16.** $625^{1/4}$ **17.** $\left(\dfrac{64}{81}\right)^{1/2}$ **18.** $\left(\dfrac{8}{27}\right)^{1/3}$

19. $(-27)^{1/3}$ **20.** $(-32)^{1/5}$ **21.** $(-144)^{1/2}$ **22.** $(-36)^{1/2}$

● 23. $100^{3/2}$ **24.** $64^{3/2}$ **25.** $81^{3/4}$ **26.** $216^{2/3}$

27. $-16^{5/2}$ **28.** $-32^{3/5}$ **29.** $(-8)^{4/3}$ **30.** $(-243)^{2/5}$

● 31. $32^{-3/5}$ **32.** $27^{-4/3}$ **33.** $64^{-3/2}$ **34.** $81^{-3/2}$

35. $\left(\dfrac{125}{27}\right)^{-2/3}$ **36.** $\left(\dfrac{64}{125}\right)^{-2/3}$ **37.** $\left(\dfrac{16}{81}\right)^{-3/4}$ **38.** $\left(\dfrac{729}{64}\right)^{-5/6}$

Write with radicals. Assume that all variables represent positive real numbers. ***See Example 4.***

● 39. $10^{1/2}$ **40.** $3^{1/2}$ **41.** $8^{3/4}$

42. $7^{2/3}$ **43.** $(9q)^{5/8} - (2x)^{2/3}$ **44.** $(3p)^{3/4} + (4x)^{1/3}$

45. $(2m)^{-3/2}$ **46.** $(5y)^{-3/5}$ **47.** $(2y + x)^{2/3}$

48. $(r + 2z)^{3/2}$ **49.** $(3m^4 + 2k^2)^{-2/3}$ **50.** $(5x^2 + 3z^3)^{-5/6}$

Simplify by first converting to rational exponents. Assume that all variables represent positive real numbers. ***See Example 4.***

51. $\sqrt{2^{12}}$ **52.** $\sqrt{5^{10}}$ **53.** $\sqrt[3]{4^9}$ **54.** $\sqrt[4]{6^8}$ **55.** $\sqrt{x^{20}}$

56. $\sqrt{r^{50}}$ **57.** $\sqrt[3]{x} \cdot \sqrt{x}$ **58.** $\sqrt[4]{y} \cdot \sqrt[5]{y^2}$ **59.** $\dfrac{\sqrt[3]{t^4}}{\sqrt[5]{t^4}}$ **60.** $\dfrac{\sqrt[4]{w^3}}{\sqrt[6]{w}}$

Simplify each expression. Write all answers with positive exponents. Assume that all variables represent positive real numbers. **See Example 5.**

61. $3^{1/2} \cdot 3^{3/2}$

62. $6^{4/3} \cdot 6^{2/3}$

63. $\dfrac{64^{5/3}}{64^{4/3}}$

64. $\dfrac{125^{7/3}}{125^{5/3}}$

65. $y^{7/3} \cdot y^{-4/3}$

66. $r^{-8/9} \cdot r^{17/9}$

67. $x^{2/3} \cdot x^{-1/4}$

68. $x^{2/5} \cdot x^{-1/3}$

69. $\dfrac{k^{1/3}}{k^{2/3} \cdot k^{-1}}$

70. $\dfrac{z^{3/4}}{z^{5/4} \cdot z^{-2}}$

71. $\dfrac{(x^{1/4}y^{2/5})^{20}}{x^2}$

72. $\dfrac{(r^{1/5}s^{2/3})^{15}}{r^2}$

73. $\dfrac{(x^{2/3})^2}{(x^2)^{7/3}}$

74. $\dfrac{(p^3)^{1/4}}{(p^{5/4})^2}$

75. $\dfrac{m^{3/4}n^{-1/4}}{(m^2n)^{1/2}}$

76. $\dfrac{(a^2b^5)^{-1/4}}{(a^{-3}b^2)^{1/6}}$

77. $\dfrac{p^{1/5}p^{7/10}p^{1/2}}{(p^3)^{-1/5}}$

78. $\dfrac{z^{1/3}z^{-2/3}z^{1/6}}{(z^{-1/6})^3}$

79. $\left(\dfrac{b^{-3/2}}{c^{-5/3}}\right)^2 (b^{-1/4}c^{-1/3})^{-1}$

80. $\left(\dfrac{m^{-2/3}}{a^{-3/4}}\right)^4 (m^{-3/8}a^{1/4})^{-2}$

81. $\left(\dfrac{p^{-1/4}q^{-3/2}}{3^{-1}p^{-2}q^{-2/3}}\right)^{-2}$

82. $\left(\dfrac{2^{-2}w^{-3/4}x^{-5/8}}{w^{3/4}x^{-1/2}}\right)^{-3}$

83. $p^{2/3}(p^{1/3} + 2p^{4/3})$

84. $z^{5/8}(3z^{5/8} + 5z^{11/8})$

85. $k^{1/4}(k^{3/2} - k^{1/2})$

86. $r^{3/5}(r^{1/2} + r^{3/4})$

87. $6a^{7/4}(a^{-7/4} + 3a^{-3/4})$

88. $4m^{5/3}(m^{-2/3} - 4m^{-5/3})$

89. $-5x^{7/6}(x^{5/6} - x^{-1/6})$

90. $-8y^{11/7}(y^{3/7} - y^{-4/7})$

Write with rational exponents, and then apply the properties of exponents. Assume that all radicands represent positive real numbers. Give answers in exponential form. **See Example 6.**

91. $\sqrt[5]{x^3} \cdot \sqrt[4]{x}$

92. $\sqrt[6]{y^5} \cdot \sqrt[3]{y^2}$

93. $\dfrac{\sqrt{x^5}}{\sqrt{x^8}}$

94. $\dfrac{\sqrt[3]{k^5}}{\sqrt[3]{k^7}}$

95. $\sqrt{y} \cdot \sqrt[3]{yz}$

96. $\sqrt[3]{xz} \cdot \sqrt{z}$

97. $\sqrt[4]{\sqrt[3]{m}}$

98. $\sqrt[3]{\sqrt{k}}$

99. $\sqrt{\sqrt{\sqrt{x}}}$

100. $\sqrt{\sqrt{\sqrt{\sqrt{x}}}}$

101. $\sqrt{\sqrt[3]{\sqrt[4]{x}}}$

102. $\sqrt[3]{\sqrt[5]{\sqrt{y}}}$

103. Show that, in general, $\sqrt{a^2 + b^2} \neq a + b$ by replacing a with 3 and b with 4.

104. Suppose someone claims that $\sqrt[n]{a^n + b^n}$ must equal $a + b$, since, when $a = 1$ and $b = 0$, a true statement results:

$$\sqrt[n]{a^n + b^n} = \sqrt[n]{1^n + 0^n} = \sqrt[n]{1^n} = 1 = 1 + 0 = a + b.$$

Explain why this is faulty reasoning.

Solve each problem.

105. Meteorologists can determine the duration of a storm by using the function defined by

$$T(D) = 0.07D^{3/2},$$

where D is the diameter of the storm in miles and T is the time in hours. Find the duration of a storm with a diameter of 16 mi. Round your answer to the nearest tenth of an hour.

106. The threshold weight T, in pounds, for a person is the weight above which the risk of death increases greatly. The threshold weight in pounds for men aged 40–49 is related to height h in inches by the function defined by

$$h(T) = (1860.867T)^{1/3}.$$

What height corresponds to a threshold weight of 200 lb for a 46-yr-old man? Round your answer to the nearest inch and then to the nearest tenth of a foot.

The **windchill factor** is a measure of the cooling effect that the wind has on a person's skin. It calculates the equivalent cooling temperature if there were no wind. The National Weather Service uses the formula

$$\text{Windchill temperature} = 35.74 + 0.6215T - 35.75V^{4/25} + 0.4275TV^{4/25},$$

where T is the temperature in °F and V is the wind speed in miles per hour, to calculate windchill. The chart gives the windchill factor for various wind speeds and temperatures at which frostbite is a risk, and how quickly it may occur.

					Temperature (°F)				
Calm	40	30	20	10	0	−10	−20	−30	−40
5	36	25	13	1	−11	−22	−34	−46	−57
10	34	21	9	−4	−16	−28	−41	−53	−66
15	32	19	6	−7	−19	−32	−45	−58	−71
20	30	17	4	−9	−22	−35	−48	−61	−74
25	29	16	3	−11	−24	−37	−51	−64	−78
30	28	15	1	−12	−26	−39	−53	−67	−80
35	28	14	0	−14	−27	−41	−55	−69	−82
40	27	13	−1	−15	−29	−43	−57	−71	−84

Wind speed (mph)

Frostbites times: ☐ 30 minutes ☐ 10 minutes ☐ 5 minutes

Source: National Oceanic and Atmospheric Administration, National Weather Service.

Use the formula and a calculator to determine the windchill to the nearest tenth of a degree, given the following conditions. Compare your answers with the appropriate entries in the table.

107. 30°F, 15-mph wind

108. 10°F, 30-mph wind

109. 20°F, 20-mph wind

110. 40°F, 10-mph wind

PREVIEW EXERCISES

*Simplify each pair of expressions, and then compare the results. **See Section 8.1.***

111. $\sqrt{25} \cdot \sqrt{36}, \quad \sqrt{25 \cdot 36}$

112. $\dfrac{\sqrt[3]{27}}{\sqrt[3]{729}}, \quad \sqrt[3]{\dfrac{27}{729}}$

8.3 Simplifying Radical Expressions

OBJECTIVES

1. Use the product rule for radicals.
2. Use the quotient rule for radicals.
3. Simplify radicals.
4. Simplify products and quotients of radicals with different indexes.
5. Use the Pythagorean theorem.
6. Use the distance formula.

OBJECTIVE 1 **Use the product rule for radicals.** Consider the expressions $\sqrt{36 \cdot 4}$ and $\sqrt{36} \cdot \sqrt{4}$. Are they equal?

$$\sqrt{36 \cdot 4} = \sqrt{144} = 12$$
$$\sqrt{36} \cdot \sqrt{4} = 6 \cdot 2 = 12$$

The result is the same.

This is an example of the **product rule for radicals.**

Product Rule for Radicals

If $\sqrt[n]{a}$ and $\sqrt[n]{b}$ are real numbers and n is a natural number, then

$$\sqrt[n]{a} \cdot \sqrt[n]{b} = \sqrt[n]{ab}.$$

That is, the product of two nth roots is the nth root of the product.

We justify the product rule by using the rules for rational exponents. Since $\sqrt[n]{a} = a^{1/n}$ and $\sqrt[n]{b} = b^{1/n}$,

$$\sqrt[n]{a} \cdot \sqrt[n]{b} = a^{1/n} \cdot b^{1/n} = (ab)^{1/n} = \sqrt[n]{ab}.$$

⚠ **CAUTION** *Use the product rule only when the radicals have the same index.*

NOW TRY
EXERCISE 1
Multiply. Assume that all variables represent positive real numbers.

(a) $\sqrt{7} \cdot \sqrt{11}$

(b) $\sqrt{2mn} \cdot \sqrt{15}$

EXAMPLE 1 Using the Product Rule

Multiply. Assume that all variables represent positive real numbers.

(a) $\sqrt{5} \cdot \sqrt{7}$
$= \sqrt{5 \cdot 7}$
$= \sqrt{35}$

(b) $\sqrt{11} \cdot \sqrt{p}$
$= \sqrt{11p}$

(c) $\sqrt{7} \cdot \sqrt{11xyz}$
$= \sqrt{77xyz}$

NOW TRY ↻

NOW TRY
EXERCISE 2
Multiply. Assume that all variables represent positive real numbers.

(a) $\sqrt[3]{4} \cdot \sqrt[3]{5}$

(b) $\sqrt[4]{5t} \cdot \sqrt[4]{6r^3}$

(c) $\sqrt[7]{20x} \cdot \sqrt[7]{3xy^3}$

(d) $\sqrt[3]{5} \cdot \sqrt[4]{9}$

EXAMPLE 2 Using the Product Rule

Multiply. Assume that all variables represent positive real numbers.

(a) $\sqrt[3]{3} \cdot \sqrt[3]{12}$
$= \sqrt[3]{3 \cdot 12}$
$= \sqrt[3]{36}$ ← Remember to write the index.

(b) $\sqrt[4]{8y} \cdot \sqrt[4]{3r^2}$
$= \sqrt[4]{24yr^2}$

(c) $\sqrt[6]{10m^4} \cdot \sqrt[6]{5m}$
$= \sqrt[6]{50m^5}$

(d) $\sqrt[4]{2} \cdot \sqrt[5]{2}$ cannot be simplified using the product rule for radicals, because the indexes (4 and 5) are different.

NOW TRY ↻

OBJECTIVE 2 Use the quotient rule for radicals. The **quotient rule for radicals** is similar to the product rule.

Quotient Rule for Radicals

If $\sqrt[n]{a}$ and $\sqrt[n]{b}$ are real numbers, $b \neq 0$, and n is a natural number, then

$$\sqrt[n]{\frac{a}{b}} = \frac{\sqrt[n]{a}}{\sqrt[n]{b}}.$$

That is, the nth root of a quotient is the quotient of the nth roots.

EXAMPLE 3 Using the Quotient Rule

Simplify. Assume that all variables represent positive real numbers.

NOW TRY ANSWERS
1. (a) $\sqrt{77}$ (b) $\sqrt{30mn}$
2. (a) $\sqrt[3]{20}$ (b) $\sqrt[4]{30tr^3}$
   (c) $\sqrt[7]{60x^2y^3}$
   (d) This expression cannot be simplified by the product rule.

(a) $\sqrt{\frac{16}{25}} = \frac{\sqrt{16}}{\sqrt{25}} = \frac{4}{5}$

(b) $\sqrt{\frac{7}{36}} = \frac{\sqrt{7}}{\sqrt{36}} = \frac{\sqrt{7}}{6}$

(c) $\sqrt[3]{-\frac{8}{125}} = \sqrt[3]{\frac{-8}{125}} = \frac{\sqrt[3]{-8}}{\sqrt[3]{125}} = \frac{-2}{5} = -\frac{2}{5}$ $\frac{-a}{b} = -\frac{a}{b}$

NOW TRY
EXERCISE 3

Simplify. Assume that all variables represent positive real numbers.

(a) $\sqrt{\dfrac{49}{36}}$ **(b)** $\sqrt{\dfrac{5}{144}}$

(c) $\sqrt[3]{-\dfrac{27}{1000}}$ **(d)** $\sqrt[4]{\dfrac{t}{16}}$

(e) $-\sqrt[5]{\dfrac{m^{15}}{243}}$

(d) $\sqrt[3]{\dfrac{7}{216}} = \dfrac{\sqrt[3]{7}}{\sqrt[3]{216}} = \dfrac{\sqrt[3]{7}}{6}$ **(e)** $\sqrt[5]{\dfrac{x}{32}} = \dfrac{\sqrt[5]{x}}{\sqrt[5]{32}} = \dfrac{\sqrt[5]{x}}{2}$

(f) $-\sqrt[3]{\dfrac{m^6}{125}} = -\dfrac{\sqrt[3]{m^6}}{\sqrt[3]{125}} = -\dfrac{m^2}{5}$ ◁ Think: $\sqrt[3]{m^6} = m^{6/3} = m^2$

NOW TRY

OBJECTIVE 3 **Simplify radicals.** We use the product and quotient rules to simplify radicals. A radical is **simplified** if the following four conditions are met.

Conditions for a Simplified Radical

1. The radicand has no factor raised to a power greater than or equal to the index.
2. The radicand has no fractions.
3. No denominator contains a radical.
4. Exponents in the radicand and the index of the radical have greatest common factor 1.

EXAMPLE 4 **Simplifying Roots of Numbers**

Simplify.

(a) $\sqrt{24}$

Check to see whether 24 is divisible by a perfect square (the square of a natural number) such as 4, 9, 16, ... The greatest perfect square that divides into 24 is 4.

$$\sqrt{24}$$
$$= \sqrt{4 \cdot 6} \qquad \text{Factor; 4 is a perfect square.}$$
$$= \sqrt{4} \cdot \sqrt{6} \qquad \text{Product rule}$$
$$= 2\sqrt{6} \qquad \sqrt{4} = 2$$

(b) $\sqrt{108}$

As shown on the left, the number 108 is divisible by the perfect square 36. If this perfect square is not immediately clear, try factoring 108 into its prime factors, as shown on the right.

$$\sqrt{108}$$
$$= \sqrt{36 \cdot 3} \qquad \text{Factor.}$$
$$= \sqrt{36} \cdot \sqrt{3} \qquad \text{Product rule}$$
$$= 6\sqrt{3} \qquad \sqrt{36} = 6$$

$$\sqrt{108}$$
$$= \sqrt{2^2 \cdot 3^3}$$
$$= \sqrt{2^2 \cdot 3^2 \cdot 3} \qquad a^3 = a^2 \cdot a$$
$$= \sqrt{2^2} \cdot \sqrt{3^2} \cdot \sqrt{3} \qquad \text{Product rule}$$
$$= 2 \cdot 3 \cdot \sqrt{3} \qquad \sqrt{2^2} = 2, \ \sqrt{3^2} = 3$$
$$= 6\sqrt{3} \qquad \text{Multiply.}$$

(c) $\sqrt{10}$

No perfect square (other than 1) divides into 10, so $\sqrt{10}$ cannot be simplified further.

NOW TRY ANSWERS

3. **(a)** $\dfrac{7}{6}$ **(b)** $\dfrac{\sqrt{5}}{12}$ **(c)** $-\dfrac{3}{10}$

(d) $\dfrac{\sqrt[4]{t}}{2}$ **(e)** $-\dfrac{m^3}{3}$

NOW TRY
EXERCISE 4
Simplify.

(a) $\sqrt{50}$ **(b)** $\sqrt{192}$

(c) $\sqrt{42}$ **(d)** $\sqrt[3]{108}$

(e) $-\sqrt[4]{80}$

(d) $\sqrt[3]{16}$

The greatest perfect *cube* that divides into 16 is 8, so factor 16 as $8 \cdot 2$.

$$\sqrt[3]{16} \qquad \boxed{\text{Remember to write the index.}}$$

$$= \sqrt[3]{8 \cdot 2} \qquad \text{8 is a perfect cube.}$$

$$= \sqrt[3]{8} \cdot \sqrt[3]{2} \qquad \text{Product rule}$$

$$= 2\sqrt[3]{2} \qquad \sqrt[3]{8} = 2$$

(e)

$$-\sqrt[4]{162}$$

$$= -\sqrt[4]{81 \cdot 2} \qquad \text{81 is a perfect 4th power.}$$

$$\boxed{\text{Remember the negative sign in each line.}} = -\sqrt[4]{81} \cdot \sqrt[4]{2} \qquad \text{Product rule}$$

$$= -3\sqrt[4]{2} \qquad \sqrt[4]{81} = 3 \qquad \text{NOW TRY}$$

⚠ **CAUTION** *Be careful with which factors belong outside the radical sign and which belong inside.* Note in **Example 4(b)** how $2 \cdot 3$ is written outside because $\sqrt{2^2} = 2$ and $\sqrt{3^2} = 3$, while the remaining 3 is left inside the radical.

NOW TRY
EXERCISE 5
Simplify. Assume that all variables represent positive real numbers.

(a) $\sqrt{36x^5}$ **(b)** $\sqrt{32m^5n^4}$

(c) $\sqrt[3]{-125k^3p^7}$

(d) $-\sqrt[4]{162x^7y^8}$

EXAMPLE 5 Simplifying Radicals Involving Variables

Simplify. Assume that all variables represent positive real numbers.

(a) $\sqrt{16m^3}$

$$= \sqrt{16m^2 \cdot m} \qquad \text{Factor.}$$

$$= \sqrt{16m^2} \cdot \sqrt{m} \qquad \text{Product rule}$$

$$= 4m\sqrt{m} \qquad \text{Take the square root.}$$

Absolute value bars are not needed around the m in color because all the variables represent *positive* real numbers.

(b) $\sqrt{200k^7q^8}$

$$= \sqrt{10^2 \cdot 2 \cdot (k^3)^2 \cdot k \cdot (q^4)^2} \qquad \text{Factor.}$$

$$= 10k^3q^4\sqrt{2k} \qquad \text{Remove perfect square factors.}$$

(c) $\sqrt[3]{-8x^4y^5}$

$$= \sqrt[3]{(-8x^3y^3)(xy^2)} \qquad \begin{array}{l}\text{Choose } -8x^3y^3 \text{ as the perfect cube that}\\ \text{divides into } -8x^4y^5.\end{array}$$

$$= \sqrt[3]{-8x^3y^3} \cdot \sqrt[3]{xy^2} \qquad \text{Product rule}$$

$$= -2xy\sqrt[3]{xy^2} \qquad \text{Take the cube root.}$$

(d) $-\sqrt[4]{32y^9}$

$$= -\sqrt[4]{(16y^8)(2y)} \qquad 16y^8 \text{ is the greatest 4th power that divides } 32y^9.$$

$$= -\sqrt[4]{16y^8} \cdot \sqrt[4]{2y} \qquad \text{Product rule}$$

$$= -2y^2\sqrt[4]{2y} \qquad \text{Take the fourth root.} \qquad \text{NOW TRY}$$

NOW TRY ANSWERS
4. (a) $5\sqrt{2}$ **(b)** $8\sqrt{3}$
(c) $\sqrt{42}$ cannot be simplified further.
(d) $3\sqrt[3]{4}$ **(e)** $-2\sqrt[4]{5}$
5. (a) $6x^2\sqrt{x}$ **(b)** $4m^2n^2\sqrt{2m}$
(c) $-5kp^2\sqrt[3]{p}$ **(d)** $-3xy^2\sqrt[4]{2x^3}$

NOTE From **Example 5**, we see that if a variable is raised to a power with an exponent divisible by 2, it is a perfect square. If it is raised to a power with an exponent divisible by 3, it is a perfect cube. *In general, if it is raised to a power with an exponent divisible by n, it is a perfect nth power.*

The conditions for a simplified radical given earlier state that an exponent in the radicand and the index of the radical should have greatest common factor 1.

NOW TRY
EXERCISE 6
Simplify. Assume that all variables represent positive real numbers.
(a) $\sqrt[6]{7^2}$ **(b)** $\sqrt[6]{y^4}$

EXAMPLE 6 Simplifying Radicals by Using Smaller Indexes

Simplify. Assume that all variables represent positive real numbers.

(a) $\sqrt[9]{5^6}$

We write this radical by using rational exponents and then write the exponent in lowest terms. We then express the answer as a radical.

$$\sqrt[9]{5^6} = (5^6)^{1/9} = 5^{6/9} = 5^{2/3} = \sqrt[3]{5^2}, \quad \text{or} \quad \sqrt[3]{25}$$

(b) $\sqrt[4]{p^2} = (p^2)^{1/4} = p^{2/4} = p^{1/2} = \sqrt{p}$ (Recall the assumption that $p > 0$.)

NOW TRY

These examples suggest the following rule.

$\sqrt[kn]{a^{km}}$

If m is an integer, n and k are natural numbers, and all indicated roots exist, then

$$\sqrt[kn]{a^{km}} = \sqrt[n]{a^m}.$$

OBJECTIVE 4 **Simplify products and quotients of radicals with different indexes.** We multiply and divide radicals with different indexes by using rational exponents.

NOW TRY
EXERCISE 7
Simplify $\sqrt[3]{3} \cdot \sqrt{6}$.

EXAMPLE 7 Multiplying Radicals with Different Indexes

Simplify $\sqrt{7} \cdot \sqrt[3]{2}$.

Because the different indexes, 2 and 3, have a least common multiple of 6, use rational exponents to write each radical as a sixth root.

$$\sqrt{7} = 7^{1/2} = 7^{3/6} = \sqrt[6]{7^3} = \sqrt[6]{343}$$
$$\sqrt[3]{2} = 2^{1/3} = 2^{2/6} = \sqrt[6]{2^2} = \sqrt[6]{4}$$

Now we can multiply.

$$\sqrt{7} \cdot \sqrt[3]{2} = \sqrt[6]{343} \cdot \sqrt[6]{4} \quad \text{Substitute; } \sqrt{7} = \sqrt[6]{343}, \sqrt[3]{2} = \sqrt[6]{4}$$
$$= \sqrt[6]{1372} \quad \text{Product rule}$$

NOW TRY

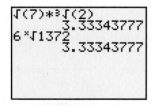

FIGURE 6

Results such as the one in **Example 7** can be supported with a calculator, as shown in **FIGURE 6**. Notice that the calculator gives the same approximation for the initial product and the final radical that we obtained.

NOW TRY ANSWERS
6. (a) $\sqrt[3]{7}$ **(b)** $\sqrt[3]{y^2}$
7. (a) $\sqrt[6]{1944}$

⚠ **CAUTION** The computation in **FIGURE 6** is not *proof* that the two expressions are equal. The algebra in **Example 7**, however, is valid proof of their equality.

OBJECTIVE 5 **Use the Pythagorean theorem.** The **Pythagorean theorem** provides an equation that relates the lengths of the three sides of a right triangle.

Pythagorean Theorem

If a and b are the lengths of the shorter sides of a right triangle and c is the length of the longest side, then

$$a^2 + b^2 = c^2.$$

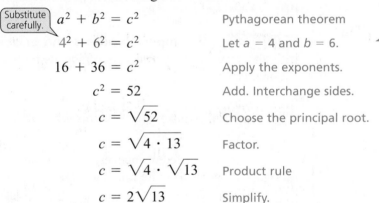

The two shorter sides are the **legs** of the triangle, and the longest side is the **hypotenuse.** The hypotenuse is the side opposite the right angle.

In **Section 9.1** we will see that an equation such as $x^2 = 7$ has two solutions: $\sqrt{7}$ (the principal, or positive, square root of 7) and $-\sqrt{7}$. Similarly, $c^2 = 52$ has two solutions, $\pm\sqrt{52} = \pm2\sqrt{13}$. In applications we often choose only the principal square root.

NOW TRY
EXERCISE 8

Find the length of the unknown side in each triangle.

(a)

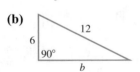

(b)

EXAMPLE 8 Using the Pythagorean Theorem

Use the Pythagorean theorem to find the length of the unknown side of the triangle in **FIGURE 7**.

$\boxed{\text{Substitute carefully.}}$ $a^2 + b^2 = c^2$ Pythagorean theorem

$\qquad 4^2 + 6^2 = c^2$ Let $a = 4$ and $b = 6$.

$\qquad 16 + 36 = c^2$ Apply the exponents.

$\qquad\qquad c^2 = 52$ Add. Interchange sides.

$\qquad\qquad c = \sqrt{52}$ Choose the principal root.

$\qquad\qquad c = \sqrt{4 \cdot 13}$ Factor.

$\qquad\qquad c = \sqrt{4} \cdot \sqrt{13}$ Product rule

$\qquad\qquad c = 2\sqrt{13}$ Simplify.

FIGURE 7

The length of the hypotenuse is $2\sqrt{13}$.   NOW TRY

⚠ **CAUTION** When substituting in the equation $a^2 + b^2 = c^2$, of the Pythagorean theorem, be sure that the length of the hypotenuse is substituted for c and that the lengths of the legs are substituted for a and b.

OBJECTIVE 6 **Use the distance formula.** The *distance formula* allows us to find the distance between two points in the coordinate plane, or the length of the line segment joining those two points.

FIGURE 8 on the next page shows the points $(3, -4)$ and $(-5, 3)$. The vertical line through $(-5, 3)$ and the horizontal line through $(3, -4)$ intersect at the point $(-5, -4)$. Thus, the point $(-5, -4)$ becomes the vertex of the right angle in a right triangle.

NOW TRY ANSWERS
8. (a) $\sqrt{89}$ **(b)** $6\sqrt{3}$

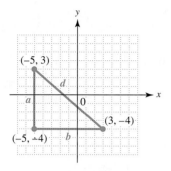

FIGURE 8

By the Pythagorean theorem, the square of the length of the hypotenuse d of the right triangle in **FIGURE 8** is equal to the sum of the squares of the lengths of the two legs a and b.

$$a^2 + b^2 = d^2$$

The length a is the difference between the y-coordinates of the endpoints. Since the x-coordinate of both points in **FIGURE 8** is -5, the side is vertical, and we can find a by finding the difference between the y-coordinates. We subtract -4 from 3 to get a positive value for a.

$$a = 3 - (-4) = 7$$

Similarly, we find b by subtracting -5 from 3.

$$b = 3 - (-5) = 8$$

Now substitute these values into the equation.

$$d^2 = a^2 + b^2$$
$$d^2 = 7^2 + 8^2 \qquad \text{Let } a = 7 \text{ and } b = 8.$$
$$d^2 = 49 + 64 \qquad \text{Apply the exponents.}$$
$$d^2 = 113 \qquad \text{Add.}$$
$$d = \sqrt{113} \qquad \text{Choose the principal root.}$$

We choose the principal root, since distance cannot be negative. Therefore, the distance between $(-5, 3)$ and $(3, -4)$ is $\sqrt{113}$.

NOTE It is customary to leave the distance in simplified radical form. Do not use a calculator to get an approximation, unless you are specifically directed to do so.

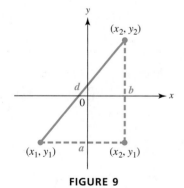

FIGURE 9

This result can be generalized. **FIGURE 9** shows the two points (x_1, y_1) and (x_2, y_2). The distance a between (x_1, y_1) and (x_2, y_1) is given by

$$a = |x_2 - x_1|,$$

and the distance b between (x_2, y_2) and (x_2, y_1) is given by

$$b = |y_2 - y_1|.$$

From the Pythagorean theorem, we obtain the following.

$$d^2 = a^2 + b^2$$
$$d^2 = (x_2 - x_1)^2 + (y_2 - y_1)^2$$

Choosing the principal square root gives the **distance formula.**

Distance Formula

The distance d between the points (x_1, y_1) and (x_2, y_2) is

$$d = \sqrt{(x_2 - x_1)^2 + (y_2 - y_1)^2}.$$

NOW TRY
EXERCISE 9
Find the distance between the points $(-4, -3)$ and $(-8, 6)$.

EXAMPLE 9 Using the Distance Formula

Find the distance between the points $(-3, 5)$ and $(6, 4)$.

Designating the points as (x_1, y_1) and (x_2, y_2) is arbitrary. We choose $(x_1, y_1) = (-3, 5)$ and $(x_2, y_2) = (6, 4)$.

$$d = \sqrt{(x_2 - x_1)^2 + (y_2 - y_1)^2}$$
$$= \sqrt{[6 - (-3)]^2 + (4 - 5)^2} \quad x_2 = 6, \ y_2 = 4, \ x_1 = -3, \ y_1 = 5$$
$$= \sqrt{9^2 + (-1)^2} \quad \boxed{\text{Substitute carefully.}}$$
$$= \sqrt{82} \quad \text{Leave in radical form.} \quad \textit{NOW TRY}$$

NOW TRY ANSWER
9. $\sqrt{97}$

8.3 EXERCISES

MyMathLab | Math XL PRACTICE | WATCH | DOWNLOAD | READ | REVIEW

⊙ *Complete solution available on the Video Resources on DVD*

Multiply, if possible, using the product rule. Assume that all variables represent positive real numbers. **See Examples 1 and 2.**

1. $\sqrt{3} \cdot \sqrt{3}$ **2.** $\sqrt{5} \cdot \sqrt{5}$ **3.** $\sqrt{18} \cdot \sqrt{2}$ **4.** $\sqrt{12} \cdot \sqrt{3}$

 5. $\sqrt{5} \cdot \sqrt{6}$ **6.** $\sqrt{10} \cdot \sqrt{3}$ **7.** $\sqrt{14} \cdot \sqrt{x}$ **8.** $\sqrt{23} \cdot \sqrt{t}$

9. $\sqrt{14} \cdot \sqrt{3pqr}$ **10.** $\sqrt{7} \cdot \sqrt{5xt}$ **11.** $\sqrt[3]{2} \cdot \sqrt[3]{5}$ **12.** $\sqrt[3]{3} \cdot \sqrt[3]{6}$

⊙ **13.** $\sqrt[3]{7x} \cdot \sqrt[3]{2y}$ **14.** $\sqrt[3]{9x} \cdot \sqrt[3]{4y}$ **15.** $\sqrt[4]{11} \cdot \sqrt[4]{3}$ **16.** $\sqrt[4]{6} \cdot \sqrt[4]{9}$

17. $\sqrt[4]{2x} \cdot \sqrt[4]{3x^2}$ **18.** $\sqrt[4]{3y^2} \cdot \sqrt[4]{6y}$ **19.** $\sqrt[3]{7} \cdot \sqrt[4]{3}$ **20.** $\sqrt[5]{8} \cdot \sqrt[6]{12}$

Simplify each radical. Assume that all variables represent positive real numbers. **See Example 3.**

⊙ **21.** $\sqrt{\dfrac{64}{121}}$ **22.** $\sqrt{\dfrac{16}{49}}$ **23.** $\sqrt{\dfrac{3}{25}}$ **24.** $\sqrt{\dfrac{13}{49}}$

25. $\sqrt{\dfrac{x}{25}}$ **26.** $\sqrt{\dfrac{k}{100}}$ **27.** $\sqrt{\dfrac{p^6}{81}}$ **28.** $\sqrt{\dfrac{w^{10}}{36}}$

29. $\sqrt[3]{-\dfrac{27}{64}}$ **30.** $\sqrt[3]{-\dfrac{216}{125}}$ **31.** $\sqrt[3]{\dfrac{r^2}{8}}$ **32.** $\sqrt[3]{\dfrac{t}{125}}$

33. $-\sqrt[4]{\dfrac{81}{x^4}}$ **34.** $-\sqrt[4]{\dfrac{625}{y^4}}$ **35.** $\sqrt[5]{\dfrac{1}{x^{15}}}$ **36.** $\sqrt[5]{\dfrac{32}{y^{20}}}$

Express each radical in simplified form. **See Example 4.**

⊙ **37.** $\sqrt{12}$ **38.** $\sqrt{18}$ **39.** $\sqrt{288}$ **40.** $\sqrt{72}$ **41.** $-\sqrt{32}$

42. $-\sqrt{48}$ **43.** $-\sqrt{28}$ **44.** $-\sqrt{24}$ **45.** $\sqrt{30}$ **46.** $\sqrt{46}$

47. $\sqrt[3]{128}$ **48.** $\sqrt[3]{24}$ **49.** $\sqrt[3]{-16}$ **50.** $\sqrt[3]{-250}$ **51.** $\sqrt[3]{40}$

52. $\sqrt[3]{375}$ **53.** $-\sqrt[4]{512}$ **54.** $-\sqrt[4]{1250}$ **55.** $\sqrt[5]{64}$ **56.** $\sqrt[5]{128}$

57. $-\sqrt[5]{486}$ **58.** $-\sqrt[5]{2048}$ **59.** $\sqrt[6]{128}$ **60.** $\sqrt[6]{1458}$

✎ **61.** A student claimed that $\sqrt[3]{14}$ is not in simplified form, since $14 = 8 + 6$, and 8 is a perfect cube. Was his reasoning correct? Why or why not?

✎ **62.** Explain in your own words why $\sqrt[3]{k^4}$ is not a simplified radical.

Express each radical in simplified form. Assume that all variables represent positive real numbers. **See Example 5.**

63. $\sqrt{72k^2}$

64. $\sqrt{18m^2}$

65. $\sqrt{144x^3y^9}$

66. $\sqrt{169s^5t^{10}}$

67. $\sqrt{121x^6}$

68. $\sqrt{256z^{12}}$

69. $-\sqrt[3]{27t^{12}}$

70. $-\sqrt[3]{64y^{18}}$

71. $-\sqrt{100m^8z^4}$

72. $-\sqrt{25t^6s^{20}}$

73. $-\sqrt[3]{-125a^6b^9c^{12}}$

74. $-\sqrt[3]{-216y^{15}x^6z^3}$

75. $\sqrt[4]{\dfrac{1}{16}r^8t^{20}}$

76. $\sqrt[4]{\dfrac{81}{256}t^{12}u^8}$

💿 **77.** $\sqrt{50x^3}$

78. $\sqrt{300z^3}$

79. $-\sqrt{500r^{11}}$

80. $-\sqrt{200p^{13}}$

81. $\sqrt{13x^7y^8}$

82. $\sqrt{23k^9p^{14}}$

83. $\sqrt[3]{8z^6w^9}$

84. $\sqrt[3]{64a^{15}b^{12}}$

85. $\sqrt[3]{-16z^5t^7}$

86. $\sqrt[3]{-81m^4n^{10}}$

87. $\sqrt[4]{81x^{12}y^{16}}$

88. $\sqrt[4]{81t^8u^{28}}$

89. $-\sqrt[4]{162r^{15}s^{10}}$

90. $-\sqrt[4]{32k^5m^{10}}$

91. $\sqrt{\dfrac{y^{11}}{36}}$

92. $\sqrt{\dfrac{v^{13}}{49}}$

93. $\sqrt[3]{\dfrac{x^{16}}{27}}$

94. $\sqrt[3]{\dfrac{y^{17}}{125}}$

Simplify each radical. Assume that $x \geq 0$. **See Example 6.**

💿 **95.** $\sqrt[4]{48^2}$

96. $\sqrt[4]{50^2}$

97. $\sqrt[4]{25}$

98. $\sqrt[6]{8}$

99. $\sqrt[10]{x^{25}}$

100. $\sqrt[12]{x^{44}}$

Simplify by first writing the radicals as radicals with the same index. Then multiply. Assume that all variables represent positive real numbers. **See Example 7.**

💿 **101.** $\sqrt[3]{4} \cdot \sqrt{3}$

102. $\sqrt[3]{5} \cdot \sqrt{6}$

103. $\sqrt[4]{3} \cdot \sqrt[3]{4}$

104. $\sqrt[5]{7} \cdot \sqrt[3]{5}$

105. $\sqrt{x} \cdot \sqrt[3]{x}$

106. $\sqrt[3]{y} \cdot \sqrt[4]{y}$

Find the unknown length in each right triangle. Simplify the answer if possible. **See Example 8.**

💿 **107.**

108.

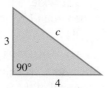

109.

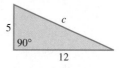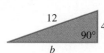

110.

111.

112.

Find the distance between each pair of points. **See Example 9.**

113. $(6, 13)$ and $(1, 1)$

114. $(8, 13)$ and $(2, 5)$

💿 **115.** $(-6, 5)$ and $(3, -4)$

116. $(-1, 5)$ and $(-7, 7)$

117. $(-8, 2)$ and $(-4, 1)$

118. $(-1, 2)$ and $(5, 3)$

119. $(4.7, 2.3)$ and $(1.7, -1.7)$

120. $(-2.9, 18.2)$ and $(2.1, 6.2)$

121. $\left(\sqrt{2}, \sqrt{6}\right)$ and $\left(-2\sqrt{2}, 4\sqrt{6}\right)$

122. $\left(\sqrt{7}, 9\sqrt{3}\right)$ and $\left(-\sqrt{7}, 4\sqrt{3}\right)$

123. $(x + y, y)$ and $(x - y, x)$

124. $(c, c - d)$ and $(d, c + d)$

Find the perimeter of each triangle. $\left(\textit{Hint: For Exercise 125, use } \sqrt{k} + \sqrt{k} = 2\sqrt{k}.\right)$

125.

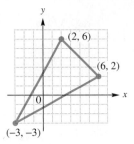

126.

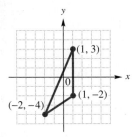

Solve each problem.

127. The following letter appeared in the column "Ask Tom Why," written by Tom Skilling of the *Chicago Tribune*:

> *Dear Tom,*
>
> *I cannot remember the formula to calculate the distance to the horizon. I have a stunning view from my 14th-floor condo, 150 ft above the ground. How far can I see?*
>
> *Ted Fleischaker; Indianapolis, Ind.*

Skilling's answer was as follows:

> To find the distance to the horizon in miles, take the square root of the height of your view in feet and multiply that result by 1.224. Your answer will be the number of miles to the horizon. (*Source: Chicago Tribune.*)

Assuming that Ted's eyes are 6 ft above the ground, the total height from the ground is $150 + 6 = 156$ ft. To the nearest tenth of a mile, how far can he see to the horizon?

128. The length of the diagonal of a box is given by

$$D = \sqrt{L^2 + W^2 + H^2},$$

where L, W, and H are, respectively, the length, width, and height of the box. Find the length of the diagonal D of a box that is 4 ft long, 2 ft wide, and 3 ft high. Give the exact value, and then round to the nearest tenth of a foot.

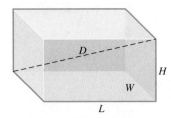

129. A Sanyo color television, model AVM-2755, has a rectangular screen with a 21.7-in. width. Its height is 16 in. What is the measure of the diagonal of the screen, to the nearest tenth of an inch? (*Source:* Actual measurements of the author's television.)

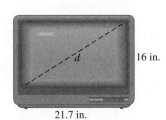

16 in.

21.7 in.

130. A formula from electronics dealing with the impedance of parallel resonant circuits is

$$I = \frac{E}{\sqrt{R^2 + \omega^2 L^2}},$$

where the variables are in appropriate units. Find I if $E = 282$, $R = 100$, $L = 264$, and $\omega = 120\pi$. Give your answer to the nearest thousandth.

131. In the study of sound, one version of the law of tensions is

$$f_1 = f_2 \sqrt{\frac{F_1}{F_2}}.$$

If $F_1 = 300$, $F_2 = 60$, and $f_2 = 260$, find f_1 to the nearest unit.

132. The illumination I, in foot-candles, produced by a light source is related to the distance d, in feet, from the light source by the equation

$$d = \sqrt{\frac{k}{I}},$$

where k is a constant. If $k = 640$, how far from the light source will the illumination be 2 foot-candles? Give the exact value, and then round to the nearest tenth of a foot.

Refer to the Chapter Opener on **page 427.** *Recall that an* **isosceles triangle** *is a triangle that has two sides of equal length.*

133. The statement made by the Scarecrow in *The Wizard of Oz* can be proved false by providing at least one situation in which it leads to a false statement. Use the isosceles triangle shown here to prove that the statement is false.

134. Use the same style of wording as the Scarecrow to state the Pythagorean theorem correctly.

The table gives data on three different solar modules available for roofing.

Model	Watts	Volts	Amps	Size (in inches)	Cost (in dollars)
MSX-77	77	16.9	4.56	44 × 26	475
MSX-83	83	17.1	4.85	44 × 24	490
MSX-60	60	17.1	3.5	44 × 20	382

Source: Solarex table in Jade Mountain catalog.

You must determine the size of frame needed to support each panel on a roof. (Note: The sides of each frame will form a right triangle, and the hypotenuse of the triangle will be the width of the panel.) In Exercises 135–136, use the Pythagorean theorem to find the dimensions of the legs for each frame under the given conditions. Round answers to the nearest tenth.

135. The legs have equal length.

136. One leg is twice the length of the other.

PREVIEW EXERCISES

*Combine like terms. **See Section 5.2.***

137. $13x^4 - 12x^3 + 9x^4 + 2x^3$

138. $-15z^3 - z^2 + 4z^4 + 12z^8$

139. $9q^2 + 2q - 5q - q^2$

140. $7m^5 - 2m^3 + 8m^5 - m^3$

8.4 Adding and Subtracting Radical Expressions

OBJECTIVE

1 Simplify radical expressions involving addition and subtraction.

OBJECTIVE 1 **Simplify radical expressions involving addition and subtraction.** Expressions such as $4\sqrt{2} + 3\sqrt{2}$ and $2\sqrt{3} - 5\sqrt{3}$ can be simplified using the distributive property.

$4\sqrt{2} + 3\sqrt{2}$
$= (4 + 3)\sqrt{2} = 7\sqrt{2}$ } This is similar to simplifying $4x + 3x$ to $7x$.

$2\sqrt{3} - 5\sqrt{3}$
$= (2 - 5)\sqrt{3} = -3\sqrt{3}$ } This is similar to simplifying $2x - 5x$ to $-3x$.

NOW TRY
EXERCISE 1

Add or subtract to simplify each radical expression.

(a) $\sqrt{12} + \sqrt{75}$

(b) $-\sqrt{63t} + 3\sqrt{28t}, \quad t \geq 0$

(c) $6\sqrt{7} - 2\sqrt{3}$

⚠ **CAUTION** *Only radical expressions with the same index and the same radicand may be combined.*

EXAMPLE 1 Adding and Subtracting Radicals

Add or subtract to simplify each radical expression.

(a) $3\sqrt{24} + \sqrt{54}$

$= 3\sqrt{4} \cdot \sqrt{6} + \sqrt{9} \cdot \sqrt{6}$ Product rule

$= 3 \cdot 2\sqrt{6} + 3\sqrt{6}$ $\sqrt{4} = 2; \sqrt{9} = 3$

$= 6\sqrt{6} + 3\sqrt{6}$ Multiply.

$= 9\sqrt{6}$ $6\sqrt{6} + 3\sqrt{6} = (6 + 3)\sqrt{6}$

(b) $2\sqrt{20x} - \sqrt{45x}, \quad x \geq 0$

$= 2\sqrt{4} \cdot \sqrt{5x} - \sqrt{9} \cdot \sqrt{5x}$ Product rule

$= 2 \cdot 2\sqrt{5x} - 3\sqrt{5x}$ $\sqrt{4} = 2; \sqrt{9} = 3$

$= 4\sqrt{5x} - 3\sqrt{5x}$ Multiply.

$= \sqrt{5x}$ Combine like terms.

(c) $2\sqrt{3} - 4\sqrt{5}$ The radicands differ and are already simplified, so $2\sqrt{3} - 4\sqrt{5}$ cannot be simplified further.   NOW TRY ⟲

⚠ **CAUTION** *The root of a sum does not equal the sum of the roots.* For example,

$$\sqrt{9 + 16} \neq \sqrt{9} + \sqrt{16}$$

since $\sqrt{9 + 16} = \sqrt{25} = 5,$ but $\sqrt{9} + \sqrt{16} = 3 + 4 = 7.$

EXAMPLE 2 Adding and Subtracting Radicals with Higher Indexes

Add or subtract to simplify each radical expression. Assume that all variables represent positive real numbers.

(a) $2\sqrt[3]{16} - 5\sqrt[3]{54}$ *Remember to write the index with each radical.*

$= 2\sqrt[3]{8 \cdot 2} - 5\sqrt[3]{27 \cdot 2}$ Factor.

$= 2\sqrt[3]{8} \cdot \sqrt[3]{2} - 5\sqrt[3]{27} \cdot \sqrt[3]{2}$ Product rule

$= 2 \cdot 2 \cdot \sqrt[3]{2} - 5 \cdot 3 \cdot \sqrt[3]{2}$ Find the cube roots.

$= 4\sqrt[3]{2} - 15\sqrt[3]{2}$ Multiply.

$= (4 - 15)\sqrt[3]{2}$ Distributive property

$= -11\sqrt[3]{2}$ Combine like terms.

(b)

$$2\sqrt[3]{x^2y} + \sqrt[3]{8x^5y^4}$$

$= 2\sqrt[3]{x^2y} + \sqrt[3]{(8x^3y^3)x^2y}$ Factor.

$= 2\sqrt[3]{x^2y} + \sqrt[3]{8x^3y^3} \cdot \sqrt[3]{x^2y}$ Product rule

$= 2\sqrt[3]{x^2y} + 2xy\sqrt[3]{x^2y}$ Find the cube root.

$= (2 + 2xy)\sqrt[3]{x^2y}$ Distributive property

This result cannot be simplified further.

NOW TRY ANSWERS

1. (a) $7\sqrt{3}$ **(b)** $3\sqrt{7t}$

   (c) The expression cannot be simplified further.

NOW TRY
EXERCISE 2

Add or subtract to simplify each radical expression. Assume that all variables represent positive real numbers.

(a) $3\sqrt[3]{2000} - 4\sqrt[3]{128}$

(b) $5\sqrt[4]{a^5b^3} + \sqrt[4]{81ab^7}$

(c) $\sqrt[3]{128t^4} - 2\sqrt{72t^3}$

> Be careful. The indexes are different.

(c) $5\sqrt{4x^3} + 3\sqrt[3]{64x^4}$

$= 5\sqrt{4x^2 \cdot x} + 3\sqrt[3]{64x^3 \cdot x}$ Factor.

$= 5\sqrt{4x^2} \cdot \sqrt{x} + 3\sqrt[3]{64x^3} \cdot \sqrt[3]{x}$ Product rule

$= 5 \cdot 2x\sqrt{x} + 3 \cdot 4x\sqrt[3]{x}$

> Keep track of the indexes.

$= 10x\sqrt{x} + 12x\sqrt[3]{x}$

The radicands are both x, but since the indexes are different, this expression cannot be simplified further.   NOW TRY

NOW TRY
EXERCISE 3

Perform the indicated operations. Assume that all variables represent positive real numbers.

(a) $5\dfrac{\sqrt{5}}{\sqrt{45}} - 4\sqrt{\dfrac{28}{9}}$

(b) $6\sqrt[3]{\dfrac{16}{x^{12}}} + 7\sqrt[3]{\dfrac{9}{x^9}}$

EXAMPLE 3   Adding and Subtracting Radicals with Fractions

Perform the indicated operations. Assume that all variables represent positive real numbers.

(a) $2\sqrt{\dfrac{75}{16}} + 4\dfrac{\sqrt{8}}{\sqrt{32}}$

$= 2\dfrac{\sqrt{25 \cdot 3}}{\sqrt{16}} + 4\dfrac{\sqrt{4 \cdot 2}}{\sqrt{16 \cdot 2}}$ Quotient rule; factor.

$= 2\left(\dfrac{5\sqrt{3}}{4}\right) + 4\left(\dfrac{2\sqrt{2}}{4\sqrt{2}}\right)$ Product rule; find the square roots.

$= \dfrac{5\sqrt{3}}{2} + 2$ Multiply; $\dfrac{\sqrt{2}}{\sqrt{2}} = 1$.

$= \dfrac{5\sqrt{3}}{2} + \dfrac{4}{2}$ Write with a common denominator.

$= \dfrac{5\sqrt{3} + 4}{2}$ $\dfrac{a}{c} + \dfrac{b}{c} = \dfrac{a + b}{c}$

(b) $10\sqrt[3]{\dfrac{5}{x^6}} - 3\sqrt[3]{\dfrac{4}{x^9}}$

$= 10\dfrac{\sqrt[3]{5}}{\sqrt[3]{x^6}} - 3\dfrac{\sqrt[3]{4}}{\sqrt[3]{x^9}}$ Quotient rule

$= \dfrac{10\sqrt[3]{5}}{x^2} - \dfrac{3\sqrt[3]{4}}{x^3}$ Simplify denominators.

$= \dfrac{10\sqrt[3]{5} \cdot x}{x^2 \cdot x} - \dfrac{3\sqrt[3]{4}}{x^3}$ Write with a common denominator.

$= \dfrac{10x\sqrt[3]{5} - 3\sqrt[3]{4}}{x^3}$ Subtract fractions.   NOW TRY

NOW TRY ANSWERS

2. (a) $14\sqrt[3]{2}$
  (b) $(5a + 3b)\sqrt[4]{ab^3}$
  (c) $4t\sqrt[3]{2t} - 12t\sqrt{2t}$

3. (a) $\dfrac{5 - 8\sqrt{7}}{3}$
  (b) $\dfrac{12\sqrt[3]{2} + 7x\sqrt[3]{9}}{x^4}$

8.4 EXERCISES

MyMathLab Math XL PRACTICE WATCH DOWNLOAD READ REVIEW

⊙ *Complete solution available on the Video Resources on DVD*

Simplify. Assume that all variables represent positive real numbers. **See Examples 1 and 2.**

1. $\sqrt{36} - \sqrt{100}$ **2.** $\sqrt{25} - \sqrt{81}$ ⊙ **3.** $-2\sqrt{48} + 3\sqrt{75}$

4. $4\sqrt{32} - 2\sqrt{8}$ **5.** $\sqrt[3]{16} + 4\sqrt[3]{54}$ **6.** $3\sqrt[3]{24} - 2\sqrt[3]{192}$

7. $\sqrt[4]{32} + 3\sqrt[4]{2}$ **8.** $\sqrt[4]{405} - 2\sqrt[4]{5}$

9. $6\sqrt{18} - \sqrt{32} + 2\sqrt{50}$ **10.** $5\sqrt{8} + 3\sqrt{72} - 3\sqrt{50}$

11. $5\sqrt{6} + 2\sqrt{10}$ **12.** $3\sqrt{11} - 5\sqrt{13}$

13. $2\sqrt{5} + 3\sqrt{20} + 4\sqrt{45}$ **14.** $5\sqrt{54} - 2\sqrt{24} - 2\sqrt{96}$

15. $\sqrt{72x} - \sqrt{8x}$ **16.** $\sqrt{18k} - \sqrt{72k}$

17. $3\sqrt{72m^2} - 5\sqrt{32m^2} - 3\sqrt{18m^2}$ **18.** $9\sqrt{27p^2} - 14\sqrt{108p^2} + 2\sqrt{48p^2}$

19. $2\sqrt[3]{16} + \sqrt[3]{54}$ **20.** $15\sqrt[3]{81} + 4\sqrt[3]{24}$

⊙ **21.** $2\sqrt[3]{27x} - 2\sqrt[3]{8x}$ **22.** $6\sqrt[3]{128m} - 3\sqrt[3]{16m}$

23. $3\sqrt[3]{x^2y} - 5\sqrt[3]{8x^2y}$ **24.** $3\sqrt[3]{x^2y^2} - 2\sqrt[3]{64x^2y^2}$

25. $3x\sqrt[3]{xy^2} - 2\sqrt[3]{8x^4y^2}$ **26.** $6q^2\sqrt[3]{5q} - 2q\sqrt[3]{40q^4}$

27. $5\sqrt[4]{32} + 3\sqrt[4]{162}$ **28.** $2\sqrt[4]{512} + 4\sqrt[4]{32}$

29. $3\sqrt[4]{x^5y} - 2x\sqrt[4]{xy}$ **30.** $2\sqrt[4]{m^9p^6} - 3m^2p\sqrt[4]{mp^2}$

31. $2\sqrt[4]{32a^3} + 5\sqrt[4]{2a^3}$ **32.** $5\sqrt[4]{243x^3} + 2\sqrt[4]{3x^3}$

33. $\sqrt[3]{64xy^2} + \sqrt[3]{27x^4y^5}$ **34.** $\sqrt[4]{625s^3t} + \sqrt[4]{81s^7t^5}$

35. $\sqrt[3]{192st^4} - \sqrt{27s^3t}$ **36.** $\sqrt{125a^5b^5} + \sqrt[3]{125a^4b^4}$

37. $2\sqrt[3]{8x^4} + 3\sqrt[4]{16x^5}$ **38.** $3\sqrt[3]{64m^4} + 5\sqrt[4]{81m^5}$

Simplify. Assume that all variables represent positive real numbers. **See Example 3.**

39. $\sqrt{8} - \dfrac{\sqrt{64}}{\sqrt{16}}$ **40.** $\sqrt{48} - \dfrac{\sqrt{81}}{\sqrt{9}}$ **41.** $\dfrac{2\sqrt{5}}{3} + \dfrac{\sqrt{5}}{6}$

42. $\dfrac{4\sqrt{3}}{3} + \dfrac{2\sqrt{3}}{9}$ **43.** $\sqrt{\dfrac{8}{9}} + \sqrt{\dfrac{18}{36}}$ **44.** $\sqrt{\dfrac{12}{16}} + \sqrt{\dfrac{48}{64}}$

45. $\dfrac{\sqrt{32}}{3} + \dfrac{2\sqrt{2}}{3} - \dfrac{\sqrt{2}}{\sqrt{9}}$ **46.** $\dfrac{\sqrt{27}}{2} - \dfrac{3\sqrt{3}}{2} + \dfrac{\sqrt{3}}{\sqrt{4}}$ ⊙ **47.** $3\sqrt{\dfrac{50}{9}} + 8\dfrac{\sqrt{2}}{\sqrt{8}}$

48. $5\sqrt{\dfrac{288}{25}} + 21\dfrac{\sqrt{2}}{\sqrt{18}}$ **49.** $\sqrt{\dfrac{25}{x^8}} + \sqrt{\dfrac{9}{x^6}}$ **50.** $\sqrt{\dfrac{100}{y^4}} + \sqrt{\dfrac{81}{y^{10}}}$

51. $3\sqrt[3]{\dfrac{m^5}{27}} - 2m\sqrt[3]{\dfrac{m^2}{64}}$ **52.** $2a\sqrt[4]{\dfrac{a}{16}} - 5a\sqrt[4]{\dfrac{a}{81}}$

53. $3\sqrt[3]{\dfrac{2}{x^6}} - 4\sqrt[3]{\dfrac{5}{x^9}}$ **54.** $-4\sqrt[3]{\dfrac{4}{t^9}} + 3\sqrt[3]{\dfrac{9}{t^{12}}}$

55. *Concept Check* Which sum could be simplified without first simplifying the individual radical expressions?

A. $\sqrt{50} + \sqrt{32}$ **B.** $3\sqrt{6} + 9\sqrt{6}$ **C.** $\sqrt[3]{32} + \sqrt[3]{108}$ **D.** $\sqrt[5]{6} + \sqrt[5]{192}$

56. *Concept Check* Let $a = 1$ and let $b = 64$.

   (a) Evaluate $\sqrt{a} + \sqrt{b}$. Then find $\sqrt{a + b}$. Are they equal?

   (b) Evaluate $\sqrt[3]{a} + \sqrt[3]{b}$. Then find $\sqrt[3]{a + b}$. Are they equal?

   (c) Complete the following: In general, $\sqrt[n]{a} + \sqrt[n]{b} \neq$ _____, based on the observations in parts (a) and (b) of this exercise.

57. Even though the root indexes of the terms are not equal, the sum $\sqrt{64} + \sqrt[3]{125} + \sqrt[4]{16}$ can be simplified quite easily. What is this sum? Why can we add these terms so easily?

58. Explain why $28 - 4\sqrt{2}$ is not equal to $24\sqrt{2}$. (This is a common error among algebra students.)

Solve each problem.

59. A rectangular yard has a length of $\sqrt{192}$ m and a width of $\sqrt{48}$ m. Choose the best estimate of its dimensions. Then estimate the perimeter.

   A. 14 m by 7 m   **B.** 5 m by 7 m   **C.** 14 m by 8 m   **D.** 15 m by 8 m

60. If the sides of a triangle are $\sqrt{65}$ in., $\sqrt{35}$ in., and $\sqrt{26}$ in., which one of the following is the best estimate of its perimeter?

   A. 20 in.   **B.** 26 in.   **C.** 19 in.   **D.** 24 in.

Solve each problem. Give answers as simplified radical expressions.

61. Find the perimeter of the triangle.

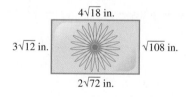

3√20 in.   2√45 in.   √75 in.

62. Find the perimeter of the rectangle.

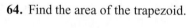

√192 m   √48 m

63. What is the perimeter of the computer graphic?

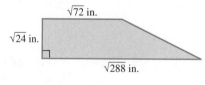

4√18 in.   3√12 in.   √108 in.   2√72 in.

64. Find the area of the trapezoid.

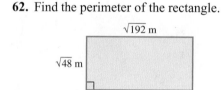

√72 in.   √24 in.   √288 in.

PREVIEW EXERCISES

Find each product. See Section 5.4.

65. $5xy(2x^2y^3 - 4x)$ **66.** $(3x + 7)(2x - 6)$ **67.** $(a^2 + b)(a^2 - b)$

68. $(2p - 7)^2$ **69.** $(4x^3 + 3)^3$ **70.** $(2 + 3y)(2 - 3y)$

Write in lowest terms. See Section 5.5.

71. $\dfrac{8x^2 - 10x}{6x^2}$ **72.** $\dfrac{15y^3 - 9y^2}{6y}$

8.5   Multiplying and Dividing Radical Expressions

OBJECTIVE 1   Multiply radical expressions. We multiply binomial expressions involving radicals by using the FOIL method from **Section 5.4.** Recall that the acronym **FOIL** refers to multiplying the **F**irst terms, **O**uter terms, **I**nner terms, and **L**ast terms of the binomials.

EXAMPLE 1   Multiplying Binomials Involving Radical Expressions

Multiply, using the FOIL method.

(a) $\left(\sqrt{5} + 3\right)\left(\sqrt{6} + 1\right)$

$$\overset{\text{First}}{\overbrace{}} \quad \overset{\text{Outer}}{\overbrace{}} \quad \overset{\text{Inner}}{\overbrace{}} \quad \overset{\text{Last}}{\overbrace{}}$$

$$= \sqrt{5} \cdot \sqrt{6} + \sqrt{5} \cdot 1 + 3 \cdot \sqrt{6} + 3 \cdot 1$$

$$= \sqrt{30} + \sqrt{5} + 3\sqrt{6} + 3 \quad \boxed{\text{This result cannot be simplified further.}}$$

(b) $\left(7 - \sqrt{3}\right)\left(\sqrt{5} + \sqrt{2}\right)$

$$\quad\quad\quad\quad \overset{\text{F}}{} \quad\quad \overset{\text{O}}{} \quad\quad \overset{\text{I}}{} \quad\quad\quad \overset{\text{L}}{}$$

$$= 7\sqrt{5} + 7\sqrt{2} - \sqrt{3} \cdot \sqrt{5} - \sqrt{3} \cdot \sqrt{2}$$

$$= 7\sqrt{5} + 7\sqrt{2} - \sqrt{15} - \sqrt{6}$$

(c) $\left(\sqrt{10} + \sqrt{3}\right)\left(\sqrt{10} - \sqrt{3}\right)$

$$= \sqrt{10} \cdot \sqrt{10} - \sqrt{10} \cdot \sqrt{3} + \sqrt{10} \cdot \sqrt{3} - \sqrt{3} \cdot \sqrt{3} \quad \text{FOIL}$$

$$= 10 - 3$$

$$= 7$$

The product $\left(\sqrt{10} + \sqrt{3}\right)\left(\sqrt{10} - \sqrt{3}\right) = \left(\sqrt{10}\right)^2 - \left(\sqrt{3}\right)^2$ is the difference of squares.

$$(x + y)(x - y) = x^2 - y^2 \quad \text{Here, } x = \sqrt{10} \text{ and } y = \sqrt{3}.$$

(d) $\left(\sqrt{7} - 3\right)^2$

$$= \left(\sqrt{7} - 3\right)\left(\sqrt{7} - 3\right)$$

$$= \sqrt{7} \cdot \sqrt{7} - 3\sqrt{7} - 3\sqrt{7} + 3 \cdot 3$$

$$= 7 - 6\sqrt{7} + 9$$

$$= 16 - 6\sqrt{7} \quad \boxed{\text{Be careful. These terms cannot be combined.}}$$

(e) $\left(5 - \sqrt[3]{3}\right)\left(5 + \sqrt[3]{3}\right)$

$$= 5 \cdot 5 + 5\sqrt[3]{3} - 5\sqrt[3]{3} - \sqrt[3]{3} \cdot \sqrt[3]{3}$$

$$= 25 - \sqrt[3]{3^2} \quad \boxed{\text{Remember to write the index 3 in each radical.}}$$

$$= 25 - \sqrt[3]{9}$$

NOW TRY
EXERCISE 1

Multiply, using the FOIL method.

(a) $\left(8 - \sqrt{5}\right)\left(9 - \sqrt{2}\right)$

(b) $\left(\sqrt{7} + \sqrt{5}\right)\left(\sqrt{7} - \sqrt{5}\right)$

(c) $\left(\sqrt{15} - 4\right)^2$

(d) $\left(8 + \sqrt[3]{5}\right)\left(8 - \sqrt[3]{5}\right)$

(e) $\left(\sqrt{m} - \sqrt{n}\right)\left(\sqrt{m} + \sqrt{n}\right)$,
$m \geq 0$ and $n \geq 0$

(f) $\left(\sqrt{k} + \sqrt{y}\right)\left(\sqrt{k} - \sqrt{y}\right)$

$= \left(\sqrt{k}\right)^2 - \left(\sqrt{y}\right)^2$ Difference of squares

$= k - y, \quad k \geq 0 \text{ and } y \geq 0$

NOW TRY

NOTE   In **Example 1(d),** we could have used the formula for the square of a binomial to obtain the same result.

$\left(\sqrt{7} - 3\right)^2$

$= \left(\sqrt{7}\right)^2 - 2\left(\sqrt{7}\right)(3) + 3^2$ $(x - y)^2 = x^2 - 2xy + y^2$

$= 7 - 6\sqrt{7} + 9$ Apply the exponents. Multiply.

$= 16 - 6\sqrt{7}$ Add.

OBJECTIVE 2   **Rationalize denominators with one radical term.** As defined earlier, a simplified radical expression has no radical in the denominator. The origin of this agreement no doubt occurred before the days of high-speed calculation, when computation was a tedious process performed by hand.

For example, consider the radical expression $\dfrac{1}{\sqrt{2}}$. To find a decimal approximation by hand, it is necessary to divide 1 by a decimal approximation for $\sqrt{2}$, such as 1.414. It is much easier if the divisor is a whole number. This can be accomplished by multiplying $\dfrac{1}{\sqrt{2}}$ by 1 in the form $\dfrac{\sqrt{2}}{\sqrt{2}}$. *Multiplying by 1 in any form does not change the value of the original expression.*

$$\frac{1}{\sqrt{2}} \cdot \frac{\sqrt{2}}{\sqrt{2}} = \frac{\sqrt{2}}{2} \qquad \text{Multiply by 1; } \frac{\sqrt{2}}{\sqrt{2}} = 1$$

Now the computation requires dividing 1.414 by 2 to obtain 0.707, a much easier task.

With current technology, either form of this fraction can be approximated with the same number of keystrokes. See **FIGURE 10**, which shows how a calculator gives the same approximation for both forms of the expression.

```
1/√(2)
        .7071067812
√(2)/2
        .7071067812
```

FIGURE 10

Rationalizing the Denominator

A common way of "standardizing" the form of a radical expression is to have the denominator contain no radicals. The process of removing radicals from a denominator so that the denominator contains only rational numbers is called **rationalizing the denominator.** This is done by multiplying by a form of 1.

EXAMPLE 2   Rationalizing Denominators with Square Roots

Rationalize each denominator.

(a) $\dfrac{3}{\sqrt{7}}$

Multiply the numerator and denominator by $\sqrt{7}$. This is, in effect, multiplying by 1.

$$\frac{3}{\sqrt{7}} = \frac{3 \cdot \sqrt{7}}{\sqrt{7} \cdot \sqrt{7}} = \frac{3\sqrt{7}}{7} \qquad \begin{array}{l}\text{In the denominator,} \\ \sqrt{7} \cdot \sqrt{7} = \sqrt{7 \cdot 7} = \sqrt{49} = 7. \\ \text{The final denominator is now a rational number.}\end{array}$$

NOW TRY ANSWERS
1. **(a)** $72 - 8\sqrt{2} - 9\sqrt{5} + \sqrt{10}$
  (b) 2   **(c)** $31 - 8\sqrt{15}$
  (d) $64 - \sqrt[3]{25}$ **(e)** $m - n$

NOW TRY
EXERCISE 2
Rationalize each denominator.

(a) $\dfrac{8}{\sqrt{13}}$ (b) $\dfrac{9\sqrt{7}}{\sqrt{3}}$

(c) $\dfrac{-10}{\sqrt{20}}$

(b) $\dfrac{5\sqrt{2}}{\sqrt{5}} = \dfrac{5\sqrt{2} \cdot \sqrt{5}}{\sqrt{5} \cdot \sqrt{5}} = \dfrac{5\sqrt{10}}{5} = \sqrt{10}$

(c) $\dfrac{-6}{\sqrt{12}}$

Less work is involved if we simplify the radical in the denominator first.

$$\dfrac{-6}{\sqrt{12}} = \dfrac{-6}{\sqrt{4 \cdot 3}} = \dfrac{-6}{2\sqrt{3}} = \dfrac{-3}{\sqrt{3}}$$

Now we rationalize the denominator.

$$\dfrac{-3}{\sqrt{3}} = \dfrac{-3 \cdot \sqrt{3}}{\sqrt{3} \cdot \sqrt{3}} = \dfrac{-3\sqrt{3}}{3} = -\sqrt{3} \qquad \text{NOW TRY} \circlearrowleft$$

NOW TRY
EXERCISE 3
Simplify each radical.

(a) $-\sqrt{\dfrac{27}{80}}$

(b) $\sqrt{\dfrac{48x^8}{y^3}}, \quad y > 0$

EXAMPLE 3 Rationalizing Denominators in Roots of Fractions

Simplify each radical. In part (b), $p > 0$.

(a) $-\sqrt{\dfrac{18}{125}}$

$= -\dfrac{\sqrt{18}}{\sqrt{125}}$ Quotient rule

$= -\dfrac{\sqrt{9 \cdot 2}}{\sqrt{25 \cdot 5}}$ Factor.

$= -\dfrac{3\sqrt{2}}{5\sqrt{5}}$ Product rule

$= -\dfrac{3\sqrt{2} \cdot \sqrt{5}}{5\sqrt{5} \cdot \sqrt{5}}$ Multiply by $\frac{\sqrt{5}}{\sqrt{5}}$.

$= -\dfrac{3\sqrt{10}}{5 \cdot 5}$ Product rule

$= -\dfrac{3\sqrt{10}}{25}$ Multiply.

(b) $\sqrt{\dfrac{50m^4}{p^5}}$

$= \dfrac{\sqrt{50m^4}}{\sqrt{p^5}}$ Quotient rule

$= \dfrac{5m^2\sqrt{2}}{p^2\sqrt{p}}$ Product rule

$= \dfrac{5m^2\sqrt{2} \cdot \sqrt{p}}{p^2\sqrt{p} \cdot \sqrt{p}}$ Multiply by $\frac{\sqrt{p}}{\sqrt{p}}$.

$= \dfrac{5m^2\sqrt{2p}}{p^2 \cdot p}$ Product rule

$= \dfrac{5m^2\sqrt{2p}}{p^3}$ Multiply.

NOW TRY $\circlearrowleft$

EXAMPLE 4 Rationalizing Denominators with Cube and Fourth Roots

Simplify.

(a) $\sqrt[3]{\dfrac{27}{16}}$

Use the quotient rule, and simplify the numerator and denominator.

NOW TRY ANSWERS

2. (a) $\dfrac{8\sqrt{13}}{13}$ (b) $3\sqrt{21}$

   (c) $-\sqrt{5}$

3. (a) $-\dfrac{3\sqrt{15}}{20}$ (b) $\dfrac{4x^4\sqrt{3y}}{y^2}$

$$\sqrt[3]{\dfrac{27}{16}} = \dfrac{\sqrt[3]{27}}{\sqrt[3]{16}} = \dfrac{3}{\sqrt[3]{8} \cdot \sqrt[3]{2}} = \dfrac{3}{2\sqrt[3]{2}}$$

Since $2 \cdot 4 = 8$, a perfect cube, multiply the numerator and denominator by $\sqrt[3]{4}$.

NOW TRY
EXERCISE 4
Simplify.

(a) $\sqrt[3]{\dfrac{8}{81}}$

(b) $\sqrt[4]{\dfrac{7x}{y}}, \quad x \geq 0, y > 0$

$$\frac{3}{2\sqrt[3]{2}} \qquad \sqrt[3]{\frac{27}{16}} = \frac{3}{2\sqrt[3]{2}} \text{ from page 460}$$

$$= \frac{3 \cdot \sqrt[3]{4}}{2\sqrt[3]{2} \cdot \sqrt[3]{4}} \qquad \begin{array}{l}\text{Multiply by } \sqrt[3]{4} \text{ in numerator} \\ \text{and denominator. This will give} \\ \sqrt[3]{8} = 2 \text{ in the denominator.}\end{array}$$

$$= \frac{3\sqrt[3]{4}}{2\sqrt[3]{8}} \qquad \text{Multiply.}$$

$$= \frac{3\sqrt[3]{4}}{2 \cdot 2} \qquad \sqrt[3]{8} = 2$$

$$= \frac{3\sqrt[3]{4}}{4} \qquad \text{Multiply.}$$

(b) $\qquad \sqrt[4]{\dfrac{5x}{z}}$

$$= \frac{\sqrt[4]{5x}}{\sqrt[4]{z}} \qquad \text{Quotient rule}$$

$$= \frac{\sqrt[4]{5x}}{\sqrt[4]{z}} \cdot \frac{\sqrt[4]{z^3}}{\sqrt[4]{z^3}} \qquad \text{Multiply by 1.}$$

> $\sqrt[4]{z} \cdot \sqrt[4]{z^3}$
> will give $\sqrt[4]{z^4}$.

$$= \frac{\sqrt[4]{5xz^3}}{\sqrt[4]{z^4}} \qquad \text{Product rule}$$

$$= \frac{\sqrt[4]{5xz^3}}{z}, \quad x \geq 0, z > 0 \qquad \text{NOW TRY}$$

⚠️ **CAUTION** In **Example 4(a)**, a typical error is to multiply the numerator and denominator by $\sqrt[3]{2}$, forgetting that $\sqrt[3]{2} \cdot \sqrt[3]{2} = \sqrt[3]{2^2}$, which does **not** equal 2. We need *three* factors of 2 to obtain 2^3 under the radical.

$$\sqrt[3]{2} \cdot \sqrt[3]{2} \cdot \sqrt[3]{2} = \sqrt[3]{2^3} \quad \text{which does equal} \quad 2.$$

OBJECTIVE 3 **Rationalize denominators with binomials involving radicals.**
Recall the special product $(x + y)(x - y) = x^2 - y^2$. To rationalize a denominator that contains a binomial expression (one that contains exactly two terms) involving radicals, such as

$$\frac{3}{1 + \sqrt{2}},$$

we must use *conjugates*. The conjugate of $1 + \sqrt{2}$ is $1 - \sqrt{2}$. In general, $x + y$ and $x - y$ are **conjugates.**

Rationalizing a Binomial Denominator

Whenever a radical expression has a sum or difference with square root radicals in the denominator, rationalize the denominator by multiplying both the numerator and denominator by the conjugate of the denominator.

NOW TRY ANSWERS
4. (a) $\dfrac{2\sqrt[3]{9}}{9}$ **(b)** $\dfrac{\sqrt[4]{7xy^3}}{y}$

NOW TRY
EXERCISE 5

Rationalize each denominator.

(a) $\dfrac{4}{1 + \sqrt{3}}$ **(b)** $\dfrac{4}{5 + \sqrt{7}}$

(c) $\dfrac{\sqrt{3} + \sqrt{7}}{\sqrt{5} - \sqrt{2}}$

(d) $\dfrac{8}{\sqrt{3x} - \sqrt{y}}$,
$3x \neq y, x > 0, y > 0$

EXAMPLE 5 Rationalizing Binomial Denominators

Rationalize each denominator.

(a) $\dfrac{3}{1 + \sqrt{2}}$

> Again, we are multiplying by a form of 1.

$= \dfrac{3(1 - \sqrt{2})}{(1 + \sqrt{2})(1 - \sqrt{2})}$ Multiply the numerator and denominator by $1 - \sqrt{2}$, the conjugate of the denominator.

$(1 + \sqrt{2})(1 - \sqrt{2})$
$= 1^2 - (\sqrt{2})^2$
$= 1 - 2,$ or -1

> The denominator is now a rational number.

$= \dfrac{3(1 - \sqrt{2})}{-1}$

$= \dfrac{3}{-1}(1 - \sqrt{2})$

$= -3(1 - \sqrt{2}),$ or $-3 + 3\sqrt{2}$ Distributive property

(b) $\dfrac{5}{4 - \sqrt{3}}$

$= \dfrac{5(4 + \sqrt{3})}{(4 - \sqrt{3})(4 + \sqrt{3})}$ Multiply the numerator and denominator by $4 + \sqrt{3}$.

$= \dfrac{5(4 + \sqrt{3})}{16 - 3}$ Multiply in the denominator.

$= \dfrac{5(4 + \sqrt{3})}{13}$ Subtract in the denominator.

Notice that the numerator is left in factored form. This makes it easier to determine whether the expression is written in lowest terms.

(c) $\dfrac{\sqrt{2} - \sqrt{3}}{\sqrt{5} + \sqrt{3}}$

$= \dfrac{(\sqrt{2} - \sqrt{3})(\sqrt{5} - \sqrt{3})}{(\sqrt{5} + \sqrt{3})(\sqrt{5} - \sqrt{3})}$ Multiply the numerator and denominator by $\sqrt{5} - \sqrt{3}$.

$= \dfrac{\sqrt{10} - \sqrt{6} - \sqrt{15} + 3}{5 - 3}$ Multiply.

$= \dfrac{\sqrt{10} - \sqrt{6} - \sqrt{15} + 3}{2}$ Subtract in the denominator.

(d) $\dfrac{3}{\sqrt{5m} - \sqrt{p}}$, $5m \neq p, m > 0, p > 0$

$= \dfrac{3(\sqrt{5m} + \sqrt{p})}{(\sqrt{5m} - \sqrt{p})(\sqrt{5m} + \sqrt{p})}$ Multiply the numerator and denominator by $\sqrt{5m} + \sqrt{p}$.

$= \dfrac{3(\sqrt{5m} + \sqrt{p})}{5m - p}$ Multiply in the denominator.

NOW TRY

NOW TRY ANSWERS

5. (a) $-2(1 - \sqrt{3})$, or $-2 + 2\sqrt{3}$

(b) $\dfrac{2(5 - \sqrt{7})}{9}$

(c) $\dfrac{\sqrt{15} + \sqrt{6} + \sqrt{35} + \sqrt{14}}{3}$

(d) $\dfrac{8(\sqrt{3x} + \sqrt{y})}{3x - y}$

OBJECTIVE 4   Write radical quotients in lowest terms.

NOW TRY
EXERCISE 6

Write each quotient in lowest terms.

(a) $\dfrac{15 - 6\sqrt{2}}{18}$

(b) $\dfrac{15k + \sqrt{50k^2}}{20k}, \quad k > 0$

EXAMPLE 6   Writing Radical Quotients in Lowest Terms

Write each quotient in lowest terms.

(a) $\dfrac{6 + 2\sqrt{5}}{4}$

$= \dfrac{2(3 + \sqrt{5})}{2 \cdot 2}$ This is a key step. Factor the numerator and denominator.

$= \dfrac{3 + \sqrt{5}}{2}$ Divide out the common factor.

Here is an alternative method for writing this expression in lowest terms.

$$\frac{6 + 2\sqrt{5}}{4} = \frac{6}{4} + \frac{2\sqrt{5}}{4} = \frac{3}{2} + \frac{\sqrt{5}}{2} = \frac{3 + \sqrt{5}}{2}$$

(b) $\dfrac{5y - \sqrt{8y^2}}{6y}, \quad y > 0$

$= \dfrac{5y - 2y\sqrt{2}}{6y}$ $\sqrt{8y^2} = \sqrt{4y^2 \cdot 2} = 2y\sqrt{2}$

$= \dfrac{y(5 - 2\sqrt{2})}{6y}$ Factor the numerator.

$= \dfrac{5 - 2\sqrt{2}}{6}$ Divide out the common factor.   NOW TRY

⚠ **CAUTION**   *Be careful to factor before writing a quotient in lowest terms.*

CONNECTIONS

In calculus, it is sometimes desirable to **rationalize the numerator.** For example, to rationalize the numerator of

$$\frac{6 - \sqrt{2}}{4},$$

we multiply the numerator and the denominator by the conjugate of the *numerator.*

$$\frac{6 - \sqrt{2}}{4} = \frac{(6 - \sqrt{2})(6 + \sqrt{2})}{4(6 + \sqrt{2})} = \frac{36 - 2}{4(6 + \sqrt{2})} = \frac{34}{4(6 + \sqrt{2})} = \frac{17}{2(6 + \sqrt{2})}$$

For Discussion or Writing

Rationalize the numerator of each expression. (*a* and *b* are nonnegative real numbers.)

1. $\dfrac{8\sqrt{5} - 1}{6}$ 2. $\dfrac{3\sqrt{a} + \sqrt{b}}{b}$ 3. $\dfrac{3\sqrt{a} + \sqrt{b}}{\sqrt{b} - \sqrt{a}} \quad (b \neq a)$

4. Rationalize the denominator of the expression in **Exercise 3,** and then describe the difference in the procedure you used from what you did in **Exercise 3.**

NOW TRY ANSWERS

6. (a) $\dfrac{5 - 2\sqrt{2}}{6}$ (b) $\dfrac{3 + \sqrt{2}}{4}$

8.5 EXERCISES

⊕ *Complete solution available on the Video Resources on DVD*

Concept Check *Match each part of a rule for a special product in Column I with the other part in Column II. Assume that A and B represent positive real numbers.*

I	II
1. $\left(A + \sqrt{B}\right)\left(A - \sqrt{B}\right)$	**A.** $A - B$
2. $\left(\sqrt{A} + B\right)\left(\sqrt{A} - B\right)$	**B.** $A + 2B\sqrt{A} + B^2$
3. $\left(\sqrt{A} + \sqrt{B}\right)\left(\sqrt{A} - \sqrt{B}\right)$	**C.** $A - B^2$
4. $\left(\sqrt{A} + \sqrt{B}\right)^2$	**D.** $A - 2\sqrt{AB} + B$
5. $\left(\sqrt{A} - \sqrt{B}\right)^2$	**E.** $A^2 - B$
6. $\left(\sqrt{A} + B\right)^2$	**F.** $A + 2\sqrt{AB} + B$

Multiply, and then simplify each product. Assume that all variables represent positive real numbers. ***See Example 1.***

7. $\sqrt{6}\left(3 + \sqrt{2}\right)$ **8.** $\sqrt{2}\left(\sqrt{32} - \sqrt{9}\right)$ **9.** $5\left(\sqrt{72} - \sqrt{8}\right)$

10. $7\left(\sqrt{50} - \sqrt{18}\right)$ **11.** $\left(\sqrt{7} + 3\right)\left(\sqrt{7} - 3\right)$ **12.** $\left(\sqrt{3} - 5\right)\left(\sqrt{3} + 5\right)$

⊕ **13.** $\left(\sqrt{2} - \sqrt{3}\right)\left(\sqrt{2} + \sqrt{3}\right)$ **14.** $\left(\sqrt{7} + \sqrt{14}\right)\left(\sqrt{7} - \sqrt{14}\right)$

15. $\left(\sqrt{8} - \sqrt{2}\right)\left(\sqrt{8} + \sqrt{2}\right)$ **16.** $\left(\sqrt{20} - \sqrt{5}\right)\left(\sqrt{20} + \sqrt{5}\right)$

17. $\left(\sqrt{2} + 1\right)\left(\sqrt{3} - 1\right)$ **18.** $\left(\sqrt{3} + 3\right)\left(\sqrt{5} - 2\right)$

19. $\left(\sqrt{11} - \sqrt{7}\right)\left(\sqrt{2} + \sqrt{5}\right)$ **20.** $\left(\sqrt{13} - \sqrt{7}\right)\left(\sqrt{3} + \sqrt{11}\right)$

21. $\left(2\sqrt{3} + \sqrt{5}\right)\left(3\sqrt{3} - 2\sqrt{5}\right)$ **22.** $\left(\sqrt{7} - \sqrt{11}\right)\left(2\sqrt{7} + 3\sqrt{11}\right)$

23. $\left(\sqrt{5} + 2\right)^2$ **24.** $\left(\sqrt{11} - 1\right)^2$

25. $\left(\sqrt{21} - \sqrt{5}\right)^2$ **26.** $\left(\sqrt{6} - \sqrt{2}\right)^2$

27. $\left(2 + \sqrt[3]{6}\right)\left(2 - \sqrt[3]{6}\right)$ **28.** $\left(\sqrt[3]{3} + 6\right)\left(\sqrt[3]{3} - 6\right)$

29. $\left(2 + \sqrt[3]{2}\right)\left(4 - 2\sqrt[3]{2} + \sqrt[3]{4}\right)$ **30.** $\left(\sqrt[3]{3} - 1\right)\left(\sqrt[3]{9} + \sqrt[3]{3} + 1\right)$

31. $\left(3\sqrt{x} - \sqrt{5}\right)\left(2\sqrt{x} + 1\right)$ **32.** $\left(4\sqrt{p} + \sqrt{7}\right)\left(\sqrt{p} - 9\right)$

33. $\left(3\sqrt{r} - \sqrt{s}\right)\left(3\sqrt{r} + \sqrt{s}\right)$ **34.** $\left(\sqrt{k} + 4\sqrt{m}\right)\left(\sqrt{k} - 4\sqrt{m}\right)$

35. $\left(\sqrt[3]{2y} - 5\right)\left(4\sqrt[3]{2y} + 1\right)$ **36.** $\left(\sqrt[3]{9z} - 2\right)\left(5\sqrt[3]{9z} + 7\right)$

37. $\left(\sqrt{3x} + 2\right)\left(\sqrt{3x} - 2\right)$ **38.** $\left(\sqrt{6y} - 4\right)\left(\sqrt{6y} + 4\right)$

39. $\left(2\sqrt{x} + \sqrt{y}\right)\left(2\sqrt{x} - \sqrt{y}\right)$ **40.** $\left(\sqrt{p} + 5\sqrt{s}\right)\left(\sqrt{p} - 5\sqrt{s}\right)$

41. $\left[\left(\sqrt{2} + \sqrt{3}\right) - \sqrt{6}\right]\left[\left(\sqrt{2} + \sqrt{3}\right) + \sqrt{6}\right]$

42. $\left[\left(\sqrt{5} - \sqrt{2}\right) - \sqrt{3}\right]\left[\left(\sqrt{5} - \sqrt{2}\right) + \sqrt{3}\right]$

Rationalize the denominator in each expression. Assume that all variables represent positive real numbers. **See Examples 2 and 3.**

43. $\dfrac{7}{\sqrt{7}}$ **44.** $\dfrac{11}{\sqrt{11}}$ **45.** $\dfrac{15}{\sqrt{3}}$ **46.** $\dfrac{12}{\sqrt{6}}$ **47.** $\dfrac{\sqrt{3}}{\sqrt{2}}$

48. $\dfrac{\sqrt{7}}{\sqrt{6}}$ **49.** $\dfrac{9\sqrt{3}}{\sqrt{5}}$ **50.** $\dfrac{3\sqrt{2}}{\sqrt{11}}$ **51.** $\dfrac{-7}{\sqrt{48}}$ **52.** $\dfrac{-5}{\sqrt{24}}$

53. $\sqrt{\dfrac{7}{2}}$ **54.** $\sqrt{\dfrac{10}{3}}$ **55.** $-\sqrt{\dfrac{7}{50}}$ **56.** $-\sqrt{\dfrac{13}{75}}$ **57.** $\sqrt{\dfrac{24}{x}}$

58. $\sqrt{\dfrac{52}{y}}$ **59.** $\dfrac{-8\sqrt{3}}{\sqrt{k}}$ **60.** $\dfrac{-4\sqrt{13}}{\sqrt{m}}$ **61.** $-\sqrt{\dfrac{150m^5}{n^3}}$ **62.** $-\sqrt{\dfrac{98r^3}{s^5}}$

63. $\sqrt{\dfrac{288x^7}{y^9}}$ **64.** $\sqrt{\dfrac{242t^9}{u^{11}}}$ **65.** $\dfrac{5\sqrt{2m}}{\sqrt{y^3}}$

66. $\dfrac{2\sqrt{5r}}{\sqrt{m^3}}$ **67.** $-\sqrt{\dfrac{48k^2}{z}}$ **68.** $-\sqrt{\dfrac{75m^3}{p}}$

Simplify. Assume that all variables represent positive real numbers. **See Example 4.**

69. $\sqrt[3]{\dfrac{2}{3}}$ **70.** $\sqrt[3]{\dfrac{4}{5}}$ **71.** $\sqrt[3]{\dfrac{4}{9}}$ **72.** $\sqrt[3]{\dfrac{5}{16}}$ **73.** $\sqrt[3]{\dfrac{9}{32}}$

74. $\sqrt[3]{\dfrac{10}{9}}$ **75.** $-\sqrt[3]{\dfrac{2p}{r^2}}$ **76.** $-\sqrt[3]{\dfrac{6x}{y^2}}$ **77.** $\sqrt[3]{\dfrac{x^6}{y}}$ **78.** $\sqrt[3]{\dfrac{m^9}{q}}$

79. $\sqrt[4]{\dfrac{16}{x}}$ **80.** $\sqrt[4]{\dfrac{81}{y}}$ **81.** $\sqrt[4]{\dfrac{2y}{z}}$ **82.** $\sqrt[4]{\dfrac{7t}{s^2}}$

Rationalize the denominator in each expression. Assume that all variables represent positive real numbers and no denominators are 0. **See Example 5.**

83. $\dfrac{3}{4+\sqrt{5}}$ **84.** $\dfrac{4}{5+\sqrt{6}}$ **85.** $\dfrac{\sqrt{8}}{3-\sqrt{2}}$

86. $\dfrac{\sqrt{27}}{3-\sqrt{3}}$ **87.** $\dfrac{2}{3\sqrt{5}+2\sqrt{3}}$ **88.** $\dfrac{-1}{3\sqrt{2}-2\sqrt{7}}$

89. $\dfrac{\sqrt{2}-\sqrt{3}}{\sqrt{6}-\sqrt{5}}$ **90.** $\dfrac{\sqrt{5}+\sqrt{6}}{\sqrt{3}-\sqrt{2}}$ **91.** $\dfrac{m-4}{\sqrt{m}+2}$

92. $\dfrac{r-9}{\sqrt{r}-3}$ **93.** $\dfrac{4}{\sqrt{x}-2\sqrt{y}}$ **94.** $\dfrac{5}{3\sqrt{r}+\sqrt{s}}$

95. $\dfrac{\sqrt{x}-\sqrt{y}}{\sqrt{x}+\sqrt{y}}$ **96.** $\dfrac{\sqrt{a}+\sqrt{b}}{\sqrt{a}-\sqrt{b}}$ **97.** $\dfrac{5\sqrt{k}}{2\sqrt{k}+\sqrt{q}}$ **98.** $\dfrac{3\sqrt{x}}{\sqrt{x}-2\sqrt{y}}$

Write each expression in lowest terms. Assume that all variables represent positive real numbers. **See Example 6.**

99. $\dfrac{30-20\sqrt{6}}{10}$ **100.** $\dfrac{24+12\sqrt{5}}{12}$ **101.** $\dfrac{3-3\sqrt{5}}{3}$ **102.** $\dfrac{-5+5\sqrt{2}}{5}$

103. $\dfrac{16-4\sqrt{8}}{12}$ **104.** $\dfrac{12-9\sqrt{72}}{18}$ **105.** $\dfrac{6p+\sqrt{24p^3}}{3p}$ **106.** $\dfrac{11y-\sqrt{242y^5}}{22y}$

Brain Busters *Rationalize each denominator. Assume that all radicals represent real numbers and no denominators are* 0.

107. $\dfrac{3}{\sqrt{x+y}}$

108. $\dfrac{5}{\sqrt{m-n}}$

109. $\dfrac{p}{\sqrt{p+2}}$

110. $\dfrac{q}{\sqrt{5+q}}$

111. The following expression occurs in a certain standard problem in trigonometry.

$$\frac{1}{\sqrt{2}} \cdot \frac{\sqrt{3}}{2} - \frac{1}{\sqrt{2}} \cdot \frac{1}{2}$$

Show that it simplifies to $\dfrac{\sqrt{6}-\sqrt{2}}{4}$. Then verify, using a calculator approximation.

112. The following expression occurs in a certain standard problem in trigonometry.

$$\frac{\sqrt{3}+1}{1-\sqrt{3}}$$

Show that it simplifies to $-2 - \sqrt{3}$. Then verify, using a calculator approximation.

Rationalize the numerator in each expression. Assume that all variables represent positive real numbers. (Hint: See the **Connections box** *following* **Example 6.)**

113. $\dfrac{6-\sqrt{3}}{8}$

114. $\dfrac{2\sqrt{5}-3}{2}$

115. $\dfrac{2\sqrt{x}-\sqrt{y}}{3x}$

116. $\dfrac{\sqrt{p}-3\sqrt{q}}{4q}$

PREVIEW EXERCISES

Solve each equation. See Sections 2.1 and 6.5.

117. $-8x + 7 = 4$

118. $3x - 7 = 12$

119. $6x^2 - 7x = 3$

120. $x(15x - 11) = -2$

SUMMARY EXERCISES on Operations with Radicals and Rational Exponents

Conditions for a Simplified Radical

1. The radicand has no factor raised to a power greater than or equal to the index.

2. The radicand has no fractions.

3. No denominator contains a radical.

4. Exponents in the radicand and the index of the radical have greatest common factor 1.

Perform all indicated operations, and express each answer in simplest form with positive exponents. Assume that all variables represent positive real numbers.

1. $6\sqrt{10} - 12\sqrt{10}$

2. $\sqrt{7}\left(\sqrt{7}-\sqrt{2}\right)$

3. $\left(1-\sqrt{3}\right)\left(2+\sqrt{6}\right)$

4. $\sqrt{50} - \sqrt{98} + \sqrt{72}$

5. $\left(3\sqrt{5}+2\sqrt{7}\right)^2$

6. $\dfrac{-3}{\sqrt{6}}$

7. $\dfrac{8}{\sqrt{7} + \sqrt{5}}$

8. $\dfrac{1 - \sqrt{2}}{1 + \sqrt{2}}$

9. $\left(\sqrt{5} + 7\right)\left(\sqrt{5} - 7\right)$

10. $\dfrac{1}{\sqrt{x} - \sqrt{5}}, \quad x \neq 5$

11. $\sqrt[3]{8a^3b^5c^9}$

12. $\dfrac{15}{\sqrt[3]{9}}$

13. $\dfrac{3}{\sqrt{5} + 2}$

14. $\sqrt{\dfrac{3}{5x}}$

15. $\dfrac{16\sqrt{3}}{5\sqrt{12}}$

16. $\dfrac{2\sqrt{25}}{8\sqrt{50}}$

17. $\dfrac{-10}{\sqrt[3]{10}}$

18. $\dfrac{\sqrt{6} + \sqrt{5}}{\sqrt{6} - \sqrt{5}}$

19. $\sqrt{12x} - \sqrt{75x}$

20. $\left(5 - 3\sqrt{3}\right)^2$

21. $\sqrt[3]{\dfrac{13}{81}}$

22. $\dfrac{\sqrt{3} + \sqrt{7}}{\sqrt{6} - \sqrt{5}}$

23. $\dfrac{6}{\sqrt[4]{3}}$

24. $\dfrac{1}{1 - \sqrt[3]{3}}$

25. $\sqrt[3]{\dfrac{x^2y}{x^{-3}y^4}}$

26. $\sqrt{12} - \sqrt{108} - \sqrt[3]{27}$

27. $\dfrac{x^{-2/3}y^{4/5}}{x^{-5/3}y^{-2/5}}$

28. $\left(\dfrac{x^{3/4}y^{2/3}}{x^{1/3}y^{5/8}}\right)^{24}$

29. $(125x^3)^{-2/3}$

30. $\dfrac{4^{1/2} + 3^{1/2}}{4^{1/2} - 3^{1/2}}$

31. $\sqrt[3]{16x^2} - \sqrt[3]{54x^2} + \sqrt[3]{128x^2}$

32. $\left(1 - \sqrt[3]{3}\right)\left(1 + \sqrt[3]{3} + \sqrt[3]{9}\right)$

Students often have trouble distinguishing between the following two types of problems:

Simplifying a Radical Involving a Square Root	**Solving an Equation Using Square Roots**
Exercise: Simplify $\sqrt{25}$. *Answer:* 5 In this situation, $\sqrt{25}$ represents the positive square root of 25, namely 5.	*Exercise:* Solve $x^2 = 25$. *Answer:* $\{-5, 5\}$ In this situation, $x^2 = 25$ has two solutions, the negative square root of 25 or the positive square root of 25: $-5, 5$.

In Exercises 33–40, provide the appropriate responses.

33. (a) Simplify $\sqrt{64}$.

    (b) Solve $x^2 = 64$.

34. (a) Simplify $\sqrt{100}$.

    (b) Solve $x^2 = 100$.

35. (a) Solve $x^2 = 16$.

    (b) Simplify $-\sqrt{16}$.

36. (a) Solve $x^2 = 25$.

    (b) Simplify $-\sqrt{25}$.

37. (a) Simplify $-\sqrt{\dfrac{81}{121}}$.

    (b) Solve $x^2 = \dfrac{81}{121}$.

38. (a) Simplify $-\sqrt{\dfrac{49}{100}}$.

    (b) Solve $x^2 = \dfrac{49}{100}$.

39. (a) Solve $x^2 = 0.04$.

    (b) Simplify $\sqrt{0.04}$.

40. (a) Solve $x^2 = 0.09$.

    (b) Simplify $\sqrt{0.09}$.

8.6
Solving Equations with Radicals

An equation that includes one or more radical expressions with a variable is called a **radical equation.**

$$\sqrt{x-4} = 8, \quad \sqrt{5x+12} = 3\sqrt{2x-1}, \quad \text{and} \quad \sqrt[3]{6+x} = 27$$

Examples of radical equations

OBJECTIVE 1 Solve radical equations by using the power rule. The equation $x = 1$ has only one solution. Its solution set is $\{1\}$. If we square both sides of this equation, we get $x^2 = 1$. This new equation has *two* solutions: -1 and 1. Notice that the solution of the original equation is also a solution of the equation following squaring. However, that equation has another solution, -1, that is *not* a solution of the original equation.

When solving equations with radicals, we use this idea of raising both sides to a power. It is an application of the **power rule.**

Power Rule for Solving an Equation with Radicals

If both sides of an equation are raised to the same power, all solutions of the original equation are also solutions of the new equation.

The power rule does not say that all solutions of the new equation are solutions of the original equation. They may or may not be. Solutions that do not satisfy the original equation are called **extraneous solutions.** They must be rejected.

⚠ **CAUTION** When the power rule is used to solve an equation, *every solution of the new equation must be checked in the original equation.*

NOW TRY EXERCISE 1

Solve $\sqrt{9x+7} = 5$.

EXAMPLE 1 Using the Power Rule

Solve $\sqrt{3x+4} = 8$.

$$\left(\sqrt{3x+4}\right)^2 = 8^2 \qquad \text{Use the power rule and square each side.}$$

$\left(\sqrt{a}\right)^2 = \sqrt{a} \cdot \sqrt{a} = a$

$$3x + 4 = 64 \qquad \text{Apply the exponents.}$$

$$3x = 60 \qquad \text{Subtract 4.}$$

$$x = 20 \qquad \text{Divide by 3.}$$

CHECK

$$\sqrt{3x+4} = 8 \qquad \text{Original equation}$$

$$\sqrt{3 \cdot 20 + 4} \overset{?}{=} 8 \qquad \text{Let } x = 20.$$

$$\sqrt{64} \overset{?}{=} 8 \qquad \text{Simplify.}$$

$$8 = 8 \checkmark \qquad \text{True}$$

Since 20 satisfies the *original* equation, the solution set is $\{20\}$.

NOW TRY ANSWER
1. $\{2\}$

NOW TRY ↻

Use the following steps to solve equations with radicals.

Solving an Equation with Radicals

Step 1 **Isolate the radical.** Make sure that one radical term is alone on one side of the equation.

Step 2 **Apply the power rule.** Raise each side of the equation to a power that is the same as the index of the radical.

Step 3 **Solve** the resulting equation. If it still contains a radical, repeat Steps 1 and 2.

Step 4 **Check** all proposed solutions in the original equation.

NOW TRY
EXERCISE 2

Solve $\sqrt{3x + 4} + 5 = 0$.

EXAMPLE 2 Using the Power Rule

Solve $\sqrt{5x - 1} + 3 = 0$.

Step 1	$\sqrt{5x - 1} = -3$	To isolate the radical on one side, subtract 3 from each side.
Step 2	$\left(\sqrt{5x - 1}\right)^2 = (-3)^2$	Square each side.
Step 3	$5x - 1 = 9$	Apply the exponents.
	$5x = 10$	Add 1.
	$x = 2$	Divide by 5.

Step 4 CHECK

$\sqrt{5x - 1} + 3 = 0$ Original equation

$\sqrt{5 \cdot 2 - 1} + 3 \overset{?}{=} 0$ Let $x = 2$.

Be sure to check the proposed solution.

$3 + 3 = 0$ False

This false result shows that the *proposed* solution 2 is *not* a solution of the original equation. It is extraneous. The solution set is $\emptyset$. NOW TRY

NOTE We could have determined after Step 1 that the equation in **Example 2** has no solution because the expression on the left cannot be negative. (Why?)

OBJECTIVE 2 **Solve radical equations that require additional steps.** The next examples involve finding the square of a binomial. Recall the rule from **Section 5.4.**

$$(x + y)^2 = x^2 + 2xy + y^2$$

EXAMPLE 3 Using the Power Rule (Squaring a Binomial)

Solve $\sqrt{4 - x} = x + 2$.

Step 1 The radical is alone on the left side of the equation.

Step 2 Square each side. The square of $x + 2$ is $(x + 2)^2 = x^2 + 2(x)(2) + 4$.

$$\left(\sqrt{4 - x}\right)^2 = (x + 2)^2$$

Remember the middle term.

$$4 - x = x^2 + 4x + 4$$

Twice the product of 2 and x

NOW TRY ANSWER
2. $\emptyset$

**NOW TRY
EXERCISE 3**
Solve $\sqrt{16-x}=x+4$.

Step 3 The new equation is quadratic, so write it in standard form.

$$4-x=x^2+4x+4 \quad \text{Equation from Step 2}$$
$$x^2+5x=0 \quad \text{Subtract 4. Add } x.$$
$$x(x+5)=0 \quad \text{Factor.}$$

Set each factor equal to 0. → $x=0 \quad \text{or} \quad x+5=0 \quad \text{Zero-factor property}$
$$x=-5 \quad \text{Solve for } x.$$

Step 4 Check each proposed solution in the original equation.

CHECK $\sqrt{4-x}=x+2$ $\qquad\qquad$ $\sqrt{4-x}=x+2$

$\sqrt{4-0}\overset{?}{=}0+2$ Let x = 0. $\qquad$ $\sqrt{4-(-5)}\overset{?}{=}-5+2$ Let x = -5.

$\sqrt{4}\overset{?}{=}2$ $\qquad\qquad\qquad$ $\sqrt{9}\overset{?}{=}-3$

$2=2$ ✓ True $\qquad\qquad$ $3=-3$ False

The solution set is $\{0\}$. The other proposed solution, -5, is extraneous. NOW TRY

**NOW TRY
EXERCISE 4**
Solve
$\sqrt{x^2-3x+18}=x+3$.

EXAMPLE 4 Using the Power Rule (Squaring a Binomial)

Solve $\sqrt{x^2-4x+9}=x-1$.

Squaring gives $(x-1)^2=x^2-2(x)(1)+1^2$ on the right.

$$\left(\sqrt{x^2-4x+9}\right)^2=(x-1)^2 \quad \boxed{\text{Remember the middle term.}}$$
$$x^2-4x+9=x^2-2x+1$$

Twice the product of x and −1

$$-2x=-8 \quad \text{Subtract } x^2 \text{ and 9. Add } 2x.$$
$$x=4 \quad \text{Divide by } -2.$$

CHECK $\qquad\qquad \sqrt{x^2-4x+9}=x-1$ Original equation

$\sqrt{4^2-4\cdot4+9}\overset{?}{=}4-1$ Let x = 4.

$3=3$ ✓ True

The solution set is $\{4\}$. $\qquad\qquad\qquad\qquad\qquad$ NOW TRY

EXAMPLE 5 Using the Power Rule (Squaring Twice)

Solve $\sqrt{5x+6}+\sqrt{3x+4}=2$.

Isolate one radical on one side of the equation by subtracting $\sqrt{3x+4}$ from each side.

$$\sqrt{5x+6}=2-\sqrt{3x+4} \quad \text{Subtract } \sqrt{3x+4}.$$
$$\left(\sqrt{5x+6}\right)^2=\left(2-\sqrt{3x+4}\right)^2 \quad \text{Square each side.}$$
$$5x+6=4-4\sqrt{3x+4}+(3x+4) \quad \boxed{\text{Be careful here.}}$$

Remember the middle term. Twice the product of 2 and $-\sqrt{3x+4}$

NOW TRY ANSWERS
3. $\{0\}$ **4.** $\{1\}$

NOW TRY
EXERCISE 5
Solve

$\sqrt{3x + 1} - \sqrt{x + 4} = 1.$

The equation still contains a radical, so isolate the radical term on the right and square both sides again.

$5x + 6 = 4 - 4\sqrt{3x + 4} + 3x + 4$	Result after squaring
$5x + 6 = 8 - 4\sqrt{3x + 4} + 3x$	Combine like terms.
$2x - 2 = -4\sqrt{3x + 4}$	Subtract 8 and 3x.
$x - 1 = -2\sqrt{3x + 4}$	Divide by 2.
$(x - 1)^2 = \left(-2\sqrt{3x + 4}\right)^2$	Square each side again.
$x^2 - 2x + 1 = (-2)^2\left(\sqrt{3x + 4}\right)^2$	On the right, $(ab)^2 = a^2b^2$.
$x^2 - 2x + 1 = 4(3x + 4)$	Apply the exponents.
$x^2 - 2x + 1 = 12x + 16$	Distributive property
$x^2 - 14x - 15 = 0$	Standard form
$(x - 15)(x + 1) = 0$	Factor.
$x - 15 = 0 \quad$ or $\quad x + 1 = 0$	Zero-factor property
$x = 15 \quad$ or $\quad x = -1$	Solve each equation.

Divide each term by 2.

CHECK		
	$\sqrt{5x + 6} + \sqrt{3x + 4} = 2$	Original equation
	$\sqrt{5(15) + 6} + \sqrt{3(15) + 4} \stackrel{?}{=} 2$	Let x = 15.
	$\sqrt{81} + \sqrt{49} \stackrel{?}{=} 2$	Simplify.
	$9 + 7 \stackrel{?}{=} 2$	Take square roots.
	$16 = 2$	False

Thus, 15 is an extraneous solution and must be rejected. Confirm that the proposed solution -1 checks, so the solution set is $\{-1\}$. NOW TRY

OBJECTIVE 3 Solve radical equations with indexes greater than 2.

NOW TRY
EXERCISE 6
Solve $\sqrt[3]{4x - 5} = \sqrt[3]{3x + 2}.$

EXAMPLE 6 Using the Power Rule for a Power Greater Than 2

Solve $\sqrt[3]{z + 5} = \sqrt[3]{2z - 6}.$

$\left(\sqrt[3]{z + 5}\right)^3 = \left(\sqrt[3]{2z - 6}\right)^3$	Cube each side.
$z + 5 = 2z - 6$	
$11 = z$	Subtract z. Add 6.

CHECK		
	$\sqrt[3]{z + 5} = \sqrt[3]{2z - 6}$	Original equation
	$\sqrt[3]{11 + 5} \stackrel{?}{=} \sqrt[3]{2 \cdot 11 - 6}$	Let z = 11.
	$\sqrt[3]{16} = \sqrt[3]{16} \checkmark$	True

The solution set is $\{11\}$. NOW TRY

NOW TRY ANSWERS
5. $\{5\}$ 6. $\{7\}$

OBJECTIVE 4   Use the power rule to solve a formula for a specified variable.

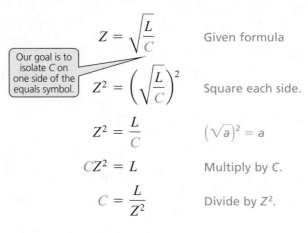

NOW TRY
EXERCISE 7

Solve the formula for a.

$$x = \sqrt{\frac{y+2}{a}}$$

EXAMPLE 7   Solving a Formula from Electronics for a Variable

An important property of a radio-frequency transmission line is its **characteristic impedance,** represented by Z and measured in ohms. If L and C are the inductance and capacitance, respectively, per unit of length of the line, then these quantities are related by the formula $Z = \sqrt{\frac{L}{C}}$. Solve this formula for C.

$$Z = \sqrt{\frac{L}{C}} \qquad \text{Given formula}$$

Our goal is to isolate C on one side of the equals symbol.

$$Z^2 = \left(\sqrt{\frac{L}{C}}\right)^2 \qquad \text{Square each side.}$$

$$Z^2 = \frac{L}{C} \qquad \left(\sqrt{a}\right)^2 = a$$

$$CZ^2 = L \qquad \text{Multiply by } C.$$

$$C = \frac{L}{Z^2} \qquad \text{Divide by } Z^2.$$

NOW TRY ANSWER

7. $a = \dfrac{y+2}{x^2}$

NOW TRY

8.6 EXERCISES

MyMathLab   Math XL PRACTICE   WATCH   DOWNLOAD   READ   REVIEW

Concept Check   *Check each equation to see if the given value for x is a solution.*

1. $\sqrt{3x + 18} - x = 0$

    (a) 6   **(b)** -3

2. $\sqrt{3x - 3} - x + 1 = 0$

    (a) 1   **(b)** 4

3. $\sqrt{x + 2} - \sqrt{9x - 2} = -2\sqrt{x - 1}$

    (a) 2   **(b)** 7

4. $\sqrt{8x - 3} - 2x = 0$

    (a) $\dfrac{3}{2}$ **(b)** $\dfrac{1}{2}$

📝 **5.** Is 9 a solution of the equation $\sqrt{x} = -3$? If not, what is the solution of this equation? Explain.

📝 **6.** Before even attempting to solve $\sqrt{3x + 18} = x$, how can you be sure that the equation cannot have a negative solution?

*Solve each equation. **See Examples 1–4.***

7. $\sqrt{x - 2} = 3$

8. $\sqrt{x + 1} = 7$

🌐 **9.** $\sqrt{6k - 1} = 1$

10. $\sqrt{7x - 3} = 6$

🌐 **11.** $\sqrt{4r + 3} + 1 = 0$

12. $\sqrt{5k - 3} + 2 = 0$

13. $\sqrt{3x + 1} - 4 = 0$

14. $\sqrt{5x + 1} - 11 = 0$

15. $4 - \sqrt{x - 2} = 0$

16. $9 - \sqrt{4x + 1} = 0$

17. $\sqrt{9x - 4} = \sqrt{8x + 1}$

18. $\sqrt{4x - 2} = \sqrt{3x + 5}$

19. $2\sqrt{x} = \sqrt{3x + 4}$

20. $2\sqrt{x} = \sqrt{5x - 16}$

21. $3\sqrt{x - 1} = 2\sqrt{2x + 2}$

22. $5\sqrt{4x + 1} = 3\sqrt{10x + 25}$

23. $x = \sqrt{x^2 + 4x - 20}$

24. $x = \sqrt{x^2 - 3x + 18}$

25. $x = \sqrt{x^2 + 3x + 9}$

26. $x = \sqrt{x^2 - 4x - 8}$

27. $\sqrt{9 - x} = x + 3$

28. $\sqrt{5 - x} = x + 1$

29. $\sqrt{k^2 + 2k + 9} = k + 3$

30. $\sqrt{x^2 - 3x + 3} = x - 1$

31. $\sqrt{x^2 + 12x - 4} = x - 4$

32. $\sqrt{x^2 - 15x + 15} = x - 5$

33. $\sqrt{r^2 + 9r + 15} - r - 4 = 0$

34. $\sqrt{m^2 + 3m + 12} - m - 2 = 0$

35. *Concept Check* In solving the equation $\sqrt{3x + 4} = 8 - x$, a student wrote the following for her first step. *WHAT WENT WRONG?* Solve the given equation correctly.

$$3x + 4 = 64 + x^2$$

36. *Concept Check* In solving the equation $\sqrt{5x + 6} - \sqrt{x + 3} = 3$, a student wrote the following for his first step. *WHAT WENT WRONG?* Solve the given equation correctly.

$$(5x + 6) + (x + 3) = 9$$

Solve each equation. ***See Examples 5 and 6.***

37. $\sqrt[3]{2x + 5} = \sqrt[3]{6x + 1}$

38. $\sqrt[3]{p + 5} = \sqrt[3]{2p - 4}$

39. $\sqrt[3]{x^2 + 5x + 1} = \sqrt[3]{x^2 + 4x}$

40. $\sqrt[3]{r^2 + 2r + 8} = \sqrt[3]{r^2 + 3r + 12}$

41. $\sqrt[3]{2m - 1} = \sqrt[3]{m + 13}$

42. $\sqrt[3]{2k - 11} = \sqrt[3]{5k + 1}$

43. $\sqrt[4]{x + 12} = \sqrt[4]{3x - 4}$

44. $\sqrt[4]{z + 11} = \sqrt[4]{2z + 6}$

45. $\sqrt[3]{x - 8} + 2 = 0$

46. $\sqrt[3]{r + 1} + 1 = 0$

47. $\sqrt[4]{2k - 5} + 4 = 0$

48. $\sqrt[4]{8z - 3} + 2 = 0$

49. $\sqrt{k + 2} - \sqrt{k - 3} = 1$

50. $\sqrt{r + 6} - \sqrt{r - 2} = 2$

51. $\sqrt{2r + 11} - \sqrt{5r + 1} = -1$

52. $\sqrt{3x - 2} - \sqrt{x + 3} = 1$

53. $\sqrt{3p + 4} - \sqrt{2p - 4} = 2$

54. $\sqrt{4x + 5} - \sqrt{2x + 2} = 1$

55. $\sqrt{3 - 3p} - 3 = \sqrt{3p + 2}$

56. $\sqrt{4x + 7} - 4 = \sqrt{4x - 1}$

57. $\sqrt{2\sqrt{x + 11}} = \sqrt{4x + 2}$

58. $\sqrt{1 + \sqrt{24 - 10x}} = \sqrt{3x + 5}$

For each equation, write the expressions with rational exponents as radical expressions, and then solve, using the procedures explained in this section.

59. $(2x - 9)^{1/2} = 2 + (x - 8)^{1/2}$

60. $(3w + 7)^{1/2} = 1 + (w + 2)^{1/2}$

61. $(2w - 1)^{2/3} - w^{1/3} = 0$

62. $(x^2 - 2x)^{1/3} - x^{1/3} = 0$

Solve each formula for the indicated variable. ***See Example 7.*** *(Source: Cooke, Nelson M., and Joseph B. Orleans,* Mathematics Essential to Electricity and Radio, *McGraw-Hill.)*

63. $Z = \sqrt{\dfrac{L}{C}}$ for L **64.** $r = \sqrt{\dfrac{\mathscr{A}}{\pi}}$ for $\mathscr{A}$ ◐ **65.** $V = \sqrt{\dfrac{2K}{m}}$ for K

66. $V = \sqrt{\dfrac{2K}{m}}$ for m **67.** $r = \sqrt{\dfrac{Mm}{F}}$ for M **68.** $r = \sqrt{\dfrac{Mm}{F}}$ for F

The formula

$$N = \frac{1}{2\pi}\sqrt{\frac{a}{r}}$$

is used to find the rotational rate N of a space station. Here, a is the acceleration and r represents the radius of the space station, in meters. To find the value of r that will make N simulate the effect of gravity on Earth, the equation must be solved for r, using the required value of N. (Source: Kastner, Bernice, Space Mathematics, *NASA.)*

69. Solve the equation for r.

70. (a) Approximate the value of r so that $N = 0.063$ rotation per sec if $a = 9.8$ m per sec^2.

     (b) Approximate the value of r so that $N = 0.04$ rotation per sec if $a = 9.8$ m per sec^2.

PREVIEW EXERCISES

Perform the indicated operations. ***See Sections 5.2 and 5.4.***

71. $(5 + 9x) + (-4 - 8x)$ **72.** $(12 + 7y) - (-3 + 2y)$ **73.** $(x + 3)(2x - 5)$

Simplify each radical. ***See Section 8.5.***

74. $\dfrac{2}{4 + \sqrt{3}}$ **75.** $\dfrac{-7}{5 - \sqrt{2}}$ **76.** $\dfrac{\sqrt{2} + \sqrt{7}}{\sqrt{5} + \sqrt{3}}$

8.7   Complex Numbers

OBJECTIVES

1 Simplify numbers of the form $\sqrt{-b}$, where $b > 0$.

2 Recognize subsets of the complex numbers.

3 Add and subtract complex numbers.

4 Multiply complex numbers.

5 Divide complex numbers.

6 Find powers of i.

OBJECTIVE 1   **Simplify numbers of the form $\sqrt{-b}$, where $b > 0$.** The equation $x^2 + 1 = 0$ has no real number solution, since any solution must be a number whose square is -1. In the set of real numbers, all squares are nonnegative numbers because the product of two positive numbers or two negative numbers is positive and $0^2 = 0$. To provide a solution of the equation $x^2 + 1 = 0$, we introduce a new number i.

Imaginary Unit i

The **imaginary unit** i is defined as

$$i = \sqrt{-1}, \quad \text{where} \quad i^2 = -1.$$

That is, i is the principal square root of -1.

We can use this definition to define any square root of a negative real number.

$$\sqrt{-b}$$

For any positive real number b, $\sqrt{-b} = i\sqrt{b}.$

NOW TRY
EXERCISE 1

Write each number as a product of a real number and i.

(a) $\sqrt{-49}$ **(b)** $-\sqrt{-121}$

(c) $\sqrt{-3}$ **(d)** $\sqrt{-32}$

EXAMPLE 1 Simplifying Square Roots of Negative Numbers

Write each number as a product of a real number and i.

(a) $\sqrt{-100} = i\sqrt{100} = 10i$ **(b)** $-\sqrt{-36} = -i\sqrt{36} = -6i$

(c) $\sqrt{-2} = i\sqrt{2}$ **(d)** $\sqrt{-54} = i\sqrt{54} = i\sqrt{9 \cdot 6} = 3i\sqrt{6}$

NOW TRY

⚠ **CAUTION** It is easy to mistake $\sqrt{2}i$ for $\sqrt{2i}$, with the i under the radical. For this reason, we usually write $\sqrt{2}i$ as $i\sqrt{2}$, as in the definition of $\sqrt{-b}$.

When finding a product such as $\sqrt{-4} \cdot \sqrt{-9}$, we cannot use the product rule for radicals because it applies only to *nonnegative* radicands. ***For this reason, we change $\sqrt{-b}$ to the form $i\sqrt{b}$ before performing any multiplications or divisions.***

NOW TRY
EXERCISE 2

Multiply.

(a) $\sqrt{-4} \cdot \sqrt{-16}$

(b) $\sqrt{-5} \cdot \sqrt{-11}$

(c) $\sqrt{-3} \cdot \sqrt{-12}$

(d) $\sqrt{13} \cdot \sqrt{-2}$

EXAMPLE 2 Multiplying Square Roots of Negative Numbers

Multiply.

(a)

$$\sqrt{-4} \cdot \sqrt{-9}$$

$= i\sqrt{4} \cdot i\sqrt{9}$ $\sqrt{-b} = i\sqrt{b}$ *First write all square roots in terms of i.*

$= i \cdot 2 \cdot i \cdot 3$ Take square roots.

$= 6i^2$ Multiply.

$= 6(-1)$ Substitute -1 for i^2.

$= -6$

(b)

$$\sqrt{-3} \cdot \sqrt{-7}$$

$= i\sqrt{3} \cdot i\sqrt{7}$ $\sqrt{-b} = i\sqrt{b}$ *First write all square roots in terms of i.*

$= i^2\sqrt{3 \cdot 7}$ Product rule

$= (-1)\sqrt{21}$ Substitute -1 for i^2.

$= -\sqrt{21}$

(c) $\sqrt{-2} \cdot \sqrt{-8}$

$= i\sqrt{2} \cdot i\sqrt{8}$ $\sqrt{-b} = i\sqrt{b}$

$= i^2\sqrt{2 \cdot 8}$ Product rule

$= (-1)\sqrt{16}$ $i^2 = -1$

$= -4$ Take the square root.

(d) $\sqrt{-5} \cdot \sqrt{6}$

$= i\sqrt{5} \cdot \sqrt{6}$

$= i\sqrt{30}$

NOW TRY

NOW TRY ANSWERS
1. **(a)** $7i$ **(b)** $-11i$
  (c) $i\sqrt{3}$ **(d)** $4i\sqrt{2}$
2. **(a)** -8 **(b)** $-\sqrt{55}$
  (c) -6 **(d)** $i\sqrt{26}$

> ⚠️ **CAUTION** Using the product rule for radicals *before* using the definition of $\sqrt{-b}$ gives a *wrong* answer. **Example 2(a)** shows that
>
> $$\sqrt{-4} \cdot \sqrt{-9} = -6, \qquad \text{Correct}$$
>
> but
>
> $$\sqrt{-4(-9)} = \sqrt{36} = 6, \qquad \text{Incorrect}$$
>
> so
>
> $$\sqrt{-4} \cdot \sqrt{-9} \neq \sqrt{-4(-9)}.$$

⟲ *NOW TRY*
EXERCISE 3
Divide.

(a) $\dfrac{\sqrt{-72}}{\sqrt{-8}}$ **(b)** $\dfrac{\sqrt{-48}}{\sqrt{3}}$

EXAMPLE 3 Dividing Square Roots of Negative Numbers

Divide.

(a) $\dfrac{\sqrt{-75}}{\sqrt{-3}}$

$= \dfrac{i\sqrt{75}}{i\sqrt{3}}$ ⟵ First write all square roots in terms of i.

$= \sqrt{\dfrac{75}{3}}$ Quotient rule

$= \sqrt{25}$ Divide.

$= 5$

(b) $\dfrac{\sqrt{-32}}{\sqrt{8}}$

$= \dfrac{i\sqrt{32}}{\sqrt{8}}$ $\sqrt{-32} = i\sqrt{32}$

$= i\sqrt{\dfrac{32}{8}}$ Quotient rule

$= i\sqrt{4}$ Divide.

$= 2i$ *NOW TRY* ⟳

OBJECTIVE 2 **Recognize subsets of the complex numbers.** A new set of numbers, the *complex numbers,* are defined as follows.

Complex Number

If a and b are real numbers, then any number of the form $a + bi$ is called a **complex number.** In the complex number $a + bi$, the number a is called the **real part** and b is called the **imaginary part.***

For a complex number $a + bi$, if $b = 0$, then $a + bi = a$, which is a real number. ***Thus, the set of real numbers is a subset of the set of complex numbers.*** If $a = 0$ and $b \neq 0$, the complex number is a **pure imaginary number.** For example, $3i$ is a pure imaginary number. A number such as $7 + 2i$ is a **nonreal complex number.** A complex number written in the form $a + bi$ is in **standard form.**

The relationships among the various sets of numbers are shown in **FIGURE 11.**

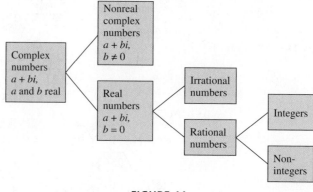

FIGURE 11

*Some texts define bi as the imaginary part of the complex number $a + bi$.

OBJECTIVE 3   **Add and subtract complex numbers.** The commutative, associative, and distributive properties for real numbers are also valid for complex numbers. *Thus, to add complex numbers, we add their real parts and add their imaginary parts.*

NOW TRY
EXERCISE 4

Add.

(a) $(-3 + 2i) + (4 + 7i)$

(b) $(5 - i) + (-3 + 3i)$
    $+ (6 - 4i)$

EXAMPLE 4   Adding Complex Numbers

Add.

(a) $(2 + 3i) + (6 + 4i)$

$\quad = (2 + 6) + (3 + 4)i$ Properties of real numbers

$\quad = 8 + 7i$ Add real parts. Add imaginary parts.

(b) $(4 + 2i) + (3 - i) + (-6 + 3i)$

$\quad = [4 + 3 + (-6)] + [2 + (-1) + 3]i$ Associative property

$\quad = 1 + 4i$ Add real parts.
Add imaginary parts.   NOW TRY

To subtract complex numbers, we subtract their real parts and subtract their imaginary parts.

NOW TRY
EXERCISE 5

Subtract.

(a) $(7 + 10i) - (3 + 5i)$

(b) $(5 - 2i) - (9 - 7i)$

(c) $(-1 + 12i) - (-1 - i)$

EXAMPLE 5   Subtracting Complex Numbers

Subtract.

(a) $(6 + 5i) - (3 + 2i)$

$\quad = (6 - 3) + (5 - 2)i$ Properties of real numbers

$\quad = 3 + 3i$ Subtract real parts. Subtract imaginary parts.

(b) $(7 - 3i) - (8 - 6i)$

$\quad = (7 - 8) + [-3 - (-6)]i$

$\quad = -1 + 3i$

(c) $(-9 + 4i) - (-9 + 8i)$

$\quad = (-9 + 9) + (4 - 8)i$

$\quad = 0 - 4i$

$\quad = -4i$ NOW TRY

OBJECTIVE 4   **Multiply complex numbers.**

EXAMPLE 6   Multiplying Complex Numbers

Multiply.

(a) $4i(2 + 3i)$

$\quad = 4i(2) + 4i(3i)$ Distributive property

$\quad = 8i + 12i^2$ Multiply.

$\quad = 8i + 12(-1)$ Substitute -1 for i^2.

$\quad = -12 + 8i$ Standard form

(b) $(3 + 5i)(4 - 2i)$

$\quad = \underbrace{3(4)}_{\text{First}} + \underbrace{3(-2i)}_{\text{Outer}} + \underbrace{5i(4)}_{\text{Inner}} + \underbrace{5i(-2i)}_{\text{Last}}$ Use the FOIL method.

$\quad = 12 - 6i + 20i - 10i^2$ Multiply.

$\quad = 12 + 14i - 10(-1)$ Combine imaginary terms; $i^2 = -1$.

$\quad = 12 + 14i + 10$ Multiply.

$\quad = 22 + 14i$ Combine real terms.

NOW TRY ANSWERS
4. **(a)** $1 + 9i$ **(b)** $8 - 2i$
5. **(a)** $4 + 5i$ **(b)** $-4 + 5i$
    (c) $13i$

NOW TRY
EXERCISE 6
Multiply.

(a) $8i(3 - 5i)$

(b) $(7 - 2i)(4 + 3i)$

(c) $(2 + 3i)(1 - 5i)$

$= 2(1) + 2(-5i) + 3i(1) + 3i(-5i)$ FOIL

$= 2 - 10i + 3i - 15i^2$ Multiply.

$= 2 - 7i - 15(-1)$ [Use parentheses around -1 to avoid errors.]

$= 2 - 7i + 15$

$= 17 - 7i$ NOW TRY

The two complex numbers $a + bi$ and $a - bi$ are called **complex conjugates,** or simply *conjugates,* of each other. ***The product of a complex number and its conjugate is always a real number,*** as shown here.

$$(a + bi)(a - bi) = a^2 - abi + abi - b^2i^2$$

$$= a^2 - b^2(-1)$$

$$(a + bi)(a - bi) = a^2 + b^2$$ [The product eliminates i.]

For example, $(3 + 7i)(3 - 7i) = 3^2 + 7^2 = 9 + 49 = 58.$

OBJECTIVE 5 **Divide complex numbers.** The quotient of two complex numbers should be a complex number. To write the quotient as a complex number, we need to eliminate i in the denominator. We use conjugates and a process similar to that for rationalizing a denominator to do this.

EXAMPLE 7 Dividing Complex Numbers

Find each quotient.

(a) $\dfrac{8 + 9i}{5 + 2i}$

Multiply both the numerator and denominator by the conjugate of the denominator. The conjugate of $5 + 2i$ is $5 - 2i$.

$$\frac{8 + 9i}{5 + 2i}$$

$$= \frac{(8 + 9i)(5 - 2i)}{(5 + 2i)(5 - 2i)} \qquad \tfrac{5 - 2i}{5 - 2i} = 1$$

$$= \frac{40 - 16i + 45i - 18i^2}{5^2 + 2^2} \qquad \begin{array}{l}\text{In the denominator,}\\ (a + bi)(a - bi) = a^2 + b^2.\end{array}$$

$$= \frac{58 + 29i}{29} \qquad \begin{array}{l}-18i^2 = -18(-1) = 18;\\ \text{Combine like terms.}\end{array}$$

$$= \frac{29(2 + i)}{29} \qquad \text{Factor the numerator.}$$

[Factor first. Then divide out the common factor.]

$$= 2 + i \qquad \text{Lowest terms}$$

NOW TRY ANSWERS
6. **(a)** $40 + 24i$ **(b)** $34 + 13i$

NOW TRY
EXERCISE 7

Find each quotient.

(a) $\dfrac{4 + 2i}{1 + 3i}$ (b) $\dfrac{5 - 4i}{i}$

(b) $\dfrac{1 + i}{i}$

$= \dfrac{(1 + i)(-i)}{i(-i)}$ Multiply numerator and denominator by $-i$, the conjugate of i.

$= \dfrac{-i - i^2}{-i^2}$ Distributive property; multiply.

$= \dfrac{-i - (-1)}{-(-1)}$ Substitute -1 for i^2.

> Use parentheses to avoid errors.

$= \dfrac{-i + 1}{1}$

$= 1 - i$ NOW TRY

OBJECTIVE 6 **Find powers of i.** Because i^2 is defined to be -1, we can find greater powers of i as shown in the following examples.

$$i^3 = i \cdot i^2 = i(-1) = -i \qquad i^6 = i^2 \cdot i^4 = (-1) \cdot 1 = -1$$
$$i^4 = i^2 \cdot i^2 = (-1)(-1) = 1 \qquad i^7 = i^3 \cdot i^4 = (-i) \cdot 1 = -i$$
$$i^5 = i \cdot i^4 = i \cdot 1 = i \qquad i^8 = i^4 \cdot i^4 = 1 \cdot 1 = 1$$

Notice that the powers of i rotate through the four numbers i, -1, $-i$, and 1. Greater powers of i can be simplified by using the fact that $i^4 = 1$.

NOW TRY
EXERCISE 8

Find each power of i.

(a) i^{16} (b) i^{21}
(c) i^{-6} (d) i^{-13}

EXAMPLE 8 Simplifying Powers of i

Find each power of i.

(a) $i^{12} = (i^4)^3 = 1^3 = 1$

(b) $i^{39} = i^{36} \cdot i^3 = (i^4)^9 \cdot i^3 = 1^9 \cdot (-i) = -i$

(c) $i^{-2} = \dfrac{1}{i^2} = \dfrac{1}{-1} = -1$

NOW TRY ANSWERS
7. (a) $1 - i$ (b) $-4 - 5i$
8. (a) 1 (b) i (c) -1 (d) $-i$

(d) $i^{-1} = \dfrac{1}{i} = \dfrac{1(-i)}{i(-i)} = \dfrac{-i}{-i^2} = \dfrac{-i}{-(-1)} = \dfrac{-i}{1} = -i$ NOW TRY

8.7 EXERCISES

 MyMathLab   Math XL PRACTICE   WATCH   DOWNLOAD   READ   REVIEW

🌐 *Complete solution available on the Video Resources on DVD*

Concept Check   *Decide whether each expression is equal to 1, -1, i, or $-i$.*

1. $\sqrt{-1}$ 2. $-\sqrt{-1}$ 3. i^2 4. $-i^2$ 5. $\dfrac{1}{i}$ 6. $(-i)^2$

Write each number as a product of a real number and i. Simplify all radical expressions. **See Example 1.**

🌐 7. $\sqrt{-169}$ 8. $\sqrt{-225}$ 9. $-\sqrt{-144}$ 10. $-\sqrt{-196}$

11. $\sqrt{-5}$ 12. $\sqrt{-21}$ 13. $\sqrt{-48}$ 14. $\sqrt{-96}$

Multiply or divide as indicated. ***See Examples 2 and 3.***

15. $\sqrt{-7} \cdot \sqrt{-15}$ **16.** $\sqrt{-3} \cdot \sqrt{-19}$ **17.** $\sqrt{-4} \cdot \sqrt{-25}$ **18.** $\sqrt{-9} \cdot \sqrt{-81}$

19. $\sqrt{-3} \cdot \sqrt{11}$ **20.** $\sqrt{-10} \cdot \sqrt{2}$ **21.** $\dfrac{\sqrt{-300}}{\sqrt{-100}}$ **22.** $\dfrac{\sqrt{-40}}{\sqrt{-10}}$

23. $\dfrac{\sqrt{-75}}{\sqrt{3}}$ **24.** $\dfrac{\sqrt{-160}}{\sqrt{10}}$ **25.** $\dfrac{-\sqrt{-64}}{\sqrt{-16}}$ **26.** $\dfrac{-\sqrt{-100}}{\sqrt{-25}}$

27. Every real number is a complex number. Explain why this is so.

28. Not every complex number is a real number. Give an example, and explain why this statement is true.

Add or subtract as indicated. Write your answers in the form a + bi. ***See Examples 4 and 5.***

29. $(3 + 2i) + (-4 + 5i)$ **30.** $(7 + 15i) + (-11 + 14i)$

31. $(5 - i) + (-5 + i)$ **32.** $(-2 + 6i) + (2 - 6i)$

33. $(4 + i) - (-3 - 2i)$ **34.** $(9 + i) - (3 + 2i)$

35. $(-3 - 4i) - (-1 - 4i)$ **36.** $(-2 - 3i) - (-5 - 3i)$

37. $(-4 + 11i) + (-2 - 4i) + (7 + 6i)$ **38.** $(-1 + i) + (2 + 5i) + (3 + 2i)$

39. $[(7 + 3i) - (4 - 2i)] + (3 + i)$ **40.** $[(7 + 2i) + (-4 - i)] - (2 + 5i)$

41. *Concept Check*   Fill in the blank with the correct response:

Because $(4 + 2i) - (3 + i) = 1 + i$, using the definition of subtraction, we can check this to find that $(1 + i) + (3 + i) =$ _____ .

42. *Concept Check*   Fill in the blank with the correct response:

Because $\dfrac{-5}{2 - i} = -2 - i$, using the definition of division, we can check this to find that $(-2 - i)(2 - i) =$ _____ .

Multiply. ***See Example 6.***

43. $(3i)(27i)$ **44.** $(5i)(125i)$ **45.** $(-8i)(-2i)$

46. $(-32i)(-2i)$ **47.** $5i(-6 + 2i)$ **48.** $3i(4 + 9i)$

49. $(4 + 3i)(1 - 2i)$ **50.** $(7 - 2i)(3 + i)$ **51.** $(4 + 5i)^2$

52. $(3 + 2i)^2$ **53.** $2i(-4 - i)^2$ **54.** $3i(-3 - i)^2$

55. $(12 + 3i)(12 - 3i)$ **56.** $(6 + 7i)(6 - 7i)$ **57.** $(4 + 9i)(4 - 9i)$

58. $(7 + 2i)(7 - 2i)$ **59.** $(1 + i)^2(1 - i)^2$ **60.** $(2 - i)^2(2 + i)^2$

61. *Concept Check*   What is the conjugate of $a + bi$?

62. *Concept Check*   If we multiply $a + bi$ by its conjugate, we get _____ , which is always a real number.

Find each quotient. ***See Example 7.***

63. $\dfrac{2}{1 - i}$ **64.** $\dfrac{2}{1 + i}$ **65.** $\dfrac{8i}{2 + 2i}$ **66.** $\dfrac{-8i}{1 + i}$

67. $\dfrac{-7 + 4i}{3 + 2i}$ **68.** $\dfrac{-38 - 8i}{7 + 3i}$ **69.** $\dfrac{2 - 3i}{2 + 3i}$ **70.** $\dfrac{-1 + 5i}{3 + 2i}$

71. $\dfrac{3 + i}{i}$ **72.** $\dfrac{5 - i}{i}$ **73.** $\dfrac{3 - i}{-i}$ **74.** $\dfrac{5 + i}{-i}$

Find each power of i. ***See Example 8.***

75. i^{18} **76.** i^{26} **77.** i^{89} **78.** i^{48} **79.** i^{38}

80. i^{102} **81.** i^{43} **82.** i^{83} **83.** i^{-5} **84.** i^{-17}

85. A student simplified i^{-18} as follows:

$$i^{-18} = i^{-18} \cdot i^{20} = i^{-18+20} = i^2 = -1.$$

Explain the mathematical justification for this correct work.

86. Explain why

$$(46 + 25i)(3 - 6i) \quad \text{and} \quad (46 + 25i)(3 - 6i)i^{12}$$

must be equal. (Do not actually perform the computation.)

Ohm's law *for the current I in a circuit with voltage E, resistance R, capacitive reactance X_c, and inductive reactance X_L is*

$$I = \frac{E}{R + (X_L - X_c)i}.$$

Use this law to work Exercises 87 and 88.

87. Find I if $E = 2 + 3i$, $R = 5$, $X_L = 4$, and $X_c = 3$.

88. Find E if $I = 1 - i$, $R = 2$, $X_L = 3$, and $X_c = 1$.

Complex numbers will appear again in this book in **Chapter 9,** *when we study quadratic equations. The following exercises examine how a complex number can be a solution of a quadratic equation.*

89. Show that $1 + 5i$ is a solution of

$$x^2 - 2x + 26 = 0.$$

Then show that its conjugate is also a solution.

90. Show that $3 + 2i$ is a solution of

$$x^2 - 6x + 13 = 0.$$

Then show that its conjugate is also a solution.

Brain Busters   *Perform the indicated operations. Give answers in standard form.*

91. $\dfrac{3}{2 - i} + \dfrac{5}{1 + i}$

92. $\dfrac{2}{3 + 4i} + \dfrac{4}{1 - i}$

93. $\left(\dfrac{2 + i}{2 - i} + \dfrac{i}{1 + i} \right)i$

94. $\left(\dfrac{4 - i}{1 + i} - \dfrac{2i}{2 + i} \right)4i$

PREVIEW EXERCISES

Solve each equation. ***See Sections 2.1 and 6.5.***

95. $6x + 13 = 0$

96. $4x - 7 = 0$

97. $x(x + 3) = 40$

98. $2x^2 - 5x - 7 = 0$

99. $5x^2 - 3x = 2$

100. $-6x^2 + 7x = -10$

STUDY SKILLS

Preparing for Your Math Final Exam

Your math final exam is likely to be a comprehensive exam, which means it will cover material from the entire term.

1. **Figure out the grade you need to earn on the final exam to get the course grade you want.** Check your course syllabus for grading policies, or ask your instructor if you are not sure.

   How many points do you need to earn on your math final exam to get the grade you want?

2. **Create a final exam week plan.** Set priorities that allow you to spend extra time studying. This may mean making adjustments, in advance, in your work schedule or enlisting extra help with family responsibilities.

   What adjustments do you need to make for final exam week?

3. **Use the following suggestions to guide your studying and reviewing.**

   ▶ **Begin reviewing several days before the final exam.** DON'T wait until the last minute.

   ▶ **Know exactly which chapters and sections will be covered on the exam.**

   ▶ **Divide up the chapters.** Decide how much you will review each day.

   ▶ **Use returned quizzes and tests to review earlier material.**

   ▶ **Practice all types of problems. Use the Cumulative Reviews** that are at the end of each chapter in your textbook. All answers, with section references, are given in the answer section.

   ▶ **Review or rewrite your notes** to create summaries of important information.

   ▶ **Make study cards for all types of problems.** Carry the cards with you, and review them whenever you have a few minutes.

   ▶ **Take plenty of short breaks to reduce physical and mental stress.** Exercising, listening to music, and enjoying a favorite activity are effective stress busters.

   Finally, *DON'T* stay up all night the night before an exam—*get a good night's sleep.*

   Select several suggestions to use as you study for your math final exam.

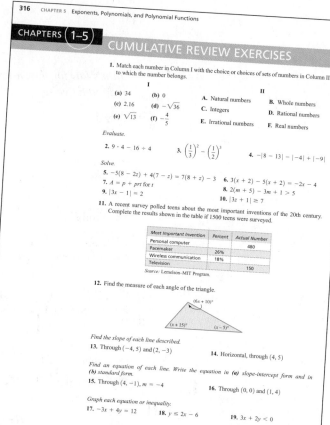

CHAPTER 8 SUMMARY

KEY TERMS

8.1
radicand
index (order)
radical
principal root
radical expression
square root function
cube root function

8.3
simplified radical

8.5
rationalizing the
denominator
conjugates

8.6
radical equation
extraneous solution

8.7
complex number
real part

imaginary part
pure imaginary number
standard form
complex conjugates

NEW SYMBOLS

$\sqrt{}$ radical symbol

$\sqrt[n]{a}$ radical; principal nth root of a

$\pm$ "positive or negative" or "plus or minus"

$\approx$ is approximately equal to

$a^{1/n}$ a to the power $\dfrac{1}{n}$

$a^{m/n}$ a to the power $\dfrac{m}{n}$

i imaginary unit

TEST YOUR WORD POWER

See how well you have learned the vocabulary in this chapter.

1. A **radicand** is
   A. the index of a radical
   B. the number or expression under the radical sign
   C. the positive root of a number
   D. the radical sign.

2. The **Pythagorean theorem** states that, in a right triangle,
   A. the sum of the measures of the angles is 180°
   B. the sum of the lengths of the two shorter sides equals the length of the longest side
   C. the longest side is opposite the right angle
   D. the square of the length of the longest side equals the sum of the squares of the lengths of the two shorter sides.

3. A **hypotenuse** is
   A. either of the two shorter sides of a triangle
   B. the shortest side of a triangle
   C. the side opposite the right angle in a triangle
   D. the longest side in any triangle.

4. **Rationalizing the denominator** is the process of
   A. eliminating fractions from a radical expression
   B. changing the denominator of a fraction from a radical to a rational number
   C. clearing a radical expression of radicals
   D. multiplying radical expressions.

5. An **extraneous solution** is a solution
   A. that does not satisfy the original equation
   B. that makes an equation true
   C. that makes an expression equal 0
   D. that checks in the original equation.

6. A **complex number** is
   A. a real number that includes a complex fraction
   B. a zero multiple of i
   C. a number of the form $a + bi$, where a and b are real numbers
   D. the square root of -1.

ANSWERS

1. B; *Example:* In $\sqrt{3xy}$, $3xy$ is the radicand.   2. D; *Example:* In a right triangle where $a = 6$, $b = 8$, and $c = 10$, $6^2 + 8^2 = 10^2$.
3. C; *Example:* In a right triangle where the sides measure 9, 12, and 15 units, the hypotenuse is the side opposite the right angle, with measure 15 units.   4. B; *Example:* To rationalize the denominator of $\dfrac{5}{\sqrt{3}+1}$, multiply both the numerator and denominator by $\sqrt{3}-1$ to get $\dfrac{5(\sqrt{3}-1)}{2}$.
5. A; *Example:* The proposed solution 2 is extraneous in $\sqrt{5x-1} + 3 = 0$.   6. C; *Examples:* -5 (or $-5 + 0i$), $7i$ (or $0 + 7i$), $\sqrt{2} - 4i$

QUICK REVIEW

CONCEPTS	EXAMPLES

8.1 Radical Expressions and Graphs

$\sqrt[n]{a} = b$ means $b^n = a$.

$\sqrt[n]{a}$ is the principal nth root of a.

$\sqrt[n]{a^n} = |a|$ if n is even $\sqrt[n]{a^n} = a$ if n is odd.

The two square roots of 64 are $\sqrt{64} = 8$ (the principal square root) and $-\sqrt{64} = -8$.

$$\sqrt[4]{(-2)^4} = |-2| = 2 \qquad \sqrt[3]{-27} = -3$$

Functions Defined by Radical Expressions

The square root function defined by $f(x) = \sqrt{x}$ and the cube root function defined by $f(x) = \sqrt[3]{x}$ are two important functions defined by radical expressions.

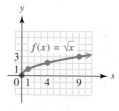

Square root function

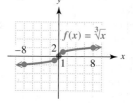

Cube root function

8.2 Rational Exponents

$a^{1/n} = \sqrt[n]{a}$ whenever $\sqrt[n]{a}$ exists.

If m and n are positive integers with $\frac{m}{n}$ in lowest terms, then $a^{m/n} = (a^{1/n})^m$, provided that $a^{1/n}$ is a real number.

All of the usual definitions and rules for exponents are valid for rational exponents.

$$81^{1/2} = \sqrt{81} = 9 \qquad -64^{1/3} = -\sqrt[3]{64} = -4$$

$$8^{5/3} = (8^{1/3})^5 = 2^5 = 32 \qquad (y^{2/5})^{10} = y^4$$

$$5^{-1/2} \cdot 5^{1/4} = 5^{-1/2+1/4} \qquad \frac{x^{-1/3}}{x^{-1/2}} = x^{-1/3-(-1/2)}$$

$$= 5^{-1/4} \qquad\qquad = x^{-1/3+1/2}$$

$$= \frac{1}{5^{1/4}} \qquad\qquad = x^{1/6}, \quad x > 0$$

8.3 Simplifying Radical Expressions

Product and Quotient Rules for Radicals

If $\sqrt[n]{a}$ and $\sqrt[n]{b}$ are real numbers and n is a natural number, then

$$\sqrt[n]{a} \cdot \sqrt[n]{b} = \sqrt[n]{ab} \quad \text{and} \quad \sqrt[n]{\frac{a}{b}} = \frac{\sqrt[n]{a}}{\sqrt[n]{b}}, \quad b \neq 0.$$

$$\sqrt{3} \cdot \sqrt{7} = \sqrt{21} \qquad \sqrt[5]{x^3y} \cdot \sqrt[5]{xy^2} = \sqrt[5]{x^4y^3}$$

$$\frac{\sqrt{x^5}}{\sqrt{x^4}} = \sqrt{\frac{x^5}{x^4}} = \sqrt{x}, \quad x > 0$$

Conditions for a Simplified Radical

1. The radicand has no factor raised to a power greater than or equal to the index.
2. The radicand has no fractions.
3. No denominator contains a radical.
4. Exponents in the radicand and the index of the radical have greatest common factor 1.

$$\sqrt{18} = \sqrt{9 \cdot 2} = 3\sqrt{2}$$

$$\sqrt[3]{54x^5y^3} = \sqrt[3]{27x^3y^3 \cdot 2x^2} = 3xy\sqrt[3]{2x^2}$$

$$\sqrt{\frac{7}{4}} = \frac{\sqrt{7}}{\sqrt{4}} = \frac{\sqrt{7}}{2}$$

$$\sqrt[9]{x^3} = x^{3/9} = x^{1/3}, \quad \text{or} \quad \sqrt[3]{x}$$

Pythagorean Theorem

If a and b are the lengths of the shorter sides of a right triangle and c is the length of the longest side, then

$$a^2 + b^2 = c^2.$$

Find b for the triangle in the figure.

$$10^2 + b^2 = \left(2\sqrt{61}\right)^2$$

$$b^2 = 4(61) - 100$$

$$b^2 = 144$$

$$b = 12$$

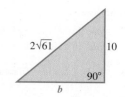

(continued)

CONCEPTS	EXAMPLES
Distance Formula The distance d between (x_1, y_1) and (x_2, y_2) is $$d = \sqrt{(x_2 - x_1)^2 + (y_2 - y_1)^2}.$$	Find the distance between $(3, -2)$ and $(-1, 1)$. $$\sqrt{(-1 - 3)^2 + [1 - (-2)]^2}$$ $$= \sqrt{(-4)^2 + 3^2}$$ $$= \sqrt{16 + 9}$$ $$= \sqrt{25}$$ $$= 5$$

8.4 Adding and Subtracting Radical Expressions

Only radical expressions with the same index and the same radicand may be combined.	$$2\sqrt{28} - 3\sqrt{63} + 8\sqrt{112}$$ $$= 2\sqrt{4 \cdot 7} - 3\sqrt{9 \cdot 7} + 8\sqrt{16 \cdot 7}$$ $$= 2 \cdot 2\sqrt{7} - 3 \cdot 3\sqrt{7} + 8 \cdot 4\sqrt{7}$$ $$= 4\sqrt{7} - 9\sqrt{7} + 32\sqrt{7}$$ $$= (4 - 9 + 32)\sqrt{7}$$ $$= 27\sqrt{7}$$ $\left.\begin{array}{l} \sqrt{15} + \sqrt{30} \\ \sqrt{3} + \sqrt[3]{9} \end{array}\right\}$ cannot be simplified further

8.5 Multiplying and Dividing Radical Expressions

Multiply binomial radical expressions by using the FOIL method. Special products from **Section 5.4** may apply.	$$\left(\sqrt{2} + \sqrt{7}\right)\left(\sqrt{3} - \sqrt{6}\right)$$ $$= \sqrt{6} - 2\sqrt{3} + \sqrt{21} - \sqrt{42} \qquad \sqrt{12} = 2\sqrt{3}$$ $$\left(\sqrt{5} - \sqrt{10}\right)\left(\sqrt{5} + \sqrt{10}\right)$$ $$= 5 - 10, \quad \text{or} \quad -5$$ $$\left(\sqrt{3} - \sqrt{2}\right)^2$$ $$= 3 - 2\sqrt{3} \cdot \sqrt{2} + 2$$ $$= 5 - 2\sqrt{6}$$
Rationalize the denominator by multiplying both the numerator and the denominator by the same expression, one that will yield a rational number in the final denominator.	$$\frac{\sqrt{7}}{\sqrt{5}} = \frac{\sqrt{7} \cdot \sqrt{5}}{\sqrt{5} \cdot \sqrt{5}} = \frac{\sqrt{35}}{5}$$ $$\frac{4}{\sqrt{5} - \sqrt{2}} = \frac{4\left(\sqrt{5} + \sqrt{2}\right)}{\left(\sqrt{5} - \sqrt{2}\right)\left(\sqrt{5} + \sqrt{2}\right)}$$ $$= \frac{4\left(\sqrt{5} + \sqrt{2}\right)}{5 - 2} = \frac{4\left(\sqrt{5} + \sqrt{2}\right)}{3}$$
To write a radical quotient in lowest terms, factor the numerator and denominator and then divide out any common factor(s).	$$\frac{5 + 15\sqrt{6}}{10} = \frac{5\left(1 + 3\sqrt{6}\right)}{5 \cdot 2} = \frac{1 + 3\sqrt{6}}{2}$$

(continued)

CONCEPTS	EXAMPLES

8.6 Solving Equations with Radicals

Solving an Equation with Radicals

Step 1 Isolate one radical on one side of the equation.

Step 2 Raise both sides of the equation to a power that is the same as the index of the radical.

Step 3 Solve the resulting equation. If it still contains a radical, repeat Steps 1 and 2.

Step 4 Check all proposed solutions in the *original* equation.

Solve $\sqrt{2x + 3} - x = 0$.

$$\sqrt{2x + 3} = x \qquad \text{Subtract } x.$$
$$\left(\sqrt{2x + 3}\right)^2 = x^2 \qquad \text{Square each side.}$$
$$2x + 3 = x^2 \qquad \text{Apply the exponents.}$$
$$x^2 - 2x - 3 = 0 \qquad \text{Standard form}$$
$$(x - 3)(x + 1) = 0 \qquad \text{Factor.}$$
$$x - 3 = 0 \quad \text{or} \quad x + 1 = 0 \qquad \text{Zero-factor property}$$
$$x = 3 \quad \text{or} \qquad x = -1 \qquad \text{Solve each equation.}$$

A check shows that 3 is a solution, but -1 is extraneous. The solution set is $\{3\}$.

8.7 Complex Numbers

$i = \sqrt{-1}$, where $i^2 = -1$.

For any positive number b, $\sqrt{-b} = i\sqrt{b}$.

To multiply radicals with negative radicands, first change each factor to the form $i\sqrt{b}$ and then multiply. The same procedure applies to quotients.

$$\sqrt{-25} = i\sqrt{25} = 5i$$

$$\sqrt{-3} \cdot \sqrt{-27}$$
$$= i\sqrt{3} \cdot i\sqrt{27} \qquad \sqrt{-b} = i\sqrt{b}$$
$$= i^2\sqrt{81}$$
$$= -1 \cdot 9 \qquad i^2 = -1$$
$$= -9$$

$$\frac{\sqrt{-18}}{\sqrt{-2}} = \frac{i\sqrt{18}}{i\sqrt{2}} = \sqrt{\frac{18}{2}} = \sqrt{9} = 3$$

Adding and Subtracting Complex Numbers

Add (or subtract) the real parts and add (or subtract) the imaginary parts.

$$\begin{array}{c|c}
(5 + 3i) + (8 - 7i) & (5 + 3i) - (8 - 7i) \\
= 13 - 4i & = -3 + 10i
\end{array}$$

Multiplying Complex Numbers

Multiply complex numbers by using the FOIL method.

$$(2 + i)(5 - 3i)$$
$$= 10 - 6i + 5i - 3i^2 \qquad \text{FOIL}$$
$$= 10 - i - 3(-1) \qquad i^2 = -1$$
$$= 10 - i + 3 \qquad \text{Multiply.}$$
$$= 13 - i \qquad \text{Combine real terms.}$$

Dividing Complex Numbers

Divide complex numbers by multiplying the numerator and the denominator by the conjugate of the denominator.

$$\frac{20}{3 + i}$$
$$= \frac{20(3 - i)}{(3 + i)(3 - i)} \qquad \text{Multiply by the conjugate.}$$
$$= \frac{20(3 - i)}{9 - i^2} \qquad (a + b)(a - b) = a^2 - b^2$$
$$= \frac{20(3 - i)}{10} \qquad i^2 = -1$$
$$= 2(3 - i), \quad \text{or} \quad 6 - 2i$$

CHAPTER **8**

REVIEW EXERCISES

8.1 *Find each root.*

1. $\sqrt{1764}$

2. $-\sqrt{289}$

3. $\sqrt[3]{216}$

4. $\sqrt[3]{-125}$

5. $-\sqrt[3]{27}$

6. $\sqrt[5]{-32}$

7. *Concept Check*   Under what conditions is $\sqrt[n]{a}$ not a real number?

8. Simplify each radical so that no radicals appear. Assume that x represents any real number.

   (a) $\sqrt{x^2}$ **(b)** $-\sqrt{x^2}$ **(c)** $\sqrt[3]{x^3}$

Use a calculator to find a decimal approximation for each number. Give the answer to the nearest thousandth.

9. $-\sqrt{47}$

10. $\sqrt[3]{-129}$

11. $\sqrt[4]{605}$

12. $\sqrt[4]{500^{-3}}$

13. $-\sqrt[3]{500^4}$

14. $-\sqrt{28^{-1}}$

Graph each function. Give the domain and range.

15. $f(x) = \sqrt{x - 1}$

16. $f(x) = \sqrt[3]{x} + 4$

17. What is the best estimate of the area of the triangle shown here?

   A. 3600   **B.** 30   **C.** 60   **D.** 360

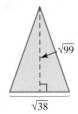

8.2

18. *Concept Check*   Fill in the blanks with the correct responses: One way to evaluate $8^{2/3}$ is to first find the _____ root of _____, which is _____. Then raise that result to the _____ power, to get an answer of _____. Therefore, $8^{2/3} = $ _____.

19. *Concept Check*   Which one of the following is a positive number?

   A. $(-27)^{2/3}$ **B.** $(-64)^{5/3}$ **C.** $(-100)^{1/2}$ **D.** $(-32)^{1/5}$

20. *Concept Check*   If a is a negative number and n is odd, then what must be true about m for $a^{m/n}$ to be   **(a)** positive   **(b)** negative?

21. *Concept Check*   If a is negative and n is even, is $a^{1/n}$ a real number?

Simplify. If the expression does not represent a real number, say so.

22. $49^{1/2}$

23. $-121^{1/2}$

24. $16^{5/4}$

25. $-8^{2/3}$

26. $-\left(\dfrac{36}{25}\right)^{3/2}$

27. $\left(-\dfrac{1}{8}\right)^{-5/3}$

28. $\left(\dfrac{81}{10,000}\right)^{-3/4}$

29. $(-16)^{3/4}$

30. *Concept Check*   Illustrate two different ways of writing $8^{2/3}$ as a radical expression.

31. Explain the relationship between the expressions $a^{m/n}$ and $\sqrt[n]{a^m}$. Give an example.

Write each expression as a radical.

32. $(m + 3n)^{1/2}$

33. $(3a + b)^{-5/3}$

Write each expression with a rational exponent.

34. $\sqrt{7^9}$

35. $\sqrt[5]{p^4}$

Use the rules for exponents to simplify each expression. Write the answer with only positive exponents. Assume that all variables represent positive real numbers.

36. $5^{1/4} \cdot 5^{7/4}$

37. $\dfrac{96^{2/3}}{96^{-1/3}}$

38. $\dfrac{(a^{1/3})^4}{a^{2/3}}$

39. $\dfrac{y^{-1/3} \cdot y^{5/6}}{y}$

40. $\left(\dfrac{z^{-1}x^{-3/5}}{2^{-2}z^{-1/2}x}\right)^{-1}$

41. $r^{-1/2}(r + r^{3/2})$

Simplify by first writing each radical in exponential form. Leave the answer in exponential form. Assume that all variables represent positive real numbers.

42. $\sqrt[8]{s^4}$

43. $\sqrt[6]{r^9}$

44. $\dfrac{\sqrt{p^5}}{p^2}$

45. $\sqrt[4]{k^3} \cdot \sqrt{k^3}$

46. $\sqrt[3]{m^5} \cdot \sqrt[3]{m^8}$

47. $\sqrt[4]{\sqrt[3]{z}}$

48. $\sqrt{\sqrt{\sqrt{x}}}$

49. $\sqrt[3]{\sqrt[5]{x}}$

50. $\sqrt{\sqrt[6]{\sqrt[3]{x}}}$

51. The product rule does not apply to $3^{1/4} \cdot 2^{1/5}$. Why?

8.3 *Simplify each radical. Assume that all variables represent positive real numbers.*

52. $\sqrt{6} \cdot \sqrt{11}$

53. $\sqrt{5} \cdot \sqrt{r}$

54. $\sqrt[3]{6} \cdot \sqrt[3]{5}$

55. $\sqrt[4]{7} \cdot \sqrt[4]{3}$

56. $\sqrt{20}$

57. $\sqrt{75}$

58. $-\sqrt{125}$

59. $\sqrt[3]{-108}$

60. $\sqrt{100y^7}$

61. $\sqrt[3]{64p^4q^6}$

62. $\sqrt[3]{108a^8b^5}$

63. $\sqrt[3]{632r^8t^4}$

64. $\sqrt{\dfrac{y^3}{144}}$

65. $\sqrt[3]{\dfrac{m^{15}}{27}}$

66. $\sqrt[3]{\dfrac{r^2}{8}}$

67. $\sqrt[4]{\dfrac{a^9}{81}}$

Simplify each radical expression.

68. $\sqrt[6]{15^3}$

69. $\sqrt[4]{p^6}$

70. $\sqrt[3]{2} \cdot \sqrt[4]{5}$

71. $\sqrt{x} \cdot \sqrt[5]{x}$

72. Find the unknown length in the right triangle. Simplify the answer if applicable.

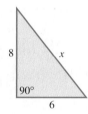

73. Find the distance between the points $(-4, 7)$ and $(10, 6)$.

8.4 *Perform the indicated operations. Assume that all variables represent positive real numbers.*

74. $2\sqrt{8} - 3\sqrt{50}$

75. $8\sqrt{80} - 3\sqrt{45}$

76. $-\sqrt{27y} + 2\sqrt{75y}$

77. $2\sqrt{54m^3} + 5\sqrt{96m^3}$

78. $3\sqrt[3]{54} + 5\sqrt[3]{16}$

79. $-6\sqrt[4]{32} + \sqrt[4]{512}$

In Exercises 80 and 81, leave answers as simplified radicals.

80. Find the perimeter of a rectangular electronic billboard having sides of lengths shown in the figure.

81. Find the perimeter of a triangular electronic highway road sign having the dimensions shown in the figure.

8.5 *Multiply.*

82. $\left(\sqrt{3} + 1\right)\left(\sqrt{3} - 2\right)$

83. $\left(\sqrt{7} + \sqrt{5}\right)\left(\sqrt{7} - \sqrt{5}\right)$

84. $\left(3\sqrt{2} + 1\right)\left(2\sqrt{2} - 3\right)$

85. $\left(\sqrt{13} - \sqrt{2}\right)^2$

86. $\left(\sqrt[3]{2} + 3\right)\left(\sqrt[3]{4} - 3\sqrt[3]{2} + 9\right)$

87. $\left(\sqrt[3]{4y} - 1\right)\left(\sqrt[3]{4y} + 3\right)$

88. Use a calculator to show that the answer to **Exercise 85,** $15 - 2\sqrt{26}$, is not equal to $13\sqrt{26}$.

89. *Concept Check* A friend tried to rationalize the denominator of $\dfrac{5}{\sqrt[3]{6}}$, by multiplying the numerator and denominator by $\sqrt[3]{6}$. **WHAT WENT WRONG?**

Rationalize each denominator. Assume that all variables represent positive real numbers.

90. $\dfrac{\sqrt{6}}{\sqrt{5}}$

91. $\dfrac{-6\sqrt{3}}{\sqrt{2}}$

92. $\dfrac{3\sqrt{7p}}{\sqrt{y}}$

93. $\sqrt{\dfrac{11}{8}}$

94. $-\sqrt[3]{\dfrac{9}{25}}$

95. $\sqrt[3]{\dfrac{108m^3}{n^5}}$

96. $\dfrac{1}{\sqrt{2} + \sqrt{7}}$

97. $\dfrac{-5}{\sqrt{6} - 3}$

Write in lowest terms.

98. $\dfrac{2 - 2\sqrt{5}}{8}$

99. $\dfrac{4 - 8\sqrt{8}}{12}$

100. $\dfrac{-18 + \sqrt{27}}{6}$

8.6 *Solve each equation.*

101. $\sqrt{8x + 9} = 5$

102. $\sqrt{2x - 3} - 3 = 0$

103. $\sqrt{3x + 1} - 2 = -3$

104. $\sqrt{7x + 1} = x + 1$

105. $3\sqrt{x} = \sqrt{10x - 9}$

106. $\sqrt{x^2 + 3x + 7} = x + 2$

107. $\sqrt{x + 2} - \sqrt{x - 3} = 1$

108. $\sqrt[3]{5x - 1} = \sqrt[3]{3x - 2}$

109. $\sqrt[3]{2x^2 + 3x - 7} = \sqrt[3]{2x^2 + 4x + 6}$

110. $\sqrt[3]{3x^2 - 4x + 6} = \sqrt[3]{3x^2 - 2x + 8}$

111. $\sqrt[3]{1 - 2x} - \sqrt[3]{-x - 13} = 0$

112. $\sqrt[3]{11 - 2x} - \sqrt[3]{-1 - 5x} = 0$

113. $\sqrt[4]{x - 1} + 2 = 0$

114. $\sqrt[4]{2x + 3} + 1 = 0$

115. $\sqrt[4]{x + 7} = \sqrt[4]{2x}$

116. $\sqrt[4]{x + 8} = \sqrt[4]{3x}$

117. Carpenters stabilize wall frames with a diagonal brace, as shown in the figure. The length of the brace is given by $L = \sqrt{H^2 + W^2}$.

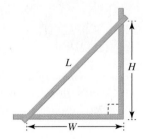

  (a) Solve this formula for H.

  (b) If the bottom of the brace is attached 9 ft from the corner and the brace is 12 ft long, how far up the corner post should it be nailed? Give your answer to the nearest tenth of a foot.

8.7 *Write each expression as a product of a real number and* i.

118. $\sqrt{-25}$

119. $\sqrt{-200}$

120. *Concept Check* If a is a positive real number, is $-\sqrt{-a}$ a real number?

Perform the indicated operations. Give answers in standard form.

121. $(-2 + 5i) + (-8 - 7i)$ **122.** $(5 + 4i) - (-9 - 3i)$ **123.** $\sqrt{-5} \cdot \sqrt{-7}$

124. $\sqrt{-25} \cdot \sqrt{-81}$ **125.** $\dfrac{\sqrt{-72}}{\sqrt{-8}}$ **126.** $(2 + 3i)(1 - i)$

127. $(6 - 2i)^2$ **128.** $\dfrac{3 - i}{2 + i}$ **129.** $\dfrac{5 + 14i}{2 + 3i}$

Find each power of i.

130. i^{11} **131.** i^{36} **132.** i^{-10} **133.** i^{-8}

MIXED REVIEW EXERCISES

Simplify. Assume that all variables represent positive real numbers.

134. $-\sqrt[4]{256}$ **135.** $1000^{-2/3}$ **136.** $\dfrac{z^{-1/5} \cdot z^{3/10}}{z^{7/10}}$

137. $\sqrt[4]{k^{24}}$ **138.** $\sqrt[3]{54z^9t^8}$ **139.** $-5\sqrt{18} + 12\sqrt{72}$

140. $\dfrac{-1}{\sqrt{12}}$ **141.** $\sqrt[3]{\dfrac{12}{25}}$ **142.** i^{-1000}

143. $\sqrt{-49}$ **144.** $(4 - 9i) + (-1 + 2i)$ **145.** $\dfrac{\sqrt{50}}{\sqrt{-2}}$

146. $\dfrac{3 + \sqrt{54}}{6}$ **147.** $(3 + 2i)^2$ **148.** $8\sqrt[3]{x^3y^2} - 2x\sqrt[3]{y^2}$

149. $9\sqrt{5} - 4\sqrt{15}$ **150.** $\left(\sqrt{5} - \sqrt{3}\right)\left(\sqrt{7} + \sqrt{3}\right)$

Solve each equation.

151. $\sqrt{x + 4} = x - 2$ **152.** $\sqrt[3]{2x - 9} = \sqrt[3]{5x + 3}$

153. $\sqrt{6 + 2x} - 1 = \sqrt{7 - 2x}$ **154.** $\sqrt{7x + 11} - 5 = 0$

155. $\sqrt{6x + 2} - \sqrt{5x + 3} = 0$ **156.** $\sqrt{3 + 5x} - \sqrt{x + 11} = 0$

157. $3\sqrt{x} = \sqrt{8x + 9}$ **158.** $6\sqrt{x} = \sqrt{30x + 24}$

159. $\sqrt{11 + 2x} + 1 = \sqrt{5x + 1}$ **160.** $\sqrt{5x + 6} - \sqrt{x + 3} = 3$

CHAPTER (8)

TEST

CHAPTER Test Prep VIDEOS

Step-by-step test solutions are found on the Chapter Test Prep Videos available via the Video Resources on DVD, in *MyMathLab*, or on You Tube (search "LialIntermediateAlg").

View the complete solutions to all Chapter Test exercises on the Video Resources on DVD.

Evaluate.

1. $-\sqrt{841}$ **2.** $\sqrt[3]{-512}$ **3.** $125^{1/3}$

4. *Concept Check* For $\sqrt{146.25}$, which choice gives the best estimate?

  A. 10 **B.** 11 **C.** 12 **D.** 13

Use a calculator to approximate each root to the nearest thousandth.

5. $\sqrt{478}$ **6.** $\sqrt[3]{-832}$

7. Graph the function defined by $f(x) = \sqrt{x + 6}$, and give the domain and range.

Simplify each expression. Assume that all variables represent positive real numbers.

8. $\left(\dfrac{16}{25}\right)^{-3/2}$

9. $(-64)^{-4/3}$

10. $\dfrac{3^{2/5}x^{-1/4}y^{2/5}}{3^{-8/5}x^{7/4}y^{1/10}}$

11. $\left(\dfrac{x^{-4}y^{-6}}{x^{-2}y^{3}}\right)^{-2/3}$

12. $7^{3/4} \cdot 7^{-1/4}$

13. $\sqrt[3]{a^4} \cdot \sqrt[3]{a^7}$

14. Use the Pythagorean theorem to find the exact length of side b in the figure.

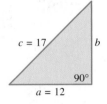

$c = 17$ b $90°$ $a = 12$

15. Find the distance between the points $(-4, 2)$ and $(2, 10)$.

Simplify each expression. Assume that all variables represent positive real numbers.

16. $\sqrt{54x^5y^6}$

17. $\sqrt[4]{32a^7b^{13}}$

18. $\sqrt{2} \cdot \sqrt[3]{5}$ (Express as a radical.)

19. $3\sqrt{20} - 5\sqrt{80} + 4\sqrt{500}$

20. $\sqrt[3]{16t^3s^5} - \sqrt[3]{54t^6s^2}$

21. $\left(7\sqrt{5} + 4\right)\left(2\sqrt{5} - 1\right)$

22. $\left(\sqrt{3} - 2\sqrt{5}\right)^2$

23. $\dfrac{-5}{\sqrt{40}}$

24. $\dfrac{2}{\sqrt[3]{5}}$

25. $\dfrac{-4}{\sqrt{7} + \sqrt{5}}$

26. Write $\dfrac{6 + \sqrt{24}}{2}$ in lowest terms.

27. The following formula is used in physics, relating the velocity V of sound to the temperature T.

$$V = \dfrac{V_0}{\sqrt{1 - kT}}$$

(a) Find an approximation of V to the nearest tenth if $V_0 = 50$, $k = 0.01$, and $T = 30$. Use a calculator.

(b) Solve the formula for T.

Solve each equation.

28. $\sqrt[3]{5x} = \sqrt[3]{2x - 3}$

29. $x + \sqrt{x + 6} = 9 - x$

30. $\sqrt{x + 4} - \sqrt{1 - x} = -1$

In Exercises 31–33, perform the indicated operations. Give the answers in standard form.

31. $(-2 + 5i) - (3 + 6i) - 7i$

32. $(1 + 5i)(3 + i)$

33. $\dfrac{7 + i}{1 - i}$

34. Simplify i^{37}.

35. *Concept Check*   Answer *true* or *false* to each of the following.

(a) $i^2 = -1$ (b) $i = \sqrt{-1}$ (c) $i = -1$ (d) $\sqrt{-3} = i\sqrt{3}$

CHAPTERS (1–8) CUMULATIVE REVIEW EXERCISES

Evaluate each expression for $a = -3$, $b = 5$, and $c = -4$.

1. $|2a^2 - 3b + c|$

2. $\dfrac{(a + b)(a + c)}{3b - 6}$

Solve each equation.

3. $3(x + 2) - 4(2x + 3) = -3x + 2$

4. $\dfrac{1}{3}x + \dfrac{1}{4}(x + 8) = x + 7$

5. $0.04x + 0.06(100 - x) = 5.88$

6. $|6x + 7| = 13$

7. Find the solution set of $-5 - 3(x - 2) < 11 - 2(x + 2)$. Write it in interval notation.

Solve each problem.

8. A piggy bank has 100 coins, all of which are nickels and quarters. The total value of the money is \$17.80. How many of each denomination are there in the bank?

9. How many liters of pure alcohol must be mixed with 40 L of 18% alcohol to obtain a 22% alcohol solution?

10. Graph the equation $4x - 3y = 12$.

11. Find the slope of the line passing through the points $(-4, 6)$ and $(2, -3)$. Then find the equation of the line and write it in the form $y = mx + b$.

12. If $f(x) = 3x - 7$, find $f(-10)$.

13. Solve by substitution or elimination.

$$3x - y = 23$$
$$2x + 3y = 8$$

14. Solve by matrix methods.

$$x + y + z = 1$$
$$x - y - z = -3$$
$$x + y - z = -1$$

Solve the problem by using a system of equations.

15. In 2010, if you had sent five 2-oz letters and three 3-oz letters by first-class mail, it would have cost you \$5.39. Sending three 2-oz letters and five 3-oz letters would have cost \$5.73. What was the 2010 postage rate for one 2-oz letter and for one 3-oz letter? (*Source:* U.S. Postal Service.)

Perform the indicated operations.

16. $(3k^3 - 5k^2 + 8k - 2) - (4k^3 + 11k + 7) + (2k^2 - 5k)$

17. $(8x - 7)(x + 3)$

18. $\dfrac{6y^4 - 3y^3 + 5y^2 + 6y - 9}{2y + 1}$

Factor each polynomial completely.

19. $2p^2 - 5pq + 3q^2$

20. $3k^4 + k^2 - 4$

21. $x^3 + 512$

Solve by factoring.

22. $2x^2 + 11x + 15 = 0$

23. $5x(x - 1) = 2(1 - x)$

24. What is the domain of $f(x) = \dfrac{4}{x^2 - 9}$?

Perform each operation and express the answer in lowest terms.

25. $\dfrac{y^2 + y - 12}{y^3 + 9y^2 + 20y} \div \dfrac{y^2 - 9}{y^3 + 3y^2}$

26. $\dfrac{1}{x + y} + \dfrac{3}{x - y}$

Simplify each complex fraction.

27. $\dfrac{\dfrac{-6}{x - 2}}{\dfrac{8}{3x - 6}}$

28. $\dfrac{\dfrac{1}{a} - \dfrac{1}{b}}{\dfrac{a}{b} - \dfrac{b}{a}}$

29. $\dfrac{x^{-1}}{y - x^{-1}}$

30. Solve the equation $\dfrac{x + 1}{x - 3} = \dfrac{4}{x - 3} + 6$.

31. Danielle can ride her bike 4 mph faster than her husband, Richard. If Danielle can ride 48 mi in the same time that Richard can ride 24 mi, what are their speeds?

Write each expression in simplest form, using only positive exponents. Assume that all variables represent positive real numbers.

32. $27^{-2/3}$

33. $\sqrt[3]{16x^2y} \cdot \sqrt[3]{3x^3y}$

34. $\sqrt{50} + \sqrt{8}$

35. $\dfrac{1}{\sqrt{10} - \sqrt{8}}$

36. Find the distance between the points $(-4, 4)$ and $(-2, 9)$.

37. Solve the equation $\sqrt{3x - 8} = x - 2$.

38. Express $\dfrac{6 - 2i}{1 - i}$ in standard form.

CHAPTER

9

Quadratic Equations, Inequalities, and Functions

The prices of food, gasoline, and other products have increased throughout the world. In particular, escalating oil prices in recent years have caused increases in transportation and shipping costs, which trickled down to affect prices of a variety of goods and services.

Although prices tend to go up over time, the rate at which they increase (the inflation rate) varies considerably. The Consumer Price Index (CPI) used by the U.S. government measures changes in prices for goods purchased by typical American families over time. In **Example 6** of **Section 9.4,** we use a *quadratic function* to model the CPI.

9.1 The Square Root Property and Completing the Square

Recall from **Section 6.5** that a *quadratic equation* is defined as follows.

> **Quadratic Equation**
>
> An equation that can be written in the form
> $$ax^2 + bx + c = 0,$$
> where a, b, and c are real numbers, with $a \neq 0$, is a **quadratic equation.** The given form is called **standard form.**

A quadratic equation is a *second-degree equation,* that is, an equation with a squared variable term and no terms of greater degree.

$$4x^2 + 4x - 5 = 0 \quad \text{and} \quad 3x^2 = 4x - 8$$

Quadratic equations (The first equation is in standard form.)

OBJECTIVE 1 **Review the zero-factor property.** In **Section 6.5** we used factoring and the zero-factor property to solve quadratic equations.

> **Zero-Factor Property**
>
> If two numbers have a product of 0, then at least one of the numbers must be 0. That is, if $ab = 0$, then $a = 0$ or $b = 0$.

NOW TRY EXERCISE 1

Use the zero-factor property to solve $2x^2 + 5x - 12 = 0$.

EXAMPLE 1 Using the Zero-Factor Property

Solve $3x^2 - 5x - 28 = 0$ by using the zero-factor property.

$$3x^2 - 5x - 28 = 0$$

$$(3x + 7)(x - 4) = 0 \qquad \text{Factor.}$$

$$3x + 7 = 0 \quad \text{or} \quad x - 4 = 0 \qquad \text{Zero-factor property}$$

$$3x = -7 \quad \text{or} \qquad x = 4 \qquad \text{Solve each equation.}$$

$$x = -\frac{7}{3}$$

To check, substitute each solution in the original equation. The solution set is $\left\{-\frac{7}{3}, 4\right\}$.

NOW TRY

OBJECTIVE 2 **Learn the square root property.** Not every quadratic equation can be solved easily by factoring. Three other methods of solving quadratic equations are based on the following property.

> **Square Root Property**
>
> If x and k are complex numbers and $x^2 = k$, then
> $$x = \sqrt{k} \quad \text{or} \quad x = -\sqrt{k}.$$

NOW TRY ANSWER
1. $\left\{-4, \frac{3}{2}\right\}$

The following steps justify the square root property.

$$x^2 = k$$
$$x^2 - k = 0 \qquad \text{Subtract } k.$$
$$\left(x - \sqrt{k}\right)\left(x + \sqrt{k}\right) = 0 \qquad \text{Factor.}$$
$$x - \sqrt{k} = 0 \quad \text{or} \quad x + \sqrt{k} = 0 \qquad \text{Zero-factor property}$$
$$\boldsymbol{x = \sqrt{k}} \quad \text{or} \qquad \boldsymbol{x = -\sqrt{k}} \qquad \text{Solve each equation.}$$

⚠ **CAUTION**   If $k \neq 0$, then using the square root property always produces *two* square roots, one positive and one negative.

**NOW TRY
EXERCISE 2**

Solve each equation.

(a) $t^2 = 10$

(b) $2x^2 - 90 = 0$

EXAMPLE 2   Using the Square Root Property

Solve each equation.

(a) $x^2 = 5$

By the square root property, if $x^2 = 5$, then

$$x = \sqrt{5} \quad \text{or} \quad x = -\sqrt{5}. \quad \boxed{\text{Don't forget the negative solution.}}$$

The solution set is $\left\{\sqrt{5}, -\sqrt{5}\right\}$.

(b)
$$4x^2 - 48 = 0$$
$$4x^2 = 48 \qquad \text{Add 48.}$$
$$x^2 = 12 \qquad \text{Divide by 4.}$$
$$x = \sqrt{12} \quad \text{or} \quad x = -\sqrt{12} \qquad \text{Square root property}$$
$$x = 2\sqrt{3} \quad \text{or} \quad x = -2\sqrt{3} \qquad \sqrt{12} = \sqrt{4} \cdot \sqrt{3} = 2\sqrt{3}$$

The solutions are $2\sqrt{3}$ and $-2\sqrt{3}$. Check each in the original equation.

CHECK
$$4x^2 - 48 = 0 \qquad \text{Original equation}$$

$$4\left(2\sqrt{3}\right)^2 - 48 \stackrel{?}{=} 0 \quad \text{Let } x = 2\sqrt{3}. \quad\Big|\quad 4\left(-2\sqrt{3}\right)^2 - 48 \stackrel{?}{=} 0 \quad \text{Let } x = -2\sqrt{3}.$$
$$4(12) - 48 \stackrel{?}{=} 0 \qquad\qquad\qquad 4(12) - 48 \stackrel{?}{=} 0$$
$$48 - 48 \stackrel{?}{=} 0 \qquad\qquad\qquad 48 - 48 \stackrel{?}{=} 0$$

$$\boxed{\begin{array}{l}\left(2\sqrt{3}\right)^2 \\ = 2^2 \cdot \left(\sqrt{3}\right)^2\end{array}}$$

$$0 = 0 \; \checkmark \; \text{True} \qquad\qquad\qquad 0 = 0 \; \checkmark \; \text{True}$$

The solution set is $\left\{2\sqrt{3}, -2\sqrt{3}\right\}$.   NOW TRY

NOTE Using the symbol $\pm$ (read "positive or negative," or "plus or minus"), the solutions in **Example 2** could be written $\pm\sqrt{5}$ and $\pm 2\sqrt{3}$.

EXAMPLE 3   Using the Square Root Property in an Application

Galileo Galilei developed a formula for freely falling objects described by

$$d = 16t^2,$$

NOW TRY ANSWERS

2. (a) $\left\{\sqrt{10}, -\sqrt{10}\right\}$

(b) $\left\{3\sqrt{5}, -3\sqrt{5}\right\}$

where d is the distance in feet that an object falls (disregarding air resistance) in t seconds, regardless of weight. Galileo dropped objects from the Leaning Tower of Pisa.

NOW TRY
EXERCISE 3

Tim is dropping roofing nails from the top of a roof 25 ft high into a large bucket on the ground. Use the formula in **Example 3** to determine how long it will take a nail dropped from 25 ft to hit the bottom of the bucket.

If the Leaning Tower is about 180 ft tall, use Galileo's formula to determine how long it would take an object dropped from the top of the tower to fall to the ground. (*Source:* www.brittanica.com)

$$d = 16t^2 \qquad \text{Galileo's formula}$$
$$180 = 16t^2 \qquad \text{Let } d = 180.$$
$$11.25 = t^2 \qquad \text{Divide by 16.}$$
$$t = \sqrt{11.25} \quad \text{or} \quad t = -\sqrt{11.25} \qquad \text{Square root property}$$

Galileo Galilei (1564–1642)

Time cannot be negative, so we discard the negative solution. Using a calculator, $\sqrt{11.25} \approx 3.4$ so $t \approx 3.4$. The object would fall to the ground in about 3.4 sec.

NOW TRY

OBJECTIVE 3 **Solve quadratic equations of the form $(ax + b)^2 = c$ by extending the square root property.** We can solve more complicated equations by extending the square root property.

NOW TRY
EXERCISE 4

Solve $(x + 3)^2 = 25$.

EXAMPLE 4 Extending the Square Root Property

Solve $(x - 5)^2 = 36$.

$$\underset{\downarrow}{x^2} \quad = \quad \underset{\downarrow}{k}$$
$$(x - 5)^2 = 36 \qquad \begin{array}{l}\text{Substitute } (x - 5)^2 \text{ for } x^2 \text{ and 36 for}\\ k \text{ in the square root property.}\end{array}$$

$$x - 5 = \sqrt{36} \quad \text{or} \quad x - 5 = -\sqrt{36}$$
$$x - 5 = 6 \qquad \text{or} \quad x - 5 = -6 \qquad \text{Simplify square roots.}$$
$$x = 11 \qquad \text{or} \qquad x = -1 \qquad \text{Add 5.}$$

CHECK
$$(x - 5)^2 = 36 \qquad \text{Original equation}$$

$$(11 - 5)^2 \overset{?}{=} 36 \quad \text{Let } x = 11. \qquad (-1 - 5)^2 \overset{?}{=} 36 \quad \text{Let } x = -1.$$
$$6^2 \overset{?}{=} 36 \qquad\qquad\qquad (-6)^2 \overset{?}{=} 36$$
$$36 = 36 \ \checkmark \ \text{True} \qquad\qquad 36 = 36 \ \checkmark \ \text{True}$$

Both solutions satisfy the original equation. The solution set is $\{-1, 11\}$.

NOW TRY

EXAMPLE 5 Extending the Square Root Property

Solve $(2x - 3)^2 = 18$.

$$2x - 3 = \sqrt{18} \qquad \text{or} \quad 2x - 3 = -\sqrt{18} \qquad \text{Square root property}$$
$$2x = 3 + \sqrt{18} \quad \text{or} \qquad 2x = 3 - \sqrt{18} \qquad \text{Add 3.}$$
$$x = \frac{3 + \sqrt{18}}{2} \quad \text{or} \qquad x = \frac{3 - \sqrt{18}}{2} \qquad \text{Divide by 2.}$$
$$x = \frac{3 + 3\sqrt{2}}{2} \quad \text{or} \qquad x = \frac{3 - 3\sqrt{2}}{2} \qquad \sqrt{18} = \sqrt{9 \cdot 2} = 3\sqrt{2}$$

NOW TRY ANSWERS
3. 1.25 sec   4. $\{-8, 2\}$

NOW TRY
EXERCISE 5
Solve $(5x - 4)^2 = 27$.

We show the check for the first solution. The check for the other solution is similar.

CHECK $(2x - 3)^2 = 18$ Original equation

$$\left[2\left(\frac{3 + 3\sqrt{2}}{2}\right) - 3\right]^2 \overset{?}{=} 18 \qquad \text{Let } x = \frac{3 + 3\sqrt{2}}{2}.$$

$$\left(3 + 3\sqrt{2} - 3\right)^2 \overset{?}{=} 18 \qquad \text{Multiply.}$$

$$\left(3\sqrt{2}\right)^2 \overset{?}{=} 18 \qquad \text{Simplify.}$$

$$18 = 18 \quad \checkmark \quad \text{True}$$

> The $\pm$ symbol denotes two solutions.

The solution set is $\left\{\dfrac{3 + 3\sqrt{2}}{2}, \dfrac{3 - 3\sqrt{2}}{2}\right\}$, abbreviated $\left\{\dfrac{3 \pm 3\sqrt{2}}{2}\right\}$.

NOW TRY

OBJECTIVE 4   Solve quadratic equations by completing the square. We can use the square root property to solve *any* quadratic equation by writing it in the form

Square of a binomial $\rightarrow (x + k)^2 = n.\leftarrow$ Constant

That is, we must write the left side of the equation as a perfect square trinomial that can be factored as $(x + k)^2$, the square of a binomial, and the right side must be a constant. This process is called **completing the square.**

Recall that the perfect square trinomial

$$x^2 + 10x + 25 \quad \text{can be factored as} \quad (x + 5)^2.$$

In the trinomial, the coefficient of x (the first-degree term) is 10 and the constant term is 25. If we take half of 10 and square it, we get the constant term, 25.

Coefficient of x Constant

$$\left[\frac{1}{2}(10)\right]^2 = 5^2 = 25$$

Similarly, in $x^2 + 12x + 36,$ $\left[\dfrac{1}{2}(12)\right]^2 = 6^2 = 36,$

and in $m^2 - 6m + 9,$ $\left[\dfrac{1}{2}(-6)\right]^2 = (-3)^2 = 9.$

This relationship is true in general and is the idea behind completing the square.

EXAMPLE 6   Solving a Quadratic Equation by Completing the Square ($a = 1$)

Solve $x^2 + 8x + 10 = 0$.

This quadratic equation cannot be solved by factoring, and it is not in the correct form to solve using the square root property. To solve it by completing the square, we need a perfect square trinomial on the left side of the equation.

$$x^2 + 8x + 10 = 0 \qquad \text{Original equation}$$

$$x^2 + 8x = -10 \qquad \text{Subtract 10.}$$

We must add a constant to get a perfect square trinomial on the left.

$$x^2 + 8x + \underline{\ \ ?\ \ }$$

Needs to be a perfect
square trinomial

NOW TRY ANSWER
5. $\left\{\dfrac{4 \pm 3\sqrt{3}}{5}\right\}$

⌒ *NOW TRY*
↪ *EXERCISE 6*
Solve $x^2 + 6x - 2 = 0$.

Take half the coefficient of the first-degree term, $8x$, and square the result.

$$\left[\frac{1}{2}(8)\right]^2 = 4^2 = 16 \longleftarrow \text{Desired constant}$$

We add this constant, 16, to *each* side of the equation and continue solving as shown.

$$x^2 + 8x + 16 = -10 + 16 \qquad \text{Add 16 to each side.}$$

> This is a key step.

$$(x + 4)^2 = 6 \qquad \begin{array}{l}\text{Factor on the left.}\\ \text{Add on the right.}\end{array}$$

$$x + 4 = \sqrt{6} \qquad \text{or} \qquad x + 4 = -\sqrt{6} \qquad \text{Square root property}$$

$$x = -4 + \sqrt{6} \quad \text{or} \qquad x = -4 - \sqrt{6} \qquad \text{Add } -4.$$

CHECK $\qquad\qquad\qquad\qquad x^2 + 8x + 10 = 0 \qquad$ Original equation

> Remember the middle term when squaring $-4 + \sqrt{6}$.

$$\left(-4 + \sqrt{6}\right)^2 + 8\left(-4 + \sqrt{6}\right) + 10 \stackrel{?}{=} 0 \qquad \text{Let } x = -4 + \sqrt{6}.$$

$$16 - 8\sqrt{6} + 6 - 32 + 8\sqrt{6} + 10 \stackrel{?}{=} 0$$

$$0 = 0 \ \checkmark \ \text{True}$$

The check of the other solution is similar. The solution set is

$$\left\{-4 + \sqrt{6}, -4 - \sqrt{6}\right\}, \quad \text{or} \quad \left\{-4 \pm \sqrt{6}\right\}. \qquad\qquad \text{NOW TRY} ⤾$$

Completing the Square

To solve $ax^2 + bx + c = 0$ $(a \neq 0)$ by completing the square, use these steps.

Step 1 **Be sure the second-degree (squared) term has coefficient 1.** If the coefficient of the second-degree term is 1, proceed to Step 2. If the coefficient of the second-degree term is not 1 but some other nonzero number a, divide each side of the equation by a.

Step 2 **Write the equation in correct form** so that terms with variables are on one side of the equals symbol and the constant is on the other side.

Step 3 **Square half the coefficient of the first-degree (linear) term.**

Step 4 **Add the square to each side.**

Step 5 **Factor the perfect square trinomial.** One side should now be a perfect square trinomial. Factor it as the square of a binomial. Simplify the other side.

Step 6 **Solve the equation.** Apply the square root property to complete the solution.

EXAMPLE 7 Solving a Quadratic Equation by Completing the Square ($a = 1$)

Solve $x^2 + 5x - 1 = 0$.

Since the coefficient of the squared term is 1, begin with Step 2.

Step 2 $\qquad\qquad\qquad\qquad x^2 + 5x = 1 \qquad$ Add 1 to each side.

Step 3 Take half the coefficient of the first-degree term and square the result.

$$\left[\frac{1}{2}(5)\right]^2 = \left(\frac{5}{2}\right)^2 = \frac{25}{4}$$

NOW TRY ANSWER
6. $\left\{-3 \pm \sqrt{11}\right\}$

**NOW TRY
EXERCISE 7**
Solve $x^2 + x - 3 = 0$.

Step 4 $\qquad x^2 + 5x + \dfrac{25}{4} = 1 + \dfrac{25}{4}$ ⟵ Add the square to each side of the equation.

Step 5 $\qquad \left(x + \dfrac{5}{2}\right)^2 = \dfrac{29}{4}$ Factor on the left. Add on the right.

Step 6 $\quad x + \dfrac{5}{2} = \sqrt{\dfrac{29}{4}} \qquad$ or $\qquad x + \dfrac{5}{2} = -\sqrt{\dfrac{29}{4}}$ Square root property

$\qquad x + \dfrac{5}{2} = \dfrac{\sqrt{29}}{2} \qquad$ or $\qquad x + \dfrac{5}{2} = -\dfrac{\sqrt{29}}{2} \qquad \sqrt{\dfrac{a}{b}} = \dfrac{\sqrt{a}}{\sqrt{b}}$

$\qquad x = -\dfrac{5}{2} + \dfrac{\sqrt{29}}{2} \qquad$ or $\qquad x = -\dfrac{5}{2} - \dfrac{\sqrt{29}}{2} \qquad$ Add $-\dfrac{5}{2}$.

$\qquad x = \dfrac{-5 + \sqrt{29}}{2} \qquad$ or $\qquad x = \dfrac{-5 - \sqrt{29}}{2} \qquad \dfrac{a}{c} \pm \dfrac{b}{c} = \dfrac{a \pm b}{c}$

Check that the solution set is $\left\{\dfrac{-5 \pm \sqrt{29}}{2}\right\}$.   NOW TRY

**NOW TRY
EXERCISE 8**
Solve $3x^2 + 12x - 5 = 0$.

EXAMPLE 8 Solving a Quadratic Equation by Completing the Square $(a \neq 1)$

Solve $2x^2 - 4x - 5 = 0$.

Divide each side by 2 to get 1 as the coefficient of the second-degree term.

$$x^2 - 2x - \dfrac{5}{2} = 0 \qquad \text{Step 1}$$

$$x^2 - 2x = \dfrac{5}{2} \qquad \text{Step 2}$$

$$\left[\dfrac{1}{2}(-2)\right]^2 = (-1)^2 = 1 \qquad \text{Step 3}$$

$$x^2 - 2x + 1 = \dfrac{5}{2} + 1 \qquad \text{Step 4}$$

$$(x - 1)^2 = \dfrac{7}{2} \qquad \text{Step 5}$$

$$x - 1 = \sqrt{\dfrac{7}{2}} \qquad \text{or} \quad x - 1 = -\sqrt{\dfrac{7}{2}} \qquad \text{Step 6}$$

$$x = 1 + \sqrt{\dfrac{7}{2}} \qquad \text{or} \qquad x = 1 - \sqrt{\dfrac{7}{2}} \qquad \text{Add 1.}$$

$$x = 1 + \dfrac{\sqrt{14}}{2} \qquad \text{or} \qquad x = 1 - \dfrac{\sqrt{14}}{2} \qquad \sqrt{\dfrac{7}{2}} = \dfrac{\sqrt{7}}{\sqrt{2}} = \dfrac{\sqrt{7}}{\sqrt{2}} \cdot \dfrac{\sqrt{2}}{\sqrt{2}} = \dfrac{\sqrt{14}}{2}$$

Add the two terms in each solution as follows.

$$1 + \dfrac{\sqrt{14}}{2} = \dfrac{2}{2} + \dfrac{\sqrt{14}}{2} = \dfrac{2 + \sqrt{14}}{2} \qquad 1 = \dfrac{2}{2}$$

$$1 - \dfrac{\sqrt{14}}{2} = \dfrac{2}{2} - \dfrac{\sqrt{14}}{2} = \dfrac{2 - \sqrt{14}}{2}$$

Check that the solution set is $\left\{\dfrac{2 \pm \sqrt{14}}{2}\right\}$.   NOW TRY

NOW TRY ANSWERS
7. $\left\{\dfrac{-1 \pm \sqrt{13}}{2}\right\}$
8. $\left\{\dfrac{-6 \pm \sqrt{51}}{3}\right\}$

OBJECTIVE 5 **Solve quadratic equations with solutions that are not real numbers.** In $x^2 = k$, if $k < 0$, there will be two nonreal complex solutions.

NOW TRY
EXERCISE 9

Solve each equation.

(a) $t^2 = -24$

(b) $(x + 4)^2 = -36$

(c) $x^2 + 8x + 21 = 0$

EXAMPLE 9 Solving for Nonreal Complex Solutions

Solve each equation.

(a)
$$x^2 = -15$$
$$x = \sqrt{-15} \quad \text{or} \quad x = -\sqrt{-15} \qquad \text{Square root property}$$
$$x = i\sqrt{15} \quad \text{or} \quad x = -i\sqrt{15} \qquad \sqrt{-1} = i$$

The solution set is $\{i\sqrt{15}, -i\sqrt{15}\}$, or $\{\pm i\sqrt{15}\}$.

(b)
$$(x + 2)^2 = -16$$
$$x + 2 = \sqrt{-16} \quad \text{or} \quad x + 2 = -\sqrt{-16} \qquad \text{Square root property}$$
$$x + 2 = 4i \quad \text{or} \quad x + 2 = -4i \qquad \sqrt{-16} = 4i$$
$$x = -2 + 4i \quad \text{or} \quad x = -2 - 4i \qquad \text{Add } -2.$$

The solution set is $\{-2 + 4i, -2 - 4i\}$, or $\{-2 \pm 4i\}$.

(c)
$$x^2 + 2x + 7 = 0$$
$$x^2 + 2x = -7 \qquad \text{Subtract 7.}$$
$$x^2 + 2x + 1 = -7 + 1 \qquad \left[\tfrac{1}{2}(2)\right]^2 = 1; \text{ Add 1 to each side.}$$
$$(x + 1)^2 = -6 \qquad \text{Factor on the left. Add on the right.}$$
$$x + 1 = \sqrt{-6} \quad \text{or} \quad x + 1 = -\sqrt{-6} \qquad \text{Square root property}$$
$$x + 1 = i\sqrt{6} \quad \text{or} \quad x + 1 = -i\sqrt{6} \qquad \sqrt{-1} = i$$
$$x = -1 + i\sqrt{6} \quad \text{or} \quad x = -1 - i\sqrt{6} \qquad \text{Add } -1.$$

The solution set is $\{-1 + i\sqrt{6}, -1 - i\sqrt{6}\}$, or $\{-1 \pm i\sqrt{6}\}$. **NOW TRY**

NOW TRY ANSWERS

9. (a) $\{\pm 2i\sqrt{6}\}$

(b) $\{-4 \pm 6i\}$

(c) $\{-4 \pm i\sqrt{5}\}$

9.1 EXERCISES **MyMathLab**
PRACTICE   WATCH   DOWNLOAD   READ   REVIEW

🌐 *Complete solution available on the Video Resources on DVD*

1. *Concept Check* Which of the following are quadratic equations?

   A. $x + 2y = 0$ **B.** $x^2 - 8x + 16 = 0$ **C.** $2t^2 - 5t = 3$ **D.** $x^3 + x^2 + 4 = 0$

2. *Concept Check* Which quadratic equation identified in **Exercise 1** is in standard form?

3. *Concept Check* A student incorrectly solved the equation $x^2 - x - 2 = 5$ as follows. **WHAT WENT WRONG?**

$$x^2 - x - 2 = 5$$
$$(x - 2)(x + 1) = 5 \qquad \text{Factor.}$$
$$x - 2 = 5 \quad \text{or} \quad x + 1 = 5 \qquad \text{Zero-factor property}$$
$$x = 7 \quad \text{or} \qquad x = 4 \qquad \text{Solve each equation.}$$

4. *Concept Check* A student was asked to solve the quadratic equation $x^2 = 16$ and did not get full credit for the solution set $\{4\}$. **WHAT WENT WRONG?**

Use the zero-factor property to solve each equation. (Hint: In Exercises 9 and 10, write the equation in standard form first.) **See Example 1.**

5. $x^2 + 3x + 2 = 0$ **6.** $x^2 + 8x + 15 = 0$ **7.** $3x^2 + 8x - 3 = 0$

8. $2x^2 + x - 6 = 0$ **9.** $2x^2 = 9x - 4$ **10.** $5x^2 = 11x - 2$

Use the square root property to solve each equation. **See Examples 2, 4, and 5.**

11. $x^2 = 81$ **12.** $x^2 = 225$ **13.** $x^2 = 17$

14. $x^2 = 19$ **15.** $x^2 = 32$ **16.** $x^2 = 54$

17. $x^2 - 20 = 0$ **18.** $p^2 - 50 = 0$ **19.** $3x^2 - 72 = 0$

20. $5z^2 - 200 = 0$ **21.** $(x + 2)^2 = 25$ **22.** $(t + 8)^2 = 9$

23. $(x - 6)^2 = 49$ **24.** $(x - 4)^2 = 64$ **25.** $(x - 4)^2 = 3$

26. $(x + 3)^2 = 11$ **27.** $(t + 5)^2 = 48$ **28.** $(m - 6)^2 = 27$

29. $(3x - 1)^2 = 7$ **30.** $(2x - 5)^2 = 10$ **31.** $(4p + 1)^2 = 24$

32. $(5t + 2)^2 = 12$ **33.** $(2 - 5t)^2 = 12$ **34.** $(1 - 4p)^2 = 24$

In Exercises 35 and 36, round answers to the nearest tenth. **See Example 3.**

35. The sculpture of American presidents at Mount Rushmore National Memorial is 500 ft above the valley floor. How long would it take a rock dropped from the top of the sculpture to fall to the ground? (*Source:* www.travelsd.com)

36. The Gateway Arch in St. Louis, Missouri, is 630 ft tall. How long would it take an object dropped from the top of the arch to fall to the ground? (*Source:* www.gatewayarch.com)

37. *Concept Check*   Which one of the two equations

$$(2x + 1)^2 = 5 \quad \text{and} \quad x^2 + 4x = 12,$$

is more suitable for solving by the square root property? Which one is more suitable for solving by completing the square?

38. Why would most students find the equation $x^2 + 4x = 20$ easier to solve by completing the square than the equation $5x^2 + 2x = 3$?

Concept Check   Decide what number must be added to make each expression a perfect square trinomial. Then factor the trinomial.

39. $x^2 + 6x + \underline{\hspace{0.5cm}}$ **40.** $x^2 + 14x + \underline{\hspace{0.5cm}}$ **41.** $p^2 - 12p + \underline{\hspace{0.5cm}}$

42. $x^2 - 20x + \underline{\hspace{0.5cm}}$ **43.** $q^2 + 9q + \underline{\hspace{0.5cm}}$ **44.** $t^2 + 13t + \underline{\hspace{0.5cm}}$

45. $x^2 + \dfrac{1}{4}x + \underline{\hspace{0.5cm}}$ **46.** $x^2 + \dfrac{1}{2}x + \underline{\hspace{0.5cm}}$ **47.** $x^2 - 0.8x + \underline{\hspace{0.5cm}}$

48. *Concept Check*   What would be the first step in solving $2x^2 + 8x = 9$ by completing the square?

Determine the number that will complete the square to solve each equation, after the constant term has been written on the right side and the coefficient of the second-degree term is 1. Do not actually solve. See Examples 6–8.

49. $x^2 + 4x - 2 = 0$ **50.** $t^2 + 2t - 1 = 0$ **51.** $x^2 + 10x + 18 = 0$

52. $x^2 + 8x + 11 = 0$ **53.** $3w^2 - w - 24 = 0$ **54.** $4z^2 - z - 39 = 0$

Solve each equation by completing the square. Use the results of Exercises 49–54 to solve Exercises 57–62. See Examples 6–8.

55. $x^2 - 2x - 24 = 0$ **56.** $m^2 - 4m - 32 = 0$ ◕ **57.** $x^2 + 4x - 2 = 0$

58. $t^2 + 2t - 1 = 0$ **59.** $x^2 + 10x + 18 = 0$ **60.** $x^2 + 8x + 11 = 0$

61. $3w^2 - w = 24$ **62.** $4z^2 - z = 39$ ◕ **63.** $x^2 + 7x - 1 = 0$

64. $x^2 + 13x - 3 = 0$ **65.** $2k^2 + 5k - 2 = 0$ **66.** $3r^2 + 2r - 2 = 0$

◕ **67.** $5x^2 - 10x + 2 = 0$ **68.** $2x^2 - 16x + 25 = 0$ **69.** $9x^2 - 24x = -13$

70. $25n^2 - 20n = 1$ **71.** $z^2 - \dfrac{4}{3}z = -\dfrac{1}{9}$ **72.** $p^2 - \dfrac{8}{3}p = -1$

73. $0.1x^2 - 0.2x - 0.1 = 0$ **74.** $0.1p^2 - 0.4p + 0.1 = 0$
(*Hint*: First clear the decimals.) (*Hint*: First clear the decimals.)

Find the nonreal complex solutions of each equation. See Example 9.

◕ **75.** $x^2 = -12$ **76.** $x^2 = -18$ **77.** $(r - 5)^2 = -4$

78. $(t + 6)^2 = -9$ **79.** $(6x - 1)^2 = -8$ **80.** $(4m - 7)^2 = -27$

81. $m^2 + 4m + 13 = 0$ **82.** $t^2 + 6t + 10 = 0$ **83.** $3r^2 + 4r + 4 = 0$

84. $4x^2 + 5x + 5 = 0$ **85.** $-m^2 - 6m - 12 = 0$ **86.** $-x^2 - 5x - 10 = 0$

RELATING CONCEPTS EXERCISES 87–92

FOR INDIVIDUAL OR GROUP WORK

The Greeks had a method of completing the square geometrically in which they literally changed a figure into a square. For example, to complete the square for $x^2 + 6x$, we begin with a square of side x, as in the figure on the left. We add three rectangles of width 1 to the right side and the bottom to get a region with area $x^2 + 6x$. To fill in the corner (complete the square), we must add nine 1-by-1 squares as shown.

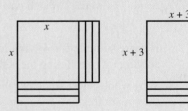

Work Exercises 87–92 in order.

87. What is the area of the original square?

88. What is the area of each strip?

89. What is the total area of the six strips?

90. What is the area of each small square in the corner of the second figure?

91. What is the total area of the small squares?

92. What is the area of the new "complete" square?

Brain Busters *Solve for x. Assume that a and b represent positive real numbers.*

93. $x^2 - b = 0$ **94.** $x^2 = 4b$ **95.** $4x^2 = b^2 + 16$

96. $9x^2 - 25a = 0$ **97.** $(5x - 2b)^2 = 3a$ **98.** $x^2 - a^2 - 36 = 0$

PREVIEW EXERCISES

Evaluate $\sqrt{b^2 - 4ac}$ *for the given values of a, b, and c.* ***See Section 1.3.***

99. $a = 3, b = 1, c = -1$ **100.** $a = 4, b = 11, c = -3$

101. $a = 6, b = 7, c = 2$ **102.** $a = 1, b = -6, c = 9$

9.2 The Quadratic Formula

OBJECTIVES

1 Derive the quadratic formula.

2 Solve quadratic equations by using the quadratic formula.

3 Use the discriminant to determine the number and type of solutions.

In this section, we complete the square to solve the general quadratic equation

$$ax^2 + bx + c = 0,$$

where a, b, and c are complex numbers and $a \neq 0$. The solution of this general equation gives a formula for finding the solution of *any* specific quadratic equation.

OBJECTIVE 1 **Derive the quadratic formula.** To solve $ax^2 + bx + c = 0$ by completing the square (assuming $a > 0$), we follow the steps given in **Section 9.1.**

$$ax^2 + bx + c = 0$$

$$x^2 + \frac{b}{a}x + \frac{c}{a} = 0 \qquad \text{Divide by } a. \text{ (Step 1)}$$

$$x^2 + \frac{b}{a}x = -\frac{c}{a} \qquad \text{Subtract } \tfrac{c}{a}. \text{ (Step 2)}$$

$$\left[\frac{1}{2}\left(\frac{b}{a}\right)\right]^2 = \left(\frac{b}{2a}\right)^2 = \frac{b^2}{4a^2} \qquad \text{(Step 3)}$$

$$x^2 + \frac{b}{a}x + \frac{b^2}{4a^2} = -\frac{c}{a} + \frac{b^2}{4a^2} \qquad \text{Add } \tfrac{b^2}{4a^2} \text{ to each side. (Step 4)}$$

$$\left(x + \frac{b}{2a}\right)^2 = \frac{b^2}{4a^2} + \frac{-c}{a} \qquad \begin{array}{l}\text{Write the left side as a perfect square.}\\ \text{Rearrange the right side. (Step 5)}\end{array}$$

$$\left(x + \frac{b}{2a}\right)^2 = \frac{b^2}{4a^2} + \frac{-4ac}{4a^2} \qquad \text{Write with a common denominator.}$$

$$\left(x + \frac{b}{2a}\right)^2 = \frac{b^2 - 4ac}{4a^2} \qquad \text{Add fractions.}$$

$$x + \frac{b}{2a} = \sqrt{\frac{b^2 - 4ac}{4a^2}} \quad \text{or} \quad x + \frac{b}{2a} = -\sqrt{\frac{b^2 - 4ac}{4a^2}} \qquad \begin{array}{l}\text{Square root property}\\ \text{(Step 6)}\end{array}$$

We can simplify $\sqrt{\dfrac{b^2 - 4ac}{4a^2}}$ as $\dfrac{\sqrt{b^2 - 4ac}}{\sqrt{4a^2}}$, or $\dfrac{\sqrt{b^2 - 4ac}}{2a}$.

The right side of each equation can be expressed as follows.

$$x + \frac{b}{2a} = \frac{\sqrt{b^2 - 4ac}}{2a} \qquad \text{or} \qquad x + \frac{b}{2a} = \frac{-\sqrt{b^2 - 4ac}}{2a}$$

$$x = \frac{-b}{2a} + \frac{\sqrt{b^2 - 4ac}}{2a} \qquad \text{or} \qquad x = \frac{-b}{2a} - \frac{\sqrt{b^2 - 4ac}}{2a}$$

| If $a < 0$, the same two solutions are obtained. |

$$x = \frac{-b + \sqrt{b^2 - 4ac}}{2a} \qquad \text{or} \qquad x = \frac{-b - \sqrt{b^2 - 4ac}}{2a}$$

The result is the **quadratic formula**, which is abbreviated as follows.

Quadratic Formula

The solutions of the equation $ax^2 + bx + c = 0$ (with $a \neq 0$) are given by

$$x = \frac{-b \pm \sqrt{b^2 - 4ac}}{2a}.$$

⚠ **CAUTION** In the quadratic formula, *the square root is added to or subtracted from the value of* $-b$ *before dividing by* $2a$.

OBJECTIVE 2 Solve quadratic equations by using the quadratic formula.

NOW TRY
EXERCISE 1
Solve $2x^2 + 3x - 20 = 0$.

EXAMPLE 1 Using the Quadratic Formula (Rational Solutions)

Solve $6x^2 - 5x - 4 = 0$.

This equation is in standard form, so we identify the values of a, b, and c. Here a, the coefficient of the second-degree term, is 6, and b, the coefficient of the first-degree term, is -5. The constant c is -4. Now substitute into the quadratic formula.

$$x = \frac{-b \pm \sqrt{b^2 - 4ac}}{2a} \qquad \text{Quadratic formula}$$

$$x = \frac{-(-5) \pm \sqrt{(-5)^2 - 4(6)(-4)}}{2(6)} \qquad a = 6,\ b = -5,\ c = -4$$

> Use parentheses and substitute carefully to avoid errors.

$$x = \frac{5 \pm \sqrt{25 + 96}}{12}$$

$$x = \frac{5 \pm \sqrt{121}}{12} \qquad \text{Simplify the radical.}$$

$$x = \frac{5 \pm 11}{12} \qquad \text{Take the square root.}$$

There are two solutions, one from the $+$ sign and one from the $-$ sign.

$$x = \frac{5 + 11}{12} = \frac{16}{12} = \frac{4}{3} \qquad \text{or} \qquad x = \frac{5 - 11}{12} = \frac{-6}{12} = -\frac{1}{2}$$

NOW TRY ANSWER
1. $\left\{-4, \frac{5}{2}\right\}$

Check each solution in the original equation. The solution set is $\left\{-\frac{1}{2}, \frac{4}{3}\right\}$.

NOW TRY

NOTE We could have used factoring to solve the equation in **Example 1.**

$$6x^2 - 5x - 4 = 0$$

$$(3x - 4)(2x + 1) = 0 \qquad \text{Factor.}$$

$$3x - 4 = 0 \quad \text{or} \quad 2x + 1 = 0 \qquad \text{Zero-factor property}$$

$$3x = 4 \quad \text{or} \qquad 2x = -1 \qquad \text{Solve each equation.}$$

$$x = \frac{4}{3} \quad \text{or} \qquad x = -\frac{1}{2} \qquad \text{Same solutions as in \textbf{Example 1}}$$

When solving quadratic equations, it is a good idea to try factoring first. If the polynomial cannot be factored or if factoring is difficult, then use the quadratic formula.

⤸ *NOW TRY*
EXERCISE 2

Solve $3x^2 + 1 = -5x$.

EXAMPLE 2 Using the Quadratic Formula (Irrational Solutions)

Solve $4x^2 = 8x - 1$.

Write the equation in standard form as $4x^2 - 8x + 1 = 0$. ⟵ This is a key step.

$$x = \frac{-b \pm \sqrt{b^2 - 4ac}}{2a} \qquad \text{Quadratic formula}$$

$$x = \frac{-(-8) \pm \sqrt{(-8)^2 - 4(4)(1)}}{2(4)} \qquad a = 4, b = -8, c = 1$$

$$x = \frac{8 \pm \sqrt{64 - 16}}{8} \qquad \text{Simplify.}$$

$$x = \frac{8 \pm \sqrt{48}}{8}$$

$$x = \frac{8 \pm 4\sqrt{3}}{8} \qquad \sqrt{48} = \sqrt{16} \cdot \sqrt{3} = 4\sqrt{3}$$

$$x = \frac{4\left(2 \pm \sqrt{3}\right)}{4(2)} \qquad \text{Factor.}$$

Factor first. Then divide out the common factor.

$$x = \frac{2 \pm \sqrt{3}}{2} \qquad \text{Lowest terms}$$

The solution set is $\left\{\dfrac{2 \pm \sqrt{3}}{2}\right\}$.

NOW TRY

⚠ **CAUTION**

1. *Every quadratic equation must be expressed in standard form* $ax^2 + bx + c = 0$ *before we begin to solve it,* whether we use factoring or the quadratic formula.

2. *When writing solutions in lowest terms, be sure to FACTOR FIRST. Then divide out the common factor,* as shown in the last two steps in **Example 2.**

NOW TRY ANSWER

2. $\left\{\dfrac{-5 \pm \sqrt{13}}{6}\right\}$

Solve $(x + 5)(x - 1) = -18$.

EXAMPLE 3 Using the Quadratic Formula (Nonreal Complex Solutions)

Solve $(9x + 3)(x - 1) = -8$.

$$(9x + 3)(x - 1) = -8$$

$$9x^2 - 6x - 3 = -8 \qquad \text{Multiply.}$$

Standard form $\rightarrow 9x^2 - 6x + 5 = 0 \qquad \text{Add 8.}$

From the equation $9x^2 - 6x + 5 = 0$, we identify $a = 9$, $b = -6$, and $c = 5$.

$$x = \frac{-b \pm \sqrt{b^2 - 4ac}}{2a} \qquad \text{Quadratic formula}$$

$$x = \frac{-(-6) \pm \sqrt{(-6)^2 - 4(9)(5)}}{2(9)} \qquad \text{Substitute.}$$

$$x = \frac{6 \pm \sqrt{-144}}{18} \qquad \text{Simplify.}$$

$$x = \frac{6 \pm 12i}{18} \qquad \sqrt{-144} = 12i$$

$$x = \frac{6(1 \pm 2i)}{6(3)} \qquad \text{Factor.}$$

$$x = \frac{1 \pm 2i}{3} \qquad \text{Lowest terms}$$

$$x = \frac{1}{3} \pm \frac{2}{3}i \qquad \begin{array}{l}\text{Standard form } a + bi \text{ for a}\\ \text{complex number}\end{array}$$

The solution set is $\left\{\frac{1}{3} \pm \frac{2}{3}i\right\}$. 　　　　　　NOW TRY

OBJECTIVE 3 Use the discriminant to determine the number and type of solutions. The solutions of the quadratic equation $ax^2 + bx + c = 0$ are given by

$$x = \frac{-b \pm \sqrt{b^2 - 4ac}}{2a}. \quad \longleftarrow \text{Discriminant}$$

If a, b, and c are integers, the type of solutions of a quadratic equation—that is, rational, irrational, or nonreal complex—is determined by the expression under the radical symbol, $b^2 - 4ac$, called the *discriminant* (because it distinguishes among the three types of solutions). By calculating the discriminant, we can predict the number and type of solutions of a quadratic equation.

Discriminant

The **discriminant** of $ax^2 + bx + c = 0$ is $b^2 - 4ac$. If a, b, and c are integers, then the number and type of solutions are determined as follows.

Discriminant	Number and Type of Solutions
Positive, and the square of an integer	Two rational solutions
Positive, but not the square of an integer	Two irrational solutions
Zero	One rational solution
Negative	Two nonreal complex solutions

Calculating the discriminant can also help you decide how to solve a quadratic equation. ***If the discriminant is a perfect square (including 0), then the equation can be solved by factoring. Otherwise, the quadratic formula should be used.***

⌒ NOW TRY
↪ EXERCISE 4

Find each discriminant. Use it to predict the number and type of solutions for each equation. Tell whether the equation can be solved by factoring or whether the quadratic formula should be used.

(a) $8x^2 - 6x - 5 = 0$

(b) $9x^2 = 24x - 16$

(c) $3x^2 + 2x = -1$

EXAMPLE 4   Using the Discriminant

Find the discriminant. Use it to predict the number and type of solutions for each equation. Tell whether the equation can be solved by factoring or whether the quadratic formula should be used.

(a) $6x^2 - x - 15 = 0$

We find the discriminant by evaluating $b^2 - 4ac$. Because $-x = -1x$, the value of b in this equation is -1.

$b^2 - 4ac$ *Use parentheses and substitute carefully.*

$= (-1)^2 - 4(6)(-15)$ $a = 6, b = -1, c = -15$

$= 1 + 360$ Apply the exponent. Multiply.

$= 361$,   or 19^2,   which is a perfect square.

Since a, b, and c are integers and the discriminant 361 is a perfect square, there will be two rational solutions. The equation can be solved by factoring.

(b) $3x^2 - 4x = 5$ Write in standard form as $3x^2 - 4x - 5 = 0$.

$b^2 - 4ac$

$= (-4)^2 - 4(3)(-5)$ $a = 3, b = -4, c = -5$

$= 16 + 60$ Apply the exponent. Multiply.

$= 76$ Add.

Because 76 is positive but *not* the square of an integer and a, b, and c are integers, the equation will have two irrational solutions and is best solved using the quadratic formula.

(c) $4x^2 + x + 1 = 0$

x = 1x, so b = 1.

$b^2 - 4ac$

$= 1^2 - 4(4)(1)$ $a = 4, b = 1, c = 1$

$= 1 - 16$ Apply the exponent. Multiply.

$= -15$ Subtract.

Because the discriminant is negative and a, b, and c are integers, this equation will have two nonreal complex solutions. The quadratic formula should be used to solve it.

(d) $4x^2 + 9 = 12x$ Write in standard form as $4x^2 - 12x + 9 = 0$.

$b^2 - 4ac$

$= (-12)^2 - 4(4)(9)$ $a = 4, b = -12, c = 9$

$= 144 - 144$ Apply the exponent. Multiply.

$= 0$ Subtract.

The discriminant is 0, so the quantity under the radical in the quadratic formula is 0, and there is only one rational solution. The equation can be solved by factoring.

NOW TRY ↩

NOW TRY ANSWERS
4. (a) 196; two rational solutions; factoring
   (b) 0; one rational solution; factoring
   (c) -8; two nonreal complex solutions; quadratic formula

NOW TRY
EXERCISE 5

Find k so that the equation will have exactly one rational solution.

$$4x^2 + kx + 25 = 0$$

EXAMPLE 5 Using the Discriminant

Find k so that $9x^2 + kx + 4 = 0$ will have exactly one rational solution.

The equation will have only one rational solution if the discriminant is 0.

$$b^2 - 4ac$$

$$= k^2 - 4(9)(4) \qquad \text{Here, } a = 9, b = k, \text{ and } c = 4.$$

$$= k^2 - 144 \longleftarrow \text{Value of the discriminant}$$

Set the discriminant equal to 0 and solve for k.

$$k^2 - 144 = 0$$

$$k^2 = 144 \qquad \text{Add 144.}$$

$$k = 12 \quad \text{or} \quad k = -12 \qquad \text{Square root property}$$

NOW TRY ANSWER
5. 20, −20

The equation will have only one rational solution if $k = 12$ or $k = -12$.

NOW TRY

9.2 EXERCISES

MyMathLab | Math XL PRACTICE | WATCH | DOWNLOAD | READ | REVIEW

🌐 *Complete solution available on the Video Resources on DVD*

Concept Check Answer each question in Exercises 1–4.

1. An early version of Microsoft *Word* for Windows included the 1.0 edition of *Equation Editor*. The documentation used the following for the quadratic formula. Was this correct? If not, correct it.

$$x = -b \pm \frac{\sqrt{b^2 - 4ac}}{2a}$$

2. The Cadillac Bar in Houston, Texas, encourages patrons to write (tasteful) messages on the walls. One person wrote the quadratic formula, as shown here. Was this correct? If not, correct it.

$$x = \frac{-b\sqrt{b^2 - 4ac}}{2a}$$

3. A student incorrectly solved $5x^2 - 5x + 1 = 0$ as follows. **WHAT WENT WRONG?**

$$x = \frac{-(-5) \pm \sqrt{(-5)^2 - 4(5)(1)}}{2(5)}$$

$$x = \frac{5 \pm \sqrt{5}}{10}$$

$$x = \frac{1}{2} \pm \sqrt{5}$$

Solution set: $\left\{ \dfrac{1}{2} \pm \sqrt{5} \right\}$

4. A student claimed that the equation $2x^2 - 5 = 0$ cannot be solved using the quadratic formula because there is no first-degree x-term. Was the student correct? If not, give the values of a, b, and c.

Use the quadratic formula to solve each equation. (All solutions for these equations are real numbers.) See Examples 1 and 2.

5. $x^2 - 8x + 15 = 0$ **6.** $x^2 + 3x - 28 = 0$ **7.** $2x^2 + 4x + 1 = 0$

8. $2x^2 + 3x - 1 = 0$ **9.** $2x^2 - 2x = 1$ **10.** $9x^2 + 6x = 1$

11. $x^2 + 18 = 10x$ **12.** $x^2 - 4 = 2x$ **13.** $4x^2 + 4x - 1 = 0$

14. $4r^2 - 4r - 19 = 0$ **15.** $2 - 2x = 3x^2$ **16.** $26r - 2 = 3r^2$

17. $\dfrac{x^2}{4} - \dfrac{x}{2} = 1$ **18.** $p^2 + \dfrac{p}{3} = \dfrac{1}{6}$ **19.** $-2t(t + 2) = -3$

20. $-3x(x + 2) = -4$ **21.** $(r - 3)(r + 5) = 2$ **22.** $(x + 1)(x - 7) = 1$

23. $(x + 2)(x - 3) = 1$ **24.** $(x - 5)(x + 2) = 6$ **25.** $p = \dfrac{5(5 - p)}{3(p + 1)}$

26. $x = \dfrac{2(x + 3)}{x + 5}$ **27.** $(2x + 1)^2 = x + 4$ **28.** $(2x - 1)^2 = x + 2$

Use the quadratic formula to solve each equation. (All solutions for these equations are non-real complex numbers.) See Example 3.

29. $x^2 - 3x + 6 = 0$ **30.** $x^2 - 5x + 20 = 0$ **31.** $r^2 - 6r + 14 = 0$

32. $t^2 + 4t + 11 = 0$ **33.** $4x^2 - 4x = -7$ **34.** $9x^2 - 6x = -7$

35. $x(3x + 4) = -2$ **36.** $z(2z + 3) = -2$

37. $(2x - 1)(8x - 4) = -1$ **38.** $(x - 1)(9x - 3) = -2$

Use the discriminant to determine whether the solutions for each equation are
    A. *two rational numbers*   **B.** *one rational number*
    C. *two irrational numbers*   **D.** *two nonreal complex numbers.*

Tell whether the equation can be solved by factoring or whether the quadratic formula should be used. Do not actually solve. See Example 4.

39. $25x^2 + 70x + 49 = 0$ **40.** $4x^2 - 28x + 49 = 0$ **41.** $x^2 + 4x + 2 = 0$

42. $9x^2 - 12x - 1 = 0$ **43.** $3x^2 = 5x + 2$ **44.** $4x^2 = 4x + 3$

45. $3m^2 - 10m + 15 = 0$ **46.** $18x^2 + 60x + 82 = 0$

*Based on your answers in **Exercises 39–46,** solve the equation given in each exercise.*

47. Exercise 39   **48. Exercise 40**   **49. Exercise 43**   **50. Exercise 44**

51. Find the discriminant for each quadratic equation. Use it to tell whether the equation can be solved by factoring or whether the quadratic formula should be used. Then solve each equation.

    (a) $3x^2 + 13x = -12$ **(b)** $2x^2 + 19 = 14x$

52. *Concept Check*   Is it possible for the solution of a quadratic equation with integer coefficients to include just one irrational number? Why or why not?

Find the value of a, b, or c so that each equation will have exactly one rational solution. See Example 5.

53. $p^2 + bp + 25 = 0$ **54.** $r^2 - br + 49 = 0$ **55.** $am^2 + 8m + 1 = 0$

56. $at^2 + 24t + 16 = 0$ **57.** $9x^2 - 30x + c = 0$ **58.** $4m^2 + 12m + c = 0$

59. One solution of $4x^2 + bx - 3 = 0$ is $-\frac{5}{2}$. Find b and the other solution.

60. One solution of $3x^2 - 7x + c = 0$ is $\frac{1}{3}$. Find c and the other solution.

Solve each equation. **See Section 2.1.**

61. $\dfrac{3}{4}x + \dfrac{1}{2}x = -10$

62. $\dfrac{x}{5} + \dfrac{3x}{4} = -19$

Solve each equation. **See Section 8.6.**

63. $\sqrt{2x + 6} = x - 1$

64. $\sqrt{2x + 1} + \sqrt{x + 3} = 0$

9.3 Equations Quadratic in Form

OBJECTIVES

1. Solve an equation with fractions by writing it in quadratic form.

2. Use quadratic equations to solve applied problems.

3. Solve an equation with radicals by writing it in quadratic form.

4. Solve an equation that is quadratic in form by substitution.

OBJECTIVE 1 **Solve an equation with fractions by writing it in quadratic form.** A variety of nonquadratic equations can be written in the form of a quadratic equation and solved by using the methods of this chapter.

EXAMPLE 1 **Solving an Equation with Fractions that Leads to a Quadratic Equation**

Solve $\dfrac{1}{x} + \dfrac{1}{x - 1} = \dfrac{7}{12}$.

Clear fractions by multiplying each term by the least common denominator, $12x(x - 1)$. (Note that the domain must be restricted to $x \neq 0, x \neq 1$.)

$$12x(x - 1)\left(\frac{1}{x} + \frac{1}{x - 1}\right) = 12x(x - 1)\left(\frac{7}{12}\right) \quad \text{Multiply by the LCD.}$$

$$12x(x - 1)\frac{1}{x} + 12x(x - 1)\frac{1}{x - 1} = 12x(x - 1)\frac{7}{12} \quad \text{Distributive property}$$

$$12(x - 1) + 12x = 7x(x - 1)$$

$$12x - 12 + 12x = 7x^2 - 7x \quad \text{Distributive property}$$

$$24x - 12 = 7x^2 - 7x \quad \text{Combine like terms.}$$

$$7x^2 - 31x + 12 = 0 \quad \text{Standard form}$$

$$(7x - 3)(x - 4) = 0 \quad \text{Factor.}$$

$$7x - 3 = 0 \quad \text{or} \quad x - 4 = 0 \quad \text{Zero-factor property}$$

$$7x = 3 \quad \text{or} \quad x = 4 \quad \text{Solve for } x.$$

$$x = \frac{3}{7}$$

The solution set is $\left\{\frac{3}{7}, 4\right\}$.

NOW TRY

OBJECTIVE 2 **Use quadratic equations to solve applied problems.** Some distance-rate-time (or motion) problems lead to quadratic equations. We continue to use the six-step problem-solving method from **Section 2.3.**

NOW TRY
EXERCISE 1

Solve $\dfrac{2}{x} + \dfrac{3}{x + 2} = 1$.

NOW TRY ANSWER
1. $\{-1, 4\}$

**NOW TRY
EXERCISE 2**

A small fishing boat averages 18 mph in still water. It takes the boat $\frac{9}{10}$ hr to travel 8 mi upstream and return. Find the rate of the current.

Current

Riverboat traveling upstream—the current slows it down.

FIGURE 1

EXAMPLE 2 Solving a Motion Problem

A riverboat for tourists averages 12 mph in still water. It takes the boat 1 hr, 4 min to go 6 mi upstream and return. Find the rate of the current.

Step 1 **Read** the problem carefully.

Step 2 **Assign a variable.** Let x = the rate of the current.

The current slows down the boat when it is going upstream, so the rate of the boat going upstream is its rate in still water *less* the rate of the current, or $(12 - x)$ mph. See **FIGURE 1**.

Similarly, the current speeds up the boat as it travels downstream, so its rate downstream is $(12 + x)$ mph. Thus,

$$12 - x = \text{the rate upstream in miles per hour,}$$

and $\qquad 12 + x = \text{the rate downstream in miles per hour.}$

	d	r	t
Upstream	6	$12 - x$	$\dfrac{6}{12 - x}$
Downstream	6	$12 + x$	$\dfrac{6}{12 + x}$

Complete a table. Use the distance formula, $d = rt$, solved for time t, $t = \frac{d}{r}$, to write expressions for t.

Times in hours

Step 3 **Write an equation.** We use the total time of 1 hr, 4 min written as a fraction.

$$1 + \frac{4}{60} = 1 + \frac{1}{15} = \frac{16}{15} \text{ hr} \qquad \text{Total time}$$

The time upstream plus the time downstream equals $\frac{16}{15}$ hr.

Time upstream	+	Time downstream	=	Total time
$\downarrow$		$\downarrow$		$\downarrow$
$\dfrac{6}{12 - x}$	$+$	$\dfrac{6}{12 + x}$	$=$	$\dfrac{16}{15}$

Step 4 **Solve** the equation. The LCD is $15(12 - x)(12 + x)$.

$$15(12 - x)(12 + x)\left(\frac{6}{12 - x} + \frac{6}{12 + x} \right)$$
$$= 15(12 - x)(12 + x)\left(\frac{16}{15} \right)$$

Multiply by the LCD.

$$15(12 + x) \cdot 6 + 15(12 - x) \cdot 6 = 16(12 - x)(12 + x)$$

Distributive property; multiply.

$$90(12 + x) + 90(12 - x) = 16(144 - x^2) \qquad \text{Multiply.}$$

$$1080 + 90x + 1080 - 90x = 2304 - 16x^2 \qquad \text{Distributive property}$$

$$2160 = 2304 - 16x^2 \qquad \text{Combine like terms.}$$

$$16x^2 = 144 \qquad \text{Add } 16x^2. \text{ Subtract 2160.}$$

$$x^2 = 9 \qquad \text{Divide by 16.}$$

$$x = 3 \quad \text{or} \quad x = -3 \qquad \text{Square root property}$$

Step 5 **State the answer.** The current rate cannot be -3, so the answer is 3 mph.

Step 6 **Check** that this value satisfies the original problem. NOW TRY

NOW TRY ANSWER
2. 2 mph

> **PROBLEM-SOLVING HINT**
>
> Recall from **Section 7.5** that a person's work rate is $\frac{1}{t}$ part of the job per hour, where t is the time in hours required to do the complete job. Thus, the part of the job the person will do in x hours is $\frac{1}{t}x$.

EXAMPLE 3 Solving a Work Problem

It takes two carpet layers 4 hr to carpet a room. If each worked alone, one of them could do the job in 1 hr less time than the other. How long would it take each carpet layer to complete the job alone?

Step 1 **Read** the problem again. There will be two answers.

Step 2 **Assign a variable.** Let x = the number of hours for the slower carpet layer to complete the job alone. Then the faster carpet layer could do the entire job in $(x - 1)$ hours. The slower person's rate is $\frac{1}{x}$, and the faster person's rate is $\frac{1}{x-1}$. Together, they do the job in 4 hr.

	Rate	Time Working Together	Fractional Part of the Job Done	
Slower Worker	$\frac{1}{x}$	4	$\frac{1}{x}(4)$	Complete a table.
Faster Worker	$\frac{1}{x-1}$	4	$\frac{1}{x-1}(4)$	Sum is 1 whole job.

Step 3 **Write an equation.**

Part done by slower worker $+$ Part done by faster worker $=$ 1 whole job

$$\frac{4}{x} \quad + \quad \frac{4}{x-1} \quad = \quad 1$$

Step 4 **Solve** the equation from Step 3.

$$x(x - 1)\left(\frac{4}{x} + \frac{4}{x - 1}\right) = x(x - 1)(1) \qquad \text{Multiply by the LCD, } x(x - 1).$$

$$4(x - 1) + 4x = x(x - 1) \qquad \text{Distributive property}$$

$$4x - 4 + 4x = x^2 - x \qquad \text{Distributive property}$$

$$x^2 - 9x + 4 = 0 \qquad \text{Standard form}$$

This equation cannot be solved by factoring, so use the quadratic formula.

$$x = \frac{-b \pm \sqrt{b^2 - 4ac}}{2a} \qquad \text{Quadratic formula}$$

$$x = \frac{-(-9) \pm \sqrt{(-9)^2 - 4(1)(4)}}{2(1)} \qquad a = 1, b = -9, c = 4$$

$$x = \frac{9 \pm \sqrt{65}}{2} \qquad \text{Simplify.}$$

$$x = \frac{9 + \sqrt{65}}{2} \approx 8.5 \quad \text{or} \quad x = \frac{9 - \sqrt{65}}{2} \approx 0.5 \qquad \text{Use a calculator.}$$

NOW TRY
EXERCISE 3

Two electricians are running wire to finish a basement. One electrician could finish the job in 2 hr less time than the other. Together, they complete the job in 6 hr. How long (to the nearest tenth) would it take the slower electrician to complete the job alone?

NOW TRY
EXERCISE 4

Solve each equation.

(a) $x = \sqrt{9x - 20}$

(b) $x + \sqrt{x} = 20$

Step 5 **State the answer.** Only the solution 8.5 makes sense in the original problem, because if $x = 0.5$, then

$$x - 1 = 0.5 - 1 = -0.5,$$

which cannot represent the time for the faster worker. The slower worker could do the job in about 8.5 hr and the faster in about $8.5 - 1 = 7.5$ hr.

Step 6 **Check** that these results satisfy the original problem.   *NOW TRY*

OBJECTIVE 3 **Solve an equation with radicals by writing it in quadratic form.**

EXAMPLE 4 **Solving Radical Equations That Lead to Quadratic Equations**

Solve each equation.

(a) $x = \sqrt{6x - 8}$

This equation is not quadratic. However, squaring each side of the equation gives a quadratic equation that can be solved by factoring.

$$
\begin{aligned}
x^2 &= \left(\sqrt{6x - 8}\right)^2 &&\text{Square each side.}\\
x^2 &= 6x - 8 &&\left(\sqrt{a}\right)^2 = a\\
x^2 - 6x + 8 &= 0 &&\text{Standard form}\\
(x - 4)(x - 2) &= 0 &&\text{Factor.}\\
x - 4 = 0 \quad &\text{or} \quad x - 2 = 0 &&\text{Zero-factor property}\\
x = 4 \quad &\text{or} \quad x = 2 &&\text{Proposed solutions}
\end{aligned}
$$

Squaring each side of an equation can introduce extraneous solutions. ***All proposed solutions must be checked in the original (not the squared) equation.***

CHECK $x = \sqrt{6x - 8}$ $\qquad\qquad$ $x = \sqrt{6x - 8}$

$4 \overset{?}{=} \sqrt{6(4) - 8}$ Let $x = 4$. $\qquad$ $2 \overset{?}{=} \sqrt{6(2) - 8}$ Let $x = 2$.

$4 \overset{?}{=} \sqrt{16}$ $\qquad\qquad\qquad$ $2 \overset{?}{=} \sqrt{4}$

$4 = 4 \ \checkmark$ $\qquad\quad$ True $\qquad$ $2 = 2 \ \checkmark$ $\qquad\quad$ True

Both solutions check, so the solution set is $\{2, 4\}$.

(b) $\qquad\qquad x + \sqrt{x} = 6$ $\boxed{(a - b)^2 = a^2 - 2ab + b^2}$

$\qquad\qquad\qquad \sqrt{x} = 6 - x$ $\qquad$ Isolate the radical on one side.

$\qquad\qquad\qquad x = 36 - 12x + x^2$ $\qquad$ Square each side.

$\qquad x^2 - 13x + 36 = 0$ $\qquad$ Standard form

$\qquad (x - 4)(x - 9) = 0$ $\qquad$ Factor.

$\qquad x - 4 = 0 \quad$ or $\quad x - 9 = 0$ $\qquad$ Zero-factor property

$\qquad\qquad x = 4 \quad$ or $\qquad x = 9$ $\qquad$ Proposed solutions

CHECK $x + \sqrt{x} = 6$ $\qquad\qquad\qquad$ $x + \sqrt{x} = 6$

$4 + \sqrt{4} \overset{?}{=} 6$ Let $x = 4$. $\qquad$ $9 + \sqrt{9} \overset{?}{=} 6$ Let $x = 9$.

$6 = 6 \ \checkmark$ True $\qquad\qquad\qquad$ $12 = 6$ False

NOW TRY ANSWERS
3. 13.1 hr
4. (a) $\{4, 5\}$ **(b)** $\{16\}$

Only the solution 4 checks, so the solution set is $\{4\}$.   *NOW TRY*

OBJECTIVE 4 **Solve an equation that is quadratic in form by substitution.**
A nonquadratic equation that can be written in the form

$$au^2 + bu + c = 0,$$

for $a \neq 0$ and an algebraic expression u, is called **quadratic in form.**

Many equations that are quadratic in form can be solved more easily by defining and substituting a "temporary" variable u for an expression involving the variable in the original equation.

> **NOW TRY**
> **EXERCISE 5**
> Define a variable u, and write each equation in the form $au^2 + bu + c = 0$.
> **(a)** $x^4 - 10x^2 + 9 = 0$
> **(b)** $6(x + 2)^2$
> $\quad - 11(x + 2) + 4 = 0$

EXAMPLE 5 Defining Substitution Variables

Define a variable u, and write each equation in the form $au^2 + bu + c = 0$.

(a) $x^4 - 13x^2 + 36 = 0$

Look at the two terms involving the variable x, ignoring their coefficients. Try to find one variable expression that is the square of the other. Since $x^4 = (x^2)^2$, we can define $u = x^2$, and rewrite the original equation as a quadratic equation.

$$u^2 - 13u + 36 = 0 \qquad \text{Here, } u = x^2.$$

(b) $2(4x - 3)^2 + 7(4x - 3) + 5 = 0$

Because this equation involves both $(4x - 3)^2$ and $(4x - 3)$, we choose $u = 4x - 3$. Substituting u for $4x - 3$ gives the quadratic equation

$$2u^2 + 7u + 5 = 0. \qquad \text{Here, } u = 4x - 3.$$

(c) $2x^{2/3} - 11x^{1/3} + 12 = 0$

We apply a power rule for exponents **(Section 5.1)**, $(a^m)^n = a^{mn}$. Because $(x^{1/3})^2 = x^{2/3}$, we define $u = x^{1/3}$. The original equation becomes

$$2u^2 - 11u + 12 = 0. \qquad \text{Here, } u = x^{1/3}. \qquad \text{NOW TRY}$$

EXAMPLE 6 Solving Equations That Are Quadratic in Form

Solve each equation.

(a) $x^4 - 13x^2 + 36 = 0$

We can write this equation in quadratic form by substituting u for x^2. (See **Example 5(a).**)

$$
\begin{array}{ll}
x^4 - 13x^2 + 36 = 0 & \\
(x^2)^2 - 13x^2 + 36 = 0 & x^4 = (x^2)^2 \\
u^2 - 13u + 36 = 0 & \text{Let } u = x^2. \\
(u - 4)(u - 9) = 0 & \text{Factor.} \\
u - 4 = 0 \quad \text{or} \quad u - 9 = 0 & \text{Zero-factor property} \\
u = 4 \quad \text{or} \quad u = 9 & \text{Solve.} \\
x^2 = 4 \quad \text{or} \quad x^2 = 9 & \text{Substitute } x^2 \text{ for } u. \\
x = \pm 2 \quad \text{or} \quad x = \pm 3 & \text{Square root property}
\end{array}
$$

(Don't stop here.) ← $u = 4$

The equation $x^4 - 13x^2 + 36 = 0$, a fourth-degree equation, has four solutions, $-3, -2, 2, 3$.* The solution set is abbreviated $\{\pm 2, \pm 3\}$. Each solution can be verified by substituting it into the original equation for x.

NOW TRY ANSWERS
5. (a) $u = x^2$; $u^2 - 10u + 9 = 0$
(b) $u = x + 2$;
$\quad 6u^2 - 11u + 4 = 0$

*In general, an equation in which an nth-degree polynomial equals 0 has n complex solutions, although some of them may be repeated.

**NOW TRY
EXERCISE 6**

Solve each equation.

(a) $x^4 - 17x^2 + 16 = 0$

(b) $x^4 + 4 = 8x^2$

(b)

$$4x^4 + 1 = 5x^2$$

$$4(x^2)^2 + 1 = 5x^2 \qquad x^4 = (x^2)^2$$

$$4u^2 + 1 = 5u \qquad \text{Let } u = x^2.$$

$$4u^2 - 5u + 1 = 0 \qquad \text{Standard form}$$

$$(4u - 1)(u - 1) = 0 \qquad \text{Factor.}$$

$$4u - 1 = 0 \quad \text{or} \quad u - 1 = 0 \qquad \text{Zero-factor property}$$

$$u = \frac{1}{4} \quad \text{or} \qquad u = 1 \qquad \text{Solve.}$$

This is a key step. → $$x^2 = \frac{1}{4} \quad \text{or} \qquad x^2 = 1 \qquad \text{Substitute } x^2 \text{ for } u.$$

$$x = \pm\frac{1}{2} \quad \text{or} \qquad x = \pm 1 \qquad \text{Square root property}$$

Check that the solution set is $\left\{\pm\frac{1}{2}, \pm 1\right\}$.

(c)
$$x^4 = 6x^2 - 3$$

$$x^4 - 6x^2 + 3 = 0 \qquad \text{Standard form}$$

$$(x^2)^2 - 6x^2 + 3 = 0 \qquad x^4 = (x^2)^2$$

$$u^2 - 6u + 3 = 0 \qquad \text{Let } u = x^2.$$

Since this equation cannot be solved by factoring, use the quadratic formula.

$$u = \frac{-(-6) \pm \sqrt{(-6)^2 - 4(1)(3)}}{2(1)} \qquad a = 1, b = -6, c = 3$$

$$u = \frac{6 \pm \sqrt{24}}{2} \qquad \text{Simplify.}$$

$$u = \frac{6 \pm 2\sqrt{6}}{2} \qquad \sqrt{24} = \sqrt{4} \cdot \sqrt{6} = 2\sqrt{6}$$

$$u = \frac{2\left(3 \pm \sqrt{6}\right)}{2} \qquad \text{Factor.}$$

$$u = 3 \pm \sqrt{6} \qquad \text{Lowest terms}$$

Find *both* square roots in each case. → $$x^2 = 3 + \sqrt{6} \quad \text{or} \quad x^2 = 3 - \sqrt{6} \qquad u = x^2$$

$$x = \pm\sqrt{3 + \sqrt{6}} \quad \text{or} \quad x = \pm\sqrt{3 - \sqrt{6}}$$

The solution set $\left\{\pm\sqrt{3 + \sqrt{6}}, \pm\sqrt{3 - \sqrt{6}}\right\}$ contains four numbers.   **NOW TRY**

NOTE Equations like those in **Examples 6(a) and (b)** can be solved by factoring.

$$x^4 - 13x^2 + 36 = 0 \qquad \textbf{Example 6(a) equation}$$

$$(x^2 - 9)(x^2 - 4) = 0 \qquad \textbf{Factor.}$$

$$(x + 3)(x - 3)(x + 2)(x - 2) = 0 \qquad \textbf{Factor again.}$$

Using the zero-factor property gives the same solutions obtained in **Example 6(a).** Equations that cannot be solved by factoring (as in **Example 6(c)**) must be solved by substitution and the quadratic formula.

NOW TRY ANSWERS

6. (a) $\{\pm 1, \pm 4\}$

 (b) $\left\{\pm\sqrt{4 + 2\sqrt{3}}, \pm\sqrt{4 - 2\sqrt{3}}\right\}$

Solving an Equation That Is Quadratic in Form by Substitution

Step 1 **Define a temporary variable u,** based on the relationship between the variable expressions in the given equation. Substitute u in the original equation and rewrite the equation in the form $au^2 + bu + c = 0$.

Step 2 **Solve the quadratic equation** obtained in **Step 1** by factoring or the quadratic formula.

Step 3 **Replace u with the expression it defined in Step 1.**

Step 4 **Solve the resulting equations for the original variable.**

Step 5 **Check** all solutions by substituting them in the original equation.

⌐ *NOW TRY*
⌐ *EXERCISE 7*

Solve each equation.

(a) $6(x - 4)^2 + 11(x - 4) - 10 = 0$

(b) $2x^{2/3} - 7x^{1/3} + 3 = 0$

EXAMPLE 7 Solving Equations That Are Quadratic in Form

Solve each equation.

(a) $2(4x - 3)^2 + 7(4x - 3) + 5 = 0$

Step 1 Because of the repeated quantity $4x - 3$, substitute u for $4x - 3$. (See **Example 5(b).**)

$$2(4x - 3)^2 + 7(4x - 3) + 5 = 0$$
$$2u^2 + 7u + 5 = 0 \qquad \text{Let } u = 4x - 3.$$

Step 2 $\qquad\qquad (2u + 5)(u + 1) = 0 \qquad \text{Factor.}$

$2u + 5 = 0 \quad$ or $\quad u + 1 = 0 \qquad$ Zero-factor property

⟨Don't stop here.⟩➤ $u = -\dfrac{5}{2} \quad$ or $\quad u = -1 \qquad$ Solve for u.

Step 3 $\quad 4x - 3 = -\dfrac{5}{2} \quad$ or $\quad 4x - 3 = -1 \qquad$ Substitute $4x - 3$ for u.

Step 4 $\qquad\quad 4x = \dfrac{1}{2} \quad$ or $\qquad 4x = 2 \qquad$ Solve for x.

$\qquad\qquad\quad x = \dfrac{1}{8} \quad$ or $\qquad\quad x = \dfrac{1}{2}$

Step 5 Check that the solution set of the original equation is $\left\{ \dfrac{1}{8}, \dfrac{1}{2} \right\}$.

(b) $2x^{2/3} - 11x^{1/3} + 12 = 0$

Substitute u for $x^{1/3}$. (See **Example 5(c).**)

$$2u^2 - 11u + 12 = 0 \qquad \text{Let } x^{1/3} = u; x^{2/3} = u^2.$$
$$(2u - 3)(u - 4) = 0 \qquad \text{Factor.}$$

$2u - 3 = 0 \qquad$ or $\quad u - 4 = 0 \qquad$ Zero-factor property

$u = \dfrac{3}{2} \qquad$ or $\qquad u = 4 \qquad$ Solve for u.

$x^{1/3} = \dfrac{3}{2} \qquad$ or $\qquad x^{1/3} = 4 \qquad u = x^{1/3}$

$(x^{1/3})^3 = \left(\dfrac{3}{2}\right)^3 \quad$ or $\quad (x^{1/3})^3 = 4^3 \qquad$ Cube each side.

$x = \dfrac{27}{8} \qquad$ or $\qquad x = 64$

NOW TRY ANSWERS

7. (a) $\left\{ \dfrac{3}{2}, \dfrac{14}{3} \right\}$ **(b)** $\left\{ \dfrac{1}{8}, 27 \right\}$

Check that the solution set is $\left\{ \dfrac{27}{8}, 64 \right\}$.

NOW TRY ↻

⚠ **CAUTION** A common error when solving problems like those in **Examples 6 and 7** is to stop too soon. *Once you have solved for u, remember to substitute and solve for the values of the original variable.*

9.3 EXERCISES

MyMathLab

Concept Check Write a sentence describing the first step you would take to solve each equation. Do not actually solve.

1. $\dfrac{14}{x} = x - 5$

2. $\sqrt{1 + x} + x = 5$

3. $(x^2 + x)^2 - 8(x^2 + x) + 12 = 0$

4. $3x = \sqrt{16 - 10x}$

5. *Concept Check* Study this incorrect "solution." *WHAT WENT WRONG?*

$$x = \sqrt{3x + 4}$$
$$x^2 = 3x + 4 \qquad \text{Square each side.}$$
$$x^2 - 3x - 4 = 0$$
$$(x - 4)(x + 1) = 0$$
$$x - 4 = 0 \quad \text{or} \quad x + 1 = 0$$
$$x = 4 \quad \text{or} \qquad x = -1$$

Solution set: $\{4, -1\}$

6. *Concept Check* Study this incorrect "solution." *WHAT WENT WRONG?*

$$2(x - 1)^2 - 3(x - 1) + 1 = 0$$
$$2u^2 - 3u + 1 = 0 \qquad \text{Let } u = x - 1.$$
$$(2u - 1)(u - 1) = 0$$
$$2u - 1 = 0 \quad \text{or} \quad u - 1 = 0$$
$$u = \dfrac{1}{2} \quad \text{or} \qquad u = 1$$

Solution set: $\left\{\dfrac{1}{2}, 1\right\}$

Solve each equation. Check your solutions. ***See Example 1.***

7. $\dfrac{14}{x} = x - 5$

8. $\dfrac{-12}{x} = x + 8$

9. $1 - \dfrac{3}{x} - \dfrac{28}{x^2} = 0$

10. $4 - \dfrac{7}{r} - \dfrac{2}{r^2} = 0$

11. $3 - \dfrac{1}{t} = \dfrac{2}{t^2}$

12. $1 + \dfrac{2}{x} = \dfrac{3}{x^2}$

🌐 **13.** $\dfrac{1}{x} + \dfrac{2}{x + 2} = \dfrac{17}{35}$

14. $\dfrac{2}{m} + \dfrac{3}{m + 9} = \dfrac{11}{4}$

15. $\dfrac{2}{x + 1} + \dfrac{3}{x + 2} = \dfrac{7}{2}$

16. $\dfrac{4}{3 - p} + \dfrac{2}{5 - p} = \dfrac{26}{15}$

17. $\dfrac{3}{2x} - \dfrac{1}{2(x + 2)} = 1$

18. $\dfrac{4}{3x} - \dfrac{1}{2(x + 1)} = 1$

19. $3 = \dfrac{1}{t + 2} + \dfrac{2}{(t + 2)^2}$

20. $1 + \dfrac{2}{3z + 2} = \dfrac{15}{(3z + 2)^2}$

21. $\dfrac{6}{p} = 2 + \dfrac{p}{p + 1}$

22. $\dfrac{x}{2 - x} + \dfrac{2}{x} = 5$

23. $1 - \dfrac{1}{2x + 1} - \dfrac{1}{(2x + 1)^2} = 0$

24. $1 - \dfrac{1}{3x - 2} - \dfrac{1}{(3x - 2)^2} = 0$

Concept Check Answer each question.

25. A boat goes 20 mph in still water, and the rate of the current is t mph.

    (a) What is the rate of the boat when it travels upstream?

    (b) What is the rate of the boat when it travels downstream?

26. (a) If it takes m hours to grade a set of papers, what is the grader's rate (in job per hour)?

    (b) How much of the job will the grader do in 2 hr?

Solve each problem. ***See Examples 2 and 3.***

27. On a windy day William Kunz found that he could go 16 mi downstream and then 4 mi back upstream at top speed in a total of 48 min. What was the top speed of William's boat if the rate of the current was 15 mph?

	d	r	t
Upstream	4	x − 15	
Downstream	16		

28. Vera Koutsoyannis flew her plane for 6 hr at a constant rate. She traveled 810 mi with the wind, then turned around and traveled 720 mi against the wind. The wind speed was a constant 15 mph. Find the rate of the plane.

	d	r	t
With Wind	810		
Against Wind	720		

29. The distance from Jackson to Lodi is about 40 mi, as is the distance from Lodi to Manteca. Adrian Iorgoni drove from Jackson to Lodi, stopped in Lodi for a high-energy drink, and then drove on to Manteca at 10 mph faster. Driving time for the entire trip was 88 min. Find the rate from Jackson to Lodi. (*Source: State Farm Road Atlas.*)

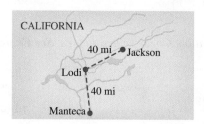

30. Medicine Hat and Cranbrook are 300 km apart. Steve Roig-Watnik rides his Harley 20 km per hr faster than Mohammad Shakil rides his Yamaha. Find Steve's average rate if he travels from Cranbrook to Medicine Hat in $1\frac{1}{4}$ hr less time than Mohammad. (*Source: State Farm Road Atlas.*)

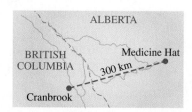

31. Working together, two people can cut a large lawn in 2 hr. One person can do the job alone in 1 hr less time than the other. How long (to the nearest tenth) would it take the faster worker to do the job? (*Hint: x is the time of the faster worker.*)

	Rate	Time Working Together	Fractional Part of the Job Done
Faster Worker	$\frac{1}{x}$	2	
Slower Worker		2	

32. Working together, two people can clean an office building in 5 hr. One person is new to the job and would take 2 hr longer than the other person to clean the building alone. How long (to the nearest tenth) would it take the new worker to clean the building alone?

	Rate	Time Working Together	Fractional Part of the Job Done
Faster Worker			
Slower Worker			

33. Rusty and Nancy Brauner are planting flats of spring flowers. Working alone, Rusty would take 2 hr longer than Nancy to plant the flowers. Working together, they do the job in 12 hr. How long (to the nearest tenth) would it have taken each person working alone?

34. Joel Spring can work through a stack of invoices in 1 hr less time than Noel White can. Working together they take $1\frac{1}{2}$ hr. How long (to the nearest tenth) would it take each person working alone?

35. Two pipes together can fill a tank in 2 hr. One of the pipes, used alone, takes 3 hr longer than the other to fill the tank. How long would each pipe take to fill the tank alone?

36. A washing machine can be filled in 6 min if both the hot and cold water taps are fully opened. Filling the washer with hot water alone takes 9 min longer than filling it with cold water alone. How long does it take to fill the washer with cold water?

Solve each equation. Check your solutions. **See Example 4.**

37. $x = \sqrt{7x - 10}$

38. $z = \sqrt{5z - 4}$

39. $2x = \sqrt{11x + 3}$

40. $4x = \sqrt{6x + 1}$

41. $3x = \sqrt{16 - 10x}$

42. $4t = \sqrt{8t + 3}$

43. $t + \sqrt{t} = 12$

44. $p - 2\sqrt{p} = 8$

45. $x = \sqrt{\dfrac{6 - 13x}{5}}$

46. $r = \sqrt{\dfrac{20 - 19r}{6}}$

47. $-x = \sqrt{\dfrac{8 - 2x}{3}}$

48. $-x = \sqrt{\dfrac{3x + 7}{4}}$

Solve each equation. Check your solutions. **See Examples 5–7.**

49. $x^4 - 29x^2 + 100 = 0$

50. $x^4 - 37x^2 + 36 = 0$

51. $4q^4 - 13q^2 + 9 = 0$

52. $9x^4 - 25x^2 + 16 = 0$

53. $x^4 + 48 = 16x^2$

54. $z^4 + 72 = 17z^2$

55. $(x + 3)^2 + 5(x + 3) + 6 = 0$

56. $(x - 4)^2 + (x - 4) - 20 = 0$

57. $3(m + 4)^2 - 8 = 2(m + 4)$

58. $(t + 5)^2 + 6 = 7(t + 5)$

59. $x^{2/3} + x^{1/3} - 2 = 0$

60. $x^{2/3} - 2x^{1/3} - 3 = 0$

61. $r^{2/3} + r^{1/3} - 12 = 0$

62. $3x^{2/3} - x^{1/3} - 24 = 0$

63. $4x^{4/3} - 13x^{2/3} + 9 = 0$

64. $9t^{4/3} - 25t^{2/3} + 16 = 0$

65. $2 + \dfrac{5}{3x - 1} = \dfrac{-2}{(3x - 1)^2}$

66. $3 - \dfrac{7}{2p + 2} = \dfrac{6}{(2p + 2)^2}$

67. $2 - 6(z - 1)^{-2} = (z - 1)^{-1}$

68. $3 - 2(x - 1)^{-1} = (x - 1)^{-2}$

The equations in Exercises 69–82 are not grouped by type. Solve each equation. Exercises 81 and 82 require knowledge of complex numbers. **See Examples 1 and 4–7.**

69. $12x^4 - 11x^2 + 2 = 0$

70. $\left(x - \dfrac{1}{2}\right)^2 + 5\left(x - \dfrac{1}{2}\right) - 4 = 0$

71. $\sqrt{2x + 3} = 2 + \sqrt{x - 2}$

72. $\sqrt{m + 1} = -1 + \sqrt{2m}$

73. $2\left(1 + \sqrt{r}\right)^2 = 13\left(1 + \sqrt{r}\right) - 6$

74. $(x^2 + x)^2 + 12 = 8(x^2 + x)$

75. $2m^6 + 11m^3 + 5 = 0$

76. $8x^6 + 513x^3 + 64 = 0$

77. $6 = 7(2w - 3)^{-1} + 3(2w - 3)^{-2}$

78. $x^6 - 10x^3 = -9$

79. $2x^4 - 9x^2 = -2$

80. $8x^4 + 1 = 11x^2$

81. $2x^4 + x^2 - 3 = 0$

82. $4x^4 + 5x^2 + 1 = 0$

PREVIEW EXERCISES

Solve each equation for the specified variable. **See Section 2.2.**

83. $P = 2L + 2W$ for W **84.** $A = \dfrac{1}{2}bh$ for h **85.** $F = \dfrac{9}{5}C + 32$ for C

SUMMARY EXERCISES on Solving Quadratic Equations

We have introduced four methods for solving quadratic equations written in standard form $ax^2 + bx + c = 0$.

Method	Advantages	Disadvantages
Factoring	This is usually the fastest method.	Not all polynomials are factorable. Some factorable polynomials are difficult to factor.
Square root property	This is the simplest method for solving equations of the form $(ax + b)^2 = c$.	Few equations are given in this form.
Completing the square	This method can always be used, although most people prefer the quadratic formula.	It requires more steps than other methods.
Quadratic formula	This method can always be used.	Sign errors are common when evaluating $\sqrt{b^2 - 4ac}$.

Concept Check   *Decide whether* factoring, *the* square root property, *or the* quadratic formula *is most appropriate for solving each quadratic equation. Do not actually solve.*

1. $(2x + 3)^2 = 4$ **2.** $4x^2 - 3x = 1$ **3.** $x^2 + 5x - 8 = 0$

4. $2x^2 + 3x = 1$ **5.** $3x^2 = 2 - 5x$ **6.** $x^2 = 5$

Solve each quadratic equation by the method of your choice.

7. $p^2 = 7$ **8.** $6x^2 - x - 15 = 0$ **9.** $n^2 + 6n + 4 = 0$

10. $(x - 3)^2 = 25$ **11.** $\dfrac{5}{x} + \dfrac{12}{x^2} = 2$ **12.** $3x^2 = 3 - 8x$

13. $2r^2 - 4r + 1 = 0$ ***14.** $x^2 = -12$ **15.** $x\sqrt{2} = \sqrt{5x - 2}$

16. $x^4 - 10x^2 + 9 = 0$ **17.** $(2x + 3)^2 = 8$ **18.** $\dfrac{2}{x} + \dfrac{1}{x - 2} = \dfrac{5}{3}$

19. $t^4 + 14 = 9t^2$ **20.** $8x^2 - 4x = 2$ ***21.** $z^2 + z + 1 = 0$

22. $5x^6 + 2x^3 - 7 = 0$ **23.** $4t^2 - 12t + 9 = 0$ **24.** $x\sqrt{3} = \sqrt{2 - x}$

25. $r^2 - 72 = 0$ **26.** $-3x^2 + 4x = -4$ **27.** $x^2 - 5x - 36 = 0$

28. $w^2 = 169$ ***29.** $3p^2 = 6p - 4$ **30.** $z = \sqrt{\dfrac{5z + 3}{2}}$

*31. $\dfrac{4}{r^2} + 3 = \dfrac{1}{r}$ **32.** $2(3x - 1)^2 + 5(3x - 1) = -2$

*This exercise requires knowledge of complex numbers.

9.4 Formulas and Further Applications

OBJECTIVE 1 Solve formulas for variables involving squares and square roots.

EXAMPLE 1 Solving for Variables Involving Squares or Square Roots

Solve each formula for the given variable. Keep $\pm$ in the answer in part (a).

(a) $w = \dfrac{kFr}{v^2}$ for v

$$w = \frac{kFr}{v^2}$$ *The goal is to isolate v on one side.*

$$v^2 w = kFr \qquad \text{Multiply by } v^2.$$

$$v^2 = \frac{kFr}{w} \qquad \text{Divide by } w.$$

$$v = \pm\sqrt{\frac{kFr}{w}} \qquad \text{Square root property}$$

$$v = \frac{\pm\sqrt{kFr}}{\sqrt{w}} \cdot \frac{\sqrt{w}}{\sqrt{w}} \qquad \text{Rationalize the denominator.}$$

$$v = \frac{\pm\sqrt{kFrw}}{w} \qquad \begin{array}{l}\sqrt{a} \cdot \sqrt{b} = \sqrt{ab};\\ \sqrt{a} \cdot \sqrt{a} = a\end{array}$$

(b) $d = \sqrt{\dfrac{4\mathscr{A}}{\pi}}$ for $\mathscr{A}$ *The goal is to isolate $\mathscr{A}$ on one side.*

$$d = \sqrt{\frac{4\mathscr{A}}{\pi}}$$

$$d^2 = \frac{4\mathscr{A}}{\pi} \qquad \text{Square both sides.}$$

$$\pi d^2 = 4\mathscr{A} \qquad \text{Multiply by } \pi.$$

$$\frac{\pi d^2}{4} = \mathscr{A}, \quad \text{or} \quad \mathscr{A} = \frac{\pi d^2}{4} \qquad \text{Divide by 4.} \qquad \text{NOW TRY}$$

NOTE In formulas like $v = \dfrac{\pm\sqrt{kFrw}}{w}$ in **Example 1(a)**, we include both positive and negative values.

**NOW TRY
EXERCISE 1**

Solve each formula for the given variable. Keep $\pm$ in the answer in part (a).

(a) $n = \dfrac{ab}{E^2}$ for E

(b) $S = \sqrt{\dfrac{pq}{n}}$ for p

EXAMPLE 2 Solving for a Variable That Appears in First- and Second-Degree Terms

Solve $s = 2t^2 + kt$ for t.

Since the given equation has terms with t^2 and t, write it in standard form $ax^2 + bx + c = 0$, with t as the variable instead of x.

$$s = 2t^2 + kt$$

$$0 = 2t^2 + kt - s \qquad \text{Subtract } s.$$

$$2t^2 + kt - s = 0 \qquad \text{Standard form}$$

NOW TRY ANSWERS

1. (a) $E = \dfrac{\pm\sqrt{abn}}{n}$

(b) $p = \dfrac{nS^2}{q}$

NOW TRY
EXERCISE 2
Solve for r.
$$r^2 + 9r = -c$$

To solve $2t^2 + kt - s = 0$, use the quadratic formula with $a = 2$, $b = k$, and $c = -s$.

$$t = \frac{-k \pm \sqrt{k^2 - 4(2)(-s)}}{2(2)} \qquad \text{Substitute.}$$

$$t = \frac{-k \pm \sqrt{k^2 + 8s}}{4} \qquad \text{Solve for } t.$$

The solutions are $t = \dfrac{-k + \sqrt{k^2 + 8s}}{4}$ and $t = \dfrac{-k - \sqrt{k^2 + 8s}}{4}$. **NOW TRY**

OBJECTIVE 2 Solve applied problems using the Pythagorean theorem. The Pythagorean theorem, represented by the equation

$$a^2 + b^2 = c^2,$$

is illustrated in **FIGURE 2** and was introduced in **Section 8.3.** It is used to solve applications involving right triangles.

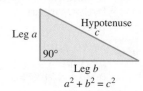

$$a^2 + b^2 = c^2$$
Pythagorean theorem
FIGURE 2

NOW TRY
EXERCISE 3
Matt Porter is building a new barn, with length 10 ft more than width. While determining the footprint of the barn, he measured the diagonal as 50 ft. What will be the dimensions of the barn?

EXAMPLE 3 Using the Pythagorean Theorem

Two cars left an intersection at the same time, one heading due north, the other due west. Some time later, they were exactly 100 mi apart. The car headed north had gone 20 mi farther than the car headed west. How far had each car traveled?

Step 1 **Read** the problem carefully.

Step 2 **Assign a variable.**

Let $x =$ the distance traveled by the car headed west.

Then $x + 20 =$ the distance traveled by the car headed north.

See **FIGURE 3**. The cars are 100 mi apart, so the hypotenuse of the right triangle equals 100.

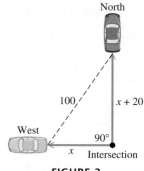

FIGURE 3

Step 3 **Write an equation.** Use the Pythagorean theorem.

$$a^2 + b^2 = c^2$$

$$x^2 + (x + 20)^2 = 100^2$$

$(x + y)^2 = x^2 + 2xy + y^2$

Step 4 **Solve.** $\quad x^2 + x^2 + 40x + 400 = 10{,}000 \qquad$ Square the binomial.

$$2x^2 + 40x - 9600 = 0 \qquad \text{Standard form}$$

$$x^2 + 20x - 4800 = 0 \qquad \text{Divide by 2.}$$

$$(x + 80)(x - 60) = 0 \qquad \text{Factor.}$$

$$x + 80 = 0 \quad \text{or} \quad x - 60 = 0 \qquad \text{Zero-factor property}$$

$$x = -80 \quad \text{or} \qquad x = 60 \qquad \text{Solve for } x.$$

NOW TRY ANSWERS
2. $r = \dfrac{-9 \pm \sqrt{81 - 4c}}{2}$
3. 30 ft by 40 ft

Step 5 **State the answer.** Since distance cannot be negative, discard the negative solution. The required distances are 60 mi and $60 + 20 = 80$ mi.

Step 6 **Check.** Since $60^2 + 80^2 = 100^2$, the answer is correct. **NOW TRY**

OBJECTIVE 3 Solve applied problems using area formulas.

NOW TRY
EXERCISE 4

A football practice field is 30 yd wide and 40 yd long. A strip of grass sod of uniform width is to be placed around the perimeter of the practice field. There is enough money budgeted for 296 sq yd of sod. How wide will the strip be?

EXAMPLE 4 Solving an Area Problem

A rectangular reflecting pool in a park is 20 ft wide and 30 ft long. The gardener wants to plant a strip of grass of uniform width around the edge of the pool. She has enough seed to cover 336 ft². How wide will the strip be?

FIGURE 4

Step 1 **Read** the problem carefully.

Step 2 **Assign a variable.** The pool is shown in **FIGURE 4**.

Let $x =$ the unknown width of the grass strip.

Then $20 + 2x =$ the width of the large rectangle (the width of the pool plus two grass strips),

and $30 + 2x =$ the length of the large rectangle.

Step 3 **Write an equation.** Refer to **FIGURE 4**.

$(30 + 2x)(20 + 2x)$ Area of large rectangle (length · width)

$30 \cdot 20,$ or 600 Area of pool (in square feet)

The area of the large rectangle minus the area of the pool should equal 336 ft², the area of the grass strip.

$$
\begin{array}{ccc}
\text{Area} & \text{Area} & \text{Area} \\
\text{of large} & - & \text{of} & = & \text{of} \\
\text{rectangle} & & \text{pool} & & \text{grass} \\
\downarrow & & \downarrow & & \downarrow
\end{array}
$$

$$(30 + 2x)(20 + 2x) - 600 = 336$$

Step 4 **Solve.**

$$600 + 100x + 4x^2 - 600 = 336 \qquad \text{Multiply.}$$
$$4x^2 + 100x - 336 = 0 \qquad \text{Standard form}$$
$$x^2 + 25x - 84 = 0 \qquad \text{Divide by 4.}$$
$$(x + 28)(x - 3) = 0 \qquad \text{Factor.}$$
$$x + 28 = 0 \quad \text{or} \quad x - 3 = 0 \qquad \text{Zero-factor property}$$
$$x = -28 \quad \text{or} \quad x = 3 \qquad \text{Solve for } x.$$

Step 5 **State the answer.** The width cannot be -28 ft, so the grass strip should be 3 ft wide.

Step 6 **Check.** If $x = 3$, we can find the area of the large rectangle (which includes the grass strip).

$$(30 + 2 \cdot 3)(20 + 2 \cdot 3) = 36 \cdot 26 = 936 \text{ ft}^2 \qquad \text{Area of pool and strip}$$

The area of the pool is $30 \cdot 20 = 600$ ft². So, the area of the grass strip is $936 - 600 = 336$ ft², as required. The answer is correct.   *NOW TRY*

OBJECTIVE 4 Solve applied problems using quadratic functions as models.

Some applied problems can be modeled by *quadratic functions,* which for real numbers a, b, and c, can be written in the form

$$f(x) = ax^2 + bx + c, \quad \text{with } a \neq 0.$$

NOW TRY
EXERCISE 5

If an object is projected upward from the top of a 120-ft building at 60 ft per sec, its position (in feet above the ground) is given by

$$s(t) = -16t^2 + 60t + 120,$$

where t is time in seconds after it was projected. When does it hit the ground (to the nearest tenth)?

NOW TRY
EXERCISE 6

Refer to **Example 6.**

(a) Use the model to approximate the CPI for 2005, to the nearest whole number.

(b) In what year did the CPI reach 500? (Round down for the year.)

NOW TRY ANSWERS
5. 5.2 sec after it is projected
6. (a) 578 **(b)** 1998

EXAMPLE 5 Solving an Applied Problem Using a Quadratic Function

If an object is projected upward from the top of a 144-ft building at 112 ft per sec, its position (in feet above the ground) is given by

$$s(t) = -16t^2 + 112t + 144,$$

where t is time in seconds after it was projected. When does it hit the ground?

When the object hits the ground, its distance above the ground is 0. We must find the value of t that makes $s(t) = 0$.

$0 = -16t^2 + 112t + 144$	Let $s(t) = 0$.
$0 = t^2 - 7t - 9$	Divide by -16.
$t = \dfrac{-(-7) \pm \sqrt{(-7)^2 - 4(1)(-9)}}{2(1)}$	Substitute into the quadratic formula.
$t = \dfrac{7 \pm \sqrt{85}}{2} \approx \dfrac{7 \pm 9.2}{2}$	Use a calculator.

The solutions are $t \approx 8.1$ or $t \approx -1.1$. Time cannot be negative, so we discard the negative solution. The object hits the ground about 8.1 sec after it is projected. **NOW TRY**

EXAMPLE 6 Using a Quadratic Function to Model the CPI

The Consumer Price Index (CPI) is used to measure trends in prices for a "basket" of goods purchased by typical American families. This index uses a base year of 1967, which means that the index number for 1967 is 100. The quadratic function defined by

$$f(x) = -0.065x^2 + 14.8x + 249$$

approximates the CPI for the years 1980–2005, where x is the number of years that have elapsed since 1980. (*Source:* Bureau of Labor Statistics.)

(a) Use the model to approximate the CPI for 1995.
For 1995, $x = 1995 - 1980 = 15$, so find $f(15)$.

$f(x) = -0.065x^2 + 14.8x + 249$	Given model
$f(15) = -0.065(15)^2 + 14.8(15) + 249$	Let $x = 15$.
$f(15) \approx 456$	Nearest whole number

The CPI for 1995 was about 456.

(b) In what year did the CPI reach 550?
Find the value of x that makes $f(x) = 550$.

$f(x) = -0.065x^2 + 14.8x + 249$	Given model
$550 = -0.065x^2 + 14.8x + 249$	Let $f(x) = 550$.
$0 = -0.065x^2 + 14.8x - 301$	Standard form
$x = \dfrac{-14.8 \pm \sqrt{14.8^2 - 4(-0.065)(-301)}}{2(-0.065)}$	Use $a = -0.065$, $b = 14.8$, and $c = -301$ in the quadratic formula.
$x \approx 22.6$ or $x \approx 205.1$	

Rounding the first solution 22.6 down, the CPI first reached 550 in $1980 + 22 = 2002$. (Reject the solution $x \approx 205.1$, as this corresponds to a year far beyond the period covered by the model.) **NOW TRY**

9.4 EXERCISES

Complete solution available on the Video Resources on DVD

Concept Check *Answer each question in Exercises 1–4.*

1. In solving a formula that has the specified variable in the denominator, what is the first step?

2. What is the first step in solving a formula like $gw^2 = 2r$ for w?

3. What is the first step in solving a formula like $gw^2 = kw + 24$ for w?

4. Why is it particularly important to check all proposed solutions to an applied problem against the information in the original problem?

In Exercises 5 and 6, solve for m in terms of the other variables $(m > 0)$.

5.

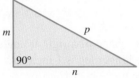

6.

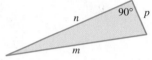

Solve each equation for the indicated variable. (Leave ± in your answers.) **See Examples 1 and 2.**

7. $d = kt^2$ for t **8.** $S = 6e^2$ for e **9.** $I = \dfrac{ks}{d^2}$ for d

10. $R = \dfrac{k}{d^2}$ for d **11.** $F = \dfrac{kA}{v^2}$ for v **12.** $L = \dfrac{kd^4}{h^2}$ for h

13. $V = \dfrac{1}{3}\pi r^2 h$ for r **14.** $V = \pi(r^2 + R^2)h$ for r **15.** $At^2 + Bt = -C$ for t

16. $S = 2\pi rh + \pi r^2$ for r **17.** $D = \sqrt{kh}$ for h **18.** $F = \dfrac{k}{\sqrt{d}}$ for d

19. $p = \sqrt{\dfrac{k\ell}{g}}$ for ℓ **20.** $p = \sqrt{\dfrac{k\ell}{g}}$ for g

21. $S = 4\pi r^2$ for r **22.** $s = kwd^2$ for d

Brain Busters *Solve each equation for the indicated variable. (Leave ± in your answers.)*

23. $p = \dfrac{E^2 R}{(r + R)^2}$ for R $(E > 0)$ **24.** $S(6S - t) = t^2$ for S

25. $10p^2 c^2 + 7pcr = 12r^2$ for r **26.** $S = vt + \dfrac{1}{2}gt^2$ for t

27. $LI^2 + RI + \dfrac{1}{c} = 0$ for I **28.** $P = EI - RI^2$ for I

Solve each problem. When appropriate, round answers to the nearest tenth. **See Example 3.**

29. Find the lengths of the sides of the triangle.

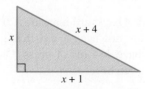

30. Find the lengths of the sides of the triangle.

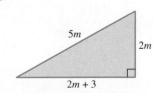

31. Two ships leave port at the same time, one heading due south and the other heading due east. Several hours later, they are 170 mi apart. If the ship traveling south traveled 70 mi farther than the other ship, how many miles did they each travel?

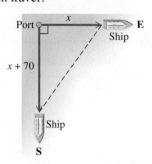

32. Deborah Israel is flying a kite that is 30 ft farther above her hand than its horizontal distance from her. The string from her hand to the kite is 150 ft long. How high is the kite?

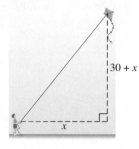

33. A game board is in the shape of a right triangle. The hypotenuse is 2 inches longer than the longer leg, and the longer leg is 1 inch less than twice as long as the shorter leg. How long is each side of the game board?

34. Manuel Bovi is planting a vegetable garden in the shape of a right triangle. The longer leg is 3 ft longer than the shorter leg, and the hypotenuse is 3 ft longer than the longer leg. Find the lengths of the three sides of the garden.

35. The diagonal of a rectangular rug measures 26 ft, and the length is 4 ft more than twice the width. Find the length and width of the rug.

36. A 13-ft ladder is leaning against a house. The distance from the bottom of the ladder to the house is 7 ft less than the distance from the top of the ladder to the ground. How far is the bottom of the ladder from the house?

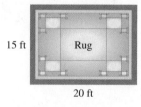

*Solve each problem. **See Example 4.***

37. A club swimming pool is 30 ft wide and 40 ft long. The club members want an exposed aggregate border in a strip of uniform width around the pool. They have enough material for 296 ft². How wide can the strip be?

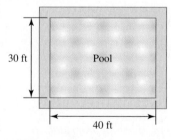

38. Lyudmila Slavina wants to buy a rug for a room that is 20 ft long and 15 ft wide. She wants to leave an even strip of flooring uncovered around the edges of the room. How wide a strip will she have if she buys a rug with an area of 234 ft²?

39. A rectangle has a length 2 m less than twice its width. When 5 m are added to the width, the resulting figure is a square with an area of 144 m². Find the dimensions of the original rectangle.

40. Mariana Coanda's backyard measures 20 m by 30 m. She wants to put a flower garden in the middle of the yard, leaving a strip of grass of uniform width around the flower garden. Mariana must have 184 m² of grass. Under these conditions, what will the length and width of the garden be?

41. A rectangular piece of sheet metal has a length that is 4 in. less than twice the width. A square piece 2 in. on a side is cut from each corner. The sides are then turned up to form an uncovered box of volume 256 in.^{3}. Find the length and width of the original piece of metal.

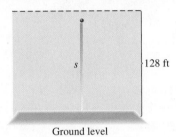

42. Another rectangular piece of sheet metal is 2 in. longer than it is wide. A square piece 3 in. on a side is cut from each corner. The sides are then turned up to form an uncovered box of volume 765 in.^{3}. Find the dimensions of the original piece of metal.

Solve each problem. When appropriate, round answers to the nearest tenth. **See Example 5.**

43. An object is projected directly upward from the ground. After t seconds its distance in feet above the ground is

$$s(t) = 144t - 16t^2.$$

After how many seconds will the object be 128 ft above the ground? (*Hint:* Look for a common factor before solving the equation.)

44. When does the object in **Exercise 43** strike the ground?

45. A ball is projected upward from the ground. Its distance in feet from the ground in t seconds is given by

$$s(t) = -16t^2 + 128t.$$

At what times will the ball be 213 ft from the ground?

46. A toy rocket is launched from ground level. Its distance in feet from the ground in t seconds is given by

$$s(t) = -16t^2 + 208t.$$

At what times will the rocket be 550 ft from the ground?

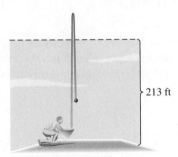

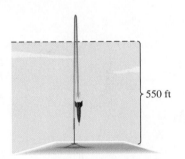

47. The function defined by

$$D(t) = 13t^2 - 100t$$

gives the distance in feet a car going approximately 68 mph will skid in t seconds. Find the time it would take for the car to skid 180 ft.

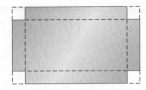

48. The function given in **Exercise 47** becomes $D(t) = 13t^2 - 73t$ for a car going 50 mph. Find the time it takes for this car to skid 218 ft.

A ball is projected upward from ground level, and its distance in feet from the ground in t seconds is given by $s(t) = -16t^2 + 160t$.

49. After how many seconds does the ball reach a height of 400 ft? How would you describe in words its position at this height?

50. After how many seconds does the ball reach a height of 425 ft? How would you interpret the mathematical result here?

Solve each problem using a quadratic equation.

51. A certain bakery has found that the daily demand for blueberry muffins is $\frac{3200}{p}$, where p is the price of a muffin in cents. The daily supply is $3p - 200$. Find the price at which supply and demand are equal.

52. In one area the demand for compact discs is $\frac{700}{P}$ per day, where P is the price in dollars per disc. The supply is $5P - 1$ per day. At what price, to the nearest cent, does supply equal demand?

53. The formula $A = P(1 + r)^2$ gives the amount A in dollars that P dollars will grow to in 2 yr at interest rate r (where r is given as a decimal), using compound interest. What interest rate will cause $2000 to grow to $2142.45 in 2 yr?

54. Use the formula $A = P(1 + r)^2$ to find the interest rate r at which a principal P of $10,000 will increase to $10,920.25 in 2 yr.

William Froude was a 19th century naval architect who used the expression

$$\frac{v^2}{g\ell}$$

in shipbuilding. This expression, known as the **Froude number,** *was also used by R. McNeill Alexander in his research on dinosaurs. (Source: "How Dinosaurs Ran,"* Scientific American, *April 1991.) In Exercises 55 and 56, find the value of* v *(in meters per second), given* $g = 9.8$ m per sec^2. *(Round to the nearest tenth.)*

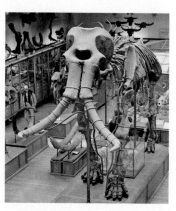

55. Rhinoceros: $\ell = 1.2$;
     Froude number = 2.57

56. Triceratops: $\ell = 2.8$;
     Froude number = 0.16

Recall that corresponding sides of similar triangles are proportional. Use this fact to find the lengths of the indicated sides of each pair of similar triangles. Check all possible solutions in both triangles. Sides of a triangle cannot be negative (and are not drawn to scale here).

57. Side AC

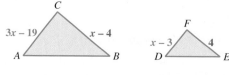

58. Side RQ

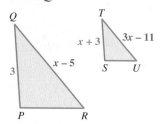

Total spending (in billions of dollars) in the United States from all sources on physician and clinical services for the years 2000–2007 are shown in the bar graph on the next page and can be modeled by the quadratic function defined by

$$f(x) = 0.3214x^2 + 25.06x + 288.2.$$

Here, $x = 0$ represents 2000, $x = 1$ represents 2001, and so on. Use the graph and the model to work Exercises 59–62. **See Example 6.**

Spending on Physician and Clinical Services

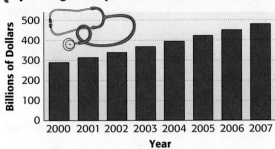

Source: U.S. Centers for Medicare and Medicaid Services.

59. (a) Use the graph to estimate spending on physician and clinical services in 2005 to the nearest $10 billion.

(b) Use the model to approximate spending to the nearest $10 billion. How does this result compare to your estimate in part (a)?

60. Based on the model, in what year did spending on physician and clinical services first exceed $350 billion? (Round down for the year.) How does this result compare to the amount of spending shown in the graph?

61. Based on the model, in what year did spending on physician and clinical services first exceed $400 billion? (Round down for the year.) How does this result compare to the amount of spending shown in the graph?

62. If these data were modeled by a *linear* function defined by $f(x) = ax + b$, would the value of a be positive or negative? Explain.

PREVIEW EXERCISES

*Find each function value. **See Section 3.6.***

63. $f(x) = x^2 + 4x - 3$. Find $f(2)$.   **64.** $f(x) = 2(x - 3)^2 + 5$. Find $f(3)$.

65. Graph $f(x) = 2x^2$. Give the domain and range. **See Section 5.3.**

9.5 Graphs of Quadratic Functions

OBJECTIVES

1 Graph a quadratic function.

2 Graph parabolas with horizontal and vertical shifts.

3 Use the coefficient of x^2 to predict the shape and direction in which a parabola opens.

4 Find a quadratic function to model data.

OBJECTIVE 1 **Graph a quadratic function.** **FIGURE 5** gives a graph of the simplest *quadratic function,* defined by $y = x^2$. This graph is called a **parabola.** (See Section 5.3.) The point $(0, 0)$, the lowest point on the curve, is the **vertex** of this parabola. The vertical line through the vertex is the **axis** of the parabola, here $x = 0$. A parabola is **symmetric about its axis**—if the graph were folded along the axis, the two portions of the curve would coincide.

As **FIGURE 5** suggests, x can be any real number, so the domain of the function defined by $y = x^2$ is $(-\infty, \infty)$. Since y is always non-negative, the range is $[0, \infty)$.

x	y
-2	4
-1	1
0	0
1	1
2	4

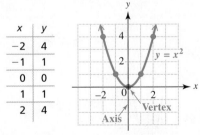

FIGURE 5

> **Quadratic Function**
>
> A function that can be written in the form
>
> $$f(x) = ax^2 + bx + c$$
>
> for real numbers a, b, and c, with $a \neq 0$, is a **quadratic function.**

The graph of any quadratic function is a parabola with a vertical axis.

NOTE We use the variable y and function notation $f(x)$ interchangeably. Although we use the letter f most often to name quadratic functions, other letters can be used. We use the capital letter F to distinguish between different parabolas graphed on the same coordinate axes.

Parabolas have a special reflecting property that makes them useful in the design of telescopes, radar equipment, solar furnaces, and automobile headlights. (See the figure.)

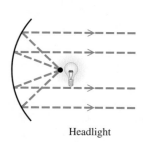

Headlight

OBJECTIVE 2 **Graph parabolas with horizontal and vertical shifts.** Parabolas need not have their vertices at the origin, as does the graph of $f(x) = x^2$.

Graph $f(x) = x^2 - 3$. Give the vertex, axis, domain, and range.

EXAMPLE 1 Graphing a Parabola (Vertical Shift)

Graph $F(x) = x^2 - 2$.

The graph of $F(x) = x^2 - 2$ has the same shape as that of $f(x) = x^2$ but is *shifted*, or *translated*, 2 units down, with vertex $(0, -2)$. Every function value is 2 less than the corresponding function value of $f(x) = x^2$. Plotting points on both sides of the vertex gives the graph in **FIGURE 6**.

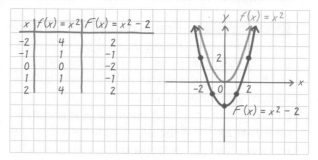

$F(x) = x^2 - 2$
Vertex: $(0, -2)$
Axis: $x = 0$
Domain: $(-\infty, \infty)$
Range: $[-2, \infty)$
The graph of $f(x) = x^2$
is shown for comparison.

FIGURE 6

This parabola is symmetric about its axis $x = 0$, so the plotted points are "mirror images" of each other. Since x can be any real number, the domain is still $(-\infty, \infty)$. The value of y (or $F(x)$) is always greater than or equal to -2, so the range is $[-2, \infty)$.

NOW TRY

NOW TRY ANSWER

1.

$f(x) = x^2 - 3$
vertex: $(0, -3)$; axis: $x = 0$;
domain: $(-\infty, \infty)$; range: $[-3, \infty)$

Vertical Shift

The graph of $F(x) = x^2 + k$ is a parabola.

- The graph has the same shape as the graph of $f(x) = x^2$.
- The parabola is shifted k units up if $k > 0$, and $|k|$ units down if $k < 0$.
- The vertex of the parabola is $(0, k)$.

NOW TRY
EXERCISE 2

Graph $f(x) = (x + 1)^2$. Give the vertex, axis, domain, and range.

EXAMPLE 2   Graphing a Parabola (Horizontal Shift)

Graph $F(x) = (x - 2)^2$.

If $x = 2$, then $F(x) = 0$, giving the vertex $(2, 0)$. The graph of $F(x) = (x - 2)^2$ has the same shape as that of $f(x) = x^2$ but is shifted 2 units to the right. Plotting points on one side of the vertex, and using symmetry about the axis $x = 2$ to find corresponding points on the other side, gives the graph in **FIGURE 7**.

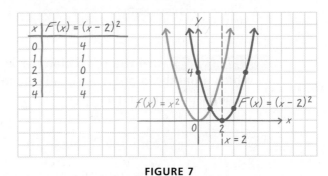

$F(x) = (x - 2)^2$
Vertex: $(2, 0)$
Axis: $x = 2$
Domain: $(-\infty, \infty)$
Range: $[0, \infty)$

FIGURE 7   NOW TRY

Horizontal Shift

The graph of $F(x) = (x - h)^2$ is a parabola.

- The graph has the same shape as the graph of $f(x) = x^2$.
- The parabola is shifted h units to the right if $h > 0$, and $|h|$ units to the left if $h < 0$.
- The vertex of the parabola is $(h, 0)$.

⚠ **CAUTION**   *Errors frequently occur when horizontal shifts are involved.* To determine the direction and magnitude of a horizontal shift, find the value that causes the expression $x - h$ to equal 0, as shown below.

$$F(x) = (x - 5)^2 \qquad\qquad\qquad F(x) = (x + 5)^2$$

Shift the graph of $F(x)$ **5 units to the right,** because $+5$ causes $x - 5$ to equal 0. | Shift the graph of $F(x)$ **5 units to the left,** because -5 causes $x + 5$ to equal 0.

NOW TRY ANSWER

2.

vertex: $(-1, 0)$; axis: $x = -1$;
domain: $(-\infty, \infty)$; range: $[0, \infty)$

EXAMPLE 3   Graphing a Parabola (Horizontal and Vertical Shifts)

Graph $F(x) = (x + 3)^2 - 2$.

This graph has the same shape as that of $f(x) = x^2$, but is shifted 3 units to the left (since $x + 3 = 0$ if $x = -3$) and 2 units down (because of the -2). See **FIGURE 8** on the next page.

**NOW TRY
EXERCISE 3**
Graph $f(x) = (x + 1)^2 - 2$.
Give the vertex, axis, domain,
and range.

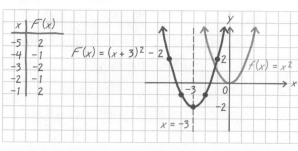

FIGURE 8

$F(x) = (x + 3)^2 - 2$
Vertex: $(-3, -2)$
Axis: $x = -3$
Domain: $(-\infty, \infty)$
Range: $[-2, \infty)$

NOW TRY

Vertex and Axis of a Parabola

The graph of $F(x) = (x - h)^2 + k$ is a parabola.

- The graph has the same shape as the graph of $f(x) = x^2$.
- The vertex of the parabola is (h, k).
- The axis is the vertical line $x = h$.

OBJECTIVE 3 **Use the coefficient of x^2 to predict the shape and direction in which a parabola opens.** Not all parabolas open up, and not all parabolas have the same shape as the graph of $f(x) = x^2$.

**NOW TRY
EXERCISE 4**
Graph $f(x) = -3x^2$. Give
the vertex, axis, domain,
and range.

EXAMPLE 4 **Graphing a Parabola That Opens Down**

Graph $f(x) = -\frac{1}{2}x^2$.

This parabola is shown in **FIGURE 9**. The coefficient $-\frac{1}{2}$ affects the shape of the graph—the $\frac{1}{2}$ makes the parabola wider $\left(\text{since the values of } \frac{1}{2}x^2 \text{ increase more slowly than those of } x^2\right)$, and the negative sign makes the parabola open down. The graph is not shifted in any direction. Unlike the parabolas graphed in **Examples 1–3**, the vertex here has the *greatest* function value of any point on the graph.

x	$f(x)$
-2	-2
-1	$-\frac{1}{2}$
0	0
1	$-\frac{1}{2}$
2	-2

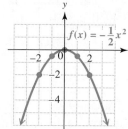

$f(x) = -\frac{1}{2}x^2$
Vertex: $(0, 0)$
Axis: $x = 0$
Domain: $(-\infty, \infty)$
Range: $(-\infty, 0]$

FIGURE 9

NOW TRY

General Characteristics of $F(x) = a(x - h)^2 + k$ $\quad(a \neq 0)$

1. The graph of the quadratic function defined by
$$F(x) = a(x - h)^2 + k, \quad \text{with } a \neq 0,$$
is a parabola with vertex (h, k) and the vertical line $x = h$ as axis.

2. The graph opens up if a is positive and down if a is negative.

3. The graph is wider than that of $f(x) = x^2$ if $0 < |a| < 1$.
The graph is narrower than that of $f(x) = x^2$ if $|a| > 1$.

NOW TRY ANSWERS

3. $f(x) = (x + 1)^2 - 2$

vertex: $(-1, -2)$; axis: $x = -1$;
domain: $(-\infty, \infty)$; range: $[-2, \infty)$

4.

vertex: $(0, 0)$; axis: $x = 0$;
domain: $(-\infty, \infty)$; range: $(-\infty, 0]$

**NOW TRY
EXERCISE 5**

Graph $f(x) = 2(x - 1)^2 + 2$.

EXAMPLE 5 Using the General Characteristics to Graph a Parabola

Graph $F(x) = -2(x + 3)^2 + 4$.

The parabola opens down (because $a < 0$) and is narrower than the graph of $f(x) = x^2$, since $|-2| = 2$ and $2 > 1$. This causes values of $F(x)$ to decrease more quickly than those of $f(x) = -x^2$. This parabola has vertex $(-3, 4)$, as shown in **FIGURE 10**. To complete the graph, we plotted the ordered pairs $(-4, 2)$ and, by symmetry, $(-2, 2)$. Symmetry can be used to find additional ordered pairs that satisfy the equation.

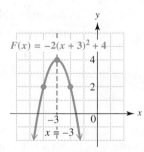

$F(x) = -2(x + 3)^2 + 4$
Vertex: $(-3, 4)$
Axis: $x = -3$
Domain: $(-\infty, \infty)$
Range: $(-\infty, 4]$

FIGURE 10

NOW TRY

OBJECTIVE 4 Find a quadratic function to model data.

EXAMPLE 6 Modeling the Number of Multiple Births

The number of higher-order multiple births (triplets or more) in the United States has declined in recent years, as shown by the data in the table. Here, x represents the number of years since 1995 and y represents the number of higher-order multiple births.

Year	x	y
1995	0	4973
1996	1	5939
1997	2	6737
1999	4	7321
2001	6	7471
2003	8	7663
2004	9	7275
2005	10	6694

Source: National Center for Health Statistics.

NOW TRY ANSWER

5.

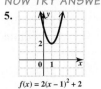

$f(x) = 2(x - 1)^2 + 2$

Find a quadratic function that models the data.

A scatter diagram of the ordered pairs (x, y) is shown in **FIGURE 11** on the next page. The general shape suggested by the scatter diagram indicates that a parabola should approximate these points, as shown by the dashed curve in **FIGURE 12**. The equation for such a parabola would have a negative coefficient for x^2 since the graph opens down.

NOW TRY
EXERCISE 6

Using the points $(0, 4973)$, $(4, 7321)$, and $(8, 7663)$, find another quadratic model for the data on higher-order multiple births in **Example 6.**

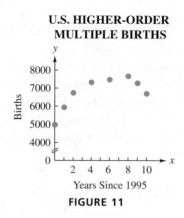

U.S. HIGHER-ORDER MULTIPLE BIRTHS

Births
Years Since 1995

FIGURE 11

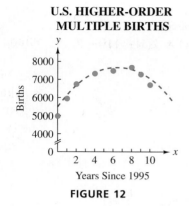

U.S. HIGHER-ORDER MULTIPLE BIRTHS

Births
Years Since 1995

FIGURE 12

To find a quadratic function of the form

$$y = ax^2 + bx + c$$

that models, or *fits*, these data, we choose three representative ordered pairs and use them to write a system of three equations. Using

$$(0, 4973), \quad (4, 7321), \quad \text{and} \quad (10, 6694),$$

we substitute the x- and y-values from the ordered pairs into the quadratic form $y = ax^2 + bx + c$ to get three equations.

$$a(0)^2 + b(0) + c = 4973 \qquad \text{or} \qquad c = 4973 \qquad (1)$$
$$a(4)^2 + b(4) + c = 7321 \qquad \text{or} \qquad 16a + 4b + c = 7321 \qquad (2)$$
$$a(10)^2 + b(10) + c = 6694 \qquad \text{or} \qquad 100a + 10b + c = 6694 \qquad (3)$$

We can find the values of a, b, and c by solving this system of three equations in three variables using the methods of **Section 4.2.** From equation (1), $c = 4973$. Substitute 4973 for c in equations (2) and (3) to obtain two equations.

$$16a + 4b + 4973 = 7321, \qquad \text{or} \qquad 16a + 4b = 2348 \qquad (4)$$
$$100a + 10b + 4973 = 6694, \qquad \text{or} \qquad 100a + 10b = 1721 \qquad (5)$$

We can eliminate b from this system of equations in two variables by multiplying equation (4) by -5 and equation (5) by 2, and adding the results.

$$120a = -8298$$
$$a = -69.15 \qquad \text{Divide by 120. Use a calculator.}$$

We substitute -69.15 for a in equation (4) or (5) to find that $b = 863.6$. Using the values we have found for a, b, and c, our model is defined by

$$y = -69.15x^2 + 863.6x + 4973. \qquad \text{NOW TRY}$$

NOTE In **Example 6,** if we had chosen three different ordered pairs of data, a slightly different model would result. The *quadratic regression* feature on a graphing calculator can also be used to generate the quadratic model that best fits given data. See your owner's manual for details.

NOW TRY ANSWER

6. $y = -62.69x^2 + 837.75x + 4973$

9.5 EXERCISES

 MyMathLab
Math XL
PRACTICE   WATCH   DOWNLOAD   READ   REVIEW

 Complete solution available on the Video Resources on DVD

1. *Concept Check*   Match each quadratic function with its graph from choices A–D.

(a) $f(x) = (x + 2)^2 - 1$ **A.** 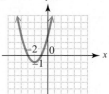   **B.**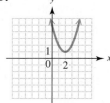

(b) $f(x) = (x + 2)^2 + 1$

(c) $f(x) = (x - 2)^2 - 1$ **C.**   **D.**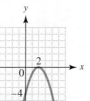

(d) $f(x) = (x - 2)^2 + 1$

2. *Concept Check*   Match each quadratic function with its graph from choices A–D.

(a) $f(x) = -x^2 + 2$ **A.** 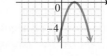   **B.**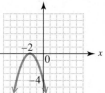

(b) $f(x) = -x^2 - 2$

(c) $f(x) = -(x + 2)^2$ **C.** 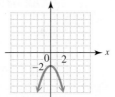   **D.**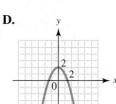

(d) $f(x) = -(x - 2)^2$

Identify the vertex of each parabola. ***See Examples 1–4.***

3. $f(x) = -3x^2$ **4.** $f(x) = \dfrac{1}{2}x^2$ **5.** $f(x) = x^2 + 4$ **6.** $f(x) = x^2 - 4$

7. $f(x) = (x - 1)^2$ **8.** $f(x) = (x + 3)^2$ **9.** $f(x) = (x + 3)^2 - 4$

10. $f(x) = (x + 5)^2 - 8$ **11.** $f(x) = -(x - 5)^2 + 6$ **12.** $f(x) = -(x - 2)^2 + 1$

For each quadratic function, tell whether the graph opens up or down and whether the graph is wider, narrower, or the same shape as the graph of $f(x) = x^2$. ***See Examples 4 and 5.***

13. $f(x) = -\dfrac{2}{5}x^2$ **14.** $f(x) = -2x^2$

15. $f(x) = 3x^2 + 1$ **16.** $f(x) = \dfrac{2}{3}x^2 - 4$

17. $f(x) = -4(x + 2)^2 + 5$ **18.** $f(x) = -\dfrac{1}{3}(x + 6)^2 + 3$

19. *Concept Check* Match each quadratic function with the description of the parabola that is its graph.

(a) $f(x) = (x - 4)^2 - 2$ **A.** Vertex $(2, -4)$, opens down

(b) $f(x) = (x - 2)^2 - 4$ **B.** Vertex $(2, -4)$, opens up

(c) $f(x) = -(x - 4)^2 - 2$ **C.** Vertex $(4, -2)$, opens down

(d) $f(x) = -(x - 2)^2 - 4$ **D.** Vertex $(4, -2)$, opens up

20. *Concept Check* For $f(x) = a(x - h)^2 + k$, in what quadrant is the vertex if

(a) $h > 0, k > 0$ (b) $h > 0, k < 0$ (c) $h < 0, k > 0$ (d) $h < 0, k < 0$?

*Graph each parabola. Plot at least two points as well as the vertex. Give the vertex, axis, domain, and range in Exercises 27–36. **See Examples 1–5.***

21. $f(x) = -2x^2$ **22.** $f(x) = -\dfrac{1}{3}x^2$ **23.** $f(x) = x^2 - 1$

24. $f(x) = x^2 + 3$ **25.** $f(x) = -x^2 + 2$ **26.** $f(x) = -x^2 - 2$

27. $f(x) = (x - 4)^2$ **28.** $f(x) = (x + 1)^2$

29. $f(x) = (x + 2)^2 - 1$ **30.** $f(x) = (x - 1)^2 + 2$

31. $f(x) = 2(x - 2)^2 - 4$ **32.** $f(x) = 3(x - 2)^2 + 1$

33. $f(x) = -\dfrac{1}{2}(x + 1)^2 + 2$ **34.** $f(x) = -\dfrac{2}{3}(x + 2)^2 + 1$

35. $f(x) = 2(x - 2)^2 - 3$ **36.** $f(x) = \dfrac{4}{3}(x - 3)^2 - 2$

Concept Check In Exercises 37–42, tell whether a linear or quadratic function would be a more appropriate model for each set of graphed data. If linear, tell whether the slope should be positive or negative. If quadratic, tell whether the coefficient a of x^2 should be positive or negative. **See Example 6.**

37. **TIME SPENT PLAYING VIDEO GAMES**

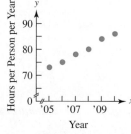

Source: Veronis Suhler Stevenson.

38. **AVERAGE DAILY VOLUME OF FIRST-CLASS MAIL**

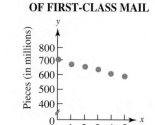

Source: General Accounting Office.

39. **FOOD ASSISTANCE SPENDING IN IOWA**

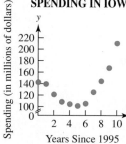

Source: Iowa Department of Human Services.

40. **PLASMA TV SALES IN U.S.**

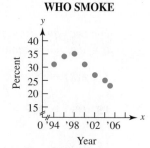

Source: Consumer Electronics Association.

41. **HIGH SCHOOL STUDENTS WHO SMOKE**

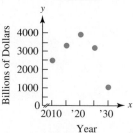

Source: www.cdc.gov

42. **SOCIAL SECURITY ASSETS***

*Projected

Source: Social Security Administration.

Solve each problem. ***See Example 6.***

43. Sales of digital cameras in the United States (in millions of dollars) between 2000 and 2006 are shown in the table. In the year column, 0 represents 2000, 1 represents 2001, and so on.

Year	Sales
0	1825
1	1972
2	2794
3	3921
4	4739
5	5611
6	7805

Source: Consumer Electronics Association.

(a) Use the ordered pairs (year, sales) to make a scatter diagram of the data.

(b) Use the scatter diagram to decide whether a linear or quadratic function would better model the data. If quadratic, should the coefficient a of x^2 be positive or negative?

(c) Use the ordered pairs $(0, 1825)$, $(3, 3921)$, and $(6, 7805)$ to find a quadratic function that models the data. Round the values of a, b, and c in your model to the nearest tenth, as necessary.

(d) Use your model from part (c) to approximate the sales of digital cameras in the United States in 2007. Round your answer to the nearest whole number (of millions).

✍ (e) Sales of digital cameras were $6517 million in 2007. Based on this, is the model valid for 2007? Explain.

44. The number (in thousands) of new, privately owned housing units started in the United States is shown in the table for the years 2002–2008. In the year column, 2 represents 2002, 3 represents 2003, and so on.

Year	Housing Starts (thousands)
2	1700
3	1850
4	1960
5	2070
6	1800
7	1360
8	910

Source: U.S. Census Bureau.

(a) Use the ordered pairs (year, housing starts) to make a scatter diagram of the data.

(b) Would a linear or quadratic function better model the data?

(c) Should the coefficient a of x^2 in a quadratic model be positive or negative?

(d) Use the ordered pairs $(2, 1700)$, $(4, 1960)$, and $(7, 1360)$ to find a quadratic function that models the data. Round the values of a, b, and c in your model to the nearest whole number, as necessary.

(e) Use your model from part (d) to approximate the number of housing starts during 2003 and 2008 to the nearest thousand. How well does the model approximate the actual data from the table?

45. In **Example 6**, we determined that the quadratic function defined by

$$y = -69.15x^2 + 863.6x + 4973$$

modeled the number of higher-order multiple births, where x represents the number of years since 1995.

(a) Use this model to approximate the number of higher-order births in 2006 to the nearest whole number.

(b) The actual number of higher-order births in 2006 was 6540. (*Source:* National Center for Health Statistics.) How does the approximation using the model compare to the actual number for 2006?

46. Should the model from **Exercise 45** be used to approximate the rate of higher-order multiple births in years after 2006? Explain.

TECHNOLOGY INSIGHTS EXERCISES 47–48

*Recall from **Section 3.3** that the x-value of the x-intercept of the graph of the line $y = mx + b$ is the solution of the linear equation $mx + b = 0$. In the same way, the x-values of the x-intercepts of the graph of the parabola $y = ax^2 + bx + c$ are the real solutions of the quadratic equation $ax^2 + bx + c = 0$.*

In Exercises 47–48, the calculator graphs show the x-values of the x-intercepts of the graph of the polynomial in the equation. Use the graphs to solve each equation.

47. $x^2 - x - 20 = 0$

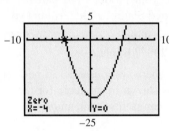 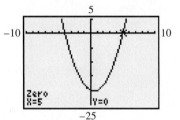

48. $x^2 + 9x + 14 = 0$

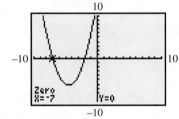 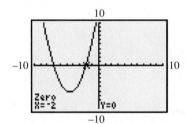

PREVIEW EXERCISES

Complete each factoring. See Section 6.1.

49. $-2x^2 + 6x = $ _____ $(x^2 - 3x)$

50. $-3x^2 - 15x = $ _____ $(x^2 + 5x)$

Solve each quadratic equation by factoring or by completing the square. See Section 9.1.

51. $x^2 + 3x - 4 = 0$

52. $x^2 - x - 6 = 0$

53. $x^2 + 6x - 3 = 0$

54. $x^2 + 8x - 4 = 0$

9.6

More About Parabolas and Their Applications

OBJECTIVES

1 Find the vertex of a vertical parabola.

2 Graph a quadratic function.

3 Use the discriminant to find the number of x-intercepts of a parabola with a vertical axis.

4 Use quadratic functions to solve problems involving maximum or minimum value.

5 Graph parabolas with horizontal axes.

OBJECTIVE 1 **Find the vertex of a vertical parabola.** When the equation of a parabola is given in the form $f(x) = ax^2 + bx + c$, there are two ways to locate the vertex.

1. Complete the square, as shown in **Examples 1 and 2,** or

2. Use a formula derived by completing the square, as shown in **Example 3.**

EXAMPLE 1 Completing the Square to Find the Vertex ($a = 1$)

Find the vertex of the graph of $f(x) = x^2 - 4x + 5$.

We can express $x^2 - 4x + 5$ in the form $(x - h)^2 + k$ by completing the square on $x^2 - 4x$, as in **Section 9.1.** The process is slightly different here because we want to keep $f(x)$ alone on one side of the equation. Instead of adding the appropriate number to each side, we *add and subtract* it on the right.

$$f(x) = x^2 - 4x + 5$$

$$= (x^2 - 4x \quad) + 5 \qquad \text{Group the variable terms.}$$

> This is equivalent to adding 0. $\qquad \left[\frac{1}{2}(-4)\right]^2 = (-2)^2 = 4$

$$= (x^2 - 4x + 4 - 4) + 5 \qquad \text{Add and subtract 4.}$$

$$= (x^2 - 4x + 4) - 4 + 5 \qquad \text{Bring } -4 \text{ outside the parentheses.}$$

$$f(x) = (x - 2)^2 + 1 \qquad \text{Factor. Combine like terms.}$$

The vertex of this parabola is $(2, 1)$. NOW TRY

NOW TRY EXERCISE 1
Find the vertex of the graph of
$f(x) = x^2 + 2x - 8$.

NOW TRY EXERCISE 2
Find the vertex of the graph of
$f(x) = -4x^2 + 16x - 10$.

EXAMPLE 2 Completing the Square to Find the Vertex ($a \neq 1$)

Find the vertex of the graph of $f(x) = -3x^2 + 6x - 1$.

Because the x^2-term has a coefficient other than 1, we factor that coefficient out of the first two terms before completing the square.

$$f(x) = -3x^2 + 6x - 1$$

$$= -3(x^2 - 2x) - 1 \qquad \text{Factor out } -3.$$

> $\left[\frac{1}{2}(-2)\right]^2 = (-1)^2 = 1$

$$= -3(x^2 - 2x + 1 - 1) - 1 \qquad \text{Add and subtract 1 within the parentheses.}$$

Now bring -1 outside the parentheses. Be sure to multiply it by -3.

$$= -3(x^2 - 2x + 1) + (-3)(-1) - 1 \qquad \text{Distributive property}$$

$$= -3(x^2 - 2x + 1) + 3 - 1 \qquad \boxed{\text{This is a key step.}}$$

$$f(x) = -3(x - 1)^2 + 2 \qquad \text{Factor. Combine like terms.}$$

NOW TRY ANSWERS
1. $(-1, -9)$ **2.** $(2, 6)$

The vertex is $(1, 2)$. NOW TRY

To derive a formula for the vertex of the graph of the quadratic function defined by $f(x) = ax^2 + bx + c$ (with $a \neq 0$), complete the square.

$$f(x) = ax^2 + bx + c \qquad \text{Standard form}$$

$$= a\left(x^2 + \frac{b}{a}x\right) + c \qquad \begin{array}{l}\text{Factor } a \text{ from the} \\ \text{first two terms.}\end{array}$$

$$\left[\tfrac{1}{2}\left(\tfrac{b}{a}\right)\right]^2 = \left(\tfrac{b}{2a}\right)^2 = \tfrac{b^2}{4a^2}$$

$$= a\left(x^2 + \frac{b}{a}x + \frac{b^2}{4a^2} - \frac{b^2}{4a^2}\right) + c \qquad \text{Add and subtract } \tfrac{b^2}{4a^2}.$$

$$= a\left(x^2 + \frac{b}{a}x + \frac{b^2}{4a^2}\right) + a\left(-\frac{b^2}{4a^2}\right) + c \qquad \text{Distributive property}$$

$$= a\left(x^2 + \frac{b}{a}x + \frac{b^2}{4a^2}\right) - \frac{b^2}{4a} + c \qquad -\tfrac{ab^2}{4a^2} = -\tfrac{b^2}{4a}$$

$$= a\left(x + \frac{b}{2a}\right)^2 + \frac{4ac - b^2}{4a} \qquad \begin{array}{l}\text{Factor. Rewrite terms with} \\ \text{a common denominator.}\end{array}$$

$$f(x) = a\underbrace{\left[x - \left(\frac{-b}{2a}\right)\right]^2}_{h} + \underbrace{\frac{4ac - b^2}{4a}}_{k} \qquad \begin{array}{l} f(x) = a(x - h)^2 + k \\ \text{The vertex } (h, k) \text{ can be ex-} \\ \text{pressed in terms of } a, b, \text{ and } c.\end{array}$$

The expression for k can be found by replacing x with $\frac{-b}{2a}$. Using function notation, if $y = f(x)$, then the y-value of the vertex is $f\left(\frac{-b}{2a}\right)$.

Vertex Formula

The graph of the quadratic function defined by $f(x) = ax^2 + bx + c$ (with $a \neq 0$) has vertex

$$\left(\frac{-b}{2a}, f\left(\frac{-b}{2a}\right)\right),$$

and the axis of the parabola is the line

$$x = \frac{-b}{2a}.$$

NOW TRY
EXERCISE 3

Use the vertex formula to find the vertex of the graph of
$$f(x) = 3x^2 - 2x + 8.$$

EXAMPLE 3 Using the Formula to Find the Vertex

Use the vertex formula to find the vertex of the graph of $f(x) = x^2 - x - 6$.

The x-coordinate of the vertex of the parabola is given by $\frac{-b}{2a}$.

$$\frac{-b}{2a} = \frac{-(-1)}{2(1)} = \frac{1}{2} \leftarrow \begin{array}{l} a = 1, b = -1, \text{ and } c = -6. \\ \text{x-coordinate of vertex} \end{array}$$

The y-coordinate is $f\left(\frac{-b}{2a}\right) = f\left(\frac{1}{2}\right)$.

$$f\left(\frac{1}{2}\right) = \left(\frac{1}{2}\right)^2 - \frac{1}{2} - 6 = \frac{1}{4} - \frac{1}{2} - 6 = -\frac{25}{4} \leftarrow \text{y-coordinate of vertex}$$

NOW TRY ANSWER
3. $\left(\frac{1}{3}, \frac{23}{3}\right)$

The vertex is $\left(\frac{1}{2}, -\frac{25}{4}\right)$.

NOW TRY

OBJECTIVE 2   Graph a quadratic function. We give a general approach.

> ### Graphing a Quadratic Function $y = f(x)$
>
> **Step 1**   **Determine whether the graph opens up or down.** If $a > 0$, the parabola opens up. If $a < 0$, it opens down.
>
> **Step 2**   **Find the vertex.** Use the vertex formula or completing the square.
>
> **Step 3**   **Find any intercepts.** To find the x-intercepts (if any), solve $f(x) = 0$. To find the y-intercept, evaluate $f(0)$.
>
> **Step 4**   **Complete the graph.** Plot the points found so far. Find and plot additional points as needed, using symmetry about the axis.

NOW TRY
EXERCISE 4

Graph the quadratic function defined by
$$f(x) = x^2 + 2x - 3.$$
Give the vertex, axis, domain, and range.

EXAMPLE 4   Graphing a Quadratic Function

Graph the quadratic function defined by $f(x) = x^2 - x - 6$.

Step 1   From the equation, $a = 1$, so the graph of the function opens up.

Step 2   The vertex, $\left(\frac{1}{2}, -\frac{25}{4}\right)$, was found in **Example 3** by using the vertex formula.

Step 3   Find any intercepts. Since the vertex, $\left(\frac{1}{2}, -\frac{25}{4}\right)$, is in quadrant IV and the graph opens up, there will be two x-intercepts. Let $f(x) = 0$ and solve.

$$f(x) = x^2 - x - 6$$

$$0 = x^2 - x - 6 \qquad \text{Let } f(x) = 0.$$

$$0 = (x - 3)(x + 2) \qquad \text{Factor.}$$

$$x - 3 = 0 \quad \text{or} \quad x + 2 = 0 \qquad \text{Zero-factor property}$$

$$x = 3 \quad \text{or} \qquad x = -2 \qquad \text{Solve each equation.}$$

The x-intercepts are $(3, 0)$ and $(-2, 0)$. Find the y-intercept by evaluating $f(0)$.

$$f(x) = x^2 - x - 6$$

$$f(0) = 0^2 - 0 - 6 \qquad \text{Let } x = 0.$$

$$f(0) = -6$$

The y-intercept is $(0, -6)$.

Step 4   Plot the points found so far and additional points as needed using symmetry about the axis, $x = \frac{1}{2}$. The graph is shown in **FIGURE 13**.

NOW TRY ANSWER

4.

$f(x) = x^2 + 2x - 3$

vertex: $(-1, -4)$; axis: $x = -1$;
domain: $(-\infty, \infty)$; range: $[-4, \infty)$

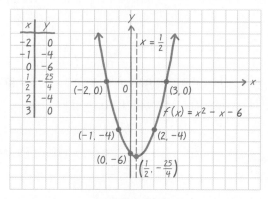

$f(x) = x^2 - x - 6$

Vertex: $\left(\frac{1}{2}, -\frac{25}{4}\right)$

Axis: $x = \frac{1}{2}$

Domain: $(-\infty, \infty)$

Range: $\left[-\frac{25}{4}, \infty\right)$

FIGURE 13

NOW TRY

OBJECTIVE 3 **Use the discriminant to find the number of x-intercepts of a parabola with a vertical axis.** Recall from **Section 9.2** that

$$b^2 - 4ac \quad \text{Discriminant}$$

is called the *discriminant* of the quadratic equation $ax^2 + bx + c = 0$ and that we can use it to determine the number of real solutions of a quadratic equation.

In a similar way, we can use the discriminant of a quadratic *function* to determine the number of x-intercepts of its graph. The three possibilities are shown in **FIGURE 14**.

1. If the discriminant is positive, the parabola will have two x-intercepts.

2. If the discriminant is 0, there will be only one x-intercept, and it will be the vertex of the parabola.

3. If the discriminant is negative, the graph will have no x-intercepts.

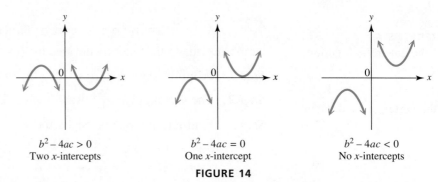

$b^2 - 4ac > 0$
Two x-intercepts

$b^2 - 4ac = 0$
One x-intercept

$b^2 - 4ac < 0$
No x-intercepts

FIGURE 14

NOW TRY
EXERCISE 5

Find the discriminant and use it to determine the number of x-intercepts of the graph of each quadratic function.

(a) $f(x) = -2x^2 + 3x - 2$

(b) $f(x) = 3x^2 + 2x - 1$

(c) $f(x) = 4x^2 - 12x + 9$

EXAMPLE 5 Using the Discriminant to Determine the Number of x-Intercepts

Find the discriminant and use it to determine the number of x-intercepts of the graph of each quadratic function.

(a) $f(x) = 2x^2 + 3x - 5$

$$b^2 - 4ac \qquad \text{Discriminant}$$
$$= 3^2 - 4(2)(-5) \qquad a = 2, b = 3, c = -5$$
$$= 9 - (-40) \qquad \text{Apply the exponent. Multiply.}$$
$$= 49 \qquad \text{Subtract.}$$

Since the discriminant is positive, the parabola has two x-intercepts.

(b) $f(x) = -3x^2 - 1$

$$b^2 - 4ac$$
$$= 0^2 - 4(-3)(-1) \qquad a = -3, b = 0, c = -1$$
$$= -12$$

The discriminant is negative, so the graph has no x-intercepts.

(c) $f(x) = 9x^2 + 6x + 1$

$$b^2 - 4ac$$
$$= 6^2 - 4(9)(1) \qquad a = 9, b = 6, c = 1$$
$$= 0$$

The parabola has only one x-intercept (its vertex).

NOW TRY

OBJECTIVE 4 **Use quadratic functions to solve problems involving maximum or minimum value.** The vertex of the graph of a quadratic function is either the highest or the lowest point on the parabola. It provides the following information.

1. The y-value of the vertex gives the maximum or minimum value of y.

2. The x-value tells where the maximum or minimum occurs.

PROBLEM-SOLVING HINT

In many applied problems we must find the greatest or least value of some quantity. When we can express that quantity in terms of a quadratic function, the value of k in the vertex (h, k) gives that optimum value.

**⌐ NOW TRY
⌐ EXERCISE 6**

Solve the problem in **Example 6** if the farmer has only 80 ft of fencing.

EXAMPLE 6 Finding the Maximum Area of a Rectangular Region

A farmer has 120 ft of fencing to enclose a rectangular area next to a building. (See **FIGURE 15**.) Find the maximum area he can enclose and the dimensions of the field when the area is maximized.

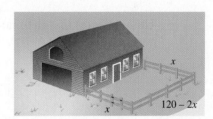

FIGURE 15

Let $x =$ the width of the field.

$$x + x + \text{length} = 120 \qquad \text{Sum of the sides is 120 ft.}$$
$$2x + \text{length} = 120 \qquad \text{Combine like terms.}$$
$$\text{length} = 120 - 2x \qquad \text{Subtract } 2x.$$

The area $\mathcal{A}(x)$ is given by the product of the length and width.

$$\mathcal{A}(x) = (120 - 2x)x \qquad \text{Area = length · width}$$
$$\mathcal{A}(x) = 120x - 2x^2 \qquad \text{Distributive property}$$

To determine the maximum area, use the vertex formula to find the vertex of the parabola given by $\mathcal{A}(x) = 120x - 2x^2$. Write the equation in standard form.

$$\mathcal{A}(x) = -2x^2 + 120x \qquad a = -2, b = 120, c = 0$$

Then $\qquad x = \dfrac{-b}{2a} = \dfrac{-120}{2(-2)} = \dfrac{-120}{-4} = 30,$

and $\qquad \mathcal{A}(30) = -2(30)^2 + 120(30) = -2(900) + 3600 = 1800.$

The graph is a parabola that opens down, and its vertex is $(30, 1800)$. Thus, the maximum area will be 1800 ft². This area will occur if x, the width of the field, is 30 ft and the length is

NOW TRY ANSWER
6. The field should be 20 ft by 40 ft with maximum area 800 ft².

$$120 - 2(30) = 60 \text{ ft.} \qquad \textit{NOW TRY}$$

⚠️ **CAUTION** *Be careful when interpreting the meanings of the coordinates of the vertex.* The first coordinate, x, gives the value for which the *function value, y* or $f(x)$, is a maximum or a minimum. Be sure to read the problem carefully to determine whether you are asked to find the value of the independent variable, the function value, or both.

⌒ *NOW TRY*
↪ *EXERCISE 7*

A stomp rocket is launched from the ground with an initial velocity of 48 ft per sec so that its distance in feet above the ground after t seconds is

$$s(t) = -16t^2 + 48t.$$

Find the maximum height attained by the rocket and the number of seconds it takes to reach that height.

EXAMPLE 7 Finding the Maximum Height Attained by a Projectile

If air resistance is neglected, a projectile on Earth shot straight upward with an initial velocity of 40 m per sec will be at a height s in meters given by

$$s(t) = -4.9t^2 + 40t,$$

where t is the number of seconds elapsed after projection. After how many seconds will it reach its maximum height, and what is this maximum height?

For this function, $a = -4.9$, $b = 40$, and $c = 0$. Use the vertex formula.

$$t = \frac{-b}{2a} = \frac{-40}{2(-4.9)} \approx 4.1 \qquad \text{Use a calculator.}$$

This indicates that the maximum height is attained at 4.1 sec. To find this maximum height, calculate $s(4.1)$.

$$s(t) = -4.9t^2 + 40t$$
$$s(4.1) = -4.9(4.1)^2 + 40(4.1) \qquad \text{Let } t = 4.1.$$
$$s(4.1) \approx 81.6 \qquad \text{Use a calculator.}$$

The projectile will attain a maximum height of approximately 81.6 m at 4.1 sec.

NOW TRY ↻

OBJECTIVE 5 **Graph parabolas with horizontal axes.** If x and y are interchanged in the equation

$$y = ax^2 + bx + c,$$

the equation becomes

$$x = ay^2 + by + c.$$

Because of the interchange of the roles of x and y, these parabolas are horizontal (with horizontal lines as axes).

Graph of a Horizontal Parabola

The graph of $x = ay^2 + by + c$ or $x = a(y - k)^2 + h$ is a parabola.

- The vertex of the parabola is (h, k).
- The axis is the horizontal line $y = k$.
- The graph opens to the right if $a > 0$ and to the left if $a < 0$.

NOW TRY ANSWER
7. 36 ft; 1.5 sec

NOW TRY
EXERCISE 8
Graph $x = (y + 2)^2 - 1$.
Give the vertex, axis, domain, and range.

EXAMPLE 8   Graphing a Horizontal Parabola ($a = 1$)

Graph $x = (y - 2)^2 - 3$. Give the vertex, axis, domain, and range.

This graph has its vertex at $(-3, 2)$, since the roles of x and y are interchanged. It opens to the right (the positive x-direction) because $a = 1$ and $1 > 0$, and has the same shape as $y = x^2$. Plotting a few additional points gives the graph shown in **FIGURE 16**.

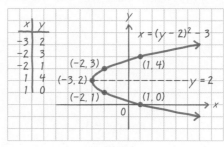

$x = (y - 2)^2 - 3$
Vertex: $(-3, 2)$
Axis: $y = 2$
Domain: $[-3, \infty)$
Range: $(-\infty, \infty)$

FIGURE 16

NOW TRY

NOW TRY
EXERCISE 9
Graph $x = -3y^2 - 6y - 5$.
Give the vertex, axis, domain, and range.

EXAMPLE 9   Completing the Square to Graph a Horizontal Parabola ($a \ne 1$)

Graph $x = -2y^2 + 4y - 3$. Give the vertex, axis, domain, and range of the relation.

$$x = -2y^2 + 4y - 3$$
$$= -2(y^2 - 2y) - 3 \qquad \text{Factor out } -2.$$
$$= -2(y^2 - 2y + 1 - 1) - 3 \qquad \begin{array}{l}\text{Complete the square within the}\\\text{parentheses. Add and subtract 1.}\end{array}$$
$$= -2(y^2 - 2y + 1) + (-2)(-1) - 3 \qquad \text{Distributive property}$$

Be careful here.

$$x = -2(y - 1)^2 - 1 \qquad \text{Factor. Simplify.}$$

Because of the negative coefficient -2 in $x = -2(y - 1)^2 - 1$, the graph opens to the left (the negative x-direction). The graph is narrower than the graph of $y = x^2$ because $|-2| > 1$. See **FIGURE 17**.

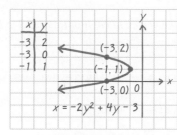

$x = -2y^2 + 4y - 3$
Vertex: $(-1, 1)$
Axis: $y = 1$
Domain: $(-\infty, -1]$
Range: $(-\infty, \infty)$

FIGURE 17

NOW TRY

NOW TRY ANSWERS
8.

$x = (y + 2)^2 - 1$

vertex: $(-1, -2)$; axis: $y = -2$;
domain: $[-1, \infty)$; range: $(-\infty, \infty)$

9.

$x = -3y^2 - 6y - 5$

vertex: $(-2, -1)$; axis: $y = -1$;
domain: $(-\infty, -2]$; range: $(-\infty, \infty)$

⚠ **CAUTION**   *Only quadratic equations solved for y (whose graphs are vertical parabolas) are examples of functions.* The horizontal parabolas in **Examples 8 and 9** are *not* graphs of functions, because they do not satisfy the conditions of the vertical line test.

In summary, the graphs of parabolas fall into the following categories.

Graphs of Parabolas

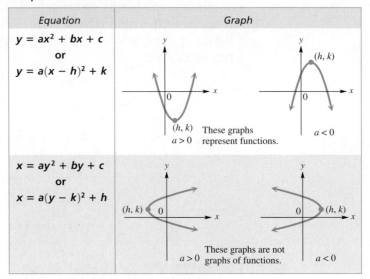

Equation	Graph
$y = ax^2 + bx + c$ or $y = a(x - h)^2 + k$	These graphs represent functions. (h, k) $a > 0$ (h, k) $a < 0$
$x = ay^2 + by + c$ or $x = a(y - k)^2 + h$	These graphs are not graphs of functions. (h, k) $a > 0$ (h, k) $a < 0$

9.6 EXERCISES

MyMathLab

Math XL PRACTICE WATCH DOWNLOAD READ REVIEW

🌐 *Complete solution available on the Video Resources on DVD*

Concept Check   In Exercises 1–4, answer each question.

1. How can you determine just by looking at the equation of a parabola whether it has a vertical or a horizontal axis?

2. Why can't the graph of a quadratic function be a parabola with a horizontal axis?

3. How can you determine the number of x-intercepts of the graph of a quadratic function without graphing the function?

4. If the vertex of the graph of a quadratic function is $(1, -3)$, and the graph opens down, how many x-intercepts does the graph have?

*Find the vertex of each parabola. **See Examples 1–3.***

🌐 **5.** $f(x) = x^2 + 8x + 10$

6. $f(x) = x^2 + 10x + 23$

🌐 **7.** $f(x) = -2x^2 + 4x - 5$

8. $f(x) = -3x^2 + 12x - 8$

🌐 **9.** $f(x) = x^2 + x - 7$

10. $f(x) = x^2 - x + 5$

*Find the vertex of each parabola. For each equation, decide whether the graph opens up, down, to the left, or to the right, and whether it is wider, narrower, or the same shape as the graph of $y = x^2$. If it is a parabola with vertical axis, find the discriminant and use it to determine the number of x-intercepts. **See Examples 1–3, 5, 8, and 9.***

🌐 **11.** $f(x) = 2x^2 + 4x + 5$

12. $f(x) = 3x^2 - 6x + 4$

13. $f(x) = -x^2 + 5x + 3$

14. $f(x) = -x^2 + 7x + 2$

15. $x = \dfrac{1}{3}y^2 + 6y + 24$

16. $x = \dfrac{1}{2}y^2 + 10y - 5$

Concept Check *Match each equation in Exercises 17–22 with its graph in choices A–F.*

17. $y = 2x^2 + 4x - 3$

18. $y = -x^2 + 3x + 5$

19. $y = -\dfrac{1}{2}x^2 - x + 1$

20. $x = y^2 + 6y + 3$

21. $x = -y^2 - 2y + 4$

22. $x = 3y^2 + 6y + 5$

A.

B.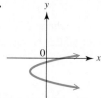

C.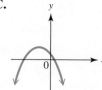

D.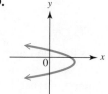

E.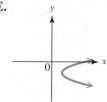

F.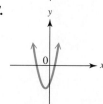

Graph each parabola. (Use the results of **Exercises 5–8** *to help graph the parabolas in Exercises 23–26.) Give the vertex, axis, domain, and range.* **See Examples 4, 8, and 9.**

◐ 23. $f(x) = x^2 + 8x + 10$

24. $f(x) = x^2 + 10x + 23$

25. $f(x) = -2x^2 + 4x - 5$

26. $f(x) = -3x^2 + 12x - 8$

◐ 27. $x = (y + 2)^2 + 1$

28. $x = (y + 3)^2 - 2$

29. $x = -\dfrac{1}{5}y^2 + 2y - 4$

30. $x = -\dfrac{1}{2}y^2 - 4y - 6$

◐ 31. $x = 3y^2 + 12y + 5$

32. $x = 4y^2 + 16y + 11$

Solve each problem. **See Examples 6 and 7.**

33. Find the pair of numbers whose sum is 40 and whose product is a maximum. (*Hint:* Let x and $40 - x$ represent the two numbers.)

34. Find the pair of numbers whose sum is 60 and whose product is a maximum.

◐ 35. Polk Community College wants to construct a rectangular parking lot on land bordered on one side by a highway. It has 280 ft of fencing that is to be used to fence off the other three sides. What should be the dimensions of the lot if the enclosed area is to be a maximum? What is the maximum area?

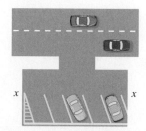

36. Bonnie Wolansky has 100 ft of fencing material to enclose a rectangular exercise run for her dog. One side of the run will border her house, so she will only need to fence three sides. What dimensions will give the enclosure the maximum area? What is the maximum area?

◐ 37. If an object on Earth is projected upward with an initial velocity of 32 ft per sec, then its height after t seconds is given by

$$s(t) = -16t^2 + 32t.$$

Find the maximum height attained by the object and the number of seconds it takes to hit the ground.

38. A projectile on Earth is fired straight upward so that its distance (in feet) above the ground t seconds after firing is given by

$$s(t) = -16t^2 + 400t.$$

Find the maximum height it reaches and the number of seconds it takes to reach that height.

39. After experimentation, two physics students from American River College find that when a bottle of California wine is shaken several times, held upright, and uncorked, its cork travels according to the function defined by

$$s(t) = -16t^2 + 64t + 1,$$

where s is its height in feet above the ground t seconds after being released. After how many seconds will it reach its maximum height? What is the maximum height?

40. Professor Barbu has found that the number of students attending his intermediate algebra class is approximated by

$$S(x) = -x^2 + 20x + 80,$$

where x is the number of hours that the Campus Center is open daily. Find the number of hours that the center should be open so that the number of students attending class is a maximum. What is this maximum number of students?

41. Klaus Loewy has a taco stand. He has found that his daily costs are approximated by

$$C(x) = x^2 - 40x + 610,$$

where $C(x)$ is the cost, in dollars, to sell x units of tacos. Find the number of units of tacos he should sell to minimize his costs. What is the minimum cost?

42. Mohammad Asghar has a frozen yogurt cart. His daily costs are approximated by

$$C(x) = x^2 - 70x + 1500,$$

where $C(x)$ is the cost, in dollars, to sell x units of frozen yogurt. Find the number of units of frozen yogurt he must sell to minimize his costs. What is the minimum cost?

43. The total receipts from individual income taxes by the U.S. Treasury in the years 2000–2007 can be modeled by the quadratic function defined by

$$f(x) = 22.88x^2 - 141.3x + 1044,$$

where $x = 0$ represents 2000, $x = 1$ represents 2001, and so on, and $f(x)$ is in billions of dollars. (*Source: World Almanac and Book of Facts.*)

  (a) Since the coefficient of x^2 given in the model is positive, the graph of this quadratic function is a parabola that opens up. Will the y-value of the vertex of this graph be a maximum or minimum?

  (b) In what year during this period were total receipts from individual taxes a minimum? (Round down for the year.) Use the actual x-value of the vertex, to the nearest tenth, to find this amount.

44. The percent of births in the United States to teenage mothers in the years 1990–2005 can be modeled by the quadratic function defined by

$$f(x) = -0.0198x^2 + 0.1054x + 12.87,$$

where $x = 0$ represents 1990, $x = 1$ represents 1991, and so on. (*Source: U.S. National Center for Health Statistics.*)

  (a) Since the coefficient of x^2 in the model is negative, the graph of this quadratic function is a parabola that opens down. Will the y-value of the vertex of this graph be a maximum or a minimum?

  (b) In what year during this period was the percent of births in the U.S. to teenage mothers a maximum? (Round down for the year.) Use the actual x-value of the vertex, to the nearest tenth, to find this percent.

45. The graph on the next page shows how Social Security trust fund assets are expected to change, and suggests that a quadratic function would be a good fit to the data. The data are approximated by the function defined by

$$f(x) = -20.57x^2 + 758.9x - 3140.$$

In the model, $x = 10$ represents 2010, $x = 15$ represents 2015, and so on, and $f(x)$ is in billions of dollars.

(a) *Concept Check* How could you have predicted this quadratic model would have a negative coefficient for x^2, based only on the graph shown?

(b) Algebraically determine the vertex of the graph, with coordinates to four significant digits.

(c) Interpret the answer to part (b) as it applies to this application.

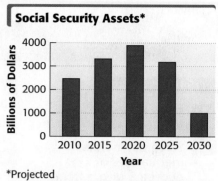

Social Security Assets*

*Projected

Source: Social Security Administration.

46. The graph shows the performance of investment portfolios with different mixtures of U.S. and foreign investments over a 25-yr period.

(a) Is this the graph of a function? Explain.

(b) What investment mixture shown on the graph appears to represent the vertex? What relative amount of risk does this point represent? What return on investment does it provide?

(c) Which point on the graph represents the riskiest investment mixture? What return on investment does it provide?

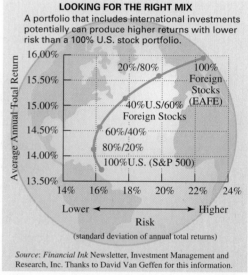

LOOKING FOR THE RIGHT MIX
A portfolio that includes international investments potentially can produce higher returns with lower risk than a 100% U.S. stock portfolio.

Source: *Financial Ink* Newsletter, Investment Management and Research, Inc. Thanks to David Van Geffen for this information.

47. A charter flight charges a fare of $200 per person, plus $4 per person for each unsold seat on the plane. If the plane holds 100 passengers and if x represents the number of unsold seats, find the following.

(a) A function defined by $R(x)$ that describes the total revenue received for the flight (*Hint:* Multiply the number of people flying, $100 - x$, by the price per ticket, $200 + 4x$.)

(b) The graph of the function from part (a)

(c) The number of unsold seats that will produce the maximum revenue

(d) The maximum revenue

48. For a trip to a resort, a charter bus company charges a fare of $48 per person, plus $2 per person for each unsold seat on the bus. If the bus has 42 seats and x represents the number of unsold seats, find the following.

(a) A function defined by $R(x)$ that describes the total revenue from the trip (*Hint:* Multiply the total number riding, $42 - x$, by the price per ticket, $48 + 2x$.)

(b) The graph of the function from part (a)

(c) The number of unsold seats that produces the maximum revenue

(d) The maximum revenue

PREVIEW EXERCISES

Graph each interval on a number line. See Section 2.5.

49. $[1, 5]$

50. $(-6, 1]$

51. $(-\infty, 1] \cup [5, \infty)$

Solve each inequality. See Section 2.5.

52. $3 - x \le 5$

53. $-2x + 1 < 4$

54. $-\dfrac{1}{2}x - 3 > 5$

9.7 Polynomial and Rational Inequalities

OBJECTIVE 1 **Solve quadratic inequalities.** Now we combine the methods of solving linear inequalities with the methods of solving quadratic equations to solve *quadratic inequalities.*

Quadratic Inequality

A **quadratic inequality** can be written in the form

$$ax^2 + bx + c < 0, \qquad ax^2 + bx + c > 0,$$

$$ax^2 + bx + c \leq 0, \qquad \text{or} \qquad ax^2 + bx + c \geq 0,$$

where a, b, and c are real numbers, with $a \neq 0$.

One way to solve a quadratic inequality is by graphing the related quadratic function.

EXAMPLE 1 Solving Quadratic Inequalities by Graphing

Solve each inequality.

(a) $x^2 - x - 12 > 0$

To solve the inequality, we graph the related quadratic function defined by $f(x) = x^2 - x - 12$. We are particularly interested in the *x*-intercepts, which are found as in **Section 9.6** by letting $f(x) = 0$ and solving the following quadratic equation.

$$x^2 - x - 12 = 0$$

$$(x - 4)(x + 3) = 0 \qquad \text{Factor.}$$

$$x - 4 = 0 \quad \text{or} \quad x + 3 = 0 \qquad \text{Zero-factor property}$$

$$x = 4 \quad \text{or} \qquad x = -3 \leftarrow \text{The } x\text{-intercepts are } (4, 0) \text{ and } (-3, 0).$$

The graph, which opens up since the coefficient of x^2 is positive, is shown in **FIGURE 18(a)**. Notice from this graph that *x*-values less than -3 or greater than 4 result in *y*-values *greater than* 0. Thus, the solution set of $x^2 - x - 12 > 0$, written in interval notation, is $(-\infty, -3) \cup (4, \infty)$.

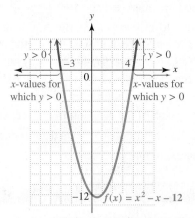

The graph is *above* the *x*-axis for $(-\infty, -3) \cup (4, \infty)$.

(a)

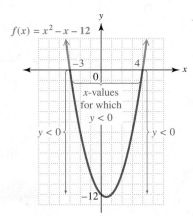

The graph is *below* the *x*-axis for $(-3, 4)$.

(b)

FIGURE 18

**NOW TRY
EXERCISE 1**
Use the graph to solve each quadratic inequality.

$f(x) = x^2 - 3x - 4$

(a) $x^2 - 3x - 4 > 0$
(b) $x^2 - 3x - 4 < 0$

(b) $x^2 - x - 12 < 0$

We want values of y that are *less than* 0. Referring to **FIGURE 18(b)**, we notice from the graph that x-values between -3 and 4 result in y-values less than 0. Thus, the solution set of $x^2 - x - 12 < 0$, written in interval notation, is $(-3, 4)$. NOW TRY

NOTE If the inequalities in **Example 1** had used $\geq$ and $\leq$, the solution sets would have included the x-values of the intercepts, which make the quadratic expression equal to 0. They would have been written in interval notation as

$$(-\infty, -3] \cup [4, \infty) \quad \text{and} \quad [-3, 4].$$

Square brackets would indicate that the endpoints -3 and 4 are *included* in the solution sets.

Another method for solving a quadratic inequality uses the basic ideas of **Example 1** without actually graphing the related quadratic function.

EXAMPLE 2 Solving a Quadratic Inequality Using Test Numbers

Solve and graph the solution set of $x^2 - x - 12 > 0$.

Solve the quadratic equation $x^2 - x - 12 = 0$ by factoring, as in **Example 1(a)**.

$$(x - 4)(x + 3) = 0$$

$$x - 4 = 0 \quad \text{or} \quad x + 3 = 0$$

$$x = 4 \quad \text{or} \quad x = -3$$

The numbers 4 and -3 divide a number line into Intervals A, B, and C, as shown in **FIGURE 19**. *Be careful to put the lesser number on the left.*

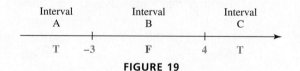

FIGURE 19

Notice the similarity between **FIGURE 19** and the x-axis with intercepts $(-3, 0)$ and $(4, 0)$ in **FIGURE 18(a)**.

The numbers 4 and -3 are the only numbers that make the quadratic expression $x^2 - x - 12$ equal to 0. All other numbers make the expression either positive or negative. The sign of the expression can change from positive to negative or from negative to positive only at a number that makes it 0. Therefore, if one number in an interval satisfies the inequality, then all the numbers in that interval will satisfy the inequality.

To see if the numbers in Interval A satisfy the inequality, choose any number from Interval A in **FIGURE 19** (that is, any number less than -3). We choose -5. Substitute this test number for x in the original inequality $x^2 - x - 12 > 0$.

$$x^2 - x - 12 > 0 \quad \text{Original inequality}$$

Use parentheses to avoid sign errors. $\rightarrow \quad (-5)^2 - (-5) - 12 \overset{?}{>} 0 \quad \text{Let } x = -5.$

$$25 + 5 - 12 \overset{?}{>} 0 \quad \text{Simplify.}$$

$$18 > 0 \quad\checkmark \text{ True}$$

NOW TRY ANSWERS
1. (a) $(-\infty, -1) \cup (4, \infty)$
  (b) $(-1, 4)$

Because -5 satisfies the inequality, *all* numbers from Interval A are solutions.

NOW TRY
EXERCISE 2
Solve and graph the solution set.

$$x^2 + 2x - 8 > 0$$

Now try 0 from Interval B.

$$x^2 - x - 12 > 0 \qquad \text{Original inequality}$$
$$0^2 - 0 - 12 \overset{?}{>} 0 \qquad \text{Let } x = 0.$$
$$-12 > 0 \qquad \text{False}$$

The numbers in Interval B are *not* solutions. Verify that the test number 5 from Interval C satisfies the inequality, so all numbers there are also solutions.

Based on these results (shown by the colored letters in **FIGURE 19**), the solution set includes the numbers in Intervals A and C, as shown on the graph in **FIGURE 20**. The solution set is written in interval notation as

$$(-\infty, -3) \cup (4, \infty).$$

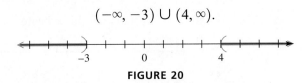

FIGURE 20

This agrees with the solution set found in **Example 1(a).**   NOW TRY

In summary, follow these steps to solve a quadratic inequality.

Solving a Quadratic Inequality

Step 1 **Write the inequality as an equation and solve it.**

Step 2 **Use the solutions from Step 1 to determine intervals.** Graph the numbers found in Step 1 on a number line. These numbers divide the number line into intervals.

Step 3 **Find the intervals that satisfy the inequality.** Substitute a test number from each interval into the original inequality to determine the intervals that satisfy the inequality. All numbers in those intervals are in the solution set. A graph of the solution set will usually look like one of these. (Square brackets might be used instead of parentheses.)

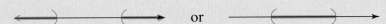

Step 4 **Consider the endpoints separately.** The numbers from Step 1 are included in the solution set if the inequality symbol is $\leq$ or $\geq$. They are not included if it is $<$ or $>$.

NOW TRY
EXERCISE 3
Solve each inequality.
(a) $(4x - 1)^2 > -3$
(b) $(4x - 1)^2 < -3$

NOW TRY ANSWERS
2. $(-\infty, -4) \cup (2, \infty)$

3. (a) $(-\infty, \infty)$ **(b)** $\emptyset$

EXAMPLE 3 Solving Special Cases

Solve each inequality.

(a) $(2x - 3)^2 > -1$

Because $(2x - 3)^2$ is never negative, it is always greater than -1. Thus, the solution set for $(2x - 3)^2 > -1$ is the set of all real numbers, $(-\infty, \infty)$.

(b) $(2x - 3)^2 < -1$

Using the same reasoning as in part (a), there is no solution for this inequality. The solution set is $\emptyset$.   NOW TRY

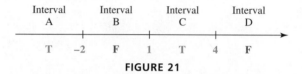

OBJECTIVE 2 Solve polynomial inequalities of degree 3 or greater.

EXAMPLE 4 Solving a Third-Degree Polynomial Inequality

Solve and graph the solution set of $(x - 1)(x + 2)(x - 4) \leq 0$.

This is a *cubic* (third-degree) inequality rather than a quadratic inequality, but it can be solved using the preceding method by extending the zero-factor property to more than two factors. (Step 1)

$$(x - 1)(x + 2)(x - 4) = 0 \qquad \text{Set the factored polynomial } equal \text{ to 0.}$$

$$x - 1 = 0 \quad \text{or} \quad x + 2 = 0 \quad \text{or} \quad x - 4 = 0 \qquad \text{Zero-factor property}$$

$$x = 1 \quad \text{or} \qquad x = -2 \quad \text{or} \qquad x = 4 \qquad \text{Solve each equation.}$$

Locate the numbers -2, 1, and 4 on a number line, as in **FIGURE 21**, to determine the Intervals A, B, C, and D. (Step 2)

	Interval A		Interval B		Interval C		Interval D

T   −2   F   1   T   4   F

FIGURE 21

Substitute a test number from each interval in the *original* inequality to determine which intervals satisfy the inequality. (Step 3)

Interval	Test Number	Test of Inequality	True or False?
A	−3	−28 ≤ 0	T
B	0	8 ≤ 0	F
C	2	−8 ≤ 0	T
D	5	28 ≤ 0	F

We use a table to organize this information. (Verify it.)

The numbers in Intervals A and C are in the solution set, which is written in interval notation as $(-\infty, -2] \cup [1, 4]$, and graphed in **FIGURE 22**. The three endpoints are included since the inequality symbol, $\leq$, includes equality. (Step 4)

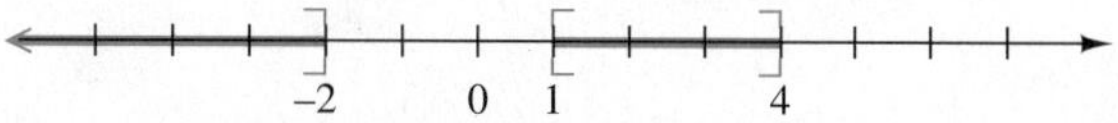

−2   0   1   4

FIGURE 22

NOW TRY

OBJECTIVE 3 Solve rational inequalities. Inequalities that involve rational expressions, called **rational inequalities,** are solved similarly using the following steps.

Solving a Rational Inequality

Step 1 **Write the inequality so that 0 is on one side** and there is a single fraction on the other side.

Step 2 **Determine the numbers that make the numerator or denominator equal to 0.**

Step 3 **Divide a number line into intervals.** Use the numbers from Step 2.

Step 4 **Find the intervals that satisfy the inequality.** Test a number from each interval by substituting it into the *original* inequality.

Step 5 **Consider the endpoints separately.** Exclude any values that make the denominator 0.

NOW TRY
EXERCISE 4

Solve and graph the solution set.

$(x + 4)(x - 3)(2x + 1) \leq 0$

NOW TRY ANSWER

4. $(-\infty, -4] \cup \left[-\frac{1}{2}, 3\right]$

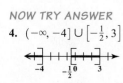

> ⚠ **CAUTION** *When solving a rational inequality, any number that makes the denominator 0 must be excluded from the solution set.*

NOW TRY
EXERCISE 5
Solve and graph the solution set.

$$\frac{3}{x+1} > 4$$

EXAMPLE 5   Solving a Rational Inequality

Solve and graph the solution set of $\dfrac{-1}{x-3} > 1$.

Write the inequality so that 0 is on one side. (Step 1)

$$\frac{-1}{x-3} - 1 > 0 \qquad \text{Subtract 1.}$$

$$\frac{-1}{x-3} - \frac{x-3}{x-3} > 0 \qquad \text{Use } x-3 \text{ as the common denominator.}$$

> Be careful with signs.

$$\frac{-1 - x + 3}{x-3} > 0 \qquad \text{Write the left side as a single fraction.}$$

$$\frac{-x + 2}{x-3} > 0 \qquad \text{Combine like terms in the numerator.}$$

The sign of $\frac{-x+2}{x-3}$ will change from positive to negative or negative to positive only at those numbers that make the numerator or denominator 0. The number 2 makes the numerator 0, and 3 makes the denominator 0. (Step 2) These two numbers, 2 and 3, divide a number line into three intervals. See **FIGURE 23**. (Step 3)

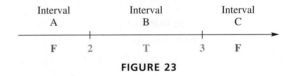

FIGURE 23

Testing a number from each interval in the *original* inequality, $\frac{-1}{x-3} > 1$, gives the results shown in the table. (Step 4)

Interval	Test Number	Test of Inequality	True or False?
A	0	$\frac{1}{3} > 1$	F
B	2.5	$2 > 1$	T
C	4	$-1 > 1$	F

The solution set is the interval $(2, 3)$. This interval does not include 3 since it would make the denominator of the original equality 0. The number 2 is not included either since the inequality symbol, $>$, does not include equality. (Step 5) See **FIGURE 24**.

NOW TRY ANSWER
5. $\left(-1, -\frac{1}{4}\right)$

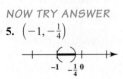

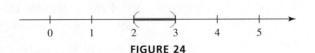

FIGURE 24

NOW TRY

NOW TRY
EXERCISE 6

Solve and graph the solution set.

$$\frac{x - 3}{x + 3} \le 2$$

EXAMPLE 6 Solving a Rational Inequality

Solve and graph the solution set of $\dfrac{x - 2}{x + 2} \le 2$.

Write the inequality so that 0 is on one side. (Step 1)

$$\frac{x - 2}{x + 2} - 2 \le 0 \qquad \text{Subtract 2.}$$

$$\frac{x - 2}{x + 2} - \frac{2(x + 2)}{x + 2} \le 0 \qquad \text{Use } x + 2 \text{ as the common denominator.}$$

Be careful with signs.
$$\frac{x - 2 - 2x - 4}{x + 2} \le 0 \qquad \text{Write as a single fraction.}$$

$$\frac{-x - 6}{x + 2} \le 0 \qquad \text{Combine like terms in the numerator.}$$

The number -6 makes the numerator 0, and -2 makes the denominator 0. (Step 2) These two numbers determine three intervals. (Step 3) Test one number from each interval (Step 4) to see that the solution set is

$$(-\infty, -6] \cup (-2, \infty).$$

The number -6 satisfies the original inequality, but -2 does not since it makes the denominator 0. (Step 5) **FIGURE 25** shows a graph of the solution set.

NOW TRY ANSWER

6. $(-\infty, -9] \cup (-3, \infty)$

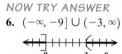

FIGURE 25

NOW TRY

9.7 EXERCISES

MyMathLab Math XL PRACTICE WATCH DOWNLOAD READ REVIEW

⊙ *Complete solution available on the Video Resources on DVD*

In Exercises 1–3, the graph of a quadratic function f is given. Use the graph to find the solution set of each equation or inequality. ***See Example 1.***

1. (a) $x^2 - 4x + 3 = 0$
   (b) $x^2 - 4x + 3 > 0$
   (c) $x^2 - 4x + 3 < 0$

2. (a) $3x^2 + 10x - 8 = 0$
   (b) $3x^2 + 10x - 8 \ge 0$
   (c) $3x^2 + 10x - 8 < 0$

3. (a) $-x^2 + 3x + 10 = 0$
   (b) $-x^2 + 3x + 10 \ge 0$
   (c) $-x^2 + 3x + 10 \le 0$

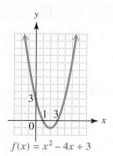

$f(x) = x^2 - 4x + 3$

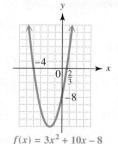

$f(x) = 3x^2 + 10x - 8$

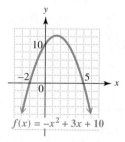

$f(x) = -x^2 + 3x + 10$

4. *Concept Check* The solution set of the inequality $x^2 + x - 12 < 0$ is the interval $(-4, 3)$. Without actually performing any work, give the solution set of the inequality $x^2 + x - 12 \ge 0$.

Solve each inequality, and graph the solution set. ***See Example 2.*** *(Hint: In Exercises 21 and 22, use the quadratic formula.)*

5. $(x + 1)(x - 5) > 0$

6. $(x + 6)(x - 2) > 0$

7. $(x + 4)(x - 6) < 0$

8. $(x + 4)(x - 8) < 0$

9. $x^2 - 4x + 3 \geq 0$

10. $x^2 - 3x - 10 \geq 0$

11. $10x^2 + 9x \geq 9$

12. $3x^2 + 10x \geq 8$

13. $4x^2 - 9 \leq 0$

14. $9x^2 - 25 \leq 0$

15. $6x^2 + x \geq 1$

16. $4x^2 + 7x \geq -3$

17. $z^2 - 4z \geq 0$

18. $x^2 + 2x < 0$

19. $3x^2 - 5x \leq 0$

20. $2z^2 + 3z > 0$

21. $x^2 - 6x + 6 \geq 0$

22. $3x^2 - 6x + 2 \leq 0$

Solve each inequality. ***See Example 3.***

23. $(4 - 3x)^2 \geq -2$

24. $(7 - 6x)^2 \geq -1$

25. $(3x + 5)^2 \leq -4$

26. $(8x + 5)^2 \leq -5$

Solve each inequality, and graph the solution set. ***See Example 4.***

27. $(x - 1)(x - 2)(x - 4) < 0$

28. $(2x + 1)(3x - 2)(4x + 7) < 0$

29. $(x - 4)(2x + 3)(3x - 1) \geq 0$

30. $(x + 2)(4x - 3)(2x + 7) \geq 0$

Solve each inequality, and graph the solution set. ***See Examples 5 and 6.***

31. $\dfrac{x - 1}{x - 4} > 0$

32. $\dfrac{x + 1}{x - 5} > 0$

33. $\dfrac{2x + 3}{x - 5} \leq 0$

34. $\dfrac{3x + 7}{x - 3} \leq 0$

35. $\dfrac{8}{x - 2} \geq 2$

36. $\dfrac{20}{x - 1} \geq 1$

37. $\dfrac{3}{2x - 1} < 2$

38. $\dfrac{6}{x - 1} < 1$

39. $\dfrac{x - 3}{x + 2} \geq 2$

40. $\dfrac{m + 4}{m + 5} \geq 2$

41. $\dfrac{x - 8}{x - 4} < 3$

42. $\dfrac{2t - 3}{t + 1} > 4$

43. $\dfrac{4k}{2k - 1} < k$

44. $\dfrac{r}{r + 2} < 2r$

45. $\dfrac{2x - 3}{x^2 + 1} \geq 0$

46. $\dfrac{9x - 8}{4x^2 + 25} < 0$

47. $\dfrac{(3x - 5)^2}{x + 2} > 0$

48. $\dfrac{(5x - 3)^2}{2x + 1} \leq 0$

PREVIEW EXERCISES

Give the domain and the range of each function. ***See Section 3.5.***

49. $\{(0, 1), (1, 2), (2, 4), (3, 8)\}$

50. $f(x) = x^2$

Decide whether each graph is that of a function. ***See Section 3.5.***

51.

52.

CHAPTER 9 SUMMARY

TEST YOUR WORD POWER

See how well you have learned the vocabulary in this chapter.

1. The **quadratic formula** is
   A. a formula to find the number of solutions of a quadratic equation
   B. a formula to find the type of solutions of a quadratic equation
   C. the standard form of a quadratic equation
   D. a general formula for solving any quadratic equation.

2. A **quadratic function** is a function that can be written in the form
   A. $f(x) = mx + b$ for real numbers m and b
   B. $f(x) = \frac{P(x)}{Q(x)}$, where $Q(x) \neq 0$
   C. $f(x) = ax^2 + bx + c$ for real numbers a, b, and c ($a \neq 0$)
   D. $f(x) = \sqrt{x}$ for $x \geq 0$.

3. A **parabola** is the graph of
   A. any equation in two variables
   B. a linear equation
   C. an equation of degree 3
   D. a quadratic equation in two variables, where one is first-degree.

4. The **vertex** of a parabola is
   A. the point where the graph intersects the y-axis
   B. the point where the graph intersects the x-axis
   C. the lowest point on a parabola that opens up or the highest point on a parabola that opens down
   D. the origin.

5. The **axis** of a parabola is
   A. either the x-axis or the y-axis
   B. the vertical line (of a vertical parabola) or the horizontal line (of a horizontal parabola) through the vertex
   C. the lowest or highest point on the graph of a parabola
   D. a line through the origin.

6. A parabola is **symmetric about its axis** since
   A. its graph is near the axis
   B. its graph is identical on each side of the axis
   C. its graph looks different on each side of the axis
   D. its graph intersects the axis.

ANSWERS

1. D; *Example:* The solutions of $ax^2 + bx + c = 0$ ($a \neq 0$) are given by $x = \dfrac{-b \pm \sqrt{b^2 - 4ac}}{2a}$.   **2.** C; *Examples:* $f(x) = x^2 - 2$, $f(x) = (x + 4)^2 + 1$, $f(x) = x^2 - 4x + 5$ **3.** D; *Examples:* See the figures in the Quick Review for **Sections 9.5 and 9.6.**   **4.** C; *Example:* The graph of $y = (x + 3)^2$ has vertex $(-3, 0)$, which is the lowest point on the graph.   **5.** B; *Example:* The axis of $y = (x + 3)^2$ is the vertical line $x = -3$.   **6.** B; *Example:* Since the graph of $y = (x + 3)^2$ is symmetric about its axis $x = -3$, the points $(-2, 1)$ and $(-4, 1)$ are on the graph.

QUICK REVIEW

CONCEPTS	EXAMPLES
9.1 The Square Root Property and Completing the Square **Square Root Property** If x and k are complex numbers and $x^2 = k$, then $$x = \sqrt{k} \quad \text{or} \quad x = -\sqrt{k}.$$	Solve $(x - 1)^2 = 8$. $x - 1 = \sqrt{8}$ or $x - 1 = -\sqrt{8}$ $x = 1 + 2\sqrt{2}$ or $x = 1 - 2\sqrt{2}$ The solution set is $\left\{1 + 2\sqrt{2}, 1 - 2\sqrt{2}\right\}$, or $\left\{1 \pm 2\sqrt{2}\right\}$.

CONCEPTS	EXAMPLES

Completing the Square

To solve $ax^2 + bx + c = 0$ (with $a \neq 0$):

Step 1 If $a \neq 1$, divide each side by a.

Step 2 Write the equation with the variable terms on one side and the constant on the other.

Step 3 Take half the coefficient of x and square it.

Step 4 Add the square to each side.

Step 5 Factor the perfect square trinomial, and write it as the square of a binomial. Simplify the other side.

Step 6 Use the square root property to complete the solution.

Solve $2x^2 - 4x - 18 = 0$.

$$x^2 - 2x - 9 = 0 \qquad \text{Divide by 2.}$$

$$x^2 - 2x = 9 \qquad \text{Add 9.}$$

$$\left[\tfrac{1}{2}(-2)\right]^2 = (-1)^2 = 1$$

$$x^2 - 2x + 1 = 9 + 1 \qquad \text{Add 1.}$$

$$(x - 1)^2 = 10 \qquad \text{Factor. Add.}$$

$$x - 1 = \sqrt{10} \quad \text{or} \quad x - 1 = -\sqrt{10} \qquad \text{Square root property}$$

$$x = 1 + \sqrt{10} \quad \text{or} \quad x = 1 - \sqrt{10}$$

The solution set is $\left\{1 + \sqrt{10}, 1 - \sqrt{10}\right\}$, or $\left\{1 \pm \sqrt{10}\right\}$

9.2 The Quadratic Formula

Quadratic Formula

The solutions of $ax^2 + bx + c = 0$ (with $a \neq 0$) are given by

$$x = \frac{-b \pm \sqrt{b^2 - 4ac}}{2a}.$$

The Discriminant

If a, b, and c are integers, then the discriminant, $b^2 - 4ac$, of $ax^2 + bx + c = 0$ determines the number and type of solutions as follows.

Discriminant	Number and Type of Solutions
Positive, the square of an integer	Two rational solutions
Positive, not the square of an integer	Two irrational solutions
Zero	One rational solution
Negative	Two nonreal complex solutions

Solve $3x^2 + 5x + 2 = 0$.

$$x = \frac{-5 \pm \sqrt{5^2 - 4(3)(2)}}{2(3)} = \frac{-5 \pm 1}{6}$$

$$x = \frac{-5 + 1}{6} = -\frac{2}{3} \quad \text{or} \quad x = \frac{-5 - 1}{6} = -1$$

The solution set is $\left\{-1, -\frac{2}{3}\right\}$.

For $x^2 + 3x - 10 = 0$, the discriminant is

$$3^2 - 4(1)(-10) = 49. \qquad \text{Two rational solutions}$$

For $4x^2 + x + 1 = 0$, the discriminant is

$$1^2 - 4(4)(1) = -15. \qquad \text{Two nonreal complex solutions}$$

9.3 Equations Quadratic in Form

A nonquadratic equation that can be written in the form

$$au^2 + bu + c = 0,$$

for $a \neq 0$ and an algebraic expression u, is called quadratic in form. Substitute u for the expression, solve for u, and then solve for the variable in the expression.

Solve $3(x + 5)^2 + 7(x + 5) + 2 = 0$.

$$3u^2 + 7u + 2 = 0 \qquad \text{Let } u = x + 5.$$

$$(3u + 1)(u + 2) = 0 \qquad \text{Factor.}$$

$$u = -\frac{1}{3} \quad \text{or} \quad u = -2$$

$$x + 5 = -\frac{1}{3} \quad \text{or} \quad x + 5 = -2 \qquad x + 5 = u$$

$$x = -\frac{16}{3} \quad \text{or} \quad x = -7 \qquad \text{Subtract 5.}$$

The solution set is $\left\{-7, -\frac{16}{3}\right\}$.

(continued)

CONCEPTS	EXAMPLES

9.4 Formulas and Further Applications

To solve a formula for a squared variable, proceed as follows.

(a) If the variable appears only to the second power: Isolate the squared variable on one side of the equation, and then use the square root property.

(b) If the variable appears to the first and second powers: Write the equation in standard form, and then use the quadratic formula.

Solve $A = \dfrac{2mp}{r^2}$ for r.

$r^2 A = 2mp$ Multiply by r^2.

$r^2 = \dfrac{2mp}{A}$ Divide by A.

$r = \pm\sqrt{\dfrac{2mp}{A}}$ Square root property

$r = \dfrac{\pm\sqrt{2mpA}}{A}$ Rationalize denominator.

Solve $x^2 + rx = t$ for x.

$x^2 + rx - t = 0$ Standard form

$x = \dfrac{-r \pm \sqrt{r^2 - 4(1)(-t)}}{2(1)}$

$a = 1,\ b = r,\ c = -t$

$x = \dfrac{-r \pm \sqrt{r^2 + 4t}}{2}$

9.5 Graphs of Quadratic Functions

1. The graph of the quadratic function defined by $F(x) = a(x - h)^2 + k$, $a \neq 0$, is a parabola with vertex at (h, k) and the vertical line $x = h$ as axis.

2. The graph opens up if a is positive and down if a is negative.

3. The graph is wider than the graph of $f(x) = x^2$ if $0 < |a| < 1$ and narrower if $|a| > 1$.

Graph $f(x) = -(x + 3)^2 + 1$.

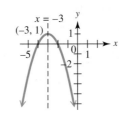

The graph opens down since $a < 0$.

Vertex: $(-3, 1)$

Axis: $x = -3$

Domain: $(-\infty, \infty)$

Range: $(-\infty, 1]$

9.6 More about Parabolas and Their Applications

The vertex of the graph of $f(x) = ax^2 + bx + c$, $a \neq 0$, may be found by completing the square.

The vertex has coordinates $\left(\dfrac{-b}{2a}, f\left(\dfrac{-b}{2a}\right)\right)$.

Graphing a Quadratic Function

Step 1   Determine whether the graph opens up or down.

Step 2   Find the vertex.

Step 3   Find the x-intercepts (if any). Find the y-intercept.

Step 4   Find and plot additional points as needed.

Horizontal Parabolas

The graph of

$$x = ay^2 + by + c \quad \text{or} \quad x = a(y - k)^2 + h$$

is a horizontal parabola with vertex (h, k) and the horizontal line $y = k$ as axis. The graph opens to the right if $a > 0$ and to the left if $a < 0$.

Horizontal parabolas do not represent functions.

Graph $f(x) = x^2 + 4x + 3$.

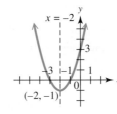

The graph opens up since $a > 0$.

Vertex: $(-2, -1)$

The solutions of $x^2 + 4x + 3 = 0$ are -1 and -3, so the x-intercepts are $(-1, 0)$ and $(-3, 0)$.

$f(0) = 3$, so the y-intercept is $(0, 3)$.

Domain: $(-\infty, \infty)$

Range: $[-1, \infty)$

Graph $x = 2y^2 + 6y + 5$.

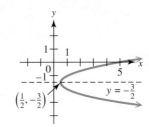

The graph opens to the right since $a > 0$.

Vertex: $\left(\dfrac{1}{2}, -\dfrac{3}{2}\right)$

Axis: $y = -\dfrac{3}{2}$

Domain: $\left[\dfrac{1}{2}, \infty\right)$

Range: $(-\infty, \infty)$

(continued)

CONCEPTS	EXAMPLES

9.7 Polynomial and Rational Inequalities

Solving a Quadratic (or Higher-Degree Polynomial) Inequality

Step 1 Write the inequality as an equation and solve.

Solve $2x^2 + 5x + 2 < 0$.
$$2x^2 + 5x + 2 = 0$$
$$(2x + 1)(x + 2) = 0$$
$$x = -\tfrac{1}{2} \quad \text{or} \quad x = -2$$

Step 2 Use the numbers found in Step 1 to divide a number line into intervals.

Intervals: $(-\infty, -2)$, $\left(-2, -\tfrac{1}{2}\right)$, $\left(-\tfrac{1}{2}, \infty\right)$

Step 3 Substitute a test number from each interval into the original inequality to determine the intervals that belong to the solution set.

Step 4 Consider the endpoints separately.

Test values: $-3, -1, 0$
$x = -3$ makes the original inequality false, $x = -1$ makes it true, and $x = 0$ makes it false. Choose the interval(s) which yield(s) a true statement. The solution set is the interval $\left(-2, -\tfrac{1}{2}\right)$.

Solving a Rational Inequality

Step 1 Write the inequality so that 0 is on one side and there is a single fraction on the other side.

Solve $\dfrac{x}{x + 2} \geq 4$.

$$\frac{x}{x + 2} - 4 \geq 0 \qquad \text{Subtract 4.}$$

$$\frac{x}{x + 2} - \frac{4(x + 2)}{x + 2} \geq 0 \qquad \text{Write with a common denominator.}$$

$$\frac{-3x - 8}{x + 2} \geq 0 \qquad \text{Subtract fractions.}$$

Step 2 Determine the numbers that make the numerator or denominator 0.

$-\tfrac{8}{3}$ makes the numerator 0, and -2 makes the denominator 0.

Step 3 Use the numbers from Step 2 to divide a number line into intervals.

Step 4 Substitute a test number from each interval into the original inequality to determine the intervals that belong to the solution set.

-4 from A makes the original inequality false, $-\tfrac{7}{3}$ from B makes it true, and 0 from C makes it false.

Step 5 Consider the endpoints separately.

The solution set is the interval $\left[-\tfrac{8}{3}, -2\right)$. The endpoint -2 is not included since it makes the denominator 0.

CHAPTER 9

REVIEW EXERCISES

9.1 *Solve each equation by using the square root property or completing the square.*

1. $t^2 = 121$ **2.** $p^2 = 3$ **3.** $(2x + 5)^2 = 100$

*4. $(3x - 2)^2 = -25$ **5.** $x^2 + 4x = 15$ **6.** $2x^2 - 3x = -1$

---

*This exercise requires knowledge of complex numbers.

7. *Concept Check* A student gave the following incorrect "solution." *WHAT WENT WRONG?*

$$x^2 = 12$$
$$x = \sqrt{12} \qquad \text{Square root property}$$
$$x = 2\sqrt{3}$$

Solution set: $\left\{2\sqrt{3}\right\}$

8. The Singapore Flyer, the world's largest Ferris wheel as of 2008, has a height of 165 m. Use the metric version of Galileo's formula,

$$d = 4.9t^2 \quad \text{(where } d \text{ is in meters),}$$

to find how long it would take a wallet dropped from the top of the Singapore Flyer to reach the ground. Round your answer to the nearest tenth of a second. (*Source:* www.singaporeflyer.com)

9.2 *Solve each equation by using the quadratic formula.*

9. $2x^2 + x - 21 = 0$ **10.** $x^2 + 5x = 7$ **11.** $(t + 3)(t - 4) = -2$

*12. $2x^2 + 3x + 4 = 0$ ***13.** $3p^2 = 2(2p - 1)$ **14.** $x(2x - 7) = 3x^2 + 3$

Use the discriminant to predict whether the solutions to each equation are

A. *two rational numbers*   **B.** *one rational number*

C. *two irrational numbers*   **D.** *two nonreal complex numbers.*

15. $x^2 + 5x + 2 = 0$ **16.** $4t^2 = 3 - 4t$

17. $4x^2 = 6x - 8$ **18.** $9z^2 + 30z + 25 = 0$

9.3 *Solve each equation. Check your solutions.*

19. $\dfrac{15}{x} = 2x - 1$ **20.** $\dfrac{1}{n} + \dfrac{2}{n + 1} = 2$

21. $-2r = \sqrt{\dfrac{48 - 20r}{2}}$ **22.** $8(3x + 5)^2 + 2(3x + 5) - 1 = 0$

23. $2x^{2/3} - x^{1/3} - 28 = 0$ **24.** $p^4 - 10p^2 + 9 = 0$

Solve each problem. Round answers to the nearest tenth, as necessary.

25. Bahaa Mourad paddled a canoe 20 mi upstream, then paddled back. If the rate of the current was 3 mph and the total trip took 7 hr, what was Bahaa's rate?

26. Carol-Ann Vassell drove 8 mi to pick up a friend, and then drove 11 mi to a mall at a rate 15 mph faster. If Carol-Ann's total travel time was 24 min, what was her rate on the trip to pick up her friend?

27. An old machine processes a batch of checks in 1 hr more time than a new one. How long would it take the old machine to process a batch of checks that the two machines together process in 2 hr?

28. Zoran Pantic can process a stack of invoices 1 hr faster than Claude Sassine can. Working together, they take 1.5 hr. How long would it take each person working alone?

9.4 *Solve each formula for the indicated variable. (Give answers with ±.)*

29. $k = \dfrac{rF}{wv^2}$ for v **30.** $p = \sqrt{\dfrac{yz}{6}}$ for y **31.** $mt^2 = 3mt + 6$ for t

---

*This exercise requires knowledge of complex numbers.

Solve each problem. Round answers to the nearest tenth, as necessary.

32. A large machine requires a part in the shape of a right triangle with a hypotenuse 9 ft less than twice the length of the longer leg. The shorter leg must be $\frac{3}{4}$ the length of the longer leg. Find the lengths of the three sides of the part.

33. A square has an area of 256 cm². If the same amount is removed from one dimension and added to the other, the resulting rectangle has an area 16 cm² less. Find the dimensions of the rectangle.

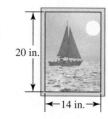

34. Allen Moser wants to buy a mat for a photograph that measures 14 in. by 20 in. He wants to have an even border around the picture when it is mounted on the mat. If the area of the mat he chooses is 352 in.², how wide will the border be?

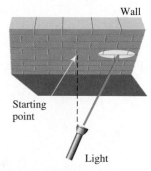

35. If a square piece of cardboard has 3-in. squares cut from its corners and then has the flaps folded up to form an open-top box, the volume of the box is given by the formula $V = 3(x - 6)^2$, where x is the length of each side of the original piece of cardboard in inches. What original length would yield a box with volume 432 in.³?

36. Wachovia Center Tower in Raleigh, North Carolina, is 400 ft high. Suppose that a ball is projected upward from the top of the tower, and its position in feet above the ground is given by the quadratic function defined by

$$f(t) = -16t^2 + 45t + 400,$$

where t is the number of seconds elapsed. How long will it take for the ball to reach a height of 200 ft above the ground? (*Source: World Almanac and Book of Facts.*)

37. A searchlight moves horizontally back and forth along a wall with the distance of the light from a starting point at t minutes given by the quadratic function defined by

$$f(t) = 100t^2 - 300t.$$

How long will it take before the light returns to the starting point?

38. Internet publishing and broadcasting revenue in the United States (in millions of dollars) for the years 2004–2007 is shown in the graph and can be modeled by the quadratic function defined by

$$f(x) = 230.5x^2 - 252.9x + 5987.$$

In the model, $x = 4$ represents 2004, $x = 5$ represents 2005, and so on.

(a) Use the model to approximate revenue from Internet publishing and broadcasting in 2007 to the nearest million dollars. How does this result compare to the number suggested by the graph?

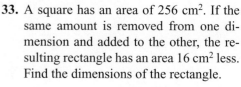

Internet Publishing and Broadcasting Revenue

Source: U.S. Census Bureau.

(b) Based on the model, in what year did the revenue from Internet publishing and broadcasting reach $14,000 million ($14 billion)? (Round down for the year.) How does this result compare to the number shown in the graph?

| 9.5–9.6 | *Identify the vertex of each parabola.*

39. $f(x) = -(x - 1)^2$ **40.** $f(x) = (x - 3)^2 + 7$

41. $x = (y - 3)^2 - 4$ **42.** $y = -3x^2 + 4x - 2$

Graph each parabola. Give the vertex, axis, domain, and range.

43. $y = 2(x - 2)^2 - 3$ **44.** $f(x) = -2x^2 + 8x - 5$

45. $x = 2(y + 3)^2 - 4$ **46.** $x = -\frac{1}{2}y^2 + 6y - 14$

Solve each problem.

47. Total consumer spending on computers, peripherals, and software in the United States for selected years is given in the table. Let $x = 0$ represent 1985, $x = 5$ represent 1990, and so on.

CONSUMER SPENDING ON COMPUTERS, PERIPHERALS, AND SOFTWARE

Year	Spending (billions of dollars)
1985	2.9
1990	8.9
1995	24.3
2000	43.8
2004	51.6
2005	56.5
2006	61.4

Source: Bureau of Economic Analysis.

  (a) Use the data for 1985, 1995, and 2005 in the quadratic form $ax^2 + bx + c = y$ to write a system of three equations.

  (b) Solve the system from part (a) to get a quadratic function f that models the data.

  (c) Use the model found in part (b) to approximate consumer spending for computers, peripherals, and software games in 2006 to the nearest tenth. How does your answer compare to the actual data from the table?

48. The height (in feet) of a projectile t seconds after being fired from Earth into the air is given by

$$f(t) = -16t^2 + 160t.$$

Find the number of seconds required for the projectile to reach maximum height. What is the maximum height?

49. Find the length and width of a rectangle having a perimeter of 200 m if the area is to be a maximum. What is the maximum area?

| 9.7 | *Solve each inequality, and graph the solution set.*

50. $(x - 4)(2x + 3) > 0$ **51.** $x^2 + x \le 12$

52. $(x + 2)(x - 3)(x + 5) \le 0$ **53.** $(4x + 3)^2 \le -4$

54. $\dfrac{6}{2z - 1} < 2$ **55.** $\dfrac{3t + 4}{t - 2} \le 1$

MIXED REVIEW EXERCISES

Solve.

56. $V = r^2 + R^2h$ for R ***57.** $3t^2 - 6t = -4$ **58.** $(3x + 11)^2 = 7$

59. $S = \dfrac{Id^2}{k}$ for d **60.** $(8x - 7)^2 \ge -1$ **61.** $2x - \sqrt{x} = 6$

*This exercise requires knowledge of complex numbers.

(continued)

62. $x^4 - 8x^2 = -1$　　　**63.** $\dfrac{-2}{x + 5} \le -5$　　　**64.** $6 + \dfrac{15}{s^2} = -\dfrac{19}{s}$

65. $(x^2 - 2x)^2 = 11(x^2 - 2x) - 24$　　　**66.** $(r - 1)(2r + 3)(r + 6) < 0$

67. *Concept Check* Match each equation in parts (a)–(f) with the figure that most closely resembles its graph in choices A–F.

(a) $g(x) = x^2 - 5$　　　**(b)** $h(x) = -x^2 + 4$　　　**(c)** $F(x) = (x - 1)^2$

(d) $G(x) = (x + 1)^2$　　　**(e)** $H(x) = (x - 1)^2 + 1$　　　**(f)** $K(x) = (x + 1)^2 + 1$

A.
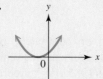

B.
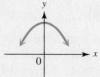

C.

D.
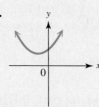

E.
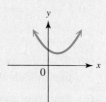

F.
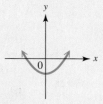

68. Graph $f(x) = 4x^2 + 4x - 2$. Give the vertex, axis, domain, and range.

69. In 4 hr, Rajeed Carriman can go 15 mi upriver and come back. The rate of the current is 5 mph. Find the rate of the boat in still water.

70. Two pieces of a large wooden puzzle fit together to form a rectangle with length 1 cm less than twice the width. The diagonal, where the two pieces meet, is 2.5 cm in length. Find the length and width of the rectangle.

CHAPTER 9 TEST

Step-by-step test solutions are found on the Chapter Test Prep Videos available via the Video Resources on DVD, in *MyMathLab*, or on YouTube (search "LialIntermediateAlg").

View the complete solutions to all Chapter Test exercises on the Video Resources on DVD.

Solve each equation by using the square root property or completing the square.

1. $t^2 = 54$　　　**2.** $(7x + 3)^2 = 25$　　　**3.** $x^2 + 2x = 4$

Solve by using the quadratic formula.

4. $2x^2 - 3x - 1 = 0$　　　***5.** $3t^2 - 4t = -5$　　　**6.** $3x = \sqrt{\dfrac{9x + 2}{2}}$

*7. *Concept Check* If k is a negative number, then which one of the following equations will have two nonreal complex solutions?

A. $x^2 = 4k$　　　**B.** $x^2 = -4k$　　　**C.** $(x + 2)^2 = -k$　　　**D.** $x^2 + k = 0$

8. What is the discriminant for $2x^2 - 8x - 3 = 0$? How many and what type of solutions does this equation have? (Do not actually solve.)

---

*This exercise requires knowledge of complex numbers.

Solve by any method.

9. $3 - \dfrac{16}{x} - \dfrac{12}{x^2} = 0$

10. $4x^2 + 7x - 3 = 0$

11. $9x^4 + 4 = 37x^2$

12. $12 = (2n + 1)^2 + (2n + 1)$

13. Solve $S = 4\pi r^2$ for r. (Leave $\pm$ in your answer.)

Solve each problem.

14. Terry and Callie do word processing. For a certain prospectus, Callie can prepare it 2 hr faster than Terry can. If they work together, they can do the entire prospectus in 5 hr. How long will it take each of them working alone to prepare the prospectus? Round your answers to the nearest tenth of an hour.

15. Qihong Shen paddled a canoe 10 mi upstream and then paddled back to the starting point. If the rate of the current was 3 mph and the entire trip took $3\frac{1}{2}$ hr, what was Qihong's rate?

16. Endre Borsos has a pool 24 ft long and 10 ft wide. He wants to construct a concrete walk around the pool. If he plans for the walk to be of uniform width and cover 152 ft², what will the width of the walk be?

17. At a point 30 m from the base of a tower, the distance to the top of the tower is 2 m more than twice the height of the tower. Find the height of the tower.

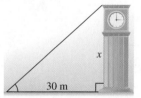

18. *Concept Check* Which one of the following figures most closely resembles the graph of $f(x) = a(x - h)^2 + k$ if $a < 0, h > 0,$ and $k < 0$?

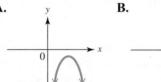

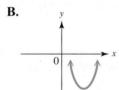

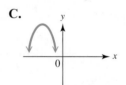

Graph each parabola. Identify the vertex, axis, domain, and range.

19. $f(x) = \dfrac{1}{2}x^2 - 2$

20. $f(x) = -x^2 + 4x - 1$

21. $x = -(y - 2)^2 + 2$

Solve each problem.

22. The total number (in millions) of civilians employed in the United States during the years 2004–2008 can be modeled by the quadratic function defined by

$$f(x) = -0.529x^2 + 8.00x + 115$$

where $x = 4$ represents 2004, $x = 5$ represents 2005, and so on. (*Source:* U.S. Bureau of Labor Statistics.)

(a) Based on this model, how many civilians, to the nearest million, were employed in the United States in 2004?

(b) In what year during this period was the maximum civilian employment? (Round down for the year.) To the nearest million, what was the total civilian employment in that year? Use the actual x-value, to the nearest tenth, to find this number.

23. Houston Community College is planning to construct a rectangular parking lot on land bordered on one side by a highway. The plan is to use 640 ft of fencing to fence off the other three sides. What should the dimensions of the lot be if the enclosed area is to be a maximum?

Solve each inequality, and graph the solution set.

24. $2x^2 + 7x > 15$

25. $\dfrac{5}{t - 4} \leq 1$

CHAPTERS **(1–9)** CUMULATIVE REVIEW EXERCISES

1. Let $S = \left\{ -\frac{7}{3}, -2, -\sqrt{3}, 0, 0.7, \sqrt{12}, \sqrt{-8}, 7, \frac{32}{3} \right\}$. List the elements of S that are elements of each set.

    (a) Integers   **(b)** Rational numbers   **(c)** Real numbers   **(d)** Complex numbers

Solve each equation or inequality.

2. $7 - (4 + 3t) + 2t = -6(t - 2) - 5$

3. $|6x - 9| = |-4x + 2|$

4. $2x = \sqrt{\dfrac{5x + 2}{3}}$

5. $\dfrac{3}{x - 3} - \dfrac{2}{x - 2} = \dfrac{3}{x^2 - 5x + 6}$

6. $(r - 5)(2r + 3) = 1$

7. $x^4 - 5x^2 + 4 = 0$

8. $-2x + 4 \leq -x + 3$

9. $|3x - 7| \leq 1$

10. $x^2 - 4x + 3 < 0$

11. $\dfrac{3}{p + 2} > 1$

Graph each relation. Tell whether or not y can be expressed as a function f of x, and if so, give its domain and range, and write using function notation.

12. $4x - 5y = 15$

13. $4x - 5y < 15$

14. $y = -2(x - 1)^2 + 3$

15. Find the slope and intercepts of the line with equation $-2x + 7y = 16$.

16. Write an equation for the specified line. Express each equation in slope-intercept form.

    (a) Through $(2, -3)$ and parallel to the line with equation $5x + 2y = 6$

    (b) Through $(-4, 1)$ and perpendicular to the line with equation $5x + 2y = 6$

Solve each system of equations.

17. $2x - 4y = 10$
    $9x + 3y = 3$

18. $x + y + 2z = 3$
    $-x + y + z = -5$
    $2x + 3y - z = -8$

19. In 2009, the two American computer software companies with the greatest revenues were Microsoft and Oracle. The two companies had combined revenues of $82.8 billion. Revenues for Microsoft were $6.8 billion less than three times those of Oracle. What were the 2009 revenues for each company? (*Source: Fortune.*)

Write with positive exponents only. Assume that variables represent positive real numbers.

20. $\left(\dfrac{x^{-3}y^2}{x^5 y^{-2}} \right)^{-1}$

21. $\dfrac{(4x^{-2})^2(2y^3)}{8x^{-3}y^5}$

Perform the indicated operations.

22. $\left(\dfrac{2}{3}t + 9\right)^2$

23. Divide $4x^3 + 2x^2 - x + 26$ by $x + 2$.

Factor completely.

24. $24m^2 + 2m - 15$

25. $8x^3 + 27y^3$

26. $9x^2 - 30xy + 25y^2$

Perform the indicated operations or simplify the complex fraction, and express each answer in lowest terms. Assume denominators are nonzero.

27. $\dfrac{5x + 2}{-6} \div \dfrac{15x + 6}{5}$

28. $\dfrac{3}{2 - x} - \dfrac{5}{x} + \dfrac{6}{x^2 - 2x}$

29. $\dfrac{\dfrac{r}{s} - \dfrac{s}{r}}{\dfrac{r}{s} + 1}$

Simplify each radical expression.

30. $\sqrt[3]{\dfrac{27}{16}}$

31. $\dfrac{2}{\sqrt{7} - \sqrt{5}}$

32. Two cars left an intersection at the same time, one heading due south and the other due east. Later they were exactly 95 mi apart. The car heading east had gone 38 mi less than twice as far as the car heading south. How far had each car traveled?

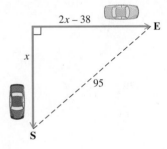

Inverse, Exponential, and Logarithmic Functions

In 2001, Apple Computer Inc., introduced the iPod. By mid-2009, the company had sold over 220 million of the popular music players, in spite of warnings by experts that listening to the devices at high volumes may put people at increased risk of hearing loss.

In **Example 4** of **Section 10.5,** we use a *logarithmic function* to calculate the volume level, in *decibels,* of an iPod.

10.1 Inverse Functions

OBJECTIVES

1 Decide whether a function is one-to-one and, if it is, find its inverse.

2 Use the horizontal line test to determine whether a function is one-to-one.

3 Find the equation of the inverse of a function.

4 Graph f^{-1} given the graph of f.

In this chapter we study two important types of functions, *exponential* and *logarithmic*. These functions are related: They are *inverses* of one another.

OBJECTIVE 1 **Decide whether a function is one-to-one and, if it is, find its inverse.** Suppose we define the function

$$G = \{(-2, 2), (-1, 1), (0, 0), (1, 3), (2, 5)\}.$$

We can form another set of ordered pairs from G by interchanging the x- and y-values of each pair in G. We can call this set F, so

$$F = \{(2, -2), (1, -1), (0, 0), (3, 1), (5, 2)\}.$$

To show that these two sets are related as just described, F is called the *inverse* of G. For a function f to have an inverse, f must be a *one-to-one function*.

One-to-One Function

In a **one-to-one function**, each x-value corresponds to only one y-value, and each y-value corresponds to only one x-value.

The function shown in **FIGURE 1(a)** is not one-to-one because the y-value 7 corresponds to *two* x-values, 2 and 3. That is, the ordered pairs $(2, 7)$ and $(3, 7)$ both belong to the function. The function in **FIGURE 1(b)** is one-to-one.

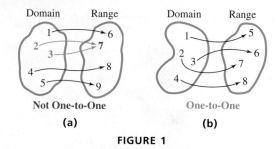

FIGURE 1

The *inverse* of any one-to-one function f is found by interchanging the components of the ordered pairs of f. The inverse of f is written f^{-1}. Read f^{-1} as **"the inverse of f"** or **"f-inverse."**

⚠ **CAUTION** The symbol $f^{-1}(x)$ does **not** represent $\dfrac{1}{f(x)}$.

The definition of the inverse of a function follows.

Inverse of a Function

The **inverse** of a one-to-one function f, written f^{-1}, is the set of all ordered pairs of the form (y, x), where (x, y) belongs to f. Since the inverse is formed by interchanging x and y, the domain of f becomes the range of f^{-1} and the range of f becomes the domain of f^{-1}.

For inverses f and f^{-1}, it follows that for all x in their domains,

$$(f \circ f^{-1})(x) = x \quad \text{and} \quad (f^{-1} \circ f)(x) = x.$$

NOW TRY
EXERCISE 1

Decide whether each function is one-to-one. If it is, find the inverse.

(a) $F = \{(-1, -2), (0, 0)$
$(1, -2), (2, -8)\}$

(b) $G = \{(0, 0), (1, 1),$
$(4, 2), (9, 3)\}$

(c) The number of stories and height of several tall buildings are given in the table.

Stories	Height
31	639
35	582
40	620
41	639
64	810

EXAMPLE 1 Finding Inverses of One-to-One Functions

Decide whether each function is one-to-one. If it is, find the inverse.

(a) $F = \{(-2, 1), (-1, 0), (0, 1), (1, 2), (2, 2)\}$

Each x-value in F corresponds to just one y-value. However, the y-value 1 corresponds to two x-values, -2 and 0. Also, the y-value 2 corresponds to both 1 and 2. Because some y-values correspond to more than one x-value, F is not one-to-one and does not have an inverse.

(b) $G = \{(3, 1), (0, 2), (2, 3), (4, 0)\}$

Every x-value in G corresponds to only one y-value, and every y-value corresponds to only one x-value, so G is a one-to-one function. The inverse function is found by interchanging the x- and y-values in each ordered pair.

$$G^{-1} = \{(1, 3), (2, 0), (3, 2), (0, 4)\}$$

The domain and range of G become the range and domain, respectively, of G^{-1}.

(c) The table shows the number of days in which the air in Connecticut exceeded the 8-hour average ground-level ozone standard for the years 1997–2006.

Year	Number of Days Exceeding Standard	Year	Number of Days Exceeding Standard
1997	27	2002	36
1998	25	2003	14
1999	33	2004	6
2000	13	2005	20
2001	26	2006	13

Source: U.S. Environmental Protection Agency.

Let f be the function defined in the table, with the years forming the domain and the numbers of days exceeding the ozone standard forming the range. Then f is not one-to-one, because in two different years (2000 and 2006), the number of days with unacceptable ozone levels was the same, 13. NOW TRY

OBJECTIVE 2 **Use the horizontal line test to determine whether a function is one-to-one.** By graphing a function and observing the graph, we can use the *horizontal line test* to tell whether the function is one-to-one.

Horizontal Line Test

A function is one-to-one if every horizontal line intersects the graph of the function at most once.

NOW TRY ANSWERS
1. (a) not one-to-one
   (b) one-to-one;
   $G^{-1} = \{(0, 0), (1, 1),$
   $(2, 4), (3, 9)\}$
   (c) not one-to-one

The horizontal line test follows from the definition of a one-to-one function. Any two points that lie on the same horizontal line have the same y-coordinate. No two ordered pairs that belong to a one-to-one function may have the same y-coordinate. Therefore, no horizontal line will intersect the graph of a one-to-one function more than once.

NOW TRY
EXERCISE 2

Use the horizontal line test to determine whether each graph is the graph of a one-to-one function.

(a)

(b)

EXAMPLE 2 Using the Horizontal Line Test

Use the horizontal line test to determine whether each graph is the graph of a one-to-one function.

(a)

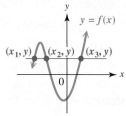

Because a horizontal line intersects the graph in more than one point (actually three points), the function is not one-to-one.

(b)

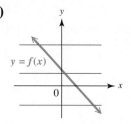

Every horizontal line will intersect the graph in exactly one point. This function is one-to-one.

NOW TRY

OBJECTIVE 3 **Find the equation of the inverse of a function.** The inverse of a one-to-one function is found by interchanging the x- and y-values of each of its ordered pairs. The equation of the inverse of a function defined by $y = f(x)$ is found in the same way.

Finding the Equation of the Inverse of $y = f(x)$

For a one-to-one function f defined by an equation $y = f(x)$, find the defining equation of the inverse as follows.

Step 1 Interchange x and y.

Step 2 Solve for y.

Step 3 Replace y with $f^{-1}(x)$.

EXAMPLE 3 Finding Equations of Inverses

Decide whether each equation defines a one-to-one function. If so, find the equation that defines the inverse.

(a) $f(x) = 2x + 5$

The graph of $y = 2x + 5$ is a nonvertical line, so by the horizontal line test, f is a one-to-one function. To find the inverse, let $y = f(x)$ and follow the steps.

$$y = 2x + 5$$

$$x = 2y + 5 \qquad \text{Interchange } x \text{ and } y. \text{ (Step 1)}$$

$$2y = x - 5 \qquad \text{Solve for } y. \text{ (Step 2)}$$

$$y = \frac{x - 5}{2}$$

$$f^{-1}(x) = \frac{x - 5}{2} \qquad \text{Replace } y \text{ with } f^{-1}(x). \text{ (Step 3)}$$

This equation can be written as follows.

NOW TRY ANSWERS
2. **(a)** one-to-one
   (b) not one-to-one

$$f^{-1}(x) = \frac{x}{2} - \frac{5}{2}, \quad \text{or} \quad f^{-1}(x) = \frac{1}{2}x - \frac{5}{2} \qquad \frac{a - b}{c} = \frac{a}{c} - \frac{b}{c}$$

NOW TRY
EXERCISE 3

Decide whether each equation defines a one-to-one function. If so, find the equation that defines the inverse.

(a) $f(x) = 5x - 7$

(b) $f(x) = (x + 1)^2$

(c) $f(x) = x^3 - 4$

Thus, f^{-1} is a linear function. In the function defined by

$$y = 2x + 5,$$

the value of y is found by starting with a value of x, multiplying by 2, and adding 5. The equation

$$f^{-1}(x) = \frac{x - 5}{2}$$

for the inverse has us *subtract* 5, and then *divide* by 2. This shows how an inverse is used to "undo" what a function does to the variable x.

(b) $y = x^2 + 2$

This equation has a vertical parabola as its graph, so some horizontal lines will intersect the graph at two points. For example, both $x = 3$ and $x = -3$ correspond to $y = 11$. Because of the x^2-term, there are many pairs of x-values that correspond to the same y-value. This means that the function defined by $y = x^2 + 2$ is not one-to-one and does not have an inverse.

Alternatively, applying the steps for finding the equation of an inverse leads to the following.

$$y = x^2 + 2$$

$$x = y^2 + 2 \qquad \text{Interchange } x \text{ and } y.$$

$$y^2 = x - 2 \qquad \text{Solve for } y.$$

$$y = \pm\sqrt{x - 2} \qquad \text{Square root property}$$

The last step shows that there are two y-values for each choice of $x > 2$, so we again see that the given function is not one-to-one. It does not have an inverse.

(c) $f(x) = (x - 2)^3$

Refer to **Section 5.3** to see that the graph of a cubing function like this is one-to-one.

$$f(x) = (x - 2)^3$$

$$y = (x - 2)^3 \qquad \text{Replace } f(x) \text{ with } y.$$

$$x = (y - 2)^3 \qquad \text{Interchange } x \text{ and } y.$$

$$\sqrt[3]{x} = \sqrt[3]{(y - 2)^3} \qquad \text{Take the cube root on each side.}$$

$$\sqrt[3]{x} = y - 2 \qquad \sqrt[3]{a^3} = a$$

$$y = \sqrt[3]{x} + 2 \qquad \text{Solve for } y.$$

$$f^{-1}(x) = \sqrt[3]{x} + 2 \qquad \text{Replace } y \text{ with } f^{-1}(x). \qquad \text{NOW TRY}$$

OBJECTIVE 4 **Graph f^{-1}, given the graph of f.** One way to graph the inverse of a function f whose equation is given is as follows.

1. Find several ordered pairs that belong to f.

2. Interchange x and y to obtain ordered pairs that belong to f^{-1}.

3. Plot those points, and sketch the graph of f^{-1} through them.

A simpler way is to select points on the graph of f and use symmetry to find corresponding points on the graph of f^{-1}.

NOW TRY ANSWERS

3. (a) one-to-one function;
   $f^{-1}(x) = \frac{x + 7}{5}$, or
   $f^{-1}(x) = \frac{1}{5}x + \frac{7}{5}$
   (b) not a one-to-one function
   (c) one-to-one function;
   $f^{-1}(x) = \sqrt[3]{x + 4}$

For example, suppose the point (a, b) shown in **FIGURE 2** belongs to a one-to-one function f. Then the point (b, a) belongs to f^{-1}. The line segment connecting (a, b) and (b, a) is perpendicular to, and cut in half by, the line $y = x$. The points (a, b) and (b, a) are "mirror images" of each other with respect to $y = x$.

We can find the graph of f^{-1} from the graph of f by locating the mirror image of each point in f with respect to the line $y = x$.

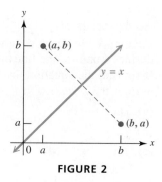

FIGURE 2

NOW TRY
EXERCISE 4

Use the given graph to graph the inverse of f.

EXAMPLE 4 Graphing the Inverse

Graph the inverses of the functions f (shown in blue) in **FIGURE 3**.

In **FIGURE 3** the graphs of two functions f are shown in blue. Their inverses are shown in red. In each case, the graph of f^{-1} is a reflection of the graph of f with respect to the line $y = x$.

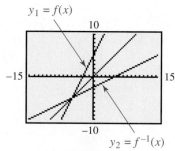

FIGURE 3

NOW TRY

CONNECTIONS

In **Example 3** we showed that the inverse of the one-to-one function defined by $f(x) = 2x + 5$ is given by $f^{-1}(x) = \frac{x - 5}{2}$. If we use a square viewing window of a graphing calculator and graph

$$y_1 = f(x) = 2x + 5, \quad y_2 = f^{-1}(x) = \frac{x - 5}{2}, \quad \text{and} \quad y_3 = x,$$

we can see how this reflection appears on the screen. See **FIGURE 4**.

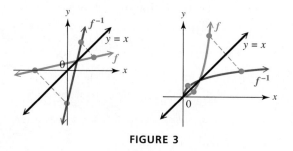

FIGURE 4

NOW TRY ANSWER

4.

For Discussion or Writing

Some graphing calculators have the capability to "draw" the inverse of a function. Use a graphing calculator to draw the graphs of $f(x) = x^3 + 2$ and its inverse in a square viewing window.

10.1 EXERCISES

Complete solution available on the Video Resources on DVD

In Exercises 1–4, write a few sentences of explanation. ***See Example 1.***

1. A study found that the trans fat content in fast-food products varied widely around the world, based on the type of frying oil used, as shown in the table. If the set of countries is the domain and the set of trans fat percentages is the range of the function consisting of the six pairs listed, is it a one-to-one function? Why or why not?

Country	Percentage of Trans Fat in McDonald's Chicken
Scotland	14
France	11
United States	11
Peru	9
Russia	5
Denmark	1

Source: New England Journal of Medicine.

2. The table shows the number of uncontrolled hazardous waste sites in 2008 that require further investigation to determine whether remedies are needed under the Superfund program. The eight states listed are ranked in the top ten on the EPA's National Priority List.

If this correspondence is considered to be a function that pairs each state with its number of uncontrolled waste sites, is it one-to-one? If not, explain why.

State	Number of Sites
New Jersey	116
California	97
Pennsylvania	96
New York	86
Michigan	67
Florida	52
Illinois	49
Texas	49

Source: U.S. Environmental Protection Agency.

3. The road mileage between Denver, Colorado, and several selected U.S. cities is shown in the table. If we consider this as a function that pairs each city with a distance, is it a one-to-one function? How could we change the answer to this question by adding 1 mile to one of the distances shown?

City	Distance to Denver (in miles)
Atlanta	1398
Dallas	781
Indianapolis	1058
Kansas City, MO	600
Los Angeles	1059
San Francisco	1235

4. Suppose you consider the set of ordered pairs (x, y) such that x represents a person in your mathematics class and y represents that person's father. Explain how this function might not be a one-to-one function.

In Exercises 5–8, choose the correct response from the given list.

5. *Concept Check* If a function is made up of ordered pairs in such a way that the same y-value appears in a correspondence with two different x-values, then

A. the function is one-to-one   **B.** the function is not one-to-one

C. its graph does not pass the vertical line test

D. it has an inverse function associated with it.

6. Which equation defines a one-to-one function? Explain why the others are not, using specific examples.

   A. $f(x) = x$ **B.** $f(x) = x^2$ **C.** $f(x) = |x|$ **D.** $f(x) = -x^2 + 2x - 1$

7. Only one of the graphs illustrates a one-to-one function. Which one is it? **(See Example 2.)**

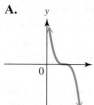 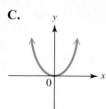 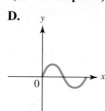

   A.   **B.**   **C.**   **D.**

8. *Concept Check*   If a function f is one-to-one and the point (p, q) lies on the graph of f, then which point *must* lie on the graph of f^{-1}?

   A. $(-p, q)$ **B.** $(-q, -p)$ **C.** $(p, -q)$ **D.** (q, p)

If the function is one-to-one, find its inverse. **See Examples 1–3.**

9. $\{(3, 6), (2, 10), (5, 12)\}$

10. $\left\{(-1, 3), (0, 5), (5, 0), \left(7, -\dfrac{1}{2}\right)\right\}$

11. $\{(-1, 3), (2, 7), (4, 3), (5, 8)\}$

12. $\{(-8, 6), (-4, 3), (0, 6), (5, 10)\}$

13. $f(x) = 2x + 4$

14. $f(x) = 3x + 1$

15. $g(x) = \sqrt{x - 3}, \quad x \geq 3$

16. $g(x) = \sqrt{x + 2}, \quad x \geq -2$

17. $f(x) = 3x^2 + 2$

18. $f(x) = 4x^2 - 1$

19. $f(x) = x^3 - 4$

20. $f(x) = x^3 + 5$

Concept Check   Let $f(x) = 2^x$. We will see in the next section that this function is one-to-one. Find each value, always working part (a) before part (b).

21. **(a)** $f(3)$ **22. (a)** $f(4)$ **23. (a)** $f(0)$ **24. (a)** $f(-2)$

   (b) $f^{-1}(8)$ **(b)** $f^{-1}(16)$ **(b)** $f^{-1}(1)$ **(b)** $f^{-1}\left(\dfrac{1}{4}\right)$

The graphs of some functions are given in Exercises 25–30. **(a)** *Use the horizontal line test to determine whether the function is one-to-one.* **(b)** *If the function is one-to-one, then graph the inverse of the function.* (*Remember that if f is one-to-one and (a, b) is on the graph of f, then (b, a) is on the graph of f^{-1}.*) **See Example 4.**

25.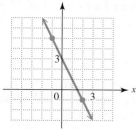

26.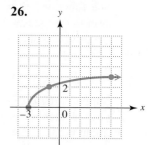

27.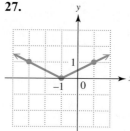

28.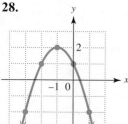

29.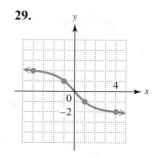

30.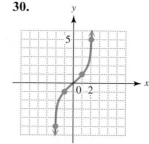

Each function defined in Exercises 31–38 is a one-to-one function. Graph the function as a solid line (or curve) and then graph its inverse on the same set of axes as a dashed line (or curve). In Exercises 35–38 complete the table so that graphing the function will be easier. **See Example 4.**

31. $f(x) = 2x - 1$ **32.** $f(x) = 2x + 3$ **33.** $g(x) = -4x$ **34.** $g(x) = -2x$

35. $f(x) = \sqrt{x},$ **36.** $f(x) = -\sqrt{x},$ **37.** $f(x) = x^3 - 2$ **38.** $f(x) = x^3 + 3$
$\quad\; x \geq 0$ $\qquad\; x \geq 0$

x	f(x)
0	
1	
4	

x	f(x)
0	
1	
4	

x	f(x)
-1	
0	
1	
2	

x	f(x)
-2	
-1	
0	
1	

RELATING CONCEPTS   EXERCISES 39–42

FOR INDIVIDUAL OR GROUP WORK

Inverse functions can be used to send and receive coded information. A simple example might use the function defined by $f(x) = 2x + 5$. (Note that it is one-to-one.) Suppose that each letter of the alphabet is assigned a numerical value according to its position, as follows.

A	1	G	7	L	12	Q	17	V	22
B	2	H	8	M	13	R	18	W	23
C	3	I	9	N	14	S	19	X	24
D	4	J	10	O	15	T	20	Y	25
E	5	K	11	P	16	U	21	Z	26
F	6								

This is an Enigma machine, used by the Germans in World War II to send coded messages.

Using the function, the word ALGEBRA would be encoded as

$$7 \quad 29 \quad 19 \quad 15 \quad 9 \quad 41 \quad 7,$$

because

$$f(A) = f(1) = 2(1) + 5 = 7, \quad f(L) = f(12) = 2(12) + 5 = 29, \quad \text{and so on.}$$

The message would then be decoded by using the inverse of f, defined by $f^{-1}(x) = \dfrac{x-5}{2}$ $\left(\text{or } f^{-1}(x) = \frac{1}{2}x - \frac{5}{2}\right)$. For example,

$$f^{-1}(7) = \frac{7-5}{2} = 1 = A, \quad f^{-1}(29) = \frac{29-5}{2} = 12 = L, \quad \text{and so on.}$$

Work Exercises 39–42 in order.

39. Suppose that you are an agent for a detective agency. Today's function for your code is defined by $f(x) = 4x - 5$. Find the rule for f^{-1} algebraically.

40. You receive the following coded message today. (Read across from left to right.)

47 95 23 67 −1 59 27 31 51 23 7 −1 43 7 79 43 −1 75 55 67

31 71 75 27 15 23 67 15 −1 75 15 71 75 75 27 31 51

23 71 31 51   7 15 71 43 31 7 15 11 3 67 15 −1 11

Use the letter/number assignment described earlier to decode the message.

41. Why is a one-to-one function essential in this encoding/decoding process?

42. Use $f(x) = x^3 + 4$ to encode your name, using the letter/number assignment described earlier.

Each function defined is one-to-one. Find the inverse algebraically, and then graph both the function and its inverse on the same graphing calculator screen. Use a square viewing window. **See the Connections box.**

43. $f(x) = 2x - 7$

44. $f(x) = -3x + 2$

45. $f(x) = x^3 + 5$

46. $f(x) = \sqrt[3]{x + 2}$

PREVIEW EXERCISES

If $f(x) = 4^x$, find each value indicated. In Exercise 50, use a calculator, and give the answer to the nearest hundredth. **See Section 3.6.**

47. $f(3)$

48. $f\left(\dfrac{1}{2}\right)$

49. $f\left(-\dfrac{1}{2}\right)$

50. $f(2.73)$

10.2 Exponential Functions

OBJECTIVES

1. Define an exponential function.
2. Graph an exponential function.
3. Solve exponential equations of the form $a^x = a^k$ for x.
4. Use exponential functions in applications involving growth or decay.

OBJECTIVE 1 Define an exponential function. In **Section 8.2** we showed how to evaluate 2^x for rational values of x.

$$2^3 = 8, \quad 2^{-1} = \frac{1}{2}, \quad 2^{1/2} = \sqrt{2}, \quad \text{and} \quad 2^{3/4} = \sqrt[4]{2^3} = \sqrt[4]{8} \qquad \text{Examples of } 2^x \text{ for rational } x$$

In more advanced courses it is shown that 2^x exists for all real number values of x, both rational and irrational. The following definition of an exponential function assumes that a^x exists for all real numbers x.

Exponential Function

For $a > 0$, $a \neq 1$, and all real numbers x,

$$f(x) = a^x$$

defines the **exponential function with base a.**

NOTE *The two restrictions on the value of a in the definition of an exponential function $f(x) = a^x$ are important.*

1. The restriction $a > 0$ is necessary so that the function can be defined for all real numbers x. Letting a be negative ($a = -2$, for instance) and letting $x = \frac{1}{2}$ would give the expression $(-2)^{1/2}$, which is not real.

2. The restriction $a \neq 1$ is necessary because 1 raised to any power is equal to 1, resulting in the linear function defined by $f(x) = 1$.

OBJECTIVE 2 Graph an exponential function. When graphing an exponential function of the form $f(x) = a^x$, pay particular attention to whether $a > 1$ or $0 < a < 1$.

NOW TRY
EXERCISE 1

Graph $y = 4^x$.

EXAMPLE 1 Graphing an Exponential Function $(a > 1)$

Graph $f(x) = 2^x$. Then compare it to the graph of $F(x) = 5^x$.

Choose some values of x, and find the corresponding values of $f(x)$. Plotting these points and drawing a smooth curve through them gives the darker graph shown in **FIGURE 5**. This graph is typical of the graphs of exponential functions of the form $F(x) = a^x$, where $a > 1$. *The larger the value of a, the faster the graph rises.* To see this, compare the graph of $F(x) = 5^x$ with the graph of $f(x) = 2^x$ in **FIGURE 5**.

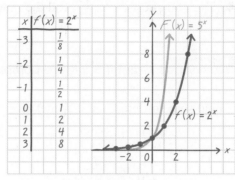

Exponential function with base $a > 1$

Domain: $(-\infty, \infty)$
Range: $(0, \infty)$

The function is one-to-one, and its graph rises from left to right.

FIGURE 5

The vertical line test assures us that the graphs in **FIGURE 5** represent functions. **FIGURE 5** also shows an important characteristic of exponential functions with $a > 1$: *As x gets larger, y increases at a faster and faster rate.*   NOW TRY

⚠️ **CAUTION** The graph of an exponential function *approaches* the x-axis, but does **not** touch it.

NOW TRY
EXERCISE 2

Graph $g(x) = \left(\dfrac{1}{10}\right)^x$.

EXAMPLE 2 Graphing an Exponential Function $(0 < a < 1)$

Graph $g(x) = \left(\dfrac{1}{2}\right)^x$.

Again, find some points on the graph. The graph, shown in **FIGURE 6**, is very similar to that of $f(x) = 2^x$ (**FIGURE 5**) except that here *as x gets larger, y decreases.* This graph is typical of the graph of a function of the form $f(x) = a^x$, where $0 < a < 1$.

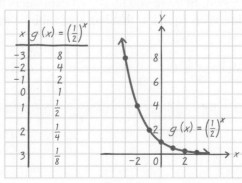

Exponential function with base $0 < a < 1$

Domain: $(-\infty, \infty)$
Range: $(0, \infty)$

The function is one-to-one, and its graph falls from left to right.

FIGURE 6   NOW TRY

NOW TRY ANSWERS

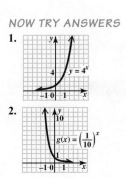

1.
2.

Characteristics of the Graph of $f(x) = a^x$

1. The graph contains the point $(0, 1)$.

2. The function is one-to-one. When $a > 1$, the graph will *rise* from left to right. (See **FIGURE 5**.) When $0 < a < 1$, the graph will *fall* from left to right. (See **FIGURE 6**.) In both cases, the graph goes from the second quadrant to the first.

3. The graph will approach the x-axis, but never touch it. (From **Section 7.4**, recall that such a line is called an **asymptote**.)

4. The domain is $(-\infty, \infty)$, and the range is $(0, \infty)$.

NOW TRY
EXERCISE 3
Graph $f(x) = 4^{2x-1}$.

EXAMPLE 3   Graphing a More Complicated Exponential Function

Graph $f(x) = 3^{2x-4}$.

Find some ordered pairs. We let $x = 0$ and $x = 2$ and find values of $f(x)$, or y.

$$y = 3^{2(0)-4} \qquad \text{Let } x = 0.$$

$$y = 3^{-4}, \quad \text{or} \quad \frac{1}{81}$$

$$y = 3^{2(2)-4} \qquad \text{Let } x = 2.$$

$$y = 3^0, \quad \text{or} \quad 1$$

These ordered pairs, $\left(0, \frac{1}{81}\right)$ and $(2, 1)$, along with the other ordered pairs shown in the table, lead to the graph in **FIGURE 7**. The graph is similar to the graph of $f(x) = 3^x$ except that it is shifted to the right and rises more rapidly.

x	y
0	$\frac{1}{81}$
1	$\frac{1}{9}$
2	1
3	9

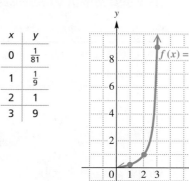

FIGURE 7   NOW TRY

OBJECTIVE 3   **Solve exponential equations of the form $a^x = a^k$ for x.** Until this chapter, we have solved only equations that had the variable as a base, like $x^2 = 8$. In these equations, all exponents have been constants. An **exponential equation** is an equation that has a variable in an exponent, such as

$$9^x = 27.$$

We can use the following property to solve certain exponential equations.

Property for Solving an Exponential Equation

For $a > 0$ and $a \neq 1$,   if $a^x = a^y$ then $x = y$.

This property would not necessarily be true if $a = 1$.

NOW TRY ANSWER
3.
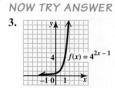

> **Solving an Exponential Equation**
>
> *Step 1*   **Each side must have the same base.** If the two sides of the equation do not have the same base, express each as a power of the same base if possible.
>
> *Step 2*   **Simplify exponents** if necessary, using the rules of exponents.
>
> *Step 3*   **Set exponents equal** using the property given in this section.
>
> *Step 4*   **Solve** the equation obtained in Step 3.

NOW TRY
EXERCISE 4
Solve the equation.

$$8^x = 16$$

EXAMPLE 4 Solving an Exponential Equation

Solve the equation $9^x = 27$.

$$9^x = 27$$

$$(3^2)^x = 3^3 \qquad \text{Write with the same base;}$$
$$\qquad\qquad\qquad 9 = 3^2 \text{ and } 27 = 3^3. \text{ (Step 1)}$$

$$3^{2x} = 3^3 \qquad \text{Power rule for exponents (Step 2)}$$

$$2x = 3 \qquad \text{If } a^x = a^y, \text{ then } x = y. \text{ (Step 3)}$$

$$x = \frac{3}{2} \qquad \text{Solve for } x. \text{ (Step 4)}$$

CHECK   Substitute $\frac{3}{2}$ for x.

$$9^x = 9^{3/2} = (9^{1/2})^3 = 3^3 = 27 \ \checkmark \ \text{True}$$

The solution set is $\left\{\frac{3}{2}\right\}$.

NOW TRY

EXAMPLE 5 Solving Exponential Equations

Solve each equation.

(a) $4^{3x-1} = 16^{x+2}$

> Be careful multiplying the exponents.

$$4^{3x-1} = (4^2)^{x+2} \qquad \text{Write with the same base; } 16 = 4^2.$$

$$4^{3x-1} = 4^{2x+4} \qquad \text{Power rule for exponents}$$

$$3x - 1 = 2x + 4 \qquad \text{Set exponents equal.}$$

$$x = 5 \qquad \text{Subtract } 2x. \text{ Add 1.}$$

Verify that the solution set is $\{5\}$.

(b) $6^x = \dfrac{1}{216}$

$$6^x = \frac{1}{6^3} \qquad 216 = 6^3$$

$$6^x = 6^{-3} \qquad \text{Write with the same base; } \frac{1}{6^3} = 6^{-3}.$$

$$x = -3 \qquad \text{Set exponents equal.}$$

CHECK $6^x = 6^{-3} = \dfrac{1}{6^3} = \dfrac{1}{216} \ \checkmark \qquad \text{Substitute } -3 \text{ for } x; \text{ true}$

NOW TRY ANSWER
4. $\left\{\frac{4}{3}\right\}$

The solution set is $\{-3\}$.

NOW TRY
EXERCISE 5

Solve each equation.

(a) $3^{2x-1} = 27^{x+4}$

(b) $5^x = \dfrac{1}{625}$

(c) $\left(\dfrac{2}{7}\right)^x = \dfrac{343}{8}$

(c) $\left(\dfrac{2}{3}\right)^x = \dfrac{9}{4}$

$\left(\dfrac{2}{3}\right)^x = \left(\dfrac{4}{9}\right)^{-1}$ $\dfrac{9}{4} = \left(\dfrac{4}{9}\right)^{-1}$

$\left(\dfrac{2}{3}\right)^x = \left[\left(\dfrac{2}{3}\right)^2\right]^{-1}$ Write with the same base.

$\left(\dfrac{2}{3}\right)^x = \left(\dfrac{2}{3}\right)^{-2}$ Power rule for exponents

$x = -2$ Set exponents equal.

Check that the solution set is $\{-2\}$.   NOW TRY

NOTE The steps used in **Examples 4 and 5** cannot be applied to an equation like

$$3^x = 12$$

because Step 1 cannot easily be done. A method for solving such exponential equations is given in **Section 10.6.**

OBJECTIVE 4 Use exponential functions in applications involving growth or decay.

EXAMPLE 6 Solving an Application Involving Exponential Growth

The graph in **FIGURE 8** shows the concentration of carbon dioxide (in parts per million) in the air. This concentration is increasing exponentially.

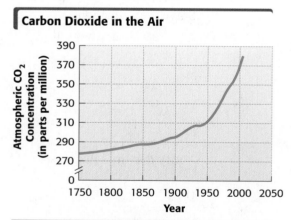

Carbon Dioxide in the Air

Source: Sacramento Bee; National Oceanic and Atmospheric Administration.

FIGURE 8

The data are approximated by the function defined by

$$f(x) = 266(1.001)^x,$$

where x is the number of years since 1750. Use this function and a calculator to approximate the concentration of carbon dioxide in parts per million, to the nearest unit, for each year.

NOW TRY
EXERCISE 6
Use the function in **Example 6** to approximate, to the nearest unit, the carbon dioxide concentration in 2000.

(a) 1900

Because x represents the number of years since 1750, $x = 1900 - 1750 = 150$.

$$f(x) = 266(1.001)^x \qquad \text{Given function}$$
$$f(150) = 266(1.001)^{150} \qquad \text{Let } x = 150.$$
$$f(150) \approx 309 \text{ parts per million} \qquad \text{Use a calculator.}$$

The concentration in 1900 was about 309 parts per million.

(b) 1950

$$f(200) = 266(1.001)^{200} \qquad x = 1950 - 1750 = 200$$
$$f(200) \approx 325 \text{ parts per million} \qquad \text{Use a calculator.}$$

The concentration in 1950 was about 325 parts per million.   NOW TRY

NOW TRY
EXERCISE 7
Use the function in **Example 7** to approximate the pressure at 6000 m. Round to the nearest unit.

EXAMPLE 7   Applying an Exponential Decay Function

The atmospheric pressure (in millibars) at a given altitude x, in meters, can be approximated by the function defined by

$$f(x) = 1038(1.000134)^{-x}, \quad \text{for values of } x \text{ between 0 and 10,000.}$$

Because the base is greater than 1 and the coefficient of x in the exponent is negative, function values decrease as x increases. This means that as altitude increases, atmospheric pressure decreases. (*Source:* Miller, A. and J. Thompson, *Elements of Meteorology*, Fourth Edition, Charles E. Merrill Publishing Company.)

(a) According to this function, what is the pressure at ground level?

$$f(0) = 1038(1.000134)^{-0} \qquad \text{Let } x = 0.$$

At ground level, $x = 0$.

$$= 1038(1) \qquad a^0 = 1$$
$$= 1038$$

The pressure is 1038 millibars.

(b) Approximate the pressure at 5000 m. Round to the nearest unit.

$$f(5000) = 1038(1.000134)^{-5000} \qquad \text{Let } x = 5000.$$
$$f(5000) \approx 531 \qquad \text{Use a calculator.}$$

NOW TRY ANSWERS
6. 342 parts per million
7. approximately 465 millibars

The pressure is approximately 531 millibars.   NOW TRY

10.2 EXERCISES   **MyMathLab**

● *Complete solution available on the Video Resources on DVD*

Concept Check   *Choose the correct response in Exercises 1–3.*

1. Which point lies on the graph of $f(x) = 3^x$?

   A. $(1, 0)$ **B.** $(3, 1)$ **C.** $(0, 1)$ **D.** $\left(\sqrt{3}, \dfrac{1}{3} \right)$

2. Which statement is true?

   A. The point $\left(\frac{1}{2}, \sqrt{5} \right)$ lies on the graph of $f(x) = 5^x$.

   B. For any $a > 1$, the graph of $f(x) = a^x$ falls from left to right.

   C. The y-intercept of the graph of $f(x) = 10^x$ is $(0, 10)$.

   D. The graph of $y = 4^x$ rises at a faster rate than the graph of $y = 10^x$.

3. The asymptote of the graph of $f(x) = a^x$

    A. is the x-axis   **B.** is the y-axis

    C. has equation $x = 1$ **D.** has equation $y = 1$.

4. In your own words, describe the characteristics of the graph of an exponential function. Use the exponential function defined by $f(x) = 3^x$ **(Exercise 5)** and the words *asymptote, domain,* and *range* in your explanation.

*Graph each exponential function. **See Examples 1–3.***

5. $f(x) = 3^x$ **6.** $f(x) = 5^x$ **7.** $g(x) = \left(\dfrac{1}{3}\right)^x$

8. $g(x) = \left(\dfrac{1}{5}\right)^x$ **9.** $y = 4^{-x}$ **10.** $y = 6^{-x}$

11. $y = 2^{2x-2}$ **12.** $y = 2^{2x+1}$

13. *Concept Check* For an exponential function defined by $f(x) = a^x$, if $a > 1$, the graph _____ from left to right. If $0 < a < 1$, the graph _____ from
                  (rises/falls)   (rises/falls)
left to right.

14. *Concept Check* Based on your answers in **Exercise 13,** make a conjecture (an educated guess) concerning whether an exponential function defined by $f(x) = a^x$ is one-to-one. Then decide whether it has an inverse based on the concepts of **Section 10.1.**

*Solve each equation. **See Examples 4 and 5.***

15. $6^x = 36$ **16.** $8^x = 64$ **17.** $100^x = 1000$

18. $8^x = 4$ **19.** $16^{2x+1} = 64^{x+3}$ **20.** $9^{2x-8} = 27^{x-4}$

21. $5^x = \dfrac{1}{125}$ **22.** $3^x = \dfrac{1}{81}$ **23.** $5^x = 0.2$

24. $10^x = 0.1$ **25.** $\left(\dfrac{3}{2}\right)^x = \dfrac{8}{27}$ **26.** $\left(\dfrac{4}{3}\right)^x = \dfrac{27}{64}$

Use the exponential key of a calculator to find an approximation to the nearest thousandth.

27. $12^{2.6}$ **28.** $13^{1.8}$ **29.** $0.5^{3.921}$

30. $0.6^{4.917}$ **31.** $2.718^{2.5}$ **32.** $2.718^{-3.1}$

A major scientific periodical published an article in 1990 dealing with the problem of global warming. The article was accompanied by a graph that illustrated two possible scenarios.

(a) *The warming might be modeled by an exponential function of the form*

$$y = (1.046 \times 10^{-38})(1.0444^x).$$

(b) *The warming might be modeled by a linear function of the form*

$$y = 0.009x - 17.67.$$

    In both cases, x represents the year, and y represents the increase in degrees Celsius due to the warming. Use these functions to approximate the increase in temperature for each of the following years.

33. 2000   **34.** 2010   **35.** 2020   **36.** 2040

Solve each problem. See Examples 6 and 7.

37. Based on figures from 1970 through 2005, the worldwide carbon dioxide emissions in millions of metric tons are approximated by the exponential function defined by

$$f(x) = 4231(1.0174)^x,$$

where $x = 0$ corresponds to 1970, $x = 5$ corresponds to 1975, and so on. (*Source:* Carbon Dioxide Information Analysis Center.) Give answers to the nearest unit.

(a) Use this model to approximate the emissions in 1980.

(b) Use this model to approximate the emissions in 1995.

(c) In 2000, the actual amount of emissions was 6735 million tons. How does this compare to the number that the model provides?

38. Based on figures from 1980 through 2007, the municipal solid waste generated in millions of tons can be approximated by the exponential function defined by

$$f(x) = 159.51(1.0186)^x,$$

where $x = 0$ corresponds to 1980, $x = 5$ corresponds to 1985, and so on. (*Source:* U.S. Environmental Protection Agency.) Give answers to the nearest hundredth.

(a) Use the model to approximate the number of tons of this waste in 1980.

(b) Use the model to approximate the number of tons of this waste in 1995.

(c) In 2007, the actual number of millions of tons of this waste was 254.1. How does this compare to the number that the model provides?

39. A small business estimates that the value $V(t)$ of a copy machine is decreasing according to the function defined by

$$V(t) = 5000(2)^{-0.15t},$$

where t is the number of years that have elapsed since the machine was purchased, and $V(t)$ is in dollars.

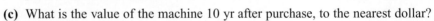

(a) What was the original value of the machine?

(b) What is the value of the machine 5 yr after purchase, to the nearest dollar?

(c) What is the value of the machine 10 yr after purchase, to the nearest dollar?

(d) Graph the function.

40. The amount of radioactive material in an ore sample is given by the function defined by

$$A(t) = 100(3.2)^{-0.5t},$$

where $A(t)$ is the amount present, in grams, of the sample t months after the initial measurement.

(a) How much was present at the initial measurement? (*Hint:* $t = 0$.)

(b) How much was present 2 months later?

(c) How much was present 10 months later?

(d) Graph the function.

41. Refer to the function in **Exercise 39.** When will the value of the machine be $2500? (*Hint:* Let $V(t) = 2500$, divide both sides by 5000, and use the method of **Example 4.**)

42. Refer to the function in **Exercise 39.** When will the value of the machine be $1250?

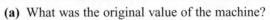

PREVIEW EXERCISES

Determine what number would have to be placed in each box for the statement to be true. See Sections 5.1 and 8.2.

43. $2^{\square} = 16$ **44.** $2^{\square} = \dfrac{1}{16}$ **45.** $2^{\square} = 1$ **46.** $2^{\square} = \sqrt{2}$

10.3 Logarithmic Functions

The graph of $y = 2^x$ is the curve shown in blue in **FIGURE 9**. Because $y = 2^x$ defines a one-to-one function, it has an inverse. Interchanging x and y gives

$$x = 2^y, \quad \text{the inverse of} \quad y = 2^x. \quad \text{Roles of } x \text{ and } y \text{ are interchanged.}$$

As we saw in **Section 10.1,** the graph of the inverse is found by reflecting the graph of $y = 2^x$ about the line $y = x$. The graph of $x = 2^y$ is shown as a red curve in **FIGURE 9**.

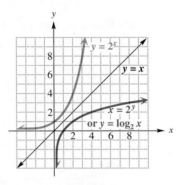

FIGURE 9

OBJECTIVE 1 **Define a logarithm.** We cannot solve the equation $x = 2^y$ for the dependent variable y with the methods presented up to now. The following definition is used to solve $x = 2^y$ for y.

Logarithm

For all positive numbers a, with $a \neq 1$, and all positive numbers x,

$$y = \log_a x \quad \text{means the same as} \quad x = a^y.$$

This key statement should be memorized. The abbreviation **log** is used for the word **logarithm.** Read $\log_a x$ as **"the logarithm of x with base a"** or **"the base a logarithm of x."** To remember the location of the base and the exponent in each form, refer to the following diagrams.

$$\text{Logarithmic form:} \quad y = \overset{\text{Exponent}}{\log_{\underset{\text{Base}}{a}} x} \qquad \text{Exponential form:} \quad x = a^{\overset{\text{Exponent}}{y}}_{\text{Base}}$$

In work with logarithmic form and exponential form, remember the following.

Meaning of $\log_a x$

A logarithm is an exponent. *The expression $\log_a x$ represents the exponent to which the base a must be raised to obtain x.*

OBJECTIVE 2 **Convert between exponential and logarithmic forms.** We can use the definition of logarithm to convert between exponential and logarithmic forms.

NOW TRY
EXERCISE 1

(a) Write $6^3 = 216$ in logarithmic form.

(b) Write $\log_{64} 4 = \frac{1}{3}$ in exponential form.

EXAMPLE 1 **Converting Between Exponential and Logarithmic Forms**

The table shows several pairs of equivalent forms.

Exponential Form	Logarithmic Form
$3^2 = 9$	$\log_3 9 = 2$
$\left(\frac{1}{5}\right)^{-2} = 25$	$\log_{1/5} 25 = -2$
$10^5 = 100{,}000$	$\log_{10} 100{,}000 = 5$
$4^{-3} = \frac{1}{64}$	$\log_4 \frac{1}{64} = -3$

$y = \log_a x$
means
$x = a^y$.

NOW TRY

OBJECTIVE 3 **Solve logarithmic equations of the form $\log_a b = k$ for a, b, or k.** A **logarithmic equation** is an equation with a logarithm in at least one term.

EXAMPLE 2 **Solving Logarithmic Equations**

Solve each equation.

(a) $\log_4 x = -2$

By the definition of logarithm, $\log_4 x = -2$ is equivalent to $x = 4^{-2}$.

$$x = 4^{-2} = \frac{1}{16}$$

The solution set is $\left\{\frac{1}{16}\right\}$.

(b)

$$\log_{1/2} (3x + 1) = 2$$

This is a key step.

$$3x + 1 = \left(\frac{1}{2}\right)^2 \qquad \text{Write in exponential form.}$$

$$3x + 1 = \frac{1}{4} \qquad \text{Apply the exponent.}$$

$$12x + 4 = 1 \qquad \text{Multiply each term by 4.}$$

$$12x = -3 \qquad \text{Subtract 4.}$$

$$x = -\frac{1}{4} \qquad \text{Divide by 12. Write in lowest terms.}$$

CHECK $\qquad \log_{1/2} \left(3\left(-\frac{1}{4}\right) + 1\right) \overset{?}{=} 2 \qquad$ Let $x = -\frac{1}{4}$.

$$\log_{1/2} \frac{1}{4} \overset{?}{=} 2 \qquad \text{Simplify within parentheses.}$$

$$\left(\frac{1}{2}\right)^2 = \frac{1}{4} \checkmark \qquad \text{Exponential form; true}$$

The solution set is $\left\{-\frac{1}{4}\right\}$.

(c) $\qquad \log_x 3 = 2$

$$x^2 = 3 \qquad \text{Write in exponential form.}$$

Be careful here.
$-\sqrt{3}$ is extraneous.

$$x = \pm\sqrt{3} \qquad \text{Take square roots.}$$

Only the *principal* square root satisfies the equation since the base must be a positive number. The solution set is $\left\{\sqrt{3}\right\}$.

NOW TRY ANSWERS
1. (a) $\log_6 216 = 3$
(b) $64^{1/3} = 4$

**NOW TRY
EXERCISE 2**
Solve each equation.

(a) $\log_2 x = -5$

(b) $\log_{3/2}(2x - 1) = 3$

(c) $\log_x 10 = 2$

(d) $\log_{125} \sqrt[3]{5} = x$

(d)

$$\log_{49} \sqrt[3]{7} = x$$

$$49^x = \sqrt[3]{7} \qquad \text{Write in exponential form.}$$

$$(7^2)^x = 7^{1/3} \qquad \text{Write with the same base.}$$

$$7^{2x} = 7^{1/3} \qquad \text{Power rule for exponents}$$

$$2x = \frac{1}{3} \qquad \text{Set exponents equal.}$$

$$x = \frac{1}{6} \qquad \begin{array}{l}\text{Divide by 2 (which is the same}\\ \text{as multiplying by } \tfrac{1}{2}\text{).}\end{array}$$

The solution set is $\left\{\tfrac{1}{6}\right\}$.

 NOW TRY

For any real number b, we know that $b^1 = b$ and for $b \neq 0$, $b^0 = 1$. Writing these statements in logarithmic form gives the following properties of logarithms.

Properties of Logarithms

For any positive real number b, with $b \neq 1$, the following are true.

$$\log_b b = 1 \qquad \text{and} \qquad \log_b 1 = 0$$

**NOW TRY
EXERCISE 3**
Evaluate each logarithm.

(a) $\log_{10} 10$

(b) $\log_8 1$

(c) $\log_{0.1} 1$

EXAMPLE 3 Using Properties of Logarithms

Evaluate each logarithm.

(a) $\log_7 7 = 1 \qquad \log_b b = 1$

(b) $\log_{\sqrt{2}} \sqrt{2} = 1$

(c) $\log_9 1 = 0 \qquad \log_b 1 = 0$

(d) $\log_{0.2} 1 = 0$ NOW TRY

OBJECTIVE 4 **Define and graph logarithmic functions.** Now we define the logarithmic function with base a.

Logarithmic Function

If a and x are positive numbers, with $a \neq 1$, then

$$g(x) = \log_a x$$

defines the **logarithmic function with base a.**

EXAMPLE 4 Graphing a Logarithmic Function ($a > 1$)

Graph $f(x) = \log_2 x$.

By writing $y = f(x) = \log_2 x$ in exponential form as $x = 2^y$, we can identify ordered pairs that satisfy the equation. It is easier to choose values for y and find the corresponding values of x. Plotting the points in the table of ordered pairs and connecting them with a smooth curve gives the graph in **FIGURE 10** on the next page. This graph is typical of logarithmic functions with base $a > 1$.

NOW TRY ANSWERS
2. **(a)** $\left\{\tfrac{1}{32}\right\}$ **(b)** $\left\{\tfrac{35}{16}\right\}$
   (c) $\left\{\sqrt{10}\right\}$ **(d)** $\left\{\tfrac{1}{9}\right\}$
3. **(a)** 1 **(b)** 0 **(c)** 0

NOW TRY
EXERCISE 4
Graph $f(x) = \log_6 x$.

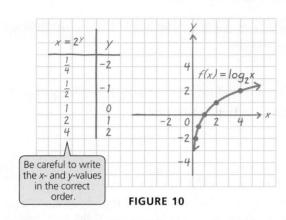

Logarithmic function with base $a > 1$

Domain: $(0, \infty)$

Range: $(-\infty, \infty)$

The function is one-to-one, and its graph rises from left to right.

Be careful to write the x- and y-values in the correct order.

FIGURE 10

NOW TRY

NOW TRY
EXERCISE 5
Graph $g(x) = \log_{1/4} x$.

EXAMPLE 5 Graphing a Logarithmic Function $(0 < a < 1)$

Graph $g(x) = \log_{1/2} x$.

We write $y = g(x) = \log_{1/2} x$ in exponential form as $x = \left(\frac{1}{2}\right)^y$, then choose values for y and find the corresponding values of x. Plotting these points and connecting them with a smooth curve gives the graph in **FIGURE 11**. This graph is typical of logarithmic functions with base $0 < a < 1$.

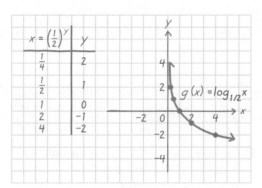

Logarithmic function with base $0 < a < 1$

Domain: $(0, \infty)$

Range: $(-\infty, \infty)$

The function is one-to-one, and its graph falls from left to right.

FIGURE 11

NOW TRY

NOTE See the box titled "Characteristics of the Graph of $f(x) = a^x$" on **page 582.** Below we give a similar set of characteristics for the graph of $g(x) = \log_a x$. Compare the four characteristics one by one to see how the concepts of inverse functions, introduced in **Section 10.1,** are illustrated by these two classes of functions.

Characteristics of the Graph of $g(x) = \log_a x$

1. The graph contains the point $(1, 0)$.

2. The function is one-to-one. When $a > 1$, the graph will *rise* from left to right, from the fourth quadrant to the first. (See **FIGURE 10.**) When $0 < a < 1$, the graph will *fall* from left to right, from the first quadrant to the fourth. (See **FIGURE 11.**)

3. The graph will approach the y-axis, but never touch it. (The y-axis is an asymptote.)

4. The domain is $(0, \infty)$, and the range is $(-\infty, \infty)$.

NOW TRY ANSWERS

4.
5.

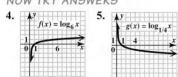

OBJECTIVE 5 Use logarithmic functions in applications involving growth or decay.

⌐ NOW TRY
↳ EXERCISE 6

Suppose the gross national product (GNP) of a small country (in millions of dollars) is approximated by

$$G(t) = 15.0 + 2.00 \log_{10} t,$$

where t is time in years since 2003. Approximate to the nearest tenth the GNP for each value of t.

(a) $t = 1$ **(b)** $t = 10$

EXAMPLE 6 Solving an Application of a Logarithmic Function

The function defined by

$$f(x) = 27 + 1.105 \log_{10}(x + 1)$$

approximates the barometric pressure in inches of mercury at a distance of x miles from the eye of a typical hurricane. (*Source:* Miller, A. and R. Anthes, *Meteorology,* Fifth Edition, Charles E. Merrill Publishing Company.) Approximate the pressure 9 mi from the eye of the hurricane.

Let $x = 9$, and find $f(9)$.

$f(9) = 27 + 1.105 \log_{10}(9 + 1)$	Let $x = 9$.
$f(9) = 27 + 1.105 \log_{10} 10$	Add inside parentheses.
$f(9) = 27 + 1.105(1)$	$\log_{10} 10 = 1$
$f(9) = 28.105$	Add.

NOW TRY ANSWERS
6. (a) $15.0 million
   (b) $17.0 million

The pressure 9 mi from the eye of the hurricane is 28.105 in.   NOW TRY ↴

10.3 EXERCISES

MyMathLab Math XL PRACTICE WATCH DOWNLOAD READ REVIEW

🌐 *Complete solution available on the Video Resources on DVD*

1. *Concept Check* Match the logarithmic equation in Column I with the corresponding exponential equation from Column II. **See Example 1.**

I	**II**
(a) $\log_{1/3} 3 = -1$	**A.** $8^{1/3} = \sqrt[3]{8}$
(b) $\log_5 1 = 0$	**B.** $\left(\dfrac{1}{3}\right)^{-1} = 3$
(c) $\log_2 \sqrt{2} = \dfrac{1}{2}$	**C.** $4^1 = 4$
(d) $\log_{10} 1000 = 3$	**D.** $2^{1/2} = \sqrt{2}$
(e) $\log_8 \sqrt[3]{8} = \dfrac{1}{3}$	**E.** $5^0 = 1$
(f) $\log_4 4 = 1$	**F.** $10^3 = 1000$

2. *Concept Check* Match the logarithm in Column I with its value in Column II. (*Example:* $\log_3 9 = 2$ because 2 is the exponent to which 3 must be raised in order to obtain 9.)

I	**II**
(a) $\log_4 16$	**A.** -2
(b) $\log_3 81$	**B.** -1
(c) $\log_3 \left(\dfrac{1}{3}\right)$	**C.** 2
(d) $\log_{10} 0.01$	**D.** 0
(e) $\log_5 \sqrt{5}$	**E.** $\dfrac{1}{2}$
(f) $\log_{13} 1$	**F.** 4

*Write in logarithmic form. **See Example 1.***

3. $4^5 = 1024$ **4.** $3^6 = 729$ **5.** $\left(\dfrac{1}{2}\right)^{-3} = 8$ **6.** $\left(\dfrac{1}{6}\right)^{-3} = 216$

7. $10^{-3} = 0.001$ **8.** $36^{1/2} = 6$ **9.** $\sqrt[4]{625} = 5$ **10.** $\sqrt[3]{343} = 7$

11. $8^{-2/3} = \dfrac{1}{4}$ **12.** $16^{-3/4} = \dfrac{1}{8}$ **13.** $5^0 = 1$ **14.** $7^0 = 1$

Write in exponential form. **See Example 1.**

15. $\log_4 64 = 3$

16. $\log_2 512 = 9$

17. $\log_{10} \dfrac{1}{10,000} = -4$

18. $\log_{100} 100 = 1$

19. $\log_6 1 = 0$

20. $\log_\pi 1 = 0$

21. $\log_9 3 = \dfrac{1}{2}$

22. $\log_{64} 2 = \dfrac{1}{6}$

23. $\log_{1/4} \dfrac{1}{2} = \dfrac{1}{2}$

24. $\log_{1/8} \dfrac{1}{2} = \dfrac{1}{3}$

25. $\log_5 5^{-1} = -1$

26. $\log_{10} 10^{-2} = -2$

27. Match each logarithm in Column I with its value in Column II. **See Example 3.**

I	II
(a) $\log_8 8$	**A.** -1
(b) $\log_{16} 1$	**B.** 0
(c) $\log_{0.3} 1$	**C.** 1
(d) $\log_{\sqrt{7}} \sqrt{7}$	**D.** 0.1

28. When a student asked his teacher to explain how to evaluate

$$\log_9 3$$

without showing any work, his teacher told him, "Think radically." Explain what the teacher meant by this hint.

Solve each equation. **See Examples 2 and 3.**

29. $x = \log_{27} 3$

30. $x = \log_{125} 5$

31. $\log_x 9 = \dfrac{1}{2}$

32. $\log_x 5 = \dfrac{1}{2}$

33. $\log_x 125 = -3$

34. $\log_x 64 = -6$

35. $\log_{12} x = 0$

36. $\log_4 x = 0$

37. $\log_x x = 1$

38. $\log_x 1 = 0$

39. $\log_x \dfrac{1}{25} = -2$

40. $\log_x \dfrac{1}{10} = -1$

41. $\log_8 32 = x$

42. $\log_{81} 27 = x$

43. $\log_\pi \pi^4 = x$

44. $\log_{\sqrt{2}} \left(\sqrt{2}\right)^9 = x$

45. $\log_6 \sqrt{216} = x$

46. $\log_4 \sqrt{64} = x$

47. $\log_4 (2x + 4) = 3$

48. $\log_3 (2x + 7) = 4$

If (p, q) is on the graph of $f(x) = a^x$ (for $a > 0$ and $a \neq 1$), then (q, p) is on the graph of $f^{-1}(x) = \log_a x$. Use this fact, and refer to the graphs required in **Exercises 5–8 in Section 10.2** *to graph each logarithmic function.* **See Examples 4 and 5.**

49. $y = \log_3 x$

50. $y = \log_5 x$

51. $y = \log_{1/3} x$

52. $y = \log_{1/5} x$

53. Explain why 1 is not allowed as a base for a logarithmic function.

54. Compare the summary of facts about the graph of $f(x) = a^x$ in **Section 10.2** with the similar summary of facts about the graph of $g(x) = \log_a x$ in this section. Make a list of the facts that reinforce the concept that f and g are inverse functions.

55. *Concept Check* The domain of $f(x) = a^x$ is $(-\infty, \infty)$, while the range is $(0, \infty)$. Therefore, since $g(x) = \log_a x$ defines the inverse of f, the domain of g is _____, while the range of g is _____.

56. *Concept Check* The graphs of both $f(x) = 3^x$ and $g(x) = \log_3 x$ rise from left to right. Which one rises at a faster rate?

Concept Check *Use the graph at the right to predict the value of $f(t)$ for the given value of t.*

57. $t = 0$

58. $t = 10$

59. $t = 60$

60. Show that the points determined in **Exercises 57–59** lie on the graph of $f(t) = 8 \log_5 (2t + 5)$.

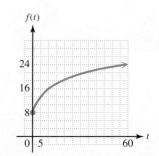

Solve each problem. See Example 6.

61. For 1981–2003, the number of billion cubic feet of natural gas gross withdrawals from crude oil wells in the United States can be approximated by the function defined by

$$f(x) = 3800 + 585 \log_2 x,$$

where $x = 1$ corresponds to 1981, $x = 2$ to 1982, and so on. (*Source:* Energy Information Administration.) Use this function to approximate, to the nearest unit, the number of cubic feet withdrawn in each of the following years.

(a) 1982   **(b)** 1988   **(c)** 1996

62. According to selected figures from the last two decades of the 20th century, the number of trillion cubic feet of dry natural gas consumed worldwide can be approximated by the function defined by

$$f(x) = 51.47 + 6.044 \log_2 x,$$

where $x = 1$ corresponds to 1980, $x = 2$ to 1981, and so on. (*Source:* Energy Information Administration.) Use the function to approximate, to the nearest hundredth, consumption in each year.

(a) 1980   **(b)** 1987   **(c)** 1995

63. Sales (in thousands of units) of a new product are approximated by the function defined by

$$S(t) = 100 + 30 \log_3 (2t + 1),$$

where t is the number of years after the product is introduced.

(a) What were the sales, to the nearest unit, after 1 yr?

(b) What were the sales, to the nearest unit, after 13 yr?

(c) Graph $y = S(t)$.

64. A study showed that the number of mice in an old abandoned house was approximated by the function defined by

$$M(t) = 6 \log_4 (2t + 4),$$

where t is measured in months and $t = 0$ corresponds to January 2008. Find the number of mice in the house in

(a) January 2008   **(b)** July 2008   **(c)** July 2010.

(d) Graph the function.

*The **Richter scale** is used to measure the intensity of earthquakes. The Richter scale rating of an earthquake of intensity x is given by*

$$R = \log_{10} \frac{x}{x_0},$$

where x_0 is the intensity of an earthquake of a certain (small) size. The figure here shows Richter scale ratings for Southern California earthquakes from 1930 to 2000 with magnitudes greater than 4.7.

65. The 1994 Northridge earthquake had a Richter scale rating of 6.7. The 1992 Landers earthquake had a rating of 7.3. How much more powerful was the Landers earthquake than the Northridge earthquake?

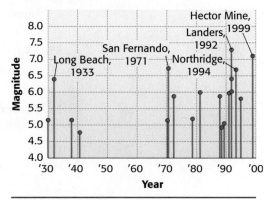

Southern California Earthquakes
(with magnitudes greater than 4.7)

Source: Caltech; U.S. Geological Survey.

66. Compare the smallest rated earthquake in the figure (at 4.8) with the Landers quake. How much more powerful was the Landers quake?

*Some graphing calculators have the capability of drawing the inverse of a function. For example, the two screens that follow show the graphs of $f(x) = 2^x$ and $g(x) = \log_2 x$. The graph of g was obtained by drawing the graph of f^{-1}, since $g(x) = f^{-1}(x)$. (Compare to **FIGURE 9** in this section.)*

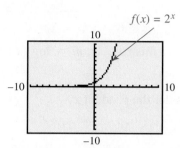 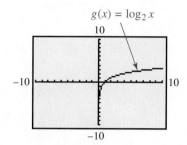

 Use a graphing calculator with the capability of drawing the inverse of a function to draw the graph of each logarithmic function. Use the standard viewing window.

67. $g(x) = \log_3 x$
   (Compare to **Exercise 49.**)

68. $g(x) = \log_5 x$
   (Compare to **Exercise 50.**)

69. $g(x) = \log_{1/3} x$
   (Compare to **Exercise 51.**)

70. $g(x) = \log_{1/5} x$
   (Compare to **Exercise 52.**)

PREVIEW EXERCISES

*Simplify each expression. Write answers using only positive exponents. **See Section 5.1.***

71. $4^7 \cdot 4^2$

72. $\dfrac{5^{-3}}{5^8}$

73. $\dfrac{7^8}{7^{-4}}$

74. $(9^3)^{-2}$

10.4   Properties of Logarithms

OBJECTIVES

1 Use the product rule for logarithms.

2 Use the quotient rule for logarithms.

3 Use the power rule for logarithms.

4 Use properties to write alternative forms of logarithmic expressions.

Logarithms were used as an aid to numerical calculation for several hundred years. Today the widespread use of calculators has made the use of logarithms for calculation obsolete. However, logarithms are still very important in applications and in further work in mathematics.

OBJECTIVE 1 **Use the product rule for logarithms.** One way in which logarithms simplify problems is by changing a problem of multiplication into one of addition. We know that $\log_2 4 = 2$, $\log_2 8 = 3$, and $\log_2 32 = 5$.

$$\log_2 32 = \log_2 4 + \log_2 8 \qquad 5 = 2 + 3$$

$$\log_2 (4 \cdot 8) = \log_2 4 + \log_2 8 \qquad 32 = 4 \cdot 8$$

This is an example of the following rule.

Product Rule for Logarithms

If x, y, and b are positive real numbers, where $b \neq 1$, then the following is true.

$$\log_b xy = \log_b x + \log_b y$$

That is, the logarithm of a product is the sum of the logarithms of the factors.

NOTE The word statement of the product rule can be restated by replacing "logarithm" with "exponent." The rule then becomes the familiar rule for multiplying exponential expressions: The *exponent* of a product is the sum of the *exponents* of the factors.

To prove this rule, let $m = \log_b x$ and $n = \log_b y$, and recall that

$$\log_b x = m \quad \text{means} \quad b^m = x \quad \text{and} \quad \log_b y = n \quad \text{means} \quad b^n = y.$$

Now consider the product xy.

$$xy = b^m \cdot b^n \qquad \text{Substitute.}$$

$$xy = b^{m+n} \qquad \text{Product rule for exponents}$$

$$\log_b xy = m + n \qquad \text{Convert to logarithmic form.}$$

$$\log_b xy = \log_b x + \log_b y \qquad \text{Substitute.}$$

The last statement is the result we wished to prove.

NOW TRY
EXERCISE 1

Use the product rule to rewrite each logarithm.

(a) $\log_{10}(7 \cdot 9)$

(b) $\log_5 11 + \log_5 8$

(c) $\log_5 (5x), \quad x > 0$

(d) $\log_2 t^3, \quad t > 0$

EXAMPLE 1   Using the Product Rule

Use the product rule to rewrite each logarithm. Assume $x > 0$.

(a) $\log_5 (6 \cdot 9)$

$\quad = \log_5 6 + \log_5 9 \qquad$ Product rule

(b) $\log_7 8 + \log_7 12$

$\quad = \log_7 (8 \cdot 12) \qquad$ Product rule

$\quad = \log_7 96 \qquad$ Multiply.

(c) $\log_3 (3x)$

$\quad = \log_3 3 + \log_3 x \qquad$ Product rule

$\quad = 1 + \log_3 x \qquad \log_3 3 = 1$

(d) $\log_4 x^3$

$\quad = \log_4 (x \cdot x \cdot x) \qquad x^3 = x \cdot x \cdot x$

$\quad = \log_4 x + \log_4 x + \log_4 x \qquad$ Product rule

$\quad = 3 \log_4 x \qquad$ Combine like terms.   **NOW TRY**

OBJECTIVE 2   **Use the quotient rule for logarithms.** The rule for division is similar to the rule for multiplication.

Quotient Rule for Logarithms

If x, y, and b are positive real numbers, where $b \neq 1$, then the following is true.

$$\log_b \frac{x}{y} = \log_b x - \log_b y$$

That is, the logarithm of a quotient is the difference between the logarithm of the numerator and the logarithm of the denominator.

NOW TRY ANSWERS
1. (a) $\log_{10} 7 + \log_{10} 9$
   (b) $\log_5 88$
   (c) $1 + \log_5 x$
   (d) $3 \log_2 t$

The proof of this rule is similar to the proof of the product rule.

NOW TRY
EXERCISE 2

Use the quotient rule to rewrite each logarithm.

(a) $\log_{10} \dfrac{7}{9}$

(b) $\log_4 x - \log_4 12, \quad x > 0$

(c) $\log_5 \dfrac{25}{27}$

EXAMPLE 2   Using the Quotient Rule

Use the quotient rule to rewrite each logarithm. Assume $x > 0$.

(a) $\log_4 \dfrac{7}{9}$

$= \log_4 7 - \log_4 9$ Quotient rule

(b) $\log_5 6 - \log_5 x$

$= \log_5 \dfrac{6}{x}$ Quotient rule

(c) $\log_3 \dfrac{27}{5}$

$= \log_3 27 - \log_3 5$ Quotient rule

$= 3 - \log_3 5$ $\log_3 27 = 3$

NOW TRY

⚠ **CAUTION**   *There is no property of logarithms to rewrite the logarithm of a sum or difference.* For example, we *cannot* write $\log_b (x + y)$ in terms of $\log_b x$ and $\log_b y$. Also,

$$\log_b \frac{x}{y} \neq \frac{\log_b x}{\log_b y}.$$

OBJECTIVE 3   **Use the power rule for logarithms.**   An exponential expression such as 2^3 means $2 \cdot 2 \cdot 2$. The base is used as a factor 3 times. Similarly, the product rule can be extended to rewrite the logarithm of a power as the product of the exponent and the logarithm of the base.

$\log_5 2^3$

$= \log_5 (2 \cdot 2 \cdot 2)$

$= \log_5 2 + \log_5 2 + \log_5 2$

$= 3 \log_5 2$

$\log_2 7^4$

$= \log_2 (7 \cdot 7 \cdot 7 \cdot 7)$

$= \log_2 7 + \log_2 7 + \log_2 7 + \log_2 7$

$= 4 \log_2 7$

Furthermore, we saw in **Example 1(d)** that $\log_4 x^3 = 3 \log_4 x$. These examples suggest the following rule.

Power Rule for Logarithms

If x and b are positive real numbers, where $b \neq 1$, and if r is any real number, then the following is true.

$$\log_b x^r = r \log_b x$$

That is, the logarithm of a number to a power equals the exponent times the logarithm of the number.

NOW TRY ANSWERS

2. (a) $\log_{10} 7 - \log_{10} 9$

(b) $\log_4 \frac{x}{12}$

(c) $2 - \log_5 27$

As further examples of this rule,

$$\log_b m^5 = 5 \log_b m \quad \text{and} \quad \log_3 5^4 = 4 \log_3 5.$$

To prove the power rule, let $\log_b x = m$.

$$b^m = x \qquad \text{Convert to exponential form.}$$
$$(b^m)^r = x^r \qquad \text{Raise to the power } r.$$
$$b^{mr} = x^r \qquad \text{Power rule for exponents}$$
$$\log_b x^r = rm \qquad \text{Convert to logarithmic form; commutative property}$$
$$\log_b x^r = r\log_b x \qquad m = \log_b x \text{ from above}$$

This is the statement to be proved.

As a special case of the power rule, let $r = \dfrac{1}{p}$, so

$$\log_b \sqrt[p]{x} = \log_b x^{1/p} = \frac{1}{p}\log_b x.$$

For example, using this result, with $x > 0$,

$$\log_b \sqrt[5]{x} = \log_b x^{1/5} = \frac{1}{5}\log_b x \qquad \text{and} \qquad \log_b \sqrt[3]{x^4} = \log_b x^{4/3} = \frac{4}{3}\log_b x.$$

Another special case is

$$\log_b \frac{1}{x} = \log_b x^{-1} = -\log_b x.$$

NOW TRY
EXERCISE 3

Use the power rule to rewrite each logarithm. Assume $a > 0, x > 0$, and $a \neq 1$.

(a) $\log_7 5^3$ (b) $\log_a \sqrt{10}$

(c) $\log_3 \sqrt[4]{x^3}$

EXAMPLE 3 Using the Power Rule

Use the power rule to rewrite each logarithm. Assume $b > 0, x > 0$, and $b \neq 1$.

(a) $\log_5 4^2$

$= 2\log_5 4 \qquad$ Power rule

(b) $\log_b x^5$

$= 5\log_b x \qquad$ Power rule

(c) $\log_b \sqrt{7}$

$= \log_b 7^{1/2} \qquad \sqrt{x} = x^{1/2}$

$= \frac{1}{2}\log_b 7 \qquad$ Power rule

(d) $\log_2 \sqrt[5]{x^2}$

$= \log_2 x^{2/5} \qquad \sqrt[5]{x^2} = x^{2/5}$

$= \frac{2}{5}\log_2 x \qquad$ Power rule

NOW TRY

Two special properties involving both exponential and logarithmic expressions come directly from the fact that logarithmic and exponential functions are inverses of each other.

Special Properties

If $b > 0$ and $b \neq 1$, then the following are true.

$$b^{\log_b x} = x, \quad x > 0 \qquad \text{and} \qquad \log_b b^x = x$$

To prove the first statement, let $y = \log_b x$.

$$y = \log_b x$$
$$b^y = x \qquad \text{Convert to exponential form.}$$
$$b^{\log_b x} = x \qquad \text{Replace } y \text{ with } \log_b x.$$

The proof of the second statement is similar.

NOW TRY ANSWERS
3. (a) $3\log_7 5$ (b) $\frac{1}{2}\log_a 10$
(c) $\frac{3}{4}\log_3 x$

NOW TRY
EXERCISE 4

Find each value.

(a) $\log_4 4^7$

(b) $\log_{10} 10{,}000$

(c) $8^{\log_8 5}$

EXAMPLE 4   Using the Special Properties

Find each value.

(a) $\log_5 5^4 = 4$, since $\log_b b^x = x$.   **(b)** $\log_3 9 = \log_3 3^2 = 2$

(c) $4^{\log_4 10} = 10$

NOW TRY

We summarize the properties of logarithms.

Properties of Logarithms

If x, y, and b are positive real numbers, where $b \neq 1$, and r is any real number, then the following are true.

Product Rule $\log_b xy = \log_b x + \log_b y$

Quotient Rule $\log_b \dfrac{x}{y} = \log_b x - \log_b y$

Power Rule $\log_b x^r = r \log_b x$

Special Properties $b^{\log_b x} = x$ and $\log_b b^x = x$

OBJECTIVE 4 Use properties to write alternative forms of logarithmic expressions.

EXAMPLE 5   Writing Logarithms in Alternative Forms

Use the properties of logarithms to rewrite each expression if possible. Assume that all variables represent positive real numbers.

(a) $\log_4 4x^3$

$\qquad = \log_4 4 + \log_4 x^3$ Product rule

$\qquad = 1 + 3 \log_4 x$ $\log_4 4 = 1$; power rule

(b) $\log_7 \sqrt{\dfrac{m}{n}}$

$\qquad = \log_7 \left(\dfrac{m}{n}\right)^{1/2}$ Write the radical expression with a rational exponent.

$\qquad = \dfrac{1}{2} \log_7 \dfrac{m}{n}$ Power rule

$\qquad = \dfrac{1}{2} (\log_7 m - \log_7 n)$ Quotient rule

(c) $\log_5 \dfrac{a^2}{bc}$

$\qquad = \log_5 a^2 - \log_5 bc$ Quotient rule

$\qquad = 2 \log_5 a - \log_5 bc$ Power rule

$\qquad = 2 \log_5 a - (\log_5 b + \log_5 c)$ Product rule

$\qquad = 2 \log_5 a - \log_5 b - \log_5 c$

Parentheses are necessary here.

NOW TRY ANSWERS
4. (a) 7 (b) 4 (c) 5

NOW TRY
EXERCISE 5

Use properties of logarithms to rewrite each expression if possible. Assume that all variables represent positive real numbers.

(a) $\log_3 9z^4$

(b) $\log_6 \sqrt{\dfrac{n}{3m}}$

(c) $\log_2 x + 3\log_2 y - \log_2 z$

(d) $\log_5(x + 10)$
$\quad + \log_5(x - 10)$
$\quad - \dfrac{3}{5}\log_5 x, \quad x > 10$

(e) $\log_7(49 + 2x)$

(d) $4\log_b m - \log_b n, \quad b \neq 1$

$= \log_b m^4 - \log_b n \qquad$ Power rule

$= \log_b \dfrac{m^4}{n} \qquad\qquad$ Quotient rule

(e) $\log_b(x + 1) + \log_b(2x + 1) - \dfrac{2}{3}\log_b x, \quad b \neq 1$

$= \log_b(x + 1) + \log_b(2x + 1) - \log_b x^{2/3} \qquad$ Power rule

$= \log_b \dfrac{(x + 1)(2x + 1)}{x^{2/3}} \qquad$ Product and quotient rules

$= \log_b \dfrac{2x^2 + 3x + 1}{x^{2/3}} \qquad$ Multiply in the numerator.

(f) $\log_8(2p + 3r)$ cannot be rewritten using the properties of logarithms. There is no property of logarithms to rewrite the logarithm of a sum. *NOW TRY*

In the next example, we use numerical values for $\log_2 5$ and $\log_2 3$. While we use the equals symbol to give these values, they are actually just approximations since most logarithms of this type are irrational numbers. *We use = with the understanding that the values are correct to four decimal places.*

NOW TRY
EXERCISE 6

Given that $\log_2 7 = 2.8074$ and $\log_2 10 = 3.3219$, evaluate the following.

(a) $\log_2 70$ **(b)** $\log_2 0.7$

(c) $\log_2 49$

EXAMPLE 6 Using the Properties of Logarithms with Numerical Values

Given that $\log_2 5 = 2.3219$ and $\log_2 3 = 1.5850$, evaluate the following.

(a) $\log_2 15$

$= \log_2(3 \cdot 5) \qquad$ Factor 15.

$= \log_2 3 + \log_2 5 \qquad$ Product rule

$= 1.5850 + 2.3219 \qquad$ Substitute the given values.

$= 3.9069 \qquad$ Add.

(b) $\log_2 0.6$

$= \log_2 \dfrac{3}{5} \qquad 0.6 = \frac{6}{10} = \frac{3}{5}$

$= \log_2 3 - \log_2 5 \qquad$ Quotient rule

$= 1.5850 - 2.3219 \qquad$ Substitute the given values.

$= -0.7369 \qquad$ Subtract.

NOW TRY ANSWERS
5. (a) $2 + 4\log_3 z$
  (b) $\frac{1}{2}(\log_6 n - \log_6 3 - \log_6 m)$
  (c) $\log_2 \dfrac{xy^3}{z}$ **(d)** $\log_5 \dfrac{x^2 - 100}{x^{3/5}}$
  (e) cannot be rewritten
6. (a) 6.1293 **(b)** -0.5145
  (c) 5.6148

(c) $\log_2 27$

$= \log_2 3^3 \qquad$ Write 27 as a power of 3.

$= 3\log_2 3 \qquad$ Power rule

$= 3(1.5850) \qquad$ Substitute the given value.

$= 4.7550 \qquad$ Multiply. *NOW TRY*

NOW TRY
EXERCISE 7

Decide whether each statement is *true* or *false*.

(a) $\log_2 16 + \log_2 16 = \log_2 32$

(b) $(\log_2 4)(\log_3 9) = \log_6 36$

EXAMPLE 7 Deciding Whether Statements about Logarithms Are True

Decide whether each statement is *true* or *false*.

(a) $\log_2 8 - \log_2 4 = \log_2 4$

Evaluate each side.

$\log_2 8 - \log_2 4$	Left side	$\log_2 4$	Right side
$= \log_2 2^3 - \log_2 2^2$	Write 8 and 4 as powers of 2.	$= \log_2 2^2$	Write 4 as a power of 2.
$= 3 - 2$	$\log_a a^x = x$	$= 2$	$\log_a a^x = x$
$= 1$	Subtract.		

The statement is false because $1 \neq 2$.

(b) $\log_3 (\log_2 8) = \dfrac{\log_7 49}{\log_8 64}$

Evaluate each side.

$\log_3 (\log_2 8)$	Left side	$\dfrac{\log_7 49}{\log_8 64}$	Right side
$= \log_3 (\log_2 2^3)$	Write 8 as a power of 2.	$= \dfrac{\log_7 7^2}{\log_8 8^2}$	Write 49 and 64 using exponents.
$= \log_3 3$	$\log_a a^x = x$	$= \dfrac{2}{2}$	$\log_a a^x = x$
$= 1$	$3 = 3^1$	$= 1$	Simplify.

The statement is true because $1 = 1$.

NOW TRY

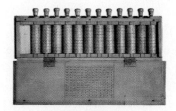

Napier's Rods

Source: IBM Corporate Archives.

CONNECTIONS

Long before the days of calculators and computers, the search for making calculations easier was an ongoing process. Machines built by Charles Babbage and Blaise Pascal, a system of "rods" used by John Napier, and slide rules were the forerunners of today's electronic marvels. The invention of logarithms by John Napier in the sixteenth century was a great breakthrough in the search for easier methods of calculation.

Since logarithms are exponents, their properties allowed users of tables of common logarithms to multiply by adding, divide by subtracting, raise to powers by multiplying, and take roots by dividing. Although logarithms are no longer used for computations, they play an important part in higher mathematics.

For Discussion or Writing

1. To multiply 458.3 by 294.6 using logarithms, we add $\log_{10} 458.3$ and $\log_{10} 294.6$, and then find 10 to this power. Perform this multiplication using the (log x) key* and the (10^x) key on your calculator. Check your answer by multiplying directly with your calculator.

2. Try division, raising to a power, and taking a root by this method.

NOW TRY ANSWERS
7. (a) false **(b)** false

*In this text, the notation log x is used to mean $\log_{10} x$. This is also the meaning of the log key on calculators.

10.4 EXERCISES

MyMathLab

PRACTICE · WATCH · DOWNLOAD · READ · REVIEW

● *Complete solution available on the Video Resources on DVD*

Use the indicated rule of logarithms to complete each equation. **See Examples 1–4.**

1. $\log_{10}(7 \cdot 8) = $ _____ (product rule)

2. $\log_{10} \dfrac{7}{8} = $ _____ (quotient rule)

● **3.** $3^{\log_3 4} = $ _____ (special property)

4. $\log_{10} 3^6 = $ _____ (power rule)

5. $\log_3 3^9 = $ _____ (special property)

6. Evaluate $\log_2(8 + 8)$. Then evaluate $\log_2 8 + \log_2 8$. Are the results the same? How could you change the operation in the first expression to make the two expressions equal?

Use the properties of logarithms to express each logarithm as a sum or difference of logarithms, or as a single number if possible. Assume that all variables represent positive real numbers. **See Examples 1–5.**

● **7.** $\log_7(4 \cdot 5)$

8. $\log_8(9 \cdot 11)$

● **9.** $\log_5 \dfrac{8}{3}$

10. $\log_3 \dfrac{7}{5}$

● **11.** $\log_4 6^2$

12. $\log_5 7^4$

● **13.** $\log_3 \dfrac{\sqrt[3]{4}}{x^2 y}$

14. $\log_7 \dfrac{\sqrt[3]{13}}{pq^2}$

15. $\log_3 \sqrt{\dfrac{xy}{5}}$

16. $\log_6 \sqrt{\dfrac{pq}{7}}$

17. $\log_2 \dfrac{\sqrt[3]{x} \cdot \sqrt[5]{y}}{r^2}$

18. $\log_4 \dfrac{\sqrt[4]{z} \cdot \sqrt[5]{w}}{s^2}$

19. *Concept Check* A student erroneously wrote $\log_a(x + y) = \log_a x + \log_a y$. When his teacher explained that this was indeed wrong, the student claimed that he had used the distributive property. **WHAT WENT WRONG?**

20. Write a few sentences explaining how the rules for multiplying and dividing powers of the same base are similar to the rules for finding logarithms of products and quotients.

Use the properties of logarithms to write each expression as a single logarithm. Assume that all variables are defined in such a way that the variable expressions are positive, and bases are positive numbers not equal to 1. **See Examples 1–5.**

21. $\log_b x + \log_b y$

22. $\log_b w + \log_b z$

23. $\log_a m - \log_a n$

24. $\log_b x - \log_b y$

25. $(\log_a r - \log_a s) + 3\log_a t$

26. $(\log_a p - \log_a q) + 2\log_a r$

27. $3\log_a 5 - 4\log_a 3$

28. $3\log_a 5 - \dfrac{1}{2}\log_a 9$

29. $\log_{10}(x + 3) + \log_{10}(x - 3)$

30. $\log_{10}(x + 4) + \log_{10}(x - 4)$

31. $3\log_p x + \dfrac{1}{2}\log_p y - \dfrac{3}{2}\log_p z - 3\log_p a$

32. $\dfrac{1}{3}\log_b x + \dfrac{2}{3}\log_b y - \dfrac{3}{4}\log_b s - \dfrac{2}{3}\log_b t$

To four decimal places, the values of $\log_{10} 2$ and $\log_{10} 9$ are

$$\log_{10} 2 = 0.3010 \quad \text{and} \quad \log_{10} 9 = 0.9542.$$

*Evaluate each logarithm by applying the appropriate rule or rules from this section. DO NOT USE A CALCULATOR. **See Example 6.***

33. $\log_{10} 18$ **34.** $\log_{10} 4$ **35.** $\log_{10} \dfrac{2}{9}$

36. $\log_{10} \dfrac{9}{2}$ **37.** $\log_{10} 36$ **38.** $\log_{10} 162$

39. $\log_{10} \sqrt[4]{9}$ **40.** $\log_{10} \sqrt[5]{2}$ **41.** $\log_{10} 3$

42. $\log_{10} \dfrac{1}{9}$ **43.** $\log_{10} 9^5$ **44.** $\log_{10} 2^{19}$

Decide whether each statement is true *or* false. ***See Example 7.***

45. $\log_2 (8 + 32) = \log_2 8 + \log_2 32$ **46.** $\log_2 (64 - 16) = \log_2 64 - \log_2 16$

47. $\log_3 7 + \log_3 7^{-1} = 0$ **48.** $\log_3 49 + \log_3 49^{-1} = 0$

49. $\log_6 60 - \log_6 10 = 1$ **50.** $\log_3 8 + \log_3 \dfrac{1}{8} = 0$

51. $\dfrac{\log_{10} 7}{\log_{10} 14} = \dfrac{1}{2}$ **52.** $\dfrac{\log_{10} 10}{\log_{10} 100} = \dfrac{1}{10}$

53. *Concept Check*   Refer to the Note following the word statement of the product rule for logarithms in this section. Now, state the quotient rule in words, replacing "logarithm" with "exponent."

54. Explain why the statement for the power rule for logarithms requires that x be a positive real number.

55. *Concept Check*   Why can't we determine a logarithm of 0? (*Hint*: Think of the definition of logarithm.)

56. *Concept Check*   Consider the following "proof" that $\log_2 16$ does not exist.

$$\log_2 16$$
$$= \log_2 (-4)(-4)$$
$$= \log_2 (-4) + \log_2 (-4)$$

Since the logarithm of a negative number is not defined, the final step cannot be evaluated, and so $\log_2 16$ does not exist. *WHAT WENT WRONG?*

PREVIEW EXERCISES

Write each exponential statement in logarithmic form. ***See Section 10.3.***

57. $10^4 = 10{,}000$ **58.** $10^{1/2} = \sqrt{10}$ **59.** $10^{-2} = 0.01$

Write each logarithmic statement in exponential form. ***See Section 10.3.***

60. $\log_{10} 0.001 = -3$ **61.** $\log_{10} 1 = 0$ **62.** $\log_{10} \sqrt[3]{10} = \dfrac{1}{3}$

10.5 Common and Natural Logarithms

Logarithms are important in many applications in biology, engineering, economics, and social science. In this section we find numerical approximations for logarithms. Traditionally, base 10 logarithms were used most often because our number system is base 10. Logarithms to base 10 are called **common logarithms,** and

$$\log_{10} x \text{ is abbreviated as } \log x,$$

where the base is understood to be 10.

OBJECTIVE 1 **Evaluate common logarithms using a calculator.** In the first example, we give the results of evaluating some common logarithms using a calculator with a (LOG) key. Consult your calculator manual to see how to use this key.

EXAMPLE 1 Evaluating Common Logarithms

Using a calculator, evaluate each logarithm to four decimal places.

(a) $\log 327.1 \approx 2.5147$ **(b)** $\log 437{,}000 \approx 5.6405$

(c) $\log 0.0615 \approx -1.2111$

FIGURE 12 shows how a graphing calculator displays these common logarithms to four decimal places.

```
log(327.1)
            2.5147
log(437000)
            5.6405
log(.0615)
           -1.2111
```

FIGURE 12

NOW TRY

In **Example 1(c),** $\log 0.0615 \approx -1.2111$, a negative result. ***The common logarithm of a number between 0 and 1 is always negative*** because the logarithm is the exponent on 10 that produces the number. In this case, we have

$$10^{-1.2111} \approx 0.0615.$$

If the exponent (the logarithm) were positive, the result would be greater than 1 because $10^0 = 1$. The graph in **FIGURE 13** illustrates these concepts.

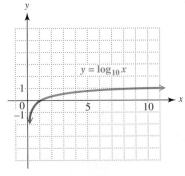

FIGURE 13

OBJECTIVE 2 **Use common logarithms in applications.** In chemistry, pH is a measure of the acidity or alkalinity of a solution. Pure water, for example, has pH 7. In general, acids have pH numbers less than 7, and alkaline solutions have pH values greater than 7, as shown in **FIGURE 14** on the next page.

C NOW TRY
EXERCISE 1

Using a calculator, evaluate each logarithm to four decimal places.

(a) $\log 115$ **(b)** $\log 0.25$

NOW TRY ANSWERS
1. **(a)** 2.0607   **(b)** -0.6021

1   7   14

Acidic   Neutral   Alkaline

FIGURE 14 pH Scale

The **pH** of a solution is defined as

$$pH = -\log [H_3O^+],$$

where $[H_3O^+]$ is the hydronium ion concentration in moles per liter. *It is customary to round pH values to the nearest tenth.*

NOW TRY
EXERCISE 2

Water taken from a wetland has a hydronium ion concentration of

$$3.4 \times 10^{-5}.$$

Find the pH value for the water and classify the wetland as a rich fen, a poor fen, or a bog.

EXAMPLE 2 Using pH in an Application

Wetlands are classified as *bogs, fens, marshes,* and *swamps,* on the basis of pH values. A pH value between 6.0 and 7.5, such as that of Summerby Swamp in Michigan's Hiawatha National Forest, indicates that the wetland is a "rich fen." When the pH is between 3.0 and 6.0, the wetland is a "poor fen," and if the pH falls to 3.0 or less, it is a "bog." (*Source:* Mohlenbrock, R., "Summerby Swamp, Michigan," *Natural History.*)

Suppose that the hydronium ion concentration of a sample of water from a wetland is 6.3×10^{-3}. How would this wetland be classified?

$pH = -\log (6.3 \times 10^{-3})$	Definition of pH
$pH = -(\log 6.3 + \log 10^{-3})$	Product rule
$pH = -[0.7993 - 3(1)]$	Use a calculator to find log 6.3.
$pH = -0.7993 + 3$	Distributive property
$pH \approx 2.2$	Add.

Since the pH is less than 3.0, the wetland is a bog. NOW TRY

NOW TRY
EXERCISE 3

Find the hydronium ion concentration of a solution with pH 2.6.

EXAMPLE 3 Finding Hydronium Ion Concentration

Find the hydronium ion concentration of drinking water with pH 6.5.

$pH = -\log [H_3O^+]$	
$6.5 = -\log [H_3O^+]$	Let pH = 6.5.
$\log [H_3O^+] = -6.5$	Multiply by −1.

Solve for $[H_3O^+]$ by writing the equation in exponential form using base 10.

$$[H_3O^+] = 10^{-6.5}$$

$$[H_3O^+] \approx 3.2 \times 10^{-7} \qquad \text{Use a calculator.}$$

NOW TRY

NOW TRY ANSWERS
2. 4.5; poor fen
3. 2.5×10^{-3}

The loudness of sound is measured in a unit called a **decibel**, abbreviated **dB.** To measure with this unit, we first assign an intensity of I_0 to a very faint sound, called the **threshold sound.** If a particular sound has intensity I, then the decibel level of this louder sound is

$$D = 10 \log \left(\frac{I}{I_0} \right).$$

The table gives average decibel levels for some common sounds. Any sound over 85 dB exceeds what hearing experts consider safe. Permanent hearing damage can be suffered at levels above 150 dB.

Decibel Level	Example
60	Normal conversation
90	Rush hour traffic, lawn mower
100	Garbage truck, chain saw, pneumatic drill
120	Rock concert, thunderclap
140	Gunshot blast, jet engine
180	Rocket launching pad

Source: Deafness Research Foundation.

NOW TRY
EXERCISE 4
Find the decibel level to the nearest whole number of the sound from a jet engine with intensity I of

$$6.312 \times 10^{13} I_0.$$

EXAMPLE 4 Measuring the Loudness of Sound

If music delivered through the earphones of an iPod has intensity I of $3.162 \times 10^9 I_0$, find the average decibel level.

$$D = 10 \log \left(\frac{I}{I_0} \right)$$

Substitute the given value for I.

$$D = 10 \log \left(\frac{3.162 \times 10^9 I_0}{I_0} \right) \quad \text{Substitute the given value for } I.$$

$$D = 10 \log (3.162 \times 10^9)$$

$$D \approx 95 \qquad \text{Use a calculator. Round to the nearest unit.}$$

NOW TRY

OBJECTIVE 3 Evaluate natural logarithms using a calculator. Logarithms used in applications are often **natural logarithms,** which have as base the number e. The number e, like π, is a **universal constant.** The letter e was chosen to honor Leonhard Euler, who published extensive results on the number in 1748. Since it is an irrational number, its decimal expansion never terminates and never repeats.

e

$$e \approx 2.718281828$$

A calculator with an $\boxed{e^x}$ key can approximate powers of e.

$$e^2 \approx 7.389056099, \quad e^3 \approx 20.08553692, \quad e^{0.6} \approx 1.8221188 \qquad \text{Powers of } e$$

Logarithms with base e are called natural logarithms because they occur in natural situations that involve growth or decay. The base e logarithm of x is written **ln x** (read "el en x"). The graph of $y = \ln x$ is given in **FIGURE 15** on the next page.

NOW TRY ANSWER
4. 138 dB

FIGURE 15

A calculator key labeled $\boxed{\ln x}$ is used to evaluate natural logarithms. Consult your calculator manual to see how to use this key.

NOW TRY
EXERCISE 5

Using a calculator, evaluate each logarithm to four decimal places.

(a) ln 0.26   (b) ln 12

(c) ln 150

EXAMPLE 5 Evaluating Natural Logarithms

Using a calculator, evaluate each logarithm to four decimal places.

(a) $\ln 0.5841 \approx -0.5377$

As with common logarithms, *a number between 0 and 1 has a negative natural logarithm.*

(b) $\ln 192.7 \approx 5.2611$ (c) $\ln 10.84 \approx 2.3832$

See **FIGURE 16**.

```
ln(.5841)
              -.5377
ln(192.7)
              5.2611
ln(10.84)
              2.3832
```

FIGURE 16

NOW TRY

OBJECTIVE 4 Use natural logarithms in applications.

NOW TRY
EXERCISE 6

Use the logarithmic function in **Example 6** to approximate the altitude when atmospheric pressure is 600 millibars. Round to the nearest hundred.

EXAMPLE 6 Applying a Natural Logarithmic Function

The altitude in meters that corresponds to an atmospheric pressure of x millibars is given by the logarithmic function defined by

$$f(x) = 51,600 - 7457 \ln x.$$

(*Source:* Miller, A. and J. Thompson, *Elements of Meteorology,* Fourth Edition, Charles E. Merrill Publishing Company.) Use this function to find the altitude when atmospheric pressure is 400 millibars. Round to the nearest hundred.

Let $x = 400$ and substitute in the expression for $f(x)$.

$$f(400) = 51,600 - 7457 \ln 400 \qquad \text{Let } x = 400.$$

$$f(400) \approx 6900 \qquad \text{Use a calculator.}$$

Atmospheric pressure is 400 millibars at approximately 6900 m.   NOW TRY

NOW TRY ANSWERS

5. (a) -1.3471 (b) 2.4849
     (c) 5.0106
6. approximately 3900 m

NOTE In **Example 6,** the final answer was obtained using a calculator *without* rounding the intermediate values. In general, it is best to wait until the final step to round the answer. Round-offs in intermediate steps can lead to a buildup of round-off error, which may cause the final answer to have an incorrect final decimal place digit or digits.

Leonhard Euler (1707–1783)

The number e is named after Euler.

OBJECTIVE 5 **Use the change-of-base rule.** We have used a calculator to approximate the values of common logarithms (base 10) and natural logarithms (base e). However, some applications involve logarithms with other bases. For example, the amount of crude oil (in millions of barrels) imported into the United States during the years 1990–2008 can be approximated by the function

$$f(x) = 2014 + 384.7 \log_2 x,$$

where $x = 1$ represents 1990, $x = 2$ represents 1991, and so on. (*Source:* U.S. Energy Information Administration.) To use this function, we need to find a base 2 logarithm. The following rule is used to convert logarithms from one base to another.

Change-of-Base Rule

If $a > 0$, $a \neq 1$, $b > 0$, $b \neq 1$, and $x > 0$, then the following is true.

$$\log_a x = \frac{\log_b x}{\log_b a}$$

NOTE Any positive number other than 1 can be used for base b in the change-of-base rule. Usually the only practical bases are e and 10 because calculators generally give logarithms only for these two bases.

To derive the change-of-base rule, let $\log_a x = m$.

$$\log_a x = m$$

$$a^m = x \qquad \text{Change to exponential form.}$$

Since logarithmic functions are one-to-one, if all variables are positive and if $x = y$, then $\log_b x = \log_b y$.

$$\log_b (a^m) = \log_b x \qquad \text{Take the logarithm on each side.}$$

$$m \log_b a = \log_b x \qquad \text{Power rule}$$

$$(\log_a x)(\log_b a) = \log_b x \qquad \text{Substitute for } m.$$

$$\log_a x = \frac{\log_b x}{\log_b a} \qquad \text{Divide by } \log_b a.$$

⌐ *NOW TRY*
↳ *EXERCISE 7*

Evaluate $\log_8 60$ to four decimal places.

EXAMPLE 7 Using the Change-of-Base Rule

Evaluate $\log_5 12$ to four decimal places.

Use common logarithms and the change-of-base rule.

$$\log_5 12 = \frac{\log 12}{\log 5} \approx 1.5440 \qquad \text{Use a calculator.}$$

NOW TRY ↻

NOTE Either common or natural logarithms can be used when applying the change-of-base rule. Verify that the same value is found in **Example 7** if natural logarithms are used.

NOW TRY ANSWER
7. 1.9690

**NOW TRY
EXERCISE 8**
Use the model in **Example 8** to estimate total crude oil imports into the United States in 2002. Compare this to the actual amount of 3336 million barrels.

EXAMPLE 8 Using the Change-of-Base Rule in an Application

Use natural logarithms in the change-of-base rule and the function defined by

$$f(x) = 2014 + 384.7 \log_2 x$$

(given earlier) to estimate total crude oil imports (in millions of barrels) into the United States in 2006. Compare this to the actual amount of 3685 million barrels. In the equation, $x = 1$ represents 1990.

$$f(x) = 2014 + 384.7 \log_2 x$$

$$f(17) = 2014 + 384.7 \log_2 17 \qquad \text{For 2006, } x = 17.$$

$$= 2014 + 384.7 \left(\frac{\ln 17}{\ln 2} \right) \qquad \text{Change-of-base rule}$$

$$\approx 3586 \qquad \text{Use a calculator.}$$

NOW TRY ANSWER
8. 3438 million barrels; This is greater than the actual amount.

The model gives about 3586 million barrels for 2006, which is less than the actual amount. NOW TRY

10.5 EXERCISES

 MyMathLab | Math XL PRACTICE | WATCH | DOWNLOAD | READ | REVIEW

● *Complete solution available on the Video Resources on DVD*

Concept Check *Choose the correct response in Exercises 1–4.*

1. What is the base in the expression $\log x$?

    A. e **B.** 1   **C.** 10   **D.** x

2. What is the base in the expression $\ln x$?

    A. e **B.** 1   **C.** 10   **D.** x

3. Since $10^0 = 1$ and $10^1 = 10$, between what two consecutive integers is the value of $\log 6.3$?

    A. 6 and 7   **B.** 10 and 11   **C.** 0 and 1   **D.** -1 and 0

4. Since $e^1 \approx 2.718$ and $e^2 \approx 7.389$, between what two consecutive integers is the value of $\ln 6.3$?

    A. 6 and 7   **B.** 2 and 3   **C.** 1 and 2   **D.** 0 and 1

5. *Concept Check* Without using a calculator, give the value of $\log 10^{31.6}$.

6. *Concept Check* Without using a calculator, give the value of $\ln e^{\sqrt{3}}$.

You will need a calculator for the remaining exercises in this set.

Find each logarithm. Give approximations to four decimal places. ***See Examples 1 and 5.***

 7. $\log 43$ **8.** $\log 98$ **9.** $\log 328.4$

10. $\log 457.2$ **11.** $\log 0.0326$ **12.** $\log 0.1741$

13. $\log (4.76 \times 10^9)$ **14.** $\log (2.13 \times 10^4)$ ● **15.** $\ln 7.84$

16. $\ln 8.32$ **17.** $\ln 0.0556$ **18.** $\ln 0.0217$

19. $\ln 388.1$ **20.** $\ln 942.6$ **21.** $\ln (8.59 \times e^2)$

22. $\ln (7.46 \times e^3)$ **23.** $\ln 10$ **24.** $\log e$

25. Use your calculator to find approximations of the following logarithms.

   (a) log 356.8   **(b)** log 35.68   **(c)** log 3.568

   (d) Observe your answers and make a conjecture concerning the decimal values of the common logarithms of numbers greater than 1 that have the same digits.

26. Let k represent the number of letters in your last name.

   (a) Use your calculator to find log k.

   (b) Raise 10 to the power indicated by the number in part (a). What is your result?

   (c) Use the concepts of **Section 10.1** to explain why you obtained the answer you found in part (b). Would it matter what number you used for k to observe the same result?

Suppose that water from a wetland area is sampled and found to have the given hydronium ion concentration. Is the wetland a rich fen, a poor fen, or a bog? See Example 2.

27. 3.1×10^{-5} **28.** 2.5×10^{-5} **29.** 2.5×10^{-2}

30. 3.6×10^{-2} **31.** 2.7×10^{-7} **32.** 2.5×10^{-7}

Find the pH of the substance with the given hydronium ion concentration. See Example 2.

33. Ammonia, 2.5×10^{-12} **34.** Sodium bicarbonate, 4.0×10^{-9}

35. Grapes, 5.0×10^{-5} **36.** Tuna, 1.3×10^{-6}

Find the hydronium ion concentration of the substance with the given pH. See Example 3.

37. Human blood plasma, 7.4   **38.** Human gastric contents, 2.0

39. Spinach, 5.4   **40.** Bananas, 4.6

Solve each problem. See Examples 4 and 6.

41. Consumers can now enjoy movies at home in elaborate home-theater systems. Find the average decibel level

$$D = 10 \log \left(\frac{I}{I_0} \right)$$

for each movie with the given intensity I.

   (a) *Avatar;* $5.012 \times 10^{10} I_0$

   (b) *Iron Man 2;* $10^{10} I_0$

   (c) *Clash of the Titans;* $6{,}310{,}000{,}000 \, I_0$

42. The time t in years for an amount increasing at a rate of r (in decimal form) to double is given by

$$t(r) = \frac{\ln 2}{\ln (1 + r)}.$$

This is called **doubling time.** Find the doubling time to the nearest tenth for an investment at each interest rate.

   (a) 2% (or 0.02)   **(b)** 5% (or 0.05)   **(c)** 8% (or 0.08)

43. The number of years, $N(r)$, since two independently evolving languages split off from a common ancestral language is approximated by

$$N(r) = -5000 \ln r,$$

where r is the percent of words (in decimal form) from the ancestral language common to both languages now. Find the number of years (to the nearest hundred years) since the split for each percent of common words.

   (a) 85% (or 0.85)   **(b)** 35% (or 0.35)   **(c)** 10% (or 0.10)

44. The concentration of a drug injected into the bloodstream decreases with time. The intervals of time T when the drug should be administered are given by

$$T = \frac{1}{k} \ln \frac{C_2}{C_1},$$

where k is a constant determined by the drug in use, C_2 is the concentration at which the drug is harmful, and C_1 is the concentration below which the drug is ineffective. (*Source:* Horelick, Brindell and Sinan Koont, "Applications of Calculus to Medicine: Prescribing Safe and Effective Dosage," *UMAP Module 202.*) Thus, if $T = 4$, the drug should be administered every 4 hr. For a certain drug, $k = \frac{1}{3}$, $C_2 = 5$, and $C_1 = 2$. How often should the drug be administered? (*Hint:* Round down.)

45. The growth of outpatient surgeries as a percent of total surgeries at hospitals is approximated by

$$f(x) = -1317 + 304 \ln x,$$

where x is the number of years since 1900. (*Source:* American Hospital Association.)

(a) What does this function predict for the percent of outpatient surgeries in 1998?

(b) When did outpatient surgeries reach 50%? (*Hint:* Substitute for y, then write the equation in exponential form to solve it.)

46. In the central Sierra Nevada of California, the percent of moisture that falls as snow rather than rain is approximated reasonably well by

$$f(x) = 86.3 \ln x - 680,$$

where x is the altitude in feet.

(a) What percent of the moisture at 5000 ft falls as snow?

(b) What percent at 7500 ft falls as snow?

47. The **cost-benefit equation**

$$T = -0.642 - 189 \ln (1 - p)$$

describes the approximate tax T, in dollars per ton, that would result in a $p\%$ (in decimal form) reduction in carbon dioxide emissions.

(a) What tax will reduce emissions 25%?

✎ **(b)** Explain why the equation is not valid for $p = 0$ or $p = 1$.

48. The age in years of a female blue whale of length L in feet is approximated by

$$t = -2.57 \ln \left(\frac{87 - L}{63} \right).$$

(a) How old is a female blue whale that measures 80 ft?

✎ **(b)** The equation that defines t has domain $24 < L < 87$. Explain why.

*Use the change-of-base rule (with either common or natural logarithms) to find each logarithm to four decimal places. **See Example 7.***

🌐 **49.** $\log_3 12$ **50.** $\log_4 18$ **51.** $\log_5 3$

52. $\log_7 4$ **53.** $\log_3 \sqrt{2}$ **54.** $\log_6 \sqrt[3]{5}$

55. $\log_\pi e$ **56.** $\log_\pi 10$ **57.** $\log_e 12$

58. To solve the equation $5^x = 7$, we must find the exponent to which 5 must be raised in order to obtain 7. This is $\log_5 7$.

(a) Use the change-of-base rule and your calculator to find $\log_5 7$.

(b) Raise 5 to the number you found in part (a). What is your result?

(c) Using as many decimal places as your calculator gives, write the solution set of $5^x = 7$. (Equations of this type will be studied in more detail in **Section 10.6.**)

59. Let m be the number of letters in your first name, and let n be the number of letters in your last name.

(a) In your own words, explain what $\log_m n$ means.

(b) Use your calculator to find $\log_m n$.

(c) Raise m to the power indicated by the number found in part (b). What is your result?

60. The value of e can be expressed as

$$e = 1 + \frac{1}{1} + \frac{1}{1 \cdot 2} + \frac{1}{1 \cdot 2 \cdot 3} + \frac{1}{1 \cdot 2 \cdot 3 \cdot 4} + \cdots.$$

Approximate e using two terms of this expression, then three terms, four terms, five terms, and six terms. How close is the approximation to the value of $e \approx 2.718281828$ with six terms? Does this infinite sum approach the value of e very quickly?

*Solve each application of a logarithmic function (from **Exercises 61 and 62** of **Section 10.3**).*

61. For 1981–2003, the number of billion cubic feet of natural gas gross withdrawals from crude oil wells in the United States can be approximated by the function defined by

$$f(x) = 3800 + 585 \log_2 x,$$

where $x = 1$ represents 1981, $x = 2$ represents 1982, and so on. (*Source:* Energy Information Administration.) Use this function to approximate the number of cubic feet withdrawn in 2003, to the nearest unit.

62. According to selected figures from the last two decades of the 20th century, the number of trillion cubic feet of dry natural gas consumed worldwide can be approximated by the function defined by

$$f(x) = 51.47 + 6.044 \log_2 x,$$

where $x = 1$ represents 1980, $x = 2$ represents 1981, and so on. (*Source:* Energy Information Administration.) Use this function to approximate consumption in 2003, to the nearest hundredth.

PREVIEW EXERCISES

*Solve each equation. **See Sections 10.2 and 10.3.***

63. $4^{2x} = 8^{3x+1}$

64. $2^{5x} = \left(\frac{1}{16}\right)^{x+3}$

65. $\log_3 (x + 4) = 2$

66. $\log_x 64 = 2$

67. $\log_{1/2} 8 = x$

68. $\log_a 1 = 0$

*Write as a single logarithm. Assume $x > 0$. **See Section 10.4.***

69. $\log (x + 2) + \log (x + 3)$

70. $\log_4 (x + 4) - 2 \log_4 (3x + 1)$

10.6 Exponential and Logarithmic Equations; Further Applications

OBJECTIVES

1. Solve equations involving variables in the exponents.
2. Solve equations involving logarithms.
3. Solve applications of compound interest.
4. Solve applications involving base e exponential growth and decay.

We solved exponential and logarithmic equations in **Sections 10.2 and 10.3.** General methods for solving these equations depend on the following properties.

Properties for Solving Exponential and Logarithmic Equations

For all real numbers $b > 0$, $b \neq 1$, and any real numbers x and y, the following are true.

1. If $x = y$, then $b^x = b^y$.
2. If $b^x = b^y$, then $x = y$.
3. If $x = y$, and $x > 0$, $y > 0$, then $\log_b x = \log_b y$.
4. If $x > 0$, $y > 0$, and $\log_b x = \log_b y$, then $x = y$.

We used Property 2 to solve exponential equations in **Section 10.2.**

OBJECTIVE 1 Solve equations involving variables in the exponents. In **Examples 1 and 2,** we use Property 3.

**NOW TRY
EXERCISE 1**

Solve the equation. Approximate the solution to three decimal places.

$$5^x = 20$$

EXAMPLE 1 Solving an Exponential Equation

Solve $3^x = 12$. Approximate the solution to three decimal places.

$$3^x = 12$$
$$\log 3^x = \log 12 \qquad \text{Property 3 (common logs)}$$
$$x \log 3 = \log 12 \qquad \text{Power rule}$$
$$\text{Exact solution} \longrightarrow x = \frac{\log 12}{\log 3} \qquad \text{Divide by log 3.}$$
$$\text{Decimal approximation} \longrightarrow x \approx 2.262 \qquad \text{Use a calculator.}$$

CHECK $\quad 3^x = 3^{2.262} \approx 12$ ✓ $\quad$ Use a calculator; true

The solution set is $\{2.262\}$. $\qquad\qquad$ *NOW TRY*

⚠ **CAUTION** Be careful: $\frac{\log 12}{\log 3}$ is **not** equal to log 4. Check to see that

$$\log 4 \approx 0.6021, \quad \text{but} \quad \frac{\log 12}{\log 3} \approx 2.262.$$

When an exponential equation has e as the base, as in the next example, it is easiest to use base e logarithms.

NOW TRY ANSWER
1. $\{1.861\}$

NOW TRY
EXERCISE 2
Solve $e^{0.12x} = 10$. Approximate the solution to three decimal places.

EXAMPLE 2 Solving an Exponential Equation with Base e

Solve $e^{0.003x} = 40$. Approximate the solution to three decimal places.

$$\ln e^{0.003x} = \ln 40 \qquad \text{Property 3 (natural logs)}$$

$$0.003x \ln e = \ln 40 \qquad \text{Power rule}$$

$$0.003x = \ln 40 \qquad \ln e = \ln e^1 = 1$$

$$x = \frac{\ln 40}{0.003} \qquad \text{Divide by 0.003.}$$

$$x \approx 1229.626 \qquad \text{Use a calculator.}$$

The solution set is $\{1229.626\}$. Check that $e^{0.003(1229.626)} \approx 40$. **NOW TRY**

General Method for Solving an Exponential Equation

Take logarithms to the same base on both sides and then use the power rule of logarithms or the special property $\log_b b^x = x$. (See **Examples 1 and 2.**)

As a special case, if both sides can be written as exponentials with the same base, do so, and set the exponents equal. (See **Section 10.2.**)

OBJECTIVE 2 Solve equations involving logarithms. We use the definition of logarithm and the properties of logarithms to change equations to exponential form.

NOW TRY
EXERCISE 3
Solve $\log_5 (x - 1)^3 = 2$. Give the exact solution.

EXAMPLE 3 Solving a Logarithmic Equation

Solve $\log_2 (x + 5)^3 = 4$. Give the exact solution.

$$\log_2 (x + 5)^3 = 4$$

$$(x + 5)^3 = 2^4 \qquad \text{Convert to exponential form.}$$

$$(x + 5)^3 = 16 \qquad 2^4 = 16$$

$$x + 5 = \sqrt[3]{16} \qquad \text{Take the cube root on each side.}$$

$$x = -5 + \sqrt[3]{16} \qquad \text{Add } -5.$$

$$x = -5 + 2\sqrt[3]{2} \qquad \sqrt[3]{16} = \sqrt[3]{8 \cdot 2} = \sqrt[3]{8} \cdot \sqrt[3]{2} = 2\sqrt[3]{2}$$

$$\text{CHECK} \qquad \log_2 (x + 5)^3 = 4 \qquad \text{Original equation}$$

$$\log_2 (-5 + 2\sqrt[3]{2} + 5)^3 \stackrel{?}{=} 4 \qquad \text{Let } x = -5 + 2\sqrt[3]{2}.$$

$$\log_2 (2\sqrt[3]{2})^3 \stackrel{?}{=} 4 \qquad \text{Work inside the parentheses.}$$

$$\log_2 16 \stackrel{?}{=} 4 \qquad (2\sqrt[3]{2})^3 = 2^3(\sqrt[3]{2})^3 = 8 \cdot 2 = 16$$

$$2^4 \stackrel{?}{=} 16 \qquad \text{Write in exponential form.}$$

$$16 = 16 \checkmark \qquad \text{True}$$

A true statement results, so the solution set is $\left\{-5 + 2\sqrt[3]{2}\right\}$. **NOW TRY**

⚠️ **CAUTION** Recall that the domain of $y = \log_b x$ is $(0, \infty)$. *For this reason, always check that each proposed solution of an equation with logarithms yields only logarithms of positive numbers in the original equation.*

NOW TRY ANSWERS
2. $\{19.188\}$ **3.** $\left\{1 + \sqrt[3]{25}\right\}$

NOW TRY
EXERCISE 4
Solve.

$\log_4 (2x + 13) - \log_4 (x + 1)$
$= \log_4 10$

EXAMPLE 4 Solving a Logarithmic Equation

Solve $\log_2 (x + 1) - \log_2 x = \log_2 7$.

$$\log_2 (x + 1) - \log_2 x = \log_2 7$$

> Transform the left side to an expresssion with only *one* logarithm.

$$\log_2 \frac{x + 1}{x} = \log_2 7 \qquad \text{Quotient rule}$$

$$\frac{x + 1}{x} = 7 \qquad \text{Property 4}$$

$$x + 1 = 7x \qquad \text{Multiply by } x.$$

$$1 = 6x \qquad \text{Subtract } x.$$

> This proposed solution must be checked.

$$\frac{1}{6} = x \qquad \text{Divide by 6.}$$

Since we cannot take the logarithm of a *nonpositive* number, both $x + 1$ and x must be positive here. If $x = \frac{1}{6}$, then this condition is satisfied.

CHECK
$$\log_2 (x + 1) - \log_2 x = \log_2 7 \qquad \text{Original equation}$$

$$\log_2 \left(\frac{1}{6} + 1 \right) - \log_2 \frac{1}{6} \stackrel{?}{=} \log_2 7 \qquad \text{Let } x = \tfrac{1}{6}.$$

$$\log_2 \frac{7}{6} - \log_2 \frac{1}{6} \stackrel{?}{=} \log_2 7 \qquad \text{Add.}$$

$$\log_2 \frac{\frac{7}{6}}{\frac{1}{6}} \stackrel{?}{=} \log_2 7 \qquad \text{Quotient rule}$$

$$\boxed{\tfrac{\frac{7}{6}}{\frac{1}{6}} = \tfrac{7}{6} \div \tfrac{1}{6} = \tfrac{7}{6} \cdot \tfrac{6}{1} = 7}$$
$$\log_2 7 = \log_2 7 \quad \checkmark \text{ True}$$

A true statement results, so the solution set is $\left\{ \frac{1}{6} \right\}$.   NOW TRY

NOW TRY
EXERCISE 5
Solve.

$\log_4 (x + 2) + \log_4 2x = 2$

EXAMPLE 5 Solving a Logarithmic Equation

Solve $\log x + \log (x - 21) = 2$.

$$\log x + \log (x - 21) = 2$$

$$\log x(x - 21) = 2 \qquad \text{Product rule}$$

> The base is 10.

$$x(x - 21) = 10^2 \qquad \text{Write in exponential form.}$$

$$x^2 - 21x = 100 \qquad \text{Distributive property; multiply.}$$

$$x^2 - 21x - 100 = 0 \qquad \text{Standard form}$$

$$(x - 25)(x + 4) = 0 \qquad \text{Factor.}$$

$$x - 25 = 0 \quad \text{or} \quad x + 4 = 0 \qquad \text{Zero-factor property}$$

$$x = 25 \quad \text{or} \qquad x = -4 \qquad \text{Proposed solutions}$$

The value -4 must be rejected as a solution since it leads to the logarithm of a negative number in the original equation.

$$\log (-4) + \log (-4 - 21) = 2 \qquad \text{The left side is undefined.}$$

Check that the only solution is 25, so the solution set is $\{25\}$.   NOW TRY

NOW TRY ANSWERS
4. $\left\{ \frac{3}{8} \right\}$ **5.** $\{2\}$

> ⚠ **CAUTION** *Do not reject a potential solution just because it is nonpositive. Reject any value that leads to the logarithm of a nonpositive number.*

Solving a Logarithmic Equation

Step 1 **Transform the equation so that a single logarithm appears on one side.** Use the product rule or quotient rule of logarithms to do this.

Step 2 (a) **Use Property 4.** If $\log_b x = \log_b y$, then $x = y$. (See **Example 4.**)

(b) **Write the equation in exponential form.** If $\log_b x = k$, then $x = b^k$. (See **Examples 3 and 5.**)

OBJECTIVE 3 **Solve applications of compound interest.** We have solved simple interest problems using the formula

$$I = prt. \qquad \text{Simple interest formula}$$

In most cases, interest paid or charged is **compound interest** (interest paid on both principal and interest). The formula for compound interest is an application of exponential functions. In this book, monetary amounts are given to the nearest cent.

Compound Interest Formula (for a Finite Number of Periods)

If a principal of P dollars is deposited at an annual rate of interest r compounded (paid) n times per year, then the account will contain

$$A = P\left(1 + \frac{r}{n}\right)^{nt}$$

dollars after t years. (In this formula, r is expressed as a decimal.)

NOW TRY
EXERCISE 6

How much money will there be in an account at the end of 10 yr if $10,000 is deposited at 2.5% compounded monthly?

EXAMPLE 6 Solving a Compound Interest Problem for A

How much money will there be in an account at the end of 5 yr if $1000 is deposited at 3% compounded quarterly? (Assume no withdrawals are made.)

Because interest is compounded quarterly, $n = 4$. The other given values are $P = 1000$, $r = 0.03$ (because 3% = 0.03), and $t = 5$.

$$A = P\left(1 + \frac{r}{n}\right)^{nt} \qquad \text{Compound interest formula}$$

$$A = 1000\left(1 + \frac{0.03}{4}\right)^{4 \cdot 5} \qquad \text{Substitute the given values.}$$

$$A = 1000(1.0075)^{20} \qquad \text{Simplify.}$$

$$A = 1161.18 \qquad \begin{array}{l}\text{Use a calculator.}\\ \text{Round to the nearest cent.}\end{array}$$

The account will contain $1161.18. (The actual amount of interest earned is $1161.18 − $1000 = $161.18. Why?)

NOW TRY

NOW TRY ANSWER
6. $12,836.92

 *NOW TRY*
EXERCISE 7
Approximate the time it would take for money deposited in an account paying 4% interest compounded quarterly to double. Round to the nearest hundredth.

EXAMPLE 7   Solving a Compound Interest Problem for t

Suppose inflation is averaging 3% per year. Approximate the time it will take for prices to double. Round to the nearest hundredth.

We want the number of years t for P dollars to grow to $2P$ dollars at a rate of 3% per year. In the compound interest formula, we substitute $2P$ for A, and let $r = 0.03$ and $n = 1$.

$$2P = P\left(1 + \frac{0.03}{1}\right)^{1t}$$ Substitute in the compound interest formula.

$$2 = (1.03)^t$$ Divide by P. Simplify.

$$\log 2 = \log (1.03)^t$$ Property 3

$$\log 2 = t \log (1.03)$$ Power rule

$$t = \frac{\log 2}{\log 1.03}$$ Interchange sides. Divide by log 1.03.

$$t \approx 23.45$$ Use a calculator.

Prices will double in about 23.45 yr. (This is called the **doubling time** of the money.) To check, verify that $1.03^{23.45} \approx 2$.   *NOW TRY*

Interest can be compounded annually, semiannually, quarterly, daily, and so on. The number of compounding periods can get larger and larger. If the value of n is allowed to approach infinity, we have an example of **continuous compounding.** The formula for continuous compounding is derived in advanced courses, and is an example of exponential growth involving the number e.

Continuous Compound Interest Formula

If a principal of P dollars is deposited at an annual rate of interest r compounded continuously for t years, the final amount A on deposit is given by

$$A = Pe^{rt}.$$

EXAMPLE 8   Solving a Continuous Compound Interest Problem

In **Example 6** we found that $1000 invested for 5 yr at 3% interest compounded quarterly would grow to $1161.18.

(a) How much would this same investment grow to if interest were compounded continuously?

$$A = Pe^{rt}$$ Continuous compounding formula

$$A = 1000e^{0.03(5)}$$ Let $P = 1000$, $r = 0.03$, and $t = 5$.

$$A = 1000e^{0.15}$$ Multiply in the exponent.

$$A = 1161.83$$ Use a calculator. Round to the nearest cent.

Continuous compounding would cause the investment to grow to $1161.83. This is $0.65 more than the amount the investment grew to in **Example 6,** when interest was compounded quarterly.

NOW TRY ANSWER
7. 17.42 yr

NOW TRY
EXERCISE 8

Suppose that \$4000 is invested at 3% interest for 2 yr.

(a) How much will the investment grow to if it is compounded continuously?

(b) Approximate the time it would take for the amount to double. Round to the nearest tenth.

(b) Approximate the time it would take for the initial investment to triple its original amount. Round to the nearest tenth.

We must find the value of t that will cause A to be $3(\$1000) = \$3000$.

$A = Pe^{rt}$	Continuous compounding formula
$3000 = 1000e^{0.03t}$	Let $A = 3P = 3000$, $P = 1000$, $r = 0.03$.
$3 = e^{0.03t}$	Divide by 1000.
$\ln 3 = \ln e^{0.03t}$	Take natural logarithms.
$\ln 3 = 0.03t$	$\ln e^k = k$
$t = \dfrac{\ln 3}{0.03}$	Divide by 0.03.
$t \approx 36.6$	Use a calculator.

It would take about 36.6 yr for the original investment to triple. NOW TRY

OBJECTIVE 4 **Solve applications involving base e exponential growth and decay.** When situations involve growth or decay of a population, the amount or number of some quantity present at time t can be approximated by

$$y = y_0 e^{kt}.$$

In this equation, y_0 is the amount or number present at time $t = 0$ and k is a constant.

The continuous compounding of money is an example of exponential growth. In **Example 9,** we investigate exponential decay.

NOW TRY
EXERCISE 9

Radium 226 decays according to the function defined by

$$y = y_0 e^{-0.00043t},$$

where t is time in years.

(a) If an initial sample contains $y_0 = 4.5$ g of radium 226, how many grams, to the nearest tenth, will be present after 150 yr?

(b) Approximate the half-life of radium 226. Round to the nearest unit.

EXAMPLE 9 Solving an Application Involving Exponential Decay

Carbon 14 is a radioactive form of carbon that is found in all living plants and animals. After a plant or animal dies, the radioactive carbon 14 disintegrates according to the function defined by

$$y = y_0 e^{-0.000121t},$$

where t is time in years, y is the amount of the sample at time t, and y_0 is the initial amount present at $t = 0$.

(a) If an initial sample contains $y_0 = 10$ g of carbon 14, how many grams, to the nearest tenth, will be present after 3000 yr?

Let $y_0 = 10$ and $t = 3000$ in the formula, and use a calculator.

$$y = 10e^{-0.000121(3000)} \approx 6.96 \text{ g}$$

(b) About how long would it take for the initial sample to decay to half of its original amount? (This is called the **half-life.**) Round to the nearest unit.

Let $y = \frac{1}{2}(10) = 5$, and solve for t.

$5 = 10e^{-0.000121t}$	Substitute in $y = y_0 e^{kt}$.
$\dfrac{1}{2} = e^{-0.000121t}$	Divide by 10.
$\ln \dfrac{1}{2} = -0.000121t$	Take natural logarithms; $\ln e^k = k$.
$t = \dfrac{\ln \frac{1}{2}}{-0.000121}$	Interchange sides. Divide by -0.000121.
$t \approx 5728$	Use a calculator.

The half-life is about 5728 yr. NOW TRY

NOW TRY ANSWERS
8. (a) \$4247.35 **(b)** 23.1 yr
9. (a) 4.2 g **(b)** 1612 yr

Recall that the x-intercepts of the graph of a function f correspond to the real solutions of the equation $f(x) = 0$. In **Example 1,** we solved the equation $3^x = 12$ algebraically using rules for logarithms and found the solution set to be $\{2.262\}$. This can be supported graphically by showing that the x-intercept of the graph of the function defined by $y = 3^x - 12$ corresponds to this solution. See **FIGURE 17**.

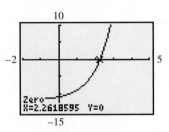

FIGURE 17

For Discussion or Writing

In **Example 5,** we solved $\log x + \log (x - 21) = 2$ to find the solution set $\{25\}$. (We rejected the proposed solution -4 since it led to the logarithm of a negative number.) Show that the x-intercept of the graph of the function defined by $y = \log x + \log (x - 21) - 2$ supports this result.

10.6 EXERCISES

Many of the problems in these exercises require a scientific calculator.

*Solve each equation. Give solutions to three decimal places. See **Example 1.***

1. $7^x = 5$

2. $4^x = 3$

3. $9^{-x+2} = 13$

4. $6^{-x+1} = 22$

5. $3^{2x} = 14$

6. $5^{0.3x} = 11$

7. $2^{x+3} = 5^x$

8. $6^{x+3} = 4^x$

9. $2^{x+3} = 3^{x-4}$

10. $4^{x-2} = 5^{3x+2}$

11. $4^{2x+3} = 6^{x-1}$

12. $3^{2x+1} = 5^{x-1}$

*Solve each equation. Use natural logarithms. When appropriate, give solutions to three decimal places. See **Example 2.***

13. $e^{0.012x} = 23$

14. $e^{0.006x} = 30$

15. $e^{-0.205x} = 9$

16. $e^{-0.103x} = 7$

17. $\ln e^{3x} = 9$

18. $\ln e^{2x} = 4$

19. $\ln e^{0.45x} = \sqrt{7}$

20. $\ln e^{0.04x} = \sqrt{3}$

21. $\ln e^{-x} = \pi$

22. $\ln e^{2x} = \pi$

23. $e^{\ln 2x} = e^{\ln(x+1)}$

24. $e^{\ln(6-x)} = e^{\ln(4+2x)}$

25. Solve one of the equations in **Exercises 13–16** using common logarithms rather than natural logarithms. (You should get the same solution.) Explain why using natural logarithms is a better choice.

26. *Concept Check*   If you were asked to solve

$$10^{0.0025x} = 75,$$

would natural or common logarithms be a better choice? Why?

Complete solution available on the Video Resources on DVD

Solve each equation. Give the exact solution. ***See Example 3.***

27. $\log_3 (6x + 5) = 2$

28. $\log_5 (12x - 8) = 3$

29. $\log_2 (2x - 1) = 5$

30. $\log_6 (4x + 2) = 2$

31. $\log_7 (x + 1)^3 = 2$

32. $\log_4 (x - 3)^3 = 4$

33. *Concept Check* Suppose that in solving a logarithmic equation having the term $\log (x - 3)$, you obtain a proposed solution of 2. All algebraic work is correct. Why must you reject 2 as a solution of the equation?

34. *Concept Check* Suppose that in solving a logarithmic equation having the term $\log (3 - x)$, you obtain a proposed solution of -4. All algebraic work is correct. Should you reject -4 as a solution of the equation? Why or why not?

Solve each equation. Give exact solutions. ***See Examples 4 and 5.***

35. $\log (6x + 1) = \log 3$

36. $\log (7 - 2x) = \log 4$

37. $\log_5 (3t + 2) - \log_5 t = \log_5 4$

38. $\log_2 (x + 5) - \log_2 (x - 1) = \log_2 3$

39. $\log 4x - \log (x - 3) = \log 2$

40. $\log (-x) + \log 3 = \log (2x - 15)$

41. $\log_2 x + \log_2 (x - 7) = 3$

42. $\log (2x - 1) + \log 10x = \log 10$

43. $\log 5x - \log (2x - 1) = \log 4$

44. $\log_3 x + \log_3 (2x + 5) = 1$

45. $\log_2 x + \log_2 (x - 6) = 4$

46. $\log_2 x + \log_2 (x + 4) = 5$

Solve each problem. ***See Examples 6–8.***

47. (a) How much money will there be in an account at the end of 6 yr if $2000 is deposited at 4% compounded quarterly? (Assume no withdrawals are made.)

(b) To one decimal place, how long will it take for the account to grow to $3000?

48. (a) How much money will there be in an account at the end of 7 yr if $3000 is deposited at 3.5% compounded quarterly? (Assume no withdrawals are made.)

(b) To one decimal place, when will the account grow to $5000?

49. (a) What will be the amount A in an account with initial principal $4000 if interest is compounded continuously at an annual rate of 3.5% for 6 yr?

(b) To one decimal place, how long will it take for the initial amount to double?

50. Refer to **Exercise 48(a).** Does the money grow to a greater value under those conditions, or when invested for 7 yr at 3% compounded continuously?

51. Find the amount of money in an account after 12 yr if $5000 is deposited at 7% annual interest compounded as follows.

(a) Annually   **(b)** Semiannually   **(c)** Quarterly

(d) Daily (Use $n = 365$.)   **(e)** Continuously

52. How much money will be in an account at the end of 8 yr if $4500 is deposited at 6% annual interest compounded as follows?

(a) Annually   **(b)** Semiannually   **(c)** Quarterly

(d) Daily (Use $n = 365$.)   **(e)** Continuously

53. How much money must be deposited today to amount to $1850 in 40 yr at 6.5% compounded continuously?

54. How much money must be deposited today to amount to $1000 in 10 yr at 5% compounded continuously?

Solve each problem. See Example 9.

55. The total volume in millions of tons of materials recovered from municipal solid waste collections in the United States during the period 1980–2007 can be approximated by the function defined by

$$f(x) = 15.94e^{0.0656x},$$

where $x = 0$ corresponds to 1980, $x = 1$ to 1981, and so on. Approximate, to the nearest tenth, the volume recovered each year. (*Source:* U.S. Environmental Protection Agency.)

(a) 1980   **(b)** 1990   **(c)** 2000   **(d)** 2007

56. Worldwide emissions in millions of metric tons of the greenhouse gas carbon dioxide from fossil fuel consumption during the period 1990–2006 can be modeled by the function defined by

$$f(x) = 20{,}761e^{0.01882x},$$

where $x = 0$ corresponds to 1990, $x = 1$ to 1991, and so on. Approximate, to the nearest unit, the emissions for each year. (*Source:* U.S. Department of Energy.)

(a) 1990   **(b)** 1995   **(c)** 2000   **(d)** 2006

57. Revenues of software publishers in the United States for the years 2004–2007 can be modeled by the function defined by

$$S(x) = 112{,}047e^{0.0827x},$$

where $x = 0$ represents 2004, $x = 1$ represents 2005, and so on, and $S(x)$ is in millions of dollars. Approximate, to the nearest unit, consumer expenditures for 2007. (*Source:* U.S. Census Bureau.)

58. Based on selected figures obtained during the years 1980–2007, the total number of bachelor's degrees earned in the United States can be modeled by the function defined by

$$D(x) = 900{,}584e^{0.0185x},$$

where $x = 0$ corresponds to 1980, $x = 10$ corresponds to 1990, and so on. Approximate, to the nearest unit, the number of bachelor's degrees earned in 2005. (*Source:* U.S. National Center for Education Statistics.)

59. Suppose that the amount, in grams, of plutonium 241 present in a given sample is determined by the function defined by

$$A(t) = 2.00e^{-0.053t},$$

where t is measured in years. Approximate the amount present, to the nearest hundredth, in the sample after the given number of years.

(a) 4   **(b)** 10   **(c)** 20   **(d)** What was the initial amount present?

60. Suppose that the amount, in grams, of radium 226 present in a given sample is determined by the function defined by

$$A(t) = 3.25e^{-0.00043t},$$

where t is measured in years. Approximate the amount present, to the nearest hundredth, in the sample after the given number of years.

(a) 20   **(b)** 100   **(c)** 500   **(d)** What was the initial amount present?

61. A sample of 400 g of lead 210 decays to polonium 210 according to the function defined by

$$A(t) = 400e^{-0.032t},$$

where t is time in years. Approximate answers to the nearest hundredth.

(a) How much lead will be left in the sample after 25 yr?

(b) How long will it take the initial sample to decay to half of its original amount?

62. The concentration of a drug in a person's system decreases according to the function defined by

$$C(t) = 2e^{-0.125t},$$

where $C(t)$ is in appropriate units, and t is in hours. Approximate answers to the nearest hundredth.

(a) How much of the drug will be in the system after 1 hr?

(b) Approximate the time it will take for the concentration to be half of its original amount.

63. Refer to **Exercise 55.** Assuming that the function continued to apply past 2007, in what year can we expect the volume of materials recovered to reach 130 million tons? (*Source: Environmental Protection Agency.*)

64. Refer to **Exercise 56.** Assuming that the function continued to apply past 2006, in what year can we expect worldwide carbon dioxide emissions from fossil fuel consumption to reach 34,000 million metric tons? (*Source: U.S. Department of Energy.*)

TECHNOLOGY INSIGHTS EXERCISES 65–66

65. The function defined by

$$A(x) = 3.25e^{-0.00043x},$$

with $x = t$, described in **Exercise 60**, is graphed on the screen at the right. Interpret the meanings of X and Y in the display at the bottom of the screen in the context of **Exercise 60**.

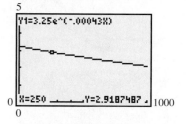

66. The screen shows a table of selected values for the function defined by $Y_1 = \left(1 + \frac{1}{X}\right)^X$.

(a) Why is there an error message for X = 0?

(b) What number does the function value seem to approach as X takes on larger and larger values?

(c) Use a calculator to evaluate this function for X = 1,000,000. What value do you get? Now evaluate $e = e^1$. How close are these two values?

(d) Make a conjecture: As the values of x approach infinity, the value of $\left(1 + \frac{1}{x}\right)^x$ approaches _____ .

PREVIEW EXERCISES

Graph each function. **See Section 9.5.**

67. $f(x) = 2x^2$

68. $f(x) = x^2 - 1$

69. $f(x) = (x + 1)^2$

70. $f(x) = (x - 1)^2 + 2$

KEY TERMS

10.1

one-to-one function
inverse of a function

10.2

exponential function
asymptote
exponential equation

10.3

logarithm
logarithmic equation
logarithmic function
   with base a

10.5

common logarithm
natural logarithm
universal constant

10.6

compound interest
continuous compounding

NEW SYMBOLS

$f^{-1}(x)$ the inverse of $f(x)$

$\log_a x$ the logarithm of x
      with base a

$\log x$ common (base 10)
     logarithm of x

$\ln x$ natural (base e)
    logarithm of x

e a constant,
   approximately
   2.718281828

TEST YOUR WORD POWER

See how well you have learned the vocabulary in this chapter.

1. In a **one-to-one function**
  A. each x-value corresponds to only one y-value
  B. each x-value corresponds to one or more y-values
  C. each x-value is the same as each y-value
  D. each x-value corresponds to only one y-value and each y-value corresponds to only one x-value.

2. If f is a one-to-one function, then the **inverse** of f is
  A. the set of all solutions of f
  B. the set of all ordered pairs formed by interchanging the coordinates of the ordered pairs of f
  C. the set of all ordered pairs that are the opposite (negative) of the coordinates of the ordered pairs of f

  D. an equation involving an exponential expression.

3. An **exponential function** is a function defined by an expression of the form
  A. $f(x) = ax^2 + bx + c$ for real numbers a, b, c $(a \neq 0)$
  B. $f(x) = \log_a x$ for positive numbers a and x $(a \neq 1)$
  C. $f(x) = a^x$ for all real numbers x $(a > 0, a \neq 1)$
  D. $f(x) = \sqrt{x}$ for $x \geq 0$.

4. An **asymptote** is
  A. a line that a graph intersects just once
  B. a line that the graph of a function more and more closely approaches as the x-values increase or decrease

  C. the x-axis or y-axis
  D. a line about which a graph is symmetric.

5. A **logarithm** is
  A. an exponent
  B. a base
  C. an equation
  D. a polynomial.

6. A **logarithmic function** is a function that is defined by an expression of the form
  A. $f(x) = ax^2 + bx + c$ for real numbers a, b, c $(a \neq 0)$
  B. $f(x) = \log_a x$ for positive numbers a and x $(a \neq 1)$
  C. $f(x) = a^x$ for all real numbers x $(a > 0, a \neq 1)$
  D. $f(x) = \sqrt{x}$ for $x \geq 0$.

ANSWERS

1. D; *Example:* The function $f = \{(0, 2), (1, -1), (3, 5), (-2, 3)\}$ is one-to-one.   **2.** B; *Example:* The inverse of the one-to-one function f defined in Answer 1 is $f^{-1} = \{(2, 0), (-1, 1), (5, 3), (3, -2)\}$.   **3.** C; *Examples:* $f(x) = 4^x, g(x) = \left(\frac{1}{2}\right)^x, h(x) = 2^{-x+3}$ **4.** B; *Example:* The graph of $f(x) = 2^x$ has the x-axis ($y = 0$) as an asymptote.   **5.** A; *Example:* $\log_a x$ is the exponent to which a must be raised to obtain x; $\log_3 9 = 2$ since $3^2 = 9$.   **6.** B; *Examples:* $y = \log_3 x, y = \log_{1/3} x$

CONCEPTS	EXAMPLES

10.1 Inverse Functions

Horizontal Line Test

A function is one-to-one if every horizontal line intersects the graph of the function at most once.

Find f^{-1} if $f(x) = 2x - 3$.

The graph of f is a non-horizontal straight line, so f is one-to-one by the horizontal line test.

Inverse Functions

For a one-to-one function f defined by an equation $y = f(x)$, the equation that defines the inverse function f^{-1} is found by interchanging x and y, solving for y, and replacing y with $f^{-1}(x)$.

To find $f^{-1}(x)$, interchange x and y in the equation $y = 2x - 3$.

$$x = 2y - 3$$

Solve for y to get $\qquad y = \dfrac{x + 3}{2}.$

Therefore, $\qquad f^{-1}(x) = \dfrac{x + 3}{2}$, or $f^{-1}(x) = \dfrac{1}{2}x + \dfrac{3}{2}.$

In general, the graph of f^{-1} is the mirror image of the graph of f with respect to the line $y = x$.

The graphs of a function f and its inverse f^{-1} are shown here.

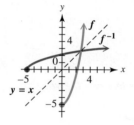

10.2 Exponential Functions

For $a > 0$, $a \neq 1$, $f(x) = a^x$ defines the exponential function with base a.

Graph of $f(x) = a^x$

1. The graph contains the point $(0, 1)$.
2. When $a > 1$, the graph rises from left to right. When $0 < a < 1$, the graph falls from left to right.
3. The x-axis is an asymptote.
4. The domain is $(-\infty, \infty)$, and the range is $(0, \infty)$.

$f(x) = 3^x$ defines the exponential function with base 3.

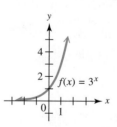

10.3 Logarithmic Functions

$y = \log_a x$ means $x = a^y.$

For $b > 0, b \neq 1$, $\log_b b = 1$ and $\log_b 1 = 0.$

$y = \log_2 x$ means $x = 2^y.$

$$\log_3 3 = 1 \qquad \log_5 1 = 0$$

CONCEPTS	EXAMPLES
For $a > 0, a \neq 1, x > 0$, $g(x) = \log_a x$ defines the logarithmic function with base a. **Graph of $g(x) = \log_a x$** 1. The graph contains the point $(1, 0)$. 2. When $a > 1$, the graph rises from left to right. When $0 < a < 1$, the graph falls from left to right. 3. The y-axis is an asymptote. 4. The domain is $(0, \infty)$, and the range is $(-\infty, \infty)$.	$g(x) = \log_3 x$ defines the logarithmic function with base 3. 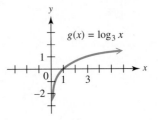

10.4 Properties of Logarithms

Product Rule $\log_a xy = \log_a x + \log_a y$	$\log_2 3m = \log_2 3 + \log_2 m$ Product rule
Quotient Rule $\log_a \dfrac{x}{y} = \log_a x - \log_a y$	$\log_5 \dfrac{9}{4} = \log_5 9 - \log_5 4$ Quotient rule
Power Rule $\log_a x^r = r \log_a x$	$\log_{10} 2^3 = 3 \log_{10} 2$ Power rule
Special Properties $\quad b^{\log_b x} = x$ and $\log_b b^x = x$	$6^{\log_6 10} = 10$ $\log_3 3^4 = 4$ Special properties

10.5 Common and Natural Logarithms

Common logarithms (base 10) are used in applications such as pH, sound level, and intensity of an earthquake.	Use the formula $\text{pH} = -\log\,[H_3O^+]$ to find the pH (to one decimal place) of grapes with hydronium ion concentration 5.0×10^{-5}. $\begin{aligned} \text{pH} &= -\log\,(5.0 \times 10^{-5}) & \text{Substitute.}\\ &= -(\log 5.0 + \log 10^{-5}) & \text{Property of logarithms}\\ &\approx 4.3 & \text{Evaluate with a calculator.} \end{aligned}$
Natural logarithms (base e) are often found in formulas for applications of growth and decay, such as time for money invested to double, decay of chemical compounds, and biological growth.	Use the formula for doubling time (in years) $t(r) = \dfrac{\ln 2}{\ln(1 + r)}$ to find the doubling time to the nearest tenth at an interest rate of 4%. $\begin{aligned} t(0.04) &= \dfrac{\ln 2}{\ln(1 + 0.04)} & \text{Substitute.}\\ &\approx 17.7 & \text{Evaluate with a calculator.} \end{aligned}$ The doubling time is about 17.7 yr.
Change-of-Base Rule If $a > 0, a \neq 1, b > 0, b \neq 1, x > 0$, then $$\log_a x = \frac{\log_b x}{\log_b a}.$$	$\log_3 17 = \dfrac{\ln 17}{\ln 3} = \dfrac{\log 17}{\log 3} \approx 2.5789$

10.6 Exponential and Logarithmic Equations; Further Applications

To solve exponential equations, use these properties $(b > 0, b \neq 1)$. 1. If $b^x = b^y$, then $x = y$.	Solve. $2^{3x} = 2^5$ $\quad\quad 3x = 5$ Set exponents equal. $\quad\quad x = \dfrac{5}{3}$ Divide by 3. The solution set is $\left\{\dfrac{5}{3}\right\}$.

(continued)

CONCEPTS	EXAMPLES
2. If $x = y$, $x > 0$, $y > 0$, then $\log_b x = \log_b y$.	Solve. $\quad 5^m = 8$
	$\log 5^m = \log 8 \qquad$ Take common logarithms.
	$m \log 5 = \log 8 \qquad$ Power rule
	$m = \dfrac{\log 8}{\log 5} \approx 1.2920 \qquad$ Divide by log 5.
	The solution set is $\{1.2920\}$.
To solve logarithmic equations, use these properties, where $b > 0$, $b \neq 1$, $x > 0$, $y > 0$. First use the properties of **Section 10.4**, if necessary, to write the equation in the proper form.	
1. If $\log_b x = \log_b y$, then $x = y$.	Solve. $\qquad \log_3 2x = \log_3 (x + 1)$
	$2x = x + 1$
	$x = 1 \qquad$ Subtract x.
	This value checks, so the solution set is $\{1\}$.
2. If $\log_b x = y$, then $b^y = x$.	Solve. $\quad \log_2 (3x - 1) = 4$
	$3x - 1 = 2^4 \qquad$ Exponential form
	$3x - 1 = 16 \qquad$ Apply the exponent.
	$3x = 17 \qquad$ Add 1.
	$x = \dfrac{17}{3} \qquad$ Divide by 3.
Always check proposed solutions in logarithmic equations.	This value checks, so the solution set is $\left\{\frac{17}{3}\right\}$.

CHAPTER **10**

REVIEW EXERCISES

10.1 *Determine whether each graph is the graph of a one-to-one function.*

1.

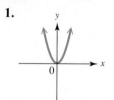

2.

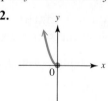

Determine whether each function is one-to-one. If it is, find its inverse.

3. $f(x) = -3x + 7$ **4.** $f(x) = \sqrt[3]{6x - 4}$ **5.** $f(x) = -x^2 + 3$

6. The table lists caffeine amounts in several popular 12-oz sodas. If the set of sodas is the domain and the set of caffeine amounts is the range of the function consisting of the six pairs listed, is it a one-to-one function? Why or why not?

Soda	Caffeine (mg)
Mountain Dew	55
Diet Coke	45
Dr. Pepper	41
Sunkist Orange Soda	41
Diet Pepsi-Cola	36
Coca-Cola Classic	34

Source: National Soft Drink Association.

Each function graphed is one-to-one. Graph its inverse.

7.

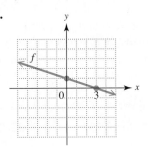

8.

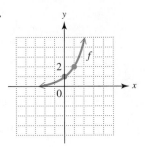

10.2   *Graph each function.*

9. $f(x) = 3^x$

10. $f(x) = \left(\dfrac{1}{3}\right)^x$

11. $y = 2^{2x+3}$

Solve each equation.

12. $5^{2x+1} = 25$

13. $4^{3x} = 8^{x+4}$

14. $\left(\dfrac{1}{27}\right)^{x-1} = 9^{2x}$

15. Sulfur dioxide emissions in the United States, in millions of tons, from 1970 through 2007 can be approximated by the exponential function defined by

$$S(x) = 33.07(1.0241)^{-x},$$

where $x = 0$ corresponds to 1970, $x = 5$ to 1975, and so on. Use this function to approximate, to the nearest tenth, the amounts for each year. (*Source:* U.S. Environmental Protection Agency.)

(a) 1975   (b) 1995   (c) 2005

10.3   *Graph each function.*

16. $g(x) = \log_3 x$ (*Hint:* See **Exercise 9.**)   17. $g(x) = \log_{1/3} x$ (*Hint:* See **Exercise 10.**)

Solve each equation.

18. $\log_8 64 = x$

19. $\log_2 \sqrt{8} = x$

20. $\log_x \left(\dfrac{1}{49}\right) = -2$

21. $\log_4 x = \dfrac{3}{2}$

22. $\log_k 4 = 1$

23. $\log_b b^2 = 2$

24. In your own words, explain the meaning of $\log_b a$.

25. *Concept Check* Based on the meaning of $\log_b a$, what is the simplest form of $b^{\log_b a}$?

26. A company has found that total sales, in thousands of dollars, are given by the function defined by

$$S(x) = 100 \log_2(x + 2),$$

where x is the number of weeks after a major advertising campaign was introduced.

(a) What were the total sales 6 weeks after the campaign was introduced?

(b) Graph the function.

10.4 *Apply the properties of logarithms to express each logarithm as a sum or difference of logarithms. Assume that all variables represent positive real numbers.*

27. $\log_2 3xy^2$

28. $\log_4 \dfrac{\sqrt{x} \cdot w^2}{z}$

Apply the properties of logarithms to write each expression as a single logarithm. Assume that all variables represent positive real numbers, $b \neq 1$.

29. $\log_b 3 + \log_b x - 2\log_b y$

30. $\log_3(x + 7) - \log_3(4x + 6)$

10.5 *Evaluate each logarithm. Give approximations to four decimal places.*

31. $\log 28.9$ **32.** $\log 0.257$ **33.** $\ln 28.9$ **34.** $\ln 0.257$

Use the change-of-base rule (with either common or natural logarithms) to find each logarithm. Give approximations to four decimal places.

35. $\log_{16} 13$

36. $\log_4 12$

Use the formula $\text{pH} = -\log[\text{H}_3\text{O}^+]$ *to find the pH of each substance with the given hydronium ion concentration.*

37. Milk, 4.0×10^{-7}

38. Crackers, 3.8×10^{-9}

39. If orange juice has pH 4.6, what is its hydronium ion concentration?

40. The magnitude of a star is defined by the equation

$$M = 6 - 2.5 \log \frac{I}{I_0},$$

where I_0 is the measure of the faintest star and I is the actual intensity of the star being measured. The dimmest stars are of magnitude 6, and the brightest are of magnitude 1. Determine the ratio of intensities between stars of magnitude 1 and 3.

41. Section 10.5, Exercise 42 introduced the doubling function defined by

$$t(r) = \frac{\ln 2}{\ln(1 + r)},$$

that gives the number of years required to double your money when it is invested at interest rate r (in decimal form) compounded annually. How long does it take to double your money at each rate? Round answers to the nearest year.

(a) 4% **(b)** 6% **(c)** 10% **(d)** 12%

(e) Compare each answer in parts (a)–(d) with the following numbers. What do you find?

$$\frac{72}{4}, \frac{72}{6}, \frac{72}{10}, \frac{72}{12}$$

10.6   *Solve each equation. Give solutions to three decimal places.*

42. $3^x = 9.42$ **43.** $2^{x-1} = 15$ **44.** $e^{0.06x} = 3$

Solve each equation. Give exact solutions.

45. $\log_3 (9x + 8) = 2$

46. $\log_5 (x + 6)^3 = 2$

47. $\log_3 (x + 2) - \log_3 x = \log_3 2$

48. $\log (2x + 3) = 1 + \log x$

49. $\log_4 x + \log_4 (8 - x) = 2$

50. $\log_2 x + \log_2 (x + 15) = \log_2 16$

51. *Concept Check*   Consider the following "solution" of the equation $\log x^2 = 2$. **WHAT WENT WRONG?** Give the correct solution set.

$\log x^2 = 2$	Original equation
$2 \log x = 2$	Power rule for logarithms
$\log x = 1$	Divide each side by 2.
$x = 10^1$	Write in exponential form.
$x = 10$	$10^1 = 10$

Solution set: $\{10\}$

Solve each problem. Use a calculator as necessary.

52. If \$20,000 is deposited at 4% annual interest compounded quarterly, how much will be in the account after 5 yr, assuming no withdrawals are made?

53. How much will \$10,000 compounded continuously at 3.75% annual interest amount to in 3 yr?

54. Which is a better plan?

Plan A: Invest \$1000 at 4% compounded quarterly for 3 yr
Plan B: Invest \$1000 at 3.9% compounded monthly for 3 yr

55. What is the half-life of a radioactive substance that decays according to the function

$$Q(t) = A_0 e^{-0.05t}, \quad \text{where } t \text{ is in days?}$$

56. A machine purchased for business use **depreciates,** or loses value, over a period of years. The value of the machine at the end of its useful life is called its **scrap value.** By one method of depreciation (where it is assumed a constant percentage of the value depreciates annually), the scrap value, S, is given by

$$S = C(1 - r)^n,$$

where C is the original cost, n is the useful life in years, and r is the constant percent of depreciation.

(a) Find the scrap value of a machine costing \$30,000, having a useful life of 12 yr and a constant annual rate of depreciation of 15%.

(b) A machine has a "half-life" of 6 yr. Find the constant annual rate of depreciation.

57. Recall from **Exercise 43** in **Section 10.5** that the number of years, $N(r)$, since two independently evolving languages split off from a common ancestral language is approximated by

$$N(r) = -5000 \ln r,$$

where r is the percent of words from the ancestral language common to both languages now. Find r if the split occurred 2000 yr ago.

58. *Concept Check*   Which one is *not* a representation of the solution of $7^x = 23$?

A. $\dfrac{\log 23}{\log 7}$ **B.** $\dfrac{\ln 23}{\ln 7}$ **C.** $\log_7 23$ **D.** $\log_{23} 7$

MIXED REVIEW EXERCISES

Evaluate.

59. $\log_2 128$

60. $5^{\log_5 36}$

61. $e^{\ln 4}$

62. $10^{\log e}$

63. $\log_3 3^{-5}$

64. $\ln e^{5.4}$

Solve.

65. $\log_3(x + 9) = 4$

66. $\ln e^x = 3$

67. $\log_x \frac{1}{81} = 2$

68. $27^x = 81$

69. $2^{2x-3} = 8$

70. $5^{x+2} = 25^{2x+1}$

71. $\log_3(x + 1) - \log_3 x = 2$

72. $\log(3x - 1) = \log 10$

73. $\ln(x^2 + 3x + 4) = \ln 2$

74. Consider the logarithmic equation

$$\log(2x + 3) = \log x + 1.$$

(a) Solve the equation using properties of logarithms.

(b) If $Y_1 = \log(2X + 3)$ and $Y_2 = \log X + 1$, then the graph of $Y_1 - Y_2$ looks like that shown. Explain how the display at the bottom of the screen confirms the solution set found in part (a).

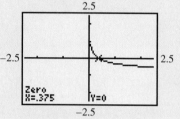

75. Based on selected figures from 1980 through 2007, the fractional part (as a decimal) of the generation of municipal solid waste recovered can be approximated by the function defined by

$$R(x) = 0.0997(e^{0.0470x}),$$

where $x = 0$ corresponds to 1980, $x = 10$ to 1990, and so on. Based on this model, approximate the percent, to the nearest hundredth, of municipal solid waste recovered in 2005. (*Source:* U.S. Environmental Protection Agency.)

76. One measure of the diversity of the species in an ecological community is the **index of diversity,** the logarithmic expression

$$-(p_1 \ln p_1 + p_2 \ln p_2 + \cdots + p_n \ln p_n),$$

where $p_1, p_2, \ldots, p_n$ are the proportions of a sample belonging to each of n species in the sample. (*Source:* Ludwig, John and James Reynolds, *Statistical Ecology: A Primer on Methods and Computing,* New York, John Wiley and Sons.) Approximate the index of diversity to the nearest thousandth if a sample of 100 from a community produces the following numbers.

(a) 90 of one species, 10 of another

(b) 60 of one species, 40 of another

CHAPTER 10 TEST

CHAPTER **Test Prep** VIDEOS

Step-by-step test solutions are found on the Chapter Test Prep Videos available via the Video Resources on DVD, in *MyMathLab*, or on YouTube (search "LialIntermediateAlg").

View the complete solutions to all Chapter Test exercises on the Video Resources on DVD.

1. Decide whether each function is one-to-one.

(a) $f(x) = x^2 + 9$

(b)

2. Find $f^{-1}(x)$ for the one-to-one function defined by $f(x) = \sqrt[3]{x + 7}$.

3. Graph the inverse of f, given the graph of f.

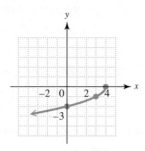

Graph each function.

4. $f(x) = 6^x$

5. $g(x) = \log_6 x$

6. Explain how the graph of the function in **Exercise 5** can be obtained from the graph of the function in **Exercise 4**.

Solve each equation. Give the exact solution.

7. $5^x = \dfrac{1}{625}$

8. $2^{3x-7} = 8^{2x+2}$

9. A 2008 report predicted that the U.S. Hispanic population will increase from 46.9 million in 2008 to 132.8 million in 2050. (*Source:* U.S. Census Bureau.) Assuming an exponential growth pattern, the population is approximated by

$$f(x) = 46.9e^{0.0247x},$$

where x represents the number of years since 2008. Use this function to approximate, to the nearest tenth, the Hispanic population in each year.

(**a**) 2015   (**b**) 2030

10. Write in logarithmic form: $4^{-2} = 0.0625$.

11. Write in exponential form: $\log_7 49 = 2$.

Solve each equation.

12. $\log_{1/2} x = -5$

13. $x = \log_9 3$

14. $\log_x 16 = 4$

15. *Concept Check* Fill in the blanks with the correct responses: The value of $\log_2 32$ is _____. This means that if we raise _____ to the _____ power, the result is _____.

Use properties of logarithms to write each expression as a sum or difference of logarithms. Assume that variables represent positive real numbers.

16. $\log_3 x^2 y$

17. $\log_5\left(\dfrac{\sqrt{x}}{yz}\right)$

Use properties of logarithms to write each expression as a single logarithm. Assume that variables represent positive real numbers, $b \neq 1$.

18. $3 \log_b s - \log_b t$

19. $\dfrac{1}{4} \log_b r + 2 \log_b s - \dfrac{2}{3} \log_b t$

20. Use a calculator to approximate each logarithm to four decimal places.

(**a**) $\log 23.1$ (**b**) $\ln 0.82$

21. Use the change-of-base rule to express $\log_3 19$

    (a) in terms of common logarithms   **(b)** in terms of natural logarithms

    (c) correct to four decimal places.

22. Solve $3^x = 78$, giving the solution to three decimal places.

23. Solve $\log_8 (x + 5) + \log_8 (x - 2) = 1$.

24. Suppose that $10,000 is invested at 4.5% annual interest, compounded quarterly. How much will be in the account in 5 yr if no money is withdrawn?

25. Suppose that $15,000 is invested at 5% annual interest, compounded continuously.

    (a) How much will be in the account in 5 yr if no money is withdrawn?

    (b) How long will it take for the initial principal to double?

CHAPTERS (1–10) CUMULATIVE REVIEW EXERCISES

Let $S = \left\{ -\frac{9}{4}, -2, -\sqrt{2}, 0, 0.6, \sqrt{11}, \sqrt{-8}, 6, \frac{30}{3} \right\}$. List the elements of S that are members of each set.

    1. Integers   **2.** Rational numbers   **3.** Irrational numbers

Simplify each expression.

    4. $|-8| + 6 - |-2| - (-6 + 2)$ **5.** $2(-5) + (-8)(4) - (-3)$

Solve each equation or inequality.

    6. $7 - (3 + 4x) + 2x = -5(x - 1) - 3$ **7.** $2x + 2 \leq 5x - 1$

    8. $|2x - 5| = 9$ **9.** $|4x + 2| > 10$

Graph.

    10. $5x + 2y = 10$ **11.** $-4x + y \leq 5$

12. The graph indicates that the number of international travelers to the United States increased from 41,218 thousand in 2003 to 57,949 thousand in 2008.

    (a) Is this the graph of a function?

    (b) What is the slope, to the nearest tenth, of the line in the graph? Interpret the slope in the context of U.S. travelers to foreign countries.

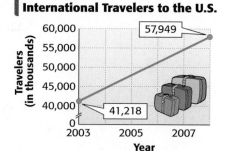

International Travelers to the U.S.

Source: U.S. Department of Commerce.

13. Find an equation of the line through $(5, -1)$ and parallel to the line with equation $3x - 4y = 12$. Write the equation in slope-intercept form.

Solve each system.

    14. $5x - 3y = 14$
            $2x + 5y = 18$

    15. $x + 2y + 3z = 11$
           $3x - y + z = 8$
           $2x + 2y - 3z = -12$

16. Candy worth $1.00 per lb is to be mixed with 10 lb of candy worth $1.96 per lb to get a mixture that will be sold for $1.60 per lb. How many pounds of the $1.00 candy should be used?

Number of Pounds	Price per Pound	Value
x	$1.00	1x
	$1.60	

Perform the indicated operations.

17. $(2p + 3)(3p - 1)$ **18.** $(4k - 3)^2$

19. $(3m^3 + 2m^2 - 5m) - (8m^3 + 2m - 4)$

20. Divide $6t^4 + 17t^3 - 4t^2 + 9t + 4$ by $3t + 1$.

Factor.

21. $8x + x^3$ **22.** $24y^2 - 7y - 6$ **23.** $5z^3 - 19z^2 - 4z$

24. $16a^2 - 25b^4$ **25.** $8c^3 + d^3$ **26.** $16r^2 + 56rq + 49q^2$

Perform the indicated operations.

27. $\dfrac{(5p^3)^4(-3p^7)}{2p^2(4p^4)}$ **28.** $\dfrac{x^2 - 9}{x^2 + 7x + 12} \div \dfrac{x - 3}{x + 5}$ **29.** $\dfrac{2}{k + 3} - \dfrac{5}{k - 2}$

Simplify.

30. $\sqrt{288}$ **31.** $2\sqrt{32} - 5\sqrt{98}$

32. Solve $\sqrt{2x + 1} - \sqrt{x} = 1$. **33.** Multiply $(5 + 4i)(5 - 4i)$.

Solve each equation or inequality.

34. $3x^2 - x - 1 = 0$ **35.** $x^2 + 2x - 8 > 0$ **36.** $x^4 - 5x^2 + 4 = 0$

37. Graph $f(x) = \dfrac{1}{3}(x - 1)^2 + 2$. **38.** Graph $f(x) = 2^x$.

39. Solve $5^{x+3} = \left(\dfrac{1}{25}\right)^{3x+2}$. **40.** Graph $f(x) = \log_3 x$.

41. Rewrite the following using the product, quotient, and power properties of logarithms.

$$\log \frac{x^3\sqrt{y}}{z}$$

42. Let the number of bacteria present in a certain culture be given by

$$B(t) = 25{,}000e^{0.2t},$$

where t is time measured in hours, and $t = 0$ corresponds to noon. Approximate, to the nearest hundred, the number of bacteria present at each time.

(a) noon **(b)** 1 P.M. **(c)** 2 P.M.

(d) When will the population double?

Nonlinear Functions, Conic Sections, and Nonlinear Systems

In this chapter, we study a group of curves known as *conic sections*. One conic section, the *ellipse*, has a special reflecting property responsible for "whispering galleries." In a whispering gallery, a person whispering at a certain point in the room can be heard clearly at another point across the room.

The Old House Chamber of the U.S. Capitol, now called Statuary Hall, is a whispering gallery. History has it that John Quincy Adams, whose desk was positioned at exactly the right point beneath the ellipsoidal ceiling, often pretended to sleep there as he listened to political opponents whispering strategies across the room. (*Source:* Aikman, Lonnelle, *We, the People, The Story of the United States Capitol.*)

In **Section 11.2,** we investigate ellipses.

11.1 Additional Graphs of Functions

OBJECTIVES

1 Recognize the graphs of the elementary functions defined by $|x|$, $\frac{1}{x}$, and $\sqrt{x}$, and graph their translations.

2 Recognize and graph step functions.

OBJECTIVE 1   Recognize the graphs of the elementary functions defined by $|x|$, $\frac{1}{x}$, and $\sqrt{x}$, and graph their translations. Earlier, we introduced the **squaring function** defined by $f(x) = x^2$. Another elementary function, defined by $f(x) = |x|$, is the **absolute value function.** This function pairs each real number with its absolute value. Its graph is shown in **FIGURE 1**.

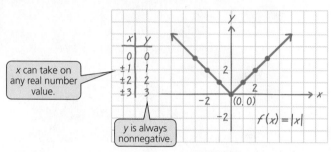

Absolute value function

$$f(x) = |x|$$

Domain: $(-\infty, \infty)$
Range: $[0, \infty)$

FIGURE 1

The **reciprocal function,** defined by $f(x) = \frac{1}{x}$ and introduced in **Section 7.4,** is a rational function. Its graph is shown in **FIGURE 2**. Since x can never equal 0, as x gets closer and closer to 0, $\frac{1}{x}$ approaches either ∞ or $-\infty$. Also, $\frac{1}{x}$ can never equal 0, and as x approaches ∞ or $-\infty$, $\frac{1}{x}$ approaches 0. The axes are called **asymptotes** for the function.

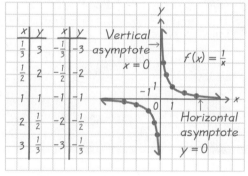

Reciprocal function

$$f(x) = \frac{1}{x}$$

Domain: $(-\infty, 0) \cup (0, \infty)$
Range: $(-\infty, 0) \cup (0, \infty)$

FIGURE 2

The **square root function,** defined by $f(x) = \sqrt{x}$ and introduced in **Section 8.1,** is shown in **FIGURE 3**.

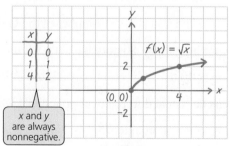

Square root function

$$f(x) = \sqrt{x}$$

Domain: $[0, \infty)$
Range: $[0, \infty)$

FIGURE 3

The graphs of these elementary functions can be shifted, or translated, just as we did with the graph of $f(x) = x^2$ in **Section 9.5.**

NOW TRY
EXERCISE 1

Graph $f(x) = \frac{1}{x+3}$. Give the domain and range.

EXAMPLE 1 Applying a Horizontal Shift

Graph $f(x) = |x - 2|$. Give the domain and range.

The graph of $y = (x - 2)^2$ is obtained by shifting the graph of $y = x^2$ two units to the right. In a similar manner, the graph of $f(x) = |x - 2|$ is found by shifting the graph of $y = |x|$ two units to the right, as shown in **FIGURE 4**.

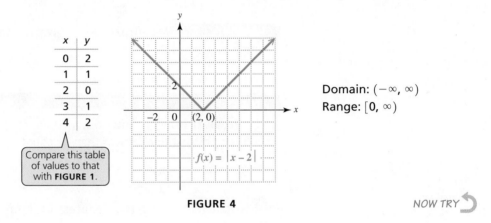

x	y
0	2
1	1
2	0
3	1
4	2

Compare this table of values to that with **FIGURE 1**.

Domain: $(-\infty, \infty)$
Range: $[0, \infty)$

FIGURE 4

NOW TRY

NOW TRY
EXERCISE 2

Graph $f(x) = \sqrt{x} + 2$. Give the domain and range.

EXAMPLE 2 Applying a Vertical Shift

Graph $f(x) = \frac{1}{x} + 3$. Give the domain and range.

The graph is found by shifting the graph of $y = \frac{1}{x}$ three units up. See **FIGURE 5**.

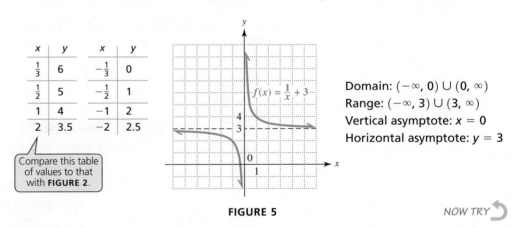

x	y	x	y
$\frac{1}{3}$	6	$-\frac{1}{3}$	0
$\frac{1}{2}$	5	$-\frac{1}{2}$	1
1	4	-1	2
2	3.5	-2	2.5

Compare this table of values to that with **FIGURE 2**.

Domain: $(-\infty, 0) \cup (0, \infty)$
Range: $(-\infty, 3) \cup (3, \infty)$
Vertical asymptote: $x = 0$
Horizontal asymptote: $y = 3$

FIGURE 5

NOW TRY

NOW TRY ANSWERS

1.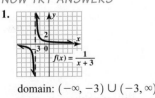

domain: $(-\infty, -3) \cup (-3, \infty)$;
range: $(-\infty, 0) \cup (0, \infty)$

2.

domain: $[0, \infty)$; range: $[2, \infty)$

EXAMPLE 3 Applying Both Horizontal and Vertical Shifts

Graph $f(x) = \sqrt{x + 1} - 4$. Give the domain and range.

The graph of $y = (x + 1)^2 - 4$ is obtained by shifting the graph of $y = x^2$ one unit to the left and four units down. Following this pattern, we shift the graph of $y = \sqrt{x}$ one unit to the left and four units down to get the graph of $f(x) = \sqrt{x + 1} - 4$. See **FIGURE 6** on the next page.

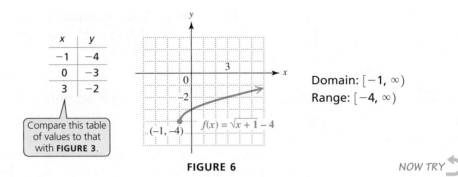

FIGURE 6

NOW TRY

NOW TRY
EXERCISE 3
Graph $f(x) = |x + 1| - 3$.
Give the domain and range.

OBJECTIVE 2 **Recognize and graph step functions.** The greatest integer function is defined as follows.

$f(x) = [\![x]\!]$

The **greatest integer function,** written $f(x) = [\![x]\!]$, pairs every real number x with the greatest integer less than or equal to x.

NOW TRY
EXERCISE 4
Evaluate each expression.
(a) $[\![5]\!]$ **(b)** $[\![-6]\!]$
(c) $[\![3.5]\!]$ **(d)** $[\![-4.1]\!]$

EXAMPLE 4 Finding the Greatest Integer

Evaluate each expression.

(a) $[\![8]\!] = 8$ **(b)** $[\![-1]\!] = -1$ **(c)** $[\![0]\!] = 0$

(d) $[\![7.45]\!] = 7$ The greatest integer *less than or equal to* 7.45 is 7.

(e) $[\![-2.6]\!] = -3$

Think of a number line with -2.6 graphed on it. Since -3 is to the *left of* (and is, therefore, *less than*) -2.6, the greatest integer less than or equal to -2.6 is -3, *not* -2.

NOW TRY

EXAMPLE 5 Graphing the Greatest Integer Function

Graph $f(x) = [\![x]\!]$. Give the domain and range.

$$\text{For } [\![x]\!], \quad \text{if } -1 \le x < 0, \quad \text{then} \quad [\![x]\!] = -1;$$
$$\text{if } \quad 0 \le x < 1, \quad \text{then} \quad [\![x]\!] = 0;$$
$$\text{if } \quad 1 \le x < 2, \quad \text{then} \quad [\![x]\!] = 1;$$
$$\text{if } \quad 2 \le x < 3, \quad \text{then} \quad [\![x]\!] = 2;$$
$$\text{if } \quad 3 \le x < 4, \quad \text{then} \quad [\![x]\!] = 3, \quad \text{and so on.}$$

Thus, the graph, as shown in **FIGURE 7** on the next page, consists of a series of horizontal line segments. In each one, the left endpoint is included and the right endpoint is excluded. These segments continue infinitely following this pattern to the left and right. The appearance of the graph is the reason that this function is called a **step function.**

NOW TRY ANSWERS
3.

domain: $(-\infty, \infty)$; range: $[-3, \infty)$
4. (a) 5 **(b)** -6 **(c)** 3 **(d)** -5

⟲ *NOW TRY*
EXERCISE 5
Graph $f(x) = [\![x - 1]\!]$. Give the domain and range.

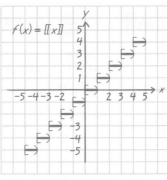

FIGURE 7

Greatest integer function

$$f(x) = [\![x]\!]$$

Domain: $(-\infty, \infty)$

Range: $\{\ldots, -3, -2, -1, 0, 1, 2, 3, \ldots\}$
(the set of integers)

The graph of a step function also may be shifted. For example, the graph of

$$h(x) = [\![x - 2]\!]$$

is the same as the graph of $f(x) = [\![x]\!]$ shifted two units to the right. Similarly, the graph of

$$g(x) = [\![x]\!] + 2$$

is the graph of $f(x)$ shifted two units up.   *NOW TRY* ⟲

⟲ *NOW TRY*
EXERCISE 6
The cost of parking a car at an airport hourly parking lot is $4 for the first hour and $2 for each additional hour or fraction thereof. Let $f(x) =$ the cost of parking a car for x hours. Graph $f(x)$ for x in the interval $(0, 5]$.

EXAMPLE 6   Applying a Greatest Integer Function

An overnight delivery service charges $25 for a package weighing up to 2 lb. For each additional pound or fraction of a pound there is an additional charge of $3. Let $D(x)$, or y, represent the cost to send a package weighing x pounds. Graph $D(x)$ for x in the interval $(0, 6]$.

For x in the interval $(0, 2]$, $y = 25$.

For x in the interval $(2, 3]$, $y = 25 + 3 = 28$.

For x in the interval $(3, 4]$, $y = 28 + 3 = 31$.

For x in the interval $(4, 5]$, $y = 31 + 3 = 34$.

For x in the interval $(5, 6]$, $y = 34 + 3 = 37$.

The graph, which is that of a step function, is shown in **FIGURE 8**.

NOW TRY ANSWERS

5.

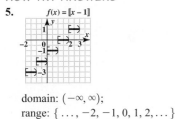

domain: $(-\infty, \infty)$;
range: $\{\ldots, -2, -1, 0, 1, 2, \ldots\}$

6.

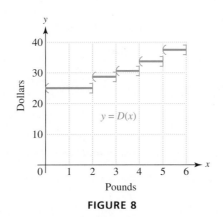

FIGURE 8

NOW TRY ⟲

11.1 EXERCISES

○ *Complete solution available on the Video Resources on DVD*

Concept Check For Exercises 1–6, refer to the basic graphs in A–F.

A.
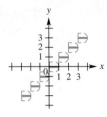

B.
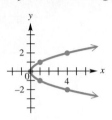

C.
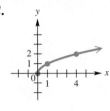

D.
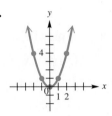

E.
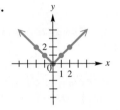

F.
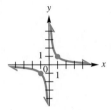

1. Which is the graph of $f(x) = |x|$? The lowest point on its graph has coordinates (____, ____).

2. Which is the graph of $f(x) = x^2$? Give the domain and range.

3. Which is the graph of $f(x) = [\![x]\!]$? Give the domain and range.

4. Which is the graph of $f(x) = \sqrt{x}$? Give the domain and range.

5. Which is not the graph of a function? Why?

6. Which is the graph of $f(x) = \frac{1}{x}$? The lines with equations $x = 0$ and $y = 0$ are called its _____.

Concept Check *Without actually plotting points, match each function defined by the absolute value expression with its graph.*

7. $f(x) = |x - 2| + 2$

8. $f(x) = |x + 2| + 2$

A.

B.

9. $f(x) = |x - 2| - 2$

10. $f(x) = |x + 2| - 2$

C.
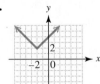

D.
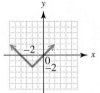

Graph each function. Give the domain and range. ***See Examples 1–3.***

○ 11. $f(x) = |x + 1|$

12. $f(x) = |x - 1|$

○ 13. $f(x) = \frac{1}{x} + 1$

14. $f(x) = \frac{1}{x} - 1$

15. $f(x) = \sqrt{x - 2}$

16. $f(x) = \sqrt{x + 5}$

17. $f(x) = \dfrac{1}{x - 2}$ **18.** $f(x) = \dfrac{1}{x + 2}$ 🌐 **19.** $f(x) = \sqrt{x + 3} - 3$

20. $f(x) = \sqrt{x - 2} + 2$ **21.** $f(x) = |x - 3| + 1$ **22.** $f(x) = |x + 1| - 4$

23. *Concept Check* How is the graph of $f(x) = \frac{1}{x - 3} + 2$ obtained from the graph of $g(x) = \frac{1}{x}$?

24. *Concept Check* How is the graph of $f(x) = \frac{1}{x + 5} - 3$ obtained from the graph of $g(x) = \frac{1}{x}$?

*Evaulate each expression. See **Example 4**.*

25. $[\![3]\!]$ **26.** $[\![18]\!]$ **27.** $[\![4.5]\!]$ **28.** $[\![8.7]\!]$ 🌐 **29.** $\left[\!\!\left[\dfrac{1}{2} \right]\!\!\right]$

30. $\left[\!\!\left[\dfrac{3}{4} \right]\!\!\right]$ **31.** $[\![-14]\!]$ **32.** $[\![-5]\!]$ 🌐 **33.** $[\![-10.1]\!]$ **34.** $[\![-6.9]\!]$

*Graph each step function. See **Examples 5 and 6**.*

35. $f(x) = [\![x]\!] - 1$ **36.** $f(x) = [\![x]\!] + 1$

🌐 **37.** $f(x) = [\![x - 3]\!]$ **38.** $f(x) = [\![x + 2]\!]$

39. Assume that postage rates are 44¢ for the first ounce, plus 17¢ for each additional ounce, and that each letter carries one 44¢ stamp and as many 17¢ stamps as necessary. Graph the function defined by

$$y = p(x) = \text{the number of stamps}$$

on a letter weighing x ounces. Use the interval $(0, 5]$.

40. The cost of parking a car at an airport hourly parking lot is $3 for the first half-hour and $2 for each additional half-hour or fraction thereof. Graph the function defined by $y = f(x) = $ the cost of parking a car for x hours. Use the interval $(0, 2]$.

41. A certain long-distance carrier provides service between Podunk and Nowhereville. If x represents the number of minutes for the call, where $x > 0$, then the function f defined by

$$f(x) = 0.40[\![x]\!] + 0.75$$

gives the total cost of the call in dollars. Find the cost of a 5.5-minute call.

42. **See Exercise 41.** Find the cost of a 20.75-minute call.

PREVIEW EXERCISES

*Find the distance between each pair of points. See **Section 8.3**.*

43. $(2, -1)$ and $(4, 3)$ **44.** (x, y) and $(-2, 5)$ **45.** (x, y) and (h, k)

11.2 The Circle and the Ellipse

OBJECTIVES

1 Find an equation of a circle given the center and radius.

2 Determine the center and radius of a circle given its equation.

3 Recognize an equation of an ellipse.

4 Graph ellipses.

When an infinite cone is intersected by a plane, the resulting figure is called a **conic section.** The parabola is one example of a conic section. Circles, ellipses, and hyperbolas may also result. See **FIGURE 9**.

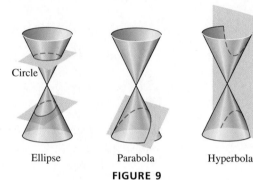

Circle

Ellipse   Parabola   Hyperbola

FIGURE 9

OBJECTIVE 1 **Find an equation of a circle given the center and radius.** A **circle** is the set of all points in a plane that lie a fixed distance from a fixed point. The fixed point is called the **center,** and the fixed distance is called the **radius.** We use the distance formula from **Section 8.3** to find an equation of a circle.

> **NOW TRY**
> **EXERCISE 1**
>
> Find an equation of the circle with radius 6 and center at $(0, 0)$, and graph it.

EXAMPLE 1 Finding an Equation of a Circle and Graphing It

Find an equation of the circle with radius 3 and center at $(0, 0)$, and graph it.

If the point (x, y) is on the circle, then the distance from (x, y) to the center $(0, 0)$ is 3.

$$\sqrt{(x_2 - x_1)^2 + (y_2 - y_1)^2} = d \quad \text{Distance formula}$$

$$\sqrt{(x - 0)^2 + (y - 0)^2} = 3 \quad \text{Let } x_1 = 0, y_1 = 0, \text{ and } d = 3.$$

$$x^2 + y^2 = 9 \quad \text{Square each side.}$$

An equation of this circle is $x^2 + y^2 = 9$. The graph is shown in **FIGURE 10**.

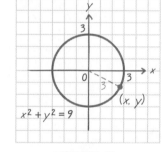

FIGURE 10

NOW TRY

A circle may not be centered at the origin, as seen in the next example.

NOW TRY ANSWER

1. $x^2 + y^2 = 36$

EXAMPLE 2 Finding an Equation of a Circle and Graphing It

Find an equation of the circle with center at $(4, -3)$ and radius 5, and graph it.

$$\sqrt{(x - 4)^2 + [y - (-3)]^2} = 5 \quad \text{Let } x_1 = 4, y_1 = -3, \text{ and } d = 5 \text{ in the distance formula.}$$

$$(x - 4)^2 + (y + 3)^2 = 25 \quad \text{Square each side.}$$

NOW TRY
EXERCISE 2
Find an equation of the circle with center at $(-2, 2)$ and radius 3, and graph it.

To graph the circle, plot the center $(4, -3)$, then move 5 units right, left, up, and down from the center, plotting the points

$$(9, -3), \quad (-1, -3), \quad (4, 2), \quad \text{and} \quad (4, -8).$$

Draw a smooth curve through these four points, sketching one quarter of the circle at a time. See **FIGURE 11**.

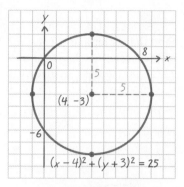

FIGURE 11

NOW TRY

Examples 1 and 2 suggest the form of an equation of a circle with radius r and center at (h, k). If (x, y) is a point on the circle, then the distance from the center (h, k) to the point (x, y) is r. By the distance formula,

$$\sqrt{(x - h)^2 + (y - k)^2} = r.$$

Squaring both sides gives the **center-radius form** of the equation of a circle.

> ### Equation of a Circle (Center-Radius Form)
>
> An equation of a circle with radius r and center (h, k) is
>
> $$(x - h)^2 + (y - k)^2 = r^2.$$

NOW TRY
EXERCISE 3
Find an equation of the circle with center at $(-5, 4)$ and radius $\sqrt{6}$.

EXAMPLE 3 Using the Center-Radius Form of the Equation of a Circle

Find an equation of the circle with center at $(-1, 2)$ and radius $\sqrt{7}$.

$$(x - h)^2 + (y - k)^2 = r^2 \qquad \text{Center-radius form}$$
$$[x - (-1)]^2 + (y - 2)^2 = \left(\sqrt{7}\right)^2 \qquad \text{Let } h = -1, k = 2, \text{ and } r = \sqrt{7}.$$

Pay attention to signs here.
$$(x + 1)^2 + (y - 2)^2 = 7 \qquad \text{Simplify; } \left(\sqrt{a}\right)^2 = a$$

NOW TRY

NOTE If a circle has its center at the origin $(0, 0)$, then its equation becomes

$$(x - 0)^2 + (y - 0)^2 = r^2 \qquad \text{Let } h = 0, k = 0 \text{ in the center-radius form.}$$
$$x^2 + y^2 = r^2. \qquad \text{See Example 1.}$$

OBJECTIVE 2 Determine the center and radius of a circle given its equation.

In the equation found in **Example 2**, multiplying out $(x - 4)^2$ and $(y + 3)^2$ gives

$$(x - 4)^2 + (y + 3)^2 = 25$$
$$x^2 - 8x + 16 + y^2 + 6y + 9 = 25 \qquad \text{Square each binomial.}$$
$$x^2 + y^2 - 8x + 6y = 0. \qquad \text{Subtract 25.}$$

This general form suggests that an equation with both x^2- and y^2-terms with equal coefficients may represent a circle.

NOW TRY ANSWERS
2. $(x + 2)^2 + (y - 2)^2 = 9$

3. $(x + 5)^2 + (y - 4)^2 = 6$

**NOW TRY
EXERCISE 4**
Find the center and radius
of the circle.

$$x^2 + y^2 - 8x + 10y - 8 = 0$$

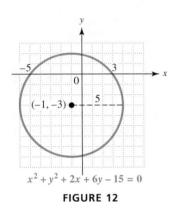

$$x^2 + y^2 + 2x + 6y - 15 = 0$$

FIGURE 12

EXAMPLE 4 Completing the Square to Find the Center and Radius

Find the center and radius of the circle $x^2 + y^2 + 2x + 6y - 15 = 0$, and graph it.

Since the equation has x^2- and y^2-terms with equal coefficients, its graph might be that of a circle. To find the center and radius, complete the squares on x and y.

$$x^2 + y^2 + 2x + 6y = 15 \qquad \text{Transform so that the constant is on the right.}$$

$$(x^2 + 2x \qquad) + (y^2 + 6y \qquad) = 15 \qquad \text{Write in anticipation of completing the square.}$$

$$\left[\frac{1}{2}(2)\right]^2 = 1 \qquad \left[\frac{1}{2}(6)\right]^2 = 9 \qquad \text{Square half the coefficient of each middle term.}$$

$$(x^2 + 2x + 1) + (y^2 + 6y + 9) = 15 + 1 + 9 \qquad \text{Complete the squares on both } x \text{ and } y.$$

> Add 1 and 9 on
> *both* sides of
> the equation.

$$(x + 1)^2 + (y + 3)^2 = 25 \qquad \text{Factor on the left. Add on the right.}$$

$$[x - (-1)]^2 + [y - (-3)]^2 = 5^2 \qquad \text{Center-radius form}$$

The final equation shows that the graph is a circle with center at $(-1, -3)$ and radius 5, as shown in **FIGURE 12**.

NOW TRY

NOTE Consider the following.

1. If the procedure of **Example 4** leads to an equation of the form

$$(x - h)^2 + (y - k)^2 = 0,$$

then the graph is the single point (h, k).

2. If the constant on the right side is *negative*, then the equation has *no graph*.

OBJECTIVE 3 **Recognize an equation of an ellipse.** An **ellipse** is the set of all points in a plane the *sum* of whose distances from two fixed points is constant. These fixed points are called **foci** (singular: *focus*). The ellipse in **FIGURE 13** has foci $(c, 0)$ and $(-c, 0)$, with x-intercepts $(a, 0)$ and $(-a, 0)$ and y-intercepts $(0, b)$ and $(0, -b)$. It is shown in more advanced courses that $c^2 = a^2 - b^2$ for an ellipse of this type. The origin is the **center** of the ellipse.

An ellipse has the following equation.

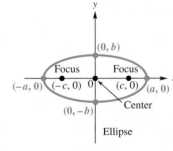

FIGURE 13

Equation of an Ellipse

The ellipse whose x-intercepts are $(a, 0)$ and $(-a, 0)$ and whose y-intercepts are $(0, b)$ and $(0, -b)$ has an equation of the form

$$\frac{x^2}{a^2} + \frac{y^2}{b^2} = 1.$$

NOW TRY ANSWER
4. center: $(4, -5)$; radius: 7

NOTE A circle is a special case of an ellipse, where $a^2 = b^2$.

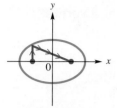

Reflecting property
of an ellipse

FIGURE 14

FIGURE 15

⌐ NOW TRY
⌐ *EXERCISE 5*

Graph $\dfrac{x^2}{16} + \dfrac{y^2}{25} = 1$.

When a ray of light or sound emanating from one focus of an ellipse bounces off the ellipse, it passes through the other focus. See **FIGURE 14**. As mentioned in the chapter introduction, this reflecting property is responsible for whispering galleries. John Quincy Adams was able to listen in on his opponents' conversations because his desk was positioned at one of the foci beneath the ellipsoidal ceiling and his opponents were located across the room at the other focus.

Elliptical bicycle gears are designed to respond to the legs' natural strengths and weaknesses. At the top and bottom of the powerstroke, where the legs have the least leverage, the gear offers little resistance, but as the gear rotates, the resistance increases. This allows the legs to apply more power where it is most naturally available. See **FIGURE 15**.

OBJECTIVE 4   Graph ellipses.

EXAMPLE 5   Graphing Ellipses

Graph each ellipse.

(a) $\dfrac{x^2}{49} + \dfrac{y^2}{36} = 1$

Here, $a^2 = 49$, so $a = 7$, and the x-intercepts are $(7, 0)$ and $(-7, 0)$. Similarly, $b^2 = 36$, so $b = 6$, and the y-intercepts are $(0, 6)$ and $(0, -6)$. Plotting the intercepts and sketching the ellipse through them gives the graph in **FIGURE 16**.

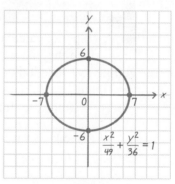

FIGURE 16

(b) $\dfrac{x^2}{36} + \dfrac{y^2}{121} = 1$

The x-intercepts are $(6, 0)$ and $(-6, 0)$, and the y-intercepts are $(0, 11)$ and $(0, -11)$. Join these with the smooth curve of an ellipse. See **FIGURE 17**.

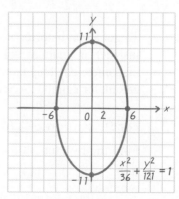

FIGURE 17

NOW TRY ANSWER

5.

$\dfrac{x^2}{16} + \dfrac{y^2}{25} = 1$

NOW TRY ⤵

NOW TRY
EXERCISE 6
Graph
$$\frac{(x-3)^2}{36} + \frac{(y-4)^2}{4} = 1.$$

EXAMPLE 6 Graphing an Ellipse Shifted Horizontally and Vertically

Graph $\dfrac{(x-2)^2}{25} + \dfrac{(y+3)^2}{49} = 1$.

Just as $(x-2)^2$ and $(y+3)^2$ would indicate that the center of a circle would be $(2,-3)$, so it is with this ellipse. **FIGURE 18** shows that the graph goes through the four points

$$(2, 4), \quad (7, -3), \quad (2, -10),$$
$$\text{and} \quad (-3, -3).$$

The x-values of these points are found by adding $\pm a = \pm 5$ to 2, and the y-values come from adding $\pm b = \pm 7$ to -3.

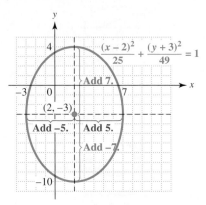

FIGURE 18

NOW TRY

NOTE *Graphs of circles and ellipses are not graphs of functions.* The only conic section whose graph represents a function is the vertical parabola with equation $f(x) = ax^2 + bx + c$.

CONNECTIONS

A graphing calculator in function mode cannot directly graph a circle or an ellipse, since they do not represent functions. We must first solve the equation for y, getting two functions y_1 and y_2. The union of these two graphs is the graph of the entire figure.

For example, to graph $(x+3)^2 + (y+2)^2 = 25$, begin by solving for y.

$$(x+3)^2 + (y+2)^2 = 25$$

$$(y+2)^2 = 25 - (x+3)^2 \qquad \text{Subtract } (x+3)^2.$$

$$y + 2 = \pm\sqrt{25 - (x+3)^2} \qquad \text{Take square roots.}$$

Remember both roots. $\quad y = -2 \pm \sqrt{25 - (x+3)^2} \qquad \text{Add } -2.$

The two functions to be graphed are

$$y_1 = -2 + \sqrt{25 - (x+3)^2} \qquad \text{and} \qquad y_2 = -2 - \sqrt{25 - (x+3)^2}.$$

To get an undistorted screen, a **square viewing window** must be used. (Refer to your instruction manual for details.) See **FIGURE 19**. The two semicircles seem to be disconnected. This is because the graphs are nearly vertical at those points, and the calculator cannot show a true picture of the behavior there.

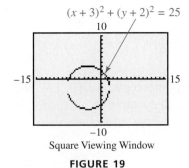

$(x+3)^2 + (y+2)^2 = 25$

Square Viewing Window

FIGURE 19

For Discussion or Writing

Find the two functions y_1 and y_2 to use to obtain the graph of the circle with equation $(x-3)^2 + (y+1)^2 = 36$. Then graph the circle using a square viewing window.

NOW TRY ANSWER
6.
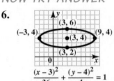

 Complete solution available on the Video Resources on DVD

 1. See Example 1. Consider the circle whose equation is $x^2 + y^2 = 25$.

  (a) What are the coordinates of its center?   **(b)** What is its radius?

  (c) Sketch its graph.

2. Why does a set of points defined by a circle *not* satisfy the definition of a function?

Concept Check *Match each equation with the correct graph.*

3. $(x - 3)^2 + (y - 2)^2 = 25$ **A.**

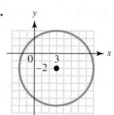

4. $(x - 3)^2 + (y + 2)^2 = 25$

B.

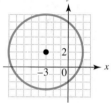

5. $(x + 3)^2 + (y - 2)^2 = 25$ **C.**

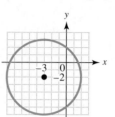

6. $(x + 3)^2 + (y + 2)^2 = 25$

D.

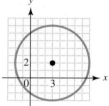

Find the equation of a circle satisfying the given conditions. **See Examples 2 and 3.**

 7. Center: $(-4, 3)$; radius: 2   **8.** Center: $(5, -2)$; radius: 4

 9. Center: $(-8, -5)$; radius: $\sqrt{5}$ **10.** Center: $(-12, 13)$; radius: $\sqrt{7}$

Find the center and radius of each circle. (Hint: In Exercises 15 and 16, divide each side by a common factor.) **See Example 4.**

 11. $x^2 + y^2 + 4x + 6y + 9 = 0$ **12.** $x^2 + y^2 - 8x - 12y + 3 = 0$

13. $x^2 + y^2 + 10x - 14y - 7 = 0$ **14.** $x^2 + y^2 - 2x + 4y - 4 = 0$

15. $3x^2 + 3y^2 - 12x - 24y + 12 = 0$ **16.** $2x^2 + 2y^2 + 20x + 16y + 10 = 0$

Graph each circle. Identify the center if it is not at the origin. **See Examples 1, 2, and 4.**

17. $x^2 + y^2 = 9$ **18.** $x^2 + y^2 = 4$

19. $2y^2 = 10 - 2x^2$ **20.** $3x^2 = 48 - 3y^2$

21. $(x + 3)^2 + (y - 2)^2 = 9$ **22.** $(x - 1)^2 + (y + 3)^2 = 16$

23. $x^2 + y^2 - 4x - 6y + 9 = 0$ **24.** $x^2 + y^2 + 8x + 2y - 8 = 0$

25. $x^2 + y^2 + 6x - 6y + 9 = 0$ **26.** $x^2 + y^2 - 4x + 10y + 20 = 0$

27. A circle can be drawn on a piece of posterboard by fastening one end of a string with a thumbtack, pulling the string taut with a pencil, and tracing a curve, as shown in the figure. Explain why this method works.

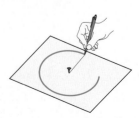

28. An ellipse can be drawn on a piece of posterboard by fastening two ends of a length of string with thumbtacks, pulling the string taut with a pencil, and tracing a curve, as shown in the figure. Explain why this method works.

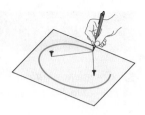

Graph each ellipse. **See Examples 5 and 6.**

29. $\dfrac{x^2}{9} + \dfrac{y^2}{25} = 1$

30. $\dfrac{x^2}{9} + \dfrac{y^2}{16} = 1$

31. $\dfrac{x^2}{36} + \dfrac{y^2}{16} = 1$

32. $\dfrac{x^2}{9} + \dfrac{y^2}{4} = 1$

33. $\dfrac{x^2}{16} + \dfrac{y^2}{4} = 1$

34. $\dfrac{x^2}{49} + \dfrac{y^2}{81} = 1$

35. $\dfrac{y^2}{25} = 1 - \dfrac{x^2}{49}$

36. $\dfrac{y^2}{9} = 1 - \dfrac{x^2}{16}$

37. $\dfrac{(x+1)^2}{64} + \dfrac{(y-2)^2}{49} = 1$

38. $\dfrac{(x-4)^2}{9} + \dfrac{(y+2)^2}{4} = 1$

39. $\dfrac{(x-2)^2}{16} + \dfrac{(y-1)^2}{9} = 1$

40. $\dfrac{(x+3)^2}{25} + \dfrac{(y+2)^2}{36} = 1$

41. Explain why a set of ordered pairs whose graph forms an ellipse does not satisfy the definition of a function.

42. (a) How many points are there on the graph of $(x-4)^2 + (y-1)^2 = 0$? Explain.

(b) How many points are there on the graph of $(x-4)^2 + (y-1)^2 = -1$? Explain.

TECHNOLOGY INSIGHTS   **EXERCISES 43 AND 44**

43. The circle shown in the calculator graph was created using function mode, with a square viewing window. It is the graph of

$$(x+2)^2 + (y-4)^2 = 16.$$

What are the two functions y_1 and y_2 that were used to obtain this graph?

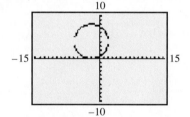

44. The ellipse shown in the calculator graph was graphed using function mode, with a square viewing window. It is the graph of

$$\dfrac{x^2}{4} + \dfrac{y^2}{9} = 1.$$

What are the two functions y_1 and y_2 that were used to obtain this graph?

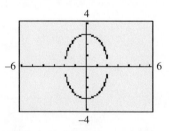

Use a graphing calculator in function mode to graph each circle or ellipse. Use a square viewing window. **See the Connections box.**

45. $x^2 + y^2 = 36$

46. $(x-2)^2 + y^2 = 49$

47. $\dfrac{x^2}{16} + \dfrac{y^2}{4} = 1$

48. $\dfrac{(x-3)^2}{25} + \dfrac{y^2}{9} = 1$

*A **lithotripter** is a machine used to crush kidney stones using shock waves. The patient is placed in an elliptical tub with the kidney stone at one focus of the ellipse. A beam is projected from the other focus to the tub, so that it reflects to hit the kidney stone. See the figure.*

49. Suppose a lithotripter is based on the ellipse with equation

$$\frac{x^2}{36} + \frac{y^2}{9} = 1.$$

How far from the center of the ellipse must the kidney stone and the source of the beam be placed? (*Hint:* Use the fact that $c^2 = a^2 - b^2$, since $a > b$ here.)

50. Rework **Exercise 49** if the equation of the ellipse is

$$9x^2 + 4y^2 = 36.$$

(*Hint:* Write the equation in fractional form by dividing each term by 36, and use $c^2 = b^2 - a^2$, since $b > a$ here.)

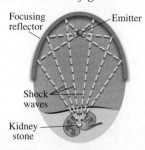

The top of an ellipse is illustrated in this depiction of how a lithotripter crushes a kidney stone.

Solve each problem.

51. An arch has the shape of half an ellipse. The equation of the ellipse is

$$100x^2 + 324y^2 = 32{,}400,$$

where x and y are in meters.

(a) How high is the center of the arch?

(b) How wide is the arch across the bottom?

NOT TO SCALE

52. A one-way street passes under an overpass, which is in the form of the top half of an ellipse, as shown in the figure. Suppose that a truck 12 ft wide passes directly under the overpass. What is the maximum possible height of this truck?

NOT TO SCALE

*In Exercises 53 and 54, see **FIGURE 13** and use the fact that $c^2 = a^2 - b^2$, where $a^2 > b^2$.*

53. The orbit of Mars is an ellipse with the sun at one focus. For x and y in millions of miles, the equation of the orbit is

$$\frac{x^2}{141.7^2} + \frac{y^2}{141.1^2} = 1.$$

(*Source:* Kaler, James B., *Astronomy!*, Addison-Wesley.)

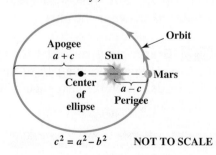

$c^2 = a^2 - b^2$ **NOT TO SCALE**

(a) Find the greatest distance (the **apogee**) from Mars to the sun.

(b) Find the least distance (the **perigee**) from Mars to the sun.

54. The orbit of Venus around the sun (one of the foci) is an ellipse with equation

$$\frac{x^2}{5013} + \frac{y^2}{4970} = 1,$$

where x and y are measured in millions of miles. (*Source:* Kaler, James B., *Astronomy!*, Addison-Wesley.)

(a) Find the greatest distance between Venus and the sun.

(b) Find the least distance between Venus and the sun.

PREVIEW EXERCISES

For Exercises 55–57, see Section 3.1.

55. Plot the points $(3, 4)$, $(-3, 4)$, $(3, -4)$, and $(-3, -4)$.

56. Sketch the graphs of $y = \frac{4}{3}x$ and $y = -\frac{4}{3}x$ on the same axes.

57. Find the x- and y-intercepts of the graph of $4x + 3y = 12$.

58. Solve the equation $x^2 = 121$. **See Section 9.1.**

11.3 The Hyperbola and Functions Defined by Radicals

OBJECTIVES

1. Recognize the equation of a hyperbola.
2. Graph hyperbolas by using asymptotes.
3. Identify conic sections by their equations.
4. Graph certain square root functions.

OBJECTIVE 1 **Recognize the equation of a hyperbola.** A **hyperbola** is the set of all points in a plane such that the absolute value of the *difference* of the distances from two fixed points (the *foci*) is constant. The graph of a hyperbola has two parts, called *branches,* and two intercepts (or *vertices*) that lie on its axis, called the **transverse axis.** The hyperbola in **FIGURE 20** has a horizontal transverse axis, with foci $(c, 0)$ and $(-c, 0)$ and x-intercepts $(a, 0)$ and $(-a, 0)$. (A hyperbola with vertical transverse axis would have its intercepts on the y-axis.)

A hyperbola centered at the origin has one of the following equations. It is shown in more advanced courses that for a hyperbola, $c^2 = a^2 + b^2$.

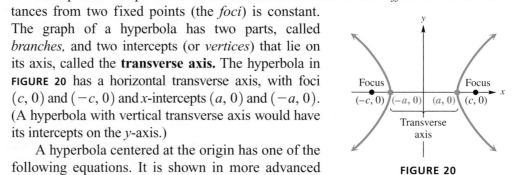

FIGURE 20

Equations of Hyperbolas

A hyperbola with x-intercepts $(a, 0)$ and $(-a, 0)$ has an equation of the form

$$\frac{x^2}{a^2} - \frac{y^2}{b^2} = 1. \qquad \text{Transverse axis on } x\text{-axis}$$

A hyperbola with y-intercepts $(0, b)$ and $(0, -b)$ has an equation of the form

$$\frac{y^2}{b^2} - \frac{x^2}{a^2} = 1. \qquad \text{Transverse axis on } y\text{-axis}$$

If we were to throw two stones into a pond, the ensuing concentric ripples would be shaped like a hyperbola. A cross-section of the cooling towers for a nuclear power plant is hyperbolic, as shown in the photo.

OBJECTIVE 2 **Graph hyperbolas by using asymptotes.** The two branches of the graph of a hyperbola approach a pair of intersecting straight lines, which are its asymptotes. See **FIGURE 21** on the next page. The asymptotes are useful for sketching the graph of the hyperbola.

Asymptotes of Hyperbolas

The extended diagonals of the rectangle with vertices (corners) at the points (a, b), $(-a, b)$, $(-a, -b)$, and $(a, -b)$ are the **asymptotes** of the hyperbolas

$$\frac{x^2}{a^2} - \frac{y^2}{b^2} = 1 \qquad \text{and} \qquad \frac{y^2}{b^2} - \frac{x^2}{a^2} = 1.$$

This rectangle is called the **fundamental rectangle.** Using the methods of **Chapter 3,** we could show that the equations of these asymptotes are

$$y = \frac{b}{a}x \qquad \text{and} \qquad y = -\frac{b}{a}x.$$ Equations of the asymptotes of a hyperbola

To graph hyperbolas, follow these steps.

Graphing a Hyperbola

Step 1   **Find the intercepts.** Locate the intercepts at $(a, 0)$ and $(-a, 0)$ if the x^2-term has a positive coefficient, or at $(0, b)$ and $(0, -b)$ if the y^2-term has a positive coefficient.

Step 2   **Find the fundamental rectangle.** Locate the vertices of the fundamental rectangle at (a, b), $(-a, b)$, $(-a, -b)$, and $(a, -b)$.

Step 3   **Sketch the asymptotes.** The extended diagonals of the rectangle are the asymptotes of the hyperbola, and they have equations $y = \pm\frac{b}{a}x$.

Step 4   **Draw the graph.** Sketch each branch of the hyperbola through an intercept and approaching (but not touching) the asymptotes.

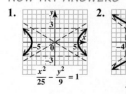

NOW TRY
EXERCISE 1

Graph $\dfrac{x^2}{25} - \dfrac{y^2}{9} = 1$.

EXAMPLE 1   Graphing a Horizontal Hyperbola

Graph $\dfrac{x^2}{16} - \dfrac{y^2}{25} = 1$.

Step 1   Here $a = 4$ and $b = 5$. The x-intercepts are $(4, 0)$ and $(-4, 0)$.

Step 2   The four points $(4, 5)$, $(-4, 5)$, $(-4, -5)$, and $(4, -5)$ are the vertices of the fundamental rectangle, as shown in **FIGURE 21** below.

Steps 3 and 4   The equations of the asymptotes are $y = \pm\frac{5}{4}x$, and the hyperbola approaches these lines as x and y get larger and larger in absolute value.   NOW TRY

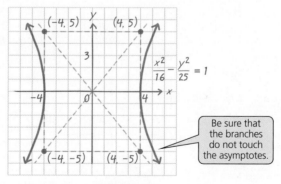

FIGURE 21

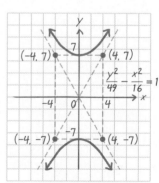

FIGURE 22

NOW TRY
EXERCISE 2

Graph $\dfrac{y^2}{9} - \dfrac{x^2}{16} = 1$.

NOW TRY ANSWERS

1.

$\dfrac{x^2}{25} - \dfrac{y^2}{9} = 1$

2.

$\dfrac{y^2}{9} - \dfrac{x^2}{16} = 1$

EXAMPLE 2   Graphing a Vertical Hyperbola

Graph $\dfrac{y^2}{49} - \dfrac{x^2}{16} = 1$.

This hyperbola has y-intercepts $(0, 7)$ and $(0, -7)$. The asymptotes are the extended diagonals of the rectangle with vertices at $(4, 7)$, $(-4, 7)$, $(-4, -7)$, and $(4, -7)$. Their equations are $y = \pm\frac{7}{4}x$. See **FIGURE 22** above.   NOW TRY

NOTE As with circles and ellipses, hyperbolas are graphed with a graphing calculator by first writing the equations of two functions whose union is equivalent to the equation of the hyperbola. A square window gives a truer shape for hyperbolas, too.

SUMMARY OF CONIC SECTIONS

Equation	Graph	Description	Identification
$y = ax^2 + bx + c$ **or** $y = a(x - h)^2 + k$	 Parabola	It opens up if $a > 0$, down if $a < 0$. The vertex is (h, k).	It has an x^2-term. y is not squared.
$x = ay^2 + by + c$ **or** $x = a(y - k)^2 + h$	 Parabola	It opens to the right if $a > 0$, to the left if $a < 0$. The vertex is (h, k).	It has a y^2-term. x is not squared.
$(x - h)^2 +$ $(y - k)^2 = r^2$	 Circle	The center is (h, k), and the radius is r.	x^2- and y^2-terms have the same positive coefficient.
$\dfrac{x^2}{a^2} + \dfrac{y^2}{b^2} = 1$	 Ellipse	The x-intercepts are $(a, 0)$ and $(-a, 0)$. The y-intercepts are $(0, b)$ and $(0, -b)$.	x^2- and y^2-terms have different positive coefficients.
$\dfrac{x^2}{a^2} - \dfrac{y^2}{b^2} = 1$	 Hyperbola	The x-intercepts are $(a, 0)$ and $(-a, 0)$. The asymptotes are found from (a, b), $(a, -b)$, $(-a, -b)$, and $(-a, b)$.	x^2 has a positive coefficient. y^2 has a negative coefficient.
$\dfrac{y^2}{b^2} - \dfrac{x^2}{a^2} = 1$	 Hyperbola	The y-intercepts are $(0, b)$ and $(0, -b)$. The asymptotes are found from (a, b), $(a, -b)$, $(-a, -b)$, and $(-a, b)$.	y^2 has a positive coefficient. x^2 has a negative coefficient.

OBJECTIVE 3 **Identify conic sections by their equations.** Rewriting a second-degree equation in one of the forms given for ellipses, hyperbolas, circles, or parabolas makes it possible to identify the graph of the equation.

⌒NOW TRY
↪ EXERCISE 3
Identify the graph of each equation.
(a) $y^2 - 10 = -x^2$
(b) $y - 2x^2 = 8$
(c) $3x^2 + y^2 = 4$

EXAMPLE 3 Identifying the Graphs of Equations

Identify the graph of each equation.

(a) $9x^2 = 108 + 12y^2$

Both variables are squared, so the graph is either an ellipse or a hyperbola. (This situation also occurs for a circle, which is a special case of an ellipse.) Rewrite the equation so that the x^2- and y^2-terms are on one side of the equation and 1 is on the other.

$$9x^2 - 12y^2 = 108 \qquad \text{Subtract } 12y^2.$$

$$\frac{x^2}{12} - \frac{y^2}{9} = 1 \qquad \text{Divide by 108.}$$

The graph of this equation is a hyperbola.

(b) $x^2 = y - 3$

Only one of the two variables, x, is squared, so this is the vertical parabola $y = x^2 + 3$.

(c) $x^2 = 9 - y^2$

Write the variable terms on the same side of the equation.

$$x^2 + y^2 = 9 \qquad \text{Add } y^2.$$

The graph of this equation is a circle with center at the origin and radius 3.

NOW TRY⤷

OBJECTIVE 4 **Graph certain square root functions.** Recall from the vertical line test that no vertical line will intersect the graph of a function in more than one point. Thus, the graphs of horizontal parabolas, all circles and ellipses, and most hyperbolas discussed in this chapter do not satisfy the conditions of a function. However, by considering only a part of each graph, we have the graph of a function, as seen in **FIGURE 23**.

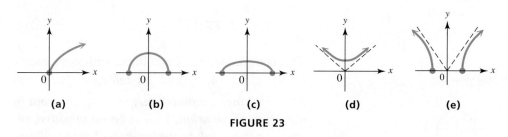

(a) (b) (c) (d) (e)

FIGURE 23

In parts (a)–(d) of **FIGURE 23**, the top portion of a conic section is shown (parabola, circle, ellipse, and hyperbola, respectively). In part (e), the top two portions of a hyperbola are shown. In each case, the graph is that of a function since the graph satisfies the conditions of the vertical line test.

In **Sections 8.1 and 11.1,** we observed the square root function defined by $f(x) = \sqrt{x}$. To find equations for the types of graphs shown in **FIGURE 23**, we extend its definition.

NOW TRY ANSWERS
3. (a) circle (b) parabola
   (c) ellipse

> ### Generalized Square Root Function
>
> For an algebraic expression in x defined by u, with $u \geq 0$, a function of the form
>
> $$f(x) = \sqrt{u}$$
>
> is a **generalized square root function.**

⤷ *NOW TRY*
   EXERCISE 4

Graph $f(x) = \sqrt{64 - x^2}$.
Give the domain and range.

EXAMPLE 4 Graphing a Semicircle

Graph $f(x) = \sqrt{25 - x^2}$. Give the domain and range.

$$f(x) = \sqrt{25 - x^2} \qquad \text{Given function}$$

$$\boxed{(\sqrt{a})^2 = a} \quad y = \sqrt{25 - x^2} \qquad \text{Replace } f(x) \text{ with } y.$$

$$y^2 = 25 - x^2 \qquad \text{Square each side.}$$

$$x^2 + y^2 = 25 \qquad \text{Add } x^2.$$

This is the graph of a circle with center at $(0, 0)$ and radius 5. Since $f(x)$, or y, represents a principal square root in the original equation, $f(x)$ must be nonnegative. This restricts the graph to the upper half of the circle, as shown in **FIGURE 24**. The domain is $[-5, 5]$, and the range is $[0, 5]$.

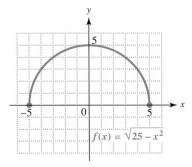

FIGURE 24

NOW TRY ⤹

⤷ *NOW TRY*
   EXERCISE 5

Graph $\dfrac{y}{4} = -\sqrt{1 - \dfrac{x^2}{9}}$.
Give the domain and range.

EXAMPLE 5 Graphing a Portion of an Ellipse

Graph $\dfrac{y}{6} = -\sqrt{1 - \dfrac{x^2}{16}}$. Give the domain and range.

Square each side to get an equation whose form is known.

$$\left(\frac{y}{6}\right)^2 = \left(-\sqrt{1 - \frac{x^2}{16}}\right)^2 \qquad \text{Square each side.}$$

$$\frac{y^2}{36} = 1 - \frac{x^2}{16} \qquad \text{Apply the exponents.}$$

$$\frac{x^2}{16} + \frac{y^2}{36} = 1 \qquad \text{Add } \frac{x^2}{16}.$$

This is the equation of an ellipse with x-intercepts $(4, 0)$ and $(-4, 0)$ and y-intercepts $(0, 6)$ and $(0, -6)$. *Since $\frac{y}{6}$ equals a negative square root in the original equation, y must be nonpositive, restricting the graph to the lower half of the ellipse,* as shown in **FIGURE 25**. The domain is $[-4, 4]$, and the range is $[-6, 0]$.

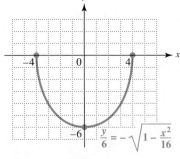

FIGURE 25

NOW TRY ⤹

NOW TRY ANSWERS

4.

domain: $[-8, 8]$; range: $[0, 8]$

5.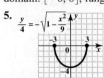

domain: $[-3, 3]$; range: $[-4, 0]$

NOTE Root functions like those graphed in **FIGURES 24 and 25**, can be entered and graphed directly with a graphing calculator.

11.3 EXERCISES   *MyMathLab*

● *Complete solution available on the Video Resources on DVD*

Concept Check   *Based on the discussions of ellipses in the previous section and of hyperbolas in this section, match each equation with its graph.*

1. $\dfrac{x^2}{25} + \dfrac{y^2}{9} = 1$

A.

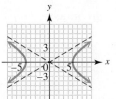

B.

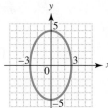

2. $\dfrac{x^2}{9} + \dfrac{y^2}{25} = 1$

3. $\dfrac{x^2}{9} - \dfrac{y^2}{25} = 1$

C.

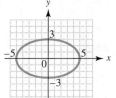

D.

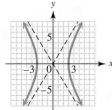

4. $\dfrac{x^2}{25} - \dfrac{y^2}{9} = 1$

Graph each hyperbola. ***See Examples 1 and 2.***

● **5.** $\dfrac{x^2}{16} - \dfrac{y^2}{9} = 1$ **6.** $\dfrac{x^2}{25} - \dfrac{y^2}{9} = 1$ ● **7.** $\dfrac{y^2}{4} - \dfrac{x^2}{25} = 1$

8. $\dfrac{y^2}{9} - \dfrac{x^2}{4} = 1$ **9.** $\dfrac{x^2}{25} - \dfrac{y^2}{36} = 1$ **10.** $\dfrac{x^2}{49} - \dfrac{y^2}{16} = 1$

11. $\dfrac{y^2}{16} - \dfrac{x^2}{16} = 1$ **12.** $\dfrac{y^2}{9} - \dfrac{x^2}{9} = 1$

Identify the graph of each equation as a parabola, circle, ellipse, *or* hyperbola, *and then sketch the graph.* ***See Example 3.***

13. $x^2 - y^2 = 16$ **14.** $x^2 + y^2 = 16$ ● **15.** $4x^2 + y^2 = 16$

16. $9x^2 = 144 + 16y^2$ **17.** $y^2 = 36 - x^2$ **18.** $9x^2 + 25y^2 = 225$

19. $x^2 - 2y = 0$ **20.** $x^2 + 9y^2 = 9$ **21.** $y^2 = 4 + x^2$

✎ **22.** State in your own words the major difference between the definitions of *ellipse* and *hyperbola*.

Graph each generalized square root function. Give the domain and range. ***See Examples 4 and 5.***

● **23.** $f(x) = \sqrt{16 - x^2}$ **24.** $f(x) = \sqrt{9 - x^2}$ ● **25.** $f(x) = -\sqrt{36 - x^2}$

26. $f(x) = -\sqrt{25 - x^2}$ **27.** $y = -2\sqrt{1 - \dfrac{x^2}{9}}$ **28.** $y = -3\sqrt{1 - \dfrac{x^2}{25}}$

29. $\dfrac{y}{3} = \sqrt{1 + \dfrac{x^2}{9}}$ **30.** $\dfrac{y}{2} = \sqrt{1 + \dfrac{x^2}{4}}$

*In **Section 11.2, Example 6,** we saw that the center of an ellipse may be shifted away from the origin. The same process applies to hyperbolas. For example, the hyperbola shown at the right,*

$$\frac{(x + 5)^2}{4} - \frac{(y - 2)^2}{9} = 1,$$

has the same graph as

$$\frac{x^2}{4} - \frac{y^2}{9} = 1,$$

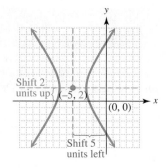

but it is centered at $(-5, 2)$. Graph each hyperbola with center shifted away from the origin.

31. $\dfrac{(x - 2)^2}{4} - \dfrac{(y + 1)^2}{9} = 1$

32. $\dfrac{(x + 3)^2}{16} - \dfrac{(y - 2)^2}{25} = 1$

33. $\dfrac{y^2}{36} - \dfrac{(x - 2)^2}{49} = 1$

34. $\dfrac{(y - 5)^2}{9} - \dfrac{x^2}{25} = 1$

Solve each problem.

35. Two buildings in a sports complex are shaped and positioned like a portion of the branches of the hyperbola with equation

$$400x^2 - 625y^2 = 250,000,$$

where x and y are in meters.

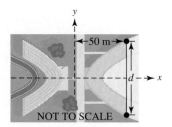

NOT TO SCALE

(a) How far apart are the buildings at their closest point?

(b) Find the distance d in the figure.

36. In rugby, after a *try* (similar to a touchdown in American football) the scoring team attempts a kick for extra points. The ball must be kicked from directly behind the point where the try was scored. The kicker can choose the distance but cannot move the ball sideways. It can be shown that the kicker's best choice is on the hyperbola with equation

$$\frac{x^2}{g^2} - \frac{y^2}{g^2} = 1,$$

where $2g$ is the distance between the goal posts. Since the hyperbola approaches its asymptotes, it is easier for the kicker to estimate points on the asymptotes instead of on the hyperbola. What are the asymptotes of this hyperbola? Why is it relatively easy to estimate them? (*Source:* Isaksen, Daniel C., "How to Kick a Field Goal," *The College Mathematics Journal.*)

37. The hyperbola shown in the figure was graphed in function mode, with a square viewing window. It is the graph of $\frac{x^2}{9} - y^2 = 1$. What are the two functions y_1 and y_2 that were used to obtain this graph?

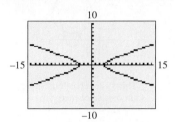

38. Repeat **Exercise 37** for the graph of $\frac{y^2}{9} - x^2 = 1$, shown in the figure.

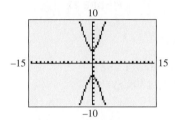

 Use a graphing calculator in function mode to graph each hyperbola. Use a square viewing window.

39. $\dfrac{x^2}{25} - \dfrac{y^2}{49} = 1$ **40.** $\dfrac{x^2}{4} - \dfrac{y^2}{16} = 1$ **41.** $y^2 - 9x^2 = 9$ **42.** $y^2 - 9x^2 = 36$

PREVIEW EXERCISES

*Solve each system. **See Section 4.1.***

43. $2x + y = 13$
$\quad\ \ y = 3x + 3$

44. $9x + 2y = 10$
$\quad\ \ x - y = -5$

45. $4x - 3y = -10$
$\quad\ \ 4x + 6y = 8$

46. $5x + 7y = 6$
$\quad\ 10x - 3y = 46$

*Solve each equation. **See Section 9.3.***

47. $2x^4 - 5x^2 - 3 = 0$

48. $x^4 - 7x^2 + 12 = 0$

11.4 Nonlinear Systems of Equations

OBJECTIVES

1 Solve a nonlinear system by substitution.

2 Solve a nonlinear system by elimination.

3 Solve a nonlinear system that requires a combination of methods.

An equation in which some terms have more than one variable or a variable of degree 2 or greater is called a **nonlinear equation.** A **nonlinear system of equations** includes at least one nonlinear equation.

When solving a nonlinear system, it helps to visualize the types of graphs of the equations of the system to determine the possible number of points of intersection. For example, if a system includes two equations where the graph of one is a circle and the graph of the other is a line, then there may be zero, one, or two points of intersection, as illustrated in **FIGURE 26.**

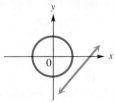

No points of intersection

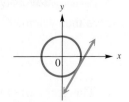

One point of intersection

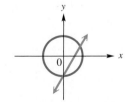

Two points of intersection

FIGURE 26

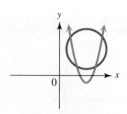

This system has four
solutions, since there are
four points of intersection.

FIGURE 27

↰ *NOW TRY*
↳ *EXERCISE 1*
Solve the system.
$$4x^2 + y^2 = 36$$
$$x - y = 3$$

If a system consists of two second-degree equations, then there may be zero, one, two, three, or four solutions. **FIGURE 27** shows a case where a system consisting of a circle and a parabola has four solutions, all made up of ordered pairs of real numbers.

OBJECTIVE 1 **Solve a nonlinear system by substitution.** We can usually solve a nonlinear system by the substitution method (**Section 4.1**) when one equation is linear.

EXAMPLE 1 Solving a Nonlinear System by Substitution

Solve the system.
$$x^2 + y^2 = 9 \qquad (1)$$
$$2x - y = 3 \qquad (2)$$

The graph of (1) is a circle and the graph of (2) is a line, so the graphs could intersect in zero, one, or two points, as in **FIGURE 26** on the preceding page. We solve the linear equation (2) for one of the two variables and then substitute the resulting expression into the nonlinear equation.

$$2x - y = 3 \qquad (2)$$
$$y = 2x - 3 \qquad \text{Solve for } y. \quad (3)$$

Substitute $2x - 3$ for y in equation (1).

$$x^2 + y^2 = 9 \qquad (1)$$
$$x^2 + (2x - 3)^2 = 9 \qquad \text{Let } y = 2x - 3.$$
$$x^2 + 4x^2 - 12x + 9 = 9 \qquad \text{Square } 2x - 3.$$
$$5x^2 - 12x = 0 \qquad \text{Combine like terms. Subtract 9.}$$
$$x(5x - 12) = 0 \qquad \text{Factor. The GCF is } x.$$
$$x = 0 \quad \text{or} \quad 5x - 12 = 0 \qquad \text{Zero-factor property}$$

> Set *both* factors equal to 0.

$$x = \frac{12}{5}$$

Let $x = 0$ in equation (3) to get $y = -3$. If $x = \frac{12}{5}$, then $y = \frac{9}{5}$. The solution set of the system is

$$\left\{ (0, -3), \left(\frac{12}{5}, \frac{9}{5} \right) \right\}.$$

See the graph in **FIGURE 28**.

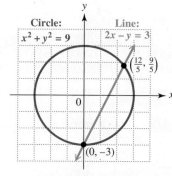

FIGURE 28

NOW TRY ↻

EXAMPLE 2 Solving a Nonlinear System by Substitution

Solve the system.
$$6x - y = 5 \qquad (1)$$
$$xy = 4 \qquad (2)$$

The graph of (1) is a line. It can be shown by plotting points that the graph of (2) is a hyperbola. Visualizing a line and a hyperbola indicates that there may be zero, one, or two points of intersection.

NOW TRY ANSWER
1. $\left\{ (3, 0), \left(-\frac{9}{5}, -\frac{24}{5} \right) \right\}$

**NOW TRY
EXERCISE 2**

Solve the system.

$$xy = 2$$
$$x - 3y = 1$$

Since neither equation has a squared term, we can solve either equation for one of the variables and then substitute the result into the other equation. Solving $xy = 4$ for x gives $x = \frac{4}{y}$. We substitute $\frac{4}{y}$ for x in equation (1).

$$6x - y = 5 \qquad (1)$$

$$6\left(\frac{4}{y}\right) - y = 5 \qquad \text{Let } x = \tfrac{4}{y}.$$

$$\frac{24}{y} - y = 5 \qquad \text{Multiply.}$$

$$24 - y^2 = 5y \qquad \text{Multiply by } y,\ y \neq 0.$$

$$y^2 + 5y - 24 = 0 \qquad \text{Standard form}$$

$$(y - 3)(y + 8) = 0 \qquad \text{Factor.}$$

$$y = 3 \quad \text{or} \quad y = -8 \qquad \text{Zero-factor property}$$

We substitute these results into $x = \frac{4}{y}$ to obtain the corresponding values of x.

If $y = 3$, then $x = \dfrac{4}{3}$.

If $y = -8$, then $x = -\dfrac{1}{2}$.

The solution set of the system is

$$\left\{ \left(\frac{4}{3}, 3\right), \left(-\frac{1}{2}, -8\right) \right\}.$$

See the graph in **FIGURE 29**.

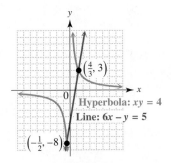

FIGURE 29

NOW TRY

OBJECTIVE 2 **Solve a nonlinear system by elimination.** We can often use the elimination method **(Section 4.1)** when both equations of a nonlinear system are second degree.

EXAMPLE 3 **Solving a Nonlinear System by Elimination**

Solve the system.

$$x^2 + y^2 = 9 \qquad (1)$$

$$2x^2 - y^2 = -6 \qquad (2)$$

The graph of (1) is a circle, while the graph of (2) is a hyperbola. By analyzing the possibilities, we conclude that there may be zero, one, two, three, or four points of intersection. Adding the two equations will eliminate y.

$$
\begin{array}{lll}
x^2 + y^2 = 9 & \quad & (1) \\
2x^2 - y^2 = -6 & \quad & (2) \\
\hline
3x^2 = 3 & \quad & \text{Add.} \\
x^2 = 1 & \quad & \text{Divide by 3.} \\
x = 1 \quad \text{or} \quad x = -1 & \quad & \text{Square root property}
\end{array}
$$

NOW TRY ANSWER

2. $\left\{(-2, -1), \left(3, \tfrac{2}{3}\right)\right\}$

⌐ NOW TRY
  ↳ EXERCISE 3
Solve the system.

$$x^2 + y^2 = 16$$
$$4x^2 + 13y^2 = 100$$

Each value of x gives corresponding values for y when substituted into one of the original equations. Using equation (1) gives the following.

$x^2 + y^2 = 9$ (1)	$x^2 + y^2 = 9$ (1)
$1^2 + y^2 = 9$ Let $x = 1$.	$(-1)^2 + y^2 = 9$ Let $x = -1$.
$y^2 = 8$	$y^2 = 8$
$y = \sqrt{8}$ or $y = -\sqrt{8}$	$y = 2\sqrt{2}$ or $y = -2\sqrt{2}$
$y = 2\sqrt{2}$ or $y = -2\sqrt{2}$	

The solution set is

$$\left\{ \left(1, 2\sqrt{2}\right), \left(1, -2\sqrt{2}\right), \right.$$
$$\left. \left(-1, 2\sqrt{2}\right), \left(-1, -2\sqrt{2}\right) \right\}.$$

FIGURE 30 shows the four points of intersection.

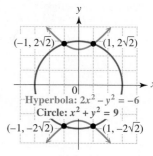

Hyperbola: $2x^2 - y^2 = -6$
Circle: $x^2 + y^2 = 9$

FIGURE 30

NOW TRY ↻

OBJECTIVE 3 **Solve a nonlinear system that requires a combination of methods.**

EXAMPLE 4 Solving a Nonlinear System by a Combination of Methods

Solve the system.

$$x^2 + 2xy - y^2 = 7 \quad (1)$$
$$x^2 - y^2 = 3 \quad (2)$$

While we have not graphed equations like (1), its graph is a hyperbola. The graph of (2) is also a hyperbola. Two hyperbolas may have zero, one, two, three, or four points of intersection. We use the elimination method here in combination with the substitution method.

$$x^2 + 2xy - y^2 = 7 \quad (1)$$
$$\underline{-x^2 + y^2 = -3} \quad \text{Multiply (2) by } -1.$$
$$ 2xy = 4 \quad \text{Add.}$$

The x^2- and y^2-terms were eliminated.

Next, we solve $2xy = 4$ for one of the variables. We choose y.

$$2xy = 4$$
$$y = \frac{2}{x} \quad \text{Divide by } 2x. \quad (3)$$

Now, we substitute $y = \frac{2}{x}$ into one of the original equations.

$$x^2 - y^2 = 3 \quad \text{The substitution is easier in (2).}$$
$$x^2 - \left(\frac{2}{x}\right)^2 = 3 \quad \text{Let } y = \frac{2}{x}.$$
$$x^2 - \frac{4}{x^2} = 3 \quad \text{Square } \frac{2}{x}.$$
$$x^4 - 4 = 3x^2 \quad \text{Multiply by } x^2, x \neq 0.$$

NOW TRY ANSWER
3. $\left\{ \left(2\sqrt{3}, 2\right), \left(2\sqrt{3}, -2\right), \right.$
$\left. \left(-2\sqrt{3}, 2\right), \left(-2\sqrt{3}, -2\right) \right\}$

NOW TRY
EXERCISE 4
Solve the system.

$$x^2 + 3xy - y^2 = 23$$
$$x^2 - y^2 = 5$$

$$x^4 - 3x^2 - 4 = 0 \qquad\qquad \text{Subtract } 3x^2.$$

$$(x^2 - 4)(x^2 + 1) = 0 \qquad\qquad \text{Factor.}$$

$$x^2 - 4 = 0 \quad \text{or} \quad x^2 + 1 = 0 \qquad \text{Zero-factor property}$$

$$x^2 = 4 \quad \text{or} \qquad x^2 = -1 \qquad \text{Solve each equation.}$$

$$x = 2 \quad \text{or} \quad x = -2 \quad \text{or} \quad x = i \quad \text{or} \quad x = -i$$

Substituting these four values into $y = \frac{2}{x}$ (equation (3)) gives the corresponding values for y.

$$\text{If } x = 2, \quad \text{then } y = \frac{2}{2} = 1.$$

$$\text{If } x = -2, \quad \text{then } y = \frac{2}{-2} = -1.$$

> Multiply by the complex conjugate of the denominator. $i(-i) = 1$

$$\text{If } x = i, \quad \text{then } y = \frac{2}{i} = \frac{2}{i} \cdot \frac{-i}{-i} = -2i.$$

$$\text{If } x = -i, \quad \text{then } y = \frac{2}{-i} = \frac{2}{-i} \cdot \frac{i}{i} = 2i.$$

If we substitute the x-values we found into equation (1) or (2) instead of into equation (3), we get extraneous solutions. *It is always wise to check all solutions in both of the given equations.* There are four ordered pairs in the solution set, two with real values and two with pure imaginary values. The solution set is

$$\{(2, 1), (-2, -1), (i, -2i), (-i, 2i)\}.$$

The graph of the system, shown in **FIGURE 31**, shows only the two real intersection points because the graph is in the real number plane. In general, if solutions contain nonreal complex numbers as components, they do not appear on the graph

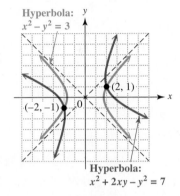

Hyperbola: $x^2 - y^2 = 3$

Hyperbola: $x^2 + 2xy - y^2 = 7$

FIGURE 31

NOW TRY

NOTE It is not essential to visualize the number of points of intersection of the graphs in order to solve a nonlinear system. Sometimes we are unfamiliar with the graphs or, as in **Example 4,** there are nonreal complex solutions that do not appear as points of intersection in the real plane. Visualizing the geometry of the graphs is only an aid to solving these systems.

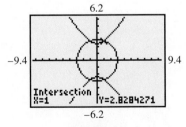

FIGURE 32

NOW TRY ANSWER
4. $\{(3, 2), (-3, -2),$
$(2i, -3i), (-2i, 3i)\}$

CONNECTIONS

If the equations in a nonlinear system can be solved for y, then we can graph the equations of the system with a graphing calculator and use the capabilities of the calculator to identify all intersection points.

For instance, the two equations in **Example 3** would require graphing four separate functions.

$$Y_1 = \sqrt{9 - X^2}, \quad Y_2 = -\sqrt{9 - X^2}, \quad Y_3 = \sqrt{2X^2 + 6}, \quad \text{and} \quad Y_4 = -\sqrt{2X^2 + 6}$$

FIGURE 32 indicates the coordinates of one of the points of intersection.

11.4 EXERCISES

⊕ *Complete solution available on the Video Resources on DVD*

Concept Check Each sketch represents the graphs of a pair of equations in a system. How many points are in each solution set?

1.

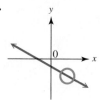

2.

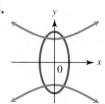

3.

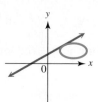

4.

Concept Check Suppose that a nonlinear system is composed of equations whose graphs are those described, and the number of points of intersection of the two graphs is as given. Make a sketch satisfying these conditions. (There may be more than one way to do this.)

5. A line and a circle; no points

6. A line and a circle; one point

7. A line and a hyperbola; one point

8. A line and an ellipse; no points

9. A circle and an ellipse; four points

10. A parabola and an ellipse; one point

11. A parabola and an ellipse; four points

12. A parabola and a hyperbola; two points

Solve each system by the substitution method. ***See Examples 1 and 2.***

13. $y = 4x^2 - x$
$y = x$

14. $y = x^2 + 6x$
$3y = 12x$

15. $y = x^2 + 6x + 9$
$x + y = 3$

16. $y = x^2 + 8x + 16$
$x - y = -4$

⊕ **17.** $x^2 + y^2 = 2$
$2x + y = 1$

18. $2x^2 + 4y^2 = 4$
$x = 4y$

⊕ **19.** $xy = 4$
$3x + 2y = -10$

20. $xy = -5$
$2x + y = 3$

21. $xy = -3$
$x + y = -2$

22. $xy = 12$
$x + y = 8$

23. $y = 3x^2 + 6x$
$y = x^2 - x - 6$

24. $y = 2x^2 + 1$
$y = 5x^2 + 2x - 7$

25. $2x^2 - y^2 = 6$
$y = x^2 - 3$

26. $x^2 + y^2 = 4$
$y = x^2 - 2$

27. $x^2 - xy + y^2 = 0$
$x - 2y = 1$

28. $x^2 - 3x + y^2 = 4$
$2x - y = 3$

Solve each system by the elimination method or a combination of the elimination and substitution methods. See Examples 3 and 4.

29. $3x^2 + 2y^2 = 12$
$x^2 + 2y^2 = 4$

30. $5x^2 - 2y^2 = -13$
$3x^2 + 4y^2 = 39$

31. $2x^2 + 3y^2 = 6$
$x^2 + 3y^2 = 3$

32. $6x^2 + y^2 = 9$
$3x^2 + 4y^2 = 36$

33. $2x^2 + y^2 = 28$
$4x^2 - 5y^2 = 28$

34. $x^2 + 6y^2 = 9$
$4x^2 + 3y^2 = 36$

35. $2x^2 = 8 - 2y^2$
$3x^2 = 24 - 4y^2$

36. $5x^2 = 20 - 5y^2$
$2y^2 = 2 - x^2$

37. $x^2 + xy + y^2 = 15$
$x^2 + y^2 = 10$

38. $2x^2 + 3xy + 2y^2 = 21$
$x^2 + y^2 = 6$

39. $3x^2 + 2xy - 3y^2 = 5$
$-x^2 - 3xy + y^2 = 3$

40. $-2x^2 + 7xy - 3y^2 = 4$
$2x^2 - 3xy + 3y^2 = 4$

Use a graphing calculator to solve each system. Then confirm your answer algebraically.

41. $xy = -6$
$x + y = -1$

42. $y = 2x^2 + 4x$
$y = -x^2 - 1$

Solve each problem by using a nonlinear system.

43. The area of a rectangular rug is 84 ft² and its perimeter is 38 ft. Find the length and width of the rug.

44. Find the length and width of a rectangular room whose perimeter is 50 m and whose area is 100 m².

45. A company has found that the price p (in dollars) of its scientific calculator is related to the supply x (in thousands) by the equation

$$px = 16.$$

The price is related to the demand x (in thousands) for the calculator by the equation

$$p = 10x + 12.$$

The **equilibrium price** is the value of p where demand equals supply. Find the equilibrium price and the supply/demand at that price. (*Hint:* Demand, price, and supply must all be positive.)

46. The calculator company in **Exercise 45** has determined that the cost y to make x (thousand) calculators is

$$y = 4x^2 + 36x + 20,$$

while the revenue y from the sale of x (thousand) calculators is

$$36x^2 - 3y = 0.$$

Find the **break-even point,** where cost equals revenue.

PREVIEW EXERCISES

Graph each inequality. See Section 3.4.

47. $2x - y \le 4$

48. $-x + 3y > 9$

11.5 Second-Degree Inequalities and Systems of Inequalities

OBJECTIVES

1 Graph second-degree inequalities.

2 Graph the solution set of a system of inequalities.

OBJECTIVE 1 **Graph second-degree inequalities.** A **second-degree inequality** is an inequality with at least one variable of degree 2 and no variable with degree greater than 2.

EXAMPLE 1 Graphing a Second-Degree Inequality

Graph $x^2 + y^2 \leq 36$.

The boundary of the inequality $x^2 + y^2 \leq 36$ is the graph of the equation $x^2 + y^2 = 36$, a circle with radius 6 and center at the origin, as shown in **FIGURE 33**.

The inequality $x^2 + y^2 \leq 36$ will include either the points outside the circle or the points inside the circle, as well as the boundary. To decide which region to shade, we substitute any test point not on the circle into the original inequality.

$$x^2 + y^2 \leq 36 \qquad \text{Original inequality}$$
$$0^2 + 0^2 \stackrel{?}{\leq} 36 \qquad \text{Use } (0, 0) \text{ as a test point.}$$
$$0 \leq 36 \checkmark \quad \text{True}$$

Since a true statement results, the original inequality includes the points *inside* the circle, the shaded region in **FIGURE 33**, and the boundary.

NOW TRY
EXERCISE 1
Graph $x^2 + y^2 \geq 9$.

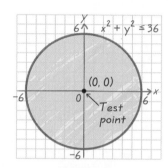

FIGURE 33

NOW TRY

NOTE Since the substitution is easy, the origin is the test point of choice unless the graph actually passes through $(0, 0)$.

NOW TRY
EXERCISE 2
Graph $y \geq -(x + 2)^2 + 1$.

EXAMPLE 2 Graphing a Second-Degree Inequality

Graph $y < -2(x - 4)^2 - 3$.

The boundary, $y = -2(x - 4)^2 - 3$, is a parabola that opens down with vertex at $(4, -3)$.

$$y < -2(x - 4)^2 - 3 \qquad \text{Original inequality}$$
$$0 \stackrel{?}{<} -2(0 - 4)^2 - 3 \qquad \text{Use } (0, 0) \text{ as a test point.}$$
$$0 \stackrel{?}{<} -32 - 3 \qquad \text{Simplify.}$$
$$0 < -35 \qquad \text{False}$$

Because the final inequality is a false statement, the points in the region containing $(0, 0)$ do not satisfy the inequality. In **FIGURE 34** the parabola is drawn as a dashed curve since the points of the parabola itself do not satisfy the inequality, and the region inside (or below) the parabola is shaded.

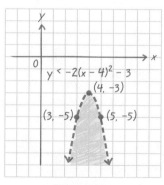

FIGURE 34

NOW TRY

NOW TRY ANSWERS

1.   2.

NOW TRY
EXERCISE 3
Graph $25x^2 - 16y^2 > 400$.

EXAMPLE 3 Graphing a Second-Degree Inequality

Graph $16y^2 \leq 144 + 9x^2$.

$$16y^2 - 9x^2 \leq 144 \qquad \text{Subtract } 9x^2.$$

$$\frac{y^2}{9} - \frac{x^2}{16} \leq 1 \qquad \text{Divide by 144.}$$

This form shows that the boundary is the hyperbola given by

$$\frac{y^2}{9} - \frac{x^2}{16} = 1.$$

Since the graph is a vertical hyperbola, the desired region will be either the region between the branches or the regions above the top branch and below the bottom branch. Choose $(0, 0)$ as a test point. Substituting into the original inequality leads to $0 \leq 144$, a true statement, so the region between the branches containing $(0, 0)$ is shaded, as shown in **FIGURE 35**.

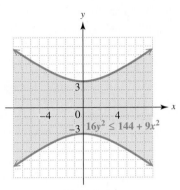

FIGURE 35

NOW TRY

OBJECTIVE 2 **Graph the solution set of a system of inequalities.** If two or more inequalities are considered at the same time, we have a **system of inequalities.** To find the solution set of the system, we find the intersection of the graphs (solution sets) of the inequalities in the system.

NOW TRY
EXERCISE 4
Graph the solution set of the system.

$$x^2 + y^2 > 9$$
$$y > x^2 - 1$$

EXAMPLE 4 Graphing a System of Two Inequalities

Graph the solution set of the system.

$$2x + 3y > 6$$
$$x^2 + y^2 < 16$$

Begin by graphing the solution set of $2x + 3y > 6$. The boundary line is the graph of $2x + 3y = 6$ and is a dashed line because of the symbol $>$. The test point $(0, 0)$ leads to a false statement in the inequality $2x + 3y > 6$, so shade the region above the line, as shown in **FIGURE 36**.

The graph of $x^2 + y^2 < 16$ is the interior of a dashed circle centered at the origin with radius 4. This is shown in **FIGURE 37**.

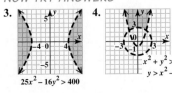

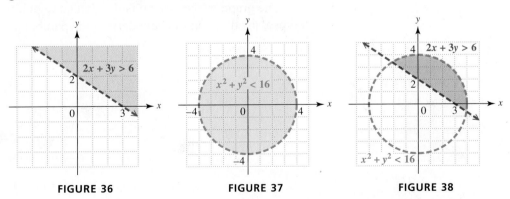

FIGURE 36　　　　**FIGURE 37**　　　　**FIGURE 38**

The graph of the solution set of the system is the intersection of the graphs of the two inequalities. The overlapping region in **FIGURE 38** is the solution set. NOW TRY

NOW TRY
EXERCISE 5

Graph the solution set of the system.

$$3x + 2y > 6$$

$$y \geq \frac{1}{2}x - 2$$

$$x \geq 0$$

EXAMPLE 5　Graphing a Linear System of Three Inequalities

Graph the solution set of the system.

$$x + y < 1$$

$$y \leq 2x + 3$$

$$y \geq -2$$

Graph each inequality separately, on the same axes. The graph of $x + y < 1$ consists of all points that lie below the dashed line $x + y = 1$. The graph of $y \leq 2x + 3$ is the region that lies below the solid line $y = 2x + 3$. Finally, the graph of $y \geq -2$ is the region above the solid horizontal line $y = -2$.

The graph of the system, the intersection of these three graphs, is the triangular region enclosed by the three boundary lines in **FIGURE 39**, including two of its boundaries.

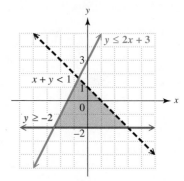

FIGURE 39

NOW TRY

NOW TRY
EXERCISE 6

Graph the solution set of the system.

$$\frac{x^2}{4} + \frac{y^2}{16} \leq 1$$

$$y \leq x^2 - 2$$

$$y + 3 > 0$$

EXAMPLE 6　Graphing a System of Three Inequalities

Graph the solution set of the system.

$$y \geq x^2 - 2x + 1$$

$$2x^2 + y^2 > 4$$

$$y < 4$$

The graph of $y = x^2 - 2x + 1$ is a parabola with vertex at $(1, 0)$. Those points above (or in the interior of) the parabola satisfy the condition $y > x^2 - 2x + 1$. Thus, the solution set of $y \geq x^2 - 2x + 1$ includes points on the parabola or in the interior.

The graph of the equation $2x^2 + y^2 = 4$ is an ellipse. We draw it as a dashed curve. To satisfy the inequality $2x^2 + y^2 > 4$, a point must lie outside the ellipse. The graph of $y < 4$ includes all points below the dashed line $y = 4$.

The graph of the system is the shaded region in **FIGURE 40**, which lies outside the ellipse, inside or on the boundary of the parabola, and below the line $y = 4$.

NOW TRY ANSWERS

5.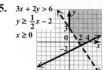

6.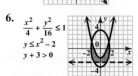

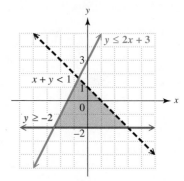

FIGURE 40

NOW TRY

11.5 EXERCISES **MyMathLab** Math XL
PRACTICE   WATCH   DOWNLOAD   READ   REVIEW

● *Complete solution available on the Video Resources on DVD*

1. *Concept Check*   Which one of the following is a description of the graph of the solution set of the following system?

$$x^2 + y^2 < 25$$
$$y > -2$$

A. All points outside the circle $x^2 + y^2 = 25$ and above the line $y = -2$

B. All points outside the circle $x^2 + y^2 = 25$ and below the line $y = -2$

C. All points inside the circle $x^2 + y^2 = 25$ and above the line $y = -2$

D. All points inside the circle $x^2 + y^2 = 25$ and below the line $y = -2$

2. *Concept Check*   Fill in each blank with the appropriate response. The graph of the system

$$y > x^2 + 1$$
$$\frac{x^2}{9} + \frac{y^2}{4} > 1$$
$$y < 5$$

consists of all points _____ the parabola $y = x^2 + 1$, _____ the
(above/below)   (inside/outside)

ellipse $\frac{x^2}{9} + \frac{y^2}{4} = 1$, and _____ the line $y = 5$.
(above/below)

Concept Check   Match each nonlinear inequality with its graph.

3. $y \geq x^2 + 4$ **4.** $y \leq x^2 + 4$ **5.** $y < x^2 + 4$ **6.** $y > x^2 + 4$

A. 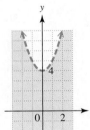   **B.** 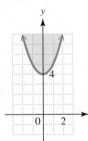   **C.** 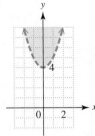   **D.**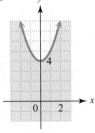

Graph each nonlinear inequality. See Examples 1–3.

7. $y^2 > 4 + x^2$ **8.** $y^2 \leq 4 - 2x^2$

● **9.** $y \geq x^2 - 2$ **10.** $x^2 \leq 16 - y^2$

11. $2y^2 \geq 8 - x^2$ **12.** $x^2 \leq 16 + 4y^2$

13. $y \leq x^2 + 4x + 2$ **14.** $9x^2 < 16y^2 - 144$

● **15.** $9x^2 > 16y^2 + 144$ **16.** $4y^2 \leq 36 - 9x^2$

17. $x^2 - 4 \geq -4y^2$ **18.** $x \geq y^2 - 8y + 14$

19. $x \leq -y^2 + 6y - 7$ **20.** $y^2 - 16x^2 \leq 16$

Graph each system of inequalities. ***See Examples 4–6.***

21. $2x + 5y < 10$
   $x - 2y < 4$

22. $3x - y > -6$
   $4x + 3y > 12$

23. $5x - 3y \leq 15$
   $4x + y \geq 4$

24. $4x - 3y \leq 0$
   $x + y \leq 5$

25. $x \leq 5$
   $y \leq 4$

26. $x \geq -2$
   $y \leq 4$

27. $y > x^2 - 4$
   $y < -x^2 + 3$

28. $x^2 - y^2 \geq 9$
   $\dfrac{x^2}{16} + \dfrac{y^2}{9} \leq 1$

29. $x^2 + y^2 \geq 4$
   $x + y \leq 5$
   $x \geq 0$
   $y \geq 0$

30. $y^2 - x^2 \geq 4$
   $-5 \leq y \leq 5$

31. $y \leq -x^2$
   $y \geq x - 3$
   $y \leq -1$
   $x < 1$

32. $y < x^2$
   $y > -2$
   $x + y < 3$
   $3x - 2y > -6$

For each nonlinear inequality in Exercises 33–40, a restriction is placed on one or both variables. For example, the inequality

$$x^2 + y^2 \leq 4, \quad x \geq 0$$

is graphed in the figure. Only the right half of the interior of the circle and its boundary is shaded, because of the restriction that x must be nonnegative. Graph each nonlinear inequality with the given restrictions.

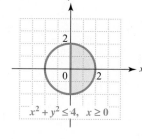

$x^2 + y^2 \leq 4, \quad x \geq 0$

33. $x^2 + y^2 > 36, \quad x \geq 0$

34. $4x^2 + 25y^2 < 100, \quad y < 0$

35. $x < y^2 - 3, \quad x < 0$

36. $x^2 - y^2 < 4, \quad x < 0$

37. $4x^2 - y^2 > 16, \quad x < 0$

38. $x^2 + y^2 > 4, \quad y < 0$

39. $x^2 + 4y^2 \geq 1, \quad x \geq 0, y \geq 0$

40. $2x^2 - 32y^2 \leq 8, \quad x \leq 0, y \geq 0$

Use the shading feature of a graphing calculator to graph each system.

41. $y \geq x - 3$
   $y \leq -x + 4$

42. $y \geq -x^2 + 5$
   $y \leq x^2 - 3$

43. $y < x^2 + 4x + 4$
   $y > -3$

44. $y > (x - 4)^2 - 3$
   $y < 5$

PREVIEW EXERCISES

Evaluate each expression for ***(a)*** *$n = 1$,* ***(b)*** *$n = 2$,* ***(c)*** *$n = 3$, and* ***(d)*** *$n = 4$.* ***See Section 1.3.***

45. $\dfrac{n + 5}{n}$

46. $\dfrac{n - 1}{n + 1}$

47. $n^2 - n$

48. $n(n - 3)$

CHAPTER 11 SUMMARY

KEY TERMS

11.1
squaring function
absolute value function
reciprocal function
asymptotes
square root function
greatest integer function
step function

11.2
conic section
circle
center (of circle)
radius
center-radius form
ellipse
foci (singular: focus)
center (of ellipse)

11.3
hyperbola
transverse axis
asymptotes of a
   hyperbola
fundamental rectangle
generalized square root
   function

11.4
nonlinear equation
nonlinear system of
   equations

11.5
second-degree inequality
system of inequalities

NEW SYMBOLS

$[\![x]\!]$ greatest integer less than or equal to x

TEST YOUR WORD POWER

See how well you have learned the vocabulary in this chapter.

1. **Conic sections** are
   A. graphs of first-degree equations
   B. the result of two or more intersecting planes
   C. graphs of first-degree inequalities
   D. figures that result from the intersection of an infinite cone with a plane.

2. A **circle** is the set of all points in a plane
   A. such that the absolute value of the difference of the distances from two fixed points is constant
   B. that lie a fixed distance from a fixed point
   C. the sum of whose distances from two fixed points is constant
   D. that make up the graph of any second-degree equation.

3. An **ellipse** is the set of all points in a plane
   A. such that the absolute value of the difference of the distances from two fixed points is constant
   B. that lie a fixed distance from a fixed point
   C. the sum of whose distances from two fixed points is constant
   D. that make up the graph of any second-degree equation.

4. A **hyperbola** is the set of all points in a plane
   A. such that the absolute value of the difference of the distances from two fixed points is constant
   B. that lie a fixed distance from a fixed point
   C. the sum of whose distances from two fixed points is constant
   D. that make up the graph of any second-degree equation.

5. A **nonlinear equation** is an equation
   A. in which some terms have more than one variable or a variable of degree 2 or greater
   B. in which the terms have only one variable
   C. of degree 1
   D. of a linear function.

6. A **nonlinear system of equations** is a system
   A. with at least one linear equation
   B. with two or more inequalities
   C. with at least one nonlinear equation
   D. with at least two linear equations.

ANSWERS

1. D; *Example:* Parabolas, circles, ellipses, and hyperbolas are conic sections. **2.** B; *Example:* See the graph of $x^2 + y^2 = 9$ in **FIGURE 10** of
Section 11.2. **3.** C; *Example:* See the graph of $\dfrac{x^2}{49} + \dfrac{y^2}{36} = 1$ in **FIGURE 16** of **Section 11.2.** **4.** A; *Example:* See the graph of $\dfrac{x^2}{16} - \dfrac{y^2}{25} = 1$
in **FIGURE 21** of **Section 11.3.** **5.** A; *Examples:* $y = x^2 + 8x + 16$, $xy = 5$, $2x^2 - y^2 = 6$ **6.** C; *Example:* $x^2 + y^2 = 2$
$2x + y = 1$

CONCEPTS	EXAMPLES

11.1 Additional Graphs of Functions

Other Functions

In addition to the squaring function, some other elementary functions include the following:

- Absolute value function, defined by $f(x) = |x|$
- Reciprocal function, defined by $f(x) = \frac{1}{x}$
- Square root function, defined by $f(x) = \sqrt{x}$
- Greatest integer function, defined by $f(x) = [\![x]\!]$, which is a step function.

Their graphs can be translated, as shown in the first three examples at the right.

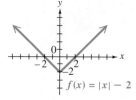
$f(x) = |x| - 2$

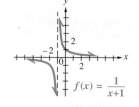
$f(x) = \frac{1}{x+1}$

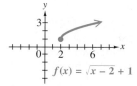
$f(x) = \sqrt{x-2} + 1$

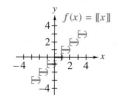
$f(x) = [\![x]\!]$

11.2 The Circle and the Ellipse

Circle

The circle with radius r and center at (h, k) has an equation of the form

$$(x - h)^2 + (y - k)^2 = r^2.$$

The circle with equation $(x + 2)^2 + (y - 3)^2 = 25$, which can be written $[x - (-2)]^2 + (y - 3)^2 = 5^2$, has center $(-2, 3)$ and radius 5.

$(x + 2)^2 + (y - 3)^2 = 25$

Ellipse

The ellipse whose x-intercepts are $(a, 0)$ and $(-a, 0)$ and whose y-intercepts are $(0, b)$ and $(0, -b)$ has an equation of the form

$$\frac{x^2}{a^2} + \frac{y^2}{b^2} = 1.$$

Graph $\dfrac{x^2}{9} + \dfrac{y^2}{4} = 1$.

$\dfrac{x^2}{9} + \dfrac{y^2}{4} = 1$

11.3 The Hyperbola and Functions Defined by Radicals

Hyperbola

A hyperbola with x-intercepts $(a, 0)$ and $(-a, 0)$ has an equation of the form

$$\frac{x^2}{a^2} - \frac{y^2}{b^2} = 1,$$

and a hyperbola with y-intercepts $(0, b)$ and $(0, -b)$ has an equation of the form

$$\frac{y^2}{b^2} - \frac{x^2}{a^2} = 1.$$

The extended diagonals of the fundamental rectangle with vertices at the points (a, b), $(-a, b)$, $(-a, -b)$, and $(a, -b)$ are the asymptotes of these hyperbolas.

Graph $\dfrac{x^2}{4} - \dfrac{y^2}{4} = 1$.

The graph has x-intercepts $(2, 0)$ and $(-2, 0)$.

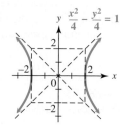
$\dfrac{x^2}{4} - \dfrac{y^2}{4} = 1$

The fundamental rectangle has vertices at $(2, 2)$, $(-2, 2)$, $(-2, -2)$, and $(2, -2)$.

(continued)

CONCEPTS	EXAMPLES
Graphing a Generalized Square Root Function To graph a generalized square root function defined by $$f(x) = \sqrt{u}$$ for an algebraic expression u, with $u \geq 0$, square each side so that the equation can be easily recognized. Then graph only the part indicated by the original equation.	Graph $y = -\sqrt{4 - x^2}$. Square each side and rearrange terms to get $$x^2 + y^2 = 4.$$ This equation has a circle as its graph. However, graph only the lower half of the circle, since the original equation indicates that y cannot be positive. 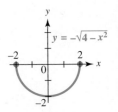

11.4 Nonlinear Systems of Equations

Solving a Nonlinear System A nonlinear system can be solved by the substitution method, the elimination method, or a combination of the two.	Solve the system. $$x^2 + 2xy - y^2 = 14 \quad (1)$$ $$x^2 - y^2 = -16 \quad (2)$$ Multiply equation (2) by -1 and use elimination. $$\begin{aligned} x^2 + 2xy - y^2 &= 14 \\ -x^2 \qquad\quad + y^2 &= 16 \\ \hline 2xy \qquad\quad &= 30 \\ xy &= 15 \end{aligned}$$ Solve $xy = 15$ for y to obtain $y = \frac{15}{x}$, and substitute into equation (2). $$x^2 - y^2 = -16 \quad (2)$$ $$x^2 - \left(\frac{15}{x}\right)^2 = -16 \qquad \text{Let } y = \frac{15}{x}.$$ $$x^2 - \frac{225}{x^2} = -16 \qquad \text{Apply the exponent.}$$ $$x^4 + 16x^2 - 225 = 0 \qquad \text{Multiply by } x^2. \text{ Add } 16x^2.$$ $$(x^2 - 9)(x^2 + 25) = 0 \qquad \text{Factor.}$$ $$x = \pm 3 \quad \text{or} \quad x = \pm 5i \qquad \text{Zero-factor property}$$ Find corresponding y-values to get the solution set $$\{(3, 5), (-3, -5), (5i, -3i), (-5i, 3i)\}.$$

11.5 Second-Degree Inequalities and Systems of Inequalities

Graphing a Second-Degree Inequality To graph a second-degree inequality, graph the corresponding equation as a boundary and use test points to determine which region(s) form the solution set. Shade the appropriate region(s). **Graphing a System of Inequalities** The solution set of a system of inequalities is the intersection of the solution sets of the individual inequalities.	Graph $$y \geq x^2 - 2x + 3.$$ 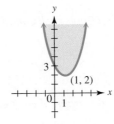 Graph the solution set of the system $$3x - 5y > -15$$ $$x^2 + y^2 \leq 25.$$ 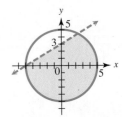

CHAPTER (11)

REVIEW EXERCISES

11.1 *Graph each function.*

1. $f(x) = |x + 4|$

2. $f(x) = \dfrac{1}{x - 4}$

3. $f(x) = \sqrt{x} + 3$

4. $f(x) = [\![x]\!] - 2$

11.2 *Write an equation for each circle.*

5. Center $(-2, 4)$, $r = 3$

6. Center $(-1, -3)$, $r = 5$

7. Center $(4, 2)$, $r = 6$

Find the center and radius of each circle.

8. $x^2 + y^2 + 6x - 4y - 3 = 0$

9. $x^2 + y^2 - 8x - 2y + 13 = 0$

10. $2x^2 + 2y^2 + 4x + 20y = -34$

11. $4x^2 + 4y^2 - 24x + 16y = 48$

Graph each equation.

12. $x^2 + y^2 = 16$

13. $\dfrac{x^2}{16} + \dfrac{y^2}{9} = 1$

14. $\dfrac{x^2}{49} + \dfrac{y^2}{25} = 1$

15. A satellite is in an elliptical orbit around Earth with perigee altitude of 160 km and apogee altitude of 16,000 km. See the figure. (*Source:* Kastner, Bernice, *Space Mathematics,* NASA.) Find the equation of the ellipse. (*Hint:* Use the fact that $c^2 = a^2 - b^2$ here.)

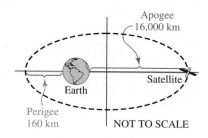

16. (a) The Roman Colosseum is an ellipse with $a = 310$ ft and $b = \dfrac{513}{2}$ ft. Find the distance, to the nearest tenth, between the foci of this ellipse.

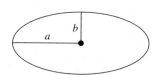

(b) The approximate perimeter of an ellipse is given by

$$P \approx 2\pi \sqrt{\dfrac{a^2 + b^2}{2}},$$

where a and b are the lengths given in part (a). Use this formula to find the approximate perimeter, to the nearest tenth, of the Roman Colosseum.

11.3 *Graph each equation.*

17. $\dfrac{x^2}{16} - \dfrac{y^2}{25} = 1$

18. $\dfrac{y^2}{25} - \dfrac{x^2}{4} = 1$

19. $f(x) = -\sqrt{16 - x^2}$

Identify the graph of each equation as a parabola, circle, ellipse, *or* hyperbola.

20. $x^2 + y^2 = 64$

21. $y = 2x^2 - 3$

22. $y^2 = 2x^2 - 8$

23. $y^2 = 8 - 2x^2$

24. $x = y^2 + 4$

25. $x^2 - y^2 = 64$

26. Ships and planes often use a location-finding system called LORAN. With this system, a radio transmitter at M sends out a series of pulses. When each pulse is received at transmitter S, it then sends out a pulse. A ship at P receives pulses from both M and S. A receiver on the ship measures the difference in the arrival times of the pulses. A special map gives hyperbolas that correspond to the differences in arrival times (which give the distances d_1 and d_2 in the figure.) The ship can then be located as lying on a branch of a particular hyperbola.

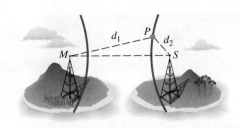

Suppose $d_1 = 80$ mi and $d_2 = 30$ mi, and the distance between transmitters M and S is 100 mi. Use the definition to find an equation of the hyperbola on which the ship is located.

11.4   *Solve each system.*

27. $2y = 3x - x^2$
    $x + 2y = -12$

28. $y + 1 = x^2 + 2x$
    $y + 2x = 4$

29. $x^2 + 3y^2 = 28$
    $y - x = -2$

30. $xy = 8$
    $x - 2y = 6$

31. $x^2 + y^2 = 6$
    $x^2 - 2y^2 = -6$

32. $3x^2 - 2y^2 = 12$
    $x^2 + 4y^2 = 18$

33. *Concept Check*   How many solutions are possible for a system of two equations whose graphs are a circle and a line?

34. *Concept Check*   How many solutions are possible for a system of two equations whose graphs are a parabola and a hyperbola?

11.5   *Graph each inequality.*

35. $9x^2 \geq 16y^2 + 144$

36. $4x^2 + y^2 \geq 16$

37. $y < -(x + 2)^2 + 1$

Graph each system of inequalities.

38. $2x + 5y \leq 10$
    $3x - y \leq 6$

39. $|x| \leq 2$
    $|y| > 1$
    $4x^2 + 9y^2 \leq 36$

40. $9x^2 \leq 4y^2 + 36$
    $x^2 + y^2 \leq 16$

MIXED REVIEW EXERCISES

Graph.

41. $\dfrac{y^2}{4} - 1 = \dfrac{x^2}{9}$

42. $x^2 + y^2 = 25$

43. $x^2 + 9y^2 = 9$

44. $x^2 - 9y^2 = 9$

45. $f(x) = \sqrt{4 - x}$

46. $4y > 3x - 12$
    $x^2 < 16 - y^2$

CHAPTER 11

TEST

CHAPTER
Test Prep
VIDEOS

Step-by-step test solutions are found on the Chapter Test Prep Videos available via the Video Resources on DVD, in *MyMathLab* , or on YouTube (search "LialIntermediateAlg").

View the complete solutions to all Chapter Test exercises on the Video Resources on DVD.

Concept Check   Fill in each blank with the correct response.

1. For the reciprocal function defined by $f(x) = \frac{1}{x}$, _____ is the only real number not in the domain.

2. The range of the square root function, given by $f(x) = \sqrt{x}$, is _____.

3. The range of $f(x) = [\![x]\!]$, the greatest integer function, is _____.

4. *Concept Check*   Match each function with its graph from choices A–D.

(a) $f(x) = \sqrt{x} - 2$

(b) $f(x) = \sqrt{x + 2}$

(c) $f(x) = \sqrt{x} + 2$

(d) $f(x) = \sqrt{x - 2}$

A.

B.

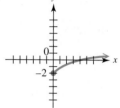

C.

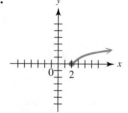

D.

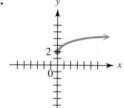

5. Sketch the graph of $f(x) = |x - 3| + 4$. Give the domain and range.

6. Find the center and radius of the circle whose equation is $(x - 2)^2 + (y + 3)^2 = 16$. Sketch the graph.

7. Find the center and radius of the circle whose equation is $x^2 + y^2 + 8x - 2y = 8$.

Graph.

8. $f(x) = \sqrt{9 - x^2}$

9. $4x^2 + 9y^2 = 36$

10. $16y^2 - 4x^2 = 64$

11. $\frac{y}{2} = -\sqrt{1 - \frac{x^2}{9}}$

Identify the graph of each equation as a parabola, hyperbola, ellipse, *or* circle.

12. $6x^2 + 4y^2 = 12$

13. $16x^2 = 144 + 9y^2$

14. $y^2 = 20 - x^2$

15. $4y^2 + 4x = 9$

Solve each system.

16. $2x - y = 9$
    $xy = 5$

17. $x - 4 = 3y$
    $x^2 + y^2 = 8$

18. $x^2 + y^2 = 25$
    $x^2 - 2y^2 = 16$

19. Graph the inequality $y < x^2 - 2$.

20. Graph the system $x^2 + 25y^2 \le 25$
    $x^2 + y^2 \le 9$.

CHAPTERS (1–11)

CUMULATIVE REVIEW EXERCISES

Solve.

1. $4 - (2x + 3) + x = 5x - 3$

2. $-4x + 7 \geq 6x + 1$

3. $|5x| - 6 = 14$

4. $|2p - 5| > 15$

5. Find the slope of the line through $(2, 5)$ and $(-4, 1)$.

6. Find the equation of the line through the point $(-3, -2)$ and perpendicular to the graph of $2x - 3y = 7$.

Solve each system.

7. $3x - y = 12$
$2x + 3y = -3$

8. $x + y - 2z = 9$
$2x + y + z = 7$
$3x - y - z = 13$

9. $xy = -5$
$2x + y = 3$

10. Al and Bev traveled from their apartment to a picnic 20 mi away. Al traveled on his bike while Bev, who left later, took her car. Al's average rate was half of Bev's average rate. The trip took Al $\frac{1}{2}$ hr longer than Bev. What was Bev's average rate?

Perform the indicated operations.

11. $(5y - 3)^2$

12. $\dfrac{8x^4 - 4x^3 + 2x^2 + 13x + 8}{2x + 1}$

Factor.

13. $12x^2 - 7x - 10$

14. $z^4 - 1$

15. $a^3 - 27b^3$

Perform the indicated operations.

16. $\dfrac{y^2 - 4}{y^2 - y - 6} \div \dfrac{y^2 - 2y}{y - 1}$

17. $\dfrac{5}{c + 5} - \dfrac{2}{c + 3}$

18. $\dfrac{p}{p^2 + p} + \dfrac{1}{p^2 + p}$

19. Henry Harris and Lawrence Hawkins want to clean their office. Henry can do the job alone in 3 hr, while Lawrence can do it alone in 2 hr. How long will it take them if they work together?

Simplify. Assume all variables represent positive real numbers.

20. $\dfrac{(2a)^{-2}a^4}{a^{-3}}$

21. $4\sqrt[3]{16} - 2\sqrt[3]{54}$

22. $\dfrac{3\sqrt{5x}}{\sqrt{2x}}$

23. $\dfrac{5 + 3i}{2 - i}$

Solve for real solutions.

24. $2\sqrt{x} = \sqrt{5x + 3}$

25. $10q^2 + 13q = 3$

26. $3x^2 - 3x - 2 = 0$

27. $2(x^2 - 3)^2 - 5(x^2 - 3) = 12$

28. $\log (x + 2) + \log (x - 1) = 1$

29. Solve $F = \dfrac{kwv^2}{r}$ for v.

30. If $f(x) = x^3 + 4$, find $f^{-1}(x)$.

31. Evaluate.   **(a)** $3^{\log_3 4}$ **(b)** $e^{\ln 7}$

32. Use properties of logarithms to write $2 \log (3x + 7) - \log 4$ as a single logarithm.

33. The bar graph shows online U.S. retail sales (in billions of dollars).

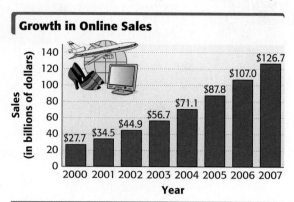

Growth in Online Sales

Source: U.S. Census Bureau.

A reasonable model for sales *y* in billions of dollars is the exponential function defined by

$$y = 28.43(1.25)^x,$$

where *x* is the number of years since 2000.

(a) Use the model to estimate sales in 2005. (*Hint:* Let $x = 5$.)

(b) Use the model to estimate sales in 2008.

34. Give the domain and range of the function defined by $f(x) = |x - 3|$.

Graph.

35. $f(x) = -3x + 5$

36. $f(x) = -2(x - 1)^2 + 3$

37. $\dfrac{x^2}{25} + \dfrac{y^2}{16} \le 1$

38. $f(x) = \sqrt{x - 2}$

39. $\dfrac{x^2}{4} - \dfrac{y^2}{16} = 1$

40. $f(x) = 3^x$

Sequences and Series

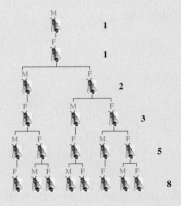

The male honeybee hatches from an unfertilized egg, while the female hatches from a fertilized one. The "family tree" of a male honeybee is shown at the left, where M represents male and F represents female. Starting with the male honeybee at the top, and counting the number of bees in each generation, we obtain the following numbers in the order shown.

$$1, 1, 2, 3, 5, 8$$

Notice the pattern. After the first two terms (1 and 1), each successive term is obtained by adding the two previous terms. This sequence of numbers is called the *Fibonacci sequence*.

In this chapter, we study *sequences* and sums of terms of sequences, known as *series*.

12.1 Sequences and Series

In the Palace of the Alhambra, residence of the Moorish rulers of Granada, Spain, the Sultana's quarters feature an interesting architectural pattern:

There are 2 matched marble slabs inlaid in the floor, 4 walls, an octagon (8-sided) ceiling, 16 windows, 32 arches, and so on.

If this pattern is continued indefinitely, the set of numbers forms an *infinite sequence* whose *terms* are powers of 2.

> **Sequence**
>
> An **infinite sequence** is a function with the set of all positive integers as the domain. A **finite sequence** is a function with domain of the form $\{1, 2, 3, \ldots, n\}$, where n is a positive integer.

OBJECTIVE 1 **Find the terms of a sequence, given the general term.** For any positive integer n, the function value of a sequence is written as a_n (read "a **sub-***n*"). The function values $a_1, a_2, a_3, \ldots$, written in order, are the **terms** of the sequence, with a_1 the first term, a_2 the second term, and so on. The expression a_n, which defines the sequence, is called the **general term** of the sequence.

In the Palace of the Alhambra example, the first five terms of the sequence are

$$a_1 = 2, \quad a_2 = 4, \quad a_3 = 8, \quad a_4 = 16, \quad \text{and} \quad a_5 = 32.$$

The general term for this sequence is $a_n = 2^n$.

NOW TRY
EXERCISE 1
Given an infinite sequence with $a_n = 5 - 3n$, find a_3.

EXAMPLE 1 **Writing the Terms of Sequences from the General Term**

Given an infinite sequence with $a_n = n + \frac{1}{n}$, find the following.

(a) The second term of the sequence

$$a_2 = 2 + \frac{1}{2} = \frac{5}{2} \qquad \text{Replace } n \text{ with 2.}$$

(b) $a_{10} = 10 + \frac{1}{10} = \frac{101}{10} \qquad 10 = \frac{100}{10}$ **(c)** $a_{12} = 12 + \frac{1}{12} = \frac{145}{12} \qquad 12 = \frac{144}{12}$

NOW TRY

Graphing calculators can be used to generate and graph sequences, as shown in **FIGURE 1** on the next page. The calculator must be in dot mode, so that the discrete points on the graph are not connected. *Remember that the domain of a sequence consists only of positive integers.*

NOW TRY ANSWER
1. $a_3 = -4$

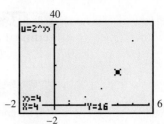

The first five terms of
the sequence $a_n = 2^n$

(a)

The first five terms of $a_n = 2^n$ are graphed
here. The display indicates that the fourth
term is 16; that is, $a_4 = 2^4 = 16$.

(b)

FIGURE 1

OBJECTIVE 2   **Find the general term of a sequence.**   Sometimes we need to
find a general term to fit the first few terms of a given sequence.

NOW TRY
EXERCISE 2

Find an expression for the general term a_n of the sequence.

$$-3, 9, -27, 81, \ldots$$

EXAMPLE 2   Finding the General Term of a Sequence

Determine an expression for the general term a_n of the sequence.

$$5, 10, 15, 20, 25, \ldots$$

Notice that the terms are all multiples of 5. The first term is $5(1)$, the second is
$5(2)$, and so on. The general term

$$a_n = 5n$$

will produce the given first five terms.   NOW TRY

⚠ **CAUTION**   Remember that when determining a general term, as in **Example 2,**
there may be more than one way to express it.

OBJECTIVE 3   **Use sequences to solve applied problems.**   Practical problems
may involve *finite* sequences.

NOW TRY
EXERCISE 3

Chase borrows $8000 and
agrees to pay $400 monthly,
plus interest of 2% on the unpaid balance from the beginning of the first month. Find
the payments for the first four
months and the remaining
debt at the end of that period.

EXAMPLE 3   Using a Sequence in an Application

Saad Alarachi borrows $5000 and agrees to pay $500 monthly, plus interest of 1% on
the unpaid balance from the beginning of the first month. Find the payments for the
first four months and the remaining debt at the end of that period.

The payments and remaining balances are calculated as follows.

First month	Payment:	$500 + 0.01($5000) = $550
	Balance:	$5000 − $500 = $4500
Second month	Payment:	$500 + 0.01($4500) = $545
	Balance:	$5000 − 2 · $500 = $4000
Third month	Payment:	$500 + 0.01($4000) = $540
	Balance:	$5000 − 3 · $500 = $3500
Fourth month	Payment:	$500 + 0.01($3500) = $535
	Balance:	$5000 − 4 · $500 = $3000

The payments for the first four months are

$$\$550, \$545, \$540, \$535$$

and the remaining debt at the end of the period is $3000.   NOW TRY

NOW TRY ANSWERS
2. $a_n = (-3)^n$
3. payments: $560, $552, $544,
  $536; balance: $6400

OBJECTIVE 4 **Use summation notation to evaluate a series.** By adding the terms of a sequence, we obtain a *series*.

Series

The indicated sum of the terms of a sequence is called a **series.**

For example, if we consider the sum of the payments listed in **Example 3,** namely,

$$550 + 545 + 540 + 535,$$

we have a series that represents the total payments for the first four months. Since a sequence can be finite or infinite, there are both finite and infinite series.

We use a compact notation, called **summation notation,** to write a series from the general term of the corresponding sequence. In mathematics, the Greek letter Σ **(sigma)** is used to denote summation. For example, the sum of the first six terms of the sequence with general term $a_n = 3n + 2$ is written as

$$\sum_{i=1}^{6} (3i + 2).$$

The letter i is called the **index of summation.** We read this as "the sum from $i = 1$ to 6 of $3i + 2$." To find this sum, we replace the letter i in $3i + 2$ with 1, 2, 3, 4, 5, and 6, and add the resulting terms.

⚠ **CAUTION** This use of i as the index of summation has no connection with the complex number i.

EXAMPLE 4 **Evaluating Series Written in Summation Notation**

Write out the terms and evaluate each series.

(a) $\displaystyle\sum_{i=1}^{6} (3i + 2)$ *Multiply and then add.*

$$= (3 \cdot 1 + 2) + (3 \cdot 2 + 2) + (3 \cdot 3 + 2)$$
$$+ (3 \cdot 4 + 2) + (3 \cdot 5 + 2) + (3 \cdot 6 + 2)$$ Replace i with 1, 2, 3, 4, 5, 6.

$$= 5 + 8 + 11 + 14 + 17 + 20$$ Work inside the parentheses.

$$= 75$$ Add.

(b) $\displaystyle\sum_{i=1}^{5} (i - 4)$

$$= (1 - 4) + (2 - 4) + (3 - 4) + (4 - 4) + (5 - 4)$$ $i = 1, 2, 3, 4, 5$

$$= -3 - 2 - 1 + 0 + 1$$ Subtract.

$$= -5$$ Simplify.

NOW TRY
EXERCISE 4

Write out the terms and evaluate the series.

$$\sum_{i=1}^{5} (i^2 - 4)$$

(c) $\displaystyle\sum_{i=3}^{7} 3i^2$

$= 3(3)^2 + 3(4)^2 + 3(5)^2 + 3(6)^2 + 3(7)^2$ $i = 3, 4, 5, 6, 7$

$= 27 + 48 + 75 + 108 + 147$ Square, and then multiply.

$= 405$ Add.   NOW TRY

OBJECTIVE 5 **Write a series with summation notation.** In **Example 4,** we started with summation notation and wrote each series using $+$ signs. Given a series, we can write it with summation notation by observing a pattern in the terms and writing the general term accordingly.

NOW TRY
EXERCISE 5

Write each sum with summation notation.

(a) $3 + 5 + 7 + 9 + 11$

(b) $-1 - 4 - 9 - 16 - 25$

EXAMPLE 5 Writing Series with Summation Notation

Write each sum with summation notation.

(a) $2 + 5 + 8 + 11$

First, find a general term a_n that will give these four terms for $a_1, a_2, a_3,$ and $a_4,$ respectively. Each term is one less than a multiple of 3, so try $3i - 1$ as the general term.

$$3(1) - 1 = 2 \quad i = 1$$
$$3(2) - 1 = 5 \quad i = 2$$
$$3(3) - 1 = 8 \quad i = 3$$
$$3(4) - 1 = 11 \quad i = 4$$

(Remember, there may be other expressions that also work.) Since i ranges from 1 to 4,

$$2 + 5 + 8 + 11 = \sum_{i=1}^{4} (3i - 1).$$

(b) $8 + 27 + 64 + 125 + 216$

These numbers are the cubes of 2, 3, 4, 5, and 6, so the general term is i^3.

$$8 + 27 + 64 + 125 + 216 = \sum_{i=2}^{6} i^3 \qquad \text{NOW TRY}$$

OBJECTIVE 6 **Find the arithmetic mean (average) of a group of numbers.**

Arithmetic Mean or Average

The **arithmetic mean,** or **average,** of a group of numbers is symbolized $\bar{x}$ and is found by dividing their sum by the number of numbers. That is,

$$\bar{x} = \frac{\displaystyle\sum_{i=1}^{n} x_i}{n}.$$

The values of x_i represent the individual numbers in the group, and n represents the number of numbers.

NOW TRY ANSWERS

4. $-3 + 0 + 5 + 12 + 21 = 35$

5. (a) $\displaystyle\sum_{i=1}^{5} (2i + 1)$ **(b)** $\displaystyle\sum_{i=1}^{5} -i^2$

NOW TRY
EXERCISE 6
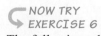

The following table shows the top 5 American Quarter Horse States in 2009 based on the total number of registered Quarter Horses. To the nearest whole number, what is the average number of Quarter Horses registered per state in these top five states?

State	Number of Registered Quarter Horses
Texas	461,054
Oklahoma	188,381
California	136,583
Missouri	107,630
Colorado	93,958

Source: American Quarter Horse Association.

EXAMPLE 6 Finding the Arithmetic Mean, or Average

The following table shows the number of FDIC-insured financial institutions for each year during the period from 2002 through 2008. What was the average number of institutions per year for this 7-yr period?

Year	Number of Institutions
2002	9369
2003	9194
2004	8988
2005	8845
2006	8691
2007	8544
2008	8314

Source: U.S. Federal Deposit Insurance Corporation.

$$\bar{x} = \frac{\sum_{i=1}^{7} x_i}{7} \qquad \text{Let } x_1 = 9369, x_2 = 9194, \text{ and so on. There are 7 numbers in the group, so } n = 7.$$

$$= \frac{9369 + 9194 + 8988 + 8845 + 8691 + 8544 + 8314}{7}$$

$$= 8849 \quad \text{(rounded to the nearest unit)}$$

The average number of institutions per year for this 7-yr period was 8849.

NOW TRY

NOW TRY ANSWER
6. 197,521

12.1 EXERCISES

MyMathLab | Math XL PRACTICE | WATCH | DOWNLOAD | READ | REVIEW

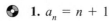 *Complete solution available on the Video Resources on DVD*

*Write out the first five terms of each sequence. **See Example 1.***

1. $a_n = n + 1$ **2.** $a_n = n + 4$ **3.** $a_n = \frac{n + 3}{n}$

4. $a_n = \frac{n + 2}{n}$ **5.** $a_n = 3^n$ **6.** $a_n = 2^n$

7. $a_n = \frac{1}{n^2}$ **8.** $a_n = \frac{-2}{n^2}$ **9.** $a_n = 5(-1)^{n-1}$

10. $a_n = 6(-1)^{n+1}$ **11.** $a_n = n - \frac{1}{n}$ **12.** $a_n = n + \frac{4}{n}$

*Find the indicated term for each sequence. **See Example 1.***

13. $a_n = -9n + 2; \quad a_8$ **14.** $a_n = 3n - 7; \quad a_{12}$

15. $a_n = \frac{3n + 7}{2n - 5}; \quad a_{14}$ **16.** $a_n = \frac{5n - 9}{3n + 8}; \quad a_{16}$

17. $a_n = (n + 1)(2n + 3); \quad a_8$ **18.** $a_n = (5n - 2)(3n + 1); \quad a_{10}$

Find a general term a_n for the given terms of each sequence. ***See Example 2.***

🌐 **19.** 4, 8, 12, 16, . . .

20. 7, 14, 21, 28, . . .

21. −8, −16, −24, −32, . . .

22. −10, −20, −30, −40, . . .

23. $\dfrac{1}{3}, \dfrac{1}{9}, \dfrac{1}{27}, \dfrac{1}{81}, \ldots$

24. $\dfrac{2}{5}, \dfrac{2}{25}, \dfrac{2}{125}, \dfrac{2}{625}, \ldots$

25. $\dfrac{2}{5}, \dfrac{3}{6}, \dfrac{4}{7}, \dfrac{5}{8}, \ldots$

26. $\dfrac{1}{2}, \dfrac{2}{3}, \dfrac{3}{4}, \dfrac{4}{5}, \ldots$

Solve each applied problem by writing the first few terms of a sequence. ***See Example 3.***

🌐 **27.** Horacio Loschak borrows $1000 and agrees to pay $100 plus interest of 1% on the unpaid balance each month. Find the payments for the first six months and the remaining debt at the end of that period.

28. Leslie Maruri is offered a new modeling job with a salary of $20,000 + 2500n$ dollars per year at the end of the nth year. Write a sequence showing her salary at the end of each of the first 5 yr. If she continues in this way, what will her salary be at the end of the tenth year?

29. Suppose that an automobile loses $\frac{1}{5}$ of its value each year; that is, at the end of any given year, the value is $\frac{4}{5}$ of the value at the beginning of that year. If a car costs $20,000 new, what is its value at the end of 5 yr, to the nearest dollar?

30. A certain car loses $\frac{1}{2}$ of its value each year. If this car cost $40,000 new, what is its value at the end of 6 yr?

Write out each series and evaluate it. ***See Example 4.***

🌐 **31.** $\displaystyle\sum_{i=1}^{5} (i + 3)$

32. $\displaystyle\sum_{i=1}^{6} (i + 9)$

33. $\displaystyle\sum_{i=1}^{3} (i^2 + 2)$

34. $\displaystyle\sum_{i=1}^{4} (i^3 + 3)$

35. $\displaystyle\sum_{i=1}^{6} (-1)^i$

36. $\displaystyle\sum_{i=1}^{5} (-1)^i \cdot i$

37. $\displaystyle\sum_{i=3}^{7} (i - 3)(i + 2)$

38. $\displaystyle\sum_{i=2}^{6} (i + 3)(i - 4)$

Write each series with summation notation. ***See Example 5.***

🌐 **39.** 3 + 4 + 5 + 6 + 7

40. 7 + 8 + 9 + 10 + 11

41. −2 + 4 − 8 + 16 − 32

42. −1 + 2 − 3 + 4 − 5 + 6

43. 1 + 4 + 9 + 16

44. 1 + 16 + 81 + 256

✎ **45.** Explain the basic difference between a sequence and a series.

46. *Concept Check*   Consider the following statement.   **WHAT WENT WRONG?**

For the sequence defined by $a_n = 2n + 4$, find $a_{1/2}$.

Find the arithmetic mean for each collection of numbers. ***See Example 6.***

47. 8, 11, 14, 9, 7, 6, 8

48. 10, 12, 8, 19, 23, 12

49. 5, 9, 8, 2, 4, 7, 3, 2, 0

50. 2, 1, 4, 8, 3, 7, 10, 8, 0

Solve each problem. ***See Example 6.***

51. The number of mutual funds operating in the United States available to investors each year during the period 2004 through 2008 is given in the table.

Year	Number of Funds Available
2004	8041
2005	7975
2006	8117
2007	8024
2008	8022

Source: Investment Company Institute.

To the nearest whole number, what was the average number of funds available per year during the given period?

52. The total assets of mutual funds operating in the United States, in billions of dollars, for each year during the period 2004 through 2008 are shown in the table. What were the average assets per year during this period?

Year	Assets (in billions of dollars)
2004	8107
2005	8905
2006	10,397
2007	12,000
2008	9601

Source: Investment Company Institute.

PREVIEW EXERCISES

Find the values of a and d by solving each system. ***See Section 4.1.***

53. $a + 3d = 12$
$a + 8d = 22$

54. $a + 7d = 12$
$a + 2d = 7$

55. Evaluate $a + (n - 1)d$ for $a = -2$, $n = 5$, and $d = 3$. **See Section 1.3.**

12.2 Arithmetic Sequences

OBJECTIVES

1. Find the common difference of an arithmetic sequence.
2. Find the general term of an arithmetic sequence.
3. Use an arithmetic sequence in an application.
4. Find any specified term or the number of terms of an arithmetic sequence.
5. Find the sum of a specified number of terms of an arithmetic sequence.

OBJECTIVE 1 **Find the common difference of an arithmetic sequence.** In this section, we introduce a special type of sequence that has many applications.

Arithmetic Sequence

An **arithmetic sequence,** or **arithmetic progression,** is a sequence in which each term after the first is found by adding a constant number to the preceding term.

For example, the sequence

$$6, 11, 16, 21, 26, \ldots \quad \text{Arithmetic sequence}$$

is an arithmetic sequence, since the difference between any two adjacent terms is always 5. The number 5 is called the **common difference** of the arithmetic sequence. The common difference, d, is found by subtracting a_n from a_{n+1} in any such pair of terms.

$$d = a_{n+1} - a_n \quad \text{Common difference}$$

NOW TRY
EXERCISE 1

Determine the common difference d for the arithmetic sequence.

$$-4, -13, -22, -31, -40, \ldots$$

NOW TRY
EXERCISE 2

Write the first five terms of the arithmetic sequence with first term 10 and common difference -8.

EXAMPLE 1 Finding the Common Difference

Determine the common difference d for the arithmetic sequence.

$$-11, -4, 3, 10, 17, 24, \ldots$$

Since the sequence is arithmetic, d is the difference between any two adjacent terms: $a_{n+1} - a_n$. We arbitrarily choose the terms 10 and 17.

$$d = 17 - 10, \quad \text{or} \quad 7$$

Verify that *any* two adjacent terms would give the same result. **NOW TRY**

EXAMPLE 2 Writing the Terms of a Sequence from the First Term and the Common Difference

Write the first five terms of the arithmetic sequence with first term 3 and common difference -2.

The second term is found by adding -2 to the first term 3, getting 1. For the next term, add -2 to 1, and so on. The first five terms are

$$3, 1, -1, -3, -5.$$ **NOW TRY**

OBJECTIVE 2 **Find the general term of an arithmetic sequence.** Generalizing from **Example 2**, if we know the first term a_1 and the common difference d of an arithmetic sequence, then the sequence is completely defined as

$$a_1, \quad a_2 = a_1 + d, \quad a_3 = a_1 + 2d, \quad a_4 = a_1 + 3d, \ldots.$$

Writing the terms of the sequence in this way suggests the following formula for a_n.

General Term of an Arithmetic Sequence

The general term of an arithmetic sequence with first term a_1 and common difference d is

$$a_n = a_1 + (n - 1)d.$$

Since $a_n = a_1 + (n - 1)d = dn + (a_1 - d)$ is a linear function in n, any linear expression of the form $kn + c$, where k and c are real numbers, defines an arithmetic sequence.

EXAMPLE 3 Finding the General Term of an Arithmetic Sequence

Determine the general term of the arithmetic sequence.

$$-9, -6, -3, 0, 3, 6, \ldots$$

Then use the general term to find a_{20}.

The first term is $a_1 = -9$.

$$d = -3 - (-6), \quad \text{or} \quad 3. \qquad \text{Let } d = a_3 - a_2.$$

Now find a_n.

$$
\begin{array}{ll}
a_n = a_1 + (n - 1)d & \text{Formula for } a_n \\
a_n = -9 + (n - 1)(3) & \text{Let } a_1 = -9, d = 3. \\
a_n = -9 + 3n - 3 & \text{Distributive property} \\
a_n = 3n - 12 & \text{Combine like terms.}
\end{array}
$$

NOW TRY ANSWERS
1. $d = -9$
2. $10, 2, -6, -14, -22$

NOW TRY
EXERCISE 3

Determine the general term of the arithmetic sequence.

$$-5, 0, 5, 10, 15, \dots$$

Then use the general term to find a_{20}.

NOW TRY
EXERCISE 4

Ginny Tiller is saving money for her son's college education. She makes an initial contribution of $1000 and deposits an additional $120 each month for the next 96 months. Disregarding interest, how much money will be in the account after 96 months?

The general term is $a_n = 3n - 12$. Now find a_{20}.

$$a_{20} = 3(20) - 12 \qquad \text{Let } n = 20.$$
$$= 60 - 12 \qquad \text{Multiply.}$$
$$= 48 \qquad \text{Subtract.} \qquad \textit{NOW TRY}$$

OBJECTIVE 3 Use an arithmetic sequence in an application.

EXAMPLE 4 Applying an Arithmetic Sequence

Leonid Bekker's uncle decides to start a fund for Leonid's education. He makes an initial contribution of $3000 and deposits an additional $500 each month. Thus, after one month the fund will have $3000 + $500 = $3500. How much will it have after 24 months? (Disregard any interest.)

After n months, the fund will contain

$$a_n = 3000 + 500n \text{ dollars.} \qquad \text{Use an arithmetic sequence.}$$

To find the amount in the fund after 24 months, find a_{24}.

$$a_{24} = 3000 + 500(24) \qquad \text{Let } n = 24.$$
$$= 3000 + 12{,}000 \qquad \text{Multiply.}$$
$$= 15{,}000 \qquad \text{Add.}$$

The account will contain $15,000 (disregarding interest) after 24 months.

NOW TRY

OBJECTIVE 4 Find any specified term or the number of terms of an arithmetic sequence. The formula for the general term of an arithmetic sequence has four variables: a_n, a_1, n, and d. If we know any three of these, the formula can be used to find the value of the fourth variable.

EXAMPLE 5 Finding Specified Terms in Sequences

Evaluate the indicated term for each arithmetic sequence.

(a) $a_1 = -6$, $d = 12$; $\quad a_{15}$

$$a_n = a_1 + (n - 1)d \qquad \text{Formula for } a_n$$
$$a_{15} = a_1 + (15 - 1)d \qquad \text{Let } n = 15.$$
$$= -6 + 14(12) \qquad \text{Let } a_1 = -6, d = 12.$$
$$= 162 \qquad \text{Multiply, and then add.}$$

(b) $a_5 = 2$ and $a_{11} = -10$; $\quad a_{17}$

Any term can be found if a_1 and d are known. Use the formula for a_n.

$$a_5 = a_1 + (5 - 1)d \qquad\qquad a_{11} = a_1 + (11 - 1)d$$
$$a_5 = a_1 + 4d \qquad\qquad\quad a_{11} = a_1 + 10d$$
$$2 = a_1 + 4d \quad {\scriptstyle a_5 = 2} \qquad -10 = a_1 + 10d \quad {\scriptstyle a_{11} = -10}$$

This gives a system of two equations in two variables, a_1 and d.

$$a_1 + 4d = 2 \qquad (1)$$
$$a_1 + 10d = -10 \qquad (2)$$

NOW TRY
EXERCISE 5

Evaluate the indicated term for each arithmetic sequence.

(a) $a_1 = 21$ and $d = -3$; a_{22}

(b) $a_7 = 25$ and $a_{12} = 40$; a_{19}

Multiply equation (2) by -1 and add to equation (1) to eliminate a_1.

$$
\begin{array}{ll}
a_1 + 4d = 2 & \text{(1)} \\
\underline{-a_1 - 10d = 10} & -1 \text{ times (2)} \\
-6d = 12 & \text{Add.} \\
d = -2 & \text{Divide by } -6.
\end{array}
$$

Now find a_1 by substituting -2 for d into either equation.

$$
\begin{array}{ll}
a_1 + 10(-2) = -10 & \text{Let } d = -2 \text{ in (2).} \\
a_1 - 20 = -10 & \text{Multiply.} \\
a_1 = 10 & \text{Add 20.}
\end{array}
$$

Use the formula for a_n to find a_{17}.

$$
\begin{array}{ll}
a_{17} = a_1 + (17 - 1)d & \text{Let } n = 17. \\
= a_1 + 16d & \text{Subtract.} \\
= 10 + 16(-2) & \text{Let } a_1 = 10, d = -2. \\
= -22 & \text{Simplify.}
\end{array}
$$

Multiply and then add.

NOW TRY

NOW TRY
EXERCISE 6

Evaluate the number of terms in the arithmetic sequence.

$$1, \frac{4}{3}, \frac{5}{3}, 2, \dots, 11$$

EXAMPLE 6   Finding the Number of Terms in a Sequence

Evaluate the number of terms in the arithmetic sequence.

$$-8, -2, 4, 10, \dots, 52$$

Let n represent the number of terms in the sequence. Since $a_n = 52$, $a_1 = -8$, and $d = -2 - (-8) = 6$, use the formula for a_n to find n.

$$
\begin{array}{ll}
a_n = a_1 + (n - 1)d & \text{Formula for } a_n \\
52 = -8 + (n - 1)(6) & \text{Let } a_n = 52, a_1 = -8, d = 6. \\
52 = -8 + 6n - 6 & \text{Distributive property} \\
66 = 6n & \text{Simplify.} \\
n = 11 & \text{Divide by 6.}
\end{array}
$$

The sequence has 11 terms.

NOW TRY

OBJECTIVE 5   **Find the sum of a specified number of terms of an arithmetic sequence.** To find a formula for the sum S_n of the first n terms of a given arithmetic sequence, we can write out the terms in two ways. We start with the first term, and then with the last term. Then we add the terms in columns.

$$
\begin{array}{l}
S_n = a_1 + (a_1 + d) + (a_1 + 2d) + \cdots + [a_1 + (n - 1)d] \\
\underline{S_n = a_n + (a_n - d) + (a_n - 2d) + \cdots + [a_n - (n - 1)d]} \\
2S_n = (a_1 + a_n) + (a_1 + a_n) + (a_1 + a_n) + \cdots + (a_1 + a_n)
\end{array}
$$

The right-hand side of this expression contains n terms, each equal to $a_1 + a_n$.

$$2S_n = n(a_1 + a_n)$$

Formula for S_n

$$S_n = \frac{n}{2}(a_1 + a_n)$$ Divide by 2.

NOW TRY ANSWERS
5. **(a)** -42 **(b)** 61
6. 31

NOW TRY
EXERCISE 7
Evaluate the sum of the first
seven terms of the arithmetic
sequence in which
$a_n = 5n - 7$.

EXAMPLE 7 Finding the Sum of the First n Terms of an Arithmetic Sequence

Evaluate the sum of the first five terms of the arithmetic sequence in which $a_n = 2n - 5$.

Begin by evaluating a_1 and a_5.

$$a_1 = 2(1) - 5 \qquad\qquad a_5 = 2(5) - 5$$
$$= -3 \qquad\qquad\qquad = 5$$

Now evaluate the sum using $a_1 = -3$, $a_5 = 5$, and $n = 5$.

$$S_n = \frac{n}{2}(a_1 + a_n) \qquad \text{Formula for } S_n$$

$$S_5 = \frac{5}{2}(-3 + 5) \qquad \text{Substitute.}$$

$$= \frac{5}{2}(2) \qquad\qquad \text{Add.}$$

$$= 5 \qquad\qquad\qquad \text{Multiply.} \qquad\qquad \text{NOW TRY}$$

It is possible to express the sum S_n of an arithmetic sequence in terms of a_1 and d, the quantities that define the sequence. Since

$$S_n = \frac{n}{2}(a_1 + a_n) \qquad \text{and} \qquad a_n = a_1 + (n-1)d,$$

by substituting the expression for a_n into the expression for S_n we obtain

$$S_n = \frac{n}{2}(a_1 + [a_1 + (n-1)d]) \qquad \text{Substitute for } a_n.$$

$$S_n = \frac{n}{2}[2a_1 + (n-1)d]. \qquad \text{Combine like terms.}$$

The summary box gives both of the alternative forms that may be used to find the sum of the first n terms of an arithmetic sequence.

Sum of the First n Terms of an Arithmetic Sequence

The sum of the first n terms of the arithmetic sequence with first term a_1, nth term a_n, and common difference d is given by either formula.

$$S_n = \frac{n}{2}(a_1 + a_n) \qquad \text{or} \qquad S_n = \frac{n}{2}[2a_1 + (n-1)d]$$

EXAMPLE 8 Finding the Sum of the First n Terms of an Arithmetic Sequence

Evaluate the sum of the first eight terms of the arithmetic sequence having first term 3 and common difference -2.

Since the known values, $a_1 = 3$, $d = -2$, and $n = 8$, appear in the second formula for S_n, we use it.

NOW TRY ANSWER
7. 91

NOW TRY
EXERCISE 8

Evaluate the sum of the first nine terms of the arithmetic sequence having first term -8 and common difference -5.

$$S_n = \frac{n}{2}[2a_1 + (n-1)d] \qquad \text{Second formula for } S_n$$

$$S_8 = \frac{8}{2}[2(3) + (8-1)(-2)] \qquad \text{Let } a_1 = 3, d = -2, n = 8.$$

$$= 4[6 - 14] \qquad \text{Work inside the brackets.}$$

$$= -32 \qquad \text{Subtract and then multiply.}$$

NOW TRY

As mentioned earlier, linear expressions of the form $kn + c$, where k and c are real numbers, define an arithmetic sequence. For example, the sequences defined by $a_n = 2n + 5$ and $a_n = n - 3$ are arithmetic sequences. For this reason,

$$\sum_{i=1}^{n} (ki + c)$$

represents the sum of the first n terms of an arithmetic sequence having first term $a_1 = k(1) + c = k + c$ and general term $a_n = k(n) + c = kn + c$. We can find this sum with the first formula for S_n, as shown in the next example.

NOW TRY
EXERCISE 9

Evaluate $\displaystyle\sum_{i=1}^{11} (5i - 7)$.

EXAMPLE 9 Using S_n to Evaluate a Summation

Evaluate $\displaystyle\sum_{i=1}^{12} (2i - 1)$.

This is the sum of the first 12 terms of the arithmetic sequence having $a_n = 2n - 1$. This sum, S_{12}, is found with the first formula for S_n.

$$S_n = \frac{n}{2}(a_1 + a_n) \qquad \text{First formula for } S_n$$

$$S_{12} = \frac{12}{2}[(2(1) - 1) + (2(12) - 1)] \qquad \text{Let } n = 12.$$

$$= 6(1 + 23) \qquad \text{Evaluate } a_1 \text{ and } a_{12}.$$

$$= 6(24) \qquad \text{Add.}$$

NOW TRY ANSWERS

8. -252 **9.** 253

$$= 144 \qquad \text{Multiply.}$$

NOW TRY

12.2 EXERCISES

MyMathLab   Math XL PRACTICE   WATCH   DOWNLOAD   READ   REVIEW

🌐 *Complete solution available on the Video Resources on DVD*

If the given sequence is arithmetic, find the common difference d. If the sequence is not arithmetic, say so. **See Example 1.**

1. $1, 2, 3, 4, 5, \ldots$

2. $2, 5, 8, 11, \ldots$

3. $2, -4, 6, -8, 10, -12, \ldots$

4. $1, 2, 4, 7, 11, 16, \ldots$

5. $10, 5, 0, -5, -10, \ldots$

6. $-6, -10, -14, -18, \ldots$

Write the first five terms of each arithmetic sequence. **See Example 2.**

7. $a_1 = 5, d = 4$ **8.** $a_1 = 6, d = 7$

9. $a_1 = -2, d = -4$ **10.** $a_1 = -3, d = -5$

Use the formula for a_n to find the general term of each arithmetic sequence. **See Example 3.**

11. $a_1 = 2, d = 5$ **12.** $a_1 = 5, d = 3$ **13.** $3, \dfrac{15}{4}, \dfrac{9}{2}, \dfrac{21}{4}, \ldots$

14. $1, \dfrac{5}{3}, \dfrac{7}{3}, 3, \ldots$ **15.** $-3, 0, 3, \ldots$ **16.** $-10, -5, 0, \ldots$

Evaluate the indicated term for each arithmetic sequence. **See Examples 3 and 5.**

17. $a_1 = 4, d = 3;\quad a_{25}$ **18.** $a_1 = 1, d = -3;\quad a_{12}$

19. $2, 4, 6, \ldots;\quad a_{24}$ **20.** $1, 5, 9, \ldots;\quad a_{50}$

21. $a_{12} = -45, a_{10} = -37;\quad a_1$ **22.** $a_{10} = -2, a_{15} = -8;\quad a_3$

Evaluate the number of terms in each arithmetic sequence. **See Example 6.**

23. $3, 5, 7, \ldots, 33$ **24.** $4, 1, -2, \ldots, -32$

25. $\dfrac{3}{4}, 3, \dfrac{21}{4}, \ldots, 12$ **26.** $2, \dfrac{3}{2}, 1, \dfrac{1}{2}, \ldots, -5$

27. *Concept Check* In the formula for S_n, what does n represent?

28. Explain when you would use each of the two formulas for S_n.

Evaluate S_6 for each arithmetic sequence. **See Examples 7 and 8.**

29. $a_1 = 6, d = 3$ **30.** $a_1 = 5, d = 4$ **31.** $a_1 = 7, d = -3$

32. $a_1 = -5, d = -4$ **33.** $a_n = 4 + 3n$ **34.** $a_n = 9 + 5n$

Use a formula for S_n to evaluate each series. **See Example 9.**

35. $\displaystyle\sum_{i=1}^{10} (8i - 5)$ **36.** $\displaystyle\sum_{i=1}^{17} (3i - 1)$ **37.** $\displaystyle\sum_{i=1}^{20} \left(\dfrac{3}{2}i + 4\right)$

38. $\displaystyle\sum_{i=1}^{11} \left(\dfrac{1}{2}i - 1\right)$ **39.** $\displaystyle\sum_{i=1}^{250} i$ **40.** $\displaystyle\sum_{i=1}^{2000} i$

Solve each problem. (Hint: Immediately after reading the problem, determine whether you need to find a specific term of a sequence or the sum of the terms of a sequence.) **See Examples 4, 7, 8, and 9.**

41. Nancy Bondy's aunt has promised to deposit $1 in her account on the first day of her birthday month, $2 on the second day, $3 on the third day, and so on for 30 days. How much will this amount to over the entire month?

42. Repeat **Exercise 41**, but assume that the deposits are $2, $4, $6, and so on, and that the month is February of a leap year.

43. Suppose that Cherian Mathew is offered a job at $1600 per month with a guaranteed increase of $50 every six months for 5 yr. What will Cherian's salary be at the end of that time?

44. Repeat **Exercise 43**, but assume that the starting salary is $2000 per month and the guaranteed increase is $100 every four months for 3 yr.

45. A seating section in a theater-in-the-round has 20 seats in the first row, 22 in the second row, 24 in the third row, and so on for 25 rows. How many seats are there in the last row? How many seats are there in the section?

46. Constantin Arne has started on a fitness program. He plans to jog 10 min per day for the first week and then add 10 min per day each week until he is jogging an hour each day. In which week will this occur? What is the total number of minutes he will run during the first four weeks?

47. A child builds with blocks, placing 35 blocks in the first row, 31 in the second row, 27 in the third row, and so on. Continuing this pattern, can she end with a row containing exactly 1 block? If not, how many blocks will the last row contain? How many rows can she build this way?

48. A stack of firewood has 28 pieces on the bottom, 24 on top of those, then 20, and so on. If there are 108 pieces of wood, how many rows are there? (*Hint: n ≤ 7.*)

PREVIEW EXERCISES

Evaluate ar^n for the given values of a, r, and n. **See Section 5.1.**

49. $a = 2, r = 3, n = 2$

50. $a = 3, r = 2, n = 4$

51. $a = 4, r = \dfrac{1}{2}, n = 3$

52. $a = 5, r = \dfrac{1}{4}, n = 2$

(12.3) Geometric Sequences

OBJECTIVES

1 Find the common ratio of a geometric sequence.

2 Find the general term of a geometric sequence.

3 Find any specified term of a geometric sequence.

4 Find the sum of a specified number of terms of a geometric sequence.

5 Apply the formula for the future value of an ordinary annuity.

6 Find the sum of an infinite number of terms of certain geometric sequences.

In an arithmetic sequence, each term after the first is found by *adding* a fixed number to the previous term. A *geometric sequence* is defined as follows.

Geometric Sequence

A **geometric sequence,** or **geometric progression,** is a sequence in which each term after the first is found by multiplying the preceding term by a nonzero constant.

OBJECTIVE 1 **Find the common ratio of a geometric sequence.** We find the constant multiplier, called the **common ratio,** by dividing any term a_{n+1} by the preceding term, a_n.

$$r = \frac{a_{n+1}}{a_n} \qquad \text{Common ratio}$$

For example,

$$2, 6, 18, 54, 162, \dots \qquad \text{Geometric sequence}$$

is a geometric sequence in which the first term, a_1, is 2 and the common ratio is

$$r = \frac{6}{2} = \frac{18}{6} = \frac{54}{18} = \frac{162}{54} = 3. \quad \leftarrow \frac{a_{n+1}}{a_n} = 3 \text{ for all } n.$$

NOW TRY
EXERCISE 1

Determine r for the geometric sequence.

$$\frac{1}{4}, -1, 4, -16, 64, \ldots$$

EXAMPLE 1 Finding the Common Ratio

Determine the common ratio r for the geometric sequence.

$$15, \frac{15}{2}, \frac{15}{4}, \frac{15}{8}, \ldots$$

To find r, choose any two successive terms and divide the second one by the first. We choose the second and third terms of the sequence.

$$r = \frac{a_3}{a_2}$$

$$= \frac{\frac{15}{4}}{\frac{15}{2}} \qquad \text{Substitute.}$$

$$= \frac{15}{4} \div \frac{15}{2} \qquad \text{Write as division.}$$

$$= \frac{15}{4} \cdot \frac{2}{15} \qquad \text{Definition of division}$$

$$= \frac{1}{2} \qquad \text{Multiply. Write in lowest terms.}$$

Any other two successive terms could have been used to find r. Additional terms of the sequence can be found by multiplying each successive term by $\frac{1}{2}$. *NOW TRY*

OBJECTIVE 2 Find the general term of a geometric sequence. The general term a_n of a geometric sequence $a_1, a_2, a_3, \ldots$ is expressed in terms of a_1 and r by writing the first few terms as

$$a_1, \quad a_2 = a_1 r, \quad a_3 = a_1 r^2, \quad a_4 = a_1 r^3, \ldots,$$

which suggests the next rule.

General Term of a Geometric Sequence

The general term of the geometric sequence with first term a_1 and common ratio r is

$$a_n = a_1 r^{n-1}.$$

⚠ **CAUTION** In finding $a_1 r^{n-1}$, be careful to use the correct order of operations. The value of r^{n-1} must be found first. Then multiply the result by a_1.

NOW TRY
EXERCISE 2

Determine the general term of the sequence.

$$\frac{1}{4}, -1, 4, -16, 64, \ldots$$

NOW TRY ANSWERS

1. -4 **2.** $a_n = \frac{1}{4}(-4)^{n-1}$

EXAMPLE 2 Finding the General Term of a Geometric Sequence

Determine the general term of the sequence in **Example 1.**
The first term is $a_1 = 15$ and the common ratio is $r = \frac{1}{2}$.

$$a_n = a_1 r^{n-1} = 15\left(\frac{1}{2}\right)^{n-1} \qquad \text{Substitute into the formula for } a_n.$$

It is not possible to simplify further, because the exponent must be applied before the multiplication can be done. *NOW TRY*

OBJECTIVE 3 Find any specified term of a geometric sequence. We can use the formula for the general term to find any particular term.

NOW TRY
EXERCISE 3
Evaluate the indicated term for each geometric sequence.

(a) $a_1 = 3, r = 2;\quad a_8$

(b) $10, 2, \frac{2}{5}, \frac{2}{25}, \ldots;\quad a_7$

EXAMPLE 3 Finding Specified Terms in Sequences

Evaluate the indicated term for each geometric sequence.

(a) $a_1 = 4, r = -3;\quad a_6$

Use the formula for the general term.

$$a_n = a_1 r^{n-1} \qquad \text{Formula for } a_n$$

$$a_6 = a_1 \cdot r^{6-1} \qquad \text{Let } n = 6.$$

$$= 4 \cdot (-3)^5 \qquad \text{Let } a_1 = 4, r = -3.$$

Evaluate $(-3)^5$ and then multiply.

$$= -972 \qquad \text{Simplify.}$$

(b) $\frac{3}{4}, \frac{3}{8}, \frac{3}{16}, \ldots;\quad a_7$

$$a_7 = \frac{3}{4} \cdot \left(\frac{1}{2}\right)^{7-1} \qquad \text{Let } a_1 = \frac{3}{4}, r = \frac{1}{2}, n = 7.$$

$$= \frac{3}{4} \cdot \frac{1}{64} \qquad \text{Apply the exponent.}$$

$$= \frac{3}{256} \qquad \text{Multiply.} \qquad \text{NOW TRY}$$

NOW TRY
EXERCISE 4
Write the first five terms of the geometric sequence whose first term is 25 and whose common ratio is $-\frac{1}{5}$.

EXAMPLE 4 Writing the Terms of a Sequence

Write the first five terms of the geometric sequence whose first term is 5 and whose common ratio is $\frac{1}{2}$.

$$a_1 = 5, \quad a_2 = 5\left(\frac{1}{2}\right) = \frac{5}{2}, \quad a_3 = 5\left(\frac{1}{2}\right)^2 = \frac{5}{4},$$

$$a_4 = 5\left(\frac{1}{2}\right)^3 = \frac{5}{8}, \quad a_5 = 5\left(\frac{1}{2}\right)^4 = \frac{5}{16}$$

Use $a_n = a_1 r^{n-1}$, with $a_1 = 5, r = \frac{1}{2}$, and $n = 1, 2, 3, 4, 5$.

NOW TRY

OBJECTIVE 4 Find the sum of a specified number of terms of a geometric sequence. It is convenient to have a formula for the sum S_n of the first n terms of a geometric sequence. We can develop a formula by first writing out S_n.

$$S_n = a_1 + a_1 r + a_1 r^2 + a_1 r^3 + \cdots + a_1 r^{n-1}$$

Next, we multiply both sides by $-r$.

$$-rS_n = -a_1 r - a_1 r^2 - a_1 r^3 - a_1 r^4 - \cdots - a_1 r^n$$

Now add.

$$S_n = a_1 + a_1 r + a_1 r^2 + a_1 r^3 + \cdots + a_1 r^{n-1}$$

$$-rS_n = \quad\ -a_1 r - a_1 r^2 - a_1 r^3 - \cdots - a_1 r^{n-1} - a_1 r^n$$

$$\overline{S_n - rS_n = a_1 \qquad\qquad\qquad\qquad\qquad\qquad\ - a_1 r^n}$$

$$S_n(1 - r) = a_1 - a_1 r^n \qquad \text{Factor on the left.}$$

$$S_n = \frac{a_1(1 - r^n)}{1 - r} \qquad \begin{array}{l}\text{Factor on the right.}\\ \text{Divide each side by } 1 - r.\end{array}$$

NOW TRY ANSWERS
3. **(a)** $3(2)^7 = 384$

(b) $10\left(\frac{1}{5}\right)^6 = \frac{2}{3125}$

4. $a_1 = 25, a_2 = -5, a_3 = 1,$

$a_4 = -\frac{1}{5}, a_5 = \frac{1}{25}$

Sum of the First *n* Terms of a Geometric Sequence

The sum of the first *n* terms of the geometric sequence with first term a_1 and common ratio *r* is

$$S_n = \frac{a_1(1 - r^n)}{1 - r} \quad (r \neq 1).$$

If $r = 1$, then $S_n = a_1 + a_1 + a_1 + \cdots + a_1 = na_1$.

Multiplying the formula for S_n by $\frac{-1}{-1}$ gives an alternative form.

$$S_n = \frac{a_1(1 - r^n)}{1 - r} \cdot \frac{-1}{-1} = \frac{a_1(r^n - 1)}{r - 1} \qquad \text{Alternative form}$$

NOW TRY
EXERCISE 5

Evaluate the sum of the first six terms of the geometric sequence with first term 4 and common ratio 2.

EXAMPLE 5 Finding the Sum of the First *n* Terms of a Geometric Sequence

Evaluate the sum of the first six terms of the geometric sequence with first term -2 and common ratio 3.

$$S_n = \frac{a_1(1 - r^n)}{1 - r} \qquad \text{Formula for } S_n$$

$$S_6 = \frac{-2(1 - 3^6)}{1 - 3} \qquad \text{Let } n = 6, a_1 = -2, r = 3.$$

$$= \frac{-2(1 - 729)}{-2} \qquad \begin{array}{l}\text{Evaluate } 3^6. \text{ Subtract in}\\ \text{the denominator.}\end{array}$$

$$= -728 \qquad \text{Simplify.} \qquad \text{NOW TRY}$$

A series of the form

$$\sum_{i=1}^{n} a \cdot b^i$$

represents the sum of the first *n* terms of a geometric sequence having first term $a_1 = a \cdot b^1 = ab$ and common ratio *b*. The next example illustrates this form.

NOW TRY
EXERCISE 6

Evaluate $\displaystyle\sum_{i=1}^{5} 8\left(\frac{1}{2}\right)^i$.

EXAMPLE 6 Using the Formula for S_n to Find a Summation

Evaluate $\displaystyle\sum_{i=1}^{4} 3 \cdot 2^i$.

Since the series is in the form $\displaystyle\sum_{i=1}^{n} a \cdot b^i$, it represents the sum of the first *n* terms of the geometric sequence with $a_1 = a \cdot b^1$ and $r = b$.

$$S_n = \frac{a_1(1 - r^n)}{1 - r} \qquad \text{Formula for } S_n$$

$$S_4 = \frac{6(1 - 2^4)}{1 - 2} \qquad \text{Let } n = 4, a_1 = 6, r = 2.$$

$$= \frac{6(1 - 16)}{-1} \qquad \begin{array}{l}\text{Evaluate } 2^4. \text{ Subtract in}\\ \text{the denominator.}\end{array}$$

$$= 90 \qquad \text{Simplify.} \qquad \text{NOW TRY}$$

NOW TRY ANSWERS
5. 252 **6.** 7.75, or $\frac{31}{4}$

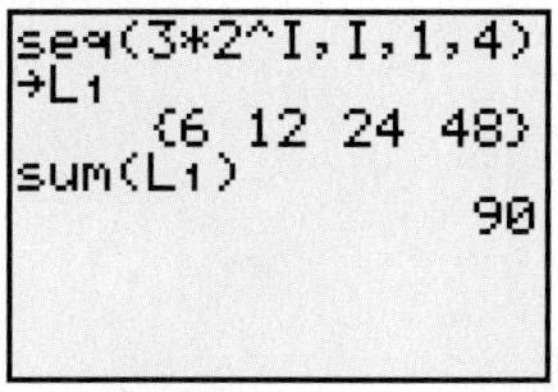

FIGURE 2

FIGURE 2 shows how a graphing calculator can store the terms in a list and then find the sum of these terms. The figure supports the result of **Example 6.**

OBJECTIVE 5 Apply the formula for the future value of an ordinary annuity. A sequence of equal payments made over equal periods is called an **annuity.** If the payments are made at the end of the period, and if the frequency of payments is the same as the frequency of compounding, the annuity is called an **ordinary annuity.** The time between payments is the **payment period,** and the time from the beginning of the first payment period to the end of the last is called the **term of the annuity.** The **future value of the annuity,** the final sum on deposit, is defined as the sum of the compound amounts of all the payments, compounded to the end of the term.

We state the following formula without proof.

> ### Future Value of an Ordinary Annuity
>
> The future value of an ordinary annuity is
>
> $$S = R\left[\frac{(1 + i)^n - 1}{i}\right],$$
>
> where S is the future value,
> R is the payment at the end of each period,
> i is the interest rate per period, and
> n is the number of periods.

**NOW TRY
EXERCISE 7**

(a) Billy Harmon deposits $600 at the end of each year into an account paying 2.5% per yr, compounded annually. Find the total amount on deposit after 18 yr.

(b) How much will be in Billy Harmon's account after 18 yr if he deposits $100 at the end of each month at 3% interest compounded monthly?

EXAMPLE 7 Applying the Formula for the Future Value of an Annuity

(a) Igor Kalugin is an athlete who believes that his playing career will last 7 yr. He deposits $22,000 at the end of each year for 7 yr in an account paying 6% compounded annually. How much will he have on deposit after 7 yr?

Igor's payments form an ordinary annuity with $R = 22,000$, $n = 7$, and $i = 0.06$. The future value of this annuity (from the formula) is

$$S = 22,000\left[\frac{(1.06)^7 - 1}{0.06}\right]$$

$$= 184,664.43, \quad \text{or} \quad \$184,664.43. \quad \text{Use a calculator.}$$

(b) Amy Loschak has decided to deposit $200 at the end of each month in an account that pays interest of 4.8% compounded monthly for retirement in 20 yr. How much will be in the account at that time?

Because the interest is compounded monthly, $i = \frac{0.048}{12}$. Also, $R = 200$ and $n = 12(20)$. The future value is

$$S = 200\left[\frac{\left(1 + \dfrac{0.048}{12}\right)^{12(20)} - 1}{\dfrac{0.048}{12}}\right] = 80,335.01, \quad \text{or} \quad \$80,335.01.$$

NOW TRY

OBJECTIVE 6 Find the sum of an infinite number of terms of certain geometric sequences. Consider an infinite geometric sequence such as

NOW TRY ANSWERS
7. (a) $13,431.81   **(b)** $28,594.03

$$\frac{1}{3}, \frac{1}{6}, \frac{1}{12}, \frac{1}{24}, \frac{1}{48}, \dots .$$

The sum of the first two terms is

$$S_2 = \frac{1}{3} + \frac{1}{6} = \frac{1}{2} = 0.5.$$

In a similar manner, we can find additional "partial sums."

$$S_3 = S_2 + \frac{1}{12} = \frac{1}{2} + \frac{1}{12} = \frac{7}{12} \approx 0.583, \quad S_4 = S_3 + \frac{1}{24} = \frac{7}{12} + \frac{1}{24} = \frac{15}{24} = 0.625,$$

$$S_5 = \frac{31}{48} \approx 0.64583, \quad S_6 = \frac{21}{32} = 0.65625, \quad S_7 = \frac{127}{192} \approx 0.6614583.$$

Each term of the geometric sequence is less than the preceding one, so each additional term is contributing less and less to the partial sum. In decimal form (to the nearest thousandth), the first 7 terms and the 10th term are given in the table.

Term	a_1	a_2	a_3	a_4	a_5	a_6	a_7	a_{10}
Value	0.333	0.167	0.083	0.042	0.021	0.010	0.005	0.001

As the table suggests, the value of a term gets closer and closer to 0 as the number of the term increases. To express this idea, we say that as n increases without bound (written $n \to \infty$), the limit of the term a_n is 0, written

$$\lim_{n \to \infty} a_n = 0.$$

A number that can be defined as the sum of an infinite number of terms of a geometric sequence is found by starting with the expression for the sum of a finite number of terms.

$$S_n = \frac{a_1(1 - r^n)}{1 - r}$$

If $|r| < 1$, then as n increases without bound, the value of r^n gets closer and closer to 0. As r^n approaches 0, $1 - r^n$ approaches $1 - 0 = 1$, and S_n approaches the quotient $\frac{a_1}{1 - r}$.

$$\lim_{r^n \to 0} S_n = \lim_{r^n \to 0} \frac{a_1(1 - r^n)}{1 - r} = \frac{a_1(1 - 0)}{1 - r} = \frac{a_1}{1 - r}$$

This limit is defined to be the sum of the infinite geometric sequence.

$$a_1 + a_1 r + a_1 r^2 + a_1 r^3 + \cdots = \frac{a_1}{1 - r}, \quad \text{if } |r| < 1$$

Sum of the Terms of an Infinite Geometric Sequence

The sum S of the terms of an infinite geometric sequence with first term a_1 and common ratio r, where $|r| < 1$, is

$$S = \frac{a_1}{1 - r}.$$

If $|r| \geq 1$, then the sum does not exist.

Now consider $|r| > 1$. For example, suppose the sequence is

$$6, 12, 24, \ldots, 3(2)^n, \ldots.$$

In this kind of sequence, as n increases, the value of r^n also increases and so does the sum S_n. Since each new term adds a greater and greater amount to the sum, there is no limit to the value of S_n. The sum S does not exist. A similar situation exists if $r = 1$.

NOW TRY
EXERCISE 8
Evaluate the sum of the terms of the infinite geometric sequence with $a_1 = -4$ and $r = \frac{2}{3}$.

EXAMPLE 8 Finding the Sum of the Terms of an Infinite Geometric Sequence

Evaluate the sum of the terms of the infinite geometric sequence with $a_1 = 3$ and $r = -\frac{1}{3}$.

Substitute into the formula.

$$S = \frac{a_1}{1 - r} \qquad \text{Infinite sum formula}$$

$$= \frac{3}{1 - \left(-\frac{1}{3}\right)} \qquad \text{Let } a_1 = 3, r = -\frac{1}{3}.$$

$$= \frac{3}{\frac{4}{3}} \qquad \text{Simplify the denominator.}$$

$$= 3 \div \frac{4}{3} \qquad \text{Write as division.}$$

$$= 3 \cdot \frac{3}{4} \qquad \text{Definition of division}$$

$$= \frac{9}{4} \qquad \text{Multiply.} \qquad \text{NOW TRY}$$

In summation notation, the sum of an infinite geometric sequence is written as

$$\sum_{i=1}^{\infty} a_i.$$

For instance, the sum in **Example 8** would be written

$$\sum_{i=1}^{\infty} 3\left(-\frac{1}{3}\right)^{i-1}.$$

NOW TRY
EXERCISE 9
Evaluate $\displaystyle\sum_{i=1}^{\infty} \left(\frac{5}{8}\right)\left(\frac{3}{4}\right)^i$.

EXAMPLE 9 Finding the Sum of the Terms of an Infinite Geometric Series

Evaluate $\displaystyle\sum_{i=1}^{\infty} \left(\frac{1}{2}\right)^i$.

This is the infinite geometric series

$$\frac{1}{2} + \frac{1}{4} + \frac{1}{8} + \cdots,$$

with $a_1 = \frac{1}{2}$ and $r = \frac{1}{2}$. Since $|r| < 1$, we find the sum as follows.

$$S = \frac{a_1}{1 - r}$$

$$= \frac{\frac{1}{2}}{1 - \frac{1}{2}} \qquad \text{Let } a_1 = \frac{1}{2}, r = \frac{1}{2}.$$

$$= \frac{\frac{1}{2}}{\frac{1}{2}} \qquad \text{Simplify the denominator.}$$

$$= 1 \qquad \text{Divide.} \qquad \text{NOW TRY}$$

NOW TRY ANSWERS
8. -12 9. $\frac{15}{8}$

● *Complete solution available on the Video Resources on DVD*

If the given sequence is geometric, find the common ratio r. If the sequence is not geometric, say so. See Example 1.

● **1.** $4, 8, 16, 32, \ldots$

2. $5, 15, 45, 135, \ldots$

3. $\dfrac{1}{3}, \dfrac{2}{3}, \dfrac{3}{3}, \dfrac{4}{3}, \ldots$

4. $\dfrac{5}{7}, \dfrac{8}{7}, \dfrac{11}{7}, 2, \ldots$

5. $1, -3, 9, -27, 81, \ldots$

6. $2, -8, 32, -128, \ldots$

7. $1, -\dfrac{1}{2}, \dfrac{1}{4}, -\dfrac{1}{8}, \ldots$

8. $\dfrac{2}{3}, -\dfrac{2}{15}, \dfrac{2}{75}, -\dfrac{2}{375}, \ldots$

Find a general term for each geometric sequence. See Example 2.

9. $-5, -10, -20, \ldots$

10. $-2, -6, -18, \ldots$

11. $-2, \dfrac{2}{3}, -\dfrac{2}{9}, \ldots$

12. $-3, \dfrac{3}{2}, -\dfrac{3}{4}, \ldots$

13. $10, -2, \dfrac{2}{5}, \ldots$

14. $8, -2, \dfrac{1}{2}, \ldots$

Evaluate the indicated term for each geometric sequence. See Example 3.

● **15.** $a_1 = 2, r = 5; \quad a_{10}$

16. $a_1 = 1, r = 3; \quad a_{15}$

17. $\dfrac{1}{2}, \dfrac{1}{6}, \dfrac{1}{18}, \ldots; \quad a_{12}$

18. $\dfrac{2}{3}, \dfrac{1}{3}, \dfrac{1}{6}, \ldots; \quad a_{18}$

19. $a_3 = \dfrac{1}{2}, a_7 = \dfrac{1}{32}; \quad a_{25}$

20. $a_5 = 48, a_8 = -384; \quad a_{10}$

Write the first five terms of each geometric sequence. See Example 4.

● **21.** $a_1 = 2, r = 3$

22. $a_1 = 4, r = 2$

23. $a_1 = 5, r = -\dfrac{1}{5}$

24. $a_1 = 6, r = -\dfrac{1}{3}$

Use the formula for S_n to determine the sum of the terms of each geometric sequence. See Examples 5 and 6. In Exercises 27–32, give the answer to the nearest thousandth.

● **25.** $\dfrac{1}{3}, \dfrac{1}{9}, \dfrac{1}{27}, \dfrac{1}{81}, \dfrac{1}{243}$

26. $\dfrac{4}{3}, \dfrac{8}{3}, \dfrac{16}{3}, \dfrac{32}{3}, \dfrac{64}{3}, \dfrac{128}{3}$

27. $-\dfrac{4}{3}, -\dfrac{4}{9}, -\dfrac{4}{27}, -\dfrac{4}{81}, -\dfrac{4}{243}, -\dfrac{4}{729}$

28. $\dfrac{5}{16}, -\dfrac{5}{32}, \dfrac{5}{64}, -\dfrac{5}{128}, \dfrac{5}{256}$

● **29.** $\displaystyle\sum_{i=1}^{7} 4\left(\dfrac{2}{5}\right)^i$

30. $\displaystyle\sum_{i=1}^{8} 5\left(\dfrac{2}{3}\right)^i$

31. $\displaystyle\sum_{i=1}^{10} (-2)\left(\dfrac{3}{5}\right)^i$

32. $\displaystyle\sum_{i=1}^{6} (-2)\left(-\dfrac{1}{2}\right)^i$

Solve each problem involving an ordinary annuity. See Example 7.

33. A father opened a savings account for his daughter on her first birthday, depositing $1000. Each year on her birthday he deposits another $1000, making the last deposit on her 21st birthday. If the account pays 4.4% interest compounded annually, how much is in the account at the end of the day on the daughter's 21st birthday?

34. B. G. Thompson puts $1000 in a retirement account at the end of each quarter $\left(\frac{1}{4}\text{ of a year}\right)$ for 15 yr. If the account pays 4% annual interest compounded quarterly, how much will be in the account at that time?

35. At the end of each quarter, a 50-year-old woman puts $1200 in a retirement account that pays 5% interest compounded quarterly. When she reaches age 60, she withdraws the entire amount and places it in a mutual fund that pays 6% interest compounded monthly. From then on, she deposits $300 in the mutual fund at the end of each month. How much is in the account when she reaches age 65?

36. Derrick Ruffin deposits $10,000 at the end of each year for 12 yr in an account paying 5% compounded annually. He then puts the total amount on deposit in another account paying 6% compounded semiannually for another 9 yr. Find the final amount on deposit after the entire 21-yr period.

Find the sum, if it exists, of the terms of each infinite geometric sequence. **See Examples 8 and 9.**

37. $a_1 = 6, r = \dfrac{1}{3}$

38. $a_1 = 10, r = \dfrac{1}{5}$

39. $a_1 = 1000, r = -\dfrac{1}{10}$

40. $a_1 = 8800, r = -\dfrac{3}{5}$

41. $\displaystyle\sum_{i=1}^{\infty} \dfrac{9}{8}\left(-\dfrac{2}{3}\right)^i$

42. $\displaystyle\sum_{i=1}^{\infty} \dfrac{3}{5}\left(\dfrac{5}{6}\right)^i$

43. $\displaystyle\sum_{i=1}^{\infty} \dfrac{12}{5}\left(\dfrac{5}{4}\right)^i$

44. $\displaystyle\sum_{i=1}^{\infty} \left(-\dfrac{16}{3}\right)\left(-\dfrac{9}{8}\right)^i$

Solve each application. (Hint: Immediately after reading the problem, determine whether you need to find a specific term of a sequence or the sum of the terms of a sequence.)

45. When dropped from a certain height, a ball rebounds $\frac{3}{5}$ of the original height. How high will the ball rebound after the fourth bounce if it was dropped from a height of 10 ft?

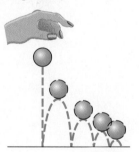

46. A fully wound yo-yo has a string 40 in. long. It is allowed to drop, and on its first rebound it returns to a height 15 in. lower than its original height. Assuming that this "rebound ratio" remains constant until the yo-yo comes to rest, how far does it travel on its third trip up the string?

47. A particular substance decays in such a way that it loses half its weight each day. In how many days will 256 g of the substance be reduced to 32 g? How much of the substance is left after 10 days?

48. A tracer dye is injected into a system with an ingestion and an excretion. After 1 hr, $\frac{2}{3}$ of the dye is left. At the end of the second hour, $\frac{2}{3}$ of the remaining dye is left, and so on. If one unit of the dye is injected, how much is left after 6 hr?

49. In a certain community, the consumption of electricity has increased about 6% per yr.

  (a) If a community uses 1.1 billion units of electricity now, how much will it use 5 yr from now?

  (b) Find the number of years it will take for the consumption to double.

50. Suppose the community in **Exercise 49** reduces its increase in consumption to 2% per yr.

  (a) How much will it use 5 yr from now?

  (b) Find the number of years it will take for the consumption to double.

51. A machine depreciates by $\frac{1}{4}$ of its value each year. If it cost $50,000 new, what is its value after 8 yr?

52. Refer to **Exercise 46.** Theoretically, how far does the yo-yo travel before coming to rest?

RELATING CONCEPTS EXERCISES 53–58

FOR INDIVIDUAL OR GROUP WORK

In **Chapter 1,** we learned that any repeating decimal is a rational number; that is, it can be expressed as a quotient of integers. Thus, the repeating decimal

$$0.99999\ldots,$$

with an endless string of 9s, must be a rational number. **Work Exercises 53–58 in order,** to discover the surprising simplest form of this rational number.

53. Use long division to write a repeating decimal representation for $\frac{1}{3}$.

54. Use long division to write a repeating decimal representation for $\frac{2}{3}$.

55. Because $\frac{1}{3} + \frac{2}{3} = 1$, the sum of the decimal representations in **Exercises 53 and 54** must also equal 1. Line up the decimals in the usual vertical method for addition, and obtain the repeating decimal result. The value of this decimal is exactly 1.

56. The repeating decimal $0.99999\ldots$ can be written as the sum of the terms of a geometric sequence with $a_1 = 0.9$ and $r = 0.1$.

$$0.99999\ldots = 0.9 + 0.9(0.1) + 0.9(0.1)^2 + 0.9(0.1)^3 + 0.9(0.1)^4 + 0.9(0.1)^5 + \cdots$$

Since $|0.1| < 1$, this sum can be found from the formula $S = \dfrac{a_1}{1-r}$. Use this formula to support the result you found another way in **Exercises 53–55.**

57. Which one of the following is true, based on your results in **Exercises 55 and 56?**

  A. $0.99999\ldots < 1$ **B.** $0.99999\ldots = 1$ **C.** $0.99999\ldots \approx 1$

58. Show that $0.49999\ldots = \frac{1}{2}$.

PREVIEW EXERCISES

Multiply. **See Section 5.4.**

59. $(3x + 2y)^2$

60. $(4x - 3y)^2$

61. $(a - b)^3$

62. $(x + y)^4$

12.4 The Binomial Theorem

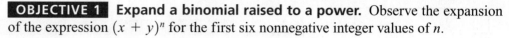

OBJECTIVE 1 **Expand a binomial raised to a power.** Observe the expansion of the expression $(x + y)^n$ for the first six nonnegative integer values of n.

$$(x + y)^0 = 1,$$
$$(x + y)^1 = x + y, \qquad \text{Expansions of } (x + y)^n$$
$$(x + y)^2 = x^2 + 2xy + y^2,$$
$$(x + y)^3 = x^3 + 3x^2y + 3xy^2 + y^3,$$
$$(x + y)^4 = x^4 + 4x^3y + 6x^2y^2 + 4xy^3 + y^4,$$
$$(x + y)^5 = x^5 + 5x^4y + 10x^3y^2 + 10x^2y^3 + 5xy^4 + y^5$$

By identifying patterns, we can write a general expansion for $(x + y)^n$.

First, if n is a positive integer, each expansion after $(x + y)^0$ begins with x raised to the same power to which the binomial is raised. That is, the expansion of $(x + y)^1$ has a first term of x^1, the expansion of $(x + y)^2$ has a first term of x^2, and so on. Also, the last term in each expansion is y to this same power, so the expansion of $(x + y)^n$ should begin with the term x^n and end with the term y^n.

The exponents on x decrease by 1 in each term after the first, while the exponents on y, beginning with y in the second term, increase by 1 in each succeeding term. Thus, the *variables* in the expansion of $(x + y)^n$ have the following pattern.

$$x^n, \quad x^{n-1}y, \quad x^{n-2}y^2, \quad x^{n-3}y^3, \quad \ldots, \quad xy^{n-1}, \quad y^n$$

This pattern suggests that the sum of the exponents on x and y in each term is n. For example, in the third term shown, the variable part is $x^{n-2}y^2$ and the sum of the exponents, $n - 2$ and 2, is n.

Now examine the pattern for the *coefficients* of the terms of the preceding expansions. Writing the coefficients alone in a triangular pattern gives **Pascal's triangle,** named in honor of the 17th-century mathematician Blaise Pascal.

Blaise Pascal (1623–1662)

			Pascal's Triangle			
			1			
		1		1		
	1		2		1	
1		3		3		1
1	4		6		4	1
1	5	10		10	5	1

and so on

The first and last terms of each row are 1. Each number in the interior of the triangle is the sum of the two numbers just above it (one to the right and one to the left). For example, in the fifth row from the top, 4 is the sum of 1 and 3, 6 is the sum of 3 and 3, and so on.

To obtain the coefficients for $(x + y)^6$, we attach the seventh row to the table by starting and ending with 1, and adding pairs of numbers from the sixth row.

$$1 \quad 6 \quad 15 \quad 20 \quad 15 \quad 6 \quad 1 \qquad \text{Seventh row}$$

We then use these coefficients to expand $(x + y)^6$ as

$$(x + y)^6 = x^6 + 6x^5y + 15x^4y^2 + 20x^3y^3 + 15x^2y^4 + 6xy^5 + y^6.$$

Although it is possible to use Pascal's triangle to find the coefficients in $(x + y)^n$ for any positive integer value of n, it is impractical for large values of n. A more efficient way to determine these coefficients uses the symbol **$n!$** (read **"n factorial"**), defined as follows.

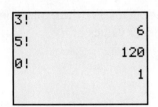

FIGURE 3

n Factorial ($n!$)

For any positive integer n,
$$n! = n(n - 1)(n - 2)(n - 3) \cdots (2)(1).$$
By definition, **$0! = 1$.**

NOW TRY
EXERCISE 1

Evaluate.

$$7!$$

EXAMPLE 1 Evaluating Factorials

Evaluate each factorial.

(a) $3! = 3 \cdot 2 \cdot 1 = 6$ **(b)** $5! = 5 \cdot 4 \cdot 3 \cdot 2 \cdot 1 = 120$

(c) $0! = 1$ 0! is defined to be 1.

FIGURE 3 shows how a graphing calculator computes factorials.   NOW TRY

NOW TRY
EXERCISE 2

Find the value of each expression.

(a) $\dfrac{8!}{6!2!}$ **(b)** $\dfrac{8!}{5!3!}$

(c) $\dfrac{6!}{6!0!}$ **(d)** $\dfrac{6!}{5!1!}$

EXAMPLE 2 Evaluating Expressions Involving Factorials

Find the value of each expression.

(a) $\dfrac{5!}{4!1!} = \dfrac{5 \cdot 4 \cdot 3 \cdot 2 \cdot 1}{(4 \cdot 3 \cdot 2 \cdot 1)(1)} = 5$

(b) $\dfrac{5!}{3!2!} = \dfrac{5 \cdot 4 \cdot 3 \cdot 2 \cdot 1}{(3 \cdot 2 \cdot 1)(2 \cdot 1)} = \dfrac{5 \cdot 4}{2 \cdot 1} = 10$

(c) $\dfrac{6!}{3!3!} = \dfrac{6 \cdot 5 \cdot 4 \cdot 3 \cdot 2 \cdot 1}{(3 \cdot 2 \cdot 1)(3 \cdot 2 \cdot 1)} = \dfrac{6 \cdot 5 \cdot 4}{3 \cdot 2 \cdot 1} = 20$

(d) $\dfrac{4!}{4!0!} = \dfrac{4 \cdot 3 \cdot 2 \cdot 1}{(4 \cdot 3 \cdot 2 \cdot 1)(1)} = 1$ NOW TRY

Now look again at the coefficients of the expansion
$$(x + y)^5 = x^5 + 5x^4y + 10x^3y^2 + 10x^2y^3 + 5xy^4 + y^5.$$

The coefficient of the second term is 5, and the exponents on the variables in that term are 4 and 1. From **Example 2(a),** $\frac{5!}{4!1!} = 5$. The coefficient of the third term is 10, and the exponents are 3 and 2. From **Example 2(b),** $\frac{5!}{3!2!} = 10$. Similar results are true for the remaining terms. The first term can be written as $1x^5y^0$, and the last term can be written as $1x^0y^5$. Then the coefficient of the first term should be $\frac{5!}{5!0!} = 1$, and the coefficient of the last term would be $\frac{5!}{0!5!} = 1$.

The coefficient of a term in $(x + y)^n$ in which the variable part is x^ry^{n-r} is

$$\frac{n!}{r!(n - r)!}.$$ This is called a **binomial coefficient.**

The binomial coefficient $\frac{n!}{r!(n - r)!}$ is often represented by the symbol $_nC_r$. This notation comes from the fact that if we choose *combinations* of n things taken r at a time, the result is given by that expression. We read $_nC_r$ as **"combinations of n things taken r at a time."** Another common representation is $\binom{n}{r}$.

NOW TRY ANSWERS
1. 5040
2. (a) 28 (b) 56 (c) 1 (d) 6

> **Formula for the Binomial Coefficient $_nC_r$**
>
> For nonnegative integers n and r, where $r \leq n$,
>
> $$_nC_r = \frac{n!}{r!(n-r)!}.$$

NOW TRY
EXERCISE 3
Evaluate $_7C_2$.

EXAMPLE 3 Evaluating Binomial Coefficients

Evaluate each binomial coefficient.

(a) $_5C_4 = \dfrac{5!}{4!(5-4)!}$ Let $n = 5, r = 4$.

$= \dfrac{5!}{4!1!}$ Subtract.

$= \dfrac{5 \cdot 4 \cdot 3 \cdot 2 \cdot 1}{4 \cdot 3 \cdot 2 \cdot 1 \cdot 1}$ Definition of n factorial

$= 5$ Lowest terms

Binomial coefficients will always be whole numbers.

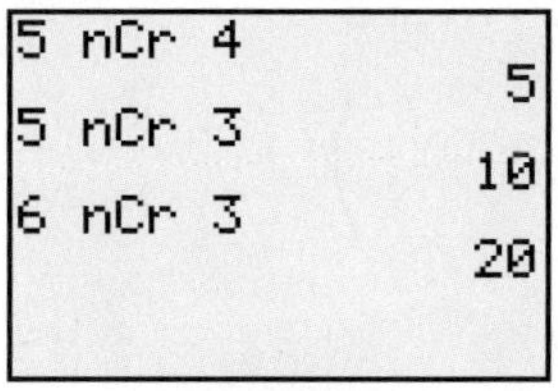

(b) $_5C_3 = \dfrac{5!}{3!(5-3)!} = \dfrac{5!}{3!2!} = \dfrac{5 \cdot 4 \cdot 3 \cdot 2 \cdot 1}{3 \cdot 2 \cdot 1 \cdot 2 \cdot 1} = 10$

(c) $_6C_3 = \dfrac{6!}{3!(6-3)!} = \dfrac{6!}{3!3!} = \dfrac{6 \cdot 5 \cdot 4 \cdot 3 \cdot 2 \cdot 1}{3 \cdot 2 \cdot 1 \cdot 3 \cdot 2 \cdot 1} = 20$

```
5 nCr 4
              5
5 nCr 3
             10
6 nCr 3
             20
```

FIGURE 4

FIGURE 4 shows how a graphing calculator displays the binomial coefficients computed here.

NOW TRY

We now state the **binomial theorem,** or the **general binomial expansion.**

> **Binomial Theorem**
>
> For any positive integer n,
>
> $$(x + y)^n = x^n + \frac{n!}{1!(n-1)!}x^{n-1}y + \frac{n!}{2!(n-2)!}x^{n-2}y^2$$
>
> $$+ \frac{n!}{3!(n-3)!}x^{n-3}y^3 + \cdots + \frac{n!}{(n-1)!1!}xy^{n-1} + y^n.$$

The binomial theorem can be written in summation notation as

$$(x + y)^n = \sum_{k=0}^{n} \frac{n!}{k!(n-k)!}x^{n-k}y^k.$$

NOTE We used the letter k as the summation index letter in the statement just given. This is customary notation in mathematics.

NOW TRY ANSWER
3. 21

NOW TRY
EXERCISE 4
Expand $(a + 3b)^5$.

EXAMPLE 4 Using the Binomial Theorem

Expand $(2m + 3)^4$.

$$(2m + 3)^4$$

$$= (2m)^4 + \frac{4!}{1!3!}(2m)^3(3) + \frac{4!}{2!2!}(2m)^2(3)^2 + \frac{4!}{3!1!}(2m)(3)^3 + 3^4$$

Remember:
$(ab)^m = a^m b^m.$

$$= 16m^4 + 4(8m^3)(3) + 6(4m^2)(9) + 4(2m)(27) + 81$$

$$= 16m^4 + 96m^3 + 216m^2 + 216m + 81 \qquad \text{NOW TRY}$$

NOW TRY
EXERCISE 5
Expand $\left(\dfrac{x}{3} - 2y\right)^4$.

EXAMPLE 5 Using the Binomial Theorem

Expand $\left(a - \dfrac{b}{2}\right)^5$.

$$\left(a - \frac{b}{2}\right)^5$$

$$= a^5 + \frac{5!}{1!4!}a^4\left(-\frac{b}{2}\right) + \frac{5!}{2!3!}a^3\left(-\frac{b}{2}\right)^2 + \frac{5!}{3!2!}a^2\left(-\frac{b}{2}\right)^3$$

$$+ \frac{5!}{4!1!}a\left(-\frac{b}{2}\right)^4 + \left(-\frac{b}{2}\right)^5$$

$$= a^5 + 5a^4\left(-\frac{b}{2}\right) + 10a^3\left(\frac{b^2}{4}\right) + 10a^2\left(-\frac{b^3}{8}\right)$$

$$+ 5a\left(\frac{b^4}{16}\right) + \left(-\frac{b^5}{32}\right)$$

Notice that signs alternate positive and negative.

$$= a^5 - \frac{5}{2}a^4b + \frac{5}{2}a^3b^2 - \frac{5}{4}a^2b^3 + \frac{5}{16}ab^4 - \frac{1}{32}b^5 \quad \text{NOW TRY}$$

⚠ **CAUTION** When the binomial is the *difference* of two terms, as in **Example 5,** the signs of the terms in the expansion will alternate. Those terms with odd exponents on the second variable expression $\left(-\frac{b}{2}\right.$ in **Example 5**$\left.\right)$ will be negative, while those with even exponents on the second variable expression will be positive.

OBJECTIVE 2 Find any specified term of the expansion of a binomial. Any single term of a binomial expansion can be determined without writing out the whole expansion. For example, if $n \geq 10$, then the 10th term of $(x + y)^n$ has y raised to the ninth power (since y has the power of 1 in the second term, the power of 2 in the third term, and so on). Since the exponents on x and y in any term must have a sum of n, the exponent on x in the 10th term is $n - 9$. The quantities 9 and $n - 9$ determine the factorials in the denominator of the coefficient. Thus, the 10th term of $(x + y)^n$ is

$$\frac{n!}{9!(n - 9)!}x^{n-9}y^9.$$

NOW TRY ANSWERS
4. $a^5 + 15a^4b + 90a^3b^2 +$
   $270a^2b^3 + 405ab^4 + 243b^5$

5. $\dfrac{x^4}{81} - \dfrac{8x^3y}{27} + \dfrac{8x^2y^2}{3} - \dfrac{32xy^3}{3} +$
   $16y^4$

rth Term of the Binomial Expansion

If $n \geq r - 1$, then the *r*th term of the expansion of $(x + y)^n$ is

$$\frac{n!}{(r - 1)![n - (r - 1)]!}x^{n-(r-1)}y^{r-1}.$$

In this general expression, remember to start with the exponent on y, which is 1 less than the term number r. Then subtract that exponent from n to get the exponent on x: $n - (r - 1)$. The two exponents are then used as the factorials in the denominator of the coefficient.

> **NOW TRY**
> **EXERCISE 6**
> Find the sixth term of the expansion of $(2m - n^2)^8$.

EXAMPLE 6 Finding a Single Term of a Binomial Expansion

Find the fourth term of the expansion of $(a + 2b)^{10}$.

In the fourth term, $2b$ has an exponent of $4 - 1 = 3$ and a has an exponent of $10 - 3 = 7$. The fourth term is determined as follows.

$$\frac{10!}{3!7!}(a^7)(2b)^3 \quad \boxed{\text{Parentheses MUST be used for } 2b.}$$

$$= \frac{10 \cdot 9 \cdot 8}{3 \cdot 2 \cdot 1}(a^7)(8b^3) \qquad \text{Let } n = 10, x = a, y = 2b, r = 4.$$

$$= 120a^7(8b^3) \qquad \text{Simplify the factorials.}$$

$$= 960a^7b^3 \qquad \text{Multiply.} \qquad \textit{NOW TRY}$$

NOW TRY ANSWER
6. $-448m^3n^{10}$

12.4 EXERCISES

MyMathLab   Math XL PRACTICE   WATCH   DOWNLOAD   READ   REVIEW

⊙ *Complete solution available on the Video Resources on DVD*

Evaluate each expression. ***See Examples 1–3.***

⊙ **1.** $6!$ **2.** $4!$ **3.** $8!$ **4.** $9!$ ⊙ **5.** $\dfrac{6!}{4!2!}$

6. $\dfrac{7!}{3!4!}$ **7.** $\dfrac{4!}{0!4!}$ **8.** $\dfrac{5!}{5!0!}$ **9.** $4! \cdot 5$ **10.** $6! \cdot 7$

⊙ **11.** $_6C_2$ **12.** $_7C_4$ **13.** $_{13}C_{11}$ **14.** $_{13}C_2$

Use the binomial theorem to expand each expression. ***See Examples 4 and 5.***

15. $(m + n)^4$ **16.** $(x + r)^5$ **17.** $(a - b)^5$ **18.** $(p - q)^4$

⊙ **19.** $(2x + 3)^3$ **20.** $(4x + 2)^3$ ⊙ **21.** $\left(\dfrac{x}{2} - y\right)^4$ **22.** $\left(\dfrac{x}{3} - 2y\right)^5$

23. $(x^2 + 1)^4$ **24.** $(y^3 + 2)^4$ **25.** $(3x^2 - y^2)^3$ **26.** $(2p^2 - q^2)^3$

Write the first four terms of each binomial expansion. ***See Examples 4 and 5.***

27. $(r + 2s)^{12}$ **28.** $(m + 3n)^{20}$ **29.** $(3x - y)^{14}$

30. $(2p - 3q)^{11}$ **31.** $(t^2 + u^2)^{10}$ **32.** $(x^2 + y^2)^{15}$

Find the indicated term of each binomial expansion. ***See Example 6.***

⊙ **33.** $(2m + n)^{10}$; fourth term   **34.** $(a + 3b)^{12}$; fifth term

35. $\left(x + \dfrac{y}{2}\right)^8$; seventh term   **36.** $\left(a + \dfrac{b}{3}\right)^{15}$; eighth term

37. $(k - 1)^9$; third term   **38.** $(r - 4)^{11}$; fourth term

39. The middle term of $(x^2 + 2y)^6$ **40.** The middle term of $(m^3 + 2y)^8$

41. The term with x^9y^4 in $(3x^3 - 4y^2)^5$ **42.** The term with x^8y^2 in $(2x^2 + 3y)^6$

CHAPTER 12 SUMMARY

KEY TERMS

12.1

infinite sequence
finite sequence
terms of a sequence
general term
series
summation notation

index of summation
arithmetic mean (average)

12.2

arithmetic sequence
(arithmetic progression)
common difference

12.3

geometric sequence
(geometric progression)
common ratio
annuity
ordinary annuity
payment period

future value of an annuity
term of an annuity

12.4

Pascal's triangle
binomial theorem (general
binomial expansion)

NEW SYMBOLS

a_n nth term of a
sequence

$\displaystyle\sum_{i=1}^{n} a_i$ summation notation

S_n sum of first n terms
of a sequence

$\displaystyle\lim_{n\to\infty} a_n$ limit of a_n as n gets
larger and larger

$\displaystyle\sum_{i=1}^{\infty} a_i$ sum of an infinite
number of terms

$n!$ n factorial

$_nC_r$ binomial coefficient
(combinations of n
things taken r at a time)

TEST YOUR WORD POWER

See how well you have learned the vocabulary in this chapter.

1. An **infinite sequence** is
   A. the values of a function
   B. a function whose domain is the
      set of positive integers
   C. the sum of the terms of a function
   D. the average of a group of
      numbers.

2. A **series** is
   A. the sum of the terms of a
      sequence
   B. the product of the terms of a
      sequence
   C. the average of the terms of a
      sequence
   D. the function values of a sequence.

3. An **arithmetic sequence** is a
   sequence in which
   A. each term after the first is a
      constant multiple of the
      preceding term

   B. the numbers are written in a
      triangular array
   C. the terms are added
   D. each term after the first differs
      from the preceding term by a
      common amount.

4. A **geometric sequence** is a sequence
   in which
   A. each term after the first is a
      constant multiple of the
      preceding term
   B. the numbers are written in a
      triangular array
   C. the terms are multiplied
   D. each term after the first differs
      from the preceding term by a
      common amount.

5. The **common difference** is
   A. the average of the terms in a
      sequence

   B. the constant multiplier in a
      geometric sequence
   C. the difference between any two
      adjacent terms in an arithmetic
      sequence
   D. the sum of the terms of an
      arithmetic sequence.

6. The **common ratio** is
   A. the average of the terms in a
      sequence
   B. the constant multiplier in a
      geometric sequence
   C. the difference between any two
      adjacent terms in an arithmetic
      sequence
   D. the product of the terms of a
      geometric sequence.

ANSWERS

1. B; *Example:* The ordered list of numbers 3, 6, 9, 12, 15, . . . is an infinite sequence.

2. A; *Example:* $3 + 6 + 9 + 12 + 15$, written in summation notation as $\displaystyle\sum_{i=1}^{5} 3i$, is a series.

3. D; *Example:* The sequence $-3, 2, 7, 12, 17, \ldots$ is arithmetic.

4. A; *Example:* The sequence 1, 4, 16, 64, 256, . . . is geometric.

5. C; *Example:* The common difference of the arithmetic sequence in Answer 3 is 5, since $2 - (-3) = 5, 7 - 2 = 5$, and so on.

6. B; *Example:* The common ratio of the geometric sequence in Answer 4 is 4, since $\frac{4}{1} = \frac{16}{4} = \frac{64}{16} = \frac{256}{64} = 4$.

QUICK REVIEW

CONCEPTS	EXAMPLES

12.1 Sequences and Series

A finite sequence is a function with domain
$$\{1, 2, 3, \ldots n\},$$
while an infinite sequence has domain
$$\{1, 2, 3, \ldots\}.$$
The nth term of a sequence is symbolized a_n. A series is an indicated sum of the terms of a sequence.

$1, \dfrac{1}{2}, \dfrac{1}{3}, \dfrac{1}{4}, \ldots, \dfrac{1}{n}$ has general term $a_n = \dfrac{1}{n}$.

The corresponding series is the *sum*

$$1 + \frac{1}{2} + \frac{1}{3} + \frac{1}{4} + \cdots + \frac{1}{n} = \sum_{i=1}^{n} \frac{1}{i}.$$

12.2 Arithmetic Sequences

Assume that a_1 is the first term, a_n is the nth term, and d is the common difference.

Common Difference
$$d = a_{n+1} - a_n$$

Consider the arithmetic sequence
$$2, 5, 8, 11, \ldots.$$

$a_1 = 2$ a_1 is the first term.

$d = 5 - 2 = 3$ Use $a_2 - a_1$.

(Any two successive terms could have been used.)

nth Term
$$a_n = a_1 + (n - 1)d$$

The tenth term is
$$a_{10} = 2 + (10 - 1)(3) \quad \text{Let } n = 10.$$
$$= 2 + 9 \cdot 3, \quad \text{or} \quad 29.$$

Sum of the First n Terms
$$S_n = \frac{n}{2}(a_1 + a_n)$$

or $\qquad S_n = \dfrac{n}{2}[2a_1 + (n - 1)d]$

The sum of the first ten terms can be found in either way.

$$S_{10} = \frac{10}{2}(2 + a_{10})$$
$$= 5(2 + 29)$$
$$= 5(31)$$
$$= 155$$

$$S_{10} = \frac{10}{2}[2(2) + (10 - 1)(3)]$$
$$= 5(4 + 9 \cdot 3)$$
$$= 5(4 + 27)$$
$$= 5(31)$$
$$= 155$$

12.3 Geometric Sequences

Assume that a_1 is the first term, a_n is the nth term, and r is the common ratio.

Common Ratio
$$r = \frac{a_{n+1}}{a_n}$$

Consider the geometric sequence
$$1, 2, 4, 8, \ldots.$$

$a_1 = 1$ a_1 is the first term.

$r = \dfrac{8}{4} = 2$ Use $\dfrac{a_4}{a_3}$.

(Any two successive terms could have been used.)

nth Term
$$a_n = a_1 r^{n-1}$$

The sixth term is
$$a_6 = (1)(2)^{6-1} = 1(2)^5 = 32. \quad \text{Let } n = 6.$$

Sum of the First n Terms
$$S_n = \frac{a_1(1 - r^n)}{1 - r} \quad (r \neq 1)$$

The sum of the first six terms is
$$S_6 = \frac{1(1 - 2^6)}{1 - 2} = \frac{1 - 64}{-1} = 63.$$

(continued)

CONCEPTS	EXAMPLES

Future Value of an Ordinary Annuity

$$S = R\left[\frac{(1 + i)^n - 1}{i}\right],$$

where S is the future value, R is the payment at the end of each period, i is the interest rate per period, and n is the number of periods.

If \$5800 is deposited into an ordinary annuity at the end of each quarter for 4 yr and interest is earned at 2.4% compounded quarterly, then

$$R = \$5800, \quad i = \frac{0.024}{4} = 0.006, \quad n = 4(4) = 16,$$

and $\quad S = 5800\left[\frac{(1 + 0.006)^{16} - 1}{0.006}\right] = \$97{,}095.24.$

Sum of the Terms of an Infinite Geometric Sequence with $|r| < 1$

$$S = \frac{a_1}{1 - r}$$

The sum S of the terms of an infinite geometric sequence with $a_1 = 1$ and $r = \frac{1}{2}$ is

$$S = \frac{1}{1 - \frac{1}{2}} = \frac{1}{\frac{1}{2}} = 1 \cdot \frac{2}{1} = 2.$$

12.4 The Binomial Theorem

Factorials

For any positive integer n,

$$n! = n(n - 1)(n - 2) \cdots (2)(1).$$

By definition, $\qquad 0! = 1.$

$$4! = 4 \cdot 3 \cdot 2 \cdot 1 = 24$$

Binomial Coefficient

$$_nC_r = \frac{n!}{r!(n - r)!}, \quad r \le n$$

$$_5C_3 = \frac{5!}{3!(5 - 3)!} = \frac{5!}{3!2!} = \frac{5 \cdot 4 \cdot 3 \cdot 2 \cdot 1}{3 \cdot 2 \cdot 1 \cdot 2 \cdot 1} = 10$$

General Binomial Expansion

For any positive integer n,

$(x + y)^n$

$$= x^n + \frac{n!}{1!(n - 1)!}x^{n-1}y + \frac{n!}{2!(n - 2)!}x^{n-2}y^2$$

$$+ \frac{n!}{3!(n - 3)!}x^{n-3}y^3 + \cdots + \frac{n!}{(n - 1)!1!}xy^{n-1}$$

$$+ y^n.$$

$(2x - 3)^4$

$$= (2x)^4 + \frac{4!}{1!3!}(2x)^3(-3) + \frac{4!}{2!2!}(2x)^2(-3)^2 +$$

$$\frac{4!}{3!1!}(2x)(-3)^3 + (-3)^4$$

$$= 2^4x^4 - 4(2)^3x^3(3) + 6(2)^2x^2(9) - 4(2x)(27) + 81$$

$$= 16x^4 - 12(8)x^3 + 54(4)x^2 - 216x + 81$$

$$= 16x^4 - 96x^3 + 216x^2 - 216x + 81$$

rth Term of the Binomial Expansion of $(x + y)^n$

$$\frac{n!}{(r - 1)![n - (r - 1)]!}x^{n-(r-1)}y^{r-1}$$

The eighth term of $(a - 2b)^{10}$ is

$$\frac{10!}{7!3!}a^3(-2b)^7$$

$$= \frac{10 \cdot 9 \cdot 8}{3 \cdot 2 \cdot 1}a^3(-2)^7b^7 \qquad \begin{array}{l} n = 10, \; x = a, \\ y = -2b, \, r = 8 \end{array}$$

$$= 120(-128)a^3b^7 \qquad \text{Simplify.}$$

$$= -15{,}360a^3b^7. \qquad \text{Multiply.}$$

CHAPTER **12**

REVIEW EXERCISES

12.1 *Write out the first four terms of each sequence.*

1. $a_n = 2n - 3$

2. $a_n = \dfrac{n - 1}{n}$

3. $a_n = n^2$

4. $a_n = \left(\dfrac{1}{2}\right)^n$

5. $a_n = (n + 1)(n - 1)$

6. $a_n = n(-1)^{n-1}$

Write each series as a sum of terms.

7. $\displaystyle\sum_{i=1}^{5} i^2$

8. $\displaystyle\sum_{i=1}^{6} (i + 1)$

9. $\displaystyle\sum_{i=3}^{6} (5i - 4)$

Evaluate each series.

10. $\displaystyle\sum_{i=1}^{4} (i + 2)$

11. $\displaystyle\sum_{i=1}^{6} 2^i$

12. $\displaystyle\sum_{i=4}^{7} \dfrac{i}{i + 1}$

13. Find the arithmetic mean, or average, of the total retirement assets of Americans for the years 2004 through 2008 shown in the table. Round to the nearest unit (in billions).

Year	Assets (in billions of dollars)
2004	13,778
2005	14,862
2006	16,680
2007	17,916
2008	13,985

Source: Investment Company Institute.

12.2–12.3 *Decide whether each sequence is* arithmetic, geometric, *or* neither. *If the sequence is arithmetic, find the common difference d. If it is geometric, find the common ratio r.*

14. $2, 5, 8, 11, \ldots$

15. $-6, -2, 2, 6, 10, \ldots$

16. $\dfrac{2}{3}, -\dfrac{1}{3}, \dfrac{1}{6}, -\dfrac{1}{12}, \ldots$

17. $-1, 1, -1, 1, -1, \ldots$

18. $64, 32, 8, \dfrac{1}{2}, \ldots$

19. $64, 32, 16, 8, \ldots$

20. *Concept Check* Refer to the Chapter Opener on **page 677**. What is the eleventh term of the Fibonacci sequence?

12.2 *Determine the indicated term of each arithmetic sequence.*

21. $a_1 = -2, d = 5; \quad a_{16}$

22. $a_6 = 12, a_8 = 18; \quad a_{25}$

Determine the general term of each arithmetic sequence.

23. $a_1 = -4, d = -5$

24. $6, 3, 0, -3, \ldots$

Determine the number of terms in each arithmetic sequence.

25. $7, 10, 13, \ldots, 49$

26. $5, 1, -3, \ldots, -79$

Evaluate S_8 for each arithmetic sequence.

27. $a_1 = -2, d = 6$

28. $a_n = -2 + 5n$

12.3 *Determine the general term for each geometric sequence.*

29. $-1, -4, -16, \ldots$

30. $\dfrac{2}{3}, \dfrac{2}{15}, \dfrac{2}{75}, \ldots$

Determine the indicated term for each geometric sequence.

31. $2, -6, 18, \ldots;\quad a_{11}$

32. $a_3 = 20, a_5 = 80;\quad a_{10}$

Evaluate each sum if it exists.

33. $\displaystyle\sum_{i=1}^{5} \left(\frac{1}{4}\right)^i$

34. $\displaystyle\sum_{i=1}^{8} \frac{3}{4}(-1)^i$

35. $\displaystyle\sum_{i=1}^{\infty} 4\left(\frac{1}{5}\right)^i$

36. $\displaystyle\sum_{i=1}^{\infty} 2(3)^i$

12.4 *Use the binomial theorem to expand each binomial.*

37. $(2p - q)^5$

38. $(x^2 + 3y)^4$

39. $(3t^3 - s^2)^4$

40. Write the fourth term of the expansion of $(3a + 2b)^{19}$.

MIXED REVIEW EXERCISES

Determine the indicated term and evaluate S_{10} for each sequence.

41. a_{10}: geometric; $-3, 6, -12, \ldots$

42. a_{40}: arithmetic; $1, 7, 13, \ldots$

43. a_{15}: arithmetic; $a_1 = -4,\quad d = 3$

44. a_9: geometric; $a_1 = 1,\quad r = -3$

Determine the general term for each arithmetic or geometric sequence.

45. $2, 8, 32, \ldots$

46. $2, 7, 12, \ldots$

47. $12, 9, 6, \ldots$

48. $27, 9, 3, \ldots$

Solve each problem.

49. When Faith's sled goes down the hill near her home, she covers 3 ft in the first second. Then, for each second after that, she goes 4 ft more than in the preceding second. If the distance she covers going down is 210 ft, how long does it take her to reach the bottom?

50. An ordinary annuity is set up so that $672 is deposited at the end of each quarter for 7 yr. The money earns 4.5% annual interest compounded quarterly. What is the future value of the annuity?

51. The school population in Middleton has been dropping 3% per yr. The current population is 50,000. If this trend continues, what will the population be in 6 yr?

52. A pump removes $\frac{1}{2}$ of the liquid in a container with each stroke. What fraction of the liquid is left in the container after seven strokes?

53. Consider the repeating decimal number $0.55555\ldots$.

(a) Write it as the sum of the terms of an infinite geometric sequence.

(b) What is r for this sequence?

(c) Find this infinite sum if it exists, and write it as a common fraction in lowest terms.

54. Can the sum of the terms of the infinite geometric sequence defined by $a_n = 5(2)^n$ be found? Explain.

CHAPTER
Test Prep
VIDEOS

Step-by-step test solutions are found on the Chapter Test Prep Videos available via the Video Resources on DVD, in *MyMathLab* , or on YouTube (search "LialIntermediateAlg").

View the complete solutions to all Chapter Test exercises on the Video Resources on DVD.

Write the first five terms of each sequence described.

1. $a_n = (-1)^n + 1$

2. arithmetic, with $a_1 = 4$ and $d = 2$

3. geometric, with $a_4 = 6$ and $r = \frac{1}{2}$

Determine a_4 for each sequence described.

4. arithmetic, with $a_1 = 6$ and $d = -2$

5. geometric, with $a_5 = 16$ and $a_7 = 9$

Evaluate S_5 for each sequence described.

6. arithmetic, with $a_2 = 12$ and $a_3 = 15$

7. geometric, with $a_5 = 4$ and $a_7 = 1$

8. The numbers of commercial bank offices (main offices and branches) in the United States for the years 2004 through 2008 are given in the table. What was the average number of banks per year for that period? Round to the nearest unit.

Year	Number
2004	78,473
2005	80,967
2006	83,860
2007	86,150
2008	97,103

Source: U.S. Federal Deposit Insurance Corporation.

9. If $4000 is deposited in an ordinary annuity at the end of each quarter for 7 yr and earns 6% interest compounded quarterly, how much will be in the account at the end of this term?

10. *Concept Check*   Under what conditions does an infinite geometric series have a sum?

Determine each sum that exists.

11. $\displaystyle\sum_{i=1}^{5} (2i + 8)$

12. $\displaystyle\sum_{i=1}^{6} (3i - 5)$

13. $\displaystyle\sum_{i=1}^{500} i$

14. $\displaystyle\sum_{i=1}^{3} \frac{1}{2}(4^i)$

15. $\displaystyle\sum_{i=1}^{\infty} \left(\frac{1}{4}\right)^i$

16. $\displaystyle\sum_{i=1}^{\infty} 6\left(\frac{3}{2}\right)^i$

Evaluate.

17. $8!$

18. $0!$

19. $\dfrac{6!}{4!2!}$

20. $_{12}C_{10}$

21. Expand $(3k - 5)^4$.

22. Write the fifth term of the expansion of $\left(2x - \dfrac{y}{3}\right)^{12}$.

Solve each problem.

23. Christian Sabau bought a new dishwasher for $300. He agreed to pay $20 per month for 15 months, plus interest of 1% each month, on the unpaid balance. Find the total cost of the machine.

24. During the summer months, the population of a certain insect colony triples each week. If there are 20 insects in the colony at the end of the first week in July, how many are present by the end of September? (Assume exactly four weeks in a month.)

Simplify each expression.

1. $|-7| + 6 - |-10| - (-8 + 3)$ **2.** $4(-6) + (-8)(5) - (-9)$

Let $P = \left\{-\frac{8}{3}, 10, 0, \sqrt{13}, -\sqrt{3}, \frac{45}{15}, \sqrt{-7}, 0.82, -3\right\}$. List the elements of P that are members of each set.

3. Rational numbers   **4.** Irrational numbers

Solve each equation or inequality.

5. $9 - (5 + 3x) + 5x = -4(x - 3) - 7$ **6.** $7x + 18 \le 9x - 2$

7. $|4x - 3| = 21$ **8.** $\dfrac{x + 3}{12} - \dfrac{x - 3}{6} = 0$

9. $2x > 8$ or $-3x > 9$ **10.** $|2x - 5| \ge 11$

11. Find the slope of the line through $(4, -5)$ and $(-12, -17)$.

12. Find the standard form of the equation of the line through $(-2, 10)$ and parallel to the line with equation $3x + y = 7$.

Graph.

13. $x - 3y = 6$ **14.** $4x - y < 4$

15. Consider the set of ordered pairs.

$$\{(-3, 2), (-2, 6), (0, 4), (1, 2), (2, 6)\}$$

    (a) Is this a function?   **(b)** What is its domain?   **(c)** What is its range?

Solve each system of equations.

16. $y = 5x + 3$
$2x + 3y = -8$

17. $x + 2y + z = 8$
$2x - y + 3z = 15$
$-x + 3y - 3z = -11$

18. Nuts worth \$3 per lb are to be mixed with 8 lb of nuts worth \$4.25 per lb to obtain a mixture that will be sold for \$4 per lb. How many pounds of the \$3 nuts should be used?

Perform the indicated operations.

19. $(4p + 2)(5p - 3)$ **20.** $(3k - 7)^2$

21. $(2m^3 - 3m^2 + 8m) - (7m^3 + 5m - 8)$

22. Divide $6t^4 + 5t^3 - 18t^2 + 14t - 1$ by $3t - 2$.

Factor.

23. $6z^3 + 5z^2 - 4z$ **24.** $49a^4 - 9b^2$ **25.** $c^3 + 27d^3$

Solve each equation or inequality.

26. $2x^2 + x = 10$ **27.** $x^2 - x - 6 \le 0$

Simplify.

28. $\left(\dfrac{2}{3}\right)^{-2}$ **29.** $\dfrac{(3p^2)^3(-2p^6)}{4p^3(5p^7)}$

Simplify.

30. $\dfrac{x^2 - 16}{x^2 + 2x - 8} \div \dfrac{x - 4}{x + 7}$

31. $\dfrac{5}{p^2 + 3p} - \dfrac{2}{p^2 - 4p}$

Solve.

32. $\dfrac{4}{x - 3} - \dfrac{6}{x + 3} = \dfrac{24}{x^2 - 9}$

33. $6x^2 + 5x = 8$

34. Simplify $5\sqrt{72} - 4\sqrt{50}$.

35. Multiply $(8 + 3i)(8 - 3i)$.

36. Find $f^{-1}(x)$ if $f(x) = 9x + 5$.

37. Graph $g(x) = \left(\dfrac{1}{3}\right)^x$.

38. Solve $3^{2x-1} = 81$.

39. Graph $y = \log_{1/3} x$.

40. Solve $\log_8 x + \log_8 (x + 2) = 1$.

Graph.

41. $f(x) = 2(x - 2)^2 - 3$ **42.** $\dfrac{x^2}{9} + \dfrac{y^2}{25} = 1$ **43.** $x^2 - y^2 = 9$

44. Solve the system $\begin{aligned} xy &= -5 \\ 2x + y &= 3. \end{aligned}$

45. Find the equation of a circle with center at $(-5, 12)$ and radius 9.

46. Write the first five terms of the sequence defined by $a_n = 5n - 12$.

47. Find each sum.

    (a) The sum of the first six terms of the arithmetic sequence with $a_1 = 8$ and $d = 2$

    (b) The sum of the geometric series $15 - 6 + \frac{12}{5} - \frac{24}{25} + \cdots$

48. Find the sum $\displaystyle\sum_{i=1}^{4} 3i$.

49. Use the binomial theorem to expand $(2a - 1)^5$.

50. What is the fourth term in the expansion of $\left(3x^4 - \frac{1}{2}y^2\right)^5$?

Determinants and Cramer's Rule

Recall from **Section 4.4** that an ordered array of numbers within square brackets is called a *matrix* (plural *matrices*). Matrices are named according to the number of rows and columns they contain. A *square matrix* has the same number of rows and columns.

Columns

Rows $\begin{bmatrix} 2 & 3 & 5 \\ 7 & 1 & 2 \end{bmatrix}$ 2 × 3 Matrix $\begin{bmatrix} -1 & 0 \\ 1 & -2 \end{bmatrix}$ 2 × 2 square matrix

Associated with every *square matrix* is a real number called the **determinant** of the matrix. A determinant is symbolized by the entries of the matrix placed between two vertical lines.

$\begin{vmatrix} 2 & 3 \\ 7 & 1 \end{vmatrix}$ 2 × 2 determinant $\begin{vmatrix} 7 & 4 & 3 \\ 0 & 1 & 5 \\ 6 & 0 & 1 \end{vmatrix}$ 3 × 3 determinant

Like matrices, determinants are named according to the number of rows and columns they contain.

OBJECTIVE 1 Evaluate 2 × 2 determinants. As mentioned above, the value of a determinant is a *real number*. We use the following rule to evaluate a 2 × 2 determinant.

Value of a 2 × 2 Determinant

$$\begin{vmatrix} a & b \\ c & d \end{vmatrix} = ad - bc$$

**NOW TRY
EXERCISE 1**

Evaluate the determinant.

$$\begin{vmatrix} 6 & 3 \\ 4 & -5 \end{vmatrix}$$

EXAMPLE 1 Evaluating a 2 × 2 Determinant

Evaluate the determinant.

$$\begin{vmatrix} -1 & -3 \\ 4 & -2 \end{vmatrix}$$

Here $a = -1$, $b = -3$, $c = 4$, and $d = -2$, so

$$\begin{vmatrix} -1 & -3 \\ 4 & -2 \end{vmatrix} = -1(-2) - (-3)4 = 2 + 12 = 14. \qquad \text{NOW TRY}$$

NOW TRY ANSWER
1. -42

A 3 × 3 determinant is evaluated in a similar way.

Value of a 3 × 3 Determinant

$$\begin{vmatrix} a_1 & b_1 & c_1 \\ a_2 & b_2 & c_2 \\ a_3 & b_3 & c_3 \end{vmatrix} = (a_1 b_2 c_3 + b_1 c_2 a_3 + c_1 a_2 b_3) \\ - (a_3 b_2 c_1 + b_3 c_2 a_1 + c_3 a_2 b_1)$$

To calculate a 3 × 3 determinant, we rearrange terms using the distributive property.

$$\begin{vmatrix} a_1 & b_1 & c_1 \\ a_2 & b_2 & c_2 \\ a_3 & b_3 & c_3 \end{vmatrix} = a_1(b_2 c_3 - b_3 c_2) - a_2(b_1 c_3 - b_3 c_1) + a_3(b_1 c_2 - b_2 c_1) \qquad (1)$$

Each quantity in parentheses represents a 2 × 2 determinant that is the part of the 3 × 3 determinant remaining when the row and column of the multiplier are eliminated, as shown below.

$$a_1(b_2 c_3 - b_3 c_2) \qquad \begin{vmatrix} a_1 & b_1 & c_1 \\ a_2 & b_2 & c_2 \\ a_3 & b_3 & c_3 \end{vmatrix} \qquad \text{Eliminate the 1st row and 1st column.}$$

$$a_2(b_1 c_3 - b_3 c_1) \qquad \begin{vmatrix} a_1 & b_1 & c_1 \\ a_2 & b_2 & c_2 \\ a_3 & b_3 & c_3 \end{vmatrix} \qquad \text{Eliminate the 2nd row and 1st column.}$$

$$a_3(b_1 c_2 - b_2 c_1) \qquad \begin{vmatrix} a_1 & b_1 & c_1 \\ a_2 & b_2 & c_2 \\ a_3 & b_3 & c_3 \end{vmatrix} \qquad \text{Eliminate the 3rd row and 1st column.}$$

These 2 × 2 determinants are called **minors** of the elements in the 3 × 3 determinant. In the determinant above, the minors of a_1, a_2, and a_3 are, respectively,

$$\begin{vmatrix} b_2 & c_2 \\ b_3 & c_3 \end{vmatrix}, \qquad \begin{vmatrix} b_1 & c_1 \\ b_3 & c_3 \end{vmatrix}, \qquad \text{and} \qquad \begin{vmatrix} b_1 & c_1 \\ b_2 & c_2 \end{vmatrix}. \qquad \text{Minors}$$

OBJECTIVE 2 Use expansion by minors to evaluate 3 × 3 determinants.
We evaluate a 3 × 3 determinant by multiplying each element in the first column by its minor and combining the products as indicated in equation (1). This procedure is called **expansion of the determinant by minors** about the first column.

EXAMPLE 2 Evaluating a 3 × 3 Determinant
Evaluate the determinant using expansion by minors about the first column.

$$\begin{vmatrix} 1 & 3 & -2 \\ -1 & -2 & -3 \\ 1 & 1 & 2 \end{vmatrix}$$

NOW TRY
EXERCISE 2

Evaluate the determinant by expansion by minors about the first column.

$$\begin{vmatrix} 0 & -2 & 3 \\ 4 & 1 & -5 \\ 6 & -1 & 5 \end{vmatrix}$$

In this determinant, $a_1 = 1$, $a_2 = -1$, and $a_3 = 1$. Multiply each of these numbers by its minor, and combine the three terms using the definition. Notice that the second term in the definition is *subtracted*.

$$\begin{vmatrix} 1 & 3 & -2 \\ -1 & -2 & -3 \\ 1 & 1 & 2 \end{vmatrix} = 1\begin{vmatrix} -2 & -3 \\ 1 & 2 \end{vmatrix} - (-1)\begin{vmatrix} 3 & -2 \\ 1 & 2 \end{vmatrix} + 1\begin{vmatrix} 3 & -2 \\ -2 & -3 \end{vmatrix}$$

Use parentheses and brackets to avoid errors.

$$= 1[-2(2) - (-3)1] + 1[3(2) - (-2)1]$$
$$\quad + 1[3(-3) - (-2)(-2)]$$
$$= 1(-1) + 1(8) + 1(-13)$$
$$= -1 + 8 - 13$$
$$= -6 \qquad \text{NOW TRY}$$

Array of Signs for a 3 × 3 Determinant

$$\begin{array}{ccc} + & - & + \\ - & + & - \\ + & - & + \end{array}$$

To obtain equation (1), we could have rearranged terms in the definition of the determinant and used the distributive property to factor out the three elements of the second or third column or of any of the three rows. **Expanding by minors about any row or any column results in the same value for a 3 × 3 determinant.**

To determine the correct signs for the terms of other expansions, the **array of signs** in the margin is helpful. The signs alternate for each row and column beginning with a $+$ in the first row, first column position. For example, if the expansion is to be about the second column, the first term would have a minus sign associated with it, the second term a plus sign, and the third term a minus sign.

NOW TRY
EXERCISE 3

Evaluate the determinant by expansion by minors about the second column.

$$\begin{vmatrix} 0 & -2 & 3 \\ 4 & 1 & -5 \\ 6 & -1 & 5 \end{vmatrix}$$

EXAMPLE 3 Evaluating a 3 × 3 Determinant

Evaluate the determinant of **Example 2** using expansion by minors about the second column.

$$\begin{vmatrix} 1 & 3 & -2 \\ -1 & -2 & -3 \\ 1 & 1 & 2 \end{vmatrix} = -3\begin{vmatrix} -1 & -3 \\ 1 & 2 \end{vmatrix} + (-2)\begin{vmatrix} 1 & -2 \\ 1 & 2 \end{vmatrix} - 1\begin{vmatrix} 1 & -2 \\ -1 & -3 \end{vmatrix}$$

$$= -3(1) - 2(4) - 1(-5)$$
$$= -3 - 8 + 5$$
$$= -6 \qquad \text{The result is the same as in Example 2.}$$

NOW TRY

OBJECTIVE 3 **Understand the derivation of Cramer's rule.** We can use determinants to solve a system of equations of the form

$$a_1x + b_1y = c_1 \qquad (1)$$
$$a_2x + b_2y = c_2. \qquad (2)$$

The result will be a formula that can be used to solve any system of two equations with two variables.

$$a_1b_2x + b_1b_2y = c_1b_2 \qquad \text{Multiply equation (1) by } b_2.$$
$$\underline{-a_2b_1x - b_1b_2y = -c_2b_1} \qquad \text{Multiply equation (2) by } -b_1.$$
$$(a_1b_2 - a_2b_1)x = c_1b_2 - c_2b_1 \qquad \text{Add.}$$

Solve for x.
$$x = \frac{c_1b_2 - c_2b_1}{a_1b_2 - a_2b_1} \qquad (\text{if } a_1b_2 - a_2b_1 \neq 0)$$

NOW TRY ANSWERS
2. 70   3. 70

To solve for y, we multiply each side of equation (1) by $-a_2$ and each side of equation (2) by a_1 and add.

$$-a_1a_2x - a_2b_1y = -a_2c_1 \qquad \text{Multiply equation (1) by } -a_2.$$
$$\underline{a_1a_2x + a_1b_2y = a_1c_2} \qquad \text{Multiply equation (2) by } a_1.$$
$$(a_1b_2 - a_2b_1)y = a_1c_2 - a_2c_1 \qquad \text{Add.}$$
$$y = \frac{a_1c_2 - a_2c_1}{a_1b_2 - a_2b_1} \qquad (\text{if } a_1b_2 - a_2b_1 \neq 0)$$

We can write both numerators and the common denominator of these values for x and y as determinants because

$$a_1c_2 - a_2c_1 = \begin{vmatrix} a_1 & c_1 \\ a_2 & c_2 \end{vmatrix}, \quad c_1b_2 - c_2b_1 = \begin{vmatrix} c_1 & b_1 \\ c_2 & b_2 \end{vmatrix}, \quad \text{and} \quad a_1b_2 - a_2b_1 = \begin{vmatrix} a_1 & b_1 \\ a_2 & b_2 \end{vmatrix}.$$

Using these results, the solutions for x and y become

$$x = \frac{\begin{vmatrix} c_1 & b_1 \\ c_2 & b_2 \end{vmatrix}}{\begin{vmatrix} a_1 & b_1 \\ a_2 & b_2 \end{vmatrix}} \quad \text{and} \quad y = \frac{\begin{vmatrix} a_1 & c_1 \\ a_2 & c_2 \end{vmatrix}}{\begin{vmatrix} a_1 & b_1 \\ a_2 & b_2 \end{vmatrix}}, \quad \text{where} \quad \begin{vmatrix} a_1 & b_1 \\ a_2 & b_2 \end{vmatrix} \neq 0.$$

For convenience, we denote the three determinants in the solution as

$$\begin{vmatrix} a_1 & b_1 \\ a_2 & b_2 \end{vmatrix} = D, \quad \begin{vmatrix} c_1 & b_1 \\ c_2 & b_2 \end{vmatrix} = D_x, \quad \text{and} \quad \begin{vmatrix} a_1 & c_1 \\ a_2 & c_2 \end{vmatrix} = D_y.$$

Notice that the elements of D are the four coefficients of the variables in the given system. The elements of D_x are obtained by replacing the coefficients of x by the respective constants. Similarly, the elements of D_y are obtained by replacing the coefficients of y by the respective constants. These results are summarized as **Cramer's rule.**

Cramer's Rule for 2 × 2 Systems

For the system $\begin{aligned} a_1x + b_1y &= c_1 \\ a_2x + b_2y &= c_2 \end{aligned}$ with $a_1b_2 - a_2b_1 = D \neq 0$, the values of x and y are given by

$$x = \frac{\begin{vmatrix} c_1 & b_1 \\ c_2 & b_2 \end{vmatrix}}{\begin{vmatrix} a_1 & b_1 \\ a_2 & b_2 \end{vmatrix}} = \frac{D_x}{D} \quad \text{and} \quad y = \frac{\begin{vmatrix} a_1 & c_1 \\ a_2 & c_2 \end{vmatrix}}{\begin{vmatrix} a_1 & b_1 \\ a_2 & b_2 \end{vmatrix}} = \frac{D_y}{D}.$$

OBJECTIVE 4 **Apply Cramer's rule to solve linear systems.** To use Cramer's rule to solve a system of equations, we find the three determinants, D, D_x, and D_y, and then write the necessary quotients for x and y.

⚠ **CAUTION** As indicated in the box, *Cramer's rule does not apply if* $D = a_1b_2 - a_2b_1 = 0$. When $D = 0$, the system is inconsistent or has dependent equations. For this reason, it is a good idea to evaluate D first.

NOW TRY
EXERCISE 4
Use Cramer's rule to solve the system.
$$3x - 2y = -33$$
$$2x + 3y = -9$$

EXAMPLE 4   Using Cramer's Rule to Solve a 2 × 2 System

Use Cramer's rule to solve the system.
$$5x + 7y = -1$$
$$6x + 8y = 1$$

By Cramer's rule, $x = \dfrac{D_x}{D}$ and $y = \dfrac{D_y}{D}$. If $D \neq 0$, then we find D_x and D_y.

$$D = \begin{vmatrix} 5 & 7 \\ 6 & 8 \end{vmatrix} = 5(8) - 7(6) = -2$$

$$D_x = \begin{vmatrix} -1 & 7 \\ 1 & 8 \end{vmatrix} = -1(8) - 7(1) = -15$$

$$D_y = \begin{vmatrix} 5 & -1 \\ 6 & 1 \end{vmatrix} = 5(1) - (-1)6 = 11$$

From Cramer's rule,

$$x = \frac{D_x}{D} = \frac{-15}{-2} = \frac{15}{2} \quad \text{and} \quad y = \frac{D_y}{D} = \frac{11}{-2} = -\frac{11}{2}.$$

The solution set is $\left\{ \left(\frac{15}{2}, -\frac{11}{2} \right) \right\}$, as can be verified by checking in the given system.

NOW TRY

Cramer's Rule for 3 × 3 Systems

For the system

$$a_1 x + b_1 y + c_1 z = d_1$$
$$a_2 x + b_2 y + c_2 z = d_2$$
$$a_3 x + b_3 y + c_3 z = d_3$$

with

$$D_x = \begin{vmatrix} d_1 & b_1 & c_1 \\ d_2 & b_2 & c_2 \\ d_3 & b_3 & c_3 \end{vmatrix}, \qquad D_y = \begin{vmatrix} a_1 & d_1 & c_1 \\ a_2 & d_2 & c_2 \\ a_3 & d_3 & c_3 \end{vmatrix},$$

$$D_z = \begin{vmatrix} a_1 & b_1 & d_1 \\ a_2 & b_2 & d_2 \\ a_3 & b_3 & d_3 \end{vmatrix}, \qquad D = \begin{vmatrix} a_1 & b_1 & c_1 \\ a_2 & b_2 & c_2 \\ a_3 & b_3 & c_3 \end{vmatrix} \neq 0,$$

the values of x, y, and z are given by

$$x = \frac{D_x}{D}, \qquad y = \frac{D_y}{D}, \qquad \text{and} \qquad z = \frac{D_z}{D}.$$

EXAMPLE 5   Using Cramer's Rule to Solve a 3 × 3 System

Use Cramer's rule to solve the system.

$$x + y - z = -2$$
$$2x - y + z = -5$$
$$x - 2y + 3z = 4$$

NOW TRY ANSWER
4. $\{(-9, 3)\}$

NOW TRY
EXERCISE 5
Use Cramer's rule to solve the system.

$$4x + 2y + z = 15$$
$$-2x + 5y - 2z = -6$$
$$x - 3y + 4z = 0$$

We expand by minors about row 1 to find D.

$$D = \begin{vmatrix} 1 & 1 & -1 \\ 2 & -1 & 1 \\ 1 & -2 & 3 \end{vmatrix}$$

Given system:
$$x + y - z = -2$$
$$2x - y + z = -5$$
$$x - 2y + 3z = 4$$

$$= 1 \begin{vmatrix} -1 & 1 \\ -2 & 3 \end{vmatrix} - 1 \begin{vmatrix} 2 & 1 \\ 1 & 3 \end{vmatrix} + (-1) \begin{vmatrix} 2 & -1 \\ 1 & -2 \end{vmatrix}$$

$$= 1(-1) - 1(5) - 1(-3)$$

$$= -3$$

Expand D_x by minors about row 1.

$$D_x = \begin{vmatrix} -2 & 1 & -1 \\ -5 & -1 & 1 \\ 4 & -2 & 3 \end{vmatrix}$$

$$= -2 \begin{vmatrix} -1 & 1 \\ -2 & 3 \end{vmatrix} - 1 \begin{vmatrix} -5 & 1 \\ 4 & 3 \end{vmatrix} + (-1) \begin{vmatrix} -5 & -1 \\ 4 & -2 \end{vmatrix}$$

$$= -2(-1) - 1(-19) - 1(14)$$

$$= 7$$

Verify that $D_y = -22$ and $D_z = -21$. Thus,

$$x = \frac{D_x}{D} = \frac{7}{-3} = -\frac{7}{3}, \qquad y = \frac{D_y}{D} = \frac{-22}{-3} = \frac{22}{3}, \qquad z = \frac{D_z}{D} = \frac{-21}{-3} = 7.$$

Check that the solution set is $\left\{ \left(-\frac{7}{3}, \frac{22}{3}, 7 \right) \right\}$.   NOW TRY

NOW TRY
EXERCISE 6
Use Cramer's rule, if possible, to solve the system.

$$5x + 3y + z = 1$$
$$x - 2y + 3z = 6$$
$$10x + 6y + 2z = 3$$

> **EXAMPLE 6**   Determining When Cramer's Rule Does Not Apply

Use Cramer's rule, if possible, to solve the system.

$$2x - 3y + 4z = 8$$
$$6x - 9y + 12z = 24$$
$$x + 2y - 3z = 5$$

First, find D.

$$D = \begin{vmatrix} 2 & -3 & 4 \\ 6 & -9 & 12 \\ 1 & 2 & -3 \end{vmatrix}$$

$$= 2 \begin{vmatrix} -9 & 12 \\ 2 & -3 \end{vmatrix} - 6 \begin{vmatrix} -3 & 4 \\ 2 & -3 \end{vmatrix} + 1 \begin{vmatrix} -3 & 4 \\ -9 & 12 \end{vmatrix}$$

$$= 2(3) - 6(1) + 1(0)$$

$$= 0$$

NOW TRY ANSWERS

5. $\{(4, 0, -1)\}$

6. Cramer's rule does not apply since $D = 0$.

Since $D = 0$ here, Cramer's rule does not apply and we must use another method to solve the system. Multiplying each side of the first equation by 3 shows that the first two equations have the same solution set, so this system has dependent equations and an infinite solution set.   NOW TRY

EXERCISES   *MyMathLab*

1. *Concept Check*   Decide whether each statement is *true* or *false*. If a statement is false, explain why.

   (a) A matrix is an array of numbers, while a determinant is a single number.

   (b) A square matrix has the same number of rows as columns.

   (c) The determinant $\begin{vmatrix} a & b \\ c & d \end{vmatrix}$ is equal to $ad + bc$.

   (d) The value of $\begin{vmatrix} 0 & 0 \\ x & y \end{vmatrix}$ is 0 for any replacements for x and y.

2. *Concept Check*   Which one of the following is the expression for the determinant

$$\begin{vmatrix} -2 & -3 \\ 4 & -6 \end{vmatrix}?$$

   A. $-2(-6) + (-3)4$ **B.** $-2(-6) - 3(4)$

   C. $-3(4) - (-2)(-6)$ **D.** $-2(-6) - (-3)4$

Evaluate each determinant. **See Example 1.**

3. $\begin{vmatrix} -2 & 5 \\ -1 & 4 \end{vmatrix}$ 4. $\begin{vmatrix} 3 & -6 \\ 2 & -2 \end{vmatrix}$ 5. $\begin{vmatrix} 1 & -2 \\ 7 & 0 \end{vmatrix}$

6. $\begin{vmatrix} -5 & -1 \\ 1 & 0 \end{vmatrix}$ 7. $\begin{vmatrix} 0 & 4 \\ 0 & 4 \end{vmatrix}$ 8. $\begin{vmatrix} 8 & -3 \\ 0 & 0 \end{vmatrix}$

Evaluate each determinant by expansion by minors about the first column. **See Example 2.**

9. $\begin{vmatrix} -1 & 2 & 4 \\ -3 & -2 & -3 \\ 2 & -1 & 5 \end{vmatrix}$ 10. $\begin{vmatrix} 2 & -3 & -5 \\ 1 & 2 & 2 \\ 5 & 3 & -1 \end{vmatrix}$

11. $\begin{vmatrix} 1 & 0 & -2 \\ 0 & 2 & 3 \\ 1 & 0 & 5 \end{vmatrix}$ 12. $\begin{vmatrix} 2 & -1 & 0 \\ 0 & -1 & 1 \\ 1 & 2 & 0 \end{vmatrix}$

Evaluate each determinant by expansion by minors about any row or column. (Hint: The work is easier if you choose a row or a column with 0s.) **See Example 3.**

13. $\begin{vmatrix} 3 & -1 & 2 \\ 1 & 5 & -2 \\ 0 & 2 & 0 \end{vmatrix}$ 14. $\begin{vmatrix} 4 & 4 & 2 \\ 1 & -1 & -2 \\ 1 & 0 & 2 \end{vmatrix}$ 15. $\begin{vmatrix} 0 & 0 & 3 \\ 4 & 0 & -2 \\ 2 & -1 & 3 \end{vmatrix}$

16. $\begin{vmatrix} 3 & 5 & -2 \\ 1 & -4 & 1 \\ 3 & 1 & -2 \end{vmatrix}$ 17. $\begin{vmatrix} 1 & 1 & 2 \\ 5 & 5 & 7 \\ 3 & 3 & 1 \end{vmatrix}$ 18. $\begin{vmatrix} 3 & 0 & -2 \\ 1 & -4 & 1 \\ 3 & 1 & -2 \end{vmatrix}$

19. *Concept Check*   For the system

$$x + 3y - 6z = 7$$
$$2x - y + z = 1$$
$$x + 2y + 2z = -1,$$

$D = -43, D_x = -43, D_y = 0,$ and $D_z = 43.$ What is the solution set of the system?

20. *Concept Check* Consider this system.

$$4x + 3y - 2z = 1$$
$$7x - 4y + 3z = 2$$
$$-2x + y - 8z = 0$$

Match each determinant in parts (a)–(d) with its correct representation from choices A–D.

(a) D **(b)** D_x **(c)** D_y **(d)** D_z

A. $\begin{vmatrix} 1 & 3 & -2 \\ 2 & -4 & 3 \\ 0 & 1 & -8 \end{vmatrix}$ **B.** $\begin{vmatrix} 4 & 3 & 1 \\ 7 & -4 & 2 \\ -2 & 1 & 0 \end{vmatrix}$

C. $\begin{vmatrix} 4 & 1 & -2 \\ 7 & 2 & 3 \\ -2 & 0 & -8 \end{vmatrix}$ **D.** $\begin{vmatrix} 4 & 3 & -2 \\ 7 & -4 & 3 \\ -2 & 1 & -8 \end{vmatrix}$

Use Cramer's rule to solve each linear system. ***See Example 4.***

21. $5x + 2y = -3$
   $4x - 3y = -30$

22. $3x + 5y = -5$
   $-2x + 3y = 16$

23. $3x - y = 9$
   $2x + 5y = 8$

24. $8x + 3y = 1$
   $6x - 5y = 2$

25. $4x + 5y = 6$
   $7x + 8y = 9$

26. $2x + 3y = 4$
   $5x + 6y = 7$

Use Cramer's rule where applicable to solve each linear system. ***See Examples 5 and 6.***

27. $x - y + 6z = 19$
   $3x + 3y - z = 1$
   $x + 9y + 2z = -19$

28. $2x + 3y + 2z = 15$
   $x - y + 2z = 5$
   $x + 2y - 6z = -26$

29. $7x + y - z = 4$
   $2x - 3y + z = 2$
   $-6x + 9y - 3z = -6$

30. $2x - 3y + 4z = 8$
   $6x - 9y + 12z = 24$
   $-4x + 6y - 8z = -16$

31. $-x + 2y = 4$
   $3x + y = -5$
   $2x + z = -1$

32. $3x + 5z = 0$
   $2x + 3y = 1$
   $-y + 2z = -11$

33. $-5x - y = -10$
   $3x + 2y + z = -3$
   $-y - 2z = -13$

34. $x - 3y = 13$
   $2y + z = 5$
   $-x + z = -7$

Solve each equation by finding an expression for the determinant on the left, and then solving using the methods of ***Chapter 2.***

35. $\begin{vmatrix} 4 & x \\ 2 & 3 \end{vmatrix} = 8$ **36.** $\begin{vmatrix} -2 & 10 \\ x & 6 \end{vmatrix} = 0$ **37.** $\begin{vmatrix} x & 4 \\ x & -3 \end{vmatrix} = 0$ **38.** $\begin{vmatrix} 5 & 3 \\ x & x \end{vmatrix} = 20$

Synthetic Division

OBJECTIVES

1 Use synthetic division to divide by a polynomial of the form $x - k$.

2 Use the remainder theorem to evaluate a polynomial.

3 Decide whether a given number is a solution of an equation.

OBJECTIVE 1 **Use synthetic division to divide by a polynomial of the form $x - k$.** If a polynomial in x is divided by a binomial of the form $x - k$, a shortcut method can be used. For an illustration, look at the division on the left below.

$$\begin{array}{r} 3x^2 + 9x + 25 \\ x - 3\overline{)3x^3 + 0x^2 - 2x + 5} \\ \underline{3x^3 - 9x^2} \\ 9x^2 - 2x \\ \underline{9x^2 - 27x} \\ 25x + 5 \\ \underline{25x - 75} \\ 80 \end{array} \qquad \begin{array}{r} 3 \quad 9 \quad 25 \\ 1 - 3\overline{)3 \quad 0 \quad -2 \quad 5} \\ 3 \quad -9 \\ 9 \quad -2 \\ 9 \quad -27 \\ 25 \quad 5 \\ 25 \quad -75 \\ 80 \end{array}$$

On the right above, exactly the same division is shown written without the variables. This is why it is *essential* to use 0 as a placeholder in synthetic division. All the numbers in color on the right are repetitions of the numbers directly above them, so we omit them, as shown on the left below.

$$\begin{array}{r} 3 \quad 9 \quad 25 \\ 1 - 3\overline{)3 \quad 0 \quad -2 \quad 5} \\ -9 \\ 9 \quad -2 \\ -27 \\ 25 \quad 5 \\ -75 \\ 80 \end{array} \qquad \begin{array}{r} 3 \quad 9 \quad 25 \\ 1 - 3\overline{)3 \quad 0 \quad -2 \quad 5} \\ -9 \\ 9 \\ -27 \\ 25 \\ -75 \\ 80 \end{array}$$

The numbers in color on the left are again repetitions of the numbers directly above them. They too are omitted, as shown on the right above. If we bring the 3 in the dividend down to the beginning of the bottom row, the top row can be omitted, since it duplicates the bottom row.

$$\begin{array}{r} 1 - 3\overline{)3 \quad 0 \quad -2 \quad 5} \\ -9 \quad -27 \quad -75 \\ \hline 3 \quad 9 \quad 25 \quad 80 \end{array}$$

We omit the 1 at the upper left, since it represents $1x$, which will always be the first term in the divisor. Also, to simplify the arithmetic, we replace subtraction in the second row by addition. To compensate for this, we change the -3 at the upper left to its additive inverse, 3.

$$\text{Additive inverse} \rightarrow 3\overline{)3 \quad 0 \quad -2 \quad 5}$$
$$\phantom{3\overline{)}} 9 \quad 27 \quad 75 \leftarrow \text{Signs changed}$$
$$\phantom{3\overline{)}} 3 \quad 9 \quad 25 \quad 80 \leftarrow \text{Remainder}$$

The quotient is read from the bottom row.

$$3x^2 + 9x + 25 + \frac{80}{x - 3}$$

The first three numbers in the bottom row are the coefficients of the quotient polynomial with degree 1 less than the degree of the dividend. The last number gives the remainder.

This shortcut procedure is called **synthetic division**. *It is used only when dividing a polynomial by a binomial of the form x − k.*

NOW TRY
EXERCISE 1

Use synthetic division to divide.

$$\frac{4x^3 + 18x^2 + 19x + 7}{x + 3}$$

EXAMPLE 1 Using Synthetic Division

Use synthetic division to divide $5x^2 + 16x + 15$ by $x + 2$.

We change $x + 2$ into the form $x - k$ by writing it as

$$x + 2 = x - (-2), \quad \text{where } k = -2.$$

Now write the coefficients of $5x^2 + 16x + 15$, placing -2 to the left.

$$x + 2 \text{ leads to } -2. \longrightarrow -2\overline{)5 \quad 16 \quad 15} \leftarrow \text{Coefficients}$$

$$-2\overline{)5 \quad 16 \quad 15}$$
$$ -10$$
$$ 5$$

Bring down the 5, and multiply: $-2 \cdot 5 = -10$.

$$-2\overline{)5 \quad 16 \quad 15}$$
$$ -10 \quad -12$$
$$ 5 \quad 6$$

Add 16 and -10, getting 6, and multiply 6 and -2 to get -12.

$$-2\overline{)5 \quad 16 \quad 15}$$
$$ -10 \quad -12$$
$$ 5 \quad 6 \quad 3 \leftarrow \text{Remainder}$$

Add 15 and -12, getting 3.

The result is read from the bottom row.

$$\frac{5x^2 + 16x + 15}{x + 2} = 5x + 6 + \frac{3}{x + 2}$$

NOW TRY

NOW TRY
EXERCISE 2

Use synthetic division to divide.

$$\frac{-3x^4 + 13x^3 - 6x^2 + 31}{x - 4}$$

EXAMPLE 2 Using Synthetic Division with a Missing Term

Use synthetic division to find $(-4x^5 + x^4 + 6x^3 + 2x^2 + 50) \div (x - 2)$.

$$2\overline{)-4 \quad 1 \quad 6 \quad 2 \quad 0 \quad 50}$$
$$ -8 \quad -14 \quad -16 \quad -28 \quad -56$$
$$ -4 \quad -7 \quad -8 \quad -14 \quad -28 \quad -6$$

Use the steps given above, first inserting a 0 for the missing x-term.

Read the result from the bottom row.

$$\frac{-4x^5 + x^4 + 6x^3 + 2x^2 + 50}{x - 2} = -4x^4 - 7x^3 - 8x^2 - 14x - 28 + \frac{-6}{x - 2}$$

NOW TRY

NOW TRY ANSWERS

1. $4x^2 + 6x + 1 + \frac{4}{x + 3}$

2. $-3x^3 + x^2 - 2x - 8 + \frac{-1}{x - 4}$

OBJECTIVE 2 **Use the remainder theorem to evaluate a polynomial.** We can use synthetic division to evaluate polynomials. For example, in the synthetic division of **Example 2,** where the polynomial was divided by $x - 2$, the remainder was -6.

Replacing x in the polynomial with 2 gives

$$-4x^5 + x^4 + 6x^3 + 2x^2 + 50$$
$$= -4 \cdot 2^5 + 2^4 + 6 \cdot 2^3 + 2 \cdot 2^2 + 50 \qquad \text{Replace } x \text{ with 2.}$$
$$= -4 \cdot 32 + 16 + 6 \cdot 8 + 2 \cdot 4 + 50 \qquad \text{Evaluate the powers.}$$
$$= -128 + 16 + 48 + 8 + 50 \qquad \text{Multiply.}$$
$$= -6, \qquad \text{Add.}$$

the same number as the remainder. Dividing by $x - 2$ produced a remainder equal to the result when x is replaced with 2. This always happens, as the following **remainder theorem** states. This result is proved in more advanced courses.

Remainder Theorem

If the polynomial $P(x)$ is divided by $x - k$, then the remainder is equal to $P(k)$.

NOW TRY
EXERCISE 3
Let $P(x) = 3x^3 - 2x^2 + 5x + 30$. Use synthetic division to evaluate $P(-2)$.

EXAMPLE 3 Using the Remainder Theorem

Let $P(x) = 2x^3 - 5x^2 - 3x + 11$. Use synthetic division to evaluate $P(-2)$.

Use the remainder theorem, and divide $P(x)$ by $x - (-2)$.

$$\text{Value of } k \rightarrow -2 \overline{\smash{)}\begin{array}{rrrr} 2 & -5 & -3 & 11 \\ & -4 & 18 & -30 \\ \hline 2 & -9 & 15 & -19 \end{array}} \leftarrow \text{Remainder}$$

Thus, $P(-2) = -19$.

NOW TRY

OBJECTIVE 3 **Decide whether a given number is a solution of an equation.** We can also use the remainder theorem to do this.

NOW TRY
EXERCISE 4
Use synthetic division to decide whether -4 is a solution of the equation.
$$5x^3 + 19x^2 - 2x + 8 = 0$$

EXAMPLE 4 Using the Remainder Theorem

Use synthetic division to decide whether -5 is a solution of the equation.
$$2x^4 + 12x^3 + 6x^2 - 5x + 75 = 0$$

If synthetic division gives a remainder of 0, then -5 is a solution. Otherwise, it is not.

$$\text{Proposed solution} \rightarrow -5 \overline{\smash{)}\begin{array}{rrrrr} 2 & 12 & 6 & -5 & 75 \\ & -10 & -10 & 20 & -75 \\ \hline 2 & 2 & -4 & 15 & 0 \end{array}} \leftarrow \text{Remainder}$$

Since the remainder is 0, the polynomial has value 0 when $k = -5$. So -5 is a solution of the given equation.

NOW TRY

The synthetic division in **Example 4** shows that $x - (-5)$ divides the polynomial with 0 remainder. Thus $x - (-5) = x + 5$ is a *factor* of the polynomial and

$$2x^4 + 12x^3 + 6x^2 - 5x + 75 \quad \text{factors as} \quad (x + 5)(2x^3 + 2x^2 - 4x + 15).$$

NOW TRY ANSWERS
3. -12 4. yes

The second factor is the quotient polynomial found in the last row of the synthetic division.

EXERCISES   **MyMathLab** PRACTICE   WATCH   DOWNLOAD   READ   REVIEW

Use synthetic division to find each quotient. **See Examples 1 and 2.**

1. $\dfrac{x^2 - 6x + 5}{x - 1}$

2. $\dfrac{x^2 - 4x - 21}{x + 3}$

3. $\dfrac{4m^2 + 19m - 5}{m + 5}$

4. $\dfrac{3x^2 - 5x - 12}{x - 3}$

5. $\dfrac{2a^2 + 8a + 13}{a + 2}$

6. $\dfrac{4y^2 - 5y - 20}{y - 4}$

7. $(p^2 - 3p + 5) \div (p + 1)$

8. $(z^2 + 4z - 6) \div (z - 5)$

9. $\dfrac{4a^3 - 3a^2 + 2a - 3}{a - 1}$

10. $\dfrac{5p^3 - 6p^2 + 3p + 14}{p + 1}$

11. $(x^5 - 2x^3 + 3x^2 - 4x - 2) \div (x - 2)$

12. $(2y^5 - 5y^4 - 3y^2 - 6y - 23) \div (y - 3)$

13. $(-4r^6 - 3r^5 - 3r^4 + 5r^3 - 6r^2 + 3r + 3) \div (r - 1)$

14. $(2t^6 - 3t^5 + 2t^4 - 5t^3 + 6t^2 - 3t - 2) \div (t - 2)$

15. $(-3y^5 + 2y^4 - 5y^3 - 6y^2 - 1) \div (y + 2)$

16. $(m^6 + 2m^4 - 5m + 11) \div (m - 2)$

Use the remainder theorem to find $P(k)$. **See Example 3.**

17. $P(x) = 2x^3 - 4x^2 + 5x - 3; k = 2$

18. $P(x) = x^3 + 3x^2 - x + 5; k = -1$

19. $P(x) = -x^3 - 5x^2 - 4x - 2; k = -4$

20. $P(x) = -x^3 + 5x^2 - 3x + 4; k = 3$

21. $P(x) = 2x^3 - 4x^2 + 5x - 33; k = 3$

22. $P(x) = x^3 - 3x^2 + 4x - 4; k = 2$

23. Explain why a 0 remainder in synthetic division of $P(x)$ by $x - k$ indicates that k is a solution of the equation $P(x) = 0$.

24. Explain why it is important to insert 0s as placeholders for missing terms before performing synthetic division.

Use synthetic division to decide whether the given number is a solution of the equation. **See Example 4.**

25. $x^3 - 2x^2 - 3x + 10 = 0; x = -2$

26. $x^3 - 3x^2 - x + 10 = 0; x = -2$

27. $3x^3 + 2x^2 - 2x + 11 = 0; x = -2$

28. $3x^3 + 10x^2 + 3x - 9 = 0; x = -2$

29. $2x^3 - x^2 - 13x + 24 = 0; x = -3$

30. $5x^3 + 22x^2 + x - 28 = 0; x = -4$

31. $x^4 + 2x^3 - 3x^2 + 8x - 8 = 0; x = -2$

32. $x^4 - x^3 - 6x^2 + 5x + 10 = 0; x = -2$

RELATING CONCEPTS   EXERCISES 33–38

FOR INDIVIDUAL OR GROUP WORK

We can show a connection between dividing one polynomial by another and factoring the first polynomial. Let $P(x) = 2x^2 + 5x - 12$. **Work Exercises 33–38 in order.**

33. Factor $P(x)$.

34. Solve $P(x) = 0$.

35. Evaluate $P(-4)$.

36. Evaluate $P\left(\frac{3}{2}\right)$.

37. Complete the following sentence: If $P(a) = 0$, then $x -$ _____ is a factor of $P(x)$.

38. Use the conclusion reached in **Exercise 37** to decide whether $x - 3$ is a factor of $Q(x) = 3x^3 - 4x^2 - 17x + 6$. Factor $Q(x)$ completely.

Answers to Selected Exercises

In this section, we provide the answers that we think most students will obtain when they work the exercises using the methods explained in the text. If your answer does not look exactly like the one given here, it is not necessarily wrong. In many cases, there are equivalent forms of the answer that are correct. For example, if the answer section shows $\frac{3}{4}$ and your answer is 0.75, you have obtained the right answer, but written it in a different (yet equivalent) form. Unless the directions specify otherwise, 0.75 is just as valid an answer as $\frac{3}{4}$.

In general, if your answer does not agree with the one given in the text, see whether it can be transformed into the other form. If it can, then it is the correct answer. If you still have doubts, talk with your instructor. You might also want to obtain a copy of the *Student's Solutions Manual* that goes with this book. Your college bookstore either has this manual or can order it for you.

1 REVIEW OF THE REAL NUMBER SYSTEM

Section 1.1 (pages 11–14)

1. $\{1, 2, 3, 4, 5\}$ **3.** $\{5, 6, 7, 8, \ldots\}$ **5.** $\{\ldots, -1, 0, 1, 2, 3, 4\}$ **7.** $\{10, 12, 14, 16, \ldots\}$ **9.** $\emptyset$ **11.** $\{-4, 4\}$

In Exercises 13 and 15, we give one possible answer.

13. $\{x \mid x$ is an even natural number less than or equal to $8\}$

15. $\{x \mid x$ is a multiple of 4 greater than $0\}$

17. (number line) $-6\ -4\ -2\ 0\ 2\ 4\ 6$ **19.** (number line) $-1\ 0\ 1\ 2\ 3\ 4\ 5$

21. (a) $8, 13, \frac{75}{5}$ (or 15) (b) $0, 8, 13, \frac{75}{5}$ (c) $-9, 0, 8, 13, \frac{75}{5}$
(d) $-9, -0.7, 0, \frac{6}{7}, 4.\overline{6}, 8, \frac{21}{2}, 13, \frac{75}{5}$ (e) $-\sqrt{6}, \sqrt{7}$ (f) All are real numbers. **23.** yes **25.** false; Some are whole numbers, but negative integers are not. **27.** false; No irrational number is an integer.
29. true **31.** true **33.** true **35.** (a) A (b) A (c) B (d) B
37. (a) -6 (b) 6 **39.** (a) 12 (b) 12 **41.** (a) $-\frac{6}{5}$ (b) $\frac{6}{5}$ **43.** 8
45. $\frac{3}{2}$ **47.** -5 **49.** -2 **51.** -4.5 **53.** 5 **55.** 6 **57.** 0
59. (a) Las Vegas; The population increased by 35.6%. (b) Detroit; The population decreased by 0.6%. **61.** Pacific Ocean, Indian Ocean, Caribbean Sea, South China Sea, Gulf of California **63.** true **65.** true
67. false **69.** true **71.** true **73.** $2 < 6$ **75.** $4 > -9$
77. $-10 < -5$ **79.** $x > 0$ **81.** $7 > y$ **83.** $5 \geq 5$
85. $3t - 4 \leq 10$ **87.** $5x + 3 \neq 0$ **89.** $-3 < t < 5$
91. $-3 \leq 3x < 4$ **93.** $-6 < 10$; true **95.** $10 \geq 10$; true
97. $-3 \geq -3$; true **99.** $-8 > -6$; false
101. $(-1, \infty)$ (number line) $-1\ 0$ **103.** $(-\infty, 6]$ (number line) $0\ 6$
105. $(0, 3.5)$ (number line) $0\ 3.5$ **107.** $[2, 7]$ (number line) $2\ 7$

109. $(-4, 3]$ (number line) $-4\ 0\ 3$ **111.** $(0, 3]$ (number line) $0\ 3$
113. Iowa (IA), Ohio (OH), Pennsylvania (PA) **115.** $x < y$

Section 1.2 (pages 20–23)

1. additive inverses; $4 + (-4) = 0$ **3.** negative; $-7 + (-21) = -28$
5. greater; $15 + (-2) = 13$ **7.** the number with lesser absolute value is subtracted from the one with greater absolute value;
$-15 - (-3) = -12$ **9.** negative; $-5(15) = -75$ **11.** -19
13. 9 **15.** $-\frac{19}{12}$ **17.** -1.85 **19.** -11 **21.** 21 **23.** -13
25. -10.18 **27.** $\frac{67}{30}$ **29.** 14 **31.** -5 **33.** -6 **35.** -11 **37.** 16
39. -4 **41.** 8 **43.** 3.018 **45.** $-\frac{7}{4}$ **47.** $-\frac{7}{8}$ **49.** 1 **51.** 6
53. $\frac{13}{2}$, or $6\frac{1}{2}$ **55.** It is true for multiplication (and division). It is false for addition and subtraction when the number to be subtracted has the lesser absolute value. A more precise statement is, "The product or quotient of two negative numbers is positive." **57.** -35 **59.** 40 **61.** 2
63. -12 **65.** $\frac{6}{5}$ **67.** 1 **69.** 5.88 **71.** -10.676 **73.** -7 **75.** 6
77. -4 **79.** 0 **81.** undefined **83.** $\frac{25}{102}$ **85.** $-\frac{9}{13}$ **87.** -2.1
89. 10,000 **91.** $\frac{17}{18}$ **93.** $\frac{17}{36}$ **95.** $-\frac{19}{24}$ **97.** $-\frac{22}{45}$ **99.** $-\frac{2}{15}$ **101.** $\frac{3}{5}$
103. $-\frac{35}{27}$ **105.** $-\frac{4}{9}$ **107.** -12.351 **109.** -15.876 **111.** -4.14
113. 4800 **115.** 51.495 **117.** $112°F$ **119.** $30.13
121. (a) $466.02 (b) $190.68 **123.** (a) $-$475 thousand
(b) $262 thousand (c) $-$83 thousand **125.** (a) 2000: $129 billion;
2010: $206 billion; 2020: $74 billion; 2030: $-$501 billion (b) The cost of Social Security will exceed revenue in 2030 by $501 billion.

Section 1.3 (pages 29–31)

1. false; $-7^6 = -(7^6)$ **3.** true **5.** true **7.** true **9.** false; The base is 8. **11.** (a) 64 (b) -64 (c) 64 (d) -64 **13.** 10^4 **15.** $\left(\frac{3}{4}\right)^5$
17. $(-9)^3$ **19.** z^7 **21.** 16 **23.** 0.021952 **25.** $\frac{1}{125}$ **27.** $\frac{256}{625}$
29. -125 **31.** 256 **33.** -729 **35.** -4096 **37.** 9 **39.** 13
41. -20 **43.** $\frac{10}{11}$ **45.** -0.7 **47.** not a real number **49.** (a) B
(b) C (c) A **51.** negative **53.** 24 **55.** 15 **57.** 55 **59.** -91
61. -8 **63.** -48 **65.** -2 **67.** -79 **69.** -10 **71.** 2
73. -2 **75.** undefined **77.** -7 **79.** -1 **81.** 17 **83.** -96
85. $-\frac{15}{238}$ **87.** 8 **89.** $-\frac{5}{16}$ **91.** -2.75 **93.** $-\frac{3}{16}$ **95.** $1572
97. $3296 **99.** 0.035 **101.** (a) $19.4 billion (b) $30.8 billion
(c) $42.3 billion (d) The amount spent on pets more than doubled from 1996 to 2008.

Section 1.4 (pages 37–38)

1. B **3.** A **5.** product; 0 **7.** grouping **9.** like **11.** $2m + 2p$
13. $-12x + 12y$ **15.** $8k$ **17.** $-2r$ **19.** cannot be simplified
21. $8a$ **23.** $-2d + f$ **25.** $x + y$ **27.** $-6y + 3$ **29.** $p + 11$

31. $-2k + 15$ **33.** $-3m + 2$ **35.** -1 **37.** $2p + 7$
39. $-6z - 39$ **41.** $(5 + 8)x = 13x$ **43.** $(5 \cdot 9)r = 45r$
45. $9y + 5x$ **47.** 7 **49.** 0 **51.** $8(-4) + 8x = -32 + 8x$ **53.** 0
55. Answers will vary. One example of commutativity is washing your face and brushing your teeth. An example of non-commutativity is putting on your socks and putting on your shoes. **57.** 1900 **59.** 75
61. 431 **63.** associative property **64.** associative property
65. commutative property **66.** associative property **67.** distributive property **68.** arithmetic facts **69.** No. One example is
$7 + (5 \cdot 3) = (7 + 5)(7 + 3)$, which is false.

Chapter 1 Review Exercises (pages 42–44)

1. (number line) **2.** (number line) **3.** 16 **4.** -8
5. 5 **6.** $0, \frac{12}{3}$ (or 4) **7.** $-9, -\sqrt{4}$ (or -2), $0, \frac{12}{3}$ (or 4) **8.** $-9, -\frac{4}{3},$
$-\sqrt{4}$ (or -2), $-0.25, 0, 0.\overline{35}, \frac{5}{3}, \frac{12}{3}$ (or 4) **9.** All are real numbers except $\sqrt{-9}$. **10.** $\{4, 5, 6, 7, 8\}$ **11.** $\{0, 1, 2, 3\}$ **12.** true
13. false **14.** true **15.** Subaru; 45.9% **16.** Toyota; 2.14%
17. false **18.** true **19.** $(-\infty, -5)$ (number line)
20. $(-2, 3]$ (number line) **21.** $\frac{41}{24}$ **22.** $-\frac{1}{2}$ **23.** -3
24. -16.99 **25.** -39 **26.** 0 **27.** $\frac{23}{20}$ **28.** -35 **29.** 11,331 ft
30. -90 **31.** $\frac{2}{3}$ **32.** -15 **33.** 3.21 **34.** $\frac{5}{7 - 7}$ **35.** 10,000
36. $\frac{27}{343}$ **37.** -125 **38.** -125 **39.** 20 **40.** $\frac{8}{11}$ **41.** -0.9
42. not a real number **43.** -4 **44.** 44 **45.** -2 **46.** -30
47. -30 **48.** $-\frac{8}{51}$ **49. (a)** 26 **(b)** Answers will vary. **50.** $20q$
51. $-4z$ **52.** $3m$ **53.** $4p$ **54.** $-2k - 6$ **55.** $6r + 18$
56. $18m + 27n$ **57.** $-p - 3q$ **58.** $y + 1$ **59.** 0 **60.** $-18m$
61. $(2 + 3)x = 5x$ **62.** -5 **63.** $(2 \cdot 4)x = 8x$
64. $13 + (-3) = 10$ **65.** 0 **66.** $6x + 6z$ **67.** 7 **68.** 1
69. 732 million; negative **70.** 1096 million; positive
71. 799 million; positive **72.** $\frac{256}{625}$ **73.** 25 **74.** 31 **75.** 9 **76.** 0
77. -5 **78.** $\frac{4}{3}$ **79.** -6.16 **80.** -9 **81.** 2 **82.** 2 **83.** not a real number **84.** $-3k + 6h$ **85.** -11.408 **86.** 24 **87.** $-6x + 4$
88. $-\frac{5}{18}$ **89. (a)** -116 **(b)** $-\frac{9}{4}$ **90.** Work within the parentheses first.

Chapter 1 Test (pages 44–45)

[1.1] **1.** (number line) **2.** $0, 3, \sqrt{25}$ (or 5), $\frac{24}{2}$ (or 12)
3. $-1, 0, 3, \sqrt{25}$ (or 5), $\frac{24}{2}$ (or 12) **4.** $-1, -0.5, 0, 3, \sqrt{25}$ (or 5),
$7.5, \frac{24}{2}$ (or 12) **5.** All are real numbers except $\sqrt{-4}$.
6. $(-\infty, -3)$ (number line) **7.** $(-4, 2]$ (number line)
[1.2] **8.** 0 [1.3] **9.** -26 **10.** 19 **11.** 1 **12.** $\frac{16}{7}$ **13.** $\frac{11}{23}$
[1.2] **14.** 50,395 ft **15.** 37,486 ft **16.** 1345 ft [1.3] **17.** 14

18. -15 **19.** not a real number **20. (a)** a must be positive.
(b) a must be negative. **(c)** a must be 0. **21.** $-\frac{6}{23}$
[1.4] **22.** $10k - 10$ **23.** It changes the sign of each term. The simplified form is $7r + 2$. **24.** B **25.** D **26.** A **27.** F **28.** C
29. C **30.** E

2 LINEAR EQUATIONS, INEQUALITIES, AND APPLICATIONS

Section 2.1 (pages 54–55)

1. A and C **3.** Both sides are evaluated as 30, so 6 is a solution.
5. equation **7.** expression **9.** equation **11.** $\{-1\}$ **13.** $\{-4\}$
15. $\{-7\}$ **17.** $\{0\}$ **19.** $\{4\}$ **21.** $\left\{-\frac{7}{8}\right\}$ **23.** $\emptyset$; contradiction
25. $\left\{-\frac{5}{3}\right\}$ **27.** $\left\{-\frac{1}{2}\right\}$ **29.** $\{2\}$ **31.** $\{-2\}$ **33.** {all real numbers};
identity **35.** $\{-1\}$ **37.** $\{7\}$ **39.** $\{2\}$ **41.** {all real numbers};
identity **43.** $\left\{\frac{3}{2}\right\}$ **45.** 12 **47. (a)** 10^2, or 100 **(b)** 10^3, or 1000
49. $\left\{-\frac{18}{5}\right\}$ **51.** $\left\{-\frac{5}{6}\right\}$ **53.** $\{6\}$ **55.** $\{4\}$ **57.** $\{3\}$ **59.** $\{3\}$
61. $\{0\}$ **63.** $\{2000\}$ **65.** $\{25\}$ **67.** $\{40\}$ **69.** $\{3\}$ **71.** 36
73. 72 **75.** 50

Section 2.2 (pages 62–67)

1. $r = \frac{I}{pt}$ **3.** $L = \frac{P - 2W}{2}$, or $L = \frac{P}{2} - W$ **5. (a)** $W = \frac{V}{LH}$
(b) $H = \frac{V}{LW}$ **7.** $r = \frac{C}{2\pi}$ **9. (a)** $h = \frac{2\mathcal{A}}{b + B}$ **(b)** $B = \frac{2\mathcal{A}}{h} - b$, or
$B = \frac{2\mathcal{A} - hb}{h}$ **11.** $C = \frac{5}{9}(F - 32)$ **13.** D **15.** $y = \frac{11 - 4x}{9}$
17. $y = \frac{5 + 3x}{2}$ **19.** $y = \frac{7 - 6x}{-5}$, or $y = \frac{6x - 7}{5}$ **21.** 3.275 hr
23. 52 mph **25.** 113°F **27.** 230 m **29.** radius: 240 in.; diameter:
480 in. **31.** 2 in. **33.** 75% water; 25% alcohol **35.** 3% **37.** $10.51
39. $45.66 **41. (a)** .586 **(b)** .519 **(c)** .463 **(d)** .395 **43.** 54%
45. 101.1 million **47.** $70,781 **49.** $35,390 **51.** 8% **53.** 3.8%
55. 47.5% **57.** $\{12\}$ **59.** $\{-6\}$ **61.** -3 **63.** 6

Section 2.3 (pages 74–80)

1. (a) $x + 15$ **(b)** $15 > x$ **3. (a)** $x - 8$ **(b)** $8 < x$ **5.** D
7. $2x - 13$ **9.** $12 + 4x$ **11.** $8(x - 16)$ **13.** $\frac{3x}{10}$ **15.** $x + 6 = -31$;
-37 **17.** $x - (-4x) = x + 9; \frac{9}{4}$ **19.** $14 - \frac{2}{3}x = 10$; 6
21. expression; $-11x + 63$ **23.** equation; $\left\{\frac{51}{11}\right\}$ **25.** expression;
$\frac{1}{3}x - \frac{13}{2}$ **27.** *Step 1:* the number of patents each corporation secured;
Step 2: patents that Samsung secured; *Step 3:* $x; x - 667$; *Step 4:* 4169;
Step 5: 4169; 3502; *Step 6:* 667; IBM patents; 3502; 7671 **29.** width:
165 ft; length: 265 ft **31.** 24.34 in. by 29.88 in. **33.** 850 mi; 925 mi;
1300 mi **35.** Exxon Mobil: $442.9 billion; Wal-Mart: $405.6 billion
37. Eiffel Tower: 1063 ft; Leaning Tower: 183 ft **39.** Obama:
365 votes; McCain: 173 votes **41.** 39.0% **43.** $7028 **45.** $44.60
47. $225 **49.** $4000 at 3%; $8000 at 4% **51.** $10,000 at 4.5%;
$19,000 at 3% **53.** $24,000 **55.** 5 L **57.** 4 L **59.** 1 gal **61.** 150 lb

63. We cannot expect the final mixture to be worth more than the more expensive of the two ingredients.   **65. (a)** $800 - x$ **(b)** $800 - y$
66. (a) $0.05x$; $0.10(800 - x)$ **(b)** $0.05y$; $0.10(800 - y)$
67. (a) $0.05x + 0.10(800 - x) = 800(0.0875)$
(b) $0.05y + 0.10(800 - y) = 800(0.0875)$ **68. (a)** \$200 at 5%;
\$600 at 10%   **(b)** 200 L of 5% acid; 600 L of 10% acid
(c) The processes are the same. The amounts of money in Problem A correspond to the amounts of solution in Problem B.   **69.** 200   **71.** 19

Section 2.4 (pages 84–89)

1. \$5.40   **3.** 30 mph   **5.** The problem asks for the *distance* to the workplace. To find this distance, we must multiply the rate, 10 mph, by the time, $\frac{3}{4}$ hr.   **7.** No, the answers must be whole numbers because they represent the number of coins.   **9.** 17 pennies; 17 dimes; 10 quarters
11. 23 loonies; 14 toonies   **13.** 28 \$10 coins; 13 \$20 coins
15. 872 adult tickets   **17.** 7.97 m per sec   **19.** 8.47 m per sec
21. $2\frac{1}{2}$ hr   **23.** 7:50 P.M.   **25.** 45 mph   **27.** $\frac{1}{2}$ hr   **29.** 60°, 60°, 60°
31. 40°, 45°, 95°   **33.** 40°, 80°   **34.** 120°   **35.** The sum is equal to the measure of the angle found in **Exercise 34.**   **36.** The sum of the measures of angles ① and ② is equal to the measure of angle ③.
37. Both measure 122°.   **39.** 64°, 26°   **41.** 19, 20, 21
43. 61 yr old   **45.** 28, 30, 32   **47.** 21, 23, 25
49. [number line]   **51.** [number line]

Summary Exercises on Solving Applied Problems (pages 89–90)

1. length: 8 in.; width: 5 in.   **2.** length: 60 m; width: 30 m
3. \$86.98   **4.** \$425   **5.** \$800 at 4%; \$1600 at 5%   **6.** \$12,000 at 3%;
\$14,000 at 4%   **7.** James: 2250; Wade: 2386   **8.** *Titanic:*
\$600.8 million; *The Dark Knight:* \$533.3 million   **9.** 5 hr
10. Merga: 12.27 mph; Kosgei: 10.37 mph   **11.** $13\frac{1}{3}$ L   **12.** $53\frac{1}{3}$ kg
13. fives: 84; tens: 42   **14.** 1650 tickets at \$9; 810 tickets at \$7
15. 20°, 30°, 130°   **16.** 107°, 73°   **17.** 31, 32, 33   **18.** 9, 11
19. 6 in., 12 in., 16 in.   **20.** 23 in.

Section 2.5 (pages 99–102)

1. D   **3.** B   **5.** F   **7.** Since $4 > 0$, the student should not have reversed the direction of the inequality symbol when dividing by 4. We reverse the symbol only when multiplying or dividing by a *negative* number. The solution set is $[-16, \infty)$.

9. $[16, \infty)$ [number line]   **11.** $(7, \infty)$ [number line]
13. $(-\infty, -4)$ [number line]
15. $(-\infty, -40]$ [number line]
17. $(-\infty, 4]$ [number line]

19. $(-\infty, -10]$ [number line]
21. $(-\infty, 14)$ [number line]
23. $\left(-\infty, -\frac{15}{2}\right)$ [number line]   **25.** $\left[\frac{1}{2}, \infty\right)$ [number line]
27. $[2, \infty)$ [number line]   **29.** $(3, \infty)$ [number line]
31. $(-\infty, 4)$ [number line]
33. $\left(-\infty, \frac{23}{6}\right]$ [number line]
35. $\left(-\infty, \frac{76}{11}\right)$ [number line]
37. $(-\infty, \infty)$ [number line]
39. $\emptyset$ **41.** $\{-9\}$ [number line]
42. $(-9, \infty)$ [number line]
43. $(-\infty, -9)$ [number line]
44. the set of all real numbers [number line]   **45.** $(-\infty, -3)$
47. $(1, 11)$ [number line]   **49.** $[-14, 10]$ [number line]
51. $[-5, 6]$ [number line]   **53.** $\left[-\frac{14}{3}, 2\right]$ [number line]
55. $\left[-\frac{1}{2}, \frac{35}{2}\right]$ [number line]   **57.** $\left(-\frac{1}{3}, \frac{1}{9}\right]$ [number line]
59. $(-2, 2)$ **61.** $[3, \infty)$ **63.** $[-9, \infty)$ **65.** at least 80
67. 26 months   **69.** 26 DVDs   **71. (a)** 140 to 184 lb
(b) 107 to 141 lb   **(c)** Answers will vary.
73. (a) [number line]   **(b)** [number line]
(c) All numbers greater than 4 and less than 5 (that is, those satisfying $4 < x < 5$) belong to both sets.

Section 2.6 (pages 108–111)

1. true   **3.** false; The union is $(-\infty, 7) \cup (7, \infty)$.   **5.** false;
The intersection is $\emptyset$.   **7.** $\{1, 3, 5\}$, or B **9.** $\{4\}$, or D **11.** $\emptyset$
13. $\{1, 2, 3, 4, 5, 6\}$, or A **15.** [number line]
17. [number line]   **19.** $(-3, 2)$ [number line]
21. $(-\infty, 2]$ [number line]   **23.** $\emptyset$
25. $[5, 9]$ [number line]   **27.** $(-3, -1)$ [number line]

29. $(-\infty, 4]$ **31.**

33. **35.** $(-\infty, 8]$

37. $[-2, \infty)$ **39.** $(-\infty, \infty)$

41. $(-\infty, -5) \cup (5, \infty)$

43. $(-\infty, -1) \cup (2, \infty)$

45. $(-\infty, \infty)$ **47.** $[-4, -1]$ **49.** $[-9, -6]$

51. $(-\infty, 3)$ **53.** $[3, 9)$

55. intersection; $(-5, -1)$

57. union; $(-\infty, 4)$

59. union; $(-\infty, 0] \cup [2, \infty)$

61. intersection; $[4, 12]$ **63.** {Tuition and fees}

65. {Tuition and fees, Board rates, Dormitory charges} **67.** Maria, Joe
68. none of them **69.** none of them **70.** Luigi, Than **71.** Maria, Joe
72. all of them **73.** $[-6, \infty)$ **75.** $(-3, 2)$ **77.** -21 **79.** false

Connections **(page 117)** The filled carton may contain between 30.4 and 33.6 oz, inclusive.

Section 2.7 (pages 118–120)

1. E; C; D; B; A **3. (a)** one **(b)** two **(c)** none **5.** $\{-12, 12\}$
7. $\{-5, 5\}$ **9.** $\{-6, 12\}$ **11.** $\{-5, 6\}$ **13.** $\{-3, \frac{11}{2}\}$
15. $\{-\frac{19}{2}, \frac{9}{2}\}$ **17.** $\{-10, -2\}$ **19.** $\{-\frac{32}{3}, 8\}$ **21.** $\{-75, 175\}$
23. $(-\infty, -3) \cup (3, \infty)$

25. $(-\infty, -4] \cup [4, \infty)$

27. $(-\infty, -25] \cup [15, \infty)$

29. $(-\infty, -12) \cup (8, \infty)$

31. $(-\infty, -2) \cup (8, \infty)$

33. $\left(-\infty, -\frac{9}{5}\right] \cup [3, \infty)$

35. (a) **(b)**

37. $[-3, 3]$ **39.** $(-4, 4)$

41. $(-25, 15)$

43. $[-12, 8]$

45. $[-2, 8]$ **47.** $\left(-\frac{9}{5}, 3\right)$

49. $(-\infty, -5) \cup (13, \infty)$

51. $(-\infty, -25) \cup (15, \infty)$

53. $\{-6, -1\}$

55. $\left[-\frac{10}{3}, 4\right]$

57. $\left[-\frac{7}{6}, -\frac{5}{6}\right]$

59. $(-\infty, -3] \cup [4, \infty)$ **61.** $\{-5, 1\}$

63. $\{3, 9\}$ **65.** $\{0, 20\}$ **67.** $\{-5, 5\}$ **69.** $\{-5, -3\}$
71. $(-\infty, -3) \cup (2, \infty)$ **73.** $[-10, 0]$ **75.** $\left\{-\frac{5}{3}, \frac{1}{3}\right\}$
77. $(-\infty, 20] \cup [30, \infty)$ **79.** $\{-1, 3\}$ **81.** $\left\{-3, \frac{5}{3}\right\}$ **83.** $\left\{-\frac{1}{3}, -\frac{1}{15}\right\}$
85. $\left\{-\frac{5}{4}\right\}$ **87.** $(-\infty, \infty)$ **89.** $\emptyset$ **91.** $\left\{-\frac{1}{4}\right\}$ **93.** $\emptyset$ **95.** $(-\infty, \infty)$
97. $\left\{-\frac{3}{7}\right\}$ **99.** $\left\{\frac{2}{5}\right\}$ **101.** $(-\infty, \infty)$ **103.** $\emptyset$
105. $|x - 1000| \le 100$; $900 \le x \le 1100$ **107.** 810.5 ft
108. Bank of America Center, Texaco Heritage Plaza
109. Williams Tower, Bank of America Center, Texaco Heritage Plaza, Enterprise Plaza, Centerpoint Energy Plaza, Continental Center I, Fulbright Tower **110. (a)** $|x - 810.5| \ge 95$
(b) $x \ge 905.5$ or $x \le 715.5$ **(c)** JPMorgan Chase Tower, Wells Fargo Plaza, One Shell Plaza **(d)** It makes sense because it includes all buildings *not* listed in the answer to **Exercise 109.**
111. (a) 12 **(b)** 4 **113. (a)** 0 **(b)** $\frac{36}{5}$

Summary Exercises on Solving Linear and Absolute Value Equations and Inequalities (page 121)

1. $\{12\}$ **2.** $\{-5, 7\}$ **3.** $\{7\}$ **4.** $\left\{-\frac{2}{5}\right\}$ **5.** $\emptyset$ **6.** $(-\infty, -1)$
7. $\left[-\frac{2}{3}, \infty\right)$ **8.** $\{-1\}$ **9.** $\{-3\}$ **10.** $\left\{1, \frac{11}{3}\right\}$ **11.** $(-\infty, 5]$
12. $(-\infty, \infty)$ **13.** $\{2\}$ **14.** $(-\infty, -8] \cup [8, \infty)$ **15.** $\emptyset$ **16.** $(-\infty, \infty)$
17. $(-5.5, 5.5)$ **18.** $\left\{\frac{13}{3}\right\}$ **19.** $\left\{-\frac{96}{5}\right\}$ **20.** $(-\infty, 32]$
21. $(-\infty, -24)$ **22.** $\left\{\frac{3}{8}\right\}$ **23.** $\left\{\frac{7}{2}\right\}$ **24.** $(-6, 8)$
25. {all real numbers} **26.** $(-\infty, 5)$ **27.** $(-\infty, -4) \cup (7, \infty)$
28. $\{24\}$ **29.** $\left\{-\frac{1}{5}\right\}$ **30.** $\left(-\infty, -\frac{5}{2}\right]$ **31.** $\left[-\frac{1}{3}, 3\right]$ **32.** $[1, 7]$
33. $\left\{-\frac{1}{6}, 2\right\}$ **34.** $\{-3\}$ **35.** $(-\infty, -1] \cup \left[\frac{5}{3}, \infty\right)$ **36.** $\left[\frac{3}{4}, \frac{15}{8}\right]$
37. $\left\{-\frac{5}{2}\right\}$ **38.** $\{60\}$ **39.** $\left[-\frac{9}{2}, \frac{15}{2}\right]$ **40.** $(1, 9)$ **41.** $(-\infty, \infty)$
42. $\left\{\frac{1}{3}, 9\right\}$ **43.** {all real numbers} **44.** $\left\{-\frac{10}{9}\right\}$ **45.** $\{-2\}$ **46.** $\emptyset$
47. $(-\infty, -1) \cup (2, \infty)$ **48.** $[-3, -2]$

Chapter 2 Review Exercises (pages 127–131)

1. $\left\{-\frac{9}{5}\right\}$ **2.** $\{16\}$ **3.** $\left\{-\frac{7}{5}\right\}$ **4.** $\emptyset$ **5.** {all real numbers}; identity
6. $\emptyset$; contradiction **7.** $\{0\}$; conditional **8.** $L = \dfrac{V}{HW}$

9. $b = \dfrac{2\mathcal{A} - Bh}{h}$, or $b = \dfrac{2\mathcal{A}}{h} - B$ **10.** $x = -4M - 3y$

11. $x = \frac{4}{3}(P + 12)$, or $x = \frac{4}{3}P + 16$ **12.** Begin by subtracting 5 from each side. Then divide each side by -2.   **13.** 6 ft   **14.** 19.0%

15. 6.5%   **16.** 25°   **17.** $617 billion   **18.** $92.4 billion   **19.** $9 - \frac{1}{3}x$

20. $\dfrac{4x}{x + 9}$ **21.** length: 13 m; width: 8 m   **22.** 17 in., 17 in., 19 in.

23. 12 kg   **24.** 30 L   **25.** 10 L   **26.** $10,000 at 6%; $6000 at 4%

27. 15 dimes; 8 quarters   **28.** 7 nickels; 12 dimes   **29.** A

30. (a) 530 mi   **(b)** 328 mi   **31.** 2.2 hr   **32.** 50 km per hr; 65 km per hr   **33.** 1 hr   **34.** 46 mph   **35.** 40°, 45°, 95°

36. 150°, 30°   **37.** $(-9, \infty)$ **38.** $(-\infty, -3]$ **39.** $\left(\frac{3}{2}, \infty\right)$

40. $[-3, \infty)$ **41.** $[3, 5)$ **42.** $\left(\frac{59}{31}, \infty\right)$ **43.** 38 m or less

44. 34 tickets or less (but at least 15)   **45.** any score greater than or equal to 61   **46.** Because the statement $-8 < -13$ is *false*, the inequality has no solution.   **47.** $\{a, c\}$ **48.** $\{a\}$

49. $\{a, c, e, f, g\}$ **50.** $\{a, b, c, d, e, f, g\}$

51. $(6, 9)$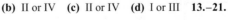

52. $(8, 14)$

53. $(-\infty, -3] \cup (5, \infty)$

54. $(-\infty, \infty)$ **55.** $\emptyset$

56. $(-\infty, -2] \cup [7, \infty)$

57. $(-3, 4)$ **58.** $(-\infty, 2)$ **59.** $(4, \infty)$ **60.** $(1, \infty)$ **61.** $\{-7, 7\}$

62. $\{-11, 7\}$ **63.** $\left\{-\frac{1}{3}, 5\right\}$ **64.** $\emptyset$ **65.** $\{0, 7\}$ **66.** $\left\{-\frac{3}{2}, \frac{1}{2}\right\}$

67. $\left\{-\frac{3}{4}, \frac{1}{2}\right\}$ **68.** $\left\{-\frac{1}{2}\right\}$ **69.** $(-14, 14)$ **70.** $[-1, 13]$

71. $[-3, -2]$ **72.** $(-\infty, \infty)$ **73.** $(-2, \infty)$ **74.** $k = \dfrac{6r - bt}{a}$

75. $[-2, 3)$ **76.** $\{0\}$ **77.** $(-\infty, \infty)$ **78.** $(-\infty, 2]$ **79.** 10 ft

80. 46, 47, 48   **81.** $\left\{-\frac{7}{3}, 1\right\}$ **82.** $\{300\}$ **83.** $[-16, 10]$

84. $\left(-\infty, \frac{14}{17}\right)$ **85.** $\left(-3, \frac{7}{2}\right)$ **86.** $(-\infty, 1]$ **87.** 80°   **88.** any amount greater than or equal to $1100   **89.** $\left(-\infty, -\frac{13}{5}\right) \cup (3, \infty)$ **90.** $(-\infty, \infty)$

91. 5 L   **92.** $\left\{-4, -\frac{2}{3}\right\}$ **93.** $\{30\}$ **94.** $[-4, -2]$ **95.** $\left\{1, \frac{11}{3}\right\}$

96. $\emptyset$ **97. (a)** $\emptyset$ **(b)** $(-\infty, \infty)$ **(c)** $\emptyset$

98.

99.

100. (a) $\{$Illinois$\}$ **(b)** $\{$Illinois, Maine, North Carolina, Oregon, Utah$\}$ **(c)** $\emptyset$

Chapter 2 Test (pages 131–132)

[2.1] **1.** $\{-19\}$ **2.** $\{5\}$ **3.** $\{$all real numbers$\}$ **4. (a)** $\emptyset$; contradiction   **(b)** $\{$all real numbers$\}$; identity   **(c)** $\{0\}$; conditional equation   [2.2] **5.** $h = \dfrac{3V}{b}$ **6.** $v = \dfrac{S + 16t^2}{t}$ [2.3, 2.4] **7.** 3.326 hr

8. 6.25%   **9.** 74.2%   **10.** $8000 at 3%; $20,000 at 5%   **11.** faster car: 60 mph; slower car: 45 mph   **12.** 40°, 40°, 100°

[2.5] **13.** $[1, \infty)$

14. $(-\infty, 28)$ **15.** $(1, 2)$

16. $[-3, 3]$ **17.** C   **18.** 82

[2.6] **19. (a)** $\{1, 5\}$ **(b)** $\{1, 2, 5, 7, 9, 12\}$ **20.** $[2, 9)$

21. $(-\infty, 3) \cup [6, \infty)$ [2.7] **22.** $\left[-\frac{5}{2}, 1\right]$ **23.** $\left(-\infty, -\frac{7}{6}\right) \cup \left(\frac{17}{6}, \infty\right)$

24. $\emptyset$ **25.** $\left(\frac{1}{3}, \frac{7}{3}\right)$ **26.** $\left\{-\frac{5}{3}, 3\right\}$ **27.** $\left\{-\frac{5}{7}, \frac{11}{3}\right\}$ **28. (a)** $\emptyset$

(b) $(-\infty, \infty)$ **(c)** $\emptyset$

Chapters 1–2 Cumulative Review Exercises (pages 133–134)

[1.1] **1.** 9, 6   **2.** 0, 9, 6   **3.** $-8, 0, 9, 6$ **4.** $-8, -\frac{2}{3}, 0, \frac{4}{5}, 9, 6$

5. $-\sqrt{6}$ **6.** All are real numbers.   [1.2] **7.** $-\frac{22}{21}$ **8.** 8   [1.3] **9.** 8

10. 0   **11.** -243 **12.** $\frac{216}{343}$ **13.** $-\frac{8}{27}$ **14.** -4096 **15.** -16

16. 184   **17.** $\frac{27}{16}$ [1.4] **18.** $-20r + 17$ **19.** $13k + 42$

20. commutative property   **21.** distributive property   [2.1] **22.** $\{5\}$

23. $\{30\}$ **24.** $\{15\}$ [2.2] **25.** $b = P - a - c$ [2.1] **26.** $\emptyset$

27. $\{$all real numbers$\}$

[2.5] **28.** $[-14, \infty)$

29. $\left[\frac{5}{3}, 3\right)$

[2.6] **30.** $(-\infty, 0) \cup (2, \infty)$

[2.7] **31.** $\left(-\infty, -\frac{1}{7}\right] \cup [1, \infty)$

[2.3] **32.** $5000 at 5%; $7000 at 6%   **33.** 2 L   [2.2, 2.3] **34.** 44 mg

35. (a) 348   **(b)** 19.8%

---

3   **GRAPHS, LINEAR EQUATIONS, AND FUNCTIONS**

Connections   **(page 143)**   **1.** x-intercept: $(-2, 0)$; y-intercept: $(0, 3)$
For Problems 2 and 3, we give each equation solved for y. Graphs are not included.   **2.** $y = 4x + 3$ **3.** $y = -0.5x$

Section 3.1 (pages 143–147)

1. (a) x represents the year; y represents the higher education aid in billions of dollars.   **(b)** about $150 billion   **(c)** $(2007, 150)$ **(d)** In 1997, higher education aid was about $75 billion.   **3.** origin   **5.** y; x; x; y **7.** two

9. (a) I   **(b)** III   **(c)** II   **(d)** IV   **(e)** none   **(f)** none   **11. (a)** I or III   **(b)** II or IV   **(c)** II or IV   **(d)** I or III   **13.–21.**

23. (a) $-4; -3; -2; -1; 0$ **25. (a)** $-3; 3; 2; -1$

(b)   **(b)**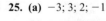

27. (a) $\frac{5}{2}$; 5; $\frac{3}{2}$; 1

(b)

29. (a) -4; 5; $-\frac{12}{5}$; $\frac{5}{4}$

(b)

31. (a) 3; 1; -1; -3

(b)

33. (a) 1   **(b)** 2   **(c)** For every increase in x by 1 unit, y increases by 2 units.

35. $(6, 0)$; $(0, 4)$

37. $(6, 0)$; $(0, -2)$

39. $(-2, 0)$; $\left(0, -\frac{5}{3}\right)$

41. $\left(\frac{21}{2}, 0\right)$; $\left(0, -\frac{7}{3}\right)$

43. none; $(0, 5)$

45. $(2, 0)$; none

47. $(-4, 0)$; none

49. none; $(0, -2)$

51. $(0, 0)$; $(0, 0)$

53. $(0, 0)$; $(0, 0)$

55. $(0, 0)$; $(0, 0)$

57. $(-5, -1)$

59. $\left(\frac{9}{2}, -\frac{3}{2}\right)$

61. $\left(0, \frac{11}{2}\right)$

63. $(2.1, 0.9)$

65. $(1, 1)$

67. $\left(-\frac{5}{12}, \frac{5}{28}\right)$ **69.** $Q(11, -4)$ **71.** $Q(4.5, 0.75)$ **73.** B

75. Window B is more useful because it shows the intercepts.

77. $y = -2.5x - 5$ **79.** 2   **81.** 0

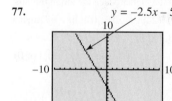

83. $y = 4x - 10$

85. $y = \dfrac{12 - 3x}{4}$, or $y = -\dfrac{3}{4}x + 3$

Section 3.2 (pages 155–161)

1. A, B, D, G   **3.** 5 ft   **5. (a)** C   **(b)** A   **(c)** D   **(d)** B   **7.** 2
9. undefined   **11.** 1   **13.** -1 **15.** 2   **17.** $\frac{5}{2}$ **19.** 0   **21.** undefined
23. (a) B **(b)** C **(c)** A **(d)** D **25. (a)** 8   **(b)** rises   **27. (a)** $\frac{5}{6}$
(b) rises   **29. (a)** 0   **(b)** horizontal   **31. (a)** $-\frac{1}{2}$ **(b)** falls

33. (a) undefined   **(b)** vertical   **35. (a)** -1 **(b)** falls   **37.** 6   **39.** -3
41. $-\frac{5}{2}$ **43.** undefined   **45.** $-\frac{1}{2}$

47. $\frac{5}{2}$

49. 4

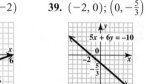

51. undefined

53. 0

55. 0

57.

59.

61.

63.

65.

67. $-\frac{4}{9}$; $\frac{9}{4}$ **69.** parallel   **71.** perpendicular
73. neither   **75.** parallel   **77.** neither
79. perpendicular   **81.** $-\$4000$ per yr; The value of the machine is decreasing \$4000 each year during these years.

83. 0% per yr (or no change); The percent of pay raise is not changing— it is 3% each year during these years.   **85.** 19.5 ft   **87. (a)** 21.2
(b) The number of subscribers increased by an average of 21.2 million each year from 2005 to 2008.   **89. (a)** -5 theaters per yr   **(b)** The negative slope means that the number of drive-in theaters decreased by an average of 5 each year from 2000 to 2007.   **91.** \$1371.67 million per yr; Sales of plasma TVs increased by an average of \$1371.67 million each year from 2003 to 2006.   **93.** Since the slopes of both pairs of opposite sides are equal, the figure is a parallelogram.   **95.** $\frac{1}{3}$ **96.** $\frac{1}{3}$ **97.** $\frac{1}{3}$
98. $\frac{1}{3} = \frac{1}{3} = \frac{1}{3}$ is true.   **99.** collinear   **100.** not collinear
101. $y = -\frac{3}{2}x + 4$ **103.** $y = 4x + 14$ **105.** $x + 2y = -4$

Connections   **(page 169)**   **1.** $\{0\}$ **2.** $\{1\}$ **3.** $\{-0.5\}$ **4.** $\{3\}$

Section 3.3 (pages 169–174)

1. A   **3.** A   **5.** $3x + y = 10$ **7.** A   **9.** C   **11.** H   **13.** B
15. $y = 5x + 15$ **17.** $y = -\frac{2}{3}x + \frac{4}{5}$ **19.** $y = x - 1$
21. $y = \frac{2}{5}x + 5$ **23.** $y = \frac{2}{3}x + 1$ **25.** $y = -x - 2$
27. (a) $y = x + 4$ **(b)** 1   **(c)** $(0, 4)$ **(d)**

29. (a) $y = -\frac{6}{5}x + 6$ **(b)** $-\frac{6}{5}$ **(c)** $(0, 6)$ **(d)**

31. (a) $y = \frac{4}{5}x - 4$ **(b)** $\frac{4}{5}$ **(c)** $(0, -4)$ **(d)**

33. (a) $y = -\frac{1}{2}x - 2$ **(b)** $-\frac{1}{2}$ **(c)** $(0, -2)$ **(d)**

35. (a) $2x + y = 18$ **(b)** $y = -2x + 18$ **37. (a)** $3x + 4y = 10$
(b) $y = -\frac{3}{4}x + \frac{5}{2}$ **39. (a)** $x - 2y = -13$ **(b)** $y = \frac{1}{2}x + \frac{13}{2}$
41. (a) $4x - y = 12$ **(b)** $y = 4x - 12$ **43. (a)** $7x - 5y = -20$
(b) $y = 1.4x + 4$ **45.** $y = 5$ **47.** $x = 9$ **49.** $y = -\frac{3}{2}$ **51.** $y = 8$
53. $x = 0.5$ **55. (a)** $2x - y = 2$ **(b)** $y = 2x - 2$
57. (a) $x + 2y = 8$ **(b)** $y = -\frac{1}{2}x + 4$ **59. (a)** $y = 5$ **(b)** $y = 5$
61. (a) $x = 7$ **(b)** not possible **63. (a)** $y = -3$ **(b)** $y = -3$
65. (a) $2x - 13y = -6$ **(b)** $y = \frac{2}{13}x + \frac{6}{13}$ **67. (a)** $y = 3x - 19$
(b) $3x - y = 19$ **69. (a)** $y = \frac{1}{2}x - 1$ **(b)** $x - 2y = 2$
71. (a) $y = -\frac{1}{2}x + 9$ **(b)** $x + 2y = 18$ **73. (a)** $y = 7$ **(b)** $y = 7$
75. $y = 45x$; $(0, 0), (5, 225), (10, 450)$ **77.** $y = 3.10x$; $(0, 0)$,
$(5, 15.50), (10, 31.00)$ **79.** $y = 111x$; $(0, 0), (5, 555), (10, 1110)$
81. (a) $y = 112.50x + 12$ **(b)** $(5, 574.50)$; The cost for 5 tickets
and a parking pass is $574.50. **(c)** $237 **83. (a)** $y = 41x + 99$
(b) $(5, 304)$; The cost for a 5-month membership is $304. **(c)** $591
85. (a) $y = 60x + 36$ **(b)** $(5, 336)$; The cost of the plan for 5 months
is $336. **(c)** $756 **87. (a)** $y = 6x + 30$ **(b)** $(5, 60)$; It costs $60
to rent the saw for 5 days. **(c)** 18 days **89. (a)** $y = 1294.7x + 3921$;
Sales of digital cameras in the United States increased by
$1294.7 million per yr from 2003 to 2006. **(b)** $9099.8 million
91. (a) $y = 5.25x + 22.25$ **(b)** $48.5 billion; It is greater than the
actual value. **93.** $32; 212$ **94.** $(0, 32)$ and $(100, 212)$ **95.** $\frac{9}{5}$
96. $F = \frac{9}{5}C + 32$ **97.** $C = \frac{5}{9}(F - 32)$ **98.** $86°$ **99.** $10°$
100. $-40°$ **101.** $(-\infty, 2)$ **103.** $\left(-\infty, -\frac{4}{3}\right]$

Summary Exercises on Slopes and Equations of Lines (page 174)

1. $-\frac{3}{5}$ **2.** $-\frac{4}{7}$ **3.** 2 **4.** $\frac{5}{2}$ **5.** undefined **6.** 0 **7. (a)** $y = -\frac{5}{6}x + \frac{13}{3}$
(b) $5x + 6y = 26$ **8. (a)** $y = 3x + 11$ **(b)** $3x - y = -11$
9. (a) $y = -\frac{5}{2}x$ **(b)** $5x + 2y = 0$ **10. (a)** $y = -8$ **(b)** $y = -8$
11. (a) $y = -\frac{7}{9}$ **(b)** $9y = -7$ **12. (a)** $y = -3x + 10$
(b) $3x + y = 10$ **13. (a)** $y = \frac{2}{3}x + \frac{14}{3}$ **(b)** $2x - 3y = -14$
14. (a) $y = 2x - 10$ **(b)** $2x - y = 10$ **15. (a)** $y = -\frac{5}{2}x + 2$
(b) $5x + 2y = 4$ **16. (a)** $y = \frac{2}{3}x + 8$ **(b)** $2x - 3y = -24$
17. (a) $y = -7x + 3$ **(b)** $7x + y = 3$ **18. (a)** B **(b)** D **(c)** A
(d) C **(e)** E

Connections **(page 179)** We include a calculator graph and supporting
explanation only with the answer to Problem 1.

1. (a) $\{-0.6\}$; The graph of $y_1 = 5x + 3$ has x-intercept $(-0.6, 0)$.

(b) $(-0.6, \infty)$; The graph of y_1 lies
above the x-axis for values of
x greater than -0.6.
(c) $(-\infty, -0.6)$; The graph of y_1 lies
below the x-axis for values of
x less than -0.6.

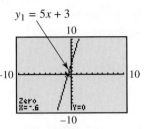

2. (a) $\{-0.5\}$ **(b)** $(-0.5, \infty)$ **(c)** $(-\infty, -0.5)$ **3. (a)** $\{-1.2\}$
(b) $(-\infty, -1.2]$ **(c)** $[-1.2, \infty)$ **4. (a)** $\{-3\}$ **(b)** $(-\infty, -3]$
(c) $[-3, \infty)$

Section 3.4 (pages 179–181)

1. solid; below **3.** dashed; above **5.** The graph of $Ax + By = C$
divides the plane into two regions. In one of the regions, the ordered
pairs satisfy $Ax + By < C$. In the other, they satisfy $Ax + By > C$.

7. **9.** **11.**

13. **15.** **17.**

19. **21.** **23.**

25. **27.** **29.** $-3 < x < 3$

31. $-2 < x + 1 < 2$ **33.** 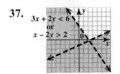 **35.**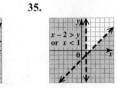
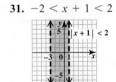

37. **39.** C **41.** A **43. (a)** $\{-4\}$ **(b)** $(-\infty, -4)$
(c) $(-4, \infty)$ **45. (a)** $\{3.5\}$ **(b)** $(3.5, \infty)$
(c) $(-\infty, 3.5)$ **47.** $x \le 200, x \ge 100$,
$y \ge 3000$

48. **49.** $C = 50x + 100y$ **50.** Some examples are
$(100, 5000), (150, 3000)$, and $(150, 5000)$.
The corner points are $(100, 3000)$ and
$(200, 3000)$. **51.** The least value occurs when
$x = 100$ and $y = 3000$.

52. The company should use 100 workers and manufacture 3000 units to achieve the least possible cost.   **53.** $[0, \infty)$ **55.** $(-\infty, 1) \cup (1, \infty)$

Section 3.5 (pages 187–189)

1. Answers will vary. A function is a set of ordered pairs in which each first component corresponds to exactly one second component. For example, $\{(0, 1), (1, 2), (2, 3), (3, 4)\}$ is a function.
3. independent variable

In Exercises 5 and 7, answers will vary.

5.   **7.**

x	y
−3	−4
−3	1
2	0

9. function; domain: $\{5, 3, 4, 7\}$; range: $\{1, 2, 9, 6\}$ **11.** not a function; domain: $\{2, 0\}$; range: $\{4, 2, 5\}$ **13.** function; domain: $\{-3, 4, -2\}$; range: $\{1, 7\}$ **15.** not a function; domain: $\{1, 0, 2\}$; range: $\{1, -1, 0, 4, -4\}$ **17.** function; domain: $\{2, 5, 11, 17, 3\}$; range: $\{1, 7, 20\}$ **19.** not a function; domain: $\{1\}$; range: $\{5, 2, -1, -4\}$ **21.** function; domain: $\{4, 2, 0, -2\}$; range: $\{-3\}$ **23.** function; domain: $\{-2, 0, 3\}$; range: $\{2, 3\}$ **25.** function; domain: $(-\infty, \infty)$; range: $(-\infty, \infty)$ **27.** not a function; domain: $(-\infty, 0]$; range: $(-\infty, \infty)$ **29.** function; domain: $(-\infty, \infty)$; range: $(-\infty, 4]$ **31.** not a function; domain: $[-4, 4]$; range: $[-3, 3]$ **33.** function; $(-\infty, \infty)$ **35.** function; $(-\infty, \infty)$ **37.** function; $(-\infty, \infty)$ **39.** not a function; $[0, \infty)$ **41.** not a function; $(-\infty, \infty)$ **43.** function; $[0, \infty)$ **45.** function; $[3, \infty)$ **47.** function; $\left[-\frac{1}{2}, \infty\right)$ **49.** function; $(-\infty, \infty)$ **51.** function; $(-\infty, 0) \cup (0, \infty)$ **53.** function; $(-\infty, 4) \cup (4, \infty)$ **55.** function; $(-\infty, 0) \cup (0, \infty)$ **57. (a)** yes   **(b)** domain: $\{2004, 2005, 2006, 2007, 2008\}$; range: $\{42.3, 42.8, 43.7, 43.8\}$ **(c)** Answers will vary. Two possible answers are $(2005, 42.3)$ and $(2008, 43.8)$.
59. -9 **61.** 1   **63.** $y = \frac{1}{2}x - \frac{7}{4}$

Section 3.6 (pages 194–197)

1. B   **3.** 4   **5.** -11 **7.** 3   **9.** 2.75   **11.** $-3p + 4$ **13.** $3x + 4$ **15.** $-3x - 2$ **17.** $-\pi^2 + 4\pi + 1$ **19.** $-3x - 3h + 4$ **21.** -9 **23. (a)** -1 **(b)** -1 **25. (a)** 2   **(b)** 3   **27. (a)** 15   **(b)** 10   **29. (a)** 4   **(b)** 1   **31. (a)** 3   **(b)** -3 **33. (a)** -3 **(b)** 2   **35. (a)** 2   **(b)** 0   **(c)** -1 **37. (a)** $f(x) = -\frac{1}{3}x + 4$ **(b)** 3   **39. (a)** $f(x) = 3 - 2x^2$ **(b)** -15 **41. (a)** $f(x) = \frac{4}{3}x - \frac{8}{3}$ **(b)** $\frac{4}{3}$
43. line; $-2; -2x + 4; -2; 3; -2$
45. domain: $(-\infty, \infty)$;   **47.** domain: $(-\infty, \infty)$;
range: $(-\infty, \infty)$ range: $(-\infty, \infty)$

49. domain: $(-\infty, \infty)$;   **51.** domain: $(-\infty, \infty)$;
range: $(-\infty, \infty)$ range: $\{-4\}$

53. domain: $(-\infty, \infty)$;   **55.** x-axis   **57. (a)** \$11.25   **(b)** 3 is
range: $\{0\}$ the value of the independent variable,
which represents a package weight of 3 lb;

$f(3)$ is the value of the dependent variable, representing the cost to mail a 3-lb package.   **(c)** \$18.75; $f(5) = 18.75$
59. (a) 194.53 cm   **(b)** 177.29 cm   **(c)** 177.41 cm   **(d)** 163.65 cm
61. (a) $f(x) = 12x + 100$ **(b)** 1600; The cost to print 125 t-shirts is \$1600.   **(c)** 75; $f(75) = 1000$; The cost to print 75 t-shirts is \$1000.
63. (a) 1.1   **(b)** 5   **(c)** -1.2 **(d)** $(0, 3.5)$ **(e)** $f(x) = -1.2x + 3.5$
65. (a) $[0, 100]$; $[0, 3000]$ **(b)** 25 hr; 25 hr   **(c)** 2000 gal
(d) $f(0) = 0$; The pool is empty at time 0.   **(e)** $f(25) = 3000$;
After 25 hr, there are 3000 gal of water in the pool.   **67.** 4   **69.** 4
71. $\{2\}$

Chapter 3 Review Exercises (pages 202–204)

1.

x	y
0	5
$\frac{10}{3}$	0
2	2
$\frac{14}{3}$	-2

2.

x	y
2	-6
5	-3
3	-5
6	-2

3. $(3, 0)$; $(0, -4)$

4. $\left(\frac{28}{5}, 0\right)$; $(0, 4)$

5. $(10, 0)$; $(0, 4)$

6. $(8, 0)$; $(0, -2)$

7. $(0, 2)$ **8.** $\left(-\frac{9}{2}, \frac{3}{2}\right)$ **9.** $-\frac{7}{5}$ **10.** $-\frac{1}{2}$ **11.** 2   **12.** $\frac{3}{4}$ **13.** undefined
14. $\frac{2}{3}$ **15.** $-\frac{1}{3}$ **16.** undefined   **17.** $-\frac{1}{3}$ **18.** -1 **19.** positive
20. negative   **21.** undefined   **22.** 0   **23.** 12 ft   **24.** \$1496 per yr
25. (a) $y = -\frac{1}{3}x - 1$ **(b)** $x + 3y = -3$ **26. (a)** $y = -2$
(b) $y = -2$ **27. (a)** $y = -\frac{4}{3}x + \frac{29}{3}$ **(b)** $4x + 3y = 29$
28. (a) $y = 3x + 7$ **(b)** $3x - y = -7$ **29. (a)** not possible
(b) $x = 2$ **30. (a)** $y = -9x + 13$ **(b)** $9x + y = 13$
31. (a) $y = \frac{7}{5}x + \frac{16}{5}$ **(b)** $7x - 5y = -16$ **32. (a)** $y = -x + 2$
(b) $x + y = 2$ **33. (a)** $y = 4x - 29$ **(b)** $4x - y = 29$
34. (a) $y = -\frac{5}{2}x + 13$ **(b)** $5x + 2y = 26$ **35. (a)** $y = 57x + 159$;
\$843   **(b)** $y = 47x + 159$; \$723   **36. (a)** $y = 143.75x + 1407.75$;
The revenue from skiing facilities increased by an average of \$143.75 million each year from 2003 to 2007.   **(b)** \$2558 million

37.   **38.**   **39.**

40.   **41.** D **42.** domain: $\{-4, 1\}$; range: $\{2, -2, 5, -5\}$; not a function **43.** domain: $\{9, 11, 4, 17, 25\}$; range: $\{32, 47, 69, 14\}$; function **44.** domain: $[-4, 4]$; range: $[0, 2]$; function

45. domain: $(-\infty, 0]$; range: $(-\infty, \infty)$; not a function **46.** function; domain: $(-\infty, \infty)$; linear function **47.** not a function; domain: $(-\infty, \infty)$ **48.** function; domain: $(-\infty, \infty)$ **49.** function; domain: $\left[-\frac{7}{4}, \infty\right)$ **50.** not a function; domain: $[0, \infty)$ **51.** function; domain: $(-\infty, 6) \cup (6, \infty)$ **52.** -6 **53.** -8.52 **54.** -8 **55.** $-2k^2 + 3k - 6$ **56.** $f(x) = 2x^2; 18$ **57.** C **58.** (a) yes (b) domain: $\{1960, 1970, 1980, 1990, 2000, 2009\}$; range: $\{69.7, 70.8, 73.7, 75.4, 76.8, 78.1\}$ (c) Answers will vary. Two possible answers are $(1960, 69.7)$ and $(2009, 78.1)$. (d) 73.7; In 1980, life expectancy at birth was 73.7 yr. (e) 2000 **59.** Because it falls from left to right, the slope is negative. **60.** $-\frac{3}{2}$ **61.** $-\frac{3}{2}; \frac{2}{3}$ **62.** $\left(\frac{7}{3}, 0\right)$ **63.** $\left(0, \frac{7}{2}\right)$ **64.** $f(x) = -\frac{3}{2}x + \frac{7}{2}$ **65.** $-\frac{17}{2}$ **66.** $x = \frac{23}{3}$

67. 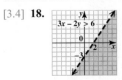 **68.** $\left\{\frac{7}{3}\right\}$ **69.** $\left(\frac{7}{3}, \infty\right)$ **70.** $\left(-\infty, \frac{7}{3}\right)$

Chapter 3 Test (pages 205–206)

[3.1] **1.** $-\frac{10}{3}; -2; 0$ **2.** $\left(\frac{20}{3}, 0\right); (0, -10)$

3. none; $(0, 5)$ **4.** $(2, 0)$; none

[3.2] **5.** $\frac{1}{2}$ **6.** It is a vertical line. **7.** perpendicular **8.** neither **9.** -929 farms per yr; The number of farms decreased, on the average, by about 929 each year from 1980 to 2008. [3.3] **10.** (a) $y = -5x + 19$ (b) $5x + y = 19$ **11.** (a) $y = 14$ (b) $y = 14$ **12.** (a) $y = -\frac{1}{2}x + 2$ (b) $x + 2y = 4$ **13.** (a) not possible (b) $x = 5$ **14.** (a) $y = -\frac{3}{5}x - \frac{11}{5}$ (b) $3x + 5y = -11$ **15.** (a) $y = -\frac{1}{2}x - \frac{3}{2}$ (b) $x + 2y = -3$ **16.** B **17.** (a) $y = 2078x + 51,557$ (b) $\$61,947$; It is more than the actual value.

[3.4] **18.**   **19.**

[3.5] **20.** D **21.** D **22.** domain: $[0, \infty)$; range: $(-\infty, \infty)$ **23.** domain: $\{0, -2, 4\}$; range: $\{1, 3, 8\}$ [3.6] **24.** (a) 0 (b) $-a^2 + 2a - 1$

25. domain: $(-\infty, \infty)$; range: $(-\infty, \infty)$

Chapters 1–3 Cumulative Review Exercises (pages 206–207)

[1.1] **1.** always true **2.** never true **3.** sometimes true; For example, $3 + (-3) = 0$, but $3 + (-1) = 2 \neq 0$. [1.2] **4.** 4 [1.3] **5.** 0.64 **6.** not a real number [1.4] **7.** $4m - 3$ **8.** $2x^2 + 5x + 4$ [1.3] **9.** $-\frac{19}{2}$ [1.1] **10.** $(-3, 5]$ [1.3] **11.** -39 **12.** undefined [2.1] **13.** $\left\{\frac{7}{6}\right\}$ **14.** $\{-1\}$ [2.4] **15.** 6 in. **16.** 2 hr

[2.5] **17.** $\left(-3, \frac{7}{2}\right)$

18. $(-\infty, 1]$

[2.6] **19.** $(6, 8)$

20. $(-\infty, -2] \cup (7, \infty)$

[2.7] **21.** $\{0, 7\}$ **22.** $(-\infty, \infty)$

[3.1] **23.** x-intercept: $(4, 0)$; y-intercept: $\left(0, \frac{12}{5}\right)$

[3.2] **24.** (a) $-\frac{6}{5}$ (b) $\frac{5}{6}$ [3.4] **25.**

[3.3] **26.** (a) $y = -\frac{3}{4}x - 1$ (b) $3x + 4y = -4$ **27.** (a) $y = -\frac{4}{3}x + \frac{7}{3}$ (b) $4x + 3y = 7$ [3.5] **28.** domain: $\{14, 91, 75, 23\}$; range: $\{9, 70, 56, 5\}$; not a function; 75 in the domain is paired with two different values, 70 and 56, in the range. [3.6] **29.** (a) domain: $(-\infty, \infty)$; range: $(-\infty, \infty)$ (b) 22 (c) 1 [3.2] **30.** -2.02; The per capita consumption of potatoes in the United States decreased by an average of 2.02 lb per yr from 2003 to 2008.

4   SYSTEMS OF LINEAR EQUATIONS

Connections   (page 219) (1) $y = \frac{2}{3}x - 1$; (2) $y = -x + 4$; $\{(3, 1)\}$

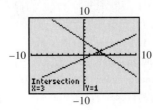

Section 4.1 (pages 219–224)

1. $4; -3$ **3.** $\emptyset$ **5.** 0 **7.** D; The ordered-pair solution must be in quadrant IV, since that is where the graphs of the equations intersect.
9. (a) B **(b)** C **(c)** A **(d)** D **11.** yes **13.** no
15. $\{(-2, -3)\}$ 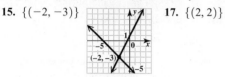 **17.** $\{(2, 2)\}$

19. $\{(1, 2)\}$ **21.** $\{(2, 3)\}$ **23.** $\left\{\left(\frac{22}{9}, \frac{22}{3}\right)\right\}$ **25.** $\{(5, 4)\}$
27. $\{(1, 3)\}$ **29.** $\left\{\left(-5, -\frac{10}{3}\right)\right\}$ **31.** $\{(2, 6)\}$
33. $\{(x, y) \mid 2x - y = 0\}$; dependent equations **35.** $\emptyset$; inconsistent system **37.** $\{(4, 2)\}$ **39.** $(0, 0)$ **41.** $\{(2, -4)\}$ **43.** $\{(3, -1)\}$
45. $\{(2, -3)\}$ **47.** $\{(x, y) \mid 7x + 2y = 6\}$; dependent equations
49. $\left\{\left(\frac{3}{2}, -\frac{3}{2}\right)\right\}$ **51.** $\emptyset$; inconsistent system **53.** $\{(0, 0)\}$
55. $\{(0, -4)\}$ **57.** $\left\{\left(6, -\frac{5}{6}\right)\right\}$ **59.** $y = -\frac{3}{7}x + \frac{4}{7}; y = -\frac{3}{7}x + \frac{3}{14}$;
no solution **61.** Both are $y = -\frac{2}{3}x + \frac{1}{3}$; infinitely many
solutions **63.** $\{(-3, 2)\}$ **65.** $\left\{\left(\frac{1}{3}, \frac{1}{2}\right)\right\}$ **67.** $\{(-4, 6)\}$
69. $\{(x, y) \mid 4x - y = -2\}$ **71.** $\left\{\left(1, \frac{1}{2}\right)\right\}$ **73.** $(3, -4)$ **75.** A
77. (a) $\{(5, 5)\}$ **(b)**

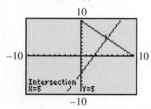

79. (a) \$4 **(b)** 300 half-gallons **(c)** supply: 200 half-gallons;
demand: 400 half-gallons **81. (a)** 2004–2008 **(b)** 2005 and 2006
(c) between 2005 and 2006; about \$4000 million **(d)** (2005, 4000)
(e) Sales of front projection displays were fairly constant. Sales of plasma
flat panel displays increased from 2003 to 2006, but then declined. Sales
of LCD flat panel displays increased over the whole period, fairly slowly
at first, and then very rapidly. **83.** 2000, 2001, first half of 2002
85. (1.4, 2675.4) (Values may vary slightly based on the method of
solution used.) **87.** $\{(2, 4)\}$ **89.** $\left\{\left(\frac{1}{2}, 2\right)\right\}$ **91.** $\left\{\left(\frac{c}{a}, 0\right)\right\}$
93. $\left\{\left(-\frac{1}{a}, -5\right)\right\}$ **95.** $8x - 12y + 4z = 20$ **97.** 4 **99.** -3

Section 4.2 (pages 231–233)

1. B **3.** $\{(3, 2, 1)\}$ **5.** $\{(1, 4, -3)\}$ **7.** $\{(0, 2, -5)\}$
9. $\{(1, 0, 3)\}$ **11.** $\left\{\left(1, \frac{3}{10}, \frac{2}{5}\right)\right\}$ **13.** $\left\{\left(-\frac{7}{3}, \frac{22}{3}, 7\right)\right\}$
15. $\{(-12, 18, 0)\}$ **17.** $\{(0.8, -1.5, 2.3)\}$ **19.** $\{(4, 5, 3)\}$
21. $\{(2, 2, 2)\}$ **23.** $\left\{\left(\frac{8}{3}, \frac{2}{3}, 3\right)\right\}$ **25.** $\{(-1, 0, 0)\}$
27. $\{(-4, 6, 2)\}$ **29.** $\{(-3, 5, -6)\}$ **31.** $\emptyset$; inconsistent system
33. $\{(x, y, z) \mid x - y + 4z = 8\}$; dependent equations **35.** $\{(3, 0, 2)\}$
37. $\{(x, y, z) \mid 2x + y - z = 6\}$; dependent equations **39.** $\{(0, 0, 0)\}$
41. $\emptyset$; inconsistent system **43.** $\{(2, 1, 5, 3)\}$ **45.** $\{(-2, 0, 1, 4)\}$
47. 100 in., 103 in., 120 in. **49.** $-4, 8, 12$

Section 4.3 (pages 241–246)

1. wins: 95; losses: 67 **3.** length: 78 ft; width: 36 ft **5.** AT&T:
\$124.0 billion; Verizon: \$97.4 billion **7.** $x = 40$ and $y = 50$, so the
angles measure 40° and 50°. **9.** NHL: \$288.23; NBA: \$291.93
11. Junior Roast Beef: \$2.09; Big Montana: \$4.39 **13. (a)** 6 oz
(b) 15 oz **(c)** 24 oz **(d)** 30 oz **15.** \$2.29x **17.** 6 gal of 25%;
14 gal of 35% **19.** pure acid: 6 L; 10% acid: 48 L **21.** nuts: 14 kg;
cereal: 16 kg **23.** \$1000 at 2%; \$2000 at 4% **25. (a)** $(10 - x)$ mph
(b) $(10 + x)$ mph **27.** train: 60 km per hr; plane: 160 km per hr
29. boat: 21 mph; current: 3 mph **31.** \$0.75-per-lb candy: 5.22 lb;
\$1.25-per-lb candy: 3.78 lb **33.** general admission: 76; with student
ID: 108 **35.** 8 for a citron; 5 for a wood apple **37.** $x + y + z = 180$;
angle measures: 70°, 30°, 80° **39.** first: 20°; second: 70°; third: 90°
41. shortest: 12 cm; middle: 25 cm; longest: 33 cm **43.** gold: 23;
silver: 21; bronze: 28 **45.** \$16 tickets: 1170; \$23 tickets: 985;
\$40 tickets: 130 **47.** bookstore A: 140; bookstore B: 280;
bookstore C: 380 **49.** first chemical: 50 kg; second chemical: 400 kg;
third chemical: 300 kg **51.** wins: 53; losses: 19; overtime losses: 10
53. (a) 6 **(b)** $-\frac{1}{6}$ **55. (a)** $-\frac{7}{8}$ **(b)** $\frac{8}{7}$

Section 4.4 (pages 252–253)

1. (a) $0, 5, -3$ **(b)** $1, -3, 8$ **(c)** yes; The number of rows is
the same as the number of columns (three). **(d)** $\begin{bmatrix} 1 & 4 & 8 \\ 0 & 5 & -3 \\ -2 & 3 & 1 \end{bmatrix}$

(e) $\begin{bmatrix} 1 & -\frac{3}{2} & -\frac{1}{2} \\ 0 & 5 & -3 \\ 1 & 4 & 8 \end{bmatrix}$ **(f)** $\begin{bmatrix} 1 & 15 & 25 \\ 0 & 5 & -3 \\ 1 & 4 & 8 \end{bmatrix}$ **3.** 3×2 **5.** 4×2

7. $\{(4, 1)\}$ **9.** $\{(1, 1)\}$ **11.** $\{(-1, 4)\}$ **13.** $\emptyset$
15. $\{(x, y) \mid 2x + y = 4\}$ **17.** $\{(0, 0)\}$ **19.** $\{(4, 0, 1)\}$
21. $\{(-1, 23, 16)\}$ **23.** $\{(3, 2, -4)\}$
25. $\{(x, y, z) \mid x - 2y + z = 4\}$ **27.** $\emptyset$ **29.** $\{(1, 1)\}$
31. $\{(-1, 2, 1)\}$ **33.** $\{(1, 7, -4)\}$ **35.** 64 **37.** 625 **39.** $\frac{81}{256}$

Chapter 4 Review Exercises (pages 257–259)

1. (a) 1980 and 1985 **(b)** just less than 500,000
2. $\{(2, 2)\}$ **3.** D

4. Answers will vary.

(a) **(b)** **(c)**

5. $\left\{\left(-\frac{8}{9}, -\frac{4}{3}\right)\right\}$ **6.** $\{(0, 4)\}$ **7.** $\{(2, 4)\}$ **8.** $\{(2, 2)\}$ **9.** $\{(0, 1)\}$
10. $\{(-1, 2)\}$ **11.** $\{(-6, 3)\}$ **12.** $\{(x, y) \mid 3x - y = -6\}$;
dependent equations **13.** $\emptyset$; inconsistent system

14. The two lines have the same slope, 3, but the y-intercepts, $(0, 2)$ and $(0, -4)$, are different. Therefore, the lines are parallel, do not intersect, and have no common solution. **15.** $\{(1, -5, 3)\}$ **16.** $\{(1, 2, 3)\}$ **17.** $\emptyset$; inconsistent system **18.** length: 200 ft; width: 85 ft **19.** New York Yankees: \$72.97; Boston Red Sox: \$50.24 **20.** plane: 300 mph; wind: 20 mph **21.** \$2-per-lb nuts: 30 lb; \$1-per-lb candy: 70 lb **22.** 85°, 60°, 35° **23.** \$40,000 at 10%; \$100,000 at 6%; \$140,000 at 5% **24.** 5 L of 8%; 3 L of 20%; none of 10% **25.** Mantle: 54; Maris: 61; Berra: 22 **26.** $\{(3, -2)\}$ **27.** $\{(-1, 5)\}$ **28.** $\{(0, 0, -1)\}$ **29.** $\{(1, 2, -1)\}$ **30.** B; The second equation is already solved for y. **31.** $\{(12, 9)\}$ **32.** $\emptyset$ **33.** $\{(3, -1)\}$ **34.** $\{(5, 3)\}$ **35.** $\{(0, 4)\}$ **36.** $\left\{\left(\frac{82}{23}, -\frac{4}{23}\right)\right\}$ **37.** 20 L **38.** U.S.: 37; Germany: 30; Canada: 26

Chapter 4 Test (pages 260–261)

[4.1] **1. (a)** Houston, Phoenix, Dallas **(b)** Philadelphia **(c)** Dallas, Phoenix, Philadelphia, Houston **2. (a)** 2010; 1.45 million **(b)** $(2025, 2.8)$ **3.** $\{(6, 1)\}$ **4.** $\{(6, -4)\}$

5. $\left\{\left(-\frac{9}{4}, \frac{5}{4}\right)\right\}$ **6.** $\{(x, y) \mid 12x - 5y = 8\}$; dependent equations **7.** $\{(3, 3)\}$ **8.** $\{(0, -2)\}$ **9.** $\emptyset$; inconsistent system [4.2] **10.** $\left\{\left(-\frac{2}{3}, \frac{4}{5}, 0\right)\right\}$ **11.** $\{(3, -2, 1)\}$ [4.3] **12.** *Star Wars Episode IV: A New Hope:* \$461.0 million; *Indiana Jones and the Kingdom of the Crystal Skull:* \$317.0 million **13.** 45 mph, 75 mph **14.** 20% solution: 4 L; 50% solution: 8 L **15.** AC adaptor: \$8; rechargeable flashlight: \$15 **16.** Orange Pekoe: 60 oz; Irish Breakfast: 30 oz; Earl Grey: 10 oz [4.4] **17.** $\left\{\left(\frac{2}{5}, \frac{7}{5}\right)\right\}$ **18.** $\{(-1, 2, 3)\}$

Chapters 1–4 Cumulative Review Exercises (pages 261–262)

[1.2] **1.** $-\frac{23}{20}$ **2.** $-\frac{2}{9}$ [1.3] **3.** 81 **4.** -81 **5.** -81 **6.** 0.7 **7.** -0.7 **8.** It is not a real number. **9.** -199 **10.** 455 [1.4] **11.** commutative property [2.1] **12.** $\left\{-\frac{15}{4}\right\}$ [2.7] **13.** $\left\{\frac{2}{3}, 2\right\}$ [2.2] **14.** $x = \dfrac{d - by}{a}$ [2.1] **15.** $\{11\}$ [2.5] **16.** $\left(-\infty, \frac{240}{13}\right]$ [2.7] **17.** $\left[-2, \frac{2}{3}\right]$ **18.** $(-\infty, \infty)$ [2.6] **19.** $(-\infty, \infty)$ [2.2] **20.** 2010; 1813; 62.8%; 57.2% [2.3, 2.4] **21.** pennies: 35; nickels: 29; dimes: 30 **22.** 46°, 46°, 88° [3.1] **23.** $y = 6$ **24.** $x = 4$ [3.2] **25.** $-\frac{4}{3}$ **26.** $\frac{3}{4}$ [3.3] **27.** $4x + 3y = 10$ [3.2] **28.** [3.4] **29.**

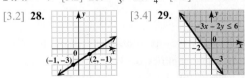

[3.6] **30. (a)** -6 **(b)** $a^2 + 3a - 6$ [4.1, 4.4] **31.** $\{(3, -3)\}$ **32.** $\{(x, y) \mid x - 3y = 7\}$ [4.2, 4.4] **33.** $\{(5, 3, 2)\}$ [4.3] **34.** Tickle Me Elmo: \$27.63; T.M.X.: \$40.00 [4.1] **35. (a)** $x = 8$, or 800 items; \$3000 **(b)** about \$400

5 **EXPONENTS, POLYNOMIALS, AND POLYNOMIAL FUNCTIONS**

Section 5.1 (pages 274–278)

1. incorrect; $(ab)^2 = a^2 b^2$ **3.** incorrect; $\left(\dfrac{3}{a}\right)^4 = \dfrac{3^4}{a^4}$ **5.** correct **7.** Do not multiply the bases. $4^5 \cdot 4^2 = 4^7$ **9.** 13^{12} **11.** 8^{10} **13.** x^{17} **15.** $-27w^8$ **17.** $18x^3 y^8$ **19.** The product rule does not apply. **21. (a)** B **(b)** C **(c)** B **(d)** C **23.** 1 **25.** -1 **27.** 1 **29.** 2 **31.** 0 **33.** -2 **35. (a)** B **(b)** D **(c)** B **(d)** D **37.** $\dfrac{1}{5^4}$, or $\dfrac{1}{625}$ **39.** $\frac{1}{9}$ **41.** $\dfrac{1}{16x^2}$ **43.** $\dfrac{4}{x^2}$ **45.** $-\dfrac{1}{a^3}$ **47.** $\dfrac{1}{a^4}$ **49.** $\frac{11}{30}$ **51.** $-\frac{5}{24}$ **53.** 16 **55.** $\frac{27}{4}$ **57.** $\frac{27}{8}$ **59.** $\frac{25}{16}$ **61. (a)** B **(b)** D **(c)** D **(d)** B **63.** 4^2, or 16 **65.** x^4 **67.** $\dfrac{1}{r^3}$ **69.** 6^6 **71.** $\dfrac{1}{6^{10}}$ **73.** 7^2, or 49 **75.** r^3 **77.** The quotient rule does not apply. **79.** x^{18} **81.** $\frac{27}{125}$ **83.** $64t^3$ **85.** $-216x^6$ **87.** $-\dfrac{64m^6}{t^3}$ **89.** $\dfrac{s^{12}}{t^{20}}$ **91.** $\frac{1}{3}$ **93.** $\dfrac{1}{a^5}$ **95.** $\dfrac{1}{k^2}$ **97.** $-4r^6$ **99.** $\dfrac{625}{a^{10}}$ **101.** $\dfrac{z^4}{x^3}$ **103.** $-14k^3$ **105.** $\dfrac{p^4}{5}$ **107.** $\dfrac{1}{2pq}$ **109.** $\dfrac{4}{a^2}$ **111.** $\dfrac{1}{6y^{13}}$ **113.** $\dfrac{4k^5}{m^2}$ **115.** $\dfrac{4k^{17}}{125}$ **117.** $\dfrac{2k^5}{3}$ **119.** $\dfrac{8}{3pq^{10}}$ **121.** $\dfrac{y^9}{8}$ **123.** $\dfrac{n^{10}}{25m^{18}}$ **125.** $-\dfrac{125y^3}{x^{30}}$ **127.** $-\dfrac{3}{32m^8 p^4}$ **129.** $\dfrac{2}{3y^4}$ **131.** $\dfrac{3p^8}{16q^{14}}$ **133.** 5.3×10^2 **135.** 8.3×10^{-1} **137.** 6.92×10^{-6} **139.** -3.85×10^4 **141.** 72,000 **143.** 0.00254 **145.** $-60,000$ **147.** 0.000012 **149.** 0.06 **151.** 0.0000025 **153.** 200,000 **155.** 3000 **157.** $\$1 \times 10^9$; $\$1 \times 10^{12}$ (or \$10^{12}); \$3.1 × 10^{12}; 2.10385×10^5 **159. (a)** 3.084×10^8 **(b)** $\$1 \times 10^{12}$ (or \$10^{12}) **(c)** \$3243 **161.** \$37,459 **163.** 300 sec **165.** approximately 5.87×10^{12} mi **167.** 998 mi² **169.** 7.5×10^9 **171.** 4×10^{17} **173.** $9x$ **175.** $3 + 5q$

Section 5.2 (pages 282–283)

1. 7; 1 **3.** -15; 2 **5.** 1; 4 **7.** $\frac{1}{6}$; 1 **9.** 8; 0 **11.** -1; 3 **13.** $2x^3 - 3x^2 + x + 4$; $2x^3$; 2 **15.** $p^7 - 8p^5 + 4p^3$; p^7; 1 **17.** $-3m^4 - m^3 + 10$; $-3m^4$; -3 **19.** monomial; 0 **21.** binomial; 1 **23.** binomial; 8 **25.** monomial; 6 **27.** trinomial; 3 **29.** none of these; 5 **31.** A **33.** $8z^4$ **35.** $7m^3$ **37.** $5x$ **39.** already simplified **41.** $-t + 13s$ **43.** $8k^2 + 2k - 7$ **45.** $-2n^4 - n^3 + n^2$ **47.** $-2ab^2 + 20a^2 b$ **49.** $3m + 11$ **51.** $-p - 4$ **53.** $8x^2 + x - 2$ **55.** $-t^4 + 2t^2 - t + 5$ **57.** $5y^3 - 3y^2 + 5y + 1$ **59.** $r + 13$ **61.** $-2a^2 - 2a - 7$ **63.** $-3z^5 + z^2 + 7z$ **65.** $12p - 4$ **67.** $-9p^2 + 11p - 9$ **69.** $5a + 18$ **71.** $14m^2 - 13m + 6$ **73.** $13z^2 + 10z - 3$ **75.** $10y^3 - 7y^2 + 5y + 8$ **77.** $-5a^4 - 6a^3 + 9a^2 - 11$ **79.** $3y^2 - 4y + 2$ **81.** $-4m^2 + 4n^2 - 7n$ **83.** $y^4 - 4y^2 - 4$ **85.** $10z^2 - 16z$ **87.** function; domain: $(-\infty, \infty)$; range: $(-\infty, \infty)$ **89.** not a function; domain: $[0, \infty)$; range: $(-\infty, \infty)$ **91. (a)** 3 **(b)** 6

Section 5.3 (pages 290–292)

1. (a) -10 **(b)** 8 **3. (a)** 8 **(b)** 2 **5. (a)** 8 **(b)** 74

7. (a) -11 **(b)** 4 **9. (a)** 4300 thousand lb **(b)** 19,371 thousand lb

(c) 64,449 thousand lb **11. (a)** \$28.2 billion **(b)** \$79.1 billion

(c) \$105.9 billion **13. (a)** $8x - 3$ **(b)** $2x - 17$

15. (a) $-x^2 + 12x - 12$ **(b)** $9x^2 + 4x + 6$ **17.** $x^2 + 2x - 9$

19. 6 **21.** $x^2 - x - 6$ **23.** 6 **25.** -33 **27.** 0 **29.** $-\frac{9}{4}$ **31.** $-\frac{9}{2}$

33. For example, let $f(x) = 2x^3 + 3x^2 + x + 4$ and $g(x) = 2x^4 +$
$3x^3 - 9x^2 + 2x - 4$. For these functions, $(f - g)(x) = -2x^4 - x^3 +$
$12x^2 - x + 8$, and $(g - f)(x) = 2x^4 + x^3 - 12x^2 + x - 8$.
Because the two differences are not equal, subtraction of functions is not
commutative. **35.** 6 **37.** 83 **39.** 53 **41.** 13 **43.** $2x^2 + 11$

45. $2x - 2$ **47.** $\frac{97}{4}$ **49.** 8 **51.** $(f \circ g)(x) = 63{,}360x$; It computes
the number of inches in x miles. **53.** $(\mathcal{A} \circ r)(t) = 4\pi t^2$; This is the
area of the circular layer as a function of time.

55. **57.** **59.**

$f(x) = -2x + 1$ $f(x) = -3x^2$ $f(x) = x^3 + 1$

domain: $(-\infty, \infty)$; domain: $(-\infty, \infty)$; domain: $(-\infty, \infty)$;
range: $(-\infty, \infty)$ range: $(-\infty, 0]$ range: $(-\infty, \infty)$

61. $12m^5$ **63.** $-6a^3b^9$ **65.** $60x^3y^4$

Section 5.4 (pages 299–301)

1. C **3.** D **5.** $-24m^5$ **7.** $-28x^7y^4$ **9.** $-6x^2 + 15x$

11. $-2q^3 - 3q^4$ **13.** $18k^4 + 12k^3 + 6k^2$ **15.** $6t^3 + t^2 - 14t - 3$

17. $m^3 - 3m^2 - 40m$ **19.** $24z^3 - 20z^2 - 16z$

21. $4x^5 - 4x^4 - 24x^3$ **23.** $6y^2 + y - 12$

25. $-2b^3 + 2b^2 + 18b + 12$ **27.** $25m^2 - 9n^2$

29. $8z^4 - 14z^3 + 17z^2 + 20z - 3$ **31.** $6p^4 + p^3 + 4p^2 - 27p - 6$

33. $m^2 - 3m - 40$ **35.** $12k^2 + k - 6$ **37.** $3z^2 + zw - 4w^2$

39. $12c^2 + 16cd - 3d^2$ **41.** The product of two binomials is the
sum of the product of the first terms, the product of the outer terms,
the product of the inner terms, and the product of the last terms.

43. $x^2 - 81$ **45.** $4p^2 - 9$ **47.** $25m^2 - 1$ **49.** $9a^2 - 4c^2$

51. $16m^2 - 49n^4$ **53.** $75y^7 - 12y$ **55.** $y^2 - 10y + 25$

57. $x^2 + 2x + 1$ **59.** $4p^2 + 28p + 49$ **61.** $16n^2 - 24nm + 9m^2$

63. Write 101 as $100 + 1$ and 99 as $100 - 1$. Then $101 \cdot 99 =$
$(100 + 1)(100 - 1) = 100^2 - 1^2 = 10{,}000 - 1 = 9999$.

65. $0.1x^2 + 0.63x - 0.13$ **67.** $3w^2 - \frac{23}{4}wz - \frac{1}{2}z^2$ **69.** $16x^2 - \frac{4}{9}$

71. $k^2 - \frac{10}{7}kp + \frac{25}{49}p^2$ **73.** $0.04x^2 - 0.56xy + 1.96y^2$

75. $25x^2 + 10x + 1 + 60xy + 12y + 36y^2$

77. $4a^2 + 4ab + b^2 - 12a - 6b + 9$ **79.** $4a^2 + 4ab + b^2 - 9$

81. $4h^2 - 4hk + k^2 - j^2$ **83.** $y^3 + 6y^2 + 12y + 8$

85. $125r^3 - 75r^2s + 15rs^2 - s^3$ **87.** $q^4 - 8q^3 + 24q^2 - 32q + 16$

89. $6a^3 + 7a^2b + 4ab^2 + b^3$ **91.** $4z^4 - 17z^3x + 12z^2x^2 - 6zx^3 + x^4$

93. $m^4 - 4m^2p^2 + 4mp^3 - p^4$ **95.** $a^4b - 7a^2b^3 - 6ab^4$

97. $49; 25; 49 \neq 25$ **99.** $2401; 337; 2401 \neq 337$ **101.** $\frac{9}{2}x^2 - 2y^2$

103. $15x^2 - 2x - 24$ **105.** $a - b$ **106.** $\mathcal{A} = s^2; (a - b)^2$

107. $(a - b)b$, or $ab - b^2; 2ab - 2b^2$ **108.** b^2 **109.** $a^2; a$

110. $a^2 - (2ab - 2b^2) - b^2 = a^2 - 2ab + b^2$ **111. (a)** They must
be equal to each other. **(b)** $(a - b)^2 = a^2 - 2ab + b^2$

112.

Area: a^2	Area: ab
Area: ab	Area: b^2

The large square is made up of two smaller
squares and two congruent rectangles. The sum of
the areas is $a^2 + 2ab + b^2$. Since $(a + b)^2$ must
represent the same quantity, they must be equal.
Thus, $(a + b)^2 = a^2 + 2ab + b^2$.

113. $10x^2 - 2x$ **115.** $2x^2 - x - 3$ **117.** $8x^3 - 27$ **119.** $2x^3 - 18x$

121. -20 **123.** $2x^2 - 6x$ **125.** 36 **127.** $\frac{35}{4}$ **129.** $\frac{1859}{64}$ **131.** $2p^4$

133. $\dfrac{-4b^6}{3a^2}$ **135.** $-8a^2 + a + 4$

Section 5.5 (pages 306–308)

1. quotient; exponents **3.** 0 **5.** $3x^3 - 2x^2 + 1$ **7.** $3y + 4 - \dfrac{5}{y}$

9. $3 + \dfrac{5}{m} + \dfrac{6}{m^2}$ **11.** $\dfrac{2}{7n} - \dfrac{3}{2m} + \dfrac{9}{7mn}$ **13.** $\dfrac{2y}{x} + \dfrac{3}{4} + \dfrac{3w}{x}$

15. $r^2 - 7r + 6$ **17.** $y - 4$ **19.** $q + 8$ **21.** $t + 5$

23. $p - 4 + \dfrac{44}{p + 6}$ **25.** $m^2 + 2m - 1$ **27.** $m^2 + m + 3$

29. $z^2 + 3$ **31.** $x^2 + 2x - 3 + \dfrac{6}{4x + 1}$ **33.** $2x - 5 + \dfrac{-4x + 5}{3x^2 - 2x + 4}$

35. $x^2 + x + 3$ **37.** $2x^2 - x - 5$ **39.** $3x^2 + 6x + 11 + \dfrac{26}{x - 2}$

41. $2k^2 + 3k - 1$ **43.** $2y^2 + 2$ **45.** $x^2 - 4x + 2 + \dfrac{9x - 4}{x^2 + 3}$

47. $p^2 + \dfrac{5}{2}p + 2 + \dfrac{-1}{2p + 2}$ **49.** $\dfrac{3}{2}a - 10 + \dfrac{77}{2a + 6}$

51. $p^2 + p + 1$ **53.** $\dfrac{2}{3}x - 1$ **55.** $\dfrac{3}{4}a - 2 + \dfrac{1}{4a + 3}$ **57.** $2p + 7$

59. $-13; -13$; They are the same, which suggests that when $P(x)$ is
divided by $x - r$, the result is $P(r)$. Here, $r = -1$. **61.** $5x - 1; 0$

63. $2x - 3; -1$ **65.** $4x^2 + 6x + 9; \frac{3}{2}$ **67.** $\dfrac{x^2 - 9}{2x}$, $x \neq 0$ **69.** $-\frac{5}{4}$

71. $\dfrac{x - 3}{2x}$, $x \neq 0$ **73.** 0 **75.** $-\frac{35}{4}$ **77.** $\frac{7}{2}$ **79.** $8y - 40$

81. $8p^2 + 4p$ **83.** $7(2x - 3z)$

Chapter 5 Review Exercises (pages 311–314)

1. 64 **2.** $\frac{1}{81}$ **3.** -125 **4.** 18 **5.** $\frac{81}{16}$ **6.** $\frac{16}{25}$ **7.** $\frac{1}{30}$ **8.** $\frac{3}{4}$ **9.** 0

10. $\dfrac{1}{3^8}$ **11.** x^8 **12.** $\dfrac{y^6}{x^2}$ **13.** $\dfrac{1}{z^{15}}$ **14.** $\dfrac{25}{m^{18}}$ **15.** $\dfrac{r^{17}}{9}$ **16.** $\dfrac{25}{z^4}$

17. $\dfrac{1}{96m^7}$ **18.** $\dfrac{2025}{8r^4}$ **19.** $-12x^2y^8$ **20.** $\dfrac{-2n}{m^5}$ **21.** $\dfrac{10p^8}{q^7}$

22. In $(-6)^0$, the base is -6 and the expression simplifies to 1. In -6^0,
the base is 6 and the expression simplifies to -1. **23.** yes **24.** No.
$(ab)^{-1} \neq ab^{-1}$ for all a, $b \neq 0$. For example, let $a = 3$ and $b = 4$. Then
$(ab)^{-1} = (3 \cdot 4)^{-1} = 12^{-1} = \frac{1}{12}$, while $ab^{-1} = 3 \cdot 4^{-1} = 3 \cdot \frac{1}{4} = \frac{3}{4}$;
$\frac{1}{12} \neq \frac{3}{4}$. **25.** 1.345×10^4 **26.** 7.65×10^{-8}

27. 1.38×10^{-1} **28.** $3.0406 \times 10^8; 9.2 \times 10^4; 1 \times 10^2$

29. 1,210,000   **30.** 0.0058   **31.** 2×10^{-4}; 0.0002   **32.** 1.5×10^3; 1500   **33.** 4.1×10^{-5}; 0.000041   **34.** 2.7×10^{-2}; 0.027
35. (a) 20,000 hr   **(b)** 833 days   **36.** 14   **37.** -1 **38.** $\frac{1}{10}$
39. 504   **40. (a)** $11k^3 - 3k^2 + 9k$ **(b)** trinomial   **(c)** 3
41. (a) $9m^7 + 14m^6$ **(b)** binomial   **(c)** 7
42. (a) $-5y^4 + 3y^3 + 7y^2 - 2y$ **(b)** none of these   **(c)** 4
43. (a) $-7q^5r^3$ **(b)** monomial   **(c)** 8
44. One example is $x^5 + 2x^4 - x^2 + x + 2$.   **45.** $-x^2 - 3x + 1$
46. $-5y^3 - 4y^2 + 6y - 12$ **47.** $6a^3 - 4a^2 - 16a + 15$
48. $8y^2 - 9y + 5$ **49.** $12x^2 + 8x + 5$ **50. (a)** -11 **(b)** 4
51. (a) $5x^2 - x + 5$ **(b)** $-5x^2 + 5x + 1$ **(c)** 11   **(d)** -9
52. (a) 167   **(b)** 1495   **(c)** 20   **(d)** 42   **(e)** $75x^2 + 220x + 160$
(f) $15x^2 + 10x + 2$ **53. (a)** 94,319   **(b)** 117,552   **(c)** 136,204
54.   **55.**   **56.**

57. $-12k^3 - 42k$ **58.** $15m^2 - 7m - 2$ **59.** $6w^2 - 13wt + 6t^2$
60. $10p^4 + 30p^3 - 8p^2 - 24p$ **61.** $3q^3 - 13q^2 - 14q + 20$
62. $36r^4 - 1$ **63.** $16m^2 + 24m + 9$ **64.** $9t^3 + 12t^2 + 4t$
65. $y^2 - 3y + \frac{5}{4}$ **66.** $x^2 - 4x + 6$ **67.** $p^2 + 6p + 9 + \frac{54}{2p - 3}$
68. $p^2 + 3p - 6$ **69. (a)** A   **(b)** G   **(c)** C   **(d)** C   **(e)** A   **(f)** E
(g) B   **(h)** H   **(i)** F   **(j)** I   **70.** $\frac{y^4}{36}$ **71.** $\frac{1}{125}$ **72.** -9
73. $21p^9 + 7p^8 + 14p^7$ **74.** $4x^2 - 36x + 81$ **75.** $-\frac{1}{5z^9}$ **76.** $\frac{1}{16y^{18}}$
77. $8x + 1 + \frac{5}{x - 3}$ **78.** $\frac{1250z^7x^6}{9}$ **79.** $9m^2 - 30mn + 25n^2 - p^2$
80. $2y^2x + \frac{3y^3}{2x} + \frac{5x^2}{2}$ **81.** $-3k^2 + 4k - 7$ **82.** 103,371 mi^2

Chapter 5 Test (pages 314–315)

[5.1] 1. (a) C   **(b)** A   **(c)** D   **(d)** A   **(e)** E   **(f)** F   **(g)** B   **(h)** G
(i) I   **(j)** C   **2.** $\frac{4x^7}{9y^{10}}$ **3.** $\frac{6}{r^{14}}$ **4.** $\frac{16}{9p^{10}q^{28}}$ **5.** $\frac{16}{x^6y^{16}}$ **6.** 0.00000091
7. 3×10^{-4}; 0.0003   **[5.3] 8. (a)** -18 **(b)** $-2x^2 + 12x - 9$
(c) $-2x^2 - 2x - 3$ **(d)** -7 **9. (a)** 23   **(b)** $3x^2 + 11$
(c) $9x^2 + 30x + 27$ **10.**   **11.**

12. (a) 446 thousand   **(b)** 710 thousand   **(c)** 907 thousand
[5.2] 13. $x^3 - 2x^2 - 10x - 13$ **[5.4] 14.** $10x^2 - x - 3$
15. $6m^3 - 7m^2 - 30m + 25$ **16.** $36x^2 - y^2$ **17.** $9k^2 + 6kq + q^2$
18. $4y^2 - 9z^2 + 6zx - x^2$ **[5.5] 19.** $4p - 8 + \frac{6}{p}$ **20.** $x^2 + 4x + 4$
[5.4] 21. (a) $x^3 + 4x^2 + 5x + 2$ **(b)** 0
[5.5] 22. (a) $x + 2$, $x \neq -1$ **(b)** 0

Chapters 1–5 Cumulative Review Exercises (pages 316–317)

[1.1] 1. (a) A, B, C, D, F   **(b)** B, C, D, F   **(c)** D, F   **(d)** C, D, F
(e) E, F   **(f)** D, F   **[1.3] 2.** 32   **3.** $-\frac{1}{72}$ **4.** 0   **[2.1] 5.** $\{-65\}$
6. {all real numbers}   **[2.2] 7.** $t = \dfrac{A - p}{pr}$ **[2.5] 8.** $(-\infty, 6)$
[2.7] 9. $\left\{-\frac{1}{3}, 1\right\}$ **10.** $\left(-\infty, -\frac{8}{3}\right] \cup [2, \infty)$ **[2.2, 2.4] 11.** 32%; 390; 270; 10%   **12.** 15°, 35°, 130°   **[3.2] 13.** $-\frac{4}{3}$ **14.** 0
[3.3] 15. (a) $y = -4x + 15$ **(b)** $4x + y = 15$
16. (a) $y = 4x$ **(b)** $4x - y = 0$
[3.1] 17.   **[3.4] 18.**   **19.**

[3.2, 3.3] 20. (a) $-12{,}272.8$ thousand lb per yr; The number of pounds of shrimp caught decreased an average of 12,272.8 thousand lb per yr.
(b) $y = -12{,}272.8x + 322{,}486$ **(c)** 285,668 thousand lb
[3.5] 21. domain: $\{-4, -1, 2, 5\}$; range: $\{-2, 0, 2\}$; function
[3.6] 22. -9 **[4.1] 23.** $\{(3, 2)\}$ **24.** $\varnothing$ **[4.2] 25.** $\{(1, 0, -1)\}$
[4.3] 26. length: 42 ft; width: 30 ft   **27.** 15% solution: 6 L; 30% solution: 3 L   **[5.1] 28.** $\dfrac{8m^9n^3}{p^6}$ **29.** $\dfrac{y^7}{x^{13}z^2}$ **30.** $\dfrac{m^6}{8n^9}$
[5.2] 31. $2x^2 - 5x + 10$ **[5.4] 32.** $x^3 + 8y^3$ **33.** $15x^2 + 7xy - 2y^2$
[5.5] 34. $4xy^4 - 2y + \dfrac{1}{x^2y}$ **35.** $m^2 - 2m + 3$

6   FACTORING

Section 6.1 (pages 324–325)

1. $12(m - 5)$ **3.** $4(1 + 5z)$ **5.** cannot be factored   **7.** $8k(k^2 + 3)$
9. $-2p^2q^4(2p + q)$ **11.** $7x^3(1 + 5x - 2x^2)$ **13.** $2t^3(5t^2 - 1 - 2t)$
15. $5ac(3ac^2 - 5c + 1)$ **17.** $16zn^3(zn^3 + 4n^4 - 2z^2)$
19. $7ab(2a^2b + a - 3a^4b^2 + 6b^3)$ **21.** $(m - 4)(2m + 5)$
23. $11(2z - 1)$ **25.** $(2 - x)^2(1 + 2x)$ **27.** $(3 - x)(6 + 2x - x^2)$
29. $20z(2z + 1)(3z + 4)$ **31.** $5(m + p)^2(m + p - 2 - 3m^2 - 6mp - 3p^2)$ **33.** $r(-r^2 + 3r + 5)$; $-r(r^2 - 3r - 5)$
35. $12s^4(-s + 4)$; $-12s^4(s - 4)$ **37.** $2x^2(-x^3 + 3x + 2)$; $-2x^2(x^3 - 3x - 2)$ **39.** $(m + q)(x + y)$ **41.** $(5m + n)(2 + k)$
43. $(2 - q)(2 - 3p)$ **45.** $(p + q)(p - 4z)$ **47.** $(2x + 3)(y + 1)$
49. $(m + 4)(m^2 - 6)$ **51.** $(a^2 + b^2)(-3a + 2b)$
53. $(y - 2)(x - 2)$ **55.** $(3y - 2)(3y^3 - 4)$ **57.** $(1 - a)(1 - b)$
59. $m^{-5}(3 + m^2)$ **61.** $p^{-3}(3 + 2p)$ **63.** The directions said that the student was to factor the polynomial *completely*. The completely factored form is $4xy^3(xy^2 - 2)$.   **65.** C   **67.** $k^2 + 6k - 7$ **69.** $25x^2 - 4t^2$
71. $6y^6 + y^3 - 12$

Section 6.2 (pages 332–333)

1. D **3.** B **5.** $(y - 3)(y + 10)$ **7.** $(p + 8)(p + 7)$ **9.** prime
11. $(a + 5b)(a - 7b)$ **13.** $(a - 6b)(a - 3b)$ **15.** prime
17. $-(6m - 5)(m + 3)$ **19.** $(5x - 6)(2x + 3)$
21. $(4k + 3)(5k + 8)$ **23.** $(3a - 2b)(5a - 4b)$ **25.** $(6m - 5)^2$
27. prime **29.** $(2xz - 1)(3xz + 4)$ **31.** $3(4x + 5)(2x + 1)$
33. $-5(a + 6)(3a - 4)$ **35.** $-11x(x - 6)(x - 4)$
37. $2xy^3(x - 12y)^2$ **39.** $6a(a - 3)(a + 5)$
41. $13y(y + 4)(y - 1)$ **43.** $3p(2p - 1)^2$ **45.** She did not factor
the polynomial *completely*. The factor $(4x + 10)$ can be factored
further into $2(2x + 5)$, giving the final form as $2(2x + 5)(x - 2)$.
47. $(6p^3 - r)(2p^3 - 5r)$ **49.** $(5k + 4)(2k + 1)$
51. $(3m + 3p + 5)(m + p - 4)$ **53.** $(a + b)^2(a - 3b)(a + 2b)$
55. $(p + q)^2(p + 3q)$ **57.** $(z - x)^2(z + 2x)$
59. $(p^2 - 8)(p^2 - 2)$ **61.** $(2x^2 + 3)(x^2 - 6)$
63. $(4x^2 + 3)(4x^2 + 1)$ **65.** $9x^2 - 25$ **67.** $p^2 + 6pq + 9q^2$
69. $y^3 + 27$

Section 6.3 (pages 337–339)

1. A, D **3.** B, C **5.** The sum of two squares can be factored
if the binomial terms have a common factor greater than 1.
7. $(p + 4)(p - 4)$ **9.** $(5x + 2)(5x - 2)$
11. $2(3a + 7b)(3a - 7b)$ **13.** $4(4m^2 + y^2)(2m + y)(2m - y)$
15. $(y + z + 9)(y + z - 9)$ **17.** $(4 + x + 3y)(4 - x - 3y)$
19. $(p^2 + 16)(p + 4)(p - 4)$ **21.** $(k - 3)^2$ **23.** $(2z + w)^2$
25. $(4m - 1 + n)(4m - 1 - n)$ **27.** $(2r - 3 + s)(2r - 3 - s)$
29. $(x + y - 1)(x - y + 1)$ **31.** $2(7m + 3n)^2$ **33.** $(p + q + 1)^2$
35. $(a - b + 4)^2$ **37.** $(x - 3)(x^2 + 3x + 9)$
39. $(6 - t)(36 + 6t + t^2)$ **41.** $(x + 4)(x^2 - 4x + 16)$
43. $(10 + y)(100 - 10y + y^2)$ **45.** $(2x + 1)(4x^2 - 2x + 1)$
47. $(5x - 6)(25x^2 + 30x + 36)$ **49.** $(x - 2y)(x^2 + 2xy + 4y^2)$
51. $(4g - 3h)(16g^2 + 12gh + 9h^2)$
53. $(7p + 5q)(49p^2 - 35pq + 25q^2)$
55. $3(2n + 3p)(4n^2 - 6np + 9p^2)$
57. $(y + z + 4)(y^2 + 2yz + z^2 - 4y - 4z + 16)$
59. $(m^2 - 5)(m^4 + 5m^2 + 25)$ **61.** $(3 - 10x^3)(9 + 30x^3 + 100x^6)$
63. $(5y^2 + z)(25y^4 - 5y^2z + z^2)$ **65.** $(x^3 - y^3)(x^3 + y^3)$;
$(x - y)(x^2 + xy + y^2)(x + y)(x^2 - xy + y^2)$
66. $(x^2 + xy + y^2)(x^2 - xy + y^2)$
67. $(x^2 - y^2)(x^4 + x^2y^2 + y^4)$; $(x - y)(x + y)(x^4 + x^2y^2 + y^4)$
68. $x^4 + x^2y^2 + y^4$ **69.** The product must equal $x^4 + x^2y^2 + y^4$.
Multiply $(x^2 + xy + y^2)(x^2 - xy + y^2)$ to verify this.
70. Start by factoring as a difference of squares.
71. $(5p + 2q)(25p^2 - 10pq + 4q^2 + 5p - 2q)$
73. $(3a - 4b)(9a^2 + 12ab + 16b^2 + 5)$
75. $(t - 3)(2t + 1)(4t^2 - 2t + 1)$
77. $(8m - 9n)(8m + 9n - 64m^2 - 72mn - 81n^2)$
79. $(2x + y)(a - b)$ **81.** $(p + 7)(p - 3)$

Section 6.4 (pages 342–343)

1. $(10a + 3b)(10a - 3b)$ **3.** $3p^2(p - 6)(p + 5)$
5. $3pq(a + 6b)(a - 5b)$ **7.** prime **9.** $(6b + 1)(b - 3)$
11. $(x - 10)(x^2 + 10x + 100)$ **13.** $(p + 2)(4 + m)$
15. $9m(m - 5 + 2m^2)$ **17.** $2(3m - 10)(9m^2 + 30m + 100)$
19. $(3m - 5n)^2$ **21.** $(k - 9)(q + r)$ **23.** $16z^2x(zx - 2)$
25. $(x + 7)(x - 5)$ **27.** $(25 + x^2)(5 - x)(5 + x)$
29. $(p + 1)(p^2 - p + 1)$ **31.** $(8m + 25)(8m - 25)$
33. $6z(2z^2 - z + 3)$ **35.** $16(4b + 5c)(4b - 5c)$
37. $8(4 + 5z)(16 - 20z + 25z^2)$ **39.** $(5r - s)(2r + 5s)$
41. $4pq(2p + q)(3p + 5q)$ **43.** $3(4k^2 + 9)(2k + 3)(2k - 3)$
45. $(m - n)(m^2 + mn + n^2 + m + n)$
47. $(x - 2m - n)(x + 2m + n)$ **49.** $6p^3(3p^2 - 4 + 2p^3)$
51. $2(x + 4)(x - 5)$ **53.** $8mn$ **55.** $2(5p + 9)(5p - 9)$
57. $4rx(3m^2 + mn + 10n^2)$ **59.** $(7a - 4b)(3a + b)$ **61.** prime
63. $(p + 8q - 5)^2$ **65.** $(7m^2 + 1)(3m^2 - 5)$
67. $(2r - t)(r^2 - rt + 19t^2)$ **69.** $(x + 3)(x^2 + 1)(x + 1)(x - 1)$
71. $(m + n - 5)(m - n + 1)$ **73.** $\left\{-\frac{2}{3}\right\}$ **75.** $\{0\}$ **77.** $\{-10\}$

Connections **(page 349)** **1.** $\{-1, 7\}$ **2.** $\{3\}$ **3.** $\{-2, 2\}$

Section 6.5 (pages 350–354)

1. First rewrite the equation so that one side is 0. Factor the other side
and set each factor equal to 0. The solutions of these linear equations are
solutions of the quadratic equation. **3.** $\{-10, 5\}$ **5.** $\left\{-\frac{8}{3}, \frac{5}{2}\right\}$
7. $\{-2, 5\}$ **9.** $\{-6, -3\}$ **11.** $\left\{-\frac{1}{2}, 4\right\}$ **13.** $\left\{-\frac{1}{3}, \frac{4}{5}\right\}$
15. $\{-3, 4\}$ **17.** $\left\{-5, -\frac{1}{5}\right\}$ **19.** $\{-4, 0\}$ **21.** $\{0, 6\}$
23. $\{-2, 2\}$ **25.** $\{-3, 3\}$ **27.** $\{3\}$ **29.** $\left\{-\frac{4}{3}\right\}$ **31.** $\{-4, 2\}$
33. $\left\{-\frac{1}{2}, 6\right\}$ **35.** $\{1, 6\}$ **37.** $\left\{-\frac{1}{2}, 0, 5\right\}$ **39.** $\{-1, 0, 3\}$
41. $\left\{-\frac{4}{3}, 0, \frac{4}{3}\right\}$ **43.** $\left\{-\frac{5}{2}, -1, 1\right\}$ **45.** $\{-3, 3, 6\}$
47. By dividing each side by a variable expression, she "lost" the
solution 0. The solution set is $\left\{-\frac{4}{3}, 0, \frac{4}{3}\right\}$. **49.** $\left\{-\frac{1}{2}, 6\right\}$
51. $\left\{-\frac{2}{3}, \frac{4}{15}\right\}$ **53.** $\left\{-\frac{3}{2}, \frac{1}{2}\right\}$ **55.** width: 16 ft; length: 20 ft
57. base: 12 ft; height: 5 ft **59.** 50 ft by 100 ft **61.** -6 and -5 or
5 and 6 **63.** length: 15 in.; width: 9 in. **65.** 5 sec **67.** $6\frac{1}{4}$ sec
69. $r = \dfrac{-2k - 3y}{a - 1}$, or $r = \dfrac{2k + 3y}{1 - a}$ **71.** $y = \dfrac{-x}{w - 3}$, or $y = \dfrac{x}{3 - w}$
73. L appears on both sides of the equation. **75.** $\{-0.5, 4\}$
77. $f(x) = -16x^2 + 80x + 100$
78.
$f(x) = -16x^2 + 80x + 100$

79. 200 ft; 2.5 sec **80.** about 6 sec
81. in the approximate interval
$(0.7, 4.3)$ **82.** $f(x) = -16x^2 + 100x$
83. $4p$ **85.** $-\dfrac{3}{4m^4n^3}$ **87.** $\dfrac{36}{75}$

Chapter 6 Review Exercises (pages 356–358)

1. $6p(2p - 1)$ **2.** $7x(3x + 5)$ **3.** $4qb(3q + 2b - 5q^2b)$

4. $6rt(r^2 - 5rt + 3t^2)$ **5.** $(x + 3)(x - 3)$ **6.** $(z + 1)(3z - 1)$

7. $(m + q)(4 + n)$ **8.** $(x + y)(x + 5)$ **9.** $(m + 3)(2 - a)$

10. $(x + 3)(x - y)$ **11.** $(3p - 4)(p + 1)$ **12.** $(3k - 2)(2k + 5)$

13. $(3r + 1)(4r - 3)$ **14.** $(2m + 5)(5m + 6)$

15. $(2k - h)(5k - 3h)$ **16.** prime **17.** $2x(4 + x)(3 - x)$

18. $3b(2b - 5)(b + 1)$ **19.** $(y^2 + 4)(y^2 - 2)$

20. $(2k^2 + 1)(k^2 - 3)$ **21.** $(p + 2)^2(p + 3)(p - 2)$

22. $(3r + 16)(r + 1)$ **23.** It is not factored because there are two terms: $x^2(y^2 - 6)$ and $5(y^2 - 6)$. The correct answer is $(y^2 - 6)(x^2 + 5)$.

24. $p + 1$ **25.** $(4x + 5)(4x - 5)$ **26.** $(3t + 7)(3t - 7)$

27. $(6m - 5n)(6m + 5n)$ **28.** $(x + 7)^2$ **29.** $(3k - 2)^2$

30. $(r + 3)(r^2 - 3r + 9)$ **31.** $(5x - 1)(25x^2 + 5x + 1)$

32. $(m + 1)(m^2 - m + 1)(m - 1)(m^2 + m + 1)$

33. $(x^4 + 1)(x^2 + 1)(x + 1)(x - 1)$

34. $(x + 3 + 5y)(x + 3 - 5y)$ **35.** $2b(3a^2 + b^2)$

36. $(x + 1)(x - 1)(x - 2)(x^2 + 2x + 4)$ **37.** $\{4\}$

38. $\left\{-1, -\frac{2}{5}\right\}$ **39.** $\{2, 3\}$ **40.** $\{-4, 2\}$ **41.** $\left\{-\frac{5}{2}, \frac{10}{3}\right\}$

42. $\left\{-\frac{3}{2}, \frac{1}{3}\right\}$ **43.** $\left\{-\frac{3}{2}, -\frac{1}{4}\right\}$ **44.** $\{-3, 3\}$ **45.** $\left\{-\frac{3}{2}, 0\right\}$

46. $\left\{\frac{1}{2}, 1\right\}$ **47.** $\{4\}$ **48.** $\left\{-\frac{7}{2}, 0, 4\right\}$ **49.** $\{-3, -2, 2\}$

50. $\left\{-2, -\frac{6}{5}, 3\right\}$ **51.** 3 ft **52.** length: 60 ft; width: 40 ft

53. 16 sec **54.** 1 sec and 15 sec **55.** The rock reaches a height of 240 ft once on its way up and once on its way down.

56. 8 sec **57.** $k = \dfrac{-3s - 2t}{b - 1}$, or $k = \dfrac{3s + 2t}{1 - b}$

58. $w = \dfrac{7}{z - 3}$, or $w = \dfrac{-7}{3 - z}$ **59.** $(4 + 9k)(4 - 9k)$

60. $a(6 - m)(5 + m)$ **61.** prime **62.** $(2 - a)(4 + 2a + a^2)$

63. $(5z - 3m)^2$ **64.** $5y^2(3y + 4)$ **65.** $\left\{-\frac{3}{5}, 4\right\}$ **66.** $\{-1, 0, 1\}$

67. 6 in. **68.** width: 25 ft; length: 110 ft

Chapter 6 Test (page 358)

[6.1–6.4] 1. $11z(z - 4)$ **2.** $5x^2y^3(2y^2 - 1 - 5x^3)$

3. $(x + y)(3 + b)$ **4.** $-(2x + 9)(x - 4)$ **5.** $(3x - 5)(2x + 7)$

6. $(4p - q)(p + q)$ **7.** $(4a + 5b)^2$ **8.** $(x + 1 + 2z)(x + 1 - 2z)$

9. $(a + b)(a - b)(a + 2)$ **10.** $(3k + 11j)(3k - 11j)$

11. $(y - 6)(y^2 + 6y + 36)$ **12.** $(2k^2 - 5)(3k^2 + 7)$

13. $(3x^2 + 1)(9x^4 - 3x^2 + 1)$ **[6.2] 14.** D **[6.5] 15.** $\left\{-2, -\frac{2}{3}\right\}$

16. $\left\{0, \frac{5}{3}\right\}$ **17.** $\left\{-\frac{2}{5}, 1\right\}$ **18.** $r = \dfrac{-2 - 6t}{a - 3}$, or $r = \dfrac{2 + 6t}{3 - a}$

19. length: 8 in.; width: 5 in. **20.** 2 sec and 4 sec

Chapters 1– 6 Cumulative Review Exercises (pages 359–360)

[1.4] 1. $-2m + 6$ **2.** $2x^2 + 5x + 4$ **[1.3] 3.** 10 **4.** undefined

[2.1] 5. $\left\{\frac{7}{6}\right\}$ **6.** $\{-1\}$ **[2.5] 7.** $\left(-\frac{1}{2}, \infty\right)$ **[2.6] 8.** $(2, 3)$

9. $(-\infty, 2) \cup (3, \infty)$ **[2.7] 10.** $\left\{-\frac{16}{5}, 2\right\}$ **11.** $(-11, 7)$

12. $(-\infty, -2] \cup [7, \infty)$ **[2.4] 13.** 2 hr

[3.1] 14.

[3.2] 15. -1 **16.** 0 **[3.6] 17.** -1

18. $\left(-\frac{7}{2}, 0\right)$ **19.** $(0, 7)$

[4.1] 20. $\{(1, 5)\}$ **[4.2] 21.** $\{(1, 1, 0)\}$

[5.1] 22. $\dfrac{y}{18x}$

[5.4] 23. $49x^2 + 42xy + 9y^2$ **[5.2] 24.** $x^3 + 12x^2 - 3x - 7$

[6.1–6.4] 25. $(2w + 7z)(8w - 3z)$ **26.** $(2x - 1 + y)(2x - 1 - y)$

27. $(10x^2 + 9)(10x^2 - 9)$ **28.** $(2p + 3)(4p^2 - 6p + 9)$

[6.5] 29. $\left\{-4, -\frac{3}{2}, 1\right\}$ **30.** $\left\{\frac{1}{3}\right\}$ **31.** 4 ft

32. longer sides: 18 in.; distance between: 16 in.

7 RATIONAL EXPRESSIONS AND FUNCTIONS

Section 7.1 (pages 368–371)

1. $7; \{x \mid x \neq 7\}$ **3.** $-\frac{1}{7}; \left\{x \mid x \neq -\frac{1}{7}\right\}$ **5.** $0; \{x \mid x \neq 0\}$

7. $-2, \frac{3}{2}; \left\{x \mid x \neq -2, \frac{3}{2}\right\}$ **9.** none; $\{x \mid x \text{ is a real number}\}$

11. none; $\{x \mid x \text{ is a real number}\}$ **13.** $\frac{2}{15}$ **15.** $\frac{9}{10}$ **17.** $\frac{3}{4}$

19. (a) C **(b)** A **(c)** D **(d)** B **(e)** E **(f)** F **21.** numerator: x^2, $4x$; denominator: $x, 4; x$ **23.** B **25.** x **27.** $\dfrac{x - 3}{x + 5}$ **29.** $\dfrac{x + 3}{2x(x - 3)}$

31. It is already in lowest terms. **33.** $\frac{6}{7}$ **35.** $\dfrac{z}{6}$ **37.** $\dfrac{t - 3}{3}$

39. $\dfrac{2}{t - 3}$ **41.** $\dfrac{x - 3}{x + 1}$ **43.** $\dfrac{4x + 1}{4x + 3}$ **45.** $a^2 - ab + b^2$

47. $\dfrac{c + 6d}{c - d}$ **49.** $\dfrac{a + b}{a - b}$ **51.** -1

In Exercises 53 and 55, there are other acceptable ways to express each answer.

53. $-(x + y)$ **55.** $-\dfrac{x + y}{x - y}$ **57.** $-\frac{1}{2}$ **59.** It is already in lowest terms.

61. $\dfrac{3y}{x^2}$ **63.** $\dfrac{3a^3b^2}{4}$ **65.** $\dfrac{27}{2mn^7}$ **67.** $\dfrac{x + 4}{x - 2}$ **69.** $\dfrac{2x + 3}{x + 2}$ **71.** $\dfrac{7x}{6}$

73. $-\dfrac{p + 5}{2p}$ (There are other ways.) **75.** $\frac{35}{4}$ **77.** $-(z + 1)$, or $-z - 1$

79. $\dfrac{14x^2}{5}$ **81.** -2 **83.** $\dfrac{x + 4}{x - 4}$ **85.** $\dfrac{a^2 + ab + b^2}{a - b}$ **87.** $\dfrac{2x - 3}{2(x - 3)}$

89. $\dfrac{a^2 + 2ab + 4b^2}{a + 2b}$ **91.** $\dfrac{2x + 3}{2x - 3}$ **93.** $\dfrac{k + 5p}{2k + 5p}$

95. $(k - 1)(k - 2)$ **97.** $-\frac{2}{3}$ **99.** $\frac{17}{42}$

Section 7.2 (pages 377–379)

1. $\frac{4}{5}$ **3.** $-\frac{1}{18}$ **5.** $\frac{31}{36}$ **7.** $\dfrac{9}{t}$ **9.** $\dfrac{6x + y}{7}$ **11.** $\dfrac{2}{x}$ **13.** $-\dfrac{2}{x^3}$ **15.** 1

17. $x - 5$ **19.** $\dfrac{5}{p + 3}$ **21.** $a - b$ **23.** $72x^4y^5$ **25.** $z(z - 2)$

27. $2(y + 4)$ **29.** $(x + 9)^2(x - 9)$ **31.** $(m + n)(m - n)$

33. $x(x - 4)(x + 1)$ **35.** $(t + 5)(t - 2)(2t - 3)$

37. $2y(y + 3)(y - 3)$ **39.** $2(x + 2)^2(x - 3)$ **41.** The expression $\dfrac{x - 4x - 1}{x + 2}$ is incorrect. The third term in the numerator should be $+1$, since the $-$ sign should be distributed over both $4x$ and -1. The answer should be $\dfrac{-3x + 1}{x + 2}$. **43.** $\dfrac{31}{3t}$ **45.** $\dfrac{5 - 22x}{12x^2y}$ **47.** $\dfrac{16b + 9a^2}{60a^4b^6}$

49. $\dfrac{4pr + 3sq^3}{14p^4q^4}$ **51.** $\dfrac{a^2b^5 - 2ab^6 + 3}{a^5b^7}$ **53.** $\dfrac{1}{x(x-1)}$

55. $\dfrac{5a^2 - 7a}{(a+1)(a-3)}$ **57.** 3 **59.** 4 **61.** $\dfrac{3}{x-4}$, or $\dfrac{-3}{4-x}$

63. $\dfrac{w+z}{w-z}$, or $\dfrac{-w-z}{z-w}$ **65.** $\dfrac{-2}{(x+1)(x-1)}$ **67.** $\dfrac{2(2x-1)}{x-1}$ **69.** $\dfrac{7}{y}$

71. $\dfrac{6}{x-2}$ **73.** $\dfrac{3x-2}{x-1}$ **75.** $\dfrac{4x-7}{x^2-x+1}$ **77.** $\dfrac{2x+1}{x}$

79. $\dfrac{4p^2 - 21p + 29}{(p-2)^2}$ **81.** $\dfrac{x}{(x-2)^2(x-3)}$

83. $\dfrac{2x(x+12y)}{(x+2y)(x-y)(x+6y)}$ **85.** $\dfrac{2x^2 + 21xy - 10y^2}{(x+2y)(x-y)(x+6y)}$

87. $\dfrac{3r-2s}{(2r-s)(3r-s)}$ **89.** $\dfrac{10x+23}{(x+2)^2(x+3)}$

91. (a) $c(x) = \dfrac{10x}{49(101-x)}$ **(b)** approximately 3.23 thousand dollars

93. $\frac{1}{6}$ **95.** $\frac{9}{5}$

Section 7.3 (pages 384–386)

1. $\frac{4}{15}$ **3.** $\frac{7}{17}$ **5.** $\dfrac{2x}{x-1}$ **7.** $\dfrac{2(k+1)}{3k-1}$ **9.** $\dfrac{5x^2}{9z^3}$ **11.** $\dfrac{6x+1}{7x-3}$

13. $\dfrac{y+x}{y-x}$ **15.** $4x$ **17.** $x+4y$ **19.** $\dfrac{y+4}{2}$ **21.** $\dfrac{a+b}{ab}$ **23.** xy

25. $\dfrac{3y}{2}$ **27.** $\dfrac{x^2+5x+4}{x^2+5x+10}$ **29.** $\dfrac{m^2+6m-4}{m(m-1)}$ **30.** $\dfrac{m^2-m-2}{m(m-1)}$

31. $\dfrac{m^2+6m-4}{m^2-m-2}$ **32.** $m(m-1)$ **33.** $\dfrac{m^2+6m-4}{m^2-m-2}$

34. Answers will vary. **35.** $\dfrac{x^2y^2}{y^2+x^2}$ **37.** $\dfrac{y^2+x^2}{xy^2+x^2y}$, or $\dfrac{y^2+x^2}{xy(y+x)}$

39. $\dfrac{1}{2xy}$ **41. (a)** $\dfrac{\dfrac{3}{mp} - \dfrac{4}{p} + \dfrac{8}{m}}{\dfrac{2}{m} - \dfrac{3}{p}}$ **(b)** In the denominator, $2m^{-1} = \dfrac{2}{m}$,

not $\dfrac{1}{2m}$, and $3p^{-1} = \dfrac{3}{p}$, not $\dfrac{1}{3p}$. **(c)** $\dfrac{3 - 4m + 8p}{2p - 3m}$ **43.** $\{-12\}$

45. $\{16\}$ **47.** 0; $\{x \mid x \neq 0\}$

Connections (page 391) **1.** $\{-1\}$ **2.** $\{0.5\}$

Section 7.4 (pages 391–394)

1. (a) 0 **(b)** $\{x \mid x \neq 0\}$ **3. (a)** $-1, 2$ **(b)** $\{x \mid x \neq -1, 2\}$

5. (a) $-4, 4$ **(b)** $\{x \mid x \neq \pm 4\}$ **7. (a)** $0, 1, -3, 2$

(b) $\{x \mid x \neq 0, 1, -3, 2\}$ **9. (a)** $-\frac{7}{4}, 0, \frac{13}{6}$ **(b)** $\{x \mid x \neq -\frac{7}{4}, 0, \frac{13}{6}\}$

11. (a) $4, \frac{7}{2}$ **(b)** $\{x \mid x \neq 4, \frac{7}{2}\}$ **13.** No, there is no possibility that the

proposed solution will be rejected, because there are no variables in the

denominators in the original equation. **15.** $\{1\}$ **17.** $\{-6, 4\}$

19. $\{-7, 3\}$ **21.** $\{-\frac{7}{12}\}$ **23.** $\emptyset$ **25.** $\{-3\}$ **27.** $\{0\}$ **29.** $\emptyset$

31. $\{5\}$ **33.** $\emptyset$ **35.** $\{\frac{27}{56}\}$ **37.** $\{-3, -1\}$ **39.** $\emptyset$ **41.** $\{-10\}$

43. $\{-1\}$ **45.** $\{13\}$ **47.** $\{x \mid x \neq \pm 3\}$

49. $x = 0; y = 0$ **51.** $x = 0; y = 0$ **53.** $x = 2; y = 0$

55. (a) 0 **(b)** 1.6 **(c)** 4.1 **(d)** The waiting time also increases.

57. (a) 500 ft **(b)** It decreases. **59.** four **61.** $t = \dfrac{d}{r}$

63. $c = P - a - b$

Summary Exercises on Rational Expressions and Equations (page 395)

1. equation; $\{20\}$ **2.** expression; $\dfrac{2(x+5)}{5}$ **3.** expression; $-\dfrac{22}{7x}$

4. expression; $\dfrac{y+x}{y-x}$ **5.** equation; $\{\frac{1}{2}\}$ **6.** equation; $\{7\}$

7. expression; $\dfrac{43}{24x}$ **8.** equation; $\{1\}$ **9.** expression; $\dfrac{5x-1}{-2x+2}$,

or $\dfrac{5x-1}{-2(x-1)}$ **10.** expression; $\dfrac{25}{4(r+2)}$ **11.** expression;

$\dfrac{x^2+xy+2y^2}{(x+y)(x-y)}$ **12.** expression; $\dfrac{24p}{p+2}$ **13.** expression; $-\frac{5}{36}$

14. equation; $\{0\}$ **15.** expression; $\dfrac{b+3}{3}$ **16.** expression; $\dfrac{5}{3z}$

17. expression; $\dfrac{2x+10}{x(x-2)(x+2)}$ **18.** equation; $\{2\}$ **19.** expression;

$\dfrac{-x}{3x+5y}$ **20.** equation; $\{-13\}$ **21.** expression; $\dfrac{3y+2}{y+3}$

22. equation; $\{\frac{5}{4}\}$ **23.** equation; $\emptyset$ **24.** expression; $\dfrac{2z-3}{2z+3}$

25. expression; $\dfrac{-1}{x-3}$, or $\dfrac{1}{3-x}$ **26.** expression; $\dfrac{t-2}{8}$

27. equation; $\{-10\}$ **28.** expression; $\dfrac{13x+28}{2x(x+4)(x-4)}$

29. equation; $\emptyset$ **30.** expression; $\dfrac{k(2k^2-2k+5)}{(k-1)(3k^2-2)}$

Section 7.5 (pages 402–406)

1. A **3.** D **5.** 24 **7.** $\frac{25}{4}$ **9.** $G = \dfrac{Fd^2}{Mm}$ **11.** $a = \dfrac{bc}{c+b}$

13. $v = \dfrac{PVt}{pT}$ **15.** $r = \dfrac{nE - IR}{In}$, or $r = \dfrac{IR - nE}{-In}$ **17.** $b = \dfrac{2\mathcal{A}}{h} - B$,

or $b = \dfrac{2\mathcal{A} - hB}{h}$ **19.** $r = \dfrac{eR}{E - e}$ **21.** Multiply each side by $a - b$.

23. 21 girls, 7 boys **25.** $\frac{1}{3}$ job per hr **27.** 1.75 in. **29.** 5.4 in.

31. 7.6 in. **33.** 56 teachers **35.** 210 deer **37.** 25,000 fish

39. 6.6 more gallons **41.** $x = \frac{7}{2}$; $AC = 8$; $DF = 12$ **43.** 2.4 mL

45. 3 mph **47.** 1020 mi **49.** 1750 mi **51.** 190 mi **53.** $6\frac{2}{3}$ min

55. 30 hr **57.** $2\frac{1}{3}$ hr **59.** 20 hr **61.** $2\frac{4}{5}$ hr **63.** $\frac{1}{3}$ **65.** 3

Section 7.6 (pages 412–415)

1. direct **3.** direct **5.** inverse **7.** inverse **9.** inverse **11.** direct

13. joint **15.** combined **17.** increases; decreases **19.** The perimeter

of a square varies directly as the length of its side. **21.** The surface

area of a sphere varies directly as the square of its radius.

23. The area of a triangle varies jointly as the length of its base and

height. **25.** 4; 2; 4π; $\frac{4}{3}\pi$; $\frac{1}{2}$; $\frac{1}{3}\pi$ **27.** 36 **29.** $\frac{16}{9}$ **31.** 0.625 **33.** $\frac{16}{5}$

35. $222\frac{2}{9}$ **37.** $2.919, or \$2.91\frac{9}{10}$ **39.** 8 lb **41.** about 450 cm^3

43. 256 ft **45.** $106\frac{2}{3}$ mph **47.** 100 cycles per sec **49.** $21\frac{1}{3}$ foot-

candles **51.** \$420 **53.** about 11.8 lb **55.** about 448.1 lb

57. about 68,600 calls **59.** Answers will vary. **61.** 8 **63.** -4
65. not a real number

Chapter 7 Review Exercises (pages 420–423)

1. (a) -6 (b) $\{x \mid x \neq -6\}$ **2.** (a) $2, 5$ (b) $\{x \mid x \neq 2, 5\}$
3. (a) 9 (b) $\{x \mid x \neq 9\}$ **4.** $\dfrac{x}{2}$ **5.** $\dfrac{5m + n}{5m - n}$ **6.** $\dfrac{-1}{2 + r}$

7. The reciprocal of a rational expression is another rational expression such that the two rational expressions have a product of 1.
8. $\dfrac{3y^2(2y + 3)}{2y - 3}$ **9.** $\dfrac{-3(w + 4)}{w}$ **10.** $\dfrac{z(z + 2)}{z + 5}$ **11.** 1 **12.** $96b^5$
13. $9r^2(3r + 1)$ **14.** $(3x - 1)(2x + 5)(3x + 4)$
15. $3(x - 4)^2(x + 2)$ **16.** $\dfrac{15y^2 - 8x^2}{9x^6y^7}$ **17.** 12 **18.** $\dfrac{71}{30(a + 2)}$
19. $\dfrac{13r^2 + 5rs}{(5r + s)(2r - s)(r + s)}$ **20.** Both students got the correct answer. The two expressions obtained are equivalent. **21.** $\dfrac{3 + 2t}{4 - 7t}$
22. -2 **23.** $\dfrac{1}{3q + 2p}$ **24.** $\dfrac{y + x}{xy}$ **25.** $\{-3\}$ **26.** $\{-2\}$ **27.** $\{0\}$
28. $\emptyset$ **29.** (a) $\{x \mid x \neq -3\}$ (b) -3 is not in the domain. **30.** In simplifying the expression, we are combining terms to get a single fraction with a denominator of $6x$. In solving the equation, we are finding a value for x that makes the equation true. **31.** C; $x = 0; y = 0$ **32.** $\dfrac{15}{2}$
33. $m = \dfrac{Fd^2}{GM}$ **34.** $M = \dfrac{m\mu}{v - \mu}$ **35.** 6000 passenger-km per day
36. 16 km per hr **37.** $4\frac{4}{5}$ min **38.** $3\frac{3}{5}$ hr **39.** C **40.** 430 mm
41. 5.59 vibrations per sec **42.** 22.5 ft³ **43.** $\dfrac{1}{x - 2y}$ **44.** $\dfrac{x + 5}{x + 2}$
45. $\dfrac{6m + 5}{3m^2}$ **46.** $\dfrac{11}{3 - x}$, or $\dfrac{-11}{x - 3}$ **47.** $\dfrac{x^2 - 6}{2(2x + 1)}$ **48.** $\dfrac{3 - 5x}{6x + 1}$
49. $\frac{1}{3}$ **50.** $\dfrac{s^2 + t^2}{st(s - t)}$ **51.** $\dfrac{k - 3}{36k^2 + 6k + 1}$ **52.** $\dfrac{x(9x + 1)}{3x + 1}$
53. $\dfrac{5a^2 + 4ab + 12b^2}{(a + 3b)(a - 2b)(a + b)}$ **54.** $\dfrac{acd + b^2d + bc^2}{bcd}$ **55.** $\{\frac{1}{3}\}$
56. $r = \dfrac{AR}{R - A}$, or $r = \dfrac{-AR}{A - R}$ **57.** $\{1, 4\}$ **58.** $\{-\frac{14}{3}\}$
59. (a) about 8.32 mm (b) about 44.9 diopters **60.** $8\frac{4}{7}$ min
61. \$36.27 **62.** 12 ft² **63.** $4\frac{1}{2}$ mi **64.** 480 mi **65.** 150 mi

Chapter 7 Test (pages 423–424)

[7.1] **1.** $-2, \frac{4}{3}; \{x \mid x \neq -2, \frac{4}{3}\}$ **2.** $\dfrac{2x - 5}{x(3x - 1)}$ **3.** $\dfrac{3(x + 3)}{4}$
4. $\dfrac{y + 4}{y - 5}$ **5.** -2 **6.** $\dfrac{x + 5}{x}$ [7.2] **7.** $t^2(t + 3)(t - 2)$
8. $\dfrac{7 - 2t}{6t^2}$ **9.** $\dfrac{9a + 5b}{21a^5b^3}$ **10.** $\dfrac{11x + 21}{(x - 3)^2(x + 3)}$ **11.** $\dfrac{4}{x + 2}$
[7.3] **12.** $\frac{72}{11}$ **13.** $-\dfrac{1}{a + b}$ **14.** $\dfrac{2y^2 + x^2}{xy(y - x)}$
[7.4] **15.** (a) expression; $\dfrac{11(x - 6)}{12}$ (b) equation; $\{6\}$
16. $\{\frac{1}{2}\}$ **17.** $\{5\}$ **18.** $\ell = \dfrac{2S}{n} - a$, or $\ell = \dfrac{2S - na}{n}$

19. $x = -1; y = 0$ [7.5] **20.** $3\frac{3}{14}$ hr
21. 15 mph **22.** 48,000 fish
23. (a) 3 units (b) 0
[7.6] **24.** 200 amps **25.** 0.8 lb

Chapters 1–7 Cumulative Review Exercises (pages 425–426)

[1.3] **1.** -199 [2.1] **2.** $\{-\frac{15}{4}\}$ [2.7] **3.** $\{\frac{2}{3}, 2\}$
[2.5] **4.** $\left(-\infty, \frac{240}{13}\right]$ [2.7] **5.** $(-\infty, -2] \cup [\frac{2}{3}, \infty)$
[2.3] **6.** \$4000 at 4%; \$8000 at 3% **7.** 6 m
[3.1] **8.** x-intercept: $(-2, 0)$; y-intercept: $(0, 4)$

[3.2] **9.** $-\frac{3}{2}$ **10.** $-\frac{3}{4}$ [3.3] **11.** $y = -\frac{3}{2}x + \frac{1}{2}$
[3.4] **12.**   **13.**
$x - y \geq 3$ and
$3x + 4y \leq 12$
[3.5] **14.** function; domain: $[-2, \infty)$; range: $(-\infty, 0]$
[3.6] **15.** (a) $\dfrac{5x - 8}{3}$, or $\dfrac{5}{3}x - \dfrac{8}{3}$ (b) -1 [4.1, 4.4] **16.** $\{(-1, 3)\}$
[4.2, 4.4] **17.** $\{(-2, 3, 1)\}$ **18.** $\emptyset$ [4.3] **19.** automobile: 42 km per hr; airplane: 600 km per hr [5.1] **20.** $\dfrac{m}{n}$ [5.2] **21.** $4y^2 - 7y - 6$
[5.4] **22.** $12f^2 + 5f - 3$ **23.** $\frac{1}{16}x^2 + \frac{5}{2}x + 25$
[5.5] **24.** $x^2 + 4x - 7$ [5.3] **25.** (a) $2x^3 - 2x^2 + 6x - 4$
(b) $2x^3 - 4x^2 + 2x + 2$ (c) -14 (d) $x^4 + 2x^2 - 3$
[6.2] **26.** $(2x + 5)(x - 9)$ [6.3] **27.** $25(2t^2 + 1)(2t^2 - 1)$
28. $(2p + 5)(4p^2 - 10p + 25)$ [6.5] **29.** $\{-\frac{7}{3}, 1\}$ [7.1] **30.** $\dfrac{y + 4}{y - 4}$
31. $\dfrac{a(a - b)}{2(a + b)}$ **32.** $\dfrac{2(x + 3)}{(x + 2)(x^2 + 3x + 9)}$ [7.2] **33.** 3
[7.4] **34.** $\{-4\}$ [7.5] **35.** $\frac{6}{5}$ hr [7.6] **36.** \$9.92

8   ROOTS, RADICALS, AND ROOT FUNCTIONS

Section 8.1 (pages 433–435)

1. E **3.** D **5.** A **7.** -9 **9.** 6 **11.** -4 **13.** -8 **15.** 6 **17.** -2
19. It is not a real number. **21.** 2 **23.** It is not a real number. **25.** $\frac{8}{9}$
27. $\frac{4}{3}$ **29.** $-\frac{1}{2}$ **31.** 3 **33.** 0.5 **35.** -0.7 **37.** 0.1
39. (a) It is not a real number. (b) negative (c) 0
In Exercises 41–47, we give the domain and then the range.
41. $[-3, \infty); [0, \infty)$ **43.** $[0, \infty); [-2, \infty)$

45. $(-\infty, \infty)$; $(-\infty, \infty)$ **47.** $(-\infty, \infty)$; $(-\infty, \infty)$

49. 12 **51.** 10 **53.** 2 **55.** -9 **57.** -5 **59.** $|x|$ **61.** $|z|$ **63.** x

65. x^5 **67.** $|x|^5$ (or $|x^5|$) **69.** C **71.** 97.381 **73.** 16.863

75. -9.055 **77.** 7.507 **79.** 3.162 **81.** 1.885 **83.** A

85. 1,183,000 cycles per sec **87.** 10 mi **89.** 392,000 mi^2

91. (a) 1.732 amps **(b)** 2.236 amps **93.** x^1, or x **95.** $\dfrac{3^3}{2^3}$, or $\dfrac{27}{8}$

Section 8.2 (pages 441–443)

1. C **3.** A **5.** H **7.** B **9.** D **11.** 13 **13.** 9 **15.** 2 **17.** $\frac{8}{9}$

19. -3 **21.** It is not a real number. **23.** 1000 **25.** 27 **27.** -1024

29. 16 **31.** $\frac{1}{8}$ **33.** $\frac{1}{512}$ **35.** $\frac{9}{25}$ **37.** $\frac{27}{8}$ **39.** $\sqrt{10}$ **41.** $\left(\sqrt[8]{8}\right)^3$

43. $\left(\sqrt[8]{9q}\right)^5 - \left(\sqrt[3]{2x}\right)^2$ **45.** $\dfrac{1}{\left(\sqrt{2m}\right)^3}$ **47.** $\left(\sqrt[3]{2y+x}\right)^2$

49. $\dfrac{1}{\left(\sqrt[3]{3m^4 + 2k^2}\right)^2}$ **51.** 64 **53.** 64 **55.** x^{10} **57.** $\sqrt[6]{x^5}$

59. $\sqrt[15]{t^8}$ **61.** 9 **63.** 4 **65.** y **67.** $x^{5/12}$ **69.** $k^{2/3}$

71. $x^3 y^8$ **73.** $\dfrac{1}{x^{10/3}}$ **75.** $\dfrac{1}{m^{1/4}n^{3/4}}$ **77.** p^2 **79.** $\dfrac{c^{11/3}}{b^{11/4}}$ **81.** $\dfrac{q^{5/3}}{9p^{7/2}}$

83. $p + 2p^2$ **85.** $k^{7/4} - k^{3/4}$ **87.** $6 + 18a$ **89.** $-5x^2 + 5x$

91. $x^{17/20}$ **93.** $\dfrac{1}{x^{3/2}}$ **95.** $y^{5/6}z^{1/3}$ **97.** $m^{1/12}$ **99.** $x^{1/8}$ **101.** $x^{1/24}$

103. $\sqrt{a^2 + b^2} = \sqrt{3^2 + 4^2} = 5$; $a + b = 3 + 4 = 7$; $5 \neq 7$

105. 4.5 hr **107.** 19.0°; The table gives 19°. **109.** 4.2°; The table gives 4°. **111.** 30; 30; They are the same.

Section 8.3 (pages 450–453)

1. $\sqrt{9}$, or 3 **3.** $\sqrt{36}$, or 6 **5.** $\sqrt{30}$ **7.** $\sqrt{14x}$ **9.** $\sqrt{42pqr}$

11. $\sqrt[3]{10}$ **13.** $\sqrt[3]{14xy}$ **15.** $\sqrt[4]{33}$ **17.** $\sqrt[4]{6x^3}$ **19.** This expression cannot be simplified by the product rule. **21.** $\frac{8}{11}$ **23.** $\dfrac{\sqrt{3}}{5}$ **25.** $\dfrac{\sqrt{x}}{5}$

27. $\dfrac{p^3}{9}$ **29.** $-\frac{3}{4}$ **31.** $\dfrac{\sqrt[3]{r^2}}{2}$ **33.** $-\dfrac{3}{x}$ **35.** $\dfrac{1}{x^3}$ **37.** $2\sqrt{3}$ **39.** $12\sqrt{2}$

41. $-4\sqrt{2}$ **43.** $-2\sqrt{7}$ **45.** This radical cannot be simplified further.

47. $4\sqrt[3]{2}$ **49.** $-2\sqrt[3]{2}$ **51.** $2\sqrt[3]{5}$ **53.** $-4\sqrt[4]{2}$ **55.** $2\sqrt[5]{2}$

57. $-3\sqrt[5]{2}$ **59.** $2\sqrt[6]{2}$ **61.** His reasoning was incorrect. Here, 8 is a term, not a factor. **63.** $6k\sqrt{2}$ **65.** $12xy^4\sqrt{xy}$ **67.** $11x^3$ **69.** $-3t^4$

71. $-10m^4 z^2$ **73.** $5a^2 b^3 c^4$ **75.** $\frac{1}{2}r^2 t^5$ **77.** $5x\sqrt{2x}$ **79.** $-10r^5\sqrt{5r}$

81. $x^3 y^4 \sqrt{13x}$ **83.** $2z^2 w^3$ **85.** $-2zt^2 \sqrt[3]{2z^2 t}$ **87.** $3x^3 y^4$

89. $-3r^3 s^2 \sqrt[4]{2r^3 s^2}$ **91.** $\dfrac{y^5 \sqrt{y}}{6}$ **93.** $\dfrac{x^5 \sqrt[3]{x}}{3}$ **95.** $4\sqrt{3}$

97. $\sqrt{5}$ **99.** $x^2\sqrt{x}$ **101.** $\sqrt[12]{432}$ **103.** $\sqrt[12]{6912}$ **105.** $\sqrt[6]{x^5}$

107. 5 **109.** $8\sqrt{2}$ **111.** $2\sqrt{14}$ **113.** 13 **115.** $9\sqrt{2}$

117. $\sqrt{17}$ **119.** 5 **121.** $6\sqrt{2}$ **123.** $\sqrt{5y^2 - 2xy + x^2}$

125. $2\sqrt{106} + 4\sqrt{2}$ **127.** 15.3 mi **129.** 27.0 in. **131.** 581

133. $\sqrt{9} + \sqrt{9} = 3 + 3 = 6$, and $\sqrt{4} = 2$; $6 \neq 2$, so the statement is false. **135.** MSX-77: 18.4 in.; MSX-83: 17.0 in.; MSX-60: 14.1 in.

137. $22x^4 - 10x^3$ **139.** $8q^2 - 3q$

Section 8.4 (pages 456–457)

1. -4 **3.** $7\sqrt{3}$ **5.** $14\sqrt[3]{2}$ **7.** $5\sqrt[4]{2}$ **9.** $24\sqrt{2}$ **11.** The expression cannot be simplified further. **13.** $20\sqrt{5}$ **15.** $4\sqrt{2x}$

17. $-11m\sqrt{2}$ **19.** $7\sqrt[3]{2}$ **21.** $2\sqrt[3]{x}$ **23.** $-7\sqrt[3]{x^2 y}$ **25.** $-x\sqrt[3]{xy^2}$

27. $19\sqrt[4]{2}$ **29.** $x\sqrt[4]{xy}$ **31.** $9\sqrt[3]{2a^3}$ **33.** $(4 + 3xy)\sqrt[3]{xy^2}$

35. $4t\sqrt[3]{3st} - 3s\sqrt{3st}$ **37.** $4x\sqrt[3]{x} + 6x\sqrt[4]{x}$ **39.** $2\sqrt{2} - 2$

41. $\dfrac{5\sqrt{5}}{6}$ **43.** $\dfrac{7\sqrt{2}}{6}$ **45.** $\dfrac{5\sqrt{2}}{3}$ **47.** $5\sqrt{2} + 4$ **49.** $\dfrac{5 + 3x}{x^4}$

51. $\dfrac{m\sqrt[3]{m^2}}{2}$ **53.** $\dfrac{3x\sqrt[3]{2} - 4\sqrt[3]{5}}{x^3}$ **55.** B

57. 15; Each radical expression simplifies to a whole number.

59. A; 42 m **61.** $\left(12\sqrt{5} + 5\sqrt{3}\right)$ in. **63.** $\left(24\sqrt{2} + 12\sqrt{3}\right)$ in.

65. $10x^3 y^4 - 20x^2 y$ **67.** $a^4 - b^2$ **69.** $64x^9 + 144x^6 + 108x^3 + 27$

71. $\dfrac{4x - 5}{3x}$

Connections **(page 463)** **1.** $\dfrac{319}{6(8\sqrt{5} + 1)}$ **2.** $\dfrac{9a - b}{b(3\sqrt{a} - \sqrt{b})}$

3. $\dfrac{9a - b}{\left(\sqrt{b} - \sqrt{a}\right)\left(3\sqrt{a} - \sqrt{b}\right)}$ **4.** $\dfrac{\left(3\sqrt{a} + \sqrt{b}\right)\left(\sqrt{b} + \sqrt{a}\right)}{b - a}$;

Instead of multiplying by the conjugate of the numerator, we use the conjugate of the denominator.

Section 8.5 (pages 464–466)

1. E **3.** A **5.** D **7.** $3\sqrt{6} + 2\sqrt{3}$ **9.** $20\sqrt{2}$ **11.** -2

13. -1 **15.** 6 **17.** $\sqrt{6} - \sqrt{2} + \sqrt{3} - 1$

19. $\sqrt{22} + \sqrt{55} - \sqrt{14} - \sqrt{35}$ **21.** $8 - \sqrt{15}$ **23.** $9 + 4\sqrt{5}$

25. $26 - 2\sqrt{105}$ **27.** $4 - \sqrt[3]{36}$ **29.** 10

31. $6x + 3\sqrt{x} - 2\sqrt{5x} - \sqrt{5}$ **33.** $9r - s$

35. $4\sqrt[3]{4y^2} - 19\sqrt[3]{2y} - 5$ **37.** $3x - 4$ **39.** $4x - y$ **41.** $2\sqrt{6} - 1$

43. $\sqrt{7}$ **45.** $5\sqrt{3}$ **47.** $\dfrac{\sqrt{6}}{2}$ **49.** $\dfrac{9\sqrt{15}}{5}$ **51.** $-\dfrac{7\sqrt{3}}{12}$ **53.** $\dfrac{\sqrt{14}}{2}$

55. $-\dfrac{\sqrt{14}}{10}$ **57.** $\dfrac{2\sqrt{6x}}{x}$ **59.** $\dfrac{-8\sqrt{3k}}{k}$ **61.** $\dfrac{-5m^2\sqrt{6mn}}{n^2}$

63. $\dfrac{12x^3\sqrt{2xy}}{y^5}$ **65.** $\dfrac{5\sqrt{2my}}{y^2}$ **67.** $-\dfrac{4k\sqrt{3z}}{z}$ **69.** $\dfrac{\sqrt[3]{18}}{3}$ **71.** $\dfrac{\sqrt[3]{12}}{3}$

73. $\dfrac{\sqrt[3]{18}}{4}$ **75.** $-\dfrac{\sqrt[3]{2pr}}{r}$ **77.** $\dfrac{x^2\sqrt[3]{y^2}}{y}$ **79.** $\dfrac{2\sqrt[4]{x^3}}{x}$ **81.** $\dfrac{\sqrt[4]{2yz^3}}{z}$

83. $\dfrac{3(4 - \sqrt{5})}{11}$ **85.** $\dfrac{6\sqrt{2} + 4}{7}$ **87.** $\dfrac{2(3\sqrt{5} - 2\sqrt{3})}{33}$

89. $2\sqrt{3} + \sqrt{10} - 3\sqrt{2} - \sqrt{15}$ **91.** $\sqrt{m} - 2$

93. $\dfrac{4(\sqrt{x} + 2\sqrt{y})}{x - 4y}$ **95.** $\dfrac{x - 2\sqrt{xy} + y}{x - y}$ **97.** $\dfrac{5\sqrt{k}(2\sqrt{k} - \sqrt{q})}{4k - q}$

99. $3 - 2\sqrt{6}$ **101.** $1 - \sqrt{5}$ **103.** $\dfrac{4 - 2\sqrt{2}}{3}$ **105.** $\dfrac{6 + 2\sqrt{6p}}{3}$

107. $\dfrac{3\sqrt{x + y}}{x + y}$ **109.** $\dfrac{p\sqrt{p + 2}}{p + 2}$ **111.** Each expression is approximately equal to 0.2588190451. **113.** $\dfrac{33}{8(6 + \sqrt{3})}$ **115.** $\dfrac{4x - y}{3x(2\sqrt{x} + \sqrt{y})}$

117. $\left\{\frac{3}{8}\right\}$ **119.** $\left\{-\frac{1}{3}, \frac{3}{2}\right\}$

Summary Exercises on Operations with Radicals and Rational Exponents (pages 466–467)

1. $-6\sqrt{10}$ **2.** $7 - \sqrt{14}$ **3.** $2 + \sqrt{6} - 2\sqrt{3} - 3\sqrt{2}$ **4.** $4\sqrt{2}$

5. $73 + 12\sqrt{35}$ **6.** $\dfrac{-\sqrt{6}}{2}$ **7.** $4(\sqrt{7} - \sqrt{5})$ **8.** $-3 + 2\sqrt{2}$

9. -44 **10.** $\dfrac{\sqrt{x} + \sqrt{5}}{x - 5}$ **11.** $2abc^3\sqrt[3]{b^2}$ **12.** $5\sqrt[3]{3}$

13. $3(\sqrt{5} - 2)$ **14.** $\dfrac{\sqrt{15x}}{5x}$ **15.** $\frac{8}{5}$ **16.** $\dfrac{\sqrt{2}}{8}$ **17.** $-\sqrt[3]{100}$

18. $11 + 2\sqrt{30}$ **19.** $-3\sqrt{3x}$ **20.** $52 - 30\sqrt{3}$ **21.** $\dfrac{\sqrt[3]{117}}{9}$

22. $3\sqrt{2} + \sqrt{15} + \sqrt{42} + \sqrt{35}$ **23.** $2\sqrt[4]{27}$ **24.** $\dfrac{1 + \sqrt[3]{3} + \sqrt[3]{9}}{-2}$

25. $\dfrac{x\sqrt[3]{x^2}}{y}$ **26.** $-4\sqrt{3} - 3$ **27.** $xy^{6/5}$ **28.** $x^{10}y$ **29.** $\dfrac{1}{25x^2}$

30. $7 + 4 \cdot 3^{1/2}$, or $7 + 4\sqrt{3}$ **31.** $3\sqrt[3]{2x^2}$ **32.** -2

33. (a) 8 **(b)** $\{-8, 8\}$ **34. (a)** 10 **(b)** $\{-10, 10\}$

35. (a) $\{-4, 4\}$ **(b)** -4 **36. (a)** $\{-5, 5\}$ **(b)** -5

37. (a) $-\frac{9}{11}$ **(b)** $\left\{-\frac{9}{11}, \frac{9}{11}\right\}$ **38. (a)** $-\frac{7}{10}$ **(b)** $\left\{-\frac{7}{10}, \frac{7}{10}\right\}$

39. (a) $\{-0.2, 0.2\}$ **(b)** 0.2 **40. (a)** $\{-0.3, 0.3\}$ **(b)** 0.3

Section 8.6 (pages 472–474)

1. (a) yes **(b)** no **3. (a)** yes **(b)** no **5.** No. There is no solution. The radical expression, which is positive, cannot equal a negative number. **7.** $\{11\}$ **9.** $\left\{\frac{1}{3}\right\}$ **11.** $\emptyset$ **13.** $\{5\}$ **15.** $\{18\}$ **17.** $\{5\}$ **19.** $\{4\}$ **21.** $\{17\}$ **23.** $\{5\}$ **25.** $\emptyset$ **27.** $\{0\}$ **29.** $\{0\}$ **31.** $\emptyset$ **33.** $\{1\}$ **35.** It is incorrect to just square each term. The right side should be $(8 - x)^2 = 64 - 16x + x^2$. The correct first step is $3x + 4 = 64 - 16x + x^2$, and the solution set is $\{4\}$. **37.** $\{1\}$ **39.** $\{-1\}$ **41.** $\{14\}$ **43.** $\{8\}$ **45.** $\{0\}$ **47.** $\emptyset$ **49.** $\{7\}$ **51.** $\{7\}$ **53.** $\{4, 20\}$ **55.** $\emptyset$ **57.** $\left\{\frac{5}{4}\right\}$ **59.** $\{9, 17\}$ **61.** $\left\{\frac{1}{4}, 1\right\}$

63. $L = CZ^2$ **65.** $K = \dfrac{V^2 m}{2}$ **67.** $M = \dfrac{r^2 F}{m}$ **69.** $r = \dfrac{a}{4\pi^2 N^2}$

71. $1 + x$ **73.** $2x^2 + x - 15$ **75.** $\dfrac{-7(5 + \sqrt{2})}{23}$

Section 8.7 (pages 479–481)

1. i **3.** -1 **5.** $-i$ **7.** $13i$ **9.** $-12i$ **11.** $i\sqrt{5}$ **13.** $4i\sqrt{3}$ **15.** $-\sqrt{105}$ **17.** -10 **19.** $i\sqrt{33}$ **21.** $\sqrt{3}$ **23.** $5i$ **25.** -2 **27.** Any real number a can be written as $a + 0i$, a complex number with imaginary part 0. **29.** $-1 + 7i$ **31.** 0 **33.** $7 + 3i$ **35.** -2

37. $1 + 13i$ **39.** $6 + 6i$ **41.** $4 + 2i$ **43.** -81 **45.** -16 **47.** $-10 - 30i$ **49.** $10 - 5i$ **51.** $-9 + 40i$ **53.** $-16 + 30i$ **55.** 153 **57.** 97 **59.** 4 **61.** $a - bi$ **63.** $1 + i$ **65.** $2 + 2i$ **67.** $-1 + 2i$ **69.** $-\frac{5}{13} - \frac{12}{13}i$ **71.** $1 - 3i$ **73.** $1 + 3i$ **75.** -1 **77.** i **79.** -1 **81.** $-i$ **83.** $-i$ **85.** Since $i^{20} = (i^4)^5 = 1^5 = 1$, the student multiplied by 1, which is justified by the identity property for multiplication. **87.** $\frac{1}{2} + \frac{1}{2}i$ **89.** Substitute both $1 + 5i$ and $1 - 5i$ for x, and show that the result is $0 = 0$ in each case. **91.** $\frac{37}{10} - \frac{19}{10}i$ **93.** $-\frac{13}{10} + \frac{11}{10}i$ **95.** $\left\{-\frac{13}{6}\right\}$ **97.** $\{-8, 5\}$ **99.** $\left\{-\frac{2}{5}, 1\right\}$

Chapter 8 Review Exercises (pages 487–490)

1. 42 **2.** -17 **3.** 6 **4.** -5 **5.** -3 **6.** -2 **7.** $\sqrt[n]{a}$ is not a real number if n is even and a is negative. **8. (a)** $|x|$ **(b)** $-|x|$ **(c)** x **9.** -6.856 **10.** -5.053 **11.** 4.960 **12.** 0.009 **13.** -3968.503 **14.** -0.189

15. domain: $[1, \infty)$; range: $[0, \infty)$

16. domain: $(-\infty, \infty)$; range: $(-\infty, \infty)$

17. B **18.** cube (third); 8; 2; second; 4; 4 **19.** A **20. (a)** m must be even. **(b)** m must be odd. **21.** no **22.** 7 **23.** -11 **24.** 32 **25.** -4 **26.** $-\frac{216}{125}$ **27.** -32 **28.** $\frac{1000}{27}$ **29.** It is not a real number. **30.** $(\sqrt[3]{8})^2$; $\sqrt[3]{8^2}$ **31.** The radical $\sqrt[n]{a^m}$ is equivalent to $a^{m/n}$. For example, $\sqrt[3]{8^2} = \sqrt[3]{64} = 4$, and $8^{2/3} = (8^{1/3})^2 = 2^2 = 4$.

32. $\sqrt{m + 3n}$ **33.** $\dfrac{1}{(\sqrt[3]{3a + b})^5}$, or $\dfrac{1}{\sqrt[3]{(3a + b)^5}}$ **34.** $7^{9/2}$

35. $p^{4/5}$ **36.** 5^2, or 25 **37.** 96 **38.** $a^{2/3}$ **39.** $\dfrac{1}{y^{1/2}}$ **40.** $\dfrac{z^{1/2}x^{8/5}}{4}$

41. $r^{1/2} + r$ **42.** $s^{1/2}$ **43.** $r^{3/2}$ **44.** $p^{1/2}$ **45.** $k^{9/4}$ **46.** $m^{13/3}$ **47.** $z^{1/12}$ **48.** $x^{1/8}$ **49.** $x^{1/15}$ **50.** $x^{1/36}$ **51.** The product rule for exponents applies only if the bases are the same. **52.** $\sqrt{66}$ **53.** $\sqrt{5r}$ **54.** $\sqrt[3]{30}$ **55.** $\sqrt[4]{21}$ **56.** $2\sqrt{5}$ **57.** $5\sqrt{3}$ **58.** $-5\sqrt{5}$ **59.** $-3\sqrt[3]{4}$ **60.** $10y^3\sqrt{y}$ **61.** $4pq^2\sqrt[3]{p}$ **62.** $3a^2b\sqrt[3]{4a^2b^2}$ **63.** $2r^2t\sqrt[3]{79r^2t}$ **64.** $\dfrac{y\sqrt{y}}{12}$ **65.** $\dfrac{m^5}{3}$ **66.** $\dfrac{\sqrt[3]{r^2}}{2}$ **67.** $\dfrac{a^2\sqrt[4]{a}}{3}$ **68.** $\sqrt{15}$ **69.** $p\sqrt{p}$ **70.** $\sqrt[12]{2000}$ **71.** $\sqrt[10]{x^7}$ **72.** 10 **73.** $\sqrt{197}$ **74.** $-11\sqrt{2}$ **75.** $23\sqrt{5}$ **76.** $7\sqrt{3y}$ **77.** $26m\sqrt{6m}$ **78.** $19\sqrt[3]{2}$ **79.** $-8\sqrt[4]{2}$ **80.** $(16\sqrt{2} + 24\sqrt{3})$ ft **81.** $(12\sqrt{3} + 5\sqrt{2})$ ft **82.** $1 - \sqrt{3}$ **83.** 2 **84.** $9 - 7\sqrt{2}$ **85.** $15 - 2\sqrt{26}$ **86.** 29 **87.** $2\sqrt[3]{2y^2} + 2\sqrt[3]{4y} - 3$ **88.** $4.801960973 \neq 66.28725368$ **89.** The denominator would become $\sqrt[3]{6^2} = \sqrt[3]{36}$, which is not rational.

90. $\dfrac{\sqrt{30}}{5}$ **91.** $-3\sqrt{6}$ **92.** $\dfrac{3\sqrt{7py}}{y}$ **93.** $\dfrac{\sqrt{22}}{4}$ **94.** $-\dfrac{\sqrt[3]{45}}{5}$

95. $\dfrac{3m\sqrt[3]{4n}}{n^2}$ **96.** $\dfrac{\sqrt{2} - \sqrt{7}}{-5}$ **97.** $\dfrac{5(\sqrt{6} + 3)}{3}$ **98.** $\dfrac{1 - \sqrt{5}}{4}$

99. $\dfrac{1 - 4\sqrt{2}}{3}$ **100.** $\dfrac{-6 + \sqrt{3}}{2}$ **101.** $\{2\}$ **102.** $\{6\}$ **103.** $\emptyset$

104. $\{0, 5\}$ **105.** $\{9\}$ **106.** $\{3\}$ **107.** $\{7\}$ **108.** $\left\{-\frac{1}{2}\right\}$
109. $\{-13\}$ **110.** $\{-1\}$ **111.** $\{14\}$ **112.** $\{-4\}$ **113.** $\emptyset$
114. $\emptyset$ **115.** $\{7\}$ **116.** $\{4\}$ **117. (a)** $H = \sqrt{L^2 - W^2}$ **(b)** 7.9 ft
118. $5i$ **119.** $10i\sqrt{2}$ **120.** no **121.** $-10 - 2i$ **122.** $14 + 7i$
123. $-\sqrt{35}$ **124.** -45 **125.** 3 **126.** $5 + i$ **127.** $32 - 24i$
128. $1 - i$ **129.** $4 + i$ **130.** $-i$ **131.** 1 **132.** -1 **133.** 1
134. -4 **135.** $\frac{1}{100}$ **136.** $\frac{1}{z^{3/5}}$ **137.** k^6 **138.** $3z^3t^2\sqrt[3]{2t^2}$
139. $57\sqrt{2}$ **140.** $-\frac{\sqrt{3}}{6}$ **141.** $\frac{\sqrt[3]{60}}{5}$ **142.** 1 **143.** $7i$
144. $3 - 7i$ **145.** $-5i$ **146.** $\frac{1 + \sqrt{6}}{2}$ **147.** $5 + 12i$ **148.** $6x\sqrt[3]{y^2}$
149. The expression cannot be simplified further.
150. $\sqrt{35} + \sqrt{15} - \sqrt{21} - 3$ **151.** $\{5\}$ **152.** $\{-4\}$ **153.** $\left\{\frac{3}{2}\right\}$
154. $\{2\}$ **155.** $\{1\}$ **156.** $\{2\}$ **157.** $\{9\}$ **158.** $\{4\}$ **159.** $\{7\}$
160. $\{6\}$

Chapter 8 Test (pages 490–491)

[8.1] **1.** -29 **2.** -8 [8.2] **3.** 5 [8.1] **4.** C **5.** 21.863 **6.** -9.405

7.

domain: $[-6, \infty)$;
range: $[0, \infty)$

[8.2] **8.** $\frac{125}{64}$ **9.** $\frac{1}{256}$ **10.** $\frac{9y^{3/10}}{x^2}$ **11.** $x^{4/3}y^6$ **12.** $7^{1/2}$, or $\sqrt{7}$

[8.3] **13.** $a^3\sqrt[3]{a^2}$, or $a^{11/3}$ **14.** $\sqrt{145}$ **15.** 10 **16.** $3x^2y^3\sqrt{6x}$
17. $2ab^3\sqrt[4]{2a^3b}$ **18.** $\sqrt[6]{200}$ [8.4] **19.** $26\sqrt{5}$ **20.** $(2ts - 3t^2)\sqrt[3]{2s^2}$
[8.5] **21.** $66 + \sqrt{5}$ **22.** $23 - 4\sqrt{15}$ **23.** $-\frac{\sqrt{10}}{4}$ **24.** $\frac{2\sqrt[3]{25}}{5}$
25. $-2(\sqrt{7} - \sqrt{5})$ **26.** $3 + \sqrt{6}$ [8.6] **27. (a)** 59.8
(b) $T = \frac{V_0^2 - V^2}{-V^2k}$, or $T = \frac{V^2 - V_0^2}{V^2k}$ **28.** $\{-1\}$ **29.** $\{3\}$
30. $\{-3\}$ [8.7] **31.** $-5 - 8i$ **32.** $-2 + 16i$ **33.** $3 + 4i$ **34.** i
35. (a) true **(b)** true **(c)** false **(d)** true

Chapters 1–8 Cumulative Review Exercises (pages 492–493)

[1.3] **1.** 1 **2.** $-\frac{14}{9}$ [2.1] **3.** $\{-4\}$ **4.** $\{-12\}$ **5.** $\{6\}$
[2.7] **6.** $\left\{-\frac{10}{3}, 1\right\}$ [2.5] **7.** $(-6, \infty)$ [2.3] **8.** 36 nickels; 64 quarters
9. $2\frac{2}{39}$ L [3.1] **10.**
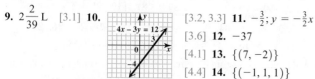
[3.2, 3.3] **11.** $-\frac{3}{2}$; $y = -\frac{3}{2}x$
[3.6] **12.** -37
[4.1] **13.** $\{(7, -2)\}$
[4.4] **14.** $\{(-1, 1, 1)\}$

[4.3] **15.** 2-oz letter: $0.61; 3-oz letter: $0.78
[5.2] **16.** $-k^3 - 3k^2 - 8k - 9$ [5.4] **17.** $8x^2 + 17x - 21$
[5.5] **18.** $3y^3 - 3y^2 + 4y + 1 + \frac{-10}{2y + 1}$
[6.2] **19.** $(2p - 3q)(p - q)$ [6.3] **20.** $(3k^2 + 4)(k - 1)(k + 1)$

21. $(x + 8)(x^2 - 8x + 64)$ [6.5] **22.** $\left\{-3, -\frac{5}{2}\right\}$ **23.** $\left\{-\frac{2}{5}, 1\right\}$
[7.1] **24.** $\{x \mid x \neq \pm 3\}$ **25.** $\frac{y}{y + 5}$ [7.2] **26.** $\frac{4x + 2y}{(x + y)(x - y)}$
[7.3] **27.** $-\frac{9}{4}$ **28.** $\frac{-1}{a + b}$ **29.** $\frac{1}{xy - 1}$ [7.4] **30.** $\emptyset$
[7.5] **31.** Danielle: 8 mph; Richard: 4 mph [8.2] **32.** $\frac{1}{9}$
[8.3] **33.** $2x\sqrt[3]{6x^2y^2}$ [8.4] **34.** $7\sqrt{2}$ [8.5] **35.** $\frac{\sqrt{10} + 2\sqrt{2}}{2}$
[8.3] **36.** $\sqrt{29}$ [8.6] **37.** $\{3, 4\}$ [8.7] **38.** $4 + 2i$

9 QUADRATIC EQUATIONS, INEQUALITIES, AND FUNCTIONS

Section 9.1 (pages 502–505)

1. B, C **3.** The zero-factor property requires a product equal to 0. The first step should have been to rewrite the equation with 0 on one side.
5. $\{-2, -1\}$ **7.** $\left\{-3, \frac{1}{3}\right\}$ **9.** $\left\{\frac{1}{2}, 4\right\}$ **11.** $\{\pm 9\}$ **13.** $\{\pm\sqrt{17}\}$
15. $\{\pm 4\sqrt{2}\}$ **17.** $\{\pm 2\sqrt{5}\}$ **19.** $\{\pm 2\sqrt{6}\}$ **21.** $\{-7, 3\}$
23. $\{-1, 13\}$ **25.** $\{4 \pm \sqrt{3}\}$ **27.** $\{-5 \pm 4\sqrt{3}\}$
29. $\left\{\frac{1 \pm \sqrt{7}}{3}\right\}$ **31.** $\left\{\frac{-1 \pm 2\sqrt{6}}{4}\right\}$ **33.** $\left\{\frac{2 \pm 2\sqrt{3}}{5}\right\}$
35. 5.6 sec **37.** Solve $(2x + 1)^2 = 5$ by the square root property. Solve $x^2 + 4x = 12$ by completing the square. **39.** 9; $(x + 3)^2$
41. 36; $(p - 6)^2$ **43.** $\frac{81}{4}$; $\left(q + \frac{9}{2}\right)^2$ **45.** $\frac{1}{64}$; $\left(x + \frac{1}{8}\right)^2$
47. 0.16; $(x - 0.4)^2$ **49.** 4 **51.** 25 **53.** $\frac{1}{36}$ **55.** $\{-4, 6\}$
57. $\{-2 \pm \sqrt{6}\}$ **59.** $\{-5 \pm \sqrt{7}\}$ **61.** $\left\{-\frac{8}{3}, 3\right\}$
63. $\left\{\frac{-7 \pm \sqrt{53}}{2}\right\}$ **65.** $\left\{\frac{-5 \pm \sqrt{41}}{4}\right\}$ **67.** $\left\{\frac{5 \pm \sqrt{15}}{5}\right\}$
69. $\left\{\frac{4 \pm \sqrt{3}}{3}\right\}$ **71.** $\left\{\frac{2 \pm \sqrt{3}}{3}\right\}$ **73.** $\{1 \pm \sqrt{2}\}$
75. $\{\pm 2i\sqrt{3}\}$ **77.** $\{5 \pm 2i\}$ **79.** $\left\{\frac{1}{6} \pm \frac{\sqrt{2}}{3}i\right\}$ **81.** $\{-2 \pm 3i\}$
83. $\left\{-\frac{2}{3} \pm \frac{2\sqrt{2}}{3}i\right\}$ **85.** $\{-3 \pm i\sqrt{3}\}$ **87.** x^2 **88.** x **89.** $6x$
90. 1 **91.** 9 **92.** $(x + 3)^2$, or $x^2 + 6x + 9$ **93.** $\{\pm\sqrt{b}\}$
95. $\left\{\pm\frac{\sqrt{b^2 + 16}}{2}\right\}$ **97.** $\left\{\frac{2b \pm \sqrt{3a}}{5}\right\}$ **99.** $\sqrt{13}$ **101.** 1

Section 9.2 (pages 510–512)

1. The documentation was incorrect, since the fraction bar should extend under the term $-b$. The correct formula is $x = \frac{-b \pm \sqrt{b^2 - 4ac}}{2a}$.
3. The last step is wrong. Because 5 is not a common factor in the numerator, the fraction cannot be simplified. The solution set is
$\left\{\frac{5 \pm \sqrt{5}}{10}\right\}$. **5.** $\{3, 5\}$ **7.** $\left\{\frac{-2 \pm \sqrt{2}}{2}\right\}$ **9.** $\left\{\frac{1 \pm \sqrt{3}}{2}\right\}$

11. $\left\{5 \pm \sqrt{7}\right\}$ **13.** $\left\{\dfrac{-1 \pm \sqrt{2}}{2}\right\}$ **15.** $\left\{\dfrac{-1 \pm \sqrt{7}}{3}\right\}$

17. $\left\{1 \pm \sqrt{5}\right\}$ **19.** $\left\{\dfrac{-2 \pm \sqrt{10}}{2}\right\}$ **21.** $\left\{-1 \pm 3\sqrt{2}\right\}$

23. $\left\{\dfrac{1 \pm \sqrt{29}}{2}\right\}$ **25.** $\left\{\dfrac{-4 \pm \sqrt{91}}{3}\right\}$ **27.** $\left\{\dfrac{-3 \pm \sqrt{57}}{8}\right\}$

29. $\left\{\dfrac{3}{2} \pm \dfrac{\sqrt{15}}{2}i\right\}$ **31.** $\left\{3 \pm i\sqrt{5}\right\}$ **33.** $\left\{\dfrac{1}{2} \pm \dfrac{\sqrt{6}}{2}i\right\}$

35. $\left\{-\dfrac{2}{3} \pm \dfrac{\sqrt{2}}{3}i\right\}$ **37.** $\left\{\dfrac{1}{2} \pm \dfrac{1}{4}i\right\}$ **39.** B; factoring

41. C; quadratic formula   **43.** A; factoring   **45.** D; quadratic formula

47. $\left\{-\dfrac{7}{5}\right\}$ **49.** $\left\{-\dfrac{1}{3}, 2\right\}$ **51. (a)** Discriminant is 25, or 5^2; solve by factoring; $\left\{-3, -\dfrac{4}{3}\right\}$ **(b)** Discriminant is 44; use the quadratic formula; $\left\{\dfrac{7 \pm \sqrt{11}}{2}\right\}$ **53.** -10 or 10 **55.** 16 **57.** 25

59. $b = \dfrac{44}{5}; \dfrac{3}{10}$ **61.** $\{-8\}$ **63.** $\{5\}$

Section 9.3 (pages 519–522)

1. Multiply by the LCD, x.   **3.** Substitute a variable for $x^2 + x$.
5. The proposed solution -1 does not check. The solution set is $\{4\}$.

7. $\{-2, 7\}$ **9.** $\{-4, 7\}$ **11.** $\left\{-\dfrac{2}{3}, 1\right\}$ **13.** $\left\{-\dfrac{14}{17}, 5\right\}$

15. $\left\{-\dfrac{11}{7}, 0\right\}$ **17.** $\left\{\dfrac{-1 \pm \sqrt{13}}{2}\right\}$ **19.** $\left\{-\dfrac{8}{3}, -1\right\}$

21. $\left\{\dfrac{2 \pm \sqrt{22}}{3}\right\}$ **23.** $\left\{\dfrac{-1 \pm \sqrt{5}}{4}\right\}$ **25. (a)** $(20 - t)$ mph

(b) $(20 + t)$ mph   **27.** 25 mph   **29.** 50 mph   **31.** 3.6 hr   **33.** Rusty: 25.0 hr; Nancy: 23.0 hr   **35.** 3 hr; 6 hr   **37.** $\{2, 5\}$ **39.** $\{3\}$

41. $\left\{\dfrac{8}{9}\right\}$ **43.** $\{9\}$ **45.** $\left\{\dfrac{2}{5}\right\}$ **47.** $\{-2\}$ **49.** $\{\pm 2, \pm 5\}$

51. $\left\{\pm 1, \pm \dfrac{3}{2}\right\}$ **53.** $\left\{\pm 2, \pm 2\sqrt{3}\right\}$ **55.** $\{-6, -5\}$

57. $\left\{-\dfrac{16}{3}, -2\right\}$ **59.** $\{-8, 1\}$ **61.** $\{-64, 27\}$ **63.** $\left\{\pm 1, \pm \dfrac{27}{8}\right\}$

65. $\left\{-\dfrac{1}{3}, \dfrac{1}{6}\right\}$ **67.** $\left\{-\dfrac{1}{2}, 3\right\}$ **69.** $\left\{\pm \dfrac{\sqrt{6}}{3}, \pm \dfrac{1}{2}\right\}$ **71.** $\{3, 11\}$

73. $\{25\}$ **75.** $\left\{-\sqrt[3]{5}, -\dfrac{\sqrt[3]{4}}{2}\right\}$ **77.** $\left\{\dfrac{4}{3}, \dfrac{9}{4}\right\}$

79. $\left\{\pm \dfrac{\sqrt{9 + \sqrt{65}}}{2}, \pm \dfrac{\sqrt{9 - \sqrt{65}}}{2}\right\}$ **81.** $\left\{\pm 1, \pm \dfrac{\sqrt{6}}{2}i\right\}$

83. $W = \dfrac{P - 2L}{2}$, or $W = \dfrac{P}{2} - L$ **85.** $C = \dfrac{5}{9}(F - 32)$

Summary Exercises on Solving Quadratic Equations (page 522)

1. square root property   **2.** factoring   **3.** quadratic formula
4. quadratic formula   **5.** factoring   **6.** square root property

7. $\left\{\pm \sqrt{7}\right\}$ **8.** $\left\{-\dfrac{3}{2}, \dfrac{5}{3}\right\}$ **9.** $\left\{-3 \pm \sqrt{5}\right\}$ **10.** $\{-2, 8\}$

11. $\left\{-\dfrac{3}{2}, 4\right\}$ **12.** $\left\{-3, \dfrac{1}{3}\right\}$ **13.** $\left\{\dfrac{2 \pm \sqrt{2}}{2}\right\}$ **14.** $\left\{\pm 2i\sqrt{3}\right\}$

15. $\left\{\dfrac{1}{2}, 2\right\}$ **16.** $\{\pm 1, \pm 3\}$ **17.** $\left\{\dfrac{-3 \pm 2\sqrt{2}}{2}\right\}$ **18.** $\left\{\dfrac{4}{5}, 3\right\}$

19. $\left\{\pm \sqrt{2}, \pm \sqrt{7}\right\}$ **20.** $\left\{\dfrac{1 \pm \sqrt{5}}{4}\right\}$ **21.** $\left\{-\dfrac{1}{2} \pm \dfrac{\sqrt{3}}{2}i\right\}$

22. $\left\{-\dfrac{\sqrt[3]{175}}{5}, 1\right\}$ **23.** $\left\{\dfrac{3}{2}\right\}$ **24.** $\left\{\dfrac{2}{3}\right\}$ **25.** $\left\{\pm 6\sqrt{2}\right\}$

26. $\left\{-\dfrac{2}{3}, 2\right\}$ **27.** $\{-4, 9\}$ **28.** $\{\pm 13\}$ **29.** $\left\{1 \pm \dfrac{\sqrt{3}}{3}i\right\}$

30. $\{3\}$ **31.** $\left\{\dfrac{1}{6} \pm \dfrac{\sqrt{47}}{6}i\right\}$ **32.** $\left\{-\dfrac{1}{3}, \dfrac{1}{6}\right\}$

Section 9.4 (pages 527–531)

1. Find a common denominator, and then multiply both sides by the common denominator.   **3.** Write it in standard form (with 0 on one side, in decreasing powers of w).   **5.** $m = \sqrt{p^2 - n^2}$

7. $t = \dfrac{\pm\sqrt{dk}}{k}$ **9.** $d = \dfrac{\pm\sqrt{skI}}{I}$ **11.** $v = \dfrac{\pm\sqrt{kAF}}{F}$

13. $r = \dfrac{\pm\sqrt{3\pi Vh}}{\pi h}$ **15.** $t = \dfrac{-B \pm \sqrt{B^2 - 4AC}}{2A}$ **17.** $h = \dfrac{D^2}{k}$

19. $\ell = \dfrac{p^2 g}{k}$ **21.** $r = \dfrac{\pm\sqrt{S\pi}}{2\pi}$ **23.** $R = \dfrac{E^2 - 2pr \pm E\sqrt{E^2 - 4pr}}{2p}$

25. $r = \dfrac{5pc}{4}$ or $r = -\dfrac{2pc}{3}$ **27.** $I = \dfrac{-cR \pm \sqrt{c^2R^2 - 4cL}}{2cL}$

29. 7.9, 8.9, 11.9   **31.** eastbound ship: 80 mi; southbound ship: 150 mi   **33.** 8 in., 15 in., 17 in.   **35.** length: 24 ft; width: 10 ft   **37.** 2 ft   **39.** 7 m by 12 m   **41.** 20 in. by 12 in.   **43.** 1 sec and 8 sec   **45.** 2.4 sec and 5.6 sec   **47.** 9.2 sec   **49.** It reaches its *maximum* height at 5 sec because this is the only time it reaches 400 ft.   **51.** \$0.80   **53.** 0.035, or 3.5%   **55.** 5.5 m per sec   **57.** 5 or 14   **59. (a)** \$420 billion   **(b)** \$420 billion; They are the same.   **61.** 2004; The graph indicates that spending first exceeded \$400 billion in 2005.   **63.** 9

65. domain: $(-\infty, \infty)$; range: $[0, \infty)$

Section 9.5 (pages 537–540)

1. (a) B   **(b)** C   **(c)** A   **(d)** D   **3.** $(0, 0)$ **5.** $(0, 4)$ **7.** $(1, 0)$

9. $(-3, -4)$ **11.** $(5, 6)$ **13.** down; wider   **15.** up; narrower

17. down; narrower   **19. (a)** D   **(b)** B   **(c)** C   **(d)** A

21. 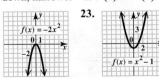   **23.**   **25.**

27. vertex: $(4, 0)$;
axis: $x = 4$;
domain: $(-\infty, \infty)$;
range: $[0, \infty)$

$f(x) = (x-4)^2$

29. vertex: $(-2, -1)$;
axis: $x = -2$;
domain: $(-\infty, \infty)$;
range: $[-1, \infty)$

$f(x) = (x+2)^2 - 1$

31. vertex: $(2, -4)$;
axis: $x = 2$;
domain: $(-\infty, \infty)$;
range: $[-4, \infty)$

$f(x) = 2(x-2)^2 - 4$

33. vertex: $(-1, 2)$;
axis: $x = -1$;
domain: $(-\infty, \infty)$;
range: $(-\infty, 2]$

$f(x) = -\frac{1}{2}(x+1)^2 + 2$

35. vertex: $(2, -3)$;
axis: $x = 2$;
domain: $(-\infty, \infty)$;
range: $[-3, \infty)$

$f(x) = 2(x-2)^2 - 3$

37. linear; positive
39. quadratic; positive
41. quadratic; negative
43. (a)

Sales (in millions of dollars) vs. Years Since 2000

(b) quadratic; positive **(c)** $f(x) = 99.3x^2 + 400.7x + 1825$
(d) \$9496 million **(e)** No. The number of digital cameras sold in 2007 is far below the number approximated by the model. Rather than continuing to increase, sales of digital cameras fell in 2007. **45. (a)** 6105
(b) The approximation using the model is low. **47.** $\{-4, 5\}$
49. -2 **51.** $\{-4, 1\}$ **53.** $\{-3 \pm 2\sqrt{3}\}$

Section 9.6 (pages 548–551)

1. If x is squared, it has a vertical axis. If y is squared, it has a horizontal axis. **3.** Use the discriminant of the function. If it is positive, there are two x-intercepts. If it is 0, there is one x-intercept (at the vertex), and if it is negative, there is no x-intercept. **5.** $(-4, -6)$ **7.** $(1, -3)$
9. $\left(-\frac{1}{2}, -\frac{29}{4}\right)$ **11.** $(-1, 3)$; up; narrower; no x-intercepts
13. $\left(\frac{5}{2}, \frac{37}{4}\right)$; down; same; two x-intercepts **15.** $(-3, -9)$; to the right; wider **17.** F **19.** C **21.** D

23. vertex: $(-4, -6)$;
axis: $x = -4$;
domain: $(-\infty, \infty)$;
range: $[-6, \infty)$

$f(x) = x^2 + 8x + 10$

25. vertex: $(1, -3)$;
axis: $x = 1$;
domain: $(-\infty, \infty)$;
range: $(-\infty, -3]$

$f(x) = -2x^2 + 4x - 5$

27. vertex: $(1, -2)$;
axis: $y = -2$;
domain: $[1, \infty)$;
range: $(-\infty, \infty)$

$x = (y+2)^2 + 1$

29. vertex: $(1, 5)$; axis: $y = 5$;
domain: $(-\infty, 1]$;
range: $(-\infty, \infty)$

$x = -\frac{1}{5}y^2 + 2y - 4$

31. vertex: $(-7, -2)$;
axis: $y = -2$; domain: $[-7, \infty)$;
range: $(-\infty, \infty)$

$x = 3y^2 + 12y + 5$

33. 20 and 20 **35.** 140 ft by 70 ft; 9800 ft^2 **37.** 16 ft; 2 sec
39. 2 sec; 65 ft **41.** 20 units; \$210 **43. (a)** minimum
(b) 2003; \$825.8 billion **45. (a)** The coefficient of x^2 is negative because a parabola that models the data must open down.
(b) $(18.45, 3860)$ **(c)** In 2018 Social Security assets will reach their maximum value of \$3860 billion.
47. (a) $R(x) = (100 - x)(200 + 4x) = 20,000 + 200x - 4x^2$
(b)

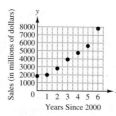

$R(x) = (100 - x)(200 + 4x)$
$(25, 22,500)$

(c) 25 **(d)** \$22,500
49.
51.
53. $\left(-\frac{3}{2}, \infty\right)$

Section 9.7 (pages 557–558)

1. (a) $\{1, 3\}$ **(b)** $(-\infty, 1) \cup (3, \infty)$ **(c)** $(1, 3)$
3. (a) $\{-2, 5\}$ **(b)** $[-2, 5]$ **(c)** $(-\infty, -2] \cup [5, \infty)$
5. $(-\infty, -1) \cup (5, \infty)$
7. $(-4, 6)$
9. $(-\infty, 1] \cup [3, \infty)$
11. $\left(-\infty, -\frac{3}{2}\right] \cup \left[\frac{3}{5}, \infty\right)$
13. $\left[-\frac{3}{2}, \frac{3}{2}\right]$
15. $\left(-\infty, -\frac{1}{2}\right] \cup \left[\frac{1}{3}, \infty\right)$
17. $(-\infty, 0] \cup [4, \infty)$
19. $\left[0, \frac{5}{3}\right]$
21. $\left(-\infty, 3 - \sqrt{3}\right) \cup \left[3 + \sqrt{3}, \infty\right)$
23. $(-\infty, \infty)$ **25.** $\emptyset$ **27.** $(-\infty, 1) \cup (2, 4)$
29. $\left[-\frac{3}{2}, \frac{1}{3}\right] \cup [4, \infty)$
31. $(-\infty, 1) \cup (4, \infty)$
33. $\left[-\frac{3}{2}, 5\right)$ **35.** $(2, 6]$
37. $\left(-\infty, \frac{1}{2}\right) \cup \left(\frac{5}{4}, \infty\right)$
39. $[-7, -2)$
41. $(-\infty, 2) \cup (4, \infty)$

43. $\left(0, \frac{1}{2}\right) \cup \left(\frac{5}{2}, \infty\right)$

45. $\left[\frac{3}{2}, \infty\right)$

47. $\left(-2, \frac{5}{3}\right) \cup \left(\frac{5}{3}, \infty\right)$

49. domain: $\{0, 1, 2, 3\}$; range: $\{1, 2, 4, 8\}$ **51.** function

Chapter 9 Review Exercises (pages 562–566)

1. $\{\pm 11\}$ **2.** $\{\pm \sqrt{3}\}$ **3.** $\left\{-\frac{15}{2}, \frac{5}{2}\right\}$ **4.** $\left\{\frac{2}{3} \pm \frac{5}{3}i\right\}$

5. $\left\{-2 \pm \sqrt{19}\right\}$ **6.** $\left\{\frac{1}{2}, 1\right\}$ **7.** By the square root property, the

first step should be $x = \sqrt{12}$ or $x = -\sqrt{12}$. The solution set is

$\left\{\pm 2\sqrt{3}\right\}$.   **8.** 5.8 sec   **9.** $\left\{-\frac{7}{2}, 3\right\}$ **10.** $\left\{\frac{-5 \pm \sqrt{53}}{2}\right\}$

11. $\left\{\frac{1 \pm \sqrt{41}}{2}\right\}$ **12.** $\left\{-\frac{3}{4} \pm \frac{\sqrt{23}}{4}i\right\}$ **13.** $\left\{\frac{2}{3} \pm \frac{\sqrt{2}}{3}i\right\}$

14. $\left\{\frac{-7 \pm \sqrt{37}}{2}\right\}$ **15.** C   **16.** A   **17.** D   **18.** B   **19.** $\left\{-\frac{5}{2}, 3\right\}$

20. $\left\{-\frac{1}{2}, 1\right\}$ **21.** $\{-4\}$ **22.** $\left\{-\frac{11}{6}, -\frac{19}{12}\right\}$ **23.** $\left\{-\frac{343}{8}, 64\right\}$

24. $\{\pm 1, \pm 3\}$ **25.** 7 mph   **26.** 40 mph   **27.** 4.6 hr

28. Zoran: 2.6 hr; Claude: 3.6 hr   **29.** $v = \frac{\pm \sqrt{rFkw}}{kw}$ **30.** $y = \frac{6p^2}{z}$

31. $t = \frac{3m \pm \sqrt{9m^2 + 24m}}{2m}$ **32.** 9 ft, 12 ft, 15 ft

33. 12 cm by 20 cm   **34.** 1 in.   **35.** 18 in.   **36.** 5.2 sec   **37.** 3 min

38. (a) \$15,511 million; It is close to the number suggested by the graph.

(b) $x = 6$, which represents 2006; Based on the graph, the revenue in

2006 was closer to \$13,000 million than \$14,000 million.   **39.** $(1, 0)$

40. $(3, 7)$ **41.** $(-4, 3)$ **42.** $\left(\frac{2}{3}, -\frac{2}{3}\right)$

43. vertex: $(2, -3)$;   **44.** vertex: $(2, 3)$;   **45.** vertex: $(-4, -3)$;
axis: $x = 2$;   axis: $x = 2$;   axis: $y = -3$;
domain: $(-\infty, \infty)$;   domain: $(-\infty, \infty)$;   domain: $[-4, \infty)$;
range: $[-3, \infty)$ range: $(-\infty, 3]$ range: $(-\infty, \infty)$

$y = 2(x-2)^2 - 3$

$f(x) = -2x^2 + 8x - 5$

$x = 2(y+3)^2 - 4$

46. vertex: $(4, 6)$;
axis: $y = 6$;
domain: $(-\infty, 4]$;
range: $(-\infty, \infty)$

$x = -\frac{1}{2}y^2 + 6y - 14$

47. (a) $c = 2.9$
$100a + 10b + c = 24.3$
$400a + 20b + c = 56.5$
(b) $f(x) = 0.054x^2 + 1.6x + 2.9$
(c) \$60.3 billion; The result using
the model is close, but slightly low.

48. 5 sec; 400 ft   **49.** length: 50 m; width: 50 m; maximum area: 2500 m²

50. $\left(-\infty, -\frac{3}{2}\right) \cup (4, \infty)$

51. $[-4, 3]$

52. $(-\infty, -5] \cup [-2, 3]$ **53.** $\varnothing$

54. $\left(-\infty, \frac{1}{2}\right) \cup (2, \infty)$

55. $[-3, 2)$ **56.** $R = \frac{\pm \sqrt{Vh - r^2 h}}{h}$

57. $\left\{1 \pm \frac{\sqrt{3}}{3}i\right\}$ **58.** $\left\{\frac{-11 \pm \sqrt{7}}{3}\right\}$ **59.** $d = \frac{\pm \sqrt{SkI}}{I}$

60. $(-\infty, \infty)$ **61.** $\{4\}$ **62.** $\left\{\pm \sqrt{4 + \sqrt{15}}, \pm \sqrt{4 - \sqrt{15}}\right\}$

63. $\left(-5, -\frac{23}{5}\right]$ **64.** $\left\{-\frac{5}{3}, -\frac{3}{2}\right\}$ **65.** $\{-2, -1, 3, 4\}$

66. $(-\infty, -6) \cup \left(-\frac{3}{2}, 1\right)$ **67. (a)** F   **(b)** B   **(c)** C
(d) A   **(e)** E   **(f)** D

68. vertex: $\left(-\frac{1}{2}, -3\right)$; axis: $x = -\frac{1}{2}$;
domain: $(-\infty, \infty)$; range: $[-3, \infty)$

$f(x) = 4x^2 + 4x - 2$

69. 10 mph   **70.** length: 2 cm; width: 1.5 cm

Chapter 9 Test (pages 566–568)

[9.1] **1.** $\left\{\pm 3\sqrt{6}\right\}$ **2.** $\left\{-\frac{8}{7}, \frac{2}{7}\right\}$ **3.** $\left\{-1 \pm \sqrt{5}\right\}$

[9.2] **4.** $\left\{\frac{3 \pm \sqrt{17}}{4}\right\}$ **5.** $\left\{\frac{2}{3} \pm \frac{\sqrt{11}}{3}i\right\}$ [9.3] **6.** $\left\{\frac{2}{3}\right\}$

[9.1] **7.** A   [9.2] **8.** discriminant: 88; There are two irrational solutions.

[9.1–9.3] **9.** $\left\{-\frac{2}{3}, 6\right\}$ **10.** $\left\{\frac{-7 \pm \sqrt{97}}{8}\right\}$ **11.** $\left\{\pm \frac{1}{3}, \pm 2\right\}$

12. $\left\{-\frac{5}{2}, 1\right\}$ [9.4] **13.** $r = \frac{\pm \sqrt{\pi S}}{2\pi}$ [9.3] **14.** Terry: 11.1 hr;

Callie: 9.1 hr   **15.** 7 mph   [9.4] **16.** 2 ft   **17.** 16 m   [9.5] **18.** A

19. vertex: $(0, -2)$; axis: $x = 0$;   [9.6] **20.** vertex: $(2, 3)$; axis: $x = 2$;
domain: $(-\infty, \infty)$;   domain: $(-\infty, \infty)$;
range: $[-2, \infty)$ range: $(-\infty, 3]$

$f(x) = \frac{1}{2}x^2 - 2$

$f(x) = -x^2 + 4x - 1$

21. vertex: $(2, 2)$; axis: $y = 2$;
domain: $(-\infty, 2]$;
range: $(-\infty, \infty)$

$x = -(y - 2)^2 + 2$

22. (a) 139 million   **(b)** 2007;
145 million   **23.** 160 ft by 320 ft

[9.7] **24.** $(-\infty, -5) \cup \left(\frac{3}{2}, \infty\right)$

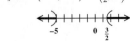

25. $(-\infty, 4) \cup [9, \infty)$

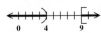

Chapters 1–9 Cumulative Review Exercises (pages 568–569)

[1.1, 8.7] **1. (a)** $-2, 0, 7$ **(b)** $-\frac{7}{3}, -2, 0, 0.7, 7, \frac{32}{3}$
(c) All are real except $\sqrt{-8}$. **(c)** All are complex numbers.

[2.1] **2.** $\left\{\frac{4}{5}\right\}$ [2.7] **3.** $\left\{\frac{11}{10}, \frac{7}{2}\right\}$ [8.6] **4.** $\left\{\frac{2}{3}\right\}$ [7.4] **5.** $\varnothing$

[9.1, 9.2] **6.** $\left\{\dfrac{7 \pm \sqrt{177}}{4}\right\}$ [9.3] **7.** $\{\pm 1, \pm 2\}$ [2.5] **8.** $[1, \infty)$

[2.7] **9.** $\left[2, \frac{8}{3}\right]$ [9.7] **10.** $(1, 3)$ **11.** $(-2, 1)$

[3.1, 3.5] **12.** function;
domain: $(-\infty, \infty)$;
range: $(-\infty, \infty)$;
$f(x) = \frac{4}{5}x - 3$

[3.4, 3.5] **13.** not a function

[9.5] **14.** function;
domain: $(-\infty, \infty)$;
range: $(-\infty, 3]$;
$f(x) = -2(x - 1)^2 + 3$

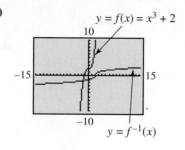

[3.2] **15.** $m = \frac{2}{7}$; x-intercept: $(-8, 0)$; y-intercept: $\left(0, \frac{16}{7}\right)$

[3.3] **16. (a)** $y = -\frac{5}{2}x + 2$
(b) $y = \frac{2}{5}x + \frac{13}{5}$

[4.1] **17.** $\{(1, -2)\}$
[4.2] **18.** $\{(3, -4, 2)\}$

[4.3] **19.** Microsoft: $60.4 billion; Oracle: $22.4 billion

[5.1] **20.** $\dfrac{x^8}{y^4}$ **21.** $\dfrac{4}{xy^2}$ [5.4] **22.** $\frac{4}{9}t^2 + 12t + 81$

[5.5] **23.** $4x^2 - 6x + 11 + \dfrac{4}{x + 2}$ [6.1–6.3] **24.** $(4m - 3)(6m + 5)$

25. $(2x + 3y)(4x^2 - 6xy + 9y^2)$ **26.** $(3x - 5y)^2$ [7.1] **27.** $-\frac{5}{18}$

[7.2] **28.** $-\dfrac{8}{x}$ [7.3] **29.** $\dfrac{r - s}{r}$ [8.3] **30.** $\dfrac{3\sqrt[3]{4}}{4}$

[8.5] **31.** $\sqrt{7} + \sqrt{5}$ [9.4] **32.** southbound car: 57 mi;
eastbound car: 76 mi

10 INVERSE, EXPONENTIAL, AND LOGARITHMIC FUNCTIONS

Connections **(page 576)**

$y = f(x) = x^3 + 2$

$y = f^{-1}(x)$

Section 10.1 (pages 577–580)

1. This function is not one-to-one because both France and the United States are paired with the same trans fat percentage, 11.
3. Yes. By adding 1 to 1058, two distances would be the same, so the function would not be one-to-one. **5.** B **7.** A
9. $\{(6, 3), (10, 2), (12, 5)\}$ **11.** not one-to-one
13. $f^{-1}(x) = \dfrac{x - 4}{2}$, or $f^{-1}(x) = \frac{1}{2}x - 2$ **15.** $g^{-1}(x) = x^2 + 3$,
$x \geq 0$ **17.** not one-to-one **19.** $f^{-1}(x) = \sqrt[3]{x + 4}$
21. (a) 8 **(b)** 3 **23. (a)** 1 **(b)** 0
25. (a) one-to-one **(b)**   **27. (a)** not one-to-one

29. (a) one-to-one **(b)**   **31.**

33.   **35.**

x	$f(x)$
0	0
1	1
4	2

37.

x	$f(x)$
-1	-3
0	-2
1	-1
2	6

39. $f^{-1}(x) = \dfrac{x + 5}{4}$, or
$f^{-1}(x) = \frac{1}{4}x + \frac{5}{4}$

40. MY GRAPHING CALCULATOR IS THE GREATEST THING SINCE SLICED BREAD. **41.** If the function were not one-to-one, there would be ambiguity in some of the characters, as they could represent more than one letter. **42.** Answers will vary. For example, Jane Doe is 1004 5 2748 129 68 3379 129.

43. $f^{-1}(x) = \dfrac{x + 7}{2}$, or $f^{-1}(x) = \frac{1}{2}x + \frac{7}{2}$ **45.** $f^{-1}(x) = \sqrt[3]{x - 5}$

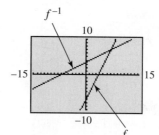

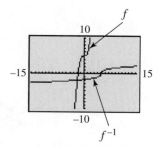

47. 64 **49.** $\frac{1}{2}$

Section 10.2 (pages 585–587)

1. C **3.** A **5.**

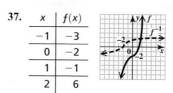

7.

$g(x) = \left(\frac{1}{3}\right)^x$

9.

$y = 4^{-x}$

11.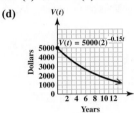

$y = 2^{2x-2}$

13. rises; falls
15. $\{2\}$ **17.** $\left\{\frac{3}{2}\right\}$
19. $\{7\}$ **21.** $\{-3\}$
23. $\{-1\}$ **25.** $\{-3\}$
27. 639.545   **29.** 0.066

31. 12.179   **33. (a)** 0.6°C   **(b)** 0.3°C   **35. (a)** 1.4°C   **(b)** 0.5°C

37. (a) 5028 million tons   **(b)** 6512 million tons

(c) It is less than what the model provides (7099 million tons).

39. (a) $5000   **(b)** $2973   **(c)** $1768

(d)

$V(t) = 5000(2)^{-0.15t}$

41. 6.67 yr after it was purchased

43. 4   **45.** 0

Section 10.3 (pages 592–595)

1. (a) B   **(b)** E   **(c)** D   **(d)** F   **(e)** A   **(f)** C   **3.** $\log_4 1024 = 5$

5. $\log_{1/2} 8 = -3$ **7.** $\log_{10} 0.001 = -3$ **9.** $\log_{625} 5 = \frac{1}{4}$

11. $\log_8 \frac{1}{4} = -\frac{2}{3}$ **13.** $\log_5 1 = 0$ **15.** $4^3 = 64$ **17.** $10^{-4} = \frac{1}{10,000}$

19. $6^0 = 1$ **21.** $9^{1/2} = 3$ **23.** $\left(\frac{1}{4}\right)^{1/2} = \frac{1}{2}$ **25.** $5^{-1} = 5^{-1}$

27. (a) C   **(b)** B   **(c)** B   **(d)** C   **29.** $\left\{\frac{1}{3}\right\}$ **31.** $\{81\}$ **33.** $\left\{\frac{1}{5}\right\}$

35. $\{1\}$ **37.** $\{x \mid x > 0, x \neq 1\}$ **39.** $\{5\}$ **41.** $\left\{\frac{5}{3}\right\}$ **43.** $\{4\}$

45. $\left\{\frac{3}{2}\right\}$ **47.** $\{30\}$ **49.**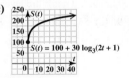

$y = \log_3 x$

51.

$y = \log_{1/3} x$

53. Every power of 1 is equal to 1, and thus it cannot be used as a base.

55. $(0, \infty)$; $(-\infty, \infty)$ **57.** 8   **59.** 24   **61. (a)** 4385 billion ft³

(b) 5555 billion ft³   **(c)** 6140 billion ft³   **63. (a)** 130 thousand units

(b) 190 thousand units   **(c)**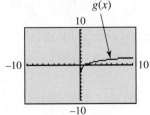

$S(t) = 100 + 30 \log_3(2t + 1)$

65. about 4 times as powerful

67.

$g(x)$

69.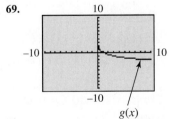

$g(x)$

71. 4^9 **73.** 7^{12}

Connections   **(page 601)**   **1.**

$$\log_{10} 458.3 \approx 2.661149857$$
$$\underline{+\log_{10} 294.6 \approx 2.469232743}$$
$$\approx 5.130382600$$
$$10^{5.130382600} \approx 135{,}015.18$$

A calculator gives

$$(458.3)(294.6) = 135{,}015.18.$$

2. Answers will vary.

Section 10.4 (pages 602–603)

1. $\log_{10} 7 + \log_{10} 8$ **3.** 4   **5.** 9   **7.** $\log_7 4 + \log_7 5$

9. $\log_5 8 - \log_5 3$ **11.** $2 \log_4 6$ **13.** $\frac{1}{3} \log_3 4 - 2 \log_3 x - \log_3 y$

15. $\frac{1}{2} \log_3 x + \frac{1}{2} \log_3 y - \frac{1}{2} \log_3 5$ **17.** $\frac{1}{3} \log_2 x + \frac{1}{5} \log_2 y - 2 \log_2 r$

19. In the notation $\log_a (x + y)$, the parentheses do not indicate multiplication. They indicate that $x + y$ is the result of raising a to some power.

21. $\log_b xy$ **23.** $\log_a \frac{m}{n}$ **25.** $\log_a \frac{rt^3}{s}$ **27.** $\log_a \frac{125}{81}$

29. $\log_{10} (x^2 - 9)$ **31.** $\log_p \frac{x^3 y^{1/2}}{z^{3/2} a^3}$ **33.** 1.2552   **35.** -0.6532

37. 1.5562   **39.** 0.2386   **41.** 0.4771   **43.** 4.7710   **45.** false

47. true   **49.** true   **51.** false   **53.** The exponent of a quotient is the difference between the exponent of the numerator and the exponent of the denominator.   **55.** No number allowed as a logarithmic base can be raised to a power with a result of 0.   **57.** $\log_{10} 10{,}000 = 4$

59. $\log_{10} 0.01 = -2$ **61.** $10^0 = 1$

Section 10.5 (pages 609–612)

1. C   **3.** C   **5.** 31.6   **7.** 1.6335   **9.** 2.5164   **11.** -1.4868

13. 9.6776   **15.** 2.0592   **17.** -2.8896 **19.** 5.9613   **21.** 4.1506

23. 2.3026   **25. (a)** 2.552424846   **(b)** 1.552424846

(c) 0.552424846   **(d)** The whole number parts will vary, but the decimal parts are the same.   **27.** poor fen   **29.** bog   **31.** rich fen   **33.** 11.6

35. 4.3   **37.** 4.0×10^{-8} **39.** 4.0×10^{-6} **41. (a)** 107 dB

(b) 100 dB   **(c)** 98 dB   **43. (a)** 800 yr   **(b)** 5200 yr   **(c)** 11,500 yr

45. (a) 77%   **(b)** 1989   **47. (a)** $54 per ton   **(b)** If $p = 0$, then

$\ln (1 - p) = \ln 1 = 0$, so T would be negative. If $p = 1$, then

$\ln (1 - p) = \ln 0$, but the domain of $\ln x$ is $(0, \infty)$.   **49.** 2.2619

51. 0.6826   **53.** 0.3155   **55.** 0.8736   **57.** 2.4849   **59.** Answers will vary. Suppose the name is Jeffery Cole, with $m = 7$ and $n = 4$.

(a) $\log_7 4$ is the exponent to which 7 must be raised to obtain 4.

(b) 0.7124143742   **(c)** 4   **61.** 6446 billion ft³   **63.** $\left\{-\frac{3}{5}\right\}$

65. $\{5\}$ **67.** $\{-3\}$ **69.** $\log (x + 2)(x + 3)$, or $\log (x^2 + 5x + 6)$

Connections   **(page 619)**

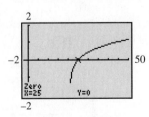

Section 10.6 (pages 619–622)

1. $\{0.827\}$ **3.** $\{0.833\}$ **5.** $\{1.201\}$ **7.** $\{2.269\}$ **9.** $\{15.967\}$
11. $\{-6.067\}$ **13.** $\{261.291\}$ **15.** $\{-10.718\}$ **17.** $\{3\}$
19. $\{5.879\}$ **21.** $\{-\pi\}$, or $\{-3.142\}$ **23.** $\{1\}$ **25.** Natural
logarithms are a better choice because e is the base.   **27.** $\left\{\frac{2}{3}\right\}$
29. $\left\{\frac{33}{2}\right\}$ **31.** $\left\{-1 + \sqrt[3]{49}\right\}$ **33.** 2 cannot be a solution because
$\log(2 - 3) = \log(-1)$, and -1 is not in the domain of $\log x$.
35. $\left\{\frac{1}{3}\right\}$ **37.** $\{2\}$ **39.** $\emptyset$ **41.** $\{8\}$ **43.** $\left\{\frac{4}{3}\right\}$ **45.** $\{8\}$
47. (a) $2539.47   **(b)** 10.2 yr   **49. (a)** $4934.71   **(b)** 19.8 yr
51. (a) $11,260.96   **(b)** $11,416.64   **(c)** $11,497.99   **(d)** $11,580.90
(e) $11,581.83   **53.** $137.41   **55. (a)** 15.9 million tons
(b) 30.7 million tons   **(c)** 59.2 million tons   **(d)** 93.7 million tons
57. $143,598 million   **59. (a)** 1.62 g   **(b)** 1.18 g   **(c)** 0.69 g
(d) 2.00 g   **61. (a)** 179.73 g   **(b)** 21.66 yr   **63.** 2012   **65.** It means
that after 250 yr, approximately 2.9 g of the original sample remain.
67.   **69.**

$f(x) = 2x^2$

$f(x) = (x + 1)^2$

Chapter 10 Review Exercises (pages 626–630)

1. not one-to-one   **2.** one-to-one   **3.** $f^{-1}(x) = \dfrac{x - 7}{-3}$,
or $f^{-1}(x) = -\dfrac{1}{3}x + \dfrac{7}{3}$ **4.** $f^{-1}(x) = \dfrac{x^3 + 4}{6}$ **5.** not one-to-one
6. This function is not one-to-one because two sodas in the list have
41 mg of caffeine.
7.   **8.**   **9.**

$f(x) = 3^x$

10.   **11.**   **12.** $\left\{\frac{1}{2}\right\}$ **13.** $\{4\}$

$f(x) = \left(\frac{1}{3}\right)^x$

$y = 2^{2x + 3}$

14. $\left\{\frac{3}{7}\right\}$

15. (a) 29.4 million tons   **(b)** 18.2 million tons   **(c)** 14.4 million tons
16.   **17.**   **18.** $\{2\}$ **19.** $\left\{\frac{3}{2}\right\}$

$g(x) = \log_3 x$

$g(x) = \log_{1/3} x$

20. $\{7\}$ **21.** $\{8\}$
22. $\{4\}$

23. $\left\{b \mid b > 0, b \neq 1\right\}$ **24.** $\log_b a$ is the exponent to which
b must be raised to obtain a.   **25.** a
26. (a) $300,000   **(b)**
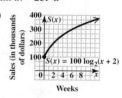

$S(x) = 100\log_2(x + 2)$

27. $\log_2 3 + \log_2 x + 2\log_2 y$ **28.** $\frac{1}{2}\log_4 x + 2\log_4 w - \log_4 z$
29. $\log_b \dfrac{3x}{y^2}$ **30.** $\log_3\left(\dfrac{x + 7}{4x + 6}\right)$ **31.** 1.4609   **32.** -0.5901
33. 3.3638   **34.** -1.3587 **35.** 0.9251   **36.** 1.7925   **37.** 6.4
38. 8.4   **39.** 2.5×10^{-5} **40.** Magnitude 1 is about 6.3 times as
intense as magnitude 3.   **41. (a)** 18 yr   **(b)** 12 yr   **(c)** 7 yr
(d) 6 yr   **(e)** Each comparison shows approximately the same number.
For example, in part (a) the doubling time is 18 yr (rounded) and $\frac{72}{4} = 18$.
Thus, the formula $t = \dfrac{72}{100\,r}$ (called the *rule of 72*) is an excellent
approximation of the doubling time formula.   **42.** $\{2.042\}$
43. $\{4.907\}$ **44.** $\{18.310\}$ **45.** $\left\{\frac{1}{9}\right\}$ **46.** $\left\{-6 + \sqrt[3]{25}\right\}$
47. $\{2\}$ **48.** $\left\{\frac{3}{8}\right\}$ **49.** $\{4\}$ **50.** $\{1\}$ **51.** When the power rule
was applied in the second step, the domain was changed from $\{x \mid x \neq 0\}$
to $\{x \mid x > 0\}$. The valid solution -10 was "lost." The solution set is
$\{\pm 10\}$.   **52.** $24,403.80   **53.** $11,190.72   **54.** Plan A is better, since
it would pay $2.92 more.   **55.** about 13.9 days   **56. (a)** about $4267
(b) about 11%   **57.** about 67%   **58.** D   **59.** 7   **60.** 36   **61.** 4
62. e **63.** -5 **64.** 5.4   **65.** $\{72\}$ **66.** $\{3\}$ **67.** $\left\{\frac{1}{9}\right\}$ **68.** $\left\{\frac{4}{3}\right\}$
69. $\{3\}$ **70.** $\{0\}$ **71.** $\left\{\frac{1}{8}\right\}$ **72.** $\left\{\frac{11}{3}\right\}$ **73.** $\{-2, -1\}$ **74. (a)** $\left\{\frac{3}{8}\right\}$
(b) The x-value of the x-intercept is 0.375, the decimal equivalent of $\frac{3}{8}$.
75. about 32.28%   **76. (a)** 0.325   **(c)** 0.673

Chapter 10 Test (pages 630–632)

[10.1] **1. (a)** not one-to-one   **(b)** one-to-one   **2.** $f^{-1}(x) = x^3 - 7$
3.   [10.2] **4.**   [10.3] **5.**

$f(x) = 6^x$

$g(x) = \log_6 x$

[10.1–10.3] **6.** Once the graph of $f(x) = 6^x$ is sketched, interchange the
x- and y-values of its ordered pairs. The resulting points will be on the
graph of $g(x) = \log_6 x$ since f and g are inverses.   [10.2] **7.** $\{-4\}$
8. $\left\{-\frac{13}{3}\right\}$ [10.5] **9. (a)** 55.8 million   **(b)** 80.8 million
[10.3] **10.** $\log_4 0.0625 = -2$ **11.** $7^2 = 49$ **12.** $\{32\}$ **13.** $\left\{\frac{1}{2}\right\}$
14. $\{2\}$ **15.** 5; 2; fifth; 32   [10.4] **16.** $2\log_3 x + \log_3 y$
17. $\frac{1}{2}\log_5 x - \log_5 y - \log_5 z$ **18.** $\log_b \dfrac{s^3}{t}$ **19.** $\log_b \dfrac{r^{1/4}s^2}{t^{2/3}}$
[10.5] **20. (a)** 1.3636   **(b)** -0.1985 **21. (a)** $\dfrac{\log 19}{\log 3}$ **(b)** $\dfrac{\ln 19}{\ln 3}$
(c) 2.6801   [10.6] **22.** $\{3.966\}$ **23.** $\{3\}$ **24.** $12,507.51
25. (a) $19,260.38   **(b)** approximately 13.9 yr

Chapters 1–10 Cumulative Review Exercises
(pages 632–633)

[1.1] **1.** $-2, 0, 6, \frac{30}{3}$ (or 10)   **2.** $-\frac{9}{4}, -2, 0, 0.6, 6, \frac{30}{3}$ (or 10)
3. $-\sqrt{2}, \sqrt{11}$ [1.2, 1.3] **4.** 16   **5.** -39 [2.1] **6.** $\left\{-\frac{2}{3}\right\}$
[2.5] **7.** $[1, \infty)$ [2.7] **8.** $\{-2, 7\}$ **9.** $(-\infty, -3) \cup (2, \infty)$

[3.1] **10.**   [3.4] **11.**

[3.2, 3.5] **12. (a)** yes   **(b)** 3346.2; The number of travelers increased by an average of 3346.2 thousand per year during 2003–2008.

[3.3] **13.** $y = \frac{3}{4}x - \frac{19}{4}$ [4.1, 4.4] **14.** $\{(4, 2)\}$

[4.2, 4.4] **15.** $\{(1, -1, 4)\}$ [4.3] **16.** 6 lb   [5.4] **17.** $6p^2 + 7p - 3$

18. $16k^2 - 24k + 9$ [5.2] **19.** $-5m^3 + 2m^2 - 7m + 4$

[5.5] **20.** $2t^3 + 5t^2 - 3t + 4$ [6.1] **21.** $x(8 + x^2)$

[6.2] **22.** $(3y - 2)(8y + 3)$ **23.** $z(5z + 1)(z - 4)$

[6.3] **24.** $(4a + 5b^2)(4a - 5b^2)$ **25.** $(2c + d)(4c^2 - 2cd + d^2)$

26. $(4r + 7q)^2$ [5.1] **27.** $-\dfrac{1875p^{13}}{8}$ [7.1] **28.** $\dfrac{x + 5}{x + 4}$

[7.2] **29.** $\dfrac{-3k - 19}{(k + 3)(k - 2)}$ [8.3] **30.** $12\sqrt{2}$ [8.4] **31.** $-27\sqrt{2}$

[8.6] **32.** $\{0, 4\}$ [8.7] **33.** 41   [9.1, 9.2] **34.** $\left\{\dfrac{1 \pm \sqrt{13}}{6}\right\}$

[9.7] **35.** $(-\infty, -4) \cup (2, \infty)$ [9.3] **36.** $\{\pm 1, \pm 2\}$

[9.5] **37.** $f(x) = \frac{1}{3}(x - 1)^2 + 2$ [10.2] **38.**   **39.** $\{-1\}$

[10.3] **40.** 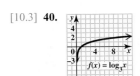   [10.4] **41.** $3 \log x + \frac{1}{2} \log y - \log z$

[10.6] **42. (a)** 25,000   **(b)** 30,500   **(c)** 37,300   **(d)** in about 3.5 hr, or at about 3:30 P.M.

---

11   NONLINEAR FUNCTIONS, CONIC SECTIONS, AND NONLINEAR SYSTEMS

Section 11.1 (pages 640–641)

1. E; 0; 0   **3.** A; $(-\infty, \infty)$; $\{\dots, -2, -1, 0, 1, 2, \dots\}$ **5.** B; It does not satisfy the conditions of the vertical line test.   **7.** B   **9.** A

11. domain: $(-\infty, \infty)$; range: $[0, \infty)$

13. domain: $(-\infty, 0) \cup (0, \infty)$; range: $(-\infty, 1) \cup (1, \infty)$

15. domain: $[2, \infty)$; range: $[0, \infty)$

17. domain: $(-\infty, 2) \cup (2, \infty)$; range: $(-\infty, 0) \cup (0, \infty)$

19. domain: $[-3, \infty)$; range: $[-3, \infty)$

21. domain: $(-\infty, \infty)$; range: $[1, \infty)$

23. Shift the graph of $g(x) = \dfrac{1}{x}$ to the right 3 units and up 2 units.

25. 3   **27.** 4   **29.** 0   **31.** -14 **33.** -11

35.   **37.**   **39.**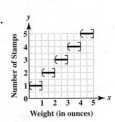

41. \$2.75   **43.** $2\sqrt{5}$ **45.** $\sqrt{(x - h)^2 + (y - k)^2}$

Connections   **(page 646)** $y_1 = -1 + \sqrt{36 - (x - 3)^2}$, $y_2 = -1 - \sqrt{36 - (x - 3)^2}$

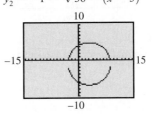

Section 11.2 (pages 647–650)

1. (a) $(0, 0)$ **(b)** 5   **(c)**   **3.** B   **5.** D

7. $(x + 4)^2 + (y - 3)^2 = 4$ **9.** $(x + 8)^2 + (y + 5)^2 = 5$

11. center: $(-2, -3)$; $r = 2$ **13.** center: $(-5, 7)$; $r = 9$

15. center: $(2, 4)$; $r = 4$

17.   **19.**   **21.** center: $(-3, 2)$

23. center: $(2, 3)$ **25.** center: $(-3, 3)$

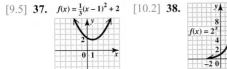

27. The thumbtack acts as the center and the length of the string acts as the radius.

29.

$$\frac{x^2}{9} + \frac{y^2}{25} = 1$$

31.

$$\frac{x^2}{36} + \frac{y^2}{16} = 1$$

33.

$$\frac{x^2}{16} + \frac{y^2}{4} = 1$$

35.

$$\frac{y^2}{25} = 1 - \frac{x^2}{49}$$

37.

$$\frac{(x+1)^2}{64} + \frac{(y-2)^2}{49} = 1$$

39.

$$\frac{(x-2)^2}{16} + \frac{(y-1)^2}{9} = 1$$

41. By the vertical line test the set is not a function, because a vertical line may intersect the graph of an ellipse in two points.

43. $y_1 = 4 + \sqrt{16 - (x+2)^2},\ y_2 = 4 - \sqrt{16 - (x+2)^2}$

45. 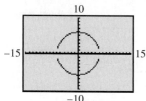   **47.**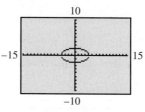

49. $3\sqrt{3}$ units   **51. (a)** 10 m   **(b)** 36 m

53. (a) 154.7 million mi   **(b)** 128.7 million mi (Answers are rounded.)

55.   **57.** $(3, 0);\ (0, 4)$

Section 11.3 (pages 655–657)

1. C   **3.** D

5.

$$\frac{x^2}{16} - \frac{y^2}{9} = 1$$

7.

$$\frac{y^2}{4} - \frac{x^2}{25} = 1$$

9.

$$\frac{x^2}{25} - \frac{y^2}{36} = 1$$

11.

$$\frac{y^2}{16} - \frac{x^2}{16} = 1$$

13. hyperbola   **15.** ellipse

$$x^2 - y^2 = 16$$ $$4x^2 + y^2 = 16$$

17. circle   **19.** parabola   **21.** hyperbola

$$y^2 = 36 - x^2$$ $$x^2 - 2y = 0$$ $$y^2 = 4 + x^2$$

23. domain: $[-4, 4]$;   **25.** domain: $[-6, 6]$;   **27.** domain: $[-3, 3]$;
range: $[0, 4]$ range: $[-6, 0]$ range: $[-2, 0]$

   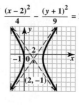   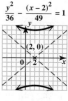

$f(x) = \sqrt{16 - x^2}$ $f(x) = -\sqrt{36 - x^2}$ $y = -2\sqrt{1 - \frac{x^2}{9}}$

29. domain: $(-\infty, \infty)$;   **31.** $\frac{(x-2)^2}{4} - \frac{(y+1)^2}{9} = 1$ **33.** $\frac{y^2}{36} - \frac{(x-2)^2}{49} = 1$
range: $[3, \infty)$

$$\frac{y}{3} = \sqrt{1 + \frac{x^2}{9}}$$

35. (a) 50 m   **(b)** 69.3 m   **37.** $y_1 = \sqrt{\frac{x^2}{9} - 1},\ y_2 = -\sqrt{\frac{x^2}{9} - 1}$

39. 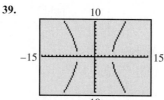   **41.**

43. $\{(2, 9)\}$ **45.** $\{(-1, 2)\}$ **47.** $\left\{\pm\sqrt{3}, \pm\frac{\sqrt{2}}{2}i\right\}$

Section 11.4 (pages 662–663)

1. one   **3.** none

5.   **7.**   **9.**   **11.**

13. $\left\{(0, 0), \left(\frac{1}{2}, \frac{1}{2}\right)\right\}$ **15.** $\{(-6, 9), (-1, 4)\}$

17. $\left\{\left(-\frac{1}{5}, \frac{7}{5}\right), (1, -1)\right\}$ **19.** $\left\{(-2, -2), \left(-\frac{4}{3}, -3\right)\right\}$

21. $\{(-3, 1), (1, -3)\}$ **23.** $\left\{\left(-\frac{3}{2}, -\frac{9}{4}\right), (-2, 0)\right\}$

25. $\left\{(-\sqrt{3}, 0), (\sqrt{3}, 0), (-\sqrt{5}, 2), (\sqrt{5}, 2)\right\}$

27. $\left\{\left(\frac{\sqrt{3}}{3}i, -\frac{1}{2} + \frac{\sqrt{3}}{6}i\right), \left(-\frac{\sqrt{3}}{3}i, -\frac{1}{2} - \frac{\sqrt{3}}{6}i\right)\right\}$

29. $\{(-2, 0), (2, 0)\}$ **31.** $\left\{(\sqrt{3}, 0), (-\sqrt{3}, 0)\right\}$

33. $\left\{(-2\sqrt{3}, -2), (-2\sqrt{3}, 2), (2\sqrt{3}, -2), (2\sqrt{3}, 2)\right\}$

35. $\left\{(-2i\sqrt{2}, -2\sqrt{3}), (-2i\sqrt{2}, 2\sqrt{3}), (2i\sqrt{2}, -2\sqrt{3}), (2i\sqrt{2}, 2\sqrt{3})\right\}$ **37.** $\left\{(-\sqrt{5}, -\sqrt{5}), (\sqrt{5}, \sqrt{5})\right\}$

39. $\{(i, 2i), (-i, -2i), (2, -1), (-2, 1)\}$

41. $\{(2, -3), (-3, 2)\}$

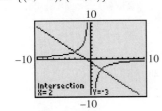

43. length: 12 ft; width: 7 ft   **45.** $20; $\frac{4}{5}$ thousand or 800 calculators

47.

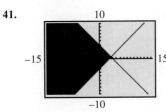

Section 11.5 (pages 667–668)

1. C   **3.** B   **5.** A

7.

9.

11.

13.

15.

17.

19.

21.

23.

25.
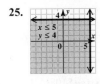

27.

29.

31.
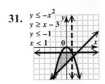

33.

35.

37.
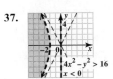

39.

41.
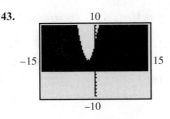

43.

Chapter 11 Review Exercises (pages 672–673)

1.

2.

3.

4.
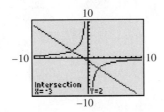

5. $(x + 2)^2 + (y - 4)^2 = 9$
6. $(x + 1)^2 + (y + 3)^2 = 25$
7. $(x - 4)^2 + (y - 2)^2 = 36$
8. center: $(-3, 2)$; $r = 4$

9. center: $(4, 1)$; $r = 2$ **10.** center: $(-1, -5)$; $r = 3$

11. center: $(3, -2)$; $r = 5$

12.

13.

14.
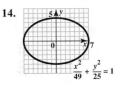

15. $\dfrac{x^2}{65,286,400} + \dfrac{y^2}{2,560,000} = 1$ **16. (a)** 348.2 ft   **(b)** 1787.6 ft

17.

18.

19.

20. circle   **21.** parabola   **22.** hyperbola   **23.** ellipse   **24.** parabola

25. hyperbola   **26.** $\dfrac{x^2}{625} - \dfrac{y^2}{1875} = 1$ **27.** $\{(6, -9), (-2, -5)\}$

28. $\{(1, 2), (-5, 14)\}$ **29.** $\{(4, 2), (-1, -3)\}$

30. $\{(-2, -4), (8, 1)\}$ **31.** $\{(-\sqrt{2}, 2), (-\sqrt{2}, -2),$
$(\sqrt{2}, -2), (\sqrt{2}, 2)\}$ **32.** $\{(-\sqrt{6}, -\sqrt{3}), (-\sqrt{6}, \sqrt{3}),$
$(\sqrt{6}, -\sqrt{3}), (\sqrt{6}, \sqrt{3})\}$ **33.** 0, 1, or 2   **34.** 0, 1, 2, 3, or 4

35.

36.

37.
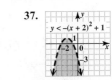

38.

39.
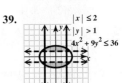

40.

41.

42.

43.
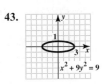

44.

$x^2 - 9y^2 = 9$

45.

$f(x) = \sqrt{4-x}$

46.

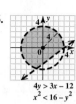

$4y > 3x - 12$
$x^2 < 16 - y^2$

Chapter 11 Test (page 674)

[11.1] **1.** 0 **2.** $[0, \infty)$ **3.** $\{\ldots, -2, -1, 0, 1, 2, \ldots\}$

4. (a) C **(b)** A **(c)** D **(d)** B

5. domain: $(-\infty, \infty)$; [11.2] **6.** center: $(2, -3)$;
range: $[4, \infty)$ radius: 4

$f(x) = |x - 3| + 4$

$(x - 2)^2 + (y + 3)^2 = 16$

7. center: $(-4, 1)$; radius: 5 [11.3] **8.**

$f(x) = \sqrt{9 - x^2}$

[11.2] **9.**

$4x^2 + 9y^2 = 36$

[11.3] **10.**

$16y^2 - 4x^2 = 64$

11.

$\frac{y}{2} = -\sqrt{1 - \frac{x^2}{9}}$

12. ellipse **13.** hyperbola **14.** circle
15. parabola [11.4] **16.** $\left\{\left(-\frac{1}{2}, -10\right), (5, 1)\right\}$
17. $\left\{(-2, -2), \left(\frac{14}{5}, -\frac{2}{5}\right)\right\}$
18. $\left\{\left(-\sqrt{22}, -\sqrt{3}\right), \left(-\sqrt{22}, \sqrt{3}\right),\right.$
$\left.\left(\sqrt{22}, -\sqrt{3}\right), \left(\sqrt{22}, \sqrt{3}\right)\right\}$

[11.5] **19.**

$y < x^2 - 2$

20.

$x^2 + 25y^2 \le 25$
$x^2 + y^2 \le 9$

Chapters 1–11 Cumulative Review Exercises (pages 675–676)

[2.1] **1.** $\left\{\frac{2}{3}\right\}$ [2.5] **2.** $\left(-\infty, \frac{3}{5}\right]$ [2.7] **3.** $\{-4, 4\}$

4. $(-\infty, -5) \cup (10, \infty)$ [3.2] **5.** $\frac{2}{3}$ [3.3] **6.** $3x + 2y = -13$

[4.1] **7.** $\{(3, -3)\}$ [4.2] **8.** $\{(4, 1, -2)\}$

[11.4] **9.** $\left\{(-1, 5), \left(\frac{5}{2}, -2\right)\right\}$ [4.3] **10.** 40 mph

[5.4] **11.** $25y^2 - 30y + 9$ [5.5] **12.** $4x^3 - 4x^2 + 3x + 5 + \dfrac{3}{2x + 1}$

[6.2] **13.** $(3x + 2)(4x - 5)$ [6.3] **14.** $(z^2 + 1)(z + 1)(z - 1)$

15. $(a - 3b)(a^2 + 3ab + 9b^2)$ [7.1] **16.** $\dfrac{y - 1}{y(y - 3)}$

[7.2] **17.** $\dfrac{3c + 5}{(c + 5)(c + 3)}$ **18.** $\dfrac{1}{p}$ [7.5] **19.** $1\frac{1}{5}$ hr [5.1] **20.** $\dfrac{a^5}{4}$

[8.4] **21.** $2\sqrt[3]{2}$ [8.5] **22.** $\dfrac{3\sqrt{10}}{2}$ [8.7] **23.** $\frac{7}{5} + \frac{11}{5}i$ [8.6] **24.** $\emptyset$

[6.5] **25.** $\left\{\frac{1}{5}, -\frac{3}{2}\right\}$ [9.1, 9.2] **26.** $\left\{\dfrac{3 \pm \sqrt{33}}{6}\right\}$

[9.3] **27.** $\left\{\pm\dfrac{\sqrt{6}}{2}, \pm\sqrt{7}\right\}$ [10.6] **28.** $\{3\}$

[9.4] **29.** $v = \dfrac{\pm\sqrt{rFkw}}{kw}$ [10.1] **30.** $f^{-1}(x) = \sqrt[3]{x - 4}$

[10.4, 10.5] **31. (a)** 4 **(b)** 7 [10.4] **32.** $\log\dfrac{(3x + 7)^2}{4}$

[10.2] **33. (a)** \$86.8 billion **(b)** \$169.5 billion

[11.1] **34.** domain: $(-\infty, \infty)$; range: $[0, \infty)$

[3.6] **35.**

$f(x) = -3x + 5$

[9.5] **36.**

$f(x) = -2(x - 1)^2 + 3$

[11.5] **37.**

$\dfrac{x^2}{25} + \dfrac{y^2}{16} \le 1$

[11.1] **38.**

$f(x) = \sqrt{x - 2}$

[11.3] **39.**

$\dfrac{x^2}{4} - \dfrac{y^2}{16} = 1$

[10.2] **40.**

$f(x) = 3^x$

12 SEQUENCES AND SERIES

Section 12.1 (pages 682–684)

1. 2, 3, 4, 5, 6 **3.** $4, \frac{5}{2}, 2, \frac{7}{4}, \frac{8}{5}$ **5.** 3, 9, 27, 81, 243 **7.** $1, \frac{1}{4}, \frac{1}{9}, \frac{1}{16}, \frac{1}{25}$

9. 5, -5, 5, -5, 5 **11.** $0, \frac{3}{2}, \frac{8}{3}, \frac{15}{4}, \frac{24}{5}$ **13.** -70 **15.** $\frac{49}{23}$ **17.** 171

19. $4n$ **21.** $-8n$ **23.** $\dfrac{1}{3^n}$ **25.** $\dfrac{n + 1}{n + 4}$ **27.** \$110, \$109, \$108, \$107,

\$106, \$105; \$400 **29.** \$6554 **31.** $4 + 5 + 6 + 7 + 8 = 30$

33. $3 + 6 + 11 = 20$ **35.** $-1 + 1 - 1 + 1 - 1 + 1 = 0$

37. $0 + 6 + 14 + 24 + 36 = 80$

Answers may vary for Exercises 39–43.

39. $\displaystyle\sum_{i=1}^{5}(i + 2)$ **41.** $\displaystyle\sum_{i=1}^{5} 2^i(-1)^i$ **43.** $\displaystyle\sum_{i=1}^{4} i^2$ **45.** A sequence is a list

of terms in a specific order, while a series is the indicated sum of the

terms of a sequence. **47.** 9 **49.** $\frac{40}{9}$ **51.** 8036 **53.** $a = 6, d = 2$

55. 10

Section 12.2 (pages 689–691)

1. $d = 1$ **3.** not arithmetic **5.** $d = -5$ **7.** 5, 9, 13, 17, 21

9. $-2, -6, -10, -14, -18$ **11.** $a_n = 5n - 3$ **13.** $a_n = \frac{3}{4}n + \frac{9}{4}$

15. $a_n = 3n - 6$ **17.** 76 **19.** 48 **21.** -1 **23.** 16 **25.** 6

27. n represents the number of terms. **29.** 81 **31.** -3 **33.** 87

35. 390   **37.** 395   **39.** 31,375   **41.** $465   **43.** $2100 per month
45. 68; 1100   **47.** no; 3; 9   **49.** 18   **51.** $\frac{1}{2}$

Section 12.3 (pages 698–700)

1. $r = 2$ **3.** not geometric   **5.** $r = -3$ **7.** $r = -\frac{1}{2}$
There are alternative forms of the answers in Exercises 9–13.

9. $a_n = -5(2)^{n-1}$ **11.** $a_n = -2\left(-\frac{1}{3}\right)^{n-1}$ **13.** $a_n = 10\left(-\frac{1}{5}\right)^{n-1}$

15. $2(5)^9 = 3{,}906{,}250$ **17.** $\frac{1}{2}\left(\frac{1}{3}\right)^{11}$, or $\frac{1}{354{,}294}$ **19.** $2\left(\frac{1}{2}\right)^{24} = \frac{1}{2^{23}}$

21. 2, 6, 18, 54, 162   **23.** $5, -1, \frac{1}{5}, -\frac{1}{25}, \frac{1}{125}$ **25.** $\frac{121}{243}$ **27.** -1.997

29. 2.662   **31.** -2.982 **33.** $33,410.84   **35.** $104,273.05   **37.** 9

39. $\frac{10{,}000}{11}$ **41.** $-\frac{9}{20}$ **43.** The sum does not exist.   **45.** $10\left(\frac{3}{5}\right)^4 \approx 1.3$ ft

47. 3 days; $\frac{1}{4}$ g   **49. (a)** $1.1(1.06)^5 \approx 1.5$ billion units

(b) approximately 12 yr   **51.** $50{,}000\left(\frac{3}{4}\right)^8 \approx \$5005.65$

53. $0.33333\ldots$ **54.** $0.66666\ldots$ **55.** $0.99999\ldots$

56. $\dfrac{a_1}{1-r} = \dfrac{0.9}{1-0.1} = \dfrac{0.9}{0.9} = 1$; Therefore, $0.99999\ldots = 1$
57. B   **58.** $0.49999\ldots = 0.4 + 0.09999\ldots = \frac{4}{10} + \frac{1}{10}(0.9999\ldots) =$
$\frac{4}{10} + \frac{1}{10}(1) = \frac{5}{10} = \frac{1}{2}$ **59.** $9x^2 + 12xy + 4y^2$
61. $a^3 - 3a^2b + 3ab^2 - b^3$

Section 12.4 (page 705)

1. 720   **3.** 40,320   **5.** 15   **7.** 1   **9.** 120   **11.** 15   **13.** 78
15. $m^4 + 4m^3n + 6m^2n^2 + 4mn^3 + n^4$ **17.** $a^5 - 5a^4b + 10a^3b^2 -$
$10a^2b^3 + 5ab^4 - b^5$ **19.** $8x^3 + 36x^2 + 54x + 27$

21. $\dfrac{x^4}{16} - \dfrac{x^3y}{2} + \dfrac{3x^2y^2}{2} - 2xy^3 + y^4$ **23.** $x^8 + 4x^6 + 6x^4 + 4x^2 + 1$

25. $27x^6 - 27x^4y^2 + 9x^2y^4 - y^6$ **27.** $r^{12} + 24r^{11}s + 264r^{10}s^2 +$
$1760r^9s^3$ **29.** $3^{14}x^{14} - 14(3^{13})x^{13}y + 91(3^{12})x^{12}y^2 - 364(3^{11})x^{11}y^3$
31. $t^{20} + 10t^{18}u^2 + 45t^{16}u^4 + 120t^{14}u^6$ **33.** $120(2^7)m^7n^3$

35. $\dfrac{7x^2y^6}{16}$ **37.** $36k^7$ **39.** $160x^6y^3$ **41.** $4320x^9y^4$

Chapter 12 Review Exercises (pages 709–710)

1. $-1, 1, 3, 5$ **2.** $0, \frac{1}{2}, \frac{2}{3}, \frac{3}{4}$ **3.** $1, 4, 9, 16$ **4.** $\frac{1}{2}, \frac{1}{4}, \frac{1}{8}, \frac{1}{16}$
5. $0, 3, 8, 15$ **6.** $1, -2, 3, -4$ **7.** $1 + 4 + 9 + 16 + 25$
8. $2 + 3 + 4 + 5 + 6 + 7$ **9.** $11 + 16 + 21 + 26$ **10.** 18
11. 126   **12.** $\frac{2827}{840}$ **13.** $15,444 billion   **14.** arithmetic; $d = 3$
15. arithmetic; $d = 4$ **16.** geometric; $r = -\frac{1}{2}$ **17.** geometric; $r = -1$
18. neither   **19.** geometric; $r = \frac{1}{2}$ **20.** 89   **21.** 73   **22.** 69
23. $a_n = -5n + 1$ **24.** $a_n = -3n + 9$ **25.** 15   **26.** 22   **27.** 152
28. 164   **29.** $a_n = -1(4)^{n-1}$ **30.** $a_n = \frac{2}{3}\left(\frac{1}{5}\right)^{n-1}$
31. $2(-3)^{10} = 118{,}098$ **32.** $5(2)^9 = 2560$ or $5(-2)^9 = -2560$
33. $\frac{341}{1024}$ **34.** 0   **35.** 1   **36.** The sum does not exist.
37. $32p^5 - 80p^4q + 80p^3q^2 - 40p^2q^3 + 10pq^4 - q^5$

38. $x^8 + 12x^6y + 54x^4y^2 + 108x^2y^3 + 81y^4$
39. $81t^{12} - 108t^9s^2 + 54t^6s^4 - 12t^3s^6 + s^8$ **40.** $7752(3)^{16}a^{16}b^3$
41. $a_{10} = 1536; S_{10} = 1023$ **42.** $a_{40} = 235; S_{10} = 280$
43. $a_{15} = 38; S_{10} = 95$ **44.** $a_9 = 6561; S_{10} = -14{,}762$
45. $a_n = 2(4)^{n-1}$ **46.** $a_n = 5n - 3$ **47.** $a_n = -3n + 15$
48. $a_n = 27\left(\frac{1}{3}\right)^{n-1}$ **49.** 10 sec   **50.** $21,973.00

51. approximately 42,000   **52.** $\frac{1}{128}$ **53. (a)** $\dfrac{5}{10} + \dfrac{5}{10}\left(\dfrac{1}{10}\right) +$

$\dfrac{5}{10}\left(\dfrac{1}{10}\right)^2 + \dfrac{5}{10}\left(\dfrac{1}{10}\right)^3 + \cdots$ **(b)** $\frac{1}{10}$ **(c)** $\frac{5}{9}$

54. No, the sum cannot be found, because $r = 2$. This value of
r does not satisfy $|r| < 1$.

Chapter 12 Test (page 711)

[12.1] **1.** 0, 2, 0, 2, 0   [12.2] **2.** 4, 6, 8, 10, 12   [12.3] **3.** 48, 24, 12, 6,
3   [12.2] **4.** 0   [12.3] **5.** $\frac{64}{3}$ or $-\frac{64}{3}$ [12.2] **6.** 75
[12.3] **7.** 124 or 44   [12.1] **8.** 85,311   [12.3] **9.** $137,925.91
10. It has a sum if $|r| < 1$.   [12.2] **11.** 70   **12.** 33   **13.** 125,250
[12.3] **14.** 42   **15.** $\frac{1}{3}$ **16.** The sum does not exist.   [12.4] **17.** 40,320
18. 1   **19.** 15   **20.** 66   **21.** $81k^4 - 540k^3 + 1350k^2 - 1500k + 625$
22. $\dfrac{14{,}080x^8y^4}{9}$ [12.1] **23.** $324   [12.3] **24.** $20(3^{11}) = 3{,}542{,}940$

Chapters 1–12 Cumulative Review Exercises (pages 712–713)

[1.2, 1.3] **1.** 8   **2.** -55 [1.1] **3.** $-\frac{8}{3}, 10, 0, \frac{45}{15}$ (or 3), 0.82, -3
4. $\sqrt{13}, -\sqrt{3}$ [2.1] **5.** $\left\{\frac{1}{6}\right\}$ [2.5] **6.** $[10, \infty)$ [2.7] **7.** $\left\{-\frac{9}{2}, 6\right\}$
[2.1] **8.** $\{9\}$ [2.6] **9.** $(-\infty, -3) \cup (4, \infty)$
[2.7] **10.** $(-\infty, -3] \cup [8, \infty)$ [3.2] **11.** $\frac{3}{4}$ [3.3] **12.** $3x + y = 4$
[3.1] **13.**   [3.4] **14.**

[3.5] **15. (a)** yes   **(b)** $\{-3, -2, 0, 1, 2\}$ **(c)** $\{2, 6, 4\}$
[4.1, 4.4] **16.** $\{(-1, -2)\}$ [4.2, 4.4] **17.** $\{(2, 1, 4)\}$
[4.3] **18.** 2 lb   [5.4] **19.** $20p^2 - 2p - 6$ **20.** $9k^2 - 42k + 49$
[5.2] **21.** $-5m^3 - 3m^2 + 3m + 8$ [5.5] **22.** $2t^3 + 3t^2 - 4t + 2 +$
$\dfrac{3}{3t - 2}$ [6.2] **23.** $z(3z + 4)(2z - 1)$ [6.3] **24.** $(7a^2 + 3b)(7a^2 - 3b)$
25. $(c + 3d)(c^2 - 3cd + 9d^2)$ [6.5] **26.** $\left\{-\frac{5}{2}, 2\right\}$ [9.7] **27.** $[-2, 3]$
[5.1] **28.** $\frac{9}{4}$ **29.** $-\dfrac{27p^2}{10}$ [7.1] **30.** $\dfrac{x + 7}{x - 2}$ [7.2] **31.** $\dfrac{3p - 26}{p(p + 3)(p - 4)}$
[7.4] **32.** $\varnothing$ [9.2] **33.** $\left\{\dfrac{-5 \pm \sqrt{217}}{12}\right\}$ [8.4] **34.** $10\sqrt{2}$
[8.7] **35.** 73   [10.1] **36.** $f^{-1}(x) = \dfrac{x - 5}{9}$, or $f^{-1}(x) = \dfrac{1}{9}x - \dfrac{5}{9}$

[10.2] **37.**

$g(x) = \left(\frac{1}{3}\right)^x$

38. $\left\{\frac{5}{2}\right\}$

[10.3] **39.**

$y = \log_{1/3} x$

[10.6] **40.** $\{2\}$

[9.5] **41.**

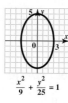

$f(x) = 2(x - 2)^2 - 3$

[11.2] **42.**

$\dfrac{x^2}{9} + \dfrac{y^2}{25} = 1$

[11.3] **43.**

$x^2 - y^2 = 9$

[11.4] **44.** $\left\{(-1, 5), \left(\frac{5}{2}, -2\right)\right\}$

[11.2] **45.** $(x + 5)^2 + (y - 12)^2 = 81$ [12.1] **46.** $-7, -2, 3, 8, 13$

[12.2, 12.3] **47. (a)** 78 **(b)** $\frac{75}{7}$ [12.2] **48.** 30

[12.4] **49.** $32a^5 - 80a^4 + 80a^3 - 40a^2 + 10a - 1$ **50.** $-\dfrac{45x^8 y^6}{4}$

APPENDIX A

Appendix A (pages 721–722)

1. (a) true **(b)** true **(c)** false; The determinant equals $ad - bc$.
(d) true **3.** -3 **5.** 14 **7.** 0 **9.** 59 **11.** 14 **13.** 16
15. -12 **17.** 0 **19.** $\{(1, 0, -1)\}$ **21.** $\{(-3, 6)\}$ **23.** $\left\{\left(\frac{53}{17}, \frac{6}{17}\right)\right\}$
25. $\{(-1, 2)\}$ **27.** $\{(4, -3, 2)\}$ **29.** Cramer's rule does not apply.
31. $\{(-2, 1, 3)\}$ **33.** $\left\{\left(\frac{49}{9}, -\frac{155}{9}, \frac{136}{9}\right)\right\}$ **35.** $\{2\}$ **37.** $\{0\}$

APPENDIX B

Appendix B (page 726)

1. $x - 5$ **3.** $4m - 1$ **5.** $2a + 4 + \dfrac{5}{a + 2}$ **7.** $p - 4 + \dfrac{9}{p + 1}$

9. $4a^2 + a + 3$ **11.** $x^4 + 2x^3 + 2x^2 + 7x + 10 + \dfrac{18}{x - 2}$

13. $-4r^5 - 7r^4 - 10r^3 - 5r^2 - 11r - 8 + \dfrac{-5}{r - 1}$

15. $-3y^4 + 8y^3 - 21y^2 + 36y - 72 + \dfrac{143}{y + 2}$ **17.** 7 **19.** -2

21. 0 **23.** By the remainder theorem, a 0 remainder means that $P(k) = 0$.
That is, k is a number that makes $P(x) = 0$. **25.** yes **27.** no **29.** yes
31. no **33.** $(2x - 3)(x + 4)$ **34.** $\left\{-4, \frac{3}{2}\right\}$ **35.** 0 **36.** 0 **37.** a
38. Yes, $x - 3$ is a factor. $Q(x) = (x - 3)(3x - 1)(x + 2)$

Glossary

For a more complete discussion, see the section(s) in parentheses.

A

absolute value The absolute value of a number is the distance between 0 and the number on a number line. (Section 1.1)

absolute value equation An absolute value equation is an equation that involves the absolute value of a variable expression. (Section 2.7)

absolute value function The function defined by $f(x) = |x|$ with a graph that includes portions of two lines is called the absolute value function. (Section 11.1)

absolute value inequality An absolute value inequality is an inequality that involves the absolute value of a variable expression. (Section 2.7)

addition property of equality The addition property of equality states that the same number can be added to (or subtracted from) both sides of an equation to obtain an equivalent equation. (Section 2.1)

addition property of inequality The addition property of inequality states that the same number can be added to (or subtracted from) both sides of an inequality to obtain an equivalent inequality. (Section 2.5)

additive inverse (negative, opposite) The additive inverse of a number x, symbolized $-x$, is the number that is the same distance from 0 on the number line as x, but on the opposite side of 0. The number 0 is its own additive inverse. For all real numbers x, $x + (-x) = (-x) + x = 0$. (Section 1.1)

algebraic expression Any collection of numbers or variables joined by the basic operations of addition, subtraction, multiplication, or division (except by 0), or the operations of raising to powers or taking roots, formed according to the rules of algebra, is called an algebraic expression. (Section 1.3)

annuity An annuity is a sequence of equal payments made at equal periods of time. (Section 12.3)

arithmetic mean (average) The arithmetic mean of a group of numbers is the sum of all the numbers divided by the number of numbers. (Section 12.1)

arithmetic sequence (arithmetic progression) An arithmetic sequence is a sequence in which each term after the first differs from the preceding term by a constant difference. (Section 12.2)

array of signs An array of signs is used when evaluating a determinant using expansion by minors. The signs alternate for each row and column, beginning with + in the first row, first column position. (Appendix A)

associative property of addition The associative property of addition states that the grouping of terms in a sum does not affect the sum. (Section 1.4)

associative property of multiplication The associative property of multiplication states that the grouping of factors in a product does not affect the product. (Section 1.4)

asymptote A line that a graph more and more closely approaches as the graph gets farther away from the origin is called an asymptote of the graph. (Section 7.4)

asymptotes of a hyperbola The two intersecting straight lines that the branches of a hyperbola approach are called asymptotes of the hyperbola. (Section 11.3)

augmented matrix An augmented matrix is a matrix that has a vertical bar that separates the columns of the matrix into two groups, separating the coefficients from the constants of the corresponding system of equations. (Section 4.4)

axis (axis of symmetry) The axis of a parabola is the vertical or horizontal line (depending on the orientation of the graph) through the vertex of the parabola. (Section 9.5)

B

base The base in an exponential expression is the expression that is the repeated factor. In b^x, b is the base. (Sections 1.3, 5.1)

binomial A binomial is a polynomial consisting of exactly two terms. (Section 5.2)

binomial theorem (general binomial expansion) The binomial theorem provides a formula used to expand a binomial raised to a power. (Section 12.4)

boundary line In the graph of an inequality, the boundary line separates the region that satisfies the inequality from the region that does not satisfy the inequality. (Sections 3.4, 11.5)

C

center of a circle The fixed point that is a fixed distance from all the points that form a circle is the center of the circle. (Section 11.2)

center of an ellipse The center of an ellipse is the fixed point located exactly halfway between the two foci. (Section 11.2)

center-radius form of the equation of a circle The center-radius form of the equation of a circle with center (h, k) and radius r is $(x - h)^2 + (y - k)^2 = r^2$. (Section 11.2)

circle A circle is the set of all points in a plane that lie a fixed distance from a fixed point. (Section 11.2)

coefficient (See **numerical coefficient.**)

column of a matrix A column of a matrix is a group of elements that are read vertically. (Section 4.4, Appendix A)

combined variation A relationship among variables that involves both direct and inverse variation is called combined variation. (Section 7.6)

combining like terms Combining like terms is a method of adding or subtracting terms having exactly the same variable factors by using the properties of real numbers. (Section 1.4)

common difference The common difference d is the difference between any two adjacent terms of an arithmetic sequence. (Section 12.2)

common logarithm A common logarithm is a logarithm having base 10. (Section 10.5)

common ratio The common ratio r is the constant multiplier between adjacent terms in a geometric sequence. (Section 12.3)

commutative property of addition The commutative property of addition states that the order of the terms in a sum does not affect the sum. (Section 1.4)

commutative property of multiplication The commutative property of multiplication states that the order of the factors in a product does not affect the product. (Section 1.4)

complementary angles (complements) Complementary angles are two angles whose measures have a sum of 90°. (Section 2.4 Exercises)

completing the square The process of adding to a binomial the expression that makes it a perfect square trinomial is called completing the square. (Section 9.1)

complex conjugate The complex conjugate of $a + bi$ is $a - bi$. (Section 8.7)

complex fraction A complex fraction is a quotient with one or more fractions in the numerator, denominator, or both. (Section 7.3)

complex number A complex number is any number that can be written in the form $a + bi$, where a and b are real numbers and i is the imaginary unit. (Section 8.7)

components In an ordered pair (x, y), x and y are called the components of the ordered pair. (Section 3.1)

composite function If g is a function of x, and f is a function of $g(x)$, then $f(g(x))$ defines the composite function of f and g. It is symbolized $(f \circ g)(x)$. (Section 5.3)

composition of functions The process of finding a composite function is called composition of functions. (Section 5.3)

compound inequality A compound inequality consists of two inequalities linked by a connective word such as *and* or *or*. (Section 2.6)

conditional equation A conditional equation is true for some replacements of the variable and false for others. (Section 2.1)

conic section When a plane intersects an infinite cone at different angles, the figures formed by the intersections are called conic sections. (Section 11.2)

conjugate The conjugate of $a + b$ is $a - b$. (Section 8.5)

consecutive integers Two integers that differ by one are called consecutive integers. (Section 2.4 Exercises)

consistent system A system of equations with a solution is called a consistent system. (Section 4.1)

constant function A linear function of the form $f(x) = b$, where b is a constant, is called a constant function. (Section 3.6)

constant of variation In the variation equations $y = kx$, $y = \frac{k}{x}$, or $y = kxz$, the nonzero real number k is called the constant of variation. (Section 7.6)

contradiction A contradiction is an equation that is never true. It has no solution. (Section 2.1)

coordinate on a number line Every point on a number line is associated with a unique real number, called the coordinate of the point. (Section 1.1)

coordinates of a point The numbers in an ordered pair are called the coordinates of the corresponding point in the plane. (Section 3.1)

Cramer's rule Cramer's rule uses determinants to solve systems of linear equations. (Appendix A)

cube root function The function defined by $f(x) = \sqrt[3]{x}$ is called the cube root function. (Section 8.1)

cubing function The polynomial function defined by $f(x) = x^3$ is called the cubing function. (Section 5.3)

D

degree of a polynomial The degree of a polynomial is the greatest degree of any of the terms in the polynomial. (Section 5.2)

degree of a term The degree of a term is the sum of the exponents on the variables in the term. (Section 5.2)

dependent equations Equations of a system that have the same graph (because they are different forms of the same equation) are called dependent equations. (Section 4.1)

dependent variable In an equation relating x and y, if the value of the variable y depends on the value of the variable x, then y is called the dependent variable. (Section 3.5)

descending powers A polynomial in one variable is written in descending powers of the variable if the exponents on the variables of the terms of the polynomial decrease from left to right. (Section 5.2)

determinant Associated with every square matrix is a real number called the determinant of the matrix, symbolized by the entries of the matrix placed between two vertical lines. (Appendix A)

difference The answer to a subtraction problem is called the difference. (Section 1.2)

difference of cubes The difference of cubes, $x^3 - y^3$, can be factored as $x^3 - y^3 = (x - y)(x^2 + xy + y^2)$. (Section 6.3)

difference of squares The difference of squares, $x^2 - y^2$, can be factored as the product of the sum and difference of two terms, or $x^2 - y^2 = (x + y)(x - y)$. (Section 6.3)

direct variation y varies directly as x if there exists a nonzero real number (constant) k such that $y = kx$. (Section 7.6)

discriminant The discriminant of $ax^2 + bx + c = 0$ is the quantity $b^2 - 4ac$ under the radical in the quadratic formula. (Section 9.2)

distributive property of multiplication with respect to addition (distributive property) For any real numbers a, b, and c, the distributive property states that $a(b + c) = ab + ac$ and $(b + c)a = ba + ca$. (Section 1.4)

distance The distance between two points on a number line is the absolute value of the difference between the two numbers. (Section 1.2)

domain The set of all first components (x-values) in the ordered pairs of a relation is called the domain. (Section 3.5)

domain of a rational equation The domain of a rational equation is the intersection of the domains of the rational expressions in the equation. (Section 7.4)

E

element of a matrix The numbers in a matrix are called the elements of the matrix. (Section 4.4)

elements (members) of a set The elements (members) of a set are the objects that belong to the set. (Section 1.1)

elimination method The elimination method is an algebraic method used to solve a system of equations in which the equations of the system are combined in order to eliminate one or more variables. (Section 4.1)

ellipse An ellipse is the set of all points in a plane such that the sum of the distances from two fixed points is constant. (Section 11.2)

empty set (null set) The empty set, denoted by $\{ \ \}$ or $\emptyset$, is the set containing no elements. (Section 1.1)

equation An equation is a statement that two algebraic expressions are equal. (Section 1.1)

equivalent equations Equivalent equations are equations that have the same solution set. (Section 2.1)

equivalent inequalities Equivalent inequalities are inequalities that have the same solution set. (Section 2.5)

expansion by minors A method of evaluating a 3×3 or larger determinant is called expansion by minors. (Appendix A)

exponent (power) An exponent, or power, is a number that indicates how many times its base is used as a factor. In b^x, x is the exponent. (Sections 1.3, 5.1)

exponential equation An exponential equation is an equation that has a variable in at least one exponent. (Section 10.2)

exponential expression A number or letter (variable) written with an exponent is an exponential expression. (Section 1.3)

exponential function with base a An exponential function with base a is a function of the form $f(x) = a^x$, where $a > 0$ and $a \neq 1$ for all real numbers x. (Section 10.2)

extraneous solution A proposed solution to an equation, following any of several procedures in the solution process, that does not satisfy the original equation is called an extraneous solution. (Section 8.6)

F

factor If a, b, and c represent numbers and $a \cdot b = c$, then a and b are factors of c. (Section 1.3)

factoring Writing a polynomial as the product of two or more simpler polynomials is called factoring. (Section 6.1)

factoring by grouping Factoring by grouping is a method of grouping the terms of a polynomial in such a way that the polynomial can be factored. It is used when the greatest common factor of the terms of the polynomial is 1. (Section 6.1)

factoring out the greatest common factor Factoring out the greatest common factor is the process of using the distributive property to write a polynomial as a product of the greatest common factor and a simpler polynomial. (Section 6.1)

finite sequence A finite sequence has a domain that includes only the first n positive integers. (Section 12.1)

first-degree equation A first-degree (linear) equation has no term with the variable to a power other than 1. (Section 2.1)

foci (singular, **focus**) Foci are fixed points used to determine the points that form a parabola, an ellipse, or a hyperbola. (Sections 11.2, 11.3)

FOIL FOIL is a mnemonic device which represents a method for multiplying two binomials $(a + b)(c + d)$. Multiply **F**irst terms ac, **O**uter terms ad, **I**nner terms bc, and **L**ast terms bd. Then combine like terms. (Section 5.4)

formula A formula is an equation in which variables are used to describe a relationship among several quantities. (Section 2.2)

function A function is a set of ordered pairs (x, y) in which each value of the first component x corresponds to exactly one value of the second component y. (Section 3.5)

function notation If a function is denoted by f, the notation $f(x)$ is called function notation. Here, $y = f(x)$ represents the value of the function at x. (Section 3.6)

fundamental rectangle The asymptotes of a hyperbola are the extended diagonals of its fundamental rectangle, with corners at the points (a, b), $(-a, b)$, $(-a, -b)$, and $(a, -b)$. (Section 11.3)

future value of an annuity The future value of an annuity is the sum of the compound amounts of all the payments, compounded to the end of the term. (Section 12.3)

G

general term of a sequence The expression a_n, which defines a sequence, is called the general term of the sequence. (Section 12.1)

geometric sequence (geometric progression) A geometric sequence is a sequence in which each term after the first is a constant multiple of the preceding term. (Section 12.3)

graph of a number The point on a number line that corresponds to a number is its graph. (Section 1.1)

graph of an equation The graph of an equation in two variables is the set of all points that correspond to all of the ordered pairs that satisfy the equation. (Section 3.1)

graph of a relation The graph of a relation is the graph of its ordered pairs. (Section 3.5)

greatest common factor (GCF) The greatest common factor of a list of integers is the largest factor of all those integers. The greatest common factor of the terms of a polynomial is the largest factor of all the terms in the polynomial. (Section 6.1)

greatest integer function The function defined by $f(x) = [\![x]\!]$, where the symbol $[\![x]\!]$ is used to represent the greatest integer less than or equal to x, is called the greatest integer function. (Section 11.1)

H

horizontal line test The horizontal line test states that a function is one-to-one if every horizontal line intersects the graph of the function at most once. (Section 10.1)

hyperbola A hyperbola is the set of all points in a plane such that the absolute value of the difference of the distances from two fixed points is constant. (Section 11.3)

hypotenuse The hypotenuse is the longest side in a right triangle. It is the side opposite the right angle. (Section 8.3)

I

identity An identity is an equation that is true for all valid replacements of the variable. It has an infinite number of solutions. (Section 2.1)

identity element for addition For all real numbers a, $a + 0 = 0 + a = a$. The number 0 is called the identity element for addition. (Section 1.4)

identity element for multiplication For all real numbers a, $a \cdot 1 = 1 \cdot a = a$. The number 1 is called the identity element for multiplication. (Section 1.4)

identity function The simplest polynomial function is the identity function, defined by $f(x) = x$. (Section 5.3)

identity property The identity property for addition states that the sum of 0 and any number equals the number. The identity property for multiplication states that the product of 1 and any number equals the number. (Section 1.4)

imaginary part The imaginary part of a complex number $a + bi$ is b. (Section 8.7)

imaginary unit The symbol i, which represents $\sqrt{-1}$, is called the imaginary unit. (Section 8.7)

inconsistent system An inconsistent system of equations is a system with no solution. (Section 4.1)

independent equations Equations of a system that have different graphs are called independent equations. (Section 4.1)

independent variable In an equation relating x and y, if the value of the variable y depends on the value of the variable x, then x is called the independent variable. (Section 3.5)

index (order) In a radical of the form $\sqrt[n]{a}$, n is called the index or order. (Section 8.1)

index of summation When using summation notation, $\sum_{i=1}^{n} f(i)$, the letter i is called the index of summation. Other letters can be used. (Section 12.1)

inequality An inequality is a statement that two expressions are not equal. (Section 1.1)

infinite sequence An infinite sequence is a function with the set of all positive integers as the domain. (Section 12.1)

integers The set of integers is $\{\ldots, -3, -2, -1, 0, 1, 2, 3, \ldots\}$. (Section 1.1)

intersection The intersection of two sets A and B, written $A \cap B$, is the set of elements that belong to *both* A *and* B. (Section 2.6)

interval An interval is a portion of a number line. (Section 1.1)

interval notation Interval notation is a simplified notation that uses parentheses ( ) and/or brackets [ ] and/or the infinity symbol ∞ to describe an interval on a number line. (Section 1.1)

inverse of a function f If f is a one-to-one function, then the inverse of f is the set of all ordered pairs of the form (y, x) where (x, y) belongs to f. (Section 10.1)

inverse property The inverse property for addition states that a number added to its opposite (additive inverse) is 0. The inverse property for multiplication states that a number multiplied by its reciprocal (multiplicative inverse) is 1. (Section 1.4)

inverse variation y varies inversely as x if there exists a nonzero real number (constant) k such that $y = \frac{k}{x}$. (Section 7.6)

irrational numbers An irrational number cannot be written as the quotient of two integers, but can be represented by a point on a number line. (Section 1.1)

J

joint variation y varies jointly as x and z if there exists a nonzero real number (constant) k such that $y = kxz$. (Section 7.6)

L

least common denominator (LCD) Given several denominators, the least multiple that is divisible by all the denominators is called the least common denominator. (Section 7.2)

legs of a right triangle The two shorter perpendicular sides of a right triangle are called the legs. (Section 8.3)

like terms Terms with exactly the same variables raised to exactly the same powers are called like terms. (Sections 1.4, 5.2)

linear equation in one variable A linear equation in one variable can be written in the form $Ax + B = C$, where A, B, and C are real numbers, with $A \neq 0$. (Section 2.1)

linear equation in two variables A linear equation in two variables is an equation that can be written in the form $Ax + By = C$, where A, B, and C are real numbers, and A and B are not both 0. (Section 3.1)

linear function A function defined by an equation of the form $f(x) = ax + b$, for real numbers a and b, is a linear function. The value of a is the slope m of the graph of the function. (Section 3.6)

linear inequality in one variable A linear inequality in one variable can be written in the form $Ax + B < C$ or $Ax + B > C$ (or with $\leq$ or $\geq$), where A, B, and C are real numbers, with $A \neq 0$. (Section 2.5)

linear inequality in two variables A linear inequality in two variables can be written in the form $Ax + By < C$ or $Ax + By > C$ (or with $\leq$ or $\geq$), where A, B, and C are real numbers, with A and B not both 0. (Section 3.4)

linear system (system of linear equations) Two or more linear equations in two or more variables form a linear system. (Section 4.1)

logarithm A logarithm is an exponent. The expression $\log_a x$ represents the exponent to which the base a must be raised to obtain x. (Section 10.3)

logarithmic equation A logarithmic equation is an equation with a logarithm of a variable expression in at least one term. (Section 10.3)

logarithmic function with base a If a and x are positive numbers with $a \neq 1$, then $f(x) = \log_a x$ defines the logarithmic function with base a. (Section 10.3)

lowest terms A fraction is in lowest terms if the greatest common factor of the numerator and denominator is 1. (Section 7.1)

M

mathematical model In a real-world problem, a mathematical model is one or more equations (or inequalities) that describe the situation. (Section 2.2)

matrix (plural, matrices) A matrix is a rectangular array of numbers consisting of horizontal rows and vertical columns. (Section 4.4, Appendix A)

minors The minor of an element in a 3×3 determinant is the 2×2 determinant remaining when a row and a column of the 3×3 determinant are eliminated. (Appendix A)

monomial A monomial is a polynomial consisting of exactly one term. (Section 5.2)

multiplication property of equality The multiplication property of equality states that the same nonzero number can be multiplied by (or divided into) both sides of an equation to obtain an equivalent equation. (Section 2.1)

multiplication property of inequality The multiplication property of inequality states that both sides of an inequality may be multiplied (or divided) by a positive number without changing the direction of the inequality symbol. Multiplying (or dividing) by a negative number reverses the direction of the inequality symbol. (Section 2.5)

multiplication property of 0 The multiplication property of 0 states that the product of any real number and 0 is 0. (Section 1.4)

multiplicative inverse (reciprocal) The multiplicative inverse (reciprocal) of a nonzero number x, symbolized $\frac{1}{x}$, is the real number which has the property that the product of the two numbers is 1. For all nonzero real numbers x, $\frac{1}{x} \cdot x = x \cdot \frac{1}{x} = 1$. (Section 1.2)

N

n-factorial ($n!$) For any positive integer n, $n(n-1)(n-2)(n-3)\cdots(2)(1) = n!$. By definition, $0! = 1$. (Section 12.4)

natural logarithm A natural logarithm is a logarithm having base e. (Section 10.5)

natural numbers (counting numbers) The set of natural numbers is the set of numbers used for counting: $\{1, 2, 3, 4, \ldots\}$. (Section 1.1)

negative of a polynomial The negative of a polynomial is that polynomial with the sign of every term changed. (Section 5.2)

nonlinear equation A nonlinear equation is an equation in which some terms have more than one variable or a variable of degree 2 or greater. (Section 11.4)

nonlinear system of equations A nonlinear system of equations consists of two or more equations to be considered at the same time, at least one of which is nonlinear. (Section 11.4)

nonlinear system of inequalities A nonlinear system of inequalities consists of two or more inequalities to be considered at the same time, at least one of which is nonlinear. (Section 11.5)

number line A line that has a point designated to correspond to the real number 0, and a standard unit chosen to represent the distance between 0 and 1, is a number line. All real numbers correspond to one and only one number on such a line. (Section 1.1)

numerical coefficient The numerical factor in a term is called the numerical coefficient, or simply, the coefficient. (Sections 1.4, 5.2)

O

one-to-one function A one-to-one function is a function in which each x-value corresponds to only one y-value and each y-value corresponds to only one x-value. (Section 10.1)

ordered pair An ordered pair is a pair of numbers written within parentheses in the form (x, y). (Section 3.1)

ordered triple An ordered triple is a triple of numbers written within parentheses in the form (x, y, z). (Section 4.2)

ordinary annuity An ordinary annuity is an annuity in which the payments are made at the end of each time period, and the frequency of payments is the same as the frequency of compounding. (Section 12.3)

origin The point at which the x-axis and y-axis of a rectangular coordinate system intersect is called the origin. (Section 3.1)

P

parabola The graph of a second-degree (quadratic) equation in two variables, with one variable first-degree, is called a parabola. It is a conic section. (Section 9.5)

parallel lines Parallel lines are two lines in the same plane that never intersect. (Section 3.2)

Pascal's triangle Pascal's triangle is a triangular array of numbers that occur as coefficients in the expansion of $(x + y)^n$, using the binomial theorem. (Section 12.4)

payment period In an annuity, the time between payments is called the payment period. (Section 12.3)

percent Percent, written with the symbol %, means "per one hundred." (Section 2.2)

perfect square trinomial A perfect square trinomial is a trinomial that can be factored as the square of a binomial. (Section 6.3)

perpendicular lines Perpendicular lines are two lines that intersect to form a right (90°) angle. (Section 3.2)

point-slope form A linear equation is written in point-slope form if it is in the form $y - y_1 = m(x - x_1)$, where m is the slope of the line and (x_1, y_1) is a point on the line. (Section 3.3)

polynomial A polynomial is a term or a finite sum of terms in which all coefficients are real, all variables have whole number exponents, and no variables appear in denominators. (Section 5.2)

polynomial function A function defined by a polynomial in one variable, consisting of one or more terms, is called a polynomial function. (Section 5.3)

polynomial in x A polynomial containing only the variable x is called a polynomial in x. (Section 5.2)

prime polynomial A prime polynomial is a polynomial that cannot be factored into factors having only integer coefficients. (Section 6.1)

principal root (principal nth root) For even indexes, the symbols $\sqrt{\ }$, $\sqrt[4]{\ }$, $\sqrt[6]{\ }$, ..., $\sqrt[n]{\ }$ are used for nonnegative roots, which are called principal roots. (Section 8.1)

product The answer to a multiplication problem is called the product. (Section 1.2)

product of the sum and difference of two terms The product of the sum and difference of two terms is the difference of the squares of the terms, or $(x + y)(x - y) = x^2 - y^2$. (Section 5.4)

proportion A proportion is a statement that two ratios are equal. (Section 7.5)

proportional If y varies directly as x and there exists some nonzero real number (constant) k such that $y = kx$, then y is said to be proportional to x. (Section 7.6)

proposed solution A value that appears as an apparent solution after a radical, rational, or logarithmic equation has been solved according to standard methods is called a proposed solution for the original equation. It may or may not be an actual solution and must be checked. (Sections 7.4, 8.6, 10.6)

pure imaginary number A complex number $a + bi$ with $a = 0$ and $b \neq 0$ is called a pure imaginary number. (Section 8.7)

Pythagorean theorem The Pythagorean theorem states that the square of the length of the hypotenuse of a right triangle equals the sum of the squares of the lengths of the two legs. (Section 8.3)

Q

quadrant A quadrant is one of the four regions in the plane determined by the axes in a rectangular coordinate system. (Section 3.1)

quadratic equation A quadratic equation is an equation that can be written in the form $ax^2 + bx + c = 0$, where a, b, and c are real numbers, with $a \neq 0$. (Sections 6.5, 9.1)

quadratic formula The quadratic formula is a general formula used to solve a quadratic equation of the form $ax^2 + bx + c = 0$, where $a \neq 0$. It is $x = \dfrac{-b \pm \sqrt{b^2 - 4ac}}{2a}$. (Section 9.2)

quadratic function A function defined by an equation of the form $f(x) = ax^2 + bx + c$, for real numbers a, b, and c, with $a \neq 0$, is a quadratic function. (Section 9.5)

quadratic inequality A quadratic inequality is an inequality that can be written in the form $ax^2 + bx + c < 0$ or $ax^2 + bx + c > 0$ (or with $\leq$ or $\geq$), where a, b, and c are real numbers, with $a \neq 0$. (Section 9.7)

quadratic in form An equation is quadratic in form if it can be written in the form $au^2 + bu + c = 0$, for $a \neq 0$ and an algebraic expression u. (Section 9.3)

quotient The answer to a division problem is called the quotient. (Section 1.2)

R

radical An expression consisting of a radical symbol, root index, and radicand is called a radical. (Section 8.1)

radical equation A radical equation is an equation with a variable in at least one radicand. (Section 8.6)

radical expression A radical expression is an algebraic expression that contains radicals. (Section 8.1)

radical symbol The symbol $\sqrt{\ }$ is called a radical symbol. (Section 1.3)

radicand The number or expression under a radical symbol is called the radicand. (Section 8.1)

radius The radius of a circle is the fixed distance between the center and any point on the circle. (Section 11.2)

range The set of all second components (y-values) in the ordered pairs of a relation is called the range. (Section 3.5)

ratio A ratio is a comparison of two quantities using a quotient. (Section 7.5)

rational expression The quotient of two polynomials with denominator not 0 is called a rational expression. (Section 7.1)

rational function A function that is defined by a quotient of polynomials is called a rational function. (Section 7.1)

rational inequality An inequality that involves rational expressions is called a rational inequality. (Section 9.7)

rationalizing the denominator The process of rewriting a radical expression so that the denominator contains no radicals is called rationalizing the denominator. (Section 8.5)

rational numbers Rational numbers can be written as the quotient of two integers, with denominator not 0. (Section 1.1)

real numbers Real numbers include all numbers that can be represented by points on the number line—that is, all rational and irrational numbers. (Section 1.1)

real part The real part of a complex number $a + bi$ is a. (Section 8.7)

reciprocal (See **multiplicative inverse**.)

reciprocal function The reciprocal function is defined by $f(x) = \frac{1}{x}$. (Sections 7.4, 11.1)

rectangular (Cartesian) coordinate system The x-axis and y-axis placed at a right angle at their zero points form a rectangular coordinate system, also called the Cartesian coordinate system. (Section 3.1)

relation A relation is a set of ordered pairs. (Section 3.5)

rise Rise refers to the vertical change between two points on a line—that is, the change in y-values. (Section 3.2)

row echelon form If a matrix is written with 1s on the diagonal from upper left to lower right and 0s below the 1s, it is said to be in row echelon form. (Section 4.4)

row of a matrix A row of a matrix is a group of elements that are read horizontally. (Section 4.4, Appendix A)

row operations Row operations are operations on a matrix that produce equivalent matrices, leading to systems that have the same solutions as the original system of equations. (Section 4.4)

run Run refers to the horizontal change between two points on a line—that is, the change in x-values. (Section 3.2)

S

scientific notation A number is written in scientific notation when it is expressed in the form $a \times 10^n$, where $1 \le |a| < 10$ and n is an integer. (Section 5.1)

second-degree inequality A second-degree inequality is an inequality with at least one variable of degree 2 and no variable with degree greater than 2. (Section 11.5)

sequence A sequence is a function whose domain is the set of natural numbers or a set of the form $\{1, 2, 3, \ldots, n\}$. (Section 12.1)

series The indicated sum of the terms of a sequence is called a series. (Section 12.1)

set A set is a collection of objects. (Section 1.1)

set-builder notation The special symbolism $\{x \mid x$ has a certain property$\}$ is called set-builder notation. It is used to describe a set of numbers without actually having to list all of the elements. (Section 1.1)

signed numbers Signed numbers are numbers that can be written with a positive or negative sign. (Section 1.1)

simplified radical A simplified radical meets four conditions:

1. The radicand has no factor (except 1) raised to a power greater than or equal to the index.

2. The radicand has no fractions.

3. No denominator contains a radical.

4. Exponents in the radicand and the index of the radical have 1 as their greatest common factor.

(Section 8.3)

slope The ratio of the change in y to the change in x for any two points on a line is called the slope of the line. (Section 3.2)

slope-intercept form A linear equation is written in slope-intercept form if it is in the form $y = mx + b$, where m is the slope and $(0, b)$ is the y-intercept. (Section 3.3)

solution of an equation A solution of an equation is any replacement for the variable that makes the equation true. (Section 2.1)

solution set The solution set of an equation is the set of all solutions of the equation. (Section 2.1)

solution set of a linear system The solution set of a linear system of equations consists of all ordered pairs that satisfy all the equations of the system at the same time. (Section 4.1)

solution set of a system of linear inequalities The solution set of a system of linear inequalities consists of all ordered pairs that make all inequalities of the system true at the same time. (Section 11.5)

square matrix A square matrix is a matrix that has the same number of rows as columns. (Section 4.4, Appendix A)

square of a binomial The square of a binomial is the sum of the square of the first term, twice the product of the two terms, and the square of the last term. That is, $(x + y)^2 = x^2 + 2xy + y^2$ and $(x - y)^2 = x^2 - 2xy + y^2$. (Section 5.4)

square root The inverse of squaring a number is called taking its square root. That is, a number a is a square root of k if $a^2 = k$. (Section 1.3)

square root function The function defined by $f(x) = \sqrt{x}$, with $x \ge 0$, is called the square root function. (Sections 8.1, 11.3)

square root property The square root property (for solving equations) states that if $x^2 = k$, then $x = \sqrt{k}$ or $x = -\sqrt{k}$. (Section 9.1)

squaring function The polynomial function defined by $f(x) = x^2$ is called the squaring function. (Section 5.3)

standard form of a complex number The standard form of a complex number is $a + bi$. (Section 8.7)

standard form of a linear equation A linear equation in two variables written in the form $Ax + By = C$, with A and B not both 0, is in standard form. (Sections 3.1, 3.3)

standard form of a quadratic equation A quadratic equation written in the form $ax^2 + bx + c = 0$, where a, b, and c are real numbers with $a \ne 0$, is in standard form. (Sections 6.5, 9.1)

step function A function that is defined using the greatest integer function and has a graph that resembles a series of steps is called a step function. (Section 11.1)

substitution method The substitution method is an algebraic method for solving a system of equations in which one equation is solved for one of the variables, and then the result is substituted into the other equation. (Section 4.1)

sum The answer to an addition problem is called the sum. (Section 1.2)

sum of cubes The sum of cubes, $x^3 + y^3$, can be factored as $x^3 + y^3 = (x + y) \cdot (x^2 - xy + y^2)$. (Section 6.3)

summation (sigma) notation Summation notation is a compact way of writing a series using the general term of the corresponding sequence. It involves the use of the Greek letter sigma, Σ. (Section 12.1)

supplementary angles (supplements) Supplementary angles are two angles whose measures have a sum of $180°$. (Section 2.4 Exercises)

synthetic division Synthetic division is a shortcut procedure for dividing a polynomial by a binomial of the form $x - k$. (Appendix B)

system of equations A system of equations consists of two or more equations to be solved at the same time. (Section 4.1)

system of inequalities A system of inequalities consists of two or more inequalities to be solved at the same time. (Section 11.5)

T

term A term is a number, a variable, or the product or quotient of a number and one or more variables raised to powers. (Sections 1.4, 5.2)

term of an annuity The time from the beginning of the first payment period to the end of the last period is called the term of an annuity. (Section 12.3)

terms of a sequence The function values in a sequence, written in order, are called terms of the sequence. (Section 12.1)

three-part inequality An inequality that says that one number is between two other numbers is called a three-part inequality. (Section 2.5)

trinomial A trinomial is a polynomial consisting of exactly three terms. (Section 5.2)

U

union The union of two sets A and B, written $A \cup B$, is the set of elements that belong to *either A or B* (or both). (Section 2.6)

universal constant The number e is called a universal constant because of its importance in many areas of mathematics. (Section 10.5)

variable A variable is a symbol, usually a letter, used to represent an unknown number. (Section 1.1)

vary directly (is directly proportional to) y varies directly as x if there exists a nonzero real number (constant) k such that $y = kx$. (Section 7.6)

vary inversely y varies inversely as x if there exists a nonzero real number (constant) k such that $y = \frac{k}{x}$. (Section 7.6)

vary jointly If one variable varies as the product of several other variables (possibly raised to powers), then the first variable is said to vary jointly as the others. (Section 7.6)

vertex The point on a parabola that has the least y-value (if the parabola opens up) or the greatest y-value (if the parabola opens down) is called the vertex of the parabola. (Section 9.5)

vertical asymptote A vertical line that a graph approaches, but never touches or intersects, is called a vertical asymptote. (Section 7.4)

vertical line test The vertical line test states that any vertical line will intersect the graph of a function in at most one point. (Section 3.5)

whole numbers The set of whole numbers is $\{0, 1, 2, 3, 4, \ldots\}$. (Section 1.1)

x-axis The horizontal number line in a rectangular coordinate system is called the x-axis. (Section 3.1)

x-intercept A point where a graph intersects the x-axis is called an x-intercept. (Section 3.1)

Y

y-axis The vertical number line in a rectangular coordinate system is called the y-axis. (Section 3.1)

y-intercept A point where a graph intersects the y-axis is called a y-intercept. (Section 3.1)

Z

zero-factor property The zero-factor property states that if two numbers have a product of 0, then at least one of the numbers must be 0. (Sections 6.5, 9.1)

Triangles and Angles

Right Triangle

Triangle has one 90° (right) angle.

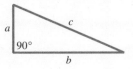

Pythagorean Theorem (for right triangles)

$a^2 + b^2 = c^2$

Right Angle

Measure is 90°.

Isosceles Triangle

Two sides are equal.

$AB = BC$

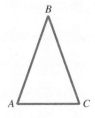

Straight Angle

Measure is 180°.

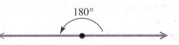

Equilateral Triangle

All sides are equal.

$AB = BC = CA$

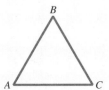

Complementary Angles

The sum of the measures of two complementary angles is 90°.

Angles ① and ② are complementary.

Sum of the Angles of Any Triangle

$A + B + C = 180°$

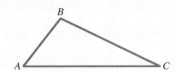

Supplementary Angles

The sum of the measures of two supplementary angles is 180°.

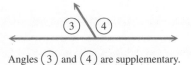

Angles ③ and ④ are supplementary.

Similar Triangles

Corresponding angles are equal. Corresponding sides are proportional.

$A = D, B = E, C = F$

$$\frac{AB}{DE} = \frac{AC}{DF} = \frac{BC}{EF}$$

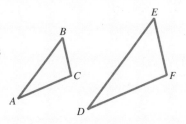

Vertical Angles

Vertical angles have equal measures.

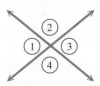

Angle ① = Angle ③

Angle ② = Angle ④

Formulas

Figure	*Formulas*	*Illustration*
Square	Perimeter: $P = 4s$ Area: $\mathcal{A} = s^2$	
Rectangle	Perimeter: $P = 2L + 2W$ Area: $\mathcal{A} = LW$	
Triangle	Perimeter: $P = a + b + c$ Area: $\mathcal{A} = \dfrac{1}{2}bh$	
Parallelogram	Perimeter: $P = 2a + 2b$ Area: $\mathcal{A} = bh$	
Trapezoid	Perimeter: $P = a + b + c + B$ Area: $\mathcal{A} = \dfrac{1}{2}h(b + B)$	
Circle	Diameter: $d = 2r$ Circumference: $C = 2\pi r$ $\qquad\qquad\qquad C = \pi d$ Area: $\mathcal{A} = \pi r^2$	